Handbook of Marine Model Organisms in Experimental Biology

Handbook of Marine Model Organisms in Experimental Biology

Established and Emerging

Edited by

Agnès Boutet and Bernd Schierwater

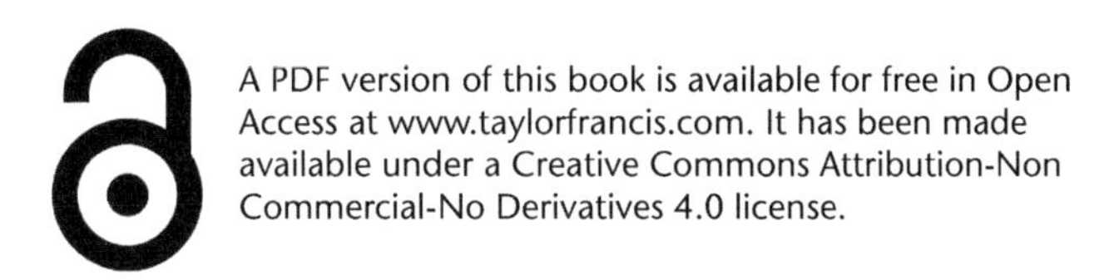

CRC Press is an imprint of the
Taylor & Francis Group, an **informa** business

First edition published 2022
by CRC Press
6000 Broken Sound Parkway NW, Suite 300, Boca Raton, FL 33487–2742

and by CRC Press
2 Park Square, Milton Park, Abingdon, Oxon, OX14 4RN

CRC Press is an imprint of Taylor & Francis Group, LLC

The Erasmus+ Digital Marine project has been funded with support from the European Commission. This publication reflects the views only of the authors, and the Commission cannot be held responsible for any use which may be made of the information contained therein.

ISBN: 978-0-367-44447-1 (hbk)
ISBN: 978-1-032-10883-4 (pbk)
ISBN: 978-1-003-21750-3 (ebk)

DOI: 10.1201/9781003217503

Typeset in Times
by Apex CoVantage, LLC

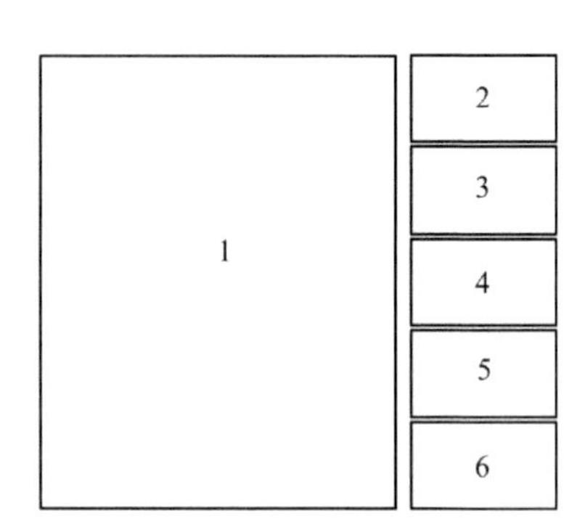

Cover artwork description

Picture 1: An illustration of the cosmopolitan marine invertebrate *Botryllus schlosseri*, a model species in the field of developmental biology, aging and allorecognition (illustrated by Oshrat Ben-Hamo).

Photos on the right (pictures 2 to 5): Courtesy of © Station Biologique de Roscoff, Wilfried THOMAS.

Photo on the right (picture 6): Courtesy of Barry Piekos & Bernd Schierwater (Yale and Hannover).

Contents

Preface

Bringing a rich diversity of living beings to the workbench is a *conditio sine qua non* to explore and understand the magical mechanisms underlying organism development and diversity. This explains why academic researchers have never ceased—and should never cease—to bring new model systems into the laboratories. In the present book, we present both the traditional and iconic marine model organisms and also some new organisms recently brought to the bench.

Marine organisms have always fertilized and nourished traditional disciplines such as neurobiology, physiology, anatomy, ontogeny or comparative zoology; they now also feed important modern fields from genomics to quantitative and computational biology.

The main purpose of this book is to provide an update on marine model organisms from two different perspectives. The first perspective focuses on the general knowledge we have so far collected from the model system; the second perspective is on the present and future importance of the organism for a given research area. To meet the goals, we have compiled 24 chapters covering some of the most important marine model organisms, from bacteria to vertebrates. All chapters are written by experts with longstanding expertise and address the following topics: history of the model, geographical distribution, life cycle, embryogenesis, anatomy, genomic data, functional approaches and challenging research questions. This layout is intended to help the reader compare marine organisms at a glance and assess to which extent they share common features or, in contrast, display specific peculiarities. Of note, several chapters contain substantial descriptive sections relating to anatomy. This is intended as a reminder that fundamental research has been emphasizing morphological descriptions as a prerequisite for pursuing molecular and functional studies.

The work of Ramón y Cajal at the end of the 19th century is a good example in this respect; his drawings are still used today to illustrate cellular and tissue morphology in review dealing with neurosciences or cancer research (Llinás 2003; López-Novoa and Nieto 2009). Remarkably, after countless tissue and cell observations and the careful restitutions with material as simple as ink and paper, Ramón y Cajal (*Histologia Del Sistema Nervioso Del Hombre Y De Los Vertebrados*, 1897–1904) was able to sketch the cellular theory of the brain parenchyma at a time when biologists were unaware of gene expression.

We hope the reader will discover or rediscover the fascination of comparing some very special marine organisms which excite biologists across disciplines. A first example is the capacity of regeneration, both at the body level (as illustrated in Chapter 4 [porifera], Chapters 7 and 8 [*Nematostella* and *Clytia*], Chapter 12 [acoela], Chapter 13 [annelids], Chapter 21 [colonial ascidians]) and organ level (such as the kidney of cartilaginous fish, Chapter 23). The organisms presented offer excellent study systems that help us understand why and how certain tissues and structures are able to renew.

Some other marine organisms are intriguing because they display particular processes that are not well understood, such as gamete formation through transdifferentiation of somatic cells (Chapter 5, *Oscarella*), the metabolic state of cryptobiosis (Chapter 15, crustaceans) or chromosome elimination during embryogenesis (Chapter 22, cyclostomes).

Although seemingly paradoxical, some marine organisms are also attractive because events as basic as embryogenesis or gametogenesis could not be described yet (example: Chapter 6, Placozoa) or because only less than a handful of species have been indexed in an entire phylum (examples: Chapter 6, Placozoa, and Chapter 14, cycliophora).

Genomic or transcriptomic data are now available for almost all marine organisms presented in this handbook. This information is crucial to develop molecular tools but also to revisit the evolution of gene families and the evolution of physiological traits. For example, the unexpected presence of endogenous glycoside hydrolase (GH) genes in the genome of the crustacean *Parhyale hawaiensis* (Chapter 16) confirms that cellulose digestion in metazoans is not necessarily fulfilled by a symbiotioc association with gut-associated bacteria and Protozoa.

Other central research questions put forward in this book include the origin of the mesoderm (Chapter 7, *Nematostella*) and of metazoan body plans (Chapter 4, Porifera; Chapter 6, Placozoa), gastrulation outside bilaterians (Chapter 8, *Clytia*), aging and longevity mechanisms, anthropogenic impact on the environment (Chapter 10 and 11, coral; Chapter 17, echinoderms), how color patterns are set up (Chapter 24, anemone fish) and which biomolecules are being considered for therapeutic or industrial applications (Chapter 1, bacteria; Chapter 5, *Oscarella*; Chapter 20, solitary ascidians; Chapter 23, cartilaginous fish). In addition, Chapter 17 gives a full measure of the complexity of biochemical mechanisms brought into play during gamete encounters.

The reader will also be able to appreciate why some marine species have served pioneering studies related to genome-wide chromatin accessibility (Chapter 19, cephalochordates) or quantitative single-cell morphology and mechanical morphogenesis modeling (Chapter 20, solitary ascidians).

The vast majority of models presented in this book are metazoans, which is not surprising considering the aforementioned biological questions. We have added some non-metazoan model systems in which similar (analogous or homologous) topics have been studied. Brown algae are the first example, as these can serve to investigate size and shape acquisition at the cellular level (Chapter 2). Unicellular holozoans and choanoflagellates are the second example, as they help us to understand how metazoans evolved (Chapter 3). The third example is marine bacteria, as they are essential to study symbiotic organisms, in our example (Chapter 1), they produce metabolites that constitute compulsory signals for jellyfish physiology and metamorphosis (Chapter 8, *Clytia* and 9, *Cassiopea*). These examples are also good illustrations of how all chapters are interconnected.

Importantly, developing new model species for experimental biology can become necessary to overcome specific disadvantages of an existing model organism and to open additional technical perspectives. For instance, until recently, producing stable genetically modified strains has not been feasible in echinoderms, because the traditional model species take several years to reach sexual maturity. Introducing a new species with a short life cycle (*Temnopleurus reevesii*) has allowed researchers to produce the first homozygous knock-out sea urchin strain (Chapter 18).

While bringing new species into the lab has always been an exciting challenge, we now face an additional question associated with our Anthropocene epoch: the conservation status of the organism we want to study. The best example for this might be the chapter dedicated to cartilaginous fish (Chapter 23), in which the reader will find a list of different species that have been used for experimental studies in this group along with their degree of vulnerability.

Having the main features of all marine model organisms presented side by side in one book will clearly be beneficial for researchers across disciplines. The reader can assess to which extent it is possible to use a specific tool and answer a specific question with one model species but not (or not as easily) with another. We thank all authors for their state-of-the-art reviews allowing the reader of this book to quickly and reliably judge the advantages and drawbacks of different model systems and pick the most appropriate one to answer his/her question.

Finally, because many disciplines within the life sciences are at crossroads between two (or more) topics (for example, mathematical modeling and biology or biophysics and cell morphogenesis), this handbook should captivate a highly diverse scientific community. Not only researchers working in developmental biology or evo-devo but also students and scientists eager to go beyond a traditional view of life sciences will find food here. We hope this handbook will find its way into all marine stations and institutes across the globe and help strengthen the network of scientists using marine organisms for their research.

This handbook was created within the Erasmus+-funded strategic partnership project DigitalMarine (2018–2021) set up to support research training on marine organisms in biology. An online distance learning platform intended for master's students is the other deliverable of this project. The combination of this platform with the Schmid Training Course, a marine biology practical course taking place in Roscoff, has been enabling the deployment of innovative teaching methods such as flipped classrooms and blended learning.

We deeply thank all the contributors for their eagerness to review and highlight the most cutting-edge research on their favorite organisms. We are also grateful to Haley Flom and David Wahnoun, respectively, educational engineer and graphic designer in the DigitalMarine project, for the help in editing and illustrations.

Agnès Boutet
Roscoff, France

Bernd Schierwater
Hannover, Germany

BIBLIOGRAPHY

Llinás RR. 2003. The contribution of Santiago Ramón y Cajal to functional neuroscience. *Nat Rev Neurosci*. 4:77–80.

López-Novoa JM, Nieto MA. 2009. Inflammation and EMT: An alliance towards organ fibrosis and cancer progression. *EMBO Mol Med*. 1:303–314.

Ramón y Cajal S. 1897–1904. *Histologia Del Sistema Nervioso Del Hombre Y De Los Vertebrados*. CSIC-Madrid.Contributors

About the Editors

Agnès Boutet has a doctorate in neurosciences from Université Paris XI (now Université Paris-Saclay). During her post-doctoral work in Spain (Angela Nieto's lab) and in France (Andreas Schedl's lab) she was interested in the role of developmental genes in the triggering of renal diseases and more generally in processes linking embryogenesis to human pathologies. In 2011, she got an academic position as a lecturer at Sorbonne Université to work at the Station Biologique de Roscoff, in France. There, she used marine organisms to conduct work in evolutionary biology to track the origin of brain asymmetries in vertebrates (in Sylvie Mazan's lab). In Roscoff, she also had the chance to continue the organization of the iconic Schmid Training Course, an international practical course on the use of marine models in biology. Her current research is still involving marine organisms, more precisely sharks as they have the property to regenerate their kidney. Her question is to decipher the molecular mechanisms underlying this incredible regenerative property. She is currently the chair of the Erasmus+-funded strategic partnership, *DigitalMarine* (2018 – 2021). This project aims to develop a hybrid training (combining self-learning through a digital platform and intense practical lab work in marine station) dedicated to the use of marine organisms in life sciences.

Bernd Schierwater is a Director ITZ and Professor of Zoology, TiHo University Hannover, Germany. He received his Ph.D. (special honors degree'summa cum laude') from Technical University Braunschweig (TUB), Germany in 1989. He was a Distinguished Sabbatical Scholar at NESCent, Duke University. He was awarded with Senior Ecologist of the Ecological Society of America (2009). His training in evolutionary and ecological genetics has arisen from running laboratories at Frankfurt University (Assistant Professor), Freiberg University (Associate Professor) and Hannover TiHo University (Full Professor) and from working as a Research Associate in different departments at Yale University and also at the AMNH New York (Rob DeSalle lab). He has developed the most primitive metazoan animals, the placozoans, into an emerging model system for next generation biodiversity and cancer research. Hans-Jürgen Osigus is at the University of Veterinary Medicine Hannover, Foundation, Institute of Animal Ecology.

Contributors

Maja Adamska
Research School of Biology
Australian National University
Canberra, Australia

Pavlopoulos Anastasios
Foundation for Research and Technology Hellas
Institute of Molecular Biology and Biotechnology
Heraklion, Greece

Xavier Bailly
Multicellular Marine Models (M3) Team
Sorbonne Université
Roscoff, France

Eldon Ball
Division of Ecology and Evolution
Australian National University
Acton, ACT, Australia

Anthony Bellantuono
Department of Biology
Florida International University
Miami, Florida, USA

Stéphanie Bertrand
Observatoire Océanologique de Banyuls sur Mer- BIOM UMR7232 CNRS/SU
Sorbonne Université
Banyuls Sur Mer, France

Laurence Besseau
CNRS – BIOM
Sorbonne University
Banyuls-Sur-Mer, France

Kilian Biasuz
Centre de Recherche de Biologie Cellulaire de Montpellier, CBRM, CNRS
Université De Montpellier
Montpellier, France

Agnès Boutet
Centre National de la Recherche Scientifique (CNRS)
Sorbonne Université
Roscoff, France

Tom Bridge
Biodiversity and Geosciences Program
Queensland Museum
Townsville, QLD, Australia

Carole Borchiellini
Aix Marseille Université
Avignon Université, CNRS, IRD, IMBE
Marseille, France

Elena Casacuberta
Functional Genomics Dept.
CSIC-University Pompeu Fabra
Barcelona, Spain

Jean-Philippe Chambon
Sorbonne Université, Paris
Paris, France

Bénédicte Charrier
Station Biologique, CNRS
Sorbonne Université
Roscoff, France

Patrick Cormier
Centre National de la Recherche Scientifique (CNRS)
Sorbonne Université
Roscoff, France

Salvatore D'aniello
Department of Biology and Evolution of Marine Organisms
Stazione Zoologica Anton Dohrn
Napoli, Italy

Justin Dalrymple
Department of Biology
Florida International University
Miami, Florida, USA

Matthew Degennaro
Department of Biology
Florida International University
Miami, Florida, USA

Renard Emmanuelle
Aix Marseille Université, Avignon Université, CNRS Ird, Imbe
Marseille, France

Alexander Ereskovsky
Aix Marseille Université, Avignon Université, CNRS Ird, Imbe
Marseille, France

William K. Fitt
Odum School of Ecology
University of Georgia
Athens, Georgia, USA

Peter Funch
Department of Biology
Aarhus University
Aarhus, Denmark

Edgar Gamero-Mora
Departamento de Zoologia
Universidade de São Paulo
São Paulo, Brasil

Brenda Gavilán
Departament de Genètica
Universitat de Barcelona,
Barcelona, Spain

Eve Gazave
Université de Paris, CNRS
Institut Jacques Monod
Paris, France

Kapai Gentian
Institute of Molecular Biology and Biotechnology - Foundation for Research and Technology Hellas
Heraklion, Greece

Régis Grimaud
Université de Pau et Des Pays de L'adour, E2S UPPA, CNRS, IPREM
Pau, France

Volker Hartenstein
Department of Molecular Cell and Developmental Biology
University of California, Los Angeles (UCLA)
Los Angeles, California, USA

Jamie Havrilak
Department of Biological Sciences
Lehigh University
Bethlehem, Pennsylvania, USA

David Hayward
Division of Biomedical Science and Biochemistry
Australian National University
Acton, ACT, Australia

Dietrich K. Hofmann
Department of Zoology & Neurobiology
Ruhr-University Bochum
Bochum, Germany

Evelyn Houliston
Laboratoire de Biologie du Développement de Villefranche-sur-Mer (LBDV)
Sorbonne Université
Villefranche-sur-Mer, France

Fabien Joux
Laboratoire D'Océanographie Microbienne (LOMIC)
Sorbonne Université
Banyuls-Sur-Mer, France

Marleen Klann
Marine Eco-Evo-Devo Unit
Okinawa Institute of Science and Technology
Japan, Okinawa

Gabriel Krasovec
National University of Ireland, Galway
Galway, Ireland

Aleksandra Kożyczkowska
Functional Genomics Department
CSIC-University Pompeu Fabra
Barcelona, Spain

Raphaël Lami
Laboratoire de Biodiversité et Biotechnologies Microbiennes (LBBM)
Sorbonne Université
Banyuls-sur-Mer, France

Michael Layden
Department of Biological Sciences
Lehigh University
Bethlehem, Pennsylvania, USA

Nicolas Rabet
UMR BOREA 7208 MNHN/UPMC/CNRS/IRD
Université Sorbonne Universités
Paris, France

Lucas Leclère
Laboratoire de Biologie du Développement de Villefranche-sur-Mer (LBDV)
Sorbonne Université
Villefranche-sur-Mer, France

Yasmine Lund-Ricard
Centre National de la Recherche Scientifique (CNRS)
Sorbonne Université
Roscoff, France

Mercader Manon
Marine Eco-Evo-Devo Unit
Okinawa Institute of Science and Technology
Japan, Okinawa

Mark Q. Martindale
Whitney Laboratory For Marine Bioscience
University of Florida
St. Augustine, Florida, USA

Pedro Martinez
Departament de Genètica, Microbiologia I Estadística, Universitat de Barcelona,
Institut Català de Recerca I Estudis Avançats (Icrea)
Barcelona, Spain

Reynaud Mathieu
Marine Eco-Evo-Devo Unit
Okinawa Institute of Science and Technology
Japan, Okinawa

Mónica Medina
Department of Biology
Pennsylvania State University
University Park, Pennsylvania, USA

David Miller
Molecular & Cell Biology
James Cook University
Townsville, Qld, Australia

Julia Morales
Centre National de la Recherche Scientifique (CNRS)
Sorbonne Université
Roscoff, France

André C. Morandini
Departamento de Zoologia
Universidade de São Paulo
São Paulo, Brasil

Roux Natacha
CNRS – BIOM
Sorbonne University
Banyuls-sur-Mer, France

Aki Ohdera
Division of Biology and Biological Engineering
California Institute of Technology
Pasadena, California, USA

Hans-Jürgen Osigus
Institute of Animal Ecology
University of Veterinary Medicine Hannover, Foundation
Hannover, Germany

Salis Pauline
CNRS – BIOM
Sorbonne University
Banyuls-sur-Mer, France

Sophie Peron
Laboratoire de Biologie du Développement de Villefranche-sur-Mer (Lbdv)
Sorbonne Université
Villefranche-sur-Mer, France

Florian Pontheaux
Centre National de la Recherche Scientifique (CNRS)
Sorbonne Université
Roscoff, France

John Rallis
Institute of Molecular Biology and Biotechnology
Foundation for Research and Technology Hellas
Heraklion, Greece

Baruch Rinkevich
Israel Oceanography and Limnological Research
National Institute of Cceanography
Haifa, Israel

Ben Hamo Rinkevich
The Department of Evolutionary and Environmental Biology
University of Haifa
Haifa, Israel

Fernando Roch
Centre National de la Recherche Scientifique (CNRS)
Sorbonne Université
Roscoff, France

Caroline Rocher
Aix Marseille Université
Avignon Université, CNRS, IRD, IMBE
Marseille, France

Iñaki Ruiz-Trillo
Functional Genomics Department
Departament de Genetica, Microbiologia I Estadistica
Universitat de Barcelona
Barcelona, Spain

Sophie Sanchez-Brosseau
CNRS, Laboratoire de Biologie Intégrative des Organismes Marins (BIOM)
Sorbonne Université
Banyuls-sur-Mer, France

Quentin Schenkelaars
Institut Jacques Monod
Université de Paris, CNRS
Paris, France

Bernd Schierwater
Institute of Animal Ecology
University of Veterinary Medicine Hannover, Foundation
Hannover, Germany

Layla Al-Shaer
Department of Biological Sciences
Lehigh University
Bethlehem, Pennsylvania, USA

Lisa Thomann
Centre de Recherche de Biologie Cellulaire de Montpellier, CRBM, CNRS
Université de Montpellier
Montpellier, France

Victoria Sharp
Department of Biology
Pennsylvania State University
University Park, Pennsylvania, USA

Dor Shefy
Department of Life Sciences, Ben-Gurion University of The Negev Eilat
Haifa, Israel

Christophe Six
Equipe Ecologie du Plancton Marin
Sorbonne Université
Roscoff, France

Simon G. Sprecher
Department of Biology
University of Fribourg
Fribourg, Switzerland

B. Steinworth
Whitney Laboratory for Marine Bioscience
University of Florida
St. Augustine, Florida, USA

Fumiaki Sugahara
Division of Biology
Hyogo College of Medicine
Riken Nishinomiya, Japan
and
Kobe, Japan

Ioannis Theodorou
Station Biologique, CNRS
Sorbonne Université
Roscoff, France

François Thomas
Station Biologique, CNRS
Sorbonne Université
Roscoff, France

Laurent Urios
Université de Pau et des Pays de L'adour, E2S UPPA, CNRS, Iprem
Pau, France

Laudet Vincent
Marine Eco-Evo-Devo Unit
Okinawa Institute of Science and Technology
Japan, Okinawa
Institute of Cellular and Organismic Biology - Lab of Marine Eco-Evo-Devo - Academia Sinica
Taipei, Taiwan

Nyree J West
Laboratoire de Biodiversité et Biotechnologies Microbiennes (LBBM)
Sorbonne Université
Banyuls-sur-Mer, France

Shunsuke Yaguchi
Shimoda Marine Research Center
University of Tsukuba
Shimoda, Shizuoka, Japan

1 Marine Bacterial Models for Experimental Biology

Raphaël Lami, Régis Grimaud, Sophie Sanchez-Brosseau, Christophe Six, François Thomas, Nyree J West, Fabien Joux and Laurent Urios

CONTENTS

DOI: 10.1201/9781003217503-1

1.1 INTRODUCTION

Bacteria are ubiquitous and abundant in the marine environment (10^5–10^6 cells.mL^{-1}), playing a multiplicity of roles in marine ecosystems that is a product of their long evolution and subsequent genetic diversification. Certain species play key roles in biogeochemical cycles, notably by contribution to primary production in the case of phototrophic *Cyanobacteria* or by the remineralization of this production by heterotrophic bacteria. Other bacterial species impact human health and the economy adversely by causing disease in humans and aquaculture facilities, whereas others interact in a coordinated fashion to form biofilms that can lead to biofouling and corrosion of marine structures. Conversely, by virtue of their wide genetic diversity, the bacterial kingdom offers a chemical and enzymatic diversity that can be exploited in many fields, for example, in the bioremediation of marine pollution or for the discovery of novel natural products for the food and medical industries. To further understanding in these diverse research domains, simple tractable bacterial model organisms are needed. In this chapter, we will briefly touch on the well-known non-marine bacterial model organisms and the criteria for a good model organism and explain some of the reasons few marine models are available despite the extraordinary reservoir of the marine environment. We will then present four different marine bacterial models applied to very different research domains, each with their own specific questions and applications but all dependent on a similar toolkit that we will develop at the end of this chapter.

1.1.1 Early Bacterial Models in Experimental Biology

One of the most famous model organisms is undoubtedly the intestinal bacterium *Escherichia coli* belonging to the *Proteobacteria* phylum that was discovered in 1885 by Theodor Escherich. With its fast growth rate in a range of inexpensive media, simple cell structure and ease of manipulation and storage, *E. coli* became the workhorse of the microbiology laboratory. With advances in molecular biology, research on *E. coli* led to a number of significant discoveries that were instrumental in developing the field of molecular genetics. A few examples of these discoveries, some of which were awarded Nobel prizes, include gene exchange between bacteria by conjugation, the elucidation of the genetic code, the mechanism of DNA replication, the organization of genes into operons and restriction enzymes (Blount 2015).

Other bacteria are also well-known models in biology, although less commonly used, and not so famous as *E. coli*. *Bacillus subtilis* is a member of the *Firmicutes* phylum and can be found in a diverse number of aquatic and terrestrial habitats and even in animal guts (Earl et al. 2008). On account of its fast growth, natural transformation, protein secretion, production of endospores and formation of biofilms, it has become an important model, notably for the food and biotechnology industries (Errington and van der Aart 2020). Despite being non-pathogenic, this bacterium has also been used to study the mechanisms of pathogenesis, as it presents some interesting features in common with pathogenic cells, including biofilm formation and sporulation. Other medically important model bacteria include *Staphylococcus* for the study of the skin microbiota and antibiotic resistance; *Bifidobacterium* for research on gut microbiota; and *Pseudomonas aeruginosa* for biofilm formation, chemotaxis and antibiotic resistance.

1.1.2 A Vast Diversity of Bacteria in the Seawater, a Reservoir of Potential Prokaryotic Models

Understandably, the best-known models mentioned previously are those organisms living in close contact with humans, as commensals or present in their immediate environment. The exploration of the oceans, combined with the molecular biology revolution, revealed a vast diversity of bacteria. Prokaryotes are incredibly abundant in seawater: their average abundance is about 5×10^5 cells per mL, and their total number in the world ocean is estimated to be about 10^{29} cells (Whitman et al. 1998). Since the 1990s, continuous and massive 16S rRNA gene sequencing of planktonic DNA has revealed the extraordinary diversity of marine prokaryotes, both for Bacteria and Archaea. An analysis of samples collected during the *Tara* research vessel's marine expeditions (https://oceans.taraexpeditions.org) has revealed 37,470 species of Bacteria and Archaea (Sunagawa et al. 2015). Analysis of sequence datasets has also revealed that we are still far from capturing the whole picture of the total prokaryotic diversity in the oceans. A considerable fraction of this diversity belongs to the "rare biosphere", an immense reservoir of species with low abundances (Overmann and Lepleux 2016). Moreover, recent studies revealed that some marine niches, like marine biofilms, are even more diverse than the pelagic waters and still constitute a substantial source of hidden diversity (Zhang et al. 2019). The recovery of large metagenomic datasets from oceanic samples has also provided evidence for the extraordinary functional diversity of marine prokaryotes; in their 193 samples, the Tara Ocean datasets revealed the presence of 39,246 different orthologous groups. The oceanic metagenomic datasets were enriched in functional categories related to transport of solutes (coenzymes, lipids, amino acids, secondary metabolites) and energy production (including photosynthesis) (Sunagawa et al. 2015). Marine bacteria are also known to produce many types of bioactive compounds that are of interest for industrial applications, including active enzymes and molecules with anticancer, antimicrobial and anti-inflammatory properties (Zeaiter et al. 2018).

1.1.3 The Need for New Marine Bacterial Models

The marine environment is potentially a very important reservoir of prokaryotic models to explore many types of biological mechanisms, either to investigate their diversity or to assess some of the particular features linked to their adaptation to marine life. We will emphasize in this chapter the diversity of biological questions that can be addressed using marine models and for which the current 'traditional' models cannot provide enough answers. Indeed, many major questions in biology and evolutionary studies cannot be fully addressed using famous bacterial models like *E. coli* or *B. subtilis*. Many of them are connected to environmental questions, and they include, for example, the ones related to molecular adaptations to environmental changes (including in ecotoxicology) or to the identification of organisms suited to develop innovative 'green' or 'blue' biotechnological applications.

1.2 EXAMPLES OF MARINE BACTERIAL MODELS

Only a few marine bacterial models currently exist, a paradox when considering the huge taxonomic and functional diversity of marine waters. In this chapter, we present a non-exhaustive collection of relevant marine models and give a snapshot of the diversity of biological mechanisms they can help us explore. We will show how *Vibrio fischeri* is a common model to examine host–symbiont interactions, bioluminescence mechanisms and cell–cell interactions; how marine cyanobacteria *Prochlorococcus* and *Synechococcus* are models to examine the mechanisms of photosynthesis and their adaptation to life in the oceans; how *Zobellia galactanivorans* allows us to study the bacterial degradation of algal biomass; and how *Marinobacter hydrocarbonoclasticus* provides us with key information on biofilm development, iron acquisition and hydrocarbon and lipid metabolism.

1.2.1 *Vibrio fischeri*, a Well-Known and Historic Marine Bacterial Model

Allivibrio fischeri (but the historical name *V. fischeri* is still widely used) is a widely known bacterial model isolated from the marine environment. We will see in this section that this bacterium serves as a model for the study of bioluminescence mechanisms, cell-to-cell communication systems and host–symbiont relationships. This first example will reveal how a marine bacterial model also serves to explore relevant mechanisms in medical sciences, biotechnology, pharmacology and many others.

1.2.1.1 Key Features of *V. fischeri*

This bacterium is a common marine *Gammaproteobacteria* that belongs to the *Vibrionaceae*. This bacterium is motile thanks to a tuft of polar flagella, which is formed by one to five flagellar filaments. The genome of *V. fischeri* has been fully sequenced and is of 4.2 Mb. It is organized in two chromosomes and usually some additional plasmids. This bacterium colonizes various marine niches, including the seawater column and marine sediments. One exceptional feature of this bacterium is its ability to colonize hosts, like the small squid *Euprymna scolopes*: when associated with its host, *V. fischeri* produces light, which makes the animal luminescent.

1.2.1.2 Bioluminescence Mechanisms in Marine Environments and Organisms

Bioluminescent marine bacteria interact with a high diversity of metazoan hosts, including squids and fishes. Like some other marine bioluminescent bacteria, *V. fischeri* exhibits a dual lifestyle, either freely floating in the water column or as a symbiont inside its host. *V. fischeri* is typically involved in symbiosis species from two families of squids as well as different families of fishes (Dunlap and Kita-Tsukamoto 2006), thus demonstrating the ubiquitous capacity of the bacterium to colonize different host types. Among the family *Sepiolidae*, the symbioses involving Mediterranean (*Sepiola*) and Pacific (*Euprymna*) squid species probably evolved independently, as they involve different *Vibrio* species (Fidopiastis et al. 1998). It is known that the light organ of *Sepiola* sp. contains a mixed population of *V. logei* and *V. fischeri* species (Fidopiastis et al. 1998), while only *V. fischeri* is strictly observed in the light organ of *Euprymna scolopes*. It appears that most of the time, the bacterial population is monospecific in a light organ (Dunlap and Urbanczyk 2013).

As for all bioluminescent organisms, the chemical reaction of bioluminescence in bacteria relies on the oxidation of a substrate (luciferin) by an enzyme (luciferase). Bacterial luciferin consists of a reduced flavin ($FMNH_2$) and an aliphatic aldehyde chain (4 to 8 carbon atoms), which serves as a cofactor. Bacterial luciferase is a flavin mono-oxygenase formed of two α (40 kDa or 355 aa) and β (37 kDa or 324 aa) subunits. The catalytic site of the enzyme consists of a TIM-type barrel (Campbell et al. 2010) located in the α subunit, while the β subunit is necessary for the stability and activity of the enzyme. In *V. fischeri*, luminescence is produced when luciferase (composed of α and β subunits) converts reduced flavin to flavin. During the dioxygen-dependent reaction, $FMNH_2$ and the aliphatic aldehyde are oxidized to flavin (FMN) and fatty acid, respectively, as follows: $FMNH_2 + O_2 + R\text{-}CHO \rightarrow FMN + R\text{-}COOH + H_2O + h\upsilon$ ($\lambda_{max} = 490$ nm). Early studies evidenced that a *V. fischeri* strain (previously also known as *Photobacterium fischeri*) was also able to emit yellow light (Ruby and Nealson 1977). This was one of the first descriptions of a bioluminescent bacterial strain emitting light in a different color than blue-green, which is the more common emission in the ocean water column. In this particular case of fluorescence associated with bioluminescence phenomenon, a yellow fluorescent protein, YFP, binds FMN and shifts the light emission from around 490 nm to 545 nm. In luminous bacteria, all products involved in the bioluminescent reaction are encoded in a *lux* operon. In *V. fischeri*, the *lux* operon comprises genes coding for different subunits of either luciferase (*luxA and luxB*), fatty acid reductase complex of the luminescence system (*luxC*, *luxD*, and *luxE*) or flavin reductase (*luxG*) (Dunlap and Kita-Tsukamoto 2006). In *V. fischeri*, the *lux* genes are cotranscribed with *luxI* (which will be defined hereafter), according to the *luxICDABEG* order, the most frequent order found in luminous bacteria.

1.2.1.3 Quorum Sensing, a Cell-to-Cell Communication System

The existence of communication between microorganisms was first suspected in *Streptococcus pneumoniae* by Alexander Tomasz in 1965. The researcher demonstrated the emission of a hormone-like based communication that controls the competent state. However, most of the observations that led to the study of communication between microorganisms and thus to the concept of "quorum sensing" were acquired from experiments conducted by marine scientists during the 1970s. During this decade, and as described, in-depth studies were conducted on *V. fischeri* strains that can colonize the light organ of the Hawaiian bobtail squid *Euprymna scolopes*, where they produce bioluminescence (Greenberg et al. 1979; Nealson et al. 1970). In particular, it has been noticed that the capacity for bioluminescence is a density-dependent phenotype. In seawater, *V. fischeri* cells are free living and scarce and do not produce light most of the time. However, in particular conditions, they can emit light when they reach high cell densities, like in laboratory cultures or when they colonize the light organ of the small squid. Since these initial studies, the concept of quorum sensing was defined in the 1990s and refers to a population density-based physiological response of bacterial cells (Fuqua et al. 1994).

After these first observations, this original system of bioluminescence regulation was fully chemically and genetically described. The diffusible signal, also named autoinducer (AI), was identified in 1981 as an acyl-homoserine lactone (AHL) and described as 3-oxo-hexanoyl-homoserine lactone (3-oxo-C6-HSL) (Eberhard et al. 1981). The genetic cluster involved in this phenomenon was then characterized as a bi-directionally transcribed operon with eight genes, named *luxA-E*, *luxG*, *luxI* and *luxR*. This genetic system has been mentioned in this chapter, except for the roles of LuxI and LuxR, which are of particular interest when focusing on quorum sensing mechanisms. LuxI is the AI synthase, while LuxR is the receptor of this diffusible signal. When the AIs reach a threshold concentration in the nearby environment of bacterial cells (reflecting the increase in cell abundance), they bind to the LuxR receptors, which act as transcription factors and activate the expression of all *lux* genes. The diffusible signal is designated as AI because it promotes its own production through the autoinduction of *luxI* (Engebrecht et al. 1983; Swartzman et al. 1990) (Figure 1.1).

After these initial discoveries and the subsequent identification of the genetic system of quorum sensing in *V. fischeri*, the study of this mechanism garnered little interest from the scientific community for more than a decade. Likely, quorum sensing appeared then to be a kind of regulation specialized for bioluminescence expressed in the *Vibrio* bacteria colonizing a small Hawaiian squid. This interest was renewed in the 1990s with the development of DNA sequencing methods and the discovery of a broad diversity of *luxI* and *luxR* homologs in many different types of bacteria: *Vibrio fisheri* has thus been little by little established as a universal model for the study of quorum sensing circuits.

Most of the scientific effort in the field of quorum sensing in the 1990s focused on strains with a medical or agronomic interest. An important reason for this interest in the medical field, among others, is that an increasing number of links were established between virulence and quorum sensing in model pathogenic bacteria, such as in *Staphylococcus* strains (Ji et al. 1995) and *Pseudomonas aeruginosa* (Pearson et al. 2000). It was only in the following decade that work began to be published about bacteria in the field of environmental sciences, including those isolated from marine waters. In 1998, one of the first reports of AIs present in the natural environment was published under the title "Quorum Sensing Autoinducers: Do They Play a Role in Natural Microbial Habitats?", which revealed some early interest in quorum sensing from the aquatic AIs in naturally occurring biofilms (Bachofen and Schenk 1998). In 2002, Gram et al. reported for the first time the production of AHLs within *Roseobacter* strains isolated from marine snow (Gram et al. 2002). Since then, a growing number of reports have focused on the nature and role of quorum sensing in marine bacteria, and large sets of culture-dependent and culture-independent studies have highlighted the importance of quorum sensing mechanisms in marine biofilms and environments (Lami 2019).

1.2.1.4 The Molecular Mechanisms of Symbiotic Associations

Nowadays, the symbiosis between *V. fischeri* and the Hawaiian bobtail squid *Euprymna scolopes* is well characterized (McFall-Ngai and Ruby 1991) and constitutes a perfect model to understand bacteria–animal interactions (McFall-Ngai 2014). The luminescence produced by the *V. fischeri* symbionts would help camouflage their host at night by eliminating its shadow within the water column ("counter-illumination"). Although this symbiosis is obligatory for the host, symbionts are horizontally transmitted as the squid host *E. scolopes* acquires its *V. fischeri* luminescent symbionts from the surrounding seawater (Wei and Young 1989). This association shows a strong species specificity initiated within hours after the juvenile squid hatches, provided that symbiotically competent *V. fischeri* cells are present in the ambient seawater (Ruby and Asato 1993; Wei and Young 1989).

Interestingly, the *E. scolopes–V. fischeri* model provided the first direct evidence of an animal host controlling the number and activity of its extracellular bacterial population as part of a circadian biological rhythm. *E. scolopes* and *Sepiola atlantica* mechanically control the emission of luminescence by periodically expelling excess *V. fischeri* symbionts, thereby adjusting bacterial density inside the light organ (Ruby and Asato 1993). As a result, the cell abundance of *V. fischeri* within the squid follow a circadian pattern. At night, *V. fischeri* cells are present at high concentrations in the crypts of the light organ (10^{10}–10^{11} cells mL^{-1}) and produce

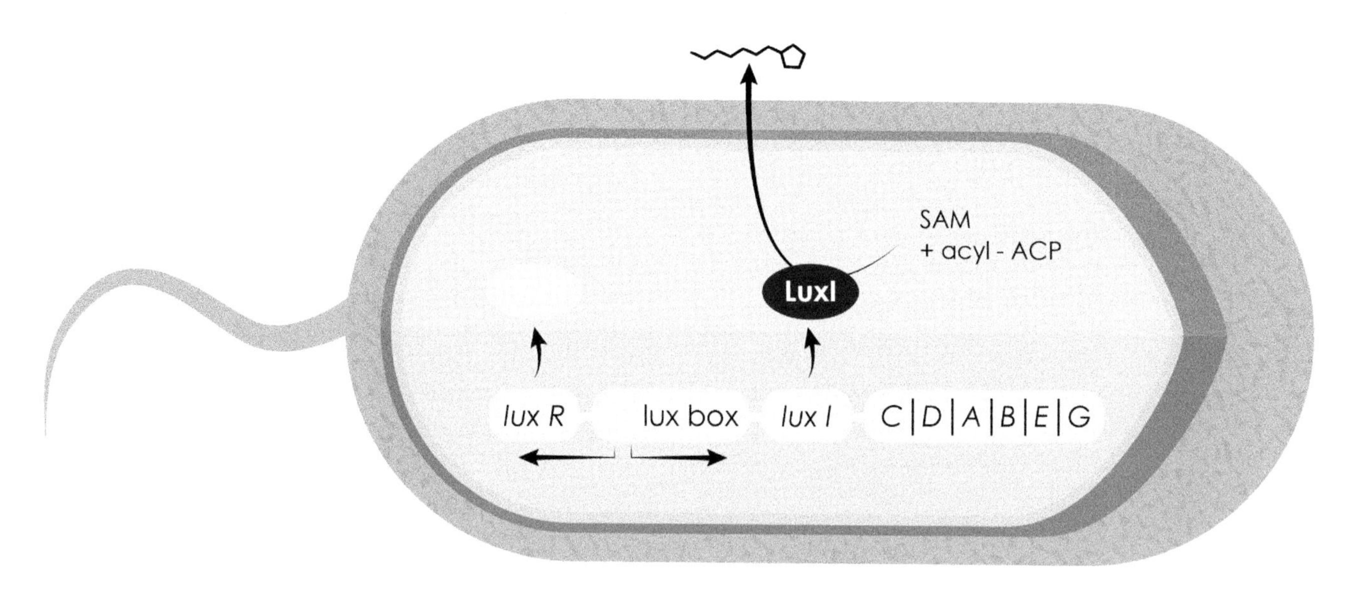

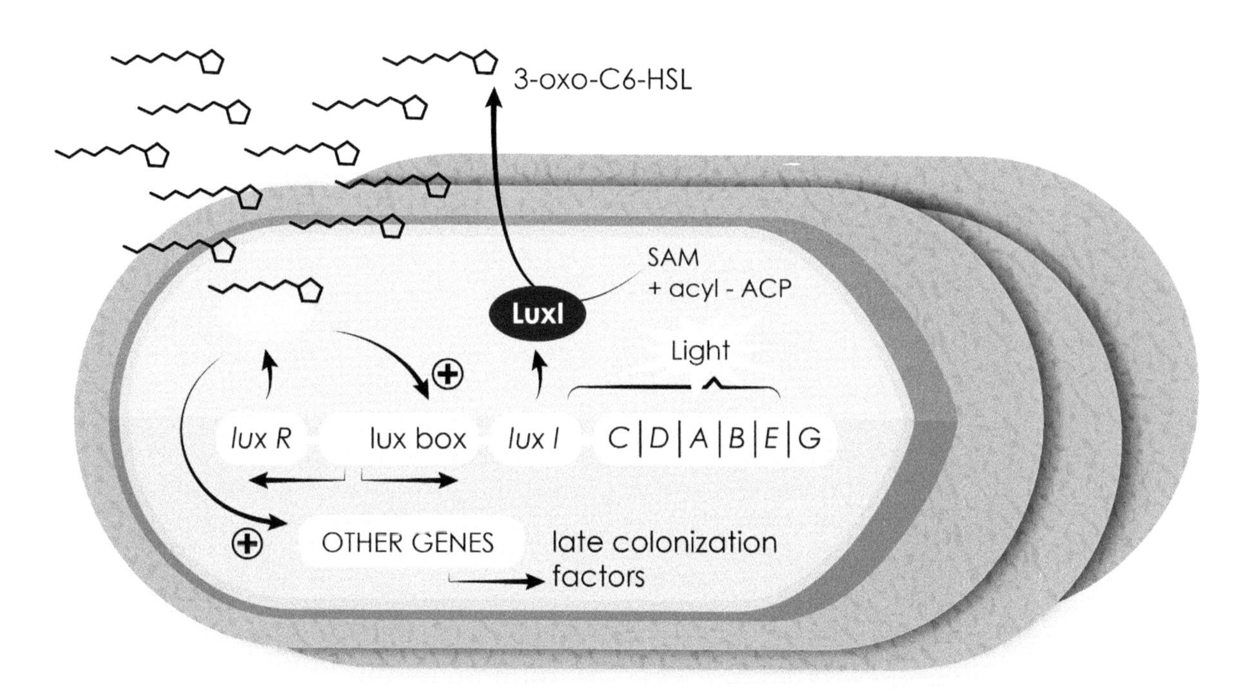

FIGURE 1.1 A schematic representation of the first discovered *luxI/luxR*-based quorum sensing system in the model species *Vibrio fischeri*, producing 3-oxo-C6-HSL as an autoinducer. Since then, a second quorum sensing system has been discovered. Based on the AinS AI synthase, it permits the liberation of C8-HSL. At low cell density, autoinducer concentration is low, while at high cell densities, autoinducers induce cytoplasmic cascades that lead to drastic genetic modifications, including transcription of genes responsible for bacterial bioluminescence.

AIs, which induce light emission (see previous paragraph). At the end of the night, most of the bacterial cells are expulsed from the light organ, leading to a dramatic reduction in bacterial concentration and of this diffusible factor. Thus, in the *V. fischeri–E. scolopes* symbiosis, the lowest production of bioluminescence is observed just before dawn to early afternoon. This coincides with the onset of environmental light (Lee and Ruby 1994). During the day, the concentrations of *V. fischeri* cells that have not been expulsed are very low, the diffusible factor is not produced and the squid does not glow.

However, this remaining population of *V. fischeri* grows steadily under favorable conditions within the squid throughout the day and at night again reaches a cell abundance that is sufficient to produce bioluminescence (Boettcher et al. 1996; Heath-Heckman et al. 2013).

A complex and specific dialog occurs between *V. fischeri* cells and the *E. scolopes* host, given that first, the *V. fischeri* cells are typically present at a concentration of less than 0.1% of the total bacterial population in the Hawaiian waters (Lee and Ruby 1994), and second, the motility of these bacterial cells is required to bring the symbionts toward the pores, the entrance of the luminescent organ in formation. Two main mechanisms were found to initiate the interaction (Visick and McFall-Ngai 2000): (i) close contact between the surfaces of the host and symbiont cells through receptor–ligand interactions and (ii) the creation of an environment in which only *V. fischeri* is viable. Receptor–ligand dynamics, often more generally referenced as microbe-associated molecular patterns (MAMPs) (Koropatnick 2004), can also be essential elements underlying the onset, maturation and persistence of mutualistic animal–microbe partnerships. Different data provided evidence that at least a portion of the host response is mediated by lipopolysaccharide-binding proteins from the LBP/BPI protein family (Chun et al. 2008; Krasity et al. 2011) and peptidoglycan-recognition proteins (PGRPs) (Troll et al. 2010). Also, studies were published concerning the complete annotated genome of *V. fischeri* (Ruby et al. 2005) and the cDNA expression libraries for colonized and uncolonized *E. scolopes* light organs (Chun et al. 2006). Numerous gene-encoding proteins known to be essential for both development and symbiosis were identified, such as reflectin, actin, myeloperoxidase, aldehyde dehydrogenase and nitric oxide synthase (Chun et al. 2008). These findings confirm the molecular dialogue between host squid and bacterial symbionts at cell surfaces. Comparison of host and symbiont population transcriptomes at four times over the day–night cycle revealed maximum expression of cytoskeleton related genes just before dawn, concordant with the daily effacement of the host epithelium and a cyclic change in the anaerobic metabolism of the symbionts (Wier et al. 2010). These host epithelium effacement and change in symbiont metabolism are clearly synchronized with the daily expulsion of most of the bacterial population (Boettcher et al. 1996; Ruby and Asato 1993). It is well known that during the colonization of the host tissue, the expression of sets of bacterial genes can be under the control of specific transcriptional regulators (Cotter and DiRita 2000), mainly described in bacteria that initiate pathogenic or benign infections (van Rhijn and Vanderleyden 1995). Interestingly, a mutant study showed that the gene *litR*, essential for the induction of luminescence, also plays a role as a transcriptional regulator in modulating the ability of *V. fischeri* to colonize juvenile squid (Fidopiastis et al. 2002).

1.2.1.5 *V. fischeri*: Conclusions

V. fischeri is now a well-known marine model in experimental biology. This first example clearly reveals how a marine bacterial strain, which at first sight appears to have a very particular mode of life (a bacterium associated with Hawaiian species), is in fact a universal model to explore mechanisms relevant to many diverse scientific fields and is at the origin of major discoveries in biology.

1.2.2 Picocyanobacteria as Models to Explore Photosynthetic Adaptations in the Oceans

Cyanobacteria are, evolutionarily speaking, very old organisms capable of producing oxygen that have significantly contributed to shape the current composition of the atmosphere. Their bioenergetic mechanisms are unique, as complex electron transfers (photosynthesis and respiration) occur in the same cell compartment. Among these organisms, the marine picocyanobacteria *Prochlorococcus* and *Synechococcus* genera provide detailed examples of photosynthetic adaptations to light conditions in the oceans. Beyond the description of unique photosynthetic mechanisms, the study of these marine cyanobacteria is key to better understanding the evolutionary origins of photosynthesis.

1.2.2.1 Key Features of *Prochlorococcus* and *Synechococcus*

The global chlorophyll biomass of oceanic ecosystems is dominated by tiny unicellular cyanobacteria of the *Prochlorococcus* and *Synechococcus* genera (~1 and 0.6 μm diameter, respectively), which are thought to account for ~25% of the global marine primary productivity (Flombaum et al. 2013). They are considered the smallest but also the numerically most abundant photosynthetic organisms on Earth, with estimations of 1.7×10^{27} cells in the World Ocean. *Prochlorococcus* and the marine *Synechococcus* diverged from a common ancestor ~150 million years ago, and the *Prochlorococcus* radiation delineates a monophyletic lineage within the complex *Synechococcus* group. Marine *Synechococcus* strains are indeed a more ancient and diverse radiation, which is usually divided into three subclusters, the major one (5.1) being subdivided into ~15 other important clades that include ~35 subclades (Farrant et al. 2016; Mazard et al. 2012). Despite their close relatedness, these two cyanobacteria have quite different ecophysiological features, as they occupy complementary though overlapping ecological niches in the ocean. *Prochlorococcus* strains are confined to the warm 45°N to 40°S latitudinal band and are very abundant in the subtropical gyres and the Mediterranean Sea but are absent from the high-latitude, colder waters. *Prochlorococcus* cell concentrations are often less important in coastal areas than offshore. By contrast, *Synechococcus* cells are detected in almost all marine environments outside of the polar circles and can be considered as the most widespread cyanobacterial genus on Earth.

Since the discovery of marine *Prochlorococcus* and *Synechococcus* only some decades ago, much progress has been made in the study of their biology. Marine picocyanobacteria have been prime targets for whole-genome sequencing projects, and more than 100 complete genomes are now available, spanning a large range of ecological niches and physiological and genetic diversity. These studies have revealed that *Prochlorococcus* is a striking example of an organism that has undergone genome "streamlining" (Dufresne et al. 2005), an evolutionary process thought to have rapidly followed the divergence from the common ancestor with *Synechococcus* and which resulted in an rapid specialization in oligotrophic marine niches. Thus, some *Prochlorococcus* isolates have a genome as small as 1.65 Mb (~1700 genes), and this cyanobacterium is often considered as approaching the near-minimal set of genes necessary for an oxygenic phototroph. The study of the *Synechococcus* genomes is more complex because of the large microdiversity of the radiation. They are on the whole bigger (2–3 Mb; 2500–3200 genes) than *Prochlorococcus* ones and, by contrast, show a relatively small range of variation in their characteristics among strains (Dufresne et al. 2008). Interestingly, the number of "unique genes", that is, the genes that are found only in one genome, is well correlated with the whole genome size. Like in *Prochlorococcus*, most of these unique genes are located in variable regions called genomic islands, whose size, position and predicted age are highly variable among genomes. This suggests that horizontal transfer of genetic material is an important process in these picocyanobacteria. Overall, the *Synechococcus* core genome includes ~70 gene families that are not present in *Prochlorococcus*, suggesting a higher diversity of metabolic processes, in line with the greater diversity of marine niches colonized by *Synechococcus* (Scanlan et al. 2009).

1.2.2.2 Different Adaptive Strategies of *Prochlorococcus* and *Synechococcus* to Light

The accumulation of (meta)genomic information has triggered the beginning of a thorough analysis of the relationships between the picocyanobacterial genotypes, phenotypes and different marine environments. In particular, the study of *Prochlorococcus* and *Synechococcus* has allowed much progress in the understanding of the selective pressures that drive the evolution of the oxygenic photosynthetic process at all scales of organization, from genes to the global ocean. Light quantity and quality are among the main drivers of photosynthesis, both showing great variability in the oceans. In tropical oligotrophic areas, the sunray angle and water transparency lead the photic zone to extend much deeper compared to higher latitudes and in turbid coastal waters. Moreover, seawater absorbs and scatters wavelengths in a selective way. Long wavelengths such as red light are absorbed within the first meters, whereas blue-green light can penetrate more deeply. In shallow coastal areas, water often carries large amounts of particulate matter that further alter the underwater light quality, inducing the presence of a green-yellow light. Successful adaptation of phototrophs to the multifaceted behavior of light in the aquatic systems notably relies on the nature and composition of the light-harvesting systems, and *Prochlorococcus* and *Synechococcus* have adopted drastically different strategies.

1.2.2.3 Adaptation of the Photosynthetic Apparatus of *Prochlorococcus*

The most-reviewed example is probably the manner by which *Prochlorococcus* modified its photosynthetic apparatus during evolution (Ting et al. 2002). Most cyanobacteria on Earth have a photosynthetic antenna consisting of a giant pigmented protein complex, called the phycobilisome. By contrast, *Prochlorococcus* is one of the rare cyanobacteria that uses membrane-intrinsic chlorophyll-binding proteins, termed Prochlorophyte-chl-binding (Pcb) proteins. Thus, most genes encoding phycobilisome components have been lost during the *Prochlorococcus* genome streamlining. As *Prochlorococcus* uses chlorophyll *b* as an accessory pigment in its atypical antenna complex, it efficiently harvests blue light, the dominant wavelength in oligotrophic and deep waters. As a result, *Prochlorococcus* populations extend deeper in the water column than almost any other phototrophs, basically defining the deepest limit of photosynthetic life in the World Ocean. The ability of *Prochlorococcus* to thrive in the entire euphotic zone also largely relies on its microdiversity, as this cyanobacteria features genetically and photophysiologically distinct populations (Biller et al. 2014). These so-called high-light and low-light ecotypes partition themselves down the water column along the light irradiance decreasing gradient. One of the main known physiological differences between *Prochlorococcus* light ecotypes is their major light-harvesting complexes, which comprise different sets of the Pcb proteins associated either with photosystem I or II, resulting in higher chl b to chl a ratio in the low-light ecotypes (Partensky and Garczarek 2010). Nevertheless, we still know very little about the differential pigmentation and function of the different Pcb proteins, especially regarding the photoprotective processes. More physiological and biochemical work is needed on this topic (Figure 1.2).

1.2.2.4 Adaptation of the Photosynthetic Apparatus of *Synechococcus*

A second interesting example is the way picocyanobacteria deal with the large variations in light spectral quality that occur along the horizontal (i.e. coastal-oceanic) gradients in the oceans. In contrast to *Prochlorococcus*, marine *Synechococcus* use phycobilisomes to harvest light, which consist of three classes of stacked phycobiliproteins. The phycobilisome core, made of allophycocyanin (APC) and connected to the photosystems, is surrounded by rods constituted of phycocyanin (PC) and/or phycoerythrin (PE). Each phycobiliprotein has a much-conserved hexameric cylindrical structure, binding one or several tetrapyrrolic chromophore (phycobilin) types: the blue phycocyanobilin (PCB), the red phycoerythrobilin (PEB), and the orange phycourobilin (PUB).

During their evolution, marine *Synechococcus* have developed an amazing variety of pigmentations by exploiting the modular nature of phycobilisomes, elaborating rods with variable pigment composition. Thus, three main pigment types can be distinguished based on the phycobiliprotein

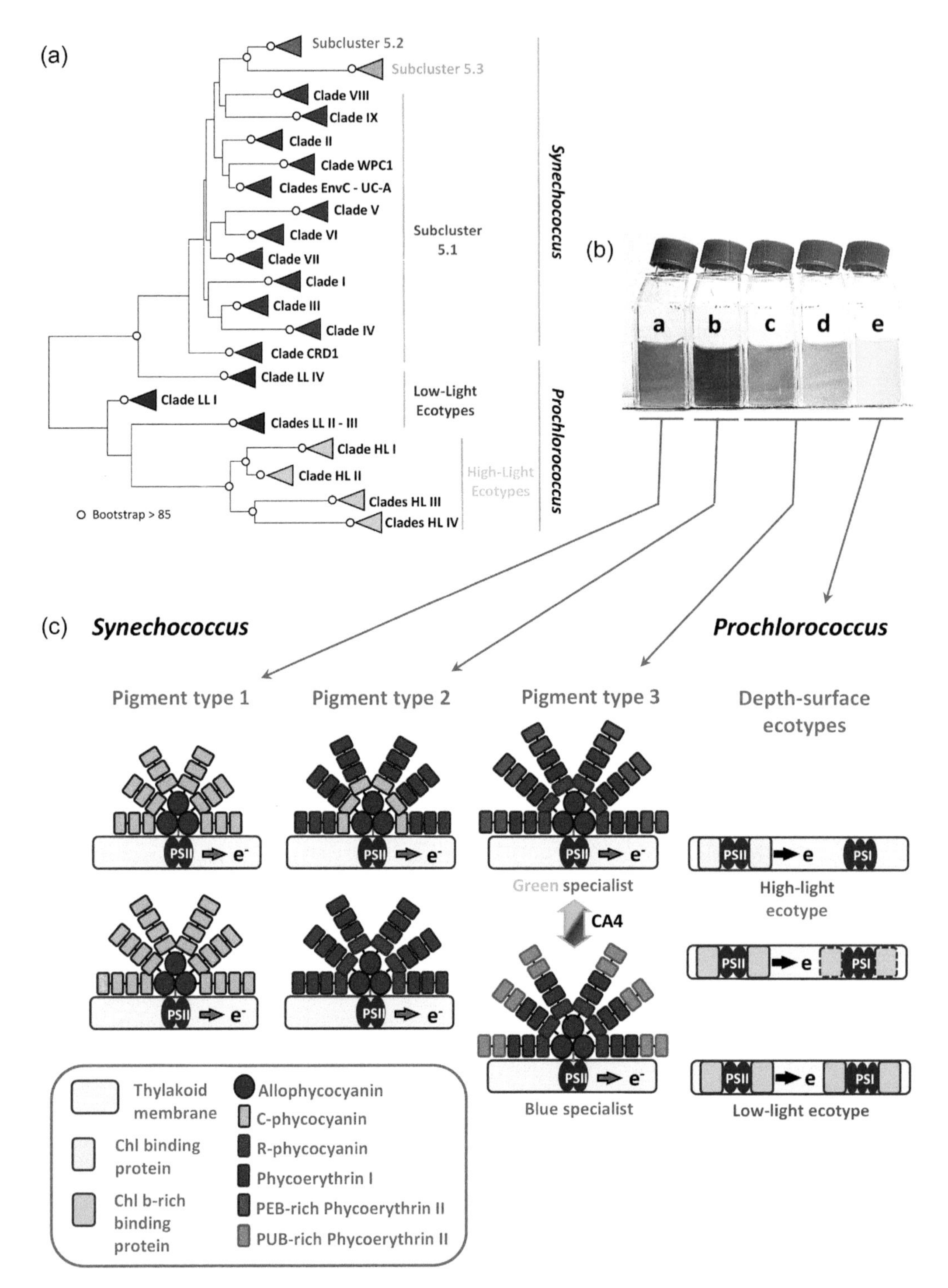

FIGURE 1.2 (a) Phylogenetic diagram (neighbor joining) showing the main marine picocyanobacterial lineages. Circled nodes are supported by bootstrap values higher than 85, and the other nodes are not well resolved; for further information. Subcluster 5.1 is the most diversified *Synechococcus* group. In contrast to *Synechococcus*, *Prochlorococcus* global phylogeny shows a microdiversification dependent on the light niche. (b) Batch cultures of *Synechococcus* spp. strains RS9917 (a), WH7805 (b), WH7803 (c), WH8102 (d) and *Prochlorococcus* sp. MED4 (e), illustrating different pigment types and their corresponding photosynthetic antenna system. (c) *Synechococcus* pigment type 1 includes C-phycocyanin rich strains with phycobilisome rods of different lengths, and pigment type 2 includes strains with one phycoerythrobilin (PEB)-rich phycoerythrin and either a C- or R-phycocyanin. *Synechococcus* pigment type 3 strains use the most sophisticated phycobilisome, including R-phycocyanin and two types of phycoerythrins with different possible proportions of PEB and phycourobilin (PUB), depending on the strain. Some strains can tune the PUB to PEB ratio through the chromatic acclimation (CA4) process. Strains of the different pigments are dispersed in the radiation and do not constitute clades, betraying the occurrence of horizontal transfer of phycobilisome related genes (see text). The represented structures of the phycobilisomes (homogeneity and number of rods, phycobiliproteins per rod, etc.) are putative. For *Prochlorococcus*, the antenna system is composed of Pcb proteins intrinsic to the thylakoidal membranes. High-light ecotypes can have a naked PSI, while low-light ecotypes may have additional Pcbs around it, sometimes inducible upon certain conditions. ([a] Mazard et al. 2012; Farrant et al. 2016.)

and phycobilin content of the phycobilisome rods. Pigment type 1 contains only phycocyanin, binding solely the orange light-absorbing phycocyanobilin (A_{MAX} = 620 nm), and is restricted to coastal, low-salinity surface waters, characterized by a high turbidity, inducing the dominance of orange wavelengths in the water. Pigment type 2 strains use PC and one type of PE binding PEB, the green-light absorbing pigment (A_{MAX} = 550 nm), and inhabit transition zones between brackish and oceanic environments with intermediate optical properties. Finally, pigment type 3 strains possess PC and two types of PEs (PE-I and PE-II), a feature specific of marine *Synechococcus* cyanobacteria. The PEs of pigment type 3 strains bind both PEB and the blue light-absorbing PUB (A_{MAX} = 495 nm) in various ratios depending on the strain, thus defining "green light specialists" (low PUB:PEB) and "blue light specialists" (high PUB:PEB) strains. Accordingly, these strains are found over large gradients from onshore mesotrophic waters, rich in green wavelengths, to offshore oligotrophic systems, where blue light is dominant. Overall, at least a dozen of optically different phycobiliproteins have been elaborated by marine *Synechococcus* during their evolution (Six et al. 2007), and there is no doubt that this is partly responsible for their global ecological success.

The genomic comparison of strains representative of these pigment types revealed that most genes involved in the biosynthesis of phycobilisome rods are located in a large (up to ~30 kb) specialized region of the genome, generally predicted to be a genomic island. The gene content and organization of this region is specific to each pigment type, independently from the strain phylogenetic position, and shows a tremendous increase in phycobilisome gene complexity from pigment types 1 to 3, the latter type being a more recent structure and the most sophisticated phycobilisome known so far. Together with the presence of phycobilisome genes in metaviriome datasets, this suggests that genes related to the phycobilisome rod region can be laterally transferred between *Synechococcus* lineages and that this might be a key mechanism facilitating adaptation of these lineages to new light niches.

Finally, there exists another particularly interesting *Synechococcus* pigment type that consists of strains capable of a unique type of chromatic acclimation (CA4), a reversible process that modifies the composition of the phycobilisomes. The strains capable of CA4 are pigment type 3 strains able to dynamically tune the PUB to PEB ratio of their phycobilisome, which becomes low under green light and high under blue light to precisely match the ambient light quality. CA is therefore predicted to increase fitness in conditions of changing light colors, allowing the harvesting of more photons than for strains with fixed pigmentation. Comparative genomic analyses of marine *Synechococcus* strains showed that the CA4 process is possible thanks to a specific small genomic island that exists in two slightly different versions, named CA4-A and CA4-B (Humily et al. 2013). The recent implementation of methods for plating and genetic manipulations such as the disruption and/or overexpression of CA4 genes in marine *Synechococcus* has allowed us to start deciphering the regulation of the CA4-A process. Thus, in the model strain *Synechococcus* sp. RS9916, isolated in the Red Sea, the CA-4 process involves chromophore switch systems at three phycoerythrin cysteines, which are regulated by the two transcription factors FciA and FciB (Sanfilippo et al. 2019). Thanks to the setup of genetic transformation methods, CA is one of the physiological processes that has been more closely studied in the laboratory in picocyanobacteria.

Using phycobiliprotein and CA4 genetic markers, the study of the extensive metagenomic Tara Oceans dataset allowed us to determine that, globally, CA4-A and CA4-B strains account for 23% and 19% of all *Synechococcus*, respectively (Grébert et al. 2018). Interestingly, CA4-A cells predominated in the nutrient-rich, temperate or cold waters found at high latitudes and in upwelling areas, while CA4-B cells were most abundant in warm, nutrient-poor waters. The reason there exist two types of CA4 genomic islands is, however, still not clear, and the functioning of the CA4-B genomic island is under investigation.

1.2.2.5 Picocyanobacterial Models: Conclusions

Picocyanobacteria (meta)genomics has greatly increased our understanding of the genomic and phenotype variations existing among these organisms, which is tightly linked to processes of niche specialization. In particular, these studies have unveiled unprecedented information on how photosynthetic complexes may drastically evolve in the oceans to fit different light niches. In this context, it is worth noting that the strength of the picocyanobacterial model is not restricted to one model organism but rather consists in a large panel of many strains that allow the understanding of the evolution of major processes like photosynthesis in the oceans. To better understand the relationships between picocyanobacterial genotypes and phenotypes, further progress requires a significant development of experimental work on the numerous picocyanobacteria strains available in culture. In this context, the development of culture axenization methods adapted to picocyanobacteria is a real necessity. Compared to other microbial models, thorough and advanced physiological studies are still scarce, and today, targeted studies of gene function should be prioritized over the overaccumulation of non-characterized genetic information. The recent development of genetic manipulation techniques on the *Synechococcus* sp. RS9916 strain gives much hope, but this will be particularly challenging for *Prochlorococcus*.

1.2.3 *Zobellia galactanivorans*, a Model for Bacterial Degradation of Macroalgal Biomass

Macroalgae and their associated microbiota provide a large diversity of enzymes, in particular involved in the degradation of many diverse types of sugars, which are also of major interest for industry. Numerous economic sectors rely on the production of efficient enzymes and are continuously searching for innovative ones. For example, alginate lyases have many applications for food and pharmaceutical companies. We will see in this section that the bacterium *Z. galactanivorans*, associated with macroalgae, is an excellent model to study the diversity and the functioning of these enzymes. Working on this bacterial model also provides interesting insights into the mechanisms of colonization of algal surfaces and degradation of macroalgal organic matter.

1.2.3.1 Key Features of *Zobellia galactanivorans*

Green, red and brown macroalgae (also known as seaweeds) are dominant primary producers in coastal regions, often locally exceeding phytoplankton and other benthic carbon fixers (Duarte et al. 2005). Seaweeds thus represent an important reservoir of organic matter and are considered a global carbon sink. The composition of macroalgal biomass is unique and consists of >50% of polysaccharides that differ from those known in terrestrial plants by the nature of their monosaccharide units and the presence of sulfated motifs and other substituents (Ficko-Blean et al. 2014). Turnover of this biomass is mostly mediated by marine heterotrophic bacteria that can colonize macroalgae and access, degrade and remineralize the algal compounds. Studies of the mechanisms underlying the interactions of these bacteria with macroalgae and their degradation pathways are therefore crucial to understanding coastal ecosystems' nutrient cycles and discovering novel enzymatic functions.

Members of the class *Flavobacteriia* (phylum *Bacteroidetes*) are recognized as key players in the degradation of marine algal polysaccharides (Thomas et al. 2011b). Among them, the cultivated species *Z. galactanivorans* has become over the past 20 years an environmentally relevant model organism to investigate macroalgal biomass degradation. Both cultivation and metagenomic approaches frequently detect members of the genus *Zobellia* in algae-dominated habitats and directly on the surface of seaweeds from different oceanic basins (Hollants et al. 2013; Nedashkovskaya et al. 2004). In particular, *Z. galactanivorans* Dsij^T was first isolated in November 1988 in Roscoff (France) from a live specimen of the red macroalga *Delesseria sanguinea* (Potin et al. 1991) and later described as the type strain of the genus *Zobellia* (Barbeyron et al. 2001). Cells are Gram-negative and rod-shaped with rounded ends (0.3–0.5 × 1.2–8.0 µm). *Z. galactanivorans* is chemoorganotroph with a strictly aerobic respiratory metabolism. Colonies on agar plates are yellow-orange due to the biosynthesis of non-diffusible flexirubin-type pigments. Cells do not possess flagella and cannot swim in liquid medium. On solid surfaces, they exhibit gliding motility at ca. 1–4 $\mu m.s^{-1}$.

1.2.3.2 An Extraordinary Set of Enzymes Made *Z. galactanivorans* a Bacterial Model for the Use of Algal Sugars

Z. galactanivorans Dsij^T has been extensively studied for its ability to use a wide array of macroalgal compounds as sole carbon and energy sources, including agars and carrageenans from red algae, as well as alginate, laminarin, mannitol and fucose-containing sulfated polysaccharides from brown algae. Recently, it was also shown to directly degrade fresh tissues of the kelp *Laminaria digitata*, corroborating its efficiency for macroalgal biomass turnover (Zhu et al. 2017). Annotation of its 5.5-Mb genome revealed that up to 9% of its gene content could be dedicated to polysaccharide utilization (Barbeyron et al. 2016). This includes genes encoding an impressive number of 142 glycoside hydrolases (GHs) and 17 polysaccharide lyases (PLs), representing 56 different functional carbohydrate active enzyme families (CAZymes), together with 37 carbohydrate-binding modules as described in the CAZy database (Lombard et al. 2014). These enzymes are accompanied by 18 carbohydrate esterases and 71 sulfatases of the S1 family, which can remove substituents from polysaccharides. These genes are often clustered in regions of the *Z. galactanivorans* genome termed polysaccharide utilization loci (PULs). PULs are frequently found in *Bacteroidetes*. They encode a suite of proteins dedicated to the utilization of a given polysaccharide, generally comprising (i) CAZYmes responsible for the breakdown of the substrate, (ii) substituent-removing enzymes, (iii) substrate-binding membrane proteins, (iv) transporters for oligosaccharides and (v) transcriptional regulators that control the PUL expression, depending on substrate availability. In particular, *Z. galactanivorans* Dsij^T harbors 71 tandems of SusC-like TonB-dependent transporter (TBDT) and SusD-like surface glycan-binding protein (SGBP) that are considered hallmarks of PUL genomic organization (Grondin et al. 2017).

Over the years, numerous biochemical and structural studies have focused on the in-depth characterization of *Z. galactanivorans* proteins dedicated to polysaccharide utilization. In September 2020, the function of 42 of these proteins was experimentally validated, and for half of them, the crystallographic 3D structure was solved. This notably includes enzymes targeting agars (Naretto et al. 2019), porphyrans (Hehemann et al. 2010), carrageenans (Matard-Mann et al. 2017), laminarin (Labourel et al. 2015), alginate (Thomas et al. 2013), mannitol (Groisillier et al. 2015) and hemicellulose (Dorival et al. 2018). In several instances, studies of *Z. galactanivorans* proteins led to the discovery of novel CAZY families [e.g. iota-carrageenases GH82 and α-1,3-L-(3,6-anhydro)-galactosidase GH117 (Rebuffet et al. 2011)] or to novel activities in existing families [e.g. exolytic α-1,3-(3,6-anhydro)-D-galactosidases in GH127 and GH129 (Ficko-Blean et al. 2017)].

Furthermore, genome-wide transcriptomes of *Z. galactanivorans* Dsij^T cells grown with different carbohydrates are publicly available, either based on microarrays (Thomas et al. 2017) or RNA-seq (Ficko-Blean et al. 2017). This is complemented by a validated reverse transcription real-time quantitative PCR (RT-qPCR) protocol to specifically target genes of interest (Thomas et al. 2011a). These transcriptomic data revealed both substrate-specific and shared responses between co-occurring polysaccharides and helped define 192 operon-like transcription units. The upregulation of 35 predicted transcriptional regulators in the presence of algal polysaccharides compared to glucose gave further insights into the regulation strategies at play to fine-tune gene expression depending on the rapidly changing glycan landscape. This was recently exemplified by the characterization of the regulator AusR, a transcriptional repressor controlling the expression of the *Z. galactanivorans* alginolytic system (Dudek et al. 2020). In addition, genetic tools were adapted for *Z. galactanivorans*, including protocols for transposon random mutagenesis, site-directed mutagenesis and complementation (Zhu et al. 2017). Integration of all these complementary tools now opens the way for functional investigations of full catabolic pathways, as illustrated by studies on *Z. galactanivorans* alginate utilization system (AUS) (Thomas et al. 2012) and carrageenan utilization system (CUS) (Ficko-Blean et al. 2017). Both systems rely on complex regulons comprising genes within and distal to a PUL and encode the full set of proteins necessary to sense the substrates, degrade polysaccharides into their monosaccharide constituents and assimilate them into the central metabolism. Interestingly, site-directed mutants of the CUS unveiled (i) the complementary functions of two α-1,3-(3,6-anhydro)-D-galactosidases that were otherwise

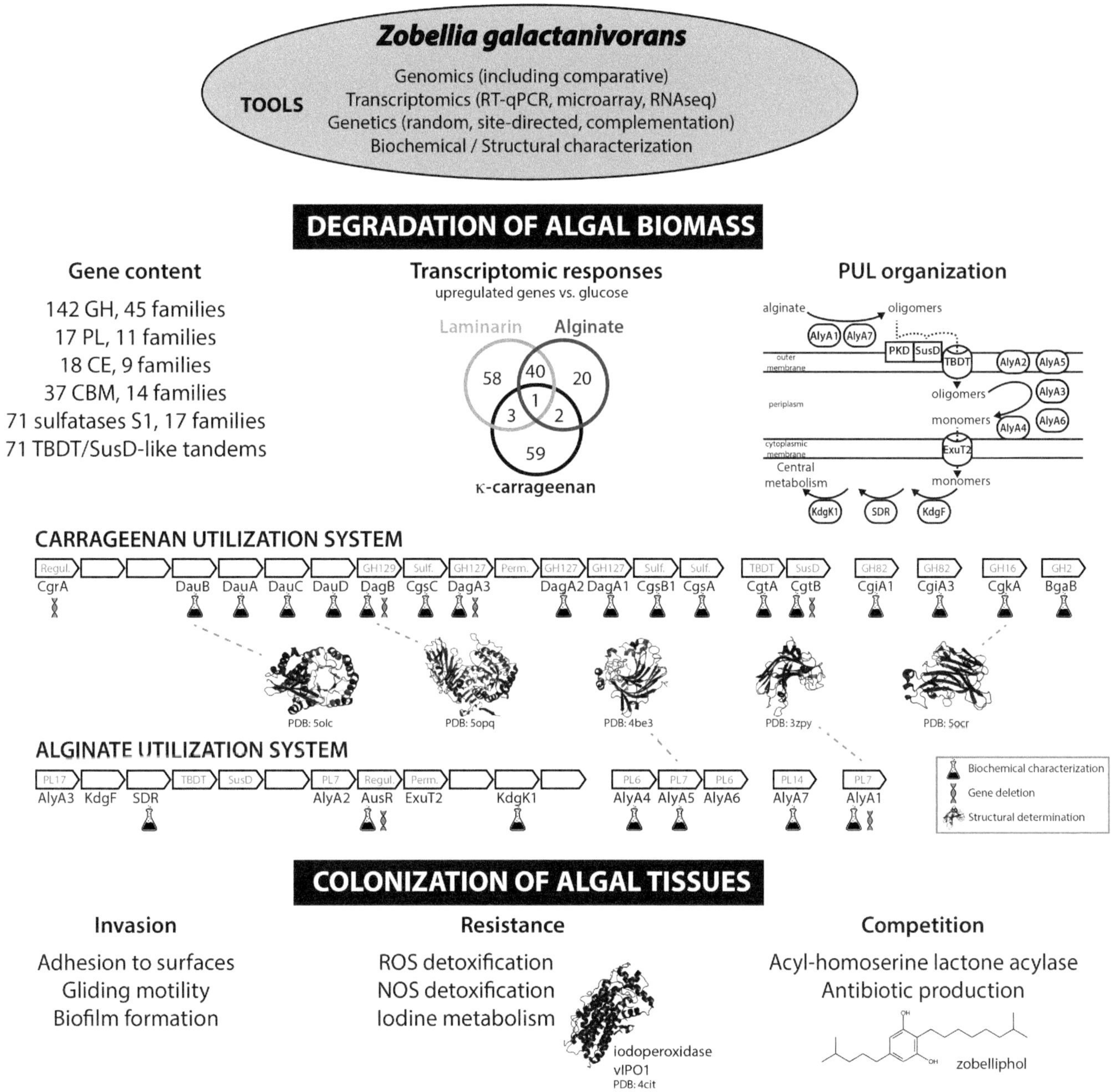

FIGURE 1.3 **Schematic view of the multifaceted model organism *Z. galactanivorans*.** The currently available experimental tools are listed, together with selected features that make *Z. galactanivorans* a useful model to investigate how marine bacteria degrade and colonize macroalgal biomass. The typical organization of polysaccharide utilization loci (PUL) is exemplified by the alginate utilization system. The genetic organization of the multi-loci carrageenan utilization system and alginate utilization system is shown, highlighting the number of proteins that have been characterized biochemically and structurally, as well as deletion mutants analyzed so far. For protein structures, the PDB accession ID is given. (Available on www.rcsb.org/.)

indistinguishable based on *in vitro* biochemical assays and (ii) the role of a distal TBDT/SusD-like tandem that was absent from the main carrageenolytic locus. These results highlight the benefit of genetic tools in a bacterial model to assess gene functions *in vivo*. Studies on *Z. galactanivorans* also provided insights into the genomic exchange of polysaccharide degradation pathways between closely and distantly related bacteria by horizontal gene transfers (HGTs). This includes acquisitions by *Z. galactanivorans* of specific genes (e.g. alginate lyase AlyA1, endoglucanase EngA) from marine *Actinobacteria* and *Firmicutes* (Zhu et al. 2017; Dorival et al. 2018) and transfers of flavobacterial PULs to marine *Proteobacteria*, as well as several iconic examples of diet-mediated HGT into gut bacteria of Asian populations (Hehemann et al. 2012) (Figure 1.3).

1.2.3.3 A Model to Study Bacterial Colonization of Algal Surfaces

Besides polysaccharide degradation, *Z. galactanivorans* is a relevant model to study other adaptations to macroalgae-associated lifestyle, such as surface colonization and resistance against algal defenses. First, its gliding motility and rapid spread on surfaces might aid in colonizing the algal thallus. Flow-cell chamber experiments showed that *Z. galactanivorans* can grow as thick biofilms (up to 90 µm), a capacity which is maintained or even increased in the presence of algal exudates (Salaün et al. 2012). Second, *Z. galactanivorans* possesses multiple enzymes predicted to cope with the reactive oxygen and nitrogen species produced by macroalgae as defense mechanisms. This includes superoxide dismutases, peroxidases, glutathione reductases, thioredoxins, thioredoxin reductases, peroxiredoxins and NO/N_2O reductases. Third, *Z. galactanivorans* features an iodotyrosine dehalogenase and biochemically active iodoperoxidases (Fournier et al. 2014) and accumulates up to 50 µM of iodine, two orders of magnitude higher than typical oceanic concentrations. This distinct iodine metabolism likely participates in the resistance against the high iodine concentration in algal cell walls and the stress-induced release of halogenated compounds. Finally, *Z. galactanivorans* strain OII3 produces a novel secondary metabolite of the dialkylresorcinol (DAR) family, named zobelliphol, with anti-microbial activity against Gram-positive bacteria (Harms et al. 2018). This compound could therefore help *Z. galactanivorans* compete with other epiphytic bacteria. It is also possible that zobelliphol acts as an antioxidant and/or signaling molecule, similar to other DAR derivatives. In line with this, *Z. galactanivorans* encodes a putative acyl-homoserine lactone acylase that might degrade communication molecules produced by competing bacteria and interfere with their quorum sensing.

1.2.3.4 *Z. galactanivorans*: Conclusions

Collectively, all these features reveal that *Z. galactanivorans* is a multifaceted model organism to investigate how marine bacteria colonize and degrade macroalgal biomass. Such studies can improve our understanding of nutrient cycles in coastal areas but also uncover novel activities with promising biotechnological applications. Considering that marine organisms represent an immense potential reservoir of bioactive compounds, such bacterial models are essential to characterize innovative molecules of interest for biotechnology but also to understand their ecological roles.

1.2.4 *Marinobacter hydrocarbonoclasticus*, a Model Bacterium for Biofilm Formation, Lipid Biodegradation and Iron Acquisition

The degradation of hydrocarbons is a bacterial activity of major industrial and environmental interest. Few microorganisms, one of which is *Marinobacter hydrocarbonoclasticus*, are able to efficiently degrade such compounds. Interestingly, and above the primary interest focused on hydrocarbon degradation, we will see in this section that this bacterium is also an excellent model to investigate the mechanisms of biofilm formation and iron acquisition, which are two universal and key features of microbial physiology.

1.2.4.1 Key Features of *Marinobacter hydrocarbonoclasticus*

Bacteria of the genus *Marinobacter*, to date composed of 57 species, are widespread in marine environments. They have been detected in the deep ocean, coastal seawater, marine sediment, hydrothermal settings, oceanic basalt, sea ice, solar salterns and oilfields, as well as in association with animal or algal hosts. These bacteria are Gram-stain-negative, rod-shaped, motile, mesophilic, halotolerant, heterotrophic and aerobic. The genus was first described with the type strain *M. hydrocarbonoclasticus* SP17 (hereafter MhSP17), which was isolated from sediments of the Mediterranean Sea near a petroleum refinery (Gauthier et al. 1992). Later, the strain *M. aquaeolei* VT8 (MhVT8) was isolated from the produced water of an offshore oil well and was recognized as a heterotypic synonym of *M. hydrocarbonoclasticus* (Huu et al. 1999; Márquez and Ventosa 2005). Since then, *M. hydrocarbonoclasticus* strains became models for studying biofilm formation on lipids and alkanes as a strategy to assimilate these insoluble substrates, production and storage of wax esters and iron acquisition through the synthesis of the siderophore petrobactin.

1.2.4.2 Biofilm Formation on Nutritive Surface and Alkane Degradation

MhSP17 exhibits a remarkable ability to grow on nearly water-insoluble compounds like long-chain alkanes (up to 32 carbons atoms), triglycerides, fatty acids and wax esters (Klein et al. 2008; Mounier et al. 2014). The water-insolubility of these substrates impairs their assimilation by bacterial cells. Growth on water-insoluble compounds can only be achieved by way of physiological and/or behavioral adaptations enabling rapid mass transfer of the substrate from the non-water-dissolved state to the cell. Biofilm formation is a widespread strategy to assimilate non-dissolved substrates, as observed, for instance, on cellulose, chitin and hydrocarbons (Sivadon et al. 2019). These biofilms develop on so-called nutritive interfaces since they play both the role of substrate and substratum. This feature distinguishes them from conventional biofilms growing on inert supports, such as minerals, metals or plastics. MhSP17 forms a biofilm at the interface between the aqueous phase and substrates that can be solid (saturated triacylglycerol, long-chain alkane, fatty acids, fatty alcohol and wax esters) or liquid (medium-chain alkane and unsaturated triacylglycerol). MhSP17 substrates also differ by the localization of their metabolism. Triglycerides must be hydrolyzed by a secreted lipase before entering the cell, whereas alkane metabolism is purely intracellular. The ability of MhSP17 to form biofilms on a variety of substrates exhibiting different physical properties or involving different metabolisms makes this bacterium a valuable model for studying biofilms on nutritive surfaces.

During biofilm formation on alkanes or triglyceride, MhSP17 cells undergo profound changes in gene expression, indicating a reshaping of the physiology of biofilm cells (Mounier et al. 2014; Vaysse et al. 2011, 2009). Interestingly, a great part of the genes modulated during biofilm formation was of unknown function, leading to potential for the discovery of new cellular functions. The role of some of these genes, like the alkane transport system AupA-AupB, has been elucidated by constructing mutants deleted of genes detected in omics analyses (Mounier et al. 2018). An extracellular matrix of biofilm developing on a nutritive surface is viewed as an external digester improving the solubilization of the substrate (Sivadon et al. 2019). This matrix function was documented in MhSP17 with the demonstration that the matrix contained extracellular factors involved in triglycerides and alkanes assimilation (Ennouri et al. 2017). A random mutational analysis led to the identification of a di-guanylate cyclase that is important for biofilm formation on alkane.

1.2.4.3 Biosynthesis and Accumulation of Wax Esters

The strains MhVT8 and, to a lesser extent, MhSP17 are also used as models for the biosynthesis of wax esters. Production and storage of neutral lipids such as wax esters and triacylglycerols are encountered in few marine bacterial genera like *Alcanivorax* and *Marinobacter*. This process is believed to be a survival strategy that allows bacteria to store energy and carbon to thrive in natural environments where nutrient availability fluctuates (Alvarez 2016; Manilla-Pérez et al. 2010). Wax esters are formed by the esterification of a fatty alcohol and an activated fatty acid. The length and desaturation degree of the fatty acid and the fatty alcohol moieties of wax esters confer on them diverse physicochemical properties that are of great interest in the industries of cosmetics, high-grade lubricants, wood coatings, antifoaming agents, printing inks, varnishes and food additives (Miklaszewska et al. 2018). Wax esters are nowadays mostly industrially produced from fossil fuels. The more sustainable production of wax esters by microbial cells from wastes is currently the object of intensive research and requires the utilization of model systems like *M. hydrocarbonoclasticus*. Strains MhVT8 and MhSP17 naturally accumulate high yields of wax esters. The two key enzymes of the biosynthesis of wax esters are the fatty acyl reductase (FAR) and the wax synthase (WS), which produce wax esters from coenzyme A (CoA) or acyl carrier protein (ACP) activated fatty acids. MhSP17 and MhVT8 possess four and five WS genes, respectively (Lenneman et al. 2013; Petronikolou and Nair 2018). Enzymatic properties of FAR and WS from *Marinobacter* strains have been extensively studied, leading to engineering efforts to alter their substrate specificity. The heterologous expression of these enzymes in hosts like *Arabidopsis thaliana* or yeasts led to the successful production of wax esters (Wenning et al. 2017; Vollheyde et al. 2020).

1.2.4.4 Iron Acquisition

In oceans, remineralization into CO_2 of the organic carbon released by marine phototrophs occurs mostly through the respiration of heterotrophic bacteria (Buchan et al. 2014). A great part of the heterotrophic activity resides in the particulate fraction of the organic carbon consisting of aggregated compounds (mostly proteins, polysaccharides and lipids) that are colonized by biofilm-forming bacteria (Benner and Amon 2015). Metal availability, particularly iron, is expected to have a strong impact on organic carbon remineralization since respiration is a highly iron-demanding process, the respiratory chain alone containing approximately 94% of the cellular iron (Tortell et al. 1999). Iron acquisition by marine heterotrophic bacteria is thus a fundamental matter to understand the recycling of organic carbon in marine environments.

MhSP17 and MhVT8 have been used as models to study iron acquisition in marine environments. MhVT8 was shown to produce three siderophores: the petrobactin and its sulfonated and disulfonated forms, while in MhSP17 culture, only petrobactin and the monosulfonated derivative were detected. The role of these sulfonations and the pathways leading to their formation are unknown. Moreover, petrobactin exhibits a typical property of marine siderophores, the photoreactivity of the ferric-complex, which causes the release of soluble Fe(II) and results in a petrobactin photoproduct that retains the capacity to complex Fe(III) (Barbeau et al. 2003). The biological significance of this photoreactivity is still not understood. Nevertheless, it might influence the iron uptake mechanism and consequently the biogeochemical cycling of iron in marine environments. It is without any doubt that the use of models that are genetically trackable will be an asset for elucidating the various mechanism facets of petrobactin and its derivatives.

1.2.4.5 Genomics and Genetics of *M. hydrocarbonoclasticus*

The genomes of MhSP17 and MhVT8 encode for 3803 and 4272 proteins, respectively. As expected for two strains from the same species, genomes of MhVT8 and MhSP17 have a great number of genes in common, their core genome consisting of 3041 genes (80% identity, 80% coverage). However, due to different sites of isolation and likely different evolutionary history, the genomes of these strains are not identical, MhVT8 and MhSP17 having 1348 and 742 strain-specific genes (80% identity, 80% coverage), respectively. In addition, MhVT8 harbors two plasmids, pMAQU01 and pMAQU02, encoding for 213 and 201 proteins, respectively, while MhSP17 does not carry any plasmid (Singer et al. 2011).

The genomic potential of *M. hydrocarbonoclasticus* strains suggests the utilization of a large variety of substrates as terminal electron acceptors, which is consistent with their occurrence in diverse environments and the multiple lifestyles, planktonic and biofilm-forming, of this species (Singer et al. 2011). One striking feature of the *M. hydrocarbonoclasticus* genomes is their high content in genes involved in the metabolism of the second messenger: bis (3'-5') cyclic dimeric guanosine monophosphate (c-di-GMP). MhSP17 and MhVT8 harbor 83 and 80 genes, respectively, encoding either diguanylate cyclases with GGDEF domains that synthesize c-di-GMP or phosphodiesterases with EAL or HD-GYP domains that hydrolyze c-di-GMP. The c-di-GMP signaling pathway controls, in particular, the switch between the sessile and biofilm mode of life. The presence of a large number of c-di-GMP-related genes suggests that *M. hydrocarbonoclasticus* lifestyles are under the control of different c-di-GMP regulatory circuits that are activated in response to multiple environmental conditions.

More details about genetic tools are provided in Section 1.3 of this chapter, but specific details about *Marinobacter* strains are provided here. In MhSP17, gene transfer has been proved successful only by conjugation, using the transfer system based on the conjugating plasmid RP4 with an *Escherichia coli* donor strain expressing the transfer functions. The *M. hydrocarbonoclasticus* receiving strain was JM1, a streptomycin-resistant derivative of MhSP17 that enables counter selection in conjugation experiments. The introduction of suicide plasmids that are unable to replicate in JM1 enables random mutagenesis using mini-Tn5 transposon and site-directed mutagenesis by allele exchange. Gene addition in MhSP17 has been achieved using the transposon vector min-Tn7 that has an integration site in the MhSP17 chromosome. This has been used to express green fluorescent protein constitutively to follow fluorescent cells under fluorescence microscopy. Plasmids with the replication origin of pBBR1 are stably maintained in JM1, and the P_{BAD} promoter was shown to be functional. This offers the possibility to introduce, maintain and express genes in MhSP17 for complementation tests or any physiological studies requiring the controlled expression of a gene (Ennouri et al. 2017; Mounier et al. 2018).

1.2.4.6 *Marinobacter hydrocarbonoclasticus*: Conclusions

The specific features of MhSP17 and MhVT8, such as biofilm formation on lipids and alkanes, accumulation of wax esters and production of siderophores, together with the availability of genetic tools, make them valuable models to study carbon and iron cycles in the ocean as well as to implement biotechnological processes for the production of lipids of industrial interest. In this sense, this example of a bacterial model has many common features with *Zobellia galactanivorans*, previously developed, and demonstrates that marine bacterial models are extremely valuable for the exploration of fundamental biological mechanisms but also for the rapidly expanding field of blue biotechnology.

1.3 THE BACTERIAL MODEL ORGANISM TOOLKIT

Although the four bacterial models presented have emerged from different labs, with the aim to answer diverse scientific questions concerning different biological mechanisms, they were developed thanks to a common toolkit consisting of optimized protocols for isolation and culture, genetic manipulation and phenotypic characterization. These key tools are under constant evolution and will be presented in this next section, beginning with the development of novel isolation methodologies essential for the discovery of original models and the establishment of strain collections. Then we will describe how classical genetic manipulation protocols allow the production of mutants to directly target key mechanisms of interest in bacterial models and present the state of the art genome editing CRISPR-Cas technology. Finally, we will see how recent omics approaches complement the characterization of bacterial models and pave the way for innovative phenotyping methods.

1.3.1 Innovative Techniques for the Isolation of New Bacterial Models: Culturing the Unculturable

The decision to develop a new bacterial model may be motivated by a lack of current models that are representative of the target species and/or a particular function they carry out in the environment. The selection of this model necessarily goes through a stage of isolation and culture in the laboratory in order to fully study its phenotype and genotype or to construct mutants. However, it is well known that isolated bacteria represent only a small fraction of the total bacterial diversity and that the culturability of environmental bacteria, is very low, ranging from less than 0.001% in seawater to about 0.3% in soils (Rappé and Giovannoni 2003). Even in the era of 'meta-omic' techniques, the objective of isolating and cultivating uncultivated bacteria remains a high priority in microbiology. This phenomenon is referred to as "the great plate count anomaly", and there are many hypotheses that could explain it: (i) some bacteria do not tolerate high concentration of nutrients; (ii) organic substrates present in culture media are inappropriate for growth; (iii) important specific vitamins or growth factors are missing in the culture media; (iv) a nutritional shock is induced by an uncontrolled production of oxygen reactive species (substrate-accelerated death); (v) growth inhibition by antagonistic interaction of other species (antibiosis); (vi) some species dependent on cell–cell communication cannot grow in the absence of chemical signals from other cells; (vii) growth of some bacteria is too slow to be detected; and (viii) unadapted pressure, O_2 concentration or inappropriate culture method (solid vs. liquid). Based on these hypotheses, different strategies can be tested to improve the isolation and cultivation of more bacterial species, especially those most abundant in the natural environment, as they could constitute interesting laboratory models.

The first strategy is to modify the culture environment and the conditions for growth. Conventional growth media are very rich in nutrients because they were originally designed for human pathogens well adapted to this type of environment. A first step is to reduce organic matter concentrations in order to favor oligotrophic species. In particular, members of Alphaproteobacteria have been shown to grow preferentially on nutrient-poor media (Senechkin et al. 2010). The reduction of the organic carbon concentration is, however, constrained: if growth is detected by observing colonies on solid media or a visible cloud in liquid media with the naked eye, a sufficient concentration of organic carbon is necessary, which would remain much higher than that of natural environments. Other studies have proposed to add peroxidase (an enzyme catalyzing the decomposition of hydrogen peroxide), to replace agar with gellan gum in solid culture media (Gelrite or Phytagel) (Tamaki et al. 2005) or to autoclave phosphate and agar in culture media separately (Kato et al. 2020, 2018). These changes could reduce the generation of hydrogen peroxide compared to conventionally prepared agar media and significantly increase the diversity of cultivable bacteria. It is also possible to complement the culture medium with components that stimulate growth, such as trace elements similar to those found in the environment, siderophores (e.g. pyoverdines-Fe, desferricoprogen), quorum sensing molecules (e.g. acylhomoserine lactone) or the supernatant of cultures of other species that stimulate the growth of others (Bruns et al. 2002; Tanaka et al. 2004). Metagenomic analysis of environmental samples can even unveil specific metabolic properties used by target non-cultivated bacteria or, inversely, the absence of genes indicating auxotrophy for certain elements that will be added to the culture medium to improve their isolation.

A second strategy is based on microculture and micromanipulation techniques. The first step consists of depositing cells from the environment on a polycarbonate membrane and then setting the membrane on a pad impregnated with nutrients or sterilized sediment. Nutrients can diffuse through the polycarbonate membrane and allow cell growth with the formation of microcolonies after a few days of incubation (Ferrari et al. 2008). Microcolonies can be observed by inverted microscopy and removed from the membrane by microdissection using ultrasound waves generated by a piezoelectric probe (Ericsson et al. 2000). Microcolonies can then be sampled using a glass capillary and transferred in tubes or microplate wells for cultivation separately from other microcolonies. This stage of microculture can then facilitate cell culture in a richer environment. This technique, however, remains tedious and requires specialized instruments.

A third strategy is to isolate single cells and try to grow them individually in order to obtain microcolonies formed of a pure culture. The separation of single cells could favour the growth of rare species, as it prevents direct competition. Obtaining microcolonies can be a first step to larger growth. Individualized cells can be grown in hundreds of diffusion chambers (called iChips) that are placed *in situ* in natural (e.g. sediments, soils) or simulated natural environments for the influx of natural compounds (Bollmann et al. 2010; Sizova et al. 2012; Van Pham and Kim 2014). These culture chambers are separated from the outside environment by semipermeable membranes of 0.03 µm, allowing fluxes of nutrients and signal molecules but preventing contamination by other microorganisms (Berdy et al. 2017; Nichols et al. 2010). This approach has been used to isolate a bacterial species producing a new antibiotic of interest (Ling et al. 2015). Another, more sophisticated approach is to encapsulate environmental bacteria into gel microdroplets (GMDs) (Liu et al. 2009), which are then placed in a chemostat fed by the nutrients extracted from the sampling environment (Zengler et al. 2002). This system also allows the transfer of communication molecules between GMDs. The GMDs in which a microcolony has developed, *a priori* consisting of a clonal culture, can then be separated by cell sorting using flow cytometry, followed by cultivation attempts. This device is attractive but expensive and complicated to implement and does not guarantee the long-term culturability of the selected cells.

A final strategy to cultivate environmental bacteria is the dilution-to-extinction technique. This approach emerged in the mid-1990s (Button et al. 1993) and was further developed in the 2000s (Connon and Giovannoni 2002; Stingl et al. 2007). It consists of performing serial dilutions of the samples using sterile natural sampling water or media on microplates or tubes to isolate one or a few cells in a single microchamber. The main benefit of this technique is to allow a slow and gradual adaptation (incubation for several weeks) of the bacterial cells in conditions that mimic the natural environment studied. Cell density is monitored by epifluorescence microscopy or flow cytometry counts, allowing even weak growth to be detected. In addition, the very low number of cells reduces the possibility of target uncultivated strains being overgrown and inhibited by opportunistic bacteria that may overgrow and inhibit slow growers of interest. The main drawback of this approach is the lack of interactions between cells of different species which could inhibit growth, as mentioned previously. While time consuming, this technique enabled the first-time isolation of many previously uncultured bacteria such as SAR11 or the oligotrophic marine gammaproteobacteria (OMG) that dominate marine ecosystems (Cho and Giovannoni 2004; Rappé et al. 2002; Stingl et al. 2007). The isolated species are mainly oligotrophic, and most of them fail to grow in a richer culture medium. Nevertheless, adaptations in the composition of the growth medium can allow for cultures to attain a fairly high biomass, as was the case for the model oligotrophic marine bacterium *Pelagibacter ubique* (Carini et al. 2013). Furthermore, additional improvements to artificial media allowed the cultivation of more than 80 new isolates belonging to abundant marine clades SAR116, OM60/NOR5, SAR92, *Roseobacter* and SAR11 (Henson et al. 2016). These authors recently expanded their collection to include members of the SAR11 LD12 and Actinobacteria acIV clades and other novel SAR11 and SAR116 strains by combining a large-scale three-year dilution to extinction campaign and modelling of taxon-specific viability variation to further refine their experimental cultivation strategy (Henson et al. 2020).

1.3.2 Genetic Manipulation of Marine Bacteria

To fully exploit a model organism, it is important to develop molecular genetics tools to be able to elucidate the functions of genes, study and modify gene expression and engineer modified organisms for biotechnological applications. The manipulation of the strain of interest may be approached using forward or reverse genetics, depending on the research question.

Forward genetics is used when researchers are interested in a particular phenotype and seek to understand the genetic basis for this phenotype and is particularly useful for genetically intractable organisms. Either natural mutants can be studied or mutations can be induced by random mutagenesis, using chemicals or UV radiation, and then the mutations are subsequently mapped to determine the genes affected. This method was used to study the process of magnetosome formation in the magnetotactic bacterium *Desulfovibrio magneticus* for which genetic tools were not available (Rahn-Lee et al. 2015). The random mutagenesis toolkit was enhanced with the discovery, in the 1940s–1950s, of mobile DNA elements known as transposons, or "jumping genes", that can insert randomly into genomes, thus creating mutations. Transposons were used for the mutagenesis of a marine archaeon (Guschinskaya et al. 2016) and marine bacteria (Ebert et al. 2013; McCarren and Brahamsha 2005; Zhu et al. 2017). This method is particularly suited to large-scale studies of genes of unknown function, as demonstrated by (Price et al. 2018), who generated thousands of mutant phenotypes from 32 species of bacteria.

In contrast to forward genetics, reverse genetics is based on modification of a target gene by deletion or insertion, for example, followed by the characterization of the mutant phenotype. Reverse genetics usually requires *a priori* knowledge of the genomic context and has been facilitated in the past 10 years owing to the increasing number of full genome sequences available (Zeaiter et al. 2018) and with the wealth of information provided by oceanic metagenomic datasets (Rusch et al. 2007; Sunagawa et al. 2015; Biller et al. 2018). Reverse genetics requires first a method to transfer foreign DNA into the target cells and then strategies for genome editing, shuttle vector and promoter design and the choice of selectable and counter-selectable markers. The toolkit can be expanded to include reporter system design to allow selection of mutated organisms or to follow gene expression. Gene inactivation is achieved primarily by homologous recombination either mediated by plasmids using the endogenous recombination machinery of the host or, more recently, by using phage recombination systems, also known as recombineering (Fels et al. 2020). Plasmid-mediated homologous recombination requires the use of traditional cloning approaches to incorporate into the plasmid vector the modified target gene with relatively long (1–2 kb) flanking homologous sequences (homology arms) that will be the site for allelic exchange for the first cross-over event. Use of a non-replicating plasmid, under antibiotic selection, forces integration of the plasmid into the host genome via a first cross-over event. However, to achieve gene replacement, a second cross-over event must occur, and these rare double-recombination events must be selected for out of the vast majority of single recombination clones that would be extremely time consuming. A strategy to promote a second cross-over event was first established with a temperature-sensitive replicon (Hamilton et al. 1989) and was later improved with the development of suicide plasmids with counter-selectable markers encoding conditional lethal genes. One of the most widely used counter-selectable markers is *sacB*, which confers sensitivity to sucrose (Gay et al. 1985) and is lethal for cells that have not undergone a second recombination to eliminate the plasmid. This strategy was used to study the role of a specific enzyme thought to be involved in alginate digestion in the model *Zobellia galactanivorans* by creating a deletion mutant of an alginase lyase gene (Zhu et al. 2017).

The more recently developed methods known as recombineering, for recombination-mediated genetic engineering, integrate linear single-stranded DNA, oligonucleotides or double-stranded DNA fragments into the target genome in cells expressing the bacteriophage λ-encoded recombination proteins (see Fels et al. 2020 for a review). Recombineering offers significant advantages over plasmid-mediated methods, since it avoids laborious *in vitro* cloning techniques, only short homology arms are required and the recombination efficiency is high. Although this method is commonly used to engineer model organisms such as *E. coli*, it has been challenging to adapt to other bacteria outside of closely related enterobacteria, since the existing phage recombination systems are not efficient in all species (Fels et al. 2020). Current research is aimed at discovering new single-stranded annealing proteins that will be able to promote recombination of ssDNA in a wider range of bacteria (Wannier et al. 2020).

For all the gene editing approaches mentioned, the final hurdle for successful genome editing is the transfer of the recombinant DNA into the target strain. DNA transfer is known to occur naturally in bacteria through transformation and conjugation (Paul et al. 1991; Chen et al. 2005) and transduction (Jiang and Paul 1998) and is the mechanism for horizontal gene transfer in bacteria. Natural competence is mediated by proteins that enable the penetration of extracellular DNA, such as type IV pili or type 2 secretion systems. For example, some cyanobacterial strains (*Synechococcus* sp. PCC 7002) and many *Vibrio* strains (including isolates related to *V. parahaemolyticus*, *V. vulnificus*, *V. fischeri*) are naturally competent (Frigaard et al. 2004; Simpson et al.

2019). In such cases, transformation protocols appear relatively simple and rely on incubation of the targeted strain with the exogenous DNA. Various factors can affect the efficiency of natural transformation, such as plasmid concentration, cell density, light conditions and pre-treatment of cells (Zang et al. 2007). For example, the natural competence of some *Vibrio* strains is induced by chitin, a biopolymer abundant in aquatic habitats, originating, for example, from crustacean exoskeletons (Meibom et al. 2005; Zeaiter et al. 2018). Tools have been developed to transform cells that are not naturally competent by artificially creating pores in the bacterial cell wall. The first artificial method to induce competency is chemical transformation, whereby treatments with salt solutions create pores in the cell membranes that allow DNA penetration into the cytoplasm. Calcium chloride, diméthylsulfoxyde, polyéthylene glycol and lysozyme are among the chemical compounds used to prepare competent cells or to improve the efficiency of other types of transformation protocols. A few positive reports of chemical transformation of marine bacteria were published. This includes transformation of *Rhodobacter sphaeroides* and *Vibrio natriegens* (Fornari and Kaplan 1982). In the latter case, it was necessary to use a *V. natriegenes* strain mutated for the chromosomal Dns endonuclease to avoid the expression of a resistance mechanism (Weinstock et al. 2016). However, several failures of chemical transformation protocols applied to marine strains were reported. For example, no transformants were obtained after testing chemical transformation protocols on 12 different *Roseobacter* strains (Piekarski et al. 2009). In general, chemical transformation does not appear to be a very efficient approach to transform marine bacteria (Zeaiter et al. 2018). The second method to induce competency is by electroporation, one of the most efficient tools to introduce DNA, particularly plasmid DNA, into a bacterial strain. This technique consists of the application of a brief electrical current to facilitate DNA uptake by a bacterial cell. Indeed, a brief pulse of 5–10 kV/cm increases cell membrane permeability and allows the production of transformants. Marine strains belonging to diverse taxonomic groups were successfully transformed using these protocols, such as strains of *Roseobacter*, *Vibrio*, *Pseudoalteromonas*, *Caulobacter*, *Halomonas* and some cyanobacteria (Zeaiter et al. 2018). However, the electric treatment applied to the cells is harsh and induces large cell mortality and many transformation failures. Indeed, many factors can influence the success of an electroporation protocol, including cell concentration, the composition of the growth medium and buffer composition, temperature, voltage of electroporation systems, plasmid size and topology (Zeaiter et al. 2018). In particular, the presence of salts is one of the most influential factors on electroporation efficiency. Therefore, careful development is needed to find the best medium for electroporation of marine strains, which require high concentrations of salts for growth.

Conjugation of bacterial strains is, together with electroporation, used more often to manipulate marine strains. Conjugation is the only method of transfer that requires cell-to-cell contact, whereby a donor (usually *E. coli*) transfers various types of mobile elements, including plasmids, transposons and integrons. One of the advantages, compared to other methods, is the capacity to transfer large amounts of genetic material. Another advantage is that conjugation involves single-strand DNA, which avoids bacterial resistance mechanisms (restriction systems) of the receptor strain. Conjugative transfer is a complex process that requires the concerted action of many gene products. The mobile element to be transferred needs to contain an origin of transfer *oriT*, and the conjugative process in itself is mediated by the transfer regions *tra*. If the donor strain possesses *tra* regions, it can directly transfer the mobile element into the recipient strain via bi-parental conjugation. When the donor strain lacks these regions, a third helper strain is needed to provide conjugative ability via tri-parental conjugation. After the conjugative transfer, donor and recipient cells will both carry the mobile element. Therefore, the selection of transconjugants is a critical step after conjugation to ensure complete removal of donor (and helper) strains. This can be achieved by using selective growth conditions (e.g. salinity, temperature) favoring the growth of marine strains over donors (usually *E. coli*). Alternatively, the mobile element can encode antibiotic resistance genes controlled by promoters that function in the recipient strain but not in the donor strain. In addition, donor strains auxotrophic for a specific compound can be used. In this case, selection occurs on a culture medium devoid of the compound.

Transduction is an efficient method of transfer of DNA from a bacteriophage to a bacterium and was successfully used to transfer genes into cultivated marine isolates and natural bacterial communities (Jiang et al. 1998). However, it is not used as widely as conjugation and transformation as a DNA delivery method, since phages generally have a limited host range, and therefore requires the careful selection of suitable phages for the target bacteria strain.

1.3.3 The Future of Gene Editing in Bacterial Models: The CRISPR-Cas Approaches

One of the most recent additions to the genetic engineering toolbox is the CRISPR-Cas technology, also known as "molecular scissors", that allows the precise cutting of DNA at specific target sites by a Cas endonuclease, guided by a short RNA sequence known as a guide RNA (sg-RNA). The CRISPR (clustered regularly interspaced short palindromic repeats)-Cas system is an adaptive immune system in prokaryotes, defending the cell against invasion by bacteriophages or extrachromosomal elements (Barrangou et al. 2007; Bolotin et al. 2005). The CRISPR loci, present in prokaryote genomes but not those of eukaryotes or viruses (Mojica et al. 2000; Jansen et al. 2002), contain short DNA repeats separated by spacer sequences, known as protospacer sequences, that correspond to fragments of the foreign DNA that are stored as a record in the CRISPR array. Although many different CRISPR-Cas systems have been discovered (Koonin et al. 2017), the most commonly used system for genome editing is based on the CRISPR-Cas9 from *Streptococcus pyogenes* and belongs to the CRISPR type II family. It functions by transcription of the repeat-spacer element to precursor CRISPR RNA (pre-crRNA), which, following base-pairing with a trans-activating cr-RNA (tracr-RNA), triggers processing of the structure to mature crRNA by RNAse III in the presence of Cas9 (Jinek et al. 2012). Site-specific cleavage of the foreign DNA by Cas9 only occurs (i) if there is complementary base-pairing between the cr-RNA and the protospacer and (ii) if the protospacer is adjacent to a short, sequence-specific region known as the protospacer adjacent motif (PAM) (Jinek et al. 2012). The sequence-specific cutting of the target DNA to create a double-stranded break (DSB) led the authors to realize the immediate potential of this mechanism for repurposing into a genome engineering tool, optimized further with the creation of single chimeric targeting RNA, a single guide RNA(sgRNA) to replace the cr-RNA:tracr-RNA duplex (Jinek et al. 2012). CRISPR technology revolutionized genome engineering in eukaryotes due to the ease of designing sgRNAs to guide the nuclease to the genome editing site, the efficiency of the Cas endonucleases and the possibility to scale up to multiple gene edits (see Hsu et al. 2014 and Pickar-Oliver and Gersbach 2019 for reviews). Whereas eukaryotes can use the error-prone non-homologous end-joining (NHEJ) system to repair DSBs, leading to small insertions or deletions, the majority of bacteria lack this pathway, making DSBs lethal. Although there are a number of hurdles to employing CRISPR in bacteria (Vento et al. 2019), CRISPR-Cas9 editing was successful in *E. coli* (Bassalo et al. 2016) and industrially important bacteria such as *Lactobacillus reuteri* (Oh and van Pijkeren 2014), *Bacillus subtilis* (Westbrook et al. 2016) and *Streptomyces* species (Alberti and Corre 2019). Considering the importance of streptomycete bacteria for the production of antimicrobials, several CRISPR plasmid toolkits have been developed for genome editing of *Streptomyces* (Alberti and Corre 2019). Examples of the application of these tools include the activation of novel transcriptionally silent biosynthetic gene clusters (BGCs) by knocking out known, preferentially or constitutively expressed BGCs (Culp et al. 2019) or the increase of their expression by "knocking in" constitutive promoters (Zhang et al. 2017). CRISPR is not limited to gene editing but can also be used to study gene repression or "knockdown" with CRISPR interference (CRISPRi). CRISPRi uses an engineered catalytically inactive (or dead) Cas9 protein (dCas9), which, instead of cutting the DNA, represses transcription of the target gene by steric interference. This approach presents several advantages, including the ease to knock down multiple genes and induction and tuning of gene repression, and requires less effort than the creation of multiple gene deletions. It has been employed for gene repression in diverse bacteria such as *Streptomyces* (Tong et al. 2015), *Synechococcus* (Knoot et al. 2020) and *B. subtilis* (Westbrook et al. 2016). More recently, the type V-A Cas protein, Cas12a (Cpf1) (Koonin et al. 2017), is showing promise for CRISPR editing (Yan et al. 2017) or interference in bacteria (Li et al. 2018) and can be a useful alternative for when Cas9 toxicity is observed, as was the case in *Streptomyces* (Li et al. 2018). Cas12a presents some advantages over Cas9, since it can enable multiplex genome editing, and the production of staggered cuts instead of blunt ends by this endonuclease promotes homology-directed repair via the provision of a repair template (Paul and Montoya 2020). And, last, an alternative CRISPR system which circumvents the difficulties of repairing DSBs carries out DSB-free single-base editing using a fusion protein of a Cas9 variant, Cas9 nickase (Cas9n). This strategy allowed efficient multiplex editing in *Streptomyces* strains that was not possible with the standard CRISPR-Cas9 system (Tong et al. 2019) and single-base editing in *Clostridium* (Li et al. 2019).

1.3.4 Phenotyping and Acquiring Knowledge on Model Strains

When the bacterial model has been isolated and preserved in appropriate conditions, and when collections of mutants have been prepared (see Section 1.3.2) to explore the role of various targeted genes and functions, the following step is to characterize in depth the model strain. Traditional phenotyping methods are still widely used in microbiology laboratories, including catabolic profiling on different nutrient sources, evaluation of growth parameters in various conditions (i.e. biofilm vs. liquid) and determination of cell shape or movements via microscopic techniques. This is especially relevant when comparing wild-type strains with mutants to evidence the role of the knocked-out genes. These traditional techniques are now complemented by the recent development of "Omics" tools providing an immense potential in model strain characterization.

First, whole-genome analysis of individual strains provides a comprehensive view of cell physiology capacities, which is an essential step when establishing a new bacterial model. Additionally, the development of genetic tools relies on thorough and precise information about gene organization and regulation in the target strain raised as a model. Accurate lists of genes, gene annotations and transcriptomic and proteomic datasets, as well as the existence of computational platforms for data integration and systems-levels analysis, are among the essential criteria to establish bacterial models (Liu and Deutschbauer 2018). An increasing quantity of genomic data for isolated strains are now available. These genomes are available in various types of databases (not specifically marine), such as that maintained by the Joint Genome Institute (JGI) Genome Portal (https://genome.jgi.doe.gov/portal/) or the one maintained by the Genoscope in France (https://mage.genoscope.cns.fr/microscope/home/index.php). For cyanobacteria, especially marine picocyanobacteria, specific databases that include genome exploration tools are available, such as Cyanobase (http://genome.kazusa.or.jp/cyanobase) and Cyanorak (http://application.sb-roscoff.fr/cyanorak/). In some databanks, one important difficulty is that many genomes are still incomplete and published as "draft genomes", which can limit their utilization in genetic approaches.

The availability of numerous complete and annotated bacterial genomes in databanks facilitates the choice of the genes to knock out when starting targeted mutagenesis approaches, which is an essential step when building an isolated strain as a model of interest. Also, the existence of many genome sequences provides potential insights into bacterial metabolic pathways: genome mining of marine strains allows the putative identification and characterization of novel biosynthetic pathways (which will have then to be confirmed by other types of experimental approaches, i.e. the preparation of collections of mutants) that are responsible for the production of bioactive compounds and the identification of physiological traits that were not suspected before. Then, comparative genomics approaches may allow the comparison of specific characteristics, even in phylogenetically closed strains. For example, comparative genomics revealed that choline metabolism is widespread among marine *Roseobacter.* Choline is an abundant organic compound in the ocean and, through its conversion to glycine betain, serves as an osmoprotectant in many marine bacteria. This molecule is also an important component of membranes (phosphatidylcholine). However, the genetic and molecular mechanisms regulating intracellular choline and glycine betaine concentrations are poorly known in marine bacteria. Following comparative genomic analysis, a targeted mutagenesis of genes involved in choline metabolism was conducted in the model bacteria *Ruegeria pomeroyi* DSS-3. The authors of this study demonstrated the key role of the *betG* gene, encoding an organic solute transporter (essential in the uptake of choline) of the *betB* gene converting choline in glycine betaine and of the *fhs* gene encoding the formyl tetrahydrofolate synthetase, essential in the oxidization of the choline methyl groups and the catabolism of glycine betaine (Lidbury et al. 2015).

While genomic analysis provides a snapshot of the physiological potential of a model strain, transcriptomics gives insights into the functions that are expressed in a given experimental condition. In the cyanobacterial *Prochlorococcus* strain AS9601, transcriptomics approaches revealed some of the mechanisms responsible for adaptation to salt stress. Under hypersaline conditions (5% w/v), 1/3 of the genome is differentially expressed compared to lower salt conditions (3.8% w/v). In hypersaline conditions, higher transcript abundance was observed for the genes involved in respiratory electron transfer, carbon fixation, osmolyte solute biosynthesis and inorganic ion transport. By contrast, a reduction of transcript abundance was noticed for the genes involved in iron transportation, heme production and photosynthesis electron transport. Such analysis thus suggests interesting mechanisms linking light utilization and salt stress in this strain of *Prochlorococcus* (Al-Hosani et al. 2015).

Proteomics is the characterization of the protein content in a cell using mass spectrometry and nuclear magnetic resonance approaches. Following the central dogma of molecular biology (DNA→RNA→proteins), focusing on protein expression allows an overall characterization of the organism's physiology in a defined experimental condition. Indeed, the function of many proteins has been described, and proteomics now provides to researchers in-depth characterization of the microbial cell physiology. Proteomics studies were conducted on various marine prokaryotes, including different cyanobacteria (*Prochlorococcus*, *Synechococcus*), *Pseudoalteromonas*, Planctomycetes, Vibrios and others (Schweder et al. 2008). For example, the planktonic/biofilm transition was investigated using proteomics in the bacterial model *Pseudoalteromonas lipolytica* TC8. This study revealed that peptidases, oxidases, transcription factors, membrane proteins and enzymes involved in histidine biosynthesis were over-expressed in biofilms. In contrast, proteins involved in heme production, nutrient assimilation, cell division and arginine/ornithine biosynthesis were over-expressed in planktonic cells (Favre et al. 2019). Collectively, all these data provide insights into the mechanisms that are expressed in bacterial cells and responsible for their adaptation to a biofilm or a planktonic way of life.

Metabolomics is now another essential approach to explore the physiology of prokaryotic models and their interactions with the environment. This approach provides global metabolite profiles under a given set of experimental conditions and a snapshot of the physiological response of prokaryotic cells. One important difficulty and technical challenge in metabolomics is the identification and dosage of thousands of molecular compounds, sometimes at very low concentrations, for which no standard is available for rapid identification. Untargeted metabolomic approaches compare the whole metabolomes in a qualitative or semi-quantitative manner and without *a priori* knowledge about the type of metabolites produced, while targeted metabolomics focuses on a particular compound. During the last decade, important improvements in the sensitivity and resolution of the analytical tools required for metabolomic analysis were achieved, including in mass spectrometry and nuclear magnetic resonance approaches (Ribeiro et al. 2019). These improvements allowed for significant progress in the characterization and identification of various compounds, including carbohydrates, alcohols, ketones, amino acids and also several types of secondary metabolites like antibiotics, pigments and infochemicals. Metabolomics is still a science in its infancy but has begun to be used to characterize the response of marine bacterial models to environmental variations. The authors of the previously mentioned study on *Pseudoalteromonas lipolytica* TC8 also used metabolomics to characterize the planktonic/biofilm transition. Interestingly, they revealed drastic modifications in the lipid composition of the membranes (Favre et al. 2019). Phosphatidylethanolamine derivatives were abundant in biofilm cells, while ornithine lipids were more present in planktonic bacteria. Thus, this study, with others, highlights the need to focus on membrane plasticity mechanisms in the planktonic-to-biofilm transition when bacteria attach to surfaces, which remains an underexplored research question in marine bacterial models.

1.4 CONCLUSIONS

This chapter reveals, through the very different selected examples (*Vibrio*, *Prochlorococcus* and *Synechococcus*, *Zobellia* and *Marinobacter*), the interest and potential as well as the difficulties to establish marine bacterial strains as models for experimental biology. The isolation of bacterial strains of interest; their full characterization; the development of genetic tools and the maintenance of strain collections; the investment in genome sequencing, including accurate gene annotation; the phenotyping of mutants relying on OMIC approaches: all these steps are crucial in the establishment of new models. Clearly, it appears from this non-exhaustive list of technical approaches as well as from the collection of examples presented in this chapter that no universal experimental approach can be applied to develop a new marine bacterial model. However, unprecedented progress has been made this last decade in synthetic biology, molecular genetic tool development, the application of omics data techniques and computational tools, which undoubtedly paves the way to the development of new bacterial models of major interest to characterize many types of biological mechanisms. The potential and the outcomes of such work are immense, and applications are found in several fields. For example, recombinant marine *Synechococcus* allowed the production of polyunsaturated acids of medical interest (Yu et al. 2000), and recombinant strains of the marine *Vibrio natriegens* species contributed to the production of melanin (Wang et al. 2020). Bacterial models can also serve as tools for biology, like the model *Vibrio fischeri*, which serves as a biosensor to detect pollutants in diverse environmental samples (Farré et al. 2002; Parvez et al. 2006; Dalzell et al. 2002) and is often reported as one of the most sensitive assays compared to others across a wide range of chemicals. Overall, new marine bacterial models have the potential to address questions which cannot be assessed by 'traditional' bacterial models. Thus, many fundamental and applied research fields would greatly benefit from investing massively in the development of new bacterial models, including research in marine sciences, marine ecology, ecotoxicology and evolutionary studies but also 'blue' biotechnology.

ACKNOWLEDGEMENTS

François Thomas acknowledges support from CNRS and the French ANR project ALGAVOR (grant agreement ANR-18-CE02-0001-01). All authors thanks Haley Flom for English grammar and spelling.

BIBLIOGRAPHY

Alberti, F., and C. Corre. 2019. Editing *Streptomycete* genomes in the CRISPR/Cas9 age. *Natural Product Reports* 36(9):1237–1248.

Al-Hosani, S., M. M. Oudah, A. Henschel, and L. F. Yousef. 2015. Global transcriptome analysis of salt acclimated *Prochlorococcus* AS9601. *Microbiological Research* 176:21–28.

Alvarez, H. M. 2016. Triacylglycerol and wax ester-accumulating machinery in Prokaryotes. *Biochimie* 120:28–39.

Bachofen, R., and A. Schenk. 1998. Quorum sensing autoinducers: Do they play a rôle in natural microbial habitats? *Microbiology Research* 153(1):61–63.

Barbeau, K., E. L. Rue, C. G. Trick, K. W. Bruland, and A. Butler. 2003. Photochemical reactivity of siderophores produced by marine heterotrophic bacteria and cyanobacteria based on characteristic Fe(III) binding groups. *Limnology and Oceanography* 48(3):1069–1078.

Barbeyron, T., S. L'Haridon, E. Corre, B. Kloareg, and P. Potin. 2001. *Zobellia galactanovorans* gen. nov., sp. nov., a marine species of *Flavobacteriaceae* isolated from a red alga, and classification of [*Cytophaga*] *uliginosa* (ZoBell and Upham 1944) Reichenbach 1989 as *Zobellia uliginosa* gen. nov., comb. nov. *International Journal of Systematic and Evolutionary Microbiology* 51(Pt 3):985–997.

Barbeyron, T., F. Thomas, V. Barbe, H. Teeling, C. Schenowitz, C. Dossat, A. Goesmann, C. Leblanc, F.-O. Glöckner, M. Czjzek, R. Amann, and G. Michel. 2016. Habitat and taxon as driving forces of carbohydrate catabolism in marine heterotrophic bacteria: Example of the model algae-associated bacterium *Zobellia galactanivorans* DsijT. *Environmental Microbiology* 18:4610–4627.

Barrangou, R., C. Fremaux, H. Deveau, M. Richards, P. Boyaval, S. Moineau, D. A. Romero, and P. Horvath. 2007. CRISPR provides acquired resistance against viruses in Prokaryotes. *Science* 315(5819):1709–1712.

Bassalo, M. C., A. D. Garst, A. A. L Halweg-Edwards, W. C. Grau, D. W. Domaille, V. K. Mutalik, A. P. Arkin, and R. T. Gill. 2016. Rapid and efficient one-step metabolic pathway integration in *E. Coli*. *ACS Synthetic Biology* 5(7):561–568.

Benner, R., and R. M. W. Amon. 2015. The size-reactivity continuum of major bioelements in the ocean. *Annual Review of Marine Science* 7(1):185–205.

Berdy, B., A. L. Spoering, L. L. Ling, and S. S. Epstein. 2017. In situ cultivation of previously uncultivable microorganisms using the ichip. *Nature Protocols* 12(10):2232–2242.

Biller, S. J., P. M. Berube, K. Dooley, M. Williams, B. M. Satinsky, T. Hackl, S. L. Hogle, A. Coe, K. Bergauer, and H. A. Bouman. 2018. Marine microbial metagenomes sampled across space and time. *Scientific Data* 5:180176.

Biller, S. J., P. M. Berube, D. Lindell, and S. W. Chisholm. 2014. *Prochlorococcus*: The structure and function of collective diversity. *Nature Reviews Microbiology* 13(1):13–27.

Blount, Z. D. 2015. The natural history of model organisms: The unexhausted potential of *E. Coli*. *Elife* 4:e05826.

Boettcher, K. J., E. G. Ruby, and M. J. McFall-Ngai. 1996. Bioluminescence in the symbiotic squid *Euprymna scolopes* is controlled by a daily biological rhythm. *Journal of Comparative Physiology A* (179):65–73.

Bollmann, A., A. V. Palumbo, K. Lewis, and S. S. Epstein. 2010. Isolation and physiology of bacteria from contaminated subsurface sediments. *Applied and Environmental Microbiology* 76(22):7413–7419.

Bolotin, A., B. Quinquis, A. Sorokin, and S. D. Ehrlich. 2005. Clustered regularly interspaced short palindrome repeats (CRISPRs) have spacers of extrachromosomal origin. *Microbiology* 151(8): 2551–2561.

Bruns, A., H. Cypionka, and J. Overmann. 2002. Cyclic AMP and acyl homoserine lactones increase the cultivation efficiency of heterotrophic bacteria from the central Baltic Sea. *Applied and Environmental Microbiology* 68(8):3978–3987.

Buchan, A., G. R. LeCleir, C. A. Gulvik, and J. M. Gonzalez. 2014. Master recyclers: Features and functions of bacteria associated with phytoplankton blooms. *Nature Reviews Microbiology* 12(10):686–698.

Button, D. K., F. Schut, P. Quang, R. Martin, and B. R. Robertson. 1993. Viability and isolation of marine bacteria by dilution culture: Theory, procedures, and initial results. *Applied and Environmental Microbiology* 59(3):881–891.

Campbell, Z. T., T. O. Baldwin, and O. Miyashita. 2010. Analysis of the bacterial luciferase mobile loop by replica-exchange molecular dynamics. *Biophysical Journal* 99(12):4012–4019.

Carini, P., L. Steindler, S. Beszteri, and S. J. Giovannoni. 2013. Nutrient requirements for growth of the extreme oligotroph 'Candidatus *Pelagibacter ubique*' HTCC1062 on a defined medium. *The ISME Journal* 7(3):592–602.

Chen, I., P. J. Christie, and D. Dubnau. 2005. The ins and outs of DNA transfer in bacteria. *Science* 310(5753):1456–1460.

Cho, J.-C., and S. J. Giovannoni. 2004. Cultivation and growth characteristics of a diverse group of oligotrophic marine Gammaproteobacteria. *Applied and Environmental Microbiology* 70(1): 432–440.

Chun, C.-K., T. E. Scheetz, MdF. Bonaldo, B. Brown, A. Clemens, W. J. Crookes-Goodson, K. Crouch, T. DeMartini, M. Eyestone, M. S. Goodson, B. Janssens, J. L. Kimbell, T. A. Koropatnick, T. Kucaba, C. Smith, J. J. Stewart, D. Tong, J. V. Troll, S. Webster, J. Winhall-Rice, C. Yap, T. L. Casavant, M. J. McFall-Ngai, and M. Bento Soares. 2006. An annotated cDNA library of juvenile *Euprymna scolopes* with and without colonization by the symbiont *Vibrio fischeri*. *BMC Genomics* 7(1):154.

Chun, C.-K., J. V. Troll, I. Koroleva, B. Brown, L. Manzella, E. Snir, H. Almabrazi, T. E. Scheetz, MdF. Bonaldo, T. L. Casavant, M. Bento Soares, E. G. Ruby, and M. J. McFall-Ngai. 2008. Effects of colonization, luminescence, and autoinducer on host transcription during development of the squid-vibrio association. *Proceedings of the National Academy of Sciences USA* 105(32):11323–11328.

Connon, S. A., and S. J. Giovannoni. 2002. High-throughput methods for culturing microorganisms in very-low-nutrient media yield diverse new marine isolates. *Applied and Environmental Microbiology* 68(8):3878–3885.

Cotter, P. A., and V. J. DiRita. 2000. Bacterial virulence gene regulation: An evolutionary perspective. *Annual Review of Microbiology* 54(1):519–565.

Culp, E. J., G. Yim, N. Waglechner, W. Wang, A. C. Pawlowski, and G. D. Wright. 2019. Hidden antibiotics in *Actinomycetes* can be identified by inactivation of gene clusters for common antibiotics. *Nature Biotechnology* 37(10):149–154.

Dalzell, D. J. B., S. Alte, E. Aspichueta, A. De la Sota, J. Etxebarria, M. Gutierrez, C. C. Hoffmann, D. Sales, U. Obst, and N. Christofi. 2002. A comparison of five rapid direct toxicity assessment methods to determine toxicity of pollutants to activated sludge. *Chemosphere* 47(5):535–545.

Dorival, J., S. Ruppert, M. Gunnoo, A. Orłowski, M. Chapelais-Baron, J. Dabin, A. Labourel, D. Thompson, G. Michel, M. Czjzek, and S. Genicot. 2018. The laterally acquired GH5 Zg EngA GH5_4 from the marine bacterium *Zobellia galactanivorans* is dedicated to hemicellulose hydrolysis. *Biochemical Journal* 475(22):3609–3628.

Duarte, C. M., J. J. Middelburg, and N. Caraco. 2005. Major role of marine vegetation on the oceanic carbon cycle. *Biogeosciences Discussions* 2:1–8.

Dudek, M., A. Dieudonné, D. Jouanneau, T. Rochat, G. Michel, B. Sarels, and F. Thomas. 2020. Regulation of alginate catabolism involves a GntR family repressor in the marine Flavobacterium *Zobellia galactanivorans* DsijT. *Nucleic Acids Research* 48(14): 7786–7800.

Dufresne, A., L. Garczarek, and F. Partensky. 2005. Accelerated evolution associated with genome reduction in a free-living prokaryote. *Genome Biology* 6(2):R14.

Dufresne, A., M. Ostrowski, D. J. Scanlan, L. Garczarek, S. Mazard, B. Palenik, I. T. Paulsen, N. Tandeau de Marsac, P. Wincker, C. Dossat, S. Ferriera, J. Johnson, A. F. Post, W. R. Hess, and F. Partensky. 2008. Unraveling the genomic mosaic of a ubiquitous genus of marine *Cyanobacteria*. *Genome Biology* 9(5):R90.

Dunlap, P. V., and K. Kita-Tsukamoto. 2006. Luminous bacteria. In *The Prokaryotes*, edited by M. Dworkin, S. Falkow, E. Rosenberg, K.-H. Schleifer, and E. Stackebrandt, 863–892. New York, NY: Springer New York.

Dunlap, P. V., and H. Urbanczyk. 2013. Luminous bacteria. *The Prokaryotes*:495–521.

Earl, A. M., R. Losick, and R. Kolter. 2008. Ecology and genomics of *Bacillus subtilis*. *Trends in Microbiology* 16(6):269–275.

Eberhard, A., A. L. Burlingame, C. Eberhard, G. L. Kenyon, K. H. Nealson, and N. J. Oppenheimer. 1981. Structural identification of autoinducer of *Photobacterium fischeri* luciferase. *Biochemistry* 20(9):2444–2449.

Ebert, M., S. Laaß, M. Burghartz, J. Petersen, S. Koßmehl, L. Wöhlbrand, R. Rabus, C. Wittmann, P. Tielen, and D. Jahn. 2013. Transposon mutagenesis identified chromosomal and plasmid genes essential for adaptation of the marine bacterium *Dinoroseobacter shibae* to anaerobic conditions. *Journal of Bacteriology* 195(20):4769–4777.

Engebrecht, J., K. Nealson, and M. Silverman. 1983. Bacterial bioluminescence: Isolation and genetic analysis of functions from *Vibrio fischeri*. *Cell* 32(3):773–781.

Ennouri, H. P. d'Abzac, F. Hakil, P. Branchu, M. Naïtali, A.-M. Lomenech, R. Oueslati, J. Desbrières, P. Sivadon, and R. Grimaud. 2017. The extracellular matrix of the oleolytic biofilms of *Marinobacter hydrocarbonoclasticus* comprises cytoplasmic proteins and T2SS effectors that promote growth on hydrocarbons and lipids. *Environmental Microbiology* 19(1):159–173.

Ericsson, M., D. Hanstorp, P. Hagberg, J. Enger, and T. Nyström. 2000. Sorting out bacterial viability with optical tweezers. *Journal of Bacteriology* 182(19):5551–5555.

Errington, J., and L. T. van der Aart. 2020. Microbe profile: *Bacillus subtilis*: Model organism for cellular development, and industrial workhorse. *Microbiology* 166(5):425–427.

Farrant, G. K., H. Doré, F. M. Cornejo-Castillo, F. Partensky, M. Ratin, M. Ostrowski, F. D. Pitt, P. Winckler, D. J. Scanlan, D. Ludicone, S. G. Acinas, and L. Garczarek. 2016. Delineating ecologically significant taxonomic units from global patterns of marine Picocyanobacteria. *Proceedings of the National Academy of Sciences USA* 113(24):E3365–E3374.

Farré, M., C. Gonçalves, S. Lacorte, D. Barcelo, and M. Alpendurada. 2002. Pesticide toxicity assessment using an electrochemical biosensor with *Pseudomonas putida* and a bioluminescence inhibition assay with *Vibrio fischeri*. *Analytical and Bioanalytical Chemistry* 373(8):696–703.

Favre, L., A. Ortalo-Magné, L. Kerloch, C. Pichereaux, B. Misson, J.-F Briand, C. Garnier, and G. Culioli. 2019. Metabolomic and proteomic changes induced by growth inhibitory concentrations of copper in the biofilm-forming marine bacterium *Pseudoalteromonas lipolytica*. *Metallomics* 11(11):1887–1899.

Fels U., K. Gevaert, and P. Van Damme. 2020. Bacterial genetic engineering by means of recombineering for reverse genetics. *Frontiers in Microbiology* 11:548410.

Ferrari, B. C., T. Winsley, M. Gillings, and S. Binnerup. 2008. Cultivating previously uncultured soil bacteria using a soil substrate membrane system. *Nature Protocols* 3(8):1261–1269.

Ficko-Blean, E., C. Hervé, and G. Michel. 2014. Sweet and sour sugars from the sea: The biosynthesis and remodeling of sulfated cell wall polysaccharides from marine macroalgae. *Perspectives in Phycology* 2(1):51–64.

Ficko-Blean, E., A. Préchoux, F. Thomas, T. Rochat, R. Larocque, Y. Zhu, M. Stam, S. Génicot, M. Jam, A. Calteau, B. Viart, D. Ropartz, D. Pérez-Pascual, G. Correc, M. Matard-Mann, K. A. Stubbs, H. Rogniaux, A. Jeudy, T. Barbeyron, C. Médigue, M. Czjzek, D. Vallenet, M. J. McBride, E. Duchaud, and G. Michel. 2017. Carrageenan catabolism is encoded by a complex regulon in marine heterotrophic bacteria. *Nature Communications* 8:1685.

Fidopiastis, P. M., S. von Boletzky, and E. G. Ruby. 1998. A new niche for *Vibrio logei*, the predominant light organ symbiont of squids in the genus *Sepiola*. *Journal of Bacteriology* 180(1):59–64.

Fidopiastis, P. M., C. M. Miyamoto, M. G. Jobling, E. A. Meighen, and E. G. Ruby. 2002. LitR, a new transcriptional activator in *Vibrio fischeri*, regulates luminescence and symbiotic light organ colonization. *Molecular Microbiology* 45(1):131–143.

Flombaum, P., J. L. Gallegos, R. A Gordillo, J. Rincon, L. L Zabala, N. Jiao, D. M. Karl, W. K. Li, M. W. Lomas, D. Veneziano, C. S. Vera, J. A. Vrugt, and A. C. Martiny. 2013. Present and future global distributions of the marine cyanobacteria *Prochlorococcus* and *Synechococcus*. *Proceedings of the National Academy of Sciences USA* 110(24):9824–9829.

Fornari, C. S., and S. Kaplan. 1982. Genetic transformation of *Rhodopseudomonas sphaeroides* by plasmid DNA. *Journal of Bacteriology* 152(1):89–97.

Fournier, J. B., E. Rebuffet, L. Delage, R. Grijol, L. Meslet-Cladière, J. Rzonca, P. Potin, G. Michel, M. Czjzek, and C. Leblanc. 2014. The vanadium iodoperoxidase from the marine *Flavobacteriaceae* species *Zobellia galactanivorans* reveals novel molecular and evolutionary features of halide specificity in the vanadium haloperoxidase enzyme family. *Applied and Environmental Microbiology* 80(24):7561–7573.

Frigaard, N.-U., S. Yumiko, and D. A. Bryant. 2004. Gene inactivation in the cyanobacterium *Synechococcus* sp. PCC7002 and the green sulfur bacterium *Chlorobium tepidum* using in vitro-made DNA constructs and natural transformation. In *Photosynthesis Research Protocols*, 325–340. Totowa, NJ: Humana Press Inc.

Fuqua, W. C., S. C Winans, and E. P. Greenberg. 1994. Quorum sensing in bacteria: The LuxR-LuxI family of cell density-responsive transcriptional regulators. *Journal of Bacteriology* 176(2):269–275.

Gauthier, M. J., B. Lafay, R. Christen, L. Fernandez, M. Acquaviva, P. Bonin, and J.-C. Bertrand. 1992. *Marinobacter hydrocarbonoclasticus* gen. nov., sp. nov., a new, extremely halotolerant, hydrocarbon-degrading marine bacterium. *International Journal of Systematic Bacteriology* 42(4):568–576.

Gay, P., D. Le Coq, M. Steinmetz, T. Berkelman, and C. I. Kado. 1985. Positive selection procedure for entrapment of insertion sequence elements in Gram-negative bacteria. *Journal of Bacteriology* 164(2):918–921.

Gram, L., H. P. Grossart, A. Schlingloff, and T. Kiorboe. 2002. Possible quorum sensing in marine snow bacteria: Production of acylated homoserine lactones by *Roseobacter* strains isolated from marine snow. *Applied and Environmental Microbiology* 68(8):4111–4116.

Grébert, T., H. Doré, F. Partensky, G. K. Farrant, E. S. Boss, and M. Picheral. 2018. Light color acclimation is a key process in the global ocean distribution of *Synechococcus* cyanobacteria. *Proceedings of the National Academy of Sciences USA* 115(9):E2010–E2019.

Greenberg, E. P., J. W. Hastings, and S. Ulitzur. 1979. Induction of luciferase synthesis in *Beneckea harveyi* by other marine bacteria. *Archives of Microbiology* 120(2):87–91.

Groisillier, A., A. Labourel, G. Michel, and T. Tonon. 2015. The mannitol utilization system of the marine bacterium *Zobellia galactanivorans*. *Applied and Environmental Microbiology* 81(5):1799–1812.

Grondin, J. M., K. Tamura, G. Déjean, D. W. Abbott, and H. Brumer. 2017. Polysaccharide utilization loci: Fuelling microbial communities. *Journal of Bacteriology* 199(15):e00860–16.

Guschinskaya, N., R. Brunel, M. Tourte, G. L. Lipscomb, M. W. W. Adams, P. Oger, and X. Charpentier. 2016. Random mutagenesis of the hyperthermophilic archaeon *Pyrococcus furiosus* using *in vitro* mariner transposition and natural transformation. *Scientific Reports* 6:36711.

Hamilton, C. M., M. Aldea, B. K. Washburn, P. Babitzke, and S. R. Kushner. 1989. New method for generating deletions and gene replacements in *Escherichia coli*. *Journal of Bacteriology* 171(9):4617–4622.

Harms, H., A. Klöckner, J. Schrör, M. Josten, S. Kehraus, M. Crüsemann, W. Hanke, T. Schneider, T. F. Schäberle, and G. M. König. 2018. Antimicrobial dialkylresorcins from marine-derived microorganisms: Insights into their mode of action and putative ecological relevance. *Planta Medica* 84(18):1363–1371.

Heath-Heckman, E. A. C., Suzanne M. Peyer, C. A. Whistler, M. A. Apicella, W. E. Goldman, and M. J. McFall-Ngai. 2013. Bacterial bioluminescence regulates expression of a host cryptochrome gene in the squid-vibrio symbiosis. *MBio* 4(2) e00167–13.

Hehemann, J.-H., G. Correc, T. Barbeyron, W. Helbert, M. Czjzek, and G. Michel. 2010. Transfer of carbohydrate-active enzymes from marine bacteria to japanese gut microbiota. *Nature* 464(7290):908–912.

Hehemann, J.-H., G. Correc, F. Thomas, T. Bernard, T. Barbeyron, M. Jam, W. Helbert, G. Michel, and M. Czjzek. 2012. Biochemical and structural characterization of the complex agarolytic enzyme system from the marine bacterium *Zobellia galactanivorans*. *The Journal of Biological Chemistry* 287(36):30571–30584.

Henson, M. W., D. M. Pitre, J. L. Weckhorst, V. C. Lanclos, A. T. Webber, and J. C. Thrash. 2016. Artificial seawater media facilitate cultivating members of the microbial majority from the Gulf of Mexico. *Msphere* 1(2):e00028–16.

Henson, M. W., V. C. Lanclos, D. M. Pitre, J. L. Weckhorst, A. M. Lucchesi, C. Cheng, . . . and J. C. Thrash. 2020. Expanding the diversity of bacterioplankton isolates and modeling isolation efficacy with large-scale dilution-to-extinction cultivation. *Applied and Environmental Microbiology* 86(17):e00943–20.

Hollants, J., F. Leliaert, O. De Clerck, and A. Willems. 2013. What we can learn from sushi: A review on seaweed-bacterial associations. *FEMS Microbiology Ecology* 83(1):1–16.

Hsu, P. D., E. S. Lander, and F. Zhang. 2014. Development and applications of CRISPR-Cas9 for genome engineering. *Cell* 157(6):1262–1278.

Humily, F., F. Partensky, C. Six, G. K. Farrant, M. Ratin, D. Marie, and L. Garczarek. 2013. A gene island with two possible configurations is involved in chromatic acclimation in marine *Synechococcus*. *PLoS One* 8(12):e84459.

Huu, N. B., E. B. M. Denner, D. T. C. Ha, G. Wanner, and H. Stan-Lotter. 1999. *Marinobacter aquaeolei* sp. nov., a halophilic bacterium isolated from a vietnamese oil-producing well. *International Journal of Systematic and Evolutionary Microbiology* 49(2):367–375.

Jansen, R., J. D. A. van Embden, W. Gaastra, and L. M. Schouls. 2002. Identification of genes that are associated with DNA repeats in prokaryotes. *Molecular Microbiology* 43(6):1565–1575.

Ji, G., R. C. Beavis, and R. P. Novick. 1995. Cell density control of *Staphylococcal* virulence mediated by an octapeptide pheromone. *Proceedings of the National Academy of Sciences USA* 92(26):12055–12059.

Jiang, S. C., and J. H. Paul. 1998. Gene transfer by transduction in the marine environment. *Applied and Environmental Microbiology* 64(8):2780–2787.

Jinek, M., K. Chylinski, I. Fonfara, M. Hauer, J. A. Doudna, and E. Charpentier. 2012. A programmable dual-RNA-guided DNA endonuclease in adaptive bacterial immunity. *Science* 337(6096): 816–821.

Kato, S., M. Terashima, A. Yama, M. Sato, W. Kitagawa, K. Kawasaki, and Y. Kamagata. 2020. Improved isolation of uncultured anaerobic bacteria using medium prepared with separate sterilization of agar and phosphate. *Microbes and Environments* 35(1):ME19060.

Kato, S., A. Yamagishi, S. Daimon, K. Kawasaki, H. Tamaki, W. Kitagawa, A. Abe, M. Tanaka, T. sone, K. Asano, and Y. Kamagata. 2018. Isolation of previously uncultured slow-growing bacteria by using a simple modification in the preparation of agar media. *Applied and Environmental Microbiology* 84(19):e00807–18.

Klein, B., V. Grossi, P. Bouriat, P. Goulas, and R. Grimaud. 2008. Cytoplasmic wax ester accumulation during biofilm-driven substrate assimilation at the alkane-water interface by *Marinobacter hydrocarbonoclasticus* SP17. *Research in Microbiology* 159(2):137–144.

Knoot, C. J., S. Biswas, and H. B. Pakrasi. 2020. Tunable repression of key photosynthetic processes using Cas12a CRISPR interference in the fast-growing cyanobacterium *Synechococcus* sp. UTEX 2973. *ACS Synthetic Biology* 9(1):132–143.

Koonin, E. V., K. S. Makarova, and F. Zhang. 2017. Diversity, classification and evolution of CRISPR-Cas systems. *Current Opinion in Microbiology* 37:67–78.

Koropatnick, T. A. 2004. Microbial factor-mediated development in a host-bacterial mutualism. *Science* 306(5699):1186–1188.

Krasity, B. C., J. V. Troll, J. P. Weiss, and M. J. McFall-Ngai. 2011. LBP/BPI proteins and their relatives: Conservation over evolution and roles in mutualism. *Biochemical Society Transactions* 39(4):1039–1044.

Labourel, A., M. Jam, L. Legentil, B. Sylla, J.-H. Hehemann, V. Ferrières, M. Czjzek, and G. Michel. 2015. Structural and biochemical characterization of the laminarinase ZgLamCGH16 from *Zobellia galactanivorans* suggests preferred recognition of branched laminarin. *Acta Crystallographica Section D Biological Crystallography* 71:173–184.

Lami, R. 2019. Chapter 3: Quorum sensing in marine biofilms and environments. In *Quorum Sensing*, edited by Giuseppina Tommonaro, 55–96. New York, NY: Academic Press.

Lee, K.-H., and E. G. Ruby. 1994. Effect of the squid host on the abundance and distribution of symbiotic *Vibrio fischeri* in nature. *Applied and Environmental Microbiology* 60(5):1565–1571.

Lenneman, E. M., J. M. Ohlert, N. P. Palani, and B. M. Barney. 2013. Fatty alcohols for wax esters in *Marinobacter aquaeolei* VT8: Two optional routes in the wax biosynthesis pathway. *Applied and Environmental Microbiology* 79(22):7055–7062.

Li, L., K. Wei, G. Zheng, X. Liu, S. Chen, W. Jiang, and Y. Lu. 2018. CRISPR-Cpf1-assisted multiplex genome editing and transcriptional repression in *Streptomyces*. *Applied and Environmental Microbiology* 84(18):e00827–18.

Li, Q., F. M. Seys, N. P. Minton, J. Yang, Y. Jiang, W. Jiang, and S. Yang. 2019. CRISPR-Cas9D10A nickase-assisted base editing in the solvent producer *Clostridium beijerinckii*. *Biotechnology and Bioengineering* 116(6):1475–1483.

Lidbury, I., G. Kimberley, D. J. Scanlan, J. C. Murrell, and Y. Chen. 2015. Comparative genomics and mutagenesis analyses of choline metabolism in the marine *Roseobacter* clade. *Environmental Microbiology* 17(12):5048–5062.

Ling, L. L., T. Schneider, A. J. Peoples, A. L. Spoering, I. Engels, B. P. Conlon, A. Mueller, T. F. Shäberle, D. E. Hughes, S. Epstein, M. Jones, L. Lazarides, V. A. Steadman, D. R. Cohen, C. R. Felix, K. A. Fetterman, W. P. Millett, A. G. Nitti, A. M. Zullo, C. Chen, and K. Lewis. 2015. A new antibiotic kills pathogens without detectable resistance. *Nature* 517(7535):455–459.

Liu, H., and A. M. Deutschbauer. 2018. Rapidly moving new bacteria to model-organism status. *Current Opinion in Biotechnology* 51:116–122.

Liu, W., H. J. Kim, E. M. Lucchetta, W. Du, and R. F. Ismagilov. 2009. Isolation, incubation, and parallel functional testing and identification by FISH of rare microbial single-copy cells from multi-species mixtures using the combination of chemistrode and stochastic confinement. *Lab on a Chip* 9(15):2153.

Lombard, V., H. G. Ramulu, E. Drula, P. M. Coutinho, and B. Henrissat. 2014. The carbohydrate-active enzymes database (CAZy) in 2013. *Nucleic Acids Research* 42:490–495.

Manilla-Pérez, E., A. B. Lange, S. Hetzler, and A. Steinbüchel. 2010. Occurrence, production, and export of lipophilic compounds by hydrocarbonoclastic marine bacteria and their potential use to produce bulk chemicals from hydrocarbons. *Applied Microbiology and Biotechnology* 86(6):1693–1706.

Márquez, M. C., and A. Ventosa. 2005. *Marinobacter hydrocarbonoclasticus* Gauthier et al. 1992 and *Marinobacter aquaeolei* Nguyen et al. 1999 are heterotypic synonyms. *International Journal of Systematic and Evolutionary Microbiology* 55(3): 1349–1351.

Matard-Mann, M., T. Bernard, C. Leroux, T. Barbeyron, R. Larocque, A. Préchoux, A. Jeudy, M. Jam, P. N. Collén, G. Michel, and M. Czjzek. 2017. Structural insights into marine carbohydrate degradation by family GH16-carrageenases. *Journal of Biological Chemistry* 292(48):19919–19934.

Mazard, S., M. Ostrowski, F. Partensky, and D. J. Scanlan. 2012. Multi-locus sequence analysis, taxonomic resolution and biogeography of marine *Synechococcus*. *Environmental Microbiology* 14(2):372–386.

McCarren, J., and B. Brahamsha. 2005. Transposon mutagenesis in a marine *Synechococcus* strain: Isolation of swimming motility mutants. *Journal of Bacteriology* 187(13):4457–4462.

McFall-Ngai, M. J. 2014. The importance of microbes in animal development: Lessons from the squid-vibrio symbiosis. *Annual Review of Microbiology* 68:177–194.

McFall-Ngai, M. J., and E. Ruby. 1991. Symbiont recognition and subsequent morphogenesis as early events in an animal-bacterial mutualism. *Science* 254(5037):1491–1494.

Meibom, K. L., M. Blokesch, N. A. Dolganov, C.-Y. Wu, and G. K. Schoolnik. 2005. Chitin induces natural competence in *Vibrio cholerae*. *Science* 310(5755):1824–1827.

Miklaszewska, M., F. Dittrich-Domergue, A. Banaś, and F. Domergue. 2018. Wax synthase MhWS2 from *Marinobacter hydrocarbonoclasticus*: Substrate specificity and biotechnological potential for wax ester production. *Applied Microbiology and Biotechnology* 102(9):4063–4074.

Mojica, F. J. M., C. Díez-Villaseñor, E. Soria, and G. Juez. 2000. Biological significance of a family of regularly spaced repeats in the genomes of Archaea, Bacteria and mitochondria. *Molecular Microbiology* 36(1):244–246.

Mounier, J., A. Camus, I. Mitteau, P.-J. Vaysse, P. Goulas, R. Grimaud, and P. Sivadon. 2014. The marine bacterium *Marinobacter hydrocarbonoclasticus* SP17 degrades a wide range of lipids and hydrocarbons through the formation of oleolytic biofilms with distinct gene expression profiles. *FEMS Microbiology Ecology* 90(3):816–831.

Mounier, J., F. Hakil, P. Branchu, M. Naïtali, P. Goulas, P. Sivadon, and R. Grimaud. 2018. AupA and AupB are outer and inner membrane proteins involved in alkane uptake in *Marinobacter hydrocarbonoclasticus* SP17. *MBio* 9(3):e00520–18.

Naretto, A., M. Fanuel, D. Ropartz, H. Rogniaux, R. Larocque, M. Czjzek, C. Tellier, and G. Michel. 2019. The agar-specific hydrolase ZgAgaC from the marine bacterium *Zobellia galactanivorans* defines a new GH16 protein subfamily. *Journal of Biological Chemistry* 294(17):6923–6939.

Nealson, K. H., T. Platt, and J. W. Hastings. 1970. Cellular control of the synthesis and activity of the bacterial luminescent system. *Journal of Bacteriology* 104(1):313–322.

Nedashkovskaya, O. I., M. Suzuki, M. Vancanneyt, I. Cleenwerck, A. M. Lysenko, V. V. Mikhailov, and J. Swings. 2004. *Zobellia amurskyensis* sp. nov., *Zobellia laminariae* sp. nov. and *Zobellia russellii* sp. nov., novel marine bacteria of the family Flavobacteriaceae. *International Journal of Systematic and Evolutionary Microbiology* 54(Pt5):1643–1648.

Nichols, D., N. Cahoon, E. M. Trakhtenberg, L. Pham, A. Mehta, A. Belanger, T. Kanigan, K. Lewis, and S. S. Epstein. 2010. Use of ichip for high-throughput in situ cultivation of uncultivable microbial species. *Applied and Environmental Microbiology* 76(8):2445–2450.

Oh, J.-H., and J.-P. van Pijkeren. 2014. CRISPR-Cas9-assisted recombineering in *Lactobacillus reuteri*. *Nucleic Acids Research* 42(17):e131.

Overmann, J., and C. Lepleux. 2016. Marine bacteria and archaea: Diversity, adaptations, and culturability. In *The Marine Microbiome*, 21–55. Switzerland: Springer.

Partensky, F., and L. Garczarek. 2010. *Prochlorococcus*: Advantages and limits of minimalism. *Annual Review of Marine Science* 2:305–331.

Parvez, S., C. Venkataraman, and S. Mukherji. 2006. A review on advantages of implementing luminescence inhibition test (*Vibrio fischeri*) for acute toxicity prediction of chemicals. *Environment International* 32(2):265–268.

Paul, B., and G. Montoya. 2020. CRISPR-Cas12a: Functional overview and applications. *Biomedical Journal* 43(1):8–17.

Paul, J. H., M. E. Frischer, and J. M. Thurmond. 1991. Gene transfer in marine water column and sediment microcosms by natural plasmid transformation. *Applied and Environmental Microbiology* 57(5):1509–1515.

Pearson, J. P., M. Feldman, B. H. Iglewski, and A. Prince. 2000. *Pseudomonas aeruginosa* cell-to-cell signaling is required for virulence in a model of acute pulmonary infection. *Infection and Immunity* 68(7):4331–4334.

Petronikolou, N., and S. K. Nair. 2018. Structural and biochemical studies of a biocatalyst for the enzymatic production of wax esters. *ACS Catalysis* 8(7):6334–6344.

Pickar-Oliver, A., and C. A. Gersbach. 2019. The next generation of CRISPR-Cas technologies and applications. *Nature Reviews Molecular Cell Biology* 20(8):490–507.

Piekarski, T., I. Buchholz, T. Drepper, M. Schobert, I. Wagner-Döbler, P. Tielen, and D. Jahn. 2009. Genetic tools for the investigation of *Roseobacter* clade bacteria. *BMC Microbiology* 9(1):265.

Potin, P., A. Sanseau, Y. Le Gall, C. Rochas, and B. Kloareg. 1991. Purification and characterization of a new kappa-carrageenase from a marine cytophaga-like bacterium. *European Journal of Biochemistry* 201(1):241–247.

Price, M. N., K. M. Wetmore, R. J. Waters, M. Callaghan, J. Ray, H. Liu, J. V. Kuehl, R. A. Melnyk, J. S. Lamson, Y. Suh, H. K. Carlson, Z. Esquivel, H. Sadeeshkumar, R. Chakraborty, G. M. Zane, B. E. Rubin, J. D. Wall, A. Visel, J. Bristow, M. J. Blow, A. P. Arkin, and A. M. Deutschbauer. 2018. Mutant phenotypes for thousands of bacterial genes of unknown function. *Nature* 557(7706):503–509.

Rahn-Lee, L., M. E. Byrne, M. Zhang, D. Le Sage, D. R. Glenn, T. Milbourne, R. L. Walsworth, H. Vali, and A. Komeili. 2015. A genetic strategy for probing the functional diversity of magnetosome formation. *PLoS Genetics* 11(1):e1004811.

Rappé, M. S., S. A. Connon, K. L. Vergin, and S. J. Giovannoni. 2002. Cultivation of the ubiquitous SAR11 marine bacterioplankton clade. *Nature* 418(6898):630–633.

Rappé, M. S., and S. J. Giovannoni. 2003. The uncultured microbial majority. *Annual Review of Microbiology* 57(1):369–394.

Rebuffet, E., A. Groisillier, A. Thompson, A. Jeudy, T. Barbeyron, M. Czjzek, and G. Michel. 2011. Discovery and structural characterization of a novel glycosidase family of marine origin. *Environmental Microbiology* 13(5):1253–1270.

Ribeiro, P. R., R. R. Barbosa, and C. P. de Almeida. 2019. Metabolomics approaches in microbial research: Current knowledge and perspective toward the understanding of microbe plasticity. In *Microbial Interventions in Agriculture and Environment*, 29–50. Singapore: Springer Nature.

Ruby, E. G., and L. M. Asato. 1993. Growth and flagellation of *Vibrio fischeri* during initiation of the sepiolid squid light organ symbiosis. *Archives of Microbiology* 159(2):160–167.

Ruby, E. G., and K. H. Nealson. 1977. A luminous bacterium that emits yellow light. *Science* 196(4288):432–434.

Ruby, E. G., M. Urbanowski, J. Campbell, A. Dunn, M. Faini, R. Gunsalus, P. Lostroh, C. Lupp, J. McCann, D. Millikan, A. Schaefer, E. Stabb, A. Stevens, K. Visick, C. Whistler, and E. P. Greenberg. 2005. Complete genome sequence of *Vibrio fischeri:* A symbiotic bacterium with pathogenic congeners. *Proceedings of the National Academy of Sciences USA* 102(8):3004–3009.

Rusch, D. B., A. L. Halpern, G. Sutton, K. B. Heidelberg, S. Williamson, S. Yooseph, D. Wu, J. A. Eisen, J. M. Hoffman, and K. Remington. 2007. The Sorcerer II global ocean sampling expedition: Northwest Atlantic through eastern tropical Pacific. *PLoS Biology* 5(3):e77.

Salaün, S., S. La Barre, M. Dos Santos-Goncalvez, P. Potin, D. Haras, and A. Bazire. 2012. Influence of exudates of the kelp *Laminaria digitata* on biofilm formation of associated and exogenous bacterial epiphytes. *Microbial Ecology* 64(2):359–369.

Sanfilippo, J. E., L. Garczarek, F. Partensky, and D. M. Kehoe. 2019. Chromatic acclimation in Cyanobacteria: A diverse and widespread process for optimizing photosynthesis. *Annual Review of Microbiology* 73(1):407–433.

Scanlan, D. J., M. Ostrowski, S. Mazard, A. Dufresne, L. Garczarek, W. R. Hess, A. F. Post, M. Hagemann, I. Paulsen, and F. Partensky. 2009. Ecological genomics of marine Picocyanobacteria. *Microbiology and Molecular Biology Reviews* 73(2):249–299.

Schweder, T., S. Markert, and M. Hecker. 2008. Proteomics of marine bacteria. *Electrophoresis* 29(12):2603–2616.

Senechkin, I. V., A. G. C. L. Speksnijder, A. M. Semenov, A. H. C. van Bruggen, and L. S. van Overbeek. 2010. Isolation and partial characterization of bacterial strains on low organic carbon medium from soils fertilized with different organic amendments. *Microbial Ecology* 60(4):829–839.

Simpson, C. A., R. Podicheti, D. B. Rusch, A. B. Dalia, and J. C. van Kessel. 2019. Diversity in natural transformation frequencies and regulation across *Vibrio* species. *MBio* 10(6):e02788–19.

Singer, E., E. A. Webb, W. C. Nelson, J. F. Heidelberg, N. Ivanova, A. Pati, and K. J. Edwards. 2011. Genomic potential of *Marinobacter aquaeolei*, a biogeochemical 'opportunitroph'. *Applied and Environmental Microbiology* 77(8):2763–2771.

Sivadon, P., C. Barnier, L. Urios, and R. Grimaud. 2019. Biofilm formation as a microbial strategy to assimilate particulate substrates. *Environmental Microbiology Reports* 11(6):749–764.

Six, C., J.-C. Thomas, L. Garczarek, M. Ostrowski, A. Dufresne, N. Blot, D. J. Scanlan, and F. Partensky. 2007. Diversity and evolution of phycobilisomes in marine *Synechococcus* spp.: A comparative genomics study. *Genome Biology* 8(12):R259.

Sizova, M. V., T. Hohmann, A. Hazen, B. J. Paster, S. R. Halem, C. M. Murphy, N. S. Panikov, and S. S. Epstein. 2012. New approaches for isolation of previously uncultivated oral bacteria. *Applied and Environmental Microbiology* 78(1):194–203.

Stingl, U., H. J. Tripp, and S. J. Giovannoni. 2007. Improvements of high-throughput culturing yielded novel SAR11 strains and other abundant marine bacteria from the oregon coast and the Bermuda Atlantic Time Series study site. *The ISME Journal* 1(4):361–371.

Sunagawa, S., L. P. Coelho, S. Chaffron, J. R. Kultima, K. Labadie, G. Salazar, B. Djahanschiri, G. Zeller, D. R. Mende, and A. Alberti. 2015. Structure and function of the global ocean microbiome. *Science* 348(6237):1261359.

Swartzman, A., S. Kapoor, A. F. Graham, and E. A. Meighen. 1990. A new *Vibrio fischeri lux* gene precedes a bidirectional termination site for the *lux* operon. *Journal of Bacteriology* 172(12):6797–6802.

Tamaki, H., Y. Sekiguchi, S. Hanada, K. Nakamura, N. Nomura, M. Matsumura, and Y. Kamagata. 2005. Comparative analysis of bacterial diversity in freshwater sediment of a shallow eutrophic lake by molecular and improved cultivation-based techniques. *Applied and Environmental Microbiology* 71(4): 2162–2169.

Tanaka, Y., S. Hanada, A. Manome, T. Tsuchida, R. Kurane, K. Nakamura, and Y. Kamagata. 2004. *Catellibacterium nectariphilum* gen. nov., sp. nov., which requires a diffusible compound from a strain related to the genus *Sphingomonas* for vigorous growth. *International Journal of Systematic and Evolutionary Microbiology* 54(3):955–959.

Thomas, F., T. Barbeyron, T. Tonon, S. Génicot, M. Czjzek, and G. Michel. 2012. Characterization of the first alginolytic operons in a marine bacterium: From their emergence in marine Flavobacteriia to their independent transfers to marine Proteobacteria and human gut Bacteroides. *Environmental Microbiology* 14(9):2379–2394.

Thomas, F., P. Bordron, D. Eveillard, and G. Michel. 2017. Gene expression analysis of *Zobellia galactanivorans* during the degradation of algal polysaccharides reveals both substrate-specific and shared transcriptome-wide responses. *Frontiers in Microbiology* 8:1808.

Thomas, F., T. Barbeyron, and G. Michel. 2011a. Evaluation of reference genes for real time quantitative PCR in the marine flavobacterium *Zobellia galactanivorans*. *Journal of Microbiological Methods* 84:61–66.

Thomas, F., J.-H. Hehemann, E. Rebuffet, M. Czjzek, and G. Michel. 2011b. Environmental and gut *Bacteroidetes*: The food connection. *Frontiers in Microbiology* 2:93.

Thomas, F., L. C. E. Lundqvist, M. Jam, A. Jeudy, T. Barbeyron, C. Sandström, G. Michel, and M. Czjzek. 2013. Comparative characterization of two marine alginate lyases from *Zobellia galactanivorans* reveals distinct modes of action and exquisite adaptation to their natural substrate. *The Journal of Biological Chemistry* 288(32):23021–23037.

Ting, C. S., G. Rocap, J. King, and S. W. Chisholm. 2002. Cyanobacterial photosynthesis in the oceans: The origins and significance of divergent light-harvesting strategies. *Trends in Microbiology* 10(3):134–142.

Tong, Y., P. Charusanti, L. Zhang, T. Weber, and S. Y. Lee. 2015. CRISPR-Cas9 based engineering of Actinomycetal genomes. *ACS Synthetic Biology* 4(9):1020–1029.

Tong, Y., C. M. Whitford, H. L. Robertsen, K. Blin, T. S. Jørgensen, A. K. Klitgaard, T. Gren, X. Jiang, T. Weber, and S. Yup Lee. 2019. Highly efficient DSB-free base editing for *Streptomycetes* with CRISPR-BEST. *Proceedings of the National Academy of Sciences* USA 116(41):20366–20375.

Tortell, P. D., M. T. Maldonado, J. Granger, and N. M. Price. 1999. Marine bacteria and biogeochemical cycling of iron in the oceans. *FEMS Microbiology Ecology* 29(1):1–11.

Troll, J. V., E. H. Bent, N. Pacquette, A. M. Wier, W. E. Goldman, N. Silverman, and M. J. McFall-Ngai. 2010. Taming the symbiont for coexistence: A host PGRP neutralizes a bacterial symbiont toxin. *Environmental Microbiology* 12(8):2190–2203.

Van Pham, H. T., and J. Kim. 2014. *Bacillus thaonhiensis* sp. nov., a new species, was isolated from the forest soil of Kyonggi University by using a modified culture method. *Current Microbiology* 68(1):88–95.

van Rhijn, P., and J. Vanderleyden. 1995. The rhizobium-plant symbiosis. *Microbiological Reviews* 59(1):124–142.

Vaysse, P.-J., L. Prat, S. Mangenot, S. Cruveiller, P. Goulas, and R. Grimaud. 2009. Proteomic analysis of *Marinobacter hydrocarbonoclasticus* SP17 biofilm formation at the alkane-water interface reveals novel proteins and cellular processes involved in hexadecane assimilation. *Research in Microbiology* 160(10): 829–837.

Vaysse, P.-J., P. Sivadon, P. Goulas, and R. Grimaud. 2011. Cells dispersed from *Marinobacter hydrocarbonoclasticus* SP17 biofilm exhibit a specific protein profile associated with a higher ability to reinitiate biofilm development at the hexadecane-water interface: *M. hydrocarbonoclasticus* biofilm-dispersed cells. *Environmental Microbiology* 13(3):737–746.

Vento, J. M., N. Crook, and C. L. Beisel. 2019. Barriers to genome editing with CRISPR in Bacteria. *Journal of Industrial Microbiology & Biotechnology* 46(9):1327–1341.

Visick, K. L., and M. J. McFall-Ngai. 2000. An exclusive contract: Specificity in the *Vibrio fischeri-Euprymna scolopes* partnership. *Journal of Bacteriology* 182(7):1779–1787.

Vollheyde, K., D. Yu, E. Hornung, C. Herrfurth, and I. Feussner. 2020. The fifth WS/DGAT enzyme of the bacterium *Marinobacter aquaeolei* VT8. *Lipids* 55(5):479–494.

Wang, Z., T. Tschirhart, Z. Schultzhaus, E. E. Kelly, A. Chen, E. Oh, O. Nag, E. R. Glaser, E. Kim, and P. F Lloyd. 2020. Melanin produced by the fast-growing marine bacterium *Vibrio natriegens* through heterologous biosynthesis: Characterization and application. *Applied and Environmental Microbiology* 86(5): e02749–19.

Wannier, T. M., A. Nyerges, H. M. Kuchwara, M. Czikkely, D. Balogh, G. T. Filsinger, N. C. Borders, C. J. Gregg, M. J. Lajoie, X. Rios, C. Pal, and G. M. Church. 2020. Improved bacterial recombineering by parallelized protein discovery. *Proceedings of the National Academy of Sciences USA* 117(24):13689–13698.

Wei, S. L., and R. E. Young. 1989. Development of symbiotic bacterial bioluminescence in a nearshore cephalopod, *Euprymna scolopes*. *Marine Biology* 103(4):541–546.

Weinstock, M. T., E. D. Hesek, C. M. Wilson, and D. G. Gibson. 2016. *Vibrio natriegens* as a fast-growing host for molecular biology. *Nature Methods* 13(10):849.

Wenning, L., T. Yu, F. David, J. Nielsen, and V. Siewers. 2017. Establishing very long-chain fatty alcohol and wax ester biosynthesis in *Saccharomyces cerevisiae*. *Biotechnology and Bioengineering* 114(5):1025–1035.

Westbrook, A. W., M. Moo-Young, and C. Perry Chou. 2016. Development of a CRISPR-Cas9 tool kit for comprehensive engineering of *Bacillus subtilis*. *Applied and Environmental Microbiology* 82(16):4876–4895.

Whitman, W. B., D. C. Coleman, and W. J. Wiebe. 1998. Prokaryotes: The unseen majority. *Proceedings of the National Academy of Sciences USA* 95(12):6578–6583.

Wier, A. M., S. V. Nyholm, M. J. Mandel, R. P. Massengo-Tiassé, A. L. Schaefer, I. Koroleva, S. Splinter-Bondurant, B. Brown, L. Manzella, E. Snir, H. Almabrazi, T. E. Scheetz, MdF. Bonaldo, T. L. Casavant, M. Bento Soares, J. E. Cronan, J. L. Reed, E. G. Ruby, and M. J. McFall-Ngai. 2010. Transcriptional patterns in both host and bacterium underlie a daily rhythm of anatomical and metabolic change in a beneficial symbiosis. *Proceedings of the National Academy of Sciences USA* 107(5):2259–2264.

Yan, M.-Y., H.-Q. Yan, G.-X. Ren, J.-P. Zhao, X.-P. Guo, and Y.-C. Sun. 2017. CRISPR-Cas12a-assisted recombineering in bacteria. *Applied and Environmental Microbiology* 83(17): e00947–17.

Yu, R., A. Yamada, K. Watanabe, K. Yazawa, H. Takeyama, T. Matsunaga, and R. Kurane. 2000. Production of eicosapentaenoic acid by a recombinant marine Cyanobacterium, *Synechococcus* sp. *Lipids* 35(10):1061–1064.

Zang, X., B. Liu, S. Liu, K. K. I. U. Arunakumara, and X. Zhang. 2007. Optimum conditions for transformation of *Synechocystis* sp. PCC 6803. *The Journal of Microbiology* 45(3):241–245.

Zeaiter, Z., F. Mapelli, E. Crotti, and S. Borin. 2018. Methods for the genetic manipulation of marine bacteria. *Electronic Journal of Biotechnology* 33:17–28.

Zengler, K., G. Toledo, M. Rappe, J. Elkins, E. J. Mathur, J. M. Short, and M. Keller. 2002. Nonlinear partial differential equations and applications: Cultivating the uncultured. *Proceedings of the National Academy of Sciences USA* 99(24):15681–15686.

Zhang, M. M., F. T. Wong, Y. Wang, S. Luo, Y. H. Lim, E. Heng, W. L. Yeo, R. E. Cobb, B. Enghiad, E. Lui Ang, and H. Zhao. 2017. CRISPR-Cas9 strategy for activation of silent *Streptomyces* biosynthetic gene clusters. *Nature Chemical Biology* 13(6):607–609.

Zhang, W., W. Ding, Y.-X. Li, C. Tam, S. Bougouffa, R. Wang, B. Pei, H. Chiang, P. Leung, and Y. Lu. 2019. Marine biofilms constitute a bank of hidden microbial diversity and functional potential. *Nature Communications* 10(1):1–10.

Zhu, Y., F. Thomas, R. Larocque, N. Li, D. Duffieux, L. Cladière, F. Souchaud, G. Michel, and M. J. McBride. 2017. Genetic analyses unravel the crucial role of a horizontally acquired alginate lyase for brown algal biomass degradation by *Zobellia galactanivorans*. *Environmental Microbiology* 19(6):2164–2181.

2 Brown Algae
Ectocarpus *and* Saccharina *as Experimental Models for Developmental Biology*

Ioannis Theodorou and Bénédicte Charrier

CONTENTS

2.1 INTRODUCTION

Brown algae (also named Phaeophyceae) are a group of eukaryotic multicellular organisms comprising ~2000 species. They are autotrophic organisms using photosynthesis to transform light into chemical energy (ATP through NADP reduction). Their evolutionary history is distinct from that of animals, fungi and plants. In the tree of life, molecular phylogeny and cytological characters position brown algae within the division of Stramenopiles (Heterokonta), diverging from the last common Stramenopile ancestor ~250 million years ago (Mya) (Kawai et al. 2015) (Figure 2.1a).

The Stramenopiles are characterized by reproductive cells that possess two flagella ("konta") of different size and structure (Derelle et al. 2016). Other photosynthetic stramenopiles

DOI: 10.1201/9781003217503-2

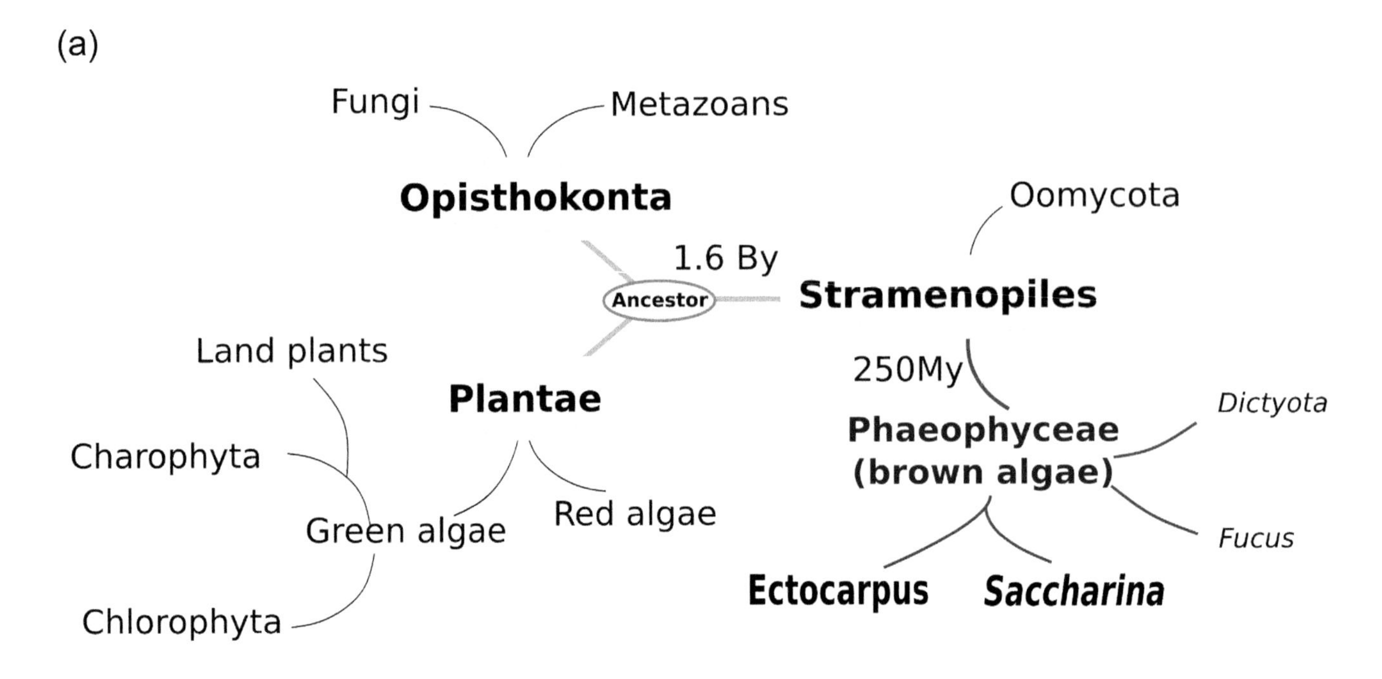

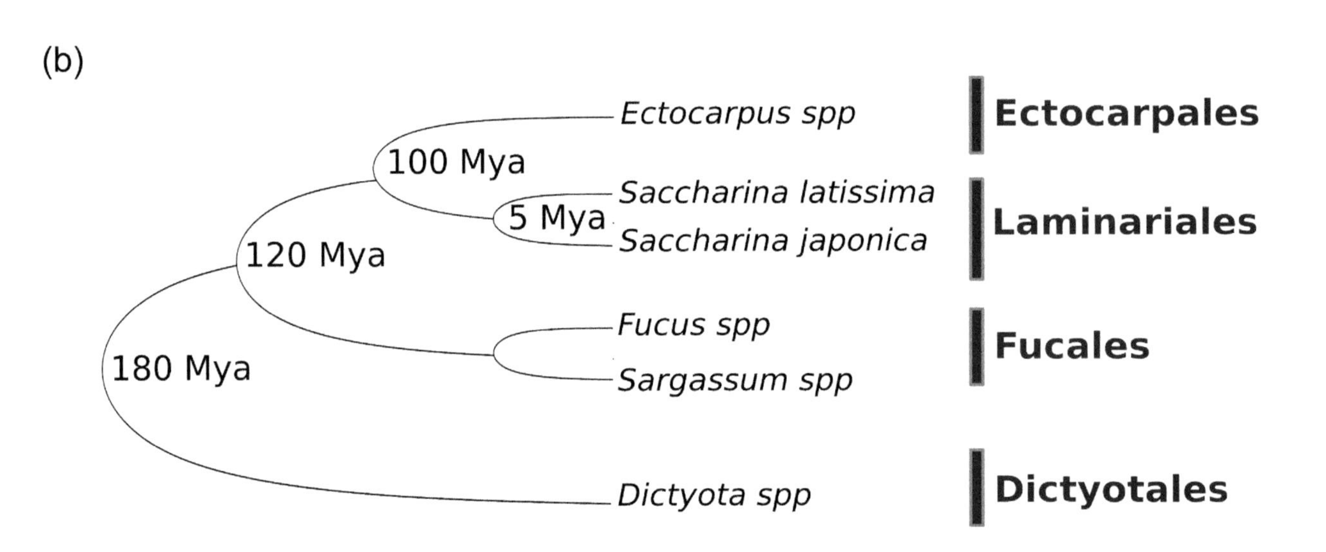

FIGURE 2.1 **Evolution of brown algae**. (a) Phylogenetic position of brown algae (Phaeophyceae) in the eukaryotic tree of life. Phaeophyceae diverged ~250 million years ago (Mya) from the last common stramenopile ancestor. Stramenopiles include multicellular organisms only (the syncitial oomycota are not considered true multicellular organisms). (b) Simplified phylogenetic tree of some brown algal genera and orders. *Ectocarpus* spp. and *Saccharina* spp. belong to closely related orders, the Ectocarpales and the Laminariales, which split ~75 Mya. Other brown algal models belonging to Fucales or Dictyotales are more distant phylogenetically (diverged 120 Mya and 180 Mya, respectively). ([a] Kawai et al. 2015; [b] Starko et al. 2019; Silberfeld et al. 2010; Kawai et al. 2015.)

(Ochrophyta) are diatoms and Xanthophyceae; however, brown algae are the only group presenting complex multicellularity. Brown algae exhibit a wide range of morphologies and a fairly high level of morphological complexity (Charrier et al. 2012). This group of algae is extremely diverse in size, ranging from just a few hundreds of micrometers to up to 40 m, for example, the kelp forests that provide shelter and feeding grounds for many marine animals. Their diversity in shape is also considerable, ranging from crusts to digitated blades, all growing attached to rocky surfaces or on other algae (epiphystism).

This chapter reports on research carried out on two very different brown algal species: the microscopic filamentous *Ectocarpus* sp., which entered the genomics and other -omics era 10 years ago, and the large laminate *Saccharina latissima*, which is currently raising increasing interest in Europe as a future source of food and derived agri-food and pharmaceutical products. These algae belong to the orders Ectocarpales and Laminariales, respectively, which diverged ~100 Mya (Silberfeld et al. 2010). Here, we present these two models in the context of studies focused primarily on development and growth.

2.2 *ECTOCARPUS* SP.

2.2.1 History of the Model and Geographical Location

Records of the occurrence of *Ectocarpus siliculosus* in the environment emerged about two centuries ago. This species was first described as *Conferva siliculosa* by Dillwyn in 1809 from material collected in England (Dillwyn 1809). Ten years later, Lyngbye recorded *Ectocarpus* sp. as *Conferva confervoides* from material collected in Denmark (Lyngbye 1819). As a result, this species is now named *Ectocarpus siliculosus* (Dillwyn) Lyngbye.

This species belongs to the order Ectocarpales, which includes most of the brown algae with a simple body architecture, mainly filamentous in habit. Due to these morphological features, *Ectocarpus* sp. was initially classified at the root of the brown algae phylogenetic tree with the Discosporangiales (e.g. *Choristocarpus* spp.), displaying similarly low morphological complexity. However, molecular markers identified in the 1980s led to more accurate phylogenetic analyses and classified the Ectocarpales as a sister group to the most morphologically complex family of brown algae, the Laminariales (kelps, see Section 2.3), far from the basal brown algal groups (Silberfeld et al. 2014) (Figure 2.1b).

Ectocarpus sp. is a tiny, filamentous brown alga, thriving in all temperate marine waters in both hemispheres. There is a recent geographical inventory of several species, together with their phylogenetic relationship (Montecinos et al. 2017). Although some *Ectocarpus* species are highly sensitive to salinity (Dittami et al. 2012; Rodriguez-Rojas et al. 2020), other species can also thrive in freshwater, particularly in rivers. The complexity of their associated microbiome may contribute to their adaptation to these environments (Dittami et al. 2016; Dittami et al. 2020). Interestingly, in contrast to other Phaeophyceae, *Ectocarpus* species have spread extensively around the world and are not confined to any specific geographical area. This wide distribution is likely due, for the most part, to the high capacity of *Ectocarpus* spp. to adhere to various artificial surfaces, such as boat hulls, ropes, and so on (biofouling), promoting their dispersal through maritime traffic (Montecinos et al. 2017).

2.2.2 Life Cycle

Ectocarpus spp. grow following a microscopic, haplodiplontic, dioicous life cycle (Figure 2.2). For some species, however, only a part of this complex life cycle can be observed in natural conditions, regardless of the ecological niche (Couceiro et al. 2015). The different stages of the life cycle and related mutants are described in Figure 2.2 and Section 2.2.6.

2.2.3 Embryogenesis and Early Development

Embryogenesis is not a term well adapted to *Ectocarpus* sp., because its early body lacks complex tissue organization and has only one growth axis. Instead, from the onset of zygote germination, *Ectocarpus* sp. develops a primary uniseriate filament along a proximo-distal axis, on which secondary filaments subsequently emerge serially (Le Bail et al. 2008). Successive and iterative branching continues and results in the development of a bushy organism of a few millimeters after 1–2 months. Interestingly, this low level of morphological complexity and slow growth (~3 µm.h-1; Rabillé et al. 2019a) endow Ectocarpus with the features of a convenient model for studying several fundamental cell growth and cell differentiation processes.

The development of the sporophyte (2n) is initiated by the emergence of a tip from the zygote (Figure 2.3a, b). The growth of this tip is indeterminate throughout the development of the organism, and it can be described by a simple and original biophysical model based on the control of the thickness of the algal cell wall in the tip area (Rabillé et al. 2019a). In this area, the cell wall is mainly composed of the two main polysaccharides identified in brown algal cell walls: alginates [combination of two types of residues: (1→4) α-L-guluronic acid (G residues) and (1→4) β-D-mannuronic acid (M residues); 40% of the cell wall] and fucans (polysaccharides containing α-L-fucosyl residues; 40%) (reviewed in Charrier et al. 2019). When sulfated, these fucans are called fucose-containing sulfated polysaccharides (FCSPs; Deniaud-Bouët et al. 2014). Although alginates may be necessary in particular for the growth of highly curved cell surfaces (Rabillé et al. 2019b), sulfated fucans may provide additional biophysical properties, for example, hygroscopy and high flexibility (Simeon et al. 2020).

In the wild type, the apical cell of each filament is a very long cylindrical cell (length > 40 µm; diameter 7 µm), but in the mutant *etoile*, the apical cell is shorter and wider. In this mutant, tip growth stops shortly after it is initiated, and cells have a thicker cell wall and an extensive Golgi apparatus (Le Bail et al. 2011).

The expansion of the tip outward is accompanied by cell division (~1 every 12 h in standard lab conditions; Nehr et al. 2011). The first cell division separating the round zygotic cell and the growing elongated cell is asymmetrical (Le Bail et al. 2011; Figure 2.3b). Once the filament has grown a few cells on one end, the initial zygotic cell germinates on the opposite end, thereby producing a filament along the same axis as the initial filament. The two processes result in the formation of a multicellular uniseriate filament made up of a series of elongated cells aligned along a single growth axis.

These cylindrical cells progressively change shape and become round (Le Bail et al. 2008) (Figure 2.3). This rounding up from a cylindrical cell to a spherical cell is reminiscent of the cell rounding that takes place in highly polarized metazoan cells before mitosis, where this process has been shown to ensure proper spindle assembly (Lancaster and Baum 2014) and equal distribution of cellular materials. In *Ectocarpus* sp., the underlying mechanisms for this cell rounding differentiation process are still unknown, but modeling has shown that local cell–cell communication between neighboring cells is likely involved, not long-range diffusion of a signaling molecule (Billoud et al. 2008).

Branching takes place primarily on maturing polarized cells and to a lesser extent on already formed round cells

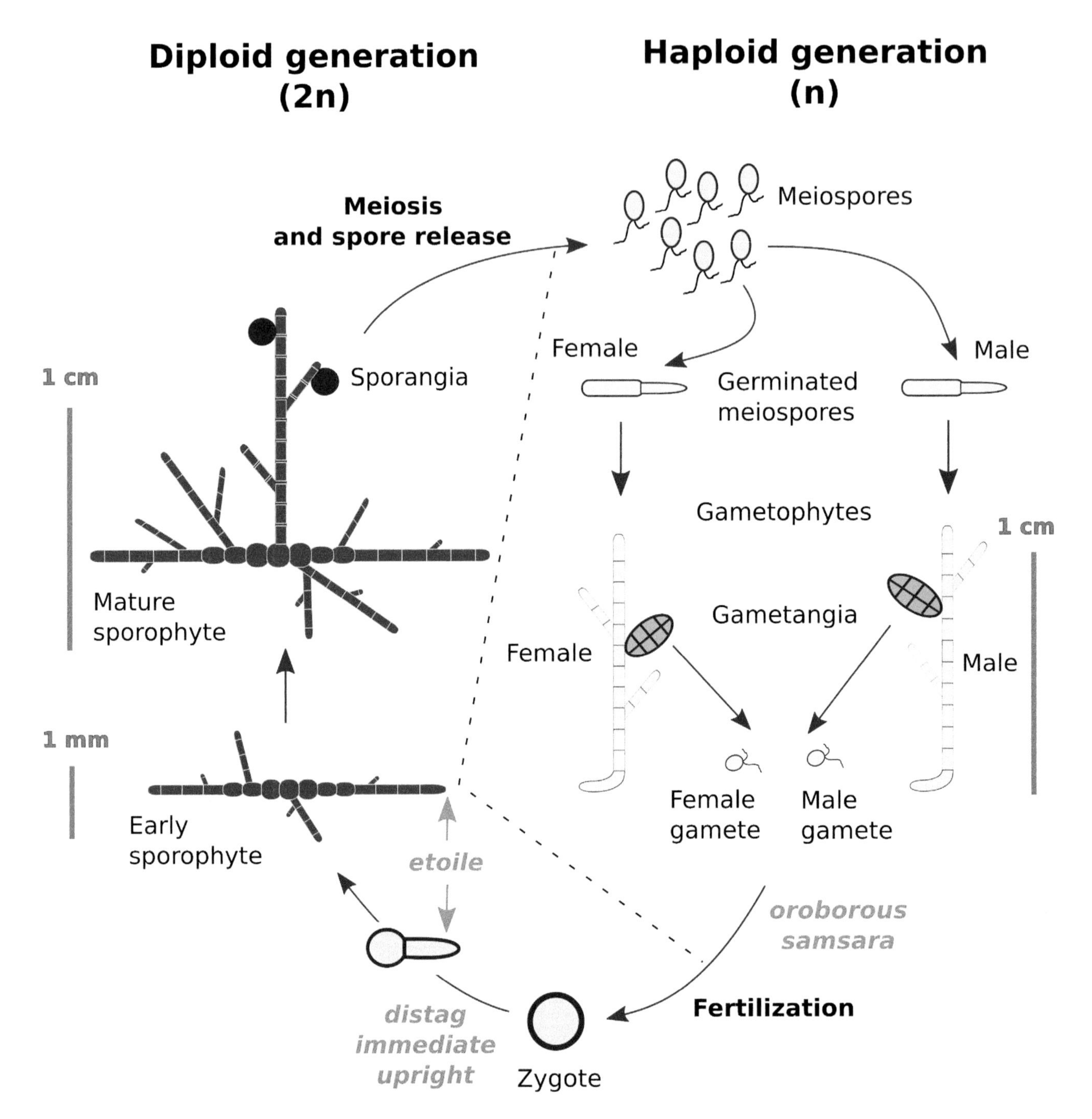

FIGURE 2.2 **Life cycle of *Ectocarpus* sp.** (summarized in Charrier et al. 2008). Diploid (brown) phase (left-hand side) is made of microscopic sporophytes composed of branched uniseriate filaments. Meiosis takes place in unilocular sporangia (dark brown circles) differentiating laterally on erect branches. Haploid (light green) phase (right-hand side, yellow shaded area) corresponds to the formation of gametophytes, which are erect branched uniseriate filaments growing from germinated meiospores (gray circles and light green cells). Male and female gametes are each released from male and female gametophytes (dioicous life cycle) and fuse freely in the external environment (seawater), producing a free zygote (orange circle). *Ectocarpus* sp. is therefore characterized by its small size, distinguishing it from most of the other brown algae (e.g. the kelp *Saccharina* sp.) (note the scale). Characterized mutants impaired in the different steps of the life cycle are indicated in light brown.

(Figure 2.3c). The detailed process is unknown. It does not seem to depend on actin filaments (although growth is) (Coudert et al. 2019) or microtubules (personal observations). A biophysical study based on the assumption that the cell wall is a poro-elastic material suggests that an increase in surface tension during the enlargement of rounding cells is sufficient to induce branching (Jia et al. 2017). Branching never occurs in the apical cell or twice in the same cell, suggesting the action of inhibitory mechanisms ensuring spacing between branches (Figure 2.3d). One potential contributor to inhibition is the phytohormone auxin, shown to accumulate at the tip of Ectocarpus filaments (Le Bail et al. 2010). Auxin may then establish a decreasing gradient along the linear filament, preventing the emergence of

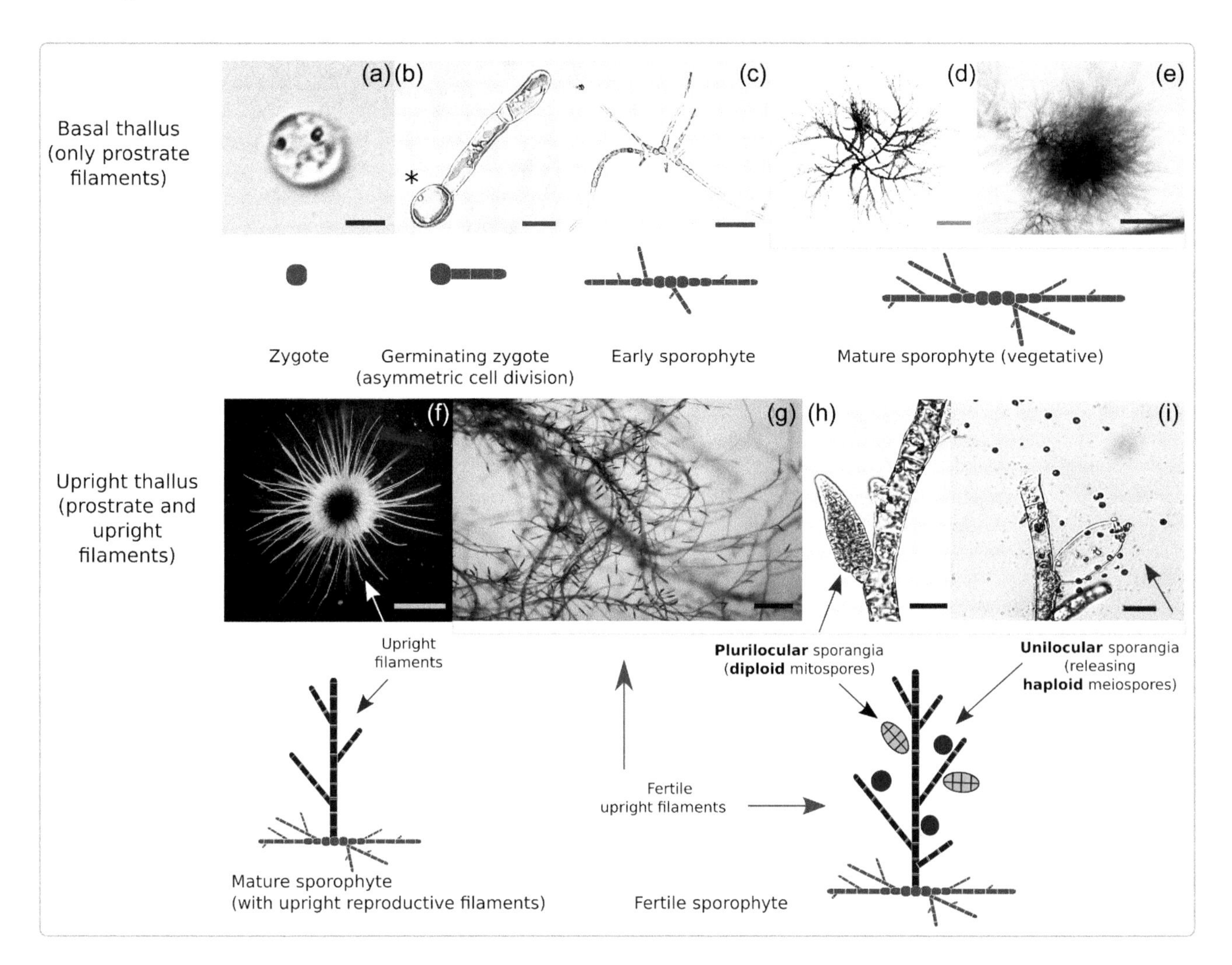

FIGURE 2.3 Developmental stages of the *Ectocarpus* sp. sporophyte. Photos and accompanying schematic representations of the different stages of sporophyte development. From top left to bottom right: the zygote (a) germinates, forming a tube, and then divides asymmetrically (b). (c) Filaments are formed by apical cell growth and cell division of the primary filament, followed by branching, leading first to a small tuft after ~20 days (d), then to a larger one after ~1 month (e). This makes up the prostrate part of the thallus (top). After ~1 month, upright filaments emerge (f, dark brown on the schematic representation), on which two kinds of reproductive organs differentiate: plurilocular sporangia (g) releasing mitospores (h, green in the schematic, not shown in Figure 2.2 for simplicity), which have the capacity to germinate as their parent, generating another sporophyte genetically and morphologically identical to its parent, and unilocular sporangia releasing meiospores after meiosis (i, brown in the schematic). These haploid spores germinate as female and male gametophytes in equal proportion (not shown). Scale bars (a, b) 5 µm, (c) 50 µm, (d, g) 100 µm, (e, f) 1 mm, (h, i) 20 µm. ([b] Le Bail et al. 2011; Billoud et al. 2008.)

branches in the most distal area of the filament and allowing branching in the more central regions. However, there must be additional mechanisms operating to explain the spacing between branches. Interestingly, during growth, growing filaments generally tend to avoid each other, following curved trajectories. This observation suggests the existence of lateral inhibition mechanisms through chemical diffusion in the environment. It is not known whether branching spacing relies on the diffusion of inhibitors in the external medium or is transported by the neighboring cells within the filaments (*Ectocarpus* sp. cells possess plasmodesmata, i.e. holes in the cell wall connecting the cytoplasms of neighboring cells; Charrier et al. 2008). Finally, branching may also be controlled by an internal clock pacing the branching process not in space but in time, ultimately resulting in an evenly spaced branching pattern in organisms growing at a regular pace (Nehr et al. 2011). Very interestingly, this cadence is maintained in the tip-growth mutant *etoile* (see previously), but the relative position of branches is not. In this mutant, branching continues at the same rate as in the wild type, but tip growth stops, leading to the formation of a compact bushy tuft (Nehr et al. 2011).

Branching results in branches with exactly the same morphology as the "parental" filament. Therefore, the reiteration of branching leads only to the addition of filaments identical to the very first one. Altogether and after ~1 month, the adult body looks like a tuft of filaments (Figure 2.3e).

Regarding the conservation of branching mechanisms on an evolutionary scale, the branching pattern observed in *Ectocarpus* sp. shares some morphological features with mosses and fungi. However, the underlying mechanism seems to be different to some extent, thereby indicating that these lineages took different evolutionary paths to develop similar, low-complexity body architectures (Coudert et al. 2019).

dis mutants lack the basal, prostrate part of the sporophyte body and are impaired in microtubule and Golgi network organization (Godfroy et al. 2017). The *DISTAG* (*DIS*) gene codes for a protein containing a TBCC domain, whose function in internal cell organization is conserved throughout the tree of life.

2.2.4 Anatomy—Later Development

Beyond the early stages of sporophyte development, *Ectocarpus* sp. develops a second type of filament (Figure 2.3f). This filament grows upright, away from the substratum surface, and differentiates into different cell types: cells are chunky and lined up on top of each other, making a straight and stout filament, on which few branches emerge. However, these filaments remain uniseriate, like the earlier, prostrate ones. Therefore, the level of complexity of the overall morphology of the *Ectocarpus* sp. sporophyte remains low. After roughly two weeks, these upright filaments allow the differentiation of lateral reproductive organs (plurilocular sporangia and unilocular sporangia; see Charrier et al. [2008] for a review; Figure 2.3g–i). The mechanisms initiating the growth of these specific filaments, and those initiating the differentiation of the reproductive organs, are completely unknown to date.

2.2.4.1 Meiosis and the Gametophytic Phase

Meiosis takes place in the unilocular sporangia borne by the upright filaments of the sporophyte (see previously). They release roughly 100 meiospores in the seawater, and each meiospore germinates into a female or male gametophyte, making this second phase of the *Ectocarpus* sp. haplodiplontic life cycle dioicous (reviewed in Charrier et al. 2008).

The first cell division in the gametophyte leads to the formation of a rhizoid and an upright filament. Upright filaments keep developing, but the rhizoid remains inconspicuous. *Dis* mutants are characterized by their lack of basal, prostrate filaments in the sporophyte (see previously) and also lack rhizoids in the gametophyte phase (Godfroy et al. 2017), suggesting that the formation of the gametophyte rhizoid and of the sporophyte prostrate filaments are controlled by the same genetic determinism.

The upright filament continues growing and produces lateral branches morphologically similar to the upright sporophyte branches, except that they never carry unilocular sporangia, and they are more densely distributed with different branching angles (Godfroy et al. 2017).

Transcriptomics studies have shown that only 0.36% of the total number of transcripts are specific to the sporophyte phase (12% are biased by a fold change ratio of at least 2), while 7.5% are specific to the gametophyte phase (23% biased) (Lipinska et al. 2015; Lipinska et al. 2019). Therefore, more than 90% of the total transcriptome identified in *Ectocarpus* sp. is shared by both generations of the life cycle. This differential expression may account for the slight morphological differences between sporophytes and gametophytes (see previously), or, more likely, to the different reproductive organs and behavior. Nevertheless, genes related to carbohydrate metabolism and small GTPase signaling processes are expressed more abundantly in sporophytes, and expression of those related to signal transduction, protein–protein interactions and microtubule and flagellum movement are enriched in gametophytes.

Ultimately, lateral buds on these gametophyte filaments differentiate into pedunculate plurilocular gametangia. Each gametangium, either female or male, releases roughly 100 flagellated gametes in the external medium. Females secrete pheromones and mediate the attraction of male gametes (e.g. ectocarpene; Müller and Schmid 1988), and specific recognition of female and male gametes is based on a glycoprotein ligand-receptor interaction (Schmid 1993; reviewed in Charrier et al. 2008).

Some *Ectocarpus* species have both sexual and asexual life cycles (Couceiro et al. 2015). In an asexual cycle, an unfertilized gamete can germinate if it does not fuse with a sexual partner, resulting in a haploid parthenosporophyte with the same morphology as the diploid sporophyte. Like the diploid sporophyte, this parthenosporophyte bears unilocular sporangia in which meiosis takes place. Endoreduplication has been shown to take place very early during growth of the parthenosporophyte or just at the onset of the sporangium emergence (Bothwell et al. 2010). The gametes of the mutant *oroborous* (*oro*) do not grow as parthenosporophytes but instead develop as gametophytes (Coelho et al. 2011). The gene *oro* codes for a homeodomain (HD) protein, which, through an heterodimer formed with the other HD protein SAMSARA, controls the sporophyte-to-gametophyte transitions, as in basal members of the Archaeplastida (Plantae) (Arun et al. 2019).

2.2.4.2 Sex Determination

The gametophyte phase is represented by female and male haploid gametophytes. Sex in *Ectocarpus* sp. is based on the UV sexual system, where female (U) and male (V) sexual traits are expressed in the haploid phase (in contrast to the XY and ZW systems in which the sexual traits are expressed in the diploid phase). Similar sexual systems are also found in green algae (e.g. the charophyte *Volvox* sp.) and in the bryophytes *Ceratodon* sp. (moss) and *Marchantia* sp. (liverwort) (Umen and Coelho 2019). In *Ectocarpus sp.*, the sex determining regions (SDRs) are relatively small genomic areas of ~0.9 Mbp (representing ~0.5% of the total genome of 214 Mbp), of similar size in females and males and framed by pseudoautosomal regions (PARs) (Ahmed et al. 2014; Bringloe et al. 2020). The SDR contains a few coding genes (15 in the female and 17 in the male) that are expressed during the haploid phase; the PAR contains genes mainly expressed during the sporophyte phase. Noteworthily, most (11) of these genes are shared by both the female and the male SDR and have homologs elsewhere in the genome (either in the PAR region or in autosomes). Therefore, the identity of the *Ectocarpus* sp. sex locus is weak compared with other species, both in the number and in the specificity of its genes. Nevertheless, these SDR loci control the expression of 753 female genes (with a -fold change [FC] > 2), representing 4.3% of the total transcripts (5.5% of the transcripts expressed in the female gametophyte), located in the rest of the genome during the haploid phase (Lipinska et al. 2015). In the male gametophyte, 1391 genes (7.9% total transcripts, 10% male-gametophyte-expressed genes) are specifically expressed with a FC > 2.

However, the role of these gametophyte genes in sex determination remains unclear, because the sexual dimorphism observed in this genus is nonexistent in vegetative gametophytes and subtle during the reproductive phase, during which male gametophytes produce more gametangia and slightly smaller gametes than female gametophytes (Lipinska et al. 2015; Luthringer et al. 2014). This slight dimorphism is reflected by the weak differential expression of sex-biased genes at these two stages of gametophyte development.

In summary, *Ectocarpus* sp. is characterized by a low level of morphological complexity: cells are aligned, and growth is one dimensional, followed by reiterated branching events producing filaments similar to the "mother" filament. The life cycle is virtually isomorphic: sporophyte and gametophyte are both filamentous, mainly made up of upright filaments, and gender traits are absent.

2.2.5 Genomic Data

The nuclear genome of *Ectocarpus* sp. (accession CCAP 1310/4) has been estimated to contain 214 Mbp, and genome sequence annotation identified 17,418 genes (Cormier et al. 2017). As the first sequence known for a brown alga at that time, it revealed unusual features. With a high GC content, genes are composed of, on average, 8 × 300 bp exons, separated by seven introns of 740 bp. Alternative splicing takes place with a frequency leading to 1.6 transcript per gene (Cormier et al. 2017), comparable to alternative splicing in metazoans and plants. Promoters have not been characterized to date, and 3'-UTR regions are particularly long (~900 bp), in contrast to most other organisms of similar genome size but similar to mammalian genomes. From this genome, several families of transposable elements, of which retrotransposons and retroposons are the most abundant, cover ~20% of the genome (Cock et al. 2010), as well as 23 microRNAs identified from a genome-based approach and whose expression has been quantified by q-RT-PCR (Billoud et al. 2014). This inventory also includes a set of 63 miR candidates identified from an RNA-seq-based approach, limited by the extent of range and level of gene expression (Tarver et al. 2015).

Interestingly, a significant proportion of *Ectocarpus* sp. genes are organized on alternating DNA strands along the chromosome, a feature specific to compact genomes.

A preliminary genetic map built with microsatellite markers was proposed in 2010 (Heesch et al. 2010), since supplemented with single nucleotide polymorphism (SNP) markers, facilitating the identification of mutated loci (Billoud et al. 2015; Cormier et al. 2017). All together, based on genetic linkage and flow cytometry data (although from another species), *Ectocarpus* sp. does not appear to have more than 28 chromosomes (Cormier et al. 2017).

2.2.6 Functional Approaches: Tools for Molecular and Cellular Analyses

Based on a solid knowledge of its biology and life cycle (reviewed in Charrier et al. 2008), *Ectocarpus* sp. was chosen as a genetic model for brown algae in the 2000s (Peters et al. 2004). Its genome was sequenced in 2010, which was a major breakthrough as the first genomic sequence for a brown alga and, what's more, the first multicellular macroalga (Cock et al. 2010). This breakthrough was accompanied by the development of a full palette of technical tools. Only techniques related to cell biology, cultivation and genetics are considered in the following.

2.2.6.1 Cultivation in the Laboratory

Ectocarpus sp. is easily grown in laboratory conditions (Le Bail and Charrier 2013). Growth speed, morphology and fertility induction depend on (white) light intensity (usually dim, <30 $\mu E.s^{-1}.m^{-2}$), photoperiod (long day or equal day: night cycle) and temperature (13–14°C). Due to its small size, optical microscopes and stereo microscopes are required to follow the different stages of the life cycle. Micromanipulation (using tweezers) is often necessary to separate the different organs of the *Ectocarpus* sp. body, such as sporangia. The adult organism is a few centimeters long, meaning that the whole life cycle can be carried out in a small recipient such as a Petri dish. Altogether, the cultivation of *Ectocarpus* sp. is amenable to rudimentary laboratory conditions and equipment. To avoid contamination with either bacteria or protozoa, *Ectocarpus* sp. is preferably handled under a sterile laminar hood.

2.2.6.2 Cell Biology and Biophysical Techniques

Transmission and scanning electronic microscopy techniques have both been used to observe *Ectocarpus* sp. cells and filaments (e.g. Le Bail et al. 2011; Tsirigoti et al. 2015), facilitated by the filamentous shape of this organism, exposing all cells to observation. However, because the cells are small (filament cell diameter, 7 µm), observation of a specific cell orientation may be difficult to handle. However, exploiting the fact that *Ectocarpus* sp. grows on surface, it is possible to make serial sections of apical filament cells in longitudinal and transversal axes, as illustrated in Rabillé et al. (2019), who measured the thickness of the cell wall along the meridional axis of the cell.

Protocols for immunocytochemistry (ICC, or immunolocalization) of cytoskeleton components have been developed in the past 20 years, inspired by protocols developed on other brown algae (reviewed in Katsaros et al. 2006). Microtubules (Coelho et al. 2012; Katsaros et al. 1992), actin filaments (Rabillé et al. 2018b) and centrin (Katsaros et al. 1991; Godfroy et al. 2017) can now be visualized in *Ectocarpus* sp. cells. These ICC protocols rely on the high conservation of these molecules, allowing the use of commercial primary antibodies raised against animal homologs. ICC using antibodies specific to *Ectocarpus* sp. has not been reported yet. However, monoclonal antibodies raised against polysaccharide components of the brown algal cell wall have been produced (Torode et al. 2016, 2015) and are now used to map specific blocks of alginates (Rabillé et al. 2019b) and fucans (Simeon et al. 2020).

A recent study on *Ectocarpus* sp. using mRNA *in situ* hybridization after an attack by a pathogen (Badstöber et al. 2020) showed mRNA in subcellular locations within the infected cell. The development of filament-wide *in situ* mRNA labeling is needed to monitor responses or cell-fate programs at the level of the whole organism.

Additional techniques, previously developed in other organisms, have been transferred to *Ectocarpus sp.* Growth of the cell surface can be monitored by loading sticky fluorescent beads on the filament surface. Recording the position of the beads as the cell expands (either during growth or in response to a stimulus) makes it possible to measure the propensity for deformation of specific cell areas. This measurement provides information on cell mechanical properties (Rabillé et al. 2018a). Mechanical properties can also be studied using atomic force microscopy, a biophysical technique that records how deep a cantilever can plunge into a cell surface and retract, according to the cell wall stiffness and adhesion (Gaboriaud and Dufrêne 2007). *Ectocarpus* sp. is particularly amenable to such approaches, because its cells are directly exposed to the cantilever (Tesson and Charrier 2014). This technique helped show that the cells along the sporophyte filament display different degrees of surface stiffness (Rabillé et al. 2019b).

2.2.6.3 Modification of Gene Expression

Attempts to genetically transform *Ectocarpus* sp. have been numerous and so far unsuccessful. Agrotransformation, electroporation, PEG-mediated protoplast or gamete transformation and micro-injection have all been tested and shown to be inefficient. A major issue is that there is little to no information on *Ectocarpus* sp. gene promoters, and heterologous promoters tested so far (e.g. diatom, Ulva, Maize or Plant virus CaMV35S) have not been shown to be functional (personal communication).

Therefore, "ready-to-use" molecules that can alter the expression of host genes without relying on the host transcription and translation machinery currently appear to be a more promising approach. Morpholinos and RNA interference have not proven to be efficient enough for routine transient knock-down experiments (personal communication; Macaisne et al. 2017).

Efforts are currently being put into the development of the CRISPR-Cas9 technology (Lino et al. 2018), shown to be a powerful tool to stably modify the genome of several marine organisms, including echinoderms (sea urchins; Lin et al. 2019) and tunicates (*Phallusia* sp.; McDougall et al. 2021). Because the expression of the guided RNA and the Cas9 protein from the host genome remains challenging, the use of pre-assembled guide RNA-Cas9 protein complex, as illustrated in Brassicaceae plants (Murovec et al. 2018), is currently considered the most promising strategy.

Several morphogenetic mutants have been generated by UV irradiation, among which some have been genetically characterized. These mutants are impaired in tip growth (Le Bail et al. 2011), cell differentiation (Godfroy et al. 2017; Macaisne et al. 2017; Le Bail et al. 2010), branching and reproductive phase change (Le Bail et al. 2010). In most mutants, several morphogenetic processes are affected, reflecting the low level of complexity of *Ectocarpus* sp. morphogenesis and suggesting an overlap in genetic functions (see transcriptomic results previously). Others are impaired in the alternation of the sporophyte and gametophyte generations (life cycle mutants: Coelho et al. 2011; Arun et al. 2019) (Figure 2.2).

2.3 *SACCHARINA LATISSIMI*

2.3.1 Nomenclature History, Evolution, Geographical Distribution and Uses

2.3.1.1 History of Its Nomenclature

Saccharina latissima (Linnaeus) C.E. Lane, C. Mayes, Druehl & G.W. Saunders 2006 is a marine photosynthetic eukaryotic organism with many different common names, including sugar kelp, sea-belt, kombu, sugar tang, poor man's weather glass and so on. Originally, in 1753, Linnaeus considered it an *Ulva* species, *Ulva latissima*, due to its sheet-like blade, common in the genus *Ulva* (Linnaeus and Salvius 1753). In 1813, Lamouroux reclassified it as *Laminaria saccharina* (Lamouroux 1813), despite its original genus name *Saccharina* given by the botanist J. Stackhouse in 1809. This genus name was resurrected in 2006 when molecular phylogenetics made it apparent that the order Laminariales should be split into two clades or families (Lane et al. 2006), which diverged ~25 Mya (Starko et al. 2019). Now, *Laminaria* spp. are assigned to the Laminariaceae family, and *Saccharina* spp. are part of the Arthrothamnaceae family (Jackson et al. 2017).

2.3.1.2 Evolution and Diversification

Classic taxonomy using morphological or physiological characteristics is useful for identifying species in the field; however, in the absence of a genetic approach, they can lead to long-lasting species confusions.

Among the brown algae, kelps are thought to have emerged ~75 Mya (Starko et al. 2019). Within the kelps (order Laminariales), *S. latissima* belongs to the so-called "complex kelps" (Starko et al. 2019) and thus shows close genetic similarity with various genera, allegedly resulting from an important upsurge in speciation beginning 31 Mya, concomitant to a massive marine species extinction due to the cooling of the Pacific Ocean during the Eocene–Oligocene boundary.

2.3.1.3 Geographical Distribution

Kelps are now almost cosmopolitan species, their presence ranging from temperate to cold waters on both sides of the Atlantic and Pacific Oceans (Bartsch et al. 2008). *Saccharina* genus appears to have initially emerged in the Northwest Pacific (North Japan, Russia) (Bolton 2010;

Luttikhuizen et al. 2018; Starko et al. 2019) and then spread further to three or four distinct regions of the globe where different lineages of *S. latissima* can be traced: in temperate to cold-temperate (sub-Arctic) waters of the Northeast Pacific, where the early diversification of Laminariales ancestors took place, and in the Northeast and Northwest Atlantic (Neiva et al. 2018; Starko et al. 2019). *S. latissima* is absent from the southern hemisphere (Bolton 2010).

Even though these populations seem to be considered as a single species (assumption supported by crosses), barcoding studies (based on the cytochrome c oxidase gene, used for) indicate high divergence between regions (Neiva et al. 2018). In combination with their morphological divergence and history of glacial vicariance (Neiva et al. 2018), these regional groups of *S. latissima* are clearly differentiating into separate species.

2.3.1.4 Uses

Individual kelp can become enormous: *S. latissima* blades can grow up to 45 m in length (Kanda 1936). As such, kelps constitute the largest coastal biomass and one of the main primary producers of the oceans. According to the FAO (2018), kelps in general, and *S. latissima* in particular, are cultivated and consumed mainly in Asia for human sustenance as well as for their alginate and iodine contents. In comparison, European consumption and production are considerably lower, and wild populations are used for various applications, mainly food and feed (Rebours et al. 2014; Barbier et al. 2019). Recent innovations aim to combine *S. latissima* cultivation with salmon aquaculture (integrated multi-trophic aquaculture) to reduce the impact of fish farms in Norway (Fossberg et al. 2018). *S. latissima* has been proposed as a source of bioethanol (Adams et al. 2008; Kraan 2016), and substances such as the sulfated polysaccharides fucoidans, laminarin and other extracts have demonstrated antitumoral effects along with anti-inflammatory and anticoagulant pharmacological properties (Cumashi et al. 2007; Mohibbullah et al. 2019; Han et al. 2019; Long et al. 2019). The number of clinical studies on mice for testing the positive effects of kelp extracts keeps increasing, as attested by a simple search in the PubMed scientific literature search engine.

2.3.2 Life Cycle

S. latissima is characterized by a highly heteromorphic haplodiplontic life cycle (Figure 2.4). Meiosis leads to the haploid gametophytic stage (or generation) of the life cycle, which, upon fertilization, gives rise to a diploid sporophyte stage. In *S. latissima*, and generally in Laminariales, the sporophyte generation is considerably different morphologically from the gametophyte generation (Kanda 1936; Fritsch 1945), in contrast to the isomorphic haplodiplontic life cycle of Ectocarpales, as seen in the previous section (Figure 2.2). The gametophyte (haploid) is microscopic and slowly grows into a prostrate filamentous thallus, and the sporophyte (diploid) is large and conspicuous. Upon favorable conditions, reproduction is initiated by a gradual differentiation of the cells of the gametophytic filaments into reproductive cells, the gametangia—antheridia (male) or oogonia (female)—a process induced by blue light (Lüning and Dring 1972). Interestingly, this induction is accompanied by changes in gene expression that are by and large common to female and male gametophytes, suggesting that the initiation of germline differentiation follows similar general mechanisms independently of gender (Pearson et al. 2019). That is, transcriptomics studies have revealed enhanced transcriptional and translational activities as well as metabolic activities (carbohydrate biosynthesis and nitrogen uptake), suggesting that gametogenesis is accompanied by an intensification of primary cellular and metabolic functions. This intensification is surprising when weighed against the fact that only one single gamete is produced by each gametangium. Yet there are differences between female and male gametogenesis proper. A small set of genes display gender-dependent induction of their expression, seemingly faster in females than in males (Pearson et al. 2019). Genes involved in basic cellular function (protein modification, nucleoplasmic transport, intron splicing), energy production and metabolic pathways and more specifically in oogenesis (reactive oxidative species metabolism) are overexpressed during female gametogenesis, in addition to prostaglandin-biosynthesis genes (Pearson et al. 2019; Monteiro et al. 2019). In turn, and as expected, male gametogenesis is accompanied mainly by the over-expression of "high mobility group" (HMG) genes, a conserved marker of male gender determination in animals, fungi and brown algae (Ahmed et al. 2014), which suppresses the development of female gender, hereby considered as set by default (Pearson et al. 2019).

In relatively high temperature conditions (20°C), the male and female gametophytes show more similar transcriptomic patterns, probably indicating a change of focus from gametogenesis-related genes to resistance to heat stress, amplified in females (Monteiro et al. 2019).

The oogonium releases an egg, leaving behind an empty apoplast, a process that is subject to the circadian rhythm and, in contrast to the formation of gametangia, is inhibited by blue light (Lüning 1981). Male gametes swim to the egg in response to female pheromones (e.g. lamoxiren; Hertweck and Boland 1997), demonstrating conspicuous chemotaxis (Maier and Müller 1986; Maier and Muller 1990; Boland 1995; Maier et al. 2001; Kinoshita et al. 2017). Upon fertilization, the early sporophyte develops as a planar embryo. In *Saccharina japonica*, the ratio of genes specific to sporophytes or gametophytes is more balanced than in *Ectocarpus* sp., with ~4% (about 700 genes) of the total number of transcripts being specifically expressed in both phase organisms (Lipinska et al. 2019). This difference in transcripts can be interpreted as a reflection of the conspicuous morphological differences between these two life cycle generations in *S. latissima*, contrasting with the near isomorphy in *Ectocarpus* sp.

The developing diploid sporophyte requires several months before reaching sexual maturity (Andersen et al. 2011; Forbord et al. 2018, 2019). Then, sori, groups of sacs (sporangia) of meiospores (swimming spores that are the

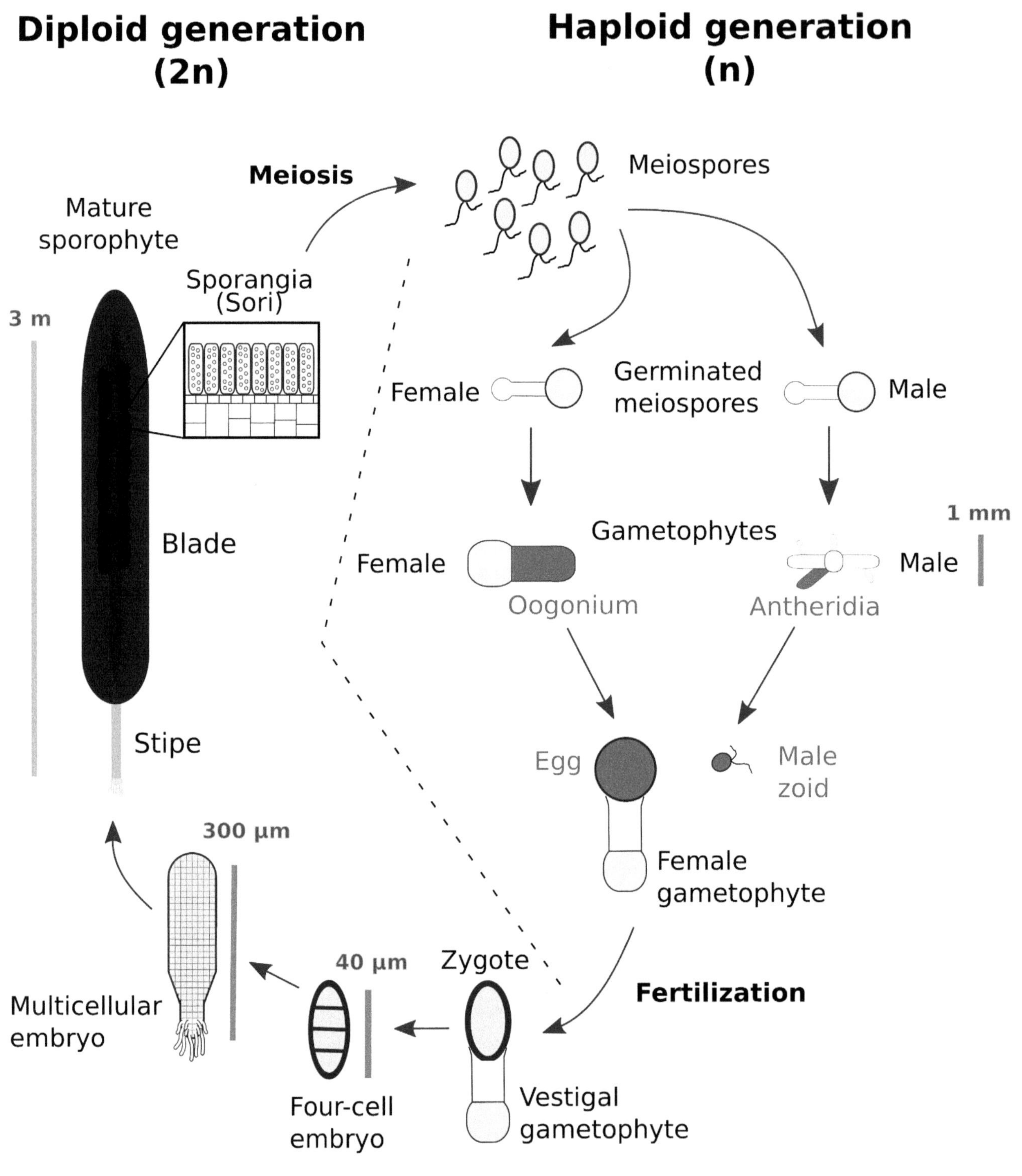

FIGURE 2.4 **The heteromorphic halplodiplontic life cycle of *Saccharina latissima*.** The large fertile sporophyte develops sporangia (located in sori) around and on the soft midrib of the blade. The released haploid meiospores germinate to female or male gametophytes. If conditions are optimal, the one- to two-celled female gametophytes and the few-celled male gametophytes produce gametes. The female gamete (egg) is retained on the empty female gametangium. Only one gamete per gametangium is produced in each sex. After fertilization, the diploid sporophyte begins to develop. After some months, a conspicuous juvenile sporophyte emerges and requires at least four to five additional months to become fertile and produce meiospores. Note the scale of the different generations and life stages.

product of meiosis) abundantly differentiate on the surface of the blade, usually on and near the midrib and far from the basal part of the blade (Drew 1910), suggesting an inhibitory control in this part of the body. In the related species *S. japonica*, the two phytohormones auxin and abscissic acid have opposite effects in the induction of sorus formation; it was hypothesized that auxin is synthesized in the basal meristem, allowing sorus differentiation only in the more distal, apical areas of the blade (reviewed in Bartsch et al. 2008). Meiospores produced from these sporangia germinate into male or female gametophytes depending on the UV sex determination type inherited from meiosis (Lipinska et al. 2017; Zhang et al. 2019). Comparison of the genome of *S. japonica* and *Ectocarpus* sp., together with other brown algal genomes, shows that the sex determining region has evolved rapidly through gene loss and gene gain, similar to organisms with an XY or ZW sex determining system (Lipinska et al. 2017).

A review of the physiological parameters controlling the whole life cycle of Laminariales can be found in Bartsch et al. (2008).

2.3.3 Embryogenesis

The development of kelps was reported in some detail in the beginning of the 20th century. Since then, the developmental and cellular data amassed during the past decades pale in comparison with the ecophysiological and biochemical studies on kelps or the bioassays on the positive effects of their extracts. Especially for *S. latissima*, the majority of our knowledge on its development and histology is restricted to studies from the 19th century. Although detailed in histology and anatomy, information regarding the development of the blade and the stipe (schematized in Figure 2.4) is scarce, particularly for the earlier stages.

The early embryo has a distinct phylloid shape shared by most kelp species (Drew 1910; Yendo 1911; Fritsch 1945). Initially, there is no visible differentiation into stipe or blade, and the embryos are made of a flat layer of cells (Figure 2.5a–f). However, the proximal ends of these phylloids are narrower in width than the rest of the flat thallus (Figure 2.5e, f). Nevertheless, cellular divisions occur throughout the phylloid tissue without any hint of a pending superficial or intercalary meristem. At a certain point, probably related to the size of the thallus, the cells of the future stipe (Figure 2.5g, red arrow) divide internally, forming the first four layers. An increased rate of anticlinal divisions of the two outer layers and slow growth of the inner layers promote the formation of a cylindrical tissue (Fritsch 1945). The peripheral layer of cells, which are considerably smaller and more actively dividing than the internal cells, defines the meristoderm. The central cells surrounded by the cortex give rise to the first medullary elements, gradually becoming thinner and elongated, while their cell walls become enriched in mucilage (Killian 1911; Smith 1939; Fritsch 1945) (schematized in Figure 2.5i). At some point, a transition zone between the lamina and the stipe becomes visible, with the former being flat (Drew 1910; Yendo 1911).

The lamina becomes progressively polystromatic (several layers of cells in width), starting first in the vicinity of the transition zone and propagating toward the more distal, apical parts of the lamina (Figure 2.5g). Therefore, gradual polystromatization is basipetal. In parallel, specific organs and tissues are formed. In the longitudinal axis, blade, stipe and haptera differentiate, resulting in a clear apico-basal, asymmetrical axis; meanwhile, in the medio-lateral axis, specific tissues differentiate, mainly in the stipe and blade (meristoderm, cortex, medulla) (Figure 2.5h, i).

2.3.4 Anatomy

The female and male gametophytes develop microscopic filamentous bodies. Only the anatomy of the sporophyte will be described here. The mature thallus of *S. latissima* is composed of three main parts, the lamina, the stipe and the holdfast (Figure 2.4). The lamina, or blade, is unserrated, flat or bullate with a potential for growth of up to several meters (~40 m, according to Kanda 1936). Damage to the lamina may be irreversible if it exceeds a certain length. Otherwise, the lamina regenerates and continues growing (Parke 1948). This process seems to be age and season dependent, with lower potential for survival and development of a new blade after the first year of growth (reviewed in Bartsch et al. 2008). The stipe is cylindrical, with a flattened zone at the top corresponding to a transition zone between the stipe and the blade (Parke 1948) (Figure 2.5g). At the opposite end, an intricate structure appears with thick branched and intermingled protrusions called haptera (pl.) (hapteron [sg.]), which progressively form the holdfast, an organ anchoring the thallus to a solid substratum of the seabed (e.g. rocks). Histological observations show high secretory activity of adhesive material coming from the epidermal meristem of the haptera (Davies et al. 1973).

Histologically, the blade and the stipe are not very different (Fritsch 1945) (Figure 2.5i). However, the blade shows a more compressed lateral arrangement of the different tissues, and the borders of the most internal tissues seem obscured: the inner cortex is often not distinguishable from the outer cortex, making the transition to the medulla sudden (Figure 2.5h).

On the surface, an epidermal tissue covers the thallus of *S. latissima*, consisting of a few layers of small isodiametric cells (Sykes 1908; Smith 1939; Fritsch 1945) (Figure 2.5h, i). This tissue demonstrates high division activity, being responsible for the thickening of the stipe and of the blade to some extent, especially in the vicinity of the transition zone. This tissue is defined as the meristoderm, as it is essentially an epidermal meristem. According to Smith (1939), the blade's superficial tissue resembles an epidermis more than a meristoderm, implying the absence of meristematic activity. In contrast, Fritsch (1945) suggests that cell divisions still occur from the meristoderm, mostly along the anticlinal plane, thereby widening the blade. However, its division ceases in distal and mature regions away above the transition zone.

At the center of the thallus is found the medulla, an intricate network of elongated filamentous cells immersed in mucilage (Figure 2.5i). This tissue raised high interest in algal histology in the past (Sykes 1908; Schmitz et al. 1972; Lüning et al. 1973; Sideman and Scheirer 1977; Schmitz and Kühn 1982), most likely because of its intriguing structure but also because of its important physiological role: it offers structural resistance and is the main transporting tissue for

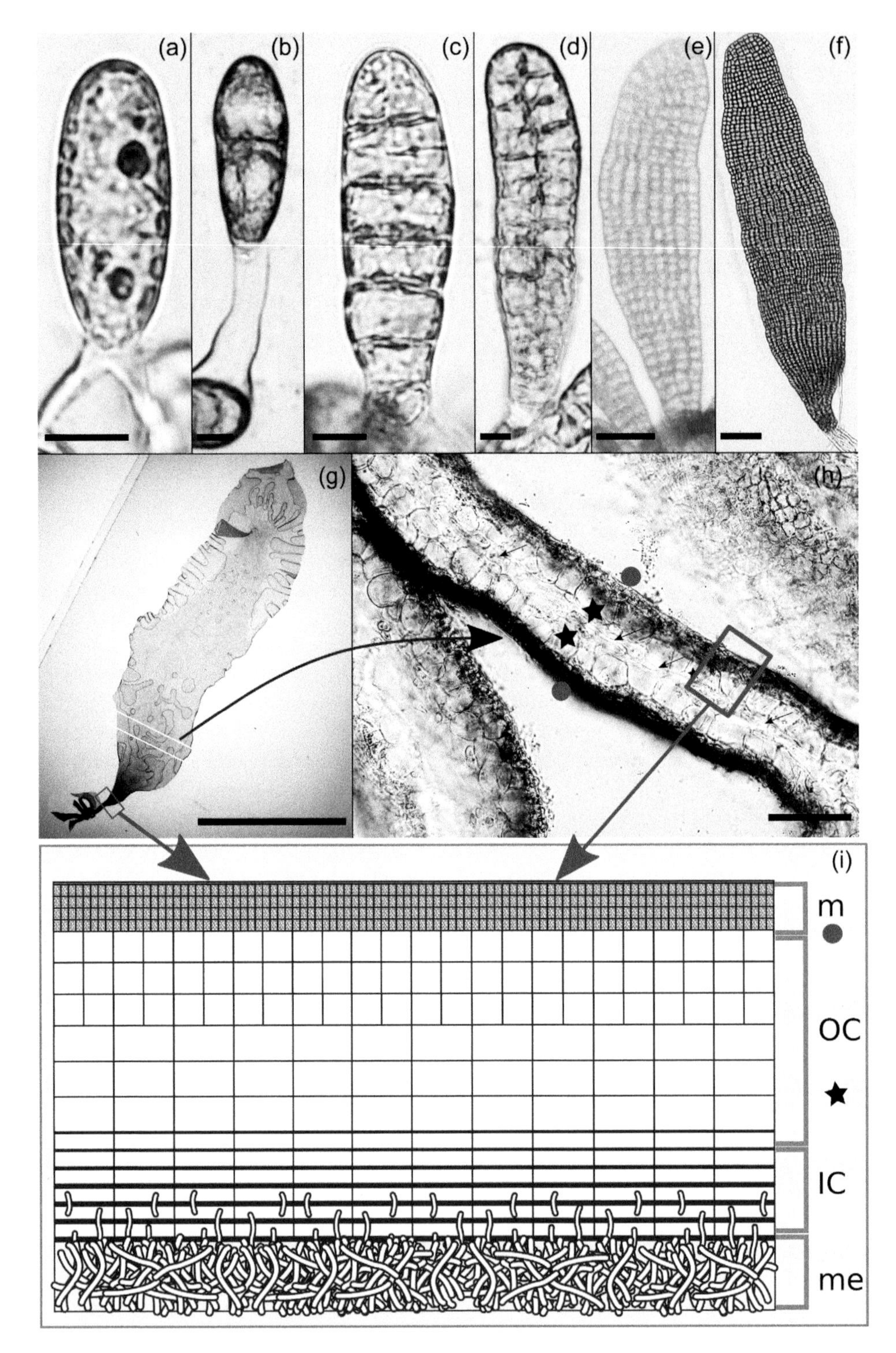

FIGURE 2.5 Developmental stages and cross-sectional histology of a *Saccharina latissima* blade. (a) A polarized zygote; (b) one- to two-day-old two-cell dividing embryo; (c) three-day-old embryo; (d) average projection from a z-stack of a four-day-old embryo; (e) average projection from a z-stack of a ten-day-old embryo; (f) focused projection from a z-stack of a three-week-old embryo; (g) a two-month-old juvenile; (h) cross-section from the middle part of the blade on (g). Red circles: meristodermal layer; black stars: cortical layer; arrows: medullary elements (hyphae-like cell protrusions); (i) schematic of the structure (cross-section) of a stipe or a blade at mature stages [older than in (g) and (h)]. Peripherally, the cylindrical stipe consists of a thin outer layer of mucilage and several layers of photosynthetic and actively dividing cells, the meristoderm (m). Inside, layers from large, opaque and highly vacuolated cells constitute the outer cortex (OC). In the inner cortex (IC), cells are thinner and elongated. The cell wall gradually thickens toward the center of the stipe; however, this is probably the result of gradual deposition of mucilage that relaxes the cell connections leading to the medulla (me). Protrusions from the innermost layers of IC already occupy the relaxed and filamentous medulla. Bars: (a-f) 10 µm, (e) 50 µm, (f,h) 100 µm, (g) 1 cm.

photoassimilates and nutrients. In recent studies on other kelp species, the medullary cells seem to have the capacity to generate turgor through the elastic properties of their cell walls (as illustrated in the kelp *Nereocystis* by Knoblauch et al. 2016a), possibly controlling the flow of the transported solutions. This and the alginate-rich extracellular matrix of the medulla make the sieve elements of kelps a study model for fluid mechanics in transport systems of plant organisms, since they are easily manipulable (Knoblauch et al. 2016b).

Between the medulla and the meristoderm resides the cortex (Figure 2.5h, i). It is divided into two parts: the outer cortex and the inner cortex. The outer cortex is easily distinguished from the meristoderm due to its sizable isodiametric opaque cells with pointy corners. The inner cortex is closer to the medulla and has elongated, thick-walled cells (closer to the medulla) with straight edges. At the transition zone and young parts of the stipe, the outer cortex cells widen and lengthen following the enlargement of the organ. A gradual change toward the more elongated cells of the inner cortex is visible. The innermost cells close to the medulla have protrusions on their most internal (proximal) longitudinal cell walls that may overlap each other, gradually resembling the shape and size of the medulla cells, as they progressively occupy this intricate mesh. At the transition zone, both the abundant mucilage deposits and the elongation of the innermost cells in combination with their growing septate protrusions "relax" the inner cortex tissue, which gradually differentiates into medullary cells (Killian 1911; Fritsch 1945). The inner cortex is supplied with cells from the outer cortex, which themselves originate from the actively dividing meristoderm.

In summary, growth of the blade and the stipe in the longitudinal axis is ensured by the transition zone, which furnishes the blade and the stipe with new tissues (Smith 1939; Fritsch 1945; Parke 1948; Steinbiss and Schmitz 1974). Therefore, the transition zone is characterized by both cell division activity in the longitudinal axis, which provides the cells for the lamina and stipe tissues, and active cell division in the peripheral meristoderm, whose role is to renew and keep providing cells to the transition zone. In this area, cell division and cell differentiation take place centripetally. Recently, transcriptomics studies confirmed an increasing meristematic activity in this location through the upregulation of ribosomal proteins and *immediate upright* genes a in the basal part of the blade (Ye et al. 2015), as in juvenile sporophytes (Shao et al. 2019).

As soon as the blade and the stipe can be identified, haptera start differentiating in the very basal part of the *S. latissima* thallus. These are outgrowths that originate from the lower end of the stipe, where a disc-like structure initially forms on top of the rhizoids (Drew 1910; Yendo 1911). Above this structure, the first haptera start developing.

While it shares the cortex and meristoderm with the stipe, the medulla of the stipe does not extend into the haptera (Yendo 1911; Smith 1939; Fritsch 1945; Davies et al. 1973). Haptera growth seems to be apical, but there is no extensive research on that matter. Haptera cells contribute to carbon fixation through photosynthesis, except when sheltered from light, resulting in cells of the haptera meristoderm displaying underdeveloped plastids with a rudimentary thylakoid membrane system (Davies et al. 1973). In addition, their endomembrane system is very well developed, with hypertrophied dictyosomes containing cell wall polysaccharides and alginate acid.

2.3.5 Genomics

The *S. latissima* genome sequence is expected to be released in 2021 (Project "Phaeoexplorer", led by FranceGenomics and the Roscoff Marine Station, www.france-genomique.org/projet/phaeoexplorer/). In the meantime, a draft genome sequence was published in 2015 for the close relative *S. japonica* (Ye et al. 2015), which diverged from *S. latissima* only ~5 Mya (Starko et al. 2019). It was enhanced by recent genome assembly work (Liu et al. 2019), leading to a genome of 580 Mbp for >35,000 genes.

The *S. japonica* genome is 2.7 times bigger than that of *Ectocarpus* sp. (Cock et al. 2010), and it contains twice as many genes; as expected, gene length is similar in the two species (Liu et al. 2019). Average exon lengths (~250 bp) are similar, but introns are less abundant (only 4.6 per gene on average in *S. japonica* vs. 7 in *Ectocarpus* sp.). Oddly, because introns are longer (1200 bp vs. 700 in *Ectocarpus* sp.), the overall exon:intron ratio per gene remains similar in *Saccharina* sp. and *Ectocarpus* sp. However, a significant difference lies in the presence of repeated sequences (46% in *S. japonica* vs. 22% in *Ectocarpus* sp.), mainly composed of class I and class II transposons and microsatellite sequences (Liu et al. 2019).

A large proportion of the gene content (85%) is distributed in gene families found in *Ectocarpus* sp. Nevertheless, detailed analysis shows interesting differences, in line with the biology of the organisms. In particular, the high capacity of *S. japonica* to accumulate iodine is reflected in the composition of its genome, which displays a very rich group of vanadium-dependent haloperoxidases (vHPOs), most likely resulting from gene expansion (Ye et al. 2015; Liu et al. 2019). Gene expansion may also have led to a significant increase in cell wall biosynthesis proteins (especially those involved in the synthesis of alginates), protein kinases and membrane-spanning receptor kinases. All together, in comparison with the *Ectocarpus* sp. genome, gene expansion would have been the genetic basis for the diversification of body plans and more generally of the complex multicellularity of Laminariales (Liu et al. 2019), which, together with the increased bioaccumulation of iodine, are the main characteristics differentiating Laminariales from Ectocarpales.

Interestingly, compared with other genomes, *Ectocarpus* sp. and *S. japonica* genomes display a significant increase in gene families (~1200) counterbalanced with a limited loss (~300), whose functions involve enzyme hydrolysis and cupin-like proteins (Ye et al. 2015). Although the functions of the gained gene families are largely unknown due to the lack of sequence conservation with other organisms, protein kinase and helix-extended-loop-helix super family domains have been identified as enriched domains in this group, suggesting a role in cell signaling and cell differentiation.

2.3.6 Functional Approaches: Tools for Molecular and Cellular Analyses

Cultivating macroalgae in laboratory conditions usually requires extensive experience and skills, because algae can be extremely sensitive to water and light parameters.

2.3.6.1 Culture Methods

2.3.6.1.1 Cultures of Gametophytes

Cultures of gametophytes can be initiated simply from fragments from an older laboratory culture or from material collected in the wild. This approach can be used for most kelps: collecting a healthy sporophyte with dark spots (sori) on the blade (schematized in Figure 2.4). Fertile blades can generally be found on the coast during the cold months. For example, in Roscoff and specifically on Perharidy beach (48°43'33.5"N 4°00'16.7"W), mature sporophytes with fully developed sori can be found from October to late April. Alternatively, fragments of large sporophytes from the intercalary meristem can be kept in short-day conditions in tanks for at least ten days to induce sporogenesis (Pang and Lüning 2004). Then, gametophytes will emerge from the released, germinated spores (Figure 2.4). More details on collecting and isolating gametophytes, as well as culture maintenance, can be found in Bartsch (2018). Care should be taken to ensure adequate temperature and light conditions while keeping the cultures under red light (Lüning and Dring 1972; Lüning 1980; Bolton and Lüning 1982; Li et al. 2020), as well as in a low concentration of chelated iron to maintain the gametophytes in a vegetative state (Lewis et al. 2013). Spontaneous gametogenesis can still be observed; however, its rate of occurrence is low and negligible. Sufficient amounts of biomass should be secured before beginning any experiments, but because *S. latissima* is a slow-growing alga, this can require several months to one year.

2.3.6.1.2 Gametogenesis

The simplest way to induce gametogenesis is to transfer the gametophytes into normal light conditions (Bartsch 2018; Forbord et al. 2018). However, if there is a high density of biomass, this may lead to reduced vegetative growth (Yabu 1965) and reproduction efficiency (Ebbing et al. 2020). Therefore, gametogenesis may be facilitated by reducing gametophytic density before transferring the cultures to normal light.

2.3.6.2 Immunochemistry and Ultrastructure Protocols

Several older studies that have examined the ultrastructure of *S. latissima* sporophytes (Davies et al. 1973; Sideman and Scheirer 1977; Schmitz and Kühn 1982), and others have employed immunochemistry on other *Saccharina* species (Motomura 1990; Motomura 1991; Klochkova et al. 2019). These studies have contributed to a better understanding of the general structure of the life cycle and histology of *Saccharina* spp. and kelps in general. However, there are no recent works focusing on the development or cytology of *S. latissima* despite its high economic and environmental interest.

These studies clearly demonstrate that *S. latissima*, as well as other brown algae, are amenable to fixation in paraformaldehyde or glutaraldehyde of various concentrations in seawater or other buffer solutions, such as microtubule stabilization buffer (Motomura 1991; Katsaros and Galatis 1992). The next step for immunochemistry is the digestion of the cell wall, which does not seem very challenging for *Saccharina angustata* when using abalone acetone powder. Because this powder has been discontinued, it has become necessary to test different cell wall digestion mixes, as shown for filamentous brown algal species (Tsirigoti et al. 2014) and green algal species (*Ulva mutabilis*) (Katsaros et al. 2011; Katsaros et al. 2017). Cell wall digestion is followed by extraction to remove most of the chlorophyll and other pigments from the cells. Triton is most commonly used, but in some cases, DMSO can be added for more efficient extraction (Rabillé et al. 2018b). This extraction step is carried out to reduce autofluorescence but also to perforate the cellular membrane to allow for the penetration of fluorescent probes. Motomura (1991) did not use an extraction step on *S. angustata* zygotes and parthenospores but noted increased autofluorescence, which can be reduced using a combination of filters during observation. The fluorescent probes, being chemical or primary and secondary antibodies, are added after the extraction step. This step can also be optimized, according to the species, because concentrations and washing steps may depend on the species and on the extraction step. The whole process can take two days of work, including observation. An antifade mounting medium, such as Vectashield or CitiFluor, can preserve the fluorescence of the samples and protect them from photobleaching. For transmission electron microscopy (TEM), there are several studies on *S. latissima* (Davies et al. 1973; Sideman and Scheirer 1977; Schmitz and Kühn 1982) that illustrate the general ultrastructure of the different cell types. In general, depending on the application, different fixatives can be chosen, and there are no cell wall digestion or extraction steps. After fixation, the specimen is post-fixed in osmium tetroxide and then dehydrated. Depending on the embedding resin, dehydration can be effected with ethanol or acetone. After embedding and polymerization of the resin, the blocks with the samples should be sectioned using an ultramicrotome. More information on the general considerations to take for TEM as well as the different protocol variations to use according to the desired application can be found in the aforementioned articles or in Raimundo et al. (2018) for a general protocol for seaweeds.

2.3.6.3 Modification of Gene Expression

To date, no genetic transformation protocol is available for *S. latissima*, but one was published for its relative *S. japonica* (formerly *Laminaria japonica*) using a biolistic approach on mature blades, showing transient expression of the GUS reporter gene (Li et al. 2009). Since then, despite demands from industry (Lin and Qin 2014; Qin et al. 2005), no additional studies have built on this technical breakthrough. Several genetic variants have been produced (reviewed in Qin et al. 2005).

2.4 CHALLENGING QUESTIONS IN BASIC AND APPLIED RESEARCH

2.4.1 Why Study Brown Algae?

2.4.1.1 Advancing Knowledge on Their Developmental Mechanisms

Brown algae make up a specific phylum of multicellular organisms. Their phylogenetic position in the eukaryotic tree (Baldauf 2008), distant from other multicellular organisms, makes them a key taxon for understanding the evolution of complex multicellularity and specific metabolic pathways. The literature abounds with biological questions and research topics positioning these organisms as essential ones to consider in future studies, and, more specifically related to this chapter, brown algae offer a wealth of candidate species to study the evolution of the formation of different body shapes. Furthermore, in contrast to the red and green algae, there is no representative unicellular species for brown algae, making the evolutionary scenario of the emergence of their diverse shapes even more intriguing.

However, the knowledge in the fields of evolution and development is very scarce compared with that on metazoans and land plants. In the following, two examples pertaining to kelp features illustrate the potential brown algae hold for leading to knowledge breakthroughs in developmental biology.

First, despite the similarities between brown algal tissues and complex histological structures in land plants, brown algal body architecture and shape remain fairly simple. Even kelps—the most complex brown algae at the morphological level—develop only a few different organs (blade, stipe and holdfast), with a limited number of specific tissues and cell types (i.e. epidermis, cortex, medulla, meristoderm, sorus [this chapter] and pneumatocysts, receptacles and conceptacles in other brown algae [reviewed in Charrier et al. 2012]). This relative simplicity provides a useful opportunity to study basic developmental mechanisms based on simple geometrical rules or morphogen gradients. Although auxin, the long-standing leading morphogen for land plants, is present in brown algae and affects morphogenesis of several morphologically simple brown algae, such as *Ectocarpus* sp. and *Dictyota* sp. (Dictyotales) (Le Bail et al. 2010; Bogaert et al. 2019), it has no conspicuous effect, nor is it specifically localized in the apex of *Sargassum* sp. (Fucales), a brown alga with relatively high morphological complexity, including the presence of an apical meristem (stem cell tissue) (Linardić and Braybrook 2017). This result casts doubt on the consistency of morphogen-mediated control mechanisms in brown algae and presages the identification of new, alternative growth control mechanisms.

The second example relates to one of the stunning characteristics of some brown algae: their size. How do cells communicate with each other over such a long distance, when it comes to organisms among the tallest on earth: kelps? The transport system in kelps is reminiscent of the vascular systems of land plants, except that the extracellular matrix (alginates) has a specific organization and distribution and contributes to the flow of photoassimilated products (Knoblauch et al. 2016a, 2016b). Cells connect with each other through pit structures where the plasmodesmata (channels or pore connecting two adjacent cells) are concentrated. These plasmodesmata are structurally similar to those in land plants (Terauchi et al. 2015), except for the absence of desmotubules and the lack of the ability to control the size of molecules transferred symplastically (Bouget et al. 1998; Terauchi et al. 2015). Although some kelps (e.g. *Macrocystis* spp.) adjust the size of their vascular tissues to the needs for photoassimilate distribution to "sink" organs (i.e. meristerm, storage tissues, sori) as land plants do, others do not, suggesting again different control mechanisms in the management of this important function (Drobnitch et al. 2015). One explanation is that larger kelp rely more heavily on an efficient transport system, especially when source and sink tissues are physically distant. Relying on a transport system would call for a regulated developmental process, as in land plants (Drobnitch et al. 2015).

2.4.1.2 Improving Aquaculture

Over the past several decades, *S. japonica* (known as "kombu") aquaculture in Asia has undergone many improvements at many different levels, because this alga has been cultivated for human consumption for several centuries. One improvement lever is breeding, and—beyond empirical approaches used in the past—genomics can now assist and speed up breeding programs (Wang et al. 2020), along with new knowledge on the control of the life cycle, reproduction and early growth steps (e.g. substrate adhesion, sensitivity to high density) (reviewed in Charrier et al. 2017). Regarding more specifically *S. latissima* cultivated in Europe, its genome has not yet been sequenced and, other than concerns on the ecological impact of seaweed aquaculture, the current bottlenecks are mainly technical and focused on scaling up production and reducing cultivation costs (reviewed in Barbier et al. 2019).

2.4.2 Biological Models: *Ectocarpus* sp., *S. latissima* or Another Brown Alga?

Because *Ectocarpus* sp. is a morphologically and sexually simple organism, it is a convenient model for cellular and molecular studies requiring microscopy, and this asset is enhanced by the availability of many additional cell biology tools (e.g. protocols for immunolocalization of the cell wall and the cytoskeleton, laser capture microdissection, *in situ* hybridization, etc.). Therefore, as illustrated in this chapter, its amenability to laboratory experimentation and its short life cycle have made it a convenient organism to explore. However, its low biomass is an impediment for biochemical research, in addition to its simple morphology, which precludes the study of complex multicellular mechanisms.

This is how *S. latissima* landed on the roadmap: based on the wealth of cultivation practice-based knowledge from applied phycology and aquaculture R&D laboratories,

TABLE 2.1
Characteristics of the Two Brown Algal Models *Ectocarpus* sp. and *Saccharina latissima* and Suitability for Lab Experiments

	Ectocarpus sp.	Saccharina latissima
Life cycle	Short, haplodiplontic, dioecious, slightly anisogamous.	Long, haplodiplontic, dioecious, strongly anisogamous/oogamous.
Amenability to lab conditions	Good.	Good, time consuming to establish a stock culture (several months). Life cycle only partially completed *in vitro*?
Size	Microscopic (100 µm–1 cm) (both sporophyte and gametophyte).	Microscopic (gametophyte: 1 mm)–macroscopic (sporophyte: up to 3 m).
Growth rate	Rapid: Spore to fertile gametophyte: two to three weeks. Zygote to fertile sporophyte: three to four weeks.	Gametophyte: extremely slow. Sporophyte: zygote → fertile sporophyte: five to six months.
Amenable to research topics in	Cell biology, developmental biology, genetics, primary and secondary metabolisms, microbiome interaction, cell wall biosynthesis.	Same. Sex determinism.
Sexual dimorphism (gametophyte phase)	Extremely low; absent in the vegetative stage; subtle on fertile organisms (gametophytes).	Significantly conspicuous in the vegetative and reproductive phases (gametophytes).
Genome	214 Mbp, ~17,000 genes, <=28 chromosomes.	Not known. In *S. japonica:* 580.5 Mbp, 35,725 encoding genes.
Genetic modification	Characterized mutants (UV irradiated). Genetic transformation: □ Stable: No. □ CRISPR: No.	Genetic transformation: □ Stable: No. □ Transient in *S. japonica* (biolistic). □ CRISPR: No.
Cell biology techniques	Immunocytochemistry. *In situ* hybridization.	Immunocytochemistry.
Phylogenetic studies	Key position, as a stramenopile, distant from metazoans and land plants.	Same + presenting complex multicellularity.
Summary	**Good for genetics and cytology, not good for biomass production.**	**Good for biomass production, cytology and all kinds of experimentation taking place at an early developmental stage (~5 cm long).**

fundamental research on *S. latissima* ramped up in the 2010s. The advent of high-throughput sequencing techniques (mainly RNA-seq) put the spotlight on this model, leading to the possibility to address biological questions specific to kelps with a new angle. Although few labs in the world work on *Ectocarpus* spp., those working on *Saccharina* spp. are numerous, driven by the potential economic benefit. However, more efforts are necessary before this model is amenable to the full range of technical tools required for comprehensive studies. Table 2.1 summarizes the main features of these two brown algal models for laboratory research.

Parallel to these avenues of research, studies have also been carried out on alternative pathways. *Dictyota* sp. (Dictyotales) has proved an excellent model for the study of early embryogenesis (Bogaert et al. 2016; Bogaert et al. 2017) and thallus dichotomy (reviewed in Bogaert et al. 2020), *Sargassum* spp. for the establishment of shoot phyllotaxis (Linardić and Braybrook 2017) and *Fucus* spp. for abundant embryogenetic studies (Brownlee et al. 2001; Corellou et al. 2001). However, these latter brown algae are relatively difficult to cultivate in the laboratory, making it impossible to address biological processes taking place later in development.

Most likely, the choice of models will continue to grow, depending on the biological features inherent to each model and on the biological question to be addressed. In the end, it is the species the most amenable to genetic transformation that will dominate the field and become the favored model.

BIBLIOGRAPHY

Adams, J.M., Gallagher, J.A. & Donnison, I.S. 2008. Fermentation study on *Saccharina latissima* for bioethanol production considering variable pre-treatments. *Journal of Applied Phycology* 21: 569.

Ahmed, S., Cock, J.M., Pessia, E., Luthringer, R., Cormier, A., Robuchon, M., Sterck, L., Peters, A.F., Dittami, S.M., Corre, E., Valero, M., Aury, J.-M., Roze, D., Van de Peer, Y., Bothwell, J., Marais, G.A.B. & Coelho, S.M. 2014. A haploid system of sex determination in the brown alga *Ectocarpus* sp. *Current Biology* 24: 1945–1957.

Andersen, S.G., Steen, H., Christie, H., Fredriksen, S. & Moy, F.E. 2011. Seasonal patterns of sporophyte growth, fertility, fouling, and mortality of *Saccharina latissima* in Skagerrak, Norway: Implications for forest recovery. *Journal of Marine Biology* 2011: e690375.

Arun, A., Coelho, S.M., Peters, A.F., Bourdareau, S., Pérès, L., Scornet, D., Strittmatter, M., Lipinska, A.P., Yao, H., Godfroy, O., Montecinos, G.J., Avia, K., Macaisne, N., Troadec, C., Bendahmane, A. & Cock, J.M. 2019. Convergent recruitment of TALE homeodomain life cycle regulators to direct

sporophyte development in land plants and brown algae. *eLife* 8: e43101.
Badstöber, J., Gachon, C.M.M., Ludwig-Müller, J., Sandbichler, A.M. & Neuhauser, S. 2020. Demystifying biotrophs: FISHing for mRNAs to decipher plant and algal pathogen-host interaction at the single cell level. *Scientic Reports* 10: 14269.
Baldauf, S.L. 2008. An overview of the phylogeny and diversity of eukaryotes. *Journal of Systematics and Evolution* 46: 263–273.
Barbier, M., Charrier, B., Araujo, R., Holdt, S.L., Jacquemin, B. & Rebours, C. 2019. *PEGASUS: PHYCOMORPH European Guidelines for a Sustainable Aquaculture of Seaweeds*, COST Action FA1406 (M. Barbier & B. Charrier, eds). Roscoff, France.
Bartsch, I. 2018. Derivation of clonal stock cultures and hybridization of kelps: A tool for strain preservation and breeding programs. In: *Protocols for Macroalgae Research*, pp. 61–78. Francis & Taylor Group, CRC Press, Boca Raton.
Bartsch, I., Wiencke, C., Bischof, K., Buchholz, C.M., Buck, B.H., Eggert, A., Feuerpfeil, P., Hanelt, D., Jacobsen, S., Karez, R., Karsten, U., Molis, M., Roleda, M.Y., Schubert, H., Schumann, R., Valentin, K., Weinberger, F. & Wiese, J. 2008. The genus *Laminaria sensu* lato: Recent insights and developments. *European Journal of Phycology* 43: 1–86.
Billoud, B., Bail, A.L. & Charrier, B. 2008. A stochastic 1D nearest-neighbour automaton models early development of the brown alga *Ectocarpus siliculosus*. *Functional Plant Biology* 35: 1014–1024.
Billoud, B., Jouanno, É., Nehr, Z., Carton, B., Rolland, É., Chenivesse, S. & Charrier, B. 2015. Localization of causal locus in the genome of the brown macroalga Ectocarpus: NGS-based mapping and positional cloning approaches. *Frontiers in Plant Sciences* 6: 68.
Billoud, B., Nehr, Z., Le Bail, A. & Charrier, B. 2014. Computational prediction and experimental validation of microRNAs in the brown alga *Ectocarpus siliculosus*. *Nucleic Acids Research* 42: 417–429.
Bogaert, K.A., Beeckman, T. & De Clerck, O. 2016. Abiotic regulation of growth and fertility in the sporophyte of *Dictyota dichotoma* (Hudson) J.V. Lamouroux (Dictyotales, Phaeophyceae). *Journal of Applied Phycology* 28: 2915–2924.
Bogaert, K.A., Beeckman, T. & De Clerck, O. 2017. Two-step cell polarization in algal zygotes. *Nature Plants* 3: 16221.
Bogaert, K.A., Blommaert, L., Ljung, K., Beeckman, T. & De Clerck, O. 2019. Auxin function in the brown alga *Dictyota dichotoma*. *Plant Physiology* 179: 280–299.
Bogaert, K.A., Delva, S. & De Clerck, O. 2020. Concise review of the genus Dictyota J.V. Lamouroux. *Journal of Applied Phycology* 32: 1521–1543. https://doi.org/10.1007/s10811-020-02121-4
Boland, W. 1995. The chemistry of gamete attraction: Chemical structures, biosynthesis, and (a)biotic degradation of algal pheromones. *Proceedings of the National Academy of Sciences U.S.A.* 92: 37–43.
Bolton, J.J. 2010. The biogeography of kelps (Laminariales, Phaeophyceae): A global analysis with new insights from recent advances in molecular phylogenetics. *Helgoland Marine Research* 64: 263–279.
Bolton, J.J. & Lüning, K. 1982. Optimal growth and maximal survival temperatures of Atlantic *Laminaria* species (Phaeophyta) in culture. *Marine Biology* 66: 89–94.
Bothwell, J.H., Marie, D., Peters, A.F., Cock, J.M. & Coelho, S.M. 2010. Role of endoreduplication and apomeiosis during parthenogenetic reproduction in the model brown alga Ectocarpus. *New Phytologist* 188: 111–121.
Bouget, F.Y., Berger, F. & Brownlee, C. 1998. Position dependent control of cell fate in the Fucus embryo: Role of intercellular communication. *Development* 125: 1999–2008.
Bringloe, T.T., Starko, S., Wade, R.M., Vieira, C., Kawai, H., De Clerck, O., Cock, J.M., Coelho, S.M., Destombe, C., Valero, M., Neiva, J., Pearson, G.A., Faugeron, S., Serrao, E.A. & Verbruggen, H. 2020. Phylogeny and evolution of the brown algae. *Critical Review of Plant Sciences* 39: 281–321.
Brownlee, C., Bouget, F.Y. & Corellou, F. 2001. Choosing sides: establishment of polarity in zygotes of fucoid algae. *Seminars in Cell & Developmental Biology* 12: 345–351. https://doi.org/10.1006/scdb.2001.0262
Charrier, B., Abreu, M.H., Araujo, R., Bruhn, A., Coates, J.C., De Clerck, O., Katsaros, C., Robaina, R.R. & Wichard, T. 2017. Furthering knowledge of seaweed growth and development to facilitate sustainable aquaculture. *New Phytologist* 216: 967–975.
Charrier, B., Coelho, S.M., Le Bail, A., Tonon, T., Michel, G., Potin, P., Kloareg, B., Boyen, C., Peters, A.F. & Cock, J.M. 2008. Development and physiology of the brown alga *Ectocarpus siliculosus*: Two centuries of research. *New Phytologist* 177: 319–332.
Charrier, B., Le Bail, A. & de Reviers, B. 2012. Plant Proteus: Brown algal morphological plasticity and underlying developmental mechanisms. *Trends in Plant Science* 17: 468–477.
Charrier, B., Rabillé, H. & Billoud, B. 2019. Gazing at cell wall expansion under a golden light. *Trends in Plant Science* 24: 130–141.
Cock, J.M., Sterck, L., Rouzé, P., Scornet, D., Allen, A.E., Amoutzias, G., Anthouard, V., Artiguenave, F., Aury, J.-M., Badger, J.H., Beszteri, B., Billiau, K., Bonnet, E., Bothwell, J.H., Bowler, C., Boyen, C., Brownlee, C., Carrano, C.J., Charrier, B., Cho, G.Y., Coelho, S.M., Collén, J., Corre, E., Da Silva, C., Delage, L., Delaroque, N., Dittami, S.M., Doulbeau, S., Elias, M., Farnham, G., Gachon, C.M.M., Gschloessl, B., Heesch, S., Jabbari, K., Jubin, C., Kawai, H., Kimura, K., Kloareg, B., Küpper, F.C., Lang, D., Le Bail, A., Leblanc, C., Lerouge, P., Lohr, M., Lopez, P.J., Martens, C., Maumus, F., Michel, G., Miranda-Saavedra, D., Morales, J., Moreau, H., Motomura, T., Nagasato, C., Napoli, C.A., Nelson, D.R., Nyvall-Collén, P., Peters, A.F., Pommier, C., Potin, P., Poulain, J., Quesneville, H., Read, B., Rensing, S.A., Ritter, A., Rousvoal, S., Samanta, M., Samson, G., Schroeder, D.C., Ségurens, B., Strittmatter, M., Tonon, T., Tregear, J.W., Valentin, K., von Dassow, P., Yamagishi, T., Van de Peer, Y. & Wincker, P. 2010. The Ectocarpus genome and the independent evolution of multicellularity in brown algae. *Nature* 465: 617–621.
Coelho, S.M., Godfroy, O., Arun, A., Le Corguillé, G., Peters, A.F. & Cock, J.M. 2011. OUROBOROS is a master regulator of the gametophyte to sporophyte life cycle transition in the brown alga Ectocarpus. *Proceedings of the National Academy of Sciences U.S.A.* 108: 11518–11523.
Coelho, S.M., Scornet, D., Rousvoal, S., Peters, N., Dartevelle, L., Peters, A.F. & Cock, J.M. 2012. Immunostaining of Ectocarpus cells. *Cold Spring Harbour Protocols* 2012: 369–372.
Corellou F, Brownlee C, Detivaud L, et al. 2001. Cell cycle in the fucus zygote parallels a somatic cell cycle but displays a unique translational regulation of cyclin-dependent kinases. *Plant Cell* 13: 585–598.
Cormier, A., Avia, K., Sterck, L., Derrien, T., Wucher, V., Andres, G., Monsoor, M., Godfroy, O., Lipinska, A., Perrineau, M.-M., Peer, Y.V.D., Hitte, C., Corre, E., Coelho, S.M. & Cock, J.M. 2017. Re-annotation, improved large-scale

assembly and establishment of a catalogue of noncoding loci for the genome of the model brown alga Ectocarpus. *New Phytologist* 214: 219–232.

Couceiro, L., Le Gac, M., Hunsperger, H.M., Mauger, S., Destombe, C., Cock, J.M., Ahmed, S., Coelho, S.M., Valero, M. & Peters, A.F. 2015. Evolution and maintenance of haploid-diploid life cycles in natural populations: The case of the marine brown alga Ectocarpus. *Evolution* 69: 1808–1822.

Coudert, Y., Harris, S. & Charrier, B. 2019. Design principles of branching morphogenesis in filamentous organisms. *Current Biology* 29: R1149–R1162.

Cumashi, A., Ushakova, N.A., Preobrazhenskaya, M.E., D'Incecco, A., Piccoli, A., Totani, L., Tinari, N., Morozevich, G.E., Berman, A.E., Bilan, M.I., Usov, A.I., Ustyuzhanina, N.E., Grachev, A.A., Sanderson, C.J., Kelly, M., Rabinovich, G.A., Iacobelli, S., Nifantiev, N.E. & Consorzio Interuniversitario Nazionale per la Bio-Oncologia, Italy. 2007. A comparative study of the anti-inflammatory, anticoagulant, antiangiogenic, and antiadhesive activities of nine different fucoidans from brown seaweeds. *Glycobiology* 17: 541–552.

Davies, J.M., Ferrier, N.C. & Johnston, C.S. 1973. The ultrastructure of the meristoderm cells of the hapteron of *Laminaria. Journal of the Marine Biological Association U-K* 53: 237–246.

Deniaud-Bouët, E., Kervarec, N., Michel, G., Tonon, T., Kloareg, B. & Hervé, C. 2014. Chemical and enzymatic fractionation of cell walls from Fucales: Insights into the structure of the extracellular matrix of brown algae. *Annals of Botany* 114: 1203–1216.

Derelle, R., López-García, P., Timpano, H. & Moreira, D. 2016. A phylogenomic framework to study the diversity and evolution of stramenopiles (=heterokonts). *Molecular Biology and Evolution* 33: 2890–2898.

Dillwyn, L.W. 1809. *British Confervae: Or Colored Figures and Descriptions of the British Plants Referred by Botanists to the Genus Conferva*. W. Phillips, London, UK.

Dittami, S.M., Duboscq-Bidot, L., Perennou, M., Gobet, A., Corre, E., Boyen, C. & Tonon, T. 2016. Host-microbe interactions as a driver of acclimation to salinity gradients in brown algal cultures. *The ISME Journal* 10: 51–63.

Dittami, S.M., Gravot, A., Goulitquer, S., Rousvoal, S., Peters, A.F., Bouchereau, A., Boyen, C. & Tonon, T. 2012. Towards deciphering dynamic changes and evolutionary mechanisms involved in the adaptation to low salinities in Ectocarpus (brown algae). *The Plant Journal* 71: 366–377.

Dittami, S.M., Peters, A.F., West, J.A., Cariou, T., KleinJan, H., Burgunter-Delamare, B., Prechoux, A., Egan, S. & Boyen, C. 2020. Revisiting Australian *Ectocarpus subulatus* (Phaeophyceae) from the Hopkins river: Distribution, abiotic environment, and associated microbiota. *Journal of Phycology* 56: 719–729.

Drew, G.H. 1910. The reproduction and early development of *Laminaria digitata* and *Laminaria saccharina*. *Annals of Botany* 24: 177–189.

Drobnitch, S.T., Jensen, K.H., Prentice, P. & Pittermann, J. 2015. Convergent evolution of vascular optimization in kelp (Laminariales). *Proceedings of Royal Society B: Biological Sciences* 282: 20151667.

Ebbing, A., Pierik, R., Bouma, T., Kromkamp, J.C. & Timmermans, K. 2020. How light and biomass density influence the reproduction of delayed *Saccharina latissima* gametophytes (Phaeophyceae). *Journal of Phycology* 56: 709–718.

FAO. 2018. The global status of seaweed production, trade and utilization. *Globefish Research Programme*. Vol. 124, p. 120. Rome.

Forbord, S., Steinhovden, K.B., Rød, K.K., Handå, A., Skjermo, J., Steinhovden, K.B., Rød, K.K., Handå, A. & Skjermo, J. 2018. Cultivation protocol for *Saccharina latissima*. In: *Protocols for Macroalgae Research*, pp. 37–60. Francis & Taylor Groups, CRC Press, Boca Raton.

Forbord, S., Steinhovden, K.B., Solvang, T., Handå, A. & Skjermo, J. 2019. Effect of seeding methods and hatchery periods on sea cultivation of *Saccharina latissima* (Phaeophyceae): A Norwegian case study. *Journal of Applied Phycology* 32: 2201–2212.

Fossberg, J., Forbord, S., Broch, O.J., Malzahn, A.M., Jansen, H., Handå, A., Førde, H., Bergvik, M., Fleddum, A.L., Skjermo, J. & Olsen, Y. 2018. The potential for upscaling Kelp (Saccharina latissima) cultivation in salmon-driven integrated multi-trophic aquaculture (IMTA). *Frontiers in Marine Science* 5: 418.

Fritsch, F.E. 1945. *The Structure and Reproduction of the Algae*. Cambridge University Press, Cambridge.

Gaboriaud, F. & Dufrêne, Y.F. 2007. Atomic force microscopy of microbial cells: Application to nanomechanical properties, surface forces and molecular recognition forces. *Colloids and Surfaces B Biointerfaces* 54: 10–19.

Godfroy, O., Uji, T., Nagasato, C., Lipinska, A.P., Scornet, D., Peters, A.F., Avia, K., Colin, S., Mignerot, L., Motomura, T., Cock, J.M. & Coelho, S.M. 2017. DISTAG/TBCCd1 is required for basal cell fate determination in Ectocarpus. *The Plant Cell* 29: 3102–3122.

Han, H., Wang, L., Liu, Y., Shi, X., Zhang, X., Li, M. & Wang, T. 2019. Combination of curcuma zedoary and kelp inhibits growth and metastasis of liver cancer in vivo and in vitro via reducing endogenous H2S levels. *Food & Function* 10: 224–234.

Heesch, S., Cho, G.Y., Peters, A.F., Le Corguillé, G., Falentin, C., Boutet, G., Coëdel, S., Jubin, C., Samson, G., Corre, E., Coelho, S.M. & Cock, J.M. 2010. A sequence-tagged genetic map for the brown alga *Ectocarpus siliculosus* provides large-scale assembly of the genome sequence. *New Phytologist* 188: 42–51.

Hertweck, C. & Boland, W. 1997. Highly efficient synthesis of (±)-lamoxirene, the gamete-releasing and gamete-attracting pheromone of the Laminariales (Phaeophyta). *Tetrahedron* 53: 14651–14654.

Jackson, C., Salomaki, E.D., Lane, C.E. & Saunders, G.W. 2017. Kelp transcriptomes provide robust support for interfamilial relationships and revision of the little known Arthrothamnaceae (Laminariales). *Journal of Phycology* 53:1–6. https://doi.org/10.1111/jpy.12465

Jia, F., Ben Amar, M., Billoud, B. & Charrier, B. 2017. Morphoelasticity in the development of brown alga *Ectocarpus siliculosus*: From cell rounding to branching. *Journal of the Royal Society Interface* 14: 20160596.

Kanda, T. 1936. On the gametophytes of some Japanese species of Laminariales I. *Scientific Papers of the Institute of Algological Research, Faculty of Science, Hokkaido Imperial University* 1: 221–260.

Katsaros, C.I. & Galatis, B. 1992. Immunofluorescence and electron microscopic studies of microtubule organization during the cell cycle of *Dictyota dichotoma* (Phaeophyta, Dictyotales). *Protoplasma* 169: 75–84.

Katsaros, C.I., Karyophyllis, D. & Galatis, B. 2006. Cytoskeleton and morphogenesis in brown algae. *Annals of Botany* 97: 679–693.

Katsaros, C.I., Kreimer, G. & Melkonian, M. 1991. Localization of tubulin and a centrin-homologue in vegetative cells and developing gametangia of *Ectocarpus siliculosus* (Dillw.) Lyngb. (Phaeophyceae, Ectocarpales). *Botanica Acta* 104: 87–92.

Katsaros, C.I., Varvarigos, V., Gachon, C.M.M., Brand, J., Motomura, T., Nagasato, C. & Küpper, F.C. 2011. Comparative immunofluorescence and ultrastructural analysis of microtubule organization in *Uronema* sp., *Klebsormidium flaccidum*, *K. subtilissimum*, *Stichococcus bacillaris* and *S. chloranthus* (Chlorophyta). *Protist* 162: 315–331.

Katsaros, C.I., Weiss, A., Llangos, I., Theodorou, I. & Wichard, T. 2017. Cell structure and microtubule organisation during gametogenesis of Ulva mutabilis Føyn (Chlorophyta). *Botanica Marina* 60: 123–135.

Kawai, H., Hanyuda, T., Draisma, S.G.A., Wilce, R.T. & Andersen, R.A. 2015. Molecular phylogeny of two unusual brown algae, *Phaeostrophion irregulare* and *Platysiphon glacialis*, proposal of the *Stschapoviales* ord. nov. and *Platysiphonaceae* fam. nov., and a re-examination of divergence times for brown algal orders. *Journal of Phycology* 51: 918–928.

Killian, K. 1911. Beiträge zur Kenntnis der Laminarien. *Zeitschr. Bot.* 3: 433–494.

Kinoshita, N., Nagasato, C. & Motomura, T. 2017. Chemotactic movement in sperm of the oogamous brown algae, *Saccharina japonica* and *Fucus distichus*. *Protoplasma* 254: 547–555.

Klochkova, T.A., Motomura, T., Nagasato, C., Klimova, A.V. & Kim, G.H. 2019. The role of egg flagella in the settlement and development of zygotes in two *Saccharina* species. *Phycologia* 58: 145–153.

Knoblauch, J., Drobnitch, S.T., Peters, W.S. & Knoblauch, M. 2016a. In situ microscopy reveals reversible cell wall swelling in kelp sieve tubes: One mechanism for turgor generation and flow control? *Plant, Cell & Environment* 39: 1727–1736.

Knoblauch, J., Peters, W.S. & Knoblauch, M. 2016b. The gelatinous extracellular matrix facilitates transport studies in kelp: Visualization of pressure-induced flow reversal across sieve plates. *Annals of Botany* 117: 599–606.

Kraan, S. 2016. Seaweed and alcohol. In: *Seaweed in Health and Disease Prevention*, pp. 169–184. Elsevier.

Lamouroux, J.V.F. 1813. Essai sur les genres de la famille des Thalassiophytes non articulées. *Annales du Muséum d'Histoire Naturelle* 20: 21–47, 115–139, 267–293, pls 7–13.

Lancaster, O.M. & Baum, B. 2014. Shaping up to divide: Coordinating actin and microtubule cytoskeletal remodelling during mitosis. *Seminars in Cell & Developmental Biology* 34: 109–115.

Lane, C.E., Mayes, C., Druehl, L.D. & Saunders, G.W. 2006. A multi-gene molecular investigation of the kelp (Laminariales, Phaeophyceae) supports substantial taxonomic re-organization. *Journal of Phycology* 42: 493–512.

Le Bail, A., Billoud, B., Kowalczyk, N., Kowalczyk, M., Gicquel, M., Le Panse, S., Stewart, S., Scornet, D., Cock, J.M., Ljung, K. & Charrier, B. 2010. Auxin metabolism and function in the multicellular brown alga *Ectocarpus siliculosus*. *Plant Physiology* 153: 128–144.

Le Bail, A., Billoud, B., Le Panse, S., Chenivesse, S. & Charrier, B. 2011. ETOILE regulates developmental patterning in the filamentous brown alga *Ectocarpus siliculosus*. *The Plant Cell* 23: 1666–1678.

Le Bail, A., Billoud, B., Maisonneuve, C., Peters, A., Cock, J.M. & Charrier, B. 2008. Initial pattern of development of the brown alga *Ectocarpus siliculosus* (Ectocarpales, Phaeophyceae) sporophyte. *Journal of Phycology* 44: 1269–1281.

Le Bail, A. & Charrier, B. 2013. Culture methods and mutant generation in the filamentous brown algae *Ectocarpus siliculosus*. In: *Plant Organogenesis* (I. De Smet, ed.), pp. 323–332. Humana Press.

Lewis, R.J., Green, M.K. & Afzal, M.E. 2013. Effects of chelated iron on oogenesis and vegetative growth of kelp gametophytes (Phaeophyceae). *Phycological Research* 61: 46–51.

Li, F., Qin, S., Jiang, P., Wu, Y. & Zhang, W. 2009. The integrative expression of GUS gene driven by FCP promoter in the seaweed *Laminaria japonica* (Phaeophyta). *Journal of Applied Phycology* 21: 287–293.

Li, H., Monteiro, C., Heinrich, S., Bartsch, I., Valentin, K., Harms, L., Glöckner, G., Corre, E. & Bischof, K. 2020. Responses of the kelp *Saccharina latissima* (Phaeophyceae) to the warming Arctic: From physiology to transcriptomics. *Physiologia Plantarum* 168: 5–26.

Lin, C.-Y., Oulhen, N., Wessel, G. & Su, Y.-H. 2019. CRISPR/Cas9-mediated genome editing in sea urchins. *Methods in Cell Biology* 151: 305–321.

Lin, H. & Qin, S. 2014. Tipping points in seaweed genetic engineering: Scaling up opportunities in the next decade. *Marine Drugs* 12: 3025–3045.

Linardić, M. & Braybrook, S.A. 2017. Towards an understanding of spiral patterning in the Sargassum muticum shoot apex. *Scientific Reports* 7: 13887.

Linne, C. von & Salvius, L. 1753. *Caroli Linnaei . . . Species plantarum: exhibentes plantas rite cognitas, ad genera relatas, cum differentiis specificis, nominibus trivialibus, synonymis selectis, locis natalibus, secundum systema sexuale digestas . . .* Impensis Laurentii Salvii, Holmiae.

Lino, C.A., Harper, J.C., Carney, J.P. & Timlin, J.A. 2018. Delivering CRISPR: A review of the challenges and approaches. *Drug Delivery* 25: 1234–1257.

Lipinska, A.P., Cormier, A., Luthringer, R., Peters, A.F., Corre, E., Gachon, C.M.M., Cock, J.M. & Coelho, S.M. 2015. Sexual dimorphism and the evolution of sex-biased gene expression in the brown alga Ectocarpus. *Molecular Biology & Evolution* 32: 1581–1597.

Lipinska, A.P., Serrano-Serrano, M.L., Cormier, A., Peters, A.F., Kogame, K., Cock, J.M. & Coelho, S.M. 2019. Rapid turnover of life-cycle-related genes in the brown algae. *Genome Biology* 20: 35.

Lipinska, A.P., Toda, N.R.T., Heesch, S., Peters, A.F., Cock, J.M. & Coelho, S.M. 2017. Multiple gene movements into and out of haploid sex chromosomes. *Genome Biology* 18: 104.

Liu, T., Wang, X., Wang, G., Jia, S., Liu, G., Shan, G., Chi, S., Zhang, J., Yu, Y., Xue, T. & Yu, J. 2019. Evolution of complex Thallus alga: Genome sequencing of *Saccharina japonica*. *Frontiers in Genetics* 10: 378.

Long, M., Li, Q.-M., Fang, Q., Pan, L.-H., Zha, X.-Q. & Luo, J.-P. 2019. Renoprotective effect of *Laminaria japonica* polysaccharide in adenine-induced chronic renal failure. *Molecules* 24: 1491.

Lüning, K. 1980. Critical levels of light and temperature regulating the gametogenesis of three *Laminaria* species (Phaeophyceae). *Journal of Phycology* 16: 1–15.

Lüning, K. 1981. Egg release in gametophytes of *Laminaria saccharina*: Induction by darkness and inhibition by blue light and u.v. *British Phycological Journal* 16: 379–393.

Lüning, K. & Dring, M.J. 1972. Reproduction induced by blue light in female gametophytes of *Laminaria saccharina*. *Planta* 104: 252–256.

Lüning, K., Schmitz, K. & Willenbrink, J. 1973. CO_2 fixation and translocation in benthic marine algae. III. Rates and ecological significance of translocation in *Laminaria hyperborea* and *Laminaria saccharina*. *Marine Biology* 23: 275–281.

Luthringer, R., Cormier, A., Ahmed, S., Peters, A.F., Cock, J.M. & Coelho, S.M. 2014. Sexual dimorphism in the brown algae. *Perspectives in Phycology* 1: 11–25.

Luttikhuizen, P.C., Heuvel, F.H.M. van den, Rebours, C., Witte, H.J., Bleijswijk, J.D.L. van & Timmermans, K. 2018. Strong population structure but no equilibrium yet: Genetic connectivity and phylogeography in the kelp *Saccharina latissima* (Laminariales, Phaeophyta). *Ecology and Evolution* 8: 4265–4277.

Lyngbye, H.C. 1819. *Tentamen hydrophytologiae danicae continens omnia hydrophyta cryptogamma Daniae, Holsatiae, Faeroae, Islandiae, Groendlandiae hucusque cognita, systematice disposita, descripta et iconibus illustrata, adjectis simul speciebus norvegicis. Opus, praemio.* Schultz. Copenhagen, Denmark.

Macaisne, N., Liu, F., Scornet, D., Peters, A.F., Lipinska, A., Perrineau, M.-M., Henry, A., Strittmatter, M., Coelho, S.M. & Cock, J.M. 2017. The Ectocarpus IMMEDIATE UPRIGHT gene encodes a member of a novel family of cysteine-rich proteins with an unusual distribution across the eukaryotes. *Development* 144: 409–418.

Maier, I., Hertweck, C. & Boland, W. 2001. Stereochemical specificity of lamoxirene, the sperm-releasing pheromone in kelp (Laminariales, Phaeophyceae). *The Biological Bulletin* 201: 121–125.

Maier, I. & Müller, D.G. 1986. Sexual pheromones in algae. *The Biological Bulletin* 170: 145–175.

Maier, I. & Muller, D.G. 1990. Chemotaxis in *Laminaria digitata* (Phaeophyceae): I. Analysis of spermatozoid movement. *Journal of Experimental Botany* 41: 869–876.

McDougall, A., Hebras, C., Gomes, I. & Dumollard, R. 2021. Gene editing in the ascidian phallusia mammillata and tail nerve cord formation. *Methods in Molecular Biology* 2219: 217–230.

Mohibbullah, M., Bashir, K.M.I., Kim, S.-K., Hong, Y.-K., Kim, A., Ku, S.-K. & Choi, J.-S. 2019. Protective effects of a mixed plant extracts derived from *Astragalus membranaceus* and *Laminaria japonica* on PTU-induced hypothyroidism and liver damages. *Journal of Food Biochemistry* 43: e12853.

Montecinos, A.E., Couceiro, L., Peters, A.F., Desrut, A., Valero, M. & Guillemin, M.-L. 2017. Species delimitation and phylogeographic analyses in the Ectocarpus subgroup siliculosi (Ectocarpales, Phaeophyceae). *Journal of Phycology* 53: 17–31.

Monteiro, C., Heinrich, S., Bartsch, I., Valentin, K., Corre, E., Collén, J., Harms, L., Glöckner, G. & Bischof, K. 2019. Temperature modulates sex-biased gene expression in the gametophytes of the kelp *Saccharina latissima*. *Frontiers in Marine Science* 6: 769.

Motomura, T. 1990. Ultrastructure of fertilization in *Laminaria angustata* (Phaeophyta, Laminariales) with emphasis on the behavior of centrioles, mitochondria and chloroplasts of the sperm. *Journal of Phycology* 26: 80–89.

Motomura, T. 1991. Immunofluorescence microscopy of fertilization and parthenogenesis in *Laminaria angustata* (Phaeophyta). *Journal of Phycology* 27: 248–257.

Müller, D.G. & Schmid, C.E. 1988. Qualitative and quantitative determination of pheromone secretion in female gametes of *Ectocarpus siliculosus* (Phaeophyceae). *Biological Chemistry* 369: 647–654.

Murovec, J., Guček, K., Bohanec, B., Avbelj, M. & Jerala, R. 2018. DNA-free genome editing of *Brassica oleracea* and *B. rapa* protoplasts using CRISPR-Cas9 Ribonucleoprotein complexes. *Frontiers in Plant Science* 9: 1594.

Nehr, Z., Billoud, B., Le Bail, A. & Charrier, B. 2011. Space-time decoupling in the branching process in the mutant étoile of the filamentous brown alga *Ectocarpus siliculosus*. *Plant Signaling & Behaviour* 6: 1889–1892.

Neiva, J., Paulino, C., Nielsen, M.M., Krause-Jensen, D., Saunders, G.W., Assis, J., Bárbara, I., Tamigneaux, É., Gouveia, L., Aires, T., Marbà, N., Bruhn, A., Pearson, G.A. & Serrão, E.A. 2018. Glacial vicariance drives phylogeographic diversification in the amphi-boreal kelp *Saccharina latissima*. *Scientific Reports* 8: 1112.

Pang, S. & Lüning, K. 2004. Photoperiodic long-day control of sporophyll and hair formation in the brown alga *Undaria pinnatifida*. *Journal of Applied Phycology* 16: 83–92.

Parke, M. 1948. Studies on British Laminariaceae. I. Growth in *Laminaria saccharina* (L.) Lamour. *Journal of the Marine Biological Association U-K* 27: 651–709.

Pearson, G.A., Martins, N., Madeira, P., Serrão, E.A. & Bartsch, I. 2019. Sex-dependent and -independent transcriptional changes during haploid phase gametogenesis in the sugar kelp *Saccharina latissima*. *PLoS One* 14: e0219723.

Peters, A.F., Marie, D., Scornet, D., Kloareg, B. & Cock, J.M. 2004. Proposal of *Ectocarpus siliculosus* (ectocarpales, Phaeophyceae) as a model organism for brown algal genetics and genomics1, 2. *Journal of Phycology* 40: 1079–1088.

Qin, S., Jiang, P. & Tseng, C. 2005. Transforming kelp into a marine bioreactor. *Trends in Biotechnology* 23: 264–268.

Rabillé, H., Billoud, B., Rolland, E. & Charrier, B. 2018a. Dynamic and microscale mapping of cell growth case of Ectocarpus filament cells. In: *Protocols for Macroalgae Research*, pp. 349–364. Taylor & Francis group, CRC Press, Boca Raton.

Rabillé, H., Billoud, B., Tesson, B., Le Panse, S., Rolland, E. & Charrier, B. 2019a. The brown algal mode of tip growth: Keeping stress under control. *PLoS Biology* 17: e2005258

Rabillé, H., Koutalianou, M., Charrier, B. & Katsaros, C. 2018b. Actin fluorescent staining in the filamentous brown alga *Ectocarpus siliculosus*. In: *Protocols for Macroalgae Research* (B. Charrier, T. Wichard, & C. Reddy, eds.), pp. 365–379. Taylor & Francis group, CRC Press, Boca Raton.

Rabillé, H., Torode, T.A., Tesson, B., Le Bail, A., Billoud, B., Rolland, E., Le Panse, S., Jam, M. & Charrier, B. 2019b. Alginates along the filament of the brown alga Ectocarpus help cells cope with stress. *Scientific Reports* 9: 12956.

Raimundo, S.C., Domozych, D.S. & Domozych, D.S. 2018. Probing the subcellular topography of seaweeds: Transmission electron microscopy, immunocytochemistry, and correlative light microscopy. In: *Protocols for Macroalgae Research*, pp. 391–410. Francis & Taylor Group, CRC Press, Boca Raton.

Rebours, C., Marinho-Soriano, E., Zertuche-González, J.A., Hayashi, L., Vásquez, J.A., Kradolfer, P., Soriano, G., Ugarte, R., Abreu, M.H., Bay-Larsen, I., Hovelsrud, G., Rødven, R. & Robledo, D. 2014. Seaweeds: An opportunity for wealth and sustainable livelihood for coastal communities. *Journal of Applied Phycology* 26: 1939–1951.

Rodriguez-Rojas, F., Lopez-Marras, A., Celis-Pla, P.S.M., Munoz, P., Garcia-Bartolomei, E., Valenzuela, F., Orrego, R., Carratala, A., Luis Sanchez-Lizaso, J. & Saez, C.A. 2020. Ecophysiological and cellular stress responses in the cosmopolitan brown macroalga Ectocarpus as biomonitoring tools for assessing desalination brine impacts. *Desalination* 489: 114527.

Schmid, C.E. 1993. Cell-cell-recognition during fertilization in *Ectocarpus siliculosus* (Phaeophyceae). In: *Fourteenth International Seaweed Symposium* (A.R.O. Chapman, M.T. Brown, & M. Lahaye, eds.), pp. 437–443. Springer Netherlands, Dordrecht.

Schmitz, K. & Kühn, R. 1982. Fine structure, distribution and frequency of plasmodesmata and pits in the cortex of *Laminaria hyperborea* and *L. saccharina*. *Planta* 154: 385–392.

Schmitz, K., Lüning, K. & Willenbrink, J. 1972. CO_2-fixierung und stofftransport in benthischen marinen algen. II. zum ferntransport 14C-markierter assimilate bei *Laminaria hyperborea* und *Laminaria saccharina*. *Zeitschrift für Pflanzenphysiologie* 67: 418–429.

Shao, Z., Zhang, P., Lu, C., Li, S., Chen, Z., Wang, X. & Duan, D. 2019. Transcriptome sequencing of *Saccharina japonica*

sporophytes during whole developmental periods reveals regulatory networks underlying alginate and mannitol biosynthesis. *BMC Genomics* 20: 975.
Sideman, E.J. & Scheirer, D.C. 1977. Some fine structural observations on developing and mature sieve elements in the brown alga *Laminaria saccharina*. *American Journal of Botany* 64: 649–657.
Silberfeld, T., Leigh, J.W., Verbruggen, H., Cruaud, C., de Reviers, B. & Rousseau, F. 2010. A multi-locus time-calibrated phylogeny of the brown algae (Heterokonta, Ochrophyta, Phaeophyceae): Investigating the evolutionary nature of the "brown algal crown radiation". *Molecular Phylogenetics and Evolution* 56: 659–674.
Silberfeld, T., Rousseau, F. & Reviers, B. de. 2014. An updated classification of brown algae (Ochrophyta, Phaeophyceae). *Cryptogamie Algologie* 35: 117–156.
Simeon, A., Kridi, S., Kloareg, B. & Herve, C. 2020. Presence of exogenous sulfate is mandatory for tip growth in the brown alga *Ectocarpus subulatus*. *Frontiers in Plant Science* 11: 1277.
Smith, A.I. 1939. The comparative histology of some of the Laminariales. *American Journal of Botany* 26: 571–585.
Starko, S., Soto Gomez, M., Darby, H., Demes, K.W., Kawai, H., Yotsukura, N., Lindstrom, S.C., Keeling, P.J., Graham, S.W. & Martone, P.T. 2019. A comprehensive kelp phylogeny sheds light on the evolution of an ecosystem. *Molecular Phylogenetics and Evolution* 136: 138–150.
Steinbiss, H.-H. & Schmitz, K. 1974. Zur Entwicklung und funktionellen Anatomie des Phylloids von Laminaria hyperborea. *Helgolander Wiss. Meeresunters* 26: 134–152.
Sykes, M.G. 1908. Anatomy and Histology of *Macrocystis pyrifera* and *Laminaria saccharina*. *Annals of Botany* 22: 291–325.
Tarver, J.E., Cormier, A., Pinzón, N., Taylor, R.S., Carré, W., Strittmatter, M., Seitz, H., Coelho, S.M. & Cock, J.M. 2015. microRNAs and the evolution of complex multicellularity: Identification of a large, diverse complement of microRNAs in the brown alga Ectocarpus. *Nucleic Acids Research* 43: 6384–6398.
Terauchi, M., Nagasato, C. & Motomura, T. 2015. Plasmodesmata of brown algae. *Journal of Plant Research* 128: 7–15.
Tesson, B. & Charrier, B. 2014. Brown algal morphogenesis: Atomic force microscopy as a tool to study the role of mechanical forces. *Frontiers in Plant Science* 5: 471.
Torode, T.A., Marcus, S.E., Jam, M., Tonon, T., Blackburn, R.S., Hervé, C. & Knox, J.P. 2015. Monoclonal antibodies directed to fucoidan preparations from brown algae. *PLoS One* 10: e0118366.
Torode, T.A., Siméon, A., Marcus, S.E., Jam, M., Le Moigne, M.-A., Duffieux, D., Knox, J.P. & Hervé, C. 2016. Dynamics of cell wall assembly during early embryogenesis in the brown alga Fucus. *Journal of Experimental Botany* 67: 6089–6100.
Tsirigoti, A., Beakes, G.W., Hervé, C., Gachon, C.M.M. & Katsaros, C. 2015. Attachment, penetration and early host defense mechanisms during the infection of filamentous brown algae by *Eurychasma dicksonii*. *Protoplasma* 252: 845–856.
Tsirigoti, A., Küpper, F.C., Gachon, C.M.M. & Katsaros, C. 2014. Cytoskeleton organisation during the infection of three brown algal species, *Ectocarpus siliculosus*, *Ectocarpus crouaniorum* and *Pylaiella littoralis*, by the intracellular marine oomycete *Eurychasma dicksonii*. *Plant Biology* 16: 272–281.
Umen, J. & Coelho, S. 2019. Algal sex determination and the evolution of anisogamy. *Annual Review of Microbiology* 73: 267–291.
Wang, X., Yao, J., Zhang, J. & Duan, D. 2020. Status of genetic studies and breeding of *Saccharina japonica* in China. *Journal of Oceanology & Limnology* 38: 1064–1079.
Yabu, H. 1965. Early development of several species of Laminariales in Hokkaido. *Memoirs of the Faculty of Fisheries Hokkaido University* 12: 1–72.
Ye, N., Zhang, X., Miao, M., Fan, X., Zheng, Y., Xu, D., Wang, J., Zhou, L., Wang, D., Gao, Y., Wang, Y., Shi, W., Ji, P., Li, D., Guan, Z., Shao, C., Zhuang, Z., Gao, Z., Qi, J. & Zhao, F. 2015. Saccharina genomes provide novel insight into kelp biology. *Nature Communications* 6: 6986.
Yendo, K.R. 1911. The development of *Costaria*, *Undaria*, and *Laminaria*. *Annals of Botany* 25: 691–716.
Zhang, L., Li, J., Wu, H. & Li, Y. 2019. Isolation and expression analysis of a candidate gametophyte sex determination gene (*sjhmg*) of kelp (*Saccharina japonica*). *Journal of Phycology* 55: 343–351.

3 Unicellular Relatives of Animals

Aleksandra Kożyczkowska, Iñaki Ruiz-Trillo and Elena Casacuberta

CONTENTS

3.1 INTRODUCTION: UNICELLULAR RELATIVES OF ANIMALS

All life on Earth has evolved from a common ancestor in a fascinating chain of events. One of the most pivotal steps in the history of life was the transition from protists into multicellular animals. However, how exactly this transition occurred remains unknown. The only way to unveil this process is by studying the unicellular relatives of animals. The Holozoa clade comprises animals and several unicellular lineages (known as unicellular Holozoa): Choanoflagellatea (King 2005), the Filasterea (Shalchian-Tabrizi et al. 2008), the Ichthyosporea (Mendoza et al. 2002) and the Corallochytrea/Pluriformea (Torruella et al. 2015; Hehenberger et al. 2017) (Figure 3.1).

The analysis of whole genomes from a wide Holozoa taxon sampling in a comparative framework has been useful to reconstruct the genetic content of their common ancestor (Sebé-Pedrós et al. 2017; Grau-Bové et al. 2017; Richter et al. 2018). These phylogenomic efforts have unveiled a unicellular ancestor of animals equipped with a much more complex genetic repertoire than previously thought. One remarkable feature of the ancestor genome is that despite of being unicellular, it already contained many genes whose function is directly related to multicellular structures. Examples of such genes are *integrins* and *cadherins*, which are directly related to cell adhesion; *tyrosine kinases* that mediate signaling in the context of cell-to-cell communication; and several transcription factors involved in development or proliferation such as *runX*, *nf-κβ* or *myc* (Abedin and King 2010; Suga

DOI: 10.1201/9781003217503-3

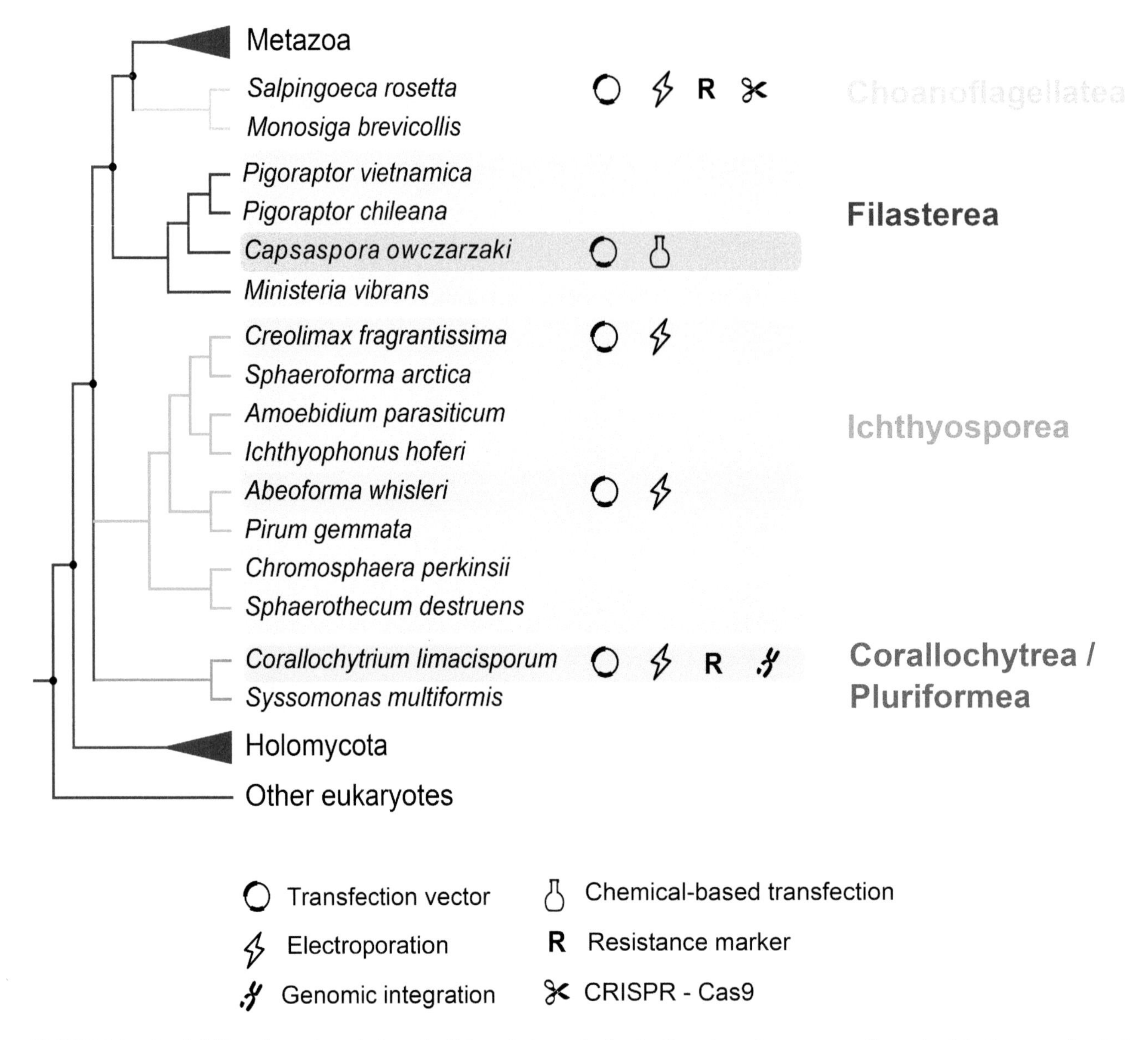

FIGURE 3.1 Availability of genetic tools for unicellular relatives of animals. Genetic tools are present for each of the lineages of unicellular Holozoa: *Salpingoeca rosetta* (Choanoflagellatea), *Capsaspora owczarzaki* (Filasterea), *Creolimax fragrantissima* and *Abeoforma whisleri* (Ichthyosporea) and *Corallochytrium limacisporum* (Corallochytrea/Pluriformea). Symbols represent transfection techniques (electroporation or chemical-based transfection), selection agent, genome editing technique (CRISPR-Cas9) and genome integration. (Phylogenetic tree adapted from Grau-Bové et al. 2017; López-Escardó et al. 2019; Hehenberger et al. 2017.)

et al. 2012; Sebé-Pedrós et al. 2017; Richter et al. 2018). After the initial studies centered in genome content, the next question was to understand if the genome of unicellular holozoans contained some of the features of the regulatory and architectural genome organization observed in Metazoa. Remarkably, genome organization and some epigenetic signatures are present in at least one filasterean, suggesting that they were already present in the genome of the unicellular ancestor (Sebé-Pedrós et al. 2016). Furthermore, since their isolation, different unicellular holozoans have been cultivated, allowing for the first observations and descriptions of some of their stages and cellular characteristics (Marshall et al. 2008; Fairclough et al. 2010; Marshall and Berbee 2011; Sebé-Pedrós et al. 2013, 2017; Torruella et al. 2015; Grau-Bové et al. 2017; Tikhonenkov et al. 2020a). From these studies, we have learned that the four unicellular holozoan lineages are diverse not only in their morphology but also in their developmental modes. Interestingly, in all lineages, there are examples of temporary "multicellular" structures during their life cycle (Figure 3.2). Choanoflagellates are able to form colonies through clonal division (Fairclough et al. 2010; Dayel et al. 2011), the filasterean *Capsaspora owczarzaki* can form cell aggregation (Sebé-Pedrós et al. 2013) and several ichthyosporeans have a multi-nucleate coenocytic stage that resembles the embryonic coenocyte of some animals (Suga and Ruiz-Trillo 2013a; Ondracka et al. 2018;

Dudin et al. 2019). Finally, *Corallochytrium limacisporum*, one of the two representatives of Corallochytrea, combines two different ways to proliferate: through binary fission or through a multi-nucleated coenocyte (Kożyczkowska et al. 2021).

The data generated so far on these unicellular relatives of animals suggest they are key to understanding the evolution from unicellular organisms to multicellular animals. However, we need to go beyond what the genomes tell us and look more particularly at functional analyses, and research efforts in this direction have begun. Genetic tools have been developed for a handful of unicellular holozoans (Figure 3.1), opening the possibility to experimentally test, in a comparative framework, some of the evolutionary hypotheses that the phylogenomic studies have put on the table. In this chapter, we provide a broad description of the general characteristics of each unicellular holozoan lineage, followed by detailed description of the taxa that have been developed into experimentally tractable organisms. We highlight, as well, their particularities and emphasize the most important optimization steps in the different protocols (Figure 3.3). The aim is to provide an updated reference for the state of the art of the methods available for the different unicellular relatives of animals.

3.2 CHOANOFLAGELLATA

Choanoflagellates are the sister-group to animals (Figure 3.1). There are around 360 species of choanoflagellates described to date, representing a considerable amount of biodiversity in life forms (King 2005). Choanoflagellates are bacterivorous, and they are commonly found in both freshwater and marine environments (Dolan and Leadbeater 2015). A typical choanoflagellate cell is composed of a single apical flagellum that is surrounded by a collar of microvilli. The currents created by the flagellum help drive bacteria into the collar, where they are phagocytized (Clark 1866; Pettitt et al. 2002). Their morphology and their feeding behavior are also found in the choanocytes, a highly specialized cell type in sponges. These similarities have historically inspired theories of a close evolutionary relationship between animals and choanoflagellates (Clark 1866; Maldonado 2004; Nielsen 2008). However, several phylogenomic analyses point to the fact that these similarities are likely the result of convergent evolution and not shared ancestry (Mah et al. 2014; Sogabe et al. 2019). Phylogenetic analyses divide choanoflagellates in two major clades, Craspedida and Acanthoecida (Carr et al. 2008; Dolan 2015; Paps et al. 2013). Accordingly, both clades show different outer morphologies. In general terms, craspedids form organic coverings which can include a thecate (a vase-like capsule) or a glycocalyx (Leadbeater et al. 2009), and acanthoecids are the species that possess an inorganic extracellular covering made of siliceous material known as the lorica (Carr et al. 2008).

Monosiga brevicollis and *Salpingoeca rosetta*, both belonging to the Craspedida, are the two better-known choanoflagellates (Figure 3.1) (King et al. 2008; Fairclough et al. 2013). The study of the genome of these two species revealed that they contain genes considered animal specific or involved in multicellular functions, as we will see for other unicellular holozoans (see next sections). Especially intriguing is the presence of synaptic proteins, even though they lack the animal-like mechanism of synapsis (Ryan and Grant 2009; Burkhardt et al. 2014). Those genomes also encode genes involved in forming multicellular structures such as the ones involved in cell adhesion and cell-to-cell communication, such as *cadherins* or *tyrosine-kinase* signaling, for example (Hoffmeyer and Burkhardt 2016; Burkhardt et al. 2014). Interestingly, these sets of genes are found in both species independently of their capacity to form multicellular structures, since *S. rosetta* is able to form colonies by clonal division (Figure 3.2a and next section), while *M. brevicollis* is unicellular throughout its life cycle.

Another important result from the study of the genome of *M. brevicolis* and *S. rosetta* is that they are evolutionarily close, show low genetic diversity and have retained the fewest ancestral gene families in comparison with the other choanoflagellate genomes now available (Richter et al. 2018).

3.2.1 *Salpingoeca rosetta*

So far, efforts to develop a choanoflagellate into an experimentally tractable system have focused on *S. rosetta*. *S. rosetta* presents several advantages among other choanoflagellates to be developed as a new model organism: it has a well-annotated genome and a colonial stage. Moreover, the mechanisms of colonial formation are well understood (Booth et al. 2018; Wetzel et al. 2018; Booth and King 2020). *Salpingoeca rosetta*, first known as *Proterospongia* sp., was isolated from a marine sample in the form of a colony (King et al. 2003). The colonies are formed by serial mitotic divisions starting from a single founding cell, which grows into a spherical multicellular structure resembling a rosette (Figure 3.2a) (Fairclough et al. 2010). Interestingly, it has been shown that inside a colony, there are differences between cells concerning their nuclei volume and conformation, the number of mitochondria or cell shapes named afterward *chili* or *carrot* cells (Naumann and Burkhardt 2019). These differences among the cells of the colony suggest that there might be spatial cell differentiation in those rosette colonies. Cells inside a rosette seem to hold to each other by cytoplasmic bridges, filopodia and extracellular matrix (ECM; Dayel et al. 2011; Laundon et al. 2019). Although, as mentioned previously, the rosette conformation was the original form in which *S. rosetta* was isolated from the ocean, soon cultured rosettes became infrequent and difficult to control under laboratory conditions, and the single cell became the main form of *S. rosetta* in *in vitro* cultures. Later, experiments of incubation of *S. rosetta* together with high densities of *Algoriphagus machipongonensis*, the bacteria with which *S. rosetta* was co-isolated from the ocean, recovered the formation of rosettes. Further investigations discovered that this phenomenon was induced by a lipid, renamed rosette inducing factor (RIF; Alegado et al. 2012; Fairclough et al. 2010; Dayel et al. 2011; Woznica et al. 2016). In parallel, a forward genetic screen for mutants unable to form rosettes allowed for the identification of a genetic factor in *S. rosetta*, which could be linked to the rosette phenotype. The recovered *rosetteless* mutant encoded a C-type lectin and was not able to develop rosettes in spite of being exposed to RIFs (Levin et al. 2014). Although it is not yet fully understood by which molecular mechanism the C-type lectin establishes the relevant interactions, it has been hypothesized that the function of the C-type lectin is related to an interaction with the ECM (Levin et al. 2014). Interestingly, colony formation is not the only stage in *S. rosetta*'s life cycle governed by bacteria. For instance, Woznica, Gerdt and collaborators discovered that the bacteria *Vibrio fischeri* was able to induce sexual behavior in *S. rosetta* through a secreted product that was conveniently labeled EroS (Woznica et al. 2017). Interestingly, EroS was biochemically identified as a chondroitin lyase. This enzyme is able to digest chondroitin sulfate and initiate mating, bearing some similarities to sperm digestion of the egg cover in animal reproduction (Miller and Ax 1990).

Under conditions promoting fast growth, *S. rosetta* is able to form yet another multicellular form different from the rosettes. Linear colonies consist of a chain of cells attached to each other and connected by intercellular bridges and ECM (Figure 3.2a) (Dayel et al. 2011). In the case of single cells, *S. rosetta* can acquire three different forms, which besides its morphology also present a specific behavior: fast swimmers, slow swimmers and thecate cells. The main difference between the different forms of single cell types is the presence of the theca in thecate cells, which consists of a vase-like capsule composed of ECM. All forms of *S. rosetta* have a flagellum that is used for swimming and orienting the colony, and fast swimmers and rosette colonies also have thin filopodia (Dayel et al. 2011).

Regardless of the availability of genetic tools, *S. rosetta* could already be considered an emerging model system because substantial information on its biology had already been obtained. The *rosetteless* mutant had been isolated by a forward genetic screen aiming to isolate defective mutants in rosette development (Levin and King 2013; Levin et al. 2014). Moreover, specific culture conditions were developed to obtain and enrich for each of the different life forms of *S. rosetta* (Dayel et al. 2011), and, finally, by the co-cultivation with specific bacteria, mating could be induced (Woznica et al. 2016; Woznica et al. 2017). Nevertheless, tools for direct genetic manipulation, which would allow us for example to fluorescently tag specific proteins to study their localization and dynamics or to knock out target genes, were missing. In recent years, Dr. Nicole King's research group has successfully developed transfection, selection and genome editing for *S. rosetta*, overcoming these limitations. In the following sections, we will briefly summarize the main steps of these achievements.

3.2.1.1 Transfection and Selection

The transfection protocol for *S. rosetta* is based on the Nucleofection technology, developed by Amaxa (Lonza Cologne AG group) (Figure 3.3a). Nucleofection is a specialized electroporation-based transfection technology engineered to transfer the DNA into the nucleus. This technique proved successful in *S. rosetta*, which can now be transiently transfected with an average efficiency of 1%, similar to what has been achieved in other protists (Janse et al. 2006; Caro et al. 2012).

In order to understand the significance of each optimization step, Booth et al. sequentially eliminated them one at a time and monitored the change in efficiency (Figure 3.3a). For example, the addition of pure and highly concentrated carrier DNA (empty plasmid, such as *pUC19*), in combination with the plasmid of interest, was key to optimize *S. rosetta* transfection, as observed in other unicellular holozoans (Faktorová et al. 2020; Kożyczkowska et al. 2021). A second key step to boost transfection in *S. rosetta* was priming the cells with a buffer that contains a combination of a protease, a reducing agent, a chelator and a chaotrope (Booth et al. 2018). This specific buffer was key in breaking down the extracellular coat and significantly improved the uptake of transfected DNA into the cell. Even though the extracellular coat is specific for this choanoflagellate, it could be of inspiration for those working on organisms that also possess an extracellular coat or wall, which usually hampers transfection efficiency.

One of the first applications of the developed transfection in *S. rosetta* by Dr. Booth and collaborators was the study of the localization of two *septin* orthologues, *SrSeptin2* and *SrSeptin6* (Booth et al. 2018). Septins are a multigenic family involved in highly conserved functions such as cell division (Neufeld and Rubin 1994) but also more specialized functions in multicellular organisms at the level of intracellular junctions and the maintenance of polarity in an epithelium (Spiliotis et al. 2008; Kim et al. 2010). The study of the involvement of *septin* orthologues of *S. rosetta* in these latter roles can help us understand the contribution of Septins in the evolution of the epithelia before the onset of animals.

Finally, at the same time as the study of Septins in *S. rosetta*, the newly developed transfection technique also proved significant for the characterization of additional rosette defective mutations (Wetzel et al. 2018). In addition, in this study, researchers went one step further by applying selection with the antibiotic puromycin. Selection is very useful in order to enrich the population in a greater proportion of transfected cells Figure 3.3a) (Wetzel et al. 2018). A public protocol for transfection and selection of *S. rosetta* is available at Protocols.io; dx.doi.org/10.17504/protocols.io.h68b9hw

3.2.1.2 Plasmids

As a first step to develop transient transfection, researchers cloned putative endogenous promoters from the *elongation factor 1*, *ef1*, β-*actin*, *act*, α-*tubulin*, *tub* and *histone H3* genes from *S. rosetta*. Two different reporter genes, *nanoluc* (monitored through a luciferase assay) and *mwassabi* (monitored through expression of green fluorescence), were chosen to test the newly cloned promoters and used to fine-tune the transfection protocol (Booth et al. 2018).

Besides the battery of transfection plasmids generated to monitor transfection carrying the previously mentioned promoters and reporter genes, researchers engineered plasmids targeting key subcellular structures for future studies on the cell biology of choanoflagellates. With this purpose, they fluorescently tagged the filopodia, cytoskeleton, endoplasmic reticulum, plasma membrane, mitochondria, cytoplasm and nuclei, using specific commercial, highly conserved peptides and protein sequences, known to localize in these cellular compartments (Booth et al. 2018).

Septin orthologues were visualized by the expression of plasmids containing *SrSeptin2* and *SrSeptin6* fused to the fluorescent reporter mTFP1 (Ai et al. 2006) under the actin promoter.

Finally, from all of the plasmids available for transfection in *S. rosetta*, we want to highlight the possibility of including the puromycin-resistant gene *pac* in order to select for puromycin-resistant cells (de la Luna S et al. 1988), since wild type *S. rosetta* shows certain susceptibility to this antibiotic (Wetzel et al. 2018).

3.2.1.3 Genome Editing; CRISPR-Cas9

Engineering genome editing from *de novo* requires not only designing the biochemical strategy that will most likely work in the chosen organism but also, and very importantly, pinpointing a good target. The ideal target should, once being edited in the transfected cells, give a phenotype that would allow further selection of those cells that have been genetically modified; antibiotic resistance or susceptibility is especially useful in this case. To illustrate this concept, we can take as an example the first attempts in genome editing in *S. rosetta* (Booth and King 2020). The first approach for using the developed CRISPR/Cas9 tools for *S. rosetta* was to introduce a mutation to the *rosetteless* gene, which had been isolated by a forward genetic screen (see previously) and encodes a C-type lectin protein that is involved in the formation of the rosette phenotype (Levin et al. 2014). The unsuccessful outcome of this first approach was likely due to a low efficiency of the genome editing procedure, which even if it worked correctly could not be detected. A solution to overcome this obstacle is to be able to select the few events of edited cells in the transfected culture by enriching successively in positively transfected cells. Booth and collaborators engineered an alternative CRISPR/Cas9 strategy to confer cycloheximide resistance as an initial step and, in this manner, optimizing the genome editing protocol in *S. rosetta*.

In terms of the molecular reagents needed for CRISPR/Cas9, the researchers decided to use a ribonucleoprotein (RNP) composed of the expressed Cas9 of *Streptomyces pyogenes* together with the *in vitro*–produced single guide RNAs, sgRNA, to direct *Sp*Cas9 to the nicking position. There is a double advantage of using an RNP instead of plasmids for the expression of the different components involved in the editing: on one hand avoiding the necessity of having an endogenous RNA polymerase III promoter in order to express the sgRNAs and on the other avoiding the possible cytotoxicity and off-target problems from uncontrolled Cas9 protein expression (Jacobs et al. 2014; Jiang et al. 2014; Shin et al. 2016; Foster et al. 2018; S. Kim et al. 2014; Liang et al. 2015; Han et al. 2020). Moreover, parallel to transfecting the RNP, a DNA repairing template should be added if the desired mutation is other than a deletion. In the case of *S. rosetta*, Booth and collaborators discovered that *S. rosetta* was able to use a variety of different templates, single and double strand. The addition of the repair template also improved genome editing efficiency. The percentage of genome editing was very similar to transfection efficiency, pinpointing the transfection technique as the limiting factor (Booth and King 2020). Nevertheless, if a good selection strategy exists, the edited cells should be efficiently recovered with this transfection rate with no difficulty.

S. rosetta is the first unicellular holozoan to be genome edited. The protocol developed by Dr. Booth and collaborators represents a technical breakthrough that will undoubtedly enhance the possibilities to perform functional studies in this organism. Needless to say, the advances in *S. rosetta* have and will keep inspiring the development of genetic tools and genome editing approaches in other closely related lineages.

3.2.2 Prospects

There is no doubt that the technical advances that we have here reported for *S. rosetta* will open new venues to functional approaches that had been hampered until now. We would also like to stress the importance of this organism beyond now being a genetically tractable organism. The importance of *S. rosetta* to address the origin of metazoans has already been broadly explained (Richter et al. 2018). Moreover, the highly organized and structured rosette colonies provide researchers with an ideal model to understand the origins of spatial cell differentiation (Naumann and Burkhardt 2019). Finally, the demonstrated influence of specific interactions with bacteria on essential life events or the transition to multicellular stages of *S. rosetta* provides a unique opportunity to study the interactions between bacteria and eukaryotes (Woznica et al. 2016, 2017).

3.3 FILASTEREA

Filasterea is one of the latest lineages of unicellular holozoans that has been described to date. Filasterea is the sister group to Choanoflagellata and Metazoa, all together forming the Filozoa clade (Shalchian-Tabrizi et al. 2008; Torruella et al. 2012, 2015) (Figure 3.1).

There are five species known to belong to Filasterea: *Capsaspora owczarzaki, Ministeria vibrans, Pigoraptor vietnamita, Pigoraptor chileana* and the recently described and potentially filasterean *Tunicaraptor* (Figure 3.1) (Owczarzak et al. 1980b; Hehenberger et al. 2017; Parra-Acero et al. 2018; Tikhonenkov et al. 2020b). Besides the endosymbiont *C. owczarzaki*, the flagellated species *Pigoraptor vietmanita* and *Pigoraptor chileana* are predatory (Hehenberger et al. 2017; Tikhonenkov et al. 2020a),

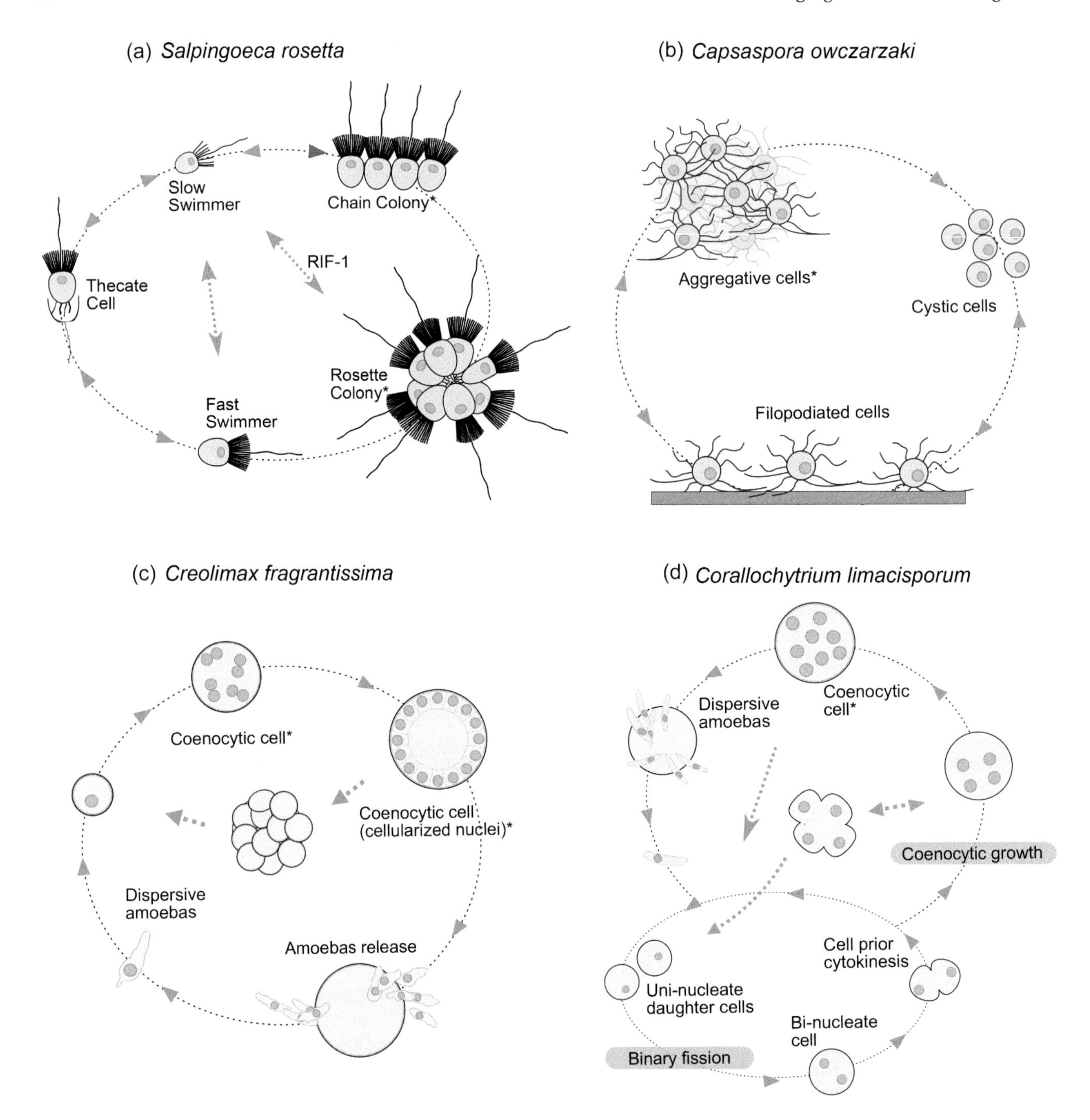

FIGURE 3.2 Models of the life cycle of unicellular relatives of animals. (a) *Salpingoeca rosetta*, (b) *Capsaspora owczarzaki*, (c) *Creolimax fragrantissima*, (d) *Corallochytrium limacisporum*. Arrows depict observed and inferred transitions between life stages partially described in the main text. Life cycles of unicellular holozoans are diverse but share an important feature: a temporary multicellular-like stage resembling those present in animals (multicellular-like stage indicated with *).

and *Ministeria vibrans* is a free-living heterotroph (Tong 1997; Cavalier-Smith and Chao 2003; Shalchian-Tabrizi et al. 2008). Filastereans have been isolated from both marine an fresh water environments. For instance, *M. vibrans* has been isolated from samples of marine coastal waters. It has been successfully grown in the laboratory but only in the presence of bacteria, making investigations more difficult. *M. vibrans* is a spherical amoeboid (aprox. 4 μm) with a stalk falgellum, surrounded by fine and long radiating arms of equal length (Torruella et al. 2015), making a characteristic vibrating movement before attaching to a substrate (Cavalier-Smith and Chao 2003). Interestingly, it has been described that this species is capable of forming aggregative cell clumps (Mylnikov et al. 2019).

Pigoraptor vietnamica and *Pigoraptor chileana* are two filasteran species isolated from freshwater environments (Hehenberger et al. 2017). Both species have an elongated-oval shape with an average size of 5–14 μm long, have predatory behavior and display a very similar life cycle. A detailed description of their complex life cycle can be found in the work done by Tikonhenko et al. (2020a). We would like to highlight that both *Pigoraptor* species can aggregate during their life cycle, as has been described for *M. vibrans* as well as for the best-studied filasterean, *C. owczarzaki* (see the following).

3.3.1 *Capsaspora owczarzaki*

First reports of *C. owczarzaki* appeared from investigations on the susceptibility of the fresh-water snail *Biomphalaria glabrata* to be infected by the parasite *Schistosoma mansoni*. Studying the possible factors underneath the resistance to infection, Stibbs and collaborators isolated a small amoeba of 3–5 µm in diameter from pericardium and mantle explants from three different strains of *B. glabrata*, two of them resistant to *Schistosoma* infection (Stibbs et al. 1979). The ability to grow *C. owczarzaki* in axenic cultures allowed researchers to test the interaction between the amoeba and the parasite. These works demonstrated that *C. owczarzaki* amoebas were able to adhere to and kill the sporocists of *S. mansoni*, resulting in a high proliferation of *C. owczarzaki*. H. Stibbs and A. Owczarzaki were the first ones to describe *C. owczarzaki* and set the initial culture conditions.

The initial stage of the life cycle of *C. owczarzaki* consists of crawling filopodiated amoebas that grow exponentially. Once the culture is saturated and nutrients become limiting, amoebas retract their filopodia and encyst in a round and compact cell, and their growth stabilizes. At this point, encysted cells can attach to each other, forming compact cell aggregates of different sizes (Figure 3.2b). *C. owczarzaki* cell aggregates can happen spontaneously or can also be induced by agitation with specific parameters (Sebé-Pedrós et al. 2013). Most importantly, electron microscopy analyses revealed that cells in the aggregates are glued together by cohesive extracellular material, which provides the aggregate with consistency but keeps cells individually separated. RNA-seq analyses demonstrated an upregulation of the expression of key genes involved in cell-to-cell communication and cell adhesion, such as the tyrosine kinase signaling pathway and the integrin adhesome (Sebé-Pedrós et al. 2013).

The study of *C. owczarzaki* has not only provided knowledge about its biology but also about the wider question of animal origins. For example, analysis of its genome revealed several genomic features previously thought to be animal specific (Suga et al. 2013; Sebé-Pedrós et al. 2017). *C. owczarzaki* contains a complete integrin adhesome necessary to mediate the interaction between the cell and the ECM (Suga et al. 2013; Parra-Acero et al. 2020). Moreover, *C. owczarzaki* also contains a set of proteins, including transcription factors (TFs), known to be involved in developmental pathways in animals; NF-κ*b*, Runx and T-box; and others involved in cell motility and proliferation such as Brachyury and MYC (Mendoza and Sebé-Pedrós 2019). Additionally, components of different signal transduction pathways have an unexpected conservation, with examples such as JAK-STAT, Notch, TGFβ or tyrosine kinases in general (RTKs) (Suga et al. 2012).

It is clear that *C. owczarzaki* was an ideal species to be developed into a genetically tractable organism in order to further investigate the different hypotheses drawn from the genomic content and signatures, as well as to plunge into the terrain of cell biology to enrich the investigations of the evolutionary path shared among holozoans.

3.3.1.1 Transfection

The first attempts to transfect a new organism fail the vast majority of times. For *C. owczarzaki*, the first protocols to be tested were based on different technologies such as electroporation, magnetofection and lipid-based transfection methods. However, these tests yielded either no positive cells or very low transfection efficiencies, hampering reproducibility (Suga and Ruiz-Trillo 2013; Ensenauer et al. 2011; Parra-Acero et al. 2018). The technology that ended up being efficient enough to be further optimized into a reliable transfection protocol was the classical calcium phosphate precipitation method (Figure 3.2b) (Graham and van der Eb 1973). Here we highlight the steps that turned out to be crucial to improve the efficiency of the transfection protocol (Parra-Acero et al. 2018). One of the factors that is important to maximize efficiency is to use cells at the exponential growth phase. The stage in which *C. owczarzaki* is growing exponentially is the adherent stage. Cells from a fresh culture at 90/95% confluence from the adherent stage were the ones with higher transfection efficiency. The size of the crystals from the DNA and the precipitates of calcium phosphate also proved important to improving the efficiency of transfection. The authors determined that the smaller the crystals, the better, as shown for other organisms such as *D. discoideum* (Jordan and Wurm 2004; Gaudet et al. 2007). In order to achieve a smaller crystal size, it is important to keep the same ratio for DNA/calcium and phosphate when preparing the DNA mix to transfect. The stability of the DNA/calcium ratio once the DNA mix was added to the media also depended on the amount of phosphate in the transfection media, which also needed to be taken into account. Similarly, the pH of the final solution should be controlled to avoid changes in the solubility of the precipitates. The last touch to further improve transfection efficiency was to expose cells to an osmotic shock, which would permeate the cell membrane for a short period of time. This technique is also used in a variety of eukaryotic cells with the application of glycerol or DMSO (10–20%) (Grosjean et al. 2006; Gaudet et al. 2007; Guo et al. 2017). In the case of *C. owczarzaki*, a 10% glycerol shock during one minute was good enough (Figure 3.3b). Finally, as in any transfection protocol, it is important to be able to identify those cells where the DNA has successfully entered the nucleus and is being expressed. The identification of transfected cells can be done by enriching the transfected population using an antibiotic or a specific drug to which wild type cells (non-transfected cells) are susceptible or by inspecting the expression of a fluorescent protein using fluorescence microscopy. Because *C. owczarzaki* seems to be resistant to different antibiotics, pesticides or cytostatic drugs that are commonly used for selection, the initial plasmids that were designed and transfected into *C. owczarzaki* contained genes encoding small fluorescent proteins. These fluorescent proteins, such as mVenus and mCherry, were expressed in the cytosol of transfected cells. Besides the microscopy observations, efficiency of transfection was also analyzed using flow cytometry by comparing the population of transfected cells with cells from a negative control population. Note that it is important to take into account the possible phenomenon of auto-fluorescence for some types of cells. Efficiency of transfection was on average around 1.132% ± 0.529 (mean ± s.d.), which might seem low for researchers working with transfection in other eukaryotic systems, but it is sufficient to efficiently further select transfected cells and proceed with downstream experiments (Parra-Acero et al. 2018).

Co-transfection is known to increase efficiency of the transfection *per se*, and it is also very useful in order to deliver two different constructs simultaneously. Dr. Parra-Acero and collaborators tested in which proportion two different plasmids were uptaken by the cells when co-transfected in order to use co-transfection to visualize simultaneously more than one subcellular structure. Co-transfection resulted, with a rate of incorporation of both constructs almost equally (72.909% ± 5.468) in *C. owczarzaki* (Parra-Acero et al. 2018).

Although stable transfection has not yet been developed in *Capsaspora*, plasmids delivered by transient transfection were shown to be expressed inside the cells for up to ten days. The life cycle of *Capsaspora* is much shorter than ten days, and therefore this protocol allows for the interrogation of the reporter expression at the different life stages of the organism.

3.3.1.2 Plasmids

The reporter plasmids (pONSY-mVenus and pONSY-mCherry) for optimizing transfection and calculating efficiency were already designed using the endogenous promoter and terminator sequences of the elongation factor 1-α gene (EF1-α) of *Capsaspora* (Parra-Acero et al. 2018). Besides the engineered plasmids to visualize the cytosol, the researchers went one step further in order to get insights into the cell biology of this species. For this reason, they designed plasmids to fluorescently label the different subcellular structures. For example, the endogenous histone 2B (H2B) gene was fused to mVenus to highlight the nucleus (pONSY-CoH2B:Venus), and the plasma membrane was visualized by cloning the N-myristoylation motif (NMM) of the endogenous Src2 tyrosine kinase gene, which is known to localize at membranes and filopodia (pONSY-CoNMM:mCherry) (Sigal et al. 1994; Parra-Acero et al. 2018). Finally, in order to visualize the cytoskeleton, a small peptide (17 amino acid) named *lifeAct* known to bind filamentous actin (Riedl et al. 2008) was fused to mCherry (pONSY-Lifeact:mCherry) to visualize the actin cytoskeleton and filopodia of transfected cells. Detailed observations using confocal microscopy of single and co-transfected *C. owczarzaki* cells with these plasmids revealed the targeted structures explaining, among others, the hollow basket structure from the actin bundles around the cell body or the dynamics of the filopodia along the different life stages (Parra-Acero et al. 2018).

3.3.2 Prospects

C. owczarzaki, in addition to its key phylogenetic position, its well-annotated genome and the number of "multicellular" genes its genome encodes, is also able to form cell aggregates during its life cycle (Figure 3.2b), making it an ideal organism to analyze the origin of animals.

Finally, the fact that this organism is able to attack and feed on *S. mansoni* sporocysts (Stibbs et al. 1979; Owczarzak et al. 1980a) also makes it a potential candidate for disease-control strategies, even though the specific interaction of *C. owczarzaki* with the snail *B. glabrata* remains unclear. Interestingly, *C. owczarzaki* exhibits high resistance to antibiotics and harsh mediums, suggesting its potential in medical applications in the case that was finally selected to control schistomiasis (Parra-Acero et al. 2018).

3.4 ICHTHYOSPOREA

Ichthyosporea is the sister-group to Corallochytrea, as well as to the Filozoa (Choanoflagellata, Filasterea and Metazoa) (Mendoza et al. 2002). All described ichthyosporeans are osmotrophs and have multiple life stages that vary greatly in shape and motility and in most cases contain a cell wall of variable composition. The developmental mode of *ichthyosporeans* is complex and contains multinucleated stages such as a coenocyte (Figures 3.1 and 3.2c).

Ichthyosporeans received this name because the early identified representatives were all parasites of fish (Cavalier-Smith 1998). Later phylogenomic analyses of rDNA with newer representatives expanded the group in two internal classes, the *Dermocystida*, which are exclusively parasites of vertebrate hosts, and the *Ichthyophonida*, which can parasitize a variety of host species (Mendoza et al. 2002; Marshall et al. 2008). In accordance with their habitat, only representatives of *Ichthyophonida* can be cultured in laboratory conditions (Jøstensen et al. 2002; Marshall et al. 2008). Interestingly, the motile representatives of *Dermocystida* are equipped with a flagellum, while the *ichthyophonids* are motile amoebas. Maybe related, it has been shown by electron microscopy studies that representatives of *Ichthyophonida* have a spindle pole body (Marshall et al. 2008), which would nicely correlate with the disappearance of centrioles and the flagellum as a consequence (Marshall and Berbee 2011). On the other hand, centrioles have been described for members of *Dermocystida* such as *Dermocystidum percae* (Pekkarinen 2003). In the coming years, further investigations on other key biological questions will be possible once experimentally tractable organisms will be developed for both subclasses. For instance, investigations on the microtubule organizing centers and the nature of the mitosis (whether it is open, closed or somewhere in between) would

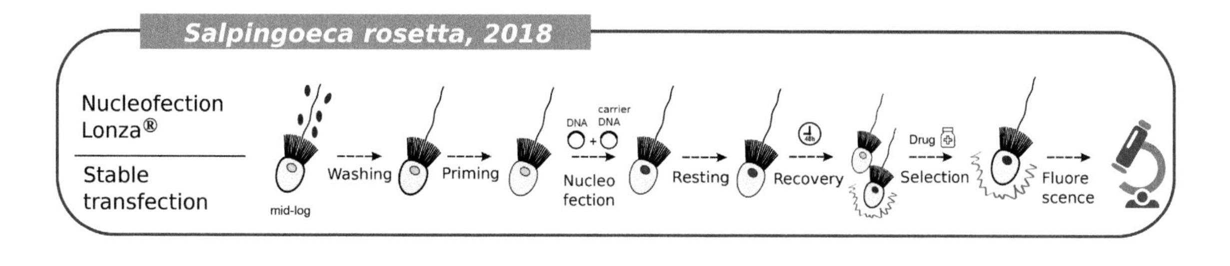

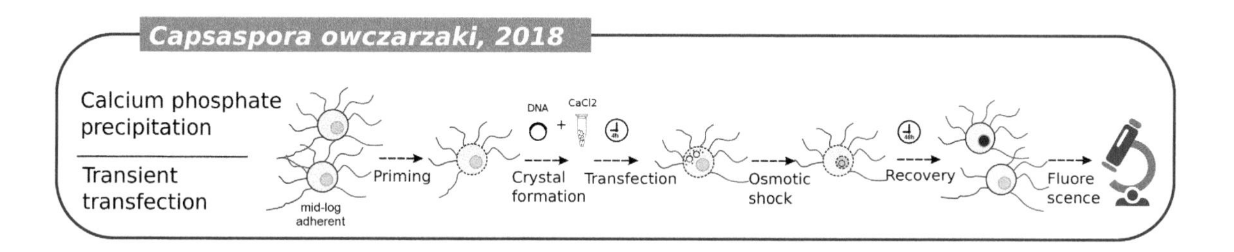

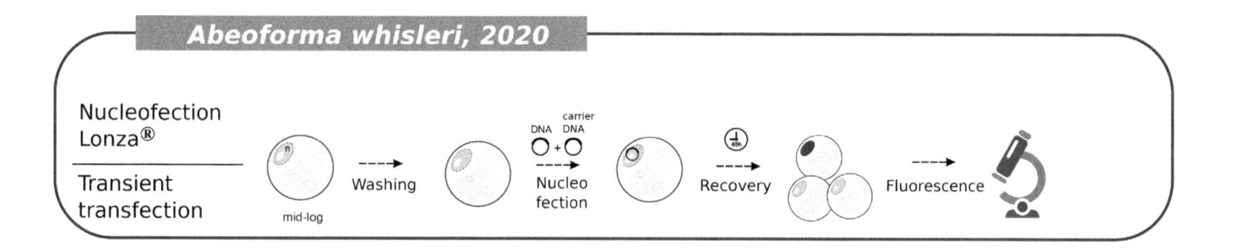

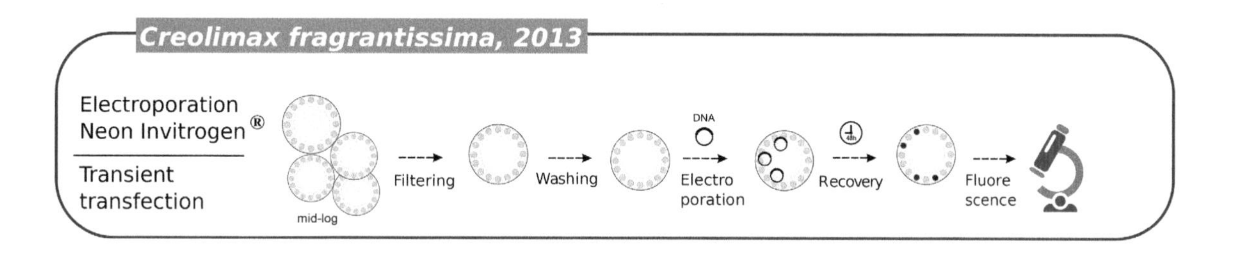

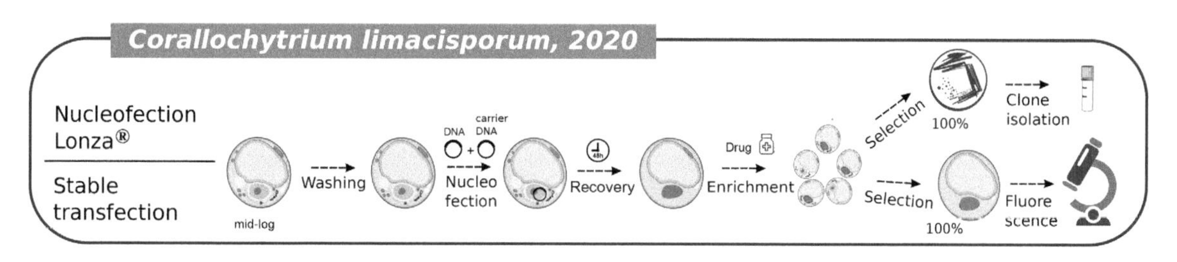

FIGURE 3.3 **Schematic diagram of transfection protocols among unicellular relatives of animals.** Basic steps have been illustrated. Key steps for electroporation-based techniques: pre-washing the remaining growth medium and addition of carrier DNA to the DNA of interest; for *S. rosetta* and *C. owczarzaki*, cells are primed for a higher membrane permeability. For calcium phosphate protocol: crystal size formation (ratio of DNA/$CaCl_2$) and an osmotic shock. For each transfection protocol, cells have been at the exponential growth phase (mid-log). Drug selection and stable transfection have been achieved in two organisms: *Salpingoeca rosetta* and *Corallochytrium limacisporum*. Additionally, *C. limacisporum* can be grown on an agar plate, allowing for single clone isolation.

be of great interest and could provide further insights on the evolutionary history of both subclasses.

3.4.1 *Abeoforma whisleri*

A. whisleri was isolated from the digestive track of the filtering mussel *Mytilus* (Figure 3.1) (Marshall and Berbee 2011). In culture, *A. whisleri* grows axenically in artificial Marine Broth (MB; GIBCO) at 13°C. Cultures can be seeded at low density 10^4/mL and reach confluence in approximately two weeks.

A. whisleri presents a vast myriad of cell shapes, which makes it difficult to reconstruct a possible life cycle from simple optical microscope observations. In a regular *A. whisleri* culture, one can observe mobile amoebas of different shapes, hypha-like stages, plasmodia cell shape, cells of different length and bigger and rounder multinucleated cells that correspond to coenocytes. Through live observations, researchers have witnessed the release of amoebas from the rounded coenocytic cells as well as vegetative reproduction, which can take place from sporadic budding of the plasmodium. For a thorough description of different cell shapes of *A. whisleri*, see Marshall and Berbee (2011).

All forms *of A. whisleri* cells are quite delicate even though it has been reported that all of them have a cell wall (Marshall and Berbee 2011). Interestingly, embedded membrane-bound microtubules (MBTs) were described for several of the morphologically different forms of *A. whisleri* cells. MBTs could be instrumental for equipping *A. whisleri* with the high membrane flexibility that it exhibits while having a cell wall. This could also be the reason behind the strong sensitivity that *A. whisleri* cells show when confronted with chemical, physical or electric shocks to create membrane pores in order to achieve transfection.

3.4.1.1 Transfection and Selection Protocol

One of the first steps toward developing genetic tools in *A. whisleri* was to test a wide battery of drugs for susceptibility in order to identify a selective agent (Faktorová et al. 2020). Puromycin resulted in the most promising acting as a cytostatic agent when assayed between 100 and 500 micrograms/mL, opening the possibility to use the resistance gene for puromycin activity (*pac*) (Luna et al. 1988) and the following protocol at Protocols.io: www.protocols.io/view/testing-selective-agents-for-the-icthyosporeans-ab-z5nf85e).

To achieve insertion of DNA inside *A. whisleri* nuclei, a battery of transfection protocols based on different methods were tested. Initially, electroporation with the Neon electroporation system (Invitrogen) was successful, but the resulting efficiency and reproducibility of this protocol did not allow for a regular establishment of transfection. During this time, researchers working on the choanoflagellate *S. rosetta* achieved promising results with another electroporation-based system, Nucleofection (Lonza), which was also more efficient and reproducible for *A. whisleri* (Figure 3.3c) (Booth et al. 2018; Faktorová et al. 2020; and Protocols.io: www.protocols.io/view/abeoforma-whisleri-transient-transfection-protocol-zexf3fn). In summary, the key steps to significantly improve efficiency and reproducibility were as follows: washing the cells with 1X PBS—which should be completely eliminated prior to re-suspension with transfection buffer—was important to maintain the low salt concentration for applying the electric shock. Small variations in this sense would make *A. whisleri* cells very susceptible to electric shock, exploding easily. On the other hand, immediate re-suspension of the cells with MB after the application of the electric current was key to obtaining the best cell recovery possible. The addition of high-concentration and high-quality carrier DNA (empty *pUC19*) was key to increasing the number of transfectants up to an order of magnitude. Finally, the best parameters for transfection were the combination of the buffer P3 in the middle of the scale of stringency and the electroporation code EN-138 (all provided by Lonza) (Figure 3.3c). After 24 h, ~1% of the culture was transformed based on the fraction of cells expressing mVFP (*venus* fluorescent protein) in the nucleus.

As an example of successful transient transfection for *A. whisleri*, Figure 3.4a shows the result of transfecting *AwH2BmVenusTer*. Several positive cells were observed with specific mVenus expression in the nuclei, demonstrating that the *AwH2BmVenusTer* plasmid was correctly delivered. Nevertheless, cells did not progress with cell division, suggesting that the expression of the fusion protein mVenus-H2B might be excessive, thus making the cells susceptible to the high levels of histone protein (Singh et al. 2010).

3.4.1.2 Plasmids

In order to deliver exogenous DNA into *A. whisleri* with the possibility to obtain transcription and protein expression, constructs with fluorescent proteins such as mCherry and mVenus (Shaner et al. 2004) (Nagai et al. 2002) were engineered using endogenous promoters to drive transcription. The actin promoter was chosen as one of the constitutive promoters widely used in molecular biology and therefore likely to work. Signatures from endogenous genes were selected in order to drive the fluorescence to a subcellular structure that could be easily identified, such as the nucleus (*AwH2BmVenusTer*) (Figure 3.4a) or the cytoskeleton (*ApmCherryTubulinaTer, ApmCherry Actina Ter*), all under the *A. whisleri* actin promoter and terminator (Faktorová et al. 2020). Moreover, a construct from which puromycin resistance could be delivered was also engineered in order to achieve stable transfected lines in the future (*ApmCherryPuromycinaTer*).

3.4.1.3 Prospects

In the near future, combined efforts to achieve stable transfection in *A. whisleri* under the effect of puromycin, together with simultaneously improving transient transfection toxicity, will be implemented. Because of the rich complexity in morphology of *A. whisleri* cells, achieving stable transfected lines with differently labeled subcellular components will be instrumental to study the sequence and diversity of its life stages and to be able to reconstruct its life cycle and the regulation of their transition.

3.4.2 *Creolimax fragrantissima*

C. fragrantissima was the first unicellular holozoan to be transiently transfected (Suga and Ruiz-Trillo 2013a), and it is so far the ichthyosporean with the greatest aptitude for being turned into a model organism (Figure 3.1). Most importantly, *C. fragrantissima* has been isolated a considerable number of times, and most of them have been successfully cultured in the laboratory. Besides having been isolated from a myriad of invertebrates belonging to four different phyla, the isolated *C. fragrantissima* strains were highly similar at both the molecular and morphological level (Marshall et al. 2008). The observed uniformity of the different strains implies relevance of the obtained results for a wide range of organisms, which is definitely desirable for a model organism.

C. fragrantissima is an osmotroph organism with an apparent asexual linear life cycle (Figure 3.2c). Cells are small and round, uni- or bi-nucleated, with a smooth cell wall and central vacuole, which pushes the nuclei to the cell periphery. There is no sign of flagella, hypha or budding behavior. The round cell grows from 6–8 mm in diameter to a mature multinucleated coenocyte of 30–70 mm in diameter, from which motile amoebas will burst from several pores of the parental coenocyte wall. Crawling uni-nucleated amoebas 12 mm long and 4.5–5 mm wide with erratic movement will become round and encyst after exploring a certain distance in various directions and finally setting, becoming round cells again, the cysts (Suga and Ruiz-Trillo 2013a; Marshall et al. 2008). The release of already round encysted cells has also been documented, as well as endospores that manage to grow without ever exiting the parental cell (Marshall et al. 2008). Fusion of cells is not observed, although clumps of cysts getting together are often found in regular cultures. The whole life cycle takes about 44 hours, where the maturation of the amoebas inside the coenocyte corresponds to 2–3 hours (Figure 3.2c).

3.4.2.1 Transfection

C. fragrantissima was the first unicellular holozoan in which transient transfection was achieved, allowing for the first investigations on its life cycle and initial characterization of life stages at the cellular level (Suga and Ruiz-Trillo 2013). Moreover, *C. fragrantissima* is the only unicellular holozoan for which morpholino RNA silencing has been successful (Suga and Ruiz-Trillo 2013).

The initial transformation protocol was based on electroporation performed inside the solution of the cell suspension using a wire-type electrode (Kim et al. 2008). With this protocol, the authors reported a remarkable transfection efficiency of 7% (Suga and Ruiz-Trillo 2013). Despite the transfection being transient, the introduced plasmid allowed for expression of the tagged protein during a two-day period. This was sufficient for the plasmid to be passed on to the next generation, enabling for the first time the description of some of the life stages of *C. fragrantissima*. The authors of the study specifically labeled the nuclei by fusing the *H2B* gene of either *C. fragrantissima* or the close relative *Sphaeroforma arctica* (Figure 3.1) with a fluorescent protein mCherry (see Figure 3.4b for an example of *C. fragrantissima* transfected with an equivalent plasmid specifically expressing mVenus in the nuclei of a coenocyte). These positively transfected cells allowed researchers to determine through time-lapse experiments the synchronicity of the nuclear divisions in the *C. fragrantissima* coenocytes.

These first transformation experiments in *C. fragrantissima* also opened the door to the possible direct manipulation of the organism by performing gene silencing. In the scenario where no transgenic organisms can be engineered, the alternative to transient gene silencing by either interfering with transcription or translation with antisense RNA matching the right targets can be an alternative functional approach. The fact that the cell wall of *C. fragrantissima* seems to be the thinnest and least complex of the known ichthyosporeans might have facilitated the success of this approach (Marshall et al. 2008). The authors chose morpholinos (i.e. synthetic small interfering RNAs, or siRNAs) to proceed with gene silencing of the transformed recombinant proteins. Because the effect of silencing was directly related to the efficiency of the transfection, an internal control needed to be established. For this reason, the authors first obtained the correlation between the intensities of the different fluorescent markers mCherry and mVenus. The transfections always proceeded with the corresponding antisense RNA targeting the gene of interest fused to mCherry together with a plasmid that expressed the cytoplasm fluorescent marker (mVenus). The decrease in mCherry fluorescence compared with the main intensity of the mVenus would give the percentage of achieved silencing. By repeating the experiments with siRNAs containing mismatches as a control, the authors were able to demonstrate that their functional RNAi approach was specific (three mismatches were enough to abolish the silencing effect on the mCherry expression). Interestingly, the authors also demonstrated that the silencing effect could be achieved by using this transfection method to block translation. In this case, the antisense RNA was directed to the 5'UTR region of one of the constructs. The results were similar, but in this case, five mismatches were necessary to lose sequence specificity (Suga and Ruiz-Trillo 2013a).

Further steps on the development of genetic tools in *C. fragrantissima* have been hampered by the lack of a suitable selective agent with a known resistance gene to achieve stable transfection. We and other researchers are working on this matter in order to be able to genetically modify *C. fragrantissima*. Previous research on this organism has unveiled a number of undoubtedly interesting avenues that will be possible to investigate after the development of more advanced genetic tools.

3.4.2.2 Plasmids

The expression cassettes reporting transfection were constructed using the endogenous *ß-tubulin* promoter of *C. fragrantissima* to drive expression of a fluorescent protein, either mCherry or mVenus. For nuclei labeling, the cassette fused the mCherry fluorescent protein to the endogenous *histone 2B* (H2B) gene of *C. fragrantissima*. Interestingly, a fusion to the *S. arctica h2B* gene was also functional in *C. fragrantissima*. For cytoplasm labeling, the authors co-transfected the H2B-mCherry construct with a vector expressing the mVenus fluorescent protein driven by the same *ß-tubulin* promoter from *C. fragrantissima* (Suga and Ruiz-Trillo 2013).

3.4.3 Prospects

Interestingly, for both *C. fragrantissima* and also for *S. arctica* (see the following), a subset of long non-coding RNAs are specifically regulated for some life stages (de Mendoza et al. 2015; Dudin et al. 2019). Being able to study this mechanism of specific gene regulation in more depth could be of relevance to elucidate the initial steps of cell specialization.

On the other hand, investigating the dynamics of cell division during the coenocytic stage of *C. fragrantissima* in depth will help us to understand the similarities and differences with the coenocytes of some animal species' embryos (Figure 3.2d) (de Mendoza et al. 2015; Ondracka et al. 2018).

As a conclusion, *C. fragrantissima* is one of the known ichthyosporeans that could be a more fruitful model organism in the near future for many reasons. First, it is easily cultivated and manipulated in laboratory conditions; second, it presents an apparently linear life cycle and a fairly good description of its different life stages, and third, it has a relatively compact and well-annotated genome, and lastly there is a reasonable availability of genetic tools. All together, this makes *C. fragrantissima* a very good candidate for the study of the evolution of the holozoa clade but also for addressing several open questions concerning the evolution toward multicellularity in animals.

3.4.4 *Sphaeroforma arctica*

Although genetic tools are yet to be developed for *Sphaeroforma arctica*, we thought it important to briefly introduce this organism in this chapter. Recently, two reports have unveiled insightful information on the cellularization and the nuclear division during the coenocytic stage of *S. arctica* (Ondracka et al. 2018; Dudin et al. 2019). These new findings will undoubtedly open new research avenues for all ichthyosporeans, and *S. arctica* will be considered a good candidate for future studies, especially those addressing questions of general interest for eukaryote biology and evolution.

S. arctica was first isolated from an artic marine amphipod, cultivated in the laboratory and described by Jøstensen and collaborators (2002). The authors also analyzed the chemical composition of its cell wall in order to find specific adaptations to cold water. Its cell wall presents a high content of polyunsaturated fatty acids (more than 70%), suggesting that they contribute to survival in cold waters (Jøstensen et al. 2002). *S. arctica* grows in laboratory conditions at 12°C in MB through a linear vegetative life cycle that is completed in approximately 48 hours. Briefly, small round newborn cells proliferate in a multinucleated coenocyte through several rounds of synchronous nuclear divisions, which cellularize at the moment of newborn cell release by bursting from the parental coenocyte (Jøstensen et al. 2002; Ondracka et al. 2018). The absence of alternative stages such as flagellated motile amoebas, budding or hyphal forms makes the *S. arctica* life cycle ideally simple for some studies. In addition, its genome and transcriptome as well as an accurate phylogenetic placement have been obtained for this species (de Mendoza et al. 2015; Torruella et al. 2015).

These features make *S. arctica* an ideal species for further investigations. Indeed, recent studies have unveiled the patterns of cellularization and control of cell division that were previously unknown outside animal lineages. The *S. arctica* cellularization process shares some mechanisms and regulatory pathways with the one present in animals, and it also presents some specific players likely shared with the rest of ichthyosporeans (Figure 3.1) (Dudin et al. 2019). Similarly, detailed studies of nuclear division in *S. arctica* cultures demonstrated that the timing of nuclear division is not affected by cell size or growth rate and is highly synchronous (Ondracka et al. 2018). This feature distinguishes *S. arctica* from filamentous fungi and more resembles the early divisions of animal embryos.

The main drawback of turning *S. arctica* into a model organism is mainly the difficulty of finding a feasible transfection method. So far, a variety of methods based both on chemical and physical approaches, such as electroporation, lipid-based methods and calcium precipitate protocols, have been tried without success (dx.doi.org/10.17504/protocols.io.z6ef9be). A hard cell wall being already present when the new generation of cells is expelled from the coenocyte is likely the main obstacle to efficiently introducing foreign DNA into the organism. Nevertheless, the fact that new model organisms are now being successfully developed using different strategies is promising for *S. arctica* to be an experimentally tractable organism in the near future.

3.5 CORALLOCHYTREA/PLURIFORMEA

The Corallochytrea clade is also known as Pluriformea because of the great variety of forms exhibited during the life cycles of the organisms composing this lineage (Hehenberger et al. 2017). Corallochytrea is the fourth clade of unicellular Holozoa, a sister-group to Ichthyosporea and in a key phylogenetic position for researchers to study the evolution from unicellular to multicellular organisms (Figure 3.1). To date, this lineage is composed of only two described species: *Corallochytrium limacisporum* and *Syssomonas multiformis* (Raghu-kumar 1987; Hehenberger et al. 2017; Tikhonenkov et al. 2020a). Intriguingly, *C. limacisporum*

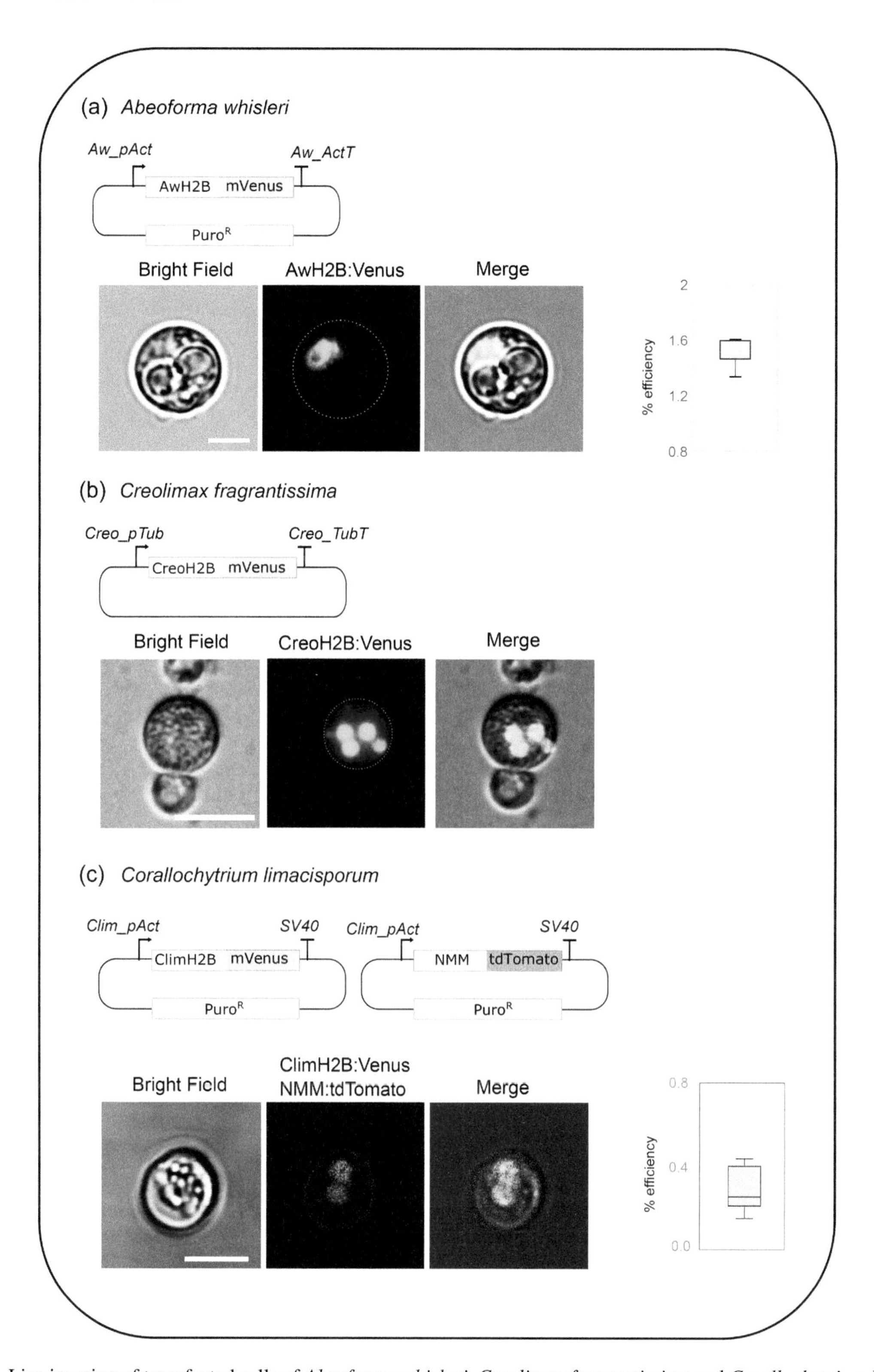

FIGURE 3.4 Live imaging of transfected cells of *Abeofroma whisleri*, *Creolimax fragrantissima* and *Corallochytrium limacisporum*. Images are complemented with diagrams of transfection cassettes. *Abeoforma whisleri*, nuclei labeling: mVenus fluorescent protein fused to endogenous Histone 2B under the actin promoter and terminator. *Creolimax fragrantissima*, nuclei labeling: mVenus fluorescent protein fused to endogenous Histone 2B under the tubulin promoter and terminator. *Corallochytrium limacisporum*, nuclei labeling: mVenus fluorescent protein fused to endogenous Histone 2B under the actin promoter and SV40 terminator. Plasma membrane labeling: tdTomato fluorescent protein fused to the endogenous N-myristoylation motif of the *src* gene (see main text) under the actin promoter and SV40 terminator. Reported transfection efficiency only for *Abeoforma whisleri* and *Coralochytrium limacisporum* from our own experiments. Scale bars (a) and (c) 5 μm, (b) 50 μm.

contains a complete flagellar toolkit (Torruella et al. 2015), but its flagellated forms occur sporadically in our culture conditions, whereas in contrast, the most commonly occurring stage of *S. multiformis* consists of flagellated forms (Tikhonenkov et al. 2020a). Both representatives of this clade show some morphological resemblance in their life cycle, *S. multiformis* being the one with a greater variety of forms. As an example, both organisms have active amoeboid forms and also present complex multicellular stages (Figure 3.2d) (Tikhonenkov et al. 2020a; Kożyczkowska et al. 2021).

In addition to its key phylogenetic position, *C. limacisporum* has many of the desirable features for an organism to be developed as genetically tractable (see next section). On the other hand, unfortunately, cultures of *S. multiformis* are no longer available, and therefore it is difficult to speculate on the possibility of this organism becoming an experimentally treatable organism.

3.5.1 *Corallochytrium limacisporum*

C. limacisporum is a small, marine, free-living corallochytrean isolated from coral reefs of India and Hawaii (Raghukumar 1987). This taxa possesses numerous features that make it an attractive candidate for further functional analysis. It grows very fast and under axenic conditions, and most importantly, it is able to grow in both liquid and agar media, allowing for easy screenings and selection of individual transformed clones. Moreover, it is the only corallochytrean with a completely sequenced and well-annotated genome (Grau-Bové et al. 2017). Finally, besides these technical advantages, *C. limacisporum* has a peculiar and understudied biology, with a complex life cycle and, as we mentioned before, some fungal-like features. For all these reasons, developing genetic tools in this fascinating unicellular organism will for sure be useful for several scientific questions/fields.

3.5.1.1 Transfection and Selection

Different antibiotics, antifungals and herbicides had been tested in *C. limacisporum*, and the antibiotic puromycin was selected as the most adequate for its efficiency and apparent low toxicity (Kożyczkowska et al. 2021). In addition to selection by antibiotics, it would be ideal to have a double selection system that would also allow us to screen transfected cells by fluorescence microscopy. Therefore, a dual selection system based on resistance to puromycin and mCherry expression was set up. Two recombinant plasmids, CAMP (***C****orallochytrium* **A**ctin **M**cherry **P**ac) and CTMP (***C****orallochytrium* **T**ubulin **M**cherry **P**ac), were used for optimizing the transfection parameters (see also "Plasmids" section).

Different methods of transfection that had worked for other protists, yeast or eukaryote cells in general based on chemical or physical methods were tested, but only electroporation was successful. Initially, positive results using an *in* electrode apparatus from Invitrogen, the Neon system, which allows modifying the electric pulse and the duration of the pulse (dx.doi.org/10.17504/protocols.io.hmwb47e), were obtained. Nevertheless, this protocol did not have enough reproducibility to carry out downstream applications, and we selected the electroporator 4D-Nucleofector from Lonza, which was being used with greater efficiency in other protists (Figure 3.3e) (Kożyczkowska et al. 2021 and Protcols.io: dx.doi.org/10.17504/protocols.io.r5ud86w; see sections for *S. rosetta, A. whisleri* and *C. fragrantissima*).

One of the important factors was the cell density and age of the starting culture to maximize efficiency. Similarly to *A. whisleri*, the cells should be washed with 1X PBS to remove the culture media. Co-transfection of highly pure and highly concentrated carrier plasmid DNA (empty *pUC19*) was another key factor that significantly increased efficiency. In general, some fluorescent cells could be observed after 24 hours post-transfection, although there was always a significant increase in positive cells after 48 hours, after which puromycin was added. In the case of *C. limacisporum*, the combination of buffer P3 and code EN-138 from the 4D-Nucleofector (Lonza) proved the most optimal for successful transfection (Figure 3.3e) (Kożyczkowska et al. 2021). Clonal lines can be obtained by plating a dilution of the cells in MB agar plates containing puromycin (Kożyczkowska et al. 2021).

As an immediate contribution from these developed genetic tools, the description of the life cycle of *C. limacisporum* and the unraveling of some unexpected traits, was possible. It has been discovered that *C. limacisporum* has two different paths for cell division, binary fission and coenocytic growth (Figure 3.2d), demonstrating that the *C. limacisporum* life cycle is non-linear and more complex than previously thought (Raghu-kumar 1987). Additionally, some particular features of *C. limacisporum* not commonly found in eukaryotes were described: first the decoupling of cytokinesis and karyokinesis in binary fission and second the observation of some examples of asynchronous nuclei divisions during coenocytic growth. The possibility to expand functional studies of these features in *C. limacisporum* will undoubtedly contribute to a better characterization of this unicellular holozoan.

3.5.1.2 Plasmids

As mentioned, a double selection system was engineered. The CAMP plasmid contained the *pac* gene to provide drug resistance (Luna et al. 1988) and the *mCherry* gene to produce fluorescence in the positively transfected cells. In order to drive transcription with endogenous promoters, the upstream non-coding sequence of the *actin* and *tubulin* genes from *C. limacisporum* and the 3'UTR terminator of the *actin* gene from the ichthyosporean *A. whisleri* were cloned in order to avoid homologous recombination at the *actin* locus. The CAMP and CTMP plasmids were indistinguishable in their phenotype, fluorescent labeling of the cytoplasm in *C. limacisporum* revealing a "crescent moon-like" shape produced by the presence of a large vacuole that occupies the 65% of the cell's volume (Kożyczkowska et al. 2021).

Progress into understanding the cell biology of *C. limacisporum* (see transfection section) was possible through the generation of constructs tagging sub-cellular components, such as the plasma membrane, cytoskeleton, cytoplasm and nucleus (Kożyczkowska et al. 2021). To construct the *pact*-NMN-tdTomato plasmid, the predicted N-myristoylation motif (NMM) from the *Src* tyrosine kinase orthologue (Gene ID Clim_evm93s153) was used. This motif has been successfully used in *C. owczarzaki* to direct the fusion protein to the plasma membrane (Parra-Acero et al. 2018). Our results show that this motif was also plasma membrane specific in *C. limacisporum* and therefore might also be useful in other organisms (Figure 3.4c) (Kożyczkowska et al. 2021). To visualize the cytoskeleton, the 17-amino acid peptide LifeAct that binds specifically to filamentous actin (ibidi) was fused to the mCherry protein *pact*-LifeAct. Finally, the construct *pact*-H2B-mVenus contains the endogenous gene of *C. limacisporum* (Gene ID Clim_evm20s1) fused to the mVenus fluorescent protein. In addition, the construct contains the actin promoter, with the dual system of puromycin resistance as well as fluorescence (Kożyczkowska et al. 2021).

3.5.2 Prospects

The development of specific recombinant plasmids together with stable transfection in *C. limacisporum* has provided insightful information about the biology of this organism while also providing the initial tools to set up functional experiments. Importantly, now *C. limacisporum* provides the opportunity to further investigate which are the factors behind different developmental routes (binary fission or coenocytic growth), as well as a promising model to study the mechanisms behind the decoupling of karyokinesis from cytokinesis and the basis of asynchronous nuclear division.

Besides the previously mentioned advances, developing CRISPR/Cas9 genome editing in *C. limacisporum* is currently ongoing. The establishment of genome editing in the future will allow us to understand, among others, the possible ancestral role of some genes related to multicellular functions in Metazoa.

3.6 CONCLUDING REMARKS

We have here described the most recent advances in the handful of model organisms available among unicellular holozoans (Figure 3.1). These model organisms belong to all four clades of unicellular relatives of animals, constituting a functional platform to experimentally address many of the hypotheses regarding the evolution of genes and cellular features along the Holozoa tree. We are eager to see how evolutionary cell biology will take advantage of all those new emerging model systems to address the function of ancestral genes and protein domains, as well as for the conservation or innovation of cell biological processes.

ACKNOWLEDGMENTS

We would like to thank Núria Ros-Rocher for critical reading of the manuscript and the Multicellgenome laboratory for ideas and discussions. We would also like to thank the European Research Council (ERC) and the Betty and Gordon Moore foundation for their strategic vision in funding the development of these organisms into experimentally tractable organisms. This work was supported by a European Research Council Consolidator Grant (ERC-2012-Co-616960) grant to I.R-T. and a Betty and Gordon Moore "New Genetic tools for Marine Protists", Grant number 4973.01 to E.C and I.R-T.

BIBLIOGRAPHY

Abedin, M., and N. King. 2010. Diverse Evolutionary Paths to Cell Adhesion. *Trends in Cell Biology* 20: 734–742.

Ai, H., Henderson, J. N., Remington, S. J., and R. E. Campbell. 2006. Directed Evolution of a Monomeric, Bright and Photostable Version of Clavularia Cyan Fluorescent Protein: Structural Characterization and Applications in Fluorescence Imaging. *The Biochemical Journal* 400: 531–540.

Alegado, R. A., Brown L. W., Cao S., Dermenjian, R. K., Zuzow, R., Fairclough, S. R., Clardy J., and Nicole King. 2012. A Bacterial Sulfonolipid Triggers Multicellular Development in the Closest Living Relatives of Animals. *ELife* 1: e00013.

Booth, D. S., and N. King. 2020. Genome Editing Enables Reverse Genetics of Multicellular Development in the Choanoflagellate Salpingoeca Rosetta. Edited by Alejandro Sánchez Alvarado, Patricia J Wittkopp, Margaret A Titus, Iñaki Ruiz-Trillo, and Matthew C Gibson. *ELife* 9: e56193.

Booth, D. S., Szmidt-Middleton, H., and N. King. 2018. Transfection of Choanoflagellates Illuminates Their Cell Biology and the Ancestry of Animal Septins. *Molecular Biology of the Cell* 29: 3026–3038.

Burkhardt, P., Grønborg, M., McDonald, K., Sulur, T., Wang, Q., and N. King. 2014. Evolutionary Insights into Premetazoan Functions of the Neuronal Protein Homer. *Molecular Biology and Evolution* 31: 2342–2355.

Caro, F., Miller, M. G., and J. L DeRisi. 2012. Plate-Based Transfection and Culturing Technique for Genetic Manipulation of *Plasmodium falciparum*. *Malaria Journal* 11: 22.

Carr, M., Leadbeater, B. S. C., Hassan, R., Nelson, M., and S. L. Baldauf. 2008. Molecular Phylogeny of Choanoflagellates, the Sister Group to Metazoa. *Proceedings of the National Academy of Sciences of the United States of America* 105: 16641–16646.

Cavalier-Smith, T. 1998. A Revised Six-Kingdom System of Life. *Biological Reviews of the Cambridge Philosophical Society* 73: 203–266.

Cavalier-Smith, T., and E. E.-Y. Chao. 2003. Phylogeny of Choanozoa, Apusozoa, and Other Protozoa and Early Eukaryote Megaevolution. *Journal of Molecular Evolution* 56: 540–563.

Clark, H. J. 1866. Conclusive Proofs of the Animality of the Ciliate Sponges, and of Their Affinities with the Infusoria Flagellata. *American Journal of Science* 42 (2): 320–324.

Dayel, M. J., Alegado, R. A., Fairclough, S. R., Levin, T. C., Nichols, S. A., McDonald, K., and N. King. 2011. Cell Differentiation and Morphogenesis in the Colony-Forming Choanoflagellate *Salpingoeca rosetta*. *Developmental Biology* 357: 73–82.

Dolan, J., and B. S. Leadbeater. 2015. *The Choanoflagellates: Evolution, Biology, and Ecology*. Cambridge University Press, Cambridge, UK. 315 pp. Hardcover, ISBN978-0-521-88444-0, $125. *Journal of Eukaryotic Microbiology*, June.

Dudin, O., Ondracka, A., Grau-Bové, X., Haraldsen, A. A. B., Toyoda, A., Suga, H., Bråte, J., and I. Ruiz-Trillo. 2019. A Unicellular Relative of Animals Generates a Layer of Polarized Cells by Actomyosin-Dependent Cellularization. *ELife* 8: e49801. https://doi.org/10.7554/eLife.49801.

Ensenauer, R., Hartl, D., Vockley, J., Roscher, A. A., and U. Fuchs. 2011. Efficient and Gentle SiRNA Delivery by Magnetofection. *Biotechnic & Histochemistry: Official Publication of the Biological Stain Commission* 86: 226–231.

Fairclough, S. R., Chen, Z., Kramer, E., Zeng, Q., Young, S., Robertson, H. M., E. Begovic, et al. 2013. Premetazoan Genome Evolution and the Regulation of Cell Differentiation in the Choanoflagellate *Salpingoeca rosetta*. *Genome Biology* 14: R15.

Fairclough, S. R., Dayel, M. J., and N. King. 2010. Multicellular Development in a Choanoflagellate. *Current Biology: CB* 20: R875–R876.

Faktorová, D., Nisbet, E. R., Fernández Robledo, J. A., Casacuberta, E., Sudek L., Allen A. E., M. Ares, et al. 2020. Genetic Tool Development in Marine Protists: Emerging Model Organisms for Experimental Cell Biology. *Nature Methods* 17: 481–494.

Foster, A. J., Martin-Urdiroz, M., Yan, X., Wright, H. S., Soanes, D. M., and N. J. Talbot. 2018. CRISPR-Cas9 Ribonucleoprotein-Mediated Co-Editing and Counterselection in the Rice Blast Fungus. *Scientific Reports* 8: 14355.

Gaudet, P., Pilcher, K. E., Fey, P., and R. L. Chisholm. 2007. Transformation of Dictyostelium Discoideum with Plasmid DNA. *Nature Protocols* 2: 1317–1324.

Graham, F. L., and A. J. van der Eb. 1973. A New Technique for the Assay of Infectivity of Human Adenovirus 5 DNA. *Virology* 52: 456–467.

Grau-Bové, X., Torruella, G., Donachie, S., Suga, H., Leonard, G., Richards, T. A., and I. Ruiz-Trillo. 2017. Dynamics of Genomic Innovation in the Unicellular Ancestry of Animals. Edited by Diethard Tautz. *ELife* 6 (July): e26036.

Grosjean, F., Bertschinger, M., Hacker, D. L., and F. M. Wurm. 2006. Multiple Glycerol Shocks Increase the Calcium Phosphate Transfection of Non-Synchronized CHO Cells. *Biotechnology Letters* 28: 1827–1833.

Guo, L., Wang, L., Yang, R., Feng, R., Li, Z., Zhou, X., Z. Dong, et al. 2017. Optimizing Conditions for Calcium Phosphate Mediated Transient Transfection. *Saudi Journal of Biological Sciences*, Computational Intelligence Research & Approaches in Bioinformatics and Biocomputing 24: 622–629.

Han, H. A., Sheng Pang, J. K., and B.-S. Soh. 2020. Mitigating Off-Target Effects in CRISPR/Cas9-Mediated In Vivo Gene Editing. *Journal of Molecular Medicine* 98: 615–632.

Hehenberger, E., Tikhonenkov, D. V., Kolisko, M., del Campo, J., Esaulov, A. S., Mylnikov, A. P., and P. J. Keeling. 2017. Novel Predators Reshape Holozoan Phylogeny and Reveal the Presence of a Two-Component Signaling System in the Ancestor of Animals. *Current Biology* 27: 2043–2050.

Hoffmeyer, T. T., and P. Burkhardt. 2016. Choanoflagellate Models: *Monosiga brevicollis* and *Salpingoeca rosetta*. *Current Opinion in Genetics & Development*, Developmental Mechanisms, Patterning and Evolution 39: 42–47.

Jacobs, J. Z., Ciccaglione, K. M., Tournier, V., and M. Zaratiegui. 2014. Implementation of the CRISPR-Cas9 System in Fission Yeast. *Nature Communications* 5: 5344.

Janse, C. J., Franke-Fayard, B., Mair, G. R., Ramesar, J., Thiel, C., Engelmann, S., Matuschewski, K., van Gemert, G. J., Sauerwein, R. W., and A. P. Waters. 2006. High Efficiency Transfection of Plasmodium Berghei Facilitates Novel Selection Procedures. *Molecular and Biochemical Parasitology* 145: 60–70.

Jiang, W., Brueggeman, A., Horken, K., Plucinak, T., and D. Weeks. 2014. Successful Transient Expression of Cas9 and Single Guide RNA Genes in *Chlamydomonas reinhardtii*. *Eukaryotic Cell* 13.

Jordan, M., and F. Wurm. 2004. Transfection of Adherent and Suspended Cells by Calcium Phosphate. *Methods (San Diego, Calif.)* 33: 136–143.

Jøstensen, J.-P., Sperstad, S., Johansen, S., and B. Landfald. 2002. Molecular-Phylogenetic, Structural and Biochemical Features of a Cold-Adapted, Marine Ichthyosporean Near the Animal-Fungal Divergence, Described from In Vitro Cultures. *European Journal of Protistology* 38: 93–104.

Kim, J. A., Cho, K., Shin, M. S., Lee, W. G., Jung, N., Chung, C., and J. Keun Chang. 2008. A Novel Electroporation Method Using a Capillary and Wire-Type Electrode. *Biosensors & Bioelectronics* 23: 1353–1360.

Kim, S. K., Kim, D., Woo Cho, S., Kim, J., and J.-S. Kim. 2014. Highly Efficient RNA-Guided Genome Editing in Human Cells via Delivery of Purified Cas9 Ribonucleoproteins. *Genome Research* 24.

Kim, S. K., Shindo, A., Park, T. J., Oh, E. C., Ghosh, S., Gray, R. S., R. A. Lewis, et al. 2010. Planar Cell Polarity Acts through Septins to Control Collective Cell Movement and Ciliogenesis. *Science (New York, N.Y.)* 329: 1337–1340.

King, N. 2005. Choanoflagellates. *Current Biology: CB* 15: R113–R114.

King, N., Hittinger, C. T., and S. B. Carroll. 2003. Evolution of Key Cell Signaling and Adhesion Protein Families Predates Animal Origins. *Science (New York, N.Y.)* 301: 361–363.

King, N., Westbrook, M. J., Young, S. L., Kuo, A., Abedin, M., Chapman, J., S. Fairclough, et al. 2008. The Genome of the Choanoflagellate *Monosiga brevicollis* and the Origin of Metazoans. *Nature* 451: 783–788.

Kożyczkowska, A., Najle, S. R., Ocaña-Pallarès, E., Aresté, C., Shabardina, V., Ara, P. S., Ruiz-Trillo, I., and E. Casacuberta. 2021. Stable Transfection in the Protist *Corallochytrium limacisporum* Allows Identification of Novel Cellular Features Among Unicellular Relatives of Animals. *Current Biology* 31: 1–7. https://doi.org/10.1016/j.cub.2021.06.061

Laundon, D., Larson, B. T., McDonald, K., King, N., and P. Burkhardt. 2019. The Architecture of Cell Differentiation in Choanoflagellates and Sponge Choanocytes. *PLoS Biology* 17 (4): e3000226.

Leadbeater, B. S. C., Yu, Q., Kent, J., and D. J. Stekel. 2009. Three-Dimensional Images of Choanoflagellate Loricae. *Proceedings of the Royal Society B: Biological Sciences* 276: 3–11.

Levin, T. C., Greaney A. J., Wetzel, L., and N. King. 2014. The Rosetteless Gene Controls Development in the Choanoflagellate *S. rosetta*. *ELife* 3 (October).

Levin, T. C., and N. King. 2013. Evidence for Sex and Recombination in the Choanoflagellate *Salpingoeca rosetta*. *Current Biology* 23: 2176–2180.

Liang, X., Potter, J., Kumar, S., Zou, Y., Quintanilla, R., Sridharan, M., J. Carte, et al. 2015. Rapid and Highly Efficient Mammalian Cell Engineering via Cas9 Protein Transfection. *Journal of Biotechnology* 208 (August): 44–53.

López-Escardó, D., Grau-Bové, X., Guillaumet-Adkins, A., Gut, M., Sieracki, M. E., and I. Ruiz-Trillo. 2019. Reconstruction of Protein Domain Evolution Using Single-Cell Amplified Genomes of Uncultured Choanoflagellates Sheds Light on the Origin of Animals. *Philosophical Transactions of the Royal Society B: Biological Sciences* 374 (1786): 20190088.

Luna, S. de la, Soria, I., Pulido, D., Ortín, J., and A. Jiménez. 1988. Efficient Transformation of Mammalian Cells with Constructs Containing a Puromycin-Resistance Marker. *Gene* 62: 121–126.

Mah, J. L., Christensen-Dalsgaard, K. K., and S. P. Leys. 2014. Choanoflagellate and Choanocyte Collar-Flagellar Systems and the Assumption of Homology. *Evolution & Development* 16: 25–37.

Maldonado, M. 2004. Choanoflagellates, Choanocytes, and Animal Multicellularity. *Invertebrate Biology* 123: 1–22.

Marshall, W. L., and M. L. Berbee. 2011. Facing Unknowns: Living Cultures (*Pirum gemmata* Gen. Nov., Sp. Nov., and *Abeoforma whisleri*, Gen. Nov., Sp. Nov.) from Invertebrate Digestive Tracts Represent an Undescribed Clade within the Unicellular Opisthokont Lineage Ichthyosporea (Mesomycetozoea). *Protist* 162: 33–57.

Marshall, W. L., Celio, G., McLaughlin, D. J., and M. L. Berbee. 2008. Multiple Isolations of a Culturable, Motile Ichthyosporean (Mesomycetozoa, Opisthokonta), *Creolimax fragrantissima* n. Gen., n. Sp., from Marine Invertebrate Digestive Tracts. *Protist* 159: 415–433.

Mendoza, A. de, and A. Sebé-Pedrós. 2019. Origin and Evolution of Eukaryotic Transcription Factors. *Current Opinion in Genetics & Development* 58–59: 25–32.

Mendoza, A. de, Suga, H., Permanyer, J., Irimia, M., and I. Ruiz-Trillo. 2015. Complex Transcriptional Regulation and Independent Evolution of Fungal-Like Traits in a Relative of Animals. Edited by Alejandro Sánchez Alvarado. *ELife* 4: e08904.

Mendoza, L., Taylor, J. W., and L. Ajello. 2002. The Class Mesomycetozoea: A Heterogeneous Group of Microorganisms at the Animal-Fungal Boundary. *Annual Review of Microbiology* 56: 315–344.

Miller, D. J., and R. L. Ax. 1990. Carbohydrates and Fertilization in Animals. *Molecular Reproduction and Development* 26: 184–198.

Mylnikov, A. P., Tikhonenkov, D. V., Karpov, S. A., and C. Wylezich. 2019. Microscopical Studies on *Ministeria vibrans* Tong, 1997 (Filasterea) Highlight the Cytoskeletal Structure of the Common Ancestor of Filasterea, Metazoa and Choanoflagellata. *Protist* 170: 385–396.

Nagai, T., Ibata, K., Sun Park, E., Kubota, M., Mikoshiba, K., and A. Miyawaki. 2002. A Variant of Yellow Fluorescent Protein with Fast and Efficient Maturation for Cell-Biological Applications. *Nature Biotechnology* 20: 87–90.

Naumann, B., and P. Burkhardt. 2019. Spatial Cell Disparity in the Colonial Choanoflagellate *Salpingoeca rosetta*. *Frontiers in Cell and Developmental Biology* 7.

Neufeld, T. P., and G. M. Rubin. 1994. The Drosophila Peanut Gene Is Required for Cytokinesis and Encodes a Protein Similar to Yeast Putative Bud Neck Filament Proteins. *Cell* 77: 371–379.

Nielsen, C. 2008. Six Major Steps in Animal Evolution: Are We Derived Sponge Larvae? *Evolution & Development* 10: 241–257.

Ondracka, A., Dudin, O., and I. Ruiz-Trillo. 2018. Decoupling of Nuclear Division Cycles and Cell Size during the Coenocytic Growth of the Ichthyosporean *Sphaeroforma arctica*. *Current Biology* 28: 1964–1969.

Owczarzak, A., Stibbs, H. H., and C. J. Bayne. 1980a. The Destruction of *Schistosoma mansoni* Mother Sporocysts In Vitro by Amoebae Isolated from *Biomphalaria glabrata*: An Ultrastructural Study. *Journal of Invertebrate Pathology* 35: 26–33.

Owczarzak, A., Stibbs, H. H., and C. J. Bayne. 1980b. The Destruction of *Schistosoma mansoni* Mother Sporocysts In Vitro by Amoebae Isolated from *Biomphalaria glabrata*: An Ultrastructural Study. *Journal of Invertebrate Pathology* 35: 26–33.

Paps, J., Medina-Chacón, L. A., Marshall, W., Suga, H., and I. Ruiz-Trillo. 2013. Molecular Phylogeny of Unikonts: New Insights into the Position of Apusomonads and Ancyromonads and the Internal Relationships of Opisthokonts. *Protist* 164: 2–12.

Parra-Acero, H., Harcet, M., Sánchez-Pons, N., Casacuberta, E., Brown, N. H., Dudin, O., and I. Ruiz-Trillo. 2020. Integrin-Mediated Adhesion in the Unicellular Holozoan *Capsaspora owczarzaki*. *Current Biology* 30: 4270–4275.

Parra-Acero, H., Ros-Rocher, N., Perez-Posada, A., Kożyczkowska, A., Sánchez-Pons, N., Nakata, A., Suga, H., Najle, S. R., and I. Ruiz-Trillo. 2018. Transfection of *Capsaspora owczarzaki*, a Close Unicellular Relative of Animals. *Development (Cambridge, England)* 145 (10).

Pekkarinen, M. 2003. Phylogenetic Position and Ultrastructure of Two Dermocystidium Species (Ichthyosporea) from the Common Perch (Perca fluviatilis). *Acta Protozoologica* 42 (4): 287–307.

Pettitt, M. E., Orme, B. A. A., Blake, J. R., and B. S. C. Leadbeater. 2002. The Hydrodynamics of Filter Feeding in Choanoflagellates. *European Journal of Protistology* 38: 313–332.

Raghu-kumar, S. 1987. Occurrence of the Thraustochytrid, Corallochytrium Limacisporum Gen. et Sp. Nov. in the Coral Reef Lagoons of the Lakshadweep Islands in the Arabian Sea. *Botanica Marina* 30 (1).

Richter, D. J., Fozouni, P., Eisen, M. B., and N. King. 2018. Gene Family Innovation, Conservation and Loss on the Animal Stem Lineage. Edited by Maximilian J Telford. *ELife* 7: e34226.

Riedl, J., Crevenna, A. H., Kessenbrock, K., Haochen Yu, J., Neukirchen, D., Bista, M., F. Bradke, et al. 2008. Lifeact: A Versatile Marker to Visualize F-Actin. *Nature Methods* 5: 605.

Ryan, T. J., and S. G. N. Grant. 2009. The Origin and Evolution of Synapses. *Nature Reviews Neuroscience* 10: 701–712.

Sebé-Pedrós, A., Ballaré, C., Parra-Acero, H., Chiva, C., Tena, J. J., Sabidó, E., Gómez-Skarmeta, J. L., Di Croce, L., and I. Ruiz-Trillo. 2016. The Dynamic Regulatory Genome of Capsaspora and the Origin of Animal Multicellularity. *Cell* 165: 1224–1237.

Sebé-Pedrós, A., Degnan, B. M., and I. Ruiz-Trillo. 2017. The Origin of Metazoa: A Unicellular Perspective. *Nature Reviews Genetics* 18: 498–512.

Sebé-Pedrós, A., Irimia, M., Del Campo, J., Parra-Acero, H., Russ, C., Nusbaum, C., Blencowe, B. J., and I. Ruiz-Trillo. 2013. Regulated Aggregative Multicellularity in a Close Unicellular Relative of Metazoa. *ELife* 2 (December): e01287.

Shalchian-Tabrizi, K., Minge, M. A., Espelund, M., Orr, R., Ruden, T., Jakobsen, K. S., and T. Cavalier-Smith. 2008. Multigene Phylogeny of Choanozoa and the Origin of Animals. *PLoS One* 3 (5).

Shaner, N. C., Campbell, R. E., Steinbach, P. A., Giepmans, B. N. G., Palmer, A. E., and R. Y. Tsien. 2004. Improved Monomeric Red, Orange and Yellow Fluorescent Proteins

Derived from *Discosoma* Sp. Red Fluorescent Protein. *Nature Biotechnology* 22: 1567–1572.

Shin, S.-E., Lim, J.-M., Koh, H. G., Kim, E. K., Kang, N. K., Jeon, S., S. Kwon, et al. 2016. CRISPR/Cas9-Induced Knockout and Knock-in Mutations in *Chlamydomonas reinhardtii*. *Scientific Reports* 6: 27810.

Sigal, C. T., Zhou, W., Buser, C. A., McLaughlin, S., and M. D. Resh. 1994. Amino-Terminal Basic Residues of Src Mediate Membrane Binding through Electrostatic Interaction with Acidic Phospholipids. *Proceedings of the National Academy of Sciences of the United States of America* 91: 12253–12257.

Singh, R. K., Liang, D., Gajjalaiahvari, U. R., Miquel Kabbaj, M.-H., Paik, J., and A. Gunjan. 2010. Excess Histone Levels Mediate Cytotoxicity via Multiple Mechanisms. *Cell Cycle* 9: 4236–4244.

Sogabe, S., Hatleberg, W. L., Kocot, K. M., Say, T. E., Stoupin, D., Roper, K. E., Fernandez-Valverde, S. L., Degnan, S. M., and B. M. Degnan. 2019. Pluripotency and the Origin of Animal Multicellularity. *Nature* 570: 519–522.

Spiliotis, E. T., Hunt, S. J., Hu, Q., Kinoshita, M., and W. J. Nelson. 2008. Epithelial Polarity Requires Septin Coupling of Vesicle Transport to Polyglutamylated Microtubules. *The Journal of Cell Biology* 180: 295–303.

Stibbs, H. H., Owczarzak, A., Bayne, C. J., and P. DeWan. 1979. Schistosome Sporocyst-Killing Amoebae Isolated from *Biomphalaria glabrata*. *Journal of Invertebrate Pathology* 33: 159–170.

Suga, H., Dacre, M., de Mendoza, A., Shalchian-Tabrizi, K., Manning, G., and I. Ruiz-Trillo. 2012. Genomic Survey of Premetazoans Shows Deep Conservation of Cytoplasmic Tyrosine Kinases and Multiple Radiations of Receptor Tyrosine Kinases. *Science Signaling* 5 (222): ra35.

Suga, H., and I. Ruiz-Trillo. 2013. Development of Ichthyosporeans Sheds Light on the Origin of Metazoan Multicellularity. *Developmental Biology* 377: 284–292.

Suga, H., Chen, Z., de Mendoza, A., Sebé-Pedrós, A., Brown, M. W., Kramer, E., M. Carr, et al. 2013. The Capsaspora Genome Reveals a Complex Unicellular Prehistory of Animals. *Nature Communications* 4 (1): 2325. https://doi.org/10.1038/ncomms3325.

Tikhonenkov, D. V., Hehenberger, E., Esaulov, A. S., Belyakova, O. I., Mazei, Y. A., Mylnikov, A.P., and P. J. Keeling. 2020a. Insights into the Origin of Metazoan Multicellularity from Predatory Unicellular Relatives of Animals. *BMC Biology* 18: 39.

Tikhonenkov, D. V., Mikhailov, K. V., Hehenberger, E., Karpov, S. A., Prokina, K. I., Esaulov, A. S., O. I. Belyakova, et al. 2020b. New Lineage of Microbial Predators Adds Complexity to Reconstructing the Evolutionary Origin of Animals. *Current Biology* 30: 4500–4509.

Tong, S. M. 1997. Heterotrophic Flagellates and Other Protists from Southampton Water, U.K. *Ophelia* 47: 71–131.

Torruella, G., de Mendoza, A., Grau-Bové, X., Antó, M., Chaplin, M. A., del Campo, J., L. Eme, et al. 2015. Phylogenomics Reveals Convergent Evolution of Lifestyles in Close Relatives of Animals and Fungi. *Current Biology* 25: 2404–2410.

Torruella, G., Derelle, R., Paps, J., Lang, B. F., Roger, A. J., Shalchian-Tabrizi, K., and I. Ruiz-Trillo. 2012. Phylogenetic Relationships within the Opisthokonta Based on Phylogenomic Analyses of Conserved Single-Copy Protein Domains. *Molecular Biology and Evolution* 29: 531–544.

Wetzel, L. A., Levin, T. C., Hulett, R. E., Chan, D., King, G. A., Aldayafleh, R., Booth, D. S., Abedin Sigg, M., and N. King. 2018. Predicted Glycosyltransferases Promote Development and Prevent Spurious Cell Clumping in the Choanoflagellate *S. rosetta*. Edited by Alejandro Sánchez Alvarado and Marianne E Bronner. *ELife* 7: e41482.

Woznica, A., Cantley, A. M., Beemelmanns, C., Freinkman, E., Clardy, J., and N. King. 2016. Bacterial Lipids Activate, Synergize, and Inhibit a Developmental Switch in Choanoflagellates. *Proceedings of the National Academy of Sciences of the United States of America* 113: 7894–7899.

Woznica, A., Gerdt, J. P., Hulett, R. E., Clardy, J., and N. King. 2017. Mating in the Closest Living Relatives of Animals Is Induced by a Bacterial Chondroitinase. *Cell* 170: 1175–1183.

4 Porifera

Maja Adamska

CONTENTS

4.1 HISTORY OF THE MODEL

Sponges (Porifera) have fascinated scientists for at least 150 years, with two key subjects of investigation remaining vibrant until today and additional areas of research emerging recently. The first of the original subjects is the relationship between sponges, other animals and protists, both in terms of their relative phylogenetic positions and the homology between body plans and cell types. The second stems from the remarkable ability of sponges to regenerate: not only by restoring lost body parts but also by completely rebuilding bodies from dissociated cells. Why can sponges do that and we cannot? While 19th- and early 20th-century biologists were equipped only with microscopes, current scientists have harnessed the power of modern genomics and gene expression analysis to address these fundamentally interesting questions. This section of the chapter sets the stage for sponges as models for biological research by (briefly) reviewing findings and opinions of 19th-century scientists on the position of sponges in the tree of life and the discoveries of sponge regenerative capacity in the early 20th century. The following sections cover modern approaches to both subjects, concluding with discussion of the most recent advances and forecasting future directions of research utilizing sponges as models.

But what are sponges, actually? Perhaps surprisingly, this simple question continues to generate heated arguments, with various answers offered (but never universally agreed on) throughout the past centuries. Are they animals of cellular grade of organization (Parazoa), with a unique body plan and independently evolved cell types? Or are they true animals, with germ layers homologous to our endoderm and ectoderm? Are they living fossils, retaining features of our distant ancestors?

When Robert Grant gave sponges the name "Porifera" (= pore bearing), he referred to the numerus tiny openings (called pores of ostia) which are present on the surface of adult sponges and which lead to (more or less complex, depending on the body plan; see Section 4.5) system of canals and chambers (Grant 1825, 1836) (Figure 4.1). The innermost surface of sponges, an epithelial layer called choanoderm, is composed of choanocytes (collar cells), which are equipped with flagella propelling water through the body. Choanocyte collars capture food particles—often bacteria—and the filtered water is then expelled through a larger opening (or openings) called osculum (plural oscula). All other surfaces of sponges (the outer, the basal and lining of the canals) are composed of flat cells called pinacocytes. In between those two epithelial layers lies the non-epithelial mesohyl layer, containing motile amoeboid cells, cells producing skeletal elements, gametes and—in viviparous sponges—embryos. With these basic building blocks, sponges form a variety of body plans, which are discussed further in Section 4.5 (Figure 4.1). Although Linnaeus listed sponges as "vegetables", Grant considered them animals.

Few decades later, the striking similarity between choanocytes and choanoflagellates, which are single-cell and colonial protists, noticed by James-Clark in 1868 and Saville-Kent in 1880, was interpreted to indicate strong affinity between sponges and protists, in effect relegating sponges from the animal kingdom. Intriguingly, all modern phylogenies place choanoflagellates as the nearest relatives (the sister group) of animals, and the majority of the genome-based phylogenies place sponges as the earliest branching animal lineage (Figure 4.1), consistent with the position of sponges as the link between protists and "true animals" (Eumetazoans).

Ernst Haeckel, considered by many the father of evolutionary developmental biology, noted similarities between body plans of sponges, in particular calcareous sponges, and cnidarians, especially coral polyps. According to his views, the sponge choanoderm was homologous to the coral

DOI: 10.1201/9781003217503-4

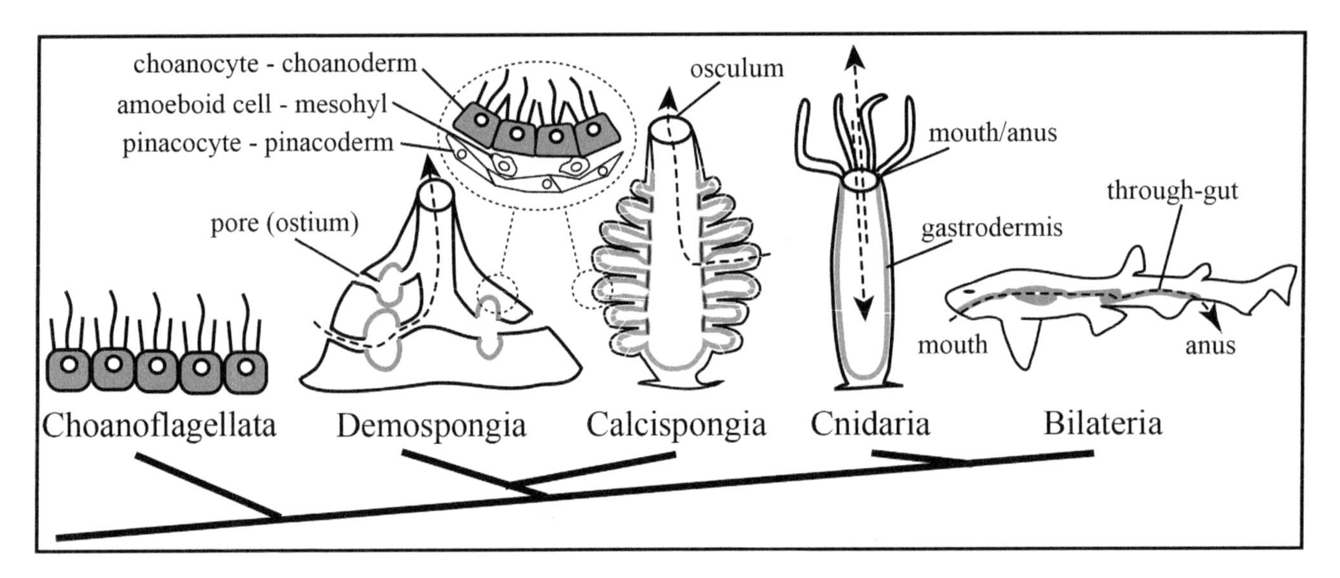

FIGURE 4.1 Phylogenetic position, major cell types and body plans of sponges. Dashed lines with arrowheads indicate direction of movement of food particles and waste products; gray color marks cells and tissues involved in food capture and digestion. (Modified from Adamska 2016.)

gastrodermis, the sponge pinacoderm to the ectoderm, and the osculum to the polyp mouth (Figure 4.1). Haeckel credited the development of the gastrea theory (stating that all animals evolved from a gastrula-like pelagic animal), and more broadly recognition of homology of germ layers, to his observations of calcareous sponges and their development (Haeckel 1870, 1874). Following the reasoning of James-Clark and Haeckel, poriferan-grade body organization appears to represent a clear transition stage between the colonial protists and complex animals. However, phylogenetic position of sponges, as well as the nature of the similarity between sponge choanocytes and choanoflagellates on the one side and the gut enterocytes on the other side of the transition (e.g. Peña et al. 2016), remain far from being settled, as discussed again in Section 4.8.

While phylogenetic position and the relationship between sponge cell types and those of other animals might be disputed (Simion et al. 2017; Whelan et al. 2017), the observations of the regenerative abilities of sponges, originally made in the early 20th century, remain as true and fascinating now as they were then. Wilson (1907), working on a marine demosponge, *Microciona prolifera*, discovered that it was capable of forming new, functional bodies after being dissociated into single cells. His experiments were soon reproduced using other sponge species, including freshwater sponges by Muller (1911a, 1911b) and the calcareous sponge *Sycon raphanus* by Huxley (1911, 1921), demonstrating that this remarkable ability is widespread among sponges. Intriguingly, it appears that the cellular mechanisms of sponge regeneration differ significantly across the phylum, and the molecular mechanisms are only beginning to be discovered. We will return to this topic, covering the intriguing recent discoveries and future research avenues, in Sections 4.7 and 4.8.

4.2 GEOGRAPHICAL LOCATION

Sponges are found in virtually all marine environments, from cold, deep waters surrounding the poles to shallow tropical environments (van Soest et al. 2012). One lineage of sponges evolved the ability to occupy freshwater environments, with species noted in lakes, rivers and creeks across the globe (Manconi and Pronzato 2002).

Sponges are notoriously difficult in lab cultivation—no sponge species can currently be reliably cultivated throughout its entire lifecycle, and the cell culture methods have only started to be established (Schippers et al. 2012; Conkling et al. 2019). This challenge in combination with interest in sponge biology resulted in proliferation of sponge models, representing all four evolutionary lineages of sponges (Figure 4.2).

From over 9,000 species of marine sponges, laboratories in Europe, North America, Asia and Australia have thus been selecting their model systems focusing attention on species which are easily accessible (abundant in shallow waters or appearing in local aquaria) and relatively robust (permitting transport to laboratories and short-term culture), in addition to possessing unique biological features making them particularly interesting or tractable. This chapter focuses on knowledge obtained using representatives of two lineages: calcareous sponges, especially those from the genus Sycon (the same that inspired Haeckel's theories), and demosponges, especially *Amphimedon queenslandica* (the first sponge to have its genome sequenced). Sponges from the relatively small (but fascinating) lineage of Hexactinellida (glass sponges, a sister group to demosponges) are generally restricted to deep waters, making them difficult to access. However, a few species, such as *Oopsacas minuta*, have been found in relatively shallow cave environments,

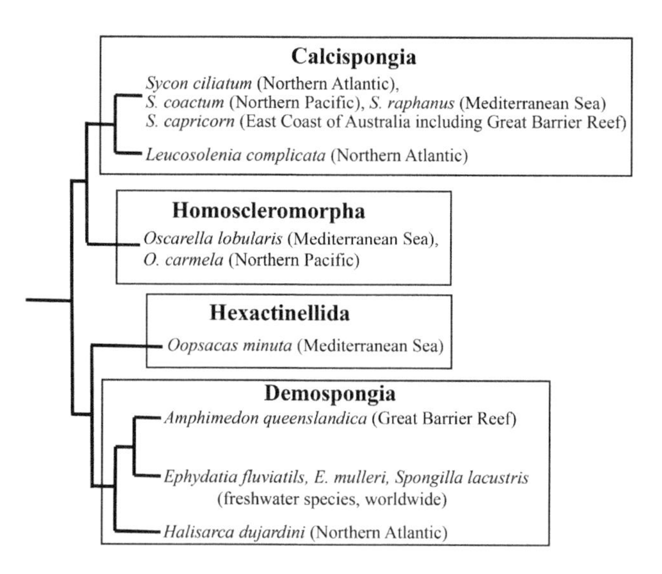

FIGURE 4.2 Phylogenetic position and geographic location of major sponge model systems.

allowing researchers to study their development leading to formation of syncytial adult body (Boury-Esnault et al. 1999; Leys et al. 2016). The highly derived genomes of Hexactinellids will be mentioned in Section 4.6. Chapter 5 focuses on Homosclermorph sponges, which are the sister group to Calcisponges.

4.3 LIFE CYCLE

Like many marine invertebrates, the majority of sponges have a biphasic life cycle, including motile, pelagic larvae and sessile, benthic adults (Figure 4.3). This lifestyle likely reflects the lifestyle of the first animals (Degnan and Degnan 2006) or perhaps even our protistan ancestors (Adamska 2016b). While very few sponge species (such as *Tetilla japonica*) secondarily lost the motile larval stage, becoming direct developers, a spectacular diversity of developmental modes and larval types has been described in sponges (Leys and Ereskovsky 2006; Ereskovsky 2010; Maldonado 2006).

Sponges can be either oviparous (that is, releasing gametes to the surrounding water, with the fertilization and subsequence development occurring in the water column) or viviparous, with embryogenesis occurring within the maternal tissues. The majority of sponge species used as models for developmental biology research are viviparous and hermaphroditic. In particular, all homoscleromorph sponges, including *Oscarella lobularis* (see Chapter 5), and all calcisponge species (including *Sycon* sp.) brood their larvae within maternal tissues (Figure 4.3c, d; see also Section 4.4); in both cases, the embryos developing in the mesohyl (the non-epithelial layer sandwiched between pinacoderm and choanoderm) are distributed across the body of the adult. In contrast, in *Amphimedon queenslandica*, the embryos develop in specialized brood chambers, generally found close to the basal region of the sponge (Figure 4.3k). In both scenarios, mature larvae (Figure 4.3e, k) leave the mother sponge through the osculum and, after a period of swimming, settle and metamorphose on suitable substrate.

During metamorphosis, larval cells undergo major rearrangement, differentiation and transdifferentiation; begin production of skeletal elements (spicules, which are built of calcite in the calcisponges, and from silica in all other sponge classes); form the first choanocyte chambers; and finally open ostia and oscula to become feeding juveniles (Figure 4.3f–h, m–o). The juvenile of calcareous sponges from the genus *Sycon* represents one of the simplest body plans found in the animal kingdom: a cup-shaped body composed of two epithelial layers, which are connected by the ostia, with a narrow mesohyl layer containing spicule-producing cells (sclerocytes) and a single apical osculum (Figure 4.3h). This body plan is referred to as the asconoid grade of organization. As development progresses, new radial chambers form to surround the original radial chamber, which becomes the atrium of the emerging syconoid body plan (Figure 4.3i; see schematic representation in Figure 4.1). Despite this substantial change of the body plan, the radial symmetry of the body, with a single osculum, is maintained in many species, including *Sycon ciliatum* (Figure 4.3b). In contrast, the juvenile form of demosponges, such as *Amphimedon queenslandica*, is of leuconoid grade (multiple choanocyte chambers connected by series of canals), with a single apical osculum (Figure 4.3o; see schematic representation in Figure 4.1). As the animal grows, the leuconoid body plan is maintained, but additional oscula are formed, disrupting the original symmetry of the body plan (compare Figure 4.3j).

The life span of sponges also varies significantly across the species. *Sycon ciliatum* can be considered an annual species in the Norwegian fjords. The larvae settling in summer grow through the autumn and resume growth in the spring before they enter the reproductive stage in late spring, with larval release and death of the majority of the post-reproductive specimens in summer (Leininger et al. 2014). In contrast, *Amphimedon queensladica* can live many years based on the apparent growth rate and the size of individuals found in nature (author's personal observations). The most extreme case of sponge longevity on record is a Hexactinellid sponge, *Monorhaphis chuni*, estimated to live 11,000 (yes, eleven thousand!) years (Jochum et al. 2012).

4.4 EMBRYOGENESIS

Sponge embryogenesis utilizes a mind-boggling array of cellular mechanisms, including individual and collective movement, differentiation and transdifferentiation, leading to development of very diverse larval types. A significant body of literature has been produced on this topic, including a dedicated book, *The Comparative Embryology of Sponges*, covering all sponge lineages in fine detail (Ereskovsky 2010). Embryonic development of *Amphimedon queenslandica*, the first sponge to have its genome sequenced, received extensive additional attention (recently summarized by Degnan et al.

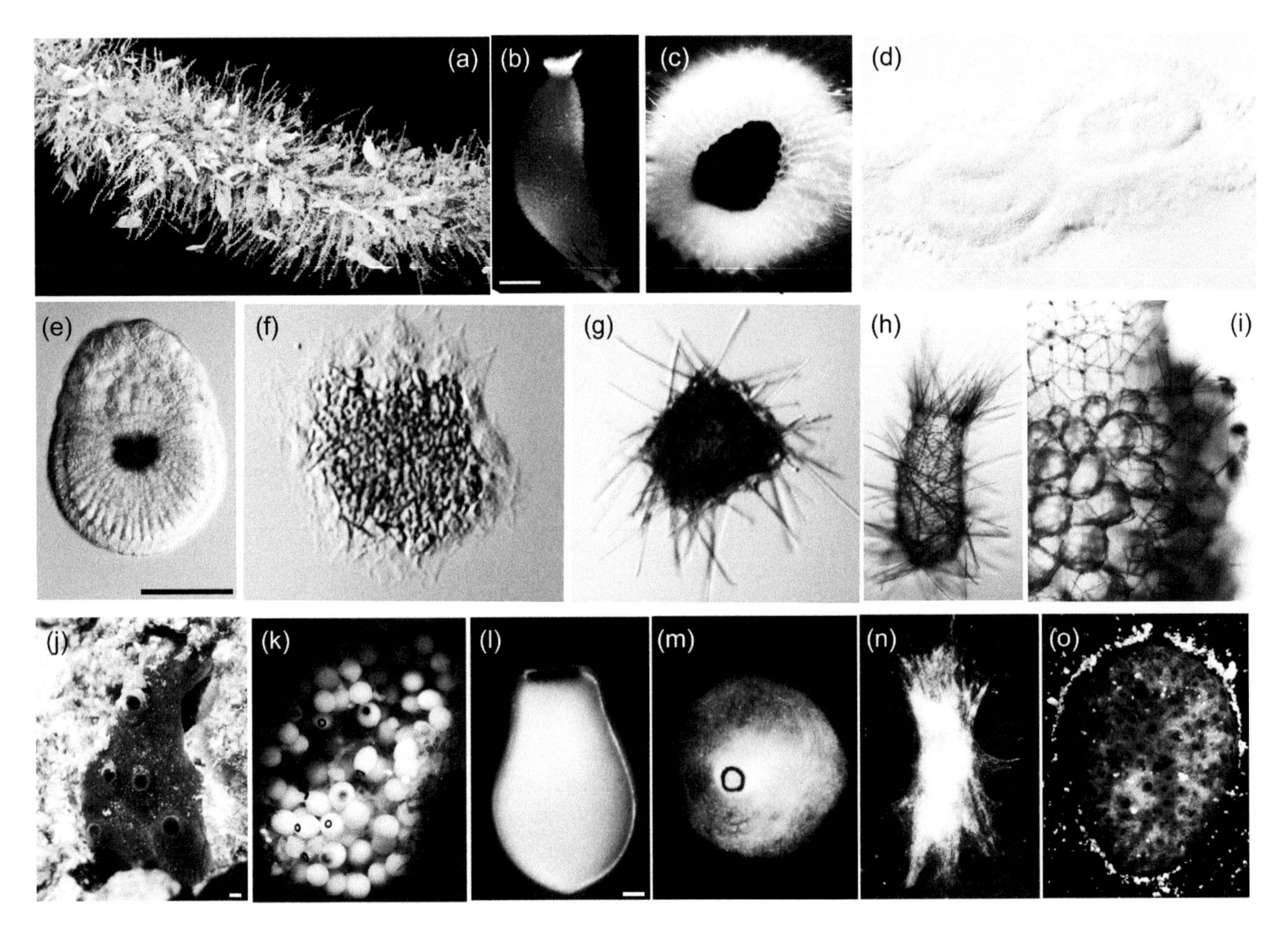

FIGURE 4.3 Sponge life cycle. Adults (a, b, j), embryos within maternal tissue (c, d, k), larvae (e, l), postlarvae (f, g, m, n) and juveniles (h, i) of two sponge model systems: the calcareous sponge *Sycon ciliatum* (a–i) and the demosponge *Amphimedon queenslandica* (j–o). (a) Multiple sponge specimens growing together on *Laminaria sp.*; (j) individual sponge on coral rubble. (d) Fixed slice of tissue with spicules removed to reveal embryos; the remaining samples are live specimens or their fragments. See text for description of embryonic development and metamorphosis. Scale bars: (b, j): 5 mm; (e, l): 50 μm. ([a–i] Reproduced from Leininger et al. 2014, [j–o] from Adamska et al. 2007.)

2015). In this species, embryonic development occurs in a brood chamber, containing a mix of embryos of all stages, from eggs to ready-to-release larvae, with the younger stages close to the edge of the chamber and more mature ones at the center (Figure 4.3k). The embryos are approximately 0.5 mm in diameter and yolky, with a cell division pattern best described as asynchronous and anarchic, leading to formation of a solid, spherical morula composed of cells of different sizes and differing by pigmentation level. Extensive cell movements result in development of a bi-layered, polarized embryo (referred to as gastrula in the original publication describing development of this species; Leys and Degnan 2002, but see Nakanishi et al. 2014 for a different view on the same process). Pigmented cells coalesce at one pole of the embryo to first form a spot and then a ring (Figure 4.3k). This ring, known to be a photosensory steering organ positioned at the posterior pole of the *Amphimedon* larva (Leys and Degnan 2001), is characteristic of parenchymella-type larvae of many other demosponges (Maldonado et al. 2006). There can be an extensive number of cell types present in mature parenchymella type larvae, including sclerocytes (cells producing spicules), archaeocytes (stem cells) and, in some cases, fully differentiated choanocytes and pinacocytes (e.g. Saller 1988).

One of the best studied of the larval types among sponges are the amphiblastula larvae of Calcaronean sponges, the lineage of calcisponges that includes *Sycon ciliatum* and related species (Franzen 1988). The other lineage of calcareous sponges, the Calcineans, has calciblastula larvae very similar to cinctoblastula found in Homoscleromorph sponges (Chapter 5), although it is not clear whether this similarity reflects shared ancestry (as Homoscleromorpha and sister group to the Calcispongiae) or is a result of convergence.

The amphiblastula larva forms through a highly stereotypic series of division followed by differentiation of only three cell types, which further undergo clear differentiation pathways upon metamorphosis. The oocytes are found uniformly distributed across the mesohyl of mature specimens. In the case of *Sycon ciliatum* in the Norwegian fjords, the development is synchronous through the local populations, with the first round of oocyte growth and fertilization occurring in the late spring (Leininger et al. 2014). Cleavage is complete, with the first two planes of division perpendicular to each other and the plane of the pinacoderm, thus dividing the zygote into four equal blastomeres. The subsequent divisions are oblique, resulting in formation of a cup-shaped embryo, with larger cells (macromeres) closer to the choanocytes and smaller cells (micromeres) facing pinacocytes (Figure 4.1). The embryonic cavity communicates with the lumen of the radial chamber, and through this opening, the embryo inverts itself so that the flagella of the micromeres (which originally form on the inner surface of the embryo) point outward.

In addition to the flagellated micromeres and larger, non-flagellated macromeres, the larva contains two other cell types: cross cells and maternal cells (Figure 4.1b). The cross cells (four in each larva) are of embryonic origin and differentiate from the outer "corners" of the four original blastomeres, with their final positions forming a cross at the equator of the larvae, conveying tetra-radial symmetry to the larva (Figure 4.1a). The function of these cells remains enigmatic, but they have been proposed to have sensory role and, consistent with this notion, express a number of genes known from other animals to be involved in specification of sensory cells and neurons (Tuzet 1973; Fortunato et al. 2014). Intriguingly, cross cells, along with maternal cells, which migrate inside of the embryo after inversion, degenerate during metamorphosis and do not contribute to formation of the juvenile body (Amano and Hori 1993).

As the larva settles on its anterior pole, the macromeres envelop the micromeres without losing epithelial character and differentiate directly to pinacocytes. The micromeres

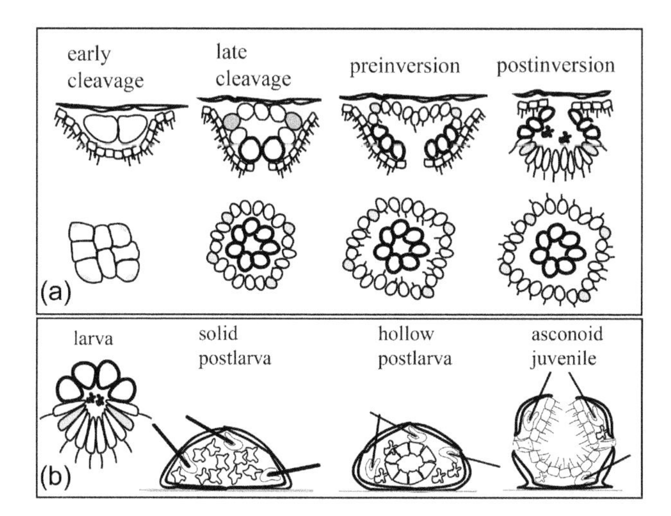

FIGURE 4.4 Schematic representation of embryonic development (a) and metamorphosis (b) in calcaronean sponges. In (a), the top row shows cross-sections of embryos surrounded by maternal tissues (pinacoderm and choanoderm); the bottom row is a top view of isolated embryos. Thick lines indicate macromeres and pinacocytes; thin lines indicate micromeres and choanocytes. Embryonic/larval cross cells and the cytoplasm of cleavage stage embryo destined to become cross cell are shaded gray.

undergo epithelial-to-mesenchymal transition and become amoeboid cells. After a period of movement (hours to days, depending on species), the micromeres differentiate into choanocytes and other juvenile cell types, including sclerocytes (spicule producing cells) (Figure 4.4b). Finally, the osculum opens at the apical pole and ostia form across the surface, resulting in formation of a functional, juvenile sponge of asconoid grade of organization. The source of porocytes is unclear, but it is likely that they differentiate from pinacocytes.

4.5 ANATOMY

All sponges (with the notable exception of carnivorous sponges, which secondarily lost choanocytes; Vacelet and Boury-Esnault 1995; Riesgo et al. 2007) are built of the same basic building blocks: choanocytes forming choanoderm of the radial chambers, the pinacoderm lining all remaining surfaces, with varying types and numbers of cells inhabiting the mesohyl. The mesohyl can be very cell poor and narrow (for example, in the Homoscleromorph sponges; see Chapter 5) or constitute most of the body of the sponge, as in many Demosponges. Traditionally, the body plans are divided into three major types. The simplest is asconoid, as described for Calcaronean juveniles (Figure 4.3h, 4.4b), with many calcisponge species retaining this body organization, with branching and anastomosing tubes forming as the body enlarges. The second type is syconoid, as in calcisponges from the genus Sycon (Figure 4.1, Figure 4.3b, c, i), with radial choanocyte chambers surrounding endopinacocyte-lined atrial cavity. The most complex, and the most common among sponges (being the typical body plan of Demosponges, the most speciose of the sponge lineages), is the leuconoid body plan composed of choanocyte chambers linked by an intricate network of endopinacyte-lined canals (Figure 4.1 and 4.3j, o). Two lesser-known sponge body plans should also be mentioned. One is the sylleibid body plan found in Homoscleromorph sponges, which can be considered a link between the syconoid and leuconoid body plans, with multiple syconoid-level units connected to the atrium. The most recently described sponge body plan, solenoid, is found in some Calcinean species and can be best described as a complex system of anastomosing tubes of the asconoid grade embedded in a thick mesohyl layer (Cavalcanti and Klautau 2011).

In the majority of sponges, the epithelial and mesenchymal layers are supported by organic and/or inorganic skeletons. The spongin-based organic skeletons of the genus *Spongia* and related species are well known as bath sponges—although, after the natural populations have been virtually exterminated by combination of harvest and pollution of the habitat, natural bath sponges have been all but replaced by artificial ones (Pronzato and Manconi 2008).

The majority of sponges produce inorganic skeletal elements, called spicules, which were traditionally the key to sponge taxonomy, given the paucity of other characters available until the advent of molecular phylogenies (Uriz 2006;

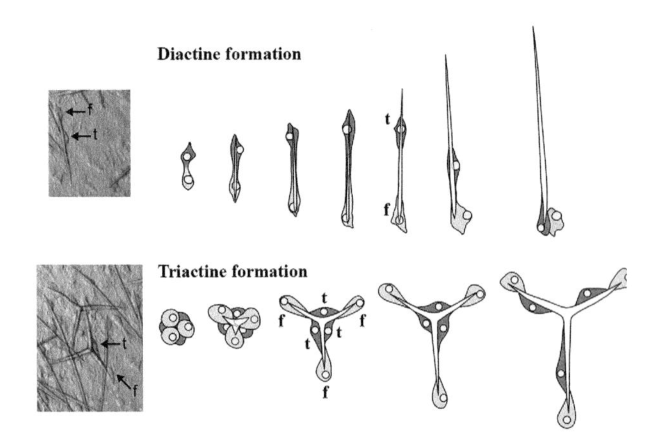

FIGURE 4.5 Spicule formation in calcareous sponges. Thickener cells (t) are dark gray, founder cells (f) are light gray. (Modified from Voigt et al. 2017, with the schematic representations re-drawn from Minchin 1908.)

van Soest et al. 2012). The spicules are of two types—built of calcite in the calcisponges and of silica in the remaining three lineages. Not only the material but also the cellular mechanism of spicule synthesis and subsequent positioning differs. The demosponge spicules are produced intracellularly, within vacuoles, and are subsequently moved to their final position by a concerted action of carrier cells (Mohri et al. 2008; Nakayama et al. 2015). In contrast, calcareous spicules are produced by groups of cells, the numbers of which depend on the type of the spicule and tend to remain in situ, without subsequent movement. For example, single-rayed spicules (diactines) are secreted by two cells, one known as the founder cell and the other as the thickener cell. On the other hand, the tri-radial triactines are produce by sextets of cells, with three founder cells and three thickener cells working together to produce one spicule (Minchin 1908; Voigt et al. 2017) (Figure 4.5). Different types of spicules form supporting structures along the body, with the long, slender diactines often found forming a crown or a collar around the osculum (Figure 4.3b).

4.6 GENOMIC DATA

The first insight into gene content of sponges was provided by transcriptome rather than genome analyses. Most significantly, the analysis of developmental regulatory genes in the transcriptome of the homoscleromorph sponge *Oscarella carmela* revealed that sponges possess multiple components of developmental signaling pathways used by animals to regulate their development (Nichols et al. 2006). However, the complete developmental regulatory gene repertoire of a sponge could only be fully appreciated by whole genome sequencing. The first sponge for which this was achieved was the demosponge *Amphimedon queenslandica*, a species inhabiting reefs fringing the Heron Island of the Great Barrier Reef (Srivastava et al. 2010). This was not only the first but also likely the last sponge genome sequenced using the traditional Sanger method. *Amphimedon* genome analysis revealed that for the overwhelming majority of developmental regulatory gene families, whether signaling molecules or transcription factors, *Amphimedon* possesses fewer family members than the more complex animals (Cnidarians and Bilaterians). This pattern, perhaps expected, was consistent with the notion that a simple animal would have a simpler regulatory gene repertoire.

It was therefore surprising when analysis of the second sponge species to be sequenced—the calcareous sponge *Sycon ciliatum*—revealed developmental gene family sizes on a par with those found in bilaterians. For example, while humans have 19 Wnt ligands and *Amphimedon* has 3 (Adamska et al. 2007), Sycon has 21 (Leininger et al. 2014). Even more strikingly—and controversially—the *Sycon* genome appears to possess a ParaHox gene, Cdx, which is clearly absent from the *Amphimedon* genome (Larroux et al. 2007; Fortunato et al. 2014). A systematic comparison of transcription factors present in *Amphimedon* and *Sycon* demonstrated that genomes of calcisponges and demosponges underwent independent events of gene loss and family expansions (Fortunato et al. 2015).

Gene content analysis of two Hexactinellids (glass sponges) revealed a different kind of surprise—it appears that neither *Oopsacas minuta* nor *Aphrocallistes vastus* possesses key components of the Wnt signaling pathway (Schenkelaars et al. 2017). As this pathway is used across the animal kingdom (including other sponges; See section 4.7) to pattern the major body axis, this finding is another key indication that insights from one lineage of sponges cannot be assumed to reflect the genome composition of all sponges—and of the last common ancestor of all animals. Instead, it thus appears that, since the divergence approximately 600 million years ago, sponge gene repertoires underwent dramatic changes, in contrast to the body plans which remained apparently stable throughout this time.

But sponge genomes can provide insight into more than just gene content: a gateway to understand evolution of genome function in animals. One of the mechanisms known to regulate gene expression in vertebrates (but not in the majority of invertebrates) is DNA methylation. However, the evolutionary history of this mechanism is not well understood. A recent study revealed that—in parallel to the differences found in gene content—sponge genomes are methylated to very different levels. While the *Amphimedon* genome is highly methylated (in striking similarly to vertebrate genomes), methylation in *Sycon* is more moderate, consistent with independent acquisition of genome methylation in sponges (de Mendoza et al. 2019).

Gaiti and colleagues (2017) used the *Amphimedon* genome find out whether two other regulatory features of animal genomes are found in sponges: the posttranslational modifications of histone H3 (linked to precise regulation of gene expression in animals) and micro-systenic units harboring distal enhancers of developmental regulatory genes.

Perhaps surprisingly, both features were found, demonstrating that they predate (and were perhaps the key to) divergence of animal lineages (Gaiti et al. 2017).

The very recent advances in genome sequencing technologies, allowing relatively cheap generation of (almost) chromosomal-level assemblies, opened the way to comparing large-scale synteny (gene order) analysis in addition to micro-synteny studied before. The first sponge genome to be assembled to this contiguity level, that of *Ephydatia mulleri*, demonstrated strong synteny conservation between this freshwater demosponge and other animals but not with choanoflagellates (Kenny et al. 2020). Time (and ongoing sequencing efforts) will tell if genomes of sponges representing other lineages also maintained this conservation or whether they hold further surprises.

4.7 FUNCTIONAL APPROACHES: TOOLS FOR MOLECULAR AND CELLULAR ANALYSES

Evolutionary genomics and developmental biology strive to go beyond cataloguing genes, attempting to reveal the links between gene expression and function. Decades of research revealed that across the animal kingdom, key developmental events, such as establishment of germ layers and polarity of embryos, as well as cell fate specification, are governed by a conserved set of regulatory genes. As soon as homologues of these genes were uncovered in sponge transcriptomes and genomes, in situ hybridization methods were developed, allowing interrogation of expression patterns of the candidate genes (Larroux et al. 2008).

One of the key examples of pan-metazoan functional conservation is the role of the Wnt pathway in specification of the primary body axis, with Wnt ligands expressed in the posterior poles of cnidarian and bilaterian embryos, as well as the apical region of cnidarian polyps. In several sponge species, Wnt ligands are expressed in the posterior pole of sponge larvae and around the osculum of sponge adults (Figure 4.6), suggesting that this role is conserved in sponges and therefore predates animal divergence (Adamska et al. 2007; Leininger et al. 2014; Borisenko et al. 2016). Similarly, genes involved in specification of animal sensory cells, such as components of the Notch pathway and the transcription factor bHLH1 (related to atonal and neurogenin in bilaterians), are expressed in the sensory cells of *Amphimedon* larvae (Richards et al. 2008).

However, gene expression patterns, while certainly suggestive, still do not demonstrate gene function. Disappointingly, functional gene expression analysis—through interference with gene function by morpholino or RNAi, or generation of transgenic animals to understand effects of gene overexpression—is still not a routine methodology in sponges. This is despite multiple efforts, some giving tantalizing results, such as successful generation of transgenic sponge cells, although with a success rate in the range of 1 in 10,000 cells (Revilla-I-Domingo et al. 2018), or downregulation genes targeted by RNAi, although with change level that required qPCR to demonstrate it (Rivera et al. 2011). Despite this limited success so far, efforts to establish robust functional genomics strategies continue in many sponge laboratories across the world. In the meantime, biologists utilize a range of other methodologies to gain functional insights into sponge development. For example, taking a drug interference approach, Windsor Reid and Leys (2010) demonstrated that the Wnt pathway is involved in specification of the main body axis of the demosponge *Ephydatia mulleri.*

4.8 CHALLENGING QUESTIONS BOTH IN ACADEMIC AND APPLIED RESEARCH

Perhaps surprisingly, the two major topics that attracted biologists to sponges in the 19th century, namely origin of the animal body plan and regeneration, continue to provide background for vibrant research programs in many laboratories—and ongoing debates in the research field. Until very recently, the relationship between sponge cell/tissue types and body plan organization was interrogated using the candidate gene approach. As discussed in Section 4.7, results of these analyses are consistent with homology of the major body axis (specified by the Wnt pathway) in sponges and cnidarians, therefore suggesting that the first animals also used the Wnt pathway to pattern their bodies (reviewed by Holstein 2012). Moreover, subsequent gene expression analyses focusing on genes involved in specification of animal endomesoderm, revealing that these genes are expressed in sponge choanocytes, are also consistent with Haeckel's idea that the sponge choanoderm is homologous to the cnidarian gastrodermis (Leininger et al. 2014; Adamska 2016a, 2016b). However, the fact that sponge cell fate specification is unusually fluid, allowing choanocytes to transdifferentiate into pinacocytes (thus apparently changing germ layer identity), makes some researchers unwilling to accept that notion (Nakanishi et al. 2014). While the question of cell type homology between sponges and other animals remains open for now, a novel approach based on expression of genes with conserved microsynteny yielded results consistent with the proposed homology of choanocytes and cells involved in cnidarian digestion (Zimmermann et al. 2019; Adamska 2019).

On the other side of the evolutionary transition leading from protists to complex animals, the similarity between choanocytes and choanoflagellates, understood to indicate homology of the collar apparatus throughout the 20th century, has become controversial again (Mah et al. 2014). Some authors take evidence of morphology, function and molecular composition of collars and flagella in choanocytes and collar cells as strong support for the proposed homology (Peña et al. 2016; Brunet and King 2017). Yet others used comparison of *Amphimedon* cell-type gene expression with cell-state gene expression data from choanoflagellates and a range of other protists to suggest that choanocyte

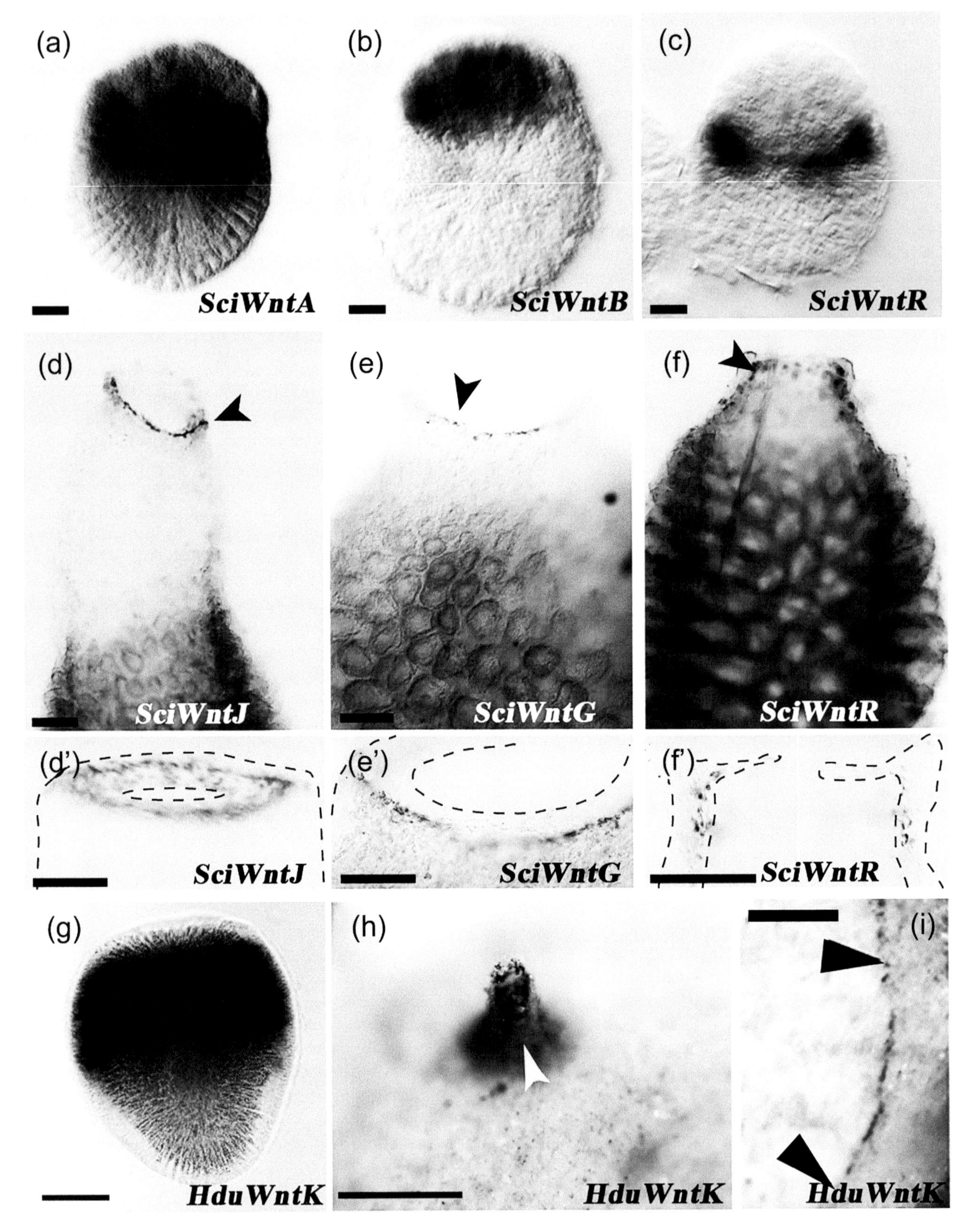

FIGURE 4.6 Expression of Wnt ligands in sponges. (a–c) Larvae of the calcareous sponge *Sycon ciliatum*. (d–f) Oscular regions of *S. ciliatum* (' indicates higher magnification; dashed lines delineate transparent tissues). (g, h, i) The demosponge, *Halisarca dujardini:* larva, the osculum and regenerating epithelium, respectively. Larval posterior and osculum are at the top of each image. Scale bars: (a–c): 10 μm, (d–f'): 100 μm, (g): 50 μm, (h–i): 3 mm. ([a–f] Reproduced from Leininger et al. 2014, [g–i] from Borisenko et al. 2016.)

morphology evolved independently from choanoflagellates (Sogabe et al. 2019). That these seemingly academic questions are also exciting to the general audience is evidenced by popular science magazines covering this debate (Cepelewicz 2019).

Less "academic", as understanding of sponge regeneration capacity might potentially be applicable to human regenerative medicine, is the question of how sponges regulate their spectacular regenerative capacities. Recent research reveals that some of the regeneration mechanisms might indeed be shared between sponges and other animals, as many of the developmental signaling pathways known to be involved in mammalian regenerations are also activated during regeneration of sponges, including re-building of bodies from dissociated cells (Soubigou et al. 2020). The most exciting aspect of sponge regeneration appears to be the capacity of sponge cells to directly transdifferentiate upon injury (Ereskovsky et al. 2015; Ereskovsky et al. 2017; reviewed by Adamska 2018). Would it be possible to utilize mechanisms involved in transdifferentiation of sponge cells to reprogram mammalian cells for therapeutic purposes?

The pharmaceutical industry has been investigating sponges as potential sources of bioactive compounds, with great success, for over 50 years. In 1969, the first sponge-derived anti-cancer drug, cytarabine (also known as Ara-C, Cytosar-U or Depocyst), originally extracted from the Caribbean demosponge *Tectitethya crypta*, was approved by the Food and Drug Administration (FDA). In 1976, the FDA also approved vidarabine (Ara-A, Vira-A) as an antiviral drug derived from the same sponge species (reviewed by Brinkmann et al. 2017). More recently, eribulin mesylate (E389, Halaven), an analog of halichondrin B isolated from Japanese demosponge *Halichondria okadai*, was approved as treatment for metastatic breast cancer (reviewed by Gerwick and Fenner 2013).

In addition to being useful, the secondary metabolites found in sponges are all the more fascinating as they are in fact produced by microbes living in close symbiosis with their poriferan hosts. The study of sponge microbiomes revealed essential roles in nutrient cycling and production of vitamins in addition to the secondary metabolites likely responsible for protection of sponges from potential predators and fouling organisms (see Reiswig 1981; Maldonado et al. 2012). It appears that the complex, species-specific assemblages of bacteria can be transmitted both horizontally (from the surrounding water) and vertically (from mother to larvae) (e.g. Schmitt et al. 2008; Webster et al. 2010). However, the molecular mechanisms involved in establishment and maintenance of these symbioses are not understood and remain an area of open and exciting investigations.

BIBLIOGRAPHY

Adamska M. 2016a. Sponges as models to study emergence of complex animals. *Curr Opin Genet Dev* 39: 21–28.

Adamska M. 2016b. Sponges as the Rosetta Stone of colonial-to-multicellular transition. In: *Multicellularity: Origins and evolution*. KJ Niklas, SA Newman (Eds.). The MIT Press, Cambridge, MA; London, England. ISBN: 978-0-262-03 415-9.

Adamska M. 2018. Differentiation and transdifferentiation of sponge cells. In: *Marine Organisms as Model Systems in Biology and Medicine*. M Kloc, JZ Kubiak (Eds.). Springer-Nature Series. Results and Problems in Cell Differentiation. Springer International Publishing. eBook ISBN: 978-3-319-92486-1. Hardcover ISBN: 978-3-319-92485-4.

Adamska M. 2019. Animal cell type diversity. *Nat Ecol Evol* 3: 1277–1278.

Adamska M, Degnan SM, Green KM, Adamski M, Craigie A, Larroux C, Degnan BM. 2007. Wnt and Tgfβ expression in the sponge *Amphimedon queenslandica* and the origin of metazoan embryonic patterning. *PLoS One* 2(10): e1031.

Amano S, Hori I. 1993. Metamorphosis of calcareous sponges. 2: Cell rearrangement and differentiation in metamorphosis. *Invert Reprod Dev* 24: 13–26.

Borisenko I, Adamski M, Ereskovsky A, Adamska M. 2016. Surprisingly rich repertoire of Wnt genes in the demosponge *Halisarca dujardini*. *BMC Evol Biol* 16: 123.

Boury-Esnault N, Efremova S, Bézac C, Vacelet J. 1999. Reproduction of a hexactinellid sponge: First description of gastrulation by cellular delamination in the Porifera. *Invertebr Reprod Dev* 35 (3): 187–201.

Brinkmann CM, Marker A, Kurtböke DI. 2017. An overview on marine sponge-symbiotic bacteria as unexhausted sources for natural product discovery diversity. *Diversity* 9: 40.

Brunet T, King N. 2017. The origin of animal multicellularity and cell differentiation. *Dev Cell* 43(2): 124–140.

Cavalcanti FF, Klautau M. 2011. Solenoid: A new aquiferous system to Porifera. *Zoomorphology* 130: 255–260.

Cepelewicz J. 2019. Scientists debate the origin of cell types in the first animals. www.quantamagazine.org/scientists-debate-the-origin-of-cell-types-in-the-first-animals-20190717/

Conkling M, Hesp K, Munroe S, et al. 2019. Breakthrough in marine invertebrate cell culture: Sponge cells divide rapidly in improved nutrient medium. *Sci Rep* 9: 17321.

Degnan BM, Adamska M, Richards GR, Larroux C, Leininger S, Bergum B, Calcino A, Maritz K, Nakanishi N, Degnan SM. 2015. Porifera. In: *Evolutionary Developmental Biology of Invertebrates*. Vol. 1. A. Wanninger (Ed.), pp. 65–106. Springer Verlag, Vien. ISBN: 978-3-7091-1861-0.

Degnan SM, Degnan BM. 2006. The origin of the pelagobenthic metazoan life cycle: What's sex got to do with it? *Integr Comp Biol* 46(6): 683–690.

de Mendoza A, Hatleberg WL, Pang K, et al. 2019. Convergent evolution of a vertebrate-like methylome in a marine sponge. *Nat Ecol Evol* 3(10): 1464–1473.

Ereskovsky AV. 2010. *The Comparative Embryology of Sponges*. Springer, New York.

Ereskovsky AV, Borisenko IE, Lapébie P, Gazave E, Tokina DB, et al. 2015. *Oscarella lobularis* (Homoscleromorpha, Porifera) regeneration: Epithelial morphogenesis and metaplasia. *PLoS One* 10(8): e0134566.

Ereskovsky AV, Lavrov AI, Bolshakov FV, Tokina DB. 2017. Regeneration in White Sea sponge *Leucosolenia complicata* (Porifera, Calcarea). *Invertebr Zool* 14(2): 108–113.

Fortunato SAV, Adamski M, Adamska M. 2015. Comparative analyses of developmental transcription factor repertoires in sponges reveal unexpected complexity of the earliest animals. *Mar Genomics* 2: 121–129.

Fortunato SAV, Adamski M, Mendivil O, Leininger S, Liu J, Ferrier DEK, and Adamska M. 2014. Calcisponges have a ParaHox gene and dynamic expression of dispersed NK homeobox genes. *Nature* 514(7524): 620–623.

Franzen W. 1988. Oogenesis and larval development of Scypha ciliata (Porifera, Calcarea). *Zoomorphology* 107: 349.

Gaiti F, Jindrich K, Fernandez-Valverde SL, Roper KE, Degnan BM, Tanurdžić M. 2017. Landscape of histone modifications in a sponge reveals the origin of animal cis-regulatory complexity. *eLife* 6: e22194.

Gerwick WH, Fenner AM. 2013. Drug discovery from marine microbes. *Microb Ecol* 65(4): 800–806.

Grant RE. 1825. Observations and experiments on the structure and functions of the sponge. *Edinburgh Phil Journ* 13: 94, 343; 14: 113–124.

Grant RE. 1836. Animal Kingdom. In: *The Cyclopaedia of Anatomy and Physiology*. Vol. 1. RB Todd (Ed.), pp. 107–118. Sherwood, Gilbert, and Piper, London.

Haeckel E. 1870. On the organization of sponges and their relationship to the corals. *Ann Mag Nat Hist* 5: 1–13, 107–120.

Haeckel E. 1874. Die Gastrae Theorie, die phylogenetische Classification des Thierreichs und die Homologie der Keimblatter. *Jena Zeitschr Naturwiss* 8: 1–55.

Holstein TW. 2012. The evolution of the Wnt pathway. *Cold Spring Harb Perspect Biol* 4: a007922.

Huxley JS. 1911. Some phenomena of regeneration in Sycon; with a note on the structure of its collar-cells. *Phil Trans R Soc B* 202: 165–189.

Huxley JS. 1921. Further studies on restitution-bodies and free tissue culture in Sycon. *Quart J Micr Sci* 65: 293–322.

James-Clark H. 1868. On the spongiae ciliatae as infusoria flagellata; or observations on the structure, animality, and relationship of *Leucosolenia botryoides*. Ann Mag Nat Hist 1: 133–142, 188–215, 250–264.

Jochum KP, Wang X, Ennemannc TW, Sinhaa B, Müller WEG. 2012. Siliceous deep-sea sponge *Monorhaphis chuni*: A potential paleoclimate archive in ancient animals. *Chemical Geology*: 300–330.

Kenny NJ, Francis WR, Rivera-Vicéns RE, et al. 2020. Tracing animal genomic evolution with the chromosomal-level assembly of the freshwater sponge *Ephydatia muelleri*. *Nat Commun* 11: 3676.

Knobloch S, Jóhannsson R, Marteinsson V. 2019. Co-cultivation of the marine sponge *Halichondria panicea* and its associated microorganisms. *Sci Rep* 9: 10403.

Larroux C, Fahey B, Adamska M, Richards GS, Gauthier M, Green K, Lovas E, Degnan BM. 2008. Whole-mount in situ hybridization in amphimedon. *CSH Protoc*. pdb.prot5096.

Larroux C, Fahey B, Degnan SM, Adamski M, Rokhsar DS, Degnan BM. 2007. The NK homeobox gene cluster predates the origin of Hox genes. *Curr Biol* Apr 17, 17(8): 706–710.

Leininger S, Adamski M, Bergum B, Guder C, Liu J, Laplante M, Bråte J, Hoffmann F, Fortunato S, Jordal S, Rapp HT, Adamska M. 2014. Developmental gene expression provides clues to relationships between sponge and eumetazoan body plans. *Nature Comm* 5: 3905.

Leys SP, Degnan BM. 2001. Cytological basis of photoresponsive behavior in a sponge larva. *Biol Bull* 201(3): 323–338.

Leys SP, Degnan BM. 2002. Embryogenesis and metamorphosis in a haplosclerid demosponge: Gastrulation and transdifferentiation of larval ciliated cells to choanocytes. *Invertebr Biol* 121: 171–189.

Leys SP, Ereskovsky AV. 2006. Embryogenesis and larval differentiation in sponges. *Can J Zool* 84(2): 262–287.

Leys SP, Kamarul Zaman A, Boury-Esnault N. 2016. Three-dimensional fate mapping of larval tissues through metamorphosis in the glass sponge *Oopsacas minuta*. *Invertebr Biol* 135: 259–272.

Mah JL, Christensen-Dalsgaard KK, Leys SP. 2014. Choanoflagellate and choanocyte collar-flagellar systems and the assumption of homology. *Evol Dev* 16(1): 25–37.

Maldonado M. 2006. The ecology of the sponge larva. *Can J Zool* 84(2): 175–194.

Maldonado M, Ribes M, van Duyl FC. 2012. Nutrient fluxes through sponges: Biology, budgets, and ecological implications. *Adv Mar Biol* 62: 113–182. Elsevier Ltd ISSN 0065-2881.

Manconi M, Pronzato R. 2002. Suborder Spongillina subord. nov.: Freshwater sponges. In: *Systema Porifera: A Guide to the Classification of Sponges*. NA John, WM Rob, Van Soest (Eds.), pp. 921–1021. Kluwer Academic/Plenum Publishers, New York.

Minchin EA. 1908. Materials for a monograph of the Ascons. II: The formation of spicules in the genus Leucosolenia, with some notes on the histology of the sponges. *Q J Microsc Sci* 52: 301–355.

Mohri K, Nakatsukasa M, Masuda Y, Agata K, Funayama N. 2008. Toward understanding the morphogenesis of siliceous spicules in freshwater sponge: Differential mRNA expression of spicule-type-specific silicatein genes in *Ephydatia fluviatilis*. *Dev Dyn* 237: 3024–3039.

Muller K. 1911a. Beobachtungen liber Reduktionsvorgange bei Spongilliden, nebst Bemerkungen zu deren ausserer Morphologie und Biologie. *Zool Anz* 37: 114–121.

Muller K. 1911b. Versuche liber die Regenerationsfahigkeit der Susswasserschwamme. *Zool Anz* 37: 83.

Nakanishi N, Sogabe S, Degnan BM. 2014. Evolutionary origin of gastrulation: Insights from sponge development. *BMC Biol* 12: 26.

Nakayama S, Arima K, Kawai K, Mohri K, Inui C, Sugano W, Koba H, Tamada K, Nakata YJ, Kishimoto K, Arai-Shindo M, Kojima C, Matsumoto T, Fujimori T, Agata K, Funayama N. 2015. Dynamic transport and cementation of skeletal elements build up the pole-and-beam structured skeleton of sponges. *Curr Biol* 25(19): 2549–2554.

Nichols SA, Dirks W, Pearse JS, King N. 2006. Early evolution of animal cell signaling and adhesion genes. *Proc Natl Acad Sci USA* 103(33): 12451–12456.

Peña JF, Alié A, Richter DJ, et al. 2016. Conserved expression of vertebrate microvillar gene homologs in choanocytes of freshwater sponges. *Evo Devo* 7: 13.

Pronzato R, Manconi R. 2008. Mediterranean commercial sponges: Over 5000 years of natural history and cultural heritage. *Mar Ecol* 29: 146–166.

Reiswig HM. 1981. Partial carbon and energy budgets of the bacteriosponge *Verongia fistularis* (Porifera: Demospongiae) in Barbados West-Indies. *Mar Biol* 2: 273–294.

Revilla-I-Domingo R, Schmidt C, Zifko C, Raible F. 2018. Establishment of transgenesis in the demosponge *Suberites domuncula*. *Genetics* 210(2): 435–443.

Richards GS, Simionato E, Perron M, Adamska M, Vervoort M, Degnan BM. 2008. Sponge genes provide new insight into the evolutionary origin of the neurogenic circuit. *Curr Biol* 18: 1156–1161.

Riesgo A, Taylor C, Leys SP. 2007. Reproduction in a carnivorous sponge: The significance of the absence of an aquiferous system to the sponge body plan. *Evol Dev* 9(6): 618–631.

Rivera AS, Hammel JU, Haen KM, Danka ES, Cieniewicz B, Winters IP, Posfai D, Wörheide G, Lavrov DV, Knight SW, Hill MS, Hill AL, Nickel M. 2011. RNA interference in marine and freshwater sponges: Actin knockdown in *Tethya wilhelma* and *Ephydatia muelleri* by ingested dsRNA expressing bacteria. *BMC Biotechnol* 11: 67.

Saller U. 1988. Oogenesis and larval development of *Ephydatia fluviatilis* (Porifera, Spongillidae). *Zoomorphology* 108: 23–28.

Saville-Kent W. 1880. *A Manual of the Infusoria: Including a Description of All Known Flagellate, Ciliate, and Tentaculiferous Protozoa, British and Foreign, and an Account of the Organization and the Affinities of the Sponges*. David Bogue, London.

Schenkelaars Q, Pratlong M, Kodjabachian L, et al. 2017. Animal multicellularity and polarity without Wnt signaling. *Sci Rep* 7: 15383.

Schippers K, Sipkema D, Osinga R, Smidt H, Pomponi S, Martens D, Wijffels R. 2012. Cultivation of sponges, sponge cells and symbionts. *Adv Mar Biol* 62: 273–337.

Schmitt S, Angermeier H, Schiller R, Lindquist N, Hentschel U. 2008. Molecular microbial diversity survey of sponge reproductive stages and mechanistic insights into vertical transmission of microbial symbionts. *Appl Environ Microbiol* 74(24): 7694–7708.

Simion P, Philippe H, Baurain D, Jager M, Richter DJ, Di Franco A, Roure B, Satoh N, Quéinnec É, Ereskovsky A, Lapébie P, Corre E, Delsuc F, King N, Wörheide G, Manuel M. 2017. A large and consistent phylogenomic dataset supports sponges as the sister group to all other animals. *Curr Biol* 27(7): 958–967.

Sogabe S, Hatleberg WL, Kocot KM, Say TE, Stoupin D, Roper KE, Fernandez-Valverde SL, Degnan SM, Degnan BM. 2019. Pluripotency and the origin of animal multicellularity. *Nature* 570(7762): 519–522.

Soubigou A, Ross EG, Touhami Y, Chrismas N, Modepalli V. 2020. Regeneration in the sponge *Sycon ciliatum* partly mimics postlarval development. *Development* 147(22): dev193714.

Srivastava M, Simakov O, Chapman J, Fahey B, Gauthier MEA, Mitros T, Richards GS, Conaco C, Dacre M, Hellsten U, Larroux C, Putnam NH, Stanke M, Adamska M, Darling A, Degnan SM, Oakley TH, Plachetzki DC, Zhai Y, Adamski M, Calcino A, Cummins SF, Goodstein DM, Harris C, Jackson DJ, Leys SP, Shu S, Woodcroft BJ, Vervoort M, Kosik KS, Manning G, Degnan BM, Rokhsar DS. 2010. The *Amphimedon queenslandica* genome and the evolution of animal complexity. *Nature* 466(7307): 720–726.

Tuzet O. 1973. Éponges calcaires. In: *Traité de Zoologie Anatomie, Systématique, Biologie Spongiaires*. PP Grassé (Ed.), pp. 27–132. Masson et Cie, Paris.

Uriz MJ. 2006. Mineral skeletogenesis in sponges. *Can J Zool* 84: 322–356.

Vacelet J, Boury-Esnault N. 1995. Carnivorous sponges. *Nature* 373: 333–335.

Van Soest RWM, Boury-Esnault N, Vacelet J, Dohrmann M, Erpenbeck D, et al. 2012. Global diversity of sponges (Porifera). *PLoS One* 7(4): e35105.

Voigt O, Adamska M, Adamski M, Kittelmann A, Wencker L, Wörheide G. 2017. Spicule formation in calcareous sponges: Coordinated expression of biomineralization genes and spicule-type specific genes. *Sci Rep* 7: 45658.

Webster NS, Taylor MW, Behnam F, Lücker S, Rattei T, Whalan S, Horn M, Wagner M. 2010. Deep sequencing reveals exceptional diversity and modes of transmission for bacterial sponge symbionts. *Environ Microbiol* 12(8): 2070–2082.

Whelan NV, Kocot KM, Moroz TP, Mukherjee K, Williams P, Paulay G, Moroz LL, Halanych KM 2017. Ctenophore relationships and their placement as the sister group to all other animals. *Nat Ecol Evol* 1(11): 1737–1746.

Wilson HV. 1907. On some phenomena of coalescence and regeneration in sponges. *J Exp Zool* 5: 245–258.

Windsor Reid P, Leys S. 2010. Wnt signaling and induction in the sponge aquiferous system: Evidence for an ancient origin of the organizer. *Evol Dev* 12: 484–493.

Zimmermann B, Robert NSM, Technau U, et al. 2019. Ancient animal genome architecture reflects cell type identities. *Nat Ecol Evol* 3: 1289–1293.

5 The Homoscleromorph Sponge, *Oscarella lobularis*

Emmanuelle Renard, Caroline Rocher, Alexander Ereskovsky and Carole Borchiellini

CONTENTS

5.1 HISTORY OF THE MODEL

Oscarella lobularis (Schmidt 1862) was first described as *Halisarca lobularis* Schmidt 1862 (Schmidt 1862). Later *Oscarella lobularis* became the type species of the genus *Oscarella* Vosmaer, 1884 (Vosmaer 1884), genus, classified until 2012 (Gazave et al. 2012) within the class Demospongiae, subclass Tetractinellida, due to the shared presence of siliceous tetractinal-like calthrops spicules (Levi 1956). Despite its reported cosmopolitan distribution (uncommon in sponges because of the low dispersal capacity of most sponge larvae) and the observation of a large variety of colors (Figure 5.1c), *O. lobularis* was long considered the only species of the genus *Oscarella*. Accordingly, all species of the *Oscarella* genus reported between 1930 and 1990 were probably wrongly assigned to *O. lobularis* (Lage et al. 2018; Pérez and Ruiz 2018).

The cosmopolitan status of *Oscarella lobularis* began to be questioned in 1992. Several color morphs assigned to the species *O. lobularis* (Schmidt 1862) living in sympatry in the west Mediterranean area were compared for the first time using a combination of characters: morphological characters, cytological characters and electric mobility of 12 protein markers. This study evidenced the presence of two

DOI: 10.1201/9781003217503-5

distinct species. The morphs with soft consistency were then referred to *O. lobularis*, while those with cartilaginous tissues were renamed as *O. tuberculata* (Boury-Esnault et al. 1992). The lack of a mineral skeleton (spicules) in the genus *Oscarella* was probably in part at the origin of species misidentification, because spicules were at that time commonly used in sponge systematics (Boury-Esnault et al. 1992). Since then, the development of multi-marker approaches (genetic, chemical, cytological, embryological characters) in conjunction with the effort deployed to explore more habitats have allowed a significant improvement in our knowledge of *Oscarella* species diversity (Bergquist and Kelly 2004; Ereskovsky 2006; Ereskovsky et al. 2009a; Ereskovsky et al. 2017b; Gazave et al. 2013; Muricy and Pearse 2004; Muricy et al. 1996; Pérez and Ruiz 2018; Pérez et al. 2011). There are so far 21 described species in the genus *Oscarella* (Table 5.1); this represents about 16% of the diversity of the Homoscleromorpha lineage (Van Soest et al. 2021).

Another major revolution in the taxonomic history of *O. lobularis* was the rise of Homoscleromorpha (previously considered a family, suborder or subclass within Demospongiae; Lévi 1973) to an upper taxonomic level. Different studies showed that Homoscleromorpha represents a fourth distinct class among Porifera (Borchiellini et al. 2004; Feuda et al. 2017; Francis and Canfield 2020; Gazave et al. 2012; Hill et al. 2013; Philippe et al. 2009; Pick et al. 2010; Pisani et al. 2015; Redmond et al. 2013; Simion et al. 2017; Thacker et al. 2013; Whelan et al. 2017; Wörheide et al. 2012) (Figure 5.1a). Homoscleromorpha is the smallest sponge class of Porifera, with only 130 exclusively marine valid species (Van Soest et al. 2021). This class is split into two families, Plakinidae Schulze, 1880, and Oscarellidae Lendenfeld, 1887 (Gazave et al. 2012) (Figure 5.1b). *Oscarella lobularis* belongs to the family Oscarellidae, a family defined by no skeleton; a variable degree of ectosome development; sylleibid-like or leuconoid organization of the aquiferous system, with eurypylous or diplodal choanocyte chambers; and the presence of the mitochondrial *tatC* gene (Gazave et al. 2010; Gazave et al. 2013; Wang and Lavrov 2007) (Figure 5.1b).

Therefore, the definition of Homoscleromorpha as a class, along with the three traditional ones Demospongiae, Hexactinellida and Calcarea (Brusca et al. 2016), shed light on homoscleromorph sponge species and evidenced the usefulness of studying and comparing these species to trace back character evolution during Poriferan evolutionary history. In accordance with the growing awareness in the evo-devo community of the need to develop studies on non-bilaterian and non-conventional animal models (Adamska

TABLE 5.1
List of *Oscarella* Species

Rank	Name	Original Description	Remarks	Geographical Location
Class	Homoscleromorpha	Bergquist (1978)	diagnosis in: Gazave et al. (2012)	Cosmopolitan
Order	Homosclerophorida	Dendy (1905)	diagnosis in: Gazave et al. (2012)	Cosmopolitan
Family	Oscarellidae	Lendenfeld (1887)	diagnosis in: Gazave et al. (2013)	Cosmopolitan
Genus	*Oscarella*	Vosmaer (1884)		Cosmopolitan
Species	*Oscarella balibaloi*	Pérez et al. (2011)		Western Mediterranean
	Oscarella bergenensis	Gazave et al. (2013)		Southern Norway
	Oscarella carmela	Muricy and Pearse (2004)		Northern California
	Oscarella cruenta	Carter (1876)		South European Atlantic Shelf
	Oscarella filipoi	Pérez and Ruiz (2018)		Eastern Caribbean
	Oscarella imperialis	Muricy et al. (1996)		Western Mediterranean
	Oscarella jarrei	Gazave et al. (2013)	accepted as *Pseudocorticium jarrei* Boury-Esnault et al. (1992)	Western Mediterranean
	Oscarella kamchatkensis	Ereskovsky et al. (2009a)		Kamchatka Shelf and Coast
	Oscarella lobularis	Schmidt (1862)		Mediterranean
	Oscarella membranacea	Hentschel (1909)		South West Australia
	Oscarella microlobata	Muricy et al. (1996)		Western Mediterranean
	Oscarella nicolae	Gazave et al. (2013)		Southern Norway
	Oscarella nigraviolacea	Bergquist and Kelly (2004)		East African
	Oscarella ochreacea	Muricy and Pearse (2004)		North east Pacific
	Oscarella pearsei	Ereskovsky et al. (2017b)		Northern California
	Oscarella rubra	Hanitsch (1890)	accepted as Aplysilla rubra (Hanitsch 1890)	Celtic seas
	Oscarella stillans	Bergquist and Kelly (2004)		North Borneo
	Oscarella tenuis	Hentschel (1909)		South West Australia
	Oscarella tuberculata	Schmidt (1868)		Mediterranean
	Oscarella viridis	Muricy et al. (1996)		Western Mediterranean
	Oscarella zoranja	Pérez and Ruiz (2018)		Eastern Caribbean

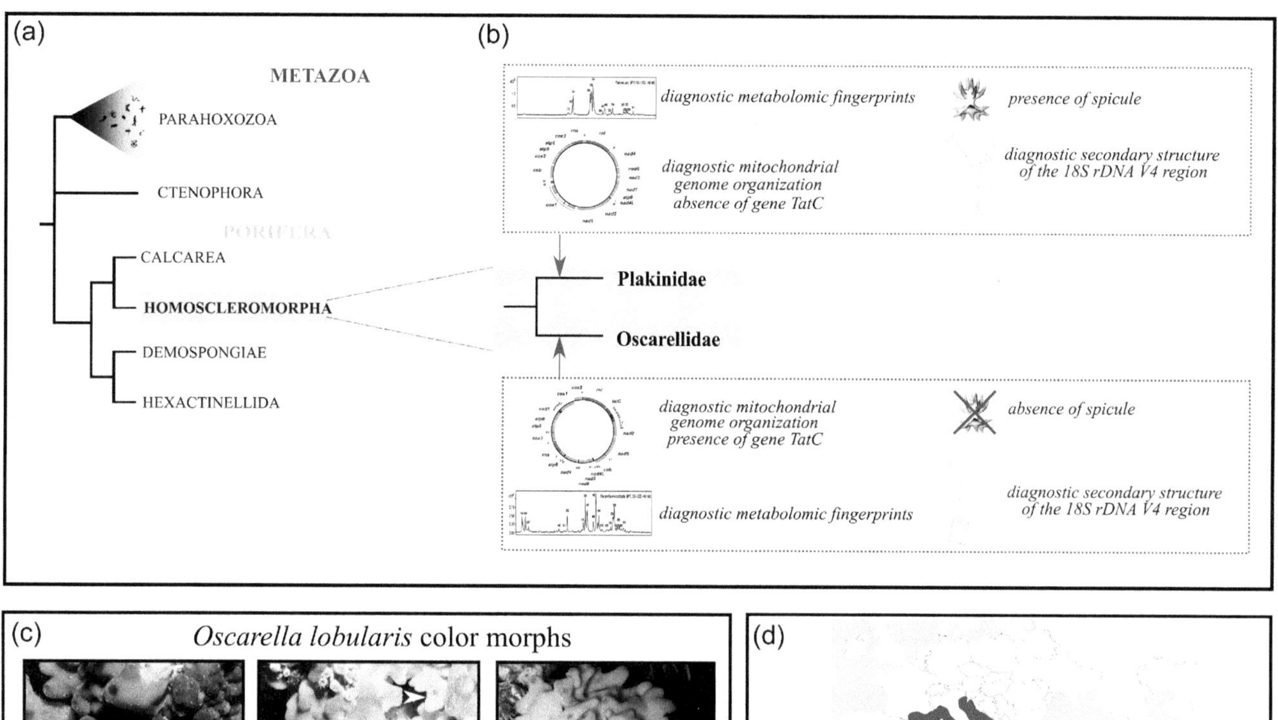

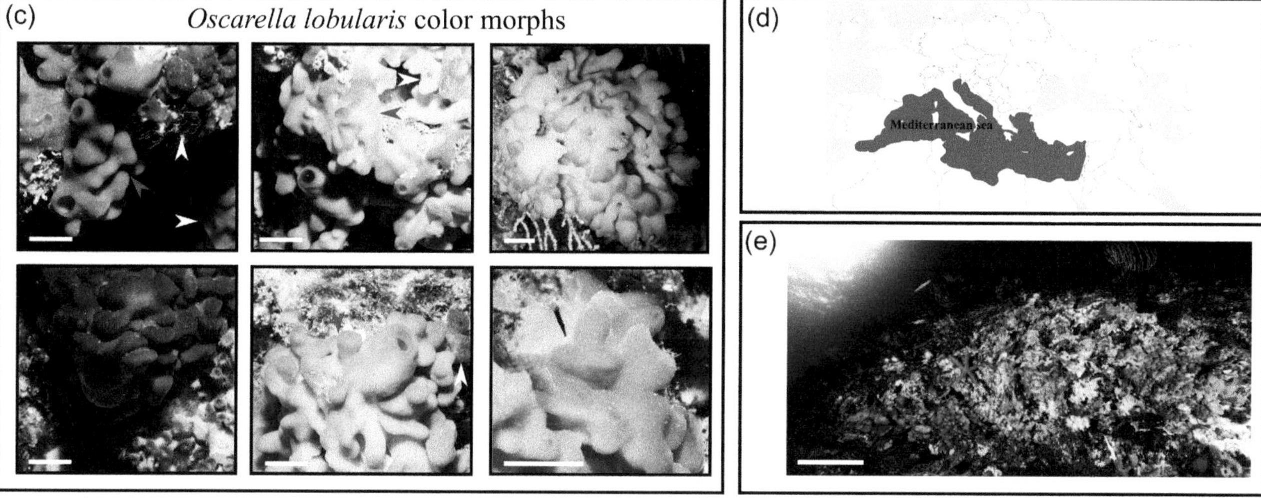

FIGURE 5.1 (a) The phylogenetic positions between Porifera and all other Metazoa and between Homoscleromorpha (to which *Oscarella lobularis* pertains) and other Poriferan classes. (b) The class Homoscleromorpha is split into Oscarellidae (to which *Oscarella lobularis* belongs) and Plakinidae, clearly distinguished by metabolomic, genetic and anatomical synapomorphies. (c) *Oscarella lobularis* harbors a high color polymorphism from yellowish to dark purple or blue; the color is unrelated to individual microbial community. *Oscarella lobularis* (red arrows) often lives in sympatry with other *Oscarella* species (white arrows), in particular its sister-species *O. tuberculata*. Scale bars represent 1 cm; photo credit: Dorian Guillemain. (d) *Oscarella lobularis* is now considered to have a geographic distribution restricted to the Mediterranean Sea. (e) *Oscarella lobularis* very often inhabits the Coralligenous habitat. Scale bar: 20 cm; photo credit: Frederic Zuberer. ([a] Borchiellini et al. 2004; Feuda et al. 2017; Francis and Canfield 2020; Gazave et al. 2012; Hill et al. 2013; Philippe et al. 2009; Pick et al. 2010; Pisani et al. 2015; Redmond et al. 2013; Simion et al. 2017; Thacker et al. 2013; Whelan et al. 2017; Wörheide et al. 2012; [b] Boury-Esnault et al. 2013; Gazave et al. 2010; Gazave et al. 2013; Ivanišević et al. 2011; [c] Gazave et al. 2012; Gloeckner et al. 2013; [d] Van Soest et al. 2021; [e] Bertolino et al. 2013.)

2016; Adamska et al. 2011; Colgren and Nichols 2019; Jenner and Wills 2007; Lanna 2015; Love and Yoshida 2019), *Oscarella lobularis* in Europe and *O. pearsei* (Ereskovsky et al. 2017b) in America began to be studied from an evo-devo perspective (Fierro-Constaín et al. 2017; Gazave et al. 2008; Gazave et al. 2009; Lapébie et al. 2009; Miller et al. 2018; Mitchell and Nichols 2019; Nichols et al. 2006; Nichols et al. 2012; Schenkelaars et al. 2015; Schenkelaars et al. 2016a).

5.2 GEOGRAPHICAL LOCATION

Homoscleromorpha, including species of the genus *Oscarella*, have a worldwide distribution, with three oceanic regions representing current hotspots of diversity (or hotspots of descriptions of new species): the Mediterranean Sea (Ereskovsky et al. 2009b; Lage et al. 2018), the tropical western Atlantic Ocean (Domingos et al. 2016; Ereskovsky et al. 2014; Pérez and Ruiz 2018; Ruiz et al. 2017; Vicente et al. 2016) and the Pacific Ocean (Bergquist and Kelly 2004; Ereskovsky 2006; Ereskovsky et al. 2009a; Lage et al. 2018; Muricy and Pearse 2004). In contrast, *O. lobularis* is found from the Gibraltar Strait to the eastern Mediterranean, including the Adriatic Sea, and is therefore presently considered a species endemic to the Mediterranean Sea (Ereskovsky et al. 2009b) (Figure 5.1d). Indeed, the other locations previously reported (for instance, Madagascar or the Manche Sea) were shown to be misidentifications (Lévi and Porte 1962; Muricy and Pearse 2004; Van Soest et al. 2007).

In Mediterranean ecosystems, sponges represent one of the main animal groups: a study by Coll *et al.* (2010) estimated that Porifera represent about 12.4% of the animal diversity (a proportion in the same range as that of vertebrate species diversity). Among the 681 poriferan species present in the Mediterranean (Coll et al. 2010), only 25 species (about 3% of the sponge species diversity) belong to Homoscleromorpha (Lage et al. 2019). Among them, *O. lobularis* is one of the most common and abundant species in some places (Ereskovsky et al. 2009b).

O. lobularis is mainly located in shallow waters from 4 to 35 m and in sciaphilic hard substratum communities including semi-dark and dark submarine caves (Ereskovsky et al. 2009b). In particular, *O. lobularis* is one of the 273 sponge species involved in coralligenous accretion (Bertolino et al. 2013) (Figure 5.1e). The infra- and circalittoral coralligenous habitats (first defined by Marion 1883) are now recognized as one of the main Mediterranean biocoenoses. In these habitats, unlike bioeroding Clionidae, *O. lobularis* usually grows on top of other sponges or on cnidarians (such as sea fans), bryozoans, annelid tubes, mollusk shells or lithophyllum; it is therefore usually considered an efficient space competitor (Garrabou and Zabala 2001).

5.3 LIFE CYCLE

Like many other sponges whose life cycles have been described (Ereskovsky 2010; Fell 1993), *Oscarella lobularis* is capable of both sexual and asexual reproduction. These types of reproduction alternate naturally during the same year (Figure 5.2).

5.3.1 Asexual Reproduction: Fragmentation and Budding

The timing and process of asexual reproduction in *Oscarella lobularis* have been described in several complementary studies (Ereskovsky 2010; Ereskovsky and Tokina 2007; Fierro-Constain 2016; Rocher et al. 2020). *O. lobularis* uses two modes of asexual reproduction: fragmentation and budding (Figure 5.2a and b).

Like sexual reproduction (see next section), fragmentation occurs once a year and often concerns most individuals of the same population. This event may be correlated with the switch to a short-day photoperiod and/or the decrease of water temperature (Fierro-Constain 2016; Rocher et al. 2020). At fall (October–November), adult individuals tend to elongate their tissues, and fragments seem to "dribble" until they separate totally (Figure 5.2b). The fate of the set-free fragments has not been monitored by any study yet, but it is supposed that these fragments can fall on a deeper substrate or be transported by the water flow; then some of them may be able to settle on rocks and develop into whole individuals.

In contrast, budding seems to occur at different periods during the year, between October and April (Figure 5.2b). This event appears not to be synchronized between individuals of the same natural population. It is therefore difficult to extrapolate the parameters triggering budding in the sea. Interestingly, budding can be triggered *in vitro* in *O. lobularis* by a mechanical stress, allowing for the monitoring and description of the whole process under laboratory conditions (Rocher et al. 2020).

The genesis and development of buds differ among sponge species (Ereskovsky et al. 2017a; Singh and Thakur 2015). In *O. lobularis*, the budding is performed in three key steps observed in a comparable manner during lab-induced budding *in vitro* and during natural budding of individuals *in situ* (Ereskovsky and Tokina 2007; Rocher et al. 2020). The budding process involves the evagination of adult tissues. The first step of budding is characterized by a transition from a smooth surface to an irregular surface. In the second step, small protrusions, responsible for this irregular aspect, grow apically to form branched finger-like structures at the surface of the adults. The third step consists of the swelling of protruding tissues and the release of free spherical buds. Once free, buds are able to float in the water flow and, *in vitro*, they have a much longer longevity than larvae: up to three months for *Oscarella* buds (Rocher et al. 2020 and for the buds of other species Maldonado and Riesgo 2008) *versus* a few days for larvae (Ereskovsky et al. 2009b; Ereskovsky et al. 2013a; Maldonado and Riesgo 2008). In standardized lab conditions, spherical buds develop outgrowths involved in the fixation to the substrate in a couple of days and an exhalant tube (osculum) in about one week, and settled juveniles can be obtained after one month (Rocher et al. 2020). These juveniles have a similar anatomy to that of juveniles resulting from sexual reproduction (Ereskovsky and Tokina 2007; Ereskovsky et al. 2007; Rocher et al. 2020) (Figure 5.2a).

We speculate that all together, the high number of buds produced by the same adult (mean 450 buds/cm^3 of adult tissue) with the floating properties of buds and their longevity (Rocher et al. 2020) make budding a crucial reproductive event in the *O. lobularis* life cycle (Fierro-Constain 2016). Asexual reproduction by budding must play an important role in the dispersion and population dynamics in natural habitats in *O. lobularis*, as proposed in demosponges (Cardone et al. 2010; Singh and Thakur 2015).

5.3.2 Sexual Reproduction, Gametogenesis and Indirect Development

Sexual reproduction takes place once a year (Figure 5.2b). A first analysis of 303 individuals of *O. lobularis* sampled monthly between 2006 and 2009 (Ereskovsky et al. 2013a) revealed that spermatogenesis occurred between June and August, differentiation of oocytes started in May and occurred until mid-August and embryogenesis occurred from mid-July to the beginning of September. A more recent study (2014–2015) based on both histological section observations and the detection of germline gene expression by *in situ* hybridization enabling a more efficient detection of earlier stages of gametogenesis allowed extension of the gametogenesis period from May–August to April–October (Fierro-Constain 2016; Fierro-Constaín et al. 2017; Rocher et al. 2020). Nevertheless, the latter study was performed on only six individuals of a population, this population being different from that considered in the previous study. This therefore does not preclude the differences observed between these studies being caused either by variations between populations or by different climatic conditions between the years considered.

Spermatogenesis and oogenesis co-occur from May to the beginning of September (Figure 5.2b), which provides an opportunity to decipher whether *O. lobularis* is a gonochoristic or hermaphroditic species. The *in situ* monitoring of localized and identified individuals in a small population suggests that *O. lobularis* is a hermaphrodite proterogyn (Fierro-Constain 2016; Fierro-Constaín et al. 2017). Both spermatocysts and oocytes were observed in the same individual as already shown in the early 20th century (Meewis 1938), and oogenesis starts earlier (April) than spermatogenesis (May). In contrast, the study of Ereskovsky et al. (2013a) suggested that this species is gonochoristic. This discrepancy may be explained by the fact that the number of oocytes and spermatocysts varies from one individual to another (Fierro-Constain 2016) and between years (Ereskovsky et al. 2013a). Nevertheless, to solve this uncertainty, we suggest that applying Fierro-Constain's approach to a higher number of individuals of different populations would be useful.

Oscarella lobularis, like all other sponges, lacks gonads as well as germ cell lineage (reviewed in Ereskovsky 2010; Leys and Ereskovsky 2006; Simpson 1984). In this context, gametes form by transdifferentiation from somatic cells with stemness properties. In *O. lobularis*, both oocytes and spermatocysts are formed by the transdifferentiation of somatic cells involved in filtration, the choanocytes (Ereskovsky 2010; Gaino et al. 1986a; Gaino et al. 1986c). It has been shown that 11 genes of the germline multipotency program (GMP) are expressed during both the spermatogenesis and oogenesis of *O. lobularis*, suggesting that the RNAs and proteins encoded by these genes are involved in gametogenesis, as described in bilaterians (Fierro-Constaín et al. 2017).

Concerning spermatogenesis, all choanocytes of the same choanocyte chamber transdifferentiate into sperm cells, and the previous choanocyte chamber becomes a spermatocyst (Figure 5.2b). Not all choanocyte chambers are concerned in the same individual, enabling the reproductive adult to continue filter feeding. Spermatocysts (size ranging from 50 to 150 µm) are randomly distributed in mesohyl and produce several asynchronous generations of male germ cells. Spermatogonia derive directly from choanocytes and will develop to produce spermatozoa by a process of centripetal differentiation, as in many other animals. During this process, spermatogonia lose morphological characteristics and histological attributes of the choanocytes (Ereskovsky 2010; Ereskovsky et al. 2013a). Spermatozoa harbor a long flagella

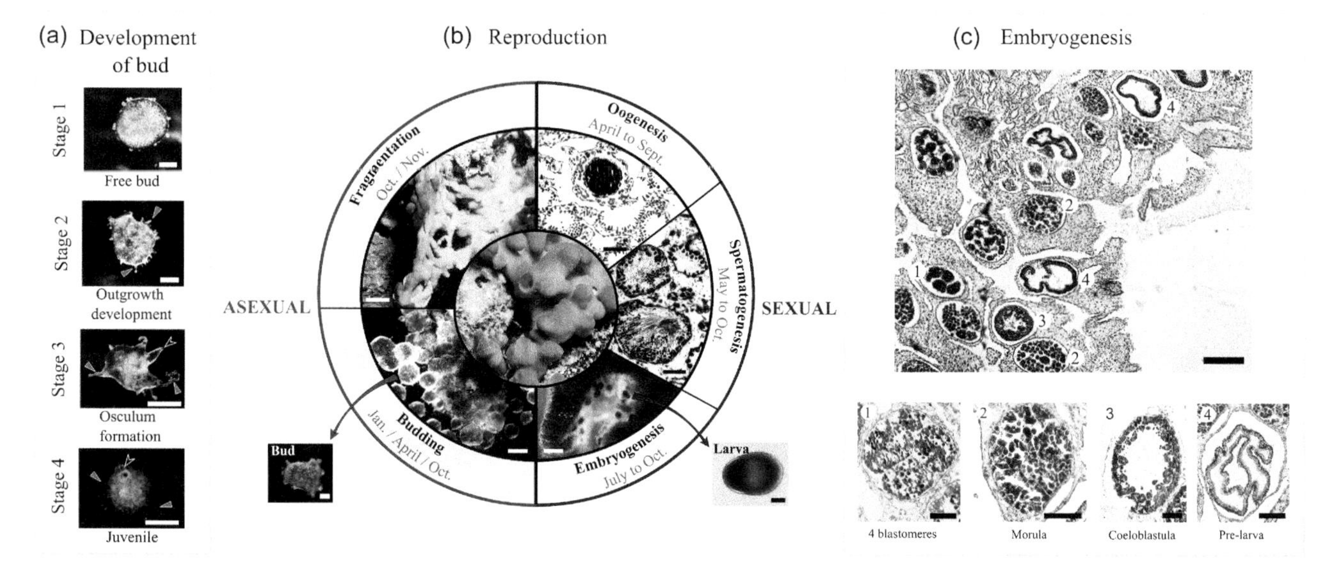

FIGURE 5.2 (A) Developmental stages from the release of free-buds to a settled juvenile (Rocher et al. 2020). Scale bars represent 500 µm (stage 1 to 4). Blue and yellow arrows indicate, respectively, outgrowths and osculum. (b) The three modes of reproduction of *Oscarella lobularis* during a year: asexual reproduction by fragmentation (scale bar: 1 cm) or budding (scale bar: 1 mm) and sexual reproduction: oogenesis (scale bar: 50 µm); spermatogenesis (scale bar: 25 µm); embryogenesis (scale bar: 1 mm). Swimming larva scale bar: 150 µm. Free bud scale bar: 200 µm. (c) Developmental stages occurring in the adult tissues from the zygote (resulting from internal fertilization) to the cinctoblastula pre-larva. Scale bar represents 200 µm. (1): Four-cell stage; (2): morula stage; (3): coeloblastula stage; (4): cinctoblastula pre-larva. Scale bars represent 50 µm (Stages 1 to 4).

and a slightly elongated head with an acrosome and a large mitochondrion (Ereskovsky 2010; Ereskovsky et al. 2013a; Gaino et al. 1986a). Spermatozoa are released into the surrounding water by the oscula via the exhalant canals.

Concerning oogenesis, a few choanocytes migrate into the mesohyl and transdifferentiate into oocytes (Figure 5.2b). The size of the young spherical oocyte corresponds to the size of one choanocyte (7–10 μm) without flagellum, microvilli and basal filopodia. This size increases significantly during vitellogenesis, although the final size of a mature oocyte is different, according to the authors (Ereskovsky 2010; Ereskovsky et al. 2013a; Fierro-Constain 2016). In this species, the great amount of vitellus (polylecithal eggs), uniformly distributed in the ooplasm (isolecithal), is produced by endogenous synthesis (Ereskovsky 2010; Ereskovsky et al. 2009b; Gaino et al. 1986b), unlike the other sponges with polylecithal oocytes in which vitellogenesis occurs by phagocytosis of somatic cells and/or bacteria (Maldonado and Riesgo 2008). Mature oocytes, located in the basal zone of the choanosome, are enclosed by endopinacoderm to form a so-called follicle. Before the closure of the follicle, maternal symbiotic bacteria and several maternal cells penetrate in the space between the oocyte and the follicle (Ereskovsky and Boury-Esnault 2002). Vertical transmission of symbionts from embryo to juvenile has been well documented in sponges (Boury-Esnault et al. 2003; Ereskovsky 2010; Ereskovsky and Boury-Esnault 2002; Ereskovsky et al. 2007; Ereskovsky et al. 2009b). Moreover, the penetration of maternal vacuolar cells inside of follicles was described in many investigated sponge species from the classes Demospongiae, Calcarea and Homoscleromorpha (Ereskovsky 2010). The oocytes remain in the adult tissue, meaning that *O. lobularis* performs internal fertilization. As fertilization *per se* has never been observed in this species, it is unknown whether it relies upon a carrier-cell system, as described in *Calcaronea* species (first described by Gatenby in 1920; reviewed in Ereskovsky 2010).

5.4 EMBRYOGENESIS

5.4.1 Cleavage and Formation of Coeloblastula

Like many sponge species described so far, *Oscarella lobularis* undergoes indirect development (Ereskovsky 2010). Additionally, as a direct consequence of internal fertilization, *O. lobularis* is a "brooding" sponge. This means that the development from a zygote to a fully developed larva (cinctoblastula) occurs within the adult tissue (Figure 5.2c): swimming larvae are then released in the surrounding water (Figure 5.2b).

The embryonic development *of O. lobularis* is similar to other species of the genus *Oscarella*. The main steps of this embryonic development have been described so far only by classical histological approaches on fixed individuals (Ereskovsky 2010; Ereskovsky and Boury-Esnault 2002; Ereskovsky et al. 2009b; Ereskovsky et al. 2013a; Ereskovsky et al. 2013b; Leys and Ereskovsky 2006). As in all Metazoa, the first developmental step consists of the cleavage of the zygote. The zygote being isolecithal (see previous section on oogenesis), this cleavage is holoblastic. The first two divisions (until the four-cell stage; Figure 5.2c) are equal and synchronous. Then the cleavage becomes irregular and asynchronous from the third division. After six divisions, the morula stage is reached: the morula is composed of 64 undifferentiated blastomeres (Ereskovsky and Boury-Esnault 2002; Ereskovsky et al. 2013a; Leys and Ereskovsky 2006) (Figure 5.2c). As cleavage progresses, the blastomeres reduce in size, and the volume of the embryo remains unchanged.

From the 64-cell morula stage, the blastomeres at the surface of the morula divide more actively, while internal blastomeres migrate to the periphery of the embryo through a process of multipolar egression (Ereskovsky 2010; Ereskovsky and Boury-Esnault 2002; Ereskovsky et al. 2013a; Leys and Ereskovsky 2006) to form a monolayered coeloblastula with a central cavity (Figure 5.2c). This central cavity has been described as containing the maternal symbiotic bacteria and maternal vacuolar cells (see previous section on oogenesis). The role and fate of these latter have not been explored and will have to be with modern molecular and cellular tools (Boury-Esnault et al. 2003; Ereskovsky and Boury-Esnault 2002; Ereskovsky et al. 2007; Ereskovsky et al. 2013a), but they seem to degenerate during metamorphosis of the larvae (personal observations).

Unlike in the three other sponge classes, the coeloblastula of *Oscarella* exhibits a monolayer columnar epithelium. This epithelium fits all classical criteria of the definition of epithelia in Bilaterians (Ereskovsky et al. 2009b; Leys and Riesgo 2012; Leys et al. 2009; Tyler 2003; Renard et al. 2021). i) Cells are highly polarized: cilia develop at the apical cell pole; ii) cells are tightened by specialized intercellular junctions, similar to adherens junctions, in the apical domain; and iii) cells are lined at their basal pole by a basement membrane consisting of collagen IV (Boute et al. 1996; Ereskovsky and Boury-Esnault 2002; Boury-Esnault et al. 2003). The establishment of this columnar epithelium at the coeloblastula stage is the first sign of cellular differentiation processes. Note that, even if the term "coeloblastula" was used in the literature because of the presence of a central cavity, this organization is not the result of the same processes (cleavage only) as in other metazoans (Boury-Esnault et al. 2003; Brusca and Brusca 2003; Ereskovsky 2010; Ereskovsky and Boury-Esnault 2002; Ereskovsky and Dondua 2006; Leys 2004; Leys and Ereskovsky 2006; Maldonado and Riesgo 2008; Wörheide et al. 2012). For this reason, some authors prefer the use of the term "prelarva" or "cinctoblastula prelarva" (Ereskovsky 2010). Unfortunately, this complex terminology makes comparison with other metazoans very difficult, and none of the embryological descriptions of embryological development available so far in sponges are based on live observations and cell tracking experiments.

5.4.2 Morphogenesis of the Cinctoblastula Larva and Larval Metamorphosis

Cells continue to divide, thus increasing the cell surface area. Because of the limited space in the follicle, the external epithelium becomes folded (Figure 5.2c). The central cavity is progressively filled by collagen fibrils, and a pronounced antero-posterior polarity is acquired: the ciliated cells contain various cytoplasmic inclusions and present a variable nucleus position according to their position along the anterior–posterior axis, unlike in coeloblastula larva of other sponges (Boury-Esnault et al. 2003; Ereskovsky 2010; Leys and Ereskovsky 2006). The cellular mechanisms by which pre-cinctoblastula larvae are transferred from the mesohyl to the exhalant canal was described in Boury-Esnault et al. (2003) and involves a fusion between the endopinacoderm forming the follicles and the endopinacoderm lining the canals. Finally, a free-swimming cinctoblastula larva is released from the adult sponge through the exhalant canals and the osculum. Larvae are uniformly flagellated (despite the presence of few scattered non-ciliated cells) and present a polarity: the anterior pole is larger than the posterior one, and the posterior pole is pigmented (pink pigments in *O. lobularis*) and rich in symbiotic bacteria and maternal vacuolar cells in the central cavity. The pigments are probably involved in the observed larval phototaxis behavior, as evidenced in the demosponge *Amphimedon queenslandica* (Degnan et al. 2015; Leys and Degnan 2001; Rivera et al. 2012).

The larva can swim in the water column for several days before settlement. The larva attaches to the substrate by the anterior pole thanks to mucus secretion, then undergoes metamorphosis (Figure 5.3a). Therefore, the A/P axis of the larva corresponds to the baso-apical axis of the juvenile sponge.

During metamorphosis, the larva undergoes radical morphological and physiological changes. The metamorphosis of the larva represents a second phase of reorganization of cell layers and corresponds to the acquisition of the typical sponge bauplan with a functional aquiferous system. The formation of the two main epithelial layers, namely the pinacoderm and the choanoderm, occurs through the transdifferentiation of the larval epithelium (fully detailed in Ereskovsky 2010; Ereskovsky et al. 2007; Ereskovsky et al. 2010).

The steps of larval metamorphosis have been described as variable and independent of environmental factors (Ereskovsky et al. 2007). However, the origin of this polyphenism is unknown. In most cases, the metamorphosis of *O. lobularis* larvae begins by a basal invagination (Figure 5.3a). In parallel to this invagination, several lateral cells ingress into the cavity. The lateral then sides fold up with the subsequent involution of marginal sides. At this stage, the future juvenile is composed of two cell layers, an external layer, from which the future exopinacoderm will originate, and an internal layer. The cells of this internal layer become flat, thereby increasing the tissue surface, which itself results in folding. This inner folded epithelium gives rise to the aquiferous system: the endopinacoderm is derived from the proximal parts of the internal cell layer, while the choanocyte chambers develop from distal parts of the internal folds (Figure 5.3a). The inhalant pores, ostia, and the exhalant pores, osculum, are formed secondarily. A settled filtering juvenile is finally formed, usually called a "rhagon".

5.4.3 Molecular Control of Development

The molecular mechanisms controlling the previously described developmental events are still unknown. As sexual reproduction occurs only once a year and embryos are not observed every year in sampled adults, and furthermore the embryos are intimately embedded in the adult tissues, their dissection and manipulation are rather tricky. Therefore, only two studies so far report gene expression patterns during embryogenesis. Due to the key role of the WNT pathway in axial patterning across the animal kingdom, several studies have investigated the pattern of *Wnt* gene expression during sponge development or during other morphogenetic processes (Adamska 2016; Adamska et al. 2007; Adamska et al. 2010; Adamska et al. 2011; Borisenko et al. 2016; Degnan et al. 2015; Lanna 2015; Leininger et al. 2014; Richards and Degnan 2009). In *Oscarella lobularis*, nine *Wnt* genes were found, as well as their target genes (Lapébie et al. 2009; Schenkelaars 2015). Even though most *Wnts* and *Fzds* genes are uniformly expressed during early stages of embryogenesis without apparent gradient or asymmetry, one *Wnt* gene is clearly localized at one pole of the embryos before any morphological polarity is observed (Schenkelaars 2015). This latter observation is in agreement with results obtained in other sponge lineages (Calcarea and Demospongiae), where WNT ligands and downstream genes are expressed in the posterior region of the embryos or larvae (Adamska 2016; Adamska et al. 2007; Adamska et al. 2010; Borisenko et al. 2016; Degnan et al. 2015; Leininger et al. 2014). These expression patterns tend to support a putative involvement of WNT pathways in patterning of the major sponge body axis. In addition, Fierro-Constaín et al. (2017) showed that 11 genes of the GMP are expressed during embryogenesis (including the most famous *piwi*, *vasa*, *nanos*, *Pl10* genes). This finding agrees with observations in other animals. Interestingly, among these genes, *nanos* harbors a highly polarized pattern in the prelarva: with a much higher expression level at the anterior pole. Such a polarized pattern was also observed in the calcarean sponge *Sycon ciliatum* (Leininger et al. 2014) and in other metazoans, but the role of this gene in axis patterning is unclear (Kanska and Frank 2013).

5.5 ANATOMY

As previously explained, developmental processes following both sexual and asexual (by budding) reproduction result

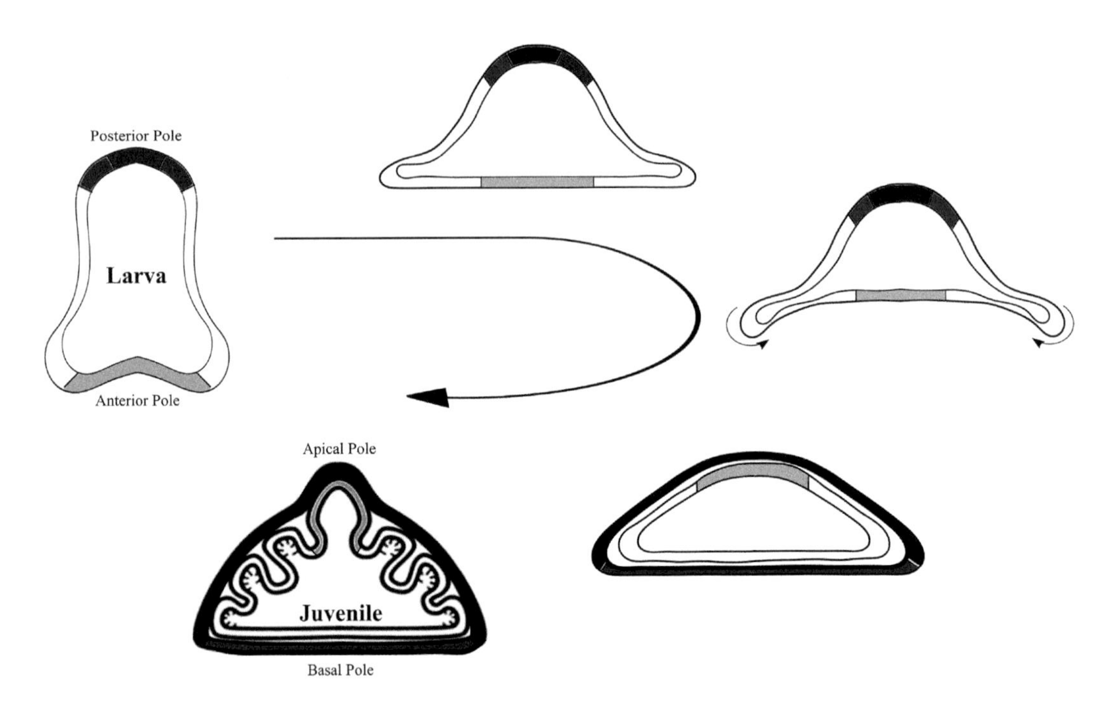

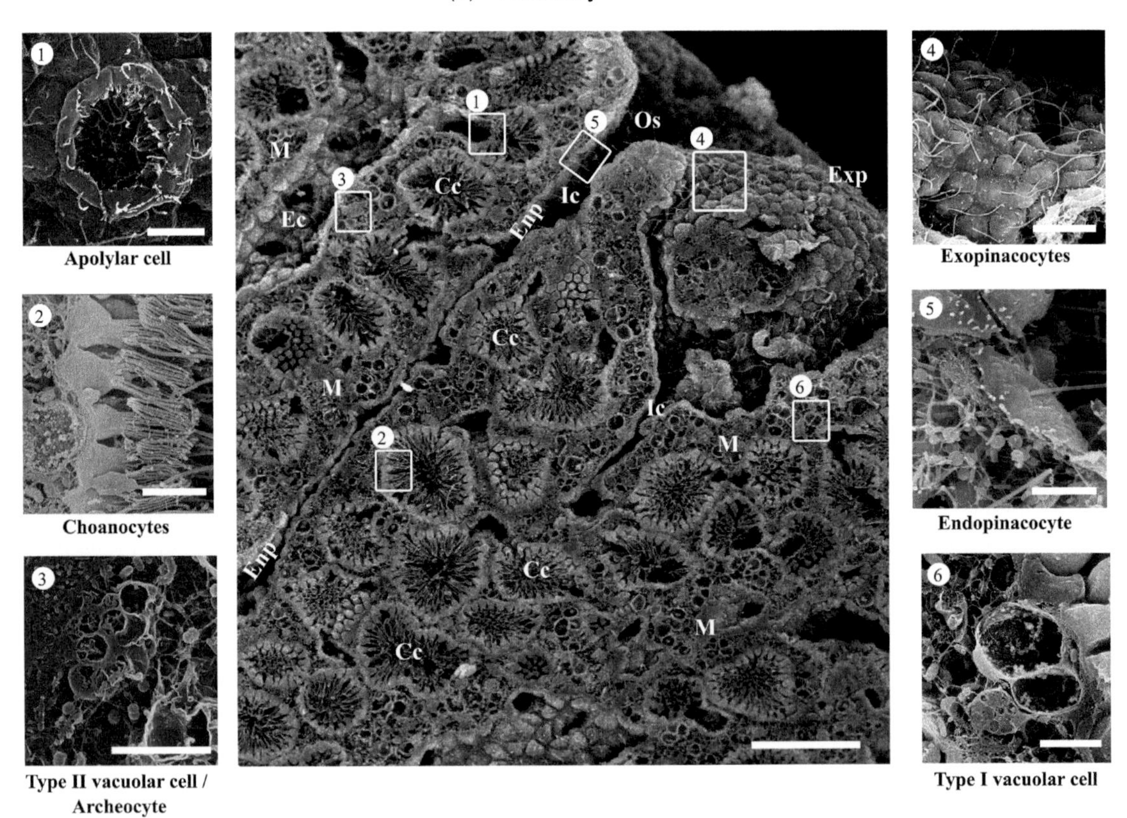

FIGURE 5.3 (a) Schematic of the steps occurring during the metamorphosis process from the free cinctoblastula larva to the settled juvenile (rhagon). The cells of the posterior pole and the posterio-lateral cells are indicated in black and dark gray. The cells of the anterior pole and the anterio-lateral cells are indicated in light gray and white. (b) Anatomy of *Oscarella lobularis* at the adult stage observed on scanning electron microscopy (SEM) sections: respective position of the mesohyl and the main parts of the aquiferous system and of the different cell types. Scale bar represents 43 μm. Scale bars: 8.6 μm (1); 5 μm (2); 7.5 μm (3); 13.6 μm (4); 4.3 μm (5, 6). Cc: choanocyte chamber; Ec: exhalant canal; Enp: endopinacoderm; Exp: exopinacoderm; Ic: inhalant canal; M: mesohyl; Os: ostium.

in the formation of sessile juveniles of *Oscarella lobularis* with a clear baso-apical polarity and a simple but functional aquiferous system. Juveniles differ in size (small, about 2 mm in length and height), color (whitish) and shape (more or less conic instead of asymmetric and multilobated) compared to adults but harbor the same main features as observed at the adult stage (Figure 5.3b).

As is the case for all other sponges, whatever their class, the adult stage of *O. lobularis* is devoid of organs, with no neuron, no muscle and no digestive cavity. *O. lobularis* adults, like most other sponges (except in the case of carnivorous demosponges; Vacelet and Boury-Esnault 1995) are sessile filter-feeders organized around a circulatory aquiferous system with a sylleibid organization (Figure 5.3b). Water flow enters through the incurrent or inhalant pores, named ostia, and is transported via the inhalant canals to the choanocyte chambers. In the choanocyte chambers, the beating of choanocyte flagella is responsible for the internal water flow, and the apical microvilli collar of choanocytes capture unicellular organisms. Trapped food particles are then phagocytized by choanocytes. The filtered water leaves choanocyte chambers via exhalant or excurrent canals and finally exits from the sponge by a large exhalant tube named the osculum.

The tissues of *Oscarella lobularis* consist of two epithelial cell layers: the pinacoderm and the choanoderm. These two layers rest on a basement membrane composed of type IV collagen (and probably of tenascin and laminin as well, as suggested in the sister species *O. tuberculata*; Humbert-David and Garrone 1993), and the epithelial cells are connected by junctions histologically similar to adherens junctions, like in the larvae (Boury-Esnault et al. 2003; Boute et al. 1996; Ereskovsky 2010; Ereskovsky et al. 2007; Ereskovsky et al. 2009b; Leys and Riesgo 2012; Leys et al. 2009) and like in buds (Rocher et al. 2020). Between these two epithelial layers, there is a loose mesenchymal layer, the mesohyl.

5.5.1 The Pinacoderm

In *Oscarella lobularis*, the pinacoderm is composed of pinacocytes organized in a monolayered squamous ciliated epithelium (Figure 5.3b). This epithelium is covered by glycocalyx and mucus layers secreted by pinacocytes. Depending on their localization, different types of pinacocytes are distinguished: the endopinacocytes line all inhalant and exhalant canals, the exopinacocytes compose the outermost layer of the body and the basopinacocytes are involved in the attachment to the substratum. According to the previously described embryology of this species, basopinacoderm and exopinacoderm originate from the same external layer of the rhagon, whereas endopinacoderm originate from the inner one.

In adults, no study has examined whether the pinacocyte cilia are motile or non-motile; in contrast, the beating of exopinacocyte cilia has been evidenced at the bud stage (Rocher et al. 2020). The authors demonstrate that a directional flow of particles (microfluorescent beads in that case) on the surface of the body is directly correlated with the exopinacocyte cilia beating. Indeed, a nocodazole treatment, well known to be a microtubule inhibitor, stops both cilia beating and the bead flow. We can extrapolate that a similar process acts at the adult stage and that the directional flow of particles (probably trapped by the external mucus) may help their convergence to the ostia and hence their absorption in the aquiferous system. Such a mechanism is akin to the ciliary-mucoid feeding process described in other suspension feeder animals (Riisgård and Larsen 2017). This hypothesis still remains to be tested by live physiological experiments.

5.5.2 The Choanoderm

In *Oscarella lobularis*, as in other sponges with leuconoid or sylleibid aquiferous systems, the choanoderm is organized in a multitude of hollow spheres named choanocyte chambers (Figure 5.3b). The choanoderm is formed by a cell type, the choanocyte, the key player of water filtration thanks to its typical microvilli collar and flagellum (whose orthology with choanoflagellate cells has been debated; Adamska 2016; Brunet and King 2017; Colgren and Nichols 2019; Dunn et al. 2015; King 2004; Laundon et al. 2019; Mah et al. 2014; Maldonado 2004; Nielsen 2008; Pozdnyakov et al. 2017; Sogabe et al. 2019). Like the pinacoderm, the choanoderm is a monolayered epithelium. In contrast to pinacocytes, choanocytes are conic cells. The filtering activity has been shown to be an active process in the bud, based on flagella beating, the arrest of beating (by nocodazole) resulting in the absence of particle absorption (Rocher et al. 2020). This observation is easily transposable to the adult stage because of previous studies in other sponges (Leys and Hill 2012; Leys et al. 2011; Ludeman et al. 2017). As in demosponges, choanocytes, even though they are a highly specialized cell type, have stemness properties: dividing activity, expression of GMP genes and capability of transdifferentiation into other cell types (Alié et al. 2015; Borisenko et al. 2015; Fierro-Constaín et al. 2017; Funayama 2013; Funayama 2018; Funayama et al. 2010; Sogabe et al. 2016).

The choanocyte chambers have large openings (eurypylous choanocyte chambers), and the opening toward exhalant canals is surrounded by a particular type of cell, named apopylar cells, which harbors an intermediate morphology between endopinacocytes and choanocytes. This cell type has been supposed to play an important role in controlling water flow in the aquiferous system (Hammel and Nickel 2014; Leys and Hill 2012).

5.5.3 The Mesohyl

The mesohyl is a mesenchymal layer. It is the inner part of the sponge body, never in direct contact with the water flow. Extracellular matrix is the main component of this layer. Extracellular bacteria are found in this internal compartment. Studies carried out by transmission electronic microscopy (TEM), by denaturating gradient gel electrophoresis (DGGE) or by 16S sequencing have shown that *O. lobularis* is a low microbial abundance (LMA) sponge (Gloeckner et al. 2013; Vishnyakov and Ereskovsky 2009). Its phylum-level microbial diversity is represented by three bacterial phyla with a large dominance (76%) of Proteobacteria. Phylogenetic analysis revealed four sequences affiliated with Verrucomicrobia, three with Gammaproteobacteria and two sequences with Bacteroidetes, and the 16 remaining sequences were affiliated with Alphaproteobacteria. Moreover, microbial diversity is neither significantly different between color morphs nor between individuals of different locations or depths (Gerçe et al. 2011; Gloeckner et al. 2013). More recently, metagenomic analyses suggest that the main bacterial symbiot of *O. lobularis* is an Alphaproteobacteria of the Rhodobacteriaceae family. This new species was named *Candidatus Rhodobacter lobularis*, it is about 20-fold more numerous than sponge cells in the mesohyl and its draft genome is available (Jourda et al. 2015). Even though no physiological studies have yet been performed to identify the mutual benefits of this association, members of the *Rhodobacter* group often perform aerobic anoxygenic photoheterotrophy (Labrenz et al. 2009; Pohlner et al. 2019; Sorokin et al. 2005); we therefore suggest that hosting such *Rhodobacter* species may supply *O. lobularis* with carbon.

In addition, several sponge cell types are present in the mesohyl (Figure 5.3b). Classically, photonic and electronic observations have defined two cell types: type I vacuolar and type II vacuolar (Boury-Esnault et al. 1992; Ereskovsky et al. 2009b). Type I vacuolar cells are characterized by two to four large empty vacuoles and a small nucleus placed laterally, and their role is unknown. Type II vacuolar cells are amoeboid cells with numerous filopodia, numerous small vacuoles and a large nucleus with a nucleolus. Because of these cytological features and the fact that these cells express 11 genes of the GMP program, they were supposed to correspond to what are defined as archaeocytes in other sponges (Fierro-Constaín et al. 2017). Comparative single-cell transcriptomic data are now awaited to establish homology between cell types between sponge species and to make clearer the sponge cell type terminology only based on cell morphology (De Vos et al. 1991; Musser et al. 2019; Rocher et al. 2020; Sogabe et al. 2019). Interestingly, the use of scanning electron microscopy and immunofluorescent techniques resulted in the identification of at least one additional cell type in the mesohyl of *O. lobularis* bud: a third vacuolar cell type (Rocher et al. 2020). Additionally, numerous, previously undescribed, tiny anucleate cell-like structures were interpreted as apoptotic extracellular vesicles (EVs) (Rocher et al. 2020). Because buds originate directly from adult tissues (see previous sections), we do not believe that these type III vacuolar cells and EVs are bud specific but rather that they were not observed on adults until now because of technical limitations. These recent findings highlight the subjectivity of cell type definition, and again, much is expected from ongoing single cell transcriptomic approaches to define cell types on the basis of a shared regulatory network (Arendt et al. 2016).

5.6 TRANSCRIPTOMIC AND GENOMIC DATA

Since the first genome of the demosponge *Amphimedon queenslandica* was published (Srivastava et al. 2010), sponge genomic resources have significantly increased (for review, see Renard et al. 2018 and references included, plus Kenny et al. 2020). These data revealed that sponges have a genome size and number of genes comparable to those of most invertebrates. In addition, these studies indicate striking genome feature differences between sponge species even within the same class: differences in predicted genome size (from 57 to 357 Mb) in agreement with very variable DNA content evidenced by old cytogenetic approaches; differences in ploidy (diploidy or probable tetraploidy in Calcarea), amount and length of non-coding regions and genes present, among others (Kenny et al. 2020; Renard et al. 2018; Santini et al. in prep).

Concerning *Oscarella lobularis*, a genome draft was sequenced with illumina technology (Belahbib et al. 2018); ongoing additional sequencing efforts are expected to improve the assembly of this genome in a near future. At present, the predicted length of the genome of *O. lobularis* is 52.34 Mb (Belahbib et al. 2018); this is even smaller than what was predicted for *O. pearsei* (57.7 Mb; Nichols et al. 2006). If confirmed when a better assembly is obtained, this genome would represent the smallest sponge genome reported so far. This genome is predicted to contain 17,885 protein-coding genes (Belahbib et al. 2018). This is surprisingly low compared to demosponges: *Ephydatia muelleri* is supposed to harbor 39,245 protein-coding genes (Kenny et al. 2020), *Amphimedon queenslandica* 40,122 (Fernandez-Valverde et al. 2015) and *Tethya wilhelma* 37,416 (Francis et al. 2017). We are expecting a better genome assembly for both *O. pearsei* and *O. lobularis* in order to be able to decipher whether these small genome sizes and low numbers of genes are due to sequencing pitfalls or represent a common feature of Oscarellidae genomes.

To date, only one study has used this genomic data to compare epithelial genes of *O. lobularis* to other sponges (Belahbib et al. 2018). All other comparative molecular studies published so far were either based on PCR approaches (Gazave et al. 2008; Lapébie et al. 2009) or on transcriptomic data obtained by 454 sequencing technology performed on a mixture of developmental stages (adult, embryos and larvae) to maximize the representativity of this transcriptome (Fierro-Constaín et al. 2017; Schenkelaars et al. 2015; Schenkelaars et al. 2016a).

These transcriptomic and genomic studies published thus far have focused on genes involved in epithelial functions, in Notch and WNT signaling and genes pertaining to the GMP. As far as the GMP and the canonical WNT pathways are concerned, genes present in *O. lobularis* are not different from what is found in other sponge classes (Fierro-Constaín et al. 2017; Lapebie 2010; Lapébie et al. 2009; Schenkelaars 2015; Schenkelaars et al. 2015). When comparisons are made at the level of gene content only, *O. lobularis*, like all other sponges, possesses all nine genes coding for proteins involved in the establishment of the CRUMBS, PAR and SCRIBBLE complexes of bilaterians needed to establish cell polarity, as well as all three genes encoding proteins needed to establish the cadherin-catenin complex (CCC) required for the formation of adherens junctions (namely alpha, beta and delta catenins as well as classical cadherin) (Belahbib et al. 2018). However, key functional domains and motif sequences are amazingly more conserved in *O. lobularis* than they are in other sponge classes. For example, PatJ protein (one of the three components of the crumbs polarity complex containing Crumbs, MPP5 and PatJ) binds MPP5 via the L27 domain: The L27 domain sequence is more conserved in *O. lobularis* compared to the other sponges (Belahbib et al. 2018). It is the same for cadherin/β-catenin/α-catenin complex. The comparison of the E-cadherin cytoplasmic tail, which contains the conserved specific binding domain for delta-catenin and β-catenin, is more conserved in *O. lobularis* than in other sponges relative to bilaterian sequences (Belahbib et al. 2018).

Concerning pathways commonly involved in epithelial patterning, it was shown that *O. lobularis* possesses all the core gene encoding for proteins needed to establish a planar cell polarity (PCP) pathway. Indeed, Strabismus (Stbm)/Van Gogh (Vang), Flamingo (Fmi), Prickle (Pk), Dishevelled (Dsh) and frizzled (Fzd) proteins are present in *O. lobularis* (Schenkelaars 2015; Schenkelaars et al. 2016a), whereas other sponges lack either one or several members of this pathway (Fmi, Fzd and/or Vang) (Schenkelaars 2015; Schenkelaars et al. 2016a) (Figure 5.4a). This finding challenged previous studies in Ctenophora and Porifera suggesting that the PCP pathway arose in the last common ancestor of Parahoxozoa (Bilateria, Cnidaria and Placozoa) (Adamska et al. 2010; Ryan et al. 2013), meaning that the PCP pathway may date back to the emergence of Metazoa. This unexpected result calls for functional studies in *O. lobularis*: Is this pathway involved in the coordination and orientation of exopinacocyte cilia (Figure 5.4b and c) in the same way it is in other animals (Devenport 2014; Schenkelaars et al. 2016a; Wallingford 2010)?

Other key genes considered absent in other sponge classes are present in *O. lobularis*. This is notably the case for the *Hes* gene belonging to group E bHLH transcription factors. To date, only *Hey* genes have been reported in Demospongiae and Calcarea (Fortunato et al. 2016; Simionato et al. 2007; Srivastava et al. 2010); in contrast, *O. pearsei* and *O. lobularis* possess *bona fide Hes* (Gazave 2010; Gazave et al. 2014) (Figure 5.4d). This means that this gene was ancestrally present in the last common ancestor of Porifera and was lost in other sponge classes. This finding offers additional possibilities to test the respective roles of canonical and non-canonical Notch signaling pathways in Metazoa and notably to explore the role of Notch signaling in animals devoid of neurons (Layden et al. 2013).

Despite a lack of neurons and conventional neurotransmitters, sponges perceive and respond to a large range of stimuli. In animals, Glutamate is the principal excitatory neurotransmitter in the central nervous system. All sponges have a number of metabotropic glutamate (mGlu) and GABA receptors, suggesting that glutamatergic signaling is common in sponges (Leys et al. 2019). In contrast, the ionotropic glutamate receptor iGluR gene is found only in calcareous sponges and homoscleromorphs (Figure 5.4e) (Ramos-Vicente et al. 2018; Renard et al. 2018; Stroebel and Paoletti 2020). However, the localization and function of these receptors remain to be identified in these animals devoid of neurons and synapses.

Much remains to explore in the transcriptome and genome of *O. lobularis*; nevertheless, according to the present knowledge, compared to other sponge classes, the homoscleromorph sponges *O. pearsei* and *O. lobularis* seem to exhibit the most complete and conserved bilaterian gene repertoire (Babonis and Martindale 2017; Fortunato et al. 2015; Gazave et al. 2014; Renard et al. 2018; Riesgo et al. 2014; Schenkelaars et al. 2016a).

5.7 FUNCTIONAL APPROACHES: TOOLS FOR MOLECULAR AND CELLULAR ANALYSES

5.7.1 Developmental and Non-Developmental Morphogenetic Contexts Accessible

Embryos and larvae are accessible only once a year between August and October, and the reproductive effort is variable from one population to another and from one year to another (Ereskovsky et al. 2013a). Therefore, because sexual reproduction cannot be triggered in the laboratory, so far, the access to embryonic developmental processes remains very limited.

To compensate for this difficulty, experimental protocols were designed to access non-developmental morphogenetic processes (Table 5.2). Wound healing experiments have already been successfully used at the adult stage (Ereskovsky et al. 2015; Fierro-Constaín et al. 2017); wound-healing and regenerative experiments are now also mastered at the bud stage: stage 3 with osculum regenerates an osculum in less than four days (Rocher et al. 2020); cell-dissociation/reaggregation experiments resulting in neo-epithelialized primmorphs (in less than three days) can be performed both on adults (unpublished data) and on buds (Rocher et al. 2020; Vernale et al. in press).

5.7.2 Polymerase Chain Reaction and Relatives

As in the case of other sponge species studied for evo-devo purposes, the first molecular studies undergone on *Oscarella lobularis* were performed using the polymerase chain reaction (PCR) technique. This resulted in the description of the phylogenetic relationships among Homoscleromorpha (described in the first section), including *O. lobularis* (Borchiellini et al. 2004; Gazave et al. 2010; Gazave et al. 2012; Gazave et al. 2013). The main pitfall faced during this simple classical PCR/cloning/sequencing was at the step of DNA extraction. For *O. lobularis*, as for other

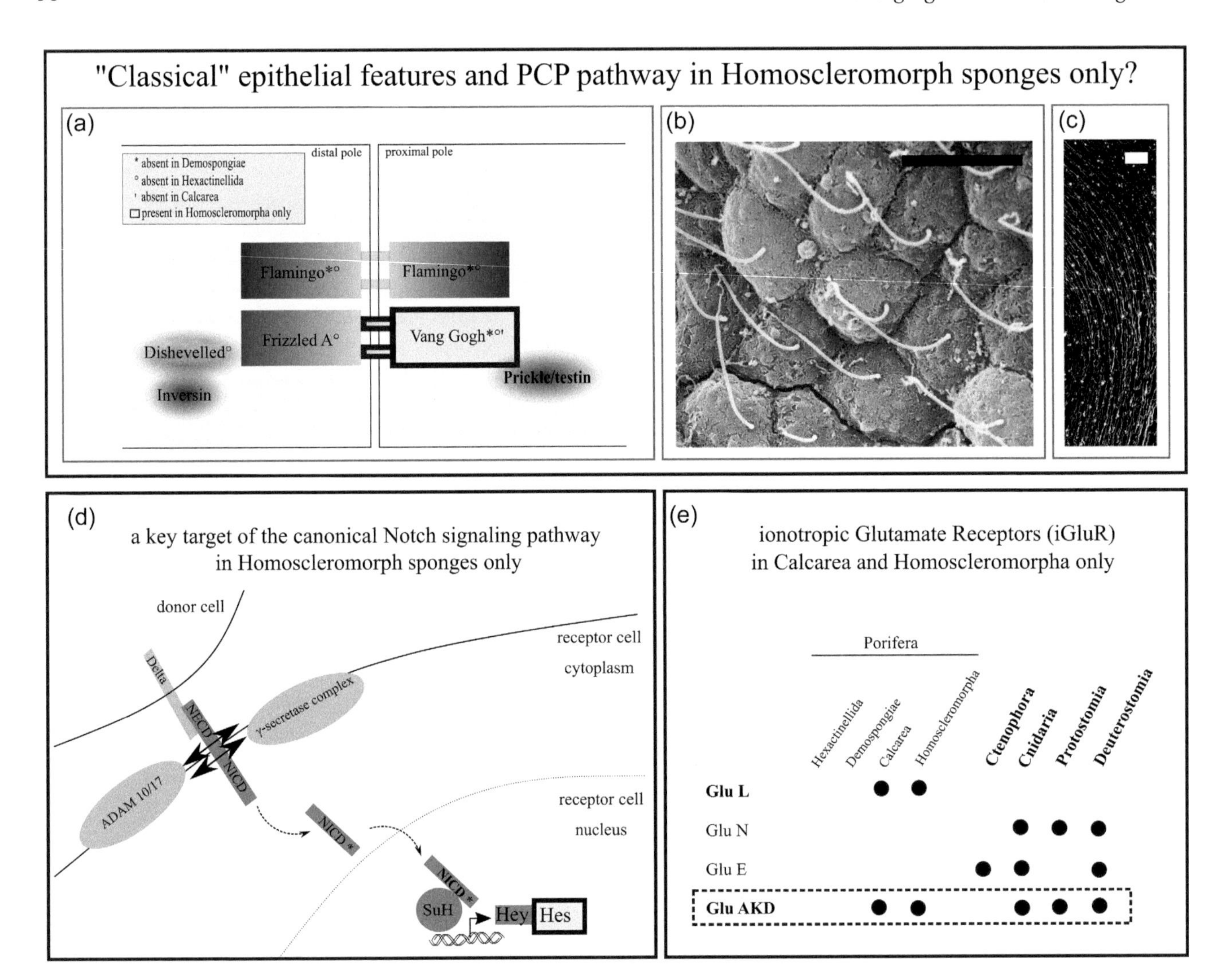

FIGURE 5.4 (a) The planar cell polarity pathway is involved in the establishment of a polarity between neighboring cells; the core members of this pathway are represented in this diagram: only Homoscleromorpha possess all these core members, whereas other sponge classes lack one to three of them. (b) Exopinacoderm of *Oscarella lobularis* showing cilia orientation. Scale bar: 10 µm. (c) The oriented beating of the cilia on the exopinacoderm was evidenced at the bud stage thanks to the monitoring of fluorescent beads. Scale bar: 50 µm. (d) Diagram of the core components of the canonical Notch signaling pathway conserved in sponges, *Hes*, was so far evidenced in the *Oscarella* genus only. (e) The ionotropic Glutamate receptors (iGluR) are split into four families (in Metazoa): the Glu L family is sponge specific, the Glu E family gathers all ctenophore *iGluRs* and genes present in cnidarians and deuterostomes and *GluN* genes are characterized in ctenophores and sponges but are found in all cnidarians and bilaterians, whereas the Glu AKD family is present from sponges to vertebrates (except ctenophores). Among sponge classes, Homoscleromorpha and Calcarea only have iGluR receptors. ([a] Schenkelaars et al. 2016a; [c] Rocher et al. 2020; [d] Fortunato et al. 2016; Gazave 2010; Gazave and Renard 2010; Gazave et al. 2009; Gazave et al. 2014; Simionato et al. 2007; [e] Stroebel and Paoletti 2020.)

Oscarella species, ethanol preservation of samples resulted in improving PCR results, probably by limiting the presence of pigments and secondary metabolites (Boury-Esnault et al. 2013; Ivanišević et al. 2011) in the tissues that might interfere with the PCR.

In parallel, a degenerated primer approach was used to search for sequences of homeobox genes encoding for transcription factors of the antennapedia (ANTP) class. This approach failed to retrieve the famous *hox* genes, as in other sponges, but *NK*-related genes were characterized (Gazave et al. 2008). Because of the usually high sequence divergence between sponge and bilaterian sequences, this PCR-based approach had low efficiency. The acquisition of expression sequence tag (EST) libraries (Lapébie et al. 2009; Philippe et al. 2009) and of a 454 transcriptome effectively made finding a candidate gene much easier (Fierro-Constain 2016; Gazave 2010; Lapebie 2010; Schenkelaars 2015). As far as PCR techniques are concerned, real-time PCR (or quantitative RTPCR [RT for reverse transcription]) was launched more recently, thereby providing the possibility of studying the expression of several genes in various conditions (Fierro-Constaín et al. 2017). For this sponge species, the mitochondrial gene *Cytochrome Oxidase subunit 1* (*CO I*) and the nuclear genes *Elongation Factor 1* (*EF1*) and *glyceraldehyde-3-phosphate dehydrogenase* (*GAPDH*) are effective reference genes, because they have stable expression during their life cycle (Fierro-Constaín et al. 2017) but also under contaminant exposure conditions (de Pao Mendonca, *unpublished data*).

5.7.3 *In Situ* Hybridization

The *in situ* hybridization (ISH) technique is also mastered in *Oscarella lobularis*, thereby allowing access to qualitative data (localization) in addition to quantitative expression gene information provided by the previously mentioned real-time PCR. The first ISH data were acquired in 2008 (Gazave et al. 2008; Lapébie et al. 2009), and the protocol was subsequently improved (Fierro-Constaín et al. 2017; Fierro-Constaín et al. 2021). The ISH can be performed at all stages (adult, bud, larvae) on sections or in whole mounts. Fluorescent ISH (FISH) is also in progress (Prünster, *unpublished data*). For colorimetric ISH, 5-brom-4-chloro-3'-indolyphosphate p-toluidine salt/nitro blue tetrazolium chloride (BCIP/NBT) was successfully used as a chromogenic phosphatase substrate for the detection of alkaline phosphatase labeled probes (with better results than BM-purple, for example). The automating of the whole mount ISH on an Intavis pro device increased the output and replicability of the technique (detailed protocols provided in Fierro-Constaín et al. 2021).

5.7.4 Fluorescent Immunolocalization

Fluorescent immunolocalization (IF) can be performed either on paraffin sections of adults and buds (*unpublished data*) or on whole mount on buds thanks to their transparency (Rocher et al. 2020) (Table 5.2). Unsurprisingly, the use of paraffin sections not only takes much longer to achieve but also can result in losing antigenic reactivity, as often observed in other tissues (Krenacs et al. 2010); for this reason, most IF experiments are performed on whole mounted buds or juveniles. The IF protocol used in *O. lobularis* buds is a classical one (Rocher et al. 2020; detailed protocol provided in Borchiellini et al. 2021). Nevertheless, the main difficulty faced is the divergence of sponge antigen sequences relative to vertebrate antigens. Most commercialized antibodies, designed against vertebrate proteins, are therefore unusable, except for highly conserved proteins. For instance, we successfully used antibodies against alpha-tubulin (Sigma) and acetylated alpha-tubulin (Sigma), phospho-histone H3 (Abcam) (Rocher et al. 2020). For other proteins, specific antibodies were raised against peptides of interest, for example, against type IV collagen (Rocher et al. 2020; Vernale et al. in press); other specific antibodies are currently under testing.

5.7.5 Cell Viability, Cell Apoptosis and Cell Proliferation Assays

During the study of morphogenetic processes or for ecotoxicological purposes, being able to measure and compare cell viability and cell activity can be useful.

Cell viability/death can be estimated very quickly (a couple of minutes) in *O. lobularis* (Table 5.2), on both dissociated cells (see next sections) or whole buds, by using propidium iodide (PI) staining on dead cell nuclei in orange and fluorescein diacetate (FA) staining on live cell cytoplasm

TABLE 5.2
Tools for Cellular and Molecular Analyses Available in *Oscarella lobularis*

Resources/Techniques	Availability in O. lobularis	References
Transcriptome	X	For review Renard et al. (2018)
Mitochondrial genome	X	Gazave et al. (2010)
Genome	IP*	Belahbib et al. (2018), Renard et al. (2018)
Single-cell transcriptome	IP*	
PCR, real-time PCR	X	Gazave et al. (2008, 2010, 2013), Fierro-Constaín et al. (2017)
In situ hybridization	X	Gazave et al. (2008), Lapébie et al. (2009), Fierro-Constaín et al. (2017), Fierro-Constaín et al. (2021)
Section 1.01 Immunolocalization	X	Boute et al. (1996), Rocher et al. (2020)
RNA interference		Rocher et al. (2020)
Morpholino		Rocher et al. (2020)
Plasmid expression		Rocher et al. (2020)
Pharmacological approach	X	Lapébie et al. (2009)
Cell proliferation assays	X	Ereskovsky et al. (2015), Rocher et al. (2020)
Cell death assays	X	Rocher et al. (2020)
Cell staining methods	X	Ereskovsky et al. (2015), Rocher et al. (2020), Borchiellini et al. (2021)
Wound healing	X	Ereskovsky et al. (2015), Rocher et al. (2020)
Regeneration	X	Rocher et al. (2020)
Cell dissociation/reaggreagation	X	Rocher et al. (2020), Vernale et al. (in press)

* IP = in progress

in green, following Sipkema's protocol (Rocher et al. 2020; Sipkema et al. 2004). As for other sponges, Trypan blue assays were not successful.

TUNEL is a classical method for detecting DNA fragmentation, used to quantify apoptotic cells, and EdU technology is also a classical way to estimate the rate of DNA synthesis (Gorczyca et al. 1992; Salic and Mitchison 2008). Both methods are now mastered on buds of *O. lobularis*, and EdU assays were also performed successfully on adult sections during wound healing and on buds of different stages (Ereskovsky et al. 2015; Rocher et al. 2020; detailed protocol in Borchiellini et al. 2021). EdU provides more readable information than antibodies against Phospho-histone H3 to estimate cell proliferation because of the low rate of cell division at that stage.

5.7.6 Cell Staining and Tracking

All embryogenetic and morphogenetic processes in *O. lobularis* were so far described on fixed samples and therefore on the interpretation of static pictures (Boury-Esnault et al. 2003; Ereskovsky and Boury-Esnault 2002; Ereskovsky et al. 2007; Ereskovsky et al. 2013a; Ereskovsky et al. 2015; Rocher et al. 2020). As mentioned in Section 5.4, this type of description results in an incomplete understanding of events occurring during the time course of the morphogenetic process. Therefore, means to stain and track cells are now under development (Table 5.2). Buds, again, because of their abundance and transparency, are suitable to test such techniques. In order to monitor epithelial morphogenesis, means to stain and track choanocytes (choanoderm epithelium) and pinacocytes (exo- and endopinacoderm epithelia) have been the subject of research. Choanocytes can be efficiently and specifically stained by using lipidic markers (CM-DiI dye), by labeled lectins (*Pha*E, *Gsl* 1 for instance) or by using their capacity of particle phagocytosis (Indian ink or fluorescent microbeads) (Ereskovsky et al. 2015; Rocher et al. 2020). Because these are non-toxic staining methods, they allow cell tracking along the time course of the process for several hours or days (Indian ink and lectins allow cell tracking after up to five days) (Ereskovsky et al. 2015; Rocher et al. 2020; Vernale et al. in press). A short incubation with wheat germ agglutinin (WGA) was also used to stain exo- and endopinacocytes (Rocher et al. 2020). Unfortunately, at present, no staining methods are available to stain embryo blastomeres or bud mesohylar cells. Because of pigmentation, an adult is much less suitable to perform live cell staining and tracking.

5.7.7 Loss-of-Function Approaches

Loss-of-function (LOF) approaches are required to study gene functions (Weiss et al. 2007; Zimmer et al. 2019). The first way to interact with gene functions was to use pharmacological approaches via small-molecule inhibition, but more recently, other knockdown (morpholino- and RNAi-mediated methods) and knockout (TALEN- and CRISPR/Cas9-mediated methods) techniques have been developed and are used successfully in various model organisms. Among Porifera, both pharmacological and RNAi techniques are so far mastered in the demosponge *Ephydatia muellieri* only (Hall et al. 2019; Rivera et al. 2011; Rivera et al. 2013; Schenkelaars et al. 2016b; Schippers et al. 2018; Windsor Reid and Leys 2010; Windsor Reid et al. 2018; http://edenrcn.com/protocols/#invertebrate), and a Crispr-Cas12 approach is recently developed in *Geodia* (Hesp et al. 2020). In *Oscarella lobularis*, pharmacological approaches were performed successfully and allowed to interfere with WNT signaling (Table 5.2). This approach showed that WNT signaling is involved in epithelial morphogenetic processes in *O. lobularis*, as is the case in other animals (Lapébie et al. 2009).

More recently, siRNA and morpholino molecules were efficiently transfected into choanocytes (Rocher et al. 2020). Nevertheless, to date, there is neither evidence of interference efficiency (with transcription and transduction, respectively) nor of phenotypic effect. This is presently the main challenging objective *O. lobularis* must reach to become a *bona fide* model organism, as is also the case for the famous marine demosponge *Amphimedon queenslandica*.

5.8 CHALLENGING QUESTIONS BOTH IN ACADEMIC AND APPLIED RESEARCH

5.8.1 Finding New Bioactive Secondary Metabolites

The pharmaceutical research field is still searching for new natural drug candidates. Among marine organisms, marine sponges represent one of the most important sources of diverse natural chemicals with potential therapeutic properties (Ancheeva et al. 2017; Genta-Jouve and Thomas 2012; Rane et al. 2014; Santhanam et al. 2018; Zhang et al. 2017). Indeed, most sponge species synthesize secondary metabolites, and this is interpreted to play a major role in these sessile animals as chemical defense against predators, overgrowth by other organisms and competition for space (Proksch 1994). Studies aiming to characterize these natural compounds therefore represent one of the main domains of applied research performed on sponges. Oscarellidae species have received less attention for this purpose until recently (Ivanišević et al. 2011). Among them, *Oscarella* species, in particular *O. lobularis*, display a high diversity of apolar compounds (Aiello et al. 1990; Aiello et al. 1991; Cimino et al. 1975; Ivanišević et al. 2011). *Oscarella* species are the most bioactive species compared to other homoscleromorph sponges: the EC50 values (measured on crude extract effect on the metabolism of the bioluminescent bacterium *Vibrio fischeri*) range from 36 to 111 µg/mL (61 µg/mL for *O. lobularis*). The authors suggest a correlation between the secondary metabolite diversity and the estimated bioactivity (Ivanišević et al. 2011). Lysophospholipids (lyso-PAF and LPE C20:2) are the major metabolites identified in *O. lobularis* (also found in its sister species *O. tuberculata*) (Aiello et al. 1990; Aiello et al. 1991; Cimino et al. 1975; Ivanišević et al. 2011). The origins (from sponge cells or bacterial cells) of these compounds and their individual bioactive properties have not been characterized yet.

5.8.2 Understanding Host–Symbiont Interactions

Thanks to molecular techniques, the microbial community of *Oscarella lobularis* is now well described (Gloeckner et al. 2013; Jourda et al. 2015). It has been described that (at least part of) this microbiont is vertically inherited (from parent to offspring) both during sexual and asexual reproduction (Boury-Esnault et al. 2003; Ereskovsky and Boury-Esnault 2002; Ereskovsky and Tokina 2007; Ereskovsky et al. 2007). But, as for many sponges, the exact nature and mutual benefits of this biotic association are not determined yet and for now remain hypothetical. Because of recent findings on the variation of the bacterial community during the life cycle in other sponges (Fieth et al. 2016), of potential metabolic complementarity between bacteria and the sponge host (Gauthier et al. 2016), evidence of bacteria–sponge horizontal gene transfers (Conaco et al. 2016) and now that metagenomic data are acquired for *O. lobularis*, we should take advantage of these data to explore by experimental approaches the ecological and physiological roles of these associations (resource partitioning/supplying between bacteria and sponge host) but also the potential impact of the microbial community on the developmental processes of the sponge as recently observed in marine cnidarians (Tivey et al. 2020; Ueda et al. 2016).

5.8.3 Deciphering the Origin and Evolution of Metazoan Epithelia

Epithelia are considered one of the four fundamental tissue types of animals (Edelblum and Turner 2015; Lowe and Anderson 2015; Yathish and Grace 2018). Epithelia cover body surfaces, organs and internal cavities, and they are essential for controlling permeability and selective exchanges between internal and external environments and between the different compartments of a body. Epithelia are patterned at the end of cleavage during embryological development (Gilbert and Barresi 2018; Tyler 2003) (see Section 5.4).

Epithelia are layers of cells defined by three main histological features, according to what is observed in bilaterians: cell polarity, lateral junctions and a basal lamina made of collagen IV (Edelblum and Turner 2015; Lowe and Anderson 2015; Tyler 2003; Renard et al. 2021). Until 1996 (Boute et al. 1996), no sponge species were known to possess all three features; sponges were therefore considered devoid of epithelia. Among sponges, Hexactinellida do not have cell layers but syncytia instead; Demospongiae and Calcarea have cell layers with cell polarity, atypical cell junctions but no basement membrane; in contrast, Homoscleromorpha possess clear cell polarity, unequivocal adherens-like junctions and obvious basement membrane. Whereas the cell layers of demosponges have similar mechanical and physiological properties like bilaterian epithelia, the epithelia of homoscleromorph sponges are the only ones that present similar histological features compared to bilaterians (Ereskovsky 2010; Ereskovsky et al. 2009b; Leys and Hill 2012; Leys and Riesgo 2012; Leys et al. 2009; Renard et al. 2021). For a while, this "true epithelium" was interpreted as a synapomorphy of Homoscleromorpha and Eumetazoa (Borchiellini et al. 2001; Sperling et al. 2007) and suggested the inclusion of Homoscleromorpha in the Epitheliozoa lineage (a clade combining Eumetazoa and Placozoa) (Sperling et al. 2009). The monophyly of Porifera, now supported by numerous phylogenomic analyses (Philippe et al. 2009; Pick et al. 2010; Pisani et al. 2015; Redmond et al. 2013; Simion et al. 2017; Thacker et al. 2013; Whelan et al. 2017; Wörheide et al. 2012), means instead that the last common ancestor of Porifera possessed all three classical features of "typical" epithelia and that some of these features were secondarily lost independently in the three other sponge classes.

Interestingly, whether species present all epithelial features or not, all sponge classes possess the same set of epithelial genes involved in the establishment of cell polarity and the composition of adherens junctions (Belahbib et al. 2018; Renard et al. 2018; Riesgo et al. 2014; Renard et al. 2021). Similar inconsistency between gene content and histological features was reported concerning the basal lamina (Fidler et al. 2017). These findings question the homology of epithelial features between sponges and other animals: Is polarity controlled by the same three polarity complexes as in bilaterians (namely Crumbs, Par and Scribble)? Are adherens junctions described in Homoscleromorpha homologous to bilaterian adherens junctions (i.e. composed of classical cadherin and alpha-beta and delta-catenins)? To answer these questions, complementary molecular and biochemical approaches are in progress in both *O. pearsei* and *O. lobularis* and in parallel in demosponges. The first results obtained suggest that the proteins involved in cell–cell and cell–matrix adhesion would be the same in demosponges and homoscleromorphs, in particular vinculin and beta-catenin (Miller et al. 2018; Mitchell and Nichols 2019; Schippers et al. 2018). To date, there is no clear information concerning the eventual implication of classical cadherins in these junctions.

5.8.4 Sponge Gastrulation and the Origin of Germ Layers

Despite the true multicellular and metazoan nature of sponges having been elucidated decades ago (reviewed in Schenkelaars et al. 2019), there is a longstanding debate in the spongiologist community on whether sponges gastrulate. Different points of view compete: i) for some authors, multipolar egression leading to the formation of the coeloblastula during embryogenesis marks the onset of polarization and regionalization processes, suggesting it may be similar to gastrulation (Maldonado and Riesgo 2008); ii) others consider that this process differs from gastrulation in that the resulting embryo apparently consists of one uniform cell layer and lacks polarity (Ereskovsky 2010; Ereskovsky and Dondua 2006) and prefer to hypothesize the gastrulation during larval metamorphosis (reviewed in Ereskovsky 2010;

Ereskovsky et al. 2013b; Lanna 2015; Leys 2004; Wörheide et al. 2012), when an "inversion of germ layers" results in the formation of the aquiferous system. In the last case, the term "inversion" means that external-most larval cells form the internal-most ("gut-like") structures of an adult sponge, namely the aquiferous system.

However, cellular tracking during the larval metamorphosis in *Amphimedon queenslandica* has shown no relation between larval and juvenile cell layers; the cells of the larvae do not have specification: all larval cell types are capable of transdifferentiating into all juvenile cell types (Nakanishi et al. 2014; Sogabe et al. 2016). This apparent lack of cell layer and fate determination and stability during metamorphosis in this sponge argues for an absence of gastrulation. In this context, the expression of the transcription factor *GATA*, a highly conserved eumetazoan endomesodermal marker, in the inner layer of *A. queenslandica* embryos, free larvae and juveniles has been interpreted to provide positional information to cells (Nakanishi et al. 2014). In contrast, in *Sycon ciliatum*, expression of the same marker in embryo/larva ciliated micromeres (at the origin of adult choanocytes) and in adult choanoderm has given rise to other conclusions (Leininger et al. 2014). Indeed, the authors suggest that the calcareous sponge choanoderm and the bilaterian endoderm are homologous structures and ciliated choanocytes are germ layers. Thus, the origin of gastrulation and germ layers is still controversial (Degnan et al. 2015; Lanna 2015). Yet the resolution of this problem is the key to comparing embryological stages between sponges and other metazoans and to discussing germ layer homology between all animal phyla.

As mentioned in the section on embryology and in the previous section, *Oscarella lobularis* (like other homoscleromorph) presents clear epithelial characteristics, and all morphogenetic processes (development, regeneration, budding) are based mainly on epithelial morphogenetic movements in contrast to demosponges (Boury-Esnault et al. 2003; Ereskovsky 2010; Ereskovsky and Tokina 2007; Ereskovsky et al. 2007; Ereskovsky et al. 2009b; Ereskovsky et al. 2013a; Ereskovsky et al. 2013b; Ereskovsky et al. 2015). This feature is expected to result in the formation of more stable cell layers during embryogenesis compared to demosponges (Ereskovsky 2010; Lanna 2015). *O. lobularis* is thus an interesting model to answer questions about the homology of embryonic morphogenesis (gastrulation) and germ layers in animals (Degnan et al. 2015; Lanna 2015). The techniques now available in this sponge species (see Section 5.7 on functional approaches) are highly significant innovations to answer this fundamental question. The main experimental limitation to do so is the difficult and limited access to embryos and larvae in this species, as sexual reproduction cannot be triggered in aquaria.

ACKNOWLEDGMENTS

The authors thank all those whose involvement made the development of this emerging model possible: Nicole Boury-Esnault and Jean Vacelet for their help in launching our studies on this species; our PhD students Pascal Lapébie, Eve Gazave, Quentin Schenkelaars, Laura Fierro-Constain, Amélie Vernale and Kassandra de Pao Mendonca; the numerous internship students who helped in performing preliminary experiments; the imaging facilities of the France Bioimaging infrastructure; the diving facilities of the Institute OSU Pytheas and divers from the IMBE lab; and the molecular biology and morphology support services of IMBE. We thank Haley Flom for English editing.

The authors acknowledge the Région Sud (Provence Alpes Côte d'Azur), the French Research ministry, the French National Center for Scientific Research (CNRS), Aix-Marseille University and the Excellence Initiative of Aix-Marseille Université–A*MIDEX for providing funds to support our fundamental research, in particular for the funding of the project for international scientific cooperation (PICS) STraS, the A*MIDEX foundation projects (n° ANR-11-IDEX-0001–02 and AMX-18-INT-021) and the LabEx INFORM (ANR-11-LABX-0054), both funded by the "Investissements d'Avenir" French Government program, managed by the French National Research Agency (ANR). The Russian Science Foundation, Grant n° 17-14-01089.

BIBLIOGRAPHY

Adamska, M. 2016. Sponges as models to study emergence of complex animals. *Curr. Opin. Genet. Dev.* 39, 21–28.

Adamska, M., Degnan, S. M., Green, K. M., Adamski, M., Craigie, A., Larroux, C. and Degnan, B. M. 2007. Wnt and TGF-beta expression in the sponge *Amphimedon queenslandica* and the origin of metazoan embryonic patterning. *PLoS One* 2, e1031.

Adamska, M., Degnan, B. M., Green, K. and Zwafink, C. 2011. What sponges can tell us about the evolution of developmental processes. *Zool. Jena Ger.* 114, 1–10.

Adamska, M., Larroux, C., Adamski, M., Green, K., Lovas, E., Koop, D., Richards, G. S., Zwafink, C. and Degnan, B. M. 2010. Structure and expression of conserved Wnt pathway components in the demosponge *Amphimedon queenslandica*. *Evol. Dev.* 12, 494–518.

Aiello, A., Fattorusso, E., Magno, S., Mayol, L. and Menna, M. 1990. Isolation of two novel 5α,6α-Epoxy-7-ketosterols from the encrusting demospongia *Oscarella lobularis*. *J. Nat. Prod.* 53, 487–491.

Aiello, A., Fattorusso, E., Magno, S. and Menna, M. 1991. Isolation of five new 5 alpha-hydroxy-6-keto-delta 7 sterols from the marine sponge *Oscarella lobularis*. *Steroids* 56, 337–340.

Alié, A., Hayashi, T., Sugimura, I., Manuel, M., Sugano, W., Mano, A., Satoh, N., Agata, K. and Funayama, N. 2015. The ancestral gene repertoire of animal stem cells. *Proc. Natl. Acad. Sci. U. S. A.* 112, E7093–E7100.

Ancheeva, E., El-Neketi, M., Song, W., Lin, W., Daletos, G., Ebrahim, W. and Proksch, P. 2017. Structurally unprecedented metabolites from marine sponges. *Curr. Org. Chem.* 21, 426–449.

Arendt, D., Musser, J. M., Baker, C. V. H., Bergman, A., Cepko, C., Erwin, D. H., Pavlicev, M., Schlosser, G., Widder, S., Laubichler, M. D., et al. 2016. The origin and evolution of cell types. *Nat. Rev. Genet.* 17, 744–757.

Babonis, L. S. and Martindale, M. Q. 2017. Phylogenetic evidence for the modular evolution of metazoan signalling pathways. *Philos. Trans. R. Soc. B Biol. Sci.* 372 (1713), 20150477.

Belahbib, H., Renard, E., Santini, S., Jourda, C., Claverie, J.-M., Borchiellini, C. and Le Bivic, A. 2018. New genomic data and analyses challenge the traditional vision of animal epithelium evolution. *BMC Genomics* 19, 393.

Bergquist, P. R. 1978. *Sponges*. Hutchinson, London; University of California Press, Berkeley & Los Angeles, CA, 268 pp.

Bergquist, P. R. and Kelly, M. 2004. Taxonomy of some Halisarcida and Homosclerophorida Porifera: Demospongiae from the Indo-Pacific. *N. Z. J. Mar. Freshw. Res*. 38, 51–66.

Bertolino, M., Cerrano, C., Bavestrello, G., Carella, M., Pansini, M. and Calcinai, B. 2013. Diversity of Porifera in the Mediterranean coralligenous accretions, with description of a new species. *ZooKeys*, 1–37.

Borchiellini, C., Degnan, S. M., Le Goff, E., Rocher, C., Vernale, A., Baghdiguian, S., Séjourné, N., Marschal, F., Le Bivic, A., Godefroy, N., et al. 2021. Staining and tracking methods for studying sponge cell dynamics. In *Developmental Biology of the Sea Urchin and Other Marine Invertebrates: Methods and Protocols*, ed. Carroll, D. J. and Stricker, S. A., pp. 81–97. Springer US, New York, NY.

Borchiellini, C., Chombard, C., Manuel, M., Alivon, E., Vacelet, J. and Boury-Esnault, N. 2004. Molecular phylogeny of Demospongiae: Implications for classification and scenarios of character evolution. *Mol. Phylogenet. Evol*. 32, 823–837.

Borchiellini, C., Manuel, M., Alivon, E., Boury-Esnault, N., Vacelet, J. and Parco, Y. L. 2001. Sponge paraphyly and the origin of Metazoa. *J. Evol. Biol*. 14, 171–179.

Borisenko, I. E., Adamski, M., Ereskovsky, A. and Adamska, M. 2016. Surprisingly rich repertoire of Wnt genes in the demosponge *Halisarca dujardini*. *BMC Evol. Biol*. 16, 123.

Borisenko, I. E., Adamska, M., Tokina, D. B. and Ereskovsky, A. V. 2015. Transdifferentiation is a driving force of regeneration in *Halisarca dujardini* Demospongiae, Porifera. *PeerJ*. 3, e1211.

Boury-Esnault, N., Ereskovsky, A., Bézac, C. and Tokina, D. 2003. Larval development in the Homoscleromorpha Porifera, Demospongiae. *Invertebr. Biol*. 122, 187–202.

Boury-Esnault, N., Lavrov, D. V., Ruiz, C. A. and Pérez, T. 2013. The integrative taxonomic approach applied to Porifera: A case study of the Homoscleromorpha. *Integr. Comp. Biol*. 53, 416–427.

Boury-Esnault, N., Sole-Cava, A. M. and Thorpe, J. P. 1992. Genetic and cytological divergence between colour morphs of the Mediterranean sponge *Oscarella lobularis* Schmidt Porifera, Demospongiae, Oscarellidae. *J. Nat. Hist*. 26, 271–284.

Boute, N., Exposito, J.-Y., Boury-Esnault, N., Vacelet, J., Noro, N., Miyazaki, K., Yoshizato, K. and Garrone, R. 1996. Type IV collagen in sponges, the missing link in basement membrane ubiquity. *Biol. Cell*. 88, 37–44.

Brunet, T. and King, N. 2017. The origin of animal multicellularity and cell differentiation. *Dev. Cell* 43, 124–140.

Brusca, R. C. and Brusca, G. J. 2003. *Invertebrates*. 2rd ed. Sinauer Associates, Sunderland, MA, USA, 936 pp.

Brusca, R. C., Moore, W. and Shuster, S. M. 2016. *Invertebrates*. 3rd ed. Sinauer Associates is an imprint of Oxford University Press, Sunderland, MA, USA, 1104 pp.

Cardone, F., Gaino, E. R. and Corriero, G. 2010. The budding process in *Tethya citrina* Sarà & Melone Porifera, Demospongiae and the incidence of post-buds in sponge population maintenance. *J. Exp. Mar. Biol. Ecol*. 389, 93–100.

Carter, H. J. 1876. Descriptions and figures of deep-sea sponges and their spicules, from the Atlantic Ocean, dredged up on board H.M.S. 'Porcupine', chiefly in 1869 (concluded). *Ann. Mag. Nat. Hist*. 18, 226–240.

Cimino, G., de Stefano, S. and Minale, L. 1975. Long alkyl chains-3-substituted pyrrole-2-aldehyde-2-carboxylic acid and methyl ester from the marine sponge *Oscarella lobularis*. *Experientia* 31, 1387–1389.

Colgren, J. and Nichols, S. A. 2019. The significance of sponges for comparative studies of developmental evolution. *Wiley Interdiscip. Rev. Dev. Biol*, e359.

Coll, M., Piroddi, C., Steenbeek, J., Kaschner, K., Lasram, F. B. R., Aguzzi, J., Ballesteros, E., Bianchi, C. N., Corbera, J., Dailianis, T., et al. 2010. The biodiversity of the Mediterranean Sea: Estimates, patterns, and threats. *PLoS One* 5, e11842.

Conaco, C., Tsoulfas, P., Sakarya, O., Dolan, A., Werren, J. and Kosik, K. S. 2016. Detection of prokaryotic genes in the *Amphimedon queenslandica* genome. *PLoS One* 11, e0151092.

Degnan, B. M., Adamska, M., Richards, G. S., Larroux, C., Leininger, S., Bergum, B., Calcino, A., Taylor, K., Nakanishi, N. and Degnan, S. M. 2015. Porifera. In *Evolutionary Developmental Biology of Invertebrates 1: Introduction, Non-Bilateria, Acoelomorpha, Xenoturbellida, Chaetognatha*, ed. Wanninger, A., pp. 65–106. Springer, Vienna.

Dendy, A. 1905. Report on the sponges collected by Professor Herdman, at Ceylon, in 1902. *In Report to the Government of Ceylon on the Pearl Oyster Fisheries of the Gulf of Manaar*, ed. Herdman, W. A., pp. 57–246. Royal Society, London. 3 (Supplement 18).

Devenport, D. 2014. The cell biology of planar cell polarity. *J. Cell Biol*. 207, 171–179.

De Vos, L., Rützler, K., Boury-Esnault, N., Donadey, C. and Vacelet, J. 1991. Atlas of sponge morphology. *Smithson. Inst. Press Wash. Lond*., 117 pp.

Domingos, C., Lage, A. and Muricy, G. 2016. Overview of the biodiversity and distribution of the class Homoscleromorpha in the tropical Western Atlantic. *J. Mar. Biol. Assoc. U. K*. 96, 379–389.

Dunn, C. W., Leys, S. P. and Haddock, S. H. D. 2015. The hidden biology of sponges and ctenophores. *Trends Ecol. Evol*. 30, 282–291.

Edelblum, K. L. and Turner, J. R. 2015. Chapter 12: Epithelial cells: Structure, transport, and barrier function. In *Mucosal Immunology*. 4th ed., ed. Mestecky, J., Strober, W., Russell, M. W., Kelsall, B. L., Cheroutre, H. and Lambrecht, B. N., pp. 187–210. Academic Press, Boston.

Ereskovsky, A. V. 2006. A new species of *Oscarella* Demospongiae: Plakinidae from the Western Sea of Japan. *Zootaxa* 1376, 37–51.

Ereskovsky, A. V. 2010. *The Comparative Embryology of Sponges*. Springer-Verlag, Dordrecht Heidelberg, London and New York, 329 pp.

Ereskovsky, A. V., Borchiellini, C., Gazave, E., Ivanisevic, J., Lapébie, P., Pérez, T., Renard, E. and Vacelet, J. 2009b. The Homoscleromorph sponge *Oscarella lobularis*, a promising sponge model in evolutionary and developmental biology. *BioEssays* 31, 89–97.

Ereskovsky, A. V., Borisenko, I. E., Lapébie, P., Gazave, E., Tokina, D. B. and Borchiellini, C. 2015. *Oscarella lobularis* Homoscleromorpha, Porifera Regeneration: Epithelial morphogenesis and metaplasia. *PLoS One* 108: e0134566.

Ereskovsky, A. V. and Boury-Esnault, N. 2002. Cleavage pattern in *Oscarella* species Porifera, Demospongiae, Homoscleromorpha: Transmission of maternal cells and symbiotic bacteria. *J. Nat. Hist*. 36, 1761–1775.

Ereskovsky, A. V., Dubois, M., Ivanišević, J., Gazave, E., Lapebie, P., Tokina, D. and Pérez, T. 2013a. Pluri-annual study of the reproduction of two Mediterranean *Oscarella* species Porifera, Homoscleromorpha: Cycle, sex-ratio, reproductive effort and phenology. *Mar. Biol.* 160, 423–438.

Ereskovsky, A. V., Geronimo, A. and Pérez, T. 2017a. Asexual and puzzling sexual reproduction of the Mediterranean sponge *Haliclona fulva* Demospongiae: Life cycle and cytological structures. *Invertebr. Biol.* 136, 403–421.

Ereskovsky, A. V. and Dondua, A. K. 2006. The problem of germ layers in sponges Porifera and some issues concerning early metazoan evolution. *Zool. Anz.: J. Comp. Zool.* 245, 65–76.

Ereskovsky, A. V., Konyukov, P. Yu. and Tokina, D. B. 2010. Morphogenesis accompanying larval metamorphosis in *Plakina trilopha* Porifera, Homoscleromorpha. *Zoomorphology* 129, 21–31.

Ereskovsky, A. V., Lavrov, D. V. and Willenz, P. 2014. Five new species of Homoscleromorpha Porifera from the Caribbean Sea and re-description of *Plakina jamaicensis*. *J. Mar. Biol. Assoc. U. K.* 94, 285–307.

Ereskovsky, A. V., Renard, E. and Borchiellini, C. 2013b. Cellular and molecular processes leading to embryo formation in sponges: Evidences for high conservation of processes throughout animal evolution. *Dev. Genes Evol.* 223, 5–22.

Ereskovsky, A. V., Richter, D. J., Lavrov, D. V., Schippers, K. J. and Nichols, S. A. 2017b. Transcriptome sequencing and delimitation of sympatric *Oscarella* species *O. carmela* and *O. pearsei* sp. nov from California, USA. *PLoS One* 129, e0183002.

Ereskovsky, A. V., Sanamyan, K. and Vishnyakov, E. 2009a. A new species of the genus *Oscarella* Porifera: Homosclerophorida: Plakinidae from the North-West Pacific. *Cah. Biol. Mar.* 50, 369–381.

Ereskovsky, A. V. and Tokina, D. B. 2007. Asexual reproduction in homoscleromorph sponges Porifera; Homoscleromorpha. *Mar. Biol.* 151, 425–434.

Ereskovsky, A. V., Tokina, D. B., Bézac, C. and Boury-Esnault, N. 2007. Metamorphosis of cinctoblastula larvae Homoscleromorpha, Porifera. *J. Morphol.* 268, 518–528.

Fell, P. 1993. Porifera. In *Asexual Propagation and Reproductive Strategies*, eds. Adiyodi, K. G. and Adiyodi, R. G., pp. 1–44. Oxford and IBH, New Delhi.

Fernandez-Valverde, S. L., Calcino, A. D. and Degnan, B. M. 2015. Deep developmental transcriptome sequencing uncovers numerous new genes and enhances gene annotation in the sponge *Amphimedon queenslandica*. *BMC Genomics* 16, 387.

Feuda, R., Dohrmann, M., Pett, W., Philippe, H., Rota-Stabelli, O., Lartillot, N., Wörheide, G. and Pisani, D. 2017. Improved modeling of compositional heterogeneity supports sponges as sister to all other animals. *Curr. Biol.* 18, 2724, 3864–3870.

Fidler, A. L., Darris, C. E., Chetyrkin, S. V., Pedchenko, V. K., Boudko, S. P., Brown, K. L., Gray Jerome, W., Hudson, J. K., Rokas, A. and Hudson, B. G. 2017. Collagen IV and basement membrane at the evolutionary dawn of metazoan tissues. *eLife* 18, 6, e24176.

Fierro-Constain, L. 2016. La reproduction chez *Oscarella lobularis* Porifera-Homoscleromorpha: gènes impliqués et effets de l'environnement. www.theses.fr.

Fierro-Constaín, L., Rocher, C., Marschal, F., Schenkelaars, Q., Séjourné, N., Borchiellini, C. and Renard, E. 2021. In situ hybridization techniques in the homoscleromorph sponge *Oscarella lobularis*. *Methods Mol. Biol. Clifton NJ* 2219, 181–194.

Fierro-Constaín, L., Schenkelaars, Q., Gazave, E., Haguenauer, A., Rocher, C., Ereskovsky, A., Borchiellini, C. and Renard, E. 2017. The conservation of the Germline multipotency program, from sponges to vertebrates: A stepping stone to understanding the somatic and germline origins. *Genome Biol. Evol.* 9, 474–488.

Fieth, R. A., Gauthier, M.-E. A., Bayes, J., Green, K. M. and Degnan, S. M. 2016. Ontogenetic changes in the bacterial symbiont community of the tropical demosponge *Amphimedon queenslandica*: Metamorphosis is a new beginning. *Front. Mar. Sci.* 3, 228.

Fortunato, S. A. V., Adamski, M. and Adamska, M. 2015. Comparative analyses of developmental transcription factor repertoires in sponges reveal unexpected complexity of the earliest animals. *Mar. Genomics* 24 (2), 121–129.

Fortunato, S. A. V., Vervoort, M., Adamski, M. and Adamska, M. 2016. Conservation and divergence of bHLH genes in the calcisponge *Sycon ciliatum*. *EvoDevo* 7, 23.

Francis, W. R. and Canfield, D. E. 2020. Very few sites can reshape the inferred phylogenetic tree. *PeerJ*. 8, e8865.

Francis, W. R., Eitel, M., R. S. V., Adamski, M., Haddock, S. H., Krebs, S., Blum, H., Erpenbeck, D. and Wörheide, G. 2017. The genome of the contractile demosponge *Tethya wilhelma* and the evolution of metazoan neural signalling pathways. *bioRxiv* 120998.

Funayama, N. 2013. The stem cell system in demosponges: Suggested involvement of two types of cells: Archeocytes active stem cells and choanocytes food-entrapping flagellated cells. *Dev. Genes Evol.* 223, 23–38.

Funayama, N. 2018. The cellular and molecular bases of the sponge stem cell systems underlying reproduction, homeostasis and regeneration. *Int. J. Dev. Biol.* 62, 513–525.

Funayama, N., Nakatsukasa, M., Mohri, K., Masuda, Y. and Agata, K. 2010. Piwi expression in archeocytes and choanocytes in demosponges: Insights into the stem cell system in demosponges. *Evol. Dev.* 12, 275–287.

Gaino, E., Burlando, B. and Buffa, P. 1986a. The vacuolar cells of *Oscarella lobularis* Porifera, Demospongiae: Ulatrastructural organization, origin, and function. *J. Morphol.* 188, 29–37.

Gaino, E., Burlando, B. and Buffa, P. 1986b. Contribution to the study of egg development and derivation in *Oscarella lobularis* Porifera, Demospongiae. *Int. J. Invertebr. Rep. Dev.* 9, 59–69.

Gaino, E., Burlando, B., Buffa, P. and Sará, M. 1986c. Ultrastructural study of spermatogenesis in *Oscarella lobularis* Porifera, Demospongiae. *Int. J. Invertebr. Rep. Dev.* 10, 297–305.

Garrabou, J. and Zabala, M. 2001. Growth dynamics in four Mediterranean demosponges. *Estuar. Coast. Shelf Sci.* 52, 293–303.

Gatenby, J. B. 1920. IX.: Further notes on the Oögenesis and fertilization of *Grantia compressa*. *J. R. Microsc. Soc.* 40, 277–282.

Gauthier, M.-E. A., Watson, J. R. and Degnan, S. M. 2016. Draft genomes shed light on the dual bacterial symbiosis that dominates the microbiome of the coral reef sponge *Amphimedon queenslandica*. *Front. Mar. Sci.* 3, 196.

Gazave, E. 2010. Etude de gène et voie de signalisation impliqués dans les processus morphogénétiques chez *Oscarella lobularis*: implications potentielles sur la compréhension de l'origine du système nerveux. www.theses.fr.

Gazave, E., Lapébie, P., Ereskovsky, A. V., Vacelet, J., Renard, E., Cárdenas, P. and Borchiellini, C. 2012. No longer Demospongiae: Homoscleromorpha formal nomination as a fourth class of Porifera. *Hydrobiologia* 687, 3–10.

Gazave, E., Lapébie, P., Renard, E., Bézac, C., Boury-Esnault, N., Vacelet, J., Pérez, T., Manuel, M. and Borchiellini, C. 2008. NK homeobox genes with choanocyte-specific expression in homoscleromorph sponges. *Dev. Genes Evol.* 218, 479–489.

Gazave, E., Lapébie, P., Richards, G. S., Brunet, F., Ereskovsky, A. V., Degnan, B. M., Borchiellini, C., Vervoort, M. and Renard, E. 2009. Origin and evolution of the Notch signalling pathway: An overview from eukaryotic genomes. *BMC Evol. Biol.* 9, 249.

Gazave, E., Lapébie, P., Renard, E., Vacelet, J., Rocher, C., Ereskovsky, A. V., Lavrov, D. V. and Borchiellini, C. 2010. Molecular phylogeny restores the supra-generic subdivision of homoscleromorph sponges Porifera, Homoscleromorpha. *PLoS One* 512, e14290.

Gazave, E., Guillou, A. and Balavoine, G. 2014. History of a prolific family: The Hes/Hey-related genes of the annelid Platynereis. *EvoDevo* 5, 29.

Gazave, E., Lavrov, D. V., Cabrol, J., Renard, E., Rocher, C., Vacelet, J., Adamska, M., Borchiellini, C. and Ereskovsky, A. V. 2013. Systematics and molecular phylogeny of the family Oscarellidae Homoscleromorpha with description of two new *Oscarella* species. *PLoS One* 8, e63976.

Gazave, E. and Renard, E. 2010. Evolution of Notch Transmembrane Receptors. In *eLS*, p. American Cancer Society.

Genta-Jouve, G. and Thomas, O. P. 2012. Sponge chemical diversity: From biosynthetic pathways to ecological roles. *Adv. Mar. Biol.* 62, 183–230.

Gerçe, B., Schwartz, T., Syldatk, C. and Hausmann, R. 2011. Differences between bacterial communities associated with the surface or tissue of Mediterranean sponge species. *Microb. Ecol.* 61, 769–782.

Gilbert, S. F. and Barresi, M. J. F. 2018. *Developmental Biology*. 11th ed. Sinauer Associates, Sunderland, MA, USA.

Gloeckner, V., Hentschel, U., Ereskovsky, A. V. and Schmitt, S. 2013. Unique and species-specific microbial communities in *Oscarella lobularis* and other Mediterranean Oscarella species Porifera: Homoscleromorpha. *Mar. Biol.* 160, 781–791.

Gorczyca, W., Bruno, S., Darzynkiewicz, R., Gong, J. and Darzynkiewicz, Z. 1992. DNA strand breaks occurring during apoptosis: Their early *in situ* detection by the terminal deoxynucleotidyl transferase and nick translation assays and prevention by serine protease inhibitors. *Int. J. Oncol.* 1, 639–648.

Hall, C., Rodriguez, M., Garcia, J., Posfai, D., DuMez, R., Wictor, E., Quintero, O. A., Hill, M. S., Rivera, A. S. and Hill, A. L. 2019. Secreted frizzled related protein is a target of PaxB and plays a role in aquiferous system development in the fresh water sponge, *Ephydatia muelleri*. *PLoS One* 14, e0212005.

Hammel, J. U. and Nickel, M. 2014. A new flow-regulating cell type in the demosponge *Tethya wilhelma*: Functional cellular anatomy of a leuconoid canal system. *PLoS One* 9, e113153.

Hanitsch, R. 1890. Third report on the Porifera of the L.M.B.C. district. *Proc. Trans. Liverpool Biol. Soc.* 4, 192–238.

Hentschel, E. 1909. Tetraxonida. I. Teil. *In Die Fauna Südwest-Australiens. Ergebnisse der Hamburger südwest-australischen Forschungsreise 1905, eds. Michaelsen, W. and* Hartmeyer, R., pp. 347–402. Jena: Fischer. 2 (21).

Hesp, K., Flores Alvarez, J. L., Alexandru, A.-M., van der Linden, J., Martens, D. E., Wijffels, R. H. and Pomponi, S. A. 2020. CRISPR/Cas12a-mediated gene editing in *Geodia barretti* sponge cell culture. *Front. Mar. Sci.* 7, 599825.

Hill, M. S., Hill, A. L., Lopez, J., Peterson, K. J., Pomponi, S., Diaz, M. C., Thacker, R. W., Adamska, M., Boury-Esnault, N., Cárdenas, P., et al. 2013. Reconstruction of family-level phylogenetic relationships within Demospongiae, Porifera using nuclear encoded housekeeping genes. *PLoS One* 8, e50437.

Humbert-David, N. and Garrone, R. 1993. A six-armed, tenascin-like protein extracted from the Porifera *Oscarella tuberculata* Homosclerophorida. *Eur. J. Biochem.* 216, 255–260.

Ivanišević, J., Thomas, O. P., Lejeusne, C., Chevaldonné, P. and Pérez, T. 2011. Metabolic fingerprinting as an indicator of biodiversity: Towards understanding inter-specific relationships among Homoscleromorpha sponges. *Metabolomics* 7, 289–304.

Jenner, R. A. and Wills, M. A. 2007. The choice of model organisms in evo-devo. *Nat. Rev. Genet.* 8, 311–314.

Jourda, C., Santini, S., Rocher, C., Le Bivic, A. and Claverie, J.-M. 2015. Draft genome sequence of an alphaproteobacterium associated with the Mediterranean sponge *Oscarella lobularis*. *Genome Announc.* 3 (5), e00977–15.

Kanska, J. and Frank, U. 2013. New roles for nanos in neural cell fate determination revealed by studies in a cnidarian. *J. Cell Sci.* 126, 3192–3203.

Kenny, N. J., Francis, W. R., Rivera-Vicéns, R. E., Juravel, K., Mendoza, A. de, Díez-Vives, C., Lister, R., Bezares-Calderon, L., Grombacher, L., Roller, M., et al. 2020. Tracing animal genomic evolution with the chromosomal-level assembly of the freshwater sponge *Ephydatia muelleri*. *bioRxiv* Feb 18, 954784.

King, N. 2004. The unicellular ancestry of animal development. *Dev. Cell* 7, 313–325.

Krenacs, L., Krenacs, T., Stelkovics, E. and Raffeld, M. 2010. Heat-induced antigen retrieval for immunohistochemical reactions in routinely processed paraffin sections. In *Immunocytochemical Methods and Protocols*, eds. Oliver, C. and Jamur, M. C., pp. 103–119. Humana Press, Totowa, NJ.

Labrenz, M., Lawson, P. A., Tindall, B. J. and Hirsch, P. 2009. *Roseibaca ekhonensis* gen. nov., sp. nov., an alkalitolerant and aerobic bacteriochlorophyll a-producing alphaproteobacterium from hypersaline Ekho Lake. *Int. J. Syst. Evol. Microbiol.* 59, 1935–1940.

Lage, A., Gerovasileiou, V., Voultsiadou, E. and Muricy, G. 2019. Taxonomy of *Plakina* Porifera: Homoscleromorpha from Aegean submarine caves, with descriptions of three new species and new characters for the genus. *Mar. Biodivers.* 49, 727–747.

Lage, A., Muricy, G., Ruiz, C. and Pérez, T. 2018. New sciaphilic plakinids Porifera, Homoscleromorpha from the Central-Western Pacific. *Zootaxa* 4466, 8–38.

Lanna, E. 2015. Evo-devo of non-bilaterian animals. *Genet. Mol. Biol.* 38, 284–300.

Lapebie, P. 2010. L'origine des morphogenèses épithéliales et leurs implications concernant l'évolution précoce des métazoaires. www.theses.fr.

Lapébie, P., Gazave, E., Ereskovsky, A., Derelle, R., Bézac, C., Renard, E., Houliston, E. and Borchiellini, C. 2009. WNT/beta-catenin signalling and epithelial patterning in the homoscleromorph sponge *Oscarella*. *PLoS One* 4, e5823.

Laundon, D., Larson, B. T., McDonald, K., King, N. and Burkhardt, P. 2019. The architecture of cell differentiation in choanoflagellates and sponge choanocytes. *PLoS Biol.* 17, e3000226.

Layden, M. J., Röttinger, E., Wolenski, F. S., Gilmore, T. D. and Martindale, M. Q. 2013. Microinjection of mRNA or morpholinos for reverse genetic analysis in the starlet sea anemone, *Nematostella vectensis*. *Nat. Protoc.* 8, 924–934.

Leininger, S., Adamski, M., Bergum, B., Guder, C., Liu, J., Laplante, M., Bråte, J., Hoffmann, F., Fortunato, S., Jordal, S., et al. 2014. Developmental gene expression provides clues to relationships between sponge and eumetazoan body plans. *Nat. Commun.* 5, ncomms4905.

Lendenfeld, R. von. 1887. On the systematic position and classification of sponges. *Proc. Zool. Soc. London.* 18, 558–662.

Lévi, C. 1956. Etude des *Hailsarca* de Roscoff. Embryologie et systematique des demosponges. *Arch. Zool. Exp. Gen.* 93, 1–184.

Lévi, C. 1973. Systématique de la classe des Demospongiaria Démosponges. Pp. 577–631. In *Traité de Zoologie. Anatomie, Systématique, Biologie*, ed. Grassé, P.-P. Spongiaires. 3(1), pp. 577–631. Masson et Cie., Paris.

Lévi, C. and Porte, A. 1962. Étude au microscope électronique de l'éponge *Oscarella lobularis* schmidt et de sa larve amphiblastula. *Cah. Biol. Mar.* 3, 307–315.

Leys, S. P. 2004. Gastrulation in sponges. *Gastrulation Cells Embryo Ed. Stern CD 2004 N. Y. Cold Spring Harb. Lab. Press*, 23–31.

Leys, S. P. and Degnan, B. M. 2001. Cytological basis of photoresponsive behavior in a sponge larva. *Biol. Bull.* 201, 323–338.

Leys, S. P. and Ereskovsky, A. V. 2006. Embryogenesis and larval differentiation in sponges. *Can. J. Zool.* 84, 262–287.

Leys, S. P. and Hill, A. 2012. The physiology and molecular biology of sponge tissues. In *Advanc. Mar. Biol.*, pp. 1–56.

Leys, S. P., Mah, J. L., McGill, P. R., Hamonic, L., De Leo, F. C. and Kahn, A. S. 2019. Sponge behavior and the chemical basis of responses: A post-genomic view. *Integr. Comp. Biol.* 59, 751–764.

Leys, S. P., Nichols, S. A. and Adams, E. D. M. 2009. Epithelia and integration in sponges. *Integr. Comp. Biol.* 49, 167–177.

Leys, S. P. and Riesgo, A. 2012. Epithelia, an evolutionary novelty of metazoans. *J. Exp. Zoolog. B Mol. Dev. Evol.* 318, 438–447.

Leys, S. P., Yahel, G., Reidenbach, M. A., Tunnicliffe, V., Shavit, U. and Reiswig, H. M. 2011. The sponge pump: The role of current induced flow in the design of the sponge body plan. *PLoS One* 6, e27787.

Love, A. C. and Yoshida, Y. 2019. Reflections on model organisms in evolutionary developmental biology. *Results Probl. Cell Differ.* 3–20.

Lowe, J. S. and Anderson, P. G. 2015. Chapter 3: Epithelial cells. In *Stevens & Lowe's Human Histology*. 4th ed., ed. Lowe, J. S. and Anderson, P. G., pp. 37–54. Mosby, Philadelphia.

Ludeman, D. A., Reidenbach, M. A. and Leys, S. P. 2017. The energetic cost of filtration by demosponges and their behavioural response to ambient currents. *J. Exp. Biol.* 220, 995–1007.

Mah, J. L., Christensen-Dalsgaard, K. K. and Leys, S. P. 2014. Choanoflagellate and choanocyte collar-flagellar systems and the assumption of homology. *Evol. Dev.* 16, 25–37.

Maldonado, M. 2004. Choanoflagellates, choanocytes, and animal multicellularity. *Invertebr. Biol.* 123, 1–22.

Maldonado, M. and Riesgo, A. 2008. Reproduction in Porifera: A synoptic overview. *Treb. Soc. Cat. Biol.* 59, 29–49.

Marion, A. F. 1883. *Esquisse d'une topographie zoologique du golfe de Marseille*. Cayer et cie.

Meewis, H. 1938. Contribution a l'étude de l'embryogenése des Myxospongiae: *Halisarca lobularis* Schmidt. *Arch. Biol. Liege* 59, 1–66.

Miller, P. W., Pokutta, S., Mitchell, J. M., Chodaparambil, J. V., Clarke, D. N., Nelson, W. J., Weis, W. I. and Nichols, S. A. 2018. Analysis of a vinculin homolog in a sponge phylum Porifera reveals that vertebrate-like cell adhesions emerged early in animal evolution. *J. Biol. Chem.* 293, 11674–11686.

Mitchell, J. M. and Nichols, S. A. 2019. Diverse cell junctions with unique molecular composition in tissues of a sponge Porifera. *EvoDevo* 10, 26.

Muricy, G., Boury-Esnault, N., Bézac, C. and Vacelet, J. 1996. Cytological evidence for cryptic speciation in Mediterranean *Oscarella* species Porifera, Homoscleromorpha. *Can. J. Zool.* 74, 881–896.

Muricy, G. and Pearse, J. S. 2004. A new species of *Oscarella* Demospongiae: Plakinidae from California. *Proc. Calif. Acad. Sci.* 55, 598–612.

Musser, J. M., Schippers, K. J., Nickel, M., Mizzon, G., Kohn, A. B., Pape, C., Hammel, J. U., Wolf, F., Liang, C., Hernández-Plaza, A., et al. 2019. Profiling cellular diversity in sponges informs animal cell type and nervous system evolution. *bioRxiv*, 758276.

Nakanishi, N., Sogabe, S. and Degnan, B. M. 2014. Evolutionary origin of gastrulation: Insights from sponge development. *BMC Biol.* 12, 26.

Nichols, S. A., Dirks, W., Pearse, J. S. and King, N. 2006. Early evolution of animal cell signaling and adhesion genes. *Proc. Natl. Acad. Sci. U. S. A.* 103, 12451–12456.

Nichols, S. A., Roberts, B. W., Richter, D. J., Fairclough, S. R. and King, N. 2012. Origin of metazoan cadherin diversity and the antiquity of the classical cadherin/β-catenin complex. *Proc. Natl. Acad. Sci. U. S. A.* 109, 13046–13051.

Nielsen, C. 2008. Six major steps in animal evolution: Are we derived sponge larvae? *Evol. Dev.* 10, 241–257.

Pérez, T., Ivanisevic, J., Dubois, M., Pedel, L., Thomas, O. P., Tokina, D. and Ereskovsky, A. V. 2011. *Oscarella balibaloi*, a new sponge species Homoscleromorpha: Plakinidae from the Western Mediterranean Sea: Cytological description, reproductive cycle and ecology. *Mar. Ecol.* 32, 174–187.

Pérez, T. and Ruiz, C. 2018. Description of the first Caribbean Oscarellidae Porifera: Homoscleromorpha. *Zootaxa* 4369, 501–514.

Philippe, H., Derelle, R., Lopez, P., Pick, K., Borchiellini, C., Boury-Esnault, N., Vacelet, J., Renard, E., Houliston, E., Quéinnec, E., et al. 2009. Phylogenomics revives traditional views on deep animal relationships. *Curr. Biol. CB* 19, 706–712.

Pick, K. S., Philippe, H., Schreiber, F., Erpenbeck, D., Jackson, D. J., Wrede, P., Wiens, M., Alié, A., Morgenstern, B., Manuel, M., et al. 2010. Improved phylogenomic taxon sampling noticeably affects nonbilaterian relationships. *Mol. Biol. Evol.* 27, 1983–1987.

Pisani, D., Pett, W., Dohrmann, M., Feuda, R., Rota-Stabelli, O., Philippe, H., Lartillot, N. and Wörheide, G. 2015. Genomic data do not support comb jellies as the sister group to all other animals. *Proc. Natl. Acad. Sci. U. S. A.* 112, 15402–15407.

Pohlner, M., Dlugosch, L., Wemheuer, B., Mills, H., Engelen, B. and Reese, B. K. 2019. The majority of active rhodobacteraceae in marine sediments belong to uncultured genera: A molecular approach to link their distribution to environmental conditions. *Front. Microbiol.* 10, 659.

Pozdnyakov, I. R., Sokolova, A. M., Ereskovsky, A. V. and Karpov, S. A. 2017. Kinetid structure of choanoflagellates and choanocytes of sponges does not support their close relationship. *Protistology* 11, 248–264.

Proksch, P. 1994. Defensive roles for secondary metabolites from marine sponges and sponge-feeding nudibranchs. *Toxicon* 32, 639–655.

Ramos-Vicente, D., Ji, J., Gratacòs-Batlle, E., Gou, G., Reig-Viader, R., Luís, J., Burguera, D., Navas-Perez, E., García-Fernández, J., Fuentes-Prior, P., et al. 2018. Metazoan evolution of glutamate receptors reveals unreported phylogenetic groups and divergent lineage-specific events. *eLife* 7, e35774.

Rane, R., Sahu, N., Shah, C. and Karpoormath, R. 2014. Marine bromopyrrole alkaloids: Synthesis and diverse medicinal applications. *Curr. Top. Med. Chem.* 14, 253–273.

Redmond, N. E., Morrow, C. C., Thacker, R. W., Diaz, M. C., Boury-Esnault, N., Cárdenas, P., Hajdu, E., Lôbo-Hajdu, G., Picton, B. E., Pomponi, S. A., et al. 2013. Phylogeny and systematics of demospongiae in light of new small-subunit ribosomal DNA 18S sequences. *Integr. Comp. Biol.* 53, 388–415.

Renard, E., Le Bivic, A. and Borchiellini, C. 2021. Origin and evolution of epithelial cell types. In *Origin and Evolution of Metazoan Cell Types*. 1st ed., eds. Leys, S. and Hejnol, A. CRC Press, Boca Raton. https://doi.org/10.1201/b21831.

Renard, E., Leys, S. P., Wörheide, G. and Borchiellini, C. 2018. Understanding animal evolution: The added value of sponge transcriptomics and genomics. *BioEssays* 409, e1700237.

Richards, G. S. and Degnan, B. M. 2009. The dawn of developmental signaling in the metazoa. *Cold Spring Harb. Symp. Quant. Biol.* 74, 81–90.

Riesgo, A., Farrar, N., Windsor, P. J., Giribet, G. and Leys, S. P. 2014. The analysis of eight transcriptomes from all Poriferan classes reveals surprising genetic complexity in sponges. *Mol. Biol. Evol.* 31, 1102–1120.

Riisgård, H. U. and Larsen, P. S. 2017. Particle capture mechanisms in suspension-feeding invertebrates. *Mar. Ecol. Prog. Ser.* 418, 255–293.

Rivera, A. S., Hammel, J. U., Haen, K. M., Danka, E. S., Cieniewicz, B., Winters, I. P., Posfai, D., Wörheide, G., Lavrov, D. V., Knight, S. W., et al. 2011. RNA interference in marine and freshwater sponges: Actin knockdown in *Tethya wilhelma* and *Ephydatia muelleri* by ingested dsRNA expressing bacteria. *BMC Biotechnol.* 11, 67.

Rivera, A. S., Ozturk, N., Fahey, B., Plachetzki, D. C., Degnan, B. M., Sancar, A. and Oakley, T. H. 2012. Blue-light-receptive cryptochrome is expressed in a sponge eye lacking neurons and opsin. *J. Exp. Biol.* 215, 1278–1286.

Rivera, A. S., Winters, I., Rued, A., Ding, S., Posfai, D., Cieniewicz, B., Cameron, K., Gentile, L. and Hill, A. 2013. The evolution and function of the Pax/Six regulatory network in sponges. *Evol. Dev.* 15, 186–196.

Rocher, C., Vernale, A., Fierro-Constain, L., Sejourne, N., Chenesseau, S., Marschal, C., Golf, E. L., Dutilleul, M., Matthews, C., Marschal, F., et al. 2020. The buds of *Oscarella lobularis* Porifera: A new convenient model for sponge cell and developmental biology. *bioRxiv* June 23, 167296.

Ruiz, C., Muricy, G., Lage, A., Domingos, C., Chenesseau, S. and Pérez, T. 2017. Descriptions of new sponge species and genus, including aspiculate Plakinidae, overturn the Homoscleromorpha classification. *Zool. J. Linn. Soc.* 179, 707–724.

Ryan, J. F., Pang, K., Schnitzler, C. E., Nguyen, A.-D., Moreland, R. T., Simmons, D. K., Koch, B. J., Francis, W. R., Havlak, P., Smith, S. A., et al. 2013. The genome of the ctenophore *Mnemiopsis leidyi* and its implications for cell type evolution. *Science* 13, 3426164, 1242592.

Salic, A. and Mitchison, T. J. 2008. A chemical method for fast and sensitive detection of DNA synthesis *in vivo*. *Proc. Natl. Acad. Sci.* 105, 2415–2420.

Santhanam, R., Ramesh, S., Sunilson, A. J., Ramesh, S. and Sunilson, A. J. 2018. *Biology and Ecology of Pharmaceutical Marine Sponges*. CRC Press. https://doi.org/10.1201/9781351132473.

Schenkelaars, Q. 2015. Origine et évolution des voies Wnt chez les métazoaires: étude comparée de diverses espèces d'éponges. www.theses.fr.

Schenkelaars, Q., Fierro-Constain, L., Renard, E. and Borchiellini, C. 2016a. Retracing the path of planar cell polarity. *BMC Evol. Biol.* 16, 69.

Schenkelaars, Q., Fierro-Constain, L., Renard, E., Hill, A. L. and Borchiellini, C. 2015. Insights into Frizzled evolution and new perspectives. *Evol. Dev.* 17, 160–169.

Schenkelaars, Q., Quintero, O., Hall, C., Fierro-Constain, L., Renard, E., Borchiellini, C. and Hill, A. L. 2016b. ROCK inhibition abolishes the establishment of the aquiferous system in *Ephydatia muelleri* Porifera, Demospongiae. *Dev. Biol.* 412, 298–310.

Schenkelaars, Q., Vernale, A., Fierro-Constaín, L., Borchiellini, C. and Renard, E. 2019. A look back over 20 years of evo-devo studies on sponges: A challenged view of Urmetazoa. In *Evolution, Origin of Life, Concepts and Methods*, ed. Pontarotti, P., pp. 135–160. Springer International Publishing, Cham, Switzerland.

Schippers, K. J., Nichols, S. A. and Wittkopp, P. 2018. Evidence of signaling and adhesion roles for β-catenin in the sponge *Ephydatia muelleri*. *Mol. Biol. Evol.* 35, 1407–1421.

Schmidt, O. 1862. *Die Spongien des adriatischen Meeres*. Wilhelm Engelmann, Leipzig, 88p.

Schmidt, O. 1868. *Die Spongien der Küste von Algier. Mit Nachträgen zu den Spongien des Adriatischen Meeres*. Wilhelm Engelmann, Leipzig, 44 pp.

Simion, P., Philippe, H., Baurain, D., Jager, M., Richter, D. J., Di Franco, A., Roure, B., Satoh, N., Quéinnec, É., Ereskovsky, A., et al. 2017. A large and consistent phylogenomic dataset supports sponges as the sister group to all other animals. *Curr. Biol. CB* 27, 958–967.

Simionato, E., Ledent, V., Richards, G., Thomas-Chollier, M., Kerner, P., Coornaert, D., Degnan, B. M. and Vervoort, M. 2007. Origin and diversification of the basic helix-loop-helix gene family in metazoans: Insights from comparative genomics. *BMC Evol. Biol.* 7, 33.

Simpson, T. L. 1984. *The Cell Biology of Sponges*. Springer-Verlag, New York.

Singh, A. and Thakur, N. L. 2015. Field and laboratory investigations of budding in the tetillid sponge *Cinachyrella cavernosa*. *Invertebr. Biol.* 134, 19–30.

Sipkema, D., Snijders, A. P. L., Schroën, C. G. P. H., Osinga, R. and Wijffels, R. H. 2004. The life and death of sponge cells. *Biotechnol. Bioeng.* 85, 239–247.

Sogabe, S., Hatleberg, W. L., Kocot, K. M., Say, T. E., Stoupin, D., Roper, K. E., Fernandez-Valverde, S. L., Degnan, S. M. and Degnan, B. M. 2019. Pluripotency and the origin of animal multicellularity. *Nature* 570, 519–522.

Sogabe, S., Nakanishi, N. and Degnan, B. M. 2016. The ontogeny of choanocyte chambers during metamorphosis in the demosponge *Amphimedon queenslandica*. *EvoDevo* 7, 6.

Sorokin, D. Y., Tourova, T. P. and Muyzer, G. 2005. *Citreicella thiooxidans* gen. nov., sp. nov., a novel lithoheterotrophic sulfur-oxidizing bacterium from the Black Sea. *Syst. Appl. Microbiol.* 28, 679–687.

Sperling, E. A., Pisani, D. and Peterson, K. J. 2007. Poriferan paraphyly and its implications for Precambrian palaeobiology. *Geol. Soc. Lond. Spec. Publ.* 286, 355–368.

Sperling, E. A., Peterson, K. J. and Pisani, D. 2009. Phylogenetic-signal dissection of nuclear housekeeping genes supports the paraphyly of sponges and the monophyly of Eumetazoa. *Mol. Biol. Evol.* 26, 2261–2274.

Srivastava, M., Simakov, O., Chapman, J., Fahey, B., Gauthier, M. E. A., Mitros, T., Richards, G. S., Conaco, C., Dacre, M., Hellsten, U., et al. 2010. The *Amphimedon queenslandica* genome and the evolution of animal complexity. *Nature* 466, 720–726.

Stroebel, D. and Paoletti, P. 2020. Architecture and function of NMDA receptors: An evolutionary perspective. *J. Physiol.* 1–24.

Thacker, R. W., Hill, A. L., Hill, M. S., Redmond, N. E., Collins, A. G., Morrow, C. C., Spicer, L., Carmack, C. A., Zappe, M. E., Pohlmann, D., et al. 2013. Nearly complete 28S rRNA gene sequences confirm new hypotheses of sponge evolution. *Integr. Comp. Biol.* 533, 373–387.

Tivey, T. R., Parkinson, J. E. and Weis, V. M. 2020. Host and symbiont cell cycle coordination is mediated by symbiotic state, nutrition, and partner identity in a model cnidarian-dinoflagellate symbiosis. *mBio* 112, e02626–19.

Tyler, S. 2003. Epithelium: The primary building block for Metazoan complexity. *Integr. Comp. Biol.* 43, 55–63.

Ueda, N., Richards, G. S., Degnan, B. M., Kranz, A., Adamska, M., Croll, R. P. and Degnan, S. M. 2016. An ancient role for nitric oxide in regulating the animal pelagobenthic life cycle: Evidence from a marine sponge. *Sci. Rep.* 6, 37546.

Vacelet, J. and Boury-Esnault, N. 1995. Carnivorous sponges. *Nature* 373, 333–335.

Van Soest, R. W. M., Boury-Esnault, N., Hooper, J. N. A., Rützler, K., de Voogd, N. J., Alvarez, B., Hajdu, E., Pisera, A. B., Manconi, R., Schönberg, C., et al. 2021. *World Porifera Database*. www.marinespecies.org/porifera on yyyy-mm-dd.

Van Soest, R. W. M., de Kluijver, M. J., van Bragt, P. H., Faasse, M., Nijland, R., Beglinger, E. J., de Weerdt, W. H. and de Voogd, N. J. 2007. Sponge invaders in Dutch coastal waters. *J. Mar. Biol. Assoc. U. K.* (876), 1733–1748.

Vernale, A., Prünster, M. M., Marchianò, F., Debost, H., Brouilly, N., Rocher, C., Massey-Harroche, D., Renard, E., Bivic, A. L., Habermann, B. H., et al. In press. Evolution of mechanisms controlling epithelial morphogenesis across animals: New insights from dissociation – Reaggregation experiments in the sponge *Oscarella lobularis*. *BMC Ecol. Evol.*

Vicente, J., Zea, S. and Hill, R. T. 2016. Sponge epizoism in the Caribbean and the discovery of new *Plakortis* and *Haliclona* species, and polymorphism of *Xestospongia deweerdtae* Porifera. *Zootaxa* 4178, 209–233.

Vishnyakov, A. E. and Ereskovsky, A. V. 2009. Bacterial symbionts as an additional cytological marker for identification of sponges without a skeleton. *Mar. Biol.* 156, 1625–1632.

Vosmaer, G. C. J. 1884. Porifera. In: Bronn, H.G., Die Klassen und Ordnungen des Thierreichs, Volume II: 65–176, Pl. III, VII-XVIII.

Wallingford, J. B. 2010. Planar cell polarity signaling, cilia and polarized ciliary beating. *Curr. Opin. Cell Biol.* 22, 597–604.

Wang, X. and Lavrov, D. V. 2007. Mitochondrial genome of the Homoscleromorph *Oscarella carmela* Porifera, Demospongiae reveals unexpected complexity in the common ancestor of sponges and other animals. *Mol. Biol. Evol.* 24, 363–373.

Weiss, W. A., Taylor, S. S. and Shokat, K. M. 2007. Recognizing and exploiting differences between RNAi and small-molecule inhibitors. *Nat. Chem. Biol.* 3, 739–744.

Whelan, N. V., Kocot, K. M., Moroz, T. P., Mukherjee, K., Williams, P., Paulay, G., Moroz, L. L. and Halanych, K. M. 2017. Ctenophore relationships and their placement as the sister group to all other animals. *Nat. Ecol. Evol.* 1, 1737–1746.

Windsor Reid, P. J. and Leys, S. P. 2010. Wnt signaling and induction in the sponge aquiferous system: Evidence for an ancient origin of the organizer. *Evol. Dev.* 12, 484–493.

Windsor Reid, P. J., Matveev, E., McClymont, A., Posfai, D., Hill, A. L. and Leys, S. P. 2018. Wnt signaling and polarity in freshwater sponges. *BMC Evol. Biol.* 18, 12.

Wörheide, G., Dohrmann, M., Erpenbeck, D., Larroux, C., Maldonado, M., Voigt, O., Borchiellini, C. and Lavrov, D. V. 2012. Deep phylogeny and evolution of sponge phylum Porifera. *Adv. Mar. Biol.* 61, 1–78.

Yathish, R. and Grace, R. 2018. Cell-cell junctions and epithelial differentiation. *J. Morphol. Anat.* 2, 111.

Zhang, H., Dong, M., Chen, J., Wang, H., Tenney, K. and Crews, P. 2017. Bioactive secondary metabolites from the marine sponge genus agelas. *Mar. Drugs* 1511, 351.

Zimmer, A. M., Pan, Y. K., Chandrapalan, T., Kwong, R. W. M. and Perry, S. F. 2019. Loss-of-function approaches in comparative physiology: Is there a future for knockdown experiments in the era of genome editing? *J. Exp. Biol.* 222 (7), jeb175737.

6 Placozoa

Bernd Schierwater and Hans-Jürgen Osigus

CONTENTS

6.1 HISTORY OF THE MODEL

More than a century ago, the simplest of all metazoan animals was discovered in a seawater aquarium and described as *Trichoplax adhaerens* (Schulze 1883). This tiny, flattened animal lacked any kind of symmetry, mouth, gut, nervous system and extra-cellular matrix and immediately stimulated inspiring discussions on the ancestral morphology of a hypothetical "urmetazoon" (for overview, see Schierwater and DeSalle 2007; Schierwater et al. 2016; Schierwater and DeSalle 2018 and references therein). For more than half a century, this important animal was completely ignored, however, because of a wrong claim that *Trichoplax* was a larva form of a hydrozoan (see Ender and Schierwater 2003; Schierwater 2005 and references therein). It was the very tedious and precise work of the German zoologist Karl Gottlieb Grell which led to the erection of its own phylum for *Trichoplax* in 1971 (Grell 1971). Just recently, two more placozoan species were described, *Hoilungia hongkongensis* and *Polyplacotoma mediterranea* (Eitel et al. 2018; Osigus et al. 2019). Genetic data suggest the presence of even more—at least several dozen—placozoan species, which might be morphologically indistinguishable, that is, cryptic species (Eitel and Schierwater 2010). A yet-undescribed species, represented by the haplotype H2 (see e.g. Kamm et al. 2018), seems to be the most robust placozoan species for culturing and manipulations in the laboratory, and we use it, for example, for gravity research on earth and in space. Most people prefer to work with the original species, *Trichoplax adhaerens*, which has been the best-studied species, since it harbors the first characterized genome (Srivastava et al. 2008).

Placozoans diverged early in metazoan history, and their morphology fits nicely into almost any of the existing urmetazoan hypotheses, no matter if we derive placozoans from an early benthic gallertoid stage or any pelagic placula or planula stage (for overview, see Syed and Schierwater 2002; Schierwater et al. 2009 and references therein). In addition, the *Trichoplax* genome resembles the best living surrogate for a metazoan ancestor genome (Srivastava et al. 2008), and almost all major gene families known from humans are already present in *Trichoplax*. Thus, it comes as no surprise that from comparative morphology to cell physiology and molecular development to cancer research, *Trichoplax* has now been used as a basic model system to answer complex questions. From the very beginning of placozoan research and also from modern integration of molecular data, many evolutionary biologists have seen compelling evidence for an early branching position of placozoans at the very root of the metazoan tree of life (e.g. Schierwater et al. 2009; Schierwater et al. 2016 for references). However, a variety of molecular trees suggests Porifera as the earliest branching metazoans (e.g. Philippe et al. 2009; Pick et al. 2010; Simion et al. 2017).

When we have been sending placozoan cultures to different laboratories worldwide, we have mostly sent benign *Trichoplax adhaerens* (the original Grell culture-strain originating from the Red Sea, haplotype H1 (Figure 6.1); see Schierwater 2005 for details) or the yet-unnamed haplotype H2 (see e.g. Eitel and Schierwater 2010; Schleicherova et al. 2017; Kamm et al. 2018). For some literature on *T. adhaerens*, it is unclear, however, which species or haplotype was actually studied. This is because of the existence of an estimated number of at least two dozen cryptic placozoan species, which under the microscope all look identical to *T. adhaerens* (e.g. Voigt et al. 2004; Eitel and Schierwater 2010; Eitel et al. 2013).

6.2 GEOGRAPHICAL LOCATION

The precise geographical and global distribution of placozoans is difficult to define, since their microscopic size and fluctuating population densities call for time-intense sampling and microscopy efforts (see Eitel et al. 2013; Voigt and Eitel 2018). Nonetheless, from available records and

DOI: 10.1201/9781003217503-6

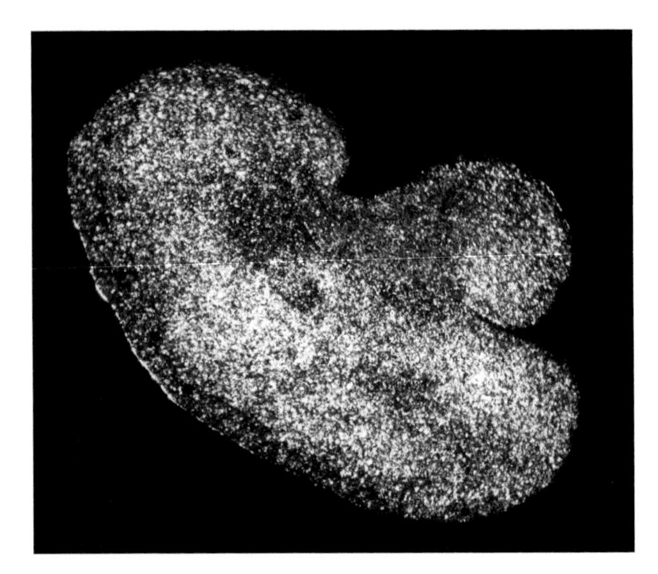

FIGURE 6.1 Life image of *Trichoplax adhaerens*. The shown animal measures about 3 mm in diameter.

FIGURE 6.2 Inferred geographic distribution of placozoans based on habitat modeling predictions. (From Paknia and Schierwater 2015.)

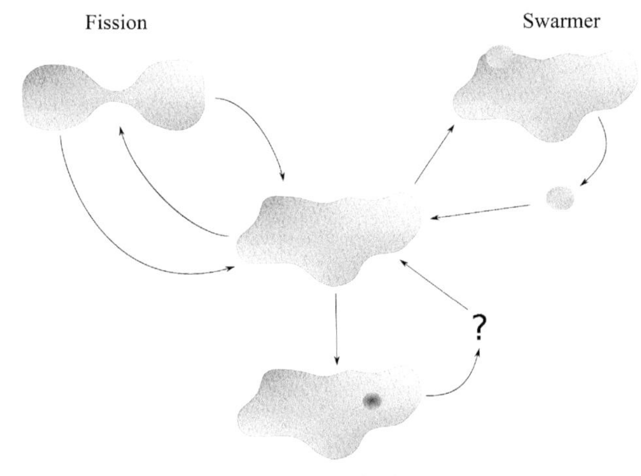

FIGURE 6.3 Schematic life cycle of placozoans. Vegetative reproduction in placozoans comprises the process of fission as well as the budding of mobile swarmer stages. Sexual development has only been recorded up to the 128-cell-stage of the embryo. (From Eitel et al. 2011.)

mathematical modelling, we conclude that placozoans are strictly marine (although they show some tolerance to brackish water, Eitel et al. 2013; Eitel et al. 2018) and are found between 55° northern and 44° southern latitude (Figure 6.2) (Paknia and Schierwater 2015). Placozoans live in all marine waters where the lowest water temperature is above 10°C (see Eitel et al. 2013).

While *Trichoplax adhaerens* (H1) is cosmopolitic and has been repeatedly found in warm oceans (see Eitel et al. 2013), the other two described placozoan species have each been found at one specific location only, but it remains to be seen whether these species are endemic (Eitel et al. 2018; Osigus et al. 2019). In general, there are clear differences between placozoan clades with respect to global distribution patterns (Eitel et al. 2013; Voigt and Eitel 2018). But, as noted before, global sampling records are highly preliminary, and reports are hard to compare because of different sampling and identification methods used. The two main sampling methods, trap-sampling and hard substrate sampling, differ substantially not only with respect to efficiency, but they also collect different life-cycle stages of placozoans (Pearse and Voigt 2007; Eitel and Schierwater 2010; Eitel et al. 2013; Miyazawa and Nakano 2018; Voigt and Eitel 2018): substrate sampling depends on the presence of a natural biofilm and mainly collects feeding adult animals, while trap sampling rather targets the planctonic placozoan swarmer stages. Thus, trap sampling methods in general shift the sampling bias toward placozoan species with higher rates of swarmer formation.

6.3 LIFE CYCLE

The complete life cycle of placozoans remains an unresolved mystery since the discovery of the first placozoan specimen in 1883 (Schulze 1883). The typical adult placozoan, that is, the benthic, disc-shaped (in one case ramified, Osigus et al. 2019) animal with no symmetry, normally reproduces by vegetative fission (see Figure 6.3), that is, by dividing into two—sometimes three—daughter individuals (Schulze 1883; Schulze 1891). Sometimes the vegetative formation of swarmers from the upper epithelium is seen in laboratory cultures (e.g. Thiemann and Ruthmann 1988). These pelagic swarmers are believed to float in the open water to eventually attach to a new substrate and this way allow dispersal if local conditions become unfavorable or population density calls for a change of location.

We know from observations in the laboratory and also from population genetics that placozoans do also reproduce sexually in the field (e.g. Eitel et al. 2011; Signorovitch et al. 2005; Kamm et al. 2018), and eggs or early embryo stages have sporadically been seen in laboratory cultures. However, a complete sexual reproductive cycle has never been reported in all the decades the animals have been kept in culture under laboratory conditions. Although eggs and early cleavage stages have been observed, the latter are cytologically anomalous and die at the 128-cell stage at the latest (Eitel et al. 2011 and references therein); neither meiosis, fertilization nor confirmed sperm cells have ever been documented. Observation of a fertilization membrane (Eitel et al. 2011) and genetic evidence for outcrossing, however, tell us that bisexual reproduction must occur in placozoans (e.g. Kamm et al. 2018). No adult sexual animals have ever been collected from the field (see also Voigt and Eitel 2018), and it remains unclear if fertilized eggs develop directly into adult placozoans or whether there is a larva or other additional life cycle stage in placozoans. We do not know if placozoans are hermaphroditic, but the genetic data do not support the idea that placozoans are using self-fertilization (Kamm et al. 2018). We have no reason to assume any derived mode of reproduction, like haploid or diploid parthenogenesis, to be present in placozoans.

6.4 EMBRYOGENESIS

As stated, only early embryogenesis has been seen in placozoans (Figure 6.4). Oocytes are built in the lower epithelium and then move into the intermediate fiber cell layer for further development, where fiber cells provide nutrition for the oocytes (Grell and Benwitz 1974; Eitel et al. 2011). One single mother animal can build up to nine oocytes simultaneously, while oocyte formation and maturation go along with the degeneration of the mother animal (Eitel et al. 2011). After an unknown fertilization process, a fertilization membrane appears around the fertilized egg (Grell and Benwitz 1974; Eitel et al. 2011). The subsequent total and equal cleavages of embryonic cells proceed to the 128-cell stage before the embryos die under laboratory conditions (Eitel et al. 2011).

6.5 ANATOMY

The general morphology of placozoans has been well known since the original description by Schulze (Schulze 1883; Schulze 1891) and the works of Karl Gottlieb Grell (e.g. Grell and Benwitz 1971). The precise ultrastructure of these organisms is still under investigation (e.g. Smith et al. 2014; Romanova et al. 2021). The general placozoan bauplan (see Figure 6.5) can be described as a three-layered disc, with an upper epithelium facing the open water, a lower (feeding) epithelium facing the substrate (see e.g. Smith et al. 2015) and a fiber cell layer (which has nothing to with an epithelium) in between.

A most remarkable and exclusive (and likely plesiomorphic) feature of the Placozoa is the lack of an extra-cellular matrix (ECM) and a basal lamina between the inner fiber cells and the enclosing epithelia (e.g. Smith et al. 2014). The reader must be aware that some textbooks (e.g. Brusca and Brusca 1990) and other publications falsely state the existence of an ECM. The interspace between the fiber cells and the epithelial cells is filled by a liquid, and both epithelia appear to be to some extent permeable for aqueous solutions (Ruthmann et al. 1986; but see also Smith and Reese 2016). The cells of the upper and lower epithelium are connected by adherens junctions, and neither tight nor septate or gap junctions have been found in *Trichoplax* (Ruthmann et al. 1986; Smith and Reese 2016).

So far, nine distinct somatic cell types have been identified in placozoans: upper and lower epithelial cells, sphere cells, crystal cells, three types of gland cells, lipophil cells and fiber cells (Schulze 1883; Smith et al. 2014; Mayorova

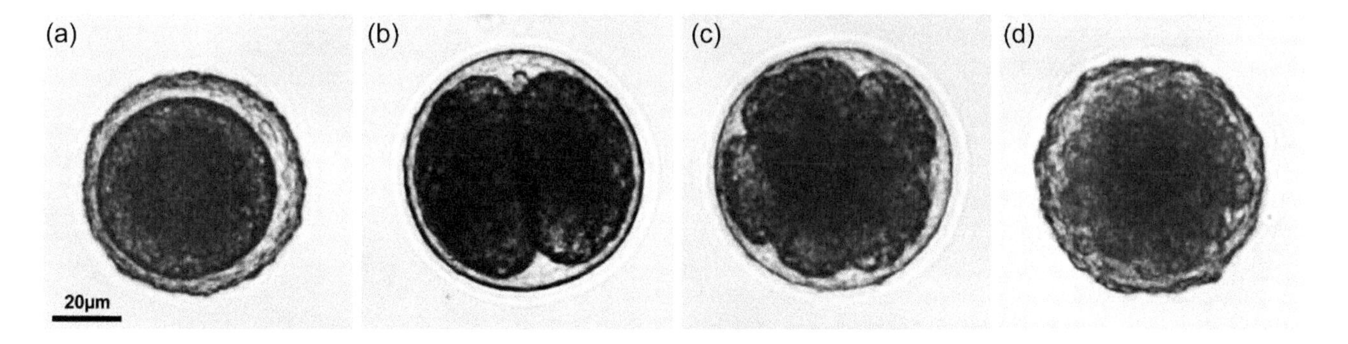

FIGURE 6.4 Early embryonic development in placozoans. A zygote is shown in (a), while (b)) to (d) show embryos at the 2-, 8- and 64-cell stage, respectively. (From Eitel et al. 2011.)

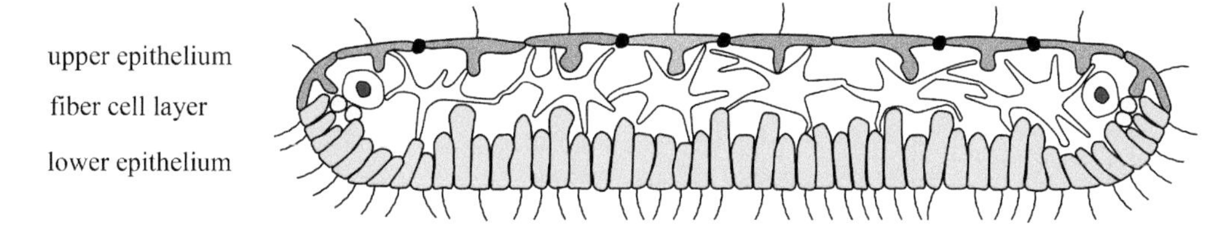

FIGURE 6.5 General anatomy of *Trichoplax adhaerens* shown as a synthesis of recent studies on the placozoan ultrastructure. The three-layered placozoan bauplan consists of an upper epithelium, a lower epithelium and a layer of fiber cells sandwiched between the two epithelia. (From Jakob et al. 2004; Guidi et al. 2011; Smith et al. 2014; and Eitel et al. 2018.)

et al. 2019; Romanova et al. 2021). The upper epithelium (consisting only of upper epithelial cells, some gland cells and sphere cells; Mayorova et al. 2019; Romanova et al. 2021) mainly has a protective function (Jackson and Buss 2009), whereas the lower epithelium (consisting of lower epithelial cells, lipophil cells and gland cells) is involved in digestion and nutrition uptake (e.g. Mayorova et al. 2019). The syncytial fiber cell layer between the two epithelia is involved in body contraction and signal transduction processes (Smith et al. 2014, Romanova et al. 2021 and references therein). The crystal cells are located at the edge of the animal and are likely involved in gravity perception (Mayorova et al. 2018). Also located close to the margin of the animal body are small undifferentiated cells, which have been regarded as omnipotent "stem" cells (Jakob et al. 2004). From comparative morphology, it is obvious that the lower epithelium resembles the entoderm and the upper epithelium the ectoderm of other metazoans (Bütschli 1884). The different lower epithelial cells use pinocytosis to take up food particles (Ruthmann et al. 1986). For this, the epithelial cells are covered with slime/mucus, allowing them to catch small food particles (Wenderoth 1986). The mucus of the lower epithelium is also involved in adhesion, movement and gliding (Mayorova et al. 2019). The upper epithelium shows lower differentiation, with the so-called 'shiny spheres' ("Glanzkugeln"; Schulze 1891; Jackson and Buss 2009), which are lipid droplets within the sphere cells (Romanova et al. 2021), as well as sporadically occuring gland cells (Mayorova et al. 2019).

6.6 GENOMIC DATA

In the last 15 years, three high-quality draft genomes have been published (Srivastava et al. 2008; Eitel et al. 2018; Kamm et al. 2018), in addition to a further three genomes of lower coverage (Laumer et al. 2018). With the genome of the haplotype H2, an additional—yet formally undescribed—*Trichoplax* species becomes available as a favorable model system (Kamm et al. 2018), which shows much higher robustness in laboratory cultures compared to other placozoans. From the available genome data, we can deduce that placozoan genomes range in size from 87–95 megabases and contain approximately 12,000 protein coding genes (Srivastava et al. 2008; Eitel et al. 2018; Kamm et al. 2018; Laumer et al. 2018). Based on the amount of conserved synteny to other metazoans like vertebrates and anthozoans (Srivastava et al. 2008), placozoans thus harbor the smallest not secondarily reduced metazoan genomes. Different placozoan species can be discriminated by a significant amount of gene sequence divergence, and less related species also show substantial differences in their gene's chromosomal arrangement (Srivastava et al. 2008; Eitel et al. 2018; Kamm et al. 2018; Laumer et al. 2018).

Compared to cnidarians and bilaterians, the complexity of the placozoan gene repertoire is lower (Schierwater et al. 2008; Srivastava et al. 2008; Alie and Manuel 2010; Eitel et al. 2018; Kamm et al. 2018; Kamm et al. 2019). Most eumetazoan gene families are present, but the expansion of several gene families, for example, homeobox genes, clearly happened after the split off of the Cnidaria (Kamm and Schierwater 2006; Kamm et al. 2006; Ryan et al. 2006; Schierwater et al. 2008). Likewise, the complexity of the gene repertoire related to cell–cell signaling (Srivastava et al. 2008), neuroendocrine function (Srivastava et al. 2008; Alie and Manuel 2010; Varoqueaux et al. 2018) or innate immunity (Kamm et al. 2019) represents a pre-cnidarian stage. On the other hand, placozoan genomes show several examples of phylum-specific gene family expansions (e.g. Eitel et al. 2018; Kamm et al. 2018; Kamm et al. 2019). These examples include genes related to innate immunity and cell death (Kamm et al. 2019) and the large group of G protein-coupled receptors (Kamm et al. 2018). The latter group of cell surface receptors also shows a high diversity within the phylum and may represent more than 6% of all genes in a species (Kamm et al. 2018). Gene duplications within such diverse gene families may thus also be a driver for speciation within the phylum (Eitel et al. 2018).

6.7 FUNCTIONAL APPROACHES: TOOLS FOR MOLECULAR AND CELLULAR ANALYSES

The simplicity of the *Trichoplax* model allows the use of the full spectrum of modern molecular methods for mapping and reconstructing fundamental cellular and organismal processes (e.g. von der Chevallerie et al. 2014; Varoqueaux et al. 2018; Popgeorgiev et al. 2020; Moroz et al. 2021 and references therein). New tools such as single-cell transcriptomics have become available and have already been tested in *Trichoplax* (Sebe-Pedros et al. 2018). So have in situ-hybridizations (Figure 6.6; see also e.g. DuBuc et al. 2019), as well as RNAi gene silencing (e.g. Jakob et al. 2004), and other modern gene knockout techniques are soon going to be established in placozoans as well.

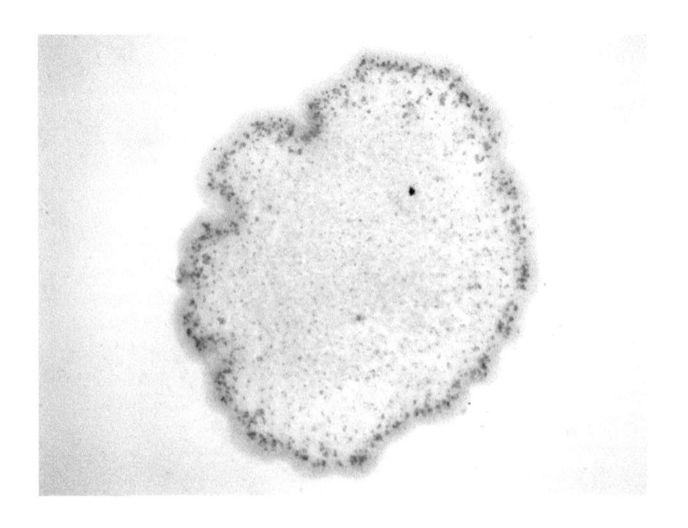

FIGURE 6.6 Whole-mount in situ hybridization reveals the typical ring-shaped expression pattern of the ParaHox gene *Trox-2* in *Trichoplax adhaerens*. (Photo by Moritz J. Schmidt and Sonja Johannsmeier.)

At the organismal level, *Trichoplax* allows the use of the *cum grano salis* full spectrum of regeneration, re-aggregation and transplantation techniques (e.g. Schwartz 1984). The size, thickness, transparency and stability of the animals make them preferred objects for traditional and modern techniques of light and high-resolution electron microscopy (e.g. Guidi et al. 2011; Smith et al. 2021). By combining these, that is, the organismal and molecular potential, placozoans offer solid prospects to answer challenging questions.

6.8 CHALLENGING QUESTIONS

While some researchers still fight over the phylogenetic position of placozoans, others have realized and accepted the outstanding importance of an early metazoan animal that harbors all the core genes for the regulation of tissue architecture in metazoans. Most regulators are highly conserved (at different levels) between *Trichoplax* and humans, and we can use a simple *Trichoplax* model to learn important details about regulatory interplays in the much more complex worm, fly and mouse models. Thus, it comes as no surprise that the current questions we are asking *Trichoplax* range from "How can symmetry be derived from polarity?" to "What is the basic genetics behind apoptosis?" to "What are the initial genetic malfunctions that start cancer growth?". And there will be many more to come.

ACKNOWLEDGMENTS

We thank Kristin Fenske and Kai Kamm for help and comments. Moritz J. Schmidt and Sonja Johannsmeier kindly provided Figure 6.6.

BIBLIOGRAPHY

Alie, A., and M. Manuel. 2010. The backbone of the post-synaptic density originated in a unicellular ancestor of choanoflagellates and metazoans. *BMC Evol Biol* 10:34.

Brusca, R. C., and G. J. Brusca. 1990. *Invertebrates*. Sunderland, MA: Sinauer Associates.

Bütschli, O. 1884. Bemerkungen zur Gastraea-Theorie. *Morphologische Jahrblatt* 9.

DuBuc, T. Q., J. F. Ryan, and M. Q. Martindale. 2019. "Dorsal-ventral" genes are part of an ancient axial patterning system: Evidence from *Trichoplax adhaerens* (Placozoa). *Mol Biol Evol* 36:966–973.

Eitel, M., W. R. Francis, F. Varoqueaux, J. Daraspe, H. J. Osigus, S. Krebs, S. Vargas, H. Blum, G. A. Williams, B. Schierwater, and G. Worheide. 2018. Comparative genomics and the nature of placozoan species. *PLoS Biol* 16:e2005359.

Eitel, M., L. Guidi, H. Hadrys, M. Balsamo, and B. Schierwater. 2011. New insights into placozoan sexual reproduction and development. *PLoS One* 6:e19639.

Eitel, M., H. J. Osigus, R. DeSalle, and B. Schierwater. 2013. Global diversity of the Placozoa. *PLoS One* 8:e57131.

Eitel, M., and B. Schierwater. 2010. The phylogeography of the Placozoa suggests a taxon-rich phylum in tropical and subtropical waters. *Mol Ecol* 19:2315–2327.

Ender, A., and B. Schierwater. 2003. Placozoa are not derived cnidarians: Evidence from molecular morphology. *Mol Biol Evol* 20:130–134.

Grell, K. G. 1971. *Trichoplax adhaerens* F.E. Schulze und die Entstehung der Metazoan. *Naturwissenschaftliche Rundschau* 24:160–161.

Grell, K. G., and G. Benwitz. 1971. Die Ultrastruktur von *Trichoplax adhaerens* F.E. Schulze. *Cytobiologie* 4:216–240.

Grell, K. G., and G. Benwitz. 1974. Elektronenmikroskopische Beobachtungen über das Wachstum der Eizelle und die Bildung der "Befruchtungsmembran" von *Trichoplax adhaerens* F.E.Schulze (Placozoa). *Zeitschrift für Morphologie der Tiere* 79:295–310.

Guidi, L., M. Eitel, E. Cesarini, B. Schierwater, and M. Balsamo. 2011. Ultrastructural analyses support different morphological lineages in the phylum Placozoa Grell, 1971. *J Morphol* 272:371–378.

Jackson, A. M., and L. W. Buss. 2009. Shiny spheres of placozoans (*Trichoplax*) function in anti-predator defense. *Invertebrate Biology* 128:205–212.

Jakob, W., S. Sagasser, S. Dellaporta, P. Holland, K. Kuhn, and B. Schierwater. 2004. The Trox-2 Hox/ParaHox gene of Trichoplax (Placozoa) marks an epithelial boundary. *Dev Genes Evol* 214:170–175.

Kamm, K., H. J. Osigus, P. F. Stadler, R. DeSalle, and B. Schierwater. 2018. Trichoplax genomes reveal profound admixture and suggest stable wild populations without bisexual reproduction. *Sci Rep* 8:11168.

Kamm, K., and B. Schierwater. 2006. Ancient complexity of the non-Hox ANTP gene complement in the anthozoan *Nematostella vectensis*: Implications for the evolution of the ANTP superclass. *J Exp Zool B Mol Dev Evol* 306:589–596.

Kamm, K., B. Schierwater, and R. DeSalle. 2019. Innate immunity in the simplest animals—placozoans. *BMC Genomics* 20:5.

Kamm, K., B. Schierwater, W. Jakob, S. L. Dellaporta, and D. J. Miller. 2006. Axial patterning and diversification in the cnidaria predate the Hox system. *Curr Biol* 16:920–926.

Laumer, C. E., H. Gruber-Vodicka, M. G. Hadfield, V. B. Pearse, A. Riesgo, J. C. Marioni, and G. Giribet. 2018. Support for a clade of Placozoa and Cnidaria in genes with minimal compositional bias. *Elife* 7.

Mayorova, T. D., K. Hammar, C. A. Winters, T. S. Reese, and C. L. Smith. 2019. The ventral epithelium of *Trichoplax adhaerens* deploys in distinct patterns cells that secrete digestive enzymes, mucus or diverse neuropeptides. *Biol Open* 8 (8).

Mayorova, T. D., C. L. Smith, K. Hammar, C. A. Winters, N. B. Pivovarova, M. A. Aronova, R. D. Leapman, and T. S. Reese. 2018. Cells containing aragonite crystals mediate responses to gravity in *Trichoplax adhaerens* (Placozoa), an animal lacking neurons and synapses. *PLoS One* 13:e0190905.

Miyazawa, H., and H. Nakano. 2018. Multiple surveys employing a new sample-processing protocol reveal the genetic diversity of placozoans in Japan. *Ecol Evol* 8:2407–2417.

Moroz, L. L., Romanova, D. Y., and A. B. Kohn. 2021. Neural versus alternative integrative systems: Molecular insights into origins of neurotransmitters. *Philos Trans R Soc Lond B Biol Sci* 376 (1821):20190762.

Osigus, H. J., S. Rolfes, R. Herzog, K. Kamm, and B. Schierwater. 2019. *Polyplacotoma mediterranea* is a new ramified placozoan species. *Curr Biol* 29:R148–R149.

Paknia, O., and B. Schierwater. 2015. Global habitat suitability and ecological niche separation in the phylum Placozoa. *PLoS One* 10 (11).

Pearse, V. B., and O. Voigt. 2007. Field biology of placozoans (Trichoplax): Distribution, diversity, biotic interactions. *Integr Comp Biol* 47:677–692.

Philippe, H., R. Derelle, P. Lopez, K. Pick, C. Borchiellini, N. Boury-Esnault, J. Vacelet, E. Renard, E. Houliston, E. Queinnec,

C. Da Silva, P. Wincker, H. Le Guyader, S. Leys, D. J. Jackson, F. Schreiber, D. Erpenbeck, B. Morgenstern, G. Worheide, and M. Manuel. 2009. Phylogenomics revives traditional views on deep animal relationships. *Curr Biol* 19:706–712.

Pick, K. S., H. Philippe, F. Schreiber, D. Erpenbeck, D. J. Jackson, P. Wrede, M. Wiens, A. Alie, B. Morgenstern, M. Manuel, and G. Worheide. 2010. Improved phylogenomic taxon sampling noticeably affects nonbilaterian relationships. *Mol Biol Evol* 27:1983–1987.

Popgeorgiev, N., J. D. Sa, L. Jabbour, S. Banjara, T. T. M. Nguyen, E. Sabet A. Akhavan, R. Gadet, N. Ralchev, S. Manon, M. G. Hinds, H. J. Osigus, B. Schierwater, P. O. Humbert, R. Rimokh, G. Gillet, and M. Kvansakul. 2020. Ancient and conserved functional interplay between Bcl-2 family proteins in the mitochondrial pathway of apoptosis. *Sci Adv* 6 (40).

Ruthmann, A., Behrendt G., and R. Wahl. 1986. The ventral epithelium of *Trichoplax adhaerens* (Placozoa). *Zoomorphology* 106:115–122.

Romanova, D. Y., Varoqueaux, F., Daraspe, J., Nikitin, M. A., Eitel, M., Fasshauer, D., and L. L. Moroz. 2021. Hidden cell diversity in Placozoa: Ultrastructural insights from *Hoilungia hongkongensis*. *Cell Tissue Res*.

Ryan, J. F., P. M. Burton, M. E. Mazza, G. K. Kwong, J. C. Mullikin, and J. R. Finnerty. 2006. The cnidarian-bilaterian ancestor possessed at least 56 homeoboxes: Evidence from the starlet sea anemone, *Nematostella vectensis*. *Genome Biol* 7:R64.

Schierwater, B. 2005. My favorite animal, *Trichoplax adhaerens*. *Bioessays* 27:1294–1302.

Schierwater, B., and R. DeSalle. 2007. Can we ever identify the Urmetazoan? *Integrative and Comparative Biology* 47: 670–676.

Schierwater, B., and R. DeSalle. 2018. Placozoa. *Curr Biol* 28: R97–R98.

Schierwater, B., M. Eitel, W. Jakob, H. J. Osigus, H. Hadrys, S. L. Dellaporta, S. O. Kolokotronis, and R. Desalle. 2009. Concatenated analysis sheds light on early metazoan evolution and fuels a modern "urmetazoon" hypothesis. *PLoS Biol* 7:e20.

Schierwater, B., K. Kamm, M. Srivastava, D. Rokhsar, R. D. Rosengarten, and S. L. Dellaporta. 2008. The early ANTP gene repertoire: Insights from the placozoan genome. *PLoS One* 3:e2457.

Schierwater, B., P. W. H. Holland, D. J. Miller, P. F. Stadler, B. M. Wiegmann, G. Wörheide, G. A. Wray, and R. DeSalle. 2016. Never ending analysis of a century old evolutionary debate: "Unringing" the urmetazoon bell. *Front Ecol Evol* 4 (5).

Schleicherova, D., K. Dulias, H. J. Osigus, O. Paknia, H. Hadrys, and B. Schierwater. 2017. The most primitive metazoan animals, the placozoans, show high sensitivity to increasing ocean temperatures and acidities. *Ecol Evol* 7:895–904.

Schulze, F. E. 1883. *Trichoplax adhaerens*, nov. gen., nov. spec. *Zoologischer Anzeiger* 6:92–97.

Schulze, F. E. 1891. Über *Trichoplax adhaerens*. In *Abhandlungen der Königlichen Preuss. Akademie der Wissenschaften zu Berlin.*, edited by G. Reimer, 1–23. Berlin: Verlag der königlichen Akademie der Wissenschaften.

Schwartz, V. 1984. The radial polar pattern of differentiation in *Trichoplax adhaerens* F.E. Schulze (Placozoa). *Zeitschrift für Naturforschung C* 39:818–832.

Sebe-Pedros, A., E. Chomsky, K. Pang, D. Lara-Astiaso, F. Gaiti, Z. Mukamel, I. Amit, A. Hejnol, B. M. Degnan, and A. Tanay. 2018. Early metazoan cell type diversity and the evolution of multicellular gene regulation. *Nat Ecol Evol* 2:1176–1188.

Signorovitch, A. Y., S. L. Dellaporta, and L. W. Buss. 2005. Molecular signatures for sex in the Placozoa. *Proc Natl Acad Sci U S A* 102:15518–15522.

Simion, P., H. Philippe, D. Baurain, M. Jager, D. J. Richter, A. Di Franco, B. Roure, N. Satoh, E. Queinnec, A. Ereskovsky, P. Lapebie, E. Corre, F. Delsuc, N. King, G. Worheide, and M. Manuel. 2017. A large and consistent phylogenomic dataset supports sponges as the sister group to all other animals. *Curr Biol* 27:958–967.

Smith, C. L., T. D. Mayorova, C. A. Winters, T. S. Reese, S. P. Leys, and A. Heyland. 2021. Microscopy studies of placozoans. *Methods Mol Biol* 2219:99–118.

Smith, C. L., N. Pivovarova, and T. S. Reese. 2015. Coordinated feeding behavior in Trichoplax, an animal without synapses. *PLoS One* 10:e0136098.

Smith, C. L., and T. S. Reese. 2016. Adherens junctions modulate diffusion between epithelial cells in *Trichoplax adhaerens*. *Biol Bull* 231:216–224.

Smith, C. L., F. Varoqueaux, M. Kittelmann, R. N. Azzam, B. Cooper, C. A. Winters, M. Eitel, D. Fasshauer, and T. S. Reese. 2014. Novel cell types, neurosecretory cells, and body plan of the early-diverging metazoan *Trichoplax adhaerens*. *Curr Biol* 24:1565–1572.

Srivastava, M., E. Begovic, J. Chapman, N. H. Putnam, U. Hellsten, T. Kawashima, A. Kuo, T. Mitros, A. Salamov, M. L. Carpenter, A. Y. Signorovitch, M. A. Moreno, K. Kamm, J. Grimwood, J. Schmutz, H. Shapiro, I. V. Grigoriev, L. W. Buss, B. Schierwater, S. L. Dellaporta, and D. S. Rokhsar. 2008. The Trichoplax genome and the nature of placozoans. *Nature* 454:955–960.

Syed, T., and B. Schierwater. 2002. The evolution of the placozoa: A new morphological model. *Senckenbergiana lethaea* 82:315–324.

Thiemann, M., and A. Ruthmann. 1988. *Trichoplax adhaerens* Schulze, F. E. (Placozoa): The formation of swarmers. *Zeitschrift für Naturforschung C* 43:955–957.

Varoqueaux, F., E. A. Williams, S. Grandemange, L. Truscello, K. Kamm, B. Schierwater, G. Jekely, and D. Fasshauer. 2018. High cell diversity and complex peptidergic signaling underlie placozoan behavior. *Curr Biol* 28:3495–3501.

Voigt, O., A. G. Collins, V. B. Pearse, J. S. Pearse, A. Ender, H. Hadrys, and B. Schierwater. 2004. Placozoa: No longer a phylum of one. *Curr Biol* 14:R944–R945.

Voigt, O., and M. Eitel. 2018. "Placozoa." In *Miscellaneous Invertebrates*, edited by A. Schmidt-Rhaesa, 41–54. Berlin, Boston: De Gruyter.

von der Chevallerie, K., S. Rolfes, and B. Schierwater. 2014. Inhibitors of the p53-Mdm2 interaction increase programmed cell death and produce abnormal phenotypes in the placozoon *Trichoplax adhaerens* (F.E. Schulze). *Dev Genes Evol* 224:79–85.

Wenderoth, H. 1986. Transepithelial cytophagy by *Trichoplax adhaerens* F.E.Schulze (Placozoa) feeding on yeast. *Zeitschrift für Naturforschung C* 41:343–347.

7 *Nematostella vectensis* as a Model System

Layla Al-Shaer, Jamie Havrilak and Michael J. Layden

CONTENTS

7.1 HISTORY OF THE MODEL

Nematostella vectensis (the starlet sea anemone) are anthozoan cnidarians. Anthozoans (e.g. corals, anemones) derive their name from the Greek *anthos*—flower—and *zōia*—animals—because their dominant polyp form shared by this class represents "a highly colored and many-petaled flower" (Figure 7.1a) (Gosse 1860). Additionally, the different morphological states of the animal can be described as the "flower" (*anthus*), when all the tentacles are extended, or the "button" (*oncus*), when the tentacles are retracted and the oral end closes in around them (Gosse 1860).

The first description of *Nematostella vectensis* was published in 1935 by Thomas A. Stephenson. Stephenson attributed the discovery of *Nematostella vectensis* and observations of their nematosomes to Ms. Gertrude F. Selwood. She found them at the Isle of Wight (England) (Figures 7.1b, 7.2) in 1929 when she was a lecturer at Municipal College, Portsmith, and sent specimens to Stephenson. Stephenson described the free-swimming nematosomes in the gastric

DOI: 10.1201/9781003217503-7

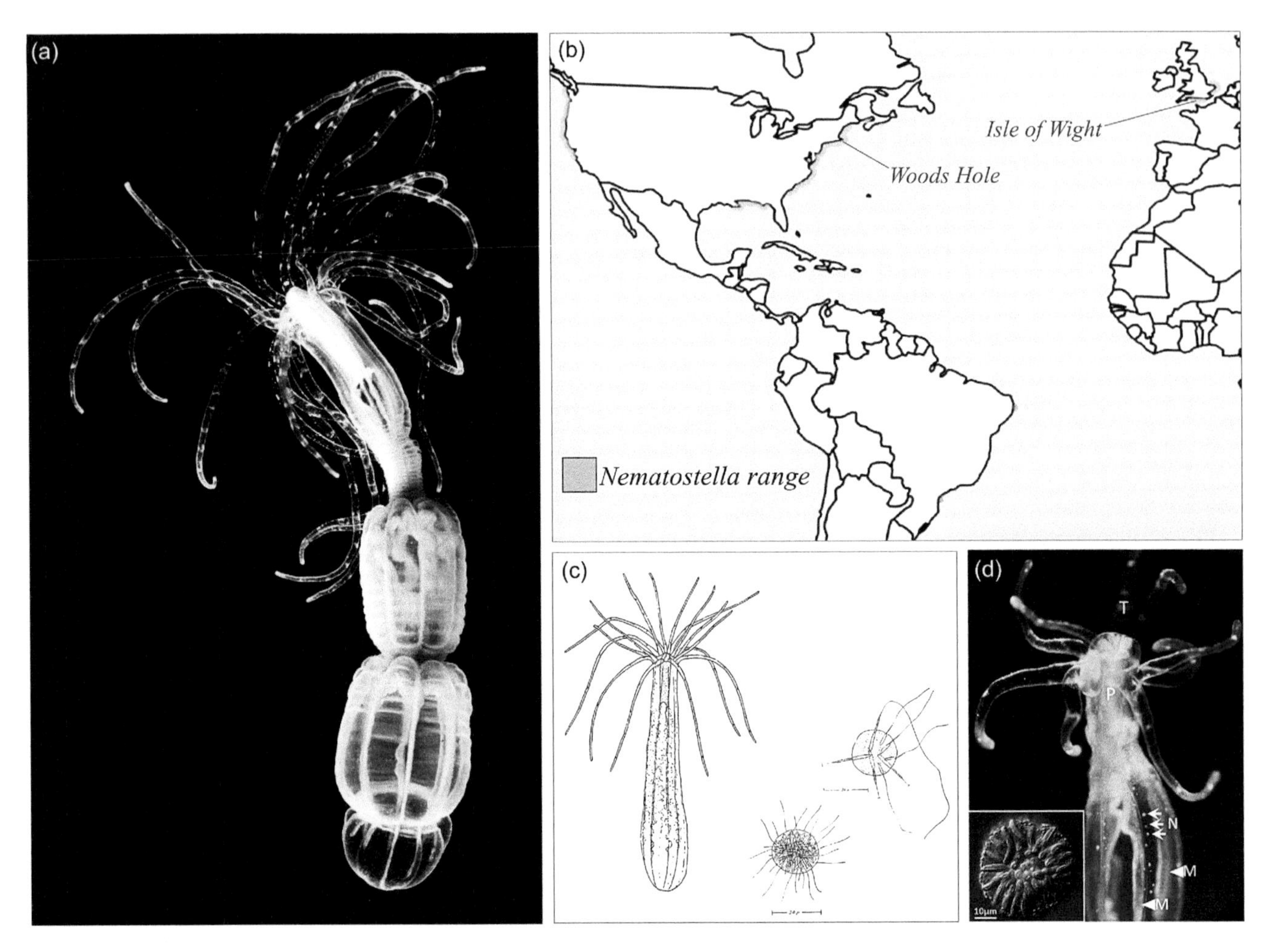

FIGURE 7.1 **Characteristics and geographical range of *Nematostella vectensis*.** (a) An adult polyp, image courtesy of Eric Röttinger. (b) Known geographical range. (c) First known illustrations of a *Nematostella* polyp and nematosomes with intact (left) and fired (right) cnidocytes. (d) Polyp showing tentacles (T) pharynx (P), mesenteries (M) and nematosomes (N). ([b] Illustrations by Sears Crowell 1946; [d] image modified from Babonis et al. 2016.)

cavity that became the characteristic feature of *Nematostella* (Figure 7.1c,d) (Stephenson 1935).

In 1939, William J. Bowden discovered *Nematostella pellucida* at Woods Hole, Massachusetts (Figure 7.1b), which was later described and published by Sears Crowell in 1946 (Figure 7.1c). *Nematostella pellucida* was initially considered distinct from *vectensis* due to color patterns on the body and the large geographical separation (Williams 1975; Williams 1976). Crowell suspected *N. vectensis* and *N. pellucida* were synonymous species, but because of the war, he was unable to get hold of *Nematostella* from the British Isles for direct comparisons (Crowell 1946). In 1957, Cadet Hand compared anemones from America and England and determined they were both *Nematostella vectensis* (Figure 7.2) (Hand 1957; Williams 1975; Williams 1976).

The various life history stages of *Nematostella* were described by several different groups from the 1940s to the 1980s (summarized in Hand and Uhlinger 1992). The potential for *Nematostella* as a laboratory model came when Cadet Hand and Kevin Uhlinger documented the ease of its culturability in the early 1990s (1992). Its ability to tolerate wide variations in salinity and temperature made it easy to maintain in laboratory cultures (Williams 1975; Williams 1976). Perhaps most importantly, *Nematostella* spawn readily with increased temperature and light. Early work established various environmental conditions that influence oogenesis, such as nutrient amount, temperature, light, density of sperm and the ideal timeframe for fertilization (Hand and Uhlinger 1992; Fritzenwanker and Technau 2002). The ability to reliably obtain thousands of embryos per spawn and close the life cycle in culture made *Nematostella* stand out as a potential cnidarian model system. Plus, due to its phylogenetic position as a basal metazoan, it is also especially well suited for evolutionary and developmental biology (evo-devo) studies.

Through the 1990s and early 2000s, studies focused on identifying the expression of known bilaterian homologues during *Nematostella* development (Figure 7.2). Initial studies focused on genes involved in axial patterning and triploblasty and provided initial insights into the origin and evolution of these genes and thus the bilaterian traits they regulate (see Darling et al. 2005). Similarly, extensive efforts

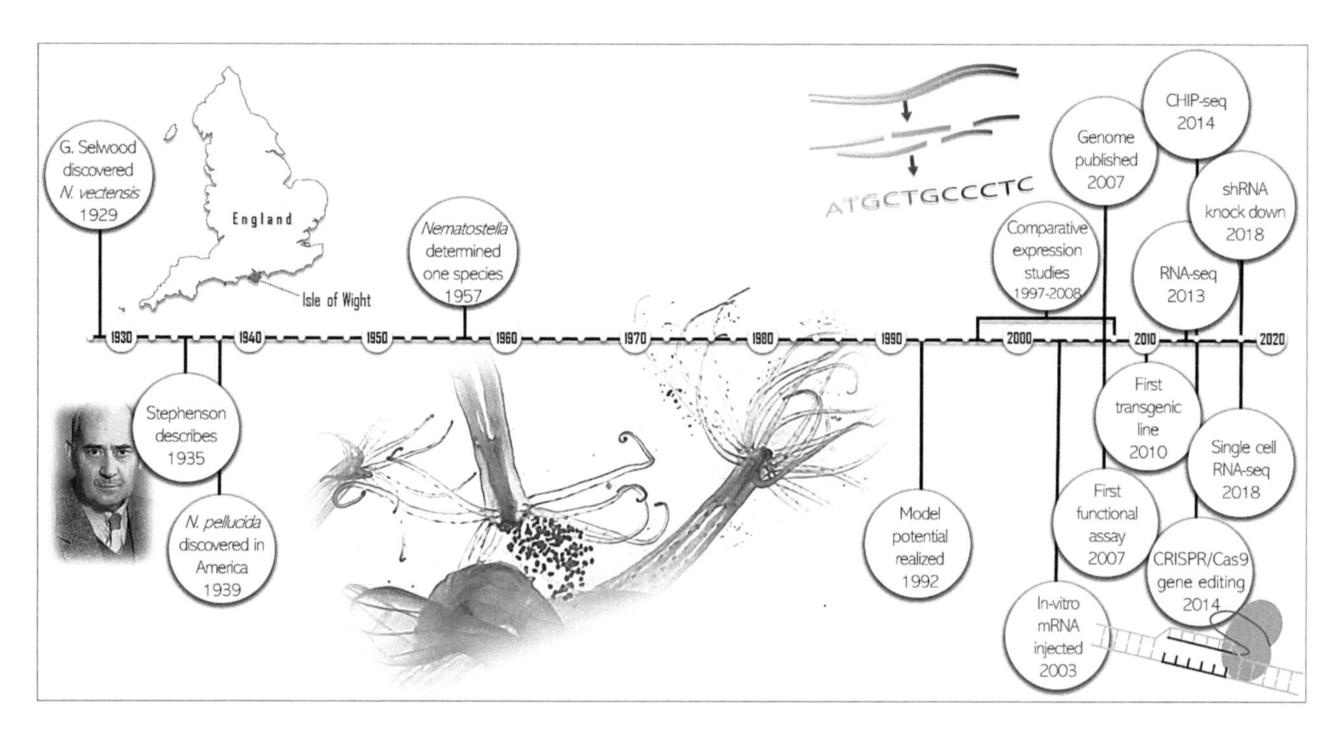

FIGURE 7.2 A timeline of major events in the *Nematostella* model system. (Picture of TA Stephenson adapted from Yonge 1962.)

focused on identifying expression of deeply conserved developmental signaling cascades. Comparative genomic studies have identified that genes involved in major families and signaling cascades were all present in the urbilaterian ancestor (e.g. Kortschak et al. 2003; Magie et al. 2005; Putnam et al. 2007). Furthermore, many cnidarian amino acid sequences are more like vertebrate sequences than other common model systems (Kortschak et al. 2003; Putnam et al. 2007) which supported the need to develop *Nematostella* as another model through the transition to more functional studies.

The application of molecular tools and reverse genetic approaches fueled the growth and use of *Nematostella* (Figure 7.2). With advances in sequencing technology and the publication of the genome in 2007 (Putnam et al. 2007), there has been a rapid increase in usage of *Nematostella* as a model organism. Morpholinos were first used successfully in *Nematostella* in 2005, and the first morphant phenotype was reported in 2008 (Magie et al. 2005; Rentzsch et al. 2008). Together these findings fueled the growth of functional studies, which have grown to include additional methods of gene knockdown and misexpression. The first transgenic line was published in 2010 with the creation of a muscle-specific reporter line, and transgenic reporter animals have also been used to identify and track specific cell types (Layden et al. 2012; Nakanishi et al. 2012).

Nematostella has repeatedly shown that it is amenable to novel and state-of-the-art molecular techniques. Genomic-level analyses were established for microarray, ChIP-seq and RNA-seq (Röttinger et al. 2012; Fritz et al. 2013; Helm et al. 2013; Tulin et al. 2013; Schwaiger et al. 2014; Sinigaglia et al. 2015). Cellular dissociation protocols and advances in single-cell sequencing technology have been successfully applied, and the use of single-cell RNA-sequencing has allowed for interrogation of the complexity of *Nematostella* cell types and characterization of gene regulatory programs (Sebé-Pedrós et al. 2018).

Their transparent body, relatively "simple" body plan, external fertilization, ease of embryo manipulation and closed life cycle in the lab make *Nematostella* amenable to a myriad of research approaches and questions, including the ability to compare development and regeneration. The application of next-generation approaches has cemented the use of *Nematostella* as a model organism. *Nematostella* joins several other cnidarian species that have become more commonly utilized laboratory models, such as *Hydractinia* and *Clytia*. *Hydra* have long been established as a model for regeneration, but they are not as amenable to developmental studies. The combination of knowledge gained from multiple cnidarian species will help to understand the ancestral toolkit in the common ancestor that gave rise to both the cnidarian and bilaterian lineages.

7.2 GEOGRAPHY AND HABITAT

Native to the Atlantic coast of North America (Hand and Uhlinger 1995; Reitzel et al. 2007), the geographic range of *Nematostella* has expanded through anthropogenic introduction to locations across at least three continents. In North America, abundant populations have been observed along the Atlantic coast from Nova Scotia to Georgia, along the Pacific coast from Washington to California and along the Gulf coast from Florida to Louisiana (Hand and Uhlinger 1992; Hand and Uhlinger 1994). In Europe, *Nematostella* occur in limited number in locations along the southern and eastern coasts of England (Stephenson 1935; Sheader et al. 1997; Pearson et al. 2002), and in South America, where they have been found in locales off the coast of Brazil (Figure 7.1b) (Silva et al. 2010; Brandão et al. 2019). Genetic and phylogeographical analyses indicate that global populations are isolated and therefore unlikely to have spread via natural dispersal mechanisms. (Hand and Uhlinger 1995; Reitzel et al. 2007). More plausible is that they were carried as hitchhikers in the ballasts of commercial seafood vessels, creating the potential for new populations to become established outside of the natural range (Sheader et al. 1997; Takahashi et al. 2008).

The successful geographic expansion of *Nematostella* can likely be attributed to their environmental plasticity, as they can inhabit a variety of coastal habitats and can tolerate fluctuating environmental conditions. They often occur burrowed in soft muddy sediments of poikilohaline lagoons, brackish mudflats, salt marshes and creeks and subtidal areas of certain estuaries and bays (Williams 1975; Williams 1976). As eurythermal animals, they can survive and adapt to a wide range of temperatures and have even been found living in habitats that approach their physiological upper limit of approximately 40°C (Williams 1975; Williams 1976). As euryhaline animals, they can contend with the spatiotemporal fluctuations in salinity common in the estuarine habitats where they are found. A testament to the remarkably flexible physiology of this anemone is that both asexual and sexual reproduction can occur under a wide range of salinities (Hand and Uhlinger 1992).

7.3 ANATOMY

Adult *Nematostella* are transparent and possess the classic polyp morphology found throughout the cnidarian lineage (Figure 7.3a). Atop the oral end of the body column is an opening that is surrounded by 4–18 long stinging tentacles, which aid in prey capture and defense but also expand the surface area of the gastric cavity (Fritz et al. 2013; Ikmi et al. 2020). This oral opening serves as both a mouth and anus by attaching to a blind-ended gut through a noticeable pharynx (Williams 1975; Williams 1976). There is also a small pore at the aboral pole (Amiel et al. 2015). The oral–aboral axis is elongated, which gives the body column a tube-like structure. Eight radially repeating body segments, which are centered around the long oral–aboral axis, give the animal what appears to be an octoradial symmetry (Figure 7.3a).

Cnidarians are generally classified as having a radially symmetric body plan, but many species have subtle bilateral differences in their anatomy that are superimposed over a general radial body plan (Martindale et al. 2002). These bilateral differences point to the presence of a secondary directive axis, which runs perpendicular to the primary oral–aboral axis (Figure 7.3b) (Berking 2007). In *Nematostella*, the presence of a directive axis is morphologically evident in adult polyps from the slit-like shape of the oral opening and pharynx, the presence of a ciliated groove (siphonoglyph) on one side of the pharynx, the asymmetric arrangement of the retractor muscles within the mesenteries and the asymmetric arrangement of the tentacles around the oral opening (Martindale et al. 2002; Berking 2007; He et al. 2018).

Nematostella are diploblastic, meaning that the entire body is composed of cells derived from two germ layers: an outer ectoderm which forms the epidermis and a bifunctional internal endoderm which forms the gastrodermis (Figure 7.3a) (Finnerty et al. 2004; Wijesena et al. 2017; but see Steinmetz et al. 2017). The epidermis covers the outside of the animal and serves as a protective barrier between the animal and its environment, while the bifunctional gastrodermis lines the coelenteron and provides both absorptive and contractile functions (Martindale et al. 2004). Separating the ectoderm and endoderm is the mesoglea, a thin extracellular matrix with no organized tissue and only a few migratory amoebocyte cells of unknown function (Tucker et al. 2011). The ectoderm contains primarily columnar cell epithelia (Magie et al. 2007), along with other differentiated cells, including stinging cells called cnidocytes (Frank and Bleakney 1976), sensory neurons, ganglion neurons (Marlow et al. 2009; Sinigaglia et al. 2015; Leclère et al. 2016), a population of myoepithelial (muscle) cells in the tentacles (Jahnel et al. 2014) and gland cells (Frank and Bleakney 1976). Ectodermal gland cells include those with exocrine and insulinergic functions (Steinmetz et al. 2017), and some produce a potent neurotoxin for both prey capture and defense (Moran et al. 2011). The endoderm possesses squamous epithelial cells (Magie et al. 2007), sensory and ganglion neurons (Marlow et al. 2009; Sinigaglia et al. 2015; Leclère et al. 2016), the majority of myoepithelial cells (Jahnel et al. 2014), gland cells (Frank and Bleakney 1976; Steinmetz et al. 2017) and gametic and absorptive cells (Layden et al. 2012; Nakanishi et al. 2012). This basic organization results in epithelial cells and differentiated cell types being scattered and intermixed with one another, as opposed to being organized into discrete organ systems.

Apart from the pharynx, the most obvious internal structures of adult *Nematostella* are the ecto- and endodermally derived lamellae known as mesenteries (Steinmetz et al. 2017). Adults have eight mesenteries, one in each body segment, that look ruffled in appearance and run the length of the body column (Figure 7.3a). Each mesentery arises from the pharynx and consists of two layers of gastrodermis epithelium separated by a layer of mesoglea (Martindale et al. 2004). Structurally, the mesenteries are important because they provide support for the pharynx, they contain muscles that allow for quick contractions of the body column (Renfer et al. 2010) and they increase the surface area of the gastrodermis. Physiologically, the mesenteries are incredibly multifunctional, as they contain absorptive cells that aid in digestion and nutrient uptake and are where gametes (Martindale et al. 2004), cnidocytes (Steinmetz 2019) and nematosomes are produced (Williams 1975, 1976). Nematosomes are the defining apomorphy of *Nematostella* (Williams 1975, 1976, 1979). They are multicellular, spherical, flagellated bodies that contain cnidocytes and can be

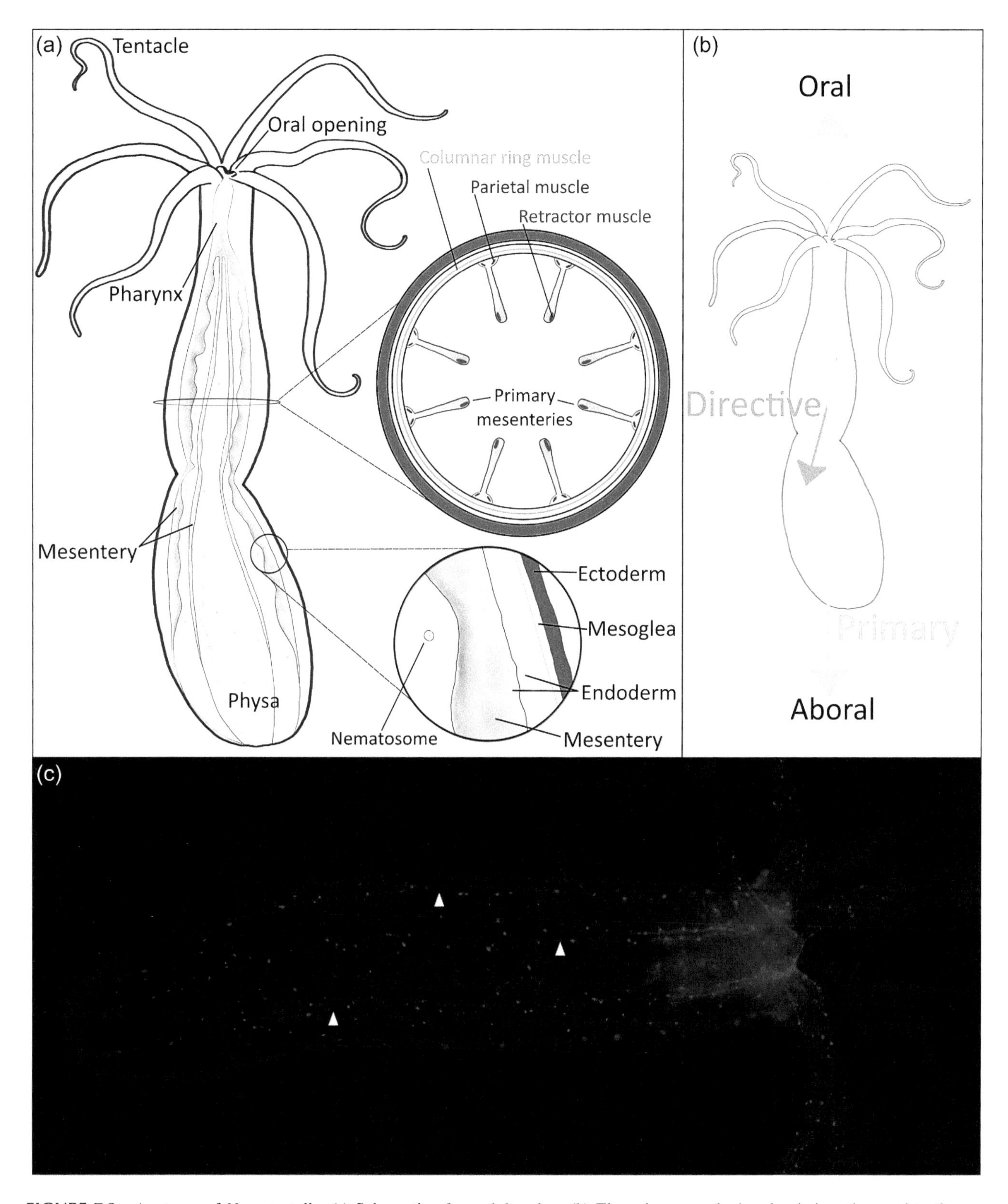

FIGURE 7.3 Anatomy of *Nematostella*. (a) Schematic of an adult polyp. (b) The primary oral–aboral axis is orthogonal to the secondary directive axis. (c) A transgenic animal is used to visualize a subset of neurons in the nerve net. Arrowheads show longitudinal neuronal tracts.

found in abundance throughout the coelenteron and packaged into egg masses (Williams 1975, 1976).

Nematostella have five functionally and morphologically distinct myoepithelial cell groups that together make up the body column and tentacular muscle systems. In the body column there are three muscle groups (Figure 7.3a). The longitudinally oriented parietal and retractor muscles are found within different regions of each mesentery and run the length of the oral–aboral axis. The columnar ring muscle group wraps around the circumference of the body wall along the oral–aboral axis (Jahnel et al. 2014). The tentacles have a similar muscle system; they contain longitudinal muscles that run the length of each tentacle, as well as ring muscles that are oriented orthogonally to the tentacular longitudinal muscles.

Nematostella possess a nerve-net nervous system—aptly named due to the way that the neurites extend from neural soma to form a diffuse interconnected web around the organism (Figure 7.3c). The nervous system is composed of both ectodermal and endodermal nerve nets (Layden et al. 2012; Nakanishi et al. 2012). Although they lack a centralized nervous system, there are distinct neural structures, including bundles of neurons that flank each mesentery within a longitudinal tract (Figure 7.3c) and condensations of neurons forming "nerve rings" around the oral opening and pharynx (Marlow et al. 2009; Sinigaglia et al. 2015; Leclère et al. 2016).

Neural cell types fall under three categories and can be found intermixed among other cell types. In the ectoderm, neural progenitor cells give rise to epithelial sensory cells (which extend an apical cilium to the body surface and neuronal processes basally) and ganglion cells (which lose their apical contacts and migrate so that their cell bodies are basally situated) (Marlow et al. 2009; Sinigaglia et al. 2015; Leclère et al. 2016). Unlike the sensory cells in the ectoderm, those in the endoderm lose their elongated appearance and become shortened along the apical–basal axis (Nakanishi et al. 2012). Cnidocytes are also considered nerve cells due to their neurophysiological properties, structure and calcium/mechanosensory-dependent exocytosis (Kass-Simon and Scappaticci 2002; Thurm et al. 2004; Galliot et al. 2009). Cnidocytes contain a unique organelle called a cnidocyst, which consists of a capsule and a harpoon-like structure that can be fired at ultra-fast speeds (Szczepanek et al. 2002). Capsules are highly specialized based on their function (e.g. feeding, defense, locomotion) and are classified based on their structure (Kass-Simon and Scappaticci 2002). *Nematostella* have three types of cnidocytes: spirocytes and two types of nematocytes. Spirocytes contain spirocyst capsules, which lack a shaft and barbs, and are found in the tentacles. Nematocytes with microbasic *p-mastigophore* capsules are found in the mesenteries and pharynx, and nematocytes with different-sized basitrichous isorhiza capsules are found mostly in the body wall, but also in the tentacles, mesenteries and pharynx (Williams 1975; Williams 1976).

Based on molecular and morphological observations, the three neural groups likely contain many subtypes that can be distinguished based on attributes including neurite number, neuropeptide profile, morphology and location (Nakanishi et al. 2012; Havrilak et al. 2017; Zang and Nakanishi 2020). Although the nerve net of *Nematostella* has been previously described as random because of its disorganized appearance (Hejnol and Rentzsch 2015), there is growing evidence that the nerve net is specifically patterned. The identification of specific neural subtypes points to a previously underappreciated complexity within the nervous system of *Nematostella*, and the presence of large neural structures suggests that neurogenesis is not random. In fact, it has also been shown that several specific neural subtypes exhibit a stereotyped developmental pattern (Havrilak et al. 2017).

7.4 LIFE HISTORY

Nematostella is a dioecious species that sexually reproduces by external fertilization, synchronously releasing eggs and sperm into the water column (Figure 7.4a) (Hand and Uhlinger 1992). Females release egg masses inside of a gelatinous sac containing nematosomes (Figure 7.4b), which are thought to provide defense to the embryos (Babonis et al. 2016). However, this does not make them immune to all predation; for example, grass shrimp will consume *Nematostella* embryos (Columbus-Shenkar et al. 2018). Embryos emerge from the protective sac as spherical, ciliated, non-feeding, planula larvae ~36–48 hours post-fertilization (Hand and Uhlinger 1992). The free-swimming planulae elongate before metamorphosing into sessile primary polyps. Metamorphosis occurs roughly six days post-fertilization and is characterized by the development of an oral opening surrounded by tentacle buds, the first two mesenteries and a loss of swimming ability that leads to larval settlement (Hand and Uhlinger 1992; Fritzenwanker et al. 2007; Fritz et al. 2013). Once settled, juvenile polyps begin to grow and mature in a nutrient-dependent manner (Ikmi et al. 2020). The polyps are opportunistic predators that feed on small estuarine invertebrates captured by their stinging tentacles (Frank and Bleakney 1978; Posey and Hines 1991). Polyps are infaunal, preferring to burrow their body column into soft substrate so that only the oral opening and tentacles are exposed (although they are sometimes found attached to vegetation) (Williams 1975; Williams 1976). Burrowing helps to protect the body column from predation and forces would-be predators to contend with their stinging tentacles first (Columbus-Shenkar et al. 2018). Sexually mature adults range in size and will grow and shrink in response to nutrient availability (Hand and Uhlinger 1994; Havrilak et al. 2021). This phenotypic plasticity allows animals to easily adapt to environmental changes and suggests that there is no set size state (Havrilak et al. 2021). In the wild, adults are typically a few centimeters in length and will reach sexual maturity in approximately six months or less (Williams 1983). In culture, this can occur in as little as ten weeks for well-fed animals (Hand and Uhlinger 1992).

Adult *Nematostella* also reproduce asexually by generating clonal individuals through two forms of transverse fission: physal pinching and polarity reversal. Physal pinching is facilitated by a deep, sustained, constriction of a site along the posterior end of the body column (physa) and results in the separation of the smaller physal fragment from the rest of the anemone (Figure 7.4a). After a few days of separation, the physal fragment will begin to generate oral structures and tentacles that will allow it to feed, ultimately resulting in a functional clone (Hand and Uhlinger 1995; Reitzel et al. 2007). Although frequent feeding can increase the amount of transverse fission that will occur in a population, there is no correlation between parent size and the size of the physal fragments they produce. Further, the number of clones generated by individuals is highly variable; in a sibling population, some will produce several clones, while others will produce none (Hand and Uhlinger 1995). Polarity reversal is like physal pinching, except that the sequence of events is different. With polarity reversal, an adult first manifests oral structures and tentacles at the aboral end of the body column, replacing the physa. A new physa will develop midway along the body column, and physal pinching will act to separate the animal into two individuals (Reitzel et al. 2007).

It is unclear which, if any, environmental conditions promote sexual versus asexual reproduction. Since *Nematostella* maintain multiple modes of reproduction, it is assumed that specific environmental and/or genetic conditions exist under which each mode would have a fitness benefit. *Nematostella* is one of only a handful of anemone species to have multiple modes of asexual reproduction (Reitzel et al. 2007), and asexual reproduction by transverse fission is rare among anthozoan cnidarians (Fautin 2002). In *Nematostella*, transverse fission by polarity reversal is less common than physal pinching and may rely on seasonal environmental cues (Frank and Bleakney 1978; Reitzel et al. 2007).

Nematostella is highly regenerative, capable of bidirectional whole body-axis regeneration and regeneration of specific structures. Although regeneration following bisection is reminiscent of physal pinching, it is markedly different because it is caused by an external factor that wounds the animal as opposed to an endogenously triggered constriction of the body column. When a complete bisection of the body column into oral and aboral fragments occurs, both fragments will regenerate missing structures, leading to the generation of two clonal individuals ~six to seven days post-amputation (Figure 7.4c) (Reitzel et al. 2007; Amiel et al. 2015; Havrilak et al. 2021).

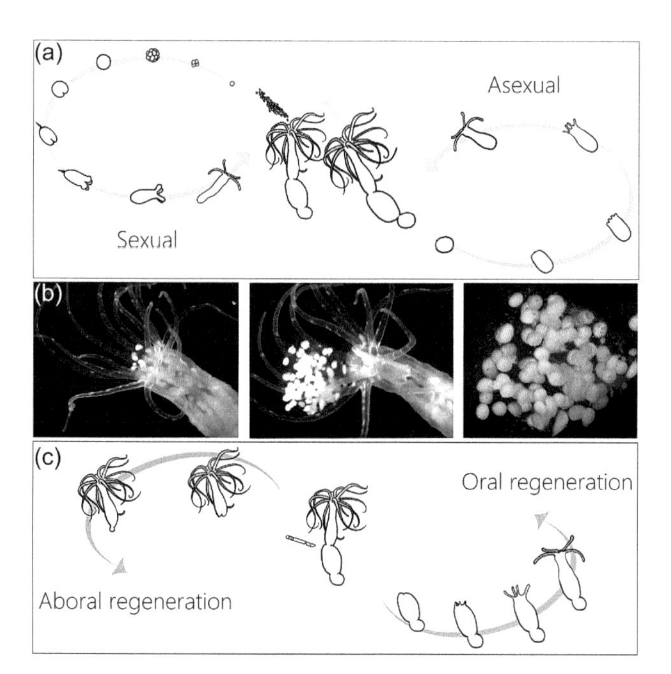

FIGURE 7.4 Reproduction and regeneration in *Nematostella*. (a) Sexual reproduction and asexual reproduction by physal pinching. (b) Spawning female releasing a clutch of eggs through the oral opening. Right panel shows the eggs shortly after being released. (c) Regeneration of oral and aboral fragments following whole-body axis bisection.

7.5 EMBRYOGENESIS

7.5.1 Process of Development

Embryogenesis can be investigated in its entirety since males and females release gametes into the water column and fertilization occurs externally (Figure 7.5). Zygote to juvenile polyp typically requires ~seven days at 22°C. However, development is temperature dependent and can be sped up or slowed down by increasing or decreasing temperatures, respectively. The first cleavage initiates ~two hours after fertilization. The first two cleavage furrows typically originate perpendicular to one another at the animal pole and progress toward the vegetal pole. Initially, cytokinesis is incomplete, and it is not until the 8-cell stage that the blastomeres become separated. While a clutch of embryos will have relatively synchronous development, there is some variability of early cleavage patterns, and odd numbers of blastomeres are occasionally observed (Fritzenwanker et al. 2007, Reitzel et al. 2007). From the 16-cell stage and on, most embryos look similar, and the blastomeres are roughly similar in size (Figure 7.5b). The blastocoel becomes visible by six hours post-fertilization, following epithelialization (Figure 7.5c), which occurs between the 16- and 32-cell stages (Fritzenwanker et al. 2007).

The 64-cell stage marks the start of a series of invagination–evagination cycles that change the shape of the embryo from spherical to a flattened "prawn chip" (characterized by having a concave side and convex side) and then back to spherical again until gastrulation. Cell divisions occur when the embryo is at its maximum flatness. This pulsing pattern continues for four to five cycles, until the onset of gastrulation (~18–22 hrs post-fertilization) (Fritzenwanker et al. 2007).

Prior to gastrulation, endodermal fates are specified by canonical Wnt/β-catenin and MAPK signaling around the animal pole forming the presumptive endoderm (Wikramanayake et al. 2003; Lee et al. 2007; Röttinger et al. 2012). Gastrulation initiates with formation of a blastopore at the animal pole as the pre-endodermal plate invaginates into the blastocoel, and the blastopore ultimately becomes the oral opening (Fritzenwanker et al. 2007; Lee et al. 2007; Magie et al. 2007). Cellular movements during gastrulation are controlled by a conserved Wnt/PCP/Stbm signaling cascade at the animal pole (Kumburegama et al. 2011) and are typified by apical constriction and weakening of cell junctions followed by invagination of the plate (Figure 7.5d) (Kraus and Technau 2006; Magie et al. 2007). Gastrulation

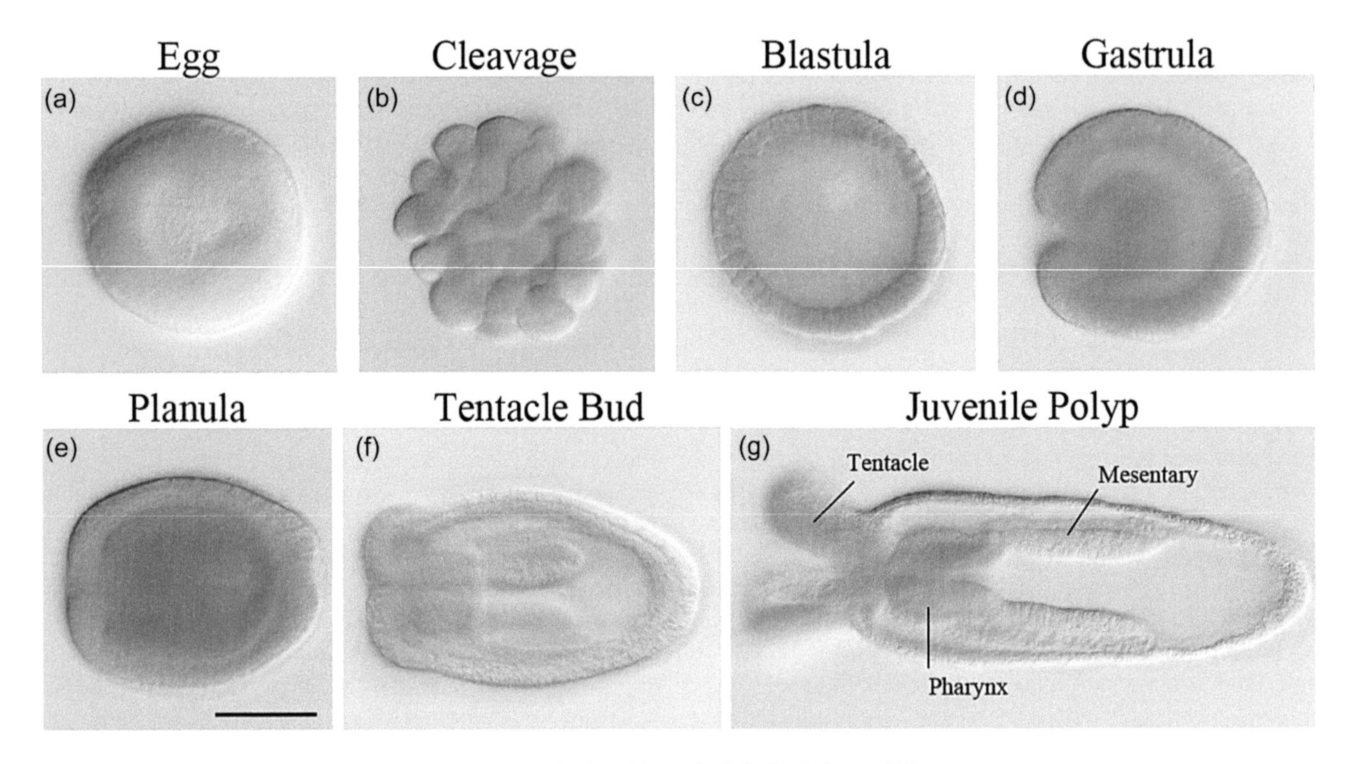

FIGURE 7.5 Developmental stages of *Nematostella*. Oral end is to the left. Scale bar = 100 μm.

completes when the ectoderm of the blastopore lip also rolls inward. This ectoderm retains its epithelial organization and gives rise to the pharynx and septal filaments at the tips of the mesenteries.

Following gastrulation, FGF activity at the aboral pole regulates formation of the apical tuft and apical organ (Rentzsch et al. 2008). The resulting planula larvae break out of the egg jelly by ~two days post-fertilization and are now free-swimming larvae (Figure 7.5e). Planula larvae initially swim in circles, but by ~three days post-fertilization they exhibit directional swimming with the apical ciliary tuft facing forward (Hand and Uhlinger 1992). The planula stage lasts three to four days, during which the planulae elongate and form the pharynx and the first two (primary) mesenteries (Figure 7.5f) (Fritzenwanker et al. 2007). A heterogeneous distribution of ectodermally derived secretory gland cells is found in the pharynx and mesenteries of the primary polyp, and gene expression studies suggest that development of these cells begins in the planula stage, as the tissues they reside in are formed (Frank and Bleakney 1976; Babonis et al. 2019). Presumptive muscle cells are detected in the early planula with F-actin staining, which becomes concentrated and oriented along the oral–aboral axis in the late planula (Jahnel et al. 2014). Besides the tentacle ring muscles, which are derived from the ectoderm, all other muscle groups are of endodermal origin, with many orthologs of genes that drive muscle development in bilaterians observed in *Nematostella* during the planula and juvenile polyp stages (Jahnel et al. 2014; Steinmetz et al. 2017). *NvMyHC1* is first detected at the mid-planula stage and is expressed in retractor muscle cells of both the tentacles and the eight mesenteries of the developing primary polyp (Renfer et al. 2010; Jahnel et al. 2014). The *NvMyHC1* transgene is further detected in retractor muscles of adult mesenteries, suggesting muscle cell differentiation in the mesenteries continues in the adult (Renfer et al. 2010).

The four tentacle buds emerge toward the end of the planula stage (~five days post-fertilization) (Figure 7.5f,e). Tentacle primordia are first identified by Fgfrb-positive cells in ring muscle around the oral opening. Stereotyped development and outgrowth of the tentacles is nutrient dependent and driven by crosstalk between TOR-mediated and FGFR signaling pathways. The tentacle buds elongate into the initial tentacles of the juvenile polyp, and *Nematostella* continue to add tentacles in the adult polyp (Stephenson 1935; Fritz et al. 2013).

Nematostella fully metamorph into juvenile polyps by 6–7 days post-fertilization (Figure 7.5h). The planula larvae settle with the aboral pole down (Rentzsch et al. 2008), then transform into a tube-shaped polyp with four tentacles around a single oral opening. Growth and maturation of the juvenile polyp into an adult is nutrient dependent, and sexual maturity can be reached in the lab in 10 weeks with regular care and feeding (Hand and Uhlinger 1992).

Neurogenesis begins with the emergence of *NvSoxB(2)* and *NvAth-like* expressing neural progenitor cells in the blastula (Richards and Rentzsch 2014, 2015) and continues throughout development. Molecular regulation of neurogenesis in *Nematostella* resembles the neurogenic cascades found in bilaterian species (Rentzsch et al. 2017), involving MEK/MAPK (Layden et al. 2016), Wnt (Marlow et al. 2013; Sinigaglia et al. 2015; Leclère et al. 2016), BMP (Watanabe et al. 2014; Saina et al. 2009) and Notch (Layden and Martindale 2014; Richards and Rentzsch 2015). *Nematostella* has both ectodermal and endodermal nerve nets (Nakanishi et al. 2012). Some neural subtypes arise at the same time as their namesake structures (e.g. tentacular neurons in the tentacles and pharyngeal neurons in the pharynx) (Havrilak et al. 2017). Cnidocyte stinging cells are also thought to be neuronal, which is supported by the fact that they require *NvSoxB(2)* and *NvPaxA* (Babonis and Martindale 2017). Cnidocyte-specific genes and proteins are detected throughout the ectoderm in the early gastrula stage to the primary polyp and in the tentacles and mesenteries of the adult (Zenkert et al. 2011; Babonis and Martindale 2014; Babonis and Martindale 2017).

7.5.2 Axial Patterning Programs

Throughout *Nematostella* development, conserved morphogen gradients and signaling cascades pattern the oral–aboral axis. Wnt/β-catenin signaling is a main driver in establishing and patterning the primary oral–aboral body axis and has a role in gastrulation, and high Wnt/β-catenin promotes oral identity (Wikramanayake et al. 2003; Kusserow et al. 2005; Kraus and Technau 2006; Lee et al. 2007; Marlow et al. 2013; Röttinger et al. 2012; Kraus et al. 2016). *Nvsix3/6* regulates the aboral domain, and its initial expression is dependent on low Wnt/β-catenin in the aboral region (Leclère et al. 2016). Further, a conserved mechanism whereby β-catenin target genes act to repress aboral gene expression in the oral region represents an ancient regulatory "logic" that may have been present in the urbilaterian ancestor (Bagaeva et al. 2020). The interaction of Wnt/β-catenin with specific *hox* genes further fine-tunes patterning along the oral–aboral axis of the *Nematostella* embryo and reflects mechanisms of patterning in bilaterians (DuBuc et al. 2018). However, much more work is needed to resolve the role that *hox* genes have in patterning the primary axis in *Nematostella*, and whether *hox* expression can be used to elucidate how the oral–aboral axis relates to the anterior–posterior axis remains a major question (Layden et al. 2016; DuBuc et al. 2018).

The secondary directive axis is established and patterned by graded BMP signaling. Following an initial radial expression in the gastrula around the blastopore, *NvBmp2/4*, *NvBmp5/8* and *NvChordin* become co-expressed on one side (Matus et al. 2006; Rentzsch et al. 2006). Active pSMAD (BMP signal transducer) is concentrated on the opposite side, suggesting a low BMP signal defines the domain and initiates transcription (Saina et al. 2009; Leclère and Rentzsch 2014). *Hox* genes also play a role in patterning the directive axis in *Nematostella*. *Hox* genes control boundary formation, which leads to the radial segmentation of the developing endoderm and positions the eight radial segments along the directive axis—thereby providing their spatial identity (He et al. 2018). Further, cross-regulatory interactions between *hox* genes occur in both bilaterians and *Nematostella* during axial patterning (Matus et al. 2006; DuBuc et al. 2018). While many of the same players are involved in patterning the secondary directive and dorsal–ventral axes in *Nematostella* and bilaterians, respectively, their positions and functions vary (see "Challenging Questions").

7.5.3 Regeneration

Many aspects of the regenerative process have been characterized at the behavioral, morphological, cellular and molecular levels (see DuBuc et al. 2014; Bossert and Thomsen 2017). The stages of oral regeneration follow a stereotypic pattern, with initial wound healing complete in ~six hours post-amputation, and complete regeneration in ~six to seven days (Figure 7.4c). In subsequent days, the mesenteries fuse, contact the wounded epithelial and then reform the pharynx as new tentacle buds elongate (Amiel et al. 2015). It is hypothesized that a population of quiescent/slow cycling stem cells in the mesenteries are necessary for regeneration (Amiel et al. 2019). Regeneration following bisection occurs at the same rate in both juvenile and adult polyps, is temperature dependent and requires both cellular proliferation and apoptosis (see DuBuc et al. 2014; Bossert and Thomsen 2017). Like what has been observed in other animals that undergo whole-body axis regeneration, some tissue remodeling may also occur during regeneration of oral structures in *Nematostella* (Amiel et al. 2015; Havrilak et al. 2021). While many of the signaling pathways necessary for *Nematostella* are redeployed during regeneration, the regulatory logic and the number of genes utilized varies, with unique gene regulatory networks utilized (Warner et al. 2019).

7.6 GENOMIC DATA

The generation of the *Nematostella* genome was a catalyst that greatly advanced the species as a model system and led to a rapid explosion of molecular techniques and publications (Figure 7.2; Table 7.1). The genome was first sequenced and assembled by the Joint Genome Institute in 2007 using a random shotgun strategy and published as a searchable database (https://mycocosm.jgi.doe.gov/Nemve) (Putnam et al. 2007). While this first genome has only partial sequence coverage and is not mapped back to chromosomes, the scaffold organization still informs researchers about syntenic relationships, gene structure and sequence. Improvements to the genome have recently been made with the publication of a second genome (Zimmermann et al. 2020). This new assembly has enhanced sequence coverage and increased chromosomal resolution (https://simrbase.stowers.org).

It was expected that the *Nematostella* genome would be relatively simple and lack many of the major gene families found in bilaterians. However, bioinformatic analysis uncovered a complex genome comparable in many ways to other animals. It turns out that the *Nematostella* genome is more like vertebrates than some popular bilaterian models such as *Caenorhabditis elegans* and *Drosophila melanogaster* (Putnam et al. 2007). The exon–intron structure of *Nematostella* is like vertebrates and other anemones, which suggests that the eumetazoan ancestor had a similar genetic organization (Putnam et al. 2007). Further, the genome includes major gene families such as *wnt* (Kusserow et al. 2005), *sox* (Magie et al. 2005), *forkhead* (Magie et al. 2005), *hedgehog* (Matus et al. 2008) and *hox* (Ryan et al. 2006). *Nematostella* utilize many major signaling pathways and possess orthologues of many effector genes and antagonists involved in signaling, revealing that the genetic components required for complete signal transduction were established in the cnidarian-bilaterian ancestor (Magie et al. 2005; Putnam et al. 2007; Galliot et al. 2009; Watanabe et al. 2009; Chapman et al. 2010).

The genome has made sequence information easy to access, analyze and manipulate, and allows for the utilization of tools for both discovery-based and comparative genomic studies of varying scales. Sophisticated gene editing is possible using TALEN and CRISPR/Cas9 systems, which can be used to induce targeted mutations and homologous-based recombination, including the generation of transgenic lines and knockout of developmental genes (Ikmi et al. 2014; Servetnick et al. 2017; He et al. 2018). Transcriptomic strategies such as ChIP-seq, RNA-seq and single-cell RNA-seq are now common practice. ChIP-seq studies have led to genome-wide predictions regarding the locations of histone modifications and have demonstrated that there is likely conservation of gene regulatory elements (such as enhancers and promoters) between *Nematostella* and bilaterians (Schwaiger et al. 2014; Technau and Schwaiger 2015; Rentzsch and Technau 2016). ChIP-seq experiments suggest acetylated histones are enriched in the 5' proximal region of gene promoters (and sometimes in the first intron) of genes they control, which facilitates identification of regulatory elements used to generate transgenic reporters. Transgenic animals have been successfully generated by capturing and cloning ~1.5–2.5 KB of the region upstream of the transcription start site (Renfer et al. 2010; Nakanishi et al. 2012; Layden et al. 2016; Renfer and Technau 2017). RNA-seq and microarrays have been used to profile gene expression levels during development and regeneration (Tulin et al. 2013; Helm et al. 2013; Fischer et al. 2014; Warner et al. 2018). The compilation of these RNA-seq studies into Nvertx, a searchable database, allows for quick comparison between timepoints and/or between the processes of development and regeneration (http://nvertx.ircan.org) (Warner et al. 2018, 2019). These databases are powerful tools because a researcher can evaluate their findings relative to this published source or can check expression profiles and make and test initial hypotheses about potential candidate genes before doing any functional studies themselves. Single-cell RNA-seq studies are now possible, and initial studies have used similarities in cellular expression profiles to generate testable hypotheses regarding cell types, their diversity and their functions (http://compgenomics.weizmann.ac.il/tanay/?page_id=724) (Sebé-Pedrós et al. 2018).

7.7 METHODS AND FUNCTIONAL APPROACHES

7.7.1 Culture and Care

Establishing and maintaining a lab population of *Nematostella* is simple and economical. Founder animals can be purchased from commercial vendors, requested from other laboratories or collected from the field using minimal equipment (see Stefanik et al. 2013). Animals can be kept in glass bowls in a cool dark room, and their husbandry only requires regular brine shrimp feeding and weekly water changes (Stephenson 1935; Williams 1983; Hand and Uhlinger 1992). They can also be maintained in modified fish aquaculture systems for large-scale cultures. Population size can be increased through sexual reproduction, and clonal lines can be developed by allowing animals to asexually reproduce or by cutting adults to create regenerates (Figure 7.4) (Hand and Uhlinger 1992; Reitzel et al. 2007; Stefanik et al. 2013). *Nematostella* spawn year-round in culture (Hand and Uhlinger 1992; Fritzenwanker and Technau 2002). Under laboratory conditions, spawning is induced by exposing animals to a light source and by increasing temperature (Niehrs 2010; Genikhovich and Technau 2017).

7.7.2 Behavioral and Ecological Approaches

The fact that these animals are found in abundance in shallow estuarine environments makes them easy to find, collect and manipulate for field studies. Due to their mostly sedentary and infaunal nature, controlled field experiments can be easily conducted without the worries of tracking individuals or animals escaping from experimental areas. Water-permeable cages allow for testing under natural conditions and provide a way to control the contents of the cage, including what can enter and exit it. For example, cages placed within natural habitats have been used to track changes within a population under different conditions, including those in which predators, food availability and abiotic environmental factors were varied (Wiltse et al. 1984; Tarrant et al. 2019). *Nematostella* can tolerate a wide range of environmental parameters and are often found living at the extremes of their tolerable ranges for temperature, salinity and oxidative stress (Williams 1983; Hand and Uhlinger 1992; Reitzel et al. 2013; Friedman et al. 2018). This remarkable environmental phenotypic plasticity makes them an intriguing indicator species and a potential model for studies of stress tolerance, effects of the environment on development, community structure and adaptive evolution. Further, existing information regarding population genetic structure, gene flow and protein-coding polymorphisms allows for studies to be placed in a broader evolutionary context (Darling et al. 2004; Reitzel et al. 2013; Friedman et al. 2018). The broad molecular toolbox available for *Nematostella* allows field researchers to take an integrative approach to experiments (Table 7.1).

Nematostella is amenable to both field and lab studies. Because it is an established laboratory model with a published genome, it is possible to determine the mechanisms of molecular, cellular and behavioral changes that occur in the wild due to environmental changes or following manipulations in a laboratory environment. Several naturally occurring behaviors have been described in *Nematostella*, including burrowing, creeping, climbing, feeding, contracting, spawning, fissioning and the propagation of peristaltic waves (Hand and Uhlinger 1992, Hand and Uhlinger 1995; Williams 2003; Faltine-Gonzalez and Layden 2019; Havrilak et al. 2020). Despite it often being difficult to observe behaviors in the field due to their small size, infaunal nature and usually low water clarity, behavioral observations can be done in the lab where video recording and magnification are easily accomplished and natural conditions can be mimicked. Besides studying behavioral observations to understand the behavioral ecology of *Nematostella*, behaviors can be used as an experimental readout due to the depth at which many behaviors have been described (e.g. Williams 2003). For instance, one can assess behaviors as a means of determining the effect of a treatment (e.g. following drug treatments, genetic manipulations) or as a measure for the completion of morphogenesis (e.g. during growth/degrowth, regeneration) (Figure 7.6) (Faltine-Gonzalez and Layden 2019; Havrilak et al. 2021).

7.7.3 Tissue Manipulation and Tracking

Classical embryological techniques, such as embryo separation, dye tracing during embryo development and tissue grafting (Lee et al. 2007; Nakanishi et al. 2012; Steinmetz et al. 2017; Warner et al. 2019), are feasible due to large transparent embryos and adults. Dissection and transplantation of fluorescent tissue from transgenic embryos into developing wild type embryos have allowed researchers to begin constructing a fate map of the germ layers, and these techniques could be useful in further constructing the *Nematostella* fate map (Steinmetz et al. 2017). Researchers have successfully cultured sheets of ectodermal tissue, which was able to transform into 3D structures and be sustained for several months (Rabinowitz et al. 2016). Cell culture techniques are being developed in *Nematostella* and are expected to be possible due to the success of tissue culture and recent

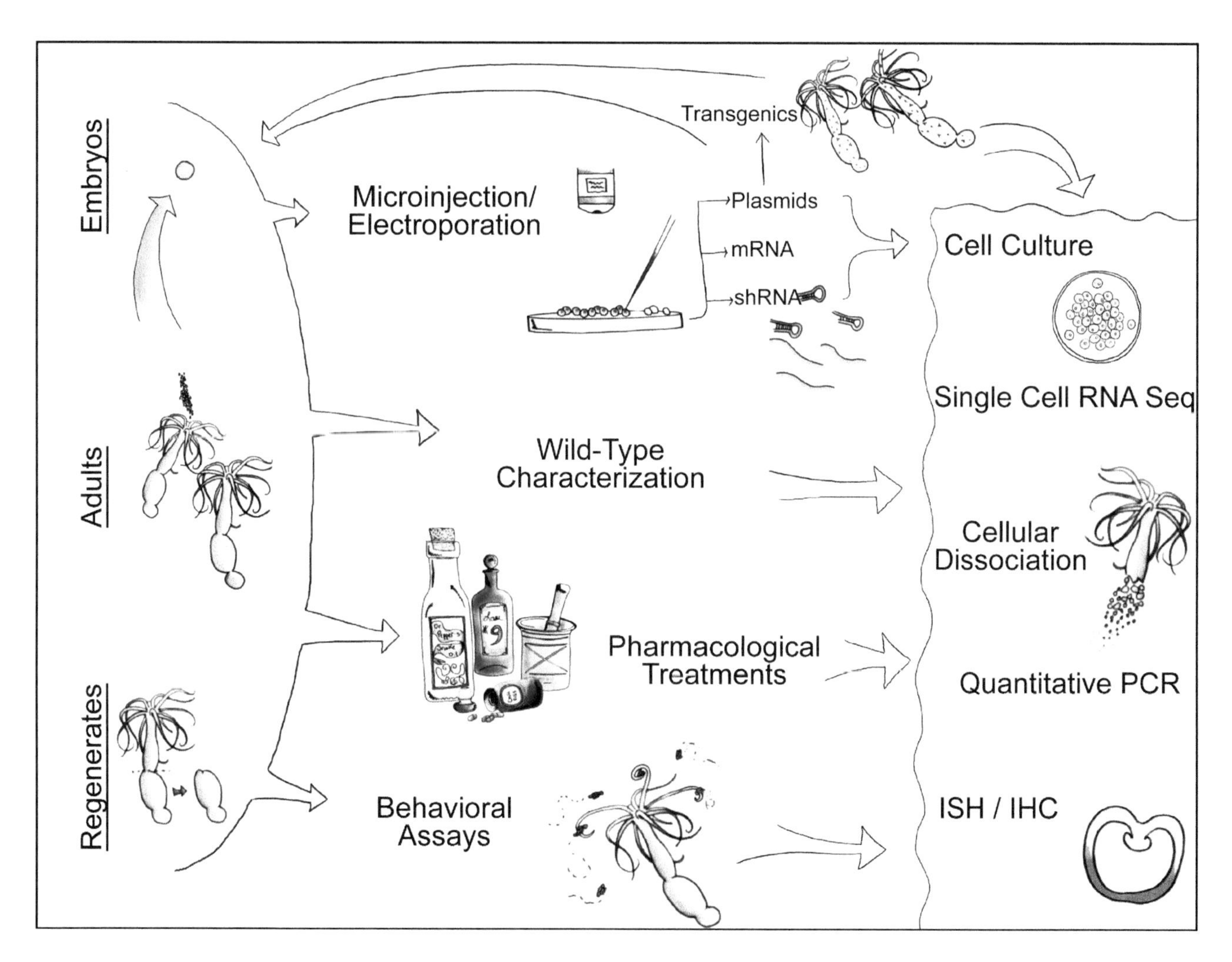

FIGURE 7.6 Potential workflow showing integration of multiple techniques using *Nematostella*.

ability to successfully dissociate animals into their cellular components (see the following paragraph). These culturing methods would allow research to be focused on a specific tissue or cell type and could negate the need to maintain an animal population due to the ability to freeze cell stocks (e.g. Fricano et al. 2020).

Dissociation of cells from transgenic and wild type adult animals has been accomplished using different combinations of enzymatic, chemical and mechanical techniques (Sebé-Pedrós et al. 2018; Clarke et al. 2019; Torres-Méndez et al. 2019; Weir et al. 2020). Cellular dissociation has also allowed for studies of cellular adhesion using the hanging drop method to study reaggregation (Clarke et al. 2019). Further, cellular dissociation has opened the door to the field of electrophysiology. For example, single-cell recordings from nematocytes have given us insight into the physiology of a novel cell type and bettered our understanding of how *Nematostella* distinguish salient environmental information to regulate cnidocyte firing (Weir et al. 2020), contributing to our understanding of cnidarian sensory systems and their stinging response.

7.7.3.1 Detection of Cellular Processes

The relatively simple body plan of *Nematostella*, consisting of only two transparent tissue layers, facilitates the use of common labeling techniques to investigate cellular processes utilized during morphogenesis and homeostasis. Standard techniques for cellular proliferation have been used by labeling animals with EdU and BrdU (Passamaneck and Martindale 2012; Richards and Rentzsch 2014; Amiel et al. 2015; Rabinowitz et al. 2016; Warner et al. 2019; Havrilak et al. 2021). TUNEL assays have been used to detect apoptotic cells during development and regeneration (Warner et al. 2019; Zang and Nakanishi 2020).

7.7.3.2 Regeneration

Inducing a regenerative response in *Nematostella* is simple, and the process has been characterized at many levels. Regeneration is induced by wounding the animal with a scalpel or probe (see DuBuc et al. 2014; Bossert and Thomsen 2017). The wound site and severity of the injury inflicted are dictated by the research question. Typically, studies have focused on whole-body axis regeneration, where live animals are bisected along the body column into oral and aboral halves and regeneration of one or both fragments is observed (Figure 7.4c) (Passamaneck and Martindale 2012; Amiel et al. 2015; Schaffer et al. 2016). However, a more acute regenerative response can be triggered following a focal injury where whole-body axis regeneration is not required (e.g. tentacle amputation, puncture wound, incomplete bisection along the body column) (Reitzel et al. 2007; DuBuc et al. 2014). This flexibility in the regeneration paradigm allows for a variety of questions to be asked. The ability to document gene expression in different regenerative paradigms, as well as to compare it to development, will continue to make this a fruitful area of research in this model. For example, many hypotheses can be tested due to comparative transcriptome analysis using RNAseq during the regeneration of oral vs. aboral fragments—which identified similarities and differences in gene expression profiles between the two halves (Schaffer et al. 2016). Methods for assessing wound closure, and detailed descriptions of key morphological landmarks that occur throughout the process of regeneration, have been described and can be used to assess the progress of the regenerative response (Bossert et al. 2013; Amiel et al. 2015). Assaying the regenerative phenotype following pharmacological or genetic manipulation could be used to understand the mechanisms of regeneration (e.g. using an inducible promoter or knockout transgenic line). Transgenic reporter lines allow for the tracking of specific cell types in live animals, including specific neural subtypes, which has made the regeneration of the nerve net tractable (Figure 7.6) (Layden et al. 2016; Havrilak et al. 2017; Sunagar et al. 2018; Havrilak et al. 2021).

7.7.4 Genetic Approaches

7.7.4.1 Microinjection and Electroporation

Molecules can be introduced into live embryos using microinjection and electroporation techniques, which facilitate the delivery of compounds such as shRNA, mRNA, morpholinos and plasmids into eggs. With microinjection, a very fine glass needle is used to penetrate an egg and deliver a small volume of the loaded injection mixture using forced air (Layden et al. 2013; Renfer and Technau 2017; Havrilak and Layden 2019). An experienced researcher can inject thousands of embryos in a single session. Microinjection offers more experimental utility due to the variety of molecular compounds that can be injected, ranging from plasmids to shRNAs. Microinjection has been used successfully for genetic knockdown and misexpression experiments, as well as for the generation of transgenic animals in *Nematostella* (Layden et al. 2013; Ikmi et al. 2014; Renfer and Technau 2017). Electroporation offers a simple and quick method for the delivery of molecules into hundreds of animals simultaneously by generating electrical pulses that create pores in the plasma membrane that allow small molecules to be taken up. So far, this method has only proved successful in the delivery of shRNA for knockdown experiments in *Nematostella* (Karabulut et al. 2019).

7.7.4.2 Gene Disruption

Tools for both gain and loss of function experiments are available (Table 7.1). Injection of *in vitro* synthesized mRNA allows for a gene of interest to be overexpressed (Wikramanayake et al. 2003), while introduction of shRNA or morpholinos facilitates genetic knockdown of a gene of interest (Magie et al. 2005; Rentzsch et al. 2008; He et al. 2018). Gene editing technologies can also be used to silence, move, knock down or overexpress a particular gene in both F0 and F1 generation mutants, and pharmacological treatments can also be performed for gain and loss of function experiments (see "Transgenics and Pharmacological Manipulations").

7.7.4.3 Transgenics

Generation of transgenic animals utilizing tissue and cell type-specific promoters driving a fluorescent tag and/or specific gene of interest has been successful using random meganuclease-assisted integration (Figure 7.3b) and site-specific CRISPR/Cas9 homologous recombination (Ikmi et al. 2014; Renfer and Technau 2017). Promoter sequences have been captured by cloning 1.5–2.5 kb of the genetic sequence upstream from the coding sequence of a gene of interest (Putnam et al. 2007; Renfer et al. 2010; Nakanishi et al. 2012; Layden et al. 2016; Renfer and Technau 2017). Transgenic lines have been made with broad expression using promoter sequences such as actin, ubiquitin and elongation factor 1α (Fritz et al. 2013; Steinmetz et al. 2017; He et al. 2018), and promoters for tissue and cell specific genes have also been utilized for restricted expression such as myosin heavy chain and *soxB(2)* (Renfer et al. 2010; Richards and Rentzsch 2014). A plasmid backbone containing I-sceI meganuclease recognition sites is available (AddGene.org: plasmid #67943) and allows for the desired construct to be swapped out using basic cloning strategies (Renfer et al. 2010). Gene editing has been achieved through homologous recombination using TALEN and more frequently CRISPR/Cas9 (Ikmi et al. 2014; Zang and Nakanishi 2020). For CRISPR/Cas9, a plasmid containing homology arms for a *Nematostella*-specific *Fp7* locus allows for expression or disruption of a desired gene of interest. Importantly, the *Fp7* locus can be disrupted without detrimental effects on the animal and allows for easy screening due to the loss of endogenous red fluorescent protein following cassette insertion. The application of conditional promoters, including an already identified heat shock promoter, opens the door for temporal control of gene expression and disruption in the future (Ikmi et al. 2014).

7.7.4.4 Visualizing Gene Expression

Several tools are available in *Nematostella* for visualizing spatial and temporal differences in gene expression. Both colorimetric and fluorescent whole mount *in situ* hybridization are widely used for determining spatial expression of mRNA at specific time points during development and regeneration (Niehrs 2010; Genikhovich and Technau 2017). Immunohistochemistry has been used to visualize protein expression (Zenkert et al. 2011; Wolenski et al. 2011;

TABLE 7.1
A List of Methods and Functional Approaches Available in *Nematostella*

	Culture, Care, and Manipulation of *Nematostella*
Culture and spawning	Hand and Uhlinger (1992), Fritzenwanker and Technau (2002), Genikhovich et al. (2009)
Inducing and staging regeneration	Bossert et al. (2013), Dubuc et al. (2014), Amiel et al. (2019)
Microinjection	Layden et al. (2013)
Field collection	Stefanik et al. (2013)
	Spatiotemporal Gene Expression
mRNA in situ	Genikhovich and Technau (2009), Wolenski et al. (2013)
Immunolocalization	Wolenski et al. (2013)
Transgenic reporters	Renfer et al. (2010), Ikmi et al. (2014)
	Gene Function
Morpholino	Magie et al. (2007), Rentzsch et al. (2008), Layden et al. (2013)
mRNA misexpression	Wikramanayake et al. (2003), Layden et al. (2013)
shRNA	He et al. (2018), Karabulut et al. (2019)
CRISPR/Cas9, TALEN/FokI	Ikmi et al. (2014)
Inducible promoters	Ikmi et al. (2014)
	Genome- and "Omics"-Level Analysis
Annotated genomes	Putnam et al. (2007), Zimmermann et al. (2020) • http://genome.jgi-psf.org/Nemve1/Nemve1.home.html • http://cnidarians.bu.edu/stellabase/index.cgi • http://metazoa.ensembl.org/Nematostella_vectensis/Info/Index • https://simrbase.stowers.org/starletseaanemone
Transcriptomes	Helm et al. (2013), Tulin et al. (2013) • http://figshare.com/articles/Nematostella_vectensis_transcriptome_and_gene_models_v2_0/807696 • http://nvertx.ircan.org/ER/ER_plotter/home
ChIP-Seq protocol	Schwaiger et al. (2014)
RNA-seq protocol	Helm et al. (2013), Tulin et al. (2013)
Micorarray approaches	Röttinger et al. (2012), Sinigaglia et al. (2015)
scRNA-seq protocol	Sebé-Pedrós et al. (2018)

Nakanishi et al. 2012; Zang and Nakanishi 2020). Transgenic reporter lines provide another means of assaying spatial and temporal protein expression and allow for live visualization and imaging. Quantitative real-time polymerase chain reaction is a quick method for determining mRNA expression levels at a given point in time and is often used experimentally in tandem with *in situ* hybridization and in the confirmation of sequencing results. Together these methods are powerful tools for characterizing wild type and transgenic expression and as a readout for gain and loss of function experiments (Figure 7.6).

7.7.4.5 Genome- and "Omics"-Level Approaches

Genomic- and transcriptomic-level experiments under varying developmental, regenerative and/or environmental paradigms are possible since the publication of the genome. ChIP-seq studies can be used to determine epigenetic protein interactions with open chromatin. For example, ChIP-seq can uncover potential genomic interactions of a protein of interest or facilitate the identification of regulatory elements for a gene of interest (Schwaiger et al. 2014; Technau and Schwaiger 2015; Rentzsch and Technau 2016). RNA-seq and single cell RNA-seq have allowed for the investigation of global gene expression levels in whole animals and single cells, respectively (Tulin et al. 2013; Helm et al. 2013; Fischer et al. 2014; Warner et al. 2018; Sebé-Pedrós et al. 2018). Ultimately, each of these methods provides different levels of resolution, and the method used will depend on the question being asked.

7.7.4.6 Pharmacological Manipulation

Pharmacological agents have been used to target specific developmental pathways, as well as to target pathways to alter the physiology of the adult animal. Administering pharmacological agents requires only introducing the desired concentration to the sea water in which treatment animals are growing. Treatments can be administered at any stage from developing embryos to mature adults. Pharmacological agents offer a quick and easy way to target pathways in a high-throughput manner. It is possible to alter basic cellular processes using drug treatments. For example, cell proliferation has been blocked with hydroxyurea (Amiel et al. 2015). Wnt/β-catenin activity can be overactivated using 1-azakenpaullone and/or alsterpaullone, and inhibited using iCRT14 (Trevino et al. 2011; Watanabe et al. 2014). The gamma secretase inhibitor DAPT can be given to effectively disrupt the Notch/Delta pathway (Layden and Martindale 2014), and the receptor tyrosine kinase inhibitor SU5402 can be used to effectively inhibit Fgf receptors (Rentzsch et al. 2008). Additionally, the mTOR pathway can be disrupted by bathing animals in rapamycin (Ikmi et al. 2020). While many of the treatments discussed previously would typically be applicable to developing animals, there are also several agents that can alter the physiology of adult *Nematostella*. For example, bathing adults in the neurotransmitter acetylcholine can induce tentacle contractions, while lidocaine can suppress these contractions (Faltine-Gonzalez and Layden 2019).

7.7.5 Integration of Approaches

While the approaches discussed here are organized into subsections, there is no hard line defining what they can be used for. The combination of various tools from field approaches to molecular, cellular and behavioral techniques can be combined to address a nearly limitless range of questions (Figure 7.6). Following the establishment of a lab population, a basic molecular biology lab setup will allow a researcher to tackle questions pertaining to the fields of molecular ecology, mechanisms of behavior, evolution, development, regeneration and so on (Figure 7.6). There is also the expectation that the *Nematostella* model will keep up with major advances in technology, since cutting-edge techniques continue to become available in this system. Advances in single-cell technologies and the application of conditional/inducible alleles will further refine the resolution and control at which experiments can be performed. Adding in the fact that field and lab comparisons and/or wild type and transgenic comparisons can be included as additional variables makes it so researchers have a high level of control, allowing them to implement experimental parameters beyond those offered by other model systems.

7.8 CHALLENGING QUESTIONS

7.8.1 Is There a Deep Evolutionary Origin for Key Bilaterian Traits?

An explosive radiation of taxa occurred within the bilaterian lineage, and it is believed to be due to the evolution of several unique characteristics (e.g. mesodermal germ layer and bilateral symmetry) that allowed them to occupy previously inaccessible niches. The evolution of these traits allowed for the evolution of larger, more complex body plans and increased specialization of structure organization and function—including cephalization and the centralization of nervous systems. Understanding the mechanisms that led to the bilaterian radiation is a longstanding evolutionary question that can only be answered by studying animals that are closely related to bilaterians in order to infer what molecular tool-kit was available to their common ancestor. Cnidarians are regarded as the sister taxon to the bilaterians (Wainright et al. 1993; Medina et al. 2001; Collins 2002), and therefore cnidarian models, such as *Nematostella*, offer an appropriate outgroup species to study the molecular basis for the origin of key bilaterian traits, such as the mesoderm and bilaterality, because they allow us to deduce the evolutionary history of these derived traits (Figure 7.7). In fact, *Nematostella* first gained momentum as a model species for its utility in uncovering the evolutionary mechanisms that led to key bilaterian features.

7.8.1.1 Origin of the Mesoderm

Thus far, studies with *Nematostella* have used comparative genetic approaches and germ layer fate mapping to form different hypotheses regarding the molecular basis of mesoderm evolution (Scholz and Technau 2003; Martindale et al. 2004; Steinmetz et al. 2017; Wijesena et al. 2017). One hypothesis suggests that the mesoderm was derived from a dual-functional endoderm originating in the diploblastic ancestor, termed the "endomesoderm", which performs both traditional endodermal and mesodermal functions within a single germ layer (Martindale et al. 2004; Wijesena et al. 2017). Expression of genes restricted to the mesoderm in bilaterians were found in the endoderm of *Nematostella*, leading to the "endomesoderm" hypothesis (Martindale et al. 2004). Further, expression of a conserved set of genes involved in the gene regulatory network driving heart field specification in bilaterian mesoderm was found to be functional in the endoderm of *Nematostella* at early developmental stages (Wijesena et al. 2017). Since they lack a closed circulatory system and other mesenchymal cell types, this begs the question: What are the functions of these heart field and other traditionally mesoderm-specific genes in *Nematostella*?

Other studies have tested the "endomesoderm" hypothesis and arrived at a different model of mesoderm evolution. Germ layer fate mapping showed that the pharynx and mesenteries are composed of cells derived from both germ layers, as opposed to being derived from only the endoderm as previously thought, and gene expression experiments suggested that these structures are also functionally partitioned. Further, the *Nematostella* endoderm has an expression profile that resembles bilaterian mesoderm (e.g. heart and gonadal genes), and the pharyngeal ectoderm expresses genes common to bilaterian endoderm (e.g. gut-specific digestion genes) (Steinmetz et al. 2017). These data point to an alternate model of germ layer homology where the cnidarian pharyngeal ectoderm is analogous to the bilaterian endoderm, and the cnidarian endoderm is analogous to the bilaterian mesoderm, supporting a proposed mechanism for bilaterian mesoderm formation where the expansion of the pharyngeal ectoderm down into the body cavity led to the formation of an internal mesodermal layer in a pre-bilaterian ancestor (Steinmetz et al. 2017; Steinmetz 2019). Support for this model requires functional studies to show that the gene expression profiles of *Nematostella* not only correspond to bilaterian germ layer profiles but also have homologous functions.

Both hypotheses propose that the cnidarian endoderm has analogous function to the bilaterian mesoderm. The main difference lies in whether the pharynx and mesenteries contain both ectodermal and endodermal tissues and function as bilaterian endoderm and mesoderm, respectively. To reconcile these different hypotheses, better resolution of gene regulatory networks in adult animals is needed in order to ascertain if mesodermal gene expression and function (such as the heart field specification network) are restricted to the endodermal portions of the bi-layered mesenteries and pharynx.

7.8.1.2 Mechanisms of Axial Patterning Leading to Bilaterality

Despite the seemingly endless variation in animal body plans, all taxa appear to have clear regimented developmental programs that set up the body axes that give rise to the

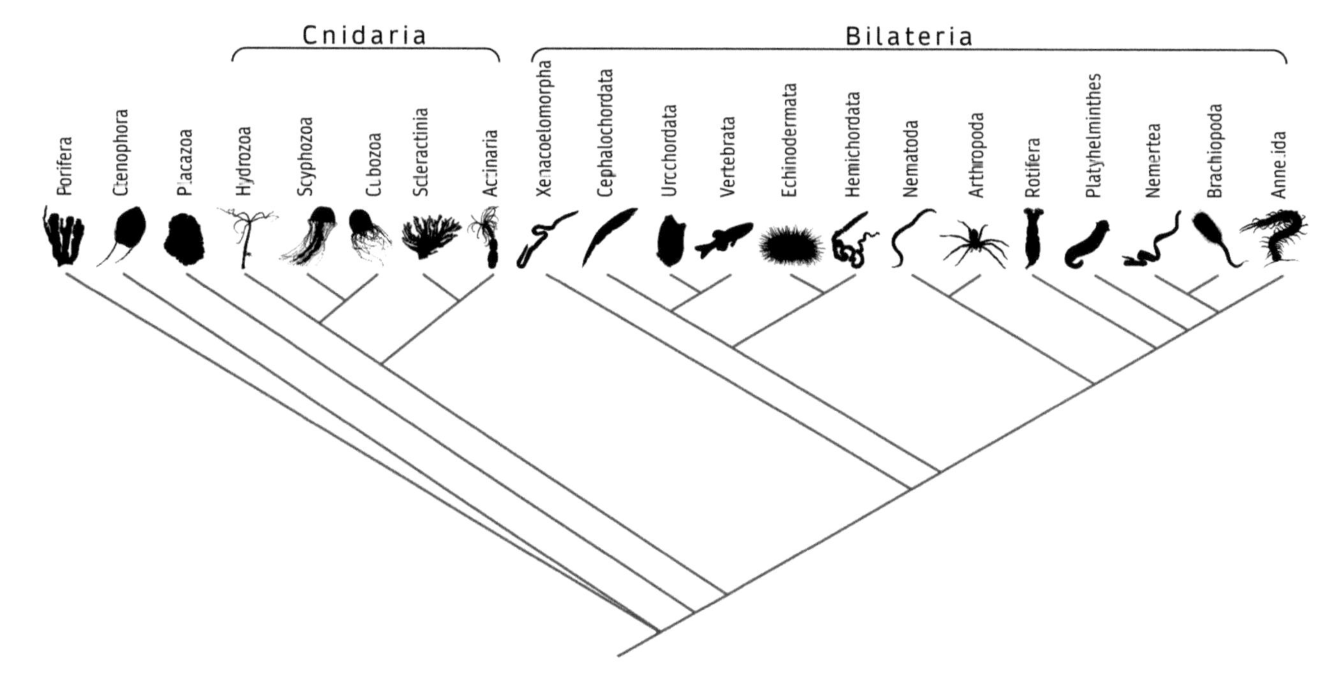

FIGURE 7.7 Phylogeny showing relationships between cnidaria, bilateria and early metazoa. *Nematostella* is an actinarian cnidarian. Cnidarians and bilaterians are sister taxa. Porifera and placazoa lineages are shown sharing a node because their phylogenetic position is unresolved.

unique morphology of each species. Many similarities in the mechanics of axial patterning have been observed between taxa. This has led to the question: Is there a conserved molecular program central to axial patterning that has been co-opted across evolutionary time?

At first glance, *Nematostella* appear radially symmetrical, with their body plan centered around a primary oral–aboral axis that runs the length of the body column. However, upon closer examination of the structural features of *Nematostella*, it is evident that they possess bilateral symmetry along a secondary "directive" axis that runs perpendicular to the primary axis (see "Anatomy"). At a molecular level, the perpendicular primary and secondary axes are derived from orthogonal morphogen gradients that work in concert to set up the body plan. In bilaterians, the orthogonal arrangement of morphogen gradients is also a fundamental aspect of body axis patterning (Niehrs 2010; Genikhovich and Technau 2017).

Major morphogen signaling pathways, with an established role in bilaterian anterior–posterior axial patterning, play a similar role in setting up domains along the oral–aboral axis in *Nematostella* (Leclère et al. 2016; Amiel et al. 2017; Bagaeva et al. 2020). Like bilaterians, a Wnt/β-catenin gradient, with a similar regulatory logic, is established along the primary axis (Marlow et al. 2013, Kraus et al. 2016, Bagaeva et al. 2020). In bilaterians, a key factor in forming the dorsal–ventral axis is the establishment of opposing gradients of bone morphogenic protein and its antagonist chordin on opposite ends of the secondary axis, perpendicular to the primary axis (Niehrs 2010, Genikhovich and Technau 2017). In contrast, expression domains of bone morphogenetic protein and chordin overlap and are on the same side of the directive axis in *Nematostella* (Matus et al. 2006; Rentzsch et al. 2006; Leclère and Rentzsch 2014). This suggests evolutionary plasticity in the BMP/Chordin systems but does not answer how they functioned and were co-opted in the establishment of diverse secondary axial patterning programs throughout evolutionary history. It is worth noting that besides these two major upstream morphogen pathways, other factors, such as *hox* genes, play critical roles as downstream effectors in shaping and refining axial patterning in *Nematostella* and bilaterians (Graham et al. 1991; Pearson et al. 2005; DuBuc et al. 2018; He et al. 2018).

The accumulation of data thus far suggests deep evolutionary roots for the morphogenetic programs governing axis patterning (Matus et al. 2006; Bagaeva et al. 2020), regardless of body plan complexity. Although the same morphogenetic pathways seem to play an important role in patterning the primary and secondary axes in *Nematostella*, there appear to be key differences in how morphogens are spatially distributed and interacting (Matus et al. 2006; Rentzsch et al. 2006; Leclère and Rentzsch 2014). Similarities in axial programming between *Nematostella* and bilaterians make them an ideal candidate for understanding if/when a general morphogenetic program was co-opted for the evolution of bilateral symmetry. Comparisons with other cnidarians and early metazoans will help to resolve how these patterning mechanisms evolved and functioned in the urbilaterian ancestor and prior to the cnidarian-bilaterian split.

7.8.2 Can *Nematostella* Be Used as a Cnidarian Model for Cnidarians?

Establishing a genetically amenable, high-throughput, cnidarian model would improve our understanding of many aspects of cnidarian biology, which has been hindered by our inability to easily access, observe and culture many species within this phylum. A major question is: Can we better understand the effects of the changing environment and inform conservation strategies by utilizing established cnidarian models that are amenable to high-throughput laboratory techniques? Although corals can be harvested and kept under laboratory conditions (provided that specific environmental parameters are met) their natural history makes it very difficult to control spawning behavior and therefore makes it so that embryos are only available up to a few times a year (Harrison et al. 1984; Baird et al. 2009; Keith et al. 2016; Craggs et al. 2017; Pollock et al. 2017; Cleves et al. 2018). In addition, there are few tools and resources available for conducting molecular, cellular or physiological research in non-model cnidarian systems (Technau and Steele 2011). A notable exception is a study that used CRISPR/Cas9 in the coral *Acropora millepora* to target a few genes of interest. However, to obtain embryos, prior knowledge of when spawning would occur was necessary so that corals could be harvested and brought into the lab just prior to their natural spawning event (Cleves et al. 2018). This exemplifies the logistical hurdles that are often present in coral and other cnidarian research.

An intriguing possibility is that *Nematostella* could be employed as a cnidarian model for cnidarians due to the repertoire of tools available and easy culture. *Nematostella* has no symbionts, and therefore it would not be useful in modeling symbiotic relationships. However, the fact that we can easily manipulate *Nematostella* at a molecular level sets it up as a good proxy to investigate fundamental molecular programs in other cnidarians. This way, hypotheses could be quickly tested in this developed model so that resources can be mobilized most efficiently in hard to study cnidarian species. An additional question is: Can *Nematostella* be used as a cnidarian model for environmental stress tolerance and adaptation? This could broaden our understanding of how imperiled cnidarians are likely to cope with ongoing environmental change. Plus, understanding the underlying mechanisms responsible for environmental plasticity in *Nematostella* could potentially be exploited in the conservation of other species.

7.8.3 How Do Novel Cell Types Evolve?

Longstanding evolutionary questions are: How does evolutionary novelty arise, and how does novelty lead to major evolutionary transitions? To investigate these questions requires a model that possesses cell types with true morphological and functional novelty. Cnidocytes are phenotypically unique stinging cells and a defining characteristic of the cnidarian phylum. Cnidocytes are one of only a handful of examples of an unequivocal evolutionary novelty and thus offer a unique opportunity to investigate the mechanisms that lead to evolutionary novelty—something that is not possible in many model systems. Using *Nematostella*, studies can be focused on the molecular basis of cnidocyte development (Babonis and Martindale 2017; Sunagar et al. 2018). This will inform how newly generated genes/proteins interact with existing biological programs, leading to the emergence of novel proteins and, in turn, cell types (Babonis and Martindale 2014; Babonis et al. 2016; Layden et al. 2016).

7.8.4 Does Regeneration Recapitulate Development?

Unraveling the molecular basis of development and regeneration is pivotal to answering the question of whether developmental programs are co-opted for regenerative processes. Complicating matters is that historically, researchers were limited by models suited to either the study of development or regeneration or those that had limited regenerative capabilities (e.g. *Ambystoma mexicanum, Danio rerio, Xenopus laevis*). Models where both processes can be examined within the same species took longer to become established (e.g. *Nematostella vectensis, Hofstenia miamia*). A distinguishing feature of *Nematostella* is that it is capable of whole-body axis regeneration. This, coupled with the fact that it is becoming a strong model for development, offers the unique ability to directly compare these two processes within the same animal. Studies in *Nematostella* and other species are gaining support for the hypothesis that regeneration is only a partial redeployment of embryonic development (e.g. Schaffer et al. 2016; Warner et al. 2019).

Moving forward, it will be necessary to study whether the same program differences arise regardless of regeneration paradigm. For example, during whole-body axis regeneration in *Nematostella*, the initial regenerative response of certain neural subtypes differs under varying regenerative paradigms, suggesting that there may be cell type differences (Havrilak et al. 2021). As we gain functional understanding of these processes, it begs the question of whether we can unlock regenerative potential in non-regenerative models and use this knowledge to develop medical therapies.

7.8.5 Other Challenging Questions

The topics addressed previously are only a small subset of the challenging questions that *Nematostella* is poised to address. For example, other questions pertaining to evolution and development, such as the centralization of the nervous system within the bilaterian lineage, are possible because of the position of cnidarians as sister taxa. Outside of its use in academia, there is also definite potential for *Nematostella* within applied research fields, such as biotechnology. Within the biotech industry, one innovative group looks to use *Nematostella* to help consumers combat the signs of ageing by harnessing the stinging action of cnidocyte cells to optimize the delivery of skin care agents deep into the skin (Toren and Gurovich 2016). Although few other examples of *Nematostella* in applied research exist, it is easy to imagine other uses for this cnidocyte-mediated injection technology throughout the beauty and medical industries, as well as many other untapped applications waiting to be uncovered.

ACKNOWLEDGMENTS

Thanks to Anna Delaney and Mark Williams for uncovering archival information on Miss. Gertrude F. Selwood from when she was at their respective universities. We thank Eric Röttinger for the photograph of the adult *Nematostella*, MingHe Chen for developmental images and Dylan Faltine-Gonzalez for insightful discussions.

BIBLIOGRAPHY

Amiel, A.R., Foucher, K., Ferreira, S., and E. Röttinger. 2019. Synergic coordination of stem cells is required to induce a regenerative response in anthozoan cnidarians. *BioRxiv*. December 31:891804.

Amiel, A.R., Hereroa J., Nedoncelle, K., Warner, J., Ferreira, S., and E. Röttinger. 2015. Characterization of morphological and cellular events underlying oral regeneration in the sea anemone. *Nematostella vectensis. International Journal of Molecular Sciences*. 16:28449–28471.

Amiel, A.R., Johnston, H., Chock, T., Dahlin, P., Iglesias, M., Layden, M., Röttinger, E., and M.Q. Martindale. 2017. A bipolar role of the transcription factor erg for cnidarian germ layer formation and apical domain patterning. *Developmental Biology*. 430:346–361.

Babonis, L.S., and M.Q. Martindale. 2014. Old cell, new trick? Cnidocytes as a model for the evolution of novelty. *Integrative and Comparative Biology*. 54:714–722.

Babonis, L.S., and M.Q. Martindale. 2017. PaxA, but not PaxC, is required for cnidocyte development in the sea anemone *Nematostella vectensis: EvoDevo*. 8:14.

Babonis, L.S., Martindale, M.Q., and J.F. Ryan. 2016. Do novel genes drive morphological novelty? An investigation of the nematosomes in the sea anemone *Nematostella vectensis*. *BMC Evolutionary Biology*. 16:114.

Babonis, L.S., Ryan, J.F., Enjolras, C., and M.Q. Martindale. 2019. Genomic analysis of the tryptome reveals molecular mechanisms of gland cell evolution. *EvoDevo*. 10:23.

Bagaeva, T., Aman, A.J., Graf, T., Niedermoser, I., Zimmermann, B., Kraus, Y., Schatka, M., Demilly, A., Technau, U., and G. Genikhovich. 2020. β-catenin dependent axial patterning in Cnidaria and Bilateria uses similar regulatory logic. *BioRxiv*. September 8:287821.

Baird, A.H., Guest, J.R., and B.L. Willis. 2009. Systematic and biogeographical patterns in the reproductive biology of Scleractinian corals. *Ecology, Evolution, and Systematics*. 40:551–571.

Berking, S. 2007. Generation of bilateral symmetry in Anthozoa: A model. *Journal of Theoretical Biology*. 246:477–490.

Bossert, P.E., Dunn, M.P., and G.H. Thomsen. 2013. A staging system for the regeneration of a polyp from the aboral physa of the anthozoan cnidarian *Nematostella vectensis*. *Developmental Dynamics*. 242:1320–1331.

Bossert, P.E., and G.H. Thomsen. 2017. Inducing complete polyp regeneration from the aboral physa of the starlet sea anemone *Nematostella vectensis*. *Journal of Visualized Experiments*. 119.

Brandão, R.A., Gusmão, L.C., and P.B. Gomes. 2019. Diversity of Edwardsiidae sea anemones (Cnidaria: Anthozoa: Actiniaria) from Brazil, with the description of a new genus and species. *Journal of the Marine Biological Association of the United Kingdom*. 99:1087–1098.

Chapman, J.A., Kirkness, E.F., Simakov, O., Hampson, S.E., Mitros, T., Weinmaier, T., Rattei, T., Balasubramanian, P.G., Borman, J., Busam, D., Disbennett, K., Pfannkoch, C., Sumin, N., Sutton, G.G., Viswanathan, L.D., Walenz, B., Goodstein, D.M., Hellsten, U., Kawashima, T., Prochnik, S.E., Putnam, N.H., Shu, S., Blumberg, B., Dana, C.E., Gee, L., Kibler, D.F., Law, L., Lindgens, D., Martinez, D.E., Peng, J., Wigge, P.A., Bertulat, B., Guder, C., Nakamura, Y., Ozbek, S., Watanabe, H., Khalturin, K., Hemmrich, G., Franke, A., Augustin, R., Fraune, S., Hayakawa, E., Hayakawa, S., Hirose, M., Hwang, J.S., Ikeo, K., Nishimiya-Fujisawa, C., Ogura, A., Takahashi, T., Steinmetz, P.R.H., Zhang, X., Aufschnaiter, R., Eder, M.K., Gorny, A.K., Salvenmoser, W., Heimberg, A.M. Wheeler, B.M., Peterson, K.J., Böttger, A., Tischler, P., Wolf, A., Gojobori, T., Remington, K.A., Strausberg, R.L., Venter, J.C., Technau, U., Hobmayer, B., Bosch, T.C.G., Holstein, T.W., Fujisawa, T., Bode, H.R., David, C.N., Rokhsar, D.S., and R.E. Steele. 2010. The dynamic genome of *Hydra*. *Nature*. 464:592–596.

Clarke, D.N., Lowe, C.J., and W.J. Nelson. 2019. The cadherin-catenin complex is necessary for cell adhesion and embryogenesis in *Nematostella vectensis*. *Developmental Biology*. 447:170–181.

Cleves, P.A., Strader, M.E., Bay, L.K., Pringle, J.R., and M.V. Matz. 2018. CRISPR/Cas9-mediated genome editing in a reef-building coral. *Proceedings of the National Academy of Sciences*. 115:201722151.

Collins, A.G. 2002. Phylogeny of Medusozoa and the evolution of cnidarian life cycles. *Journal of Evolutionary Biology*. 15:418–432.

Columbus-Shenkar, Y.Y., Sachkova, M.Y., Macrander, J., Fridrich, A., Modepalli, V., Reitzel, A.M., Sunagar, K., and Y. Moran. 2018. Dynamics of venom composition across a complex life cycle. *ELife*. 7:e35014.

Craggs, J., Guest, J.R., Davis, M., Simmons, J., Dashti, E., and M. Sweet. 2017. Inducing broadcast coral spawning ex situ: Closed system mesocosm design and husbandry protocol. *Ecology and Evolution*. 7:11066–11078.

Crowell, S. 1946. A new sea anemone from Woods Hole, Massachusetts. *Journal of the Washington Academy of Sciences*. 36:47–60.

Darling, J.A., Reitzel, A.M., Burton, P.M., Mazza, M.E., Ryan, J.F., Sullivan, J.C., and J.R. Finnerty. 2005. Rising starlet: The starlet sea anemone, *Nematostella vectensis*. *BioEssays*. 27:211–221.

Darling, J.A., Reitzel, A.M., and J.R. Finnerty. 2004. Regional population structure of a widely introduced estuarine invertebrate: *Nematostella vectensis* Stephenson in New England. *Molecular Ecology*. 13:2969–2981.

DuBuc, T.Q., Stephenson, T.B., Rock, A.Q., and M.Q. Martindale. 2018. Hox and Wnt pattern the primary body axis of an anthozoan cnidarian before gastrulation. *Nature Communications*. 9:2007.

DuBuc, T.Q., Traylor-Knowles, N., and M.Q. Martindale. 2014. Initiating a regenerative response; cellular and molecular features of wound healing in the cnidarian *Nematostella vectensis*. *BMC Biology*. 12:24.

Faltine-Gonzalez, D.Z., and M.J. Layden. 2019. Characterization of NAChRs in *Nematostella vectensis* supports neuronal and non-neuronal roles in the cnidarian–bilaterian common ancestor. *EvoDevo*. 10:27.

Fautin, D.G. 2002. Reproduction of Cnidaria. *Canadian Journal of Zoology*. 80:1735–1754.

Finnerty, J.R., Pang, K., Burton, P., Paulson, D., and M.Q. Martindale. 2004. Origins of bilateral symmetry: Hox and Dpp expression in a sea anemone. *Science*. 304:1335–1337.

Fischer, A.H.L., Mozzherin, D.A.M.E., Lans, K.D., Wilson, N., Cosentino, C., and J. Smith. 2014. SeaBase: A multispecies transcriptomic resource and platform for gene network inference. *Integrative and Comparative Biology*. 54:250–263.

Frank, P., and J.S. Bleakney. 1976. Histology and sexual reproduction of the anemone *Nematostella vectensis* Stephenson 1935. *Journal of Natural History*. 10:441–449.

Frank, P., and J.S. Bleakney. 1978. Asexual reproduction diet and anomalies of the anemone *Nematostella vectensis* in Nova Scotia Canada. *Canadian Field Naturalist*. 259–263.

Fricano, C., Röttinger, E., Furla, P., and S. Barnay-Verdier. 2020. Cnidarian cell cryopreservation: A powerful tool for cultivation and functional assays. *Cells*. 9:2541.

Friedman, L.E., Gilmore, T.D., and J.R. Finnerty. 2018. Intraspecific variation in oxidative stress tolerance in a model cnidarian: Differences in peroxide sensitivity between and within populations of *Nematostella vectensis*. *PLoS One*. 13:e0188265.

Fritz, A.E., Ikmi, A., Seidel, C., Paulson, A., and M.C. Gibson. 2013. Mechanisms of tentacle morphogenesis in the sea anemone *Nematostella vectensis*. *Development*. 140:2212–2223.

Fritzenwanker, J.H., Genikhovich, G., Kraus, Y., and U. Technau. 2007. Early development and axis specification in the sea anemone *Nematostella vectensis*. *Developmental Biology*. 310:264–279.

Fritzenwanker, J.H., and U. Technau. 2002. Induction of gametogenesis in the basal cnidarian *Nematostella vectensis* (Anthozoa). *Development Genes and Evolution*. 212:99–103.

Galliot, B., Quiquand, M., Ghila, L., de Rosa, R., Miljkovic-Licina, M., and S. Chera. 2009. Origins of neurogenesis, a Cnidarian view. *Developmental Biology*. 332:2–24.

Genikhovich, G., and U. Technau. 2009. The starlet sea anemone *Nematostella vectensis*: An anthozoan model organism for studies in comparative genomics and functional evolutionary developmental biology. *Cold Spring Harbor Protocols*. 2009:pdb.emo129.

Genikhovich, G., and U. Technau. 2017. On the evolution of bilaterality. *Development*. 144:3392–3404.

Gosse, P.H. 1860. *Actinologia britannica: A history of the British sea-anemones and corals*. London: Van Voorst.

Graham, A., Maden, M., and R. Krumlauf. 1991. The murine Hox-2 genes display dynamic dorsoventral patterns of expression during central nervous system development. *Development*. 112:255–264.

Hand, C. 1957. Another sea anemone from California and the types of certain Californian anemones. *Journal of the Washington Academy of Sciences*. 47:411–414.

Hand, C., and K.R. Uhlinger. 1992. The culture, sexual and asexual reproduction, and growth of the sea anemone *Nematostella vectensis*. *The Biological Bulletin*. 182:169–76.

Hand, C., and K.R. Uhlinger. 1994. The unique, widely distributed, estuarine sea anemone, *Nematostella vectensis*

Stephenson: A review, new facts, and questions. *Estuaries*. 17:501.

Hand, C., and K.R. Uhlinger. 1995. Asexual reproduction by transverse fission and some anomalies in the sea anemone *Nematostella vectensis*. *Invertebrate Biology*. 114:9.

Harrison, P.L., Babcock, R.C., Bull, G.D., Oliver, J.K., Wallace, C.C., and B.L. Willis. 1984. Mass spawning in tropical reef corals. *Science*. 223:1186–1189.

Havrilak, J.A., Al-Shaer, L., Baban, N., Akinci, N., and M.J. Layden. 2021. Characterization of the dynamics and variability of neuronal subtype responses during growth, degrowth, and regeneration. *BMC Biology*. 19:104.

Havrilak, J.A., Faltine-Gonzalez, D., Wen, Y., Fodera, D., Simpson, A.C., Magie, C.R., and M.J. Layden. 2017. Characterization of NvLWamide-like neurons reveals stereotypy in *Nematostella* nerve net development. *Developmental Biology*. 431:336–346.

Havrilak, J.A., and M.J. Layden. 2019. Reverse genetic approaches to investigate the neurobiology of the cnidarian sea anemone *Nematostella vectensis*. In *Methods in molecular biology*, ed. S.G. Sprecher, 25–43. New York: Humana Press.

He, S., del Viso, F., Chen, C.Y., Ikmi, A., Kroesen, A.E., and M.C. Gibson. 2018. An axial Hox code controls tissue segmentation and body patterning in *Nematostella vectensis*. *Science*. 361:1377–1380.

Hejnol, A., and F. Rentzsch. 2015. Neural nets. *Current Biology*. 25:R782–R786.

Helm, R.R., Siebert, S., Tulin, S., Smith, J., and C.W. Dunn. 2013. Characterization of differential transcript abundance through time during *Nematostella vectensis* development. *BMC Genomics*. 14:266.

Ikmi, A., McKinney, S.A., Delventhal, K.M., and M.C. Gibson. 2014. TALEN and CRISPR/Cas9-mediated genome editing in the early-branching metazoan *Nematostella vectensis*. *Nature Communications*. 5:5486.

Ikmi, A., Steenbergen, P.J., Anzo, M., McMullen, M.R. Stokkermans, A., Ellington, L.R., and M.C. Gibson. 2020. Feeding-dependent tentacle development in the sea anemone *Nematostella vectensis*. *Nature Communications*. 11:4399.

Jahnel, S.M., Walzl, M., and U. Technau. 2014. Development and epithelial organisation of muscle cells in the sea anemone *Nematostella vectensis*. *Frontiers in Zoology*. 11:44.

Karabulut, A., He, S., Chen, C.Y., McKinney, S.A., and M.C. Gibson. 2019. Electroporation of short hairpin RNAs for rapid and efficient gene knockdown in the starlet sea anemone, *Nematostella vectensis*. *Developmental Biology*. 448:7–15.

Kass-Simon, G., and A.A. Scappaticci Jr. 2002. The behavioral and developmental physiology of nematocysts. *Canadian Journal of Zoology*. 80:1772–1794.

Keith, S.A., Maynard, J.A., Edwards, A.J., Guest, J.R., Bauman, A.G., van Hooidonk, R., Heron, S.F., Berumen, M.L., Bouwmeester, J., Piromvaragorn, S., Rahbek C., and A.H. Baird. 2016. Coral mass spawning predicted by rapid seasonal rise in ocean temperature. *Proceedings of the Royal Society B: Biological Sciences*. 283:20160011.

Kortschak, R.D., Samuel, G., Saint, R., and D.J. Miller. 2003. EST analysis of the cnidarian *Acropora millepora* reveals extensive gene loss and rapid sequence divergence in the model invertebrates. *Current Biology*. 13:2190–2195.

Kraus, Y., Aman, A., Technau, U., and G. Genikhovich. 2016. Prebilaterian origin of the blastoporal axial organizer. *Nature Communications*. 7:11694.

Kraus, Y., and U. Technau. 2006. Gastrulation in the sea anemone *Nematostella vectensis* occurs by invagination and immigration: An ultrastructural study. *Development Genes and Evolution*. 216:119–132.

Kumburegama, S., Wijesena, N., Xu, R., and A.H. Wikramanayake. 2011. Strabismus-mediated primary archenteron invagination is uncoupled from Wnt/β-catenin-dependent endoderm cell fate specification in *Nematostella vectensis* (Anthozoa, Cnidaria): Implications for the evolution of gastrulation. *EvoDevo*. 2:2.

Kusserow, A., Pang, K., Sturm, C., Hrouda, M., Lentfer, J., Schmidt, H.A., Technau, U., von Haeseler, A., Hobmayer, B., Martindale M.Q., and W. Thomas. 2005. Unexpected complexity of the Wnt gene family in a sea anemone. *Nature*. 433:156–160.

Layden, M.J., Boekhout, M., and M.Q. Martindale. 2012. *Nematostella vectensis* achaete-scute homolog NvashA regulates embryonic ectodermal neurogenesis and represents an ancient component of the metazoan neural specification pathway. *Development*. 139:1013–1022.

Layden, M.J., Johnston, H., Amiel, A.R., Havrilak, J., Steinworth, B., Chock, T., Röttinger, E., and M.Q. Martindale. 2016. MAPK signaling is necessary for neurogenesis in *Nematostella vectensis*. *BMC Biology*. 14:61.

Layden, M.J., and M.Q. Martindale. 2014. Non-canonical Notch signaling represents an ancestral mechanism to regulate neural differentiation. *EvoDevo*. 5:30.

Layden, M.J., Rentzsch, F., and E. Röttinger. 2016. The rise of the starlet sea anemone *Nematostella vectensis* as a model system to investigate development and regeneration. *Wiley Interdisciplinary Reviews: Developmental Biology*. 5:408–428.

Layden, M.J., Röttinger, E., Wolenski, F.S., Gilmore, T.D., and M.Q. Martindale. 2013. Microinjection of mRNA or morpholinos for reverse genetic analysis in the starlet sea anemone, *Nematostella vectensis*. *Nature Protocols*. 8:924–934.

Leclère, L., Bause, M., Sinigaglia, C., Steger, J., and F. Rentzsch. 2016. Development of the aboral domain in *Nematostella* requires β-catenin and the opposing activities of Six3/6 and Frizzled5/8. *Development*. 143:1766–1777.

Leclère, L., and F. Rentzsch. 2014. RGM regulates BMP-mediated secondary axis formation in the sea anemone *Nematostella vectensis*. *Cell Reports*. 9:1921–1930.

Lee, P.N., Kumburegama, S., Marlow, H.Q., M.Q. Martindale, and A.H. Wikramanayake. 2007. Asymmetric developmental potential along the animal: Vegetal axis in the anthozoan cnidarian, *Nematostella vectensis*, is mediated by Dishevelled. *Developmental Biology*. 310:169–186.

Magie, C.R., Daly, M., and M.Q. Martindale. 2007. Gastrulation in the cnidarian *Nematostella vectensis* occurs via invagination not ingression. *Developmental Biology*. 305:483–497.

Magie, C.R., Pang, K., and M.Q. Martindale. 2005. Genomic inventory and expression of Sox and Fox genes in the cnidarian *Nematostella vectensis*. *Development Genes and Evolution*. 215:618–630.

Marlow, H.Q., Matus, D.Q., and M.Q. Martindale. 2013. Ectopic activation of the canonical Wnt signaling pathway affects ectodermal patterning along the primary axis during larval development in the anthozoan *Nematostella vectensis*. *Developmental Biology*. 380:324–334.

Marlow, H.Q., Srivastava, M., Matus, D.Q., Rokhsar, D., and M.Q. Martindale. 2009. Anatomy and development of the nervous system of *Nematostella vectensis*, an anthozoan cnidarian. *Developmental Neurobiology*. 69:235–254.

Martindale, M.Q., Finnerty, J.R., and J.Q. Henry. 2002. The radiata and the evolutionary origins of the bilaterian body plan. *Molecular Phylogenetics and Evolution*. 24:358–365.

Martindale, M.Q., Pang, K., and J.R. Finnerty. 2004. Investigating the origins of triploblasty: 'mesodermal' gene expression in a diploblastic animal, the sea anemone *Nematostella vectensis* (Phylum, Cnidaria; Class, Anthozoa). *Development*. 131:2463–2474.

Matus, D.Q., Magie, C.R., Pang, K., Martindale, M.Q., and G.H. Thomsen. 2008. The Hedgehog gene family of the cnidarian, *Nematostella vectensis*, and implications for understanding metazoan Hedgehog pathway evolution. *Developmental Biology*. 313:501–518.

Matus, D.Q., Pang, K., Marlow, H.Q., Dunn, C.W., Thomsen, G.H., and M.Q. Martindale. 2006. Molecular evidence for deep evolutionary roots of bilaterality in animal development. *Proceedings of the National Academy of Sciences*. 103:11195–11200.

Medina, M., Collins, A.G., Silberman, J.D., and M.L. Sogin. 2001. Evaluating hypotheses of basal animal phylogeny using complete sequences of large and small subunit rRNA. *Proceedings of the National Academy of Sciences*. 98:9707–9712.

Moran, Y., Genikhovich, G., Gordon, D., Wienkoop, S., Zenkert, C., Özbek, S., Technau, U., and M. Gurevitz. 2011. Neurotoxin localization to ectodermal gland cells uncovers an alternative mechanism of venom delivery in sea anemones. *Proceedings of the Royal Society B: Biological Sciences*. 279:1351–1358.

Nakanishi, N., Renfer, E., Technau, U., and F. Rentzsch. 2012. Nervous systems of the sea anemone *Nematostella vectensis* are generated by ectoderm and endoderm and shaped by distinct mechanisms. *Development*. 139:347–357.

Niehrs, C. 2010. On growth and form: A Cartesian coordinate system of Wnt and BMP signaling specifies bilaterian body axes. *Development*. 137:845–857.

Passamaneck, Y.J., and M.Q. Martindale. 2012. Cell proliferation is necessary for the regeneration of oral structures in the anthozoan cnidarian *Nematostella vectensis*. *BMC Developmental Biology*. 12:34.

Pearson, C.V.M., Rogers, A.D., and M. Sheader. 2002. The genetic structure of the rare lagoonal sea anemone, *Nematostella vectensis* Stephenson (Cnidaria; Anthozoa) in the United Kingdom based on RAPD analysis. *Molecular Ecology*. 11:2285–2293.

Pearson, J.C., Lemons, D., and W. McGinnis. 2005. Modulating Hox gene functions during animal body patterning. *Nature Reviews Genetics*. 6:893–904.

Pollock, F.J., Katz, S.M., van de Water, J.A.J.M., Davies, S.W., Hein, M., Torda, G., Matz, M.V., Beltran, V.H., Buerger, P., Puill-Stephan, E., Abrego, D., Bourne, D.G., and B.L. Willis. 2017. Coral larvae for restoration and research: A large-scale method for rearing *Acropora millepora* larvae, inducing settlement, and establishing symbiosis. *PeerJ*. 5:e3732.

Posey, M.H., and A.H. Hines. 1991. Complex predator-prey interactions within an estuarine benthic community. *Ecology*. 72: 2155–2169.

Putnam, N.H., Srivastava, M., Hellsten, U., Dirks, B., Chapman, J., Salamov, A., Terry, A., Shapiro, H., Lindquist, E., Kapitonov, V.V., Jurka, J., Genikhovich, G., Grigoriev, I.V., Lucas, S.M., Steele, R.E., Finnerty, J.R., Technau, U., Martindale, M.Q., and D.S. Rokhsar. 2007. Sea anemone genome reveals ancestral eumetazoan gene repertoire and genomic organization. *Science*. 317:86–94.

Rabinowitz, C., Moiseeva, E., and B. Rinkevich. 2016. In vitro cultures of ectodermal monolayers from the model sea anemone *Nematostella vectensis*. *Cell and Tissue Research*. 366:693–705.

Reitzel, A.M., Burton, P.M., Krone, C., and J.R. Finnerty. 2007. Comparison of developmental trajectories in the starlet sea anemone *Nematostella vectensis*: Embryogenesis, regeneration, and two forms of asexual fission. *Invertebrate Biology*. 126:99–112.

Reitzel, A.M., Chu, T., Edquist, S., Genovese, C., Church, C., Tarrant, A.M., and J.R. Finnerty. 2013. Physiological and developmental responses to temperature by the sea anemone *Nematostella vectensis*. *Marine Ecology Progress Series*. 484:115–130.

Reitzel, A.M., Darling, J.A., Sullivan, J.C., and J.R. Finnerty. 2007. Global population genetic structure of the starlet anemone *Nematostella vectensis*: Multiple introductions and implications for conservation policy. *Biological Invasions*. 10:1197–1213.

Reitzel, A.M., Herrera, S., Layden, M.J., Martindale, M.Q., and T.M. Shank. 2013. Going where traditional markers have not gone before: Utility of and promise for RAD sequencing in marine invertebrate phylogeography and population genomics. *Molecular Ecology*. 22:2953–2970.

Renfer, E., Amon-Hassenzahl, A., Steinmetz, P.R.H., and U. Technau. 2010. A muscle-specific transgenic reporter line of the sea anemone, *Nematostella vectensis*. *Proceedings of the National Academy of Sciences of the United States of America*. 107:104–108.

Renfer, E., and U. Technau. 2017. Meganuclease-assisted generation of stable transgenics in the sea anemone *Nematostella vectensis*. *Nature Protocols*. 12:1844–1854.

Rentzsch, F., Anton, R., Saina, M., Hammerschmidt, M., Holstein, T.W., and U. Technau. 2006. Asymmetric expression of the BMP antagonists Chordin and Gremlin in the sea anemone *Nematostella vectensis*: Implications for the evolution of axial patterning. *Developmental Biology*. 296:375–387.

Rentzsch, F., Fritzenwanker, J.H., Scholz, C.B., and U. Technau. 2008. FGF signaling controls formation of the apical sensory organ in the cnidarian *Nematostella vectensis*. *Development*. 135:1761–1769.

Rentzsch, F., Layden, M.J., and M. Manuel. 2017. The cellular and molecular basis of cnidarian neurogenesis. *Wiley Interdisciplinary Reviews: Developmental Biology*. 6:e257.

Rentzsch, F., and U. Technau. 2016. Genomics and development of *Nematostella vectensis* and other anthozoans. *Current Opinion in Genetics & Development*. 39:63–70.

Richards, G.S., and F. Rentzsch. 2014. Transgenic analysis of a SoxB gene reveals neural progenitor cells in the cnidarian *Nematostella vectensis*. *Development*. 141:4681–4689.

Richards, G.S., and F. Rentzsch. 2015. Regulation of *Nematostella* neural progenitors by Soxb, Notch and bHLH genes. *Development*. 142:3332–3342.

Röttinger, E., Dahlin, P., and M.Q. Martindale. 2012. A framework for the establishment of a cnidarian gene regulatory network for 'endomesoderm' specification: The inputs of ß-catenin/TCF signaling. *PLoS Genetics*. 8:e1003164.

Ryan, J.F., Burton, P.M., Mazza, M.E., Kwong, G.K., Mullikin, J.C., and J.R. Finnerty. 2006. The cnidarian-bilaterian ancestor possessed at least 56 homeoboxes: Evidence from the starlet sea anemone, *Nematostella vectensis*. *Genome Biology*. 7:R64.

Saina, M., Genikhovich, G., Renfer, E., and U. Technau. 2009. BMPs and Chordin regulate patterning of the directive axis in a sea anemone. *Proceedings of the National Academy of Sciences of the United States of America*. 106:18592–18597.

Schaffer, A.A., Bazarsky, M., Levy, K., Chalifa-Caspi, V., and U. Gat. 2016. A transcriptional time-course analysis of oral vs. aboral whole-body regeneration in the sea anemone *Nematostella vectensis*. *BMC Genomics*. 17:718.

Scholz, C.B., and U. Technau. 2003. The ancestral role of Brachyury: Expression of NemBra1 in the basal cnidarian *Nematostella vectensis* (Anthozoa). *Development Genes and Evolution*. 212:563–570.

Schwaiger, M., Schönauer, A., Rendeiro, A.F., Pribitzer, C., Schauer, A., Gilles, A.F., Schinko, J.B., Renfer, E., Fredman, D., and U. Technau. 2014. Evolutionary conservation of the eumetazoan gene regulatory landscape. *Genome Research*. 24:639–650.

Sebé-Pedrós, A., Saudemont, B., Chomsky, E., Plessier, F., Mailhé, M.P., Renno, J., Loe-Mie, Y., Lifshitz A., Mukamel Z., Schmutz S., Novault S., Steinmetz P.R.H., Spitz F., Tanay A., H. Marlow. 2018. Cnidarian cell type diversity and regulation revealed by whole-organism single-cell RNA-seq. *Cell*. 173:1520–1534.

Servetnick, M.D., Steinworth, B., Babonis, L.S., Simmons, D., Salinas-Saavedra, M., and M.Q. Martindale. 2017. Cas9-mediated excision of *Nematostella brachyury* disrupts endoderm development, pharynx formation, and oral-aboral patterning. *Development*. 144:dev.145839.

Sheader, M., Suwailem, A.M., and G.A. Rowe. 1997. The anemone, *Nematostella vectensis*, in Britain: Considerations for conservation management. *Aquatic Conservation: Marine and Freshwater Ecosystems*. 7:13–25.

Silva, J.F., Lima, C.A.C., Perez, C.D., and P.B. Gomes. 2010. First record of the sea anemone *Nematostella vectensis* (Actiniaria: Edwardsiidae) in southern hemisphere waters. *Zootaxa*. 2343:66–68.

Sinigaglia, C., Busengdal, H., Lerner, A., Oliveri, P., and F. Rentzsch. 2015. Molecular characterization of the apical organ of the anthozoan *Nematostella vectensis*. *Developmental Biology*. 398:120–133.

Stefanik, D.J., Friedman, L.E., and J.R. Finnerty. 2013. Collecting, rearing, spawning and inducing regeneration of the starlet sea anemone, *Nematostella vectensis*. *Nature Protocols*. 8:916–923.

Steinmetz, P.R.H. 2019. A non-bilaterian perspective on the development and evolution of animal digestive systems. *Cell and Tissue Research*. 377:321–339.

Steinmetz, P.R.H., Aman, A., Kraus, J.E.M., and U. Technau. 2017. Gut-like ectodermal tissue in a sea anemone challenges germ layer homology. *Nature Ecology & Evolution*. 1:1535–1542.

Stephenson, T.A. 1935. *The British sea anemones*. London: Ray Society.

Sunagar, K., Columbus-Shenkar, Y.Y., Fridrich, A., Gutkovich, N., Aharoni, R., and Y. Moran. 2018. Cell type-specific expression profiling unravels the development and evolution of stinging cells in sea anemone. *BMC Biology*. 16:108.

Szczepanek, S., Cikala, M., and C.N. David. 2002. Poly-gamma-glutamate synthesis during formation of nematocyst capsules in *Hydra*. *Journal of Cell Science*. 115:745–751.

Takahashi, C.K., Lourenço, N.G.G.S., Lopes, T.F., Rall, V.L.M., and C.A.M. Lopes. 2008. Ballast water: A review of the impact on the world public health. *Journal of Venomous Animals and Toxins Including Tropical Diseases*. 14:393–408.

Tarrant, A.M., Helm, R.R., Levy, O., and H.E. Rivera. 2019. Environmental entrainment demonstrates natural circadian rhythmicity in the cnidarian *Nematostella vectensis*. *The Journal of Experimental Biology*. 222: jeb205393.

Technau, U., and M. Schwaiger. 2015. Recent advances in genomics and transcriptomics of cnidarians. *Marine Genomics*. 24:31–38.

Technau, U., and R.E. Steele. 2011. Evolutionary crossroads in developmental biology: Cnidaria. *Development*. 138:1447–1458.

Thurm, U., Brinkmann, M., Golz, R., Holtmann, M., Oliver, D., and T. Sieger. 2004. Mechanoreception and synaptic transmission of hydrozoan nematocytes. *Hydrobiologia*. 530–531:97–105.

Toren, A., and M. Gurovich. 2016. An innovative method to optimize cosmeceutical delivery to skin natural, safe, painless, sustainable transdermal microinjection based on sea anemone nematocysts. *Household and Personal Care Today*. 11: 41–46.

Torres-Méndez, A., Bonnal, S., Marquez, Y., Roth, J., Iglesias, M., Permanyer, J., Almudí, I., O'Hanlon, D., Guitart, T., Soller, M., Gingras, A.C., Gebauer, F., Rentzsch, F., Blencowe, B.J., Valcárcel, J., and M. Irimia. 2019. A novel protein domain in an ancestral splicing factor drove the evolution of neural microexons. *Nature Ecology & Evolution*. 3:691–701.

Trevino, M., Stefanik, D.J., Rodriguez, R., Harmon, S., and P.M. Burton. 2011. Induction of canonical Wnt signaling by alsterpaullone is sufficient for oral tissue fate during regeneration and embryogenesis in *Nematostella vectensis*. *Developmental Dynamics*. 240:2673–2679.

Tucker, R.P., Shibata, B., and T.N. Blankenship. 2011. Ultrastructure of the mesoglea of the sea anemone *Nematostella vectensis* (Edwardsiidae). *Invertebrate Biology*. 130:11–24.

Tulin, S., Aguiar, D., Istrail, S., and J. Smith. 2013. A quantitative reference transcriptome for *Nematostella vectensis* early embryonic development: A pipeline for de novo assembly in emerging model systems. *EvoDevo*. 4:16.

Wainright, P.O., Hinkle, G., Sogin, M.L., and S.K. Stickel. 1993. Monophyletic origins of the metazoa: An evolutionary link with fungi. *Science*. 260:340–342.

Warner, J.F., Amiel, A.R., Johnston, H., and E. Röttinger. 2019. Regeneration is a partial redeployment of the embryonic gene network. *BioRxiv*.

Warner, J.F., Guerlais, V., Amiel, A.R., Johnston, H., Nedoncelle, K., and E. Röttinger. 2018. NvERTx: A gene expression database to compare embryogenesis and regeneration in the sea anemone *Nematostella vectensis*. *Development*. 145:dev.162867.

Watanabe, H., Fujisawa, T., and T.W. Holstein. 2009. Cnidarians and the evolutionary origin of the nervous system. *Development, Growth & Differentiation*. 51:167–183.

Watanabe, H., Kuhn, A., Fushiki, M., Agata, K., Özbek, S., Fujisawa, T., and T.W. Holstein. 2014. Sequential actions of β-catenin and Bmp pattern the oral nerve net in *Nematostella vectensis*. *Nature Communications*. 5:5536.

Weir, K., Dupre, C., van Giesen, L., Lee, A.S.Y., and N.W. Bellono. 2020. A molecular filter for the cnidarian stinging response. *eLife*. 9:e57578.

Wijesena, N., Simmons, D.K., and M.Q. Martindale. 2017. Antagonistic BMP–cWNT signaling in the cnidarian *Nematostella vectensis* reveals insight into the evolution of mesoderm. *Proceedings of the National Academy of Sciences*. 114:E5608–E5615.

Wikramanayake, A.H., Hong, M., Lee, P.N., Pang, K., Byrum, C.A., Bince, J.M., Xu, R., and M.Q. Martindale. 2003. An ancient role for nuclear β-catenin in the evolution of axial polarity and germ layer segregation. *Nature*. 426:446–450.

Williams, R.B. 1975. A redescription of the brackish-water sea anemone *Nematostella vectensis* Stephenson, with an appraisal of congeneric species. *Journal of Natural History*. 9:51–64.

Williams, R.B. 1976. Conservation of the sea anemone *Nematostella vectensis* in Norfolk, England and its world distribution. *Transactions of the Norfolk and Norwich Naturalists' Society*. 23:257–266.

Williams, R.B. 1979. Studies on the nematosomes of *Nematostella vectensis* Stephenson (Coelenterata: Actiniaria). *Journal of Natural History*. 13:69–80.

Williams, R.B. 1983. Starlet sea anemone: *Nematostella vectensis*. In *The IUCN invertebrate red data book*, ed. S.M. Wells, R.M. Pyle, and N.M. Collins, 43–46. Gland: IUCAN.

Williams, R.B. 2003. Locomotory behaviour and functional morphology of *Nematostella vectensis* (Anthozoa: Actiniaria: Edwardsiidae): A contribution to a comparative study of burrowing behaviour in athenarian sea anemones. *Zoologische Verhandelingen*. 345:437–84.

Wiltse, W.I., Foreman, K.H., Teal, J.M., and I. Valiela. 1984. Effects of predators and food resources on the macrobenthos of salt marsh creeks. *Journal of Marine Research*. 42:923–942.

Wolenski, F.S., Garbati, M.R., Lubinski, T.J., Traylor-Knowles, N., Dresselhaus, E., Stefanik, D.J., Goucher, H., Finnerty, J.F., and T.D. Gilmore. 2011. Characterization of the core elements of the NF-KB signaling pathway of the sea anemone *Nematostella vectensis*. *Molecular and Cellular Biology*. 31:1076–1087.

Wolenski, F.S., Layden, M.J., Martindale, M.Q., Gilmore T.D., and J.R. Finnerty. 2013. Characterizing the spatiotemporal expression of RNAs and proteins in the starlet sea anemone, *Nematostella vectensis*. *Nature Protocols*. 8:900–915.

Yonge, M. 1962. Thomas Alan Stephenson, 1898–1961. *Biographical Memoirs of Fellows of the Royal Society*. 8:136–148.

Zang, H., and N. Nakanishi. 2020. Expression analysis of cnidarian-specific neuropeptides in a sea anemone unveils an apical-organ-associated nerve net that disintegrates at metamorphosis. *Frontiers in Endocrinology*. 11:63.

Zenkert, C., Takahashi, T., Diesner, M.O., and S. Özbek. 2011. Morphological and molecular analysis of the *Nematostella vectensis* cnidom. *PLoS One*. 6:e22725.

Zimmermann, B., Robb, S.M.C., Genikhovich, G., Fropf, W.J., Weilguny, L., He, S., Chen, S., Lovegrove-Walsh, J., Hill, E.M., Ragkousi K., Praher, D., Fredman, D., Moran, Y., Gibson, M.C., and U. Technau. 2020. Sea anemone genomes reveal ancestral metazoan chromosomal macrosynteny. *BioRxiv*.

8 The Marine Jellyfish Model, *Clytia hemisphaerica*

Sophie Peron, Evelyn Houliston and Lucas Leclère

CONTENTS

DOI: 10.1201/9781003217503-8

8.1 HISTORY OF THE MODEL

Classical "model" organisms (such as mouse, drosophila, nematode, etc.) have contributed a huge amount of knowledge in biology but represent only a small fraction of the diversity of organisms. Marine animals are very diverse in term of morphology and physiology and cover a wide range of taxa. Therefore, they can make a valuable contribution to research, both for addressing biological processes and evolutionary questions.

This chapter presents *Clytia hemisphaerica*, a jellyfish with growing interest as an experimental model. This species can be cultured in the lab in reconstituted sea water, allowing use in any laboratory and a constant supply in animals. First the history of *Clytia* as an experimental model and the characteristics of *Clytia* life stages will be presented. Then diverse experimental tools, currently available and still in development, will be described, before presenting some biological questions that can be addressed using *Clytia*.

Due to their phylogenetic position as a sister group to the bilaterians, cnidarians are a valuable study group for addressing evolutionary questions. Cnidarians are divided into two main clades: the anthozoans, comprising animals only living as polyps for the adult form, and the medusozoans, characterized by the presence of the jellyfish stage in the life cycle (Collins et al. 2006). *Clytia hemisphaerica* (Linnaeus 1767) is a medusozoan species of the class Hydrozoa, the order Leptothecata (characterized by a chitinous envelop protecting the polyps and by the flat shape of the jellyfish) and the family Clytiidae (Cunha et al. 2020). Its life cycle is typical of hydrozoans, alternating between two adult forms: the free-swimming jellyfish (= medusa) and asexually propagating polyp-forming colonies.

Clytia hemisphaerica has long been recognized as a valuable research organism for studying several aspects of hydrozoan biology thanks to its ease of culture; its total transparency; and its triphasic life cycle, including a medusa stage. This last feature distinguishes it from the other main cnidarian model organisms (*Hydra*, *Nematostella* and *Hydractinia*). It is thus possible to study in *Clytia hemisphaerica* complex characters absent from the polyp-only model species, notably: striated muscles; a well-organized nervous system condensed in two nerve rings at the margin of the umbrella; and well-defined and localized organs: the gonads, the manubrium regrouping the mouth and the stomach and the tentacle bulbs.

8.1.1 Early Studies on *Clytia hemisphaerica* Anatomy and Development

Clytia hemisphaerica was referred to in earlier literature under a number of synonyms, such as *Clytia johnstoni* (in: Alder 1856), *Clytia laevis* (in: Weismann 1883), *Clytia viridicans* (in: Metchnikoff 1886), *Phialidium hemisphaericum* (in: Bodo and Bouillon 1968) or *Campanularia johnstoni* (in: Schmid and Tardent 1971; Schmid et al. 1976). *Clytia*, when used alone in this chapter, will refer to the species *Clytia hemisphaerica*.

8.1.1.1 First Descriptions of *Clytia* Embryonic Development

The first detailed description of embryogenesis in *Clytia* was conducted by Elie Metschnikoff in the late 19th century in the marine stations of Naples and Villefranche-sur-Mer (Metchnikoff 1886). In his book *Embryologische studien an Medusen* (1886), he described and compared the development and larva morphology of several medusa species from these sites, including *Clytia hemisphaerica* (= *Clytia viridicans*). Lacassagne (1961) performed histological studies, comparing planulae belonging to the family of "calyptoblastiques à gonophores" including *Clytia*. Seven years later, Bodo and Bouillon (1968) published a description of the embryonic development of five hydromedusae from Roscoff. Their study contains a detailed description of *Clytia* planulae, particularly their cell types and mode of settlement.

8.1.1.2 *Clytia* as a Model for Experimental Embryology

A distinct but closely related species, *Clytia gregaria* (= *Phialidium gregarium*), abundant on the west coast of the United States, was used extensively by the embryologist Gary Freeman and played an important part in the history of cnidarian experimental embryology (Freeman 1981a; Freeman 1981b; Freeman 2005; Freeman and Ridgway 1987; Thomas et al. 1987). Through cutting and grafting experiments using embryos and larvae from wild caught medusae, Freeman investigated the establishment of polarity in *Clytia gregaria* larvae, termed antero-posterior (AP) at that time but now commonly referred to as oral–aboral (OA). He determined i) that isolated parts of the cleaving embryo develop into normal planulae; ii) that they conserve their original antero-posterior axis (Freeman 1981a); iii) that the position of the posterior (oral) pole can be traced back to the initiation site of the first cleavage (Freeman 1980); and iv) that during gastrulation, interactions between the parts of the embryo determine the axis of the planula (Freeman 1981a). This work highlighted the precise regulation of *Clytia* embryogenesis and its flexibility, allowing the development of a correctly patterned planula even if a part of the embryo is missing.

8.1.1.3 *Clytia* Medusa Regeneration

The *Clytia* medusa, like its embryo, can cope with various types of injuries by repatterning and restoration of lost parts. This marked ability to self-repair and regenerate is another particularity that raised interest in early studies. Among cnidarians, the regenerative abilities of polyps (e.g. *Hydra*, *Hydractinia*, *Nematostella*) are well known (Amiel et al. 2015; Bradshaw et al. 2015; DuBuc et al. 2014; Galliot 2012; Schaffer et al. 2016). The huge regenerative abilities of *Hydra* were first documented in the 18th century by Trembley in an attempt to determine whether *Hydra* belonged to plants or animals (1744). In contrast, jellyfish were considered to have lesser abilities due to their greater anatomic complexity (Hargitt 1897). Compared to the literature about the regeneration abilities of the polyps, relatively few studies documented the abilities of hydrozoans and scyphozoans jellyfish (Abrams et al. 2015; Hargitt 1897; Morgan 1899; Okada 1927; Schmid and Tardent 1971; Schmid et al. 1982; Weber 1981; Zeleny 1907).

Neppi (1918) documented the regeneration abilities of wild-caught *Clytia* (*Phialidium variabile*). She concluded that fragments of the umbrella can restore their typical bell shape, and the manubrium and radial canals are restored if they are missing from the fragment, as seen also for other hydrozoan jellyfish (*Gonionemus:* Morgan 1899; and *Obelia*: Neppi 1918). More detailed studies were performed in the 1970s by Schmid and collaborators (Schmid and Tardent 1971; Schmid 1974; Schmid et al. 1976; Schneider 1975; Stidwill 1974). These researchers documented the self-repair and regeneration abilities of wild-caught *Clytia* caught near Villefranche and Banyuls marine stations (Schmid and Tardent 1971). Like Neppi in 1918, they observed that a fragment of the umbrella is able to restore the circular jellyfish shape in a quick and stereotypical process. Any missing organs (manubrium, canals and gonads) then regenerate, the manubrium being the first organ to reform. While the circular shape and missing organs are consistently restored, they found that the original tetraradial symmetry is not necessarily reestablished (Schmid and Tardent 1971). Subsequent studies focused on the mechanisms regulating manubrium regeneration (Schmid 1974; Schmid et al. 1976). They first looked for an induction/inhibition system based on morphogens, similar to that described in *Hydra*. The results of grafting experiments suggested that such diffusing molecules in the tissue are not responsible for guiding the regeneration of the manubrium in *Clytia* (Schmid 1974; Schmid et al. 1976; Stidwill 1974). An alternative hypothesis coming from this work was that tension forces generated by the muscle fibers and the underlying mesoglea are important in patterning during regeneration (Schmid et al. 1976; Schneider 1975). Further regeneration studies on jellyfish were performed on *Podocoryna carnea* and focused on the ability of its striated muscle cells to transdifferentiate after isolation from the jellyfish (Schmid et al. 1982). *Clytia* was not used further to study regeneration until its establishment as an experimental lab-cultured model species (see Section 8.8.1).

8.1.1.4 Sex Determination and the Origin of Germ Cells

In cnidarian life cycles, asexual and sexual reproduction often coexist. In medusozoans, the polyp stage ensures asexual reproduction, whereas the jellyfish is the sexual and dispersive form. Some medusae, including *Clytia mccrady*, a leptomedusa found in the Atlantic ocean and Mediterranean sea, are also able to generate medusae asexually through a budding zone, called the blastostyle, positioned in the place of the gonads (Carré et al. 1995). Carré et al. (1995) showed that, in this species, asexually reproducing jellyfish produce asexual jellyfish.

The origin of germ cells and the mode of sex determination were studied in *Clytia hemisphaerica* by Carré and Carré (2000). Medusae produced from newly established polyp colonies kept at 15°C were mostly male, whereas most of those produced at 24°C were female. However, some medusa produced at 24°C, then raised at 15°C, became male. These findings indicate that sex is not determined genetically. Carré and Carré proposed that two populations of germ cell precursors could coexist in newly released *Clytia* medusa: a dominant female population, temperature sensitive and inactivated at 15°C, and a male population, active at low temperatures (2000). In *Hydra*, it has been shown that grafting of male germ cells in a female polyp leads to the masculinization of the polyp. The male germ cells migrate into the polyp and proliferate, whereas the existing female germ cells are eliminated (Nishimiya-Fujisawa and Kobayashi 2012). In *Clytia*, the male and female germ cell populations could be competing as well, with low temperature favoring male germ cells (Siebert and Juliano 2017).

8.1.2 *Clytia* as a Model after 2000

Following the suggestion of Danielle Carré, Evelyn Houliston started in 2002 *Clytia* cultures in the marine station of Villefranche-sur-Mer, initially to study egg and embryo polarity in this transparent animal. Daily spawning of males and females and external fertilization allowed easy access to all developmental stages for microscopy and experimentation. The culture system is now standardized (Lechable et al. 2020), and different inbred lines have been established by successive self-crossing, starting from a founder colony "Z" obtained from crossing wild medusa collected in the bay of Villefranche. A male colony resulting from three successive self-crossing (Z4C)[2] was used for genome sequencing (Leclère et al. 2019). Several Z-derived male and female lines are currently used in Villefranche (Houliston et al. 2010; Leclère et al. 2019). Medusae from a given line are produced asexually from a polyp colony and therefore are genetically identical.

Clytia started as a model for developmental studies from 2005. Until 2010, it was mostly studied in two laboratories, in Villefranche-sur-Mer and Paris. The main research topics were oogenesis, embryonic patterning and polarity, evolution of developmental mechanisms, nematogenesis and gametogenesis (Amiel et al. 2009; Amiel and Houliston 2009; Chevalier et al. 2006; Chiori et al. 2009; Denker et al. 2008a; Denker et al. 2008b; Denker et al. 2008c; Derelle et al. 2010; Forêt et al. 2010; Fourrage et al. 2010; Momose et al. 2008; Momose and Houliston 2007; Philippe et al. 2009; Quiquand et al. 2009, reviewed by Houliston et al. 2010; Leclère et al. 2016). Tools have been progressively developed for imaging during embryogenesis and in the adult, and for gene function analysis in the embryo (injection of Morpholino oligonucleotides [MOs] or mRNAs into the egg or the embryo: Houliston et al. 2010) and in the adult (gene knock out with CRISPR-Cas9: Momose et al. 2018).

Clytia studies continue in Villefranche, with recently published work concerning, for instance, oocyte maturation (Quiroga Artigas et al. 2020; Quiroga Artigas et al. 2018), embryogenesis (Kraus et al. 2020; van der Sande et al. 2020) and regeneration (Sinigaglia et al. 2020). Michael Manuel and colleagues in Paris also worked extensively on *Clytia* until recently, notably focusing on the jellyfish tentacle bulb (Condamine et al. 2019; Coste et al. 2016; Denker et al. 2008c). The team of Jocelyn Malamy from Chicago University has started to work on wound healing in *Clytia*. They uncovered two healing mechanisms (actomyosin cable and lamelipods crawling) and developed a DIC microscopy system allowing visualization of individual cell movements (Kamran et al. 2017; Malamy and Shribak 2018). Other published articles on *Clytia* include the work of Ulrich Technau's group (Gur Barzilai et al. 2012; Kraus et al. 2015; Steinmetz et al. 2012), notably demonstrating the convergence of hydrozoan and bilaterian striated muscles, and from Noriyo Takeda identifying the maturation-inducing hormones (MIHs) in *Clytia* and *Cladonema* jellyfish (Takeda et al. 2018). Other groups worldwide are starting to adopt *Clytia* for their research.

8.2 GEOGRAPHICAL LOCATION

Clytia hemisphaerica is a cosmopolitan jellyfish species. Its presence has been documented in many places, including the Mediterranean sea (between September and March in Villefranche; Carré and Carré 2000), Brittany (in Roscoff in 1968, particularly during summer and fall; Bodo and Bouillon 1968), the English Channel (Lucas et al. 1995), as well as Japan (= *Clytia edwardsi*) (Kubota 1978) and the US north Pacific coast (Roosen-Runge 1962).

Clytia undergo light-dependent diel vertical migrations following a day/night cycle, like many hydrozoan jellyfish (Mills 1983). The physiological, ecological and evolutionary relevance of this daily migration remains to be studied. In laboratory conditions, *Clytia hemisphaerica* medusae spawn two hours after a dark–light transition after migrating to the surface of the tank, matching the morning spawning of local populations (Quiroga Artigas 2017). Variant spawning patterns have been reported at other locations, for instance, at dawn and dusk for *Clytia hemisphaerica* in Friday Harbor (US north Pacific coast) (Roosen-Runge 1962).

8.3 LIFE CYCLE

Clytia belongs to the hydrozoan class and exhibits the typical life cycle, alternating between a planula larva, benthic polyp and pelagic medusa (Figure 8.1).

8.3.1 From Eggs to Larva

Gametes are released daily by male and female medusae, triggered by light following a dark period (Amiel et al. 2010). The fertilized eggs develop into a torpedo-shaped planula larva, swimming by ciliary beating (Figure 8.1). Three days after fertilization, the larva settles on a substrate by the aboral pole. Metamorphosis into a primary polyp is induced by bacterial biofilms in natural conditions and can be triggered in the laboratory by the peptide GLW-amide on glass or plastic slide (Lechable et al. 2020; Piraino et al. 2011; Takahashi and Hatta 2011). During metamorphosis, the larva flattens on the substrate, and all the polyp structures are formed de novo. The oral part of the planula will give rise to the hypostome (mouth) of the polyp (Freeman 2005).

8.3.2 The Polyp Colony

The polyp colony is the asexually propagating, benthic stage of the life cycle. The body of the primary polyp is composed of a tube with a cylindrical shape, surmounted by a hypostome, surrounded by tentacles. After the first feeding, the colony starts to form by the growth of a stolon, a tubular structure spreading on the substrate, at the foot of the primary polyp. Other polyps are formed by lateral budding of the stolon, spaced by distances of 3 to 4 mm (Hale 1973). The gastrovascular system is shared between all the zooids through the stolon, allowing specialization of the zooids in two types: the gastrozooids catch and digest prey, and the gonozooids produce jellyfish by lateral budding (Figure 8.1). Well-fed and cleaned *Clytia* colonies show unlimited growing capacity, continuously extending their stolons and budding new zooids. The life span of a *Clytia* colony is unknown. In our lab culture conditions, the oldest colonies are 15 years old and show no obvious sign of aging.

8.3.3 The Swimming Medusa

Polyp colonies release hundreds of clonal and genetically identical jellyfish daily, produced by the gonozooids (Figure 8.1). Budding of the jellyfish starts with the growth of ectoderm and endoderm of the polyp wall. A group of cells then appears to delaminate from the distal ectoderm of the bud, forming the entocodon, a cell layer giving rise to the striated muscle of the medusa sub-umbrella. The ectoderm will give rise to the exumbrella, the external part of the velum and the tentacle epidermis, whereas the endoderm forms the gastrovascular system and the internal tentacular epithelium (Kraus et al. 2015). The formed jellyfish is folded inside the gonozooid and unfolds after release. The jellyfish are gonochoric. As mentioned, sex is influenced by the temperature of growth of the young polyp colony (Carré and Carré 2000). Depending on feeding, jellyfish reach sexual maturity in two to three weeks after release (Figure 8.1). *Clytia* jellyfish reach an adult size of 1 to 2 centimeters of diameter and live for up to two months.

8.3.4 Life Cycle in the Laboratory

Clytia cultures can be maintained in glass beakers containing filtered sea water, but a more convenient tank system has now been developed—see Lechable et al. (2020) for full details. Medusa and polyps are kept in kreisel tanks with circulating reconstituted sea water. Temperature and salinity are controlled. Jellyfish are fed twice a day with hatched *Artemia* nauplii. The use of artificial sea water allows culture of *Clytia* in inland labs.

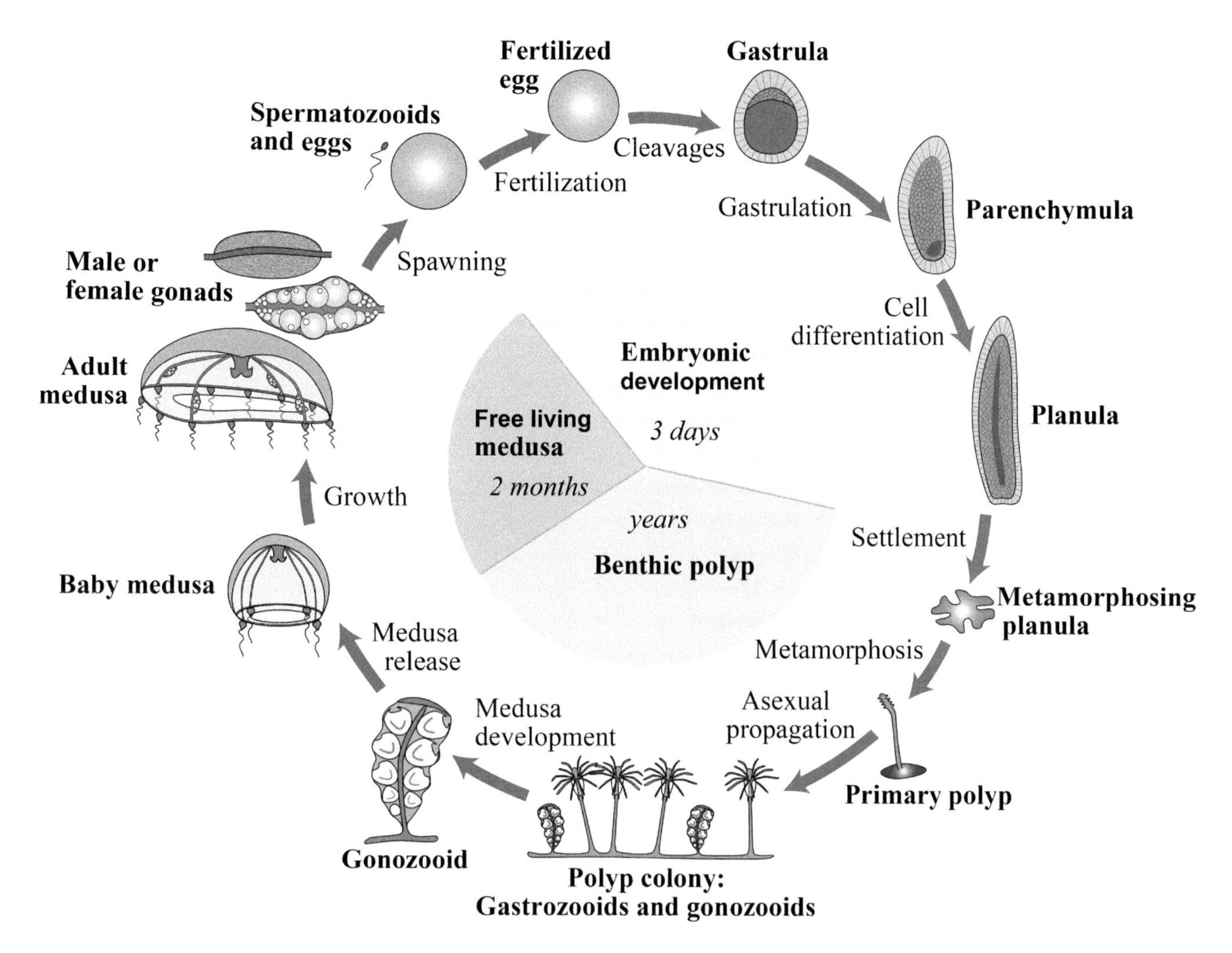

FIGURE 8.1 **The triphasic life cycle of *Clytia hemisphaerica*.** After fertilization, the embryo develops into a swimming planula larva in three days. The larva settles on a substrate and undergoes metamorphosis. Growth of the stolon from the primary polyp and budding of new zooids on the stolon lead to the formation of a colony composed of two types of polyps: gastrozooids ensure feeding of the colony, and gonozooids produce the medusae by asexual budding. Male and female medusae are mature two to three weeks after release and spawn gametes after a light cue.

8.4 EMBRYOGENESIS AND PLANULA LARVA FORMATION

8.4.1 Embryonic Development

After spawning and fertilization, the egg undergoes successive divisions until formation of a monolayered blastula (Figure 8.2A). The first division occurs 50 min after fertilization at 18°C, each following division cycle taking around 30 minutes (Kraus et al. 2020). The initiation site of the first cleavage at the animal pole of the egg marks the site of cell ingression during gastrulation and will give rise to the future oral pole of the larva (Freeman 1981b). Polarity is specified by maternal determinants localized in the oocyte: mRNAs coding for Wnt3 and Fz1 (Frizzled 1) at the animal pole which promote oral fate of the planula and for Fz3 at the vegetal pole which promote aboral fate, via the activation of the Wnt canonical pathway in the future oral territory (Momose et al. 2008; Momose and Houliston 2007).

The blastula stage begins at the 32-cell stage, with the appearance of the blastocoel. At about seven hours post-fertilization (hpf), the cells of the blastula elongate and become polarized along their apico-basal axes, forming an epithelium with apical cell–cell junctions. In parallel to this epithelialization, the diameter of the embryo reduces as the thickness of the blastoderm increases, a process called "compaction" (Kraus et al. 2020). At the late blastula stage, cilia appear on the apical surface of the embryo. Gastrulation starts at around 11–12 hpf at 18°C. Individual cells detach from the blastoderm at the future oral pole and fill the blastocoel by migrating inside, where they will form the endoderm (Figure 8.2A, B, C). This mode of gastrulation is called unipolar cell ingression (Byrum 2001). During gastrulation, the embryo elongates along the oral–aboral axis by a cell intercalation mechanism dependent on planar cell polarity (Momose et al. 2012). Gastrulation is completed at around 20–24 hpf at 18°C (Kraus et al. 2020). The resulting parenchymula has an elongated shape, but the endoderm

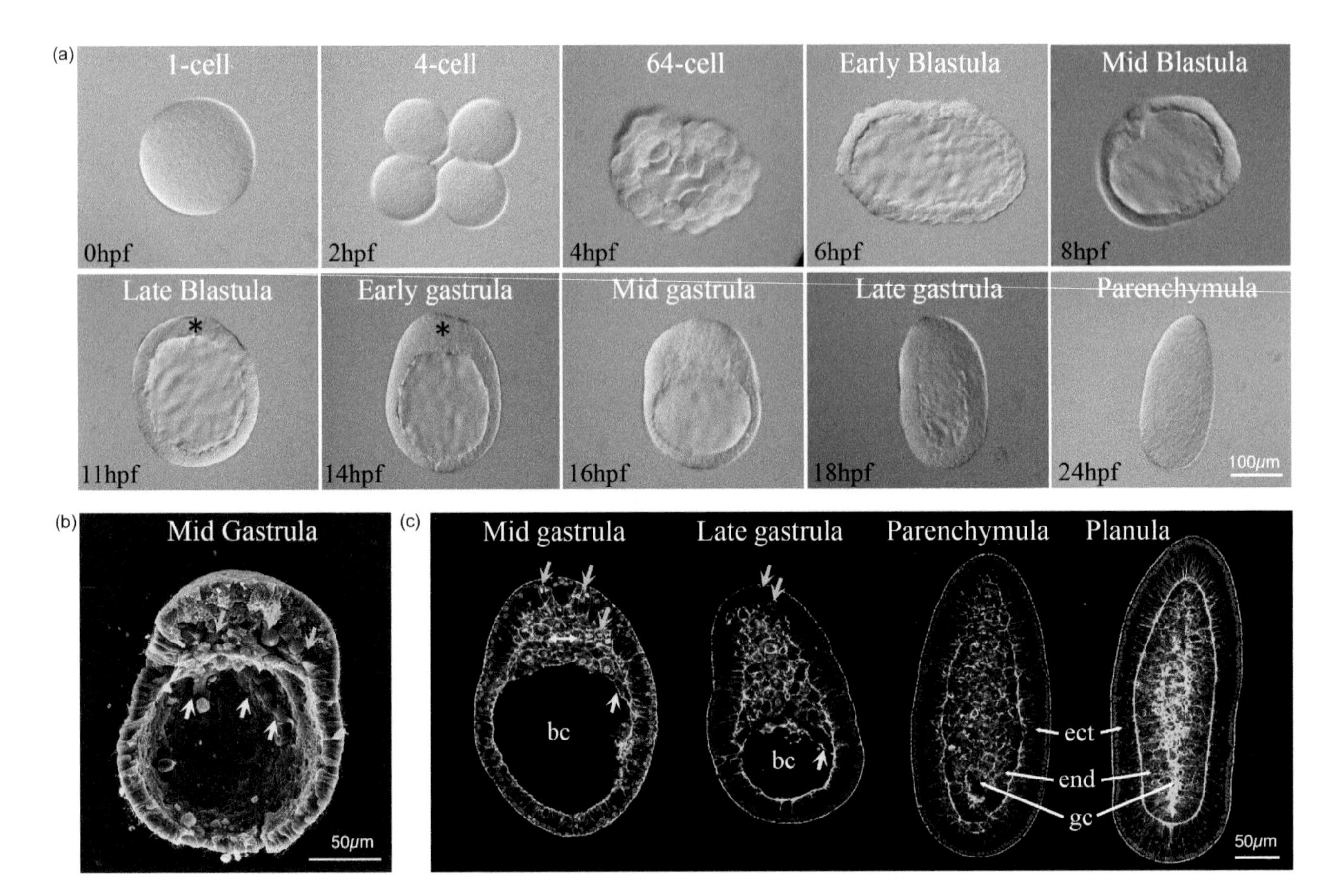

FIGURE 8.2 ***Clytia*** **embryonic development.** (a) DIC images of successive developmental stages until the end of gastrulation (Parenchymula stage). After fertilization, successive cleavage divisions increase the number of cells during the first hours, forming a hollow blastula. Between early and mid-blastula stages, epithelization of the blastoderm is accompanied by "compaction", that is, reduction in embryo diameter. The embryo oral pole is first visible as local cell layer thickening ahead of gastrulation (asterisks). Gastrulation proceeds by unipolar cell ingression from around the oral pole. Ingressed cells colonize the blastocoel, providing the future endoderm. Concomitantly, the embryo elongates. hpf = hours post-fertilization at 18°C. (b) Scanning electron micrograph of a mid-gastrula embryo split perpendicular to the oral–aboral axis to reveal the inner face of the blastocoel. Purple arrows show examples of ingressing cells at the oral pole and pink arrows ingressed cells with mesenchymal morphology migrating toward aboral pole. (c) Confocal images of embryos and planulae following staining of cell contours with phalloidin (green) and nuclei with Hoechst dye (magenta), as described in Kraus et al. (2020). Purple and pink arrows again show ingressing and migrating cells. The double-headed yellow arrow shows a region where lateral intercalation of ingressed cells is likely contributing to embryo elongation. gc: gastrocoel, ect: ectoderm, end: endoderm. (a–c) Gastrula and planulae are all oriented with the oral pole at the top. ([a] Adapted from van der Sande et al. 2020; [b] from Kraus et al. 2020.)

is not differentiated. A thin extracellular matrix layer separating the ectoderm and the endoderm (basal lamina) starts forming at the aboral pole, and a central gastric cavity progressively develops between one and two days after fertilization (Figure 8.2A, C). By two days after fertilization, the ectodermal and endodermal epithelia of the planula larva are fully developed and totally separated by the basal lamina, and the gastrocoel is complete (Figure 8.2C). Cell types continue to differentiate until the larva can metamorphose at around three days after fertilization.

8.4.2 The Planula Larva

The larva has a simple morphology. It has a torpedo shape and swims with the aboral pole in front, thanks to coordinated beating of the cilia on the ectoderm cells. Cilia orientation is coordinated by planar cell polarity along the aboral–oral axis, the protein Strabismus being located to the aboral side of each cell and Fz1 on the oral side (Momose et al. 2012).

The planula larva of *Clytia* is lecitotroph and has few cell types. The ectoderm and endoderm are composed of a typical cnidarian cell type called myoepithelial cells (epithelial cells with basal muscle fibers), nerve cells (including neurosensory and ganglion cells; Thomas et al. 1987), nematocytes (stinging cells used for prey capture and defense) (Bodo and Bouillon 1968) and interstitial stem cells called i-cells (see the following). Secretory cells and i-cells are scattered in the endoderm, with the secretory cells being also present in the aboral ectoderm (Bodo and Bouillon 1968; Leclère et al. 2012). Nematoblasts start to differentiate in the endoderm of the planula from 24 hpd before migrating to the ectoderm (Bodo and Bouillon 1968; Ruggiero 2015).

I-cells are multipotent stem cells (Bosch and David 1987), found only in hydrozoans. They are small round cells with a high nucleo-cytoplasmic ratio and are localized in the spaces between the epitheliomuscular cells. They have been well investigated in *Hydra*, where they have been shown to give rise to the nematocytes (Slautterback and Fawcett 1959), nerve cells (Davis 1974), gland cells (Bode et al. 1987) and gametes (Nishimiya-Fujisawa and Kobayashi 2012; reviewed in: Bode 1996; Bosch et al. 2010). I-cells in *Clytia* can be detected by their expression of the stem cell markers *Nanos1*, *Piwi*, *Vasa* and *PL10* (Leclère et al. 2012). These genes are also expressed in the precursors of somatic derivatives, such as nematocytes (Denker et al. 2008c), and in germ cells. In *Clytia*, i-cells appear during embryonic development (Leclère et al. 2012). Maternal mRNAs for the stem cells markers *Nanos1* and *Piwi* are concentrated in the egg next to the female pronucleus at the animal pole. During the cleavage stages, these mRNAs appear to be segregated into animal blastomeres. During gastrulation, expression of *Nanos1* and *Piwi* is taken up by cells positioned at the site of cell ingression that are internalized with the future endoderm. In the three-day-old planula, *Nanos1* and *Piwi* expressing cells are present in the endodermal layer and have typical i-cell morphology (Leclère et al. 2012). The developmental potential of i-cells in different *Clytia* life stages remain to be investigated.

8.5 ANATOMY OF THE POLYPS AND JELLYFISH

8.5.1 Anatomy of *Clytia* Polyps

The two types of polyps composing the colony have clear morphological differences linked to their specialized functions in the colony. The feeding polyps or gastrozooids are very similar to the primary polyp (described in Section 8.3.2). They are protected by a cup-shaped chitinous structure called the hydrotheca. The medusa budding polyps, or gonozooids, do not have a mouth and receive nutrients digested by the gastrozooids through the stolon network. They are completely enveloped by a chitinous gonotheca. They possess an internal structure called the gonophores, producing the medusae by lateral budding. The base of all zooids is attached to the stolon, composed from outside to inside by the perisarc (a chitinous exoskeleton), an ectodermal epithelium and an endodermal epithelium surrounding the gastric cavity that distributes nutrients throughout the whole colony.

Polyps are composed of the following cell types: myoepithelial cells (ectodermal and endodermal), nerve cells, nematocytes (only ectodermal) and gland cells. I-cells are found in the stolon. Nematocytes differentiate in the stolon and then migrate into the polyp bodies (Leclère 2008; Weiler-Stolt 1960).

8.5.2 Anatomy of the *Clytia* Jellyfish

Compared to the polyp, the jellyfish has a more complex anatomy, with well-organized smooth and striated muscle, organized nervous system, balance organs (statocysts) and well-defined organs.

8.5.2.1 Umbrella Organization

The *Clytia* jellyfish body exhibits tetraradial symmetry (Figure 8.3A, B). The oral–aboral axis is the sole axis of symmetry at the scale of the whole medusa. The bell-shaped umbrella is composed of two parts, the convex exumbrella and the concave subumbrella, separated by a thick acellular layer called the mesoglea (Figure 8.3C). The exumbrella is composed of a monolayer of epidermal cells (Kamran et al. 2017). Different cell populations are present in the subumbrella: i) an epithelium lining the mesoglea; ii) epidermal cells with myofilaments forming radial smooth muscle cover the entire subumbrella, responsible for the folding of the umbrella to bring prey to the mouth and for shock-induced protective crumpling; and iii) striated circular muscle fibers responsible for the contraction of the umbrella and the swimming movements, located between the two body layers in a band around the bell margin (Figure 8.3C, D) (Sinigaglia et al. 2020). At the periphery of the umbrella, an extension of the umbrella called the velum increases propulsion efficiency. This tissue membrane is a characteristic of hydrozoan jellyfish (Brusca et al. 2016). Medusa growth involves addition of new tissue to the peripheral region of the bell (Schmid et al. 1974).

Movements of the medusa are coordinated by a diffuse nerve net reaching all parts (Figure 8.3E, F). Two nerve rings are located at the margin of the bell. The external nerve ring integrates sensory information, while the inner nerve ring is responsible for coordinating contraction (Houliston et al. 2010; Satterlie 2002). Statocysts (balance sensory organs) located between the tentacle bulbs likely ensure orientation in the water column (Figure 8.3G). They comprise a vesicle of ectoderm with ciliated internal walls enclosing a statolith made of magnesium and calcium phosphate ($MgCaPO_4$) (Chapman 1985; Singla 1975).

8.5.2.2 A Cnidarian with Organs

From the center of the subumbrella hangs the manubrium, which is the feeding organ (Figure 8.3B, H). At its distal end is located the cross-shaped mouth, connected to the gastric cavity at the base. The outer layer of the manubrium comprises a layer of epidermal epitheliomuscular cells continuous with the subumbrella radial muscle cell layer. A distinct inner gastroderm layer lines the gastric cavity and contains both epithelial cells and populations of gland cells expressing different enzymes for extracellular digestion (Peron 2019). Four pools of i-cells positioned at the base of the manubrium likely generate the loose nerve net that lies between the gastroderm and the epiderm, as well as nematocytes mostly found concentrated on the manubrium lips.

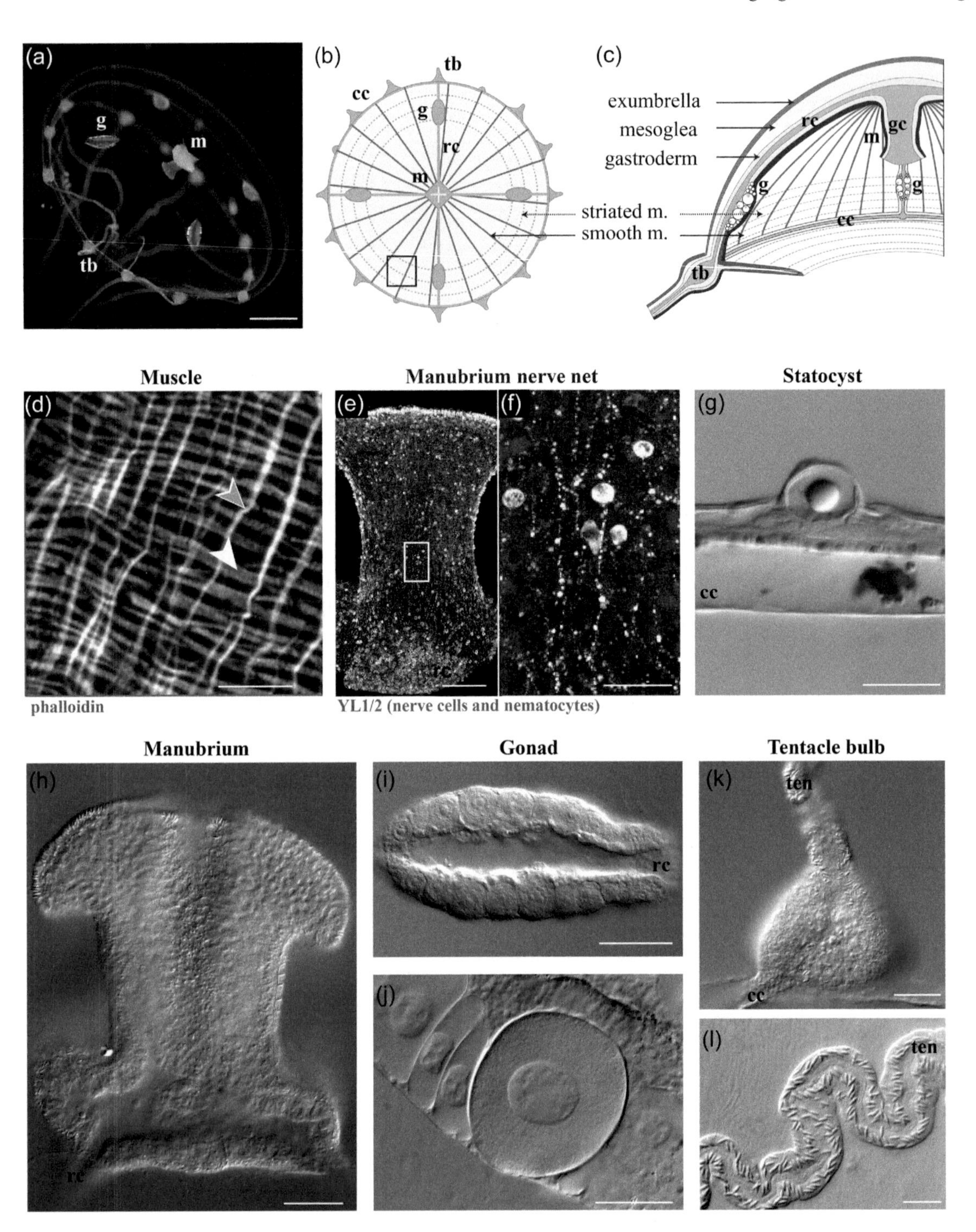

FIGURE 8.3 **Morphology of *Clytia* jellyfish**. (a) Two-week-old female jellyfish (m: manubrium, g: gonads, tb: tentacle bulbs). (b) Diagram of *Clytia* body organization: the jellyfish has a tetraradial symmetry organized around the centrally located tetraradial manubrium. Each quadrant contains a portion of the manubrium (m), a radial canal (rc) bearing a gonad (g) and up to eight tentacle bulbs (tb) located on the circular canal (cc). Two sets of muscle cells cause contractions of the umbrella: the radial smooth muscles (smooth m.) and the circular striated muscles (striated m.). (c) Tissue layers of the umbrella. The bell-shaped umbrella is composed of an epithelial exumbrella layer lying on the mesoglea and the subumbrella composed of an epithelial layer, the smooth muscle fibers and striated muscle fibers. (d) Confocal image of the muscles in the area marked with the square in (b). Gray and white arrowheads indicate, respectively, the smooth and striated muscle fibers stained with phalloidin. (e–f) Nervous system of the manubrium visualized by confocal microscopy, using YL1/2 antibody against tyrosinated tubulin. (g) DIC image of a statocyst located next to the circular canal (cc). (h–l) DIC pictures of the main organs of *Clytia*: manubrium (h) and female gonads (i–j) linked to the radial canals (rc), and tentacle bulbs (k) on the circular canal (cc), with visible nematocytes capsules on the tentacle (ten) (l). Scale bars: (a) 1 mm, (d,f) 20 µm, (e) (h–k) 100 µm, (G,L) 50 µm. ([a–c] Adapted from Sinigaglia et al. 2020.)

Nutrients are distributed to the umbrella through four radial canals, which run from the manubrium to the umbrella margin and are linked to the circular canal around the bell periphery. Four gonads are located on the radial canals and become visible as they start to swell during the growth of the medusa (Figure 8.3I, J). They become ready to release fully grown oocytes or sperm after two to three weeks. Proliferating cells, germline precursors deriving from the i-cells and growing oocytes are sandwiched between two epithelial layers: the gastroderm, continuous with the radial canal endoderm, and a thin epidermal covering (Amiel et al. 2010). Proliferating cells and early stages of differentiation are positioned closer to the bell, whereas the growing oocytes are located on the flanks of the gonad (Amiel and Houliston 2009; Jessus et al. 2020). Spawning is triggered by dark–light transitions.

The circular canal bears the tentacle bulbs, the structure producing nematocyte-rich tentacles (Figure 8.3K, L). After release from the gonozooid, the baby jellyfish has four primary tentacle bulbs located at the junction between the radial and circular canals. Additional bulbs are added during the growth of the umbrella, to a maximum of 32. Nematogenesis takes place in the ectoderm of the tentacle bulbs, which is polarized (Denker et al. 2008c). I-cells expressing *Nanos1* and *Piwi* are located in the proximal area only, while genes for the different stages of nematogenesis (*mcol3–4a, dkk, NOWA*) are expressed in a staggered way along the ectoderm of the bulb. During nematogenesis, nematoblasts are thus displaced from the proximal area of the bulb to the distal area and end up in the tentacle, forming a conveyor belt (Condamine et al. 2019; Coste et al. 2016; Denker et al. 2008c).

Cnidarians are often considered to lack true organs (e.g.: Pierobon 2012). In *Clytia* medusae, however, manubrium, gonads and tentacle bulbs can be defined as such. Indeed, they are specialized structures performing specific functions (feeding and digestion, tentacle production, oocyte production), harboring distinct cell types (gland cells, nematocytes, germ line) and i-cell populations (manubrium: Sinigaglia et al. 2020; gonads: Leclère et al. 2012; and tentacle bulbs: Denker et al. 2008c). Moreover, these three organs are still able to perform their functions for several days after isolation from the jellyfish. Isolated gonads are able to support oocyte growth, maturation and spawning (Amiel and Houliston 2009; Quiroga Artigas et al. 2018); isolated manubria will catch and digest prey (Peron 2019); and isolated tentacle bulbs will keep producing tentacles.

8.6 GENOMIC DATA

8.6.1 The *Clytia hemisphaerica* Genome

The genomes of *Nematostella vectensis* (Putnam et al. 2007) and *Hydra magnipapillata* (Chapman et al. 2010) were the first cnidarian genomes to be published. Genomes from the five main cnidarians classes are now available, with the first genomes of jellyfish species published in 2019 (Gold et al. 2019; Khalturin et al. 2019; Kim et al. 2019; Leclère et al. 2019; Ohdera et al. 2019). The sequences of the different genomes showed that cnidarians possess all the main families of signaling pathways and transcription factors regulating development found in bilaterians (reviewed in: Schnitzler 2019; Technau and Schwaiger 2015).

The genome of *Clytia*, derived from the self-crossed lab Z strains (see Section 8.1.2), was made publicly available in 2019 (Leclère et al. 2019; http://marimba.obs-vlfr.fr/home). It was the first published genome of a hydrozoan jellyfish. Sequencing was performed by the Genoscope using a whole-genome shotgun approach. The overall length of the published assembly was 445 megabases (Leclère et al. 2019); 26,727 genes and 69,083 transcripts were identified, which are distributed on 15 chromosome pairs. The frequency of polymorphism was relatively low (0.9%).

Analyses of the genome highlighted gene gain and loss in the *Clytia* lineage. Examples of horizontal gene transfer (HGT) were identified including one of two UDP-glucose 6-dehydrogenase-like genes (Leclère et al. 2019). This enzyme is used for biosynthesis of proteoglycans and known to regulate signaling pathways during embryonic development. Some examples of gene family expansion were also identified in *Clytia*, such as the Innexin gap junction genes, GFP and Clytin photoprotein genes, with 39, 14 and 18 copies, respectively (Leclère et al. 2019). The analyses also revealed extensive losses of transcription factors in the hydrozoan lineage and notably several homeobox-containing transcription factors involved in nervous system development in bilaterians, as well as genes regulating the anthozoan secondary body axis.

Comparisons of transcriptomes from life cycle stages (Leclère et al. 2019) highlighted the different gene usage at planula, polyp and medusa stages. Planula stages are enriched with GPCR signaling components, polyp and medusa stages with cell–cell and cell–matrix adhesion proteins and medusa stages with a subset of transcription factors (Leclère et al. 2019). Many of the bilaterian orthologs of transcription factors specifically expressed at the medusa play important functions in neural patterning during development. *Clytia*-specific genes, with no identifiable ortholog in any other species, were also found to be enriched in all three stages (Leclère et al. 2019).

Together, *Clytia* recently published genomic and transcriptomic data revealed that: i) the genome of *Clytia* evolved rapidly since the divergence of hydrozoans and anthozoans, ii) this rapid evolution in the hydrozoan lineage can be linked to the evolutionary acquisition of the medusa stage and to morphological simplification of the planula and polyp and iii) the medusa stage is enriched in transcription factors conserved between bilaterians and cnidarians. Since these genes are not expressed in the planula and associated with nervous structures, they are likely involved in the establishment or maintenance of neural cell types (Leclère et al. 2019).

8.6.2 Transcriptomic Data

In addition to the data included in the genome release, other transcriptomic data have been published. These focus on the gastrula stage (Lapébie et al. 2014) and tentacle bulbs (Condamine et al. 2019), as well the early stages of manubrium regeneration (Sinigaglia et al. 2020). Transcriptomes of the different tissue composing the gonad (ectoderm, endoderm, growing and fully grown oocytes) were also generated to help identify actors of oocyte maturation (Quiroga Artigas et al. 2018). About 90,000 EST and full-length sequences from cDNA libraries derived from a mix of stages (embryo, larva and medusa) are also available on NCBI dbEST (Forêt et al. 2010; Philippe et al. 2009).

8.7 FUNCTIONAL APPROACHES: TOOLS FOR MOLECULAR AND CELLULAR ANALYSES

Clytia is amenable for the development of tools for experimental biology at the cellular and molecular levels.

8.7.1 Cellular Analysis

Clytia eggs and jellyfish can be easily manipulated in a petri dish under a stereomicroscope and kept in beakers or six-well plastic plates in an incubator for further observation and manipulation. This allows pharmacological treatments for several days, as well as surgical procedures like dissections and grafts (Figure 8.4A–E) (jellyfish: Sinigaglia et al. 2020; embryos: Leclère et al. 2012; Momose and Houliston 2007). Manubriums and gonads can be easily grafted, the grafted organs connecting to the canal system of the host jellyfish (Figure 8.4A–E). The grafting approach in adult jellyfish was used to determine whether the manubrium could be a source of inductive of inhibitory signals during manubrium regeneration (Sinigaglia et al. 2020). Regeneration of the manubrium was not impaired by the grafting of an entire manubrium on the medusa subumbrella except after a graft in close proximity to the wound area, therefore excluding the hypothesis of long-range inhibition from the manubrium (Sinigaglia et al. 2020).

Embryonic stages, polyps and jellyfish are entirely transparent, making staining and imaging of different cell populations possible on fixed and living samples. Immunohistochemistry, in situ hybridization and staining using the click-it chemistry (EdU and TUNEL) are performed routinely on this species and can be combined with in situ hybridization (Figure 8.4F–H) (Sinigaglia et al. 2018). A combination of the EdU click-it staining marking proliferating cell and detection of i-cells by in situ hybridization with the probe *Nanos1* during regeneration of the manubrium demonstrated the displacement of *Nanos1*+ cells from the gonad to the regenerating manubrium to be followed (Sinigaglia et al. 2020).

8.7.2 Gene Function Analysis during Embryogenesis and Oocyte Maturation

The jellyfish used in the lab have the same genetic background, and it is easy to perform fertilizations and obtain embryo stages, facilitating gene function analyses (gain and loss of function) by injection of ARNs or MOs into the unfertilized egg (Figure 8.4I) (Momose and Houliston 2007; Momose et al. 2008). The high efficiency of loss of function by MO is likely due to low sequence polymorphism in the laboratory strains. Injection of mRNAs and MOs into the egg has helped us understand mechanisms involved in establishing polarity in *Clytia* larvae by revealing the function of maternal localized mRNAs (Wnt3, Fzl1 and Fzl3—see Section 8.4.1) (Figure 8.4J).

Clytia gonads are particularly convenient to study the molecular mechanisms underlying oogenesis. They are transparent, contain different stages of oocyte growth and continue to mature and release eggs following dark–light transition even isolated from the body of the jellyfish (Amiel et al. 2009). These characteristics were used to study the role of the Mos proteins, a conserved kinase family regulating meiosis (Amiel et al. 2009). Injection of MOs and mRNAs into the oocyte demonstrated the role of the two *Clytia* Mos homologs during oocyte maturation in regulating the formation and localization of the meiotic spindle, as well as oocyte cell cycle arrest after meiosis (Amiel et al. 2009). These functions have also been described in bilaterian species and likely represent an ancestral function of this protein family (Amiel et al. 2009).

8.7.3 Gene Function Analysis in the Adult

8.7.3.1 RNA Interference

RNA interference (RNAi) has been successfully used for downregulation of gene expression in the adult in the cnidarian *Hydractinia*, allowing, for instance, study of the role of i-cell genes during regeneration (Bradshaw et al. 2015). Gene expression perturbation through RNAi has not yet been performed in *Clytia* jellyfish; however, preliminary results indicate that the cellular machinery is present in *Clytia* larvae. Another promising avenue to explore is shRNA, also effective in both *Hydractinia* and *Nematostella* (DuBuc et al. 2020; He et al. 2018).

8.7.3.2 The Development of Mutant Lines

A robust protocol for achieving loss of gene function in *Clytia* lines by CRISPR/Cas9 has been developed (Figure 8.4K) (Momose et al. 2018). The approach was first tested on a gene involved in ciliogenesis (*CheRfx123*), whose defect leads to defect in sperm motility, and genes coding for the fluorescent protein GFP (Figure 8.4K) (double mutant *GFP1/GFP2* in F1) (Momose et al. 2018). After injection of high doses of Cas9 RNP, mutants in the F0 generation were nearly non-mosaic and already had visible phenotypes

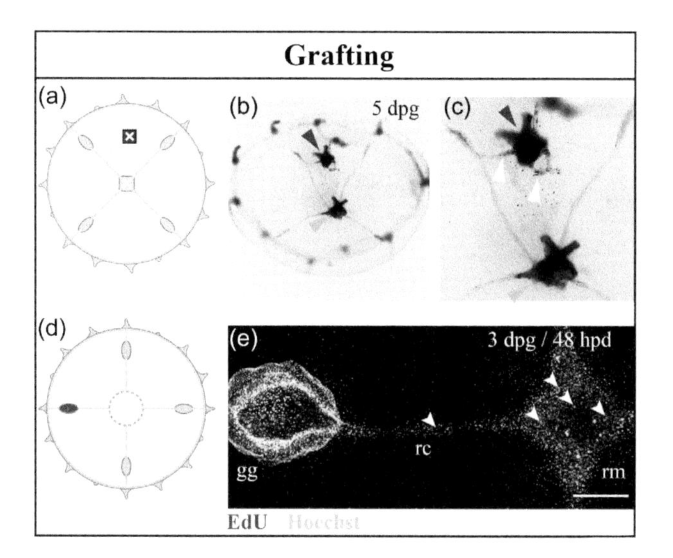

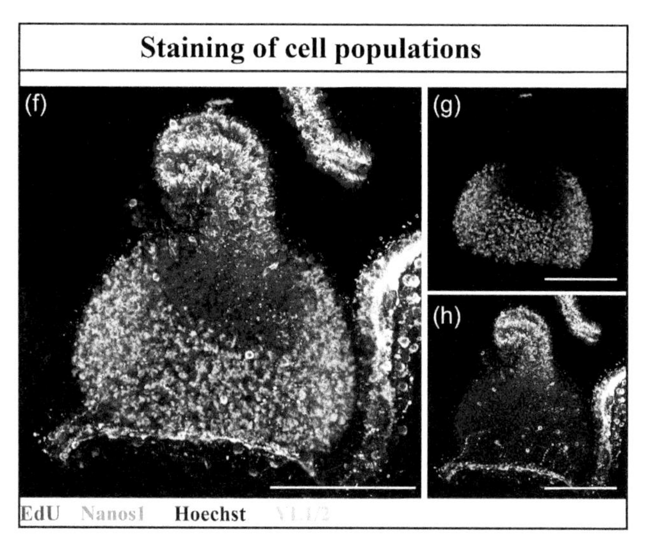

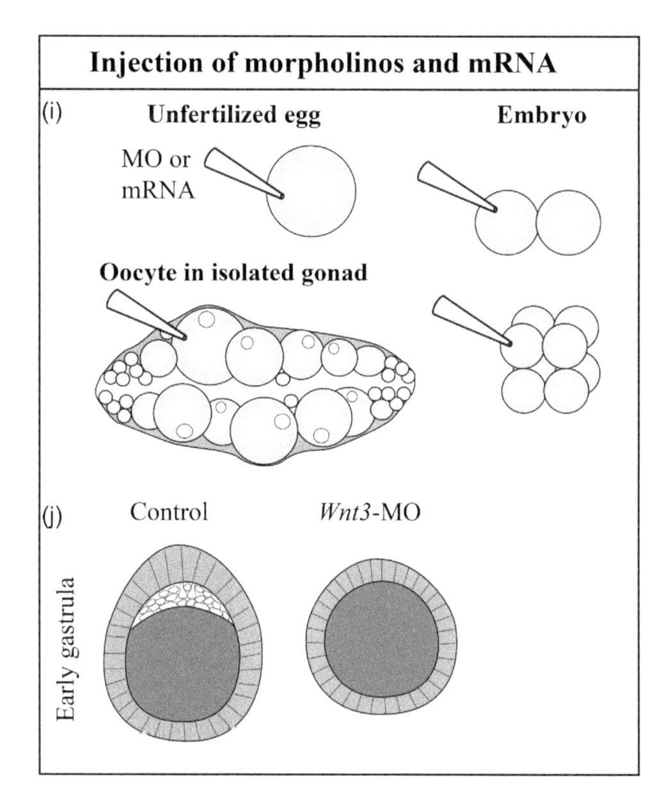

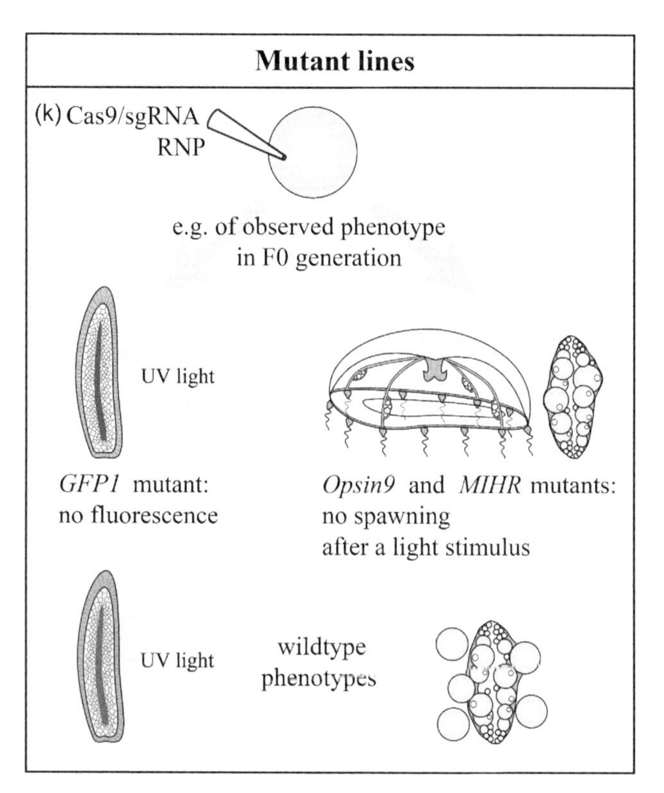

FIGURE 8.4 Tools for cellular and molecular analysis. (a–e) Organ grafting in the medusa. (a, d) Cartoons illustrating the grafting procedure: the manubrium or a gonad (both depicted in magenta) are excised from a donor medusa and placed on a host jellyfish anesthetized in menthol. After dissection, the jellyfish tissues adhere to each other. (b, c) Five days after grafting (dpg), the grafted manubrium (magenta arrowhead) has integrated the host tissue and stably coexists with the endogenous manubrium (yellow arrowhead). Both are able to catch prey and contribute to feeding; new radial canals grew from the base of the grafted manubrium (white arrowheads) and are connected to the host radial canal. (d) Donor medusa for the gonad was previously incubated in EdU, thereby marking the proliferating cells. 24 hpg, the manubrium of the host medusa was removed (dotted orange line). (e) White arrowheads indicate some EdU+ cells (magenta) from the grafted gonad (gg), which migrated into the host jellyfish through the radial canal (rc) and integrated into the regenerating manubrium (rm). (f–h) Proliferating cells (red: EdU), i-cells (green: *Nanos1* in situ hybridization), nerve cells and nematocytes (white: tyrosynated tubulin YL1/2 antibody staining) and nuclei (blue: Hoechst) were marked in the same tentacle bulb. (i) Perturbation of gene function through MO or ARNm injection in unfertilized oocytes, gonads or individual blastomeres of two- to eight-cell embryos. (j) Cartoons of embryos at the gastrula stage (15 hpf). Injection of Wnt3 MO before fertilization abolishes oral specification, delaying gastrulation and abolishing embryo elongation. (k) CRISPR/Cas9 mutagenesis allows gene function to be addressed at all life cycle stages. The diagrams illustrate examples of existing mutant lines and the associated phenotypes, published in Momose et al. 2018 (*GFP1*), Quiroga Artigas et al. 2018 (*Opsin9*) and 2020 (*MIH-R*: Maturation inducing hormone receptor). Scale bars: (e–h) 100 μm, (J) 40 μm. ([b, c] Adapted from Sinigaglia et al. 2020; [e] Chiara Sinigaglia.)

(Momose et al. 2018). The relatively short *Clytia* life cycle allows quick generation of mutant lines. The vegetatively growing polyp colonies are essentially immortal and can be kept in the aquarium for years with minimal care (daily feeding with *Artemia* larvae and regular cleaning). Moreover, mutant polyp colonies can be easily split and shared between laboratories. Those characteristics make *Clytia* a promising genetic model. Gene insertion protocols are under development.

CRISPR/Cas9-directed mutagenesis has been used to study the molecular mechanisms of oocyte maturation and spawning triggered by light cues. It was used to knock out function of an opsin photopigment candidate for light reception (Opsin9: Quiroga Artigas et al. 2018), as well as a GPCR candidate for the oocyte maturation hormone receptor (MIHR: Quiroga Artigas et al. 2020). Lines of jellyfish carrying frame-shift mutations in the Opsin9 and MIHR genes were created by CRISPR/Cas9 (Figure 8.4K). As expected, the mutant jellyfish were unable to respond to light cues, either to trigger oocyte maturation or release gametes as in control jellyfish. Specificity was validated by reversal of Opsin mutant phenotype by treatment of oocytes with the maturation-inducing hormone or in both mutants using the downstream pathway effector cAMP (Quiroga Artigas et al. 2018, 2020).

8.8 CHALLENGING QUESTIONS

With the tools currently available, *Clytia* has the potential to address many fascinating biological questions. We illustrate this with a selection of open questions related to the extensive ability of *Clytia* jellyfish to regenerate and aspects of the behavior and physiology regulated by the environment.

8.8.1 *Clytia* as a Regeneration Model

Cnidarians display huge regeneration capacities, which have been well characterized in *Hydra* and *Nematostella* (Amiel et al. 2015; DuBuc et al. 2014; Galliot 2012; Schaffer et al. 2016). In contrast, cellular and molecular mechanisms of regeneration in jellyfish have been relatively unstudied. Regeneration studies in *Clytia* were started in the 1970s by Schmid and Tardent (see 8.1.1.3). A recent study using modern tools allowed cellular mechanisms involved in repair of the umbrella and organ regeneration to be uncovered (Sinigaglia et al. 2020). This work confirmed the potential of *Clytia* laboratory strains to restore their shape after amputation (Figure 8.5A, B) and to regenerate missing organs, including the manubrium (Figure 8.5C). Two different mechanisms were identified (Figure 8.5D). Repair of a fragment of the umbrella, called remodeling, relies on a supracellular actomyosin cable lining the wound area and does not require cell proliferation. In contrast, morphogenesis of the regenerating manubrium requires cell proliferation, is fuelled by cell migration through the radial canals and depends on Wnt/β-catenin signaling (Sinigaglia et al. 2020). Moreover, the regenerating manubrium is systematically associated with the point of junction of the smooth muscle fiber (called the hub), forming as a consequence of the remodeling process and expressing *CheWnt6* before any visible sign of morphogenesis (Sinigaglia et al. 2020). These data suggest that local cues are involved in positioning the regenerating manubrium rather than a global patterning system. This study raises many questions about the regulation of regeneration in *Clytia* jellyfish.

8.8.1.1 How Is the Cellular Response Controlled during Regeneration?

Manubrium regeneration is fueled by both cell proliferation in the regeneration blastema and cell migration from distant parts of the jellyfish. At least two types of cells are mobilized: multipotent stem cells (i-cells) and differentiated digestive cells, called mobilizing gastro-digestive cells (MDG cells) (Sinigaglia et al. 2020). Cell proliferation and migration through the radial canals are necessary for regeneration of the manubrium, since regeneration is blocked at early stages in the absence of cell proliferation and if the connection to the radial canal system is interrupted (Sinigaglia et al. 2020). It is not known yet which cells are proliferating and to which extent both mechanisms of proliferation and migration contribute to the regenerating organ.

Regeneration models like planarians and the cnidarians *Hydractinia* require proliferation and migration of multipotent stem cells for regeneration of the anterior part (Bradshaw et al. 2015; Newmark and Sánchez Alvarado 2000). However, modes of regeneration are diverse, even within the same organism: *Clytia* shape restoration relies on remodeling and repatterning of existing tissues, whereas the manubrium is regenerated through cell proliferation and migration (Sinigaglia et al. 2020). Those different cell behaviors must be tightly coordinated to ensure regeneration of a correctly patterned and functional structure. Repatterning during shape restoration is controlled by tension forces generated by the actomyosin cytoskeleton. However, the mechanisms allowing fine control of cell proliferation and directing the migrating cells during organ regeneration are unknown. Elucidating the molecular control of stem cell proliferation and migration in the context of regeneration in *Clytia* will allow a better understanding of stem cell regulation systems in metazoans.

8.8.1.2 What Are the I-Cell Fates in *Clytia*?

I-cells are multipotent stem cells (see Section 8.4.2) involved in regeneration in hydrozoans (Bradshaw et al. 2015; Galliot 2013; Sinigaglia et al. 2020). The fate of i-cells has been well characterized in *Hydra* and *Hydractinia* (Gold and Jacobs 2013; Müller et al. 2004; Siebert et al. 2019). In both animals, they give rise to the gland cells, nerve cells, nematocytes and gametes. However, in *Hydractinia*, they also differentiate into the epithelial epidermal and gastrodermal cells; whereas in *Hydra*, i-cells and ectodermal and endodermal epithelial cells form three independent populations. In *Clytia*, only nematogenesis has been well characterized (Denker et al. 2008c). It is still unknown whether i-cells in *Clytia* give rise to all cell types, particularly to epithelial lineages. However, since only a small portion of *Clytia*

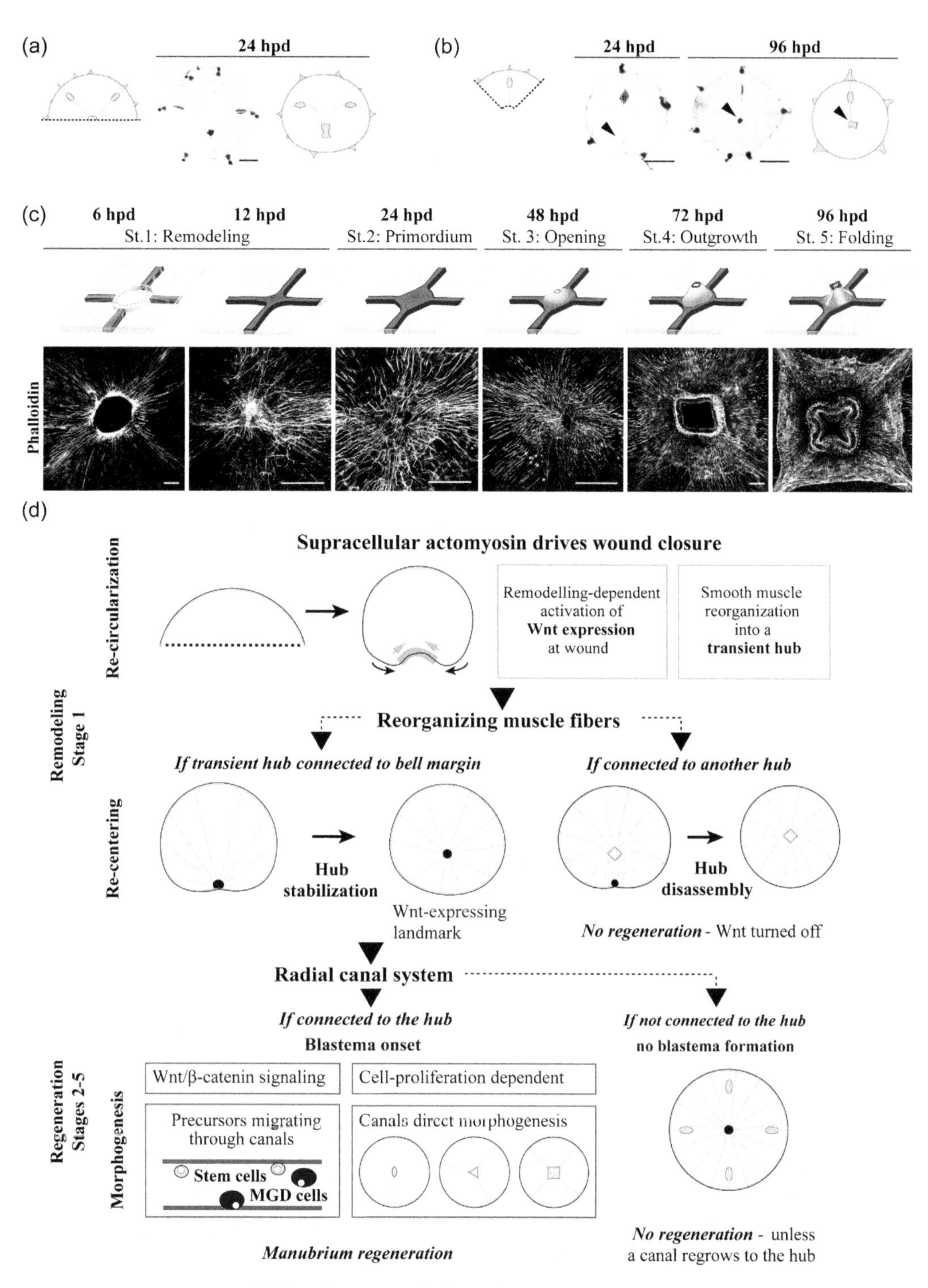

FIGURE 8.5 Regeneration of *Clytia* jellyfish. (a–b) Circular shape restoration after amputation. In the cartoon, the gray dashed line indicates the location of the cut. A half jellyfish with a half manubrium (a) and a quarter jellyfish without the manubrium (b) recover the circular jellyfish shape in 24 h. In the quarter, a manubrium blastema and a tiny regenerated manubrium are visible at 24 hpd (hours post-dissection) and 4 dpd (days post-dissection), respectively (black arrowhead). (c) Manubrium regeneration. Schematic (top line) and phalloidin staining (bottom line) of manubrium regeneration stages from 6 hpd to complete regeneration after 4 dpd. After closing of the dissection hole, a regeneration blastema forms at the junction of the radial canals. As the blastema becomes thicker, the gastric cavity opens. The regenerating manubrium first elongates, followed by the formation of four lobes. (d) Summary of the main cellular and molecular events allowing manubrium regeneration. After a cut in the umbrella, an actomyosin cable allows a rapid reestablishment of the circular jellyfish shape, affecting the organization of the smooth muscle. A new muscle hub is formed close to the former wound area. If not attached to another hub, the new hub is stabilized, as well as the associated *CheWnt6* expression. The connection to the radial canal system allows the formation of a regeneration blastema by proliferation and migration of stem cells and differentiated cells, leading to the full regeneration of the missing manubrium in only four days. Scale bars: (a–b) 1 mm, (c) 100 µm. ([a–d] Adapted from Sinigaglia et al. 2020.)

proliferating cells express *Nanos1*, the *Clytia* i-cell system is likely to be similar to *Hydra* with separated i-cells and epithelial lineages. Transgenic lines with reporters for different cell populations allowing in vivo tracing of i-cell are necessary to identify i-cell derivatives.

It is also unknown whether all *Nanos1*-expressing cells have the same potency and particularly whether some are committed to the germline. After complete ablation, the gonads regenerate, and oocyte growth resumes. This could indicate the presence of multipotent stem cell populations in the main organs, migrating through the radial canal to repopulate the regenerating gonads. *Clytia* is a promising model to study early oocyte differentiation because the gonads are fully transparent and continue to function when isolated from the jellyfish.

8.8.1.3 How Are Mechanical Cues and Signaling Pathways Integrated?

After amputation, actomyosin contractility at the wound area ensures restoration of the circular jellyfish shape. During shape restoration, the signaling molecule *CheWnt6* is expressed at the wound site. Its expression is inhibited by pharmacological inhibition of actomyosin contractility, suggesting a likely modulation of Wnt/β-catenin activity by mechanical cues (Sinigaglia et al. 2020). How mechanical cues can activate Wnt/β-catenin pathway and thus permit regeneration of the manubrium is unknown. The integration between mechanical cues and signaling pathways has been raising interest (Chiou and Collins 2018; Heisenberg and Bellaïche 2013; Urdy 2012; Vining and Mooney 2017). In *Hydra*, the actin cytoskeleton has also been proposed to influence body axis formation during regeneration (Livshits et al. 2017, Maroudas-Sacks et al. 2021) and is likely to be interacting with the Wnt/β-catenin signaling pathway, inducing hypostome formation at the oral pole (Broun 2005; Gee et al. 2010).

8.8.2 Regulation of Behavior and Physiology by Environmental Cues

Clytia life cycle and physiology of the different life stages are influenced by the environment in many ways: i) in the ocean, settlement of the planula larva occurs upon an unknown cue from bacterial biofilms; ii) growth of the polyp colony is constrained by feeding and space availability; iii) sex of the released medusa can be influenced by the temperature at which the polyp colony is growing; and iv) in the jellyfish, oocyte maturation and gamete release are triggered by a light stimulus. Gamete release is associated with light information in many cnidarian species (e.g. scyphozoans *Pelagia*: Lilley et al. 2014; *Clytia*: Amiel et al. 2010).

8.8.2.1 Which Bacterial Cues Induce Settlement of the Planula? Which Molecular Mechanisms Are Triggered?

In cnidarians, including *Clytia*, settlement of the planula larva and metamorphosis into a primary polyp is induced by bacterial biofilms (*Hydractinia echinata*: Kroiher and Berking 1999; Leitz and Wagner 1993; Seipp et al. 2007; *Acropora sp*: Negri et al. 2001; Tebben et al. 2011; Webster et al. 2004). The cellular response is mediated by neuropeptides of the GLW-amide family, secreted by sensory neurons of the planula (Takahashi and Takeda 2015). Synthetic GLW-amide neuropeptides induce settlement and metamorphosis in laboratory conditions in several planulae (Acropora: Iwao et al. 2002; *Hydractinia*: Müller and Leitz 2002; both reviewed in: Takahashi and Hatta 2011). Concerning *Clytia* planula, the synthetic peptide GLWamide2 (GNPPGLW-NH2) has been used in the laboratory to induce settlement (Momose et al. 2018; Quiroga Artigas et al. 2018). A recent study testing the efficiency of 15 other neuropeptides, derived from sequences of potential GLWamide precursors, showed that GLWamide-6 (pyro-Glu-QQAPKGLW-NH3) has an even greater efficiency (Lechable et al. 2020).

The roles of bacteria and neuropeptides in settlement have long been known. However, the signal from the bacteria inducing settlement and metamorphosis, as well as the molecular mechanisms triggering settlement and metamorphosis, are still unknown. The morphological and cellular events occurring during the metamorphosis of *Clytia* planula have been recently studied (Krasovec 2020) and provide a framework for further studies on metamorphosis.

8.8.2.2 Is There a Physiological Link between Gametogenesis and Nutrition?

In *Clytia* jellyfish, spawning and oocyte maturation occurs in males and females two hours after a light stimulus (Amiel et al. 2010). Part of the signaling cascade triggering light-induced oocyte maturation has recently been elucidated. After light reception by the photoprotein Opsin9 by neurosecretory cells of the gonad ectoderm, those cells release a maturation-inducing hormone (Quiroga Artigas et al. 2018). MIH activates in turn a GPCR, located on the oocyte surface, called the MIH-Receptor, thus triggering the rise in cAMP responsible for the initiation of oocyte maturation (Quiroga Artigas et al. 2020). Besides their function in oocyte maturation, *Clytia* MIH and MIH-R are likely to play a role in nutrition or other physiological processes. Indeed, both are expressed in the gastrovascular system and the tentacles as well as in the gonads. Moreover, MIHR is part of a superfamily of cnidarian and bilaterian GPCRs playing a role in nutrition, as well as regulation of sexual reproduction (Quiroga Artigas et al. 2020). Additional knowledge in the functions of *Clytia* MIHR could give insight in the evolution of the link between gametogenesis and nutrition.

8.8.2.3 How Does Feeding Availability Regulate Growth of Polyps and Medusa?

Some cnidarians are able to modify their size depending on feeding availability. The jellyfish *Pelagia noctiluca* and *Aurelia aurita* shrink during starvation conditions and regrow when prey are again available (Frandsen and Riisgård 1997; Hamner and Jenssen 1974; Lilley et al. 2014). In laboratory conditions, *Aurelia aurita* loses 3–5% of its mass per day without feeding and regrows after feeding to reach the original size. Starved jellyfish are not able to spawn (Frandsen and Riisgård 1997; Hamner and Jenssen 1974). Similarly to *Aurelia*, *Pelagia* loses about 7% of its mass per day and can regrow after feeding. However, egg production is maintained, with a number of eggs correlated with the size of the jellyfish (Lilley et al. 2014).

The process of shrinking in conditions of starvation is also a feature of other invertebrates. In planarians, the size depends on the feeding levels (Felix et al. 2019); in the annelid *Pristina leidyi*, feeding causes the increase and decrease of the gonads (Özpolat et al. 2016). This process has also been documented, although more rarely, in the vertebrates. The marine iguana *Amblyrhynchus cristatus* can lose up to 20% of its size after the loss of its main source of food during El Niño events (Wikelski and Thom 2000). Whether the same mechanisms are involved between metazoans still remains to be investigated.

A similar shrinking/re-growth event in case of starvation has been observed in *Clytia* jellyfish (unpublished). Moreover, the gonads also shrink and egg production declines before totally stopping. Gametogenesis resumes after feeding of the jellyfish. The recently described MDG cells, with a putative role in the distribution of nutrients, circulate more in the canals in case of starvation (Sinigaglia et al. 2020). Feeding also influences the growth of newly released jellyfish: indeed, jellyfish fed with smaller prey, and thus with a bigger food intake, grow faster than jellyfish fed with bigger prey that are harder to catch (Lechable et al. 2020).

To summarize, in *Clytia*, like in other cnidarians, the feeding levels control the rate of growth and gametogenesis. The cellular and molecular mechanisms allowing the control of growth in *Clytia* jellyfish are unknown. One level of regulation is potentially the cell cycle, since in *Hydra* and *Nematostella* polyps, the rate of cell proliferation depends on the feeding level of the animal (Campbell 1967; Otto and Campbell 1977; Passamaneck and Martindale 2012; Webster and Hamilton 1972). *Clytia* jellyfish could be used to investigate the feedback between feeding levels and cell proliferation, as well as cellular events during degrowth.

Many fascinating questions can be addressed with *Clytia*. Due to its practicality as a model organism and the tools already available and in development, *Clytia* has the potential to provide a fresh perspective on a wide range of research topics.

BIBLIOGRAPHY

Abrams, M.J., T. Basinger, W. Yuan, C.-L. Guo, and L. Goentoro. 2015. Self-repairing symmetry in jellyfish through mechanically driven reorganization. *Proceedings of the National Academy of Sciences*. 112: E3365–E3373.

Alder, J. 1856. A notice of some new genera and species of British hydroid zoophytes. *Annals and Magazine of Natural History*. 2: 12–14.

Amiel, A., P. Chang, T. Momose, and E. Houliston. 2010. *Clytia hemisphaerica*: A cnidarian model for studying oogenesis. In *Oogenesis*, ed. M.-H. Verlhac & A. Villeneuve, 81–101. John Wiley & Sons.

Amiel, A., and E. Houliston. 2009. Three distinct RNA localization mechanisms contribute to oocyte polarity establishment in the cnidarian *Clytia hemisphærica*. *Developmental Biology*. 327: 191–203.

Amiel, A.R., H.T. Johnston, K. Nedoncelle, J.F. Warner, S. Ferreira, and E. Röttinger. 2015. Characterization of morphological and cellular events underlying oral regeneration in the sea anemone, *Nematostella vectensis*. *International Journal of Molecular Sciences*. 16: 28449–28471.

Amiel, A., L. Leclère, L. Robert, S. Chevalier, and E. Houliston. 2009. Conserved functions for Mos in eumetazoan oocyte maturation revealed by studies in a cnidarian. *Current Biology*. 19: 305–311.

Bodo, F., and J. Bouillon. 1968. Etude histologique du développement embryonnaire de quelques hydromeduses de Roscoff: *Phialidium hemisphaericum* (l.), *Obelia* sp. peron et lesueur, *Sarsia eximia* (allman), *Podocoryne carnea* (sars), *Gonionemus vertens* Agassiz. *Cahiers de Biologie Marine*. 9: 69–104.

Bode, H.R. 1996. The interstitial cell lineage of *Hydra*: A stem cell system that arose early in evolution. *Journal of Cell Science*. 109: 1155–1164.

Bode, H.R., S. Heimfeld, M.A. Chow, and L.W. Huang. 1987. Gland cells arise by differentiation from interstitial cells in *Hydra attenuata*. *Developmental Biology*. 122: 577–585.

Bosch, T.C.G., F. Anton-Erxleben, G. Hemmrich, and K. Khalturin. 2010. The *Hydra* polyp: Nothing but an active stem cell community. *Development Growth and Differentiation*. 52: 15–25.

Bosch, T.C.G., and C.N. David. 1987. Stem cells of *Hydra magnipapillata* can differentiate into somatic cells and germ line cells. *Developmental Biology*. 121: 182–191.

Bradshaw, B., K. Thompson, and U. Frank. 2015. Distinct mechanisms underlie oral vs aboral regeneration in the cnidarian *Hydractinia echinata*. *ELife*. 4: 1–19.

Broun, M. 2005. Formation of the head organizer in hydra involves the canonical Wnt pathway. *Development*. 132: 2907–2916.

Brusca, R.C., W. Moore, and S.M. Shuster. 2016. *Invertebrates*. 3rd ed. Sinauer Associates Inc., Sunderland, MA.

Byrum, C.A. 2001. An analysis of hydrozoan gastrulation by unipolar ingression. *Developmental Biology*. 240: 627–640.

Campbell, R.D. 1967. Tissue dynamics of steady state growth in *Hydra littoralis*. I: Patterns of cell division. *Developmental Biology*. 15: 487–502.

Carré, D., and C. Carré. 2000. Origin of germ cells, sex determination, and sex inversion in medusae of the genus *Clytia* (Hydrozoa, Leptomedusae): The influence of temperature. *Journal of Experimental Zoology*. 287: 233–242.

Carré, D., C. Carré, F. Pagès, and J.-M. Gili. 1995. Asexual reproduction in the pelagic phase of *Clytia mccrady* (Hydrozoa, Leptomedusae). *Scientia Marina*. 59: 193–202.

Chapman, D.M. 1985. X-ray microanalysis of selected coelenterate statoliths. *Journal of the Marine Biological Association of the United Kingdom*. 65: 617–627.

Chapman, J.A., E.F. Kirkness, O. Simakov, S.E. Hampson, T. Mitros, T. Weinmaier, T. Rattei, P.G. Balasubramanian, J. Borman, D. Busam, K. Disbennett, C. Pfannkoch, N. Sumin, G.G. Sutton, L.D. Viswanathan, B. Walenz, D.M. Goodstein, U. Hellsten, T. Kawashima, et al. 2010. The dynamic genome of *Hydra*. *Nature*. 464: 592–596.

Chevalier, S., A. Martin, L. Leclère, A. Amiel, and E. Houliston. 2006. Polarised expression of FoxB and FoxQ2 genes during development of the hydrozoan *Clytia hemisphaerica*. *Development Genes and Evolution*. 216: 709–720.

Chiori, R., M. Jager, E. Denker, P. Wincker, C. Da Silva, H. Le Guyader, M. Manuel, and E. Quéinnec. 2009. Are Hox genes ancestrally involved in axial patterning? evidence from the hydrozoan *Clytia hemisphaerica* (Cnidaria). *PLoS One*. 4: e4231.

Chiou, K., and E.M.S. Collins. 2018. Why we need mechanics to understand animal regeneration. *Developmental Biology*. 433: 155–165.

Collins, A., P. Schuchert, A. Marques, T. Jankowski, M. Medina, and B. Schierwater. 2006. Medusozoan phylogeny and character evolution clarified by new large and small subunit RDNA data and an assessment of the utility of phylogenetic mixture models. *Systematic Biology*. 55: 97–115.

Condamine, T., M. Jager, L. Leclère, C. Blugeon, S. Lemoine, R.R. Copley, and M. Manuel. 2019. Molecular characterisation of a cellular conveyor belt in *Clytia* medusae. *Developmental Biology*. 456: 212–225.

Coste, A., M. Jager, J.P. Chambon, and M. Manuel. 2016. Comparative study of Hippo pathway genes in cellular conveyor belts of a ctenophore and a cnidarian. *EvoDevo*. 7: 1–19.

Cunha, A.F., A.G. Collins, and A.C. Marques. 2020. When morphometry meets taxonomy: Morphological variation and species boundaries in Proboscoida (Cnidaria: Hydrozoa). *Zoological Journal of the Linnean Society*. 190: 417–447.

Davis, L.E. 1974. Ultrastructural studies of the development of nerves in *Hydra*. *Integrative and Comparative Biology*. 14: 551–573.

Denker, E., E. Bapteste, H. Le Guyader, M. Manuel, and N. Rabet. 2008a. Horizontal gene transfer and the evolution of cnidarian stinging cells. *Current Biology*. 18: 858–859.

Denker, E., A. Chatonnet, and N. Rabet. 2008b. Acetylcholinesterase activity in *Clytia hemisphaerica* (Cnidaria). *Chemico-Biological Interactions*. 175: 125–128.

Denker, E., M. Manuel, L. Leclère, H. Le Guyader, and N. Rabet. 2008c. Ordered progression of nematogenesis from stem cells through differentiation stages in the tentacle bulb of *Clytia hemisphaerica* (Hydrozoa, Cnidaria). *Developmental Biology*. 315: 99–113.

Derelle, R., T. Momose, M. Manuel, C.D.A. Silva, P. Wincker, and E. Houliston. 2010. Convergent origins and rapid evolution of spliced leader trans-splicing in Metazoa: Insights from the Ctenophora and Hydrozoa. *RNA*. 16: 696–707.

DuBuc, T.Q., C.E. Schnitzler, E. Chrysostomou, E.T. McMahon, Febrimarsa, J.M. Gahan, T. Buggie, S.G. Gornik, S. Hanley, S.N. Barreira, P. Gonzalez, A.D. Baxevanis, and U. Frank. 2020. Transcription factor AP2 controls cnidarian germ cell induction. *Science*. 367: 757–762.

DuBuc, T.Q., N. Traylor-Knowles, and M.Q. Martindale. 2014. Initiating a regenerative response: Cellular and molecular features of wound healing in the cnidarian *Nematostella vectensis*. *BMC Biology*. 12: 1–20.

Felix, D.A., Ó. Gutiérrez-Gutiérrez, L. Espada, A. Thems, and C. González-Estévez. 2019. It is not all about regeneration: Planarians striking power to stand starvation. *Seminars in Cell and Developmental Biology*. 87: 169–181.

Forêt, S., B. Knack, E. Houliston, T. Momose, M. Manuel, E. Quéinnec, D.C. Hayward, E.E. Ball, and D.J. Miller. 2010. New tricks with old genes: The genetic bases of novel cnidarian traits. *Trends in Genetics*. 26: 154–158.

Fourrage, C., S. Chevalier, and E. Houliston. 2010. A highly conserved poc1 protein characterized in embryos of the hydrozoan *Clytia hemisphaerica*: Localization and functional studies. *PLoS One*. 5: e13994.

Frandsen, K.T., and H.U. Riisgård. 1997. Size dependent respiration and growth of jellyfish, *Aurelia aurita*. *Sarsia*. 82: 307–312.

Freeman, G. 1980. The role of cleavage in the establishment of the anterior-posterior axis of the hydrozoan embryo. In *Developmental and Cellular Biology of Coelenterates*, ed. R. Tardent & P. Tardent, 97–108. Elsevier/North Holland. Amsterdam.

Freeman, G. 1981a. The role of polarity in the development of the hydrozoan planula larva. *Wilhelm Roux's Archives of Developmental Biology*. 190: 168–184.

Freeman, G. 1981b. The cleavage initiation site establishes the posterior pole of the hydrozoan embryo. *Wilhelm Roux's Archives of Developmental Biology*. 190: 123–125.

Freeman, G. 2005. The effect of larval age on developmental changes in the polyp prepattern of a hydrozoan planula. *Zoology*. 108: 55–73.

Freeman, G., and E.B. Ridgway. 1987. Endogenous photoproteins, calcium channels and calcium transients during metamorphosis in hydrozoans. *Roux's Archives of Developmental Biology*. 196: 30–50.

Galliot, B. 2012. *Hydra*, a fruitful model system for 270 years. *International Journal of Developmental Biology*. 56: 411–423.

Galliot, B. 2013. Regeneration in *Hydra*. In *ELS*, John Wiley & Sons. Chichester.

Gee, L., J. Hartig, L. Law, J. Wittlieb, K. Khalturin, T.C.G. Bosch, and H.R. Bode. 2010. β-Catenin plays a central role in setting up the head organizer in *Hydra*. *Developmental Biology*. 340: 116–124.

Gold, D.A., and D.K. Jacobs. 2013. Stem cell dynamics in cnidaria: Are there unifying principles? *Development Genes and Evolution*. 223: 53–66.

Gold, D.A., T. Katsuki, Y. Li, X. Yan, M. Regulski, D. Ibberson, T. Holstein, R.E. Steele, D.K. Jacobs, and R.J. Greenspan. 2019. The genome of the jellyfish *Aurelia* and the evolution of animal complexity. *Nature Ecology & Evolution*. 3: 96–104.

Gur Barzilai, M., A.M. Reitzel, J.E.M. Kraus, D. Gordon, U. Technau, M. Gurevitz, and Y. Moran. 2012. Convergent evolution of sodium ion selectivity in metazoan neuronal signaling. *Cell Reports*. 2: 242–248.

Hale, L.J. 1973. The pattern of growth of *Clytia johnstoni*. *Journal of Embryology and Experimental Morphology*. 29: 283–309.

Hamner, W.M., and R.M. Jenssen. 1974. Growth, degrowth, and irreversible cell differentiation in *Aurelia aurita*. *American Zoologist*. 14: 833–849.

Hargitt, C.W. 1897. Recent experiments on regeneration. *Zoological Bulletin*. 1: 27.

He, S., F. Del Viso, C.Y. Chen, A. Ikmi, A.E. Kroesen, and M.C. Gibson. 2018. An axial Hox code controls tissue segmentation and body patterning in *Nematostella vectensis*. *Science*. 361: 1377–1380.

Heisenberg, C.P., and Y. Bellaïche. 2013. Forces in tissue morphogenesis and patterning. *Cell*. 153: 948.

Houliston, E., T. Momose, and M. Manuel. 2010. *Clytia hemisphaerica*: A jellyfish cousin joins the laboratory. *Trends in Genetics*. 26: 159–167.

Iwao, K., T. Fujisawa, and M. Hatta. 2002. A cnidarian neuropeptide of the glwamide family induces metamorphosis of reef-building corals in the genus *Acropora*. *Coral Reefs*. 21: 127–129.

Jessus, C., C. Munro, and E. Houliston. 2020. Managing the oocyte meiotic arrest—Lessons from frogs and jellyfish. *Cells*. 9: 1150–1185.

Kamran, Z., K. Zellner, H. Kyriazes, C.M. Kraus, J.B. Reynier, and J.E. Malamy. 2017. In vivo imaging of epithelial wound healing in the cnidarian *Clytia hemisphaerica* demonstrates early evolution of purse string and cell crawling closure mechanisms. *BMC Developmental Biology*. 17: 1–14.

Khalturin, K., C. Shinzato, M. Khalturina, M. Hamada, M. Fujie, R. Koyanagi, M. Kanda, H. Goto, F. Anton-Erxleben, M. Toyokawa, S. Toshino, and N. Satoh. 2019. Medusozoan genomes inform the evolution of the jellyfish body plan. *Nature Ecology and Evolution*. 3: 811–822.

Kim, H.M., J.A. Weber, N. Lee, S.G. Park, Y.S. Cho, Y. Bhak, N. Lee, Y. Jeon, S. Jeon, V. Luria, A. Karger, M.W. Kirschner, Y.J. Jo, S. Woo, K. Shin, O. Chung, J.C. Ryu, H.S. Yim, J.H. Lee, J.S. Edwards, A. Manica, J. Bhak, and S. Yum. 2019. The genome of the giant nomura's jellyfish sheds light on the early evolution of active predation. *BMC Biology*. 17: 1–12.

Krasovec, G. 2020. Compréhension du role morphogenetique de l'apoptose et de son évolution: apports de l'etude de la metamorphose de *Ciona Intestinalis* (Tunicata) et de *Clytia hemisphaerica* (Cnidaria). PhD thesis. Sorbonne Université, Paris.

Kraus, J.E., D. Fredman, W. Wang, K. Khalturin, and U. Technau. 2015. Adoption of conserved developmental genes in development and origin of the medusa body plan. *EvoDevo*. 6.

Kraus, Y., S. Chevalier, and E. Houliston. 2020. Cell shape changes during larval body plan development in *Clytia hemisphaerica*. *Developmental Biology*. 468: 59–79.

Kroiher, M., and S. Berking. 1999. On natural metamorphosis inducers of the cnidarians *Hydractinia echinata* (Hydrozoa) and *Aurelia aurita* (Scyphozoa). *Helgoland Marine Research*. 53: 118–121.

Kubota, S. 1978. The life-history of *Clytia edwardsi* (Hydrozoa; Campanulariidae) in Hokkaido, Japan. *Jour. Fac. Sci. Hokkaido Univ. Ser. VI, Zool*. 21: 317–354.

Lacassagne M. 1961. *Histologie comparée des planulas de quelques Hydraires*. Diplome d'études supérieures. Paris.

Lapébie, P., A. Ruggiero, C. Barreau, S. Chevalier, P. Chang, P. Dru, E. Houliston, and T. Momose. 2014. Differential responses to Wnt and PCP disruption predict expression and developmental function of conserved and novel genes in a cnidarian. *PLoS Genetics*. 10: e1004590.

Lechable, M., A. Jan, A. Duchene, J. Uveira, B. Weissbourd, L. Gissat, S. Collet, L. Gilletta, S. Chevalier, L. Leclère, S. Peron, C. Barreau, R. Lasbleiz, E. Houliston, and T. Momose. 2020. An improved whole life cycle culture protocol for the hydrozoan genetic model *Clytia hemisphaerica*. *Biology Open*: bio.051268.

Leclère, L. 2008. Evolution de la reproduction sexuee des hydrozoaires: aspects historiques, analyse phylogenetique et developpementale. PhD thesis. Université Pierre et Marie Curie, Paris.

Leclère, L., R.R. Copley, T. Momose, and E. Houliston. 2016. Hydrozoan insights in animal development and evolution. *Current Opinion in Genetics and Development*. 39: 157–167.

Leclère, L., C. Horin, S. Chevalier, P. Lapébie, P. Dru, S. Peron, M. Jager, T. Condamine, K. Pottin, S. Romano, J. Steger, C. Sinigaglia, C. Barreau, G. Quiroga Artigas, A. Ruggiero, C. Fourrage, J.E.M. Kraus, J. Poulain, J.M. Aury, P. Wincker, E. Quéinnec, U. Technau, M. Manuel, T. Momose, E. Houliston, and R.R. Copley. 2019. The genome of the jellyfish *Clytia hemisphaerica* and the evolution of the cnidarian life-cycle. *Nature Ecology and Evolution*. 3: 801–810.

Leclère, L., M. Jager, C. Barreau, P. Chang, H. Le Guyader, M. Manuel, and E. Houliston. 2012. Maternally localized germ plasm mRNAs and germ cell/stem cell formation in the cnidarian *Clytia*. *Developmental Biology*. 364: 236–248.

Leitz, T., and T. Wagner. 1993. The marine bacterium *Alteromonas espejiana* induces metamorphosis of the hydroid *Hydractinia Echinata*. *Marine Biology*. 115: 173–178.

Lilley, M.K.S., A. Elineau, M. Ferraris, A. Thiéry, L. Stemmann, G. Gorsky, and F. Lombard. 2014. Individual shrinking to enhance population survival: Quantifying the reproductive and metabolic expenditures of a starving jellyfish, *Pelagia noctiluca*. *Journal of Plankton Research*. 36: 1585–1597.

Linnaeus, C. 1767. *Systema naturae per regna tria naturae: Secundum classes, ordines, genera, species, cum characteribus, differentiis, synonymis, locis*. Laurentii Salvii. Stockholm.

Livshits, A., L. Shani-Zerbib, Y. Maroudas-Sacks, E. Braun, and K. Keren. 2017. Structural inheritance of the actin cytoskeletal organization determines the body axis in regenerating *Hydra*. *Cell Reports*. 18: 1410–1421.

Lucas, C.H., D.W. Williams, J.A. Wiliams, and M. Sheader. 1995. Seasonal dynamics and production of the hydromedusan *Clytia hemisphaerica* (Hydromedusa: Leptomedusa) in Southampton water. *Estuaries*. 18: 362–372.

Malamy, J.E., and M. Shribak. 2018. An orientation-independent DIC microscope allows high resolution imaging of epithelial cell migration and wound healing in a cnidarian model. *Journal of Microscopy*. 270: 290–301.

Maroudas-Sacks Y., L. Garion, L. Shani-Zerbib, A. Livshits, E. Braun, and K. Keren. 2021. Topological defects in the nematic order of actin fibres as organization centres of Hydra morphogenesis. *Nature Physics*. 17: 251–259.

Metchnikoff, E. 1886. *Embryologische Studien an Medusen: Ein Beitrag Zur Genealogie Der Primitiv-Organe*. Alfred Hölder. Wien.

Mills, C.E. 1983. Vertical migration and diel activity patterns of hydromedusae: Studies in a large tank. *Journal of Plankton Research*. 5: 619–635.

Momose, T., A. De Cian, K. Shiba, K. Inaba, C. Giovannangeli, and J.P. Concordet. 2018. High doses of CRISPR/Cas9 ribonucleoprotein efficiently induce gene knockout with low mosaicism in the hydrozoan *Clytia hemisphaerica* through microhomology-mediated deletion. *Scientific Reports*. 8: 5–10.

Momose, T., R. Derelle, and E. Houliston. 2008. A maternally localised Wnt ligand required for axial patterning in the cnidarian *Clytia hemisphaerica*. *Development*. 135: 2105–2113.

Momose, T., and E. Houliston. 2007. Two oppositely localised Frizzled RNAs as axis determinants in a cnidarian embryo. *PLoS Biology*. 5: 889–899.

Momose, T., Y. Kraus, and E. Houliston. 2012. A conserved function for Strabismus in establishing planar cell polarity in the ciliated ectoderm during cnidarian larval development. *Development*. 139: 4374–4382.

Morgan, T.H. 1899. Regeneration in the Hydromedusa, *Gonionemus vertens*. *The American Naturalist*. 33: 939–951.

Müller, W.A., and T. Leitz. 2002. Metamorphosis in the Cnidaria. *Canadian Journal of Zoology*. 80: 1755–1771.

Müller, W.A., R. Teo, and U. Frank. 2004. Totipotent migratory stem cells in a hydroid. *Developmental Biology*. 275: 215–224.

Negri, A.P., N.S. Webster, R.T. Hill, and A.J. Heyward. 2001. Metamorphosis of broadcast spawning corals in response to bacteria isolated from crustose algae. *Marine Ecology Progress Series*. 223: 121–131.

Neppi, V. 1918. Sulla rigenerazione nelle idromeduse. *Pubblicazioni Della Stazione Zoologica Di Napoli*. 2: 191–207.

Newmark, P.A., and A. Sánchez Alvarado. 2000. Bromodeoxyuridine specifically labels the regenerative stem cells of planarians. *Developmental Biology*. 220: 142–153.
Nishimiya-Fujisawa, C., and S. Kobayashi. 2012. Germline stem cells and sex determination in *Hydra*. *International Journal of Developmental Biology*. 56: 499–508.
Ohdera, A., C.L. Ames, R.B. Dikow, E. Kayal, M. Chiodin, B. Busby, S. La, S. Pirro, A.G. Collins, M. Medina, and J.F. Ryan. 2019. Box, stalked, and upside-down? Draft genomes from diverse jellyfish (Cnidaria, Acraspeda) lineages: *Alatina alata* (Cubozoa), *Calvadosia cruxmelitensis* (Staurozoa), and *Cassiopea xamachana* (Scyphozoa). *GigaScience*. 8: 1–15.
Okada, Y.K. 1927. Etudes sur la regeneration chez les coelentérés. *Archives de Zoologie Expérimentale et Générale: Histoire Naturelle*. 66: 497–551.
Otto, J.J., and R.D. Campbell. 1977. Tissue economics of *Hydra*: Regulation of cell cycle, animal size and development by controlled feeding rates. *Journal of Cell Science*. 28: 117–132.
Özpolat, B.D., E.S. Sloane, E.E. Zattara, and A.E. Bely. 2016. Plasticity and regeneration of gonads in the annelid *Pristina leidyi*. *EvoDevo*. 7: 22.
Passamaneck, Y.J., and M.Q. Martindale. 2012. Cell proliferation is necessary for the regeneration of oral structures in the anthozoan cnidarian *Nematostella vectensis*. *BMC Developmental Biology*. 12: 34.
Peron, S. 2019. Bases cellulaires et moleculaires de la regeneration chez la meduse *Clytia hemisphaerica*. PhD thesis. Sorbonne université, Paris.
Philippe, H., R. Derelle, P. Lopez, K. Pick, C. Borchiellini, N. Boury-Esnault, J. Vacelet, E. Renard, E. Houliston, E. Quéinnec, C. Da Silva, P. Wincker, H. Le Guyader, S. Leys, D.J. Jackson, F. Schreiber, D. Erpenbeck, B. Morgenstern, G. Wörheide and M. Manuel. 2009. Phylogenomics revives traditional views on deep animal relationships. *Current Biology*. 19: 706–712.
Pierobon, P. 2012. Coordinated modulation of cellular signaling through ligand-gated ion channels in *Hydra vulgaris* (Cnidaria, Hydrozoa). *International Journal of Developmental Biology*. 56: 551–565.
Piraino, S., G. Zega, C. Di Benedetto, A. Leone, A. Dell'Anna, R. Pennati, D. Candia Carnevali, V. Schmid, and H. Reichert. 2011. Complex neural architecture in the diploblastic larva of *Clava multicornis* (Hydrozoa, Cnidaria). *Journal of Comparative Neurology*. 519: 1931–1951.
Putnam, N.H., M. Srivastava, U. Hellsten, B. Dirks, J. Chapman, A. Salamov, A. Terry, H. Shapiro, E. Lindquist, V. Vladimir, J. Jurka, G. Genikhovich, I. Grigoriev, J.G.I. Sequencing, R.E. Steele, J. Finnerty, U. Technau, M.Q. Martindale, and D.S. Rokhsar. 2007. Sea anemone genome reveals the gene repertoire and genomic organization of the eumetazoan ancestor. *Science*. 317: 86–94.
Quiquand, M., N. Yanze, J. Schmich, V. Schmid, B. Galliot, and S. Piraino. 2009. More constraint on ParaHox than Hox gene families in early metazoan evolution. *Developmental Biology*. 328: 173–187.
Quiroga Artigas, G. 2017. Light-induced oocyte maturation in the hydrozoan *Clytia hemisphaerica*. PhD thesis. Université Pierre et Marie Curie, Paris.
Quiroga Artigas, G., P. Lapébie, L. Leclère, P. Bauknecht, J. Uveira, S. Chevalier, G. Jékely, T. Momose, and E. Houliston. 2020. A G protein-coupled receptor mediates neuropeptide-induced oocyte maturation in the jellyfish *Clytia*. *PLoS Biology*. 18: 1–38.
Quiroga Artigas, G., P. Lapébie, L. Leclère, N. Takeda, R. Deguchi, G. Jékely, T. Momose, and E. Houliston. 2018. A gonad-expressed opsin mediates light-induced spawning in the jellyfish *Clytia*. *ELife*. 7: 1–22.
Roosen-Runge, E. 1962. On the biology of sexual reproduction of hydromedusae, genus *Phialidium* Leuckhart. *Pacific Science*. 16: 15–31.
Ruggiero, A. 2015. Impact of Wnt signalling on multipotent stem cell dynamics during *Clytia hemisphaerica* embryonic and larval development. PhD thesis. Université Pierre et Marie Curie, Paris.
Satterlie, R.A. 2002. Neuronal control of swimming in jellyfish: A comparative story. *Canadian Journal of Zoology*. 80: 1654–1669.
Schaffer, A.A., M. Bazarsky, K. Levy, V. Chalifa-Caspi, and U. Gat. 2016. A transcriptional time-course analysis of oral vs. aboral whole-body regeneration in the sea anemone *Nematostella vectensis*. *BMC Genomics*. 17: 718.
Schmid, B., V. Schmid, and P. Tardent. 1974. The umbrellar growth process in the Leptomedusae *Phialidium hemisphaericum* (Syn. *Campanularia johnstoni*). *Experienti* 30: 1399–1400.
Schmid, V. 1974. Regeneration in medusa buds and medusae of Hydrozoa. *American Zoologist*. 14: 773–781.
Schmid, V., and P. Tardent. 1971. The reconstitutional performances of the Leptomedusa *Campanularia jonstoni*. *Marine Biology*. 8: 99–104.
Schmid, V., B. Schmid, B. Schneider, and G. Baker. 1976. Factors effecting manubrium-regeneration in Hydromedusae (Coelenterata). *Roux's Archives of Developmental Biology*. 56: 41–56.
Schmid, V., M. Wydler, and H. Alder. 1982. Transdifferentiation and regeneration in vitro. *Developmental Biology*. 92: 476–488.
Schneider, B. 1975. Die manubrium-regeneration bei der Leptomeduse *Phialidium hemisphaericum*: einfluss der ernafhrung und der gewebespannung. Diplomarbeit. Universität Zürich, Zürich.
Schnitzler, C.E. 2019. What makes a jellyfish. *Nature Ecology & Evolution*. 3: 724–725.
Seipp, S., J. Schmich, T. Kehrwald, and T. Leitz. 2007. Metamorphosis of *Hydractinia echinata*—Natural versus artificial induction and developmental plasticity. *Development Genes and Evolution*. 217: 385–394.
Siebert, S., J.A. Farrell, J.F. Cazet, Y. Abeykoon, A.S. Primack, C.E. Schnitzler, and C.E. Juliano. 2019. Stem cell differentiation trajectories in *Hydra* resolved at single-cell resolution. *Science*. 365: eaav9314.
Siebert, S., and C.E. Juliano. 2017. Sex, polyps, and medusae: Determination and maintenance of sex in cnidarians. *Molecular Reproduction and Development*. 84: 105–119.
Singla, C.L. 1975. Statocysts of Hydromedusae. *Cell and Tissue Research*. 158: 391–407.
Sinigaglia, C., S. Peron, J. Eichelbrenner, S. Chevalier, J. Steger, C. Barreau, E. Houliston, and L. Leclère. 2020. Pattern regulation in a regenerating jellyfish. *ELife*. 9: 1–33.
Sinigaglia, C., D. Thiel, A. Hejnol, E. Houliston, and L. Leclère. 2018. A safer, urea-based in situ hybridization method improves detection of gene expression in diverse animal species. *Developmental Biology*. 434: 15–23.
Slautterback, D.B., and D.W. Fawcett. 1959. The development of the cnidoblasts of *Hydra*. an electron microscope study of cell differentiation. *The Journal of Biophysical and Biochemical Cytology*. 5: 441–452.
Steinmetz, P.R.H., J.E.M. Kraus, C. Larroux, J.U. Hammel, A. Amon-Hassenzahl, E. Houliston, G. Wörheide, M. Nickel, B.M. Degnan, and U. Technau. 2012. Independent evolution of striated muscles in cnidarians and bilaterians. *Nature*. 487: 231–234.

Stidwill, R. 1974. Inhibitions-und induktionspheanomene bei der manubriumregeneration der leptomeduse *Phialidium hemisphaericum* (L.). Diplomarbeit. Universität Zürich, Zürich.

Takahashi, T., and M. Hatta. 2011. The importance of GLWamide neuropeptides in cnidarian development and physiology. *Journal of Amino Acids*. 2011: 1–8.

Takahashi, T., and N. Takeda. 2015. Insight into the molecular and functional diversity of cnidarian neuropeptides. *International Journal of Molecular Sciences*.16: 2610–2625.

Takeda, N., Y. Kon, G. Quiroga Artigas, P. Lapébie, C. Barreau, O. Koizumi, T. Kishimoto, K. Tachibana, E. Houliston, and R. Deguchi. 2018. Identification of jellyfish neuropeptides that act directly as oocyte maturation-inducing hormones. *Development*. 145: dev156786.

Tebben, J., D.M. Tapiolas, C.A. Motti, D. Abrego, A.P. Negri, L.L. Blackall, P.D. Steinberg, and T. Harder. 2011. induction of larval metamorphosis of the coral *Acropora millepora* by tetrabromopyrrole isolated from a *Pseudoalteromonas Bacterium*. *PLoS One*. 6: 1–8.

Technau, U., and M. Schwaiger. 2015. Recent advances in genomics and transcriptomics of cnidarians. *Marine Genomics* 24: 131–138.

Thomas, M.B., G. Freeman, and V.J. Martin. 1987. The embryonic origin of neurosensory cells and the role of nerve cells in metamorphosis in *Phialidium gregarium* (Cnidaria, Hydrozoa). *International Journal of Invertebrate Reproduction and Development* 11: 265–287.

Trembley, A. 1744. *Memoires pour servir a l'histoire d'un genre de polyes d'eau douce, a bras en forme de cornes*. Leide: Verbeek, Jean & Herman.

Urdy, S. 2012. On the evolution of morphogenetic models: Mechano-chemical interactions and an integrated view of cell differentiation, growth, pattern formation and morphogenesis. *Biological Reviews*. 87: 786–803.

van der Sande, M., Y. Kraus, E. Houliston, and J. Kaandorp. 2020. A cell-based boundary model of gastrulation by unipolar ingression in the hydrozoan cnidarian *Clytia hemisphaerica*. *Developmental Biology*. 460: 176–186.

Vining, K.H., and D.J. Mooney. 2017. Mechanical forces direct stem cell behaviour in development and regeneration. *Nature Reviews Molecular Cell Biology*. 18: 728–742.

Weber, C. 1981. Structure, histochemistry, ontogenetic development, and regeneration of the ocellus of *Cladonema radiatum* Dujardin (Cnidaria, Hydrozoa, Anthomedusae). *Journal of Morphology*. 167: 313–331.

Webster, B.G., and S. Hamilton. 1972. Budding in *Hydra*: The role of cell multiplication and cell movement in bud initiation. *Journal of Embryology and Experimental Morphology*. 27: 301–316.

Webster, N.S., L.D. Smith, A.J. Heyward, J.E.M. Watts, R.I. Webb, L.L. Blackall, and A.P. Negri. 2004. Metamorphosis of a scleractinian coral in response to microbial biofilms. *Applied and Environmental Microbiology*.70: 1213–1221.

Weiler-Stolt, B. 1960. Über die bedeutung der interstitiellen zellen für die entwicklung und fortpflanzung mariner hydroiden. *Roux's Archives of Developmental Biology*. 152: 398–454.

Weismann, A. 1883. *Die entstehung der sexualzellen bei den hydromedusen*. Gustav Fisher. Jena.

Wikelski, M., and C. Thom. 2000. Marine iguanas shrink to survive El Niño. *Nature*. 403: 37–38.

Zeleny, C. 1907. The effect of degree of injury, successive injury and functional activity upon regeneration in the scyphomedusan, *Cassiopea xamachana*. *Journal of Experimental Zoology*. 5: 265–274.

9 The Upside-Down Jellyfish *Cassiopea xamachana* as an Emerging Model System to Study Cnidarian–Algal Symbiosis

Mónica Medina, Victoria Sharp, Aki Ohdera, Anthony Bellantuono, Justin Dalrymple, Edgar Gamero-Mora, Bailey Steinworth, Dietrich K. Hofmann, Mark Q. Martindale, André C. Morandini, Matthew DeGennaro and William K. Fitt

CONTENTS

9.1 HISTORY OF THE MODEL

The model *Cassiopea xamachana*, also known as the upside-down jellyfish, was first described for the Caribbean (Jamaica) by Bigelow in 1892. *Cassiopea xamachana* is a tropical species belonging to the cnidarian class Scyphozoa, order Rhizostomeae, family Cassiopeidae. Substantially different from typically pelagic scyphozoan medusae, *Cassiopea* spp. jellyfish show an epibenthic lifestyle, resting upside-down with the bell turned to the substrate and the oral arms and appendages exposed upward. They preferentially occur in shallow water on soft bottom areas, often also in seagrass beds, in tropical, mangrove-sheltered lagoons.

Historically, Peter S. Pallas published the first formal description of a rhizostome medusa termed *Medusa* (now *Cassiopea*) *frondosa* in 1774, based on a preserved specimen originating from an unreported site in the Caribbean. However, Peter Forskål, a member of a Danish expedition sent to explore Arab countries in the years 1761–1767, first observed, collected and described in his data log an upside-down–type rhizostomatous medusa under the name *Medusa* (now *Cassiopea*) *andromeda* at Tôr on the southwestern coast of the Sinai Peninsula in October 1762. Tragically, Forskål and all but one participant of the expedition succumbed to disease or fatal incidents. As the only survivor, the surveyor Carsten Nibuhr wrote an account of the expedition and published *postum* only in 1775 the scientific descriptions of plants and animals Forskål had left behind. The plates depicting the described *C. andromeda* specimen were published a year later in 1776. Several

DOI: 10.1201/9781003217503-9

more forms of *Cassiopea* medusae have been described from various tropical regions of the world by 19th-century authors, either as varieties of *C. andromeda* or as separate species and varieties thereof. These descriptions were compiled and critically reviewed by Mayer (1910). For an actual listing of valid *Cassiopea* species, see Ohdera et al. (2018) and Jarms and Morandini (2019). *Cassiopea* spp. have been recorded as alien or introduced species first in the Mediterranean Sea by Maas (1903), as so-called "Lessepsian migrants" originating from the Red Sea through the Suez Canal, and in O'ahu, Hawaii, described by Cutress in Doty (1961) as most probably introduced during World War II.

In his keystone paper, Bigelow (1892) provided a detailed description of the anatomy and development of *C. xamachana* from Jamaica bearing on both the medusa and the scyphopolyp (scyphistoma). He included medusa formation by strobilation of the polyp and the asexual propagation of the polyp through the budding of ciliated, spindle-shaped propagules that settle and develop into new polyps. Sexual reproduction by the typically gonochoric medusae was assessed much later and embryonic development approached only recently (see Section 9.4). Bigelow was a pioneer in noticing the presence of green cells, or "zoanthelae", in medusae, scyphistomae and buds of this species, recognized as symbiotic unicellular algae and described much later by Freudenthal (1959). They became commonly termed "zooxanthellae". A wealth of information on *C. andromeda* from the Red Sea became available through the two monographs by Gohar and Eisawy (1960a, 1960b), closing gaps in knowledge of the life-history. In contrast, information on *C. frondosa* remained scarce (Bigelow 1893; Smith 1936; Hummelinck 1968). Providing easily collectable mature medusae from tropical and subtropical habitats almost year-round, and with scyphistomae performing asexual reproduction under relatively simple conditions in the lab, *C. xamachana* was setting out to become a versatile symbiotic scyphozoan model species.

The Carnegie Marine Biological Laboratory on Loggerhead Key in the Dry Tortugas, Gulf of Mexico, commonly called Tortugas Marine Laboratory, was founded in 1904 with Alfred Goldsborough Mayer as its first director (Stephens and Calder 2006). This lab, in fortunate association with the publication series *Papers from Tortugas Laboratory* by the Carnegie Institution, was pivotal in hosting experimental studies of *Cassiopea* spp. (Perkins 1908). Some of the research topics included *Cassiopea*'s rhythmical pulsation and its causes (Mayer 1908), the rate of regeneration in *C. xamachana* medusae (Stockard 1908), the physiology of the *C. xamachana* nervous system (Cary 1917) and the anatomy and physiology of the sympatric *C. frondosa* (Smith 1936). Mayer (1910) contributed volume III, The Scyphomedusae, of his monumental work, *Medusae of the World*. In it, he provides a detailed account of the genera *Toreuma* and *Cassiopea* in the context of history, taxonomy and biology. After those early 20th-century works, there was a slowdown in research in *Cassiopea*, with a renaissance in the 1970s. Curtis and Cowden (1972) meticulously investigated the significant regenerative capacities of *C. xamachana* scyphistomae. More recently, Hamlet et al. (2011) and Santhanakrishnan et al. (2012) introduced advanced high speed kinematic and modeling techniques to study the hydrodynamics of the conspicuous pulsation behavior of the *Cassiopea* jellyfish. Moreover, in the wake of photo-physiological studies of zooxanthellate scleractinian corals (e.g. Yonge and Nicholls 1931), the *Cassiopea–Symbiodinium* symbiosis prompted a rapidly growing number of studies bearing on the mutualistic relationship between the host and the algal symbionts in different phases of the life cycle (e.g. Ludwig 1969; Balderston and Claus 1969; Hofmann and Kremer 1981; Fitt and Trench 1983a). Contemporary work on bud-to-polyp transition by Curtis and Cowden (1971) initiated a search for extrinsic natural and synthetic factors inducing metamorphosis of planula larvae and buds and studies to elucidate their putative mode of action (see Section 9.3). In recent years, research on *C. xamachana* diversified considerably, as described in 2018 by Ohdera and a consort of co-authors. Their review exposes work on behavior, quiescence, bioinvasions and blooms, environmental monitoring and ecotoxicology, toxicology and cnidome and virology, in addition to expanding on topics that have briefly been considered here. The isolation of Hox genes by Kuhn et al. (1999) was a landmark timepoint indicating that *C. xamachana* research had entered the age of evo-devo and genomics (see Section 9.6).

9.2 GEOGRAPHICAL LOCATION

9.2.1 Species and Endemic Distributions

It is often the case that jellyfish clades include cryptic species not easily distinguished by morphological characteristics (Holland et al. 2004; Arai 2001), and this is further complicated by the fact that intraspecific morphological diversity is often quite high (Gomez-Daglio and Dawson 2017). Nine *Cassiopea* species are currently recognized by the World Register of Marine Species: *C. andromeda* (Forskål 1775), *C. depressa* (Haeckel 1880), *C. frondosa* (Pallas 1774), *C. maremetens* (Gershwin et al. 2010), *C. medusa* (Light 1914), *C. mertensi* (Brandt 1835), *C. ndrosia* (Agassiz and Mayer 1899), *C. ornata* (Haeckel 1880) and *C. xamachana* (Bigelow 1892). Additionally, *C. vanderhorsti* has been proposed as a species (Stiasny 1924) but may be a variety of *C. xamachana* (Jarms and Morandini 2019). *Cassiopea* species are distributed throughout tropical and subtropical waters all over the world, with *C. frondosa* and *C. xamachana* in the Caribbean and Gulf of Mexico; *C. andromeda* in the Red Sea, invasive in Hawaii, Brazil and the Asian-Australian sea; *C. medusa, C. mertensi, C. maremetens, C. ndrosia* and *C. ornata* in the eastern South Pacific; and *C. depressa* along the coral coast of eastern African in the Indian Ocean (Figure 9.1).

Morphological work would go on to merge *C. medusa* and *C. mertensi* into *C. andromeda* (Gohar and Eisawy 1960a)

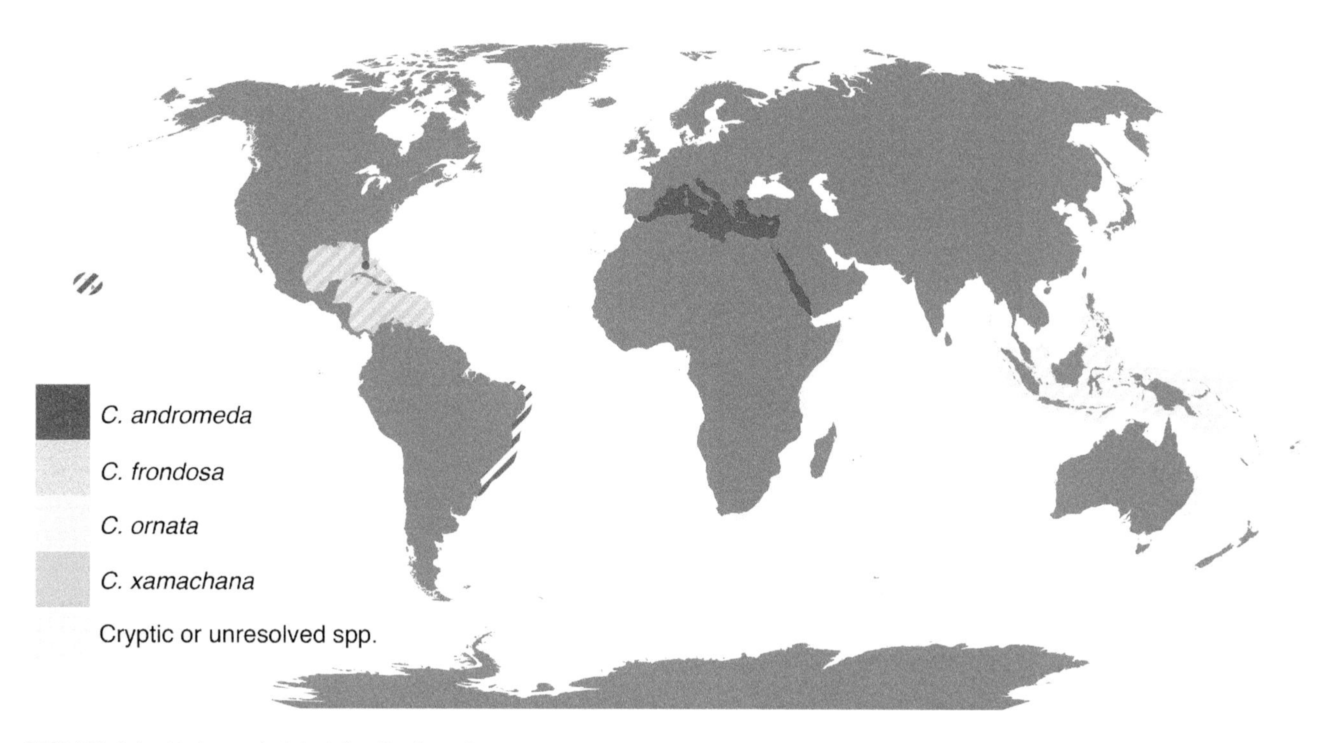

FIGURE 9.1 Estimated global distribution of *Cassiopea* species, compiled from the World Register of Marine Species. (From Holland et al. 2004, Arai et al. 2017, and Morandini et al. 2017.)

before further reorganization of the clade by molecular phylogenetic analysis. In recent years, several groups have used DNA barcoding of the mitochondrial gene cytochrome c oxidase subunit 1 (usually denoted as COI or COX1) to resolve ambiguities in the phylogeny of *Cassiopea*. Analysis of COX1 sequences from *Cassiopea* around the world by Holland et al. (2004) supports six species: *C. frondosa* in the western Atlantic; *C. andromeda* in the Red Sea, western Atlantic and Hawaii; *C. ornata* in Indonesia, Palau and Fiji; cryptic *Cassiopea* species 1 in eastern Australia; cryptic *Cassiopea* species 2 in Papua New Guinea; and cryptic *Cassiopea* species 3 in Papua New Guinea and Hawaii. The three cryptic species suggested by this analysis were previously classified as *C. andromeda*. This study also shows that specimens identified as *C. xamachana* from the Gulf of Mexico and the Caribbean are actually *C. andromeda*. Later studies by Morandini et al. (2017) and Arai et al. (2017) largely recapitulate these findings, but Arai et al. (2017) suggest three more cryptic species within *C. andromeda*, potentially bringing the total number of *Cassiopea* species to as many as nine, plus the valid morphospecies without molecular data associated with them (*C. depressa, C. maremetens, C. medusa, C. mertensi* and *C. ndrosia*). Further work remains to be done in this field, especially considering the claim that COX1 barcoding may be insufficient to distinguish between cnidarian congeners due to exceptionally low rates of mitochondrial evolution within Cnidaria (France and Hoover 2002; Shearer et al. 2002). This is possibly due to the presence of excision repair, which is absent in other animal mitochondria (Hebert et al. 2003).

9.2.2 Invasion and Human Impacts

Cassiopea jellyfish possess multiple characteristics which make them a potential invasive threat, particularly their high tolerance to both salinity (Goldfarb 1914) and thermal stress (Klein et al. 2019), as well as their capacity for thermal acclimation to 32°C (Al-jbour et al. 2017). Recent work suggests that rising seawater temperatures may increase the range of *Cassiopea* (Al-jbour et al. 2017). With cryptic life phases and potential to persist as scyphistomae (= benthic stages) for extended periods of time, *Cassiopea* have great potential to be transported as hitchhikers on ships. Additionally, proximity to human populations may enhance *Cassiopea* growth: there is some evidence from Abaco Island (Bahamas) that *Cassiopea* populations are larger in areas with high human density, presumably since high human densities are also correlated with higher levels of nutrients (Stoner et al. 2011; Thé et al. 2020).

The potential for *Cassiopea* invasion and blooms has been realized in multiple instances. Humans have a historical role in spreading *Cassiopea*, with molecular evidence suggesting that Floridian and Bermudan *Cassiopea* were spread to Brazil approximately 500 years ago—a time contemporaneous with the beginning of Portuguese shipping and colonization in the region (Morandini et al. 2017).

The relationship between human movement and *Cassiopea* range extension has also been documented more recently. The Hawaiian Islands have apparently been colonized by *Cassiopea* in the past century, as a 1902 survey by Mayer (1906) on the USS Albatross, the first purpose-built marine research ship, found no *Cassiopea* on the islands. *Cassiopea* were first reported after World War II, presumably transported to Hawaii by US naval traffic. According to reports by residents, *Cassiopea medusa* first appeared exclusively in Pearl Harbor on O'ahu between 1941 and 1945 but were observed circa 1950 in Honolulu Harbor and the Ala Wai Canal (Doty 1961). Observations in 1964 (Uchida 1970) reported *Cassiopea* in Kane'ohe Bay. These early reports of *Cassiopea* initially identified *C. medusa* and *C. mertensi*, but the taxa have since been collapsed to a single species, *Cassiopea andromeda*, due to morphological similarity (Hofmann and Hadfield 2002). Curiously, however, the *Cassiopea* found near Ala Wai Harbor exhibited hermaphroditism, though this characteristic was not stable over time (Hofmann and Hadfield 2002).

Baker's law (1955) hypothesizes that species which can reproduce with only a single hermaphroditic parent will colonize new areas more successfully than gonochoristic species. While the advantages in invasion capacity of uniparental reproduction have not been tested in cnidarians, this ability is the basis of a longstanding hypothesis in terrestrial plants (Baker 1965; Van Etten et al. 2017). The hermaphroditic capacity of some *Cassiopea* may facilitate their invasion, particularly of islands seeded by chance through human introduction, where a founding population may originate from a single scyphistoma hitchhiking on a hull or in ballast water. Indeed, Hofmann and Hadfield (2002) hypothesize that the founder of the invasive population in Ala Wai Canal may have consisted of a single clonal individual. Morandini et al. (2017) note that all 200 medusae collected in Cabo Frio (Brazil) were male and potentially the result of clonal reproduction, suggesting that asexual reproduction as scyphistomae is yet another method of uniparental reproduction that may play a part in the capacity of *Cassiopea* to expand their range. A recent study from northeastern Brazil (Ceará state) also reported only female individuals in the population (Thé et al. 2020).

The first molecular phylogenetics of *Cassiopea* indicated that the species identified as *C. andromeda* in O'ahu, Hawaii, waters in fact comprised two distinct clades representing a cryptic species (Holland et al. 2004), with one clade of Indo-Pacific origin and the other established from either the Western Atlantic or Red Sea. Arai et al. (2017) further examined the molecular phylogenetics of *Cassiopea* and also found that *C. xamachana* from the Western Atlantic and *C. andromeda* from the Red Sea fell into the same clade, indicating that these are likely the result of an introduction of *C. andromeda* into the Caribbean.

Cassiopea have recently spread even farther, with reports in the central Mediterranean originally in 2005 in the Maltese Islands (Schembri et al. 2010) and again in 2006 in the Levantine coast of Turkey (Çevik et al. 2006). Keable and Ahyong (2016) identified multiple species in coastal lakes of eastern Australia, representing the southernmost reported invasion of the genus (Figure 9.1). The growing geographic range and propensity of *Cassiopea* to form blooms further supports the need for revised systematic and taxonomic methods for the accurate classification of these organisms in order to more meaningfully categorize them and identify their origins.

9.3 LIFE CYCLE

Like the majority of scyphozoans, *C. xamachana* alternates between the asexual polyp (i.e. scyphistoma) and a sexual medusa (Figure 9.2). Planula larvae, the result of sexual reproduction, settle and metamorphose in response to bacterial cues on environmental substrates (Hofmann et al. 1996) (for early development, see Section 9.4). The resulting scyphistomae can reproduce asexually via budding or strobilation to produce either a male or female medusa. Strobilation is initiated following the establishment of symbiosis with dinoflagellates of the family Symbiodiniaceae (LaJeunesse et al. 2018). Therefore, in addition to environmental factors, life cycle completion partly involves association with two different organisms: settlement of the larvae happens in response to different bacterial cues, and strobilation occurs in response to cues associated with the establishment of symbiosis with Symbiodiniaceae.

The planula larva does not have dinoflagellate symbionts but does rely on specific bacteria such as *Vibrio* spp. (Neumann 1979; Hofmann and Brand 1987) and *Pseudoalteromonas* sp. (Ohdera, et al., in prep a) that release cues to induce their settlement and metamorphosis. The cues appear to be peptides that are either released by the bacteria or the result of biodegradation of the substrate they are on (Fleck et al. 1999). A number of artificial peptides have been identified and the mechanism of interaction with larval receptors proposed (Hofmann et al. 1996; Fleck and Hofmann 1995). The scyphistomae are frequently found on the shaded side of degraded mangrove leaves during the summer (Fleck and Fitt 1999; Fleck et al. 1999) but also settle on other leaves and hard surfaces.

Newly settled scyphistomae of *C. xamachana* exhibit horizontal transfer of symbiotic Symbiodiniaceae, meaning they collect their symbionts from the environment rather than inheriting them. Shortly after settling and metamorphosing into polyps and developing a mouth, endodermal digestive cells (i.e. gastrodermis) phagocytose Symbiodiniaceae from the water column (Colley and Trench 1983). Soon after being infected with symbiotic algae, the scyphistoma undergoes strobilation. Algae live within the symbiosome, also known as the amoebocyte, formed from the initial vacuoles which engulf the ingested symbiont cells. Amoebocytes migrate to the base of the gastrodermis by approximately day 3 after ingestion and subsequently migrate to the mesoglea by approximately day 8 post-infection (Colley and Trench 1985). When the number of Symbiodiniaceae reach 5–12,000 in large (>1 mm) scyphistomae at ≥25°C, they will strobilate a single medusa in one to three weeks depending on temperature and light levels (Hofmann et al. 1978). We have observed that scyphistomae can continue strobilating throughout the summer and fall in the Florida Keys and in culture indefinitely. *C. xamachana* has been found to establish a symbiosis with different Symbiodiniaceae species in

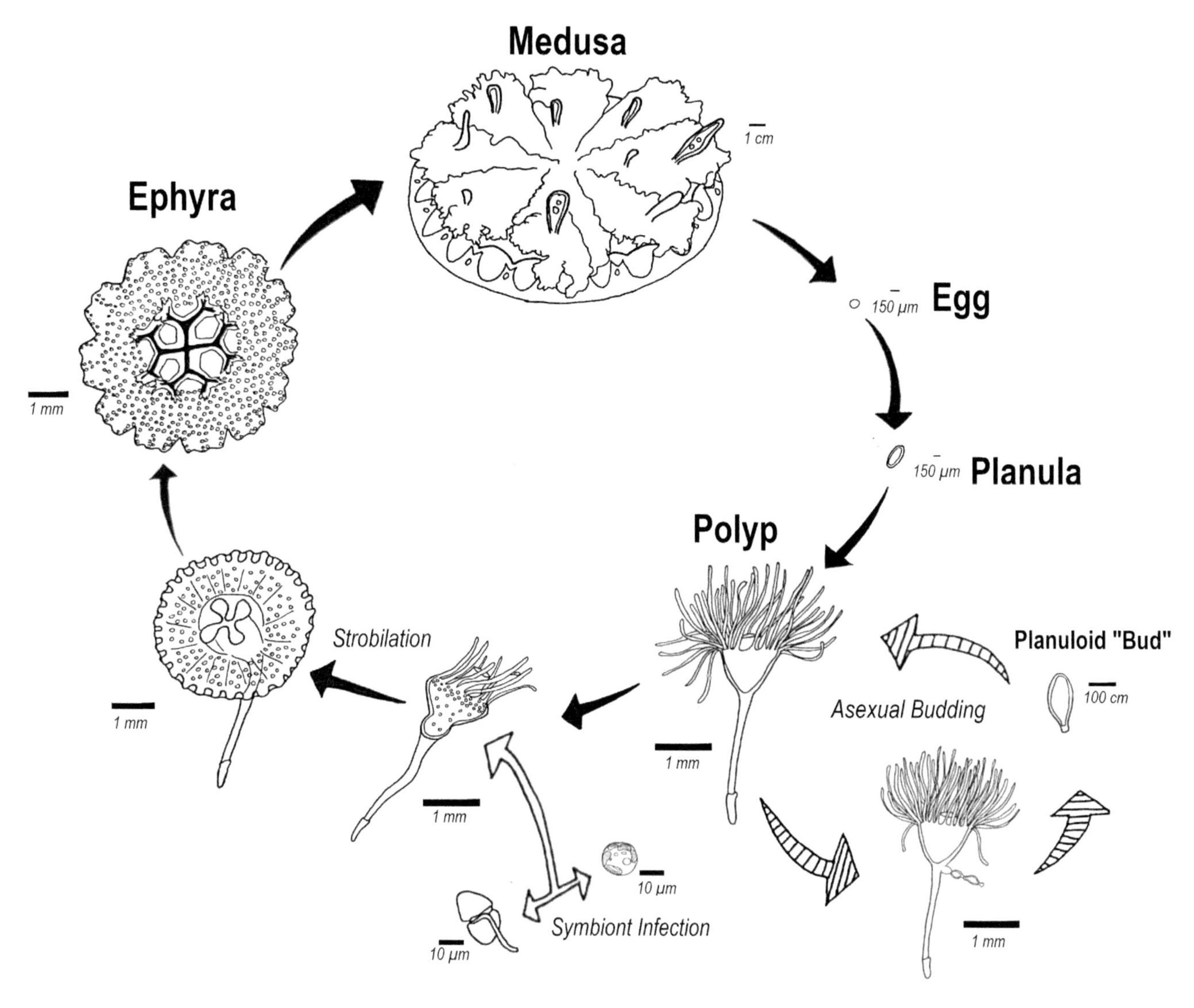

FIGURE 9.2 Life cycle of *Cassiopea xamachana* with scale bars per developmental stage. Ontogenetic stage names in bold. Non-sexual processes in italics. Black arrows; metagenic life cycle. Striped arrows; asexual "budding" reproduction. White arrows; symbiont infection and induction of strobilation.

fewer than three days while being held on the reef, back reef, seagrass bed or mangroves in the Florida Keys (Thornhill et al. 2006). If exposed to the homologous (found most frequently and at highest relative densities in *C. xamachana*) symbiont species *Symbiodinium microadriaticum*, the symbiont composition switches to *Symbiodinium microadriaticum* in a short period of time (via competitive exclusion), and the scyphistomae strobilates shortly thereafter (Thornhill et al. 2006). The role *S. microadriaticum* plays in inducing strobilation is not currently known.

The medusa and symbiotic scyphistomae are both photosynthetic and predatory. Photosynthesis occurs in the symbiotic dinoflagellates contained in digestive or ameobocytic cells, usually in direct sun in very shallow water, and is thought to provide the bulk of the fixed carbon to fulfill the energy requirements of their hosts (Verde and McCloskey 1998). However, they also use their mouth arm *digitata*, which contain the stinging organelles called nematocysts, to capture small zooplankton and other particles. Rhizostomes feed via many small mouths rather than the single mouth found in all other scyphozoans. *C. xamachana* can also shed clumps of nematocysts—dubbed cassiosomes—presumably to aid in obtaining food or as a defense from predators (Ames et al. 2020). External feeding is thought to provide the protein for growth of the jellyfish.

Temperature is a decisive factor in the life cycle of *C. xamachana*. Whereas rhizostome jellyfish typically over-winter in the scyphistomae stage, *C. xamachana* are present in the South Florida winter only as a medusa, as the polyps cannot feed themselves and disappear at temperatures ≤18°C (Fitt and Costley 1998). As the water temperature rises, planulae settle and metamorphose into scyphistomae which catch and consume food. It is not known if scyphistomae can survive winter temperatures in lower latitudes of the Caribbean Sea. *C. xamachana* begins to strobilate when temperatures are ≥25°C, thus completing the life cycle (Rahat and Adar 1980).

As temperatures increase with global climate change, populations of *C. xamachana* appear to be expanding (Morandini et al. 2005, Morandini lab unpublished) with a longer season to strobilate (Richardson et al. 2009). In addition, *C. andromeda* has become an exotic species, with populations in Australia, Hawaii, the Mediterranean and potentially the entire Caribbean (Çevik et al. 2006; Morandini et al. 2017; Holland et al. 2004; Schembri et al. 2010; Keable and Ahyong 2016), possibly partially due to higher temperatures. Whether the exotic *C. xamachana*'s recent range expansions will harm the environment remains to be seen.

9.4 EMBRYOGENESIS

9.4.1 Sexual Reproduction

Members of the genus *Cassiopea* are generally gonochoristic, though hermaphrodites have been observed in at least one population (Hofmann and Hadfield 2002). In males, appendages are homogenous across the oral disc, whereas in females, there is a region of appendages at the center of the oral disc that are specialized for brooding embryos (circled in Figure 9.3a). The precise timing of sexual maturity is not known in terms of age or diameter; however, viable gametes have been recovered from individuals as small as 7 cm in bell diameter (Hofmann and Hadfield 2002). The gonads can be accessed through the four prominent openings (subgenital pits) located between the oral arms and the bell. In the Florida Keys, the temperatures are often colder during winter cold fronts, which could reduce the number of eggs female medusae produce.

Despite the existence of separate sexes, the site of fertilization is unknown. Free spawning has never been observed. Martin and Chia (1982) claim to have performed *in vitro* fertilization: they collected gonadal material from inside the gastrovascular cavity, combined ovary and testes in seawater and observed swimming planulae. Fertilization seems to occur either within the mother, with sperm taken in from the water column, or quickly after unfertilized eggs are deposited onto the brooding tentacles.

In laboratory conditions with adult wild-caught animals, new embryos can be collected daily from the brooding region of female medusae. Spawning seems to be regulated by light. When medusae are kept on a light cycle of 12 hours of darkness and 12 hours of light at 24°C at the Whitney Lab for Marine Bioscience, zygotes can be observed among the brooding appendages of female medusae, but only if male medusae are also present. If females are maintained separately from males, no eggs (fertilized or unfertilized) are observed to be released into the brooding appendage region. Unlike some symbiotic cnidarians, eggs do not contain symbiotic dinoflagellates; symbionts are acquired horizontally via acquisition from the environment rather than vertically inherited from the mother.

Within a few hours, clusters of zygotes become encased in a stiff membrane that attaches them firmly to the brooding tentacles (Figure 9.3b). This membrane is maternally produced, as zygotes collected from the mother before the membrane appears do not develop this membrane. Eggs have already been fertilized before this membrane appears. Embryos are tightly packed within this membrane, often causing them to take on irregular shapes as development progresses. If left undisturbed, zygotes will continue to develop encased in this membrane, attached to the mother's brooding appendages, until reaching the stage when they can swim using cilia and eventually free themselves and swim away.

Observations of development have been made from embryos removed at the one-cell stage and kept at 24°C. Zygotes are 100–150 um in diameter (Figure 9.3c). Cleavage begins approximately two hours after zygotes are first observed (Figure 9.3d). Initial cell divisions are unipolar, beginning at the animal pole, and are complete, producing clear two-cell (Figure 9.3e) and four-cell (Figure 9.3f) stages. The embryo reaches the blastula stage, a hollow ball of cells with no yolk in the blastocoel, around 24 hours after the first cleavage (Figure 9.3g), and gastrulation is complete within 48 hours after the first cleavage is observed (Figure 9.3h). The exterior of the gastrula is ciliated (Figure 9.3k). Gastrulae move with a spinning motion, unlike the directed swimming later seen in the planula.

Further study is needed to fully understand the morphological details of development from zygote to planula. The mode of gastrulation is not yet known, though invagination is the most common form of gastrulation in the Scyphozoa (Morandini and da Silveira 2001; Nakanishi et al. 2008; Yuan et al. 2008; Kraus and Markov 2016). During gastrulation by invagination, the epithelium of the blastula folds inward at the future oral end while maintaining its epithelial identity. The epithelium continues to migrate inward until there are two layers of epithelium, the endoderm and ectoderm. Some cnidarians have complex patterns of gastrulation involving multiple waves of cellular movement (reviewed by Kraus and Markov 2016). While the mode of gastrulation has not been confirmed in *Cassiopea*, images of gastrulae appear to support the possibility of gastrulation by invagination (Figures 9.3h–i). Molecular studies using endomesodermal markers in other cnidarians are underway to confirm the location of presumptive endodermal precursors.

At three days old, an opening to the external sea water is still present and is located at the site of gastrulation, the blastopore (Figure 9.3i). By four days, the blastopore has closed completely, so that the inner epithelium has no connection to the outside of the embryo (Figure 9.3j). The structure of four-day-old planulae was described by Martin and Chia (1982) using transmission electron microscopy (TEM). Planulae range from 120 to 220 μm in length and 85 to 100 μm in width at the midpoint. The exterior of the planula is uniformly ciliated (Figure 9.3l), and planulae swim leading with the future aboral end ahead, but there is no apical tuft at the leading edge. Planulae contain endodermal and ectodermal epithelia separated by a thin layer of mesoglea (Martin and Chia 1982).

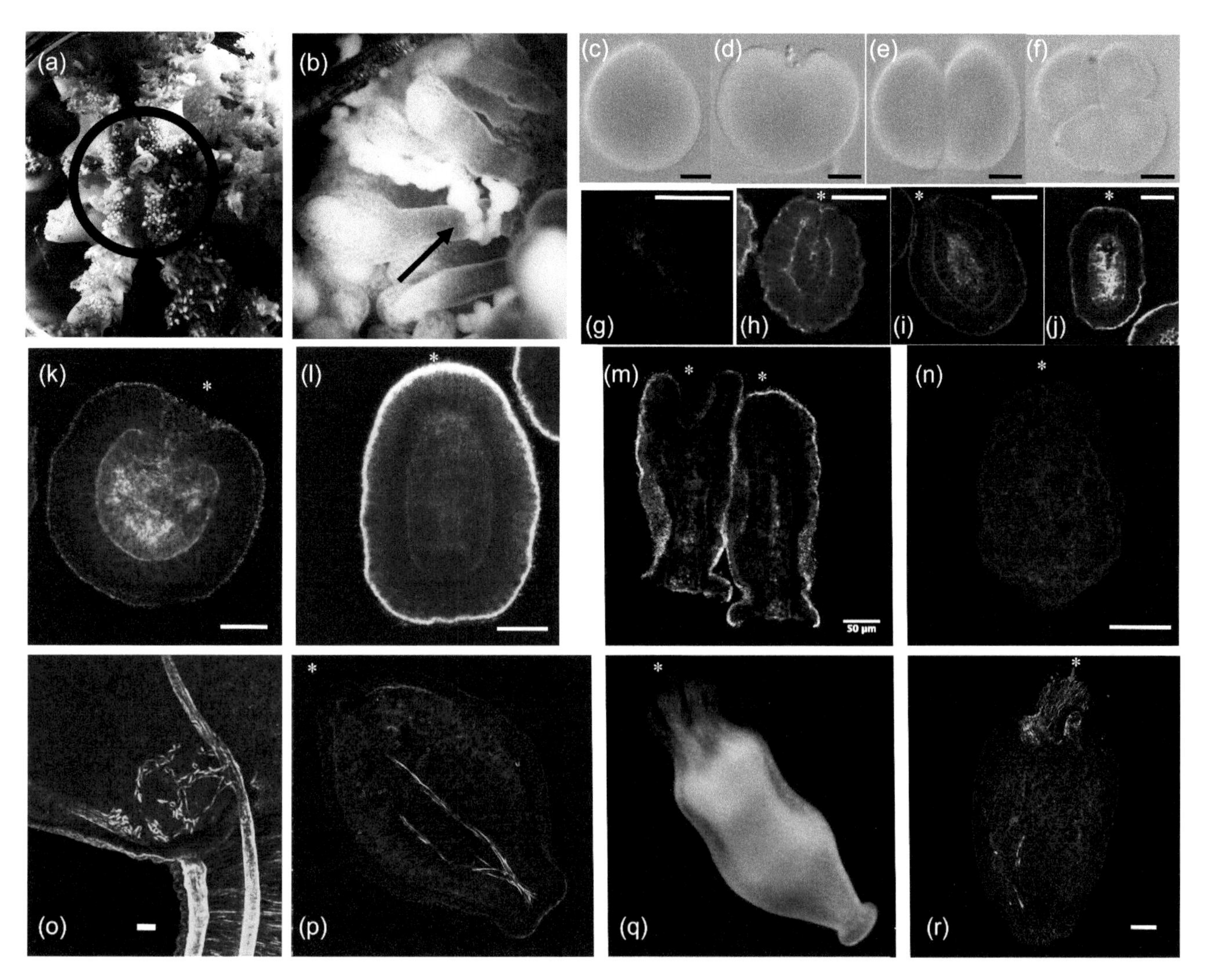

FIGURE 9.3 If female medusae (a) are kept with male medusae, zygotes can be found daily among the brooding appendages (b, circled in a) at the center of the oral disc. Zygotes (c) are packaged in a thin membrane and attached to the brooding appendages. Arrow in (b) points to attachment point where a package of embryos is wrapped around a brooding appendage. Location of fertilization is unknown. Initial cleavage (d) produces a two-cell stage (e), and each cell divides equally to produce a four-cell stage (f). Embryos reach the blastula stage (g) at approximately 24 hours after first cleavage and the gastrula stage (h) approximately 48 hours after first cleavage. At 72 hours after first cleavage (i), the blastopore can still be observed, but it is no longer observable by 96 hours (j). (g–j) Confocal slices stained to show actin. (k–l) and (o–p) Confocal slices stained to show actin (green), nuclei (blue), and cilia [magenta, no cilia stain in (o)]. Gastrulae (k) and planulae (l) are ciliated, and no mouth is observable in planulae. After attachment to a surface (m, right side), the polyp mouth forms *de novo* (m, left side). Asexually produced planuloids contain septal muscle fibers from the parent polyp (o and p) and can contain symbiotic dinoflagellates in the gastrodermis, shown by magenta autofluorescence in (p). Mouth and tentacles can form in asexually produced planuloids without attachment to a substrate (q). Both planulae and asexually produced planuloids stain with antibodies to the neural marker protein RFamide (n and r), shown here on 3D projections of confocal stacks with RFamide in magenta and actin in green. All scale bars are 50 micrometers. Asterisks indicate the future oral end of planulae and planuloids.

Four cell morphologies have been previously described in the planula: two types in the ectoderm and two in the endoderm. The ectoderm consists of support cells and cnidocytes. Ectodermal support cells extend from the mesoglea to the exterior surface. The apical surface of a support cell is covered in microvilli, and each cell has a single cilium (Martin and Chia 1982). Martin and Chia report one type of cnidocyte in the planula but do not specify what type it is; in other life stages of *Cassiopea*, different types of cnidocytes have been described (Heins et al. 2015) (see Section 9.5 for additional detail). The apical surface of a mature cnidocyte is exposed to the exterior, and the cell does not appear to extend basally to the mesoglea, based on TEM. Developing cnidocytes can be identified by their capsule and are located between support cells near the basal region of these cells; they do not connect to the exterior. The endoderm also contains two cell types: support cells and interstitial cells. Endodermal support cells extend from the mesoglea to the interior lumen of the planula and bear an apical cilium. Interstitial cells are clustered among the endodermal support cells, and their function is unknown (Martin and Chia 1982). Staining with an antibody to the neurotransmitter RFamide implies the presence of neural cells, specifically concentrated at the aboral end of the planula. The potential presence of neural cells indicates there may be additional cell types present that have not yet been described.

Planulae are competent to settle by the age of four or five days (Martin and Chia 1982). Attachment to a surface usually precedes development into a polyp, but planulae have been observed to metamorphose without attachment (Martin and Chia 1982). Planula settlement can be induced by *Vibrio alginolyticus* bacteria or by the hexapeptide Z-Gly-Pro-Gly-Gly-Pro-Ala (Hofmann and Brand 1987). The polyp mouth forms *de novo* at the site of blastopore closure (Figure 9.3m), followed by four initial tentacles surrounding it, then four additional tentacles at the spaces between those. At this point, the former planula is recognizable as a small polyp. Once the mouth has developed, polyps are capable of both eating and taking in dinoflagellates from the environment to establish symbiosis. As the polyp grows, the region of the stolon that lacks a gastrovascular cavity continues to lengthen.

9.4.2 Asexual Reproduction

In addition to sexual reproduction, polyps can reproduce asexually to form more polyps. Clonal daughter offspring bud from the side of polyps, usually at consistent spots near the base of the calyx, in the form of swimming oblongs researchers have called planuloids or planuloid buds (Khabibulina and Starunov 2019). The future oral–aboral axis of the planuloid forms at an angle to the oral–aboral axis of the parent polyp. Clonal planuloids are superficially similar to planulae produced as a result of spawning in a number of ways. Both planulae and planuloids have a uniformly ciliated exterior; both swim leading with the future aboral end of the polyp ahead, rotating about the oral–aboral axis. An oral opening is absent in both (Figures 9.3l and p) and forms during development into a polyp (Figures 9.3m and q). Additionally, antibody staining against the neural marker RFamide (Figures 9.3n and r) displays concentrated signal at the future aboral end, which is the leading pole during swimming.

There are notable differences between the morphologies of planulae and planuloids. The most obvious difference is that planuloids are much larger than planulae. Planuloids can be over 2 mm in length and 1 mm in width at their widest point. Planuloids also contain longitudinal muscle fibers running from the future oral to future aboral end (Figure 9.3p), and no such muscle fibers are present in sexually produced embryos (Figure 9.3l). Development of asexual propagules begins with an outpocketing of the body wall of the parent polyp, with the longitudinal muscle fibers of the polyp extending into the developing propagule (Figure 9.3o). However, Khabibulina and Starunov (2019) report that these muscle fibers are lost during propagule development, and the fibers observed in the propagule form *de novo*. Unlike planulae, asexual propagules regularly begin to metamorphose into polyps before attachment to a surface. Finally, asexual propagules may contain symbiotic dinoflagellates in cells of the gastrodermis if the parent polyp is inoculated with symbionts (Figure 9.3p), while planulae only acquire symbionts from the environment once they have developed a mouth in the process of becoming a polyp.

9.5 ANATOMY

The *C. xamachana* body is composed of three layers: epidermis, gastrodermis and mesoglea (Mayer 1910). Planulae are uniformly ciliated and polarized, swimming with the anterior end forward. The anterior end is the precursor to the polyp pedal disk and where settlement occurs. As previously mentioned in this chapter (see Section 9.4), planulae are aposymbiotic and additionally have cnidoblasts (precursors to cnidocytes, the cells which produce cnidocysts or "stinging cells") in their epidermis. Fully differentiated cnidocytes are present in the ectoderm (Martin and Chia 1982). A full description of *Cassiopea* cnidocysts is located at the end of this section.

After settlement, *C. xamachana* larvae develop into scyphistomae (polyps). A scyphistoma is composed of a pedal disc securing the polyp to a substrate, a stem rising to meet the head or calyx and a centrally located mouth or hypostome (Figure 9.4a) (Bigelow 1900). The calyx contains four gastric pouches separated by four septal muscles (Bigelow 1892). It has 32 total tentacles: 4 pairs of perradial, 4 pairs of interradial and 8 pairs of adradial tentacles. When fully expanded, the tentacles exceed the length of the body (Bigelow 1900) which is 3 to 4 mm long with a 1-mm-diameter calyx (head) in fully grown polyps (Figure 9.4a) (Curtis and Cowden 1974). Budding occurs at the base of the calyx in a perradial distribution (Hofmann et al. 1978). The planuloid buds have a single-layered ectoderm with three cell types, an endoderm with two cell types and a thin mesoglea separating the ectoderm from the endoderm. Cnidoblasts are located at the base of the epithelial cells, while cnidocytes are near the epithelial surface (Hofmann and Honegger 1990). While buds detach independently from the polyp, they can form budding chains where two to four buds are connected by ectodermal tubes which eventually sever when the bud detaches. The bud at the base of this chain forms a continuous endoderm with the polyp (Figure 9.4a). Buds are spindle shaped and uniformly ciliated, rotating around a longitudinal axis and swimming with the distal anterior pole forward. This anterior end eventually forms the pedal disc upon settlement (Hofmann et al. 1978).

Symbiosomes localize at the base of a host cell, away from maximum lysosomal activity (Fitt and Trench 1983b). Algae are most dense in the subtentacular region of the polyp and at lowest density in the pedal disk region. The positioning of symbionts ensures transfer of algae to the developing ephyra. Ephyra initially have four simple oral arms with a central mouth opening and develop marginal lobes and rhopalia, the sense-organs of adult *C. xamachana* (Figure 9.4b–c).

FIGURE 9.4 (a) Aposymbiotic budding scyphistoma. (b) Symbiotic polyp in beginning stages of strobilation. Tentacles have not fully retracted and brown-green algae cells visible within translucent polyp. (c) Symbiotic polyp in late stages of strobilation before ephyra has fully detached. Rhopalia labeled with white arrows. The 32 radial canals are visible on the subumbrella. (d) View of a single oral arm. Symbiont cells are seen within every oral vesicle and the oral arm as a whole. (e) Light passing through the umbrella, highlighting the muscle fibers and also the canal system within. (f–j) Adult *Cassiopea* photographed in Key Largo, Florida. Multiple color variations and oral appendage distributions seen. Key: H, hypostome; T, tentacles; C, calyx; B, bud; ET, ectodermal tube; S, stem; PD, pedal disc; OA, oral arms; ML, marginal lappets; RG, radial canals; OV, oral vesicles; D, digitata; OAP, oral appendages.

After detachment of the ephyra, the remaining polyp stem will regenerate a new calyx and tentacles and is capable of strobilating once more, and, in fact, head regeneration has been shown to begin before the strobila fully detaches from the polyp (Hofmann et al. 1978).

While adult *C. xamachana* are physically typical jellyfish, they are unique in that the bell rests on the sandy bottom of their habitats, which has given them the name "upside-down jellyfish" (Figure 9.4f–j). The adult can secure itself to a surface by using the concave shape of the exumbrella to create suction and adhere to the substrate. The average size of adults seems to vary based on habitat, although a comprehensive size range has not been created to date. Bigelow (1900) reported bell diameter sizes ranging from 6.5 to 24 cm, but Mayer (1910) reports diameters usually around 150 mm. The umbrella perimeter is composed of 80 marginal lappets with corresponding white markings (Figure 9.4g). *C. xamachana* is characterized by its white circular band on the exumbrella, though the exact pattern of these markings differs between individuals. Additionally, there are typically 16 oval-shaped white spots around the umbrella margin corresponding with the rhopalia (sense organs) (Figure 9.4g). Adult *C. xamachana* have on average 16 rhopalia, but individuals have been recorded with anywhere from 10 to 23 rhopalia (Bigelow 1900). Rhopalia are located on notches along the margin of the umbrella and are marked by a reddish-brown pigment spot (Mayer 1910).

Attached to the bell is the oral disc from which the oral arms sprout. Adults have eight oral arms formed in pairs, which are described as rounded and slender compared to those in other *Cassiopea* species (Figure 9.4f–j). Their length can be greater than the radius of the jellyfish by up to one half. The oral arms have 9 to 15 branches, which are then further branched, giving them a fluffy appearance. Many appendages (oral vesicles) are found at the base of these branches, and they greatly vary in size throughout a jellyfish (Figure 9.4f–j) (Bigelow 1900). The oral arms are also covered with paddle-shaped oral appendages, which are often highly pigmented (Figure 9.4f–g, i–j). While *C. xamachana* have reported color morphs of brown and green (Figure 9.4f–j), the morph of deep blue is the most well known and studied. The blue pigment, Cassio Blue, is found in both the oral appendages and diffused within the mesoglea (Blanquet and Phelan 1987). The green and brown morphs have not yet been studied or their pigments characterized, though adult color pattern has been found independent of symbiont species (Lampert et al. 2012).

Brachial canals attach to each pair of arms and converge within the oral disc to empty into the stomach. The stomach contains 32 radial grooves connected by a network of anastomosing branches (Figure 9.4e) (Bigelow 1900). The stomach is surrounded by four subgenital pits and four genital sacs, which are accessible from the outside via four subgenital ducts (Mayer 1910). Adults exhibit sexual dimorphism. Females have visually distinctive brooding appendages, seen as a white cluster of appendages in the center of the oral disc (for more information, see Section 9.4) (Figure 9.4h). The mesoglea makes up most of the body and contains symbiotic cells, which have highest density in the umbrella. An endodermic layer separates the subumbrellar and exumbrellar mesoglea (Bigelow 1900). Muscle fibers cover the subumbrella, and muscle activity has been connected with rhopalia signaling and activity (Mayer 1910). Adults have mostly epitheliomuscular cells with muscle fibers in sheets folded into the mesoglea (Blanquet and Riordan 1981).

Scyphozoan cnidocysts fall into three different categories: isorhizas, anisorhizas and rhopaloids. *C. xamachana* have three different types of cnidocysts, though the presence and abundance differ based on life stage. Additionally, the names of two of these cnidocysts have been reported differently in literature, and we will list both names for comprehension. Heterotrichous microbasic euryteles (Jensch and Hofmann 1997), or rhopaloids (Ames et al. 2020), are present in the both the ectoderm and endoderm of all life stages. Holotrichous α-isorhizas are also found in both the ectoderm and endoderm of the polyp and adult but have not been detected in all parts of the scyphistoma body. Finally, heterotrichous anisorhizas (Jensch and Hofmann 1997), or O-isorhizas (Ames et al. 2020), are only detected in the polyp after strobilation has begun. All three cnidocyst types are found in the adult within the ectoderm, and no cnidocysts are located within the mesoglea of any part of the life cycle. Oral vesicles and adjacent tentacle-like structures called digitata contain clusters of cnidocysts in the ectoderm (Figure 9.4d) (Jensch and Hofmann 1997). These digitata immobilize prey when the natural pulsations of the umbrella pump surrounding water against the oral arms. Additionally, *C. xamachana* ephyrae and adults release large amounts of cnidocyst-containing mucus into the surrounding water upon agitation, a response associated with defense and predation. The undeployed cnidocysts inside this mucus are termed cassiosomes and, unlike the oral arms of the adult, only contain the heterotrichous anisorhiza/O-isorhiza cnidocysts. These cnidocysts line the cassiosome periphery interspaced with ectoderm cells containing cilia, allowing temporary mobility of the unit. The interior space of a cassiosome is mostly empty but uniquely contains symbiont cells. A cassiosome ranges from 100 to 550 μm in diameter (Ames et al. 2020). *C. xamachana* had been reported as both venomous and nonvenomous in different habitats, and potency has been related to venom composition, as the cnidocyst composition is identical between these varieties. *C. xamachana* stings are described as relatively mild to humans but are capable of hemolytic, proteolytic, cardiotoxic and dermonecrotic effects (Radwan et al. 2001).

9.6 GENOMIC DATA

With renewed interest in establishing *C. xamachana* as a model to study cnidarian–dinoflagellate symbiosis, efforts have been put forth to compile genomic and transcriptomic data. The first *C. xamachana* transcriptomic dataset became publicly available in 2018, and the first *Cassiopea* genome (T1-A clonal line) was published in 2019 (Kayal et al. 2018;

Ohdera et al. 2019). The T1-A line is available from the labs of the authors in this chapter. The initial draft genome of *C. xamachana* was composed entirely of Illumina short-read data, resulting in a fragmented assembly (N50 = 15,563 Kb) compared to the recently published scyphozoan genomes employing third-generation sequencing technology (Gold et al. 2019; Khalturin et al. 2019; Kim et al. 2019; Li et al. 2020). An updated assembly is now available at the US Department of Energy's Joint Genome Institute (JGI)'s web portal, with significant improvements across all assembly statistics (N50 = 17.8 Mb) (https://mycocosm.jgi.doe.gov/Casxa1). We will continue efforts to improve the assembly and make updates available on the portal. *C. xamachana* remains the only non-anthozoan cnidarian genome available that establishes a stable symbiosis with Symbiodiniaceae, making it a highly attractive model to study the evolution and genetics of symbiosis. In addition to future resources that will become available, past studies have already begun to utilize and illuminate the genetics underlying *Cassiopea*.

In silico prediction of the genome size of *C. xamachana* suggests roughly 360 Mb, consistent with previous measurements of genome sizes for *C. ornata* and *Cassiopea* sp. (Mirsky and Ris 1951; Adachi et al. 2017; Ohdera et al. 2019). A marginally larger assembly of 393.5 Mb was obtained, in line with previous predictions. These values suggest the genus to have genome sizes comparable to other members of the order Rhizostomeae (Kim et al. 2019; Li et al. 2020), but two-fold smaller than the predicted genome size of *Aurelia* sp1. (Adachi et al. 2017; Gold et al. 2019; Khalturin et al. 2019). A genome size greater than 500 Mb appears to be the exception given the average genome sizes for the two additional *Aurelia* species sequenced, which may suggest genome size to be relatively constant within the class. Approximately 31,459 protein-coding genes have been predicted from the *C. xamachana* draft genome, similar to the currently available *Aurelia* genomes. This is in contrast to its close relatives *Nemopilema nomurai* and *Rhopilema esculentum*, which were predicted to contain 18,962 and 17,219 protein coding genes, respectively (Kim et al. 2019; Li et al. 2020). It remains to be seen whether the ancestor of the suborder Dactyliophorae experienced gene loss or a gene expansion occurred after the split of Kolpophorae.

The gene content and its similarity to bilaterians have prompted researchers to investigate the evolution of genomic organization (Hui et al. 2008; Schierwater and Kuhn 1998; Gauchat et al. 2000; Garcia-Fernàndez 2005). Cnidarians occupy a unique position as sister group to bilaterians. Early investigations into genomic architecture suggested high conservation of protein coding gene between cnidarians and humans despite the large divergence time (Schierwater and Kuhn 1998). A recent analysis of medusozoan genomes showed genetic divergence between major cnidarian lineages to be equivalent to that found in bilaterians (Khalturin et al. 2019). Humans share a remarkable number of genes with jellyfish, offering an opportunity to study the evolution of pre-bilaterian genomic architecture and gene conservation. Ohdera et al. (2018) found nearly 5,000 orthologous gene groups (orthogroups) between cnidarians and humans. *C. xamachana* in particular shared 444 unique orthogroups with humans, far more than other cnidarian classes. Similar findings were reported for the moon jelly *Aurelia aurita*, where a high degree of macrosyntenic linkage with humans was found relative to the anemone *Nematostella vectensis* (Khalturin et al. 2019), suggesting a greater genomic conservation since the cnidarian-bilaterian split. Cnidarians have thus played a crucial role in helping us understand gene family evolution and expansion in metazoans (e.g. *Hox* genes).

In cnidarians, *Hox* genes were first recovered from three species of the class Hydrozoa (Schummer et al. 1992), but *Cassiopea* was the first scyphozoan in which *Hox* genes were identified (Kuhn et al. 1999). Initial investigations explored how *Hox* genes may regulate morphological patterning considering the relatively simple body plan. *Hox* gene expression defines the anterior–posterior axis in Bilateria, and similar regulatory roles have been identified for cnidarian *Hox* genes (DuBuc et al. 2018; He et al. 2018). As with other cnidarian lineages, *Cassiopea* maintains a similar repertoire of homeobox genes (Table 9.1). The first homeobox gene identified within Scyphozoa was the *Scox1–5* of *Cassiopea* (Kuhn et al. 1999), which were grouped within two major cnidarian homeobox groups (*Cnox1, Cnox2*). While *Cnox2* has since been classified as a parahox gene, all five *Cnox* groups show highest homology to the bilaterian *Antp* class of homeobox genes. Moreover, *hox* gene orientation within clusters is not expressed as such, similar to that seen in bilaterians. In fact, *hox* expression is not conserved even between cnidarians. It remains to be seen how *homeobox* genes are involved in strobilation and body polarity. With the improvement in genome quality, investigations of genomic synteny will likely address the questions regarding genomic architecture of the ancestral genome prior to the cnidarian–bilaterian split. Previously, a syntenic linkage between a *POU* and *Hox* gene was thought to have been a pre-bilaterian ancestral feature, as it was found in both vertebrates and the hydrozoan *Eleutheria* (Kamm and Schierwater 2007). The availability of new medusozoan genomes, including *Cassiopea*, revealed the linkage may have arisen independently in the medusozoan and vertebrate ancestors (Ohdera et al. 2019).

Another aspect of cnidarian biology that has intrigued biologists is the capacity of *Cassiopea* to regenerate as well as the lack of senescence. While research has focused largely on *Hydra* and corals, chromosome specific telomere length was first investigated in *Cassiopea* (Ojimi and Hidaka 2010). *Cassiopea* exhibits unequal telomere length depending on life stage, with the bell margin of adult medusae having the longest telomeres (2,000 bp) compared to other tissue types (~1,200 bp). This is despite telomerase activity remaining relatively similar across multiple life-stages (Ojimi et al. 2009). Ojimi et al. (2010) also found the *Cassiopea* telomeres to resemble the vertebrate sequence (TTAGGG), in agreement with members of other cnidarian classes, suggesting the vertebrate telomere sequence to be ancestral at the cnidarian–bilaterian split (Grant et al. 2003).

TABLE 9.1
Repertoire of Homeobox Genes in Cnidaria

	Anthozoa		Cubozoa	Scyphozoa						Hydrozoa	
	Exaiptasia diaphana	*Nematostella vectensis*	*Morbakka virulenta*	*Aurelia sp. 1*	*Aurelia aurita*	*Chrysaora quinquecirrha*	*Cassiopea xamachana*	*Nemopilema nomural*	*Rhopilema esculentum*	*Hydra vulgaris*	*Clytia hemisphaerica*
ANTP	62	78	33	35	33	22	32	38	31	17	28
CERS	1	1	0	0	0	0	0	0	0	0	0
HNF	0	1	0	0	0	0	0	0	0	0	0
LIM	6	6	5	3	5	3	4	5	5	5	5
POU	5	6	4	5	3	3	4	3	4	3	4
PRD	36	44	25	30	29	22	28	29	20	18	17
SINE	4	6	5	4	5	5	4	6	5	2	4
TALE	8	5	4	5	7	3	3	7	5	6	5
OTHER	1	4	1	0	0	0	1	0	1	0	0
TOTAL	123	151	77	82	82	58	76	88	71	51	63

Note: Homeobox genes were classified according to the classification outlined by Zhong and Holland (2011), following the method outline by Gold et al. (2019). Protein models from each genome were initially blasted against the curated dataset used by Gold et al. (2019), combined with previously identified cnidarian hox genes from *C. xamachana* and *Aurelia* sp1. Matching hits were further assessed using Interpro (https://github.com/ebi-pf-team/interproscan) to confirm the presence of the homeodomain. Genes were further classified using homeoDB (http://homeodb.zoo.ox.ac.uk/) to generate the final counts.

As previously mentioned, species within the order Rhizostomeae are characterized by the blue pigment Cassio Blue. First isolated in *Cassiopea* and subsequently described in *Rhizostoma*, Cassio Blue likely plays a photoprotective role (Blanquet and Phelan 1987; Bulina et al. 2004). Researchers also found this chromoprotein to exhibit promiscuous metal binding properties but, strikingly, to contain domains for *Frizzled* and *Kringle*, genes involved in *wnt* signaling (Bulina et al. 2004; Phelan et al. 2006). While the function of the chromoprotein beyond its photoprotective role is unknown, the presence of the *wnt* domains has led to speculation of the protein's additional roles. Given the overlap in protein deposition and symbiont localization, Cassio Blue may be involved in regulation of symbiont density, though this remains to be examined.

The *C. xamachana* mitochondrial genome was sequenced in 2012 (Kayal et al. 2012). The *Cassiopea* mitochondrial genome is linear and approximately 17,000 kb in length (Bridge et al. 1992), with 17 conserved genes and two tRNAs and an intact gene order relative to other medusozoan mitochondrial genomes. Medusozoan mtDNA appears to be streamlined, with short intergenic regions. Scyphozoans including *Cassiopea* are characterized by a ~90 bp intergenic region capable of forming a conserved stem loop motif potentially involved in transcriptional regulation and replication. Scyphozoan mtDNAs are also characterized by the presence of a *pol-B* and *ORF314* gene at the chromosome end, a likely signature of an ancient integration of a linear plasmid and consequent linearization of the chromosome. *ORF314* may be a terminal protein involved in maintaining mtDNA integrity by binding to the short, inverted terminal repeats at the end of the mtDNA. In addition to gene organization, the COX1 gene has revealed high genetic divergence to exist within the genus. For example, a mean pairwise divergence of 20.3% was calculated for the two likely invasive species present in Hawaii. This is remarkable considering the morphological similarity between species.

Despite a significant increase in the number of available medusozoan genomes over the past several years. *C. xamachana* offers a unique position as the sole symbiotic species with a genome currently available. Researchers now have the opportunity to investigate the genetic basis of symbiosis by having access to genomes of different cnidarian lineages exhibiting photosymbiosis with different Symbiodinaceae taxa such as the scyphozoan *C. xamachana* (Ohdera et al. 2019), the sea anemone *Exaptasia diaphana* (Baumgarten et al. 2015), the octocoral *Xenia* sp. (Hu et al. 2020) and a growing number of scleractinian corals (e.g. Shinzato et al. 2011; Fuller et al. 2020; Cunning et al. 2018; Shumaker et al. 2018). While the underlying mechanism is yet unclear, the availability of the *C. xamachana* genome will provide an opportunity to study the convergent evolution of symbiosis within Cnidaria and whether *cis-* and *trans*-regulatory mechanisms underlie the evolution of symbiosis within the cnidarian lineage.

9.7 FUNCTIONAL APPROACHES: TOOLS FOR MOLECULAR AND CELLULAR ANALYSES

9.7.1 Toward a Genetic Model to Study Cnidarian Symbiosis

Genetically accessible model organisms have been crucial tools for biologists to understand the molecular underpinnings of life as we know it. Great strides have been made in the past century using genetic model systems to study gene function in other invertebrates, but some systems have not been empowered by these methods. The symbiosis between corals and their photosynthetic endosymbionts is the basis of coral reef ecosystems throughout the world, but the absence of genetic tools in a laboratory model system for the investigation of symbiotic cnidarians has prevented a mechanistic understanding of this symbiosis.

Selection of an appropriate laboratory genetic model system is critical for the implementation of genetic tools (Matthews and Vosshall 2020). Successful systems are marked by key features, namely 1) the capacity to close the life cycle in the laboratory, 2) efficient methods for mutagenesis and transgenesis and 3) germline transmission of mutations/transgenes. Reef-building corals generally spawn once annually, with development to sexual maturity requiring multiple years. Infrequent spawning and long generation time impose extreme limitations on hard coral systems for rapid progress in genetics. The anemone *Exaiptasia diaphana* has been a useful model for cell biology and physiology, but the inability to close the life cycle makes this organism, at present, an intractable system for comprehensive molecular genetic analysis (Jones et al. 2018).

C. xamachana is an apt genetic model system for the study of symbiotic cnidarians. Like reef-building corals, *Cassiopea* engage in a nutritional endosymbiosis with Symbiodiniaceae and are susceptible to thermal bleaching. However, this organism has multiple characteristics which make it an attractive laboratory system. *Cassiopea* spawns daily in aquaria (see Section 9.4), providing regular access to single-cell embryos that are necessary to genetically manipulate the organism using microinjection or electroporation (Figure 9.5a–b). The life cycle of this organism has been closed in the laboratory. Development from embryo to polyp (Figure 9.5c–f) and the subsequent formation of ephyrae spans approximately two months. Medusae require additional time to reach sexual maturity, leading to a generation time of fewer than six months. Additionally, polyps can be maintained as immortal lines in the lab, producing buds at rates associated with how much they are fed. Infected scyphistomae can also live forever under constant culture conditions, though in the field, they will be affected by seasonal conditions (e.g. in the Florida Keys, they disappear in the winter months). Medusae require additional time to reach sexual maturity, leading to a generation time of fewer than six months. Given these qualities, *Cassiopea* provides a practical and relevant model system for a more expedient genetic analysis than in corals. Here we provide some pragmatic information for those interested in using *Cassiopea* as a laboratory model.

9.7.2 Establishing a Lab Colony from Wild Collection

The ability to maintain a breeding *C. xamachana* colonies in relatively simple aquaria is a strength of this model system for cnidarian symbiosis. Reproductive adults can be readily collected from their nearshore natural habitats by snorkeling or wading in the shallow waters they inhabit. In the state of Florida, USA, *C. xamachana* can be collected under a recreational saltwater fishing license. For the purposes of lab-based spawning, medusae from 10–15 cm in bell diameter are appropriate for long-term culture in aquaria. Males and females can be readily identified via externally visible morphological characteristics, namely the presence of central brooding appendages on females (Hofmann and Hadfield 2002). While larger individuals can be kept, their higher biomass and food requirements make them less conducive to sustained culture in closed systems. Medusae can be shipped overnight and fare well when packaged inside of individual poly bags, approximately half filled with water to allow for airspace for gas exchange, shipped inside of an insulated foam box to stabilize temperature during the journey.

9.7.3 Culturing *Cassiopea* in the Lab

A stable, purpose-built aquarium system greatly facilitates the maintenance of a spawning *C. xamachana* colony. Overall, these organisms fare well with high levels of light (250–400 μE m^{-2} s^{-1}), frequent and heavy feeding (freshly hatched *Artemia* sp. Nauplii, which can be supplemented with rotifers) and low water flow. A shallow tank with a plumbed sump functions well as a foundation for a colony, with a few considerations of our organism. While relatively robust, *C. xamachana* will readily be pulled into overflows as well as powerheads and other circulation pumps. Long, shallow tanks of 15–30 cm depth provide convenient access and reduce crowding. No powerheads, pumps or other equipment should be located directly in the tank. The overflow which brings water from the tank to the sump via gravity should be covered with a protective grate constructed from polystyrene egg crate lighting diffuser. In the sump, water first passes through a filter sock or floss, which should be washed/exchanged at least every other day. The sump also contains live rock or other media to serve as biological filtration, as well as an efficient and appropriately sized protein skimmer which both removes waste and facilitates gas exchange. A temperature of 25–26°C is maintained with an aquarium heater located in the sump. As aquarium heaters are notoriously unreliable and failure in the on position may result in severe impacts to the colony, the heater should be backed up by a secondary temperature controller. Activated carbon is also located in the sump in order to remove organics that reduce water clarity; this should be kept in a filter bag or nylons and changed monthly; approximately 60 mL per 100 liters of water in the system is sufficient. The return pump delivers water back to the aquarium. This should be relatively low flow so as not to unnecessarily disturb the medusae in the main tank; approximate turnover of one to three times the volume of the aquarium is sufficient. Diffusing the water returning to the tank will also prevent the disturbance of the medusae (Widmer 2008).

Heavy feeding of freshly hatched live *Artemia* sp. one to three times daily facilitates continued, regular spawning. Though *C. xamachana* are not particularly demanding of water quality, attention to water parameters will promote the

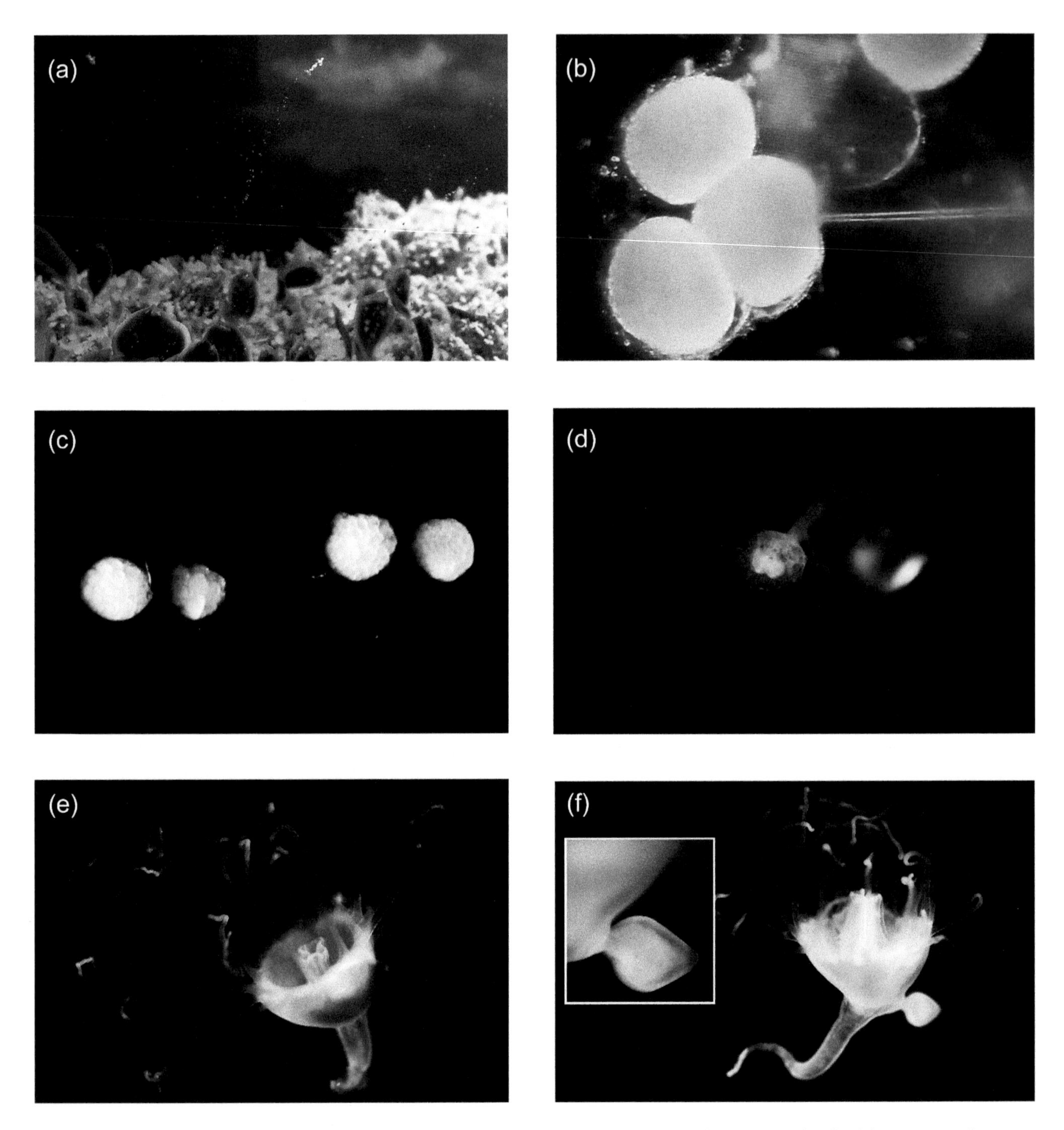

FIGURE 9.5 Spawning, injection and settlement of *Cassiopea*. (a) Daily spawning of *Cassiopea* in the laboratory environment. (b) Injection of Cas9-RNPs into single-cell embryos, with visualization aided by phenol red tracer dye. (c) Development of injected embryos, ten hours after injection. (d) Metamorphosis and settlement of injected *Cassiopea* embryo into a small polyp, ten days following injection. (e) Growth of an injected embryo into a polyp, 30 days after injection. (f) Development of asexual planuloid buds on a polyp (see inset for detail) 45 days following injection.

longevity of the culture and consistent spawning. Artificial seawater should be mixed using 0 TDS RO/DI water to a salinity of 34–36 PSU. Weekly water changes of 20% are helpful in long-term maintenance and stability. Nitrate and phosphate levels should be monitored weekly; low or high levels can be problematic. As a guideline, nitrate levels of 2–10 ppm and phosphate levels of 0.03 to 0.10 ppm have provided for consistent maintenance and spawning of broodstock. Excess nutrients can be managed by increasing the volume of water changes and implementing an algal refugium (e.g. *Chaetomorpha*) in the sump. Insufficient nutrients in the water can be ameliorated by increasing feeding, reducing skimming or with the careful dosing of sodium nitrate or sodium phosphate solutions to achieve desired levels.

As photosymbiotic organisms with spawning controlled by the daily light cycle, appropriate lighting is a critical component of *Cassiopea* husbandry. Lighting solutions designed for reef-building corals are appropriate for these shallow-water animals that require high levels of photosynthetically available radiation (PAR) to maximize the nutritional benefits from their endosymbionts. Modern high-output LEDs designed for reef tanks can be implemented to blanket the bottom of the tank with PAR levels of 250–400 μE on a 12:12 daily cycle. Light levels should be assessed with a submersible PAR meter and lighting adjusted as appropriate.

9.7.4 Microinjection of Single-Cell Embryos for the Generations of Mutants and Transgenic *Cassiopea*

The study of symbiosis in cnidarians has long sought to identify the mechanistic basis of the interactions between the animal host and intracellular algal partner. Studies comparing symbiotic and aposymbiotic hosts have been performed in numerous cnidarian taxa (Lehnert et al. 2014; Rodriguez-Lanetty et al. 2006), as well as numerous studies examining the response to heat stress and the breakdown of symbiosis (Pinzón et al. 2015; DeSalvo et al. 2010) and gene expression patterns associated with thermal tolerance (Bellantuono et al. 2012; Barshis et al. 2013). This broad body of work has resulted in the identification of numerous genes of interest, including molecular chaperones and antioxidant enzymes associated with the response to thermal stress (Császár et al. 2009; Fang et al. 1997), as well as lectins which may mediate the relationship between the host and symbiont (Kvennefors et al. 2008). However, the field has largely been missing crucial tools of genetics to robustly test these hypotheses. Microinjection of *C. xamachana* embryos opens a path to understand the molecular genetic basis of symbiosis, chemosensation and sleep in an early diverging metazoan with a decentralized nervous system (Figure 9.5b).

A basic tool of genetics is the capacity to perform loss-of-function studies such as gene knockout experiments. With the development of genome editing techniques, *C. xamachana* is an apt model system to test hypotheses of cnidarian symbiosis. Using microinjection, *C. xamachana* embryos are amenable to CRISPR-mediated mutagenesis, a technology which allows for precise, targeted mutagenesis and transgenesis using a programmable nuclease comprised of a guide RNA and the protein Cas9 (Jinek et al. 2012). CRISPR-Cas9 can be used by delivering the Cas9 protein complexed with single guide RNAs (sgRNA) which direct the nuclease to the locus of interest in the nucleus of a living cell. This Cas9-sgRNA complex cleaves the targeted DNA, resulting in endogenous DNA repair. In the absence of homologous template, non-homologous end joining (NHEJ) repair occurs (Doudna and Charpentier 2014). By injecting a Cas9-sgRNA complex into single cell embryos, mutants are generated with small insertions or deletions (indels) induced by the imperfect DNA repair mechanisms of the cell. These indels often result in frameshift mutations of the target gene, generating loss-of-function alleles. These mutagenized embryos can then be reared to polyps and induced to strobilate by exposure to an algal symbiont, generating medusae that can be used for subsequent crosses once sexually mature. As the life cycle of *Cassiopea* can be completed in the lab within four to six months, the crosses necessary to generate a homozygous mutant can be completed within 18 months. Work to establish this technology in *Cassiopea* is ongoing.

In addition to using CRISPR to generate loss-of-function alleles, this technology can also be implemented to perform gene knock-in. By providing donor DNA consisting of a transgene flanked by sequence homologous to the both sides of the cut site, CRISPR can be used to engineer knock-in at a specific locus (Barrangou and Doudna 2016). This will allow the generation of diverse molecular tools for *Cassiopea* for the study of cnidarian symbiosis, development and neuroscience in this unique model system with the future implementation of genetically encoded calcium indicators (GECIs) such as GCaMP (Nakai et al. 2001) for the real-time fluorescent readout of nervous system activity, as well as genetically encoded fluorescent redox sensors (Lukyanov and Belousov 2014) to test longstanding hypotheses regarding the role of ROS stress in cnidarian bleaching. *Cassiopea* are transparent and lack endogenous host autofluorescence, making them well suited to molecular imaging.

Spawning is timed by the daily light cycle, occurring five to six hours after artificial sunrise in aquaria. In order to collect unicellular embryos, clear selected spawning female medusae of previously extruded, multicellular embryos approximately two hours prior to spawning using a baster. Selected female medusae can then be placed in shallow black polycarbonate pans under a light source to improve the visibility of embryos at the time of release. Once released, the 80-μm embryos can be collected with a transfer pipette into small glass dishes, taking care to avoid mucus. Prior to injection, unicellular embryos are transferred and aligned in polystyrene culture dishes containing 40 PSU seawater. The increased salinity results in a slight reduction of cell volume due to osmosis and allows the cell to accommodate the volume of the injected liquid payload. Transfer and positioning of embryos is performed using an aspirator constructed from a 1-mm glass capillary fitted with a length of 1-mm ID silicone tubing. Embryos readily adhere to new, virgin polystyrene and can be arranged in a row for efficient microinjection. Dishes with tight-fitting lids are best employed to reduce evaporation, as the injection dish also houses embryos during development to planulae.

Typical injection payloads include Cas9-sgRNA ribonucleoprotein injection mixture, composed of a guide RNA complexed with Cas9 protein (with NLS), injection buffer and phenol red dye microinjected into single-cell *Cassiopea* embryos (Figure 9.5b–c). Custom needles are prepared with thin-walled 1-mm aluminosilicate glass capillaries on a P-1000 horizontal pipette puller (Sutter Instrument, CA, USA) and beveled on BV-10 micropipette beveler (Sutter) to 17°. Microinjection is performed using a Xenoworks digital injector and manipulator system (Sutter Instrument, CA, USA) under a SteREO Discovery V8 microscope (Zeiss, Germany). Current injection methods yield survival rates of up to 40%. In the three to six hours following injection, each embryo is examined to assess whether it has survived and entered the cleavage stage. Non-dividing embryos are culled and removed, and the water in the dish is carefully replaced with filtered 34 PSU artificial seawater. Planulation of viable embryos occurs approximately one week following injection, with a developmental delay often observed in comparison to uninjected embryos. *Cassiopea* larvae readily settle and metamorphose in response to a number of cues, including bacteria, degrading mangrove leaves and the previously mentioned endogenous metamorphosis-inducing peptide (Neumann 1979; Fleck and Fitt 1999; Thieme and Hofmann 2003). We have found that settlement dishes can easily be prepared by using a cotton swab to transfer biofilm from the sump of an established *Cassiopea* tank to polystyrene dishes and then covering with seawater and incubating at room temperature for three to five days. Prior to transferring planulae to settlement dishes, water should be exchanged with filtered 34 PSU artificial seawater. Planulae should be monitored regularly; once settlement occurs and nascent scyphistomae have developed tentacles, regular feeding of freshly hatched *Artemia* nauplii should begin. Daily feeding is optimal. The survival of recent settlers can be enhanced by placing a nauplius on the hypostome with forceps. In order to maintain polyps in an aposymbiotic state and prevent strobilation, polyps can be maintained in 10 µm DCMU without apparent detriment. In order to generate medusae, mature polyps can be challenged with symbionts to induce strobilation. Once released from the polyp, the ephyra will develop into a medusa. Growth is facilitated with ample feeding of *Artemia* (at least daily) and high artificial light levels (250–400 µE) on a 12:12 cycle or natural light. With regular water changes, medusae can be cultured in 1-liter beakers or polycarbonate pans to bell diameters of at least 5 cm. The generation of sexually mature medusae takes several months. Work is in progress to develop the most efficient methods to cross medusae.

9.8 CHALLENGING QUESTIONS

While a lot of emphasis has been placed on understanding the origins of the first metazoan body plans, less is known about how those early animals interacted with their surrounding microbial seas. The establishment of holobiont communities (i.e. a multicellular host and its associated microbiome) required the evolution of novel interkingdom communication. As metazoan life cycles evolved, their associated microbial communities diversified with them (McFall-Ngai et al. 2013). The study of host-microbe associations throughout an organism's life cycle is now feasible (Gilbert et al. 2015; Gilbert 2016). There is a growing interest in ontogenetic microbiomes (i.e. microbial associates over a host developmental time course) (Fieth et al. 2016; Carrier and Reitzel 2018; Vijayan et al. 2019) and how they can affect developmental phenotypes (Tran and Hadfield 2011; Thompson et al. 2015; Fieth et al. 2016; Shikuma et al. 2016; Carrier and Reitzel 2018). While a few microbes have been shown to induce larval settlement in *C. xamachana*, such as *Vibrio* spp. (Neumann 1979; Hofmann and Brand 1987) and *Pseudoalteromonas* sp. (Ohdera et al. *in prep* a), it is likely that the complex microbiomes in settlement substrates as well as developmental microbiomes acquired by the organism through ontogeny will also play critical roles in driving phenotypic and physiological traits as *C. xamachana* goes through its life cycle (Medina lab, unpublished). Our ability to infect with different Symbiodiniaceae that will in turn harbor different microbiomes as well as potentially developing axenic and gnotobiotic animals will also open doors to understand host–microbiome interactions at the developmental level (Medina lab, unpublished).

Many cnidarian taxa establish endosymbioses with Symbiodiniaceae, and this symbiosis is crucial in the maintenance of coral reef ecosystems (LaJeunesse 2020). Scleractinian corals usually establish their photosymbiosis during the larval stage (Schwarz et al. 1999; Abrego et al. 2009; Voolstra et al. 2009; McIlroy and Coffroth 2017). Mounting evidence now supports the role of Symbiodiniaceae (LaJeunesse et al. 2018) in the onset of host development (Mohamed et al. 2016; Reich et al. 2017). Coral larval manipulation experiments are challenging given the limited availability of larvae due to annual spawning events (Harrison et al. 1984; Szmant 1986; Van Woesik et al. 2006). Although the pelago-benthic transition from larva to settled polyp is partially linked to onset of photosymbiosis (Mohamed et al. 2016; Reich et al. 2017), discerning the role of photosymbionts as drivers of this developmental transition has not been clearly elucidated (Hartmann et al. 2019). *Cassiopea* therefore represents an efficient model system to study developmental symbioses.

We believe that *C. xamachana* can become an ideal system to study environmental canalization (Waddington 1942) because of the clear and easily manipulated developmental switch (i.e. onset of photosymbiosis) that we can also obviate with artificial inducers. We can alter the phenotypic outcome of strobilation by using different photosymbionts in comparative infection experiments. Once the polyp stage is infected, it can take different developmental trajectories that lead to divergent morphospaces between homologous and heterologous photosymbiotic infections (Figure 9.6). These different developmental phenotypes also likely have diverging underlying molecular regulatory mechanisms. Robert Trench had indeed already proposed that this type of photosymbiosis would be ideal for the study of cross-genome regulation (Trench 1979). In support of this idea, we have uncovered a possible role of *S. microadriaticum* photosynthetic pigments in the regulation of *C. xamachana* strobilation (Ohdera et al. in prep b).

Both the host (*C. xamachana*) (Ohdera et al. 2019) and the homologous photosymbiont (*S. microadriaticum*) (Aranda et al. 2016) are now genome enabled, facilitating any downstream molecular analysis. Establishing

FIGURE 9.6 Symbiosis-driven development in *C. xamachana*. The small white circle represents the zygote stage that follows different developmental trajectories. Strobilation can lead to different phenotypic outcomes (i.e. symbiotic *vs*. aposymbiotic strobila) driven by photosymbiosis vs environmental and/or chemical cues. The symbiotic route is the one that occurs primarily in nature. The aposymbiotic route can be lab induced and is probably environmentally induced as well. The underlying genetic network is therefore dynamic and slightly modified depending on the trigger of strobilation.

laboratory lines of both host and photosymbionts has been straightforward, and we can complete the *C. xamachana* life cycle in the lab in which aposymbiotic asexual polyps (scyphistomae) metamorphose (strobilation) into sexual medusae (ephyrae) due to onset of photosymbiosis (Figure 9.2). Cell-type specific genes have not yet been identified in *C. xamachana*; however, single-cell transcriptomics has already been successfully used for the study of other cnidarian symbiosis (Hu et al. 2020) and can therefore readily be implemented in the upside-down jellyfish. We can now also chemically induce strobilation (Cabrales-Arellano et al. 2017), providing a suitable control for the study of photosymbiosis-driven development. In addition to onset of developmental symbiosis, we are able to perform timely thermal stress (disruption of symbiosis) experiments that can shed light on the mechanism of cnidarian bleaching affecting coral reefs worldwide due to climate change (Newkirk et al. 2020).

The nervous system is a key driver of animal responses to environmental changes; *Cassiopea* and other cnidarians are likely to be no exception. The roles of circadian rhythm and sleep in a photosymbiotic animal have only begun to be characterized. *C. xamachana* is the earliest branching metazoan to exhibit sleep (Nath et al. 2017) that coincidentally is also symbiotic. Thus, of particular interest is host cellular responses to photosynthetic products from the algal symbiont (Ohdera et al. in prep b). In addition, the sensory biology of cnidarians is poorly understood. How the animal may sense heat or chemical stressors may have an impact on the maintenance of symbiosis.

Regeneration has been reported in *C. xamachana* since the turn of the 20th century (Mayer 1908; Stockard 1910; Cary 1916; Curtis and Cowden 1974; Gamero et al. 2019), but the environmental and molecular drivers of regeneration have not been tackled in this organism. Thus, it is not well known how regeneration progresses and how to successfully induce it in lab. It is still unknown whether *C. xamachana* has stem cells and, if so, what type and where they are generated. Metazoan regeneration (Li et al. 2015; Tiozzo and Copley 2015) is a burgeoning field thanks to increasingly readily available genomic tools for diverse taxa (e.g. Shao et al. 2020; Medina-Feliciano et al. 2020; Gerhke et al. 2019) and increased awareness of the importance of new relevant model systems (Sanchez-Alvarado 2004). Studies of regeneration in *C. xamachana* can provide a new perspective by being a symbiotic organism as well as basal animal that can shed light in possible shared regenerative traits in the pre-bilaterian ancestor.

As mentioned earlier in the chapter, *C. xamachana* sexual reproduction in the field and lab still needs additional research. We have yet to uncover when and what triggers male sperm release in the wild. Fertilization is internal, and it is unknown what the female attractants are and when exactly it takes place. Uncovering these aspects of sexual reproduction will yield knowledge useful in understanding gamete recognition in marine taxa, possibly understanding if hybrids can form between congeneric species and improving husbandry techniques.

Adult *C. xamachana* phenotypic plasticity in color morphotypes and variation in number and size of lappets (Figure 9.4f–j) becomes more apparent at densely populated sites. The vast variation of color morphotypes deserves investigation to understand whether coloration is inherited or environmentally driven and how much of this variation is linked to the photosymbiosis life style. These chromoproteins can potentially have biotechnological application.

In summary, there are many aspects of cnidarian and photosymbiosis biology that will be better understood with the use of *C. xamachana* as a model system. The growing *Cassiopea* scientific community holds an annual workshop at the Key Largo Marine Research Lab every year where participants can exchange ideas and perform experiments on the readily available *Cassiopea* population. Additional information about the workshop and resources can be found at http://cassiopeabase.org/. We hope this chapter offers enough information for the community to implement the use of *C. xamachana* as a model system in labs around the world.

ACKNOWLEDGMENTS

We thank Justin Wheeler for help with the design of Figure 9.6. Igor Grigoriev and Sajeet Haridas released the JGI *C. xamachana* genome assembly in time for the publication of this chapter. M. Medina was funded by NSF grants OCE 1442206 and OCE 1642311. A.C. Morandini was funded by CNPq 309440/2019–0 and FAPESP 2011/50242–5, 2015/21007–9. We thank Key Largo Marine Research Lab for the incessant hospitality over the years. D. Hoffman also extends his thanks and deep appreciation to Dr. Bill Fitt for decades of joint research on *Cassiopea*.

BIBLIOGRAPHY

Abrego, D., M. J. H. Van Oppen and B. L. Willis. 2009. Onset of algal endosymbiont specificity varies among closely related species of *Acropora* corals during early ontogeny. *Molecular Ecology* 18: 3532–3543.

Adachi, K., H. Miyake, T. Kuramochi, et al. 2017. Genome size distribution in phylum Cnidaria. *Fisheries Science* 83: 107–112.

Agassiz, A. and A. G. Mayer. 1899. Acalephs from the Fiji Islands. 33: 157–189.

Al-jbour, S. M., M. Zimmer and A. Kunzmann. 2017. Cellular respiration, oxygen consumption, and trade-offs of the jellyfish *Cassiopea* sp. in response to temperature change. *Journal of Sea Research* 128: 92–97.

Ames, C. L., A. M. L. Klompen, K. Badhiwala, et al. 2020. Cassiosomes are stinging-cell structures in the mucus of the upside-down jellyfish *Cassiopea xamachana*. *Communications Biology* 3: 67.

Arai, M. N. 2001. Pelagic coelenterates and eutrophication: A review. In: *Hydrobiologia* 451: 69–87.

Arai, Y., R. Gotoh, J. Yokoyama, et al. 2017. Phylogenetic relationships and morphological variations of upside-down jellyfishes, *Cassiopea* spp. inhabiting Palau Islands. *Biogeography* 10: 133–141.

Aranda, M., Y. Li, Y. J. Liew, et al. 2016. Genomes of coral dinoflagellate symbionts highlight evolutionary adaptations conducive to a symbiotic lifestyle. *Scientific Reports* 6: 39734.

Baker, H. G. 1955. Self-compatibility and establishment after "long-distance" dispersal. *Evolution* 9: 347–349.

Baker, H. G. 1965. Characteristics and modes of origin of weeds, In *The genetics of colonizing species*.

Balderston, L. and G. Claus 1969. Study of symbiotic relationship between *Symbiodinium microadriaticum* Freudenthal, zooxanthella and upside down jellyfish. *Nova Hedwigia* 17: 373–382.

Barrangou, R. and J. A. Doudna. 2016. Applications of CRISPR technologies in research and beyond. *Nature Biotechnology* 34: 933–941.

Barshis, D. J., J. T. Ladner, T. A. Oliver, et al. 2013. Genomic basis for coral resilience to climate change. *Proceedings of the National Academy of Sciences of the United States of America* 110: 1387–1392.

Baumgarten, S., O. Simakov, L. Y. Esherick, et al. 2015. The genome of *Aiptasia*, a sea anemone model for coral symbiosis. *Proceedings of the National Academy of Sciences* 112: 11893–11898.

Bayha, K. M. and W. M. Graham. 2014. Nonindigenous marine jellyfish: invasiveness, invasibility, and impacts. In *Jellyfish blooms*, K. A. Pitt and C. H. Lucas (Eds.), pp. 45–78. Springer.

Bellantuono, A. J., C. Granados-Cifuentes, D. J. Miller, et al. 2012. Coral thermal tolerance: Tuning gene expression to resist thermal stress. *PLoS One* 7, Nr. 11.

Bigelow, R. P. 1892. On a new species of *Cassiopea* from Jamaica. *Zoologischer Anzeiger* 15: 212–214.

Bigelow, R. P. 1893. Some observations on *Polyclonia frondosa*. Baltimore: John Hopkins University.

Bigelow, R. P. 1900. The anatomy and development of *Cassiopea xamachana*. Boston Society of Natural History.

Blanquet, R. S. and M. A. Phelan. 1987. An unusual blue mesogleal protein from the mangrove jellyfish *Cassiopea xamachana*. *Marine Biology* 94: 423–430.

Blanquet, R. S. and G. P. Riordan. 1981. An ultrastructural study of the subumbrellar musculature and desmosomal complexes of *Cassiopea xamachana* (Cnidaria: Scyphozoa). *Transactions of the American Microscopical Society* 100: 109–119.

Brandt, J. F. 1835. *Prodromus Descriptionis Animalium Ab H. Mertensio in Orbis Terrarum Circumnavigatione Observatorum*.

Brandt, J. F. and C. H. Mertens. 1838. Ausführliche Beschreibung der von CH Mertens auf seiner Weltumsegelung beobachteten Schirmquallen: Nebst allgemeinen Bemerkungen über die Schirmquallen überhaupt. 2: 237–411.

Bridge, D., C. W. Cunningham, B. Schierwater et al. 1992. Class-level relationships in the phylum Cnidaria: Evidence from mitochondrial genome structure. *Proceedings of the National Academy of Sciences of the United States of America* 89: 8750–8753.

Bulina, M. E., K. A. Lukyanov, I. V. Yampolsky, et al. 2004. New class of blue animal pigments based on Frizzled and Kringle protein domains. *Journal of Biological Chemistry* 279: 43367–43370.

Cabrales-Arellano, P., T. Islas-Flores, P. E. Thomé, et al. 2017. Indomethacin reproducibly induces metamorphosis in *Cassiopea xamachana* scyphistomae. *PeerJ*, Nr. 3.

Carrier, T. J. and A. M. Reitzel. 2018. Convergent shifts in host-associated microbial communities across environmentally elicited phenotypes. *Nature Communications* 9: 952.

Cary, L. R. 1916. The influence of the marginal sense organs on the rate of regeneration in *Cassiopea xamachana*. *Journal of Experimental Zoology* 21: 1–32.

Cary, L. R. 1917. Studies on the physiology of the nervous system of *Cassiopea xamachana*. *Carnegie Institution of Washington Plub*, Nr 251: 121–170.

Çevik, C., I. T. Erkol and B. Toklu. 2006. A new record of an alien jellyfish from the Levantine coast of Turkey—*Cassiopea andromeda* (Forskål, 1775) [Cnidaria: Scyphozoa: Rhizostomea]. *Aquatic Invasions* 1, Nr. 3.

Colley, N. J. and R. K. Trench. 1983. Selectivity in phagocytosis and persistence of symbiotic algae in the scyphistoma stage of the jellyfish *Cassiopeia xamachana*. *Proceedings of the Royal Society of London. Series B, Containing papers of a Biological character. Royal Society (Great Britain)* 219: 61–82.

Colley, N. J. and R. K. Trench. 1985. Cellular events in the reestablishment of a symbiosis between a marine dinoflagellate and a coelenterate. *Cell and Tissue Research* 239: 93–103.

Császár, N. B. M., F. O. Seneca and M. J. H. Van Oppen. 2009. Variation in antioxidant gene expression in the scleractinian coral *Acropora millepora* under laboratory thermal stress. *Marine Ecology Progress Series* 392: 93–102.

Cunning, R., R. A. Bay, P. Gillette, et al. 2018. Comparative analysis of the *Pocillopora damicornis* genome highlights role of immune system in coral evolution. *Scientific Reports* 8: 16134.

Curtis, S. K. and R. R. Cowden. 1971. Normal and experimentally modified development of buds in *Cassiopea* (phylum Coelenterata; class Scyphozoa). *Acta embryologiae experimentalis* 3: 239–259.

Curtis, S. K. and R. R. Cowden. 1972. Regenerative capacities of the scyphistoma of *Cassiopea* (Phylum, Coelenterata; class, Scyphozoa). *Acta embryologiae experimentalis* 1972: 429–454.

Curtis, S. K. and R. R. Cowden. 1974. Some aspects of regeneration in the scyphistoma of *Cassiopea* (class Scyphozoa) as

revealed by the use of antimetabolites and microspectrophotometry. *American Zoologist* 14: 851–866.
DeSalvo, M. K., S. Sunagawa, P. L. Fisher, et al. 2010. Coral host transcriptomic states are correlated with *Symbiodinium* genotypes. *Molecular Ecology* 19: 1174–1186.
Doty, M. S. 1961. Acanthophora, a possible invader of the marine flora of Hawaii. *Pacific Science* 15: 547–552.
Doudna, J. A. and E. Charpentier. 2014. The new frontier of genome engineering with CRISPR-Cas9. *Science* 346: 1258096.
Dubuc, T. Q., T. B. Stephenson, A. Q. Rock, et al. 2018. Hox and Wnt pattern the primary body axis of an anthozoan cnidarian before gastrulation. *Nature Communications* 9, Nr. 1.
Fang, L. S., S. P. Huang and K. L. Lin. 1997. High temperature induces the synthesis of heat-shock proteins and the elevation of intracellular calcium in the coral Acropora grandis. *Coral Reefs* 16: 127–131.
Fieth, R. A., M. E. A. Gauthier, J. Bayes, et al. 2016. Ontogenetic changes in the bacterial symbiont community of the tropical demosponge *Amphimedon queenslandica*: Metamorphosis is a new beginning. *Frontiers in Marine Science* 3: 228.
Fitt, W. K. 1984. The role of chemosensory behavior of *Symbiodinium microadriaticum*, intermediate hosts, and host behavior in the infection of coelenterates and molluscs with zooxanthellae. *Marine Biology* 81: 9–17.
Fitt, W. K. 1985. Effect of different strains of the zooxanthellae *Symbiodinium microadriaticum* on growth and survival of their coelenterate and molluscan hosts. *Proceedings of the 5th International Coral Reef Congress*, 6: 228.
Fitt, W. K. and R. K. Trench. 1983a. Endocytosis of the symbiotic dinoflagellate *Symbiodinium microadriaticum* Freudenthal by endodermal cells of the scyphistomae of *Cassiopeia xamachana* and resistance of the algae to host digestion. *Journal of Cell Science* 64: 195–212.
Fitt, W. K. and R. K. Trench. 1983b. Infection of coelenterate hosts with the symbiotic dinoflagellate *Symbiodinium microadriaticum*. In: *Intracellular space as oligogenetic ecosystem. Proceedings*, 2: 675–681.
Fitt, W. K. and K. Costley. 1998. The role of temperature in survival of the polyp stage of the tropical rhizostome jellyfish *Cassiopea xamachana*. *Journal of Experimental Marine Biology and Ecology* 222: 79–91.
Fitt, W. K. and D. K. Hofmann. 1985. Chemical induction of settlement and metamorphosis of planulae and buds of the reef-dwelling coelenterate *Cassiopeia andromeda*. *Proceedings of the 5th International Coral Reef Congress*, 5: 239–244, Tahiti.
Fitt, W. K., D. K. Hofmann and M. Rahat. 1987. Requirement of exogenous inducers for metamorphosis of axenic larvae and buds of *Cassiopeia andromeda* (Cnidaria: Scyphozoan). *Marine Biology* 94: 415–422.
Fleck, J. and A. Bischoff. 1993. Protein kinase C is possibly involved in chemical induction of metamorphosis in *Cassiopea* spp. (Cnidaria: Scyphozoa). *Proceedings of the 5th International Coral Reef Congress*, 456–462. Richmond RH (ed) University of Guam Press, KOG Station, Mangilao, Guam, USA.
Fleck, J., W. K. Fitt and M. G. Hahn. 1999. A proline-rich peptide originating from decomposing mangrove leaves is one natural metamorphic cue of the tropical jellyfish *Cassiopea xamachana*. *Marine Ecology Progress Series* 183: 115–124.
Fleck, J. and W. K. Fitt. 1999. Degrading mangrove leaves of Rhizophora mangle Linne provide a natural cue for settlement and metamorphosis of the upside down jellyfish *Cassiopea xamachana*. *Journal of Experimental Marine Biology and Ecology* 234: 83–94.
Fleck, J. and D. K. Hofmann. 1995. In vivo binding of a biologically active oligopeptide in vegetative buds of the scyphozoan *Cassiopea andromeda*: Demonstration of receptor-mediated induction of metamorphosis. *Marine Biology* 122: 447–451.
Fleck, J. and D. K. Hofmann. 1990. The efficiency of metamorphosis inducing oligopeptides in *Cassiopea* species (Cnidaria: Scyphozoa) depends on both primary structure and amino- and carboxy terminal substituents. *Verh Dtsch Zool Ges* 83: 452–453.
Fleck, J. 1998. Chemical fate of a metamorphic inducer in larvae-like buds of the cnidarian *Cassiopea andromeda*. *Biological Bulletin* 194: 83–91.
Forskål, P. 1775. *Desriptiones Animalium Avium, Amphibiorum, Piscium, Insectorum, Vermium; Quae in Intinere Orientali Observavit Petrus Forskål*. Copenhagen: Niebuhr, Carsten.
Forskål, P. and C. Niebuhr. 1776. *Icones rerum naturalium: quas in itinere orientali depingi*. Copenhagen.
France, S. C. and L. L. Hoover. 2002. DNA sequences of the mitochondrial COI gene have low levels of divergence among deep-sea octocorals (Cnidaria: Anthozoa). *Hydrobiologia* 471: 149–155.
Freeman, C. J., E. W. Stoner, C. G. Easson, et al. 2016. Symbiont carbon and nitrogen assimilation in the *Cassiopea-Symbiodinium* mutualism. *Marine Ecology Progress Series* 544: 281–286.
Freudenthal, H. D. 1959. Observations on the algal cells (zooxanthellae) inhabiting the anemone *Cassiopea* sp. *J. Protozool* Nr. 6: 12.
Freudenthal, H. D. 1962. *Symbiodinium* gen. nov. and *Symbiodinium microadriaticum* sp. nov., a Zooxanthella: Taxonomy, life cycle, and morphology. *J. Protozool.* 9: 45–52.
Fuller, Z. L., V. J. L. Mocellin, L. A. Morris, et al. 2020. Population genetics of the coral *Acropora millepora*: Toward genomic prediction of bleaching. *Science* 369: eaba4674.
Gamero-Mora, E., R. Halbauer, V. Bartsch, et al. 2019. Regenerative capacity of the upside-down jellyfish *Cassiopea xamachana*. *Zoological Studies* 58.
Garcia-Fernàndez, J. 2005. Hox, ParaHox, ProtoHox: Facts and guesses. *Hereditary* 94: 145–152.
Gauchat, D., F. Mazet, C. Berney, et al. 2000. Evolution of Antp-class genes and differential expression of *Hydra* Hox/para-Hox genes in anterior patterning. *Proceedings of the National Academy of Sciences of the United States of America* 97: 4493–4498.
Gehrke, A. R., E. Neverett, Y. Luo, et al. 2019. Acoel genome reveals the regulatory landscape of whole-body regeneration. *Science* 363, Nr. 6432.
Gershwin, L. A., W. Zeidler and P. J. F. Davie. 2010. Medusae (Cnidaria) of Moreton Bay, Queensland, Australia. *Memoirs of the Queensland Museum* 54: 47–108.
Gilbert, S. F. 2016. Chapter twenty-two: Developmental plasticity and developmental symbiosis: The return of eco-devo. In *Current topics in developmental biology*, hg. von Paul M. Wassarman, 116: 415–433. Essays on Developmental Biology, Part A. Academic Press, 1.
Gilbert, S. F. and D. Epel. 2015. *Ecological developmental biology: The environmental regulation of development, health, and evolution*. 2nd ed. Sunderland, MA, USA: Sinauer Associates is an imprint of Oxford University Press.
Gohar, H. A. F. and A. M. Eisawy. 1960a. The biology of *Cassiopea andromeda* (from the Red Sea) (with a note on the species

problem). *Publications of the Marine Biological Station, Ghardaqa* 11: 3–39.

Gohar, H. A. F. and M. Eisawy. 1960b. The development of *Cassiopea andromeda* (Scyphomedusae). *Publs Marine Biology Station Ghardaqa* 11: 148–190.

Gold, D. A., T. Katsuki, Y. L., Xifeng, et al. 2019. The genome of the jellyfish *Aurelia* and the evolution of animal complexity. *Nature Ecology and Evolution* 3: 96–104.

Goldfarb, A. J. 1914. Changes in salinity and their effects upon the regeneration of *Cassiopea xamachana. Papers from the Tortugas Laboratory of the Carnegie Institution* 6: 85–94.

Gómez-Daglio, L. and M. N. Dawson. 2017. Species richness of jellyfishes (Scyphozoa: Discomedusae) in the tropical Eastern Pacific: Missed taxa, molecules, and morphology match in a biodiversity hotspot. *Invertebrate Systematics* 31: 635–663.

Grant, A. J., D. A. Trautman, S. Frankland, et al. 2003. A symbiosome membrane is not required for the actions of two host signalling compounds regulating photosynthesis in symbiotic algae isolated from cnidarians. *Comparative Biochemistry and Physiology: A Molecular and Integrative Physiology* 135: 337–345.

Haeckel, E. 1880. System der Acraspeden-Zweite Hälfte des Systems der Medusen. *Denkschriften der Medizinisch—Naturwissenschaftlichen Gesellschaft zu Jena*: 82.

Hamlet, C., A. Santhanakrishnan and L. A. Miller. 2011. A numerical study of the effects of bell pulsation dynamics and oral arms on the exchange currents generated by the upside-down jellyfish *Cassiopea xamachana. Journal of Experimental Biology* 214: 1911–1921.

Harrison, P. L., R. C. Babcock, G. D. Bull, et al. 1984. Mass spawning in tropical reef corals. *Science* 223: 1186–1189.

Hartmann, A. C., K. L. Marhaver, A. Klueter, et al. 2019. Acquisition of obligate mutualist symbionts during the larval stage is not beneficial for a coral host. *Molecular Ecology* 28: 141–155.

He, S., F. Del Viso, C. Y. Chen, et al. 2018. An axial Hox code controls tissue segmentation and body patterning in *Nematostella vectensis. Science* 361: 1377–1380.

Hebert, P. D. N., S. Ratnasingham and J. R. DeWaard. 2003. Barcoding animal life: Cytochrome c oxidase subunit 1 divergences among closely related species. *Proceedings of the Royal Society B: Biological Sciences* 270: S96–S99.

Heins, A., T. Glatzel and S. Holst. 2015. Revised descriptions of the nematocysts and the asexual reproduction modes of the scyphozoan jellyfish *Cassiopea andromeda* (Forskål, 1775). *Zoomorphology* 134: 351–366.

Hofmann, D. K. and M. Gottlieb. 1991. Bud formation in the scyphozoan *Cassiopea andromeda*: Epithelial dynamics and fate map. *Hydrobiologia* 216: 53–59.

Hofmann, D. K. and M. G. Hadfield. 2002. Hermaphroditism, gonochorism, and asexual reproduction in *Cassiopea* sp.: An immigrant in the islands of Hawai'i. *Invertebrate Reproduction and Development* 41: 215–221.

Hofmann, D. K. and T. G. Honegger. 1990. Bud formation and metamorphosis in *Cassiopea andromeda* (Cnidaria: Scyphozoa): A developmental and ultrastructural study. *Marine Biology* 105: 509–518.

Hofmann, D. K. and B. P. Kremer. 1981. Carbon metabolism and strobilation in *Cassiopea andromedea* (Cnidaria: Scyphozoa): Significance of endosymbiotic dinoflagellates. *Marine Biology* 65: 25–33.

Hofmann, D. K., R. Neumann and K. Henne. 1978. Strobilation, budding and initiation of scyphistoma morphogenesis in the Rhizostome *Cassiopea andromeda* (Cnidaria: Scyphozoa). *Marine Biology* 47: 161–176.

Hofmann, D. K. and U. Brandt. 1987. Induction of metamorphosis in the symbiotic scyphozoan *Cassiopea andromeda*: Role of marine bacteria and of biochemicals. *Symbiosis* 4: 99–116.

Hofmann, D. K., W. K. Fitt and J. Fleck. 1996. Checkpoints in the life-cycle of *Cassiopea* spp.: Control of metagenesis and metamorphosis in a tropical jellyfish. *International Journal of Developmental Biology* 40: 331–338.

Hofmann, D. K. and G. Henning. 1991. Effects of axenic culture conditions on asexual reproduction and metamorphosis in the symbiotic scyphozoan *Cassiopea andromeda. Symbiosis* 10: 83–93.

Holland, B. S., M. N. Dawson, G. L. Crow, et al. 2004. Global phylogeography of *Cassiopea* (Scyphozoa: Rhizostomeae): Molecular evidence for cryptic species and multiple invasions of the Hawaiian Islands. *Marine Biology* 145: 1119–1128.

Hu, M., X. Zheng, C. M. Fan, et al. 2020. Lineage dynamics of the endosymbiotic cell type in the soft coral *Xenia. Nature* 582: 534–538.

Hui, J. H. L., P. W. H. Holland and D. E. K. Ferrier. 2008. Do cnidarians have a ParaHox cluster? Analysis of synteny around a *Nematostella* homeobox gene cluster. *Evolution and Development* 10: 725–730.

Hummelinck, P. Wagenaar. 1968. Caribbean scyphomedusae of the genus *Cassiopea. Studies on the Fauna of Curaçao and other Caribbean Islands* 25: 1–57.

Jantzen, C., C. Wild, M. Rasheed, et al. 2010. Enhanced pore-water nutrient fluxes by the upside-down jellyfish *Cassiopea* sp. in a Red Sea coral reef. *Marine Ecology Progress Series* 411: 117–125.

Jarms, G. and A. C. Morandini. 2019. *World Atlas of Jellyfish.* Dölling and Galitz Verlag.

Jensch, F. and D. K. Hofmann. 1997. The cnidomes of *Cassiopea andromeda* Forskål, 1775, and *Cassiopea xamachana* Bigelow, 1882 (Cnidaria: Scyphozoa). *Proceedings of the 6th International Conference on Coelenterate Biology*, 279–285.

Jinek, M., K. Chylinski, I. Fonfara, et al. 2012. A programmable dual-RNA-guided DNA endonuclease in adaptive bacterial immunity. *Science* 337, Nr. 6096.

Jones, V. A. S., M. Bucher, E. A. Hambleton et al. 2018. Microinjection to deliver protein, mRNA, and DNA into zygotes of the cnidarian endosymbiosis model *Aiptasia* sp. *Scientific Reports* 8: 16437.

Kamm, K. and B. Schierwater. 2007. Ancient linkage of a POU class 6 and an anterior Hox-like gene in Cnidaria: Implications for the evolution of homeobox genes. *Journal of Experimental Zoology Part B: Molecular and Developmental Evolution* 308B: 777–784.

Kayal, E., B. Bentlage, A. G. Collins, et al. 2012. Evolution of linear mitochondrial genomes in medusozoan cnidarians. *Genome Biology and Evolution* 4: 1–12.

Kayal, E., B. Bentlage, M. S. Pankey, et al. 2018. Phylogenomics provides a robust topology of the major cnidarian lineages and insights on the origins of key organismal traits. *BMC Evolutionary Biology* 18, Nr. 68.

Keable, S. J. and S. T. Ahyong. 2016. First records of the invasive "upside-down jellyfish", *Cassiopea* (Cnidaria: Scyphozoa: Rhizostomeae: Cassiopeidae), from coastal lakes of New South Wales, Australia. *Records of the Australian Museum* 68: 23–30.

Khabibulina, V. and V. Starunov. 2019. Musculature development in planuloids of *Cassiopeia xamachana* (Cnidaria: Scyphozoa). *Zoomorphology* 138: 297–306.

Khalturin, K., C. Shinzato, M. Khalturina, et al. 2019. Medusozoan genomes inform the evolution of the jellyfish body plan. *Nature Ecology and Evolution* 3: 811–822.

Kim, H. M., J. A. Weber, N. Lee, et al. 2019. The genome of the giant *Nomura*'s jellyfish sheds light on the early evolution of active predation. *BMC Biology* 17: 28.

King, J. M. 1980. Direct fission: An undescribed reproductive method in hydromedusae. *Bulletin of Marine Science* 30: 522–525.

Klein, S. G., K. A. Pitt, C. H. Lucas, et al. 2019. Night-time temperature reprieves enhance the thermal tolerance of a symbiotic cnidarian. *Frontiers in Marine Science* 6, Nr. July: 453.

Kraus, Y. A. and A. V. Markov. 2016. The gastrulation in Cnidaria: A key to understanding phylogeny or the chaos of secondary modifications? *Zhurnal Obshcheĭ Biologii* 77: 83–105.

Kuhn, K., B. Streit and Bernd Schierwater. 1999. Isolation of Hox genes from the Scyphozoan *Cassiopeia xamachana*: Implications for the early evolution of Hox genes. *Journal of Experimental Zoology* 285: 63–75.

Kvennefors, E. C. E., W. Leggat, O. Hoegh-Guldberg, et al. 2008. An ancient and variable mannose-binding lectin from the coral *Acropora millepora* binds both pathogens and symbionts. *Developmental and Comparative Immunology* 32: 1582–1592.

LaJeunesse, T. C. 2017. Validation and description of *Symbiodinium microadriaticum*, the type species of Symbiodinium (Dinophyta). *Journal of Phycology* 53: 1109–1114.

LaJeunesse, T. C. 2020. Zooxanthellae. *Current Biology* 30: R1110–R1113.

LaJeunesse, T. C., J. E. Parkinson, P. W. Gabrielson, et al. 2018. Systematic revision of Symbiodiniaceae highlights the antiquity and diversity of coral endosymbionts. *Current Biology* 28: 2570–2580.

LaJeunesse, T. C., D. T. Pettay, E. M. Sampayo, et al. 2010. Long-standing environmental conditions, geographic isolation and host-symbiont specificity influence the relative ecological dominance and genetic diversification of coral endosymbionts in the genus *Symbiodinium*. *Journal of Biogeography* 37: 785–800.

Lampert, K. P. 2016. *Cassiopea* and its zooxanthellae. In: *The Cnidaria, past, present and future: The world of Medusa and her sisters*, Stefano Goffredo and Zvy Dubinsky (Eds.). Springer International Publishing.

Lampert, K. P., P. Bürger, S. Striewski, et al. 2012. Lack of association between color morphs of the jellyfish *Cassiopea andromeda* and zooxanthella clade. *Marine Ecology* 33: 364–369.

Lehnert, E. M., M. E. Mouchka, M. S. Burriesci, et al. 2014. Extensive differences in gene expression between symbiotic and aposymbiotic cnidarians. *G3: Genes, Genomes, Genetics* 4: 277–295.

Li, Qiao, H. Y. and T. P. Zhong. 2015. Regeneration across metazoan phylogeny: Lessons from model organisms. *Journal of Genetics and Genomics* 42: 57–70.

Li, Yunfeng, L. G., Y. Pan, M. Tian, et al. 2020. Chromosome-level reference genome of the jellyfish *Rhopilema esculentum*. *GigaScience* 9, Nr. 4.

Lieshout, J., S. V. and V. J. Martin. 1992. Development of planuloid buds of *Cassiopea xamachana* (Cnidaria: Scyphozoa). *Transactions of the American Microscopical Society* 111: 89–110.

Light, S. F. 1914. Some Philippine Scyphomedusae, including two new genera, five new species, and one new variety. *Philippine Journal of Science* 9: 195–231.

Light, S. F. 1924. A new species of Scyphomedusan jellyfish in Chinese waters. *The China Journal of Science and Arts* 2: 449–450.

Ludwig, F. D. 1969. Die Zooxanthellen bei *Cassiopea andromeda* Eschscholtz 1829 (Polyp-Stadium) und ihre Bedeutung für die Strobilation. *Zoologische Jahrbücher. Abteilung für Anatomie und Ontogenie der Tiere Abteilung für Anatomie und Ontogenie der Tiere* 86: 238–277.

Lukyanov, K. A. and V. V. Belousov. 2014. Genetically encoded fluorescent redox sensors. *Biochimica et Biophysica Acta (BBA)—General Subjects* 1840: 745–756.

Maas, O. 1903. *Die Scyphomedusen der Siboga-Expedition*. Siboga-expeditie 11. Leiden: Buchhandlung und druckerei vormals E. J. Brill.

Martin, V. J. and F. S. Chia. 1982. Fine structure of a scyphozoan planula, *Cassiopeia xamachana*. *The Biological Bulletin* 163, Nr. 2.

Matthews, B. J. and L. B. Vosshall. 2020. How to turn an organism into a model organism in 10 "easy" steps. *Journal of Experimental Biology* 223: 223.

Mayer, A. G. 1908. *Rhythmical pulsation of the medusae*. 47. Aufl. Washington, DC.

Mayer, A. G. 1906. *Medusae of the Hawaiian Islands collected by the steamer Albatross in 1902*. Washington, DC: Government Printing Office.

Mayer, A. G. 1910. *Medusae of the world: Volume III the scyphomedusae*. Washington, DC: Carnegie Institution of Washington.

McFall-Ngai, M., M. G. Hadfield, T. C. G. Bosch, et al. 2013. Animals in a bacterial world, a new imperative for the life sciences. *Proceedings of the National Academy of Sciences of the United States of America* 110: 3229–3236.

McGill, C. J. and C. M. Pomory. 2008. Effects of bleaching and nutrient supplementation on wet weight in the jellyfish *Cassiopea xamachana* (Bigelow) (Cnidaria: Scyphozoa). *Marine and Freshwater Behaviour and Physiology* 41: 179–189.

McIlroy, S. E. and M. A. Coffroth. 2017. Coral ontogeny affects early symbiont acquisition in laboratory-reared recruits. *Coral Reefs* 36: 927–932.

Medina-Feliciano, J. G., S. Pirro, J. E. García-Arrarás, et al. 2020. Draft genome of the sea cucumber *Holothuria glaberrima*, a model for the study of regeneration. *bioRxiv*.

Mellas, R. E., S. E. McIlroy, W. K. Fitt et al. 2014. Variation in symbiont uptake in the early ontogeny of the upside-down jellyfish, *Cassiopea* spp. *Journal of Experimental Marine Biology and Ecology* 459: 38–44.

Mirsky, A. E. and H. Ris. 1951. The desoxyribonucleic acid content of animal cells and its evolutionary significance. *The Journal of general physiology* 34: 451–462.

Mohamed, A. R., V. Cumbo, S. Harii, et al. 2016. The transcriptomic response of the coral Acropora digitifera to a competent *Symbiodinium* strain: The symbiosome as an arrested early phagosome. *Molecular Ecology* 25: 3127–3141.

Morandini, A. C., D. Ascher, S. N. Stampar, et al. 2005. Cubozoa e Scyphozoa (Cnidaria: Medusozoa) de águas costeiras do Brasil. *Iheringia. Série Zoologia* 95: 281–294.

Morandini, A. C. and F. L. da Silveira. 2001. Sexual reproduction of *Nausithoe aurea* (Scyphozoa, Coronatae). Gametogenesis, egg release, embryonic development, and gastrulation. *Scientia Marina* 65: 139–149.

Morandini, A. C., S. N. Stampar, M. M. Maronna, et al. 2017. All non-indigenous species were introduced recently? The case study of *Cassiopea* (Cnidaria: Scyphozoa) in Brazilian waters. *Journal of the Marine Biological Association of the United Kingdom* 97: 321–328.

Mortillaro, J. M., K. A. Pitt, S. Y. Lee, et al. 2009. Light intensity influences the production and translocation of fatty acids by zooxanthellae in the jellyfish *Cassiopea* sp. *Journal of Experimental Marine Biology and Ecology* 378: 22–30.

Muscatine, L. and J. W. Porter. 1977. Reef Corals: Mutualistic symbioses adapted to nutrient-poor environments. *BioScience* 27: 454–460.

Nakai, J., M. Ohkura and K. Imoto. 2001. A high signal-to-noise Ca2+ probe composed of a single green fluorescent protein. *Nature Biotechnology* 19: 137–141.

Nakanishi, N., D. Yuan, D. K. Jacobs, et al. 2008. Early development, pattern, and reorganization of the planula nervous system in *Aurelia* (Cnidaria, Scyphozoa). *Development Genes and Evolution* 218: 511–524.

Nath, R. D., C. N. Bedbrook, M. J. Abrams, et al. 2017. The jellyfish *Cassiopea* exhibits a sleep-like state. *Current Biology* 27: 2984–2990.

Neumann, R. 1979. Bacterial induction of settlement and metamorphosis in the planula larvae of *Cassiopea andromeda* (Cnidaria: Scyphozoa, Rhizostomeae). *Marine Ecology Progress Series* 1, Nr. 1: 21–28.

Newkirk, C. R., T. K. Frazer, M. Q. Martindale, et al. 2020. Adaptation to bleaching: Are thermotolerant Symbiodiniaceae strains more successful than other strains under elevated temperatures in a model symbiotic cnidarian? *Frontiers in Microbiology* 11: 822.

Niggl, W., M. S. Naumann, U. Struck, et al. 2010. Organic matter release by the benthic upside-down jellyfish *Cassiopea* sp. fuels pelagic food webs in coral reefs. *Journal of Experimental Marine Biology and Ecology* 384: 99–106.

Ohdera, A., C. L. Ames, R. B. Dikow, et al. 2019. Box, stalked, and upside-down? Draft genomes from diverse jellyfish (Cnidaria, Acraspeda) lineages: *Alatina alata* (Cubozoa), *Calvadosia cruxmelitensis* (Staurozoa), and *Cassiopea xamachana* (Scyphozoa). *GigaScience* 8, Nr. 7.

Ohdera, A. H., M. J. Abrams, C. L. Ames et al. 2018. Upside-down but headed in the right direction: Review of the highly versatile *Cassiopea xamachana* system. *Frontiers in Ecology and Evolution* 6: 35.

Ohdera, A. H., K. Attarwala, H. Rubain, et al. In prep (a). Genomic insights of bacteria responsible for settlement and metamorphosis of *Cassiopea xamachana* larvae.

Ohdera, A. H., V. Avila-Magaña, V. Sharp et al. In prep (b). Modulation of gene expression driven by symbiosis: Strobilation mechanism in the upside-down jellyfish.

Ojimi, M. C. and M. Hidaka. 2010. Comparison of telomere length among different life cycle stages of the jellyfish *Cassiopea andromeda*. *Marine Biology* 157: 2279–2287.

Ojimi, M. C., N. Isomura and M. Hidaka. 2009. Telomerase activity is not related to life history stage in the jellyfish *Cassiopea* sp. *Comparative Biochemistry and Physiology: A Molecular and Integrative Physiology* 152: 240–244.

Pallas, Peter Simon. 1774. *Spicilegia zoologica: quibus novae imprimis et obscurae animalium species iconibus, descriptionibus atque commentariis illustrantur*. Hg. von August Lange. 10. Aufl. Berlin.

Passano, L. M. 2004. Spasm behavior and the diffuse nerve-net in *Cassiopea xamachana* (Scyphozoa: Coelenterata). In: *Hydrobiologia*, 530–531:91–96.

Perkins, H. F. 1908. Notes on the occurrence of *Cassiopea xamachana* and *Polyclonia frondosa* at the Tortugas. *Papers from the Tortugas Laboratory of the Carnegie Institution* 1.

Phelan, M. A., J. L. Matta, Y. M. Reyes, et al. 2006. Associations between metals and the blue mesogleal protein of *Cassiopea xamachana*. *Marine Biology* 149: 307–312.

Pierce, J. 2005. A system for mass culture of upside-down jellyfish *Cassiopea* spp. as a potential food item for medusivores in captivity. *International Zoo Yearbook* 39: 62–69.

Pinzón, J. H., B. Kamel, C. A. Burge, et al. 2015. Whole transcriptome analysis reveals changes in expression of immune-related genes during and after bleaching in a reef-building coral. *Royal Society Open Science* 2: 140214.

Polteya, D. G. D. Zhinidarich and A. Lui. 1985. Observations of asexual reproduction and regeneration in *Cassiopea* (Scyphozoa, Coelenterata). *Zoologicheski Zhurnal* 64:172–180.

Radwan, F. F. Y., J. W. Burnett, D. A. Bloom, et al. 2001. A comparison of the toxinological characteristics of two *Cassiopea* and *Aurelia* species. *Toxicon* 39: 245–257.

Rahat, M. and O. Adar. 1980. Effect of symbiotic zooxanthellae and temperature on budding and strobilation in *Cassiopeia andromeda* (Eschscholz). *Biological Bulletin* 159: 394–401.

Reich, H. G., D. L. Robertson and G. Goodbody-Gringley. 2017. Do the shuffle: Changes in *Symbiodinium* consortia throughout juvenile coral development. *PLoS One* 12: e0171768.

Richardson, A. J., A. Bakun, G. C. Hays, et al. 2009. The jellyfish joyride: Causes, consequences and management responses to a more gelatinous future. *Trends in Ecology & Evolution* 24: 312–322.

Rodriguez-Lanetty, M., W. S. Phillips and V. M. Weis. 2006. Transcriptome analysis of a cnidarian–dinoflagellate mutualism reveals complex modulation of host gene expression. *BMC Genomics* 7, Nr. 23.

Ruppert, E. E., R. S. Fox and R. D. Barnes. 2004. *Invertebrate zoology: A functional evolutionary approach*. Belmont, CA: Thomson-Brooks/Cole.

Sánchez-Alvarado, A. 2004. Regeneration and the need for simpler model organisms. *Philosophical Transactions of the Royal Society of London. Series B: Biological Sciences* 359: 759–763.

Santhanakrishnan, A., M. Dollinger, C. L. Hamlet, et al. 2012. Flow structure and transport characteristics of feeding and exchange currents generated by upside-down *Cassiopea* jellyfish. *Journal of Experimental Biology* 215: 2369–2381.

Schembri, P. J., A. Deidun and P. J. Vella. 2010. First record of *Cassiopea andromeda* (Scyphozoa: Rhizostomeae: Cassiopeidae) from the central Mediterranean Sea. *Marine Biodiversity Records* 3: e6.

Schierwater, B. and K. Kuhn. 1998. Homology of Hox genes and the Zootype concept in early metazoan evolution. *Molecular Phylogenetics and Evolution* 9: 375–381.

Schummer, M., I. Scheurlen, C. Schaller, et al. 1992. HOM/HOX homeobox genes are present in hydra (*Chlorohydra viridissima*) and are differentially expressed during regeneration. *EMBO Journal* 11: 1815–1823.

Schwarz, J. A., D. A. Krupp and V. M. Weis. 1999. Late larval development and onset of symbiosis in the scleractinian coral *Fungia scutaria*. *The Biological Bulletin* 196: 70–79.

Shao, Y., X. B. Wang, J. J. Zhang, et al. 2020. Genome and single-cell RNA-sequencing of the earthworm *Eisenia andrei* identifies cellular mechanisms underlying regeneration. *Nature Communications* 11: 2656.

Shearer, T. L., M. J. H. Van Oppen, S. L. Romano, et al. 2002. Slow mitochondrial DNA sequence evolution in the Anthozoa (Cnidaria). *Molecular Ecology* 11: 2475–2487.

Shikuma, N. J., I. Antoshechkin, J. M. Medeiros, et al. 2016. Stepwise metamorphosis of the tubeworm *Hydroides elegans* is mediated by a bacterial inducer and MAPK signaling.

Proceedings of the National Academy of Sciences of the United States of America 113: 10097–10102.

Shinzato, C., E. Shoguchi, T. Kawashima, et al. 2011. Using the *Acropora digitifera* genome to understand coral responses to environmental change. *Nature* 476: 320–323.

Shumaker, A., H. M. Putnam, H. Qiu et al. 2019. Genome analysis of the rice coral *Montipora capitata*. *Scientific Reports* 9: 2571.

Smith, H. G. 1936. *Contribution to the anatomy and physiology of* Cassiopea frondosa. *Papers from the Tortugas Laboratory of the Carnegie Institution* 31 p.

Stephens, L. and D. Calder. 2006. *Seafaring scientist: Alfred Goldsborough Mayor, pioneer in marine biology*. University of South Caroline Press, Columbia.

Stiasny, G. 1924. Ueber einige von Dr. CJ van der Horst bei Curaçao gesammelte Medusen. *Bijdragen tot de Dierkunde* 23: 83–92.

Stockard, C. R. 1908. Studies of tissue growth. *Archiv für Entwicklungsmechanik der Organismen* 29: 15–32.

Stoner, E. W., C. A. Layman, et al. 2011. Effects of anthropogenic disturbance on the abundance and size of epibenthic jellyfish *Cassiopea* spp. *Marine Pollution Bulletin* 62: 1109–1114.

Szmant, A. M. 1986. Reproductive ecology of Caribbean reef corals. *Coral Reefs* 5: 43–53.

Thé, J., H. S. Barroso, M. Mammone, et al. 2020. Aquaculture facilities promote populational stability throughout seasons and increase medusae size for the invasive jellyfish *Cassiopea andromeda*. *Marine Environmental Research*: 105161.

Thieme, C. and D. K. Hofmann. 2003. An endogenous peptide is involved in internal control of metamorphosis in the marine invertebrate *Cassiopea xamachana* (Cnidaria: Scyphozoa). *Development Genes and Evolution* 213: 97–101.

Thompson, J. R., H. E. Rivera, C. J. Closek, et al. 2015. Microbes in the coral holobiont: Partners through evolution, development, and ecological interactions. *Frontiers in Cellular and Infection Microbiology* 4: 176.

Thornhill, D. J., M. W. Daniel, T. C. LaJeunesse, et al. 2006. Natural infections of aposymbiotic *Cassiopea xamachana* scyphistomae from environmental pools of *Symbiodinium*. *Journal of Experimental Marine Biology and Ecology* 338: 50–56.

Tiozzo, S. and R. R. Copley. 2015. Reconsidering regeneration in metazoans: An evo-devo approach. *Frontiers in Ecology and Evolution* 3: 67.

Todd, B. D., D. J. Thornhill and W. K. Fitt. 2006. Patterns of inorganic phosphate uptake in *Cassiopea xamachana*: A bioindicator species. *Marine Pollution Bulletin* 52: 515–521.

Tran, C. and M. G. Hadfield. 2011. Larvae of *Pocillopora damicornis* (Anthozoa) settle and metamorphose in response to surface-biofilm bacteria. *Marine Ecology Progress Series* 433: 85–96.

Trench, R. K. 1979. The cell biology of plant-animal symbiosis. *Annual Review of Plant Physiology* 30: 485–531.

Uchida, T. 1970. Occurrence of a rhizostome medusa, *Cassiopea mertensii* Brandt from the Hawaiian Islands. *Annotat. Zool. Jap*. 43: 102–104.

Van Etten, M. L., J. K. Conner, S. M. Chang, et al. 2017. Not all weeds are created equal: A database approach uncovers differences in the sexual system of native and introduced weeds. *Ecology and Evolution* 7: 2636–2642.

Van Woesik, R., F. Lacharmoise and S. Köksal. 2006. Annual cycles of solar insolation predict spawning times of Caribbean corals. *Ecology Letters* 9: 390–398.

Verde, E. A. and L. R. McCloskey. 1998. Production, respiration, and photophysiology of the mangrove jellyfish *Cassiopea xamachana* symbiotic with zooxanthellae: Effect of jellyfish size and season. *Marine Ecology Progress Series* 168: 147–162.

Vijayan, N., K. A. Lema, B. T. Nedved, et al. 2019. Microbiomes of the polychaete *Hydroides elegans* (Polychaeta: Serpulidae) across its life-history stages. *Marine Biology* 166: 19.

Voolstra, C. R., J. A. Schwarz, J. Schnetzer, et al. 2009. The host transcriptome remains unaltered during the establishment of coral–algal symbioses. *Molecular Ecology* 18: 1823–1833.

Waddington, C. H. 1942. Canalization of development and the inheritance of acquired characters. *Nature* 150: 563–565.

Welsh, D. T., R. J. K. Dunn and T. Meziane. 2009. Oxygen and nutrient dynamics of the upside down jellyfish (*Cassiopea* sp.) and its influence on benthic nutrient exchanges and primary production. *Hydrobiologia* 635: 351–362.

Widmer, C. L. 2008. Life cycle of *Chrysaora fuscescens* (Cnidaria: Scyphozoa) and a key to sympatric ephyrae. *Pacific Science* 62: 71–82.

Winstead, D., A. Ohdera, M. Medina, et al. 2018. *Symbiodinium* proliferation inside a cnidarian host vessel are competitive and dynamic. *SICB 2018 Annual Meeting Abstracts*, 58: e1–e264. San Francisco, CA, März.

Wolk, M., M. Rahat, W. K. Fitt, et al. 1985. Cholera toxin and thyrotropine can replace natural inducers required for the metamorphosis of larvae and buds of the scyphozoan *Cassiopea andromeda*. *Wilhelm Roux's Archives of Developmental Biology* 194: 487–490.

Yokoyama, J., N. Hanzawa, Y. Arai, et al. 2017. Phylogenetic relationships and morphological variations of upside-down jellyfishes, *Cassiopea* spp. inhabiting Palau Islands. Phylogenetic relationships and morphological variations of upside-down jellyfishes, *Cassiopea* spp. inhabiting Palau Islands. *Biogeography* 10: 133–141.

Yonge, C. M. 1928–1929. Studies on the physiology of corals. IV: The structure, distribution and physiology of the zooxanthellae. *Scientific Reports/Great Barrier Reef Expedition* 1: 135–176.

Yonge, C.M., Nicholls, A.G. 1931. Studies on the physiology of corals. IV. The structure, distribution and physiology of the zooxanthellae. Sci. Rept. Gr. Barrier Reef Exped. 1928–1929. *Brit. Mus. (Nat. Hist.)* 1:135–176.

Yuan, D., N. Nakanishi, D. K. Jacobs, et al. 2008. Embryonic development and metamorphosis of the scyphozoan *Aurelia*. *Development Genes and Evolution* 218: 525–539.

10 *Acropora*—The Most-Studied Coral Genus

Eldon E. Ball, David C. Hayward, Tom C.L. Bridge and David J. Miller

CONTENTS

10.1 HISTORY AND TAXONOMIC STATUS OF THE GENUS

Corals belong to the phylum Cnidaria, the class Anthozoa (along with the sea anemones) and the Order Scleractinia (the stony corals). Within this order, there are two major clades, the Complexa and Robusta (Romano & Palumbi 1996). These clades, which were originally separated on the basis of 16S sRNA sequences and named on the basis of their skeletal characteristics, have been confirmed by more recent sequencing approaches that have resulted in the phylogenetic reclassification of corals at all taxonomic levels (Kitahara et al. 2016; Ying et al. 2018). The family Acroporidae, to which the genus *Acropora* belongs, falls within the complex clade and is the most speciose family of corals (Madin et al. 2016; Renema et al. 2016), as well as being responsible for much of the three-dimensional structure of modern reefs. Members of this family are commonly known as staghorn or elkhorn corals.

As summarized in Table 10.1, based on the number of mentions in Google Scholar, the genus *Acropora* is by far the most-studied genus of corals, and this has meant that we have had to be very selective in what to include in this chapter. For this, we apologize to the many authors whose excellent work we have failed to cite.

Our goal has been to provide the information required for an understanding of the basic biology of members of the genus *Acropora* and then to focus on some of the most recent findings and debates. Within the genus, the Caribbean species *Acropora palmata* (#1) and *Acropora cervicornis* (#3)

DOI: 10.1201/9781003217503-10

TABLE 10.1
Most-Studied Corals Based on Number of Mentions in Google Scholar 2020

Widely studied coral genera	
Acropora	78,500
Pocillopora	19,000
Orbicella (Montastraea)	14,990
Stylophora	12,200
Widely studied species within the genus *Acropora*	
Acropora palmata	10,800
Acropora millepora	10,300
Acropora cervicornis	8,310
Acropora digitifera	3,230
Acropora tenuis	3,190

rank highly on the scale of mentions. The five most-studied *Acropora* species, as listed in Table 10.1, are pictured in Figure 10.1a–e, while Figure 10.1f–l shows the diverse morphology of other members of the genus.

In spite of the popularity of the Caribbean species, much of the *Acropora* research of this century has focused on Indo-Pacific species, partly due to the rise of large research centers in Australia (e.g. the ARC Centre of Excellence for Coral Reef Studies, James Cook University and the Australian Institute of Marine Science, all in the Townsville area, and the University of Queensland in Brisbane), as well as the Okinawa Institute of Science and Technology in Japan. There are additional major foci of coral research at King Abdullah University of Science and Technology (KAUST) in Saudi Arabia and in Israel, although with somewhat less emphasis on *Acropora* research, perhaps reflecting the composition of the fauna.

There has been a long-standing debate over what the type specimen of the genus *Acropora* should be. The situation was summarized in 1999 by Stephen Cairns (quoted in Wallace 1999) as follows: "The largest and most important genus of hermatypic Scleractinia does not have a recognisable type species". After an extensive historical review of names, Wallace designated a neotype for *Acropora muricata* (originally described as *Millepora muricata* by Linnaeus 1758) (Wallace 1999, p. iv). The description by Linnaeus was based on a drawing of a specimen from Ambon, Indonesia, by G.E. Rumphius, and therefore did not include a type specimen, necessitating Wallace to designate a neotype. The first use of the name *Acropora* for the genus was by Oken (1815), although most nominal *Acropora* species were described as *Madrepora* until Verrill (1901) formalized the genus *Acropora* within the newly designated family Acroporidae.

The genus *Acropora* currently contains approximately 408 nominal species (Hoeksema & Cairns 2020). However, many of these nominal species were synonymized in taxonomic works based on skeletal morphology in the late 20th century, while the status of others remains unresolved (Veron & Wallace 1984; Wallace 1999). Based largely on morphological features, Wallace (1999) recognized only 114 species, leaving almost three-quarters of nominal species either synonymized or unresolved. This was followed in 2012 by a revised monograph recognizing 122 species (Wallace et al. 2012). However, this monograph was completed just as molecular phylogeny was emerging, changing many of our views on relationships throughout the animal kingdom, including among corals, where environmental factors can have a major effect on micromorphology and few taxonomically informative morphological features have been identified. The switch from a taxonomy based exclusively on morphology to one utilizing an integrated approach combining morphology with sequence data has resulted in frequently changing views of relationships within the Scleractinia. Although molecular phylogenetics has largely stabilized genus- and family-level relationships (Kitahara et al. 2016), there is still considerable uncertainty at the species level in many groups, especially in the hyperdiverse family Acroporidae. Fortunately, newly developed molecular techniques such as targeted capture of conserved loci may allow resolution of species-level relationships (Cowman et al. 2020) and, combined with comparison to type material, should allow the testing of species boundaries and identification of informative characters for delineating species. This work suggests that the diversity of the genus *Acropora* is far higher than currently appreciated and that many species are not widespread across the Indo-Pacific, but restricted to specific biogeographic regions. So, while much of the material on structure and biology in Wallace's 1999 book is still valid and useful, the taxonomy is mostly in the process of revision.

Acropora taxonomy, as traditionally practiced, was based on qualitative morphological differences which were not easily recognized by the non-specialist, a situation which is problematic in a genus with environmentally induced morphological variability. This problem is exacerbated by the issue of potential hybridization among species in the genus, as was first brought to widespread attention by J.E.N. Veron in his book *Corals in Space and Time* (1995). This book popularized the idea of reticulate evolution in corals and called into question the definition of a species. For the species, Veron suggested substituting a grouping called a syngameon, which is an interconnected group of potentially interbreeding populations. Hybridization, to the extent it exists, will make it difficult to define a species, but molecular phylogenetics is also calling into question many of the morphological characters formerly used to define species. Indeed, several studies have highlighted extensive "cryptic" species complexes within morphological species (e.g. Richards et al. 2016; Sheets et al. 2018), and at least some of the characters used to define morphological species and species groups are invalid (Cowman et al. 2020). The existence of "cryptic" species is also supported by other lines of evidence. For example, the putatively widespread species *Acropora tenuis* was chosen for detailed study of spawning patterns by Gilmour et al. (2016) specifically because it was thought to be easily recognizable in the field. However,

FIGURE 10.1 Diverse morphologies within the genus *Acropora*. (a–e) The five most-studied species: (a) *A. palmata* (Florida), (b) *A. millepora* (Magnetic Island, central Great Barrier Reef), (c) *A. cervicornis* (Florida), (d) *A.* cf *digitifera* (Kimbe Bay, New Britain, Papua New Guinea), (e) *A. tenuis* (Fiji), (f) *A.* aff. *palmerae* (Tonga), (g) *A. echinata* (Mantis Reef, northern Great Barrier Reef), (h) *A.* aff. *listeri* (Ha'apai, Tonga), (i) *A.* cf. *pacifica* (Ha'apai, Tonga), (j) *A. pichoni* (Kimbe Bay, Papua New Guinea), (k) *Acropora cf. rongelapensis* (Pohnpei, Micronesia), (l) *A. walindii* (Kimbe Bay, Papua New Guinea). Species identifications based on comparisons to type material of all nominal species using open nomenclature outlined in Cowman et al. (2020). (Photos [a,c] courtesy Peter Leahy; [b, d–l] Tom Bridge. Copyright is retained by the photographers.)

in spite of morphological similarity, the population was divided into two genetically distinct groups, as judged by microsatellites and time of spawning. In this chapter, we have retained the names used by the authors of the papers cited while noting that these identifications may be subject to future revision.

In spite of these difficulties, taxonomy is fundamental to the study of coral biology, especially for the field biologist, and no one has proposed a practical way to do without the concept of a species. Several efforts are underway to try to improve identification while maintaining the species concept. In one approach, Kitchen et al. (2019) used shallow genome sequencing to sample multiple populations of the two Caribbean acroporids, *A. palmata* and *A. cervicornis*, to establish the degree of intraspecific genomic variability and to find single nucleotide variants that allowed the two species to be distinguished. They also set up computational tools and stored workflows on the Galaxy server, to which others can add data from other *Acropora* species as these become available. A second approach uses targeted sequence capture of conserved genomic elements found in all corals to produce phylogenies that are stronger than those based on one or a few genomic loci and at a lower cost than whole genome sequencing (Cowman et al. 2020). These robust phylogenies can then be combined with other lines of evidence (e.g. morphological, ecological or geographic data) to support the delineation of species. As in other coral taxa examined using such approaches (e.g. Benzoni et al. 2010; Budd et al. 2012; Huang et al. 2014), there is evidence that morphological characters for delineating species and therefore useful for field research do exist, although they are sometimes incongruent with traditional taxonomic classification.

This integrated approach combining phylogenomics with other lines of evidence, such as spawning times and geographical partitioning, forms the basis for re-examining the taxonomy of the group. The strong evidence for extensive "cryptic" speciation within putatively widespread *Acropora* species (e.g. Richards et al. 2016) necessitates comparison of operational taxonomic units (OTUs) to the type material of all 408 nominal species, not just those accepted in recent revisions, given that many of these "cryptic" species likely represent nominal species that have been synonymized based on morphological characters.

Possible approaches to dealing with the identification problem for future workers include collection of field photos and voucher specimens, use of single nucleotide polymorphisms (which unfortunately can only be done post-hoc back in the lab) and a better understanding of phylogenetically informative morphological features which can be used to identify species in the field.

Staghorn corals are the most important contributors to the three-dimensional structure of modern reefs and are therefore vital for maintaining the biodiversity of these systems (Renema et al. 2016). Much of their success has been due to their mutualistic association with photosynthetic endosymbionts belonging to the family Symbiodinaceae, on which they depend for much of the energy needed for growth. They are therefore most common at shallow depths with good light penetration in tropical and sub-tropical regions, although some species have become specialized to mesophotic coral ecosystems. Originally all of the photosynthetic endosymbionts were treated as a single species, but they are now known to form a diverse group and are placed in different genera. They confer different physiological properties on the colonies that contain them, one of which is resistance to bleaching. The relationship between the coral and its symbionts is a very active area of research, as will be discussed in later sections.

10.2 GEOGRAPHICAL OCCURRENCE—PAST AND PRESENT

The geographical occurrence and paleontology of staghorn corals have recently been summarized by Renema et al. (2016). The earliest described *Acropora* is from the Paleocene, with 10 species known by the end of the Oligocene, 37 in the Miocene, 60 in the Pleistocene and up to 408 nominal species at present (Wallace & Rosen 2006; Santodomingo et al. 2015). However, it should be noted that because these identifications were based on morphology, they are probably conservative, because recent molecular phylogenies have suggested different relationships and will probably increase the number of species (Cowman et al. 2020). In addition, the fragile skeletons of many *Acropora* species are not well suited to fossilization, making their identification in fossil assemblages extremely difficult, particularly at the species level. In spite of their long history, staghorn corals were not dominant reef builders until approximately 1.8 million years ago at the start of a period of high amplitude sea level fluctuations which favored *Acropora* due to high growth rates and the ability to propagate by fragmentation as well as sexually (Renema et al. 2016).

The diversity of staghorn corals belonging to the genus *Acropora* is greater now than at any time in the past. As shown in Figure 10.2, they are currently found in the tropics and subtropics in all three of the world's major oceans between 30°N and 30°S, with their peak distribution in the Central Indo-Pacific. Within this range, they are found in diverse habitats, including reef flats, reef crests and slopes and down to the mesophotic zone (reviewed in Wallace 1999; Muir et al. 2015).

It appears that all species presently described as belonging to the genus *Acropora* reproduce by releasing their buoyant gametes into the water column where fertilization occurs, a process known as "broadcast spawning". Older literature (e.g. Kojis 1986a, 1986b) describes brooding in *Acropora palifera*, but all brooding species are now included in the sister genus *Isopora* (Wallace et al. 2007). In several parts of the world, most notably in northeastern Australia and in the waters around Okinawa, multiple species of *Acropora* spawn together on just a few nights of the year, in a phenomenon known as mass spawning. The term "mass spawning" is controversial (see Baird et al. 2009), but we are using it to refer to spawning on the same night by multiple species in a limited area. Once the egg has been fertilized, the resulting larva can survive for weeks or months on its stored lipid, perhaps supplemented by captured organic matter (Ball et al. 2002a). The longest documented survival time for an *Acropora* larva that we know of is 209 days (Graham et al. 2008), although in the field, much of a larval population is likely to have died long before that. This longevity is facilitated by a rapid decline in larval metabolism (Graham et al. 2013) during which larvae could theoretically be carried hundreds of kilometers by currents before settling to found colonies which could then colonize a new area by a combination of fragmentation and further mass spawning.

Although the Quaternary has seen a peak in *Acropora* abundance and diversity, populations started to shrink in the 20th century due to myriad anthropogenically induced threats to coral health. The greatest of these threats is global warming. Most corals live near their upper thermal limits, so a temperature rise of as little as 3°C for more than a few days causes them to lose the photosynthetic endosymbionts, members of the dinoflagellate family Symbiodinaceae, on which they depend for much of their energy, in a phenomenon known as coral bleaching. If bleaching is prolonged,

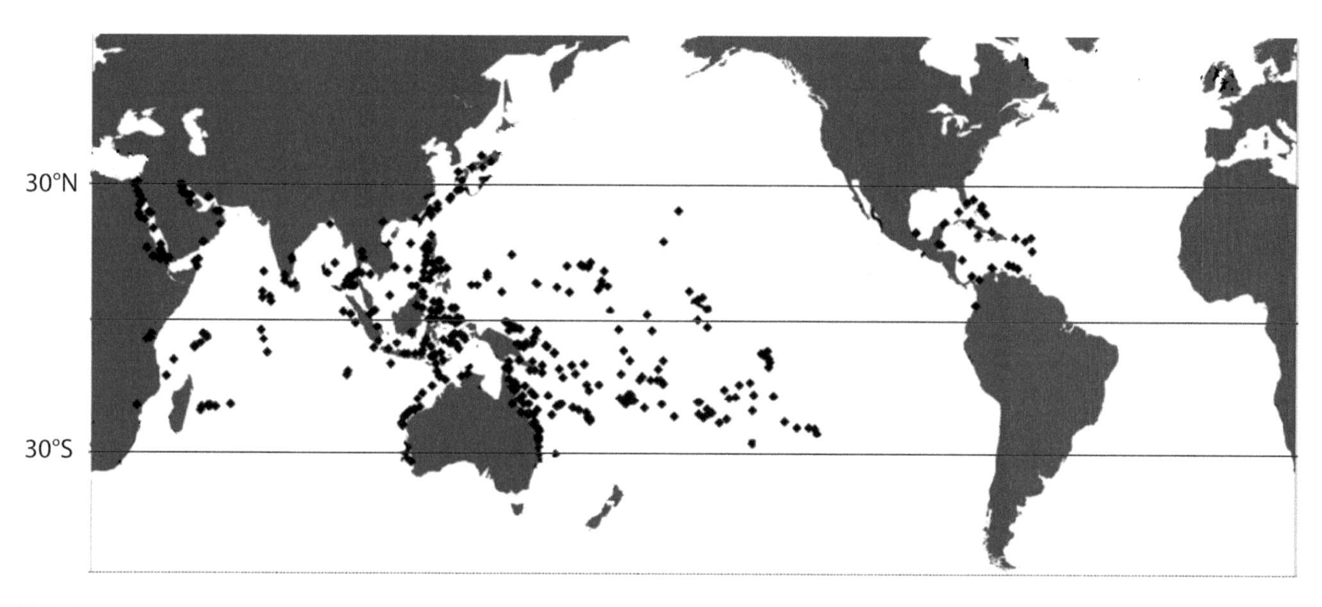

FIGURE 10.2 The worldwide distribution of *Acropora* species is essentially between 30°N and 30°S. (Modified from Wallace and Rosen 2006.)

the corals die, and members of the genus *Acropora* are particularly susceptible to bleaching. Episodes of bleaching are becoming increasingly widespread and frequent and have considerably reduced *Acropora* populations worldwide (Hughes et al. 2017, 2018). In addition to global warming, a second threat arising from rising atmospheric CO_2 levels is ocean acidification. Although a less immediate threat than bleaching, ocean acidification slows the rate of calcification and weakens coral skeletons and may therefore prove significant in the longer term. Other anthropogenic threats include severe weather events, reduced water quality, predator outbreaks (e.g. Crown of Thorns on the Great Barrier Reef), incidental damage due to fishing and diving, the aquarium trade and so on. All of these threats will result in changes to the distribution of individual species and may result in the extinction of some within this century.

10.3 LIFE CYCLE

There is a vast literature on various aspects of reproduction in *Acropora* to which we can't hope to do justice. Among the major reviews of coral reproduction which include information on *Acropora* are those of Harrison and Wallace (1990), Baird et al. (2009) and Harrison (2011), as well as a chapter specifically on reproduction in *Acropora* (Morita & Kitanobo 2020). In addition, the other references cited in this chapter contain many further references. Here we focus our discussion on the life cycle of *A. millepora*, as that is the species with which we are most familiar, but to the best of our knowledge, the life cycles of all members of the genus are very similar.

The month and day of spawning are determined mainly by seawater temperatures in the weeks before potential spawning dates and by phases of the moon, which in turn determine the tides (Keith et al. 2016). The importance of a rapid increase in temperature as a cue for spawning is evident on the central Great Barrier Reef (GBR) where corals on inshore reefs, where the water warms first, frequently spawn one month ahead of offshore reefs, although separated from the latter by only tens of kilometers. Thus, on the central GBR, inshore reefs usually spawn three to five days after the full moon in October or November, with offshore reefs a month later. The night of spawning is not totally synchronous within a population, as spawning may extend over a few nights, although peak spawning is usually restricted to a single night. Not only is there a peak night, but there is usually a peak time of the night at which each species characteristically spawns. For instance, at Magnetic Island, *A. tenuis* usually spawns approximately two hours before *A. millepora* (personal observation). For broadcast spawning corals, onset of darkness is typically the final cue determining the hour of spawning (Babcock et al. 1986). Fukami et al. (2003) describe a similar temporal separation of spawning times in sympatric acroporids in Okinawa.

In some years on the GBR, there is a split spawning, with part of the population spawning in one month and the remainder a month later. A recent modeling study using seven years of data from the GBR has combined data on the time and place of *Acropora* spawning with oceanographic data and has found that split spawning increases the robustness of coral larval supply and inter-reef connectivity due to temporal changes in the currents (Hock et al. 2019).

While the spectacular synchronous multispecies mass spawnings on the Great Barrier Reef have attracted considerable popular and scientific attention, synchrony is by no means universal, even there. In fact, in eastern Australia, synchrony is greatest at mid-latitudes and is reduced to both north and south, and populations in the north often have two spawnings per year.

A major study of *Acropora* spawning patterns was undertaken at Scott Reef (14°S) off northwestern Australia (Gilmour et al. 2016), where 13 species of *Acropora* were followed over three years (n = 1,855 colonies). Of these, seven species spawned in both autumn and spring, five only in autumn and one only in spring. However, the vast majority of individuals spawned only once a year in the same season. The most-studied species, *A. tenuis*, was divided into two genetically distinct but morphologically indistinguishable groups, one spawning in autumn and the other in spring.

On the night of spawning, egg–sperm bundles, which have been developing on the mesenteries of the individual polyps of the colony, are released from their mouths. The egg-sperm bundles contain a number of eggs, surrounding a mass of sperm. They are buoyant due to the high lipid content of the eggs, which is mainly in the form of wax esters (Harii et al. 2007). Once these bundles are released, they float to the surface, breaking up as they go and releasing the sperm. However, how synchronization between colonies is achieved is unknown. One possibility is a so-far-undescribed chemical cue, and there appears to be nothing in the literature to indicate that this has been investigated. In a mass spawning event, the eggs and sperm from one colony will join millions of others coming from diverse individuals and species, although the neighbors will often be predominantly of the same species, thus facilitating fertilization. It seems likely, just on consideration of gamete density, that the majority of fertilizations will occur within the first hour or two of gamete release, although Willis et al. (1997) report that gamete viability does not fall for six to eight hours after release. Cross-fertilization between closely related species is minimized in several ways. First, temporal separation of spawning times is important, as most eggs are apparently fertilized within a relatively short period after release. Second, according to Morita et al. (2006) *Acropora* sperm are not motile when spawned and only become so in the vicinity of conspecific eggs, first swimming in circles and then in a straight line as they get nearer to the egg. However, apparent hybridization between recognizably different morphospecies does occur, reaffirming questions about the nature of "species" in *Acropora*. Several generalizations emerged from the extensive hybridization experiments reported by Willis et al. (1997). First, self-fertilization of eggs from a colony by sperm from that same colony was rare, indicating that sperm can distinguish eggs from their own colony from those from other conspecific colonies. Second, morphologically similar "species" were more likely to hybridize than those which were dissimilar. Third, fertilization success was bimodal in *Acropora millepora*, and on closer inspection, it was found that low fertilization success was associated with differing morphologies of the parent colonies, suggesting the existence of two distinct populations (or of two separate species), one thick branched and the other thin branched. This is a particularly interesting case if the two morphs were both sympatric and spawning at similar times. Apparent cases of hybridization were recorded in more than one-third of 42 species pairs tested, but these results must be considered in light of more recent understanding of species boundaries. Hybrids survived just as well as non-hybrids. The paper of Willis et al. (1997) considers the many implications of their hybridization experiments and concludes, "The complexity in coral mating systems revealed by our experimental crosses suggest that a number of alternative speciation processes, as well as reticulate evolutionary pathways, may have contributed to shaping modern coral species". The take-home lesson for present-day workers is the need to carefully document their experimental material in every way possible, including photos, exact locality data and, if possible, molecular data to support the accurate delineation of species.

Moving on from these complications, the life cycle itself (Figure 10.3) seems to be basically similar for all of the species that have been studied. Once the egg has been fertilized, it continues to float for at least an hour before starting

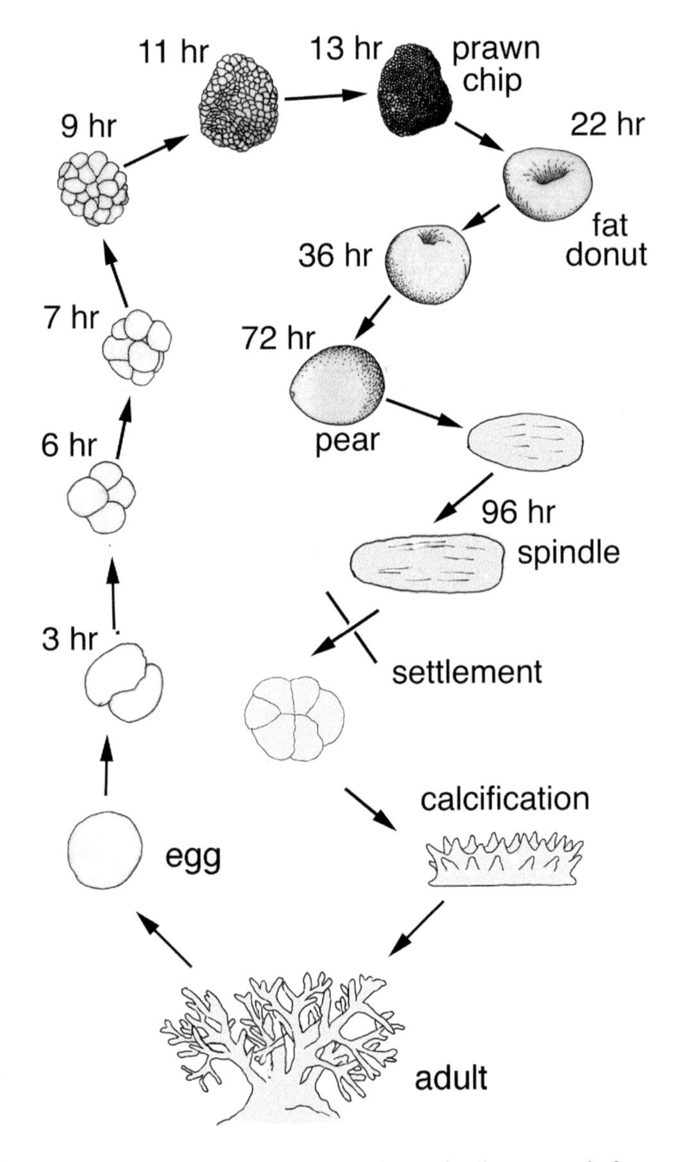

FIGURE 10.3 Life cycle of *A. millepora* in diagrammatic form. (Modified and reproduced with permission of UPV/EHU Press from Ball et al. 2002b. Coral development: from classical embryology to molecular control. *Int. J. Dev. Biol.* 46: 671–678.)

to divide. Then, once cell division has started, it progresses fairly steadily in a temperature-dependent fashion, initially resulting in a ball of cells, known as a morula (Figure 10.4h, i). This then flattens into a stage known colloquially as a prawn chip, due to its resemblance to a prawn cracker (Figure 10.4j–m). As cell division continues, this structure bends and thickens, taking on the appearance of a fat donut, with a depression in one side (Figure 10.4o). Tissue then sinks into this hole, the blastopore, which gradually closes as cells move in from the sides until a closed sphere is formed (Figure 10.4p). At about this stage, cilia appear and the sphere begins to elongate, taking on a pear shape with an oral pore at its apex (Figure 10.4q). The process of elongation continues until, at an age of four to five days, the planula larva has achieved the shape of a ciliated spindle, swimming independently through the water column. Up to this point, the population has remained relatively synchronous in its morphological development. Once elongation to a spindle has occurred, there is relatively little overt morphological change until just before settlement, although differentiation is continuing at the cellular level with an accompanying increase in the number of genes expressed. Somewhere between four and seven days in culture, the developmental synchrony breaks down, and a portion of the population shows a dramatic change in behavior, changing from horizontal swimming to corkscrew swimming into the bottom, apparently testing the substratum. By seven days post-fertilization >50% of the population studied by Strader et al. (2018) had settled and metamorphosed. The delay in settlement by part of a population occurs even in members of a single cross (Meyer et al. 2011; Strader et al. 2018), and its basis is not understood. An interesting correlate of this difference is that those larvae with higher levels of expression of red fluorescent protein are less responsive to settlement cues (Kenkel et al. 2011) and have "gene expression signatures of cell cycle arrest and decreased transcription accompanied by elevated ribosome production and heightened defenses against oxidative stress" (Strader et al. 2016). This pattern of gene expression is consistent with elevated thermal tolerance and greater dispersal potential.

For details of the settlement process, see the section on unresolved problems, but as far as the life cycle is concerned, at the time of settlement, the planula larva samples the substratum with unknown receptors on or toward its aboral end. Once it detects a favorable chemical signal, it flattens onto the substratum, and the oral end spreads to form a primary polyp. The morphology of larvae at this stage is remarkably labile, as they can appear to start to settle but then resume swimming in a matter of seconds. However, shortly after settlement, they attach themselves to the substratum and within a day or so have begun to calcify, first forming a basal plate and then starting to erect septa in a six-part symmetry corresponding to the mesenteries which divide the developing polyp into chambers. Growth is at first two-dimensional along the substratum, with additional polyps appearing in the developing tissue mass beside the first. Then the colony becomes dome shaped as polyps are added over the next few months, and finally vertical branches are sent up from the dome-shaped structure (Abrego et al. 2009). In *A. tenuis*, reproduction begins at colony diameters >10 cm, with the percentage of colonies reproducing steadily rising from there; once colony diameter is >21 cm, all are reproductively mature (Abrego et al. 2009).

10.4 EMBRYOGENESIS

The important stages in *Acropora* development were outlined in the previous section and are similar in all of the *Acropora* species studied. These include *A. hyacinthus*, *A. nasuta*, *A. florida* and *A. secale* (Hayashibara et al. 1997); *A. millepora* (Hayward et al. 2002, 2004, 2015; Okubo et al. 2016); *A. intermedia, A. solitaryensis, A. hyacinthus, A. digitifera* and *A. tenuis* (Okubo & Motokawa 2007); *A. digitifera* (Harii et al. 2009); and *A. digitifera* and *A. tenuis* (Yasuoka et al. 2016), and the embryology of several of these species has been studied in considerable detail.

As in the life cycle, we will start with release of an egg–sperm bundle by the adult coral. This consists of 4–17 eggs surrounding a tightly packed core of sperm (Hayashibara et al. 1997; Okubo & Motokawa 2007). The eggs are at first compressed into ellipsoidal shapes but round up to form a sphere (Figure 10.4a) within an hour of release. Sperm consist of an anterior head and a collar surrounding the base of a flagellum (Figure 10.5a). Ultrastructural features of the sperm are described by Harrison and Wallace (1990) and Wallace (1999). The speed at which cell division occurs varies with the temperature, but following the timetable in Figure 10.3, by three hours, the two-cell stage has been reached. The first cleavage division is equal and holoblastic and occurs by progressive furrow formation; the cleavage furrow initiates on one side of the fertilized egg and moves across to the opposite side, resulting in the formation of two equal blastomeres (Figure 10.4b–d). At this stage, the blastomeres may be parallel (Figure 10.4c) or at right angles to each other (Figure 10.4d). At the four-cell stage, the blastomeres lie in a single plane (Figure 10.4e), but as cell division continues, they form a cube (Figure 10.4f, g). With further cell division, the cube of cells becomes more rounded (Figure 10.4h). Anti-tubulin staining at this stage reveals no clear pattern in the orientation of dividing cells (Figure 10.5b). Next a depression appears in one side of the mass of dividing cells (Figure 10.4i); then the cells spread and flatten, eventually forming a bilayer (Figure 10.4j–m, 10.5c, d). At this stage, lipid is distributed evenly within the cells (Figure 10.5c, d), and DAPI staining reveals extra-nuclear bodies (Figure 10.5e, arrowheads) for which we have no explanation, unless they are mitochondria. As development continues, this bilayer thickens and rounds up, probably by a combination of cell movement and cell division (Figure 10.5f), although the relative contribution of these two processes has not been established (Figure 10.4n, o). We have described this process as gastrulation, as cells expressing

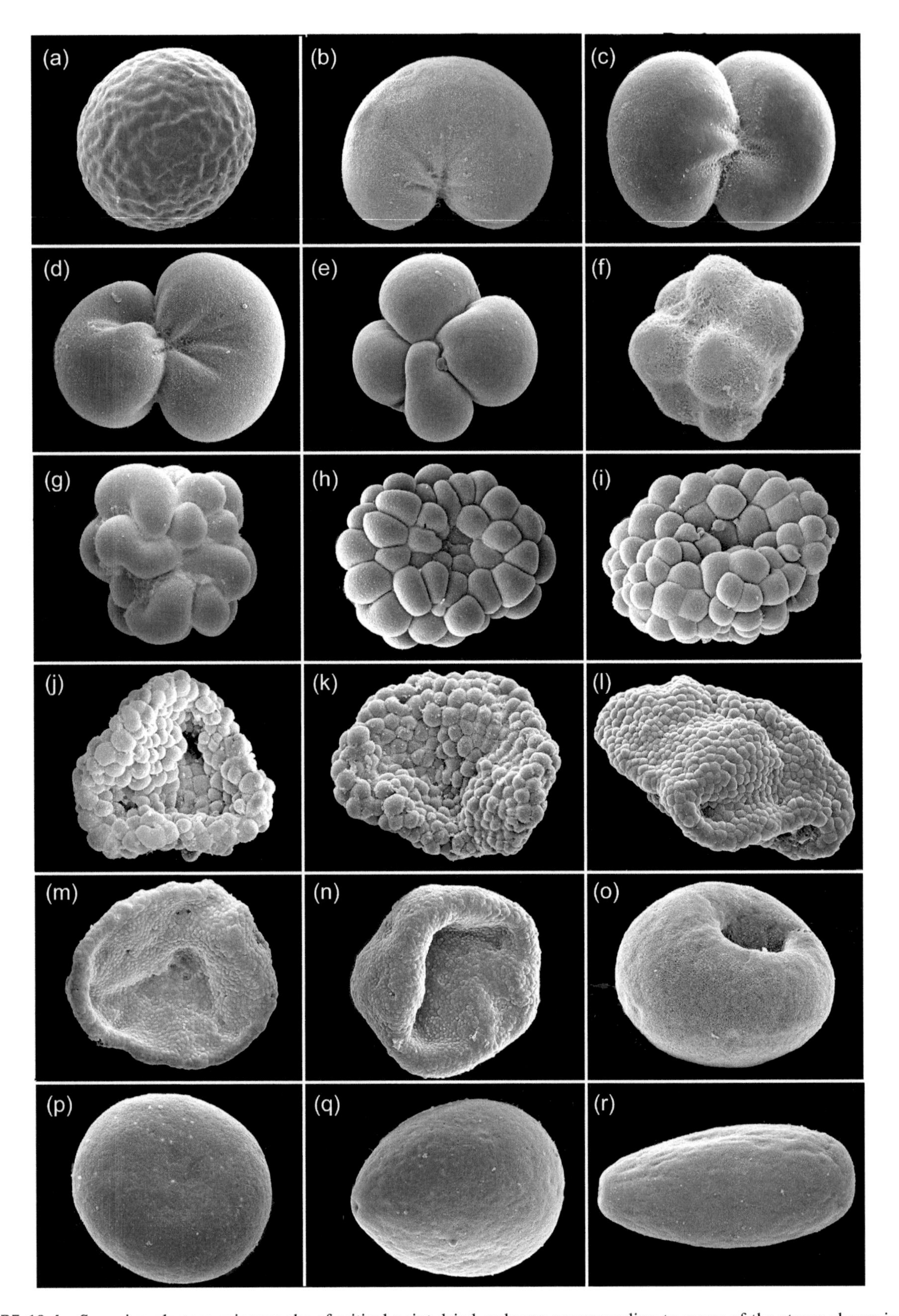

FIGURE 10.4 Scanning electron micrographs of critical point dried embryos corresponding to many of the stages shown in Figure 10.3 (life cycle). (a) Egg; (b) first cleavage division; (c) two-cell stage, blastomeres parallel; (d) two-cell stage, blastomeres at right angles; (e) four-cell stage; (f) eight-cell stage, divisions becoming asynchronous; (g) approximately 20 cells; (h,i) morula stage; (j–m) prawn chip stage, consisting of a steadily increasing number of cells; (n) the transition from prawn chip to gastrula; (o) gastrulation—cells are moving inward as the blastopore closes; (p) the blastopore has closed, and the embryo is spherical; (q) cilia have formed, and the sphere is elongating to form a pear; (r) the planula stage—this is the basic morphology until settlement, although the planula can change shape rapidly and dynamically.

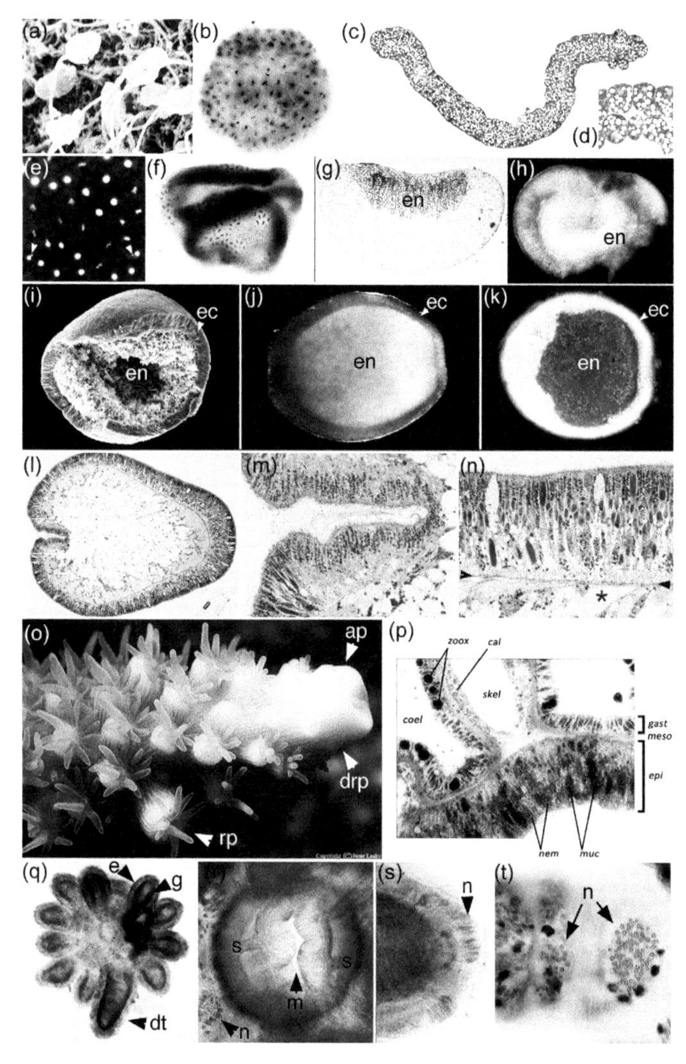

FIGURE 10.5 Aspects of *Acropora* development and anatomy visualized using different technologies. (a) Scanning electron micrograph of critical point dried *Acropora* sperm on the surface of an egg. (b) Anti-tubulin staining of mitotic spindles reveals no clearly ordered pattern of cell division at the morula stage. (c) Transverse section of a prawn chip stained with methylene blue and fuchsin, showing that it consists of a bilayer of cells containing evenly distributed droplets of lipid. (d) Higher magnification view of a portion of (c). (e) DAPI-stained whole mount of a prawn chip with mysterious extranuclear bodies (arrowheads). (f) Late prawn chip stained with anti-tubulin to reveal the patterns of cell division. (g) Section of an in situ hybridization of a bowl-shaped embryo. Tissue expressing the *snail* gene is moving inward to form the endoderm (en). (h) Section of a BMP2/4 in situ preparation reveals a well-developed endoderm at this stage. (i–k) Three embryos at the pear/planula stage examined using different technologies: (i) critical point drying reveals a clear demarcation between ectoderm (ec) and endoderm (en). Solvents used in preparation have removed lipid from the endoderm, giving it a frothy appearance. The central cavity is an artifact of the way in which the embryo fractured. (j) Light micrograph of an unstained embryo showing the highly reflective endodermal lipid (en) contrasting with the much less reflective ectoderm (ec). (k) DAPI staining of an embryo of similar age reveals the contrasting density of cells in the ectoderm (ec), as compared to the endoderm (en). This is consistent with the trichrome stained section shown in (l), in which the large, lipid-filled cells with small nuclei are apparent. (m) Blow-up of the boxed portion of the embryo shown in (l). The uniform nature and appearance of cells in this region contrast with the diversity of cell types apparent elsewhere in the ectoderm and are consistent with a possible function in extracellular digestion. (n) Trichrome staining reveals the diversity of cell types in the body wall away from the oral pore. Clearly apparent are dark-blue-staining cnidocytes (containing nematocysts) and gland cells (large empty-appearing cells). Arrowheads mark the mesoglea, beneath which lie lipid-filled cells (*), as well as smaller cells of unknown function. (o) Branch tip of *A. cervicornis*, showing the arrangement of the two types of polyps. At the tip of the branch is the large axial polyp (ap) which lacks zooxanthellae; behind it are small developing radial polyps (drp), and further proximally lie full-sized radial polyps (rp). (p) Polyp cross-section of *A. longicyathus* showing tissue layers. The coelenteron is lined with gastrodermis containing photosynthetic dinoflagellates (zoox). The calicoblastic epithelium (cal) lines areas occupied by the skeleton (skel) prior to decalcification for sectioning. The epithelium of the body wall contains mucocytes (muc) and nematocytes (nem) and is separated from gastrodermis by the acellular mesoglea (meso). (q) A radial polyp showing the longer directive tentacle (dt). The ectoderm (e), gastrodermis (g) and hollow nature of the tentacles are clearly visible. (r) The muscular mouth (m), showing the arrangement of the septa (s) and the abundant nematocysts (n) located on the oral disc. (s, t) Nematocysts (n) are abundant at the tips of the tentacles (s), particularly on their oral sides (t). (Photo in [o] courtesy Peter Leahy; photo in [p] courtesy Daniel Bucher and Peter Harrison from Bucher and Harrison 2018.)

the gene *snail* move inward through the pore (Figure 10.5g) to form a second tissue layer (Figure 10.5h). As development continues, the pore closes, forming a sphere (Figure 10.4p). Shortly thereafter, the sphere starts to elongate, becoming pear shaped (Figure 10.4q, 10.5i–k), and cilia form. As this elongation occurs, an oral pore (the future mouth of the polyp) opens at or near the site of the blastopore (Okubo & Motokawa 2007). Then, over the next 24–36 hours, cell division continues, new cell types differentiate and the pear elongates into a spindle-shaped planula larva (Figure 10.4r, 10.5l–n), a stage in which it may remain for days or weeks before settlement. Hayashibara et al. (2000) studied the development of cnidae in *Acropora nasuta* and found two types in planulae, a microbasic b-mastigophore nematocyst and a spirocyst. The appearance of cnidae in the planula at three to four days coincided with the start of settlement, and their abundance peaked at eight days, coinciding with maximum settlement. Interestingly, the number of spirocysts then fell in planulae which had failed to settle after eight days, possibly because they were used up in failed attempts to do so. These same two types of cnidae were present in the primary polyp, along with two additional types, the microbasic p-mastigophore and the holotrichous isorhiza.

10.5 ANATOMY

Before turning to anatomical details, a note on terminology relating to tissue layers is needed. The terms "endoderm" and "gastroderm/gastrodermis" are used interchangeably in the literature, as are "ectoderm" and "epithelium". Technically, the former term in each pair refers to embryonic tissue layers, while the latter is used for adult tissues, but this convention is often ignored.

There is no detailed account of what happens immediately after settlement for any one species, but by combining descriptions from several species, it is possible to put together a description that probably is correct in its general outlines for all species. The early steps in the process described in the following are shown in Figure 10.6a.

According to Goreau and Hayes (1977), working on *Porites*, the first step, once the planula larva has chosen a place to settle, is the laying down of a pad of a mucoid substance. Then, within a few hours or days of settlement, depending on species and conditions, the nature of the aboral ectoderm adjacent to the substratum undergoes a morphological change from a columnar epithelium consisting of multiple cell types to a flattened squamous epithelium consisting of a single cell type—the calicoblast cell. This process has been most studied in the genus *Pocillopora* (Vandermeulen 1975; LeTissier 1988; Clode & Marshall 2004), but those observations are consistent with what is known for *Acropora*. Hirose et al. (2008) has a series of photos showing the development of the living primary polyp, while a corresponding sequence of the early stages of skeleton formation in *A. millepora* is shown in Figure 8 of Wallace (1999). According to this sequence, by the third day after settlement, a disc-shaped basal plate has been laid down on which are 12 equally spaced protosepta radiating from the central area occupied by the polyp, like spokes of a wheel (Figure 10.6a4). By the fifth day, the inner ends of the septa have grown laterally and joined to form a circle known as a synapticular ring. The places where these lateral outgrowths meet are called nodes. By the seventh day, the nodes send projections centrally, and a second synapticular ring has formed concentric to and outside of the first (Figure 10.6a5). Further upward and outward growth occurs by addition of more synapticular rings. It is actually outgrowths from the nodes, rather than further development of the protosepta, that will form the adult septa (Piromvaragorn, cited in Wallace 1999). Once the tissue of the primary polyp has spread laterally across the substratum, secondary polyps start to appear by its side. As polyps are added, the colony becomes dome shaped. Then, once a colony consists of 15–20 polyps, some of these start to elongate, founding branches (Abrego et al. 2009).

Adult colonies of all species consist of numerous branches. The colony is organized so that the living tissue lies over of the skeleton that it is secreting (Figure 10.6d). The tissue throughout the colony is organized into two layers, an outer epidermis (or ectoderm) and an inner gastrodermis (Figure 10.5p, 10.6d). The nature of these two layers varies depending on where they are on the colony. At the tip of each branch, there is an axial polyp (Figure 10.5o, ap), while below it, on the sides of the branch, developing radial polyps are budded off (Figure 10.5o, drp) as the colony grows steadily larger. The axial polyp is the largest and fastest-growing polyp. It lacks zooxanthellae and contrasts in color with the radial polyps and the tissue covering the lower part of the branch, which contain zooxanthellae as well as often being pigmented.

Branches of *A. cervicornis* have been recorded to extend by as much as 300 um/day under favorable conditions (Gladfelter 1982). The axial and radial polyps are interconnected by a gastrovascular system of canals (Figure 10.6d–h) filled with fluid and lined with ciliated gastrodermal cells. It has been suggested that this allows sharing of photosynthate produced by zooxanthellate parts of the colony with the rapidly growing axial polyp, which lacks zooxanthellae of its own (Pearse & Muscatine 1971). Bucher and Harrison (2018) have hypothesized that the axial polyp may suppress others from forming as long as the photosynthate supply is limiting. Using time-lapse photography at six-hour intervals, Barnes and Crossland (1980) established that the peak period of daily branch extension was 1200–0600 and did not correspond to the peak period of accretion (0600–1200) as measured using ^{45}Ca. Gladfelter (1982) hypothesized that these observations could be explained by the rapidly growing axial polyp laying down a relatively flimsy framework during the first period, which is then filled in by continuing calcification behind the tip in the second. This is consistent with the observation that permeability and porosity of the

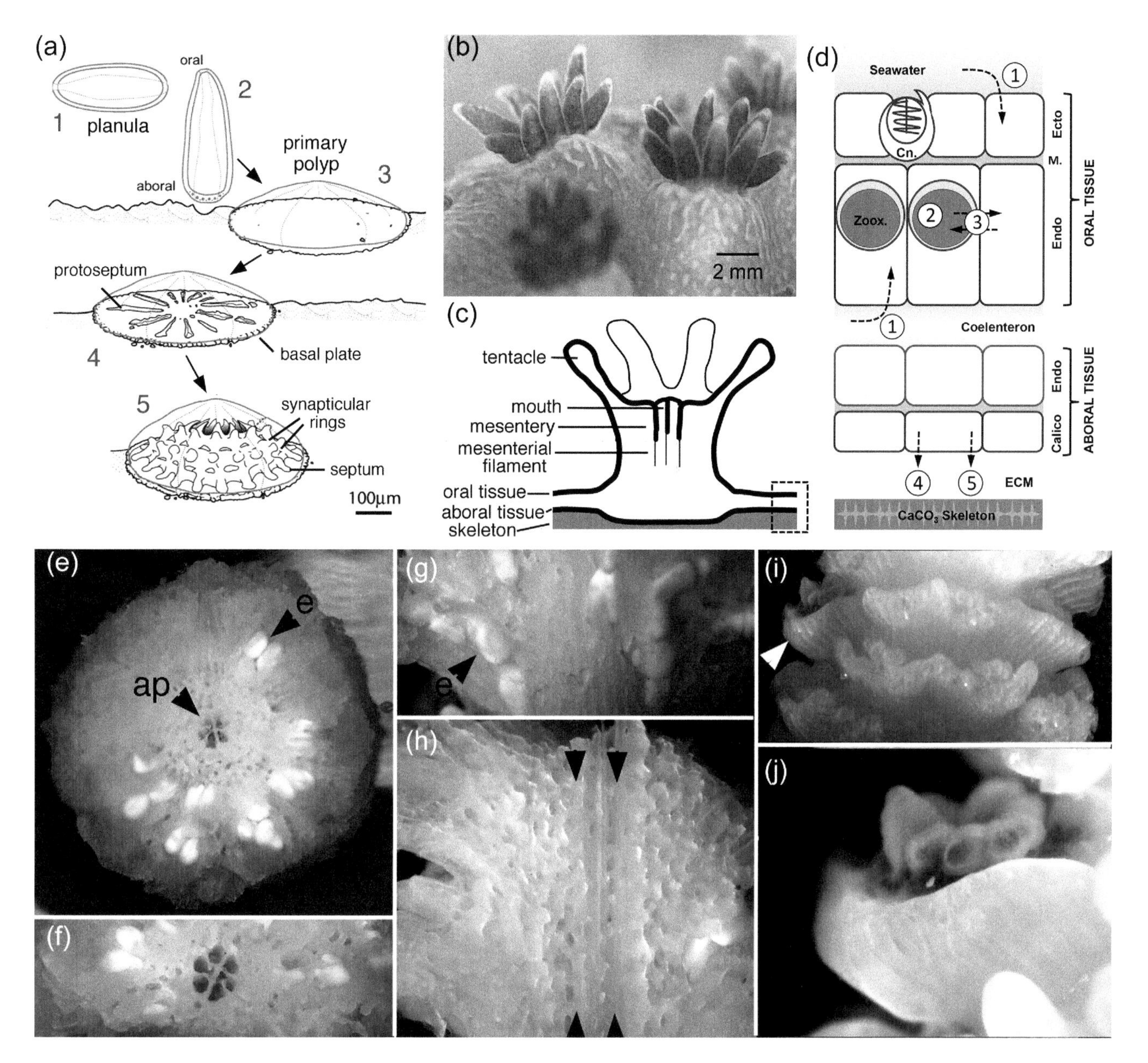

FIGURE 10.6 Anatomy. (a) Settlement, metamorphosis and the initiation of calcification. (a1) Initially, the planula larva swims horizontally well away from the bottom. (a2) When ready to settle, the planula initiates searching behavior, swimming into the bottom in a corkscrew fashion and apparently testing the substratum. (a3) Once a site is selected, the planula flattens in the oral/aboral axis and expands laterally, and a mucoid pad is laid down. (a4) Next calcification begins, first with the deposition of a calcified basal plate and then with the erection of radial protosepta on it. (a5) The protosepta are replaced by septa, which expand laterally at their inner ends to form a synapticular ring. Then more rings are added as the polyp grows. (b–d) Anatomy and function of the adult. (b) Expanded polyps of *A. digitifera*. (c) Diagrammatic view of a polyp with the parts labeled. (d) Histological organization of an area of calcifying tissue showing the relation of the tissue layers and the main metabolic pathways: (1) nutrient uptake, (2) photosynthesis, (3) nutrient exchange, (4) ion secretion, (5) organic matrix secretion. Cn, cnidocyte; M, mesoglea; ECM, extracellular matrix. (e–j) The skeleton. (e) Transverse section of a branch of *A. millepora* showing the central canal leading from the axial polyp (ap) and egg–sperm bundles (e) in canals leading from the radial polyps. (f) Blow-up of the central portion of (e). (g) Branch broken in the long axis showing the arrangement of the egg–sperm bundles in the canals leading to the radial polyps. (h) Another branch broken axially in the plane of the central canal (arrowheads). (i) Lateral view of a branch, showing the organization of the radial polyps. (j) Blow-up of the corallite arrowed in (i) showing a radial polyp with its long directive tentacle. ([a] Modified from Reyes-Bermudez et al. 2009; [b–d] modified from Bertucci et al. 2015.)

skeleton decrease with increasing distance from the branch tip (Gladfelter 1982).

The coral skeleton consists of calcium carbonate ($CaCO_3$) in the form of aragonite in an organic matrix consisting mostly of proteins, polysaccharides and lipids. The composition of the *A. millepora* (Ramos-Silva et al. 2013) and *A. digitifera* (Takeuchi et al. 2016) organic matrices has been determined, and progress has been made toward understanding basic mechanisms of calcification in other species (reviewed in Drake et al. 2019). However, how the characteristic morphology of individual species is produced is still not understood.

As the colony grows, new branches are founded by appearance of a new axial polyp somewhere along an existing branch or by conversion of a radial polyp into an axial polyp (Wallace 1999). The tentacles of the polyps are mostly in multiples of six (hence the classification of *Acropora* in the Hexacorallia), with 12 tentacles being the most frequent (Figure 10.5o, q; Figure 10.6b, c, j). The radial polyps are retractile and can withdraw into the skeleton surrounding them when disturbed. The parts of a radial polyp are shown schematically in Figure 10.6c and in greater detail in Figure 10.5q–t. One tentacle (known as the directive tentacle) is consistently longer than all of the rest and is typically unpigmented, in contrast to the others (Figure 10.5o, q;10.6i, j). The organization of a radial polyp is clearly apparent in Figure 10.5q. Each tentacle is hollow and consists of an outer layer of ectoderm surrounding an inner layer of gastroderm, which in turn surrounds a hollow cavity, connecting to the central cavity, or coelenteron, of the columnar polyp. The mouth is at the center of a flattened area known as the oral disc and is closed by a muscular sphincter (Figure 10.5r). The central cavity is partially partitioned by mesenteries from which hang mesenterial filaments, containing nematocysts which help to subdue struggling prey. Nematocysts are also abundant at the tips of the tentacles (Figure 10.5s) and particularly on their oral sides (Figure 10.5t). The ectoderm consists of diverse cell types, including cnidocytes (which produce several types of nematocyst) as well as gland cells and neurons. Gastrodermal cells are ciliated, have a digestive function and frequently contain photosynthetic dinoflagellates belonging to the family Symbiodinaceae (LaJeunesse et al. 2018).

10.6. GENOMICS

Prior to 2011, only limited transcriptomic and genomic data were available for corals (reviewed in Miller et al. 2011), but in that year, the first coral whole genome assembly was published (Shinzato et al. 2011). Fittingly, the species sequenced was *A. digitifera*—a common species that dominates reefs in many parts of Okinawa and on which Japanese biologists regularly carry out research. Comparison of the *A. digitifera* genome with that of the sea anemone *Nematostella vectensis* (the first cnidarian whole genome sequence assembly) revealed a number of differences. For example, it was suggested that the requirement for a sophisticated symbiont recognition system might underlie the observed enrichment of predicted immune receptors in the *A. digitifera* genome relative to *N. vectensis* (Shinzato et al. 2011). Another surprise was the discovery in *A. digitifera* of a suite of genes that together may enable biosynthesis of mycosporine-like amino acids, "natural sunscreen" which was previously assumed to be produced by the algal symbionts rather than the coral animal. A third key finding arising from analyses of the *A. digitifera* genome was that this coral lacked cystathionine ß-synthase (Cbs), one of the enzymes required for biosynthesis of cysteine. All *Acropora* species examined to date lack Cbs, although a Cbs homolog is present in a wide range of other corals (Shinzato et al. 2011).

The early availability of significant bodies of molecular data for *A. millepora* (e.g. Kortschak et al. 2003; Meyer et al. 2009; Moya et al. 2012) led to widespread use of this coral for experimental purposes, making this species an obvious target for whole genome sequencing. In 2019, the first genome assembly for *A. millepora* became available (Ying et al. 2019); as with the *A. digitifera* assembly, the first *A. millepora* genome was based on short-read data, but a long-read–based assembly became available shortly thereafter (Fuller et al. 2020). There has recently been a rapid increase in the number of genome assemblies available for *Acropora* species, largely carried out at the Okinawa Institute for Science and Technology (OIST)—the institution responsible for the first coral genome assembly. Mao et al. (2018) generated short-read assemblies for four additional species of *Acropora* (*A. gemmifera*, *A. echinata*, *A. subglabra* and *A. tenuis*), and Shinzato et al. (2020) analyzed the genomes of an additional 11 *Acropora* species and those of the confamilial taxa *Montipora cactus*, *M. efflorescens* and *Astreopora myriophthalma*.

Although genomes were not actually assembled, extensive genomic sequence data are also available for the Caribbean species *A. palmata* and *A. cervicornis* (Kitchen et al. 2019).

10.6.1 What Have We Learned from All of Those Genomes?

Despite early speculation on the possibility of a whole genome duplication having facilitated the evolutionary success of *Acropora* (Mao & Satoh 2019), it is now clear that such a duplication is unlikely to have occurred (Shinzato et al. 2020). Rather, many independent gene duplication events occurred in the *Acropora* lineage (Hislop et al. 2005; Shinzato et al. 2020).

The genomes of *Acropora* species vary surprisingly little. Based on short-read assemblies, Shinzato et al. (2020) estimated gene numbers across the genus to be around 22–24,000. However, gene predictions from the two long-read assemblies are significantly higher—28,000 for *A. millepora* (Fuller et al. 2020) and around 30,000 for *A. tenuis* (Cooke et al. 2020). Within the genus *Acropora*, some gene families have been dramatically expanded, interesting examples of which are those encoding the atypical two-domain caspase-X, small cysteine-rich proteins (SCRiPs) and dimethylsulfoniopropionate (DMSP)-lyases (Shinzato et al. 2020). The caspase-X proteins have both active and inactive caspase domains, the latter being likely to normally hold the protein in an inactive state in a manner resembling the interaction of caspase-8 and c-FLIP (Moya et al. 2016). SCRiPs have been implicated in a wide range of functions, including skeletogenesis (Sunagawa et al. 2009; Hayward et al. 2011) and stress responses (DeSalvo et al. 2008; Meyer et al. 2011; Moya et al. 2012), as toxins

(Jouiaei et al. 2015) and possibly also in symbiont acquisition (Mohamed et al. 2020a). *Acropora* spp. are known to produce large amounts of DMSP, which is cleaved by DMSP-lyase to dimethyl sulfate (DMS) and acrylate. As DMS is volatile and can seed cloud formation, a role in local climate moderation has been proposed (Vallina & Simó 2007). Although roles for SCRiPs and caspase-X proteins in stress responses and for DMSP-lyases in mitigating solar radiation have been interpreted as adaptations within the *Acropora* lineage to deal with environmental stressors (Shinzato et al. 2020), *Acropora* species remain among the most sensitive of reef-building corals to thermal stress, and at this stage, it is unclear whether these gene family expansions are related to that.

With the exceptions of the Fuller et al. (2020) assembly for *A. millepora* and the Cooke et al. (2020) assembly for *A. tenuis*, all of these other genomes have been based on short-read data. So, while they have provided some high-quality gene prediction datasets, they do not provide comprehensive coverage. Comparison between the Cooke et al. (2020) *A. tenuis* and the Fuller et al. (2020) *A. millepora* assemblies shows a remarkable level of macrosynteny (Cooke et al. 2020). Given that these species are highly diverged within the genus (Cowman et al. 2020), it is likely that the overall genome architecture varies little within *Acropora*—note that data from Shinzato et al. (2020) are consistent with this view.

10.6.2 How Does the *Acropora* Genome Compare with Those of Other Coral Genera?

With the caveat that, at the time of writing, data are not available for a representative range of reef-building corals, based on the long-read assemblies, at around ~480 Mb (Fuller et al. 2020; Cooke et al. 2020), the estimated size of the *Acropora* genome appears to be fairly typical of corals. Although estimates of both genome size and gene number for some members of the Robusta are much larger (Ying et al. 2018), these were based on short-read assemblies, and it is as yet unclear whether the larger genomes are consequences of higher content of repetitive elements and transposons—as in the case of several bilaterian lineages—or higher gene content. Until higher-quality genome assemblies are available for a phylogenetically representative range of corals, general evolutionary patterns will remain unclear.

10.6.3 Why Has *Acropora* Been Such an Evolutionary Success Story?

Throughout the Indo-Pacific, *Acropora* is the dominant reef-building coral and is one of the most speciose coral genera. As speculated on by Shinzato et al. (2020) and others, its evolutionary success may be due to acquisition and amplification of gene families that have enabled rapid adaptation to changing conditions. However, *Acropora* is almost always associated with one particular genus of Symbiodiniaceae, *Cladocopium*, and we speculate that this partnership may have facilitated the observed rise to dominance of this genus. Comparative transcriptomics has demonstrated the over-representation of (for example) ABC-transporters in *Cladocopium goreaui* compared to *Breviolum minutum* and *Fugacium kawagutii*—other Symbiodiniaceae associated with corals—and among the transporters known so far only in *Cladocopium*, there are components of transport systems for both cysteine and histidine (Mohamed et al. 2020b). The significance of cysteine in the case of *Acropora* was discussed previously; although members of the Robusta are capable of histidine biosynthesis, along with other Complexa and bilaterians, *Acropora* species cannot synthesize it. Hence the association between *Acropora* as host and *Cladocopium* as symbiont may be a particularly good "fit" and have contributed to the rise of the genus during the Neogene and Quaternary.

10.7 FUNCTIONAL APPROACHES: TOOLS FOR MOLECULAR AND CELLULAR ANALYSES

For many reasons, the functional approaches that have proven so fruitful in other organisms such as *Drosophila* and *Caenorhabditis* have been difficult or impossible to implement in *Acropora*. First, there is ease and cost of culture. While adult corals have been kept in aquaria for years, albeit in varying degrees of health, it is only in the past year that there has been a report in the literature of successful production of a second generation of *Acropora* in captivity (Craggs et al. 2020), and this required a sophisticated and expensive aquarium system. Second, there is the problem of generation time; it is probably at least three years before a second generation of *Acropora* would produce sufficient embryos for experimental purposes. Third, there is genome size. Compared to the best-understood "model" organism, *Drosophila melanogaster* (genome size ~140 Mb; 15,700 genes), at 400–500 Mb and with ~28–30,000 genes, the genomes of *A. millepora* and *A. tenuis*, the two *Acropora* species for which we have the best data, are relatively large. In addition, *Drosophila* has only 8 chromosomes (four pairs), while *A. millepora* has 28 (Kenyon 1997; Flot et al. 2006), as does *A. digitifera* (Supp Fig 1 in Shinzato et al. 2011). Twenty-eight chromosomes is most common in the genus, as Kenyon (1997) found this number in 16 species, but this is by no means universal, as 6 other species had 24, 30 (2 species), 42, 48 and 54.

Studies on *Acropora* also require several additional considerations that may not be relevant to other organisms. One is the taxonomic problem dealt with in Section 10.1. Molecular markers may be required in the future to be sure that one is really dealing with the same species in different parts of the world. A final difficulty is that a coral is in fact a holobiont, usually consisting of the coral itself, one or more species of photosynthetic microalgae and numerous other micro-organisms. In nature, this assemblage will vary somewhat from coral to coral and locality to locality and may have considerable effects on the health and physiology of the individual coral and therefore on experimental repeatability.

Genetic and cell biological manipulations have been done on other cnidarians, most notably on *Hydra* and *Nematostella*, in both of which gene knockdown experiments have been successful. However, culturing these species is much less demanding than for corals. Of greater relevance to studies on corals have been experiments on the sea anemone *Exaiptasia* (often under the name *Aiptasia*), which is relatively easy to culture and which shares with corals the presence of photosynthetic endosymbionts. There has been an attempt by the *Exaiptasia* community to standardize strains of anemone and endosymbionts in order to achieve a greater level of experimental consistency across the community (e.g. Cziesielski et al. 2018), but this will be difficult in the case of *Acropora*.

In spite of the challenges noted previously, there have been some successful attempts at experimental manipulation in corals. For example, lithium chloride and 1-azakenpaullone (AZ) have been used to inhibit GSK3 and activate the wnt pathway in *A. digitifera* (Yasuoka et al. 2016), resulting in the expansion of *brachyury* expression throughout the embryonic ectoderm in a dose-dependent manner. In contrast, wnt/ßcatenin signaling inhibitors (pyrvinium pamoate, IWR1 or iCRT14) reduced *Adi_bra* expression in a dose-dependent fashion, leading to the conclusion that it is positively regulated by wnt/ßcatenin signalling. In a following experiment, FITC-labeled anti-sense morpholinos were designed to bind to and inhibit *Adi_bra* RNAs, resulting in loss of function of the *brachyury* gene and a lack of pharynx formation in the morphants, although gastrulation still occurred. The authors then went on to compare bra-morphants, control morphants and uninjected embryos using RNA seq in order to identify genes downstream from *Adi_bra*.

Although morpholinos gave results which could be interpreted in the case described previously, in most studies in other organisms, they have now been replaced by CRISPR/Cas9 gene editing technology, which can result in permanent heritable genetic changes. This was first applied to corals by Cleves et al. (2018), who targeted the *A. millepora* genes encoding fibroblast growth factor 1a (FGF1a), green fluorescent protein (GFP) and red fluorescent protein (RFP) in an attempt to prove that CRISPR/Cas9 could be applied to corals. FGF1a is a single copy gene chosen for its probable role "in sensing the environment and/or in modulating gene expression during larval settlement and metamorphosis". The GFP and RFP are multicopy but were chosen for ease of assay and for their probable ecological importance as well as the ability to target multiple copies due to their sequence similarity. Sequencing of 11 mutant larvae revealed both wild type and multiple different mutant alleles of target genes, indicating that the injected sgRNA-Cas9 remained active for several cell cycles after injection and that the target gene was never knocked out biallelically (i.e. on both copies of the chromosome). While this study was a great technical success, the authors are careful to point out some of its limitations and provide recommendations for further studies using this technique. They point out that "As there is little immediate prospect of raising mutagenized animals to adulthood and generating homozygous individuals by genetic crosses, obtaining animals that have sustained early biallelic mutations will be critical to the analysis of phenotypes of interest". A further consideration, in order to avoid equivocal results, is the need to choose a single copy gene with a clear assay for whether gene knockout has been achieved.

The examples discussed previously were both carried out by injecting eggs, and it should be stressed that such experiments require a high degree of organization on the part of the experimenters because eggs from mass-spawning acroporids are only available for a few nights once or twice per year. A promising new gene knockdown technology has recently been developed using electroporation of short hairpin RNA that has been successfully used on *Nematostella* (Karabulut et al. 2019) and on the hydroid *Hydractinia symbiolongicarpus* (Quiroga-Artigas et al. 2020). This technology would mark a huge advance if it could be developed for broadcast spawning corals such as *Acropora*, as it would allow processing of hundreds of embryos, and testing of multiple genes, in the short annual time window that eggs are available.

Another recently reported innovation, which may prove important for future studies, is gel immobilization (Randall et al. 2019), in which developmental stages of corals are embedded in low-melting-point agarose. The authors used this on developmental stages of five species of corals, including *A. millepora*, and obtained good survival in all species when embedding was done after larvae had become ciliated. This technique could prove particularly valuable for experimental studies since it allows larvae to be individually tracked, manipulated and photographed.

Living *Acropora muricata* colonies were recently imaged in unprecedented detail using light sheet illumination (Laissue et al. 2020). This technique allows the study of any processes in the living coral that would be interfered with by bright light. Unfortunately, it requires a rather specialized optical setup, so it probably will not be widely available, but it may enable certain observations that would not otherwise be possible.

10.8 CHALLENGING QUESTIONS

10.8.1 How Can We Deal with Hybridization and the Species Problem?

The taxonomic problems outlined in Section 10.1 may cause issues with reproducibility and will have to be taken into consideration as possible causes of differing experimental results. For this reason, careful documentation of specimens is of the utmost importance.

10.8.2 What Is the Genomic Basis of the Differing Morphologies of Different Species of *Acropora* and Other Corals?

Presumably the answer to this question lies in gene regulation, as there are few genes involved in skeletogenesis that are species specific, especially if we limit consideration to the genus *Acropora*. So, this will be an interesting, but probably difficult-to-resolve, question.

10.8.3 What Determines the Time and Place at Which Coral Larvae Settle and Undergo Metamorphosis?

Settlement and metamorphosis in *Acropora* are obviously critical for completion of the life cycle and survival of the species but are surprisingly poorly understood. A first important question is what triggers the process of searching and settlement. Some of the temporal variability has a genetic basis, with 47% of variation due to parental effects (Kenkel et al. 2011), but what is it that sends some larvae into searching behavior (a dramatic behavioral change in which larvae go from horizontal swimming to corkscrew swimming into the bottom, apparently testing for chemical cues) in a few days, while others take weeks?

In an early effort to identify the inducer, Morse et al. (1996) surveyed the responses of ten species of Indo-Pacific *Acropora* and found that for all of them, an unidentified sulfated glycosaminoglycan emanating from crustose coralline algae (CCA) was the settlement inducer. While this compound may be the most effective settlement cue, it appears from several lines of evidence that there may be more than one cue that induces settlement and that there is a hierarchy of such cues in relation to their effectiveness in inducing the normally combined processes of settlement and metamorphosis. For instance, Negri et al. (2001) reported that it was actually inducers from the bacterium *Pseudoalteromonas* growing on the CCA that were responsible for settlement. Tebben et al. (2011) took this analysis further, establishing that it was tetrabromopyrrole (TBP) produced by the *Pseudoalteromonas* that was the critical compound for successful metamorphosis of *A. millepora*. However, 90% of the larvae induced to metamorphose by application of TBP did so in the water column and did not successfully attach to the substratum. Successful completion of the entire sequence of settlement, metamorphosis and attachment was only observed in the presence of two species of CCA, and it was determined in a later paper (Tebben et al. 2015) that in order to produce the complete normal sequence of going to the bottom, metamorphosing and attaching, the presence of CCA cell-wall–associated glycoglycerolipids and polysaccharides was required.

10.8.4 What Are the Receptor Molecules Driving Metamorphosis and How Is the Signal Transduced?

There are further related questions about how the larva receives and processes the information relating to settlement and metamorphosis. First, what is the receptor (or receptors) for the CCA compounds that stimulate settlement and metamorphosis? Second, what is the chain of transduction between this receptor and the effector molecules that produce the morphological changes of metamorphosis? There are some clues relating to the answer to the second question in that Iwao et al. (2002) tested the effect of several GLWamide peptides on larvae of *Acropora* and found that the *Hydra* peptide Hym-248 (EPLPIGLWa) induced metamorphosis in all of them but not in the other corals tested, while Erwin & Szmant (2010) found that the same peptide induced metamorphosis in *A. palmata* but not in *Orbicella (Montastrea) faveolata*. The cell bodies of cells expressing the *A. millepora* LWamide gene lie on the mesoglea but project to the surface of the planula larva (Attenborough et al. 2019), but whether these cells also contain the unknown metamorphosis receptors is unknown. A final puzzle is how the signal to metamorphose is distributed to the cells that must respond in larvae that lack a circulatory system.

10.8.5 There Are Many Questions Relating to the Symbiosis between Corals and Their Photosynthetic Dinoflagellate Endosymbionts Belonging to the Family Symbiodinaceae

The ecological success of reef-building corals in nutrient-poor tropical waters is due to their symbiosis with photosynthetic dinoflagellates belonging to the family Symbiodinaceae. These dinoflagellates are remarkable in that many or all occur in both a free-living, flagellated form and a coccoid symbiotic form, with individuals capable of switching between these forms depending on their environment. The relationship with the coral has been assumed to be a classical symbiosis (i.e. a mutualism) from which both partners benefit, with the coral receiving the energy for growth from the dinoflagellate's photosynthate, while the latter utilizes the nitrogenous and phosphate-containing waste produced by the coral, as well as obtaining what is normally a secure place to live. However, the assumption of mutualism as a general property of Symbiodiniaceae is currently being revisited (LaJeunesse et al. 2018; Liu et al. 2018; Mohamed et al. 2020b).

Understanding of the relationship between corals and their symbionts has grown explosively in the last few years, driven by the worldwide breakdown in this symbiosis reflected in widespread coral bleaching, which occurs when the symbionts leave or are expelled by the coral. Bleaching is most commonly caused by thermal stress, as most corals live very near their upper thermal limit and will die if the heating is prolonged.

Progress and problems in studying the symbiosis between cnidarians and their photosynthetic endosymbionts were summarized in a comprehensive review by Davy et al. (2012), and while considerable progress has been made in the intervening years, most of the questions raised in that review are still under investigation using newly developed molecular techniques which have opened the way to a much greater understanding of the symbiotic relationship and its complexity. So, just in the last 20 years, the field has gone from lumping all of the endosymbionts into a common basket, to recognizing a steadily increasing number of clades, to realizing that members of these clades differed in their physiology, to most recently classifying these clades into different genera (LaJeunesse et al. 2018). In the space available, it is only possible to outline some of the most active areas of research and some key literature references. These involve all aspects of the relationship between host and symbiont, including establishment, maintenance and breakdown. Unfortunately, the literature is full of apparently contradictory results which are difficult to interpret because of differing combinations of corals and their potential symbionts and differing experimental techniques. Some of the areas under most active investigation are the following. When and how is symbiosis established in *Acropora*? What is the mechanism of symbiont uptake and retention or rejection? What do the host and symbiont contribute to each other? What happens when corals bleach—does the coral evict its symbionts, or do they flee? Recent summaries of research in these areas include Morrow et al. (2018) and van Oppen and Medina (2020).

10.8.6 How Does the Coral Interact with Its Non-Dinoflagellate Endosymbionts and They with Each Other?

The coral is a metaorganism, playing host to many microorganisms in addition to the members of the Symbiodinaceae on which it is reliant for much of its energy. These include bacteria, viruses and other microbes such as apicomplexans. Recently, many techniques, including genomics and metabolomics, have been developed that facilitate study of these interactions. Deep sequencing enabled Robbins et al. (2019) to assemble "complete" metagenomes for 52 bacterial and archaeal taxa associated with in the coral *Porites lutea*, and analyses of these reveal numerous ways in which they could be contributing to the success of the metaorganism. Now it is a matter of establishing actual, as opposed to theoretical, contributions. Similarly, certain micro-organisms seem to be associated with coral diseases, but is the relationship causal, or is it just a reflection of stress? A few of the many recent reviews of this area include O'Brien et al. (2019), Matthews et al. (2020) and McIllroy et al. (2020).

10.8.7 Can Coral Reefs Be Restored, and What Is the Best Way to Accomplish This?

Due to their morphology, corals belonging to the genus *Acropora* are among the most sensitive to bleaching and death induced by global warming and, as pointed out in earlier sections, they are among the most important structural constituents of many reef systems. As a result of this, a great deal of effort is going into reef restoration, with much of it centered on *Acropora*. Three approaches which we will discuss here are assisted settlement, planting of nubbins and assisted evolution. A comprehensive summary and evaluation of reef restoration techniques is given by Boström-Einarsson et al. (2018) and Zoccola et al. (2020). In the following, we have discussed examples particularly involving *Acropora*.

10.8.7.1 Assisted Settlement

Optimal laboratory conditions have been determined for culture of larvae, induction of settlement and infection with symbiont (Pollock et al. 2017). In field applications of this technique, eggs and sperm are trapped in large floating traps, moved to enclosed rearing pens and then moved on to the desired site of settlement. This technique was pioneered in the Philippines (dela Cruz & Harrison 2017) and on the southern Great Barrier Reef by Peter Harrison and his colleagues and has now moved to a larger scale project near Cairns (https://citizensgbr.org/p/larval-restoration-project). The greatest effectiveness of this technique will almost certainly be in restoration of relatively small areas of high tourist value or for seeding source reefs for recolonization, for example, following a cyclone.

10.8.7.2 Planting of Nubbins

This technique has been attempted in several parts of the world, most notably in the Caribbean and in the waters surrounding Okinawa. There is no doubt that, although it is expensive, it can be successful, at least in limited areas, especially where reefs have suffered physical damage due to hurricanes or cyclones. However, it is difficult to judge success objectively since successes are considered newsworthy, while failures are generally ignored. Efforts over many years in the Caribbean are summarized by Calle-Triviño et al. (2020), and there are certainly examples of success. However, in Okinawa, restoration efforts seem to have been much less successful. For example, 89.2 % of the 79,487 corals transplanted in the Onna village area of Okinawa died within the first five years due to typhoons, bleaching and for "unknown reasons" (Nature Conservation Division D.o.E.A. 2017).

10.8.7.3 Assisted Evolution

These approaches, which have been championed by Madeleine van Oppen and colleagues (van Oppen et al. 2015), were nicely summarized by Zoccola et al (2020) as follows:

> The authors propose to promote resilience/resistance of coral colonies by (1) inducing laboratory stress and selecting the colonies that survive, (2) actively modifying the coral-associated microbiota, (3) applying environmental stress hardening to generate more resistant phenotypes, and (4) genetically enhancing coral host-associated microalgae by means of mutation and selection using artificial evolution. Subsequently, methods for active modification of the coral genome through approaches such as CRISPR and synthetic biology were suggested.

While these methods may have some success, they may be outrun by climate change, and selection in the lab may not be relevant to survival in the field due to fitness tradeoffs.

10.8.7.4 Conclusions

While the previous measures may have some success, economics limits their application to relatively small scales. Experiments conducted under the umbrella of "assisted evolution" will be useful in delivering basic science outcomes, but their real-world relevance has yet to be demonstrated. Technical solutions would be much closer if coral holobionts comprised "plug-and-play" components, but this is clearly not the case (see, for example, Herrera et al. 2020). Moreover, there is a real danger that by focusing attention on reef restoration efforts, perspective on the big picture is lost—ultimately, there is only one solution to the problem of coral bleaching and death, and that means dealing with the anthropogenic impacts of pollution, coastal runoff and climate change. In the meantime, conservation of genetic resources is of critical importance in ensuring the long-term survival of coral reefs in anything like their current state.

ACKNOWLEDGMENTS

The authors would like to thank the Australian Research Council for supporting our coral research over many years via the ARC Centre of Excellence for Coral Reef Research, the Centre for Molecular Genetics of Development and Discovery Grants, as well as the many students and colleagues who have participated in the research reported on here. We thank Peter Leahy for use of his photos, and we dedicate this chapter to the memory of Sylvain Forêt, an enthusiastic collaborator in our research until his untimely death.

BIBLIOGRAPHY

Abrego, D., Van Oppen, M. J., & Willis, B. L. 2009. Onset of algal endosymbiont specificity varies among closely related species of Acropora corals during early ontogeny. *Molecular Ecology* 18:3532–3543.

Attenborough, R. M., Hayward, D. C., Wiedemann, U., Forêt, S., Miller, D. J., & Ball, E. E. 2019. Expression of the neuropeptides RFamide and LWamide during development of the coral *Acropora millepora* in relation to settlement and metamorphosis. *Developmental Biology* 446:56–67.

Babcock, R. C., Bull, G. D., Harrison, P. L., Heyward, A. J., Oliver, J. K., Wallace, C. C., & Willis, B. L. 1986. Synchronous spawnings of 105 scleractinian coral species on the Great Barrier Reef. *Marine Biology* 90:379–394.

Baird, A. H., Guest, J. R., & Willis, B. L. 2009. Systematic and biogeographical patterns in the reproductive biology of scleractinian corals. *Annual Review of Ecology, Evolution, and Systematics* 40:551–571.

Ball, E. E., Hayward, D. C., Catmull, J., Reece-Hoyes, J. S., Hislop, N. R., Harrison, P. L., Samuel, G., et al. 2002a. Molecular control of development in the reef coral, *Acropora millepora*. *Proceedings of the 9th International Coral Reef Symposium (Bali, Indonesia)*. Vol. 1, 395–402.

Ball, E. E., Hayward, D. C., Reece-Hoyes, J., Hislop, N., Samuel, G., Saint, R., Harrison, P. L., et al. 2002b. Coral development: From classical embryology to molecular control. *International Journal of Developmental Biology* 46:671–678.

Barnes, D. J., & Crossland, C. J. 1980. Diurnal and seasonal variations in the growth of a staghorn coral measured by time-lapse photography. *Limnology and Oceanography* 25:1113–1117.

Benzoni, F., Stefani, F., Pichon, M., & Galli, P. 2010. The name game: Morpho-molecular species boundaries in the genus Psammocora (Cnidaria, Scleractinia). *Zoological Journal of the Linnean Society* 160:421–456.

Bertucci, A., Forêt, S., Ball, E. E., & Miller, D. J. 2015. Transcriptomic differences between day and night in *Acropora millepora* provide new insights into metabolite exchange and light-enhanced calcification in corals. *Molecular Ecology* 24:4489–4504.

Boström-Einarsson, L., Ceccarelli, D., Babcock, R. C., Bayraktarov, E., Cook, N., Harrison, P., & Hein, M. 2018. Coral restoration in a changing world: A global synthesis of methods and techniques. Report to the National Environmental Science Program. Reef and Rainforest Research Centre Ltd, Cairns (63pp.).

Bucher, D. J., & Harrison, P. L. 2018. Changes in radial polyp tissues of *Acropora longicyathus* after long-term exposure to experimentally elevated nutrient concentrations. *Frontiers in Marine Science* 5:390.

Budd, A. F., Fukami, H., Smith, N. D., & Knowlton, N. 2012. Taxonomic classification of the reef coral family Mussidae (Cnidaria: Anthozoa: Scleractinia). *Zoological Journal of the Linnean Society* 166:465–529.

Calle-Triviño, J., Rivera-Madrid, R., León-Pech, M. G., Cortés-Useche, C., Sellares-Blasco, R. I., Aguilar-Espinosa, M., & Arias-González, J. E. 2020. Assessing and genotyping threatened staghorn coral *Acropora cervicornis* nurseries during restoration in southeast Dominican Republic. *PeerJ* 8:e8863. DOI:10.7717/peerj.8863.

Cleves, P. A., Strader, M. E., Bay, L. K., Pringle, J. R., & Matz, M. V. 2018. CRISPR/Cas9-mediated genome editing in a reef-building coral. *Proceedings of the National Academy of Sciences* 115:5235–5240.

Clode, P. L., & Marshall, A. T. 2004. Calcium localisation by X-ray microanalysis and fluorescence microscopy in larvae of zooxanthellate and azooxanthellate corals. *Tissue and Cell* 36:379–390.

Cooke, I. R., Ying, H., Forêt, S., Bongaerts, P., Strugnell, J. M., Simakov, O., Zhang, J., et al. 2020. Genomic signatures in the coral holobiont reveal host adaptations driven by Holocene climate change and reef specific symbionts. *Science Advances* 6:eabc6318.

Cowman, P. F., Quattrini, A. M., Bridge, T. C., Watkins-Colwell, G. J., Fadli, N., Grinblat, M., Roberts, T. E., et al. 2020. An enhanced target-enrichment bait set for Hexacorallia provides phylogenomic resolution of the staghorn corals (Acroporidae) and close relatives. *Molecular Phylogenetics and Evolution*. DOI:10.1016/j.ympev.2020.106944.

Craggs, J., Guest, J., Davis, M., & Sweet, M. 2020. Completing the life cycle of a broadcast spawning coral in a closed mesocosm. *Invertebrate Reproduction & Development*:1–4. DOI: 10.1080/07924259.2020.1759704.

Cziesielski, M. J., Liew, Y. J., Cui, G., Schmidt-Roach, S., Campana, S., Marondedze, C., & Aranda, M. 2018. Multi-omics analysis of thermal stress response in a zooxanthellate cnidarian reveals the importance of associating with thermotolerant symbionts. *Proceedings of the Royal Society B: Biological Sciences* 285:20172654.

Davy, S. K., Allemand, D., & Weis, V. M. 2012. Cell biology of cnidarian-dinoflagellate symbiosis. *Microbiology and Molecular Biology Reviews* 76:229–261.

dela Cruz, D. W., & Harrison, P. L. 2017. Enhanced larval supply and recruitment can replenish reef corals on degraded reefs. *Scientific Reports* 7:1–13.

DeSalvo, M. K., Voolstra, C. R., Sunagawa, S., Schwarz, J. A., Stillman, J. H., Coffroth, M. A., Szmant, A. M., et al. 2008. Differential gene expression during thermal stress and bleaching in the Caribbean coral *Montastraea faveolata*. *Molecular Ecology* 17:3952–3971.

Drake, J. L., Mass, T., Stolarski, J., Von Euw, S., van de Schootbrugge, B., & Falkowski, P. G. 2019. How corals made rocks through the ages. *Global Change Biology* 26:31.

Erwin, P. M., & Szmant, A. M. 2010. Settlement induction of *Acropora palmata* planulae by a GLW-amide neuropeptide. *Coral Reefs* 29:929–939.

Flot, J. F., Ozouf-Costaz, C., Tsuchiya, M., & Van Woesik, R. 2006. Comparative coral cytogenetics. *Proceedings of the 10th International Coral Reef Symposium, Okinawa*, 4–8.

Fukami, H., Omori, M., Shimoike, K., Hayashibara, T., & Hatta, M. 2003. Ecological and genetic aspects of reproductive isolation by different spawning times in Acropora corals. *Marine Biology* 142:679–684.

Fuller, Z. L., Mocellin, V. J., Morris, L. A., Cantin, N., Shepherd, J., Sarre, L., Peng, J., et al. 2020. Population genetics of the coral *Acropora millepora*: Toward genomic prediction of bleaching. *Science* 369(6501).

Gilmour, J. P., Underwood, J. N., Howells, E. J., Gates, E., & Heyward, A. J. 2016. Biannual spawning and temporal reproductive isolation in Acropora corals. *PLoS One* 11(3):e0150916.

Gladfelter, E. H. 1982. Skeletal development in *Acropora cervicornis*. I: Patterns of calcium carbonate accretion in the axial corallite. *Coral Reefs* 1:45–51.

Goreau, N. I., & Hayes, R. L. 1977. Nucleation catalysis in coral skeletogenesis. *Proceedings of the 3rd International Coral Reef Symposium, Miami* 2:439–445.

Graham, E. M., Baird, A. H., & Connolly, S. R. 2008. Survival dynamics of scleractinian coral larvae and implications for dispersal. *Coral Reefs* 27:529–539.

Graham, E. M., Baird, A. H., Connolly, S. R., Sewell, M. A., & Willis, B. L. 2013. Rapid declines in metabolism explain extended coral larval longevity. *Coral Reefs* 32:539–549.

Harii, S., Nadaoka, K., Yamamoto, M., & Iwao, K. 2007. Temporal changes in settlement, lipid content and lipid composition of larvae of the spawning hermatypic coral *Acropora tenuis*. *Marine Ecology Progress Series* 346:89–96.

Harii, S., Yasuda, N., Rodriguez-Lanetty, M., Irie, T., & Hidaka, M. 2009. Onset of symbiosis and distribution patterns of symbiotic dinoflagellates in the larvae of scleractinian corals. *Marine Biology* 156:1203–1212.

Harrison, P. L. 2011. Sexual reproduction of scleractinian corals. In *Coral Reefs: An Ecosystem in Transition*. eds. Z. Dubinsky & N. Stambler, 59–85. Springer, Dordrecht.

Harrison, P. L., & Wallace, C. C. 1990. A review of reproduction, larval dispersal and settlement of scleractinian corals. In *Ecosystems of the World 25 Coral Reefs*. ed. Z. Dubinsky, 133–196. Elsevier, Amsterdam.

Hayashibara, T., Kimura, T., & Hatta, M. 2000. Changes of cnida composition during planula development of a reef-building coral *Acropora nasuta*. *Journal of the Japanese Coral Reef Society* 2000(2):39–42.

Hayashibara, T., Ohike, S., & Kakinuma, Y. 1997. Embryonic and larval development and planula metamorphosis of four gamete-spawning Acropora (Anthozoa, Scleractinia). *Proceedings of the 8th International Coral Reef Symposium, Panama* 2:1231–1236.

Hayward, D. C., Grasso, L. C., Saint, R., Miller, D. J., & Ball, E. E. 2015. The organizer in evolution: Gastrulation and organizer gene expression highlight the importance of Brachyury during development of the coral, *Acropora millepora*. *Developmental Biology* 399:337–347.

Hayward, D. C., Hetherington, S., Behm, C. A., Grasso, L. C., Forêt, S., Miller, D. J., & Ball, E. E. 2011. Differential gene expression at coral settlement and metamorphosis—A subtractive hybridization study. *PLoS One* 6(10):e26411.

Hayward, D. C., Miller, D. J., & Ball, E. E. 2004. Snail expression during embryonic development of the coral Acropora: Blurring the diploblast/triploblast divide? *Development Genes and Evolution* 214:257–260.

Hayward, D. C., Samuel, G., Pontynen, P. C., Catmull, J., Saint, R., Miller, D. J., & Ball, E. E. 2002. Localized expression of a dpp/BMP2/4 ortholog in a coral embryo. *Proceedings of the National Academy of Sciences* 99:8106–8111.

Herrera, M., Klein, S. G., Schmidt-Roach, S., Campana, S., Cziesielski, M. J., Chen, J. E., Duarte, C. M., et al. 2020. Unfamiliar partnerships limit cnidarian holobiont acclimation to warming. *Global Change Biology* 26:5539–5553.

Hirose, M., Yamamoto, H., & Nonaka, M. 2008. Metamorphosis and acquisition of symbiotic algae in planula larvae and primary polyps of *Acropora* spp. *Coral Reefs* 27:247–254.

Hislop, N. R., de Jong, D., Hayward, D. C., Ball, E. E., & Miller, D. J. 2005. Tandem organization of independently duplicated homeobox genes in the basal cnidarian *Acropora millepora*. *Development Genes and Evolution* 215:268–273.

Hock, K., Doropoulos, C., Gorton, R., Condie, S. A., & Mumby, P. J. 2019. Split spawning increases robustness of coral larval supply and inter-reef connectivity. *Nature Communications* 10:1–10.

Hoeksema, B. W., & Cairns, S. 2020. World list of Scleractinia. Acropora Oken, 1815. Accessed through: World Register of Marine Species. http://marinespecies.org/aphia.php?p=taxdetails&id=205469 on 2020-09-15.

Huang, D., Benzoni, F., Arrigoni, R., Baird, A. H., Berumen, M. L., Bouwmeester, J., Chou, L. M., et al. 2014. Towards a phylogenetic classification of reef corals: The Indo-Pacific genera Merulina, Goniastrea and Scapophyllia (Scleractinia, Merulinidae). *Zoologica Scripta* 43:531–548.

Hughes, T. P., Anderson, K. D., Connolly, S. R., Heron, S. F., Kerry, J. T., Lough, J. M., Baird, A. H., et al. 2018. Spatial and temporal patterns of mass bleaching of corals in the Anthropocene. *Science* 359:80–83.

Hughes, T. P., Kerry, J. T., Álvarez-Noriega, M., Álvarez-Romero, J. G., Anderson, K. D., Baird, A. H., Babcock, R. C., et al. 2017. Global warming and recurrent mass bleaching of corals. *Nature* 543:373–377.

Iwao, K., Fujisawa, T., & Hatta, M. 2002. A cnidarian neuropeptide of the GLWamide family induces metamorphosis of reef-building corals in the genus Acropora. *Coral Reefs* 2:127–129.

Jouiaei, M., Sunagar, K., Federman Gross, A., Scheib, H., Alewood, P. F., Moran, Y., & Fry, B. G. 2015. Evolution of an ancient venom: Recognition of a novel family of cnidarian toxins and the common evolutionary origin of sodium and potassium neurotoxins in sea anemone. *Molecular Biology and Evolution* 32:1598–1610.

Karabulut, A., He, S., Chen, C. Y., McKinney, S. A., & Gibson, M. C. 2019. Electroporation of short hairpin RNAs for rapid and efficient gene knockdown in the starlet sea anemone, Nematostella vectensis. *Developmental Biology* 448:7–15.

Keith, S. A., Maynard, J. A., Edwards, A. J., Guest, J. R., Bauman, A. G., Van Hooidonk, R., Heron, S. F., et al. 2016. Coral mass spawning predicted by rapid seasonal rise in ocean temperature. *Proceedings of the Royal Society B: Biological Sciences* 283:20160011.

Kenkel, C. D., Traylor, M. R., Wiedenmann, J., Salih, A., & Matz, M. V. 2011. Fluorescence of coral larvae predicts their settlement response to crustose coralline algae and reflects stress. *Proceedings of the Royal Society B: Biological Sciences* 278:2691–2697.

Kenyon, J. C. 1997. Models of reticulate evolution in the coral genus Acropora based on chromosome numbers: Parallels with plants. *Evolution* 51:756–767.

Kitahara, M. V., Fukami, H., Benzoni, F., & Huang, D. 2016. The new systematics of Scleractinia: Integrating molecular and morphological evidence. In *The Cnidaria, Past, Present and Future: The World of Medusa and Her Sisters*. eds. S. Goffredo & Z. Dubinsky, 41–59. Springer, Cham.

Kitchen, S. A., Ratan, A., Bedoya-Reina, O. C., Burhans, R., Fogarty, N. D., Miller, W., & Baums, I. B. 2019. Genomic variants among threatened Acropora corals. *G3: Genes, Genomes, Genetics* 9:1633–1646.

Kojis, B. L. 1986a. Sexual reproduction in Acropora (Isopora) species (Coelenterata: Scleractinia). I. *A. cuneata* and *A. palifera* on Heron Island reef, Great Barrier Reef. *Marine Biology* 91:291–309.

Kojis, B. L. 1986b. Sexual reproduction in Acropora (Isopora) species (Coelenterata: Scleractinia). II. Latitudinal variation in *A. palifera* from the Great Barrier Reef and Papua New Guinea. *Marine Biology* 91:311–318.

Kortschak, R. D., Samuel, G., Saint, R., & Miller, D. J. 2003. EST analysis of the cnidarian *Acropora millepora* reveals extensive gene loss and rapid sequence divergence in the model invertebrates. *Current Biology* 13:2190–2195.

Laissue, P. P., Roberson, L., Gu, Y., Qian, C., & Smith, D. J. 2020. Long-term imaging of the photosensitive, reef-building coral *Acropora muricata* using light-sheet illumination. *Scientific Reports* 10:1–12.

LaJeunesse, T. C., Parkinson, J. E., Gabrielson, P. W., Jeong, H. J., Reimer, J. D., Voolstra, C. R., & Santos, S. R. 2018. Systematic revision of Symbiodiniaceae highlights the antiquity and diversity of coral endosymbionts. *Current Biology* 28:2570–2580.

LeTissier, M. D. A. 1988. Patterns of formation and the ultrastructure of the larval skeleton of Pocillopora damicornis. *Marine Biology* 98:493–501.

Linnaeus, C. 1758. Systema naturae (ed. 10) 1:1–824. Laurentii Salvii, Holmiae.

Liu, H., Stephens, T. G., Galzalez-Pech, R., Beltran, V. H., Lapeyre, B., Bongaerts, P., Cooke, I., et al. 2018. *Symbiodinium* genomes reveal adaptive evolution of functions related to coral-dinoflagellate symbiosis. *Communications Biology* 1:95.

Madin, J. S., Hoogenboom, M. O., Connolly, S. R., Darling, E. S., Falster, D. S., Huang, D., Keith, S. A., et al. 2016. A trait-based approach to advance coral reef science. *Trends in Ecology & Evolution* 31:419–428.

Mao, Y., Economo, E. P., & Satoh, N. 2018. The roles of introgression and climate change in the rise to dominance of Acropora corals. *Current Biology* 28:3373–3382.

Mao, Y., & Satoh, N. 2019. A likely ancient genome duplication in the speciose reef-building coral genus, Acropora. *iScience* 13:20–32.

Matthews, J. L., Raina, J. B., Kahlke, T., Seymour, J. R., van Oppen, M. J., & Suggett, D. J. 2020. Symbiodiniaceae-bacteria interactions: Rethinking metabolite exchange in reef-building corals as multi-partner metabolic networks. *Environmental Microbiology* 22:1675–1687.

McIlroy, S. E., Wong, J. C., & Baker, D. M. 2020. Competitive traits of coral symbionts may alter the structure and function of the microbiome. *ISMEJ* 14:2424–2432.

Meyer, E., Aglyamova, G. V., & Matz, M. V. 2011. Profiling gene expression responses of coral larvae (*Acropora millepora*) to elevated temperature and settlement inducers using a novel RNA-Seq procedure. *Molecular Ecology* 20:3599–3616.

Meyer, E., Aglyamova, G. V., Wang, S., Buchanan-Carter, J., Abrego, D., Colbourne, J. K., Willis, B. L., et al. 2009. Sequencing and de novo analysis of a coral larval transcriptome using 454 GSFlx. *BMC Genomics* 10:219.

Miller, D. J., Ball, E. E., Forêt, S., & Satoh, N. 2011. Coral genomics and transcriptomics—Ushering in a new era in coral biology. *Journal of Experimental Marine Biology and Ecology* 408:114–119.

Mohamed, A. R., Andrade, N., Moya, A., Chan, C. X., Negri, A. P., Bourne, D. G., Ying, H., et al. 2020a. Dual RNA-seq analyses of a coral and its native symbiont during the establishment of symbiosis. *Molecular Ecology*. DOI:10.1111/mec.15612.

Mohamed, A. R., Chan, C. X., Ragan, M. A., Zhang, J., Cooke, I., Ball, E. E., & Miller, D. J. 2020b. Comparative transcriptomic analyses of Chromera and Symbiodiniaceae. *Environmental Microbiology Reports* 12:435–443.

Morita, M., & Kitanobo, S. 2020. Reproduction in the coral Acropora. In *Reproduction in Aquatic Animals*, eds. M Yoshida & J. Asturiano, 167–177. Springer, Singapore.

Morita, M., Nishikawa, A., Nakajima, A., Iguchi, A., Sakai, K., Takemura, A., & Okuno, M. 2006. Eggs regulate sperm flagellar motility initiation, chemotaxis and inhibition in the coral *Acropora digitifera*, *A. gemmifera* and *A. tenuis*. *Journal of Experimental Biology* 209:4574–4579.

Morrow, K., Muller, E., & Lesser, M. 2018. How does the coral microbiome cause, respond to, or modulate the bleaching process? In *Coral Bleaching: Patterns, Processes, Causes and Consequences*, 2nd edn, eds. M. J. H. van Oppen & J. M. Lough, 153–188. Springer, Berlin.

Morse, A. N., Iwao, K., Baba, M., Shimoike, K., Hayashibara, T., & Omori, M. 1996. An ancient chemosensory mechanism brings new life to coral reefs. *The Biological Bulletin* 191:149–154.

Moya, A., Huisman, L., Ball, E. E., Hayward, D. C., Grasso, L. C., Chua, C. M., Woo, H. N., et al. 2012. Whole transcriptome analysis of the coral *Acropora millepora* reveals complex responses to CO_2-driven acidification during the initiation of calcification. *Molecular Ecology* 21:2440–2454.

Moya, A., Sakamaki, K., Mason, B. M., Huisman, L., Forêt, S., Weiss, Y., Bull, T. E., et al. 2016. Functional conservation of the apoptotic machinery from coral to man: The diverse and complex Bcl-2 and caspase repertoires of *Acropora millepora*. *BMC Genomics* 17:62.

Muir, P., Wallace, C., Bridge, T. C., & Bongaerts, P. 2015. Diverse staghorn coral fauna on the mesophotic reefs of north-east Australia. *PLoS One* 10(2):e0117933.

Nature Conservation Division D.o.E.A., Okinawa Prefectural Government. 2017. Report of the Coral Reef Preservation and Restoration Project. p. 68.

Negri, A. P., Webster, N. S., Hill, R. T., & Heyward, A. J. 2001. Metamorphosis of broadcast spawning corals in response to bacteria isolated from crustose algae. *Marine Ecology Progress Series* 223:121–131.

O'Brien, P. A., Webster, N. S., Miller, D. J., & Bourne, D. G. 2019. Host-microbe coevolution: Applying evidence from model systems to complex marine invertebrate holobionts. *MBio* 10(1), e02241–18.

Oken, L. 1815. Steinkorallen. *Lehrbuch Naturgesch* 3:59–74.

Okubo, N., Hayward, D. C., Forêt, S., & Ball, E. E. 2016. A comparative view of early development in the corals *Favia lizardensis*, *Ctenactis echinata*, and *Acropora millepora*-morphology, transcriptome, and developmental gene expression. *BMC Evolutionary Biology* 16:48.

Okubo, N., & Motokawa, T. 2007. Embryogenesis in the reef-building coral *Acropora* spp. *Zoological Science* 24:1169–1177.

Pearse, V. B., & Muscatine, L. 1971. Role of symbiotic algae (zooxanthellae) in coral calcification. *The Biological Bulletin* 141:350–363.

Pollock, F. J., Katz, S. M., van der Water, A. J. M., Davies, S. W., Hein, M., Torda, G., Matz, M. V., et al. 2017. Coral larvae for restoration and research: A large-scale method for rearing *Acropora millepora* larvae, inducing settlement, and establishing symbiosis. *PeerJ* 5:e3732.

Quiroga-Artigas, G., Alexandrea, D., Katelyn, L., Justin, W., & Schnitzler, C. E. 2020. Gene knockdown via electroporation of short hairpin RNAs in embryos of the marine hydroid *Hydractinia symbiolongicarpus*. *Scientific Reports* 10:1.

Ramos-Silva, P., Kaandorp, J., Huisman, L., Marie, B., Zanella-Cléon, I., Guichard, N., Miller, D. J., et al. 2013. The skeletal proteome of the coral *Acropora millepora*: The evolution of calcification by co-option and domain shuffling. *Molecular Biology and Evolution* 30:2099–2112.

Randall, C. J., Giuliano, C., Mead, D., Heyward, A. J., & Negri, A. P. 2019. Immobilisation of living coral embryos and larvae. *Scientific Reports* 9:14596.

Renema, W., Pandolfi, J. M., Kiessling, W., Bosellini, F. R., Klaus, J. S., Korpanty, C., Rosen, B. R., et al. 2016. Are coral reefs victims of their own past success? *Science Advances* 2:e1500850.

Reyes-Bermudez, A., Lin, Z., Hayward, D. C., Miller, D. J., & Ball, E. E. 2009. Differential expression of three galaxin-related genes during settlement and metamorphosis in the scleractinian coral *Acropora millepora*. *BMC Evolutionary Biology* 9:1–12.

Richards, Z. T., Berry, O., & van Oppen, M. J. 2016. Cryptic genetic divergence within threatened species of Acropora coral from the Indian and Pacific Oceans. *Conservation Genetics* 17:577–591.

Robbins, S. J., Singleton, C. M., Chan, C. X., Messer, L. F., Geers, A. U., Ying, H., Baker, A., et al. 2019. A genomic view of the reef-building coral *Porites lutea* and its microbial symbionts. *Nature Microbiology* 4912:2090–2100.

Romano, S. L., & Palumbi, S. R. 1996. Evolution of scleractinian corals inferred from molecular systematics. *Science* 271:640–642.

Santodomingo, N., Novak, V., Pretković, V., Marshall, N., Di Martino, E., Capelli, E. L. G., Roesler, A., et al. 2015. A diverse patch reef from turbid habitats in the middle Miocene (East Kalimantan, Indonesia). *Palaios* 30:128–149.

Sheets, E. A., Warner, P. A., & Palumbi, S. R. 2018. Accurate population genetic measurements require cryptic species identification in corals. *Coral Reefs* 37:549–563.

Shinzato, C., Khalturin, K., Inoue, J., Zayasu, Y., Kanda, M., Kawamitsu, M., Yoshioka, Y., et al. 2020. Eighteen coral genomes reveal the evolutionary origin of Acropora strategies to accommodate environmental changes. *Molecular Biology and Evolution*. DOI:10.1093/molbev/msaa216.

Shinzato, C., Shoguchi, E., Kawashima, T., Hamada, M., Hisata, K., Tanaka, M., Fujie, M., et al. 2011. Using the *Acropora digitifera* genome to understand coral responses to environmental change. *Nature* 476:320–323.

Strader, M. E., Aglyamova, G. V., & Matz, M. V. 2016. Red fluorescence in coral larvae is associated with a diapause-like state. *Molecular Ecology*. 25:559–569.

Strader, M. E., Aglyamova, G. V., & Matz, M. V. 2018. Molecular characterization of larval development from fertilization to metamorphosis in a reef-building coral. *BMC Genomics* 19:1–17.

Sunagawa, S., DeSalvo, M. K., Voolstra, C. R., Reyes-Bermudez, A., & Medina, M. 2009. Identification and gene expression analysis of a taxonomically restricted cysteine-rich protein family in reef-building corals. *PLoS One* 4:e4865.

Takeuchi, T., Yamada, L., Shinzato, C., Sawada, H., & Satoh, N. 2016. Stepwise evolution of coral biomineralization revealed with genome-wide proteomics and transcriptomics. *PLoS One* 11:e0156424.

Tebben, J., Motti, C. A., Siboni, N., Tapiolas, D. M., Negri, A. P., Schupp, P. J., Kitamura, M., et al. 2015. Chemical mediation of coral larval settlement by crustose coralline algae. *Scientific Reports* 5(1):1–11.

Tebben, J., Tapiolas, D. M., Motti, C. A., Abrego, D., Negri, A. P., Blackall, L. L. Steinberg, P. D., et al. 2011. Induction of larval metamorphosis of the coral *Acropora millepora* by tetrabromopyrrole isolated from a Pseudoalteromonas bacterium. *PLoS One* 6(4):e19082.

Vallina, S. M., & Simó, R. 2007. Strong relationship between DMS and the solar radiation dose over the global surface ocean. *Science* 315:506–508.

van Oppen, M. J. H., & Medina, M. 2020. Coral evolutionary responses to microbial symbioses. *Philosophical Transactions of the Royal Society B* 375:20190591.

van Oppen, M. J. H., Oliver, J. K., Putnam, H. M., & Gates, R. D. 2015. Building coral reef resilience through assisted evolution. *Proceedings of the National Academy of Sciences* 112:2307–2313.

Vandermeulen, J. H. 1975. Studies on reef corals. III: Fine structural changes of calicoblast cells in *Pocillopora damicornis* during settling and calcification. *Marine Biology* 31:69–77.

Veron, J. E. N. 1995. *Corals in Space and Time: The Biogeography and Evolution of the Scleractinia*. Cornell University Press, Ithaca, NY.

Veron, J. E. N., & Wallace, C. C. 1984. Scleractinia of Eastern Australia: Part V. Family Acroporidae. *Australian Institute of Marine Science Monograph Series* 6:1–485.

Verrill, A. E. 1901. Variations and nomenclature of Bermudian, West Indian and Brazilian reef corals, with notes on various Indo-Pacific corals. *Transactions of the Connecticut Academy of Arts and Sciences* 11:63–168.

Wallace, C. C. 1999. *Staghorn corals of the world: A revision of the coral genus Acropora (Scleractinia; Astrocoeniina; Acroporidae) worldwide, with emphasis on morphology, phylogeny and biogeography*. CSIRO publishing. Collingwood, Victoria.

Wallace, C. C., Chen, C. A., Fukami, H., & Muir, P. R. 2007. Recognition of separate genera within Acropora based on new morphological, reproductive and genetic evidence from *Acropora togianensis*, and elevation of the subgenus Isopora Studer, 1878 to genus (Scleractinia: Astrocoeniidae; Acroporidae). *Coral Reefs* 26:231–239.

Wallace, C. C., Done, B. J., & Muir, P. R. 2012. Revision and catalogue of worldwide staghorn corals Acropora and Isopora (Scleractina: Acroporidae) in the Museum of Tropical Queensland. *Memoirs of the Queensland Museum—Nature* 57:1–255.

Wallace, C. C., & Rosen, B. R. 2006. Diverse staghorn corals (Acropora) in high-latitude Eocene assemblages: Implications for the evolution of modern diversity patterns of reef corals. *Proceedings of the Royal Society B: Biological Sciences* 273:975–982.

Willis, B. L., Babcock, R. C., Harrison, P. L., & Wallace, C. C. 1997. Experimental hybridization and breeding incompatibilities within the mating systems of mass spawning reef corals. *Coral Reefs* 16:S53–S65.

Yasuoka, Y., Shinzato, C., & Satoh, N. 2016. The mesoderm-forming gene Brachyury regulates ectoderm-endoderm demarcation in the coral *Acropora digitifera*. *Current Biology* 26:2885–2892.

Ying, H., Hayward, D. C., Cooke, I., Wang, W., Moya, A., Siemering, K. R., Sprungala, S., et al. 2019. The whole-genome sequence of the coral *Acropora millepora*. *Genome Biology and Evolution* 11:1374–1379.

Ying, H., Cooke, I., Sprungala, S., Wang, W., Hayward, D. C., Tang, Y., Huttley, G., et al. 2018. Comparative genomics reveals the distinct evolutionary trajectories of the robust and complex coral lineages. *Genome Biology* 19:175.

Zoccola, D., Ounais, N., Barthelemy, D., Calcagno, R., Gaill, F., Henard, S., Hoegh-Guldberg, O., et al. 2020. The World Coral Conservatory: A Noah's ark for corals to support survival of reef ecosystems. *PLoS Biology* 18(9):e3000823.

11 *Stylophora pistillata*—A Model Colonial Species in Basic and Applied Studies

Dor Shefy and Baruch Rinkevich

CONTENTS

11.1 HISTORY OF THE MODEL

Stylophora pistillata (Pocilloporidae; Scleractinia) is a common Indo-Pacific branching coral species, also known by the common name smooth cauliflower coral (Figures 11.1, 11.2, 11.3). This species was first named more than 220 years ago as *Madrepora pistillata* (Esper 1797) (Figure 11.1a), which was followed by many synonymous names in this period, until it stabilized on the current name. To the best of our knowledge, the first focused study on the biology of this species was engaged with sexual reproduction, settlement and metamorphosis in Palau's colonies (Atoda 1947a). Three decades later, Loya (1976) referred to some ecological attributes of this species and suggested that *S. pistillata* from the Red Sea is an "r strategist" species. This work was followed by a wide range of studies, with most performed on *S. pistillata* populations from the Gulf of Aqaba/Eilat (GOA/E; Red Sea) along the Israeli coast. The studies in the late 1970s and early 1980s were focused on the species' reproductive activities and the impacts of oil pollution on sexual reproduction (Loya and Rinkevich 1979; Rinkevich and Loya 1977, 1979a, 1979b, 1979c, 1985b, 1987), allorecognition and ecological interactions (Mokady et al. 1991; Edwards and Emberton 1980; Müller et al. 1984; Rinkevich and Loya 1983a, 1985a; Rinkevich et al. 1991, 1993; Rinkevich and Weissman 1987), as on basic coral physiology, pattern formation and senescence (Dubinsky et al. 1984, 1990; Falkowski and Dubinsky 1981; Falkowski et al. 1984; Loya and Rinkevich 1987; Muscatine et al. 1984, 1985, 1989; Rinkevich 1989; Weis et al. 1989; McCloskey and Muscatine 1984; Rahav et al. 1989; Rinkevich and Loya 1983b, 1983c, 1984a, 1984b, 1986). From the late 80s, more and more studies have focused on *S. pistillata* as a model species in search of a wide range of biological queries, all over the Indo-Pacific area and as an important, sometimes key, species in reef assemblages. Following the observations on coral bleaching events and the high mortality rates that have been documented globally, more attention has been devoted to *S. pistillata*'s metabolism, nutrient uptake and interaction with environmental drivers, making this species a model species for studying the complex interactions between the animal, its symbiotic algae and the

DOI: 10.1201/9781003217503-11

environment (Abramovitch-Gottlib et al. 2003; Dubinsky et al. 1990; Dubinsky and Jokiel 1994; Ferrier-Pagès et al. 2000, 2001, 2003, 2010; Franklin et al. 2004; Grover et al. 2002, 2003, 2006, 2008; Hoegh-Guldberg and Smith 1989a, 1989b; Houlbrèque et al. 2003, 2004; Lampert-Karako et al. 2008; Muscatine et al. 1989; Nakamura et al. 2003; Rahav et al. 1989; Rinkevich 1989; Shashar et al. 1993; Tchernov et al. 2004; Titlyanov et al. 2000a; Titlyanov et al. 2000b; Titlyanov et al. 2001; Weis et al. 1989). The accumulated knowledge on the species distribution and the reproductive mode of *S. pistillata* has led researchers to study population dynamics, population genetic structures, modes of reproduction and larval dispersal in a specific reef and among reefs (Ayre and Hughes 2000; Zvuloni et al. 2008; Klueter and Andreakis 2013; Douek et al. 2011; Guerrini et al. 2020; Takabayashi et al. 2003; Nishikawa et al. 2003). *S. pistillata* colonies are also often used for understanding the impacts of anthropogenic activities and climate change disturbances on coral reefs and, together with the rapid advances in technology, scientists have examined the combined effects of anthropogenic/climate change impacts on *S. pistillata*'s biological and ecological parameters (Ammar et al. 2012; Guerrini et al. 2020; Horwitz et al. 2017; Loya and Rinkevich 1979; Shefy et al. 2018; Tamir et al. 2020), physiology (Abramovitch-Gottlib et al. 2003; Banc-Prandi and Fine 2019; Bellworthy and Fine 2017; Bellworthy et al. 2019; Dias et al. 2019; Epstein et al. 2005; Fitt et al. 2009; Grinblat et al. 2018; Hall et al. 2018; Hawkins et al. 2015; Hoegh-Guldberg and Smith 1989b; Krueger et al. 2017; Reynaud et al. 2003; Rinkevich et al. 2005; Rosic et al. 2020; Sampayo et al. 2008; Sampayo et al. 2016; Saragosti et al. 2010; Shick et al. 1999; Stat et al. 2009) and gene expression patterns (Maor-Landaw and Levy 2016; Oren et al. 2010, 2013; Voolstra et al. 2017). Several studies have focused on *in vitro* approaches with *S. pistillata* cells and minute fragments for the development of novel methodologies; cell culture, nubbin and larvae usage for ecotoxicology and for reef restoration and for the elucidation of biological features, such as calcification and algal movements (Bockel and Rinkevich 2019; Danovaro et al. 2008; Downs et al. 2014; Epstein et al. 2000; Frank et al. 1994; Horoszowski-Fridman et al. 2020; Mass et al. 2012, 2017a; Raz-Bahat et al. 2006; Shafir et al. 2001, 2003, 2007, 2014); on anatomical features (Raz-Bahat et al. 2017); and on applied approaches (Rinkevich 2015a; Rinkevich and Shafir 1998; Rinkevich et al. 1999; Shafir et al. 2001). The understanding that coral reefs around the world are degrading has led, in the last two decades, to the development of an additional applied route, an active reef restoration that is based on a wide range of methodologies being tested on *S. pistillata* as a model species (Amar and Rinkevich 2007; Epstein and Rinkevich 2001 Epstein et al. 2001, 2005; Golomb et al. 2020; Horoszowski-Fridman et al. 2015, 2020; Horoszowski-Fridman and Rinkevich 2020; Linden and Rinkevich 2011, 2017; Linden et al. 2019; Rachmilovitz and Rinkevich 2017; Rinkevich 2000, 2015a, 2019a, 2019b; Shafir and Rinkevich 2008, 2010; Shafir et al. 2006a, 2009).

Here, we aim to review the knowledge about *S. pistillata*'s biological features in various scientific disciplines for the last eight decades of research.

11.2 GEOGRAPHICAL LOCATION

S. pistillata colonies are found in shallow waters and up to 70 meters deep (Fishelson 1971; Kramer et al. 2019; Muir and Pichon 2019; Veron 2000). This species has a wide geographical range in the tropical and sub-tropical Indo-Pacific Ocean; central and west Pacific; tropical Australia; South China Sea; southern Japan; central Indian Ocean; southwest and northwest Indian Ocean; Arabian/Iranian Gulf; Gulf of Aden and the Red Sea, including the gulfs of Suez and Aqaba/Eilat (Veron 2000).

11.3 ANATOMY

An *S. pistillata* colony consists of up to tens of thousands of polyps at adulthood, each about 1–2 mm in diameter, where each polyp creates a small skeletal cup (termed a corallite), the hard supporting blueprint of the polyp's tissue (Veron 2000). The external soft tissues of the polyps and their extensions that connect between the polyps (coenosarc) overlie the coral skeleton that is made of calcium carbonate (Veron 2000). The polyps are anchored to the underlying skeletons by cells called desmocytes that connect the lower ectodermic layer (the calicoblastic layer) to the perforated calcium carbonate milieu (Muscatine et al. 1997; Raz-Bahat et al. 2006; Tambutté et al. 2007). Each polyp is a hollow cylindrical blind-ended sac that resembles a sea anemone in structure with a mouth in the center of the polyp, surrounded by 12 hollow retractable tentacles (Figure 11.2d) that connect to the gastric cavity by the pharynx. This is the gateway for food particles to the coelenteron, but studies revealed further roles in chemical digestion (Raz-Bahat et al. 2017). All polyps within a colony are connected to each other via a network of cell-lined tubes (gastrovascular canals) that radiate from the gastric cavity of the polyps. The polyp's internal gastric cavity is divided by 12 partitions (mesenteries; 6 are complete) into compartments which run radially from the body wall's gastrodermis to the actinopharynx and are connected to the pharynx carrying six long extensions (mesenterial filaments; Raz-Bahat et al. 2017). Two types of mesenterial filaments exist in *S. pistillata*, distinct, as much to be known by general morphology: four short filaments with no secretory cells and two long convoluted filaments with stinging and secretory cells (Raz-Bahat et al. 2017) that penetrate the gastric cavity and into the gastrovascular canals. The compartments between the mesenteries are also the sites where male and female gonads are developed (Ammar et al. 2012; Rinkevich and Loya 1979a). As in all corals, each polyp and the connected coenosarc consist of two epithelial layers, the ectodermic and gastrodermis (endodermis), separated by the mesoglea. This non-epithelial milieu binds the two epithelial layers together throughout the colony while consisting of a gelatinous substance, with collagen fibers and some cells. The columnar ectodermic layer contains mucus gland cells, nematocytes and spyrocyte cells, and the gastrodermis layer contains the zooxanthellae (Al-Sofyani 1991; Raz-Bahat et al. 2017, Bockel and Rinkevich 2019). The tentacles that are located above the oral disk are loaded with zooxanthellae in their gastrodermis cells, while the epidermis contains nematocytes.

As mentioned, the skeleton is secreted by the calicoblastic tissue (also named calicodermis), which forms the lower ectodermal layer (Allemand et al. 2004, 2011). The calicoblastic epithelium is very thin and has only calicoblastic cells anchored to the skeleton by the desmocytes (Raz-Bahat et al. 2006; Tambutté et al. 2007). The calicoblastic epithelium secretes amorphous nano-calcium carbonate crystals into microenvironments enriched in organic material. The carbonate crystals aggregate and then crystallize to create ordered aragonitic structures (Mass et al. 2017b; Von Euw et al. 2017). On the coenosteum (skeleton secreted by the coenosarc), skeletal spines called coenosteal spines are developed, and in shallow water colonies, they have granular textures as compared to smoother textures in deeper water colonies (Malik et al. 2020).

11.4 LIFE CYCLE

11.4.1 Sexual Reproduction, Seasonality and General Reproductive Characteristics

While most of the coral species are broadcast spawners, together with other 61 species, *S. pistillata* belongs to a group of brooding coral species, where fertilization and larval development take place inside the polyps (Ammar et al. 2012; Fan and Dai 2002; Rinkevich and Loya 1979a) for an estimated duration of two weeks (Fan and Dai 2002; Shefy et al. 2018). The planula larvae are released to the water column about one to two hours after sunset (Atoda 1947a; Rinkevich and Loya 1979b)

S. pistillata is a hermaphrodite species, and male and female gonads are situated side by side within the polyp's coelenteron, extended into the body cavities and attached to the mesenteries by stalks. Along astogeny, the male gonads appear first when the colonies reach an approximate radius of 2 cm, and female gonads develop a year later (Rinkevich and Loya 1979a). A wide range of anthropogenic and natural stressors may affect gonadal development. Early studies revealed that oil pollution and sedimentation directly reduce male and female gonad numbers and significantly affect the developing planulae (Loya and Rinkevich 1979; Rinkevich and Loya 1979c). Even nutrient-enriched environments may affect gonads and larval development, and while phosphorus load may have a minor impact on the reproductive efforts (Ammar et al. 2012), particulate matter (PM) and particulate organic matter (POM) may increase the size and number of oocytes and testes (Bongiorni et al. 2003a 2003b). Yet, resident fish within coral colonies that secrete nutrients (Liberman et al. 1995) do not have impacts on fecundity, as on the colony color morph (Rinkevich 1982). In contrast, intraspecific (within the same species) and interspecific (with different species) interactions have impacts on the number of female gonads per polyp (Rinkevich and Loya 1985b).

S. pistillata's reproductive patterns, seasonality and reproductive efforts vary among bio-geographical regions. In Palau, Atoda (1947a) recorded planulae release one to two weeks after a full moon all year long. Differences in seasonality are also present in the population at Yabnu (South Res Sea) and Tarut Bay (Arabian Gulf), which are in the same latitude but in different seas. In Tarut Bay, embryos were observed for just two months a year (before seawater temperature exceeded 31°C), while in Yabnu, embryos were documented ten months a year (before temperature exceeded 29°C) (Fadlallah and Lindo 1988). In the Philippines, the reproductive season of *S. pistillata* lasts just three months, from November to January (Baird et al. 2015), while in Taiwan, documentations revealed all-year-round larval release, with no obvious lunar periodicity (Fan and Dai 2002). *S. pistillata* colonies in the southern hemisphere release planulae from August to December in the Great Barrier Reef (GBR) (Tanner 1996) and from August to May with lunar periodicity in south Australia (Villanueva et al. 2008).

The reproduction of the *S. pistillata* populations in the Gulf of Aqaba/Eilat, Red Sea, is a model case for coral reproduction for over five decades, allowing a glimpse of changes in reproduction on an extended time scale. During the 1970s and 1980s, shallow-water *S. pistillata* colonies in Eilat released planulae for seven to eight months (December–July) (Rinkevich and Loya 1979b, 1987). Recent observations revealed that seasonality of larval release during the 2010s is extended by one to two months, from December to September–October (Rinkevich and Loya 1979b, 1987; Shefy et al. 2018) and year-round recruitment (Guerrini et al. 2020). Studies also revealed a bell-shaped curve in the larval release of most *S. pistillata* populations characterized in Eilat by an increase in planulae numbers until reaching a peak and then, in the second half, a decrease in the release until the end of the season (Amar et al. 2007; Fan and Dai 2002; Rinkevich and Loya 1979a, 1979b, 1987; Shefy et al. 2018; Tanner 1996). Fecundity among different colonies (even those of the same size that are situated side by side in the reef) or within a coral colony over several reproductive seasons is portrayed by high variability (Rinkevich and Loya 1987; Shefy et al. 2018). Variation is also recorded for lunar periodicity that was assigned for some populations (Atoda 1947a; Dai et al. 1992; Fan and Dai 2002; Tanner 1996; Villanueva et al. 2008; Zakai et al. 2006) while missing in others (Linden et al. 2018; Rinkevich and Loya 1979b). Linden et al. (2018) revealed that larval release by *S. pistillata* colonies does not comply with the assumed entrainment by the lunar cycle, further documenting that the lunar cycle does not provide a strict zeitgeber and can better be classified as a circatrigintan pattern. Water temperature and solar radiation did not correlate significantly with larval release.

FIGURE 11.1 (a) The first description from 1797 of *Stylophora pistillata* (assigned the name *Madrepora pistillata*) by Eugenius Johann Christoph Esper in his book: *Fortsetzungen der Pflanzenthiere in Abbildungennach der Natur mit Farben erleuchtet nebst Beschreibungen.* (b–c) *S. pistillata* colonies representing two common color morphs (Gulf of Aqaba/Eilat). (d) The *S. danae* morphotype of *S. pistillata* (South Sinai, Red Sea; following Stefani et al. 2011). (e) Two juvenile colonies in allogeneic contact, rejecting each other (sensu Rinkevich and Loya 1983a) marked by the black arrowhead. (Photographs [b–e] courtesty of D. Shefy.)

11.4.2 Planulae, Metamorphosis and Settlement

Without an efficient sexual reproduction process and successful settlement (recruitment) of coral larvae, a coral reef will not grow and thrive. For recruitment, the planula larvae need to find suitable substrates to settle and to develop. The ball-shaped planulae are released from the polyp mouths of shallow water *S. pistillata* colonies with the oral part upward and then alter to 1–2-mm-long rod-like-shaped swimming larvae (Figure 11.2a, b; Rinkevich and Loya 1979a). Planulae from mesophotic colonies are smaller than shallow-water planulae, contain different symbiont clades and have lower GFP-like chromoprotein mRNA expression (Scucchia et al. 2020; Rinkevich and Loya 1979a; Byler et al. 2013; Lampert-Karako et al. 2008; Winters et al. 2009). Planulae are released to the water loaded with zooxanthellae inherited from the mother colony (vertical transmission) but can also acquire zooxanthellae from the water column (horizontal transmission) (Byler et al. 2013).

Similar to other Pocilloporidae species, the planulae of *S. pistillata* settle within a few hours upon release, with the majority settling in the first 48 hours upon release (Amar et al. 2007; Atoda 1947a; Atoda 1947b; Atoda 1951; Nishikawa et al. 2003; Richmond 1997; Wallace and Harrison 1990). Unlike other coral species, these planulae settle and metamorphose on any available substrate, including natural hard layers, manmade and fabricated substrates (glass, plastic, metal, concrete, etc.), such as on water upper-surface tension layers under laboratory conditions (Nishikawa et al. 2003; Putnam et al. 2008; Rinkevich and Loya 1979a), and metamorphose to primary polyps, even without the presence of crustose coralline algae (CCA) or preconditioned biofilm (Amar et al. 2007; Atoda 1947a; Baird and Morse 2004; Heyward and Negri 1999; Nishikawa et al. 2003; Putnam et al.2008; Rinkevich and Loya 1979a). In Eilat, Red Sea, year-round recruitment has recently been documented (Guerrini et al. 2020). Planulae settlement is associated with mucus secretion from aboral epidermal cells, followed by flattened larvae that form disc-like shapes and the completion of basal plates carrying 24 basal ridges toward the formation of columellas three to four days post-settlement (Baird and Babcock 2000). Planulae settle either separate from each other or in aggregates, a distribution setting that leads to allogeneic contacts between adjacent spat either to morphological fusions into coral chimeras or allogeneic rejections characterized by necrotic areas and pseudo-fusion events (Figure 11.1e; Amar et al. 2007; Frank et al. 1997; Linden and Rinkevich 2017; Raymundo and Maypa 2004; Rinkevich 2011). Aggregated settlement and chimerism have further been documented in other marine invertebrates and are claimed to benefit coral chimeras through an immediate increase in colonial size and survival rates (Amar et al. 2008; Puill-Stephan et al. 2012; Raymundo and Maypa 2004; Rinkevich 2019b).

11.4.3 Colony Formation, Growth and Survivorship

Colonial astogeny occurs through iterated polyp buddings, with an axially rod-like growth form of branches where each branch consists of numerous small polyps, with a colonial symmetry that approximates a sphere (Loya 1976), all configured by a pre-designed colonial architecture (Rinkevich 2001, 2002) and nutritional resources that provide positional information for colonial structures (Kücken et al. 2011). Settled primary polyps start to deposit calcareous skeletons from one day following metamorphosis, which bud in extra-tentacular mode, starting from one to two weeks following settlement, a process that adds up to six additional polyps as a circlet around the primary polyp, all further forming the basal plate which is the initial colonial anchor to the substrate. Growth rates of new polyps over time are highly variable among young colonies (Frank et al. 1997). At some yet-unidentified stage, branches initiate by apical growth, usually just as a single apical ramified structure from each basal plate. New upgrowing and side-growing branches are then added by dichotomous fission at a branch tip (Rinkevich 2000, 2001, 2002; Rinkevich and Loya 1985a), developing in conformity with the basic architectural rules of this species, all together forming reiterated complexes (Epstein and Rinkevich 2001; Shaish et al. 2006, 2007; Shaish and Rinkevich 2009). The colony's growth exhibits allometric ratios within the newly developing dichotomous up-growing branches that differ significantly from those of older branches, decrease in growth rates of inward-growing lateral branches and changes in growth directionality of isogeneic branches that risk contiguity (Rinkevich and Loya 1985a). In addition to that, the lack of fusion between closely growing branches within a colony and the retreat growth occasionally recorded between closely growing allogeneic branches (Rinkevich and Loya 1985b) further emphasizes the within-colony genetic background for spatial configuration (Rinkevich 2001, 2002). The deduced genetic control (Rinkevich 2001, 2002; Shaish et al. 2006, 2007; Shaish and Rinkevich 2009), internal and external transport of signals (Kücken et al. 2011; Rinkevich and Loya 1985a) and external and internal nutrients (Rinkevich 1989, 1991) may have substantial impacts on the pattern formation of *S. pistillata* colonies.

The growth of *S. pistillata* can be measured by several methodologies. Linear extension represents the increase in the length of a single branch or the diameter of a colony by units of distance (i.e. mm, cm). Aerial size represents the increase in surface area as viewed from above, in units of surface area (mm^2). Tissue surface area (including all branches) measurements can further be evaluated by wrapping all branches in aluminum foil (Marsh 1970) or by dipping the colony in wax (parafilm) and comparing the wax/aluminum foil weights with calibrated curves of mass increment vs. surface area (Stimson and Kinzie 1991), translating weighs to units of area (mm^2). The parameter of the ecological volume of a colony is the aerial size multiplied by the height and is measured by an increase of the whole space encompassed by the coral branches in mm³ (Shafir and Rinkevich 2010; Shafir et al. 2006b). Other size methods, such as 3D photography for measuring parameters of growth rates (surface areas, volumes, etc.), do not always give accurate results due to the high structural complexity of developing colonies.

S. pistillata is a fast-growing species as compared to massive and encrusting species and some other branching species. Branches can grow up to 5 cm per year, depending on the conditions and the initial fragment size (Dar and Mohamed 2017; Bockel and Rinkevich 2019; Hasan 2019; Liberman et al. 1995; Loya 1976; Shafir and Rinkevich 2010; Shafir et al. 2006b; Tamir et al. 2020), and small fragments can multiply their ecological volumes by 200 times within 8–12 months (Shafir and Rinkevich 2008). In old senescent colonies, calcification rates, as reproductive activities, decrease synchronically in all branches, and the whole colony as a single unit, new and old polyps alike, exhibits senescence concurrently, leading to accelerated degradation and colonial death within few months (Rinkevich and Loya 1986).

S. pistillata colonies that grow under improved water flows (primarily in mid-water floating nurseries) that assist the polyps in catching prey exhibit enhanced growth rates and advanced recovery from bleaching (in all parameters mentioned previously) (Bongiorni et al. 2003a; Nakamura et al. 2003; Shafir and Rinkevich 2010). In contrast to the high and fast growth rates characteristic to *S. pistillata* and although it is one of the most abundant species in the GOA/E (Shaked and Genin 2019; Shlesinger and Loya 2016), adult colonies and primarily recruits have high mortality rates (Doropoulos et al. 2015; Linden and Rinkevich 2011, 2017; Loya 1976; Shafir, Van Rijn, and Rinkevich 2006b; Shlesinger and Loya 2016; Tamir et al. 2020). Assuming 50–80% settlement rates in the wild (Amar et al. 2007; Linden and Rinkevich 2011), only a small portion of recruits will develop into gravid colonies out of tens of millions and more of planulae released during any reproduction season. Under *in-situ* aqua-culture conditions, young colonies can reach a 40–80% survival rate if protected by cages and 10–30% if not protected (Linden and Rinkevich 2017; Shafir et al. 2006b), orders of magnitude above natural figures. Nevertheless, size structure demographic models for *S. pistillata* populations in various reefs were not constructed and are not yet available, in spite of their importance for conservation and management plans (Doropoulos et al. 2015).

11.4.4 Metabolism

In the past four decades, *S. pistillata* has been used as a model species in studies on carbon and nutrients assimilation and their acquisition, allocation and uptake by coral and by symbiotic algae. Since most coral reefs thrive in oligotrophic waters, it is essential to understand nutrient recycling by reef communities, as it may shed light on coral life histories and reef-resilient. *S. pistillata* colonies, as other coral species get carbon and nutrients through two main processes: via photosynthesis, provided by the symbiotic autotrophic algae (Muscatine et al. 1981), and by feeding on particular or dissolved sources of organic carbon (Houlbrèque and Ferrier-Pagès 2009). The symbiotic dinoflagellates cannot provide all the essential carbon and organic nitrogen needed for the coral, especially under low light regimens (Falkowski et al. 1984; Muscatine et al. 1984; Tremblay et al. 2014). Yet corals may modify their algal numbers and their activities. Studies on *S. pistillata* revealed that under high light regimes, respiration and calcification rates increased (Dubinsky and Jokiel 1994), while the symbiotic algae decreased in size and numbers, further showing high respiration and lower quantum yields (Dubinsky et al. 1984). With regard to nitrogen, another limiting nutrient source for the algae (Hoegh-Guldberg and Smith 1989a), increasing concentrations of nitrogen compounds such as ammonium, urea, amino acids, nitrite and nitrate lead to an increase in the nitrogen uptake by the holobiont (Dubinsky and Jokiel 1994; Grover et al. 2002, 2003, 2006, 2008; Houlbre'que and Ferrier-page 2009; Rahav et al. 1989). The fate and path of each nitrogen source, whether consumed via water or by feeding (praying of zooplankton), is mostly determined by light intensity and photosynthetic products (Dubinsky and Jokiel 1994; Houlbrèque and Ferrier-Pagès 2009). Assuming constant low nutrient concentration in the reef, under high light intensity regimes, most of the carbon goes to respiration and growth, including calcification by the host, while under low light, the zooxanthellae use the carbon and nutrients (Dubinsky and Jokiel 1994). Feeding on zooplankton or other pico- and nano-planktonic organisms increases nutrients uptake (including phosphate) that provides the nutrients needed for coral growth and reproduction (Ferrier-Pagès et al. 2003; Houlbrèque et al. 2004; Houlbrèque et al. 2003) and enhances the numbers of zooxanthellae in the coral tissues (Dubinsky et al. 1990; Houlbrèque et al. 2003; Titlyanov et al. 2001; Titlyanov et al. 2000, 2001; Titlyanov et al. 2000). Studies on *S. pistillata*'s symbiotic relationships further revealed the translocation of photosynthates between branches and along a branch within a colony and between genotypes (Rinkevich 1991; Rinkevich and Loya 1983b, 1983c, 1984a), and were used in quest on the "light enhanced calcification" enigma (Houlbrèque et al. 2003; Moya et al. 2006; Muscatine et al. 1984; Reynaud-Vaganay et al. 2001; Rinkevich and Loya 1984b). Despite all the previous studies on *S. pistillata* symbiotic relationships, there is a need for additional studies to reveal the more intimate interactions between the holobiont participants (Ferrier-Pagès et al. 2018; Hédouin et al. 2016; Metian et al. 2015).

11.5 EMBRYOGENESIS

As a hermaphroditic brooder species, *S. pistillata* fertilization and larval development take place within the body cavities of the polyps, thus making it challenging to study embryogenesis and larval development. Rinkevich and Loya (1979a) and then Ammar et al. (2012) observed that male and female gonads, situated on small stalks, start to develop at two and five months, respectively, before the onset of larval

FIGURE 11.2 (a) Planula of *Stylophora pistillata* as a rod-like shape, the oral part facing to the left side of the picture. (b) Planula of *S. pistillata* as a ball-like shape. The planula is "enveloped" by secreted mucus, further revealing the pattern of symbiotic algae (brown dots) that also depict the mesenteries' tissues (ms). (c) A primary polyp, one day after settlement. (d) Extended polyps in *S. pistillata*, each with an open mouth (m) surrounded by 12 tentacles (tn), loaded with zooxanthellae, which give the coral its brown color. (e) The Christmas tree worm *Spirobranchus giganteus* (Polychaete) on top of an *S. pistillata* branch. (f) *Trapezia cymodoce* (Decapoda) "guarding" a juvenile *S. pistillata* colony (red arrowhead). The green arrowheads point to the coral gall crabs *Hapalocarcinus marsupialis* (Cryptochiridae) that modify the morphology of the branch. (Photographs [a–e] courtesy of D. Shefy; [f] courtesy of Y. Shmuel.)

release (reproductive season), filling up the gastric cavities of the polyps during the peak of reproduction season. At the start, 4–16 oocytes per polyp develop; some are absorbed during the development in such a way that only a single mature egg at a specific time is left (Rinkevich and Loya 1979a). The migration of the egg nuclei to the periphery signals that the eggs are ready for fertilization (Rinkevich and Loya 1979a). Larval development is assumed to take 14 days, but the whole development process was not studied (Fan and Dai 2002; Rinkevich and Loya 1979b). Planulae develop in most polyps (except for the sexually sterile branch tips), and upon the release of the larva from a specific polyp, another oocyte becomes ready for fertilization. A mature planula has an organized ectodermal epithelium and a less organized gastrodermis loaded with zooxanthellae, separated by a thin mesoglea, and has six pairs of mesenteries (Figure 11.2a, b) (Atoda 1947a; Fan and Dai 2002; Rinkevich and Loya 1979a; Scucchia et al. 2020). Further, planulae of *S. pistillata* from the Red Sea (not observed in other planulae, including of *S. pistillata* from other places) show temporary extensions from the body wall, consisting of ectodermal-mesogleal material ("filaments") and extensions containing endodermal epithelium only ("nodules") that regularly appear and absorb (Rinkevich and Loya 1979a). The developing larvae are flexible in their morphologies, and, while globular upon release, they appear as pear-like, disk-like or rod-like structures (Figure 11.2a, b) (Atoda 1947a; Rinkevich and Loya 1979a). Planulae of *S. pistillata* that are released from shallow water gravid colonies are fluorescent (Grinblat et al. 2018; Rinkevich and Loya 1979a; Scucchia et al. 2020), with a lower expression of the green fluorescence protein (GFP) gene in planulae originating from >30 m colonies (Scucchia et al. 2020). It has further been documented that planulae start to precipitate minerals in the form of small crystals that may assist in rapid calcification upon settlement (Akiva et al. 2018).

11.6 GENOMIC DATA

Advances, reduced costs of sequencing and improved technologies over the past decade enabled the recent sequencing and assembling of the *S. pistillata* genome (the full sequenced genome can be found at http://spis.reefgenomics.org/) (Banguera-Hinestroza et al. 2013; Voolstra et al. 2017). The sequenced genome enabled studies on evolutionary adaptation and origin of this species (Voolstra et al. 2017), algae–host relationships, gene expression analyses (Barott et al. 2015b; Gutner-Hoch et al. 2017; Karako-Lampert et al. 2014; Liew et al. 2014; Maor-Landaw and Levy 2016) and studies on epigenetics (Dimond and Roberts 2016; Liew et al. 2018). Results further revealed the genes involved in stressed (and not stressed) colonies as the molecular mechanisms for adaptation to global change impacts. *S. pistillata* mitochondrial DNA (mDNA) was used to investigate phylogenetic aspects, species delineation and the taxonomical status of this species (Chen et al. 2008; Flot et al. 2011; Keshavmurthy et al. 2013; Klueter and Andreakis 2013; Stefani et al. 2011), further elucidating that the origin of *S. pistillata* is from the west Indian ocean and that this species presents of up to six distinct morphs. Molecular markers such as ITS1, amplified fragment length polymorphism (AFLP) and allozymes were used to assess the genetic structure among different *S. pistillata* populations, within populations and coral recruits (Amar et al. 2008; Ayre and Hughes 2000; Douek et al. 2011; Takabayashi et al. 2003; Zvuloni et al. 2008) yet are too few to reveal clear genetic landscapes.

11.7 FUNCTIONAL APPROACHES: TOOLS FOR MOLECULAR AND CELLULAR ANALYSES

Despite the claim that *S. pistillata* is a "weedy species" (Loya 1976), the biological characteristics of this species, such as its fast growth rates, abundance and long reproductive season, made *S. pistillata* a model animal in a wide range of ecological settings and for functional approaches. It also helped that while *S. pistillata* colonies present several color morphs (Figure 11.1b, c) (Stambler and Shashar 2007), this diversity has no connection to either ecological feature studied (Rinkevich and Loya 1979b, 1985b).

11.7.1 The Use of *S. pistillata* as a Model Species in Studies on Climate Change and Anthropogenic Impacts

The decline of coral reef resilience and persistence due to anthropogenic impacts and global warming is of great concern for the future of reef ecosystems (Bindoff et al. 2019). *S. pistillata* has further served as a model species for analyzing a wide range of stressors on corals and symbionts, on various life history parameters and on coral adaption to changing environments. These studies further examined the holobiont (coral/algal) symbiotic relationships on the whole-organism level (respiration, calcification rates, survival and photosynthesis), on the cellular level (organelles, lipids, proteins and stress-related proteins) and on a molecular level (DNA damage, gene expression and symbiont identity). In these studies, *S. pistillata* colonies are often used for elucidating coral responses to thermal stress (increasing of seawater temperatures), with consequences that are determined by the specific zooxanthellae species and the coral genotype subjected to specific stress conditions (Sampayo et al. 2008), further associated with alteration in the symbiont clades toward more physiologically suited algal populations (Fitt et al. 2009; Sampayo et al. 2016).

Ex-situ and *in-situ* experiments with *S. pistillata* revealed damages to the thylakoid membranes of the symbiotic algae when colonies are exposed to elevated temperatures and increased light intensities (Tchernov et al. 2004), also following other biological and physiological stresses, all expressed with induced photoinhibition and decreased photosynthesis (Bhagooli and Hidaka 2004; Cohen and Dubinsky 2015; Falkowski and Dubinsky 1981; Franklin et al. 2004; Hawkins et al. 2015; Hoegh-Guldberg and Smith

1989b; Yakovleva et al. 2004), reduced algal density with time (Abramovitch-Gottlib et al. 2003; Biscéré et al. 2018; Cohen and Dubinsky 2015) and decreased protein concentration (Falkowski and Dubinsky 1981; Hoegh-Guldberg and Smith 1989b; Rosic et al. 2020). When evaluating the impacts on the host *S. pistillata* and its responses, studies documented that elevated temperatures increase coral respiration (Hall et al. 2018; Hoegh-Guldberg and Smith 1989b; Reynaud et al. 2003); enforced impacts on calcification rates (mixed results, increase or decrease; Abramovitch-Gottlib et al. 2003; Biscéré et al. 2018; Hall et al. 2018; Reynaud et al. 2003); decreased protein and lipid contents (Falkowski and Dubinsky 1981; Hall et al. 2018; Rosic et al. 2020); imposed fluctuations in ROS and antioxidant enzymes) such as superoxide dismutase [SOD], catalase [CAT], ascorbate peroxidase [APX], glutathione S-transferase [GST] and glutathione peroxidase [GPX]), primarily if light stress was co-involved (Hawkins et al. 2015; Saragosti et al. 2010; Yakovleva et al. 2004); and increased coral mortality rates (Dias et al. 2019). These physiological responses are further reflected in gene expression patterns, including the upregulation of key cellular processes associated with heat stress such as oxidative stress, energy metabolism, DNA repair and apoptosis (Maor-Landaw and Levy 2016). While it is a possibility that higher-latitude *S. pistillata* populations show a general improved tendency for adaptation to temperature changes (Pontasch et al. 2017), the suggestion that *S. pistillata* from the Red Sea specifically went through evolutionary adaptation to heat stress (Fine et al. 2013) made this species a model animal for experiments examining climate change impacts on corals (Bellworthy and Fine 2017; Bellworthy et al. 2019; Bellworthy et al. 2019; Hall et al. 2018; Grottoli et al. 2017; Krueger et al. 2017). Other studies examined the ecological consequences of global change, such as on allogeneic and xenogeneic interactions (Horwitz et al. 2017).

Following the results that *S. pistillata* colonies accumulate metal from seawater (Ali et al. 2011; Al-Sawalmih et al. 2017; Ferrier-Pagès et al. 2005), studies have further investigated *S. pistillata* holobiont responses to metal pollution and the combined effects with warming seas. High concentrations of copper have negative impacts on the holobiont, expressed as a decrease in photosynthesis efficiency, algal density, host respiration rate and host protein and increase in SOD activity, especially when combined with elevated temperature (Banc-Prandi and Fine 2019). Biscéré et al. (2018) further found that while manganese (Mn) enhances cellular chlorophyll concentration and photosynthesis efficiency and increases *S. pistillata* resistance to heat stress, and iron (Fe) positively affects the holobiont and symbionts (Biscéré et al. 2018; Shick et al. 2011), seawater enriched with Mn and iron decreases calcification and induces bleaching. Increased concentrations of Cobalt (Co) inflicted decreased growth rates under ambient pH conditions and in lower-pH water but had no impacts on photosynthesis under ambient pH conditions (Biscéré et al. 2015).

Numerous studies used *S. pistillata* as a model coral species to investigate the impacts of a wide range of pollutants on corals, such as oil pollution, sunscreen lotion detergents and eutrophication. Results revealed that some sunscreen ingredients might induce extensive necrosis in the coral's epidermis and gastrodermis layers (Downs et al. 2014), impair photosynthetic activity (Fel et al. 2019) and promote viral infection followed by bleaching (Danovaro et al. 2008). *In-situ* and *ex-situ* experiments showed that crude oil and its derivatives have a destructive effect on sexual reproduction in *S. pistillata* by reducing the number of female gonads per polyp (Rinkevich and Loya 1979c), by inducing the abortion of planulae (Epstein et al. 2000; Loya and Rinkevich 1979), by decreasing the settlement rate (Epstein et al. 2000), through DNA damage (Kteifan et al. 2017) and by intensifying coral and larval mortalities (Epstein et al. 2000). The same applies to detergents in seawater that impair basic *S. pistillata* biological features (Shafir et al. 2014) and anti-fouling compounds (Shafir et al. 2009). Studies also revealed that under various scenarios for nutrient-enriched environments, eutrophication even enhances *S. pistillata* performance, as colonies exhibited increased growth rates (Bongiorni et al. 2003a, 2003b), increases in host mitochondrial and protein concentrations (Kramarsky-Winter et al. 2009; Sawall et al. 2011), decreases in oxidation (Kramarsky-Winter et al. 2009) and increases in teste and egg numbers with a decrease in their size (Ammar et al. 2012; Bongiorni et al. 2003a). The healthy physiological status, in contrast to lab experiment results, suggests that the corals gain more energy through heterotrophy (increase in zooplankton) rather than autotrophy (Rinkevich 2015c).

Light has a significant role in marine invertebrates' biological clocks and is a cue in the regulation of circadian rhythms (zeitgeber) and physiological processes. Therefore, *S. pistillata* was further used as a model species for light pollution, following the observation that the coral reefs in the northern tip of the GOA/E, Red Sea, are heavily subjected to artificial light pollution at night (ALAN) (Aubrecht et al. 2008; Tamir et al. 2017). Shefy et al. (2018) postulated that changes in the length of the reproductive season in *S. pistillata* from Eilat might be the outcomes of increased ALAN in the last four decades. Further, reduced settlement rates were recorded in planulae exposed to ALAN as compared to regular light regimes, and a year upon settlement, the formerly impacted young colonies exhibited lower photosynthesis efficiency, albeit higher survival, growth and calcification rates (Tamir et al. 2020). Adult *S. pistillata* colonies as their symbionts showed increased oxidative damage in lipids and increased respiration rate and experienced loss of symbionts and enhanced photoinhibition at decreased photosynthetic rates (Levy et al. 2020).

11.7.2 Larval Collection and Settlement

As mentioned, *S. pistillata* is a brooding coral with a long reproduction season in some bio-geographical areas. By using this reproduction strategy, scientists can also use the planulae of *S. pistillata* as a model animal. In order to catch planulae easily, a planulae trap is used (Akiva et al. 2018;

FIGURE 11.3 A shallow reef in Eilat, Gulf of Aqaba/Eilat dominated by colonies of *S. pistillata*. The future reefs ("reefs of tomorrow") will be dominated by a small number of species and lower diversity but may still keep their 3D structure and substrate complexity. (Photograph courtesy of D. Shefy.)

Amar et al. 2007, 2008; Douek et al. 2011; Horoszowski-Fridman et al. 2020; Linden et al. 2018, 2019; Linden and Rinkevich 2011, 2017; Rinkevich and Loya 1979b, 1987; Scucchia et al. 2020; Shefy et al. 2018; Tamir et al. 2020; Zakai et al. 2006). This is a trap that is similar to a plankton trap but on a smaller scale, and its use is passive (no need to tow) (Amar et al. 2007; Rinkevich and Loya 1979b; Zakai et al. 2006). The planulae are released from the colony at night and have positive buoyancy in the first few hours after release. As a result, the trap should be placed slightly before sunset and picked up in the early morning or in the middle of the night. The released planulae are trapped in a jar that is located at the top of the traps. Because in some bio-geographical regions, *S. pistillata* does not reproduce according to the lunar phase, and the reproduction season is long, planulae can be collected with few limitations on dates. In contrast to *in-situ* collection with planulae traps, *ex-situ* collection of planulae does not require a trap. Nevertheless, *ex-situ* planulae collection results in a lower number of planulae per colony that do not represent the planulae yield in the field (Zakai et al. 2006). To the best of our knowledge, sexual reproduction of *S. pistillata* has never been documented in a closed-system aquarium. Large amounts of planulae during the majority of the year also enable the study of settlement or early life stages (Amar et al. 2007, 2008; Atoda 1947a; Baird and Morse 2004; Heyward and Negri 2010; Nishikawa et al. 2003; Putnam et al. 2008; Rinkevich and Loya 1979a; Tamir et al. 2020). As mentioned earlier, the planulae of *S. pistillata* are not very selective for substrate and may settle on smooth materials (like microscope slides) without the presence of red algae such as in other coral species (Atoda 1947b; Nishikawa et al. 2003; Putnam et al. 2008; Rinkevich and Loya 1979a). Planulae which settled on the water surface can be resettled (Frank et al. 1997). By using a fine small brush, one can gently move the floating primary polyps to the desired substrate.

11.7.3 Establishing Allorecognition Assays

This species is commonly used to elucidate the nature and dynamics of intraspecific interactions (between *S. pistillata* individuals) and interspecific interaction (between *S. pistillata* colonies and other species in the reef) and to elucidate "self" and "non-self" recognition. Studies clearly showed that a *S. pistillata* colony might distinguish between different neighbors and responds differentially to different allogeneic and xenogeneic challenges (Chadwick-Furman and Rinkevich 1994; Frank et al. 1997; Frank and Rinkevich 1994; Müller et al. 1984; Rinkevich 2004, 2012; Rinkevich and Loya 1985a 1985b). By detecting degraded tissues at contact areas between adjacent coral species in the field, Abelson and Loya (1999) and Rinkevich et al. (1993) defined linear and circular aggression hierarchies among coral species in the GOA/E where *S. pistillata* has emerged as one of the inferior partners in the hierarchies of interspecific interactions. Employing grafting assays, whether *in-situ* or *ex-situ* settings, gained control of the participants' identity in the interaction. Experiments with grafts were conducted by simple methodologies such as attaching allogeneic coral fragments by laundry clips. Conducting hundreds of allogenic assays, Rinkevich and Loya (1983a) and Chadwick-Furman and Rinkevich (1994) further confirmed the control of genetic background on intra- and interspecific interactions in *S. pistillata*. While allografts (interaction between different *S. pistillata* genotypes) will have an array of different responses (Figure 11.1e), iso-grafts (within the same *S. pistillata* genotype) will fuse upon direct tissue contacts (Chadwick-Furman and Rinkevich 1994; Müller et al. 1984; Rinkevich and Loya 1983a), some of which are the outcome of the secretion of isomones—unknown chemical substances that are released into the water column (Rinkevich and Loya 1985a). In *S. pistillata*, adult genotypes do not fuse, yet, in the early life stages of the coral, fusion may occur in zero- to four-month-old colonies (Amar et al. 2008; Amar and Rinkevich 2010; Frank et al. 1997). Genetic relatedness was observed to affect the fusion rates between juveniles, where young colonies that shared at least one parent (kins) had higher fusion rates than non-siblings (Amar et al. 2008; Amar and Rinkevich 2010; Frank et al. 1997; Shefy, personal communication).

11.7.4 Population Genetics

Since kin relatedness level (coefficient of relationship) may influence genetic diversity, and larval connectivity may affect the intraspecific interactions within a population and consequently shape population fitness, it is necessary to understand the population genetics in and between different reefs. A comparison of microsatellites or other genetic markers of gravid colonies and planulae among different reefs may reveal connectivity and genetic flow processes and patterns.

Elements of population genetic structures of *S. pistillata* populations were studied along the GBR, Okinawa and GOA\E, revealing a high contribution of sexual reproduction to the populations (Ayre and Hughes 2000; Takabayashi et al. 2003; Zvuloni et al. 2008). Yet significant differences in polymorphic allozyme loci diversity were recorded between populations in the same geographical region, implying low levels of connectivity but sufficient genetic diversity to maintain gene flow among reefs (Ayre and Hughes 2000). The low genetic flow among reefs is also related to the fast settlement rates of most released larvae, where the vast majority of the planulae metamorphose 24–48 hours upon release, a time scale that is varied between early and late phases of the reproduction season (Nishikawa et al. 2003; Amar et al. 2007; Rinkevich and Loya 1979a). Yet there are no detailed population genetics studies that employed highly polymorphic markers, reinforcing the need to develop additional efficient and inexpensive tools.

11.7.5 Establishing *S. pistillata* as a Model Orgnism for Reef Restoration

The accelerating climate change and its effects on the coral reefs and the recognition that passive management measures (such as the declaration of marine protected areas) are not enough to cope with climate change (Bindoff et al. 2019; Rinkevich 2008) have raised the need for active reef restoration (Rinkevich 1995, 2000, 2005, 2014, 2015a, 2015b). Much of the work published on active reef restoration has emerged as of the end of the 1990s and has considered colonies of *S. pistillata* for the research and development of new reef restoration methods and approaches. Most of the colonies that were maintained in the first constructed floating nurseries in the GOA/E, including microcolonies and 2–5-cm-long fragments of *S. pistillata*, exhibited fast growth rates and high survival rates (Epstein et al. 2001; Linden and Rinkevich 2017; Linden et al. 2018; Rinkevich 2000; Shafir and Rinkevich 2010; Shafir et al. 2001, 2003, 2006b). The same applied to transplantation acts performed in Eilat and other Indo-Pacific sites (Golomb et al. 2020; Horoszowski-Fridman et al. 2015; Horoszowski-Fridman et al. 2020). *S. pistillata* was further used in various ecological engineering approaches. To achieve higher genetic diversity, several studies (Linden and Rinkevich 2011, 2017; Linden et al. 2019) worked on *S. pistillata* planulae as source material for reef restoration. They collected planulae and reared them in two ways: (1) *in situ*, using a special designated settlement box that allowed the planulae to settle *in situ* on artificial substrates (Linden et al. 2019), and (2) *ex situ*, in outdoor aquarium systems (Linden and Rinkevich 2011), and then developing spat were moved and farmed in floating nurseries (Linden and Rinkevich 2011, 2017). Several versions of methodologies adopted various colony orientations (vertical or horizontal), protection methods against predation (in or out of cages) and locations in the nursery. These developing methods yielded high survival rates, involved minimal maintenance in the developing spat and successfully enhanced genetic diversity. By harnessing the ability of isogeneic fragments to fuse, Rachmilovitz and Rinkevich (2017) formed, within six to seven months, flat *S. pistillata* tissue plates from glued fragments on plastic tiles in the purpose of creating two-dimensional corals units (that can cover degraded substrates). Furthermore, it was shown that nursery-farmed coral colonies that had been transplanted into a degraded reef at Eilat (Dekel Beach) revealed higher fecundity (Horoszowski-Fridman et al. 2020) than native colonies, and when transplanted with other species, they attracted planulae settlement (Golomb et al. 2020). Harnessing chimerism, the fusion between different genotypes (possible during only at early life stages), has also been proposed as an active reef restoration tool to mitigate climate change impacts (Rinkevich 2019b). Chimerism can benefit the coral entity by causing increased sizes, high genotypic diversity and and enhanced phenotipic plasticity.

11.8 CHALLENGING QUESTIONS BOTH IN ACADEMIC AND APPLIED RESEARCH

Out of the many challenging topics associated with the use of *S. pistillata* as a model system for coral biology, three challenging topics are outlined in the following as being of primary importance in the biology of this species.

11.8.1 Biomineralization

The mechanisms controlling coral calcification at the molecular, cellular and entire tissue levels are still not fully understood. Over the past few decades, *S. pistillata* has been used as one of the model organisms for studying calcification in corals. Although numerous papers has been published, the calcification process remains an enigmatic biological phenomenon, as its nature, including physiochemically controlled mechanisms or its biologically mediated machinery, have not yet been resolved (Allemand et al. 2011). Within the last three decades, numerous studies have engaged with various aspects of coral calcification, while many of them have used *S. pistillata* as the model organism for corals (Allemand et al. 2004; Drake et al. 2019; Falini et al. 2015). As mentioned earlier, the calcifying tissue is the calicoblastic layer, an epithelium attached to the skeleton with desmocytes (Muscatine et al. 1997; Raz-Bahat et al. 2006; Tambutté et al. 2007), thus found in direct contact with the skeleton surface (Tambutté et al. 2007). The calicoblastic ectoderm produces the extracellular matrix (ECM) proteins that are secreted to the calcifying medium and remain preserved in the skeleton organic matrix (Allemand et al. 2011). Coral skeletal aragonite is produced within the ECM, which is secreted into semi-enclosed extracellular compartments and composed of a few nano-micrometers-thick matrix elements (Mass et al. 2017a; Sevilgen et al. 2019; Tambutté et al. 2007). The cells in the calicoblastic layer are connected through tight junctions that control the diffusion of molecules to the ECM (Barott et al. 2015a; Raz-Bahat et al. 2006; Tambutté et al. 1996, 2007, 2012; Zoccola et al. 1999, 2004).

This paracellular pathway depends on the charge and size of the molecules (Tambutté et al. 2012). Furthermore, a second path of calcium ions to the center of calcification through an intracellular pathway was proposed. By using *in vitro* primary cell cultures of *S. pistillata* and employing antibodies against ion transporters, several studies (Barott et al. 2015a; Mass et al. 2012, 2017a) showed that calcium is concentrated in intracellular pockets and is exported to the site of calcification via vesicles (Ganot et al. 2020). Dissolved inorganic carbon (DIC) can diffuse from the coral tissue to the ECM (Furla et al. 2000) or, alternatively, be transported via bicarbonate transporters from the calicoblastic cells' cytosol to the ECM (Zoccola et al. 2015). The transport of proteins and minerals to the ECM is influenced and mediated by environmental parameters such as temperature, pH, calcium saturation levels, pollutants and enzymes (Al-Sawalmih 2016; Allemand et al. 2004; Furla et al. 2000; Gattuso et al. 1998; Gutner-Hoch et al. 2017; Malik et al. 2020; Puverel et al. 2005; Zoccola et al. 1999, 2004, 2015). It is suggested that high amounts of acidic amino acids and glycine in the ECM (Puverel et al. 2005) allow the control of its chemical composition by increasing pH and DIC concentration above the surrounding water and enable the formation of aragonite (Drake et al. 2019; Venn et al. 2011). The skeletal organic matrix within the skeletal framework contains at least 60 proteins and glycosylated derivatives which remain entrapped within the crystalline units (Allemand et al. 2011; Drake et al. 2013; Mass et al. 2014; Peled et al. 2020; Puverel et al. 2007). The calicoblastic tissue secretes amorphous nano-calcium carbonate particles in the created microenvironments enriched in organic material aggregates that then crystallize to create ordered aragonitic structures (Mass et al. 2012, 2017b; Von Euw et al. 2017). *S. pistillata* colonies grow their skeletons from the centers of calcification areas of spherulitic shapes (radial distributions of acicular crystals), forming bundles of aragonite crystals (Sun et al. 2017, 2020).

11.8.2 Taxonomy

S. pistillata is considered a model organism in research and has been the focus of coral research over the past four decades. This species is widely distributed in the Indo-Pacific region and represented by numerous morphological variations (morphotypes) associated with different reef habitats, geographical regions and reef depth zones (Figures 11.1b, c, d, 11.3). Thus, for comparative studies, it is imperative to ensure its correct taxonomy and species delineation. Using molecular markers (mitochondrial and nuclear genes), aided by comparisons of morphological characteristics, enabled scientists to point toward the west Pacific and not the coral triangle, like for other corals, as the origin of *S. pistillata* (Flot et al. 2011; Stefani et al. 2011). Keshavmurthy et al. (2013) further revealed the presence of cryptic divergence and four distinct evolutionary lineages (clades) within *S. pistillata* across its distribution range: clade 1 is distributed in the Pacific Ocean (Klueter and Andreakis 2013), clade 2 is distributed over the Indian Ocean and clade 3 is found in the west Indian Ocean. The distribution of the fourth clade overlaps with clades 2 and 3, but this clade inhabits the Red Sea as well (Keshavmurthy et al. 2013). In contrast, Arrigoni et al. (2016) postulated that the different species of the genus *Stylophora* found in the Red Sea are actually ecomorphs of a single phenotypically plastic species that belong to a single molecular lineage. Further analyses are thus needed to evaluate the taxonomic status of *S. pistillata* and whether other species of *Stylophora* represent valid endemic species arising from speciation or locally emerged ecomorphs of *S. pistillata* that had been adapted to different environmental conditions (depth, temperature, etc.).

11.8.3 Aging

How long can a colony of *S. pistillata* live? Are colonies that Jacques Cousteau saw still alive? Some of the coral species attain considerable ages (>400 years), but others have a shorter life span (reviewed in Bythell et al. 2018). The life span of *S. pistillata* was never followed in detail, but studies assumed it to be in the range of 20–30 years (Rinkevich, personal communication). Before natural death, a colony exhibits a decrease in the rate of reproduction, tissue degradation and a decrease in growth (Rinkevich and Loya 1986). Aging in such colonial species is of great interest, and telomeres can be used in the research as molecular markers of aging due to the common loss of telomeres repeating in other aging multicellular organisms, including humans. Additionally, coral stem cells, which can be used as another marker for aging, are not yet known in *S. pistillata*, nor in other coral species. Decreased regeneration abilities in some colonies could also be related to stem cell aging (Y. Rinkevich et al. 2009). Hence, *S. pistillata* may be used as a model species for aging and stem cell biology research of corals in general.

11.8.4 Interactions with Associated Species That Colonize Harbors

S. pistillata is an ecologically important key species, considered an r-strategist (Loya 1976) and an ecological engineering species (Rinkevich 2020) that harbors on branches, between branches and within the skeleton a wide range of fish species and species of large invertebrates, including cryptic, boring and encrusting organisms such as sponges, bivalves, polychaetes, crabs and others (Figure 11.2e, f) (Barneah et al. 2007; Belmaker et al. 2007; Berenshtein et al. 2015; El-Damhougy et al. 2018; Mbije et al. 2019; Garcia-Herrera et al. 2017; Goldshmid et al. 2004; Kotb and Hartnoll 2002; Kuwamura et al. 1994; Limviriyakul et al. 2016; Mohammed and Yassien 2013; Mokady et al. 1991, 1993, 1994; Pratchett 2001; Rinkevich et al. 1991; Shafir et al. 2008). Some of these organisms are commensals; others are corallivores, passing organisms or symbionts. The nature of such interactions is not always explicit. Garcia-Herrera et al. (2017) found that *Dascyllus marginatus* fish that are fanning their fins keep oxygen levels high during the night hours in the inner spaces of the colony between branches, where the photosynthetic oxygen levels are decreased (Shashar et al. 1993). *Trapezia cymodoce*, a xanthid crab which lives between *S. pistillata*'s colony branches, grazes on the coral tissue (Rinkevich et al. 1991), yet colonies harboring this "parasitic" crab demonstrated higher survival rates (Glynn 1983), partly due to their aggressive behavior toward predators (Pratchett 2001). Some of the species live exclusively on/in *S. pistillata* colonies, including the gobiid fish *Paragobiodon echinocephalus* (Belmaker et al. 2007; Kuwamura et al. 1994) and the boring bivalve *Lithophaga lessepsiana* (Mokady et al. 1994). While very little is known about such biological associations, boring organisms such as bivalves and crustaceans can modify the colony morphology (Abelson et al. 1991). These associations become a challenging question, further highlighted by reef restoration acts that consider the whole reef communities and not solely the coral transplants.

BIBLIOGRAPHY

Abelson, A., B. Galil, and Y. Loya. 1991. "Skeletal modifications in stony corals caused by indwelling crabs: Hydrodynamical advantages for crab feeding." *Symbiosis* 10: 233–248.

Abelson, A., and Y. Loya. 1999. "Interspecific aggression among stony corals in Eilat, Red Sea: A hierarchy of aggression ability and related parameters." *Bulletin of Marine Science* 65 (3): 851–860.

Abramovitch-Gottlib, L., D. Katoshevski, and R. Vago. 2003. "Responses of *Stylophora pistillata* and *Millepora dichotoma* to seawater temperature elevation." *Bulletin of Marine Science* 73 (3): 745–755.

Akiva, A., M. Neder, K. Kahil, R. Gavriel, I. Pinkas, G. Goobes, and T. Mass. 2018. "Minerals in the pre-settled coral *Stylophora pistillata* crystallize via protein and ion changes." *Nature Communications* 9 (1). doi:10.1038/s41467-018-04285-7.

Al-Sawalmih, A. 2016. "Calcium composition and microstructure of coral *Stylophora pistillata* under phosphate pollution stress in the Gulf of Aqaba." *Natural Science* 8 (3): 89–95. doi:10.4236/ns.2016.83012.

Al-Sawalmih, A., F.A. Al-Horani, and S. Al-Rousan. 2017. "Elemental analysis of the coral *Stylophora pistillata* incubated along the Jordanian coast of the Gulf of Aqaba." *Fresenius Environmental Bulletin* 26 (4): 3029–3036.

Al-Sofyani, A.A. 1991. "Physiology and ecology of *Stylophora pistillata* and *Echinopora gemmacea* from the Red Sea." University of Glasgow, United Kindom.

Ali, A. hamid A.M., M.A. Hamed, and H.A. El-Azim. 2011. "Heavy metals distribution in the coral reef ecosystems of the Northern Red Sea." *Helgoland Marine Research* 65 (1): 67–80. doi:10.1007/s10152-010-0202-7.

Allemand, D., C. Ferrier-pagès, P. Furla, F. Houlbrèque, S. Puverel, S. Reynaud, É. Tambutté, S. Tambutté, and D. Zoccola. 2004. "Biomineralisation in reef-building corals: From molecular mechanisms to environmental control." *General Palaeontology (Palaeobiochemistry) Biomineralisation* 3: 453–467. doi:10.1016/j.crpv.2004.07.011.

Allemand, D., É. Tambutté, D. Zoccola, and S. Tambutté. 2011. "Coral calcification, cells to reefs." In *Coral Reefs: An Ecosystem in Transition*, eds. Dubinsky, Z. and Stambler, N., 119–150. Dordrecht: Springer Netherlands. doi:10.1007/978-94-007-0114-4_9.

Amar, K.O., N. Chadwick, and B. Rinkevich. 2007. "Coral planulae as dispersion vehicles: Biological properties of larvae released early and late in the season." *Marine Ecology Progress Series* 350 (November): 71–78. doi:10.3354/meps07125.

Amar, K.O., N.E. Chadwick, and B. Rinkevich. 2008. "Coral kin aggregations exhibit mixed allogeneic reactions and enhanced fitness during early ontogeny." *BMC Evolutionary Biology* 8 (1): 126. doi:10.1186/1471-2148-8-126.

Amar, K.O., J. Douek, C. Rabinowitz, and B. Rinkevich. 2008. "Employing of the amplified fragment length polymorphism (AFLP) methodology as an efficient population genetic tool for symbiotic cnidarians." *Marine Biotechnology* 10 (4): 350–357. doi:10.1007/s10126-007-9069-2.

Amar, K.O., and B. Rinkevich. 2007. "A floating mid-water coral nursery as larval dispersion hub: Testing an idea." *Marine Biology* 151: 713–718. doi:10.1007/s00227-006-0512-0.

Amar, K.O., and B. Rinkevich. 2010. "Mounting of erratic histo-incompatible responses in hermatypic corals: A multi-year interval comparison." *Journal of Experimental Biology* 213 (4): 535–540.

Ammar, M.S.A., A.H. Obuid-Allah, and M.A.M. Al-Hammady. 2012. "Patterns of fertility in the two Red Sea corals *Stylophora pistillata* and *Acropora humilis*." *Nusantara Bioscience* 4 (2): 62–75. doi:10.13057/nusbiosci/n040204.

Arrigoni, R., F. Benzoni, T.I. Terraneo, A. Caragnano, and M.L. Berumen. 2016. "Recent origin and semi-permeable species boundaries in the scleractinian coral genus *Stylophora* from the Red Sea." *Scientific Reports* 6 (January): 1–13. doi:10.1038/srep34612.

Atoda, K. 1947a. "The larva and postlarval development of some reef-building corals. II. *Stylophora pistillata* (Esper)." *Science Reports of the Tohoku Imperial University, 4th Series (Biology)* 18: 48–64.

Atoda, K. 1947b. "The larva and postlarval development of some reef-building corals. I. *Pocillopora damicornis cespitosa* (Dana)." *Science Reports of the Tohoku Imperial University, 4th Series (Biology)* 18: 24–47.

Atoda, K. 1951. "The larva and postlarval development of some reef-building corals. V. (*Seritopora hystrix*) Dana." *Science Reports of the Tohoku University 4th Ser. (Biology)* 19 (1): 33–39.

Aubrecht, C., C.D. Elvidge, T. Longcore, C. Rich, J. Safran, A.E. Strong, M. Eakin, et al. 2008. "A global inventory of coral reef stressors based on satellite observed nighttime lights." *Geocarto International* 23 (December 2015): 467–479. doi:10.1080/10106040802185940.

Ayre, D.J., and T.P. Hughes. 2000. "Genotypic diversity and gene flow in brooding and spawning corals along the Great Barrier Reef, Australia." *Evolution* 54 (5): 1590–1605. doi:10.1554/0014-3820(2000)054[1590:GDAGFI]2.0.CO;2.

Baird, A.H., and R.C. Babcock. 2000. "Morphological differences among three species of newly settled pocilloporid coral recruits." *Coral Reefs* 19 (2): 179–183. doi:10.1007/PL00006955.

Baird, A.H., V.R. Cumbo, S. Gudge, S.A. Keith, J.A. Maynard, C.H. Tan, and E.S. Woolsey. 2015. "Coral reproduction on the world's southernmost reef at Lord Howe Island, Australia." *Aquatic Biology* 23 (3): 275–284. doi:10.3354/ab00627.

Baird, A.H., and A.N.C. Morse. 2004. "Induction of metamorphosis in larvae of the brooding corals *Acropora palifera* and *Stylophora pistillata*." *Marine and Freshwater Research* 55 (5): 469–472. doi:10.1071/MF03121.

Banc-Prandi, G., and M. Fine. 2019. "Copper enrichment reduces thermal tolerance of the highly resistant Red Sea coral *Stylophora pistillata*." *Coral Reefs* 38 (2): 285–296. doi:10.1007/s00338-019-01774-z.

Banguera-Hinestroza, E., P. Saenz-Agudelo, T. Bayer, M.L. Berumen, and C.R. Voolstra. 2013. "Characterization of new microsatellite loci for population genetic studies in the smooth cauliflower coral (*Stylophora sp.*)." *Conservation Genetics Resources* 5 (2): 561–563. doi:10.1007/s12686-012-9852-x.

Barneah, O., I. Brickner, M. Hooge, V.M. Weis, T.C. LaJeunesse, and Y. Benayahu. 2007. "Three party symbiosis: Acoelomorph worms, corals and unicellular algal symbionts in Eilat (Red Sea)." *Marine Biology* 151 (4): 1215–1223. doi:10.1007/s00227-006-0563-2.

Barott, K.L., S.O. Perez, L.B. Linsmayer, and M. Tresguerres. 2015a. "Differential localization of ion transporters suggests distinct cellular mechanisms for calcification and photosynthesis between two coral species." *American Journal of Physiology: Regulatory Integrative and Comparative Physiology* 309 (3): R235–R246. doi:10.1152/ajpregu.00052.2015.

Barott, K.L., A.A. Venn, S.O. Perez, S. Tambutteeé, M. Tresguerres, and G.N. Somero. 2015b. "Coral host cells acidify symbiotic algal microenvironment to promote photosynthesis." *Proceedings of the National Academy of Sciences of the United States of America* 112 (2): 607–612. doi:10.1073/pnas.1413483112.

Bellworthy, J., and M. Fine. 2017. "Beyond peak summer temperatures, branching corals in the Gulf of Aqaba are resilient to thermal stress but sensitive to high light." *Coral Reefs* 36 (4). Springer Berlin Heidelberg: 1071–1082. doi:10.1007/s00338-017-1598-1.

Bellworthy, J., M. Menoud, T. Krueger, A. Meibom, and M. Fine. 2019. "Developmental carryover effects of ocean warming and acidification in corals from a potential climate refugium, the Gulf of Aqaba." *Journal of Experimental Biology* 222 (1). doi:10.1242/jeb.186940.

Bellworthy, J., J.E. Spangenberg, and M. Fine. 2019. "Feeding increases the number of offspring but decreases parental investment of Red Sea coral *Stylophora pistillata*." *Ecology and Evolution* 9 (21): 12245–12258. doi:10.1002/ece3.5712.

Belmaker, J., O. Polak, N. Shashar, and Y. Ziv. 2007. "Geographic divergence in the relationship between *Paragobiodon echinocephalus* and its obligate coral host." *Journal of Fish Biology* 71 (5): 1555–1561. doi:10.1111/j.1095-8649.2007.01619.x.

Berenshtein, I., Y. Reuben, and A. Genin. 2015. "Effect of oxygen on coral fanning by mutualistic fish." *Marine Ecology* 36 (4): 1171–1175. doi:10.1111/maec.12218.

Bhagooli, R., and M. Hidaka. 2004. "Photoinhibition, bleaching susceptibility and mortality in two scleractinian corals, *Platygyra ryukyuensis* and *Stylophora pistillata*, in response to thermal and light stresses." *Comparative Biochemistry and Physiology: A Molecular and Integrative Physiology* 137 (3): 547–555. doi:10.1016/j.cbpb.2003.11.008.

Bindoff, N.L., W.W.L. Cheung, J.G. Kairo, J. Arístegui, V.A. Guinder, R. Hallberg, N. Hilmi, et al. 2019. *Changing Ocean, Marine Ecosystems, and Dependent Communities. IPCC Special Report on the Ocean and Cryosphere in a Changing Climate [H.-O. Pörtner, D.C. Roberts, V. Masson-Delmotte, P. Zhai, M. Tignor, E. Poloczanska, K. Mintenbeck, A. Alegría, M. Nicolai, A. Okem, J. Petzold, B. Rama, N.M. Weyer (Eds.)].* www.ipcc.ch/srocc/download-report/.

Biscéré, T., C. Ferrier-Pagès, A. Gilbert, T. Pichler, and F. Houlbrèque. 2018. "Evidence for mitigation of coral bleaching by manganese." *Scientific Reports* 8 (1): 1–10. doi:10.1038/s41598-018-34994-4.

Biscéré, T., R. Rodolfo-Metalpa, A. Lorrain, L. Chauvaud, J. Thébault, J. Clavier, and F. Houlbrèque. 2015. "Responses of two scleractinian corals to cobalt pollution and ocean acidification." *PLoS One* 10 (4): 1–18. doi:10.1371/journal.pone.0122898.

Bockel, T., and B. Rinkevich. 2019. "Rapid recruitment of symbiotic algae into developing Scleractinian coral tissues." *Journal of Marine Science and Engineering* 7 (306). doi:10.3390/jmse7090306.

Bongiorni, L., S. Shafir, D. Angel, and B. Rinkevich. 2003a. "Survival, growth and gonad development of two hermatypic corals subjected to in situ fish-farm nutrient enrichment." *Marine Ecology Progress Series* 253 (2001): 137–144. doi:10.3354/meps253137.

Bongiorni, L., S. Shafir, and B. Rinkevich. 2003b. "Effects of particulate matter released by a fish farm (Eilat, Red Sea) on survival and growth of *Stylophora pistillata* coral nubbins." *Marine Pollution Bulletin* 46 (9): 1120–1124. doi:10.1016/S0025-326X(03)00240-6.

Byler, K.A., M. Carmi-Veal, M. Fine, and T.L. Goulet. 2013. "Multiple symbiont acquisition strategies as an adaptive mechanism in the coral *Stylophora pistillata*." *PLoS One* 8 (3): 1–7. doi:10.1371/journal.pone.0059596.

Bythell, J.C., B.E. Brown, and T.B.L. Kirkwood. 2018. "Do reef corals age?" *Biological Reviews* 93 (2): 1192–1202. doi:10.1111/brv.12391.

Chadwick-Furman, N., and B. Rinkevich. 1994. "A complex allorecognition system in a reef-building coral: Delayed responses, reversals and nontransitive hierarchies." *Coral Reefs* 13 (1): 57–63. doi:10.1007/BF00426436.

Chen, C., C.Y. Chiou, C.F. Dai, and C.A. Chen. 2008. "Unique mitogenomic features in the scleractinian family pocilloporidae (Scleractinia: Astrocoeniina)." *Marine Biotechnology* 10 (5): 538–553. doi:10.1007/s10126-008-9093-x.

Cohen, I., and Z. Dubinsky. 2015. "Long term photoacclimation responses of the coral *Stylophora pistillata* to reciprocal deep to shallow transplantation: Photosynthesis and calcification." *Frontiers in Marine Science* 2 (June): 1–13. doi:10.3389/fmars.2015.00045.

Dai, C.F., K. Soong, and T.Y. Fan. 1992. "Sexual reproduction of corals in northern and southern Taiwan." *7th Int'l. Coral Reef Symp.*, 1: 448–455.

Danovaro, R., L. Bongiorni, C. Corinaldesi, D. Giovannelli, E. Damiani, P. Astolfi, L. Greci, and A. Pusceddu. 2008. "Sunscreens cause coral bleaching by promoting viral infections." *Environmental Health Perspectives* 116 (4): 441–447. doi:10.1289/ehp.10966.

Dar, M.A., and T.A.A. Mohamed. 2017. "Coral growth and skeletal density relationships in some branching corals of the Red Sea, Egypt." *Journal of Environment and Earth Science* 7 (11): 66–79.

Dias, M., A. Ferreira, R. Gouveia, C. Madeira, N. Jogee, H. Cabral, M. Diniz, and C. Vinagre. 2019. "Long-term exposure to increasing temperatures on scleractinian coral fragments reveals oxidative stress." *Marine Environmental Research* 150 (January): 104758. doi:10.1016/j.marenvres.2019.104758.

Dimond, J.L., and S.B. Roberts. 2016. "Germline DNA methylation in reef corals: Patterns and potential roles in response to environmental change." *Molecular Ecology* 25 (8): 1895–1904. doi:10.1111/mec.13414.

Doropoulos, C., S. Ward, G. Roff, M. González-Rivero, and P.J. Mumby. 2015. "Linking demographic processes of juvenile corals to benthic recovery trajectories in two common reef habitats." *PLoS One* 10 (5): 1–23. doi:10.1371/journal.pone.0128535.

Douek, J., K.O. Amar, and B. Rinkevich. 2011. "Maternal-larval population genetic traits in *Stylophora pistillata*, a hermaphroditic brooding coral species." *Genetica* 139 (11–12): 1531–1542. doi:10.1007/s10709-012-9653-x.

Downs, C.A., E. Kramarsky-Winter, J.E. Fauth, R. Segal, O. Bronstein, R. Jeger, Y. Lichtenfeld, et al. 2014. "Toxicological effects of the sunscreen UV filter, benzophenone-2, on planulae and in vitro cells of the coral, *Stylophora pistillata*." *Ecotoxicology* 23 (2): 175–191. doi:10.1007/s10646-013-1161-y.

Drake, J.L., T. Mass, L. Haramaty, E. Zelzion, D. Bhattacharya, and P.G. Falkowski. 2013. "Proteomic analysis of skeletal organic matrix from the stony coral *Stylophora pistillata*." *Proceedings of the National Academy of Sciences* 110 (10): 3788–3793. doi:10.1073/pnas.1304972110.

Drake, J.L., T. Mass, J. Stolarski, S. Von Euw, B. van de Schootbrugge, and P.G. Falkowski. 2019. "How corals made rocks through the ages." *Global Change Biology* 26 (1): 31–53. doi:10.1111/gcb.14912.

Dubinsky, Z., P.G. Falkowski, J.W. Porter, and L. Muscatine. 1984. "Absorption and utilization of radiant energy by light and shade adapted colonies of the hermatypic coral *Stylophora pistillata*." *Proceeding of the Royal Society of London* 214 (1227): 203–214. https://doi.org/10.1098/rspb.1984.0059.

Dubinsky, Z., and P.L. Jokiel. 1994. "Ratio of energy and nutrient fluxes regulates symbiosis between zooxanthelae and corals." *Pacific Science* 48 (3): 313–324.

Dubinsky, Z., N. Stambler, M. Ben-Zion, L.R. McCloskey, L. Muscatine, and P.G. Falkowski. 1990. "The effect of external nutrient resources on the optical properties and photosynthetic efficiency of *Stylophora pistillata*." *Proc. R. Soc. Lond. B* 239: 231–246. doi:10.1098/rspb.1990.0015.

Edwards, A., and H. Emberton. 1980. "Crustacea associated with the scleractinian coral, *Stylophora pistillata* (Esper), in the Sudanese Red Sea." *Journal of Experimental Marine Biology and Ecology* 42: 225–240. doi:10.1016/0022-0981(80)90178-1.

El-Damhougy, K.A., E.-S.S.E. Salem, M.M.A. Fouda, and M.A.M.M. Al-Hammady. 2018. "The growth and reproductive biology of the coral gall crab, *Hapalocarcinus marsupialis* Stimpson, 1859 (Crustacea: Cryptochiridae) from Gulf of Aqaba, Red Sea, Egypt." *The Journal of Basic and Applied Zoology* 79 (1). The Journal of Basic and Applied Zoology. doi:10.1186/s41936-017-0010-6.

Epstein, N., R.P.M. Bak, and B. Rinkevich. 2000. "Toxicity of third generation dispersants and dispersed Egyptian crude oil on Red Sea coral larvae." *Marine Pollution Bulletin* 40 (6): 497–503. doi:10.1016/S0025-326X(99)00232-5.

Epstein, N., R.P.M. Bak, and B. Rinkevich. 2001. "Strategies for gardening denuded coral reef areas: The spplicability of using different types of coral material for reef restoration." *Restoration Ecology* 9 (4): 432–442.

Epstein, N., and B. Rinkevich. 2001. "From isolated ramets to coral colonies: The significance of colony pattern formation in reef restoration practices." *Basic and Applied Ecology* 2 (3): 219–222. doi:10.1078/1439-1791-00045.

Epstein, N., M.J.A. Vermeij, R.P.M. Bak, and B. Rinkevich. 2005. "Alleviating impacts of anthropogenic activities by traditional conservation measures: Can a small reef reserve be sustainedly managed?" *Biological Conservation* 121 (2): 243–255. doi:10.1016/j.biocon.2004.05.001.

Esper, E.J.C. 1797. *Fortsetzungen Der Pflanzenthiere in Abbildugen Nach Der Natur Mit Farben Erleuchtet Nebst Beschreibungen*. Vol. 1. In der Raspeschen Buchhandlung.

Fadlallah, Y.H., and R.T. Lindo. 1988. "Contrasting cycles of reproduction in *Stylophora pistillata* from the Red Sea and the Arabian Gulf, with emphasis on temperature." *Proceedings of the 6th International Coral Reef Symposium* 3: 225–230.

Falini, G., S. Fermani, and S. Goffredo. 2015. "Coral biomineralization: A focus on intra-skeletal organic matrix and calcification." *Seminars in Cell and Developmental Biology* 46: 17–26. doi:10.1016/j.semcdb.2015.09.005.

Falkowski, P.G., and Z. Dubinsky. 1981. "Light-shade adaptation of *Stylophora pistillata*, a hermatypic coral from the Gulf of Eilat." *Nature* 289 (5794): 172–174.

Falkowski, P.G., Z. Dubinsky, L. Muscatine, and J.W. Porter. 1984. "Light and the bioenergetics of a symbiotic coral." *BioScience* 34 (11): 705–709.

Fan, T., and C. Dai. 2002. "Sexual reproduction of the reef coral *Stylophora pistillata* in southern Taiwan." *Acta Oceanographica Taiwanica* 40 (2): 107–120.

Fel, J.P., C. Lacherez, A. Bensetra, S. Mezzache, E. Béraud, M. Léonard, D. Allemand, and C. Ferrier-Pagès. 2019. "Photochemical response of the scleractinian coral *Stylophora pistillata* to some sunscreen ingredients." *Coral Reefs* 38 (1): 109–122. doi:10.1007/s00338-018-01759-4.

Ferrier-Pagès, C., J.P. Gattuso, S. Dallot, and J. Jaubert. 2000. "Effect of nutrient enrichment on growth and photosynthesis of the zooxanthellate coral *Stylophora pistillata*." *Coral Reefs* 19 (2): 103–113. doi:10.1007/s003380000078.

Ferrier-Pagès, C., F. Houlbrèque, E. Wyse, C. Richard, D. Allemand, and F. Boisson. 2005. "Bioaccumulation of zinc in the scleractinian coral *Stylophora pistillata*." *Coral Reefs* 24 (4): 636–645. doi:10.1007/s00338-005-0045-x.

Ferrier-Pagès, C., C. Rottier, E. Beraud, and O. Levy. 2010. "Experimental assessment of the feeding effort of three scleractinian coral species during a thermal stress: Effect on the rates of photosynthesis." *Journal of Experimental Marine Biology and Ecology* 390 (2). Elsevier B.V.: 118–124. doi:10.1016/j.jembe.2010.05.007.

Ferrier-Pagès, C., L. Sauzéat, and V. Balter. 2018. "Coral bleaching is linked to the capacity of the animal host to supply essential metals to the symbionts." *Global Change Biology* 24 (7): 3145–3157. doi:10.1111/gcb.14141.

Ferrier-Pagès, C., V. Schoelzke, J. Jaubert, L. Muscatine, and O. Hoegh-Guldberg. 2001. "Response of a scleractinian coral, *Stylophora pistillata*, to iron and nitrate enrichment." *Journal of Experimental Marine Biology and Ecology* 259 (2): 249–261. doi:10.1016/S0022-0981(01)00241-6.

Ferrier-Pagès, C., J. Witting, E. Tambutté, and K.P. Sebens. 2003. "Effect of natural zooplankton feeding on the tissue and skeletal growth of the scleractinian coral *Stylophora pistillata*." *Coral Reefs* 22 (3): 229–240. doi:10.1007/s00338-003-0312-7.

Fine, M., H. Gildor, and A. Genin. 2013. "A coral reef refuge in the Red Sea." *Global Change Biology* 19 (12): 3640–3647. doi:10.1111/gcb.12356.

Fishelson, L. 1971. "Ecology and distribution of the benthic fauna in the shallow waters of the Red Sea." *Marine Biology: International Journal on Life in Oceans and Coastal Waters* 10 (2): 113–133. doi:10.1007/BF00354828.

Fitt, W.K., R.D. Gates, O. Hoegh-Guldberg, J.C. Bythell, A. Jatkar, A.G. Grottoli, M. Gomez, et al. 2009. "Response of two species of Indo-Pacific corals, *Porites cylindrica* and *Stylophora pistillata*, to short-term thermal stress: The host does matter in determining the tolerance of corals to bleaching." *Journal of Experimental Marine Biology and Ecology* 373 (2). Elsevier B.V.: 102–110. doi:10.1016/j.jembe.2009.03.011.

Flot, J.F., J. Blanchot, L. Charpy, C. Cruaud, W.Y. Licuanan, Y. Nakano, C. Payri, and S. Tillier. 2011. "Incongruence between morphotypes and genetically delimited species in the coral genus *Stylophora:* phenotypic plasticity, morphological convergence, morphological stasis or interspecific hybridization?" *BMC Ecology* 11 (1). BioMed Central Ltd: 22. doi:10.1186/1472-6785-11-22.

Frank, U., U. Oren, Y. Loya, and B. Rinkevich. 1997. "Alloimmune maturation in the coral *Stylophora pistillata* is achieved through three distinctive stages, 4 months post-metamorphosis." *Proceedings of the Royal Society of London: Series B: Biological Sciences* 264 (1378): 99–104.

Frank, U., C. Rabinowitz, and B. Rinkevich. 1994. "*In vitro* establishment of continuous cell cultures and cell lines from ten colonial cnidarians." *Marine Biology* 120: 491–499.

Frank, U., and B. Rinkevich. 1994. "Nontransitive patterns of historecognition phenomena in the Red Sea hydrocoral *Millepora dichotoma*." *Marine Biology* 118 (4): 723–729. doi:10.1007/BF00347521.

Franklin, D.J., O. Hoegh-Guldberg, R.J. Jones, and J.A. Berges. 2004. "Cell death and degeneration in the symbiotic dinoflagellates of the coral *Stylophora pistillata* during bleaching." *Marine Ecology Progress Series* 272: 117–130. doi:10.3354/meps272117.

Furla, P., I. Galgani, I. Durand, and D. Allemand. 2000. "Sources and mechanisms of inorganic carbon transport for coral calcification and photosynthesis." *Journal of Experimental Biology* 203 (22): 3445–3457.

Ganot, P., E. Tambutté, N. Caminiti-Segonds, G. Toullec, D. Allemand, and S. Tambutté. 2020. "Ubiquitous macropinocytosis in anthozoans." *ELife* 9: 1–25. doi:10.7554/eLife.50022.

Garcia-Herrera, N., S.C.A. Ferse, A. Kunzmann, and A. Genin. 2017. "Mutualistic damselfish induce higher photosynthetic rates in their host coral." *Journal of Experimental Biology* 220 (10): 1803–1811. doi:10.1242/jeb.152462.

Gattuso, J.P., M. Frankignoulle, I. Bourge, S. Romaine, and R.W. Buddemeier. 1998. "Effect of calcium carbonate saturation of seawater on coral calcification." *Global and Planetary Change* 18 (1–2): 37–46. doi:10.1016/S0921-8181(98)00035-6.

Glynn, P.W. 1983. "Increased survivorship on corals harbouring crustacean symbionts." *Marine Biology and Fisheries* 4 (2): 105–111.

Goldshmid, R., R. Holzman, D. Weihs, and A. Genin. 2004. "Aeration of corals by sleep-swimming fish." *Limnology and Oceanography* 49 (5): 1832–1839. doi:10.4319/lo.2004.49.5.1832.

Golomb, D., N. Shashar, and B. Rinkevich. 2020. "Coral carpets—A novel ecological engineering tool aimed at constructing coral communities on soft sand bottoms." *Ecological Engineering* 145 (February): 105743. doi:10.1016/j.ecoleng.2020.105743.

Grinblat, M., M. Fine, Y. Tikochinski, and Y. Loya. 2018. "*Stylophora pistillata* in the Red Sea demonstrate higher GFP fluorescence under ocean acidification conditions." *Coral Reefs* 37 (1). Springer Berlin Heidelberg: 309–320. doi:10.1007/s00338-018-1659-0.

Grottoli, A.G., D. Tchernov, and G. Winters. 2017. "Physiological and biogeochemical responses of super-corals to thermal stress from the northern gulf of Aqaba, Red Sea." *Frontiers in Marine Science* 4 (July): 1–12. doi:10.3389/fmars.2017.00215.

Grover, R., J.F. Maguer, D. Allemand, and C. Ferrier-page. 2006. "Urea uptake by the scleractinian coral *Stylophora pistillata*." *Journal of Experimental Marine Biology and Ecology* 332: 216–225. doi:10.1016/j.jembe.2005.11.020.

Grover, R., J.F. Maguer, D. Allemand, and C. Ferrier-Pagès. 2003. "Nitrate uptake in the scleractinian coral *Stylophora pistillata*." *Limnology and Oceanography* 48 (6): 2266–2274. doi:10.4319/lo.2003.48.6.2266.

Grover, R., J.F. Maguer, D. Allemand, and C. Ferrier-Pagès. 2008. "Uptake of dissolved free amino acids by the scleractinian coral *Stylophora pistillata*." *Journal of Experimental Biology* 211 (6): 860–865. doi:10.1242/jeb.012807.

Grover, R., J.F. Maguer, S. Reynaud-Vaganay, and C. Ferrier-Pagès. 2002. "Uptake of ammonium by the scleractinian coral *Stylophora pistillata*: Effect of feeding, light, and ammonium concentrations." *Limnology and Oceanography* 47 (3): 782–790. doi:10.4319/lo.2002.47.3.0782.

Guerrini, G., M. Yerushalmy, D. Shefy, N. Shashar, and B. Rinkevich. 2020. "Apparent recruitment failure for the vast majority of coral species at Eilat, Red Sea." *Coral Reefs*. doi:10.1007/s00338-020-01998-4.

Gutner-Hoch, E., H.W. Ben-Asher, R. Yam, A. Shemesh, and O. Levy. 2017. "Identifying genes and regulatory pathways associated with the scleractinian coral calcification process." *PeerJ* 2017 (7): 3590. doi:10.7717/peerj.3590.

Hall, E.R., E.M. Muller, T. Goulet, J. Bellworthy, K.B. Ritchie, and M. Fine. 2018. "Eutrophication may compromise the resilience of the Red Sea coral *Stylophora pistillata* to global change." *Marine Pollution Bulletin* 131 (May). Elsevier: 701–711. doi:10.1016/j.marpolbul.2018.04.067.

Hasan, M.H. 2019. "Effect of hard installation on coral community succession and growth rate at taba heights international marina at the northern Gulf of Aqaba, Egypt." *Egyptian Journal of Aquatic Biology and Fisheries* 23 (3): 225–243. doi:10.21608/ejabf.2019.41918.

Hawkins, T.D., T. Krueger, S.P. Wilkinson, P.L. Fisher, and S.K. Davy. 2015. "Antioxidant responses to heat and light stress differ with habitat in a common reef coral." *Coral Reefs* 34 (4): 1229–1241. doi:10.1007/s00338-015-1345-4.

Hédouin, L., M. Metian, J.L. Teyssié, F. Oberhänsli, C. Ferrier-Pagès, and M. Warnau. 2016. "Bioaccumulation of 63Ni in the scleractinian coral *Stylophora pistillata* and isolated *Symbiodinium* using radiotracer techniques." *Chemosphere* 156: 420–427. doi:10.1016/j.chemosphere.2016.04.097.

Heyward, A.J., and A.P. Negri. 1999. "Natural inducers for coral larval metamorphosis." *Coral Reefs* 18 (3): 273–279. doi:10.1007/s003380050193.

Heyward, A.J., and A.P. Negri. 2010. "Plasticity of larval pre-competency in response to temperature: Observations on multiple broadcast spawning coral species." *Coral Reefs* 29 (3): 631–636. doi:10.1007/s00338-009-0578-5.

Hoegh-Guldberg, O., and J.G. Smith. 1989a. "Influence of the population density of zooxanthellae and supply of ammonium on the biomass and metabolic characteristics of the reef corals *Seriatopora hystrix* and *Stylophora pistillata*." *Marine Ecology Progress Series* 57 (October): 173–186. doi:10.3354/meps057173.

Hoegh-Guldberg, O., and J.G. Smith. 1989b. "The effect of sudden changes in temperature, light and salinity on the population density and export of zooxathellae from the reef corals *Stylophora pistillata* Esper and *Seriatopora hystrix* Dana." *Genetics and Ceff Biology* 129: 279–303.

Horoszowski-Fridman, Y., J.C. Brêthes, N. Rahmani, and B. Rinkevich. 2015. "Marine silviculture: Incorporating ecosystem engineering properties into reef restoration acts." *Ecological Engineering* 82. Elsevier B.V.: 201–213. doi:10.1016/j.ecoleng.2015.04.104.

Horoszowski-Fridman, Y., I. Izhaki, and B. Rinkevich. 2020. "Long-term heightened larval production in nursery-bred coral transplants." *Basic and Applied Ecology*: 0–25. doi:10.1016/j.baae.2020.05.003.

Horoszowski-Fridman, Y., and B. Rinkevich. 2020. "Active coral reef restoration in Eilat, Israel: Reconnoitering the long-term prospectus." In *Active Coral Restoration*, ed. Vaughan, D. Platation, FL: J. Ross Publishing.

Horwitz, R., M.O. Hoogenboom, and M. Fine. 2017. "Spatial competition dynamics between reef corals under ocean acidification." *Scientific Reports* 7: 1–13. doi:10.1038/srep40288.

Houlbre'que, F., and C. Ferrier-page. 2009. "Heterotrophy in tropical scleractinian corals." *Biological Reviews* 84: 1–17. doi:10.1111/j.1469-185X.2008.00058.x.

Houlbrèque, F., E. Tambutté, D. Allemand, and C. Ferrier-Pagès. 2004. "Interactions between zooplankton feeding, photosynthesis and skeletal growth in the scleractinian coral *Stylophora pistillata*." *Journal of Experimental Biology* 207 (9): 1461–1469. doi:10.1242/jeb.00911.

Houlbrèque, F., E. Tambutté, and C. Ferrier-Pagès. 2003. "Effect of zooplankton availability on the rates of photosynthesis, and tissue and skeletal growth in the scleractinian coral *Stylophora pistillata*." *Journal of Experimental Marine Biology and Ecology* 296 (2): 145–166. doi:10.1016/S0022-0981(03)00259-4.

Karako-Lampert, S., D. Zoccola, M. Salmon-Divon, M. Katzenellenbogen, S. Tambutté, A. Bertucci, O. Hoegh-Guldberg, E. Deleury, D. Allemand, and O. Levy. 2014. "Transcriptome analysis of the scleractinian coral *Stylophora pistillata*." *PLoS One* 9 (2). doi:10.1371/journal.pone.0088615.

Keshavmurthy, S., S.Y. Yang, A. Alamaru, Y.Y. Chuang, M. Pichon, D. Obura, S. Fontana, et al. 2013. "DNA barcoding reveals the coral 'laboratory-rat', *Stylophora pistillata* encompasses multiple identities." *Scientific Reports* 3: 1–7. doi:10.1038/srep01520.

Klueter, A., and N. Andreakis. 2013. "Assessing genetic diversity in the scleractinian coral *Stylophora pistillata* (Esper 1797) from the Central Great Barrier Reef and the Coral Sea." *Systematics and Biodiversity* 11 (1): 67–76. doi:10.1080/14772000.2013.770419.

Kotb, M.M.A., and R.G. Hartnoll. 2002. "Aspects of the growth and reproduction of the coral gall crab *Hapalocarcinus marsupialis*." *Journal of Crustacean Biology* 22 (3): 558–566. doi:10.1651/0278-0372(2002)022[0558:aotgar]2.0.co;2.

Kramarsky-Winter, E., C.A. Downs, A. Downs, and Y. Loya. 2009. "Cellular responses in the coral *Stylophora pistillata* exposed to eutrophication from fish mariculture." *Evolutionary Ecology Research* 11 (3): 381–401.

Kramer, N., G. Eyal, R. Tamir, and Y. Loya. 2019. "Upper mesophotic depths in the coral reefs of Eilat, Red Sea, offer suitable refuge grounds for coral settlement." *Scientific Reports* 9 (1): 1–12. doi:10.1038/s41598-019-38795-1.

Krueger, T., N. Horwitz, J. Bodin, M.E. Giovani, S. Escrig, A. Meibom, and M. Fine. 2017. "Common reef-building coral in the northern red sea resistant to elevated temperature and acidification." *Royal Society Open Science* 4 (5). doi:10.1098/rsos.170038.

Kteifan, M., M. Wahsha, and F.A. Al-Horani. 2017. "Assessing stress response of *Stylophora pistillata* towards oil and phosphate pollution in the Gulf of Aqaba, using molecular and biochemical markers." *Chemistry and Ecology* 33 (4): 281–294. doi:10.1080/02757540.2017.1308500.

Kücken, M., B. Rinkevich, L. Shaish, and A. Deutsch. 2011. "Nutritional resources as positional information for morphogenesis in the stony coral *Stylophora pistillata*." *Journal of Theoretical Biology* 275 (1): 70–77. doi:10.1016/j.jtbi.2011.01.018.

Kuwamura, T., Y. Yogo, and Y. Nakashima. 1994. "Population dynamics of goby *Paragobiodon echinocephalus* and host coral *Stylophora pistillata*." *Marine Ecology Progress Series* 103 (1–2): 17–24. doi:10.3354/meps103017.

Lampert-Karako, S., N. Stambler, D.J. Katcoff, Y. Achituv, Z. Dubinsky, and N. Simon-Blecher. 2008. "Effects of depth and eutrophication on the zooxanthella clades of *Stylophora pistillata* from the Gulf of Eilat (Red Sea)." *Aquatic Conservetion: Marine and Fresh Water Ecosystems* 18: 1039–1045. doi:10.1002/aqc.927 Effects.

Levy, O., L. Fernandes de Barros Marangoni, J.I. Cohen, C. Rottier, E. Béraud, R. Grover, and C. Ferrier-Pagès. 2020. "Artificial light at night (ALAN) alters the physiology and biochemistry of symbiotic reef building corals." *Environmental Pollution* 266: 114987. doi:10.1016/j.envpol.2020.114987.

Liberman, T., A. Genin, and Y. Loya. 1995. "Effects on growth and reproduction of the coral *Stylophora pistillata* by the mutualistic damselfish *Dascyllus marginatus*." *Marine Biology* 121 (4): 741–746. doi:10.1007/BF00349310.

Liew, Y.J., M. Aranda, A. Carr, S. Baumgarten, D. Zoccola, S. Tambutté, D. Allemand, G. Micklem, and C.R. Voolstra. 2014. "Identification of microRNAs in the coral *Stylophora pistillata*." *PLoS One* 9 (3): 1–11. doi:10.1371/journal.pone.0091101.

Liew, Y.J., D. Zoccola, Y. Li, E. Tambutte, A.A. Venn, C.T. Michell, G. Cui, et al. 2018. "Epigenome-associated phenotypic acclimatization to ocean acidification in a reef-building coral." *Science Advances* 4 (6). doi:10.1126/sciadv.aar8028.

Limviriyakul, P., L.C. Tseng, J.S. Hwang, and T.W. Shih. 2016. "Anomuran and brachyuran symbiotic crabs in coastal areas between the Southern Ryukyu arc and the coral triangle." *Zoological Studies* 55. doi:10.6620/ZS.2016.55-07.

Linden, B., J. Huisman, and B. Rinkevich. 2018. "Circatrigintan instead of lunar periodicity of larval release in a brooding coral species." *Scientific Reports* 8 (1): 1–9. doi:10.1038/s41598-018-23274-w.

Linden, B., and B. Rinkevich. 2011. "Creating stocks of young colonies from brooding coral larvae, amenable to active reef restoration." *Journal of Experimental Marine Biology and Ecology* 398 (1–2): 40–46. doi:10.1016/j.jembe.2010.12.002.

Linden, B., and B. Rinkevich. 2017. "Elaborating an eco-engineering approach for stock enhanced sexually derived coral colonies." *Journal of Experimental Marine Biology and Ecology* 486 (October): 314–321. doi:10.1016/j.jembe.2016.10.014.

Linden, B., M.J.A. Vermeij, and B. Rinkevich. 2019. "The coral settlement box: A simple device to produce coral stock from brooded coral larvae entirely in situ." *Ecological Engineering* 132. Elsevier: 115–119. doi:10.1016/j.ecoleng.2019.04.012.

Loya, Y. 1976. "The Red Sea coral *Stylophora pistillata* is an r strategist." *Nature* 259: 478–480.

Loya, Y., and B. Rinkevich. 1979. "Abortion effect in corals induced by oil pollution." *Marine Ecology Progress Series* 1: 77–80. doi:10.3354/meps001077.

Loya, Y., and B. Rinkevich. 1987. "Effects of petroleum hydrocarbons on corals." *Human Impacts on Coral Reefs: Facts and Recommendations*. French Polynesia: Antenne Museum EPHE, 91–102.

Malik, A., S. Einbinder, S. Martinez, D. Tchernov, S. Haviv, R. Almuly, P. Zaslansky, et al. 2020. "Molecular and skeletal fingerprints of scleractinian coral biomineralization: From the sea surface to mesophotic depths." *Acta Biomaterialia*, 1–14. doi:10.1016/j.actbio.2020.01.010.

Maor-Landaw, K., and O. Levy. 2016. "Gene expression profiles during short-term heat stress; branching vs. massive Scleractinian corals of the Red Sea." *PeerJ* 4 (E1814). doi:10.7717/peerj.1814.

Marsh, J.A. 1970. "Primary productivity of reef-building calcareous red algae." *Ecological Society of America* 51 (2): 255–263.

Mass, T., J.L. Drake, L. Haramaty, Y. Rosenthal, O.M.E. Schofield, R.M. Sherrell, and P.G. Falkowski. 2012. "Aragonite precipitation by 'proto-polyps' in coral cell cultures." *PLoS One* 7 (4): 8–15. doi:10.1371/journal.pone.0035049.

Mass, T., J.L. Drake, J.M. Heddleston, and P.G. Falkowski. 2017a. "Nanoscale visualization of biomineral formation in coral proto-polyps." *Current Biology* 27 (20). Elsevier Ltd.: 3191–3196.e3. doi:10.1016/j.cub.2017.09.012.

Mass, T., J.L. Drake, E.C. Peters, W. Jiang, and P.G. Falkowski. 2014. "Immunolocalization of skeletal matrix proteins in tissue and mineral of the coral *Stylophora pistillata*." *Proceedings of the National Academy of Sciences of the United States of America* 111 (35): 12728–12733. doi:10.1073/pnas.1408621111.

Mass, T., A.J. Giuffre, C.Y. Sun, C.A. Stifler, M.J. Frazier, M. Neder, N. Tamura, C. V. Stan, M.A. Marcus, and P.U.P.A. Gilbert. 2017b. "Amorphous calcium carbonate particles form coral skeletons." *Proceedings of the National Academy of Sciences of the United States of America* 114 (37): E7670–E7678. doi:10.1073/pnas.1707890114.

Mbije, N.E.J., J. Douek, E. Spanier, and B. Rinkevich. 2019. "Population genetic parameters of the emerging corallivorous snail *Drupella cornus* in the northern Gulf of Eilat and Tanzanian coastlines based on mitochondrial COI gene sequences." *Marine Biodiversity* 49: 147–161. doi:10.1007/s12526-017-0768-2.

McCloskey, L.R., and L. Muscatine. 1984. "Production and respiration in the Red Sea coral *Stylophora pistillata* as a function of depth." *Proceedings of the Royal Society of London: Series B, Biological Sciences* 222 (1227): 215–230.

Metian, M., L. Hédouin, C. Ferrier-Pagès, J.L. Teyssié, F. Oberhansli, E. Buschiazzo, and M. Warnau. 2015. "Metal bioconcentration in the scleractinian coral *Stylophora pistillata*: Investigating the role of different components of the holobiont using radiotracers." *Environmental Monitoring and Assessment* 187 (4). doi:10.1007/s10661-015-4383-z.

Mohammed, T.A.A., and M.H. Yassien. 2013. "Assemblages of two gall crabs within coral species northern Red Sea, Egypt." *Asian Journal of Scientific Research*. doi:10.3923/ajsr.2013.98.106.

Mokady, O., D.B. Bonar, G. Arazi, and Y. Loya. 1991. "Coral host specificity in settlement and metamorphosis of the date mussel *Lithophaga lessepsiana* (Vaillant, 1865)." *Journal of Experimental Marine Biology and Ecology* 146 (2): 205–216. doi:10.1016/0022-0981(91)90026-S.

Mokady, O., D.B. Bonar, G. Arazi, and Y. Loya. 1993. "Spawning and development of three coral-associated Lithophaga species in the Red Sea." *Marine Biology* 115 (2): 245–252. doi:10.1007/BF00346341.

Mokady, O., S. Rozenblatt, D. Graur, and Y. Loya. 1994. "Coral-host specificity of Red Sea Lithophaga bivalves: Interspecific and intraspecific variation in 12S mitochondrial ribosomal RNA." *Molecular Marine Biology and Biotechnology* 3 (3): 158–164.

Moya, A., S. Tambutté, E. Tambutté, D. Zoccola, N. Caminiti, and D. Allemand. 2006. "Study of calcification during a daily cycle of the coral *Stylophora pistillata*: Implications for 'light-enhanced calcification.'" *Journal of Experimental Biology* 209 (17): 3413–3419. doi:10.1242/jeb.02382.

Muir, P.R., and M. Pichon. 2019. "Biodiversity of reef-building, scleractinian corals." In *Mesophotic Coral Ecosystems*, eds. Loya, Y., Puglise, K.A., and Bridge, T.C.L., 589–620. Cham: Springer International Publishing. doi:10.1007/978-3-319-92735-0_33.

Müller, W.E.G., I. Müller, R. Zhan, and A. Maidhof. 1984. "Intraspecific recognition system in scleractinian corals: Morphological and cytochemical description of the autolysis mechanism." *Journal of Histochemistry & Cytochemistry* 32 (3): 285–288.

Muscatine, L., P. Falkowski, J.W. Porter, and Z. Dubinsky. 1984. "Fate of photosynthetic fixed carbon in light-and shade-adapted colonies of the symbiotic coral *Stylophora pistillata*." *Society of London: Series B: Biological Sciences* 222 (1227): 181–202.

Muscatine, L., L.R. McCloskey, and Y. Loya. 1985. "A comparison of the growth rates of zooxanthellae and animal tissue in the red sea coral *Stylophora pistillata*." *Proceedings of the Fifth International Coral Reef Congress, Tahiti*, 6: 119–123.

Muscatine, L., L.R. Mccloskey, and R.E. Marian. 1981. "Estimating the daily contribution of carbon from zooxanthellae to coral animal respiration." *Limnolgy and Oceanography* 26 (4): 601–611.

Muscatine, L., J.W. Porter, and I.R. Kaplan. 1989. "Resource partitioning by reef corals as determined from stable isotope composition." *Marine Biology* 100: 185–193.

Muscatine, L., E. Tambutte, and D. Allemand. 1997. "Morphology of coral desmocytes, cells that anchor the calicoblastic epithelium to the skeleton." *Coral Reefs* 16 (4): 205–213. doi:10.1007/s003380050075.

Nakamura, T., H. Yamasaki, and R. Van Woesik. 2003. "Water flow facilitates recovery from bleaching in the coral *Stylophora pistillata*." *Marine Ecology Progress Series* 256: 287–291.

Nishikawa, A., M. Katoh, and K. Sakai. 2003. "Larval settlement rates and gene flow of broadcast-spawning (*Acropora tenuis*) and planula-brooding (*Stylophora pistillata*) corals." *Marine Ecology Progress Series* 256: 87–97. doi:10.3354/meps256087.

Oren, M., K.O. Amar, J. Douek, T. Rosenzweig, G. Paz, and B. Rinkevich. 2010. "Assembled catalog of immune-related genes from allogeneic challenged corals that unveils the participation of vWF-like transcript." *Developmental and Comparative Immunology* 34 (6): 630–637. doi:10.1016/j.dci.2010.01.007.

Oren, M., G. Paz, J. Douek, A. Rosner, K.O. Amar, and B. Rinkevich. 2013. "Marine invertebrates cross phyla comparisons reveal highly conserved immune machinery." *Immunobiology* 218 (4): 484–495. doi:10.1016/j.imbio.2012.06.004.

Peled, Y., J.L. Drake, A. Malik, R. Almuly, M. Lalzar, D. Morgenstern, and T. Mass. 2020. "Optimization of skeletal protein preparation for LC-MS/MS sequencing yields additional coral skeletal proteins in *Stylophora pistillata*." *BMC Materials* 2 (1): 8. doi:10.1186/s42833-020-00014-x.

Pontasch, S., P.L. Fisher, T. Krueger, S. Dove, O. Hoegh-Guldberg, W. Leggat, and S.K. Davy. 2017. "Photoacclimatory and photoprotective responses to cold versus heat stress in high latitude reef corals." *Journal of Phycology* 53 (2): 308–321. doi:10.1111/jpy.12492.

Pratchett, M.S. 2001. "Influence of coral symbionts on feeding preferences of crown-of-thorns starfish *Acanthaster planci* in the Western Pacific." *Marine Ecology Progress Series* 214: 111–119. doi:10.3354/meps214111.

Puill-Stephan, E., M.J.H. van Oppen, K. Pichavant-Rafini, and B.L. Willis. 2012. "High potential for formation and persistence of chimeras following aggregated larval settlement in the broadcast spawning coral, *Acropora millepora*." *Proceedings of the Royal Society of London: Series B: Biological Sciences* 279 (1729): 699–708. doi:10.1098/rspb.2011.1035.

Putnam, H.M., P.J. Edmunds, and T.-Y. Fan. 2008. "Effect of temperature on the settlement choice and photophysiology of larvae from the reef coral *Stylophora pistillata*." *The Biological Bulletin* 215 (2): 135–142. doi:10.2307/25470694.

Puverel, S., F. Houlbrèque, E. Tambutté, D. Zoccola, P. Payan, N. Caminiti, S. Tambutté, and D. Allemand. 2007. "Evidence of low molecular weight components in the organic matrix of the reef building coral, *Stylophora pistillata*." *Comparative Biochemistry and Physiology: A Molecular and Integrative Physiology* 147 (4): 850–856. doi:10.1016/j.cbpa.2006.10.045.

Puverel, S., E. Tambutté, L. Pereira-Mouriès, D. Zoccola, D. Allemand, and S. Tambutté. 2005. "Soluble organic matrix of two Scleractinian corals: Partial and comparative analysis." *Comparative Biochemistry and Physiology: B Biochemistry and Molecular Biology* 141 (4): 480–487. doi:10.1016/j.cbpc.2005.05.013.

Rachmilovitz, E.N., and B. Rinkevich. 2017. "Tiling the reef: Exploring the first step of an ecological engineering tool that may promote phase-shift reversals in coral reefs." *Ecological Engineering* 105. Elsevier B.V.: 150–161. doi:10.1016/j.ecoleng.2017.04.038.

Rahav, O., Z. Dubinsky, Y. Achituv, and P.G. Falkowski. 1989. "Ammonium metabolism in the zooxanthellate coral, *Stylophora pistillata*." *Proc. R. Soc. Lond. B* 236: 325–337.

Raymundo, L.J., and A.P. Maypa. 2004. "Getting bigger faster: Mediation of size-specific mortality via fusion in juvenile coral transplants." *Ecological Applications* 14 (1): 281–295.

Raz-Bahat, M., J. Douek, E. Moiseeva, E.C. Peters, and B. Rinkevich. 2017. "The digestive system of the stony coral *Stylophora pistillata*." *Cell and Tissue Research* 368 (2): 311–323. doi:10.1007/s00441-016-2555-y.

Raz-Bahat, M., J. Erez, and B. Rinkevich. 2006. "In vivo light-microscopic documentation for primary calcification processes in the hermatypic coral *Stylophora pistillata*." *Cell Tisue Res* 325: 361–368. doi:10.1007/s00441-006-0182-8.

Reynaud-Vaganay, S., A. Juillet-Leclerc, J. Jaubert, and J.P. Gattuso. 2001. "Effect of light on skeletal δ13C and δ18O, and interaction with photosynthesis, respiration and calcification in two zooxanthellate scleractinian corals." *Palaeogeography, Palaeoclimatology, Palaeoecology*. doi:10.1016/S0031-0182(01)00382-0.

Reynaud, S., N. Leclercq, S. Romaine-Lioud, C. Ferrier-Pagès, J. Jaubert, and J. Pierre-Gattuso. 2003. "Interacting effects of CO_2 partial pressure and temperature on photosynthesis and calcification in a scleractinian coral." *Global Change Biology* 9: 1660–1668. doi:10.1046/j.1529-8817.2003.00678.x.

Richmond, R.H. 1997. "Reproduction and recruitment in corals: Critical links in the persistence of reefs." *Life and Death of Coral Reefs*, 175–197. New York: Chapman & Hall.

Rinkevich, B. 1982. "*Stylophora pistillata*: Ecophysiological aspects in biology of an hermatypic coral." Tel-Aviv university, Tel-Aviv.

Rinkevich, B. 1989. "The contribution of photosynthetic products to coral reproduction." *Marine Biology* 101 (2): 259–263. doi:10.1007/BF00391465.

Rinkevich, B. 1991. "A long-term compartmental partitioning of photosynthetically fixed carbon in a symbiotic reef coral." *Symbiosis* 10: 175–194.

Rinkevich, B. 1995. "Restoration strategies for coral reefs damaged by recreational activities: The use of sexual and asexual recruits." *Society for Ecological Restoration* 3 (4): 241–251. doi:10.1111/j.1526-100X.1995.tb00091.x.

Rinkevich, B. 2000. "Steps towards the evaluation of coral reef restoration by using small branch fragments." *Marine Biology* 136 (5): 807–812. doi:10.1007/s002270000293.

Rinkevich, B. 2001. "Genetic regulation in the branching stony coral *Stylophora pistillata*." In *The Algorithmic Beauty of Seaweeds, Sponges, and Corals*, 62–66. Berlin: Springer.

Rinkevich, B. 2002. "The branching coral *Stylophora pistillata*: Contribution of genetics in shaping colony landscape." *Israel Journal of Zoology* 48 (1): 71–82. doi:10.1560/bcpa-um3a-mkbp-hgl2.

Rinkevich, B. 2004. "Allorecognition and xenorecognition in reef corals: A decade of interactions." *Hydrobiologia* 530–531 (March): 443–450. doi:10.1007/s10750-004-2686-0.

Rinkevich, B. 2005. "Conservation of coral reefs through active restoration measures: Recent approaches and last decade progress." *Environmental Science and Technology* 39 (12): 4333–4342. doi:10.1021/es0482583.

Rinkevich, B. 2008. "Management of coral reefs: We have gone wrong when neglecting active reef restoration." *Marine Pollution Bulletin* 56 (11). Elsevier Ltd: 1821–1824. doi:10.1016/j.marpolbul.2008.08.014.

Rinkevich, B. 2011. "Quo vadis chimerism?" *Chimerism* 2 (1): 1–5. doi:10.4161/chim.2.1.14725.

Rinkevich, B. 2012. "Neglected biological features in Cnidarians self-nonself recognition." In *Self-NonSelf Recognition*, ed. López-Larrea, C., 46–59. New York: Springer.

Rinkevich, B. 2014. "Rebuilding coral reefs: Does active reef restoration lead to sustainable reefs?" *Current Opinion in Environmental Sustainability* 7: 28–36. doi:10.1016/j.cosust.2013.11.018.

Rinkevich, B. 2015a. "Novel tradable instruments in the conservation of coral reefs, based on the coral gardening concept for reef restoration." *Journal of Environmental Management* 162: 199–205. doi:10.1016/j.jenvman.2015.07.028.

Rinkevich, B. 2015b. "Climate change and active reef restoration-ways of constructing the 'reefs of tomorrow'." *Journal of Marine Science and Engineering* 3 (1): 111–127. doi:10.3390/jmse3010111.

Rinkevich, B. 2015c. "A critique of why looks can be deceptive in judging the health of well-fed corals (related to DOI 10.1002/bies.201400074)." *BioEssays* 37 (4): 354–355. doi:10.1002/bies.201400216.

Rinkevich, B. 2019a. "The active reef restoration toolbox is a vehicle for coral resilience and adaptation in a changing world." *Journal of Marine Science and Engineering* 7: 201. doi:10.3390/jmse7070201.

Rinkevich, B. 2019b. "Coral chimerism as an evolutionary rescue mechanism to mitigate global climate change impacts." *Global Change Biology* 25 (4): 1198–1206. doi:10.1111/gcb.14576.

Rinkevich, B. 2020. "Ecological engineering approaches in coral reef restoration." *ICES Journal of Marine Science* 2100. doi:10.1093/icesjms/fsaa022.

Rinkevich, B., S. Ben-Yakir, and R. Ben-Yakir. 1999. "Regeneration of amputated avian bone by a coral skeletal implant." *Biological Bulletin* 197 (1): 11–13.

Rinkevich, B., and Y. Loya. 1977. "Harmful effects of chronic oil pollution on a Red Sea scleractinian coral population." *Proceedings of the Third Internacional Coral Reef Symposium*. Miami.

Rinkevich, B., and Y. Loya. 1979a. "The reproduction of the Red Sea coral *Stylophora pistillata*. I: Gonads and planulae." *Marine Ecology Progress Series* 1: 145–152. doi:10.3354/meps001145.

Rinkevich, B., and Y. Loya. 1979b. "The reproduction of the Red Sea coral *Stylophora pistillata*. II: Synchronization in breeding and seasonality of planulae shedding." *Marine Ecology Progress Series* 1: 133–144. doi:10.3354/meps001133.

Rinkevich, B., and Y. Loya. 1979c. "Laboratory experiments on the effects of crude oil on the Red Sea coral *Stylophora pistillata*." *Marine Pollution Bulletin* 10 (11): 328–330. doi:10.1016/0025-326X(79)90402-8.

Rinkevich, B., and Y. Loya. 1983a. "Intraspecific competitive networks in the Red Sea coral *Stylophora pistillata*." *Coral Reefs* 1 (3): 161–172. doi:10.1007/BF00571193.

Rinkevich, B., and Y. Loya. 1983b. "Oriented translocation of energy in grafted reef corals." *Coral Reefs* 1 (4): 243–247. doi:10.1007/BF00304422.

Rinkevich, B., and Y. Loya. 1983c. "Short-term fate of photosynthetic products in a hermatypic coral." *Journal of Experimental Marine Biology and Ecology* 13: 175–184.

Rinkevich, B., and Y. Loya. 1984a. "Coral illumination through an optic glass-fiber: Incorporation of 14C photosynthates." *Marine Biology* 80: 7–15. doi:10.1038/164914a0.

Rinkevich, B., and Y. Loya. 1984b. "Does light enhance calcification in hermatypic corals?" *Marine Biology* 80 (1): 1–6. doi:10.1007/BF00393120.

Rinkevich, B., and Y. Loya. 1985a. "Coral isomone: A proposed chemical signal controlling intraclonal growth patterns in a branching coral." *Bulletin of Marine Science* 36 (2): 319–324.

Rinkevich, B., and Y. Loya. 1985b. "Intraspecific competition in a reef coral: Effects on growth and reproduction." *Oecologia* 66 (1): 100–105. doi:10.1007/BF00378559.

Rinkevich, B., and Y. Loya. 1986. "Senescence and dying signals in a reef building coral." *Experientia* 42: 320–322.

Rinkevich, B., and Y. Loya. 1987. "Variability in the pattern of sexual reproduction of the coral *Stylophora pistillata* at Eilat, Red Sea: A long-term study." *Biological Bulletin* 173 (2): 335. doi:10.2307/1541546.

Rinkevich, B., A. Nanthawan, and C. Rabinowitz. 2005. "UV incites diverse levels of DNA breaks in different cellular compartments of a branching coral species." *Journal of Experimental Biology* 208 (5): 843–848. doi:10.1242/jeb.01496.

Rinkevich, B., and S. Shafir. 1998. "Ex situ culture of colonial marine ornamental invertebrates: Concepts for domestication." *Aquarium Sciences and Conservation* 2: 237–250.

Rinkevich, B., N. Shashar, and T. Liberman. 1993. "Nontransitive xenogeneic interactions between four common Red Sea sessile invertebrates." *Proceedings of the Seventh International Coral Reef Symposium* 2: 833–839.

Rinkevich, B., and I.L. Weissman. 1987. "Chimeras in colonial invertebrates: A synergistic symbiosis or somatic-cell and germ-cell parasitism?" *Symbiosis* 4 (1–3): 117–134.

Rinkevich, B., Z. Wolodarsky, and Y. Loya. 1991. "Coral-crab association: A compact domain of a multilevel trophic system." *Hydrobiologia* 216–217 (1): 279–284. doi:10.1007/BF00026475.

Rinkevich, Y., V. Matranga, and B. Rinkevich. 2009. "Stem cells in aquatic invertebrates: Common premises and emerging unique themes." In *Stem Cells in Marine Organisms*, eds. Rinkevich, B. and Matranga, V., 61–103. Dordrecht: Springer Netherlands. doi:10.1007/978-90-481-2767-2_4.

Rosic, N., C. Rémond, and M.A. Mello-Athayde. 2020. "Differential impact of heat stress on reef-building corals under different light conditions." *Marine Environmental Research* 158. doi:10.1016/j.marenvres.2020.104947.

Sampayo, E.M., T. Ridgway, P. Bongaerts, and O. Hoegh-Guldberg. 2008. "Bleaching susceptibility and mortality of corals are determined by fine-scale differences in symbiont type." *Proceedings of the National Academy of Sciences of the United States of America* 105 (30): 10444–10449. doi:10.1073/pnas.0708049105.

Sampayo, E.M., T. Ridgway, L. Franceschinis, G. Roff, and O. Hoegh-Guldberd. 2016. "Coral symbioses under prolonged environmental change: Living near tolerance range limits." *Scientific Reports* 6 (1): 1–12. doi:10.1038/srep36271.

Saragosti, E., D. Tchernov, A. Katsir, and Y. Shaked. 2010. "Extracellular production and degradation of superoxide in the coral *Stylophora pistillata* and cultured *Symbiodinium*." *PLoS One* 5 (9): 1–10. doi:10.1371/journal.pone.0012508.

Sawall, Y., M.C. Teichberg, J. Seemann, M. Litaay, J. Jompa, and C. Richter. 2011. "Nutritional status and metabolism of the coral *Stylophora subseriata* along a eutrophication gradient in Spermonde Archipelago (Indonesia)." *Coral Reefs* 30 (3): 841–853. doi:10.1007/s00338-011-0764-0.

Scucchia, F., H. Nativ, M. Neder, G. Goodbody-Gringley, and T. Mass. 2020. "Physiological characteristics of *Stylophora pistillata* larvae across a depth gradient." *Frontiers in Marine Science* 7 (January): 1–9. doi:10.3389/fmars.2020.00013.

Sevilgen, D.S., A.A. Venn, M.Y. Hu, E. Tambutté, D. De Beer, V. Planas-Bielsa, and S. Tambutté. 2019. "Full *in vivo* characterization of carbonate chemistry at the site of calcification in corals." *Science Advances* 5 (1): 1–10. doi:10.1126/sciadv.aau7447.

Shafir, S., S. Abady, and B. Rinkevich. 2009. "Improved sustainable maintenance for mid-water coral nursery by the application of an anti-fouling agent." *Journal of Experimental Marine Biology and Ecology Improv* 368: 124–128. doi:10.1016/j.jembe.2008.08.017.

Shafir, S., O. Gur, and B. Rinkevich. 2008. "A *Drupella cornus* outbreak in the northern Gulf of Eilat and changes in coral prey." *Coral Reefs* 27 (2): 379. doi:10.1007/s00338-008-0353-z.

Shafir, S., I. Halperin, and B. Rinkevich. 2014. "Toxicology of household detergents to reef corals." *Water, Air, and Soil Pollution* 225 (3). doi:10.1007/s11270-014-1890-4.

Shafir, S., and B. Rinkevich. 2008. "The underwater silviculture approach for reef restoration: An emergent aquaculture theme." In *Aquaculture Research Trends*, ed. Schwartz, S.H., 279–295. New York: Nova Science Publishers, Inc.

Shafir, S., and B. Rinkevich. 2010. "Integrated long-term mid-water coral nurseries: A management instrument evolving into a floating ecosystem." *University of Mauritius Research Journal* 16: 365–386.

Shafir, S., J. Van Rijn, and B. Rinkevich. 2001. "Nubbing of coral colonies: A novel approach for the development of inland broodstocks." *Aquarium Sciences and Conservation* 3: 183–190.

Shafir, S., J. Van Rijn, and B. Rinkevich. 2003. "The use of coral nubbins in coral reef ecotoxicology testing." *Biomolecular Engineering* 20: 401–406. doi:10.1016/S1389-0344(03)00062-5.

Shafir, S., J. Van Rijn, and B. Rinkevich. 2006a. "A mid-water coral nursery." In *Proceeding of 10th International Coral Reef Symposium*, 1674–1679. Okinawa, Japan: International Society for Reef Stusied.

Shafir, S., J. Van Rijn, and B. Rinkevich. 2006b. "Steps in the construction of underwater coral nursery, an essential component in reef restoration acts." *Marine Biology* 149 (3): 679–687. doi:10.1007/s00227-005-0236-6.

Shafir, S., J. Van Rijn, and B. Rinkevich. 2007. "Short and long term toxicity of crude oil and oil dispersants to two representative coral species." *Environmental Science and Technology* 41 (15): 5571–5574. doi:10.1021/es0704582.

Shaish, L., A. Abelson, and B. Rinkevich. 2006. "Branch to colony trajectory in a modular organism: Pattern formation in the Indo-Pacific coral *Stylophora pistillata*." *Developmental Dynamics* 235 (8). Wiley Online Library: 2111–2121.

Shaish, L., A. Abelson, and B. Rinkevich. 2007. "How plastic can phenotypic plasticity be? The branching coral *Stylophora pistillata* as a model system." *PLoS One* 2 (7): 1–9. doi:10.1371/journal.pone.0000644.

Shaish, L., and B. Rinkevich. 2009. "Critical evaluation of branch polarity and apical dominance as dictators of colony astogeny in a branching coral." *PLoS One* 4 (1). doi:10.1371/journal.pone.0004095.

Shaked, Y., and A. Genin. 2019. *The Israel National Monitoring Program in the Northern Gulf of Aqaba 2018*. Jerusalem.

Shashar, N., Y. Cohen, and Y. Loya. 1993. "Extreme diel fluctuations of oxygen in diffusive boundary layers surrounding stony corals." *The Biological Bulletin* 185 (3): 455–461. doi:10.2307/1542485.

Shefy, D., N. Shashar, and B. Rinkevich. 2018. "The reproduction of the Red Sea coral *Stylophora pistillata* from Eilat: 4-decade perspective." *Marine Biology* 165 (2): 27. doi:10.1007/s00227-017-3280-0.

Shick, J.M., S. Romaine-Lioud, C. Ferrier-Pagès, and J.P. Gattuso. 1999. "Ultraviolet-B radiation stimulates shikimate pathway-dependent accumulation of mycosporine-like amino acids in the coral *Stylophora pistillata* despite decreases in its population." *Limnology and Oceanography* 44 (7): 1667–1682.

Shick, J.M., M.L. Wells, C.G. Trick, and W.C. Dunlap. 2011. "Responces to iron limitation in two colonies of *Stylophora pistillata* exposed to high temperature: Implications for coral bleaching." *Limnology and Oceanography* 56 (3): 813–828. doi:10.4319/LO.2011.56.3.0813.

Shlesinger, T., and Y. Loya. 2016. "Recruitment, mortality, and resilience potential of scleractinian corals at Eilat, Red Sea." *Coral Reefs* 35 (4): 1–12. doi:10.1007/s00338-016-1468-2.

Stambler, N., and N. Shashar. 2007. "Variation in spectral reflectance of the hermatypic corals, *Stylophora pistillata* and *Pocillopora damicornis*." *Journal of Experimental Marine Biology and Ecology* 351 (1): 143–149.

Stat, M., W.K.W. Loh, T.C. LaJeunesse, O. Hoegh-Guldberg, and D.A. Carter. 2009. "Stability of coral-endosymbiont associations during and after a thermal stress event in the southern Great Barrier Reef." *Coral Reefs* 28 (3): 709–713. doi:10.1007/s00338-009-0509-5.

Stefani, F., F. Benzoni, S.Y. Yang, M. Pichon, P. Galli, and C.A. Chen. 2011. "Comparison of morphological and genetic analyses reveals cryptic divergence and morphological plasticity in *Stylophora* (Cnidaria, Scleractinia)." *Coral Reefs* 30 (4): 1033–1049. doi:10.1007/s00338-011-0797-4.

Stimson, J., and R.A. Kinzie. 1991. "The temporal pattern and rate of release of zooxanthellae from the reef coral *Pocillopora damicornis* (Linnaeus) under nitrogen-enrichment and control conditions." *Journal of Experimental Marine Biology and Ecology* 153: 63–67.

Sun, C.-Y., L. Gránásy, C.A. Stifler, T. Zaquin, R. V. Chopdekar, N. Tamura, J.C. Weaver, et al. 2020. "Crystal nucleation and growth of spherulites demonstrated by coral skeletons and phase-field simulations." *Acta Biomaterialia*. doi:10.1016/j.actbio.2020.06.027.

Sun, C.-Y., M.A. Marcus, M.J. Frazier, A.J. Giuffre, T. Mass, and P.U.P.A. Gilbert. 2017. "Spherulitic growth of coral skeletons and synthetic aragonite: Nature's three-dimensional printing." *ACS Nano* 11 (7): 6612–6622. doi:10.1021/acsnano.7b00127.

Takabayashi, M., D.A. Carter, J. V. Lopez, and O. Hoegh-Guldberg. 2003. "Genetic variation of the scleractinian coral *Stylophora pistillata*, from western Pacific reefs." *Coral Reefs* 22 (1): 17–22. doi:10.1007/s00338-002-0272-3.

Tambutté, É., D. Allemand, E. Mueller, and J. Jaubert. 1996. "A compartmental approach to the mechanism of calcification in hermatypic corals." *Journal of Experimental Biology* 199 (5): 1029–1041.

Tambutté, É., A. D. D. Zoccola, A. Meibom, S. Lotto, N. Caminiti, and S. Tambutté. 2007. "Observations of the tissue-skeleton interface in the scleractinian coral *Stylophora pistillata*." *Coral Reefs* 205: 517–529. doi:10.1007/s00338-007-0263-5.

Tambutté, É., S. Tambutté, N. Segonds, D. Zoccola, A. Venn, J. Erez, and D. Allemand. 2012. "Calcein labelling and electrophysiology: Insights on coral tissue permeability and calcification." *Proceedings of the Royal Society B: Biological Sciences* 279 (1726): 19–27. doi:10.1098/rspb.2011.0733.

Tamir, R., G. Eyal, I. Cohen, and Y. Loya. 2020. "Effects of light pollution on the early life stages of the most abundant northern red sea coral." *Microorganisms* 8 (2). doi:10.3390/microorganisms8020193.

Tamir, R., A. Lerner, C. Haspel, Z. Dubinsky, and D. Iluz. 2017. "The spectral and spatial distribution of light pollution in the waters of the northern Gulf of Aqaba (Eilat)." *Scientific Reports* 7: 42329. doi:10.1038/srep42329.

Tanner, J.E. 1996. "Seasonality and lunar periodicity in the reproduction of Pocilloporid corals." *Coral Reefs* 15 (1): 59–66. doi:10.1007/s003380050028.

Tchernov, D., M.Y. Gorbunov, C. De Vargas, S.N. Yadav, A.J. Milligant, M. Häggblom, and P.G. Falkowski. 2004. "Membrane lipids of symbiotic algae are diagnostic of sensitivity to thermal bleaching in corals." *Proceedings of the National Academy of Sciences of the United States of America* 101 (37): 13531–13535. doi:10.1073/pnas.0402907101.

Titlyanov, E.A., K. Bil', I. Fomina, T. Titlyanova, V. Leletkin, N. Eden, A. Malkin, and Z. Dubinsky. 2000a. "Effects of dissolved ammonium addition and host feeding with Artemia salina on photoacclimation of the hermatypic coral *Stylophora pistillata*." *Marine Biology* 137 (3): 463–472. doi:10.1007/s002270000370.

Titlyanov, E.A., T. V Titlyanova, K. Yamazato, and R. Van Woesik. 2001. "Photo-acclimation of the hermatypic coral *Stylophora pistillata* while subjected to either starvation or food provisioning." *Journal of Experimental Marine Biology and Ecology* 257 (2001): 163–181.

Titlyanov, E.A., J. Tsukahara, T. V Titlyanova, V.A. Leletkin, R. Van Woesik, and K. Yamazato. 2000b. "Zooxanthellae population density and physiological state of the coral *Stylophora pistillata* during starvation and osmotic shock." *Symbiosis* 28: 303–322.

Tremblay, P., R. Grover, J.F. Maguer, M. Hoogenboom, and C. Ferrier-Pagès. 2014. "Carbon translocation from symbiont to host depends on irradiance and food availability in the tropical coral *Stylophora pistillata*." *Coral Reefs* 33 (1): 1–13. doi:10.1007/s00338-013-1100-7.

Venn, A., E. Tambutté, M. Holcomb, D. Allemand, and S. Tambutté. 2011. "Live tissue imaging shows reef corals elevate pH under their calcifying tissue relative to seawater." *PLoS One* 6 (5). doi:10.1371/journal.pone.0020013.

Veron, J.E.N. 2000. "Corals of the World, vol. 1–3." *Australian Institute of Marine Science* 58–63. Townsville.

Villanueva, R., H. Yap, and M. Montaño. 2008. "Timing of planulation by pocilloporid corals in the northwestern Philippines." *Marine Ecology Progress Series* 370 (October): 111–119. doi:10.3354/meps07659.

Von Euw, S., Q. Zhang, V. Manichev, N. Murali, J. Gross, L.C. Feldman, T. Gustafsson, C. Flach, R. Mendelsohn, and P.G. Falkowski. 2017. "Biological control of aragonite formation in stony corals." *Science* 356 (6341): 933–938. doi:10.1126/science.aam6371.

Voolstra, C.R., Y. Li, Y.J. Liew, S. Baumgarten, D. Zoccola, J.F. Flot, S. Tambutté, D. Allemand, and M. Aranda. 2017. "Comparative analysis of the genomes of *Stylophora pistillata* and *Acropora digitifera* provides evidence for extensive differences between species of corals." *Scientific Reports* 7 (1): 1–14. doi:10.1038/s41598-017-17484-x.

Wallace, C.C., and P.L. Harrison. 1990. "Reproduction, dispersal and recruitment of scleractinian corals." In *Ecosystems of the World, 25: Coral Reefs*, ed. Dubinsky, Z., 25:133–207. Amsterdam: Coral Reefs. Elsevier Science Publishing Company, Inc.

Weis, V.M., G.J. Smith, and L. Muscatine. 1989. "A 'CO_2 supply' mechanism in zooxanthellate cnidarians: Role of carbonic anhydrase." *Marine Biology* 100: 195–202.

Winters, G., S. Beer, B. Ben Zvi, I. Brickner, and Y. Loya. 2009. "Spatial and temporal photoacclimation of *Stylophora pistillata*: Zooxanthella size, pigmentation, location and clade." *Marine Ecology Progress Series* 384: 107–119. doi:10.3354/meps08036.

Yakovleva, I., R. Bhagooli, A. Takemura, and M. Hidaka. 2004. "Differential susceptibility to oxidative stress of two scleractinian corals: Antioxidant functioning of mycosporine-glycine." *Comparative Biochemistry and Physiology: B Biochemistry and Molecular Biology* 139 (4): 721–730. doi:10.1016/j.cbpc.2004.08.016.

Zakai, D., Z. Dubinsky, A. Avishai, T. Caaras, and N. Chadwick. 2006. "Lunar periodicity of planula release in the reef-building coral *Stylophora pistillata*." *Marine Ecology Progress Series* 311: 93–102. doi:10.3354/meps311093.

Zoccola, D., P. Ganot, A. Bertucci, N. Caminiti-Segonds, N. Techer, C.R. Voolstra, M. Aranda, et al. 2015. "Bicarbonate transporters in corals point towards a key step in the evolution of cnidarian calcification." *Scientific Reports* 5: 1–11. doi:10.1038/srep09983.

Zoccola, D., E. Tambutté, E. Kulhanek, S. Puverel, J.C. Scimeca, D. Allemand, and S. Tambutté. 2004. "Molecular cloning and localization of a PMCA P-type calcium ATPase from the coral *Stylophora pistillata*." *Biochimica et Biophysica Acta—Biomembranes* 1663 (1–2): 117–126. doi:10.1016/j.bbamem.2004.02.010.

Zoccola, D., E. Tambutté, F. Sénégas-Balas, J.F. Michiels, J.P. Failla, J. Jaubert, and D. Allemand. 1999. "Cloning of a calcium channel α1 subunit from the reef-building coral, *Stylophora pistillata*." *Gene* 227 (2): 157–167. doi:10.1016/S0378-1119(98)00602-7.

Zvuloni, A., O. Mokady, M. Al-Zibdah, G. Bernardi, S.D. Gaines, and A. Abelson. 2008. "Local scale genetic structure in coral populations: A signature of selection." *Marine Pollution Bulletin* 56 (3): 430–438. doi:10.1016/j.marpolbul.2007.11.002.

12 *Symsagittifera roscoffensis* as a Model in Biology

Pedro Martinez, Volker Hartenstein, Brenda Gavilán, Simon G. Sprecher and Xavier Bailly

In memoriam of our friend and colleague Heinrich Reichert.

CONTENTS

12.1 INTRODUCTION

Lynn Margulis (1938–2011), the iconoclastic scientist who shed light on biological evolutionary mechanisms that have driven the emergence of eukaryotic cell complexity by sequences of mergers of different type of bacteria, often referred in her works to marine "sunbathing green worms" from beaches of Brittany, France (Margulis 1998). She exemplified the sometimes uncritically accepted serial endosymbiotic theory (Sagan 1967) by pointing at this photosynthetic animal, a sustainable assemblage combining a marine flatworm and a dense population of photosynthetically active green microalgae localized under its epidermis (Figure 12.1a, b). From a rhetorical standpoint, the use of an oxymoron to describe a biological system (photosynthesis is not expected to be a property of metazoan tissues) can be a crucial educational and pedagogical lever. It provides a strong illustration for introducing and promoting the holobiont paradigm, which conceives of all living beings as complex assemblages formed by different organisms that constantly communicate.

We present herein descriptions related to the history, biology and ecology of *Symsagittifera roscoffensis* which have led to the emergence of this metazoan as a marine model organism, a photosymbiotic flatworm living together with *in hospite* green microalgae in its tissues, giving the typically green color to the animals (hence the name "mint-sauce worm"). *Symsagittifera roscoffensis* became attractive for research because gravid specimens can be found abundantly on specific beaches along the Atlantic coast, and all stages of development are easily accessible in the lab. Recent zootechnical advances allow for completing the life cycle in captivity; this includes deseasonalization (bypassing the annual reproductive diapause) but above all conserving colonies for months, with very low mortality and high reproduction rate. Culture standardization is critical to provide wide access to *S. roscoffensis* as a system exhibiting various biological properties, from brain regeneration to photosymbiosis.

DOI: 10.1201/9781003217503-12

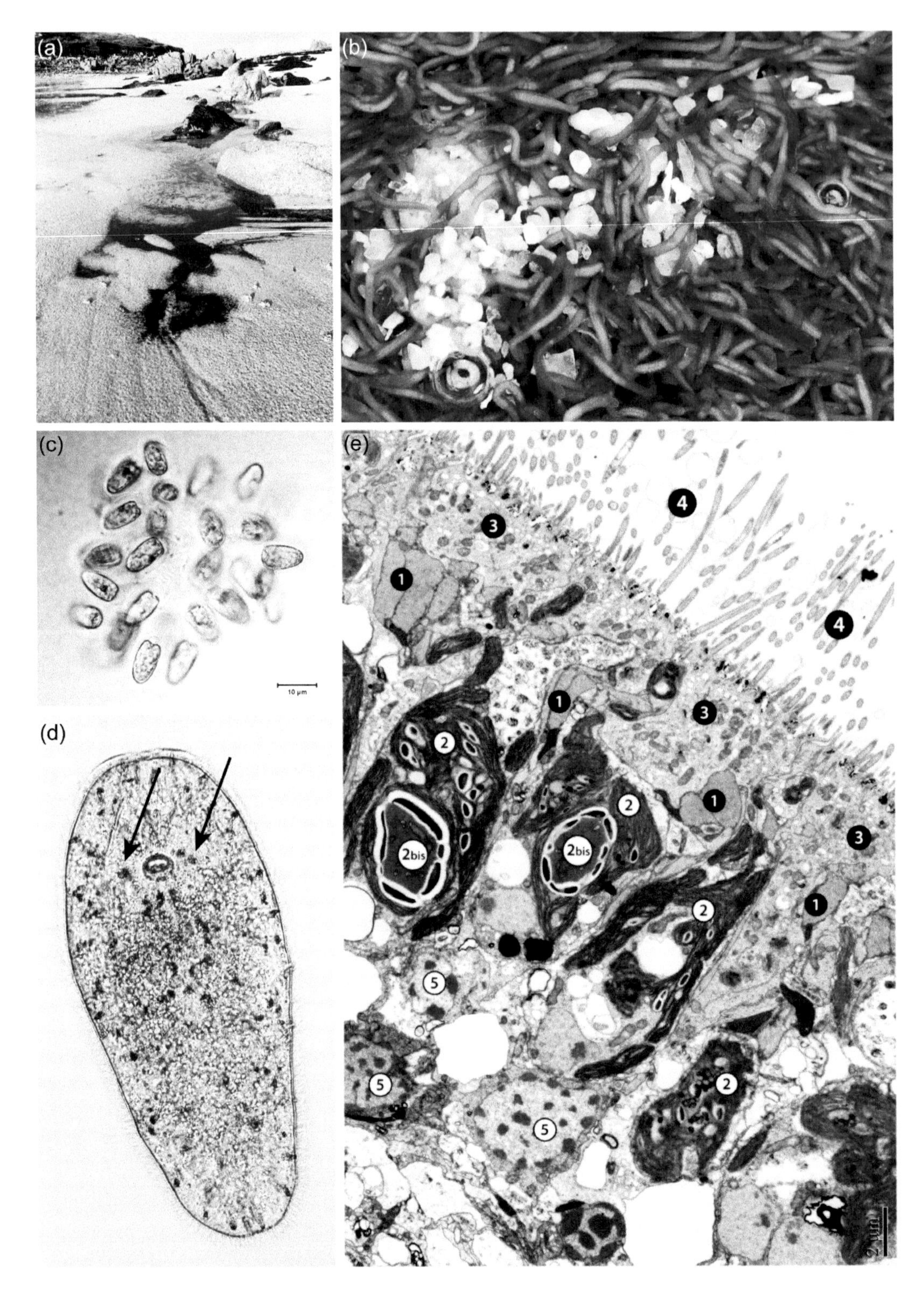

FIGURE 12.1 *S. roscoffensis* biotope and its photosymbiont. (a) At low tide, millions of *S. roscoffensis* specimens emerge from the sand and aggregate in puddles or gentle flow streams until the next high tide. The whole colony appears as a green mat. (b) Enlarged view of (a) showing high density of *S. roscoffensis*. Each adult flatworm is about 3 millimeters long. The white filaments in the middle of the body are oocytes (gravid animals). (c) Free-living algae *Tetraselmis convolutae*: The difference of phenotype between the *in hospite* microalgae and the free-living relatives are mainly noticeable by the absence of a cell wall (and the flagella) resulting from its ingestion in the animal tissues. (d) A freshly hatched, transparent juvenile of about 250 to 300 micrometers long. The brownish cells homogeneously spread along the body are rhabdites, rod-shaped, epidermal, mucus-secreting bodies (Smith et al. 1982). Two black arrows point to the photoreceptors at both sides of the statocyst (gravity sensor). (e) A transmission electron microscopy picture of the epidermal and sub-epidermal layers of the animal. Above the muscle fibers, organized as a net (1), lay the epidermal ciliated cells (3 and 4). The photosymbiont algae (2) are localized beneath the muscle layer (the closest position within the parenchyma to sense the light). Most of the microalgae cellular space is occupied by the thylakoids (lamellar-like structure = dedicated to photon harvesting) with a characteristic central structure, the pyrenoid (2bis), surrounded by the white halo (a sign of starch synthesis). Microalgae are in close contact with animal cells (5).

12.2 HISTORY OF THE MODEL AND GEOGRAPHICAL LOCATION

In the first publications, addressing the nature and origin of the "green bodies" conferring the animals' green color (Geddes 1879) and the intriguing simplicity of the body plan (Delage 1886), *S. roscoffensis* was first mistakenly referred to as *Convoluta schultzii*, a phenotypically similar species previously described from the Adriatic Sea. An accurate taxonomic description was performed by Ludwig von Graff, hosted in a marine biological laboratory outpost on the coasts of North Brittany, France, now called the *Station Biologique de Roscoff*. As a tribute to the spirit of hospitality associated with facilities provided for exploration and experimentation of the surrounding marine environment, von Graff named this species *Convoluta roscoffensis* (von Graff 1891). Since then, colonies of billions of individuals have been observed on sandy beaches, distributed all along the Atlantic coast of Europe, from Wales to Portugal. The *in hospite* enigmatic green cells in the original description were first described as chloroplasts vertically transferred as colorless leucoplasts (Graff and Haberlandt 1891). They were later isolated and identified as free-living quadri-flagellate green microalgae (Gamble and Keeble 1904), known today as *Tetraselmis convolutae* (Figure 12.1c), and formerly named *Platymonas convolutae* (Parke and Manton 1967). Revisited with molecular taxonomy tools (Kostenko and Mamkaev 1990), *Convoluta roscoffensis* was renamed *Symsagittifera roscoffensis*. Initially positioned inside the Platyhelminthes phylum as an acoel turbellarian, this species is now a member of the phylum Xenacoelomorpha (Philippe et al. 2011), whose critically—and currently unresolved—phylogenetic position in the animal tree of life is discussed further.

S. roscoffensis has initially been used in a wide range of studies as a model for deciphering the mechanisms of the setting up, specificity and trophic relationship of this photosymbiosis in the intertidal zone. Gravid adult *S. roscoffensis* lay a translucid cocoon with embryos that develop to the aposymbiotic juvenile stage within four to five days (Figure 12.1d). If juveniles, once outside the cocoon, fail to ingest the microalgae, they do not survive to maturity, indicating that this association is obligate, with the animal feeding on photosynthates transferred from the photosymbiont (Keeble 1907). The aposymbiotic *S. roscoffensis* juvenile specifically incorporates but do not digest some *Tetraselmis convolutae*. These microalgae, in comparison to other closely related species (*T. chui/subcoriformis/suecica*), exhibit a special mode of division, whereby daughter cells stay in pairs in the parent theca for a much longer period, a factor favoring ingestion by the "benthic" juvenile acoel. The *in hospite* microalgae are taken up into the digestive syncytium and undergo morphological alterations compared to the free-living state, losing their theca (cell wall), eyespot and flagella but retaining an imposing chloroplast and a specific shape with finger-like processes (Oshman 1966; Figure 12.1e). This suggests that microalgal cellular processes leading to high levels of energy consumption are drastically reduced in favor of increasing photosynthesis and production of organic molecules. Mannitol and starch (visible as grains in the chloroplast—Figure 12.1e) are the major carbohydrates in both free-living and *in hospite* microalgae (Gooday 1970). The photosynthetically fixed carbon, moving from the microalgae to the animal are mostly amino acids (Muscatine 1974). The nitrogen source for the *in hospite* algae (i.e. for amino acid synthesis) is ammonia stemming from the animal's uric acid catabolism (Boyle 1975). Both adult and aposymbiotic juvenile worms produce nitrogen waste (i.e. uric acid/ammonia) that is recycled by the algae for protein synthesis. In juveniles, uric acid crystals accumulate until photosynthesis sets in, then decline once photosynthesis is fully operational (Douglas 1983a).

According to the literature (Oshman 1966; Nozawa et al. 1972; Muscatine et al. 1974; Meyer et al. 1979), microalgal photosynthetic activity provides all of the energy and nutrients (proteins, polysaccharides, lipids) for feeding the worm. However, strict photo-autotrophy has never been formally demonstrated for this association, and one cannot rule out a mixotrophic regime: *S. roscoffensis* could indeed take up some additional organic molecules released by benthic organisms, including the environmental microbiome.

The paucity of data describing the trophic relationship between *S. roscoffensis* and *T. convolutae* prevents one from assigning a mutualistic status between these organisms, with the idea of a reciprocal benefit and egalitarian partnership, as has often been claimed. Controversially, recent surveys on photosynthetic endosymbiosis rather suggest that microalgae are exploited by their host (Kiers and West 2016; Lowe et al. 2016).

The *S. roscoffensis* biotope is localized within the upper sandy part of the intertidal zone. During high tide, animals live inside the interstitial sandy net, but as soon as the tide goes out (uncovers the sand) and until it comes in again, the animals are exposed to the sunlight in seepages or pools of seawater.

12.3 LIFE CYCLE AND REPRODUCTION

Exploring the diversity and complexity of body plans and their evolutionary and developmental basis requires that the entire life cycle of a species be accessible, from the freshly fertilized oocyte to the gravid reproducer. Controlling all the developmental steps of a species in captivity is essential to undertake necessary experimental steps, including genetic analysis and genome editing. An often-ignored obstacle is a non-negligible investment in time and expenses, a suite of trials, errors and chance findings that slow down access to many crucial stages of ontogenesis.

12.3.1 Reproductive Organs

Acoels are hermaphroditic and reproduce by internal fertilization. Sperm cells and eggs develop from neoblast-derived progenitors which divide and mature in the parenchyma in an anterior–posterior gradient (Figure 12.2a, b). Figure 12.2c shows V-shaped bundles of sperm ("sperm tracts"),

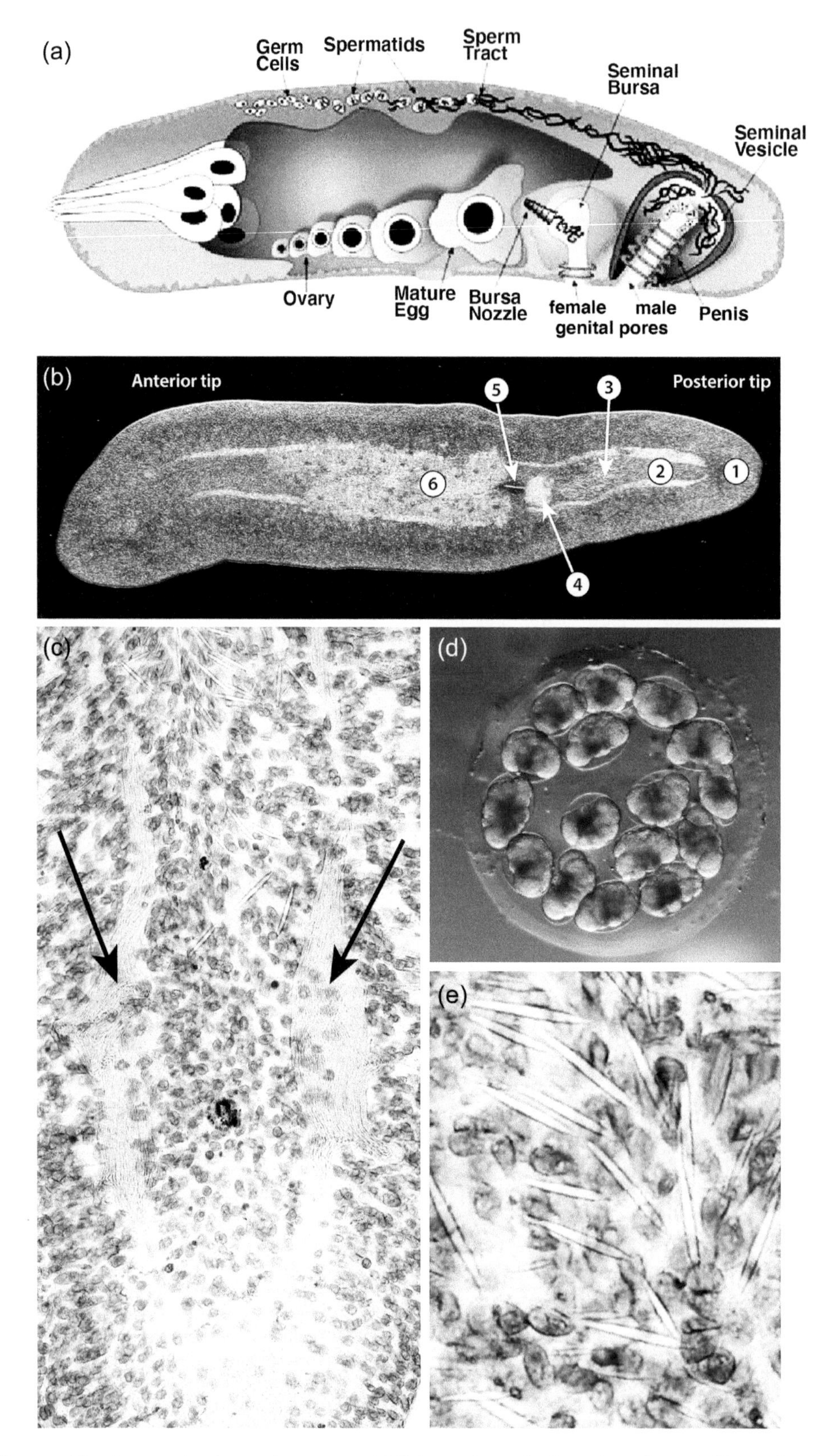

FIGURE 12.2 *S. roscoffensis* reproduction and anatomy. (a) Schematic sagittal section of acoel illustrating reproductive organs. (b) Photograph showing gravid *S. roscoffensis* reproductive organs: a male gonopore (1) is associated with bundles of mature sperm (2); flanking the gonopore area, there are an important group of saggitocysts (3). A female genital pore (not visible in the picture) gives access to the spermatheca, full of spermatozoids (4) ready to fertilize mature oocytes (6), an event mediated by a bursal nozzle (5). (c) V-shaped bundles of sperm ("sperm tracts"), localized in the posterior part of the body and converging into the male gonopore (invisible in this picture). (d) Cocoon with cluster of cleavage stage embryos. (e) Needle-like structures, the sagittocysts, are found around the genitalia at the end the body. ([a] After Kathryn Apse and Prof. Seth Tyler, University of Maine; with permission. http://turbellaria.umaine.edu/globalworming/.)

localized in the posterior part of the body and converging onto the male gonopore. Fertilization is mutual, and sperm are transferred into the seminal bursa and stored there until the eggs are ready to be fertilized (Figure 12.2a, b). Acoel egg and sperm morphologies vary among species, and their characteristics have been used for taxonomic classification (e.g. Achatz et al. 2013). Their copulatory organs are well developed and also show great morphological variety across different taxa. The members of the family Sagittiferidae, for example, develop an antrum that is turned inside out, and the bursa of many sagittiferid species lacks a muscular lining (Kostenko and Mamkaev 1990). In general, the copulatory apparatus of Sagittiferidae is considered a simplified version when compared to those of other families, such as Convolutidae (Zabotin and Golubev 2014).

Most species release the fertilized eggs through the mouth. A few species release eggs through the female genital pore (in those species that have this structure), but all species release the sperm through the male gonopores (Figure 12.2a, b). Genital pores in Acoela are by no means simple structures but have specific associated muscle systems. *Symsagittifera roscoffensis* has both male and female genital pores. The female genital pore lies in much closer proximity to the male pore than to the mouth, namely at 70% of the anterior–posterior axis, while the male genital pore is located at the 90% position (Semmler et al. 2008). The male copulatory organ presents a complex associated musculature. In the position where it is located, the regular grid of circular and longitudinal muscles of the body wall is disrupted, as also happens in the area of the female genital pore. The bursal nozzle is composed of a sclerotized lamellate stack of cells, forming a tubule. This tubiform structure on the seminal bursa is believed to behave like a sperm duct, through which allosperm are transported to the oocytes (Figure 12.2a, b).

In addition to the copulatory organs themselves, certain structures of yet-unknown function are clustered around the male gonopore. Called saggitocysts, these have a needle-like shape with a clear muscle mantle that wraps around an interior protusible filament, being located below the body's muscular grid (Figure 12.2d). Some authors have speculated that the needles might be released and be functionally relevant during copulation (e.g. Yamasu 1991).

12.3.2 Egg Deposition

In the natural environment, *S. roscoffensis* is not gravid from July to September and usually reproduces from October to June. In the lab, each gravid adult (Figure 12.2b) maintained in filtered or artificial sea-water spontaneously lays embryos. Embryos are surrounded by a viscous *mucous* layer, a cocoon or capsule (Figure 12.2e). The lack of extracellular coats around oocytes prior to capsule formation is functionally very significant, since it allows the incorporation of multiple cells per capsule (Shinn 1993). Once the cocoon with a diameter of approximately 750 micrometers is finished, the adult deposits the eggs inside it. The number of eggs inside each cocoon can reach a maximum of 30. After four to five days of development, embryos become actively moving transparent juvenile flatworms, approximately 250 micrometers long (Figure 12.1d). After some hours, the juveniles hatch from the cocoon. The absence of microalgae in the juvenile tissues indicates that the transmission of the microalgae is not vertical (i.e. transmission through the oocytes) but horizontal: the free-living microalgae live in the sand and seawater of flatworm's habitat. In the lab, without providing the free-living algae, the juvenile reared in sterile seawater do not survive more than 10 to 15 days, indicating that this partnership is obligatory with respect to the animal.

12.4 ANATOMY

12.4.1 General Architecture of Cells and Tissues

As a member of the clade Acoelomorpha, *S. roscoffensis* lacks a body cavity. A body wall, consisting of processes of epidermal cells and muscle cells, encloses a solid parenchyma whose cells serve the digestion and distribution of nutrients. Embedded in the parenchyma are the nervous system, a variety of glands and the reproductive organs (Ehlers 1985; Rieger et al. 1991).

A fundamental aspect of acoelomorph cellular architecture is the highly branched nature of virtually all cell types. Cells possess a cell body, formed by the nucleus surrounded by scant cytoplasm, and one or (more often) multiple processes which emerge from the cell body (Ehlers 1985; Rieger et al. 1991; Figure 12.3a, b). Processes display a great variety of shapes depending on the type of cell considered. There is the main, or "functional" process(es), next to one or more leaf-like ensheathing processes that many cells project around neighboring structures. Epidermal cells, for example, emit their one "connecting" process radially toward the periphery, where it spreads out to form a large (compared to the size of the cell body), flattened layer that displays the complex ultrastructural features, such as microvilli and cilia, intercellular junctional complexes and epitheliosomes (Rieger et al. 1991; Lundin 1997; Figure 12.3b, c and see subsequently). Additional branched and variably shaped processes of the epidermal cell body project horizontally and intermingle with peripheral nerves, muscle fibers and parenchymal cells (Figure 12.3b, c). Similar to epidermal cells, muscle cells give rise to connecting processes which branch out into long, slender fibers (myofibers) that contain contractile actin-myosin filaments (myofilaments; Figure 12.3b, c). Many cells, including muscle and glands, possess a third type of thin, cylindrical process that enters the neuropil of the central nervous system (see subsequently).

Their branched anatomy implies that the cell bodies of epidermal cells or muscle cells (and other cell types) are located at a distance from their "functional parts", that is, the myofibers or epithelial processes forming the body wall. Cell bodies are embedded in the parenchyma, where they are arranged as an irregular layer ("cell body domain") around

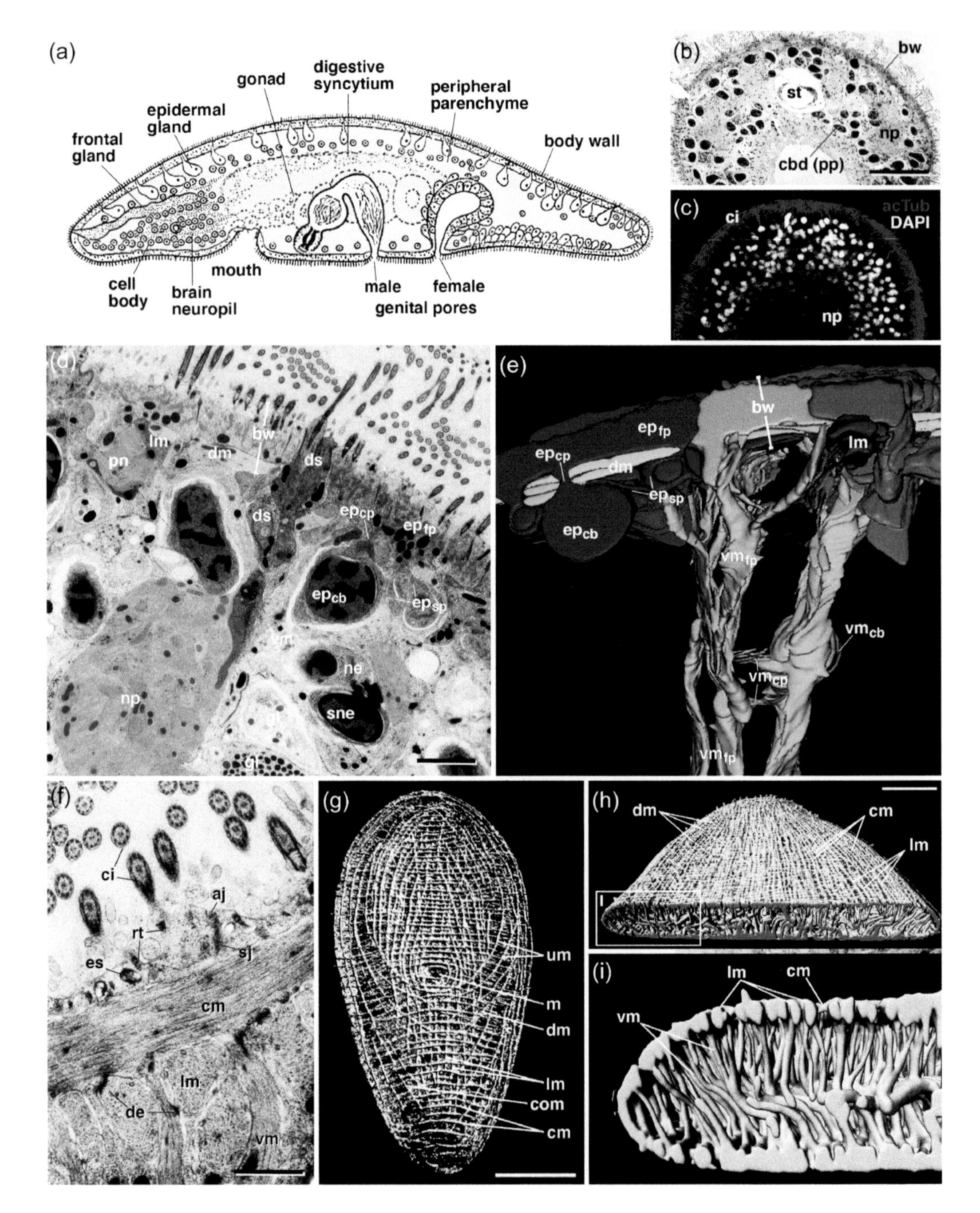

FIGURE 12.3 Anatomy of *S. roscoffensis*. (a) Schematic sagittal section of acoel (modified from Hyman 1951). (n) Ultrathin cross-section of juvenile *S. roscoffensis* at level of statocyst (st), showing body wall (bw), domain of cell bodies (cbd), sunken into peripheral parenchyma (pp) and neuropil (np). (c) Confocal section of juvenile *S. roscoffensis* labeled with anti-acetylated tubulin (acTub, red; marking epidermal cilia [ci] and neuronal fibers forming neuropil [np]). (d) Ultrathin cross-section of juvenile *S. roscoffensis*, showing structures of bodywall (bw), peripheral parenchyma/cell body domain, and neuropil (np). Different cell types are rendered in shades of blue (epidermal cells), green (muscle cells), red (neurons) and yellow (gland cells). Basic architecture of acoel cell types is shown for epidermal cell at upper right, for which cell body (ep_{cb}), connecting process (ep_{cp}), functional process (ep_{fp}) and sheath processes (ep_{sp}) are visible. Muscle cell fibers include longitudinal fibers (lm), diagonal fibers (dm) and vertical fibers (vm). A bundle of peripheral sensory dendrites (ds; shades of purple) penetrate the bodywall. (e) 3D digital model of juvenile *S. roscoffensis* bodywall, showing partial reconstructions of three epidermal cells (blue) and vertical muscle cell (green). Components of the epidermal cell on the left and of the muscle cells are indicated. Both cells are composed of a cell body (ep_{cb}, vm_{cb}), connecting process(es) (ep_{cp}, vm_{cp}), functional processes (ep_{fp}, vm_{fp}) and sheath processes (ep_{sp}; no sheath processes are formed by the muscle cell shown). (f) Electron micrograph of cross-section of body wall of juvenile *S. roscoffensis*, showing ultrastructural aspects of epidermal cells (ci: cilia; es: epitheliosome; rt: rootlet of cilium; aj: adherens junction; sj: septate junction) and body wall–associated muscle fibers (cm: circular muscle; lm: longitudinal muscle; vm: vertical muscle; de: desmosomes between muscle fibers). (g–i) 3D rendering of *S. roscoffensis* muscles labeled by phalloidin. Ventral view (g), dorso-posterior view (h), frontal view (i; digital cross-section). Other abbreviations: com: ventral cross-over muscles; m: mouth; ne: central neuron; pn: peripheral nerve; sne: sensory neuron; um: U-shaped muscles. Scale bars: 20 micrometers (b, c); 2 micrometers (d, e); 1 micrometer (f); 50 micrometers (g). ([g–i] From Semmler et al. 2008, with permission.)

an interior neuropil and digestive syncytium (Figure 12.3d, e; see subsequently). Importantly, bodies of different cell types, in particular neurons, muscle cells and gland cells, appear to be intermingled in the cell body domain rather than forming separate organs or tissues (Figure 12.3a; Arboleda et al. 2018; Gavilan et al. in prep).

The unusual cellular architecture in acoelomorphs has been related to the absence of a basement membrane, another unique character of this clade (Smith and Tyler 1985; Rieger et al. 1991; Morris 1993; Tyler and Rieger 1999). In other animals, a basement membrane, composed of robust and highly interconnected filamentous proteins including collagens and laminins, separates epidermal cells and muscle cells and surrounds internal organs such as the intestinal tube, glands and nerves. The basement membrane also provides the point of anchorage between muscles and epidermis or other epithelial tissues. As a result, cells have a more or less symmetric shape, resembling cubes or cylinders, with the cell body included within these shapes. In acoelomorphs, lacking a basement membrane, cell bodies can be extruded out from their working parts, intermingle and adopt highly irregular, branched shapes.

12.4.2 Epidermis

The squamous functional processes of epidermal cells that cover the surface of the animal are of a fairly regular polygonal shape. Epidermal cells of *S. roscoffensis* are interconnected by belt-like junctional complexes, consisting of an apical adheres junction followed proximally by a prominent septate junction (Rieger et al. 1991; Lundin 1997; Figure 12.3f). Epidermal motile cilia power locomotion of the animal. Following the ground pattern of acoelomorphs and flatworms in general, epidermal cells are multiciliated (Figure 12.3f). Cilia are anchored by vertically oriented striated rootlets, conspicuous cytoskeletal elements consisting of the conserved protein rootletin (Yang et al. 2002). Since rootlets are interconnected by evenly sized horizontal processes, cilia of each epidermal cell form a highly symmetric array. More irregularly spaced microvilli are interspersed with the cilia. Another characteristic of epidermal cells are closely packed, moderately electron-dense vesicles called epitheliosomes, or ultrarhabdites (Rieger et al. 1991). Epitheliosomes are of rounded or elongated shape and can be seen to be extruded from the apical membrane to release their presumably mucous content (Figure 12.3f).

12.4.3 Muscle System

The musculature of the acoelomorph body wall is formed by three layers of myofibers, circular fibers, diagonal fibers and longitudinal fibers (Rieger et al. 1991; Hooge 2001). In early larval *S. roscoffensis*, one finds approximately 60 circular and 30 longitudinal fibers; in adults, these numbers increase to 300 and 140, respectively (Semmler et al. 2010; Figure 12.3g–i). Note that these numbers do not necessarily reflect the number of muscle cells, since one muscle cell soma can give rise to more than one myofiber (see previously). In addition to the outer muscles, a large number of regularly spaced, short vertical muscle fibers penetrate the parenchyma and nervous system and insert at the dorsal and ventral body wall. Specialized muscle fibers surround the mouth opening (see section on digestive system). In all muscle fibers, myofilaments show a smooth architecture (Figure 12.3f), lacking the Z-discs of striated muscles found in other clades. Myofibers are typically branched near their point of attachments to each other and to epidermal cells (Figure 12.3e, f) and exhibit electron-dense junctional complexes ("maculae adherentes" or desmosomes; Tyler and Rieger 1999; Figure 12.3f).

The innervation of the musculature of *S. roscoffensis*, as with acoelomorphs in general, is mediated by thin processes branching off the myofibers and extending into peripheral nerves or the neuropil (Rieger et al. 1991). In addition, large numbers of neuronal fibers exiting neuropil and peripheral nerves terminate in close contact to myofibers, as well as epidermal and glandular processes (Gavilan et al. in prep.). The exact mechanism of neural control of muscle contraction and ciliary movement is clearly one of the research areas that needs much attention.

12.4.4 Central Nervous System

Acoelomorphs have a central nervous system consisting of an anterior brain, several and paired longitudinal nerve cords that issue from the brain (Martinez et al., 2017). Brain and nerve trunks are formed by neuronal somata that are located in the cell body domain underlying the body wall and a central neuropil enclosed within the cell body domain. The neuropil, labeled by markers such as anti-acetylated tubulin or anti-Synapsin (Bery et al. 2010; Sprecher et al. 2015; Arboleda et al. 2018), is built of stereotypically patterned elements and provides an internal scaffold to which other cells and organs can be related. In *S. roscoffensis*, one distinguishes a dorsomedial compartment, dorsolateral compartment and ventral compartment along the dorsoventral axis (Figure 12.4a, b). As described for other acoelomorph taxa (Martinez et al. 2017), the brain neuropil of *S. roscoffensis* encloses in its center the statocyst, which demarcates within each of the compartments an anterior domain (relative to the midpoint of the statocyst) and a posterior domain (Figure 12.4a, b). Three commissures connect these compartments: the ventro-anterior commissure (vac) arises from the convergence of the anterior ventral and anterior dorso-lateral compartment, the dorso-anterior commissure (dac; c1 in Bery et al. 2010) interconnects the anterior dorso-medial compartments right in front of the statocyst and the dorso-posterior commissure (dpc; c2 in Bery et al. 2010) forms a bridge between the posterior dorso-medial compartments. The nerve cords projecting posteriorly from the brain include the dorso-medial cord (dmc, originating from dorsomedial compartment), dorsolateral cord (dlc) and ventrolateral cord (vlc) (Bery et al. 2010). The cords are also interconnected by several anastomoses and commissures.

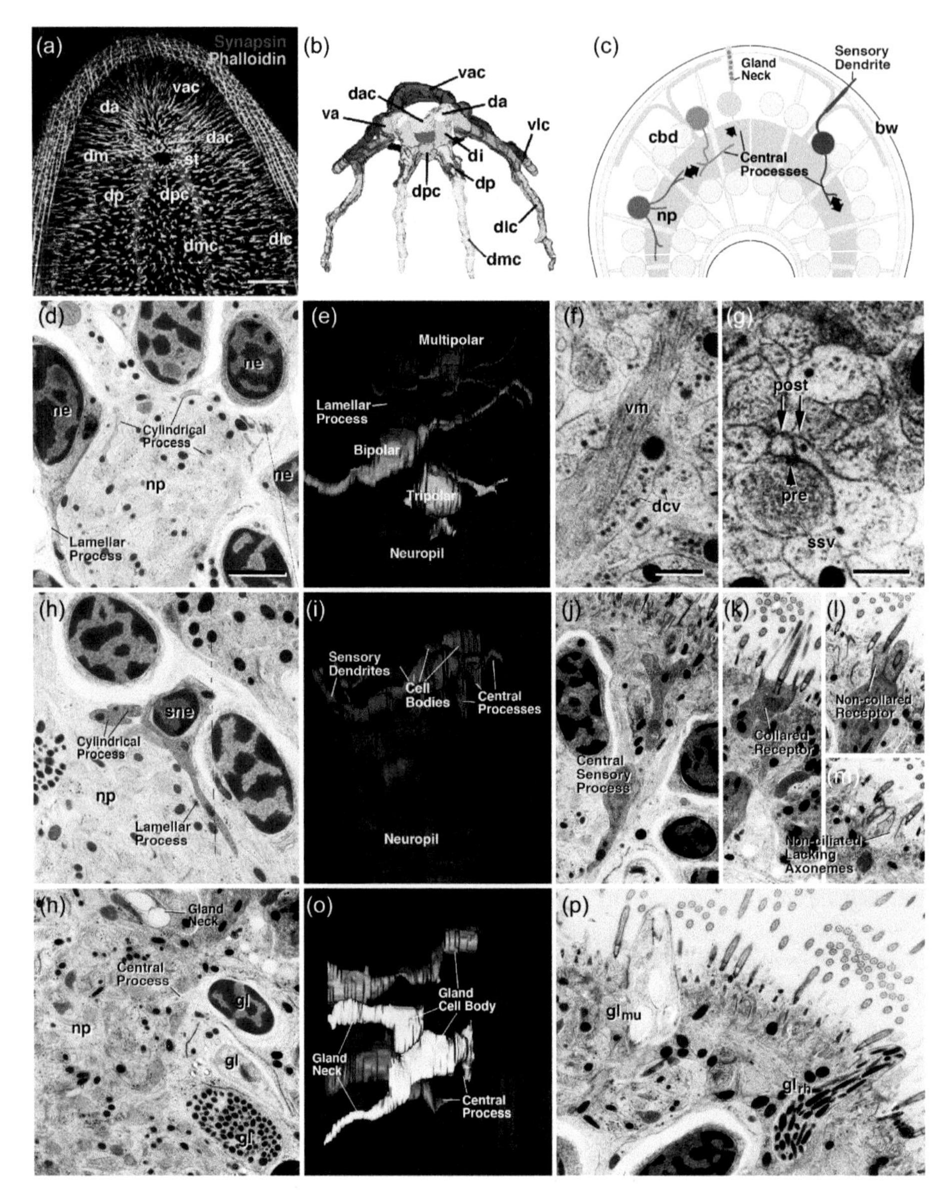

FIGURE 12.4 Anatomy of *S. roscoffensis*. (a, b) Central nervous system and neuropil. (a) A confocal section of adult *S. roscoffensis*. Muscles are labeled by phalloidin (green), central neuropil by an antibody against synapsin (red). (b) A 3D digital model of neuropil with different neuropil domains rendered in different colors. Neuropil domains visible in the dorsal view shown include dorso-anterior compartment (da), dorso-intermediate compartment (di; flanking statocyst shaded gray), dorso-posterior compartment (dp), and ventro-anterior compartment (va). The three brain commissures connecting right and left compartments are the ventro-anterior (ring) commissure (vac), dorso-anterior commissure (dac) and dorso-posterior commissure (dpc). Three pairs of nerve cords exit the brain: the dorso-medial cord (dmc), dorso-lateral cord (dlc) and ventro-lateral cord (vlc). (c) Schematic section of *S. roscoffensis*, illustrating the processes of neurons (red), sensory neurons (purple) and gland cells (yellow) in relationship to the body wall (bw), neuropil (np) and cell body domain (cbd). Thick black arrows symbolize synaptic interaction between central processes of the cells shown and elements of the neuropil. (d–g) Cytological details of central neurons. (d) Cell bodies surrounding neuropil (np; shaded blue). Three cell bodies belong to central neurons (ne; rendered in shades of red). Central neurons emit processes into neuropil. In some cases, processes exhibit particular sheath-like shapes ("lamellar processes"), aside from the cylindrical processes typical for neurons in general. (e) 3D digital model (lateral view) of four representative partially reconstructed central neurons exhibiting different shapes. (f, g) Electron microscopic sections of neuropil at high magnification. Note the high proportion of axons with dense core vesicles (dcv). Vertical muscle fibers (vm) penetrate neuropil and could receive extra-synaptic input from these axons. (g) An example of synaptic connection between large presynaptic element (pre) with small synaptic vesicles (ssv) and two small postsynaptic elements (post). (h–m) Cytological details of sensory neurons. As shown in (h), cell bodies of sensory neurons (sne) frequently lie adjacent to the neuropil and emit cylindrical or lamellar processes into the neuropil. (i) Shapes of ciliated sensory neurons (lateral view). (j) Bundle of four sensory processes linking neuropil to the body wall. (k–m) Three different types of frequently seen sensory endings, a collared receptor (k), non-collared receptor (l) and non-ciliated receptor (m). (n–p) Details of gland cell structure. In (n), cell bodies of three gland cells (rendered in shades of yellow) surround the central neuropil (np). One gland cell emits a central process into the neuropil. Digital 3D models shown in (o) illustrate representative gland cells (lateral view). (p) Section of body wall with endings of two different types of gland cells, a mucus gland cell with large electron-lucent vesicles (gl_{mu}) and a rhabdoid gland cell (g_{lrh}) with elongated, electron-dense inclusions. Scale bars: 40 micrometers (a); 2 micrometers (d, h–p); 0.5 micrometers (f, g). (From Sprecher et al. 2015, with permission.)

Neuronal cell bodies (somata) of the *S. roscoffensis* nervous system are small and have a round heterochromatin-rich nucleus (Figure 12.4d). Based on light-microscopic analysis, the larval brain contains an estimated 800 somata overall, but more precise numbers have to await serial EM analysis, since somata of neurons located in the diffuse cell body domain that surrounds the neuropil cannot be told apart with certainty from cell bodies of muscle cells or gland cells. EM reconstruction shows that many neurons are bipolar, extending an anterior process that in many cases may reach the epidermal surface to end as a sensory receptor, and one or more posterior or central process(es) that reaches into the neuropil, where it shows a modest amount of branching (Figure 12.4d, e). Along with neuronal processes, central extensions of muscle cells and gland cells also form part of the neuropil (Figure 12.4c).

Based on the types of vesicles they contain, neuronal processes of acoelomorphs were divided into four classes (Bedini and Lanfranchi 1991; Bery et al. 2010), including fibers with small clear vesicles (20–40 nm), which are associated with the "classical" transmitters acetylcholine, GABA or glutamate, and dense vesicles (70–90 nm), which resemble the dense core vesicles that, in vertebrates and many invertebrates alike, have been described to contain neuropeptides (Figure 12.4f, g). As in these other species, many neurons of *S. roscoffensis* have both types of vesicles. What stands out, however, is the large proportion of neuronal processes with dense vesicles, a finding that matches descriptions of light microscopic studies detecting peptide transmitters in large neuron populations in acoelomorphs (Reuter et al. 2001). Aside from small clear vesicles and dense vesicles, two other types with so far unknown significance and neurotransmitter content were described for acoelomorphs: another type of "dense core vesicles" (60–120 nm), containing small, dense centers surrounded by a light halo (not to be confused with the peptide-containing dense core vesicles in vertebrates or insects) and large irregularly shaped clear vesicles (20–400 nm; Bedini and Lanfranchi 1991).

Neuronal processes containing small clear vesicles in conjunction with membrane densities can be recognized as synapses (Bedini and Lanfranchi 1991; Bery et al. 2010; Figure 12.4g). However, thus defined synapses are relatively few in number, at least in the larval brain, and it is very possible that neural transmission relies heavily on extra-synaptic transmitter release. This is made all the more likely looking at the processes with dense vesicles, which fill the entire length of neurons, including the cell body, and peripheral processes. Peptide release from dense core vesicles in vertebrates has been definitively shown to occur extrasynaptically ("volume release") in many instances (Fuxe et al. 2007).

12.4.5 Peripheral Nervous System and Sensory Receptors

The peripheral nervous system consists of sensory receptors integrated in the body wall and an anastomosing meshwork of thin "nerves" that contain fibers formed by sensory receptors, muscle cells and gland cells, as well as cells effector cells ("motor neurons") that, aside from processes in the neuropil, project processes through the peripheral nerves into the periphery. Sensory neurons form part of the cell body domain surrounding the neuropil (Figure 12.4c, h, i). Their peripheral dendrites project into the body wall (Figure 12.4j), where they terminate as conspicuous elements that have been described for many flatworms, including acoels (Rieger et al. 1991). Unlike epidermal cells, sensory receptors typically contain a single cilium, aside from other apical membrane specializations. Based on these specializations, one distinguishes collared receptors from non-collared receptors (Bedini et al. 1973; Todt and Tyler 2006). In the former, a central cilium is surrounded by a ring (collar) of long, stout microvilli; this collar is lacking in the latter class. Both classes are further subdivided into several types (Todt and Tyler 2006). In *S. roscoffensis*, three types of sensory receptors have described, including non-collared receptors with a hollow ciliary rootlet containing a granulated core (Type 3 of Todt and Tyler 2006; Figure 12.4l), collared receptors with rootlets (Type 4) and collared receptors with granular body (Type 5; Figure 12.4k). Another frequently encountered type of presumed receptors are non-ciliated endings (Figure 12.4m). Receptors are distributed in characteristic patterns all at different positions (Bery et al. 2010). Nothing is known about the specific modalities and functions of sensory receptors.

Two other sensory elements, the statocyst and eyes, are surrounded by neuropil and thereby form part of the CNS (Figure 12.1c). The statocyst, thought to sense gravity, is formed by a capsule of two parietal cells enclosing a cavity that houses a specialized statolith cell (lithocyte; Ferrero 1973; Ehlers 1991). A small group of specialized muscle cells inserts at the capsule. No recognizable sensory neuronal structures are associated with the statocyst, and it has been proposed that gravity-induced displacements of the statolith could inform the CNS by affecting the muscles by which the statocyst is suspended.

The eye of convolutid acoels, including *S. roscoffensis*, is embedded into the brain on either side of the statocyst. The eye consists of a pigment cell with electron-dense granules and crystalline inclusions ("platelets") that may act as reflectors; enclosed by the pigment cell are two to three receptor cells with axons connecting to the neuropil (Yamasu 1991). Unlike most photoreceptors described for other taxa, acoel photoreceptors cells lack conspicuous microvilli or cilia.

12.4.6 Glandular System

Glands are unicellular, consisting of individual gland cells that constitute a major part of the acoelomorph body in terms of number and function. As stated for epidermal and muscle cells, gland cells consist of a cell body that forms part of the internal cell body domain and one or more elongated processes ("gland necks") that project peripherally and open to the outside (Figure 12.4c, n, o). Certain clusters of gland cells, located posteriorly of the brain, project their long necks forward through the neuropil and open at the anterior tip of the body, some of them in an acoelomorph-characteristic pore, the "frontal pore" (Pedersen 1965; Smith and Tyler 1986; Klauser et al. 1986; Ehlers 1992; Figure 12.3a). Cell bodies and gland necks contain secretory vesicles of different shape and texture by which gland cells have been divided into different classes, as summarized in the following. Gland necks carry a characteristic array of microtubules around their periphery. In addition to secretory gland necks, many gland cells appear to have central processes that invade peripheral nerves or the neuropil. These processes, like the ones formed by myofibers (see previously), may mediate the connection between nerve impulses and secretory function (Figure 12.4c, n).

Functionally and biochemically, acoelomorph gland secretions include mucus (mucopolysaccharides) that serves for locomotion, attachment and protection, as well as proteinaceous enzymes for digestion and degradation of macromolecules. Mucus-producing glands, called cyanophilic glands in the classical light microscopy literature, are structurally associated with densely packed, electron-lucent vesicles with a rounded or oval shape (Pedersen 1965; Rieger et al. 1991). Gland cells of this type open in the frontal pore but also occur all over the body surface of *S. roscoffensis*. Aside from gland cells with electron-lucent inclusions, a variety of cells with electron-dense vesicles of different sizes and shapes have been described for the acoelomorphs (Smith and Tyler 1986; Klauser et al. 1986; Todt 2009). These have been given different names (e.g. "ellipsoid" glands, "target glands", "alcian blue-positive rhabdoid glands") but cannot be assigned to specific functions. In the larva of *S. roscoffensis*, we detect glands with large, electron-lucent inclusions (mucus glands; Figure 12.4p) all over the body but preferentially anteriorly and ventrally; in addition, there are three clearly distinguishable types of gland cells with electron-dense inclusions (Gavilan et al. in prep):

1. A rare type we call a rhabdoid gland cell, with elongated inclusions of approximately 500 nm length and 100 nm diameter (Figure 12.4p).
2. Glands with pleomorphic vesicles: Inclusions are more rounded than those of rhabdoid glands and possess different diameters and electron densities (Figure 12.4n, bottom). Rhabdoid glands and pleomorphic glands are located ventro-anteriorly.
3. Glands with mixed electron-dense and electron-lucent vesicles: These are more numerous and ventro-laterally overlie the ventral nerve cord.

12.4.7 Parenchyma and Digestive Syncytium

The name-giving feature of acoels is their lack of a gut cavity. The interior of the animal is filled with a solid parenchyma that is divided into a central and peripheral domain (Smith and Tyler 1985; Gavilán et al. 2019). The central parenchyma is typically a syncytium ("digestive syncytium") formed by the merger of multiple endodermal cells; in the larva of *S. roscoffensis*, the digestive syncytium contains an estimated 6–10 nuclei (Gavilan et al. in prep.). At a mid-ventral position, the digestive syncytium is in contact with the interior through a pore ("mouth") in the epidermal covering. A pharynx, in the shape of an invagination of the ventral epidermis surrounded by specialized muscle and neural elements, is absent (Todt 2009; Semmler et al. 2010). Only a slender muscle ring from which a few fibers radiate outward marks the mouth. In addition, several ventral longitudinal muscle fibers cross over the midline right behind the mouth, giving rise to the U-shaped muscles that are the characteristic of the derived acoel clade of "Crucimusculata" to which *S. roscoffensis* belongs. It is thought that contraction of these fibers tilts the mouth forward, facilitating the uptake of food stuff.

The digestive syncytium is filled with a great diversity of organelles related to phagocytosis and digestion. In *S. roscoffensis*, symbiotic algae of the genus *Tetraselmys* are taken into the syncytium, where they lose part of their cell wall. The digestive syncytium emits processes that reach throughout the entire body, ensheathing (parts of) many cell bodies in the cell body domain and wrapping around peripheral nerves, muscle fibers and epidermal processes (Smith and Tyler 1985; Gavilan et al. in prep). One has to assume that this architecture enables the syncytium not only to digest but also distribute nutrients throughout the body. In the case of *S. roscoffensis*, algae ingested at the early larval stage multiply within vacuoles of the digestive syncytium (Oshman 1966; Douglas 1983b). In the adult, algae form a dense layer underneath the body wall, interspersed with epidermal and muscle processes (Figure 12.1d). EM analysis indicates that algae remain enclosed within the processes of the digestive syncytium (Douglas 1983b).

The peripheral parenchyma is formed by cells called "wrapping cells" (Smith and Tyler 1985) which are similar in ultrastructure to the digestive syncytium. They also form elaborate sheaths around other cells, interdigitating with processes of the digestive syncytium. It has been proposed that wrapping cells merge with the digestive syncytium, manifesting part of a dynamic process whereby newly generated cells proliferated from neoblasts (see section 12.4.8) mature, have a transient life as wrapping cells and end up as part of the central syncytium (reviewed in Gavilán et al. 2019).

12.4.8 Neoblasts (Stem Cells)

Regeneration of acoel tissues is a well-known phenomenon. This process depends on the deployment of a pool of stem-like cells called neoblasts that are present within parenchymal tissues (De Mulder et al. 2009; Srivastava et al. 2014). In all species in which neoblasts have been mapped, these cells are distributed in two lateral bands and mostly excluded from the head region. Neoblasts are easily identifiable by their intensive basophilic cytoplasm and relative scarcity of cytoplasmic organelles (Brøndsted 1955). Neoblasts are the only dividing cells in adult organisms, and they have the potential to differentiate into all, or most, cell types during regeneration (Gschwentner et al. 2001). In *Symsagittifera roscoffensis*, neoblasts have been detected using EdU labeling, and their global distribution is similar to what has been reported for other acoels (Arboleda et al. 2018). Using more detailed TEM images, these cells can be seen characteristically embedded in the parenchyma, showing the typical high nuclear/cytoplasmic ratio—a characteristic shared by all known neoblasts, including those of the distantly related Platyhelminthes phylum. After amputation of anterior structures (unpublished data), neoblasts start to proliferate immediately, in the next few hours, and are subsequently mobilized to the wound area. After this initial burst period, the number of neoblasts seems to decrease, likely due to their differentiation into newly formed tissues. Interestingly, the analysis of TEM data has shown that at least some neoblast groups (composed of three to four cells each) seem to be associated with the nerve cord and muscle fibers. This could reflect a close interaction of neoblasts with these tissues, both in regular homeostasis and in regeneration. A fraction of the cells with neoblast characteristics seem to be undergoing differentiation. The cytoplasm of these differentiating cells extends processes filled with microtubules and vesicles in between the surrounding neuronal somata or epidermal cells (Bery et al. 2010). Regeneration in Platyhelminthes and Acoela has been shown to be regulated by neural trophic factors with positional cues from musculature (Hori 1997; Hori 1999; Raz et al. 2017). This suggests that *S. roscoffensis* neoblasts may be actively receiving signals from their close environment (the niche?). A recent study from the Sprecher laboratory using single-cell technology (data not shown here) elucidates the molecular signatures characteristic of neoblasts in *Isodiametra pulchra*. These findings should enable a more detailed characterization of the regulatory factors that control the stemness state of neoblasts in acoel species and also how they make decisions to differentiate.

12.5 EMBRYOGENESIS

The embryonic development of acoels is poorly understood. Various problems, mostly practical in nature, have impaired the study of early acoel embryos. In fact, the lineage of early blastomeres has been described in detail for only one acoel species, *Neochildia fusca* (Henry et al. 2000). Later stages of development in this species have also been studied, in combination with molecular markers, by Ramachandra et al. (2002).

All acoel embryos studied thus far—including our species, *Symsagittifera roscoffensis* (Georgévitch 1899; Bresslau 1909)—appear to share the same pattern of early divisions (Georgévitch 1899; Bresslau 1909; Apelt 1969; Boyer 1971; Henry et al. 2000). Acoels' unique pattern of cleavage is termed "duet spiral" cleavage in order to differentiate it from the more common "quartet spiral" cleavage. It is important to note that although the acoel's unique form of cleavage was recognized early on by researchers such as Ernst Bresslau, it was still considered a modified version of the typical "spiral cleavage". Barbara Boyer and colleagues introduced the term "duet spiral" in 1996, after it became clear that the pattern is in fact very specific to acoels (Boyer et al. 1996).

As explained by Henry et al. (2000), the "duet" form of cleavage is characterized by the presence of a second cleavage plane oblique to the animal–vegetal axis. At the four-cell stage, the first cleavage plane corresponds to the plane of bilateral symmetry. The first two divisions give rise to four equal blastomeres, while the third division generates the first set of four micromeres in the animal half. The first division plane corresponds to the plane of bilateral symmetry, and the second cleavage always occurs in a leiotropically oblique plane relative to the animal–vegetal axis. After this second division, all remaining cleavages are symmetrical across the sagittal plane. The second sets of micromeres are given off of the macromeres. These micromeres will all give rise to the ectoderm. A fourth quartet of micromeres, plus the macromeres, will give rise to the endoderm. Finally, derivatives of some of these micromeres will give rise to the mesoderm.

The early embryos of *Symsagittifera roscoffensis* were described for the first time by Jivoïn Georgévitch in 1899, using histological sections in paraffin. He observed that the embryos are enveloped in a thick, cocoon-like membrane where they develop more or less synchronously for one week outside the animal, until hatching. The first embryonic division begins after the fertilized egg is enveloped in the cocoon membrane and outside the animal. Cleavage follows, and the embryo reaches the blastula state at the eight-blastomere stage. Here, the ectodermal cells occupy the dorsal part of the embryo, and the endodermal cells occupy the ventral part of the embryo. After a few more divisions, the embryo reaches the gastrula stage. This is achieved through the process of epiboly, in which the ectodermal cells—originally in the dorsal part—migrate downward to cover the whole embryo. No gastric cavity is observed in the gastrula, similar to Gardiner's (1895) observations in *Polychaerus*. At later stages—but before hatching—the primordia of the different tissues can be observed. Outside the embryo, the ciliary cover of the epithelial cells is clearly visible. These organ systems further mature after hatching, reaching adult-level complexity a few weeks later. The previous descriptions, while correct overall, were immediately criticized by Ernst Bresslau for inaccuracies in many details. In 1909, Bresslau published a more accurate account of each cleavage stage, from the 2-cell stage to the 32-cell stage (Figure 12.5a), Using live embryos, he was able to describe the different divisions (and their relative orientations) in great detail. Initial unequal cleavages led to a blastula at the eight-cell stage. He insisted that the changes in the configuration of the blastomeres between the 8-cell and 16-cell stages could be understood as a gastrulation process, whereby the 14 micromeres produced thus far undergo a process of epiboly that internalizes the 3A/B macromeres, the founder cells of the endo-mesoderm (Figure 12.5a). All in all, Bresslau provided the first accurate description of the first stages of development, consistent in many details with Henry et al.'s (2000) report on *Neochilida fusca* using

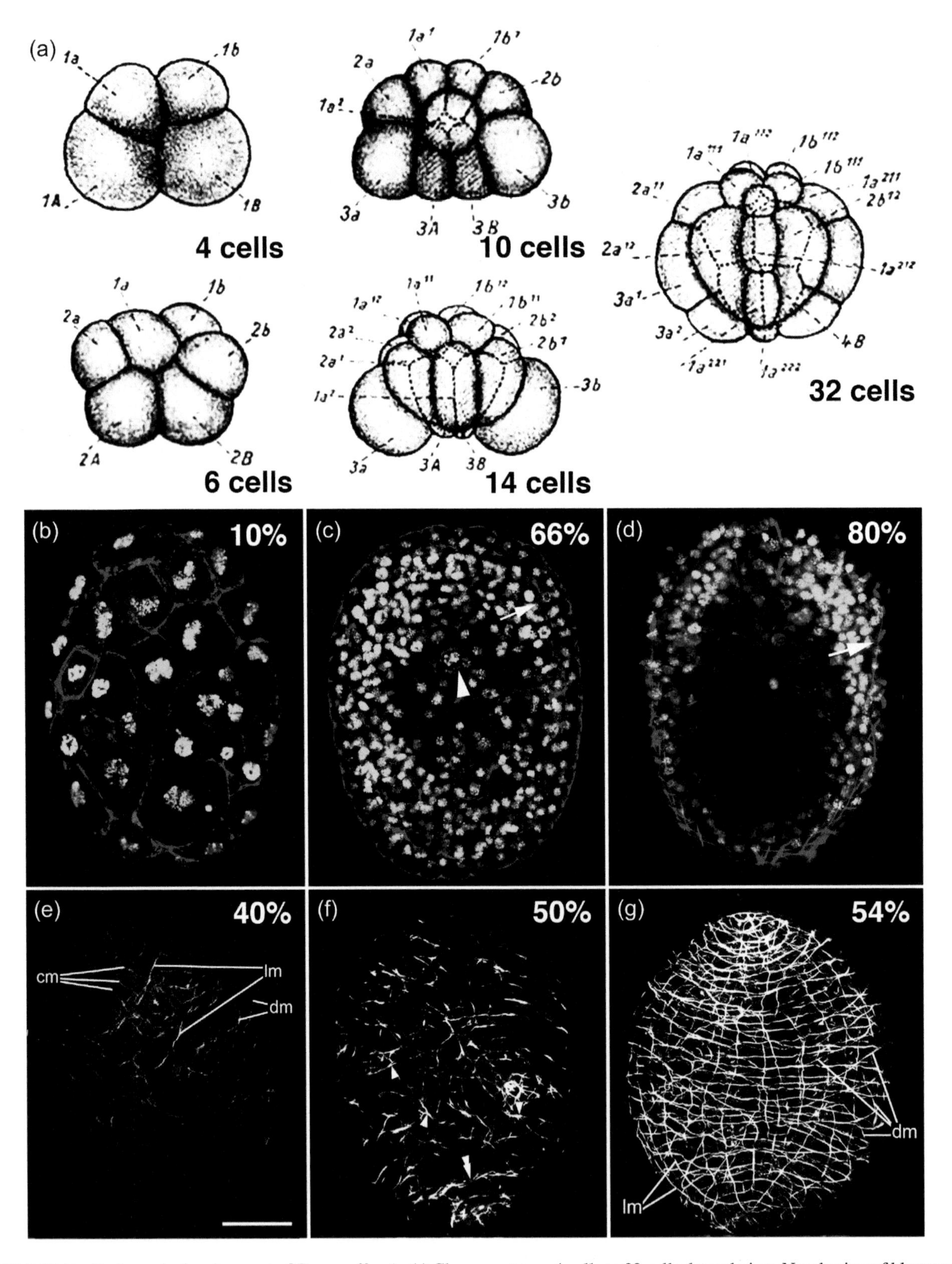

FIGURE 12.5 Embryonic development of *S. roscoffensis*. (a) Cleavage stages 4 cells to 32 cells, lateral view. Numbering of blastomeres by the author. (b–d) Horizontal confocal sections of *S. roscoffensis* at 10% development (b), 66% development (c) and 80% development (d). Nuclei are labeled with Topro (green). Phalloidin (red) labels cell membrane associated actin filaments as well as myofilaments. Arrow in (c) points at basal membranes of ectoderm cells and emerging myofilaments; note that ectodermal (epidermal) nuclei still form a layer peripherally of this boundary. Arrowhead indicates membrane around internal endodermal cells. At later stages (d), most epidermal nuclei have sunk below the level of body wall muscle fibers (arrow); endoderm cells have fused into digestive syncytium. (e–g) Emergence of muscle fibers, labeled with phalloidin (orange) between embryonic stages 40% and 54%. Z-projection, dorsal view. Abbreviations: cm: circular muscles; dm: diagonal muscles; lm: longitudinal muscles. Scale bar: 50 micrometers (b–g). ([a] From Bresslau 1909; [e-g] from Semmler et al. 2008, with permission.)

lineage tracing. Moreover, Bresslau is the first to present a lineage map of the *Convoluta* (*Symsagitifera*) embryo, an impressive feat of detailed observation at the beginning of the 20th century. Notably, the duet spiral cleavage characteristic of acoels is not present in members of the closely related Nemertodermatida order (Børve and Hejnol 2014), which exhibit a slightly different pattern of blastomere divisions during early embryonic development.

The embryological origin of tissues hasn't been thoroughly studied in *S. roscoffensis*. Following cleavage and gastrulation, the embryo forms a solid mass of cells, with an outer epithelial layer giving rise to the epidermis and an inner mass of cells to digestive cells (Figure 12.5b), parenchyma and musculature. It is not known whether, at this stage, progenitors of neurons or gland cells are already part of the inner mass or are still integrated in the epithelial outer layer. Until about 60% of development, a regular surface epithelium remains visible; subsequently, cell bodies of epidermal cells, as well as all other cells which potentially are initially at the surface, like glands or sensory neurons, sink inward (Figure 12.5c, d).

The genesis of the musculature has been observed in detail using F-actin labeling (Semmler et al. 2008). The process of myogenesis is very similar to that observed in another acoels (i.e. *Isodiametra pulchra*: Ladurner et al. 2000 or *Neochildia fusca*: Ramachandra et al. 2002). The latter study shows the initial stages of muscle formation, probably common to many acoels, with the first signs of musculature being myoblasts forming a thin layer underneath the epidermis, laterally and posteriorly to the brain. Some early muscular fibers penetrate the brain. During the very first days of *Symsagittifera* embryo development, a grid of circular and longitudinal muscles appears, with circular muscles preceding longitudinal ones. Myogenesis in the anterior part of the animal occurs first and then proceeds in an anterior–posterior progression (Figure 12.5e). Muscular circular fibers are added by a process involving the branching of previous ones (Figure 12.5f). The grid of muscles is more regular in the dorsal part of the embryo than in the ventral, probably due to the need to accommodate additional muscles in ventral structures such as the mouth and the copulatory organs (Figure 12.5g). The embryos hatch with a basic grid composed of about 30 longitudinal and 60 circular muscles (Semmler et al. 2008). During the later development, additional muscles are incorporated, including specialized muscles around the mouth and the copulatory system, plus a whole array of transversal (dorso-ventral) fibers. The adults have a total of about 300 circular muscles and 140 longitudinal ones.

The embryonic origin of the brain and the neural chords hasn't been studied in detail, but it is assumed to occur in early embryogenesis, based on early embryonic expression (bilateral lobes) of some bHLH "neurogenic" genes (Perea-Atienza et al. 2018). A better understanding of the genesis of the nervous system is derived from the study of *Neochildia fusca* embryos (Ramachandra et al. 2002). These authors documented the presence in late embryos of the brain primordia, which can be clearly distinguished at the anterior pole of the embryo and consists of an external cortex of neuronal bodies around an internal neuropil. Given the consistency of these observations with those of Perea-Atienza, and with both acoels being members of the same class, Crucimusculata, we can hypothesize that the neurogenesis is following identical, or very similar, paths. A more comprehensive analysis of gene expression patterns during *S. roscoffensis* embryogenesis is urgently needed in order to understand the mechanisms regulating embryonic development and patterning.

12.6 REGENERATION

Acoel flatworms show an enormous capacity for regeneration. The extent of this regeneration varies from species to species, with some even relying on regeneration for reproduction (Sikes and Bely 2010). Investigation of the regenerative capacity of acoels dates back to the beginning of the 20th century, when Elsa Keil (1929) described some histological aspects of regeneration in the acoel flatworm *Polychaerus caudatus*. Keil's work was a revision of even earlier data provided by Stevens and Boring (1905) and Child (1907). In the 1950s and 1960s, researchers including Steinböck (1954) and Hanson (1960, 1967) undertook a more systematic analysis of the regeneration process in some acoel "turbellarians", resulting in the creation of some now-classical monographs.

One interesting aspect of acoel regeneration is that different species have the capacity to regenerate different bodily areas. For example, *Symsagittifera roscoffensis* and *Hofstenia miamia* can regenerate the anterior area (Bailly et al. 2014; Hulett et al. 2020), while *Isodiametra pulchra* can regenerate the posterior area (De Mulder et al. 2009; Perea-Atienza et al. 2013). Many other varieties of regeneration have been described for other species (Bely and Sikes 2010). The reasons underlying these different capacities remain unknown.

Symsagittifera roscoffensis is a particularly interesting system in which to study regeneration, since this species has the capacity to regenerate the whole brain anew. This has interesting implications for understanding the mechanisms involved in the regeneration of the nervous tissue. In *Symsagittifera roscoffensis*, the regeneration of the brain anatomy after amputation takes between one week and ten days, similar to the time taken by *Hofstenia miamia*. However, some additional structures, such as the statocyst, require a few weeks for complete regeneration. The regenerative process involves the mobilization of stem cells (neoblasts) that begin actively proliferating in response to amputation and subsequently concentrate in the wound area (BG and PM, unpublished data). The active proliferation of neoblasts is followed by a differentiation of mature tissues. A clear blastemal area is missing in this process. Regeneration follows three broad and distinct steps: (1) a contraction of the anterior musculature immediately following amputation; (2) a subsequent closure of the wound area; (3) an extension of the three pairs of nerve cords into the anterior domain of the animal's body; and (4) the final connection of these nerves to form two ring-shaped, symmetrical neuronal structures with increasing numbers of mature neurons (i.e. the brain). Based on indirect observations (see Bery et al. 2010), it has been proposed that nerve chords and muscular fibers at the amputation site could somehow guide the process of tissue repair. This would be in line with indications in *Hofstenia miamia* that muscles provide positional information to regenerating tissue in acoels (Raz et al. 2017), as is also the case in platyhelminth species. The process of regeneration in *Symsagittifera roscoffensis* has not been well characterized due to a lack of studies using molecular markers. Studies of this nature have been undertaken recently in *Hofstenia miamia* (Hulett et al. 2020). However, it is important to note that in *Symsagittifera*, it has been possible to test the functional reconstruction of the brain area using various behavioral tests assessing functions such as phototaxis and geotaxis (Sprecher et al. 2015). These behaviors, though recognized for decades (Keeble 1910), are only now being studied quantitatively (Nissen et al. 2015). Sprecher and colleagues (2015) have used different paradigms to assess the behavior of amputated worms at different stages of recovery, evaluating their responses to light, vibration and settling in columns. The researchers also followed the motility of the animals over the recovery period (Sprecher et al. 2015). The functional assessment of brain activity was done in parallel with a careful analysis of nervous system anatomy by immunostaining, allowing the correlation of functional and structural aspects of the regeneration process. This study represents the first time that tests of this nature have been used to understand the physiological consequences of acoel regenerative processes (beyond the obvious characteristics like recovery of body movement). A striking finding of this study is that different sensory modalities are restored at different times. For instance, phototaxis is restored at about 20 days post-decapitation, while geotaxis takes approximately 50 days to be restored. The growing recognition that *Symsagittifera roscoffensis* is able to follow more complex behaviors (Franks et al. 2016) and even social behaviors offers further opportunities to study functional recovery in the nervous systems of these animals, once considered "simple". The use of automated tracking systems and computer simulation of individual and collective behaviors—as Franks and collaborators (2016) have done—will provide us with the necessary tools to analyze different aspects of the brain's functional recovery in detail.

12.7 PRELIMINARY GENOMIC DATA

The so-called post-genomic era has produced a flurry of papers addressing the characterization of many animal genomes and transcriptomes, information that allows us to trace the evolutionary history of animals with unprecedented detail. Among those animals for which new information has been gathered are several members of the phylum Xenacoelomorpha (an updated list appears in: Jondelius et al. 2019).

Three acoel genomes with different degrees of completeness have been produced in the last few years—those of species *Hofstenia miamia* (Gehrke et al. 2019), *Praesagittifera naikaiensis* (Arimoto et al. 2019) and *Symsagittifera roscoffensis* (Philippe et al. 2019). While the first is quite complete, that of our species is only a preliminary draft. Despite the relatively low quality of the *Symsagittifera* genome (a high-quality version is currently being generated), some basic facts can be extracted. The first is that the genome of *Symsagittifera* is quite big, around 1.4 Gb, approximately half the size of the human genome. This is supported by an independent analysis of the genome size carried out by flow cytometry. This genome is much bigger than that of *Hofstenia miamia*, which has been reported to be 950 Mb long, and *Praesagittifera naikaiensis*, which is estimated at 654 Mb. The genome of *Symsagittifera* is packed into 20 chromosomes of seemingly equal size ($2n = 20$), as determined cytochemically using chromosomal spreads (Moreno et al. 2009).

Briefly, in the case of *Symsagittifera roscoffensis*, a standard fragment Illumina library was made from a pool of symbiont-free hatchlings, which were raised in artificial seawater in the presence of antibiotics. The genome fragments were assembled with a mix of SOAPdenovo2 (–M3, –R,–d1, –K31) and the Celera assemblers, resulting in an N50 of 2,905 bp. The introduction of PacBio sequencing methodologies has recently allowed us to increase the N50 to above 100 kb (PM, unpublished data). Genome and transcriptome assemblies, including the genome of *Symsagittifera*, have been deposited in https://figshare.com/search, project number PRJNA517079. In parallel, a transcriptome was also sequenced from mixed-stage *S. roscoffensis* embryos using standard methods.

This is an A+T-rich genome with a 36% content of G+C and a high representation of repetitive elements and transposons (data not shown). Some of the transposon sequences have been mapped to specific locations in the genome, such as the neighborhood of the Hox genes (Moreno et al. 2011), a particularity that would explain their dispersion in different chromosomes by rearrangements. The draft genome and the transcriptomes have allowed for the exploration of gene families and their compositions. Families such as those containing bHLH, GPCRs, Wnts or homeobox have been explored extensively in recent years (Perea-Atienza et al. 2015; Gavilán et al. 2016; Brauchle et al. 2018). Strikingly, many of these sequences show specific patterns of divergence with respect to the putative orthologs in other bilaterian clades (i.e. Wnts), corroborating the well-known fast rate of evolution of acoels, and in particular *Symsagittifera*, genomes (Philippe et al. 2019). Moreover, these gene family characterizations provide a source of sequences necessary for the design of probes used in downstream experiments by situ hybridization (Perea-Atienza et al. 2018) or in the identification of BAC clones used in studies of chromosomal mapping (Moreno et al. 2009).

12.8 CHALLENGING QUESTIONS FOR THE FUTURE

Some challenging questions need to be addressed in this model. The lack of functional tools has been a hindrance in the analysis of *Symsagittifera* biology from both a developmental and physiological perspective. Until now, we have relied on several molecular, anatomical and biochemical techniques to analyze aspects of the anatomy, embryology and metabolic activity of these animals under different conditions. This has provided us with an enormous body of knowledge, though mostly descriptive. The development of tools for knockdown and biochemical intervention (i.e. pharmacological agents) should be a priority in the field, so that phenomena discovered observationally can be tested directly through experimental intervention. Specifically, the following are needed:

1. A deeper understanding of the embryology of *S. roscoffensis*, including lineage maps and a dissection of blastomere contributions (through ablation methodologies). Furthermore, molecular markers should be incorporated into our understanding of embryonic regulation in *S. roscoffensis*.
2. We need a better understanding of how the *S. roscoffensis* genome is organized. This is necessary not only for the identification of key features of the genome (including intron/exon boundaries, synteny conservation, non-coding RNAs, indels, etc.) but also as an alternative tool for tackling the difficult problem of phylogenetic affinities. We believe that genomic characteristics can be of critical importance for phylogenomic reconstruction, beyond the "classical" use of primary sequence data.
3. A detailed characterization of cell types and their architectural organization in tissues is still missing in *S. roscoffensis*. High-throughput TEM reconstructions aided by single-cell transcriptomics would provide ample opportunities to understand how cell types are organized in *S. roscoffensis* and their putative enrichment in different subtypes. Combinations of single-cell data plus *in situ* hybridization will be necessary to reach this goal (spatial transcriptomics).
4. *S. roscoffensis* is a unique system for the study of symbiotic relationships. The host–algae interaction provides a rich metabolic partnership and is critical to the survival of animals in their environment. It is unknown how this symbiosis is achieved and controlled at the genetic level. The fact that both the host and the algae can be independently cultivated and mixed provides us with a unique opportunity to follow, in real time, the molecular activities involved in the symbiogenic process. The use of complementary techniques, such as TEM, can also aid our understanding of the morphological changes that take place in both partners during the symbiogenic process.
5. *S. roscoffensis* exhibits complex behavior at both the individual and collective levels. Factors such light, gravity or animal crowds elicit a clear behavioral response in *S. roscoffensis*. These diverse and rich behaviors observed in a relatively "simple" animal merit a deeper investigation. Genetic intervention—and, perhaps, neuronal ablations—could provide insight into the regulation of the *S. roscoffensis* behavioral repertoire.
6. Acoels show a remarkable capacity for regeneration of body parts. *S. roscoffensis* has been identified as an ideal system to study the regeneration of the head (and brain) from scratch. Understanding how this process occurs could be of great importance beyond the domain of fundamental biology. A combination of tools including gene mapping, gene editing or gene knockout approaches (such as CRISPR/CAS9) and single-cell sequencing could give us unprecedented access to the mechanisms that regulate nervous system reconstruction.

The implications of this work for biomedicine cannot be overstated.

The availability of some of the required technologies in related acoel species should prove especially relevant. Over the last years, we have seen the incorporation of RNAi methodologies in the study of the development of *Isodiametra pulchra* (De Mulder et al. 2009; Moreno et al. 2010) and *Hofstenia miamia* (Srivastava et al. 2014). Moreover, conventional techniques such as colorimetric and fluorescent multiplex in situ hybridizations plus immunochemical tools are now regular tools used in the analysis of the species of this chapter, *S. roscoffensis*, and have been described at extenso in the chapter published by Perea-Atienza and collaborators (Perea-Atienza et al. 2018; Perea-Atienza et al. 2020). To end this short overview, note that *S. roscoffensis* is the first acoel species in which behavioral tests have been devised (Nissen et al. 2015; Sprecher et al. 2015), opening the possibility of carrying out detailed analysis of the physiological role that tissues, cells and genes have in the Acoela.

BIBLIOGRAPHY

Achatz, J. G., Chiodin, M., Salvenmoser, W., Tyler, S., and Martinez, P. 2013. The Acoela: On their kind and kinships, especially with nemertodermatids and xenoturbellids (Bilateria incertae sedis). *Org Divers Evol* 13(2):267–286.

Apelt, G. 1969. Fortpflanzungsbiologie, Entwicklungszyklen und vergleichende Frühentwicklung acoeler Turbellarien. *Marine Biol* 8:267–325.

Arboleda, E., Hartenstein, V., Martinez, P., Reichert, H., Sen, S., Sprecher, S., and Bailly, X. 2018. An emerging system to study photosymbiosis, brain regeneration, chronobiology, and behavior: The marine acoel *Symsagittifera roscoffensis*. *Bioessays* 40:e1800107.

Arimoto, A., Hikosaka-Katayama, T., Hikosaka, A., Tagawa, K., Inoue, T., Ueki, T., Yoshida, M. A., Kanda, M., Shoguchi, E., Hisata, K., and Satoh, N. 2019. A draft nuclear-genome assembly of the acoel flatworm *Praesagittifera naikaiensis*. *GigaScience* 8:giz023.

Bailly, X., Laguerre, L., Correc, G., Dupont, S., Kurth, T., Pfannkuchen, A., Entzeroth, R., Probert, I., Vinogradov, S., Lechauve, C., Garet-Delmas, M. J., Reichert, H., and Hartenstein, V. 2014. The chimerical and multifaceted marine acoel *Symsagittifera roscoffensis*: From photosymbiosis to brain regeneration. *Front Microbiol* 5:498.

Bedini, C., Ferrero, E., and Lanfranchi, A. 1973. The ultrastructure of ciliary sensory cells in two turbellaria acoela. *Tissue and Cell* 5:359–372.

Bedini, C., and Lanfranchi, A. 1991. The central and peripheral nervous system of Acoela (Plathelminthes): An electron microscopical study. *Acta Zoologica* 72:101–106.

Bery, A. E., Cardona, A., Martinez, P., and Hartenstein, V. 2010. Structure of the central nervous system of a juvenile acoel, *Symsagittifera roscoffensis*. *Development Genes and Evolution* 220:61–76.

Bely, A. E., and Sikes, J. M. 2010. Acoel and platyhelminth models for stem-cell research. *J Biol* 9:14.

Børve, A., and Hejnol, A. 2014. Development and juvenile anatomy of the nemertodermatid *Meara stichopi* (Bock) Westblad 1949 (Acoelomorpha). *Front Zool* 11:50.

Boyer, B. C. 1971. Regulative development in a spiralian embryo as shown by cell deletion experiments on the Acoel, *Childia*. *J Exp Zool* 176:97–105.

Boyer, B. C., Henry, J. Q., and Martindale, M. Q. 1996. Modified spiral cleavage: The duet cleavage pattern and early blastomere fates in the acoel turbellarian *Neochildia fusca*. *Biol Bull* 191:285–286.

Boyle, J. E., and Smith, D. C. 1975. Biochemical interactions between the symbionts of *Convoluta roscoffensis*. *Proceedings of the Royal Society of London: Series B. Biological Sciences* 189:121–135.

Brauchle, M., Bilican, A., Eyer, C., Bailly, X., Martínez, P., Ladurner, P., Bruggmann, R., and Sprecher, S. G. 2018. Xenacoelomorpha survey reveals that all 11 animal homeobox classes were present in the first bilaterians. *Genome. Biol. Evol.* 10:2205–2217.

Bresslau, E. 1909. Die Entwicklung der Acoelen. *Verh Deutsch Zoologisch Gesell* 19:314–323.

Brøndsted, H. V. 1955. Planarian regeneration. *Biol Rev* 30:65–126.

Child, C. M. 1907. The localisation of different methods of form-regulation in *Polychoerus caudatus*. *Arch Entwm Org* 23:227–248.

Delage, Y. 1886. Etudes histologiques sur les planaires rhabdocoeles acoeles (*Convoluta schultzii O. Sch.*). *Arch Zool Exp Gen* T4:109–144.

De Mulder, K., Kuales, G., Pfister, D., Willems, M., Egger, B., Salvenmoser, W., Thaler, M., Gorny, A. K., Hrouda, M., Borgonie, G., and Ladurner, P. 2009. Characterization of the stem cell system of the acoel *Isodiametra pulchra*. *BMC Dev Bio* 9:69.

Douglas, A. E. 1983a. Uric acid utilization in *Platymonas convolutae* and symbiotic *Convoluta roscoffensis*. *J Mar Biol Ass UK* 63:435–447.

Douglas, A. E. 1983b. Establishment of the symbiosis in *Convoluta roscoffensis*. *J Mar Biol Ass UK* 63:419–434.

Ehlers, U. 1985. *Das phylogenetische System der Plathelminthes*. Gustav Fischer Verlag, Stuttgart, New York.

Ehlers, U. 1991. Comparative morphology of statocysts in the Plathelminthes and the Xenoturbellida. *Hydrobiologia* 227:263–271.

Ehlers, U. 1992. Frontal glandular and sensory structures in *Nemertoderma* (Nemertodermatida) and *Paratomella* (Acoela): Ultrastructure and phylogenetic implications for the monophyly of the Euplathelminthes (Plathelminthes). *Zoomorphology* 112:227–236.

Ferrero, E. 1973. A fine structural analysis of the statocyst in *Turbellaria acoela*. *Zoologica Scripta* 2:5–16.

Franks, N. R., Worley, A., Grant, K. A., Gorman, A. R., Vizard, V., Plackett, H., Doran, C., Gamble, M. L., Stumpe, M. C., and Sendova-Franks, A. B. 2016. Social behaviour and collective motion in plant-animal worms. *Proc Biol Sci* 283.

Fuxe, K., Dahlström, A., Höistad, M., Marcellino, D., Jansson, A., Rivera, A., Diaz-Cabiale, Z., Jacobsen, K., Tinner-Staines, B., Hagman, B., Leo, G., Staines, W., Guidolin, D., Kehr, J., Genedani, S., Belluardo, N., and Agnati, L. F. 2007. From the Golgi-Cajal mapping to the transmitter-based characterization of the neuronal networks leading to two modes of brain communication: Wiring and volume transmission. *Brain Res Rev* 55:17–54.

Gamble, F. W., and Keeble, F. 1904. The bionomics of *Convoluta roscoffensis*, with special reference to its green cells. *Proceedings of the Royal Society of London* 72:93–98.

Gardiner, E. G. 1895. Early development of *Polychoerus caudatus* Mark. *J Morphol* 11:155–176.

Gavilán, B., Perea-Atienza, E., and Martinez, P. 2016. Xenacoelomorpha: A case of independent nervous system centralization? *Philos Trans R Soc Lond B Biol Sci* 371:1685.

Gavilán, B., Sprecher, S. G., Hartenstein, V., and Martinez, P. 2019. The digestive system of xenacoelomorphs. *Cell Tissue Res* 377:369–382.

Geddes, P. 1879. II: Observations on the physiology and histology of *Convoluta schultzii*. *Proceedings of the Royal Society of London* 28:449–457.

Gehrke, A. R., Neverett, E., Luo, Y. J., Brandt, A., Ricci, L., Hulett, R. E., Gompers, A., Ruby, J. G., Rokhsar, D. S., Reddien, P. W., and Srivastava, M. 2019. Acoel genome reveals the regulatory landscape of whole-body regeneration. *Science* 363.

Georgévitch, J. 1899. Etude sur le développement de la *Convoluta roscoffensis* Graff. *Arch Zool Experim* 3:343–361.

Gooday, G. W. 1970. A physiological comparison of the symbiotic alga *Platymonas convolutae and its free-living relatives. J Mar Biol Ass UK* 50:199–208.

Graff, L. 1891. Sur l'organisation des turbellariés acoeles. *Arch Zool Exp Gén* T9:1–12.

Graff, L., and Haberlandt, G. 1891. *Die Organisation der Turbellaria acoela; mit einem Anhange über den Bau und die Bedeutung der Chlorophyllzellen von Convoluta roscoffensis von Gottlieb Haberlandt*. Engelmann, Leipzig.

Gschwentner, R., Ladurner, P., Nimeth, K., and Rieger, R. 2001. Stem cells in a basal bialterian: S phase and mitotic cells in *Convolutriloba longifissura*. *Cell Tissue Res* 304:401–408.

Hanson, E. D. 1960. Asexual reproduction in acoelous Turbellaria. *Yale J Biol Med* 33:107–111.

Hanson, E. D. 1967. Regeneration in acoelous flatworms: The role of the peripheral parenchyma. *W. Roux' Archiv f. Entwicklungsmechanik* 159:298–313.

Henry, J. Q., Martindale, M. Q., and Boyer, B. C. 2000. The unique developmental program of the acoel flatworm, *Neochildia fusca*. *Dev Biol* 220:285–295.

Hooge, M. D. 2001. Evolution of body-wall musculature in the platyhelminthes (Acoelomorpha, catenulida, rhabditophora). *J Morphol* 249:171–194.

Hori, I. 1997. Cytological approach to morphogenesis in the planarian blastema: The effect of neuropeptides. *J Submicrosc Cytol Pathol* 29:91–97.

Hori, I. 1999. Cytological approach to morphogenesis in the planarian blastema: Ultrastructure and regeneration of the acoel turbellarian *Convoluta naikaiensis*. *J Submicrosc Cytol Pathol* 31:247–258.

Hulett, R. E., Potter, D., and Srivastava, M. 2020. Neural architecture and regeneration in the acoel *Hofstenia miamia*. *Proc Biol Sci* 287.

Hyman, L. H. 1951. *Invertebrates: Platyhelminthes and Rhynchocoela*, vol. 2. McGraw-Hill Inc., New York.

Jondelius, U., Raikova, O. I., and Martinez, P. 2019. Xenacoelomorpha, a key group to understand bilaterian evolution: Morphological and molecular perspectives. In P. Pontarotti (Ed.), *Evolution, Origin of Life, Concepts and Methods* (pp. 287–315). Springer Nature.

Keeble, F. 1910. *Plant-Animals: A Study in Symbiosis*. Cambridge University Press, Cambridge.

Keeble, F., and Gamble, F. W. 1907. The origin and nature of the green cells of *Convoluta roscoffensis*. *QJ Microsc. Sci* 51:167–217.

Keil, E. 1929. Regeneration in *Polychaerus caudatus* Mark. *Bio Bull* 57:223–244.

Klauser, M. D., Smith, J. P. S., and Tyler, S. 1986. Ultrastructure of the frontal organ in Convoluta and *Macrostomum* spp.: Significance for models of the turbellarian archetype. *Hydrobiologia* 132:47–52.

Kiers, E. T., and West, S. A. 2016. Evolution: Welcome to symbiont prison. *Current Biology* 26:R66–R68.

Kostenko, A. G., and Mamkaev, Y. V. 1990. The position of green convoluts IN the system of acoel turbellarians (Turbellaria, Acoela). 2. *Sagittiferidae fam. n. Zool. Zh.* (Moscow) 69:5–16.

Ladurner, P., and Rieger, R. 2000. Embryonic muscle development of *Convoluta pulchra* (Turbellaria-acoelomorpha, platyhelminthes). *Dev Biol* 222:359–375.

Lowe, C. D., Minter, E. J., Cameron, D. D., and Brockhurst, M. A. 2016. Shining a light on exploitative host control in a photosynthetic endosymbiosis. *Current Biology* 26:207–211.

Lundin, K. 1997. Comparative ultrastructure of the epidermal ciliary rootlets and associated structures in species of the Nemertodermatida and Acoela (Plathelminthes). *Zoomorphology* 117:81–92.

Margulis, L. 1998. *Symbiotic Planet: A New Look at Evolution*, Kindle edition. Basic Books, New York.

Martinez, P., Hartenstein, V., and Sprecher, S. 2017. Xenacoelomorpha Nervous Systems. In S. Murray Sherman (Ed.), *Oxford Research Encyclopedia of Neuroscience*. Oxford University Press, New York.

Meyer, H., Provasoli, L., and Meyer, F. 1979. Lipid biosynthesis in the marine flatworm *Convoluta roscoffensis* and its algal symbiont *Platymonas convoluta*. *Biochim Biophys Acta* 573:464–480.

Moreno, E., De Mulder, K., Salvenmoser, W., Ladurner, P., and Martínez, P. 2010. Inferring the ancestral function of the posterior Hox gene within the Bilateria: Controlling the maintenance of reproductive structures, the musculature and the nervous system in the acoel flatworm *Isodiametra pulchra*. *Evolution and Development* 12:258–266.

Moreno, E., Nadal, M., Baguñá, J., and Martínez, P. 2009. The origin of the bilaterian Hox patterning system: Insights from the acoel flatworm *Symsagittifera roscoffensis*. *Evolution and Development* 11:574–581.

Moreno, E., Permanyer, J., and Martinez, P. 2011. The origin of patterning systems in bilateria-insights from the Hox and ParaHox genes in Acoelomorpha. *Genomics, Proteomics and Bioinformatics* 9:65–76.

Morris, P. J. 1993. The developmental role of the extracellular matrix suggests a monophyletic origin of the kingdom Animalia. *Evolution* 47:152–165.

Muscatine, L., Boyle, J. E., and Smith, D. C. 1974. Symbiosis of the acoel flatworm *Convoluta roscoffensis* with the alga *Platymonas convolutae*. *Proceedings of the Royal Society of London: Series B. Biological Sciences* 187:221–234.

Nissen, M., Shcherbakov, D., Heyer, A. G., Brümmer, F., and Schill, R. O. 2015. Behaviour of the platyhelminth *Symsagittifera roscoffensis* under different light conditions and the consequences for the symbiotic algae *Tetraselmis convolutae*. *J Exper Biol* 218:1693–1698.

Nozawa, K., Taylor, D. L., and Provasoli, L. 1972. Respiration and photosynthesis in *Convoluta roscoffensis* Graff, infected with various symbionts. *Biol Bull* 143:420–430.

Oshman, J. 1966. Development of the symbiosis of *Convoluta voscoffensis* Graff and Platy-monas sp. *J Phycol* 2:111–116.

Parke, M., and Manton, I. 1967. The specific identity of the algal symbiont in *Convoluta roscoffensis*. *J Mar Biol Ass UK* 47:445–464.

Pedersen, K. J. 1965. Cytological and cytochemical observations on the mucous gland cells of an acoel turbellarian, *Convoluta convulta*. *Ann NY Acad Sci* 118:930–965.

Perea-Atienza, E., Botta, M., Salvenmoser, W., Gschwentner, R., Egger, B., Kristof, A., Martinez, P., and Achatz, J. G. 2013. Posterior regeneration in *Isodiametra pulchra* (Acoela, Acoelomorpha). *Front Zool* 10:64.

Perea-Atienza, E., Gavilan, B., Chiodin, M., Abril, J. F., Hoff, K. J., Poustka, A. J., and Martinez, P. 2015. The nervous system of Xenacoelomorpha: A genomic perspective. *J. Exp. Biol.* 218:618–628.

Perea-Atienza, E., Gavilán, B., Sprecher, S. G., and Martinez, P. 2020. Immunostaining and in situ hybridization of the developing acoel nervous system. In: S. Sprecher (Ed.), *Brain Development: Methods in Molecular Biology*, vol. 2047. Humana, New York, NY.

Perea-Atienza, E., Sprecher, S. G., and Martinez, P. 2018. Characterization of the bHLH family of transcriptional regulators in the acoel *S. roscoffensis* and their putative role in neurogenesis. *EvoDevo* 9:8.

Philippe, H., Brinkmann, H., Copley, R. R., Moroz, L. L., Nakano, H., Poustka, A. J., Wallberg, A., Peterson, K. J., and Telford, M. J. 2011. Acoelomorph flatworms are deuterostomes related to Xenoturbella. *Nature* 470:255–258.

Philippe, H., Poustka, A. J., Chiodin, M., Hoff, K. J., Dessimoz, C., Tomiczek, B., Schiffer, P. H., Müller, S., Domman, D., Horn, M., Kuhl, H., Timmermann, B., Satoh, N., Hikosaka-Katayama, T., Nakano, H., Rowe, M. L., Elphick, M. R., Thomas-Chollier, M., Hankeln, T., Mertes, F., Wallberg, A., Rast, J. P., Copley, R. R., Martinez, P., and Telford, M. J. 2019. Mitigating anticipated effects of systematic errors supports sister-group relationship between xenacoelomorpha and ambulacraria. *Curr Biol* 29:1818–1826.

Ramachandra, N. B., Gates, R. D., Ladurner, P., Jacobs, D. K., and Hartenstein, V. 2002. Embryonic development in the primitive bilaterian *Neochildia fusca*: Normal morphogenesis and isolation of POU genes Brn-1 and Brn-3. *Development Genes and Evolution* 212:55–69.

Raz, A. A., Srivastava, M., Salvamoser, R., and Reddien, P. W. 2017. Acoel regeneration mechanisms indicate an ancient role for muscle in regenerative patterning. *Nat Commun* 8:1260.

Reuter, M., Raikova, O. I., Jondelius, U., Gustafsson, M. K. S., Maule, A. G., and Halton, D. W. 2001. Organisation of the nervous system in the Acoela: An immunocytochemical study. *Tissue and Cell* 33:119–128.

Rieger, R. M., Tyler, S., Smith, J. P. S., III, and Rieger, G. E. 1991. Platyhelminthes: Turbellaria. In F. W. Harrison and B. J. Bogitsh (Eds.), *Microscopic Anatomy of Invertebrates, Vol. 3: Platyhelminthes and Nemertinea* (pp. 7–140). Wiley-Liss, New York.

Sagan, L. 1967. On the origin of mitosing cells. *Journal of Theoretical Biology* 14:225-IN6.

Semmler, H., Bailly, X., and Wanninger, A. 2008. Myogenesis in the basal bilaterian *Symsagittifera roscoffensis* (Acoela). *Front Zool* 5:14.

Semmler, H., Chiodin, M., Bailly, X., Martinez, P., and Wanninger, A. 2010. Steps towards a centralized nervous system in basal bilaterians: Insights from neurogenesis of the acoel *Symsagittifera roscoffensis*. *Dev Growth Differ* 52(8):701–713.

Shinn, G. 1993. Formation of egg capsules by flatworms (Phylum platyhelminthes). *Transactions of the American Microscopical Society* 112:18–34.

Sikes, J. M., and Bely, A. E. 2010. Making heads from tails: Development of a reversed anterior-posterior axis during budding in an acoel. *Dev Biol* 338(1):86–97.

Smith, J., III, and Tyler, S. 1985. The acoel turbellarians: Kingpins of metazoan evolution or a specialized offshoot? In H. M. Conway, S. Morris, J. D. George, R. Gibson, and Platt (Eds.), *The Origins and Relationships of Lower Invertebrates* (p. 123142). Clarendon Press, Oxford.

Smith, J. P. S., and Tyler, S. 1986. Frontal organs in the Acoelomorpha (Turbellaria): Ultrastructure and phylogenetic significance. *Hydrobiologia* 132:71–78.

Smith, J. P. S., Tyler, S., Thomas, M. B., and Rieger, R. M. 1982. The morphology of turbellarian rhabdites: Phylogenetic implications. *Trans Am Microsc Soc* 101(3):209–228.

Sprecher, S. G., Bernardo-Garcia, F. J., van Giesen, L., Hartenstein, V., Reichert, H., Neves, R., Bailly, X., Martinez, P., and Brauchle, M. 2015. Functional brain regeneration in the acoel worm *Symsagittifera roscoffensis*. *Biol Open* 4:1688–1695.

Srivastava, M., Mazza-Curll, K. L., van Wolfswinkel, J. C., and Reddien, P. W. 2014. Whole-body acoel regeneration is controlled by Wnt and Bmp-Admp signaling. *Curr Biol* 24:1107–1113.

Steinböck, O. 1954. Regeneration azöler Turbellarien. *Verb. Dtsch. Zool. Gesell., Tübingen, S.* 86–94.

Stevens, N. M., and Boring, A. M. 1905. Regeneration in *Polychœrus caudatus*, Part I: Observations on living material. *J. Experimental Zoology* 2:335–346.

Todt, C. 2009. Structure and evolution of the pharynx simplex in acoel flatworms (Acoela). *Journal of Morphology* 270:271–290.

Todt, C., and Tyler, S. 2006. Ciliary receptors associated with the mouth and pharynx of Acoela (Acoelomorpha): A comparative ultrastructural study. *Acta Zoologica* 88:41–58.

Tyler, S., and Rieger, R. 1999. Functional morphology of musculature in the acoelomate worm, *Convoluta pulchra* (Platyhelminthes). *Zoomorphology* 119:127–141.

Yamasu, T. 1991. Fine structure and function of ocelli and sagittocysts of acoel flatworms. *Hydrobiologia* 227:273–282.

Yang, J., Liu, X., Yue, G., Adamian, M., Bulgakov, O., and Li, T. J. 2002. Rootletin, a novel coiled-coil protein, is a structural component of the ciliary rootlet. *J Cell Biol* 159:431–440.

Zabotin, Y. I., and Golubev, A. I. 2014. Ultrastructure of oocytes and female copulatory organs of acoela. *Biology Bulletin* 41:722–735.

13 The Annelid *Platynereis dumerilii* as an Experimental Model for Evo-Devo and Regeneration Studies

Quentin Schenkelaars and Eve Gazave

CONTENTS

DOI: 10.1201/9781003217503-13

13.1 HISTORY OF THE MODEL

Annelids, also known as segmented worms, are a major group of non-vertebrate bilaterian animals. Annelid name comes from the Latin *annellus*, meaning "little ring", and refers to their segmented or metamerized body plan. Annelids represent a large number of species and an ecologically diversified animal taxon with over 18,950 described species living in various ecosystems from deep sea to rainforest canopy (Brusca and Brusca 2003). They are especially abundant in sea water but also occupy humid terrestrial and freshwater habitats. Some species are parasitic, mutualist or commensal (Rouse and Pleijel 2001; Piper 2015). Annelid species present a huge diversity of body forms coexisting with various life history strategies, being either scavengers, bioturbators, predators or filter feeders. They also harbor a multitude of (sometimes) extravagant forms of sexual and asexual reproduction (Caspers 1984; Fischer 1999; Schroeder et al. 2017). The annelid phylum, like Mollusca, Platyhelminthes, Bryozoa and more, is part of the Lophotrochozoa clade (Laumer et al. 2019), which together with Ecdyzosoa form the large group of Protostomia within Bilateria. Annelid phylogenetic relationships were, for a long time, mostly based on morphological characteristics and thus were difficult to ascertain (Weigert and Bleidorn 2016). The first classification of annelids separated them in three main groups, Polychaeta, Oligochaeta and Hirudinae (Lamarck 1818; Weigert and Bleidorn 2016). Briefly, Polychaeta, or bristle worms, referred to a large and diverse group of worms presenting numerous bristles, or chaetae, hence the name "poly-chaeta". In contrast, worms with very few or reduced chaetae were grouped together into Oligochaeta, while Hirudinae referred to worms with no chaetae and presenting a sucker. In addition, a multitude of other groups of "invertebrates" such as Sipuncula and Echiura were, at that time, considered closely related to annelids. During the 20th century, new morphology-based classifications proposed the separation of annelids into two main groups, Polychaeta and Clitellata, the latter containing the Hirudinae (Weigert and Bleidorn 2016). Polychaeta were themselves divided into two groups, the Errantia and the Sedentaria, based on worm lifestyles. Free-moving and predatory worms were encompassed in Errantia, while sessile and tube-dwelling worms formed the Sedentaria group (de Quatrefarges 1865; Fauvel 1923, 1927). This Errantia/Sedentaria separation was dismissed with the advent of morphological cladistic analysis (Rouse and Fauchald 1997). Indeed, in 1997, Rouse and Fauchald proposed to separate Polychaeta into the clades Palpata and Scolecida based on the presence or absence of palps, respectively. Over the last 20 years, with the rise of molecular biology, the phylogenetic relationships of annelids were regularly reassessed. A recent seminal phylogenomic study highlighted the division of annelids into two main subgroups, reviving the ancient Errantia and Sedentaria nomenclature, in addition to a couple of early branching lineages such as Sipuncula (Struck et al. 2011). Errantia and Sedentaria together form the Pleistoannelida (Struck 2011). Internal relationships among those two groups are quite well defined (Struck 2011; Weigert et al. 2014; Weigert and Bleidorn 2016). Notably, Sedentaria now also includes the Clitellata and the Echiura. The Polychaeta term is consequently no longer valid, as "polychaete worms" are present in both the Errantia and Sedentaria groups. In contrast, the phylogenetic affiliations of early branching annelid lineages (notably Sipuncula) are not yet stable (Struck et al. 2011; Weigert et al. 2014; Andrade et al. 2015; Weigert and Bleidorn 2016). Recent discovery of new annelid fossils and reassessment of their discrete morphological characters allowed for the reconciliation of annelid fossil records and new molecular phylogenetic relationships (Parry et al. 2016; Chen et al. 2020). Thanks to their huge diversity and rich phylogenetic and evolutionary histories, annelids represent a key source of potential model species to investigate a variety of biological questions, notably the evolution of developmental mechanisms (Ferrier 2012).

Among Errantia, the nereididae *Platynereis dumerilii* (Audouin and Milne Edwards 1833) is an important annelid model species developed by the scientific community to address key biological questions. *Platynereis dumerilii*, also named "*Néréide de Dumeril*", was discovered thanks to an oceanographic campaign around the French North coast of the English Channel (Granville, Chausey Island and Saint Malo) that occurred from 1826 to 1829. Jean Victor Audouin and Henri Milne Edwards subsequently described the type species (deposited in the La Rochelle museum, France) and named it *Nereis dumerilii* in their "*Classification des Annélides et description de celles qui habitent les côtes de la France*" book chapter containing dozens of new annelid species descriptions, especially for the Nereididae family (Audouin and Milne Edwards 1833, 1834) (Figure 13.1a). *Platynereis dumerilii* and Nereididae in general have been the subject of intense studies in the past century, especially regarding embryology, reproduction strategies and regeneration. Their fascinating nuptial dance behavior observed before reproduction (Just 1929; Boilly-Marer 1973; Zeeck et al. 1990), the influence of a brain hormone on their reproduction, their regeneration and growth processes (Hauenschild 1956, 1960; Hofmann 1976) and their oogenesis and spiral embryonic development (Fischer 1974; Dorresteijn et al. 1987; Dorresteijn 1990) were the main scientific questions addressed at that time. Those pioneer studies still provide important information for current challenging research questions (see Section 13.8). Carl Haeuenschild established the first *Platynereis* year-round laboratory culture in 1953 in Germany from a Mediterranean population (Caspers 1971). Since then, *Platynereis* culture procedures have been slightly refined, allowing them to be easily bred in a dozen of research laboratories all over the world (Kuehn et al. 2019).

13.2 GEOGRAPHICAL LOCATION

Platynereis dumerilii worms live in coastal marine waters, especially inhabiting shallow (usually between 0 and 5 meter deep), hard-bottom, algae-covered substrates. They

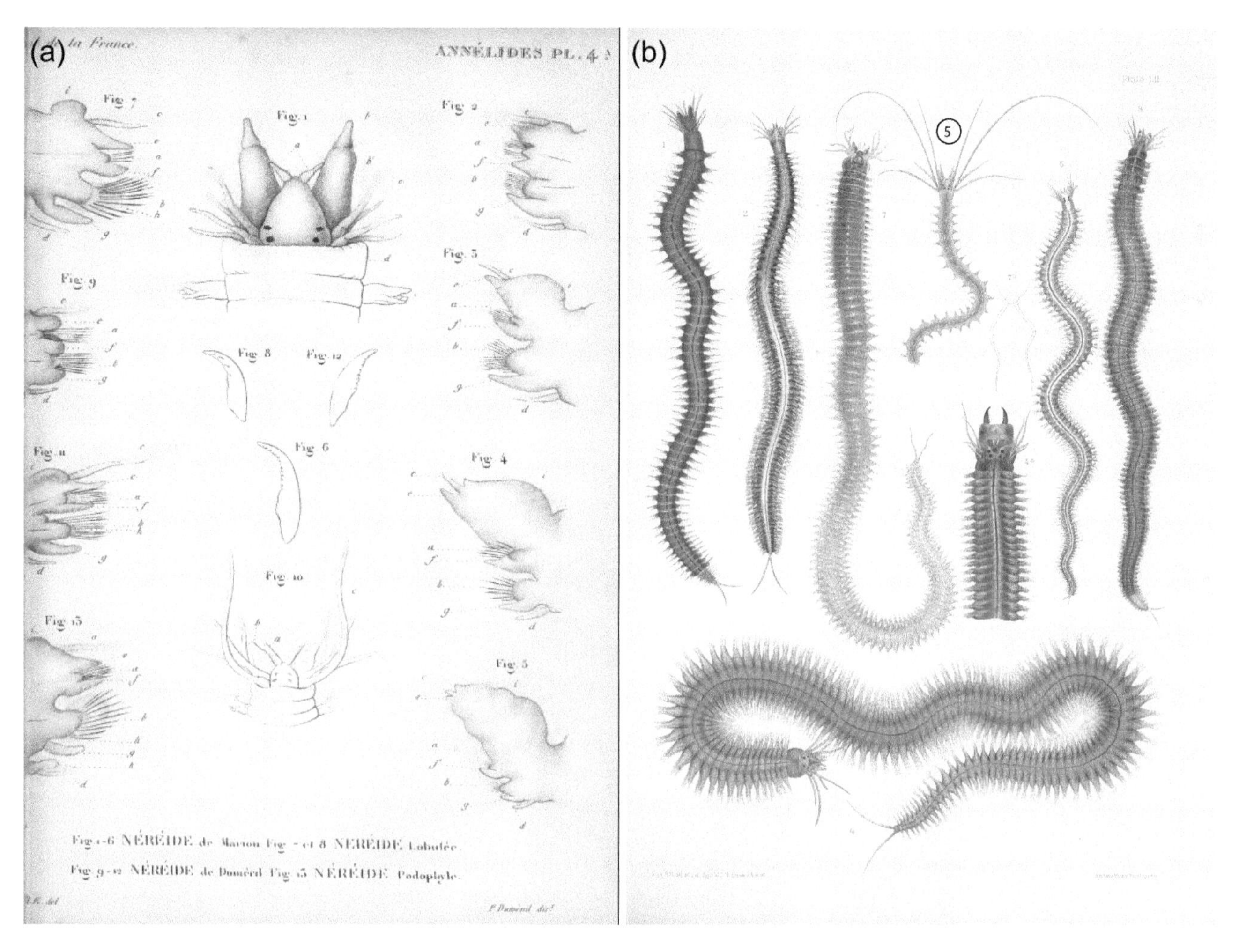

FIGURE 13.1 Original and historical drawings of *Platynereis* (initially named *Nereis*) *dumerilii*. (a) Drawings of the original description of *Platynereis dumerilii* mentioned as "*Nereide de dumeril*". (Plate 4A, drawings 9 to 12: 9 = parapodia; 10 = anterior part and head sensory structures; 11 = parapodia; 12 = denticulated jaw). (b) Drawing of annelids (5 = *Nereis dumerilli*, from a sketch drawn at Lochmaddy). ([a] From Audouin and Milne Edwards 1833, 1834; [b] from M'Intosh et al. 1910.)

can also directly live on seaweeds and marine plant leaves such as *Posidonia oceanica* and *Zostera marina* (Jacobs and Pierson 1979). As mentioned before, *Platynereis dumerilii* was first described from the French north coast of the English Channel. Surprisingly, *Platynereis dumerilii* is also found in many other locations, from temperate to tropical zones: they are often encountered throughout the Mediterranean Sea (Gambi et al. 2000) but also in the North Sea, the English Channel (Figure 13.1b), the Atlantic down to the Cape of Good Hope, the Black Sea, the Red Sea, the Persian Gulf, the Gulf of Mexico, Cuba, the Sea of Japan, the Pacific, the Kerguelen Islands and the coasts of Mozambique and South Africa (Read and Fauchald 2018; Kara et al. 2020). As a consequence of this very broad geographical distribution, *Platynereis dumerilii* is considered a cosmopolitan species (Fischer and Dorresteijn 2004; Read and Fauchald 2018). However, cosmopolitan species rarely exist, since they actually often pool together sibling species or a species complex with (nearly) identical morphologies (Knowlton 1993). As shown for many other marine non-vertebrates species, recent population genetic studies from the Mediterranean Sea (Italian coast) and South Africa revealed that *Platynereis dumerilii*, in those localities, is in fact a species complex. In Italy, *P. dumerilii* is frequently mistaken for it sibling species *P. massiliensis* (Moquin-Tandon 1869; Valvassori et al. 2015; Wäge et al. 2017). In South Africa, *P. dumerilii* lives in sympatry with *P. australis* (Schmarda 1861), another morphologically sibling species (Kara et al. 2020). Population genetic analysis of specimens found in South Africa initially identified as *P. dumerilii* are probably a new species, *P. entshonae*, highlighting the fact that only rigorous and broad-scale population genetic studies worldwide will help to uncover the real geographic distribution of *P. dumerilii* (Kara et al. 2020) and the diversity of the *Platynereis* genus that currently contains 41 valid species (Read and Fauchald 2018).

13.3 LIFE CYCLE

Platynereis's life cycle exhibits several interesting features and encompasses three phases separated by metamorphosis events (Fischer and Dorresteijn 2004) (Figure 13.2). Like many other marine animals, such as corals, sea urchins and even fishes, *Platynereis* sexual maturation and reproduction are synchronized with the natural moon phases (Bentley et al. 1999). This fascinating biological characteristic of lunar-controlled reproductive periodicity, regulated thanks to an endogenous oscillator, is called a circalunar life cycle (Tessmar-Raible et al. 2011; Raible et al. 2017). Each worm reproduce only once, and the timing of this reproduction is tightly regulated by this clock (Zantke et al. 2013). The number of animals reaching sexual maturity is maximal shortly after the new moon and minimal during the full moon (Hauenschild 1955; Zantke et al. 2013). In nature, this reproductive period occurs between May and September in the Mediterranean Sea (Giangrande et al. 2002). When *Platynereis dumerilii* worms are ready to spawn, usually at night, males and females reach the surface, start an elegant nuptial dance and synchronously release eggs and sperm in a massive spawning event. This external fertilization induces the formation of thousands of small zygotes and ultimately implies the death of the reproducing males and females (Figure 13.2). By the third day, the zygote develops into small, segmented, planktonic larva named nectochaetae (see Section 13.5). Nectochaete larvae live on their own nutritive stock and move thanks to marine currents and ciliary belts. After five to seven days of planktonic life, small juvenile worms, while still able to swim, switch to a benthic/errant life mode following metamorphosis (Fischer et al. 2010). This first metamorphosis event corresponds to the disappearance of ciliated belts. Juvenile worms then continue to grow throughout their lives at a rate that is highly dependent on food availability. At some point, a second metamorphosis occurs, inducing profound morphological modifications of the head and first segment. Additionally, worms start to produce silk, important for the building of a tube in which they will live for several months. They continue to grow, to regenerate following injury and to grow posteriorly until they initiate their last metamorphosis, corresponding to the appearance of sexual traits (Fischer et al. 2010). This sexual metamorphosis is moon dependent and implies drastic morphological changes (see Section 13.4) to allow the production of thousands of gametes. During this very short reproductive period, worms become pelagic.

This eventful life cycle can be reproduced in laboratory in culture rooms maintained at 18°C and a daily artificial illumination regime (16 hours of light/8 hours of darkness). To induce sexual maturation, a low-light lamp is used to mimic the lunar stimulus seven days per month. A couple of days after this week of artificial full moon, juvenile (or atoke) worms start sexual maturation for a two-week period, allowing the production of sexually-mature (or epitoke) worms every day (Fischer and Dorresteijn 2004; Kuehn et al. 2019; Vervoort and Gazave in press).

13.4 ANATOMY

Like many other annelid species, *Platynereis dumerilii* worms have a complex body plan with various tissues, structures and organs that are described in the following sections for both atoke and epitoke forms.

13.4.1 External Anatomy of *Platynereis dumerilii* Juvenile (Atoke) Worms

Platynereis dumerilii juvenile worm size can be up to 90 mm for around 100 segments (Figure 13.3a). Their body color is highly variable, from yellowish and reddish to greenish, and this coloration mostly relies on pigmented cells, or chromatophores (Arboleda et al. 2019), that shine at the surface of the epidermis, itself secreting a cellular cuticle. While their sex is genetically determined, at the juvenile stage, male and female animals are indistinguishable. The morphology of Nereid annelids such as *Platynereis* is often described in zoological textbooks as representative of the typical annelid body plan, composed of three main parts: (i) an anterior region, the head with a substantial cephalization; (ii) the segmented or metamerized trunk composed of many identical units called segments, with appendages named parapodia; and (iii) a post-segmental terminal part, containing the pygidium, a differentiated structure notably containing the anus (Figure 13.3a) (Fischer and Dorresteijn 2004; Fischer et al. 2010).

Platynereis's head is composed of different structures, many of them being sensory (Chartier et al. 2018). These structures ensure crucial functions for the worm's life (Purschke 2005) (Figure 13.3b and c). To begin, *Platynereis* possesses two pairs of pigmented cup brown adult eyes, in a trapezoid arrangement, only visible on the dorsal part of the worm (Figure 13.3b). These pairs of adult eyes represent a distinct type of eyes in comparison to larval eyes, as revealed by their specific developmental program (Arendt et al. 2002; Guhmann et al. 2015). They also harbor a very specific cellular structure with rhabdomeric photoreceptor extensions traversing the pigmented cell layer (Arendt et al. 2002). These eyes are localized on a specific structure of the head, named the prostomium (Figure 13.3b). The prostomium also bears a pair of highly chemosensory antennae localized at the front of the head (Chartier et al. 2018) (Figure 13.3b). A pair of sensory palps are present near the antennae; based on their cellular ultrastructure (Dorsett and Hyde 1969) and a physiological experiment (Chartier et al. 2018), they have been proposed to be chemosensory as well (Figure 13.3c). The head is also composed of four pairs of long sensory and photosensitive tentacular cirri (namely anterior/posterior dorsal/ventral tentacle cirrus) (Figure 13.3c). They are involved in the worm's "shadow reflex", a defensive behavior triggered by a decrease in illumination (Ayers et al. 2018). At the posterior dorsal margin of the prostomium are nuchal organs, a pair of ciliated cavities also considered important chemosensory structures (Schmidtberg and Dorresteijn 2010; Chartier et al. 2018)

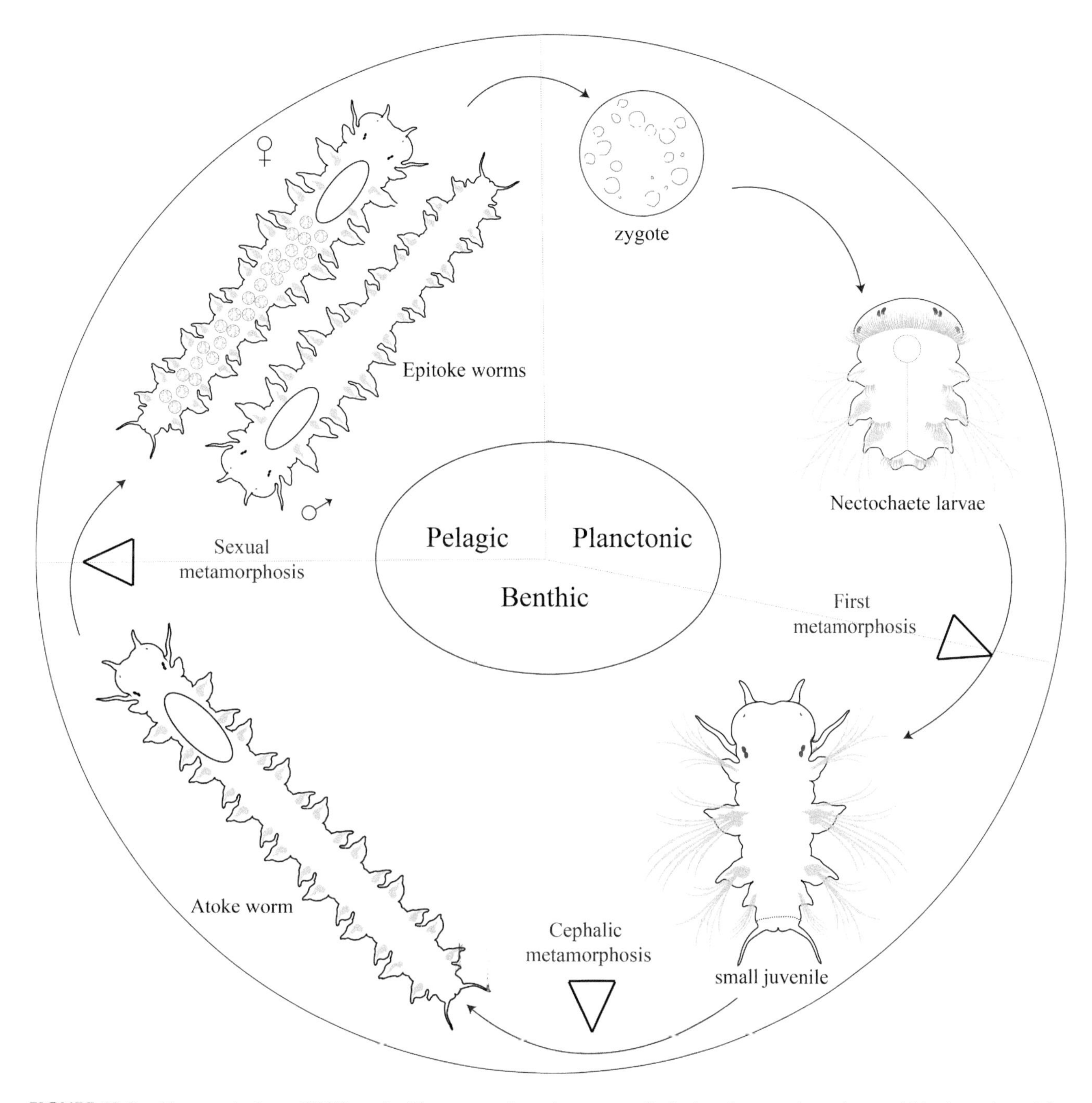

FIGURE 13.2 *Platynereis dumerilii* life cycle. The zygote gives rise to a small planktonic nectochaete larva within three days. After five to seven days post-fertilization (dpf), the small juvenile worm starts to feed and become benthic. Shortly thereafter, the small worm undergoes cephalic metamorphosis. The atoke worm lives inside its tube and grows continuously until sexual maturation. The sexually mature or epitoke worm then leaves its tube and swims into the water column until it performs mass spawning.

(Figure 13.3b). Finally, the aforementioned structures are located on a specific segment with no appendages, named the peristomium (Figure 13.3b). The peristomium also contains a large structure, visible only on the ventral side of the worm, which is the stomium or mouth (Figure 13.3c). Following the mouth, the pharynx contains a pair of chitinous and denticulated jaws, invaginated in the first segment of the worm, which are evaginated to catch food (Figure 13.3d). This eversible pharynx corresponds to the anterior part of the digestive tract (Verdonschot 2015).

Platynereis's trunk is composed of identical segments (Figure 13.3a). Its segmentation is thus named homonomous (Fischer and Dorresteijn 2004). Each segment is externally composed of an outer annulus and parapodia (Scholtz 2002). Parapodia are paired appendages, found in many annelids, and which have locomotion, respiratory and sensory functions (Figure 13.3e). They notably allow the worm to crawl and swim (Grimmel et al. 2016). Parapodia are biramous and thus composed of two parts, the notopodium in the dorsal side of the animal and the neuropodium in its ventral side

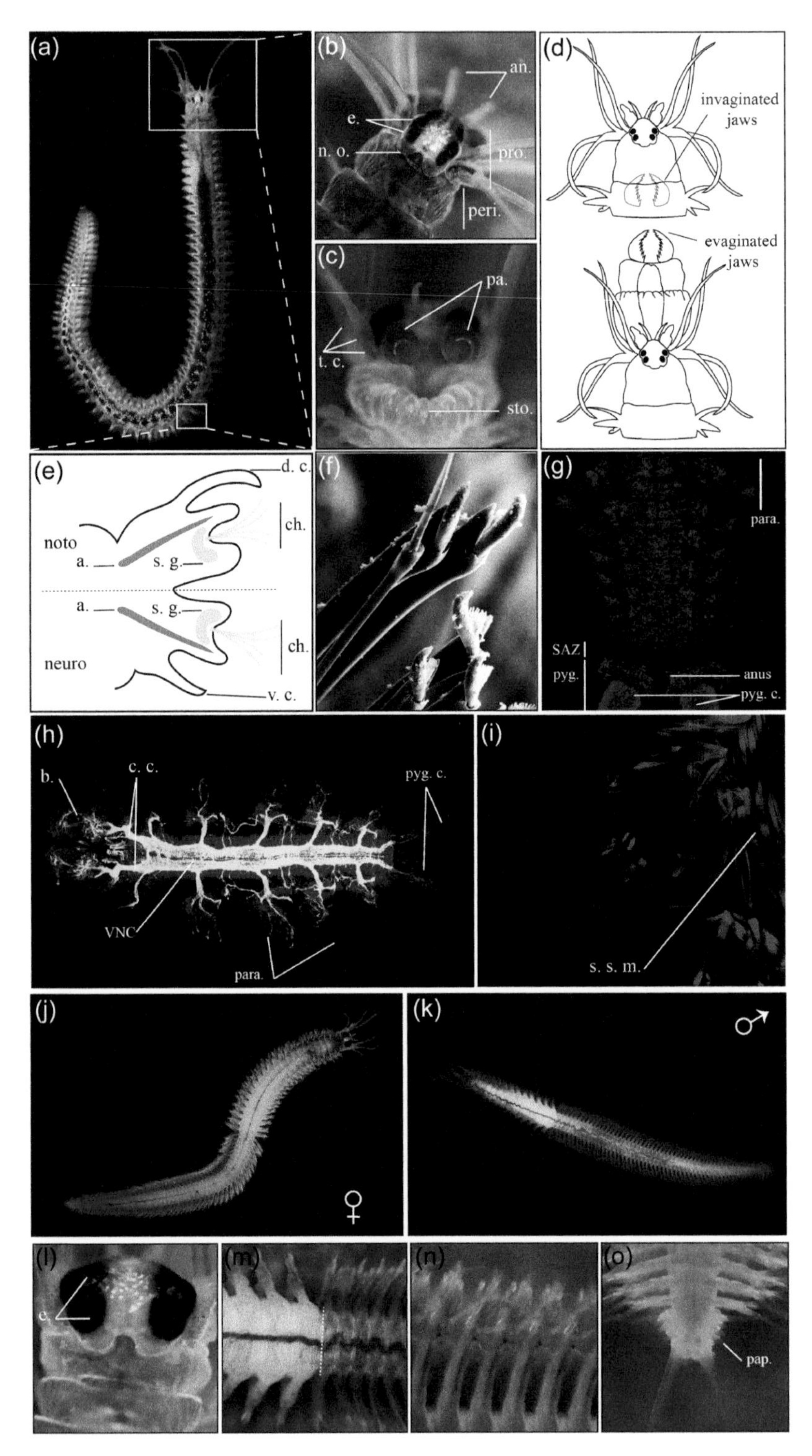

FIGURE 13.3 Anatomical features of *Platynereis dumerilii* atoke and epitoke worms. (a) Juvenile of *Platynereis dumerilii*, head (see b to d) and parapodia (see e and f) are framed. (b) Dorsal close view of the head, bearing sensory structures. (c) Ventral close view of the head, bearing the mouth or stomodeum. (d) Process of jaw evagination to catch prey. (e) Schematic representation of parapodia; the dorsal part is facing up. (f) Scanning electron microscopy (SEM) of chaetae, photo courtesy of N. Dray, PhD (CNRS). (g) Posterior part of the worm, showing the parapodia, segment addition zone, and pygidium. Hoechst nuclear staining in blue. (h) Nervous system of small juvenile worm. Ventral view; anterior is to the left. Nerves are labeled by acetylated-Tubulin antibody in green, and hoechst nuclear staining is in blue. (i) Musculature of small juvenile worm parapodia. Muscles are labeled with Phalloidin. (j) *Platynereis* mature female. (k) *Platynereis* mature male. (l) Enlarged *Platynereis* eyes during sexual maturation. (m) Boundary between anterior (left) and posterior (right) segments of mature worms. (n) Important blood network within the posterior parapodia of mature male. (o) Male pygidium presenting extra papillae. Abv.: a. = acicula, an. = antennae, b. = brain, c. c. = circumpharyngeal connectives, ch. = chaetae, d. c. = dorsal cirri, e. = eye, neuro = neuropodium; n. o. = nuchal organ, noto. = notopodium, pa. = palps, pap. = papillae, para. = parapodia, peri. = peristomium, pro. = prostomium, pyg. = pygidium, pyg. c. = pygidial cirri, SAZ = segment addition zone, s. g. = spinning gland, s. s. m. = somatic striated muscle, sto. = stomium, t. c. = tentacular cirri, v. c. = ventral cirri, VNC = ventral nerve chord.

(Figure 13.3e). Each rami is composed of cirri (dorsal and ventral cirrus), a lobe and a beam of extracellular chitineous structures named chaetae or bristles (Verdonschot 2015) (Figure 13.3f). The latter are surrounded by specific glands that secrete material for tube or cocoon synthesis and are named spinning glands (Fischer et al. 2010). These external chaetae are constantly produced by internal structures named chaetal sacs (Gazave et al. 2017). Chaetae have a terminal articulated portion at their tips (Figure 13.3f). In addition, a robust, skeletal, internal, peculiar chaeta, named the acicula, stabilizes each lobe (Figure 13.3e). Interestingly, in annelid systematics, the shape of the parapodia and the type of chaetae are informative characteristics to determine species (Zakrzewski 2011).

The terminal part of the worm is named the pygidium (Starunov et al. 2015) (Figure 13.3g). The pygidium contains the anus and presents two sensory anal or pygidial cirri in its ventral part (Ayers et al. 2018).

13.4.2 Internal Anatomy of *Platynereis dumerilii* Juvenile (Atoke) Worms

As for its external anatomy, *Platynereis*'s internal body plan is segmented and presents a repetition of internal structures within each segment. Indeed, each body unit contains a body cavity or coelom (separated from the next one by an incomplete intersegmental piece of tissue called septa), each containing a part of the (i) nervous system, (ii) circulatory system, (iii) musculature and (iv) excretory system (Verdonschot 2015). The non-metamerized digestive tract (v) runs along the antero–posterior axis of the worm.

13.4.2.1 Nervous System

Platynereis worms possess a central nervous system (CNS) and a peripheral nervous system (PNS). One main element of the CNS is the highly developed brain that resides in a dorsal position within the head (Starunov et al. 2017) (Figure 13.3h). Another important element of the CNS is a paired serial chain of spherical ganglia in a ventral position (named the ventral nerve chord or VNC) that runs all along the length of the worm's body, making a ladder-shaped structure (Figure 13.3h). The brain is connected to the VNC by circumpharyngeal connectives, which surround the pharynx (Verdonschot 2015) (Figure 13.3h). *Platynereis*'s brain contains prominent dorsal neuropile arrangements named mushroom bodies (Tomer et al. 2010), a structure shared with arthropods (Heuer et al. 2010). Interestingly, the mushroom body anlagen in *Platynereis* larvae expresses a similar molecular signature to developing mushroom bodies in *Drosophila melanogaster*, thus providing strong evidence in favor of an evolutionary relatedness of insect and annelid mushroom bodies (Tomer et al. 2010). *Platynereis*'s brain is also an important neurosecretory center. Its developing forebrain expresses the neuropeptides FMRFa and vasotocin (Tessmar-Raible et al. 2007), plus a diversity of other neuropeptides recently identified, such as somatostatin, galanin and so on (Williams et al. 2017). Interestingly, this annelid brain region shares a common molecular signature with the vertebrate hypothalamus, furthering the hypothesis of an evolutionary relationship between those two structures (Williams and Nagy 2017). In addition, *Platynereis*'s brain produces a brain hormone responsible for the switch from a growing juvenile to a sexually mature worm (Hauenschild 1956). This hormone, whose activity suppresses reproduction, was recently identified as Methylfarnesoate (Schenk et al. 2016). In addition to the brain, the VNC is also a complex structure that has been shown to harbor around 200 distinct types of neurons, expressing specific combinations of transcription factors in a small juvenile (Vergara et al. 2017). *Platynereis*'s PNS is prominent in the head, being associated with the many sensory structures it contains. The PNS also contains the parapodial and pygidial nerve extensions. Indeed, the terminal part of the worm is highly innervated with nerve projections into the pygidium and the anal cirri (Starunov et al. 2015) (Figure 13.3h).

13.4.2.2 Circulatory System

Platynereis has a closed circulatory system mainly composed of two vessels and capillary networks. The dorsal and ventral vessels are connected by a capillary network forming a ring around the intestine. The dorsal pulsatile vessel is the main pump of the circulatory system, pumping the blood anteriorly (from the tail to the head), while the ventral vessel pumps the blood posteriorly. Segmental lateral vessels irrigate the parapodia in each segment in order to ensure their respiratory function (Saudemont et al. 2008; Verdonschot 2015). A circular blood sinus is present in the pygidium (Starunov et al. 2015).

13.4.2.3 Musculature

Platynereis has two main types of muscles, smooth and striated, which together ensure precise movements of the worm's body structures (Brunet et al. 2016) (Figure 13.3i). Some somatic striated muscles run longitudinally from the head to the tail of the animal. Additional somatic striated muscles control the movements of parapodia thanks to ventral oblique and parapodial fibers (Figure 13.3i). In contrast, visceral muscles are mainly smooth muscle (with the noticeable exception of the anterior part of the gut that contains striated visceral muscles). They form a specific muscular structure, the orthogon, which is composed of both circular and longitudinal fibers (Brunet et al. 2016). Smooth muscles are also associated with the pulsatile dorsal vessel (Brunet et al. 2016). A peculiar somatic striated and longitudinal muscle, the axochord, is found between the VNC and the dorsal vessel and is proposed to be at the origin of the chordate notochord (Lauri et al. 2014; Brunet et al. 2015). The pygidium musculature is also highly complex, mainly composed of a strong array of circular muscles that plays the role of the anal sphincter (Starunov et al. 2015).

13.4.2.4 Excretory System

Platynereis atoke worms possess in each segment, except the pygidium, a pair of metanephridia that connects the coelomic compartment to the exterior to ensure the excretion of waste products (Hasse et al. 2010; Verdonschot 2015).

13.4.2.5 Digestive System

The digestive system of *Platynereis* is mainly composed of three successive elements called the foregut, the midgut and the hindgut (Fischer et al. 2010; Zidek et al. 2018). The foregut is composed of the mouth, the eversible pharynx and the jaws, in charge of collecting and grinding the food. Digestive enzymes are secreted and active in the midgut, where food absorption occurs. The last section of the digestive system is a hindgut connecting the midgut to the anus and producing digestive enzymes, too (Verdonschot 2015; Williams et al. 2015).

13.4.3 External and Internal Anatomy of *Platynereis dumerilii* Adult (Epitoke) Worms

As mentioned before, one of the main events in *Platynereis* life cycle is sexual maturation (epitoky), since it induces not only drastic morphological modifications but also changes in behavior (Fischer and Dorresteijn 2004). A striking difference between mature and juvenile worms is the difference in body color (Figure 13.3j and k): while juveniles mainly show a sex-independent brownish color, sexual dimorphism appears during epitoky, as females become bright yellow (Figure 13.3j) and males display white anterior and red posterior body regions (Figure 13.3k). During sexual maturation, worms stop food intake, their gut regresses and becomes non-functional. The trunk of the animal is progressively modified to become a "bag" full of gametes, visible through the body wall, which loses its pigmentation (Fischer and Dorresteijn 2004; Fischer et al. 2010). The yellow oocytes and white spermatozoids both contribute to the main color of the female and male anterior parts, respectively. Among other morphological changes, the eyes enlarge dramatically (Figure 13.3l), and the homonomous segmentation present in juveniles is lost. Indeed, while anterior segments are not modified, posterior segments are substantially reshaped, and a clear boundary between these two parts of the trunk becomes visible (between the 15th and 20th segments, depending on the sex; Figure 13.3m) (Schulz et al. 1989; Fischer 1999). In modified posterior segments, parapodia flatten and develop paddle-shaped chaetae in both sexes. In males, posterior parapodia show a significant increase in vascularization, conferring its red color to the posterior part (Figure 13.3n). Muscles present in juvenile worms degenerate and are replaced by new muscle fibers which are specific to sexually mature animals. This dramatic reorganization of the body enables the formerly benthic juvenile worms to swim quickly to ensure the nuptial dance required for sexual reproduction (Fischer and Dorresteijn 2004; Fischer et al. 2010). Finally, while the terminal part of the female is not modified, the male pygidium presents extra papillae, allowing the sperm to be released in many directions (Figure 13.3o) (Starunov et al. 2015). In *Platynereis*, the switch from a growing worm to its reproductive life stage is controlled by brain hormone activity. Interestingly, worm decapitation (i.e. artificial reduction of brain hormone) induces worms' sexual maturation similarly to natural conditions (Schenk et al. 2016).

13.5 EMBRYOGENESIS AND LARVAL DEVELOPMENT

More than a century ago, Edmund B. Wilson retraced an incredibly relevant and reliable cell lineage of embryo blastomeres in order to depict the origin of the germ layers in annelids (Wilson 1892). To do so, he took advantage of the transparency of *Nereis limbate* (now *Alitta succinea*) and *Nectonereis megalops* (now *Platynereis megalops*) embryos and of their stereotypic development. Indeed, as all embryos develop in exactly the same way, they provide an ideal framework to link cell division to blastomere formation and cell fate. Interestingly, since publication, his work has been reasserted by the description of *Platynereis dumerilli* embryogenesis in the early 90s (Dorresteijn 1990), and his assumptions regarding blastomere cleavage and fate remain a reference in the field of annelid development. Indeed, micro-injection of individual blastomeres at different embryonic stages with fluorescent dyes has more recently confirmed previous observations (Fischer and Dorresteijn 2004; Ackermann et al. 2005). Hereafter, we have mainly compiled the previously mentioned publications to depict the main events of embryogenesis and larval development (Fischer et al. 2010).

13.5.1 Embryo Development

13.5.1.1 Unfertilized Eggs

Unfertilized eggs are packed within the coelomic cavity of the mature female, causing their polymorphous shapes. Upon laying, the pressure is released and the eggs rapidly undergo a massive shape change to become ellipsoid (the short axis of the unfertilized egg corresponds to the future animal–vegetal axis of the zygote). At that stage, their cytoplasm is organized, in a concentric fashion, around the central nucleus which is wrapped in yolk-free cytoplasm. The latter is surrounded by a shell of yolk containing large lipid droplets (in particular in the equatorial plane where they are bigger) and a thick outer layer of cortical granules (secretory organelles found within oocytes). Finally, the egg is itself protected within a vitelline envelope. Interestingly, in *Platynereis*, eggs are in fact oocytes blocked in metaphase and, as such, the release of polar bodies occurs after fertilization.

13.5.1.2 Fertilization

Upon fertilization, the fertilizing spermatozoid sticks to the cell surface until the emission of the first polar body (a small haploid cell). As soon as this contact is established, substantial changes in the cytoplasmic organization of the oocyte occur. The cortical granules are released to form an external jelly layer (0–23 minutes post-fertilization, mnpf). As a consequence, the yolk granules are less packed within the spherical egg and more broadly distributed, while the lipid droplet pattern remains as a readout of the equatorial plan. When the vitelline envelope breaks down (18 mnpf), a small area is progressively cleared from yolk at one pole of the egg compared to the equatorial location of lipid droplets. This area marks the future animal pole where the first polar body is formed (60 mnpf, Figure 13.4a). The sperm pronucleus finally enters into the ooplasm, and a second polar body is formed (80 mnpf). Yolk granules migrate toward the vegetal pole, allowing the rapid expansion of the clear cytoplasm, the female pronucleus forms and karyogamy (fusion of the two nuclei) occurs (90–100 mnpf). Subsequently, the animal pole is completely cleared from yolk granules, and the first cleavage is initiated.

13.5.1.3 First Cleavages (120–420 mnpf)

The first cleavage is unequal, giving rise to a small AB blastomere and a large CD blastomere (73% of the volume, 100 mnpf, Figure 13.4b1). This unequal cleavage induces a new axis, perpendicular to the vegetal/animal axis, which nearly corresponds to the dorsoventral axis. The second cleavage is slightly asynchronous (Figure 13.4a), unequal in the CD blastomere (D blasomere inheriting 50% of the total egg volume) and equal in the small AB blastomere (135 mnpf; Figure 13.4a, 4b2). Each of the resulting four macromeres is the founder of a distinct quadrant (e.g. A-quadrant corresponds to the offspring of A macromere). The third cleavage is slightly asynchronous as well (Figure 13.4a) and corresponds to the first "spiral" cleavage (clockwise), producing two batches of two nearly identical micromeres (1a and 1b *versus* 1c and 1d) (170 mnpf, Figure 13.4b3). Before the fourth asynchronous cleavage (Figure 13.4a), yolk granules are segregated at the vegetal pole of each blastomere. As a result, after completion of the fourth cleavage (i.e. 16-cell stage), blastomeres $1a^1$–$1d^1$ contain less yolk than blastomeres $1a^2$–$1d^2$. The latter are called the primary trochoblasts and give rise to the equatorial ciliated belt, or prototroch, of the trochophore larva (Figure 13.4b4). The 2d blastomere is by far the largest micromere, since its size even exceeds that of the macromeres 2A-2C. After the fourth cleavage, cleavages become highly asynchronous (except in the trochoblasts), and the cleavage strategy of the D quadrant strongly differs from the others with the short cell cycle of $2d^1$ and 4d cell lines (Figure 13.4a). At the 38-cell stage, the fate of the three germ layers is established. The four macromeres (3A-3C and 4D) give rise to the endoderm, and the mesoderm mainly arises from 4d micromere (also called mesoblast or "M"), as well as the germ line. All other micromeres form the ectoderm (Figure 13.4a, 4b5 to 4b5''').

13.5.1.4 Stereoblastula/Stereogastrula/Protrochophore Larva (7–24 Hours Post-Fertilization of hpf)

After the 38-cell stage, micromeres no longer undergo spiral cleavage but rather progressively follow a bilateral symmetry. They rapidly divide and initiate their epibolic movement toward the vegetal pole, thus covering macromeres. This movement of micromeres results in the final equatorial position of trochoblasts, thus forming the prototroch. At the vegetal pole, cells arising from the cleavage of the 4d micromere submerge beneath the large cells produced by the 2d micromere and start to form the mesodermal bands ($4d^{122}$ and $4d^{222}$ lines the dorsal rim of the blastopore); gastrulation is thus initiated. In *Platynereis*, this process shows amphistome mode, meaning that the blastopore gives rise to both the mouth and the anus (Figure 13.4c) (Steinmetz et al. 2007). During this massive rearrangement of embryonic cells, the D-quadrant plays a key role, especially in the formation of trunk tissues. Indeed, the 4D blastomere participates in midgut anlage, and the 2d offspring forms the somatic plate and the entire trunk ectoderm (i.e. epidermis and nervous system). The 4d lineage provides the full trunk mesoderm, including the four quiescent putative primordial germ cells (i.e. 1mL, 1mr, 2mL and 2mr resulting from two asymmetric divisions of M-daughter cells) but also the cells composing the growth zone where new segments are added after the larval stage (Fischer and Arendt 2013; Ozpolat et al. 2017). During gastrulation and later on, the presence of four lipid droplets appears as a good readout of the proper development of the embryo. After the gastrulation stage, the embryo, often called the protrochophore larva (13–24 hpf), despite the persisting jelly, is slowly rotating within the jelly thanks to the prototroch (Figure 13.4c). It develops an apical tuft (apical ciliated organ), and the stomodeal field (i.e. the mouth anlage) starts to develop too. At around 17 hpf, the first serotonergic neuron differentiates at the posterior extremity of the protrochophore larvae (Starunov et al. 2017).

13.5.2 Larvae Development

13.5.2.1 Trochophore Larva (24–48hpf)

The trochophore larva is a phototactic swimming larva possessing two pigmented eyes that become more and more prominent (Figure 13.4c). With age, the spherical larva elongates, and three segments start to appear. Consistent with this first sign of segmentation, three pairs of ectodermal bulges develop laterally from 2d descendants to form the ventral chaetal sac pairs (Figure 13.3c'). An additional band of ciliated cells, called the telotroch, is formed at the posterior end, marking the edge between the pygidium and the rest of the trunk (26 hpf). Regarding the establishment of the digestive tract, the number of stomodeal cells slightly increases, and they start to form a ring (i.e. the stomodeal rosette). The stomodeal field progressively moves toward the anterior pole, and the rosette opens just below the prototroch to form the mouth (40 hpf). Meanwhile, the overall nervous system rapidly develops (also from 2d micromeres) in part along with the increase in ciliated structures. From 24 hpf, various nervous connections are also implemented. Indeed, the apical ganglion at the posterior pole, containing the pioneer neuron of the VNC, is linked to the prototroch nerve ring by two ventral connectives. These connectives of the

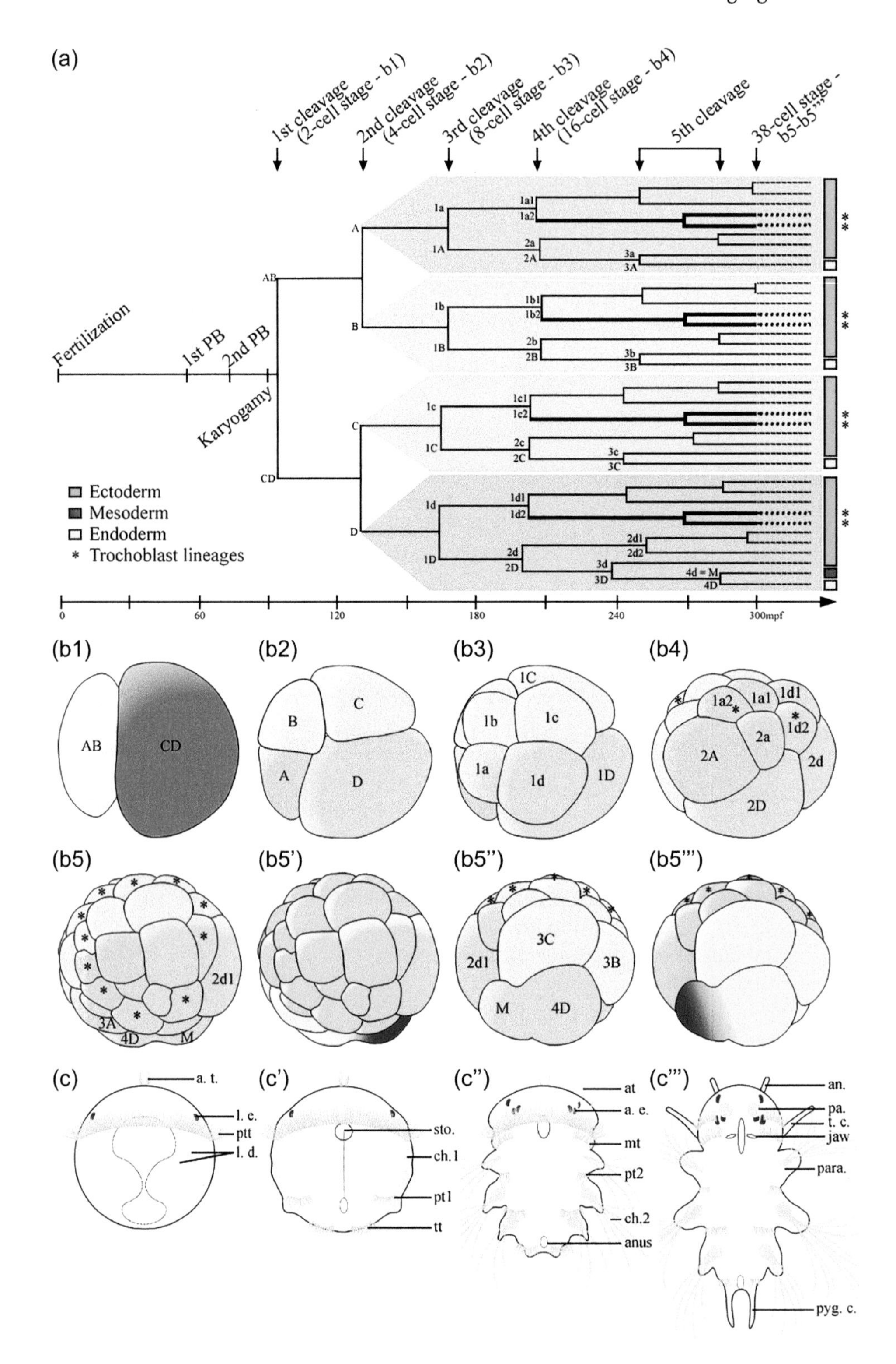

FIGURE 13.4 Embryogenesis and larval development in *Platynereis dumerilii.* (a) Dendrogram summarizing the stereotypic steps following fertilization, including the emission of the two polar bodies (PB), karyogamy and the first cleavages (see b1 to b5''') that give rise to a 38-cell embryo in which the fate of the three germ layers is established (see legend on the panel) as well as trochoblast lineages ($1a^2$–$1d^2$). Colored backgrounds represent each quadrant (i.e. A, B, C and D). Blastomere names are provided above and below nodes (in capital letters for macromeres), including the highly proliferative $2d^1$ and M micromeres. Time frame is provided below the dendrogram. (b1–b5''') Schematic representation of embryo following the five first cleavages. Color codes are similar to those in (a). (b1–b3, b5, b5') Animal views. (b4) Animal pole at the top. (b5'', b5''') Vegetal views. Based on (Dorresteijn 1990). (c–c''') Schematic representation of larval development. Ventral view of (c) 24-hour post-fertilization (hpf) larva, (c') 48-hpf larva, (c'') 72-hpf larva and (c''') 6-day post-fertilization larva. Abv: a. e. = adult eye, an. = antenna, a. t. = apical tuft, at = akrotroch, ch.1/2 = chaeta within/outside the body wall, l. e. = larval eye, l. d. = lipid droplets, mnpf = minutes post-fertilization, mt = metatroch, para = functional parapodia, pa. = palpa, pt1/2 = paratroch 1 and 2, ptt = prototroch, pyg.c. = pygidial cirrus, sto. = stomodeum, t.c. = tentacular cirrus, tt = telotroch. (Based on Dorresteijn 1990.)

VNC represent the two first axons of the brain. Immediately thereafter, the dorsal root of the circumesophagial connectives develop as well, followed by the ventral root (26 hpf) to connect the VNC to the brain. At the same time, the single asymmetric unpaired dorsal axon and the first cerebral commissures appear. Later on, three additional serotoninergic cells arise at the apical part (30–34 hpf), as well as one pair at the first ventral commissure (40 hpf). Finally, the second ventral commissure appears (44 hpf) (Starunov et al. 2017). Similarly, muscles appear and develop during the trochophore stage. The dorsal longitudinal muscles develop first (28 hpf), followed by ventral longitudinal muscles (32 hpf), while the oblique and parapodial muscles start to be visible at the late trochophore stage (46–48 hpf). The excretory system appears also at the trochophore stage, with the emergence of small, lateral, non-ciliated tubules (Hasse et al. 2010).

13.5.2.2 Metatrochophore Larva (48–66 hpf)

The metatrochophore larval stage is marked by the appearance of the two adult eye pairs (Figure 13.4c''). In addition, the three first segments appear more defined due to the formation of non-functional parapodia and the significant growth of the chaetae outside the body wall (these segments are so called the chaetigerous segments). In addition, the first paratroch appears between the second and the third chaetigerous segments (48 hpf), and a second one is visible later on between the first and the second chaetigerous segments (56 hpf), thus participating in segment delimitation. Then, above the prototroch, an additional ciliated structure progressively develops—the akrotroch—close to the apical tuft (60 hpf). The stomodeal rosette size increases with an additional ring of cells (52 hpf). The stomodeum invaginates, resulting in the larval foregut that elongates toward the posterior part. The nervous system also rapidly develops. A third commissure appears (48 hpf), and all commissures thicken (54 hpf). Axon projections from the VNC are observed laterally and redirected ventrally toward the surface. The circumesophagial connectives get closer to each other, and the prototroch nerve ring moves toward the brain. All these phenomena participate in the formation and growth of the brain (52 hpf). Additionally, the number of serotonergic cells along the ventral nerve cord increases with the occurrence of three additional pairs. Finally, the ventral medial longitudinal muscle appears (56 hpf) and elongates up to the posterior border of the third segment. Similarly, oblique and parapodial muscles also elongate (Figure 13.4c''). Excretory system development continues as non-ciliated tubules elongate laterally toward the developing stomodeum (Hasse et al. 2010).

13.5.2.3 Nectochaete Larva (66 hpf–5 dpf)

The nectochaete larva corresponds to a major lifestyle transition. Indeed, the pelago-benthic larva starts crawling on the substrate thanks to functional parapodia and starts to eat. The sensory organs, including antenna, palps, tentacular antero-dorsal cirri and anal cirri, appear and develop (75 hpf) (Figure 13.4c'''). The trunk continues to elongate, providing a worm-like shape to the larva, and a constriction distinguishes the trunk from the head. The two adult eyes found on both sides of the head increase in size and become extremely close. Lipid droplets progressively move toward the posterior part. Ciliogenesis progresses with the establishment of the metatroch, an additional line of ciliated cells that develops below the prototroch and fuses with this latter on the lateral sides. The midgut forms, as well as the proctodeum (anal region), and the stomodeum/foregut continues to elongate toward the posterior part, resulting in a fully functional digestive tract (75 hpf–4 dpf). Furthermore, the jaws develop within the foregut (4 dpf) and a pair of primary teeth appears (5 dpf). Meanwhile, the brain continuously grows, the convergence of the circumesophagial connective roots progresses, axon numbers increase in connectives and commissures and additional serotonergic cells arise both in the ventral nerve cord (66–72 hpf) and in the brain (4–5 dpf). The overall musculature develops as well, especially around the stomodeum, to form the pharynx. Additionally, muscles and nerves associated with the development of antennae and cirri increase. Seventy-two-hpf larvae possess a pair of anterio-lateral non-ciliated tubules named "head kidneys" located close to the episphere. These larval structures are transitory, since they disappear before 96 hpf. In parallel, larval nephridia or protonephridia, formed from ciliated tubules and localized between segments, start to appear (Hasse et al. 2010).

13.5.2.4 Young Errant Juvenile

At this stage, the development of animals is no longer synchronous. Very young worms start to sequentially produce additional segments through posterior elongation, a process relying on a thin row of cells (presumably stem cells) that forms the segment addition zone (SAZ) in front of the pygidium (Gazave et al. 2013). Worms also lose several larva-specific features such as the prototroch, the apical tuft, larval eyes and lipid droplets. The excretory system is composed of segmented protonephridia until the worms reach the size of 20 segments, at which stage metanephridia appear (Hasse et al. 2010). In addition, the first chaetigerous segment fuses with the head. This important morphological transition, called cephalization, consists of the transformation of the first pair of parapodia into tentacular posterior–dorsal cirri and the progressive loss of chaetae. Finally, spinning glands develop and produce mucus, allowing worms to build their first cocoon network.

13.6 GENOMIC DATA

As in many animals, counting chromosomes during metaphase revealed that *Platynereis dumerilii* is diploid ($2n = 28$) (Jha et al. 1995). More precisely, the *Platynereis* karyotype encompasses seven chromosome pairs showing a median arm ratio, while the seven other pairs show a sub-median ratio (Figure 13.5). Different regular staining techniques were used to further characterize chromosome pairs. For instance, *Chromosome 2* shows a clear C-band-positive

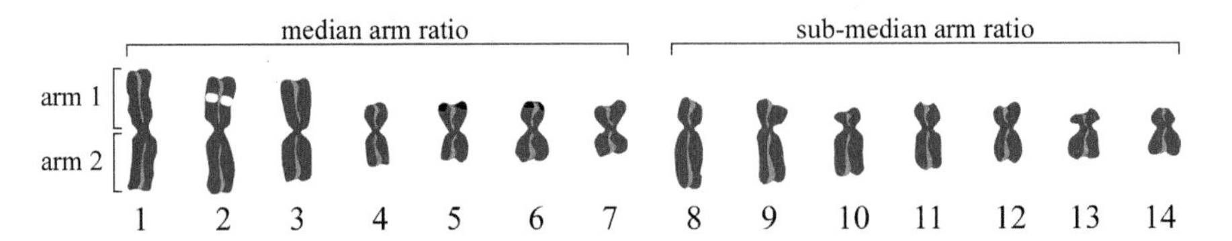

FIGURE 13.5 Schematic representation of *Platynereis* chromosomes. *Platynereis* possesses $2n = 28$ chromosomes, five pairs showing a median position of centromeres (*Chromosome 1* to *7*). Black and white areas represent heterochromatin (C-band-positive band) and ribosomal RNA genes (NOR staining), respectively.

region, revealing a constitutive heterochromatin region not localized at the centromic region. Nuclear organizer regions (NORs) are found at the terminal positions of *Chromosomes 5* and *6*, thus revealing the localization of genes coding for ribosomal RNA.

The precise genetic content of these chromosomes is in the course of being uncovered. Indeed, the *Platynereis dumerilii* genome has recently been sequenced by the D. Arendt laboratory (EMBL, Germany), notably from sperm. Although this genome is currently being refined with the aim of obtaining a chromosome-level assembly, a high-quality draft version is already available, upon request, for the whole community working on *Platynereis*. Preliminary data revealed that *Platynereis* genome appears less compact than in other annelids (~1 Gpb) (Zantke et al. 2014), and a previous analysis comparing bacterial artificial chromosome (BAC) sequencing and expressed sequence tags (ESTs) on a subset of 30 randomly detected genes suggested that *Platynereis* genes are intron rich, surprisingly, with two-thirds of introns shared between *Platynereis* and human orthologs (Raible et al. 2005). Various additional transcriptomic databases have been acquired during the past years (Table 13.1), including bulk RNA-seq data for all key stages of embryonic and larval development, juveniles of different ages and adults (Conzelmann et al. 2013; Chou et al. 2016). These data have been grouped together and are now publicly available on the *Pdumbase* website (Chou et al. 2018) (http://140.109.48.81/platynereis/controller.php?action=home). *Platynereis* is being actively studied by a scientific community, notably in the field of evolution and developmental biology, and as such, additional transcriptomic databases are constantly produced. For instance, Achim and collaborators shed light on the transcriptomic landscape of cell diversity in 48 hpf-larvae using a single-cell RNA-seq (scRNA-seq) approach (Achim et al. 2015; Williams et al. 2017; Achim et al. 2018). In addition, bulk RNA-seq were acquired to unravel the dynamic of gene expression during circalunar-dependent sexual maturation (Schenk et al. 2019) and posterior regeneration (Vervoort's Lab, unpublished data).

Finally, in addition to the significant *Platynereis* resources acquired during the past decade, the availability of genome sequences of the Sedentaria *Capitella teleta*, *Helobdella robusta* (Simakov et al. 2013), *Spirobranchus lamarcki* (Kenny et al. 2015), *Lamellibrachia luymesi* (Li et al. 2019), *Eisenia Andrei* (Shao et al. 2020) and *Eisenia fetida* (Bhambri et al. 2018) as well as the Dinophiliformia (sister group to Sedentaria + Errantia) *Dimorphilus gyrociliatus* (Martin-Duran et al. 2021) allow for comparative analyses within annelids.

13.7 FUNCTIONAL APPROACHES: TOOLS FOR MOLECULAR AND CELLULAR ANALYSES

In addition to its scientific relevance and its easy maintenance in laboratory, the success of *Platynereis* as a new model system also strongly relies on the efforts that have been undertaken to develop a large panel of molecular and cellular tools to successfully tackle interesting biological questions in evolutionary and developmental biology (Backfisch et al. 2014; Williams and Jekely 2016).

13.7.1 Descriptive Approaches

13.7.1.1 Detection of mRNA: Whole-Mount *In Situ* Hybridization

As mentioned in the genomic data section, several high-quality bulk RNA-seq and scRNA-seq were recently used to investigate modulations in gene expression during various processes in *Platynereis*. Nevertheless, bulk RNA-seq average information from various cell populations and scRNA-seq remains expensive, and their interpretation relies on a comprehensive description of cell populations *in vivo*. Accordingly, despite important breakthroughs in sequencing technologies, whole-mount *in situ* hybridization (WMISH) remains an indispensable molecular approach to localize gene expression. WMISH has been established in *Platynereis* to investigate gene expression during early embryonic/larval stages (Arendt et al. 2001), posterior elongation (Prud'homme et al. 2003, Gazave et al., 2013), regeneration (Planques et al. 2019) and the adult stage (Backfisch et al. 2013) using the regular NBT/BCIP colorimetric staining (Figure 13.6a and a'). Similarly, fluorescent *in situ* hybridization (FISH, Figure 13.6b and b') has been established (Tessmar-Raible et al. 2005), while current efforts are now also dedicated to implement hybridization chain reactions (HCRs) (Choi et al. 2018), thus allowing multiple transcript detection to be required for co-expression analysis. Finally, the stereotypic development of embryo and larva coupled with *in situ* hybridizations allows for image registration (Figure 13.6c), which consists of a virtual atlas of expression patterns for their systematic comparison (Tomer et al. 2010; Asadulina et al. 2012).

TABLE 13.1
***Platynereis* genomic (BAC) and transcriptomic (EST and RNA-seq) databases**

	Stage	Sequencing information		Repository	References
BAC	Sperm of mature Male	Sanger (shotgun)	15 contigs	Genbank: CT030666 - CT030681	Raible et al., 2005
EST	Larvae (48hpf)	Sanger (3730xl)	1,484 expressed sequence tags	Genbank: CT032248 - CT033731	Raible et al., 2005
EST	Larvae and juvenile stages	Sanger + 454 Roche	77,419 expressed sequence tags	Genbank: JZ391525 - JZ468943	Conzelmann et al., 2013
Bulk RNA-seq	Fertilized eggs, larvae (24, 36, 48, 72hpf and 4dpf), juveniles (10, 15dpf, 1, 3mpf) and adults (males and females)	Illumina (HiSeq 2000)	351,625 reads, 87,686 contigs (>500bp), 28,067 (>1000bp), 51,767 ORFs (>120aa)	Supp. Data	Conzelmann et al., 2013
Bulk RNA-seq	Embryonic development (2, 4, 6, 8, 10, 12hpf) and larvae (14hpf)	Illumina (HiSeq)	273,087 contigs, 51,260 ORFs (>100aa)	https://github.com/hsienchao/pdu_sqs/find/master	Chou et al., 2016
Bulk RNA-seq	Head samples under various circalunar conditions and maturation stages	Illumina (HiSeq 2000)	52,059 contigs (>500bp)	ENA repository: PRJEB27496	Schenk et al., 2019
scRNA-seq	47hpf-larva epispheres	Fluidigm C1 Single-Cell Auto Prep System / Illumina (HiSeq 2000)		ArrayExpress: E-MTAB-2865	Achim et al., 2015
scRNA-seq	48hpf-larvae	Fluidigm C1 Single-Cell Auto Prep System / Illumina (HiSeq 2000)		ArrayExpress: E-MTAB-2865 and E-MTAB-5953	Achim et al., 2018

BAC = bacterial artificial chromosome sequencing; EST = expressed sequences tags

13.7.1.2 Detection of Proteins: Immunohistochemistry and Western Blot

The *in vivo* detection of proteins has been developed as well. The first detection of proteins in *Platynereis* dates back to the early 90s with the visualization of the nervous system, ciliated cells and the entire epidermis during early development using various antibodies (Abs) raised against *Nereis diversicolor* (Annelida) proteins, *Drosophila* Engrailed and Antennapedia, respectively (Dorresteijn et al. 1993). Since then, antibodies such as those against acetylated-Tubulin Abs are now routinely used to depict the *Platynereis* nervous system (Figure 13.6d). In contrast to WMISH that can be performed on virtually all genes, immunohistochemistry (IHC) suffers from the lack of appropriate Abs developed against *Platynereis* proteins (or proteins from closely related species). Accordingly, WMISH remains the preferred approach used as a proxy of protein location, while IHC is often restricted to highly conserved proteins (e.g. proteins from the cytoskeleton and histones). Nevertheless, IHC against other proteins such as MIP peptides, β-catenin or neuropeptides have also proven successful (Schneider and Bowerman 2007; Conzelmann et al. 2011; Williams et al. 2015; Gazave et al. 2017). Western blots (WBs) have been also developed from whole cell extract (Schneider and Bowerman 2007) and nuclear extracts (Figure 13.6e, unpublished data Vervoort's Lab), thus allowing for the quantification of specific proteins in different tissues or upon various conditions.

13.7.1.3 Tracking Cell, Cell Components and Monitoring Key Cellular Processes

Staining approaches: Various staining using commercially available dyes were used to study, for instance, muscles (phalloidin) or chaetae (wheat germ agglutinin) or to stain cell membranes (mCLING–ATTO 647N, FM-464) either on fixed or live animals, depending on the dye used (Lauri et al. 2014; Williams et al. 2015; Gazave et al. 2017; Chartier et al. 2018). Staining to monitor key cellular processes has also been developed in *Platynereis*. For instance, EdU (5-ethynyl-2'deoxyuridine, Figure 13.6f and f') and BrdU (Bromo-desoxyuridine) incorporations followed by chasing are used to highlight proliferative cells and their progenies, a key approach to characterize putative stem cells / progenitors and their lineage during early development (Rebscher et al. 2012; Demilly et al. 2013), posterior elongation (Gazave et al. 2013) and regeneration (Planques et al. 2019). Cell death can be assessed as well, using real-time apoptosis detection (TUNEL) (Demilly et al. 2013; Lauri et al. 2014; Zidek et al. 2018).

Microinjection of dyes: As reported in the "Embryogenesis and Larval Development" section, Ackermann and colleagues injected *Platynereis* embryos at the two-, four- and eight-cell stages with fluorescent dyes (e.g. FITC-dextrane) to trace blastomere lineages and their respective contribution to tissue in young worms (Fischer and Dorresteijn 2004; Ackermann et al. 2005).

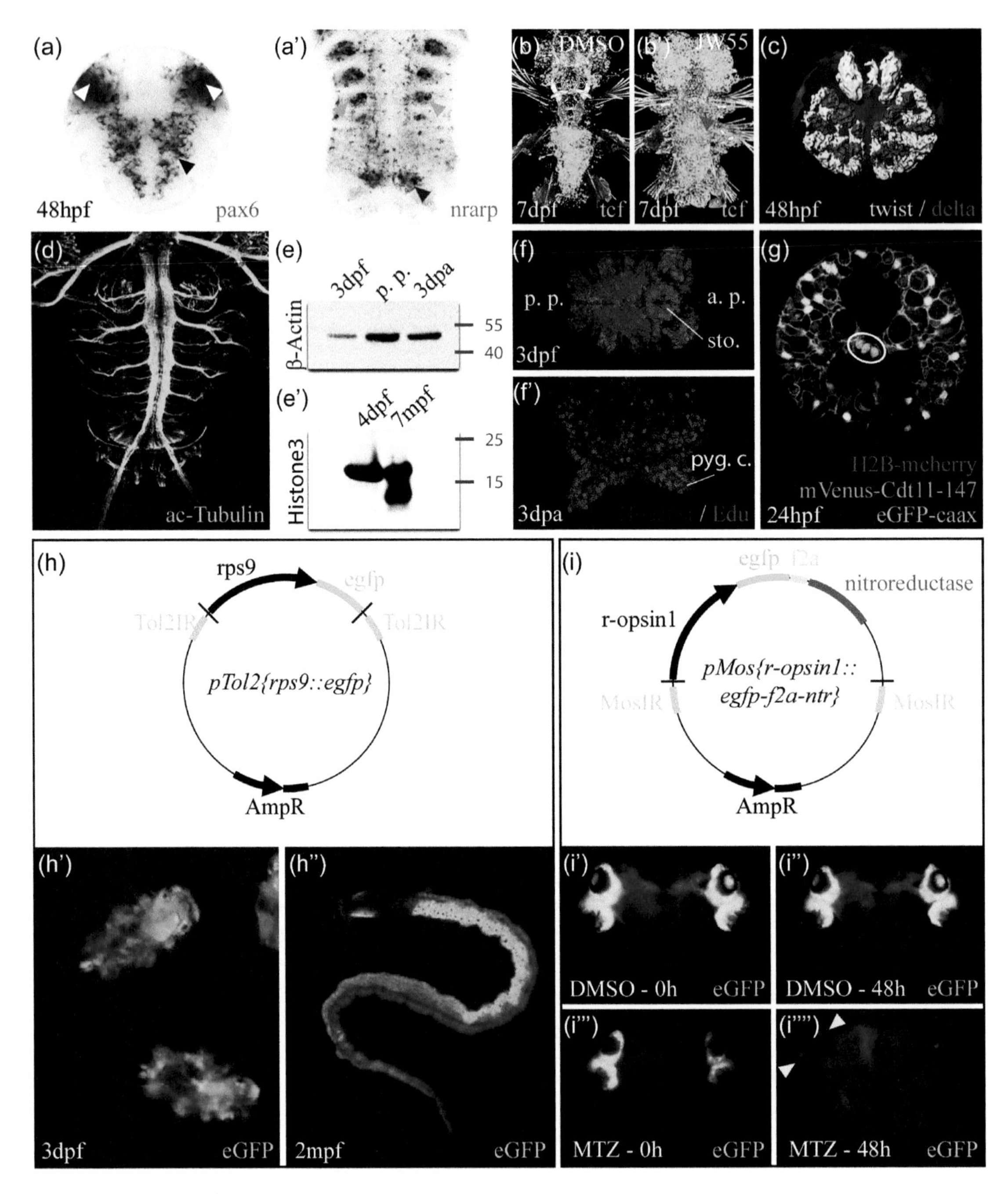

FIGURE 13.6 Molecular and cellular tools for functional approaches in *Platynereis*. (a–a') Whole-mount *in situ* hybridizations using NBT/BCIP colorimetric staining showing (a) the expression pattern of *pax6* in brain hemispheres (white arrowheads) and ventral neurectoderm (black arrowhead) and (a') *nrarp* expression in chaetal sacs (blue arrowheads) and cells of growth zone (purple arrowhead) during posterior elongation. (b–b') Fluorescent *in situ* hybridization showing the effect of Wnt/β-catenin pathway inhibition on *tcf* expression. Upon JW55 treatment (Axin2 stabilization), *tcf* expression is extended to other tissues (red arrowhead) in addition to its regular expression in brain ganglia and midgut. (c) Image registration showing *Platynereis twist* and *delta* expressions in mesoderm and chaetal sacs, respectively. Ventral view. (d) Acetylated-Tubulin immunohistochemistry revealing the ventral nervous system in posterior part. (e–e') Western blot of *Platynereis* using (e) whole cell extracts (β-Actin) and (e') nuclear extracts (Histone 3). (f–f') EdU staining to investigate proliferative cells (f) in larva and (f') posterior part regeneration. (g) Co-injection of *H2B-mcherry* (nuclear marker), *mvenus-cdt11–147* (cell cycle biosensor) and *egfp-caax* (membrane marker) mRNAs in fertilized embryo used to follow cell cycle progression during embryo development. The green staining of putative primordial germ cells (white circle) suggests that they no longer divide (Ozpolat et al. 2017). (h–h'') Tol2 transposase system for transient transgenesis using the promoter of the ribosomal protein Rps9 (*rps9*) to ubiquitously express *enhanced green fluorescent protein* (*egfp*) (h') in larvae and (h'') young worms. (i–i'''') Mos transposase system for heritable transgenesis using the promoter of *r-opsin1* to co-express *egfp* and bacterial *nitroreductase* in (i') adult eyes. (i'–i'''') Upon metronidazole 48h-treatment (MTZ), Nitroreductase converts MTZ into a toxic compound leading to the death of positive cells (yellow arrowheads). Abv: a. p. = anterior part, dpa = days post-amputation, hpf/dpf/mpf = hours/days/month post-fertilization, p. p. = posterior part, pyg. c. = pygidial cirrus, sto = stomodeum. ([b–b'] Zidek et al. 2018; [h''] Backfisch et al. 2014; [i'–i''''] Backfisch et al. 2014.)

Microinjection of mRNA: mRNA were also successfully injected into fertilized embryos to induce the expression of different fluorescent proteins such as the photoactivable mCherry (PAmCherryl) and the photoconvertible Kikume Green-Red (KIKGR) protein for cell tracking (Lauri et al. 2014; Veraszto et al. 2017), biomarkers to mark nucleus (*H2A-mCherry; H2B-eGFP*) and cell membranes (*egfp-caax; mYfp*) (Lauri et al. 2014; Ozpolat et al. 2017; Kuehn et al. 2019) or biosensors to monitor cell cycle progression (mVenus-cdt1$^{aa1-147}$) (Ozpolat et al. 2017) (Figure 13.6g) and neuronal activity (GCaMP6, calcium imaging) (Veraszto et al. 2017; Chartier et al. 2018) in live animals.

Transposon-based transgenesis of reporter cassettes: While mRNA represent an incredible useful technique for biomarker and biosensor expression during early development, transgenic animals allow for a tight control of gene expression. In *Platynereis*, two transposon-mediated systems (i.e. Tol2 and Mos-based constructs) were efficiently developed for both transient and stable transgenesis (Backfisch et al. 2014). To implement this approach, the promoter of the ribosomal protein Rps9 has been used to drive ubiquitous expression of the *enhanced green fluorescent protein* (*egfp*) (Figure 13.6h to h''). Interestingly, comparison of Tol2- and Mos-based systems using similar constructs [i.e. *pTol2(rps9::egfp)* and *pMos(rps9::egfp)*] revealed that whereas embryos injected with Tol2-based plasmids tend to show a higher frequency of genome integration than those injected with Mos-based plasmids, transgenes are heritable to progeny only through Mos-mediated transgenesis (Backfisch et al. 2014). Additional promoters to *rps9* have been developed to target specific cell populations such as the *r-opsin1* promoter for adult eye cells and their neuronal projections (Backfisch et al. 2013; Veedin-Rajan et al. 2013) (Figure 13.6i to i'), a specific *alpha-tubulin* promoter (*tuba*) for cells with motile cilia in larvae, a *maf* promoter for a subtype of nerve cells in the larval brain (Backfisch et al. 2014) and a *guanylyl cyclase-β* promoter for the cholinergic motorneurons (Veraszto et al. 2017).

Serial section transmission electron microscopy (ssTEM): By imaging and assembling numerous serial sections (around 1,700 sections for a head and the first chaetigerous segment or 5,000 sections for a full individual) and manually tracing all neurons, researchers were able to reconstruct a comprehensive three-dimensional cell atlas of the visual neuronal circuit in 72 hpf larvae, including 106 neurons (i.e. photoreceptor cells, interneurons and motoneurons) and their synaptic connectivity (Randel et al. 2014; Randel et al. 2015). This sophisticated approach has been more recently extended to other circuits such as the neurosecretory connectome (Williams et al. 2017).

13.7.2 Functional Approaches

13.7.2.1 Gene Knock-Down: Translation-Blocking Morpholinos

Although they are used infrequently, morpholinos (MOs) represent an interesting knock-down approach to assess gene functions during early development. In a study aiming to show the implication of myoinhibitory (MIP) peptides on larval settlement, two MIP-receptor MOs were successfully used. Indeed, in embryos injected with MOs, MIP treatment-induced settlement was no longer observed (Conzelmann et al. 2013).

13.7.2.2 Protein Inhibition/Activation: Pharmacological and Peptide Treatments

Although concerns regarding putative off-target effects have been raised with the pharmacological approaches, often addressed by the use of different molecules in parallel, the treatment using inhibitors is an easy approach to assess the function of specific proteins in live animals, especially in water-dwelling animals such as *Platynereis*. In addition, this approach allows researchers to interfere with proteins at specific timepoints and during processes that cannot be reached using MOs (e.g. post-larval and regeneration processes). Accordingly, a broad range of studies has developed this approach, for instance, to investigate the function of key signaling pathways such as Wnt/β-catenin (Schneider and Bowerman 2007; Steinmetz et al. 2007; Demilly et al. 2013; Marlow et al. 2014; Zidek et al. 2018) (Figure 13.6b and b'), Planar cell polarity (Steinmetz et al. 2007), Notch (Gazave et al. 2017) or Hedgehog (Dray et al. 2010) or to assess the role of key cellular processes such as cell proliferation (Planques et al. 2019). Similarly, successful results were obtained by incubating *Platynereis* larvae with zebrafish BMP4 peptides (Denes et al. 2007), *Platynereis* synthetic neuropeptide (Conzelmann et al. 2011) or *Platynereis* synthetic MIB peptides (Conzelmann et al. 2013; Williams et al. 2015).

13.7.2.3 Genome Editing

Transgenesis: Transgenesis in *Platynereis* has so far mainly been used to monitor gene expression and to study specific cell populations (see previously). However, this technique now opens a broad range of subsequent functional approaches, including conditional knock-down and ectopic expression. In *Platynereis*, transgenesis has been used for effective targeted cell ablation. Indeed, the use of *r-opsin1* promoter allowed the expression of the bacterial nitroreductase enzyme (Ntr) in *Platynereis* adult eyes (Veedin-Rajan et al. 2013). This enzyme converts metronidazole (MTZ) into a toxic product that induces the death of the corresponding cells (Figure 13.6i to i''''). Thus, transgenic animals expressing nitroreductase represent a great alternative to laser ablation to specifically remove a subset of cells.

Transcriptional activator-like nuclease (TALEN): In *Platynereis*, TALEN has been established as an efficient tool to induce heritable mutagenesis (Bannister et al. 2014), and this approach has been recently used to highlight the involvement of gonadotropin-releasing hormone (GnRH, known to integrate environmental stimuli for vertebrate sexual maturation and breeding) in the regulation of growth and sexual maturation by lunar phases. Indeed, maturation, growth and regeneration were reduced in animals where mutations leading to *corazonin1/gnrhl1* knock-outs were performed (Andreatta et al. 2020).

CRISPR/Cas9: CRISPR/Cas9 also has recently been used in *Platynereis*. In planktonic larvae, the startle response is mediated by collar receptor neurons expressing *polycystin* genes (*PKD1-1* and *PKD2-1*). Interestingly, this freezing response is abolished in both *PKD1-1* and *PKD2-1* mutants (Bezares-Calderon et al. 2018).

13.8 CHALLENGING QUESTIONS

Platynereis has been successfully developed as a powerful marine model thanks to the development of many tools (see Section 13.7), allowing researchers to address a variety of biological questions, mostly related to evolutionary developmental biology (Ferrier 2012). Several of these questions have already been raised earlier in this chapter and have been the subject of detailed recent reviews, notably (i) biological rhythms and clocks (Tessmar-Raible et al. 2011; Raible and Falciatore 2014; Raible et al. 2017; Andreatta and Tessmar-Raible 2020), (ii) neuronal connectomics and plankton behavior (Jekely et al. 2018; Williams and Jekely 2019; Bezares-Calderon et al. 2020; Marinkovic et al. 2020) and (iii) cell type evolution (Brunet et al. 2015; Arendt, Musser et al. 2016; Arendt, Tosches et al. 2016; Arendt 2018; Nielsen et al. 2018; Arendt et al. 2019). Here, we choose to introduce two additional lines of research that are currently (re)emerging: the regeneration processes and epigenetic modifications during embryonic and post-embryonic development.

13.8.1 Regeneration

Animal regeneration is defined as the ability to restore a lost or damaged body part (Poss 2010). This fascinating process has intrigued scientists for centuries, and we recently observed a strong re-emergence of the regeneration field thanks to the availability of new tools for less conventional models (Gazave and Rottinger 2021). Injury-induced regeneration is a widespread phenomenon harbored by species of all the major lineages of Metazoa. In addition, the extent of what can be regenerated after an injury greatly varies among animals (Grillo et al. 2016; Bideau et al. 2021). The origin and evolution of animal regeneration is a long-standing debate, and the questions of why and how regeneration abilities evolved are still poorly understood (Bely 2010). Annelids show amazing regenerative capabilities, as most species are able to regenerate the posterior part of their body and their parapodia following an amputation, as well as, for some species, their anterior part (including the head) (Ozpolat and Bely 2016). Experimental and descriptive morphological studies of annelid regeneration have provided important knowledge (Boilly 1969a, 1969b) (for recent reviews, see Kostyuchenko and Kozin 2020; Nikanorova et al. 2020). Nowadays, some cellular and molecular aspects of these processes have been addressed in a limited number of models (Myohara 2012; Sugio et al. 2012; de Jong and Seaver 2018; Ribeiro et al. 2019), notably *Platynereis* (Planques et al. 2019).

Platynereis is able to regenerate its posterior part as well as various body outgrowths, such as tentacles and parapodia, but not its head. Its posterior regeneration was recently carefully described at the morphological, cellular and molecular levels (Planques et al. 2019). After amputation of the posterior part of their body (segments, growth zone and pygidium), *Platynereis* worms rapidly regenerate both the posterior-most part of the body, the pygidium and the stem cell-rich growth zone, the latter then producing new segments through posterior elongation (Gazave et al. 2013). Interestingly, both complex differentiated structures and stem cell populations are regenerated during this event (Gazave et al. 2013). In precise conditions of worm age/size and a specific amputation procedure, *Platynereis* posterior regeneration follows five well-defined stages, which correspond to particular timepoints after amputation. Briefly, (i) wound healing is achieved one day post-amputation (1 dpa); (ii) a proliferating blastema appears around 2 dpa; (iii) at 3 dpa, this blastema shows a conspicuous antero-posterior and dorso-ventral organization; (iv) a well-differentiated pygidium is formed at 4 dpa; and (v) from 5 dpa, new morphologically visible segments are produced by the growth zone (Planques et al. 2019). While several parameters such as the size of the worms, the position of amputation, and the realization of serial amputations affect the timing of the process, posterior regeneration is always successful (except when the amputation is performed close to the pharynx and in sexually mature animals). Further characterization of posterior regeneration using various labelings and *in situ* hybridizations for tissue patterning genes indicates that regeneration is a rapid process: important cell and tissue differentiation starts at 3 dpa, and at this stage, the growth zone is already re-established and starts to produce segments. Thanks to EdU incorporations, cell cycle marker labelings and the use of an inhibitor of cell divisions, it has been also shown that cell proliferation is strictly required for regeneration (Planques et al. 2019). These findings pave the way for a better understanding of *Platynereis* posterior regeneration, while many pressing questions remain unanswered.

An important question in the regeneration field concerns the initiation and control of regeneration (Ricci and Srivastava 2018). Recent studies have suggested that cell death could be a crucial event by triggering cell proliferation (Perez-Garijo and Steller 2015). Cell death seems to be itself stimulated by the production of reactive oxygen species (ROS), essential for regeneration in several models (*Hydra*, *Drosophila* and so on) through the activation of various signaling pathways (Vriz et al. 2014). Whether the cascade ROS $\rightarrow$ apoptosis $\rightarrow$ proliferation may represent a general principle of regeneration is, however, not known. In annelids, this question has not been addressed yet, but preliminary data for *Platynereis* strongly suggest the occurrence of cell death at 1 and 2 dpa, concomitantly with a peak of cell proliferation (unpublished data).

Thanks to recently developed tools for molecular and cellular analyses in *Platynereis*, it is now possible to characterize the *in vivo* distribution of apoptotic cells and to detect the ROS production cells using fluorescent dyes or genetically encoded biosensors (Vullien et al. 2021). This in-depth description of the processes at play, combined with functional tools and transcriptomic analysis, will certainly in the near future uncover the initiation and control mechanisms of *Platynereis* posterior regeneration.

Another key question is to determine the origin and fate of blastema cells, which give rise to the regenerated structures (Morgan 1901; Tanaka 2016). They can derive from pre-existing stem cells present in the body before the amputation and/or being produced by dedifferentiation of cells at the amputation site (Tanaka and Reddien 2011). These cells could be pluripotent stem cells and/or more tissue-restricted progenitor cells with limited potency. In annelids, the source of cells involved in posterior and anterior regeneration has been partially addressed in a couple of species, such as *Nereis diversicolor* (Boilly 1969c), *Enchytraeus japonensis* (Myohara 2012; Sugio et al. 2012) or *Capitella teleta* (de Jong and Seaver 2018). During *Platynereis* posterior regeneration, EdU pulse and chase experiments strongly support the idea that blastema cells mostly derive from dedifferentiation of cells coming from the segment abutting the amputation plane, with the notable exception of the gut, which probably regenerates from pre-existing gut stem cells (Planques et al. 2019). In addition, blastema cells from very early stages express a collection of genes belonging to the GMP signature (Juliano et al. 2010), whose orthologs in other species are expressed in pluripotent/multipotent somatic stem cells and primordial germ cells. This suggests that blastema may contain multi- or pluripotent progenitors/stem cells, even if this needs to be confirmed. To better assess the origin and fate of blastema cells, it would be highly valuable to perform blastema cell lineage tracing experiments. This would allow us to clearly define the respective contribution of resident stem cells and local dedifferentiation events to blastema formation in *Platynereis* as well as the fate of blastema cells.

13.8.2 Epigenetic Modifications during Embryonic/Larval Development and Regeneration

Development and regeneration are highly dynamic processes both requiring important changes in gene expression to handle the establishment of various cell populations (Gerber et al. 2018; Cao et al. 2019; Pijuan-Sala et al. 2019; Shao et al. 2020). This cell fate trajectory, allowing cells to progressively acquire their molecular and functional identities, implies dynamic modulations of epigenetic marks. Nowadays, in developmental biology and cell biology, epigenetics includes any alteration of gene expression that is not associated with changes in the DNA sequence but is due to other molecular mechanisms such as changes in the chromatin structure, histone post-translational modifications and non-coding RNAs (Nicoglou and Merlin 2017). By revealing how each locus is activated or downregulated, epigenetics represents a tremendous step forward by allowing comprehensive overviews of biological processes. Among epigenetic marks, DNA methylation (5-methyl-cytosine, 5mC) appears to be the most extensively studied one (Greenberg and Bourc'his 2019). Basically, two different DNA methylation patterns exist, both occurring at CpG sites (CG motif in the DNA sequence) (Zemach et al. 2010). On the one hand, high levels of methylation at CpG islands (DNA regions where CpG sites are abundant) of promoter regions tends to be associated with low gene expression, while low methylation corresponds to active genes. Although this regulatory-promoter methylation is well identified in vertebrates, only few cases have been reported in non-vertebrates so far (de Mendoza et al. 2019). On the other hand, gene body methylation (GBM, i.e. methylation on coding regions, exons and introns) is found in vertebrates, non-vertebrate animals and other multicellular organisms (Suzuki and Bird 2008; Zemach et al. 2010). However, the function of this type of methylation remains largely unknown. Beyond DNA methylation, epigenetics also strongly relies on Histone mark modifications (e.g. acetylation, methylation, phosphorylation, ubiquitination). For instance, the study of Histone methylation and acetylation in vertebrates allowed researchers to describe specific marks of active and inactive genes (Karlic et al. 2010; Dai and Wang 2014). Among them, Histone 3 (H3) tri-methylation (me3) at lysine 4 (H3K4me3), H3K36me and H3 acetylation at K27 (H3K27ac) coincide with gene activation during embryonic development in sponges, cnidarians, planarians and vertebrates, while H3K9me3 and H3K27me3 represent repressive marks (Karlic et al. 2010; Schwaiger et al. 2014; Cunliffe 2016; Gaiti et al. 2017; Dattani et al. 2018). Accordingly, epigenetics represents one of the most active domains in biology, especially in the context of biological phenomena such as cell differentiation and development. However, epigenetics is often restricted to vertebrates and a few non-vertebrate organisms (e.g. cnidarians and poriferans), while no data have been acquired for other lineages such as annelids, thus calling for comparative studies. In *Platynereis*, gene coding for orthologous proteins of all main actors of 5mC DNA methylation/demethylation machinery were found (Planques et al. 2021). In addition, computational analyses (CpG observed/expected) and assays with methylation-sensitive restriction enzymes revealed a high level of DNA methylation during embryonic and larval development. Interestingly, treatment with a hypomethylating agent (Decitabine/5-aza-2'deoxycytidine) during larval development impairs parapodia, chaetae and pygidium formation and eventually leads to the death of juvenile worms, suggesting a fundamental role of DNA methylation during larval development. Similarly, Decitabine greatly delays worm regeneration and sometimes leads to abnormal posterior elongation (i.e. no or reduced number of new segments, abnormal parapodia and cirri) after drug removal. This suggests that the regenerated growth zone is affected by Decitabine-mediated hypomethylation, leading to persistent defects of its function thereafter. Now, additional data are required to assess the precise methylation patterns in *Platynereis* (e.g. genome-wide bisulfite sequencing) and the link between modulations in methylation patterns and changes in gene expression. Furthermore, extending research to other epigenetic mechanisms such as the role of post-translational Histone marks and non-coding RNA would bring additional clues to questions on the tight mechanisms controlling cell fate trajectories during dynamic processes, especially in non-vertebrate animals and during regeneration, for which studies remain highly scarce.

ACKNOWLEDGMENTS

Work in our team is supported by funding from the Labex "Who Am I" laboratory of excellence (No. ANR-11-LABX-0071) funded by the French government through its "Investments for the Future" program operated by the Agence Nationale de la Recherche under grant No. ANR-11-IDEX-0005–01, the Centre National de la Recherche Scientifique, the INSB department (grant "Diversity of Biological Mechanisms"), the Agence Nationale de la Recherche (grant TELOBLAST no. ANR-16-CE91–0007 and grant STEM no. ANR-19-CE27-0027-02), the "Association pour la Recherche sur le Cancer" (grant PJA 20191209482) and the "Ligue Nationale Contre le Cancer" (grant RS20/75–20). QS is a fellow of the labex "Who Am I" and the "Paris Region Fellowship Programme" 2021. We thank Dr. Nicolas Dray and Loïc Bideau for providing pictures. We are grateful to Haley Flom for critical reading of the manuscript. The authors warmly thank all current and past members of the 'Stem cells, Development and Evolution' team at the Institut Jacques Monod, Paris, France, especially Prof. Michel Vervoort for his valuable comments on this chapter.

BIBLIOGRAPHY

Achim, K., Eling, N., Vergara, H. M., Bertucci, P. Y., Musser, J., Vopalensky, P., Brunet, T., Collier, P., Benes, V., Marioni, J. C., and Arendt, D. 2018. Whole-body single-cell sequencing reveals transcriptional domains in the annelid larval body. *Mol Biol Evol* 35 (5):1047–1062.

Achim, K., Pettit, J. B., Saraiva, L. R., Gavriouchkina, D., Larsson, T., Arendt, D., and Marioni, J. C. 2015. High-throughput spatial mapping of single-cell RNA-seq data to tissue of origin. *Nature Biotechnology* 33 (5):503-U215.

Ackermann, C., Dorresteijn, A., and Fischer, A. 2005. Clonal domains in postlarval *Platynereis dumerilii* (Annelida: Polychaeta). *J Morphol* 266 (3):258–280.

Andrade, S. C., Novo, M., Kawauchi, G. Y., Worsaae, K., Pleijel, F., Giribet, G., and Rouse, G. W. 2015. Articulating "archiannelids": Phylogenomics and annelid relationships, with emphasis on meiofaunal taxa. *Mol Biol Evol* 32 (11):2860–2875.

Andreatta, G., Broyart, C., Borghgraef, C., Vadiwala, K., Kozin, V., Polo, A., Bileck, A., Beets, I., Schoofs, L., Gerner, C., and Raible, F. 2020. Corazonin signaling integrates energy homeostasis and lunar phase to regulate aspects of growth and sexual maturation in Platynereis. *Proc Natl Acad Sci U S A* 117 (2):1097–1106.

Andreatta, G., and Tessmar-Raible, K. 2020. The still dark side of the moon: Molecular mechanisms of lunar-controlled rhythms and clocks. *J Mol Biol* 432 (12):3525–3546.

Arboleda, E., Zurl, M., Waldherr, M., and Tessmar-Raible, K. 2019. Differential impacts of the head on *Platynereis dumerilii* peripheral circadian rhythms. *Front Physiol* 10:900.

Arendt, D. 2018. Animal evolution: Convergent nerve cords? *Curr Biol* 28 (5):R225–R227.

Arendt, D., Bertucci, P. Y., Achim, K., and Musser, J. M. 2019. Evolution of neuronal types and families. *Curr Opin Neurobiol* 56:144–152.

Arendt, D., Musser, J. M., Baker, C. V. H., Bergman, A., Cepko, C., Erwin, D. H., Pavlicev, M., Schlosser, G., Widder, S., Laubichler, M. D., and Wagner, G. P. 2016. The origin and evolution of cell types. *Nat Rev Genet* 17 (12):744–757.

Arendt, D., Technau, U., and Wittbrodt, J. 2001. Evolution of the bilaterian larval foregut. *Nature* 409 (6816):81–85.

Arendt, D., Tessmar, K., de Campos-Baptista, M. I., Dorresteijn, A., and Wittbrodt, J. 2002. Development of pigment-cup eyes in the polychaete *Platynereis dumerilii* and evolutionary conservation of larval eyes in Bilateria. *Development* 129 (5):1143–1154.

Arendt, D., Tosches, M. A., and Marlow, H. 2016. From nerve net to nerve ring, nerve cord and brain-evolution of the nervous system. *Nat Rev Neurosci* 17 (1):61–72.

Asadulina, A., Panzera, A., Veraszto, C., Liebig, C., and Jekely, G. 2012. Whole-body gene expression pattern registration in Platynereis larvae. *Evodevo* 3 (1):27.

Audouin, J. V., and Milne Edwards, H. 1833. Classification des annélides, et description de celles qui habitent les côtes de la France. In *Annales des Sciences Naturelles*, 195–269.

Audouin, J. V., and Milne Edwards, H. 1834. *Recherches pour servir à l'histoire naturelle du littoral de la France* Vol. 2, *Recueil de mémoires sur l'anatomie, la physiologie, la classification et les mœurs des animaux de nos côtes*. Paris.

Ayers, T., Tsukamoto, H., Guhmann, M., Veedin Rajan, V. B., and Tessmar-Raible, K. 2018. A Go-type opsin mediates the shadow reflex in the annelid *Platynereis dumerilii. BMC Biol* 16 (1):41.

Backfisch, B., Kozin, V. V., Kirchmaier, S., Tessmar-Raible, K., and Raible, F. 2014. Tools for gene-regulatory analyses in the marine annelid *Platynereis dumerilii. PLoS One* 9 (4):e93076.

Backfisch, B., Veedin Rajan, V. B., Fischer, R. M., Lohs, C., Arboleda, E., Tessmar-Raible, K., and Raible, F. 2013. Stable transgenesis in the marine annelid *Platynereis dumerilii* sheds new light on photoreceptor evolution. *Proc Natl Acad Sci U S A* 110 (1):193–198.

Bannister, S., Antonova, O., Polo, A., Lohs, C., Hallay, N., Valinciute, A., Raible, F., and Tessmar-Raible, K. 2014. TALENs mediate efficient and heritable mutation of endogenous genes in the marine annelid *Platynereis dumerilii. Genetics* 197 (1):77–89.

Bely, A. E. 2010. Evolutionary loss of animal regeneration: Pattern and process. *Integr Comp Biol* 50 (4):515–527.

Bentley, M. G., Olive, P. J. W., and Last, K. 1999. Sexual satellites, moonlight and the nuptial dances of worms: The influence of the moon on the reproduction of marine animals. *Earth, Moon, and Planets* 85/86 (0):67–84.

Bezares-Calderon, L. A., Berger, J., Jasek, S., Veraszto, C., Mendes, S., Guhmann, M., Almeda, R., Shahidi, R., and Jekely, G. 2018. Neural circuitry of a polycystin-mediated hydrodynamic startle response for predator avoidance. *Elife* 7.

Bezares-Calderon, L. A., Berger, J., and Jekely, G. 2020. Diversity of cilia-based mechanosensory systems and their functions in marine animal behaviour. *Philos Trans R Soc Lond B Biol Sci* 375 (1792):20190376.

Bhambri, A., Dhaunta, N., Patel, S. S., Hardikar, M., Bhatt, A., Srikakulam, N., Shridhar, S., Vellarikkal, S., Pandey, R., Jayarajan, R., Verma, A., Kumar, V., Gautam, P., Khanna, Y., Khan, J. A., Fromm, B., Peterson, K. J., Scaria, V., Sivasubbu,

S., and Pillai, B. 2018. Large scale changes in the transcriptome of *Eisenia fetida* during regeneration. *PLoS One* 13 (9):e0204234.

Bideau, L., Kerner, P., Hui, J., Vervoort, M., and Gazave, E. 2021. Animal regeneration in the era of transcriptomics. *Cell Mol Life Sci* 78:3941–3956.

Boilly, B. 1969a. Sur l'origine des cellules régénératrices chez les annélides polychètes. *Archives de Zoologie expérimentale et générale* 110 (1):127–143.

Boilly, B. 1969b. Experimental study of the localization, by relation to the amputation plan, of the source of mesodermal regeneration cells in an annelid polychaete (*Syllis amica* Quatrefages). *J Embryol Exp Morphol* 21 (1):193–206.

Boilly, B. 1969c. Origine des cellules régénératrices chez Nereis diversicolor O. F. Müller (Annélide Polychète). *Wilhelm Roux'Archiv für Entwicklungsmechanik der Organismen* 162 (3):286–305.

Boilly-Marer, Y. 1973. Etude expérimentale du comportement nuptial de *Platynereis dumerilii* (Annélide Polychète): chémoreception, émission des produits génitaux. *Marine Biology* 24:167–179.

Brunet, T., Fischer, A. H., Steinmetz, P. R., Lauri, A., Bertucci, P., and Arendt, D. 2016. The evolutionary origin of bilaterian smooth and striated myocytes. *Elife* 5.

Brunet, T., Lauri, A., and Arendt, D. 2015. Did the notochord evolve from an ancient axial muscle? The axochord hypothesis. *Bioessays* 37 (8):836–850.

Brusca, R. C., and Brusca, G. J., eds. 2003. *Invertebrates*, second edition. Sunderland: Sinauer Associates, Inc.

Cao, J., Spielmann, M., Qiu, X., Huang, X., Ibrahim, D. M., Hill, A. J., Zhang, F., Mundlos, S., Christiansen, L., Steemers, F. J., Trapnell, C., and Shendure, J. 2019. The single-cell transcriptional landscape of mammalian organogenesis. *Nature* 566 (7745):496–502.

Caspers, H. 1971. C. Hauenschild und A. Fischer: *Platynereis dumerilii*. Mikroskopische Anatomie, Fortpflanzung, Entwicklung.—Großes Zoologisches Praktikum Heft 10b. Mit 37 Abb., Stuttgart: Gustav Fischer Verlag 1969. 55 S. DM 26,—. *Internationale Revue der gesamten Hydrobiologie und Hydrographie* 56 (2):326–326.

Caspers, H. 1984. Spawning periodicity and habitat of the palolo worm eunice-viridis (Polychaeta, eunicidae) in the Samoan Islands. *Marine Biology* 79 (3):229–236.

Chartier, T. F., Deschamps, J., Durichen, W., Jekely, G., and Arendt, D. 2018. Whole-head recording of chemosensory activity in the marine annelid *Platynereis dumerilii*. *Open Biol* 8 (10).

Chen, H., Parry, L. A., Vinther, J., Zhai, D., Hou, X., and Ma, X. 2020. A Cambrian crown annelid reconciles phylogenomics and the fossil record. *Nature* 583 (7815):249–252.

Choi, H. M. T., Schwarzkopf, M., Fornace, M. E., Acharya, A., Artavanis, G., Stegmaier, J., Cunha, A., and Pierce, N. A. 2018. Third-generation in situ hybridization chain reaction: Multiplexed, quantitative, sensitive, versatile, robust. *Development* 145 (12).

Chou, H. C., Acevedo-Luna, N., Kuhlman, J. A., and Schneider, S. Q. 2018. PdumBase: A transcriptome database and research tool for *Platynereis dumerilii* and early development of other metazoans. *BMC Genomics* 19 (1):618.

Chou, H. C., Pruitt, M. M., Bastin, B. R., and Schneider, S. Q. 2016. A transcriptional blueprint for a spiral-cleaving embryo. *BMC Genomics* 17:552.

Conzelmann, M., Offenburger, S. L., Asadulina, A., Keller, T., Munch, T. A., and Jekely, G. 2011. Neuropeptides regulate swimming depth of Platynereis larvae. *Proc Natl Acad Sci U S A* 108 (46):E1174–E1183.

Conzelmann, M., Williams, E. A., Tunaru, S., Randel, N., Shahidi, R., Asadulina, A., Berger, J., Offermanns, S., and Jekely, G. 2013. Conserved MIP receptor-ligand pair regulates Platynereis larval settlement. *Proc Natl Acad Sci U S A* 110 (20):8224–8229.

Cunliffe, V. T. 2016. Histone modifications in zebrafish development. *Methods Cell Biol* 135:361–385.

Dai, H., and Wang, Z. 2014. Histone modification patterns and their responses to environment. *Current Environmental Health Reports* 1 (1):11–21.

Dattani, A., Kao, D., Mihaylova, Y., Abnave, P., Hughes, S., Lai, A., Sahu, S., and Aboobaker, A. A. 2018. Epigenetic analyses of planarian stem cells demonstrate conservation of bivalent histone modifications in animal stem cells. *Genome Res* 28 (10):1543–1554.

de Jong, D. M., and Seaver, E. C. 2018. Investigation into the cellular origins of posterior regeneration in the annelid *Capitella teleta*. *Regeneration (Oxf)* 5 (1):61–77.

de Mendoza, A., Hatleberg, W. L., Pang, K., Leininger, S., Bogdanovic, O., Pflueger, J., Buckberry, S., Technau, U., Hejnol, A., Adamska, M., Degnan, B. M., Degnan, S. M., and Lister, R. 2019. Convergent evolution of a vertebrate-like methylome in a marine sponge. *Nat Ecol Evol* 3 (10):1464–1473.

Demilly, A., Steinmetz, P., Gazave, E., Marchand, L., and Vervoort, M. 2013. Involvement of the Wnt/beta-catenin pathway in neurectoderm architecture in *Platynereis dumerilii*. *Nat Commun* 4:1915.

Denes, A. S., Jekely, G., Steinmetz, P. R., Raible, F., Snyman, H., Prud'homme, B., Ferrier, D. E., Balavoine, G., and Arendt, D. 2007. Molecular architecture of annelid nerve cord supports common origin of nervous system centralization in bilateria. *Cell* 129 (2):277–288.

de Quatrefarges, A. 1865. Note sur la classification des Annélides. *Comptes rendus hebdomadaires des séances de l'Académie des sciences* 60:586–600.

Dorresteijn, A. W. 1990. Quantitative analysis of cellular differentiation during early embryogenesis of *Platynereis dumerilii*. *Rouxs Arch Dev Biol* 199 (1):14–30.

Dorresteijn, A. W., Bornewasser, H., and Fischer, A. 1987. A correlative study of experimentally changed first cleavage and Janus development in the trunk of *Platynereis dumerilii* (Annelida, Polychaeta). *Rouxs Arch Dev Biol* 196 (1):51–58.

Dorresteijn, A. W., O'Grady, B., Fischer, A., Porchet-Hennere, E., and Boilly-Marer, Y. 1993. Molecular specification of cell lines in the embryo of Platynereis (Annelida). *Rouxs Arch Dev Biol* 202 (5):260–269.

Dorsett, D. A., and Hyde, R. 1969. The fine structure of the compound sense organs on the cirri of *Nereis diversicolor*. *Z Zellforsch Mikrosk Anat* 97 (4):512–527.

Dray, N., Tessmar-Raible, K., Le Gouar, M., Vibert, L., Christodoulou, F., Schipany, K., Guillou, A., Zantke, J., Snyman, H., Behague, J., Vervoort, M., Arendt, D., and Balavoine, G. 2010. Hedgehog signaling regulates segment formation in the annelid Platynereis. *Science* 329 (5989):339–342.

Fauvel, P. 1923. Polychètes errantes. *Faune de France* 5:1–488.

Fauvel, P. 1927. Polychètes Sédentaires. Addenda aux Errantes, Archiannélides, Myzostomaires. *Faune de France* 16:1–494.

Ferrier, D. E. 2012. Evolutionary crossroads in developmental biology: Annelids. *Development* 139 (15):2643–2653.

Fischer, A. 1974. Stages and stage distribution in early oogenesis in the Annelid, *Platynereis dumerilii*. *Cell Tissue Res* 156 (1):35–45.

Fischer, A. 1999. Reproductive and developmental phenomena in annelids: A source of exemplary research problems. In

Reproductive Strategies and Developmental Patterns in Annelids, edited by W. Westheide and A. W. C. Dorresteijn, 1–20. Dordrecht: Springer Netherlands.

Fischer, A., and Dorresteijn, A. 2004. The polychaete *Platynereis dumerilii* (Annelida): A laboratory animal with spiralian cleavage, lifelong segment proliferation and a mixed benthic/pelagic life cycle. *Bioessays* 26 (3):314–325.

Fischer, A. H., Henrich, T., and Arendt, D. 2010. The normal development of *Platynereis dumerilii* (Nereididae, Annelida). *Front Zool* 7:31.

Fischer, A. H. L., and Arendt, D. 2013. Mesoteloblast-like mesodermal stem cells in the polychaete annelid *Platynereis dumerilii* (Nereididae). *Journal of Experimental Zoology Part B-Molecular and Developmental Evolution* 320b (2):94–104.

Gaiti, F., Jindrich, K., Fernandez-Valverde, S. L., Roper, K. E., Degnan, B. M., and Tanurdzic, M. 2017. Landscape of histone modifications in a sponge reveals the origin of animal cis-regulatory complexity. *Elife* 6.

Gambi, M. C., Zupo, V., Buia, M. C., and Mazzella, L. 2000. Feeding Ecology of *Platynereis dumerilii* (Audouin & Milne-Edwards) in the Seagrass *Posidonia Oceanica* System: The Role of the Epiphytic Flora (Polychaeta, Nereididae). *Ophelia* 12 (53):189–202.

Gazave, E., Behague, J., Laplane, L., Guillou, A., Preau, L., Demilly, A., Balavoine, G., and Vervoort, M. 2013. Posterior elongation in the annelid *Platynereis dumerilii* involves stem cells molecularly related to primordial germ cells. *Dev Biol* 382 (1):246–267.

Gazave, E., Lemaitre, Q. I. B., and Balavoine, G. 2017. The Notch pathway in the annelid Platynereis: Insights into chaetogenesis and neurogenesis processes. *Open Biology* 7 (2).

Gazave, E., and Rottinger, E. 2021. 7th Euro Evo Devo meeting: Report on the "evolution of regeneration in Metazoa" symposium. *J Exp Zool (Mol Dev Evol)* 336:89–93.

Gerber, T., Murawala, P., Knapp, D., Masselink, W., Schuez, M., Hermann, S., Gac-Santel, M., Nowoshilow, S., Kageyama, J., Khattak, S., Currie, J. D., Camp, J. G., Tanaka, E. M., and Treutlein, B. 2018. Single-cell analysis uncovers convergence of cell identities during axolotl limb regeneration. *Science* 362 (6413).

Giangrande, A., Fraschetti, S., and Terlizzi, A. 2002. Local recruitment differences in *Platynereis dumerilii* (Polychaeta, Nereididae) and their consequences for population structure. *Italian Journal of Zoology* 69 (2):133–139.

Greenberg, M. V. C., and Bourc'his, D. 2019. The diverse roles of DNA methylation in mammalian development and disease. *Nat Rev Mol Cell Biol* 20 (10):590–607.

Grillo, M., Konstantinides, N., and Averof, M. 2016. Old questions, new models: Unraveling complex organ regeneration with new experimental approaches. *Curr Opin Genet Dev* 40:23–31.

Grimmel, J., Dorresteijn, A. W., and Frobius, A. C. 2016. Formation of body appendages during caudal regeneration in *Platynereis dumerilii*: Adaptation of conserved molecular toolsets. *Evodevo* 7:10.

Guhmann, M., Jia, H., Randel, N., Veraszto, C., Bezares-Calderon, L. A., Michiels, N. K., Yokoyama, S., and Jekely, G. 2015. Spectral tuning of phototaxis by a go-opsin in the rhabdomeric eyes of Platynereis. *Curr Biol* 25 (17):2265–2271.

Hasse, C., Rebscher, N., Reiher, W., Sobjinski, K., Moerschel, E., Beck, L., Tessmar-Raible, K., Arendt, D., and Hassel, M. 2010. Three consecutive generations of nephridia occur during development of *Platynereis dumerilii* (Annelida, Polychaeta). *Dev Dyn* 239 (7):1967–1976.

Hauenschild, C. 1955. Photoperiodizität als ursache des von der mondphase abhangigen metamorphose-rhythmus bei dem polychaeten *Platynereis dumerilii. Z. Naturforsch. B.* 10:658–662.

Hauenschild, C. 1956. Hormonale Hemmung der Geschlechtsreife und Metamorphose bei dem Polychaeten *Platynereis dumerilii. Zeitschrift für Naturforschung B* 11 (3):125–132.

Hauenschild, C. 1960. Abhängigkeit der Regenerationsleistung von der inneren Sekretion im Prostomium bei *Platynereis dumerilii. Zeitschrift für Naturforschung B* 15 (1):52–55.

Heuer, C. M., Muller, C. H., Todt, C., and Loesel, R. 2010. Comparative neuroanatomy suggests repeated reduction of neuroarchitectural complexity in Annelida. *Front Zool* 7:13.

Hofmann, D. K. 1976. Regeneration and endocrinology in the polychaete *Platynereis dumerilii*: An experimental and structural study. *Wilehm Roux Arch Dev Biol* 180 (1):47–71.

Jacobs, R. P. W. M., and Pierson, E. S. 1979. *Zostera marina* spathes as a habitat for *Platynereis dumerilii* (Audouin and Milne-Edwards, 1834). *Aquatic Botany* 6:403–406.

Jekely, G., Melzer, S., Beets, I., Kadow, I. C. G., Koene, J., Haddad, S., and Holden-Dye, L. 2018. The long and the short of it: A perspective on peptidergic regulation of circuits and behaviour. *J Exp Biol* 221 (Pt 3).

Jha, A. N., Hutchinson, T. H., Mackay, J. M., Elliott, B. M., Pascoe, P. L., and Dixon, D. R. 1995. The chromosomes of *Platynereis dumerilii* (Polychaeta, Nereidae). *Journal of the Marine Biological Association of the United Kingdom* 75 (3):551–562.

Juliano, C. E., Swartz, S. Z., and Wessel, G. M. 2010. A conserved germline multipotency program. *Development* 137 (24):4113–4126.

Just, E. E. 1929. Breeding habits of *Nereis dumerilii* at Naples. *Biol Bull* 57 (5):307–310.

Kara, J., Santos, C. S. G., Macdonald, Angus, H. H., and Simon, C. A. 2020. Resolving the taxonomic identities and genetic structure of two cryptic *Platynereis* Kinberg species from South Africa. *Invertebrate Systematics* 34 (6):618–636, 619.

Karlic, R., Chung, H. R., Lasserre, J., Vlahovicek, K., and Vingron, M. 2010. Histone modification levels are predictive for gene expression. *Proc Natl Acad Sci U S A* 107 (7):2926–2931.

Kenny, N. J., Namigai, E. K., Marletaz, F., Hui, J. H., and Shimeld, S. M. 2015. Draft genome assemblies and predicted microRNA complements of the intertidal lophotrochozoans *Patella vulgata* (Mollusca, Patellogastropoda) and *Spirobranchus* (Pomatoceros) *lamarcki* (Annelida, Serpulida). *Mar Genomics* 24 (Pt 2):139–146.

Knowlton, N. 1993. Sibling species in the sea. *Annual Review of Ecology and Systematics* 24 (1):189–216.

Kostyuchenko, R. P., and Kozin, V. V. 2020. Morphallaxis versus epimorphosis? Cellular and molecular aspects of regeneration and asexual reproduction in annelids. *Biology Bulletin* 47 (3):237–246.

Kuehn, E., Stockinger, A. W., Girard, J., Raible, F., and Ozpolat, B. D. 2019. A scalable culturing system for the marine annelid *Platynereis dumerilii. PLoS One* 14 (12):e0226156.

Lamarck, J. B. 1818. *Histoire naturelle des animaux sans vertèbres*. Vol. 5. Paris: Deterville & Verdiere.

Laumer, C. E., Fernandez, R., Lemer, S., Combosch, D., Kocot, K. M., Riesgo, A., Andrade, S. C. S., Sterrer, W., Sorensen, M. V., and Giribet, G. 2019. Revisiting metazoan phylogeny with genomic sampling of all phyla. *Proc Biol Sci* 286 (1906):20190831.

Lauri, A., Brunet, T., Handberg-Thorsager, M., Fischer, A. H., Simakov, O., Steinmetz, P. R., Tomer, R., Keller, P. J., and Arendt, D. 2014. Development of the annelid axochord: Insights into notochord evolution. *Science* 345 (6202):1365–1368.

Li, Y., Tassia, M. G., Waits, D. S., Bogantes, V. E., David, K. T., and Halanych, K. M. 2019. Genomic adaptations to chemosymbiosis in the deep-sea seep-dwelling tubeworm Lamellibrachia luymesi. *BMC Biol* 17 (1):91.

Marinkovic, M., Berger, J., and Jekely, G. 2020. Neuronal coordination of motile cilia in locomotion and feeding. *Philos Trans R Soc Lond B Biol Sci* 375 (1792):20190165.

Marlow, H., Tosches, M. A., Tomer, R., Steinmetz, P. R., Lauri, A., Larsson, T., and Arendt, D. 2014. Larval body patterning and apical organs are conserved in animal evolution. *BMC Biol* 12:7.

Martin-Duran, J. M., Vellutini, B. C., Marletaz, F., Cetrangolo, V., Cvetesic, N., Thiel, D., Henriet, S., Grau-Bové, X., Carillo-Baltodano, A. M., Gu, W., Kerbl, A., Marquez, Y., Bekkouche, N., Chourrout, D., Gomez-Skarmeta, J. L., Irimia, M., Lenhard, B., Worsaae, K., and Hejnol, A. 2021. Conservative route to genome compaction in a miniature annelid. *Nature Ecology & Evolution* 5:231–242.

M'Intosh, W. C., Ford, G. H., McIntosh, R., and Walker, A. H. 1910. *A Monograph of the British Marine Annelids*. Vol. 2:pt.2 (1910). London: The Ray Society.

Moquin-Tandon, G. 1869. Note sur une nouvelle annelide chetopode hermaphrodite (*Nereis massiliensis*). *Annales des Sciences Naturelles* 9:97–106.

Morgan, T. H. 1901. *Regeneration*. New York: Macmillan.

Myohara, M. 2012. What role do annelid neoblasts play? A comparison of the regeneration patterns in a neoblast-bearing and a neoblast-lacking enchytraeid oligochaete. *PLoS One* 7 (5):e37319.

Nicoglou, A., and Merlin, F. 2017. Epigenetics: A way to bridge the gap between biological fields. *Stud Hist Philos Biol Biomed Sci* 66:73–82.

Nielsen, C., Brunet, T., and Arendt, D. 2018. Evolution of the bilaterian mouth and anus. *Nat Ecol Evol* 2 (9):1358–1376.

Nikanorova, D. D., Kupriashova, E. E., and Kostyuchenko, R. P. 2020. Regeneration in annelids: Cell sources, tissue remodeling, and differential gene expression. *Russian Journal of Developmental Biology* 51 (3):148–161.

Ozpolat, B. D., and Bely, A. E. 2016. Developmental and molecular biology of annelid regeneration: A comparative review of recent studies. *Curr Opin Genet Dev* 40:144–153.

Ozpolat, B. D., Handberg-Thorsager, M., Vervoort, M., and Balavoine, G. 2017. Cell lineage and cell cycling analyses of the 4d micromere using live imaging in the marine annelid *Platynereis dumerilii*. *Elife* 6.

Parry, L. A., Edgecombe, G. D., Eibye-Jacobsen, D., and Vinther, J. 2016. The impact of fossil data on annelid phylogeny inferred from discrete morphological characters. *Proc Biol Sci* 283 (1837).

Perez-Garijo, A., and Steller, H. 2015. Spreading the word: Non-autonomous effects of apoptosis during development, regeneration and disease. *Development* 142 (19): 3253–3262.

Pijuan-Sala, B., Griffiths, J. A., Guibentif, C., Hiscock, T. W., Jawaid, W., Calero-Nieto, F. J., Mulas, C., Ibarra-Soria, X., Tyser, R. C. V., Ho, D. L. L., Reik, W., Srinivas, S., Simons, B. D., Nichols, J., Marioni, J. C., and Gottgens, B. 2019. A single-cell molecular map of mouse gastrulation and early organogenesis. *Nature* 566 (7745):490–495.

Piper, R. 2015. *Animal Earth: The Amazing Diversity of Living Creatures*. London: Thames & Hudson, Ltd.

Planques, A., Kerner, P., Ferry, L., Grunau, C., Gazave, E., and Vervoort, M. 2021. DNA methylation atlas and machinery in the developing and regenerating annelid *Platynereis dumerilii*. *BMC Bio*. 19: 148.

Planques, A., Malem, J., Parapar, J., Vervoort, M., and Gazave, E. 2019. Morphological, cellular and molecular characterization of posterior regeneration in the marine annelid *Platynereis dumerilii*. *Dev Biol* 445 (2):189–210.

Poss, K. D. 2010. Advances in understanding tissue regenerative capacity and mechanisms in animals. *Nat Rev Genet* 11 (10):710–722.

Prud'homme, B., de Rosa, R., Arendt, D., Julien, J. F., Pajaziti, R., Dorresteijn, A. W., Adoutte, A., Wittbrodt, J., and Balavoine, G. 2003. Arthropod-like expression patterns of engrailed and wingless in the annelid *Platynereis dumerilii* suggest a role in segment formation. *Curr Biol* 13 (21):1876–1881.

Purschke, G. 2005. Sense organs in polychaetes (Annelida). *Hydrobiologia* 535:53–78.

Raible, F., and Falciatore, A. 2014. It's about time: Rhythms as a new dimension of molecular marine research. *Mar Genomics* 14:1–2.

Raible, F., Takekata, H., and Tessmar-Raible, K. 2017. An overview of monthly rhythms and clocks. *Front Neurol* 8:189.

Raible, F., Tessmar-Raible, K., Osoegawa, K., Wincker, P., Jubin, C., Balavoine, G., Ferrier, D., Benes, V., de Jong, P., Weissenbach, J., Bork, P., and Arendt, D. 2005. Vertebrate-type intron-rich genes in the marine annelid *Platynereis dumerilii*. *Science* 310 (5752):1325–1326.

Randel, N., Asadulina, A., Bezares-Calderon, L. A., Veraszto, C., Williams, E. A., Conzelmann, M., Shahidi, R., and Jekely, G. 2014. Neuronal connectome of a sensory-motor circuit for visual navigation. *Elife* 3.

Randel, N., Shahidi, R., Veraszto, C., Bezares-Calderon, L. A., Schmidt, S., and Jekely, G. 2015. Inter-individual stereotypy of the Platynereis larval visual connectome. *Elife* 4:e08069.

Read, G., and Fauchald, K. 2018. World Polychaeta database. *Platynereis dumerilii* (Audouin & Milne Edwards, 1833).

Rebscher, N., Lidke, A. K., and Ackermann, C. F. 2012. Hidden in the crowd: Primordial germ cells and somatic stem cells in the mesodermal posterior growth zone of the polychaete *Platynereis dumerilii* are two distinct cell populations. *Evodevo* 3:9.

Ribeiro, R. P., Ponz-Segrelles, G., Bleidorn, C., and Aguado, M. T. 2019. Comparative transcriptomics in Syllidae (Annelida) indicates that posterior regeneration and regular growth are comparable, while anterior regeneration is a distinct process. *BMC Genomics* 20 (1).

Ricci, L., and Srivastava, M. 2018. Wound-induced cell proliferation during animal regeneration. *Wiley Interdiscip Rev Dev Biol*: e321.

Rouse, G. W., and Fauchald, K. 1997. Cladistics and polychaetes. *Zoologica Scripta* 26 (2):139–204.

Rouse, G. W., and Pleijel, F. 2001. *Polychaetes*. New York: Oxford University Press.

Saudemont, A., Dray, N., Hudry, B., Le Gouar, M., Vervoort, M., and Balavoine, G. 2008. Complementary striped expression patterns of NK homeobox genes during segment formation in the annelid Platynereis. *Dev Biol* 317 (2):430–443.

Schenk, S., Bannister, S. C., Sedlazeck, F. J., Anrather, D., Minh, B. Q., Bileck, A., Hartl, M., von Haeseler, A., Gerner, C., Raible, F., and Tessmar-Raible, K. 2019. Combined transcriptome and proteome profiling reveals specific molecular brain signatures for sex, maturation and circalunar clock phase. *Elife* 8.

Schenk, S., Krauditsch, C., Fruhauf, P., Gerner, C., and Raible, F. 2016. Discovery of methylfarnesoate as the annelid brain hormone reveals an ancient role of sesquiterpenoids in reproduction. *Elife* 5.

Schmarda, L. K. 1861. Neue Wirbellose Thiere: Beobachted und Gesammelt auf einer Reise um die Erdr 1853 bis 1857. In *In Turbellarien, Rotatorien und Anneliden*, edited by Zweite Hälfte Erster Band. Leipzig: Verlag von Wilhelm Engelmann.

Schmidtberg, H., and Dorresteijn, A. W. C. 2010. Ultrastructure of the nuchal organs in the polychaete *Platynereis dumerilii* (Annelida, Nereididae). *Invertebrate Biology* 129 (3):252–265.

Schneider, S. Q., and Bowerman, B. 2007. Beta-catenin asymmetries after all animal/vegetal-oriented cell divisions in *Platynereis dumerilii* embryos mediate binary cell-fate specification. *Dev Cell* 13 (1):73–86.

Scholtz, G. 2002. The Articulata hypothesis: Or what is a segment? *Organisms Diversity & Evolution* 2 (3):197–215.

Schroeder, P. C., Aguado, M. T., Malpartida, A., and Glasby, C. J. 2017. New observations on reproduction in the branching polychaetes *Ramisyllis multicaudata* and *Syllis ramosa* (Annelida: Syllidae: Syllinae). *Journal of the Marine Biological Association of the United Kingdom* 97 (5):1167–1175.

Schulz, G., Ulbrich, K. P., Hauenschild, C., and Pfannenstiel, H. D. 1989. The atokous-epitokous border is determined before the onset of heteronereid development in *Platynereis dumerilii* (Annelida, Polychaeta). *Rouxs Arch Dev Biol* 198 (1):29–33.

Schwaiger, M., Schonauer, A., Rendeiro, A. F., Pribitzer, C., Schauer, A., Gilles, A. F., Schinko, J. B., Renfer, E., Fredman, D., and Technau, U. 2014. Evolutionary conservation of the eumetazoan gene regulatory landscape. *Genome Res* 24 (4):639–650.

Shao, Y., Wang, X. B., Zhang, J. J., Li, M. L., Wu, S. S., Ma, X. Y., Wang, X., Zhao, H. F., Li, Y., Zhu, H. H., Irwin, D. M., Wang, D. P., Zhang, G. J., Ruan, J., and Wu, D. D. 2020. Genome and single-cell RNA-sequencing of the earthworm *Eisenia andrei* identifies cellular mechanisms underlying regeneration. *Nat Commun* 11 (1):2656.

Simakov, O., Marletaz, F., Cho, S. J., Edsinger-Gonzales, E., Havlak, P., Hellsten, U., Kuo, D. H., Larsson, T., Lv, J., Arendt, D., Savage, R., Osoegawa, K., de Jong, P., Grimwood, J., Chapman, J. A., Shapiro, H., Aerts, A., Otillar, R. P., Terry, A. Y., Boore, J. L., Grigoriev, I. V., Lindberg, D. R., Seaver, E. C., Weisblat, D. A., Putnam, N. H., and Rokhsar, D. S. 2013. Insights into bilaterian evolution from three spiralian genomes. *Nature* 493 (7433):526–531.

Starunov, V. V., Dray, N., Belikova, E. V., Kerner, P., Vervoort, M., and Balavoine, G. 2015. A metameric origin for the annelid pygidium? *BMC Evol Biol* 15:25.

Starunov, V. V., Voronezhskaya, E. E., and Nezlin, L. P. 2017. Development of the nervous system in *Platynereis dumerilii* (Nereididae, Annelida). *Front Zool* 14:27.

Steinmetz, P. R., Zelada-Gonzales, F., Burgtorf, C., Wittbrodt, J., and Arendt, D. 2007. Polychaete trunk neuroectoderm converges and extends by mediolateral cell intercalation. *Proc Natl Acad Sci U S A* 104 (8):2727–2732.

Struck, T. H. 2011. Direction of evolution within Annelida and the definition of Pleistoannelida. *Journal of Zoological Systematics and Evolutionary Research* 49 (4):340–345.

Struck, T. H., Paul, C., Hill, N., Hartmann, S., Hosel, C., Kube, M., Lieb, B., Meyer, A., Tiedemann, R., Purschke, G., and Bleidorn, C. 2011. Phylogenomic analyses unravel annelid evolution. *Nature* 471 (7336):95–98.

Sugio, M., Yoshida-Noro, C., Ozawa, K., and Tochinai, S. 2012. Stem cells in asexual reproduction of *Enchytraeus japonensis* (Oligochaeta, Annelid): Proliferation and migration of neoblasts. *Development Growth & Differentiation* 54 (4):439–450.

Suzuki, M. M., and Bird, A. 2008. DNA methylation landscapes: Provocative insights from epigenomics. *Nat Rev Genet* 9 (6):465–476.

Tanaka, E. M. 2016. The molecular and cellular choreography of appendage regeneration. *Cell* 165 (7):1598–1608.

Tanaka, E. M., and Reddien, P. W. 2011. The cellular basis for animal regeneration. *Dev Cell* 21 (1):172–185.

Tessmar-Raible, K., Raible, F., and Arboleda, E. 2011. Another place, another timer: Marine species and the rhythms of life. *Bioessays* 33 (3):165–172.

Tessmar-Raible, K., Raible, F., Christodoulou, F., Guy, K., Rembold, M., Hausen, H., and Arendt, D. 2007. Conserved sensory-neurosecretory cell types in annelid and fish forebrain: Insights into hypothalamus evolution. *Cell* 129 (7):1389–1400.

Tessmar-Raible, K., Steinmetz, P. R., Snyman, H., Hassel, M., and Arendt, D. 2005. Fluorescent two-color whole mount in situ hybridization in *Platynereis dumerilii* (Polychaeta, Annelida), an emerging marine molecular model for evolution and development. *Biotechniques* 39 (4):460, 462, 464.

Tomer, R., Denes, A. S., Tessmar-Raible, K., and Arendt, D. 2010. Profiling by image registration reveals common origin of annelid mushroom bodies and vertebrate pallium. *Cell* 142 (5):800–809.

Valvassori, G., Massa-Gallucci, A., and Gambi, M. C. 2015. Reappraisal of *Platynereis massiliensis* (Moquin-Tandon) (Annelida, Nereididae), a neglected sibling species of *Platynereis dumerilii* (Audouin & Milne Edwards). *Biologia Marina Mediterranea* 22 (1):113–116.

Veedin-Rajan, V. B., Fischer, R. M., Raible, F., and Tessmar-Raible, K. 2013. Conditional and specific cell ablation in the marine annelid *Platynereis dumerilii*. *PLoS One* 8 (9):e75811.

Veraszto, C., Ueda, N., Bezares-Calderon, L. A., Panzera, A., Williams, E. A., Shahidi, R., and Jekely, G. 2017. Ciliomotor circuitry underlying whole-body coordination of ciliary activity in the Platynereis larva. *Elife* 6.

Verdonschot, P. F. M. 2015. Introduction to annelida and the class polychaeta. In *Thorp and Covich's Freshwater Invertebrates*, edited by James H. Thorp and Christopher Rogers, 509–528. Elsevier Inc.; London: Academic Press.

Vergara, H. M., Bertucci, P. Y., Hantz, P., Tosches, M. A., Achim, K., Vopalensky, P., and Arendt, D. 2017. Whole-organism cellular gene-expression atlas reveals conserved cell types in the ventral nerve cord of *Platynereis dumerilii*. *Proc Natl Acad Sci U S A* 114 (23):5878–5885.

Vervoort, M., and Gazave, E. in press. Zoological and molecular methods to study Annelida regeneration using *Platynereis dumerilii*. In *Methods in Molecular Biology*, edited by Stephen Stricker David Carroll. Springer.

Vriz, S., Reiter, S., and Galliot, B. 2014. Cell death: A program to regenerate. *Curr Top Dev Biol* 108:121–151.

Vullien, A., Röttinger, E., Vervoort, M., and Gazave, E. 2021. A trio of mechanisms involved in regeneration initiation in animals. *Médecine/Sciences* (in French) 37 (4):349–358.

Wäge, J., Valvassori, G., Hardege, J. D., Schulze, A., and Gambi, M. C. 2017. The sibling polychaetes *Platynereis dumerilii* and *Platynereis massiliensis* in the Mediterranean Sea: Are phylogeographic patterns related to exposure to ocean acidification? *Marine Biology* 164 (10):164–199.

Weigert, A., and Bleidorn, C. 2016. Current status of annelid phylogeny. *Organisms Diversity & Evolution* 16 (2):345–362.

Weigert, A., Helm, C., Meyer, M., Nickel, B., Arendt, D., Hausdorf, B., Santos, S. R., Halanych, K. M., Purschke, G., Bleidorn, C., and Struck, T. H. 2014. Illuminating the base of the annelid tree using transcriptomics. *Mol Biol Evol* 31 (6):1391–1401.

Williams, E. A., Conzelmann, M., and Jekely, G. 2015. Myoinhibitory peptide regulates feeding in the marine annelid Platynereis. *Front Zool* 12 (1):1.

Williams, E. A., and Jekely, G. 2016. Towards a systems-level understanding of development in the marine annelid *Platynereis dumerilii*. *Curr Opin Genet Dev* 39:175–181.

Williams, E. A., and Jekely, G. 2019. Neuronal cell types in the annelid *Platynereis dumerilii*. *Curr Opin Neurobiol* 56:106–116.

Williams, E. A., Veraszto, C., Jasek, S., Conzelmann, M., Shahidi, R., Bauknecht, P., Mirabeau, O., and Jekely, G. 2017. Synaptic and peptidergic connectome of a neurosecretory center in the annelid brain. *Elife* 6.

Williams, T. A., and Nagy, L. M. 2017. Linking gene regulation to cell behaviors in the posterior growth zone of sequentially segmenting arthropods. *Arthropod Struct Dev* 46 (3):380–394.

Wilson, E. B. 1892. The cell-lineage of Nereis: A contribution to the cytogeny of the annelid body. *J Morphol* 6 (3):361–480.

Zakrzewski, A. C. 2011. Molecular Characterization of Chaetae Formation in Annelida and Other Lophotrochozoa. Doktors der Naturwissenschaften, Fachbereich Biologie, Chemie und Pharmazie der Freien, Berlin.

Zantke, J., Bannister, S., Rajan, V. B., Raible, F., and Tessmar-Raible, K. 2014. Genetic and genomic tools for the marine annelid *Platynereis dumerilii*. *Genetics* 197 (1):19–31.

Zantke, J., Ishikawa-Fujiwara, T., Arboleda, E., Lohs, C., Schipany, K., Hallay, N., Straw, A. D., Todo, T., and Tessmar-Raible, K. 2013. Circadian and circalunar clock interactions in a marine annelid. *Cell Rep* 5 (1):99–113.

Zeeck, E., Hardege, J., and Bartels-Hardege, H. 1990. Sex pheromones and reproductive isolation in two nereid species, *Nereis succinea* and *Platynereis dumerilii*. *Marine Ecology Progress Series* 67:183–188.

Zemach, A., McDaniel, I. E., Silva, P., and Zilberman, D. 2010. Genome-wide evolutionary analysis of eukaryotic DNA methylation. *Science* 328 (5980):916–919.

Zidek, R., Machon, O., and Kozmik, Z. 2018. Wnt/beta-catenin signalling is necessary for gut differentiation in a marine annelid, *Platynereis dumerilii*. *Evodevo* 9:14.

14 Cycliophora—An Emergent Model Organism for Life Cycle Studies

Peter Funch

CONTENTS

14.1 HISTORY OF THE MODEL

Cycliophora is a phylum of marine, microscopic, solitary epizoans found on the mouthparts of three common species of commercially exploited lobsters (Decapoda, Nephropidae) (Figures 14.1 and 14.2). Surprisingly, they were described as late as in 1995, but they were noticed already in the 1960s by Profs. Tom Fenchel and José Bresciani (Funch and Kristensen 1995; Funch and Kristensen 1997; Kristensen 2002; Funch and Neves 2019). At that time, cycliophorans were regarded as aberrant rotifers and got the nickname "Mysticus enigmaticus". Prof. Claus Nielsen at the Zoological Museum in Copenhagen then collected mouth parts from *N. norvegicus* with cycliophorans and prepared this material for ultrastructural studies. He kindly handed the embedded material to the author to be included in his master thesis project (Andersen 1992). Transmission electron microscopy of this material revealed that the cycliophorans had a well-developed cuticle very different from the syncytial integument with an intracytoplasmic lamina known form rotifers (Clément and Wurdak 1991). This observation lead to more extensive studies of ultrastructure, life cycle and host range (Andersen 1992).

To date, only two species have been formally described. The first studies showed that cycliophorans have an elaborate life cycle with a number of morphologically distinct stages that involve alternations between attached and free stages and asexual and sexual cycles (Funch and Kristensen 1995; Funch and Kristensen 1997) (Figure 14.3). The first species, *Symbion pandora* (Funch and Kristensen 1995), was described from the Norway lobster, *Nephrops norvegicus,* from Scandinavian waters, but before this description, a similar epibiont, still undescribed, was found on the mouthparts of the European lobster, *Homarus gammarus* (Andersen 1992; Funch and Kristensen 1997). The second described cycliophoran species, *Symbion americanus*, occurs on the American lobster, *Homarus americanus* (Obst et al. 2006), but cycliophorans from this host species are more genetically diverse due to the presence of at least three cryptic lineages (Obst et al. 2005; Baker et al. 2007; Baker and Giribet 2007). A study on *S. pandora* on *N. norvegicus* showed that this epizoan species is an obligatory commensal that depends on microscopic food particles generated during host feeding (Funch et al. 2008). Cycliophorans have also been found attached to harpacticoid copepods in a study of cycliophorans from European lobsters (Neves et al. 2014), but how common this association is and if it has any role in assisted migration of the cycliophorans is unclear. The integument including gills and mouth parts of a broader range of crustaceans—for example, *Cancer pagurus*, *Carcinus maenas*, *Pagurus bernhardus*, *Geryon trispinosus*, *Galathea* sp., *Hyas* sp. and *Munida* sp.—were examined for Cycliophora but did not reveal any (Andersen 1992). Also, a survey on a broader range of crustaceans from museum material only recovered cycliophorans on nephropid hosts (Funch and Kristensen 1997; Plaza 2012).

14.2 GEOGRAPHICAL LOCATION

Thus far, cycliophorans are known from coastal areas of the North Atlantic Ocean and the Mediterranean Sea where their decapod hosts also occur. The first known observations of cycliophorans from the 1960s were from mouthparts of *Nephrops norvegicus* from Kattegat, Denmark, and later the Gulf of Naples, Italy (pers. comm. Tom Fenchel and José Bresciani). The type locality for *Symbion pandora* is

DOI: 10.1201/9781003217503-14

FIGURE 14.1 Sessile feeding stages of *Symbion pandora* on the setae of *Nephrops norvegicus*. DIC.

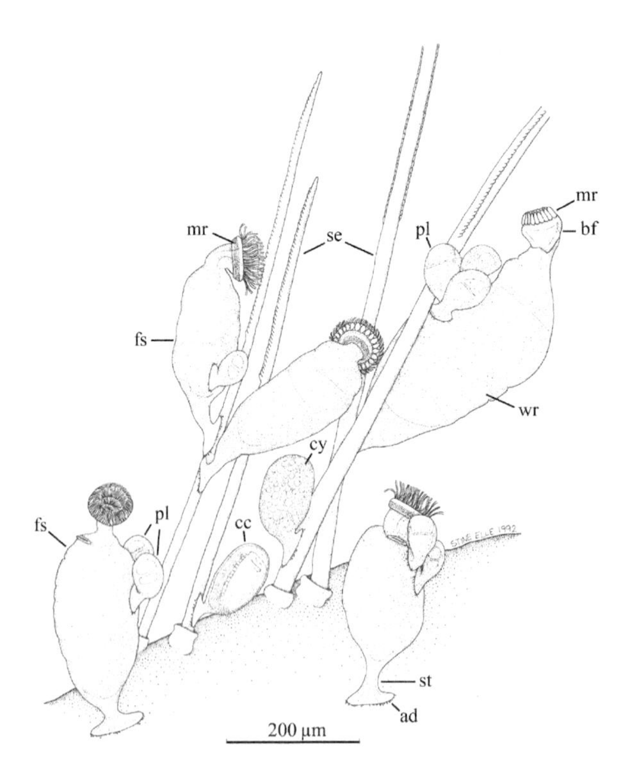

FIGURE 14.2 Various attached life cycle stages of *Symbion pandora*, type material. The cycliophorans are attached to the endopod of the first maxilla of *Nephrops norvegicus*. In the front are two feeding stages (fs) with an open mouth ring (mr) in feeding position and two Prometheus larvae (pl) attached on the trunk. Two cyst-like stages are attached to the bases of the setae. The one to the left is a chordoid cyst (cc) that contains a chordoid larva with ventral ciliation and a chordoid organ. The cyst-like stage (cy) to the right contains undifferentiated cells. Distally on the same seta is a larger feeding stage with a closed mouth ring and three attached Prometheus larvae. On the rightmost seta are two feeding stages with open mouth ring. The one to the left has an attached Prometheus larva—the right one has none. ad, adhesive disc; bf, buccal funnel; se, seta from the host; st, stalk; wr, wrinkles. (Reproduced with permission from Andersen 1992.)

NW Kattegat, Denmark at 20–40 m of depth. In 1992, the known geographic range was extended to the coastal areas around the Faroe Islands (on fixed material collected in 1990 in Kaldbak Fjord), Orkney Islands and Southern Norway, and the host range was extended to include *Homarus gammarus* with an undescribed cycliophoran species (Andersen 1992). Nedved (2004) also reported the occurrence of cycliophorans on *Homarus gammarus* from the Adriatic Sea. The third cycliophoran species, *Symbion americanus*, was described from the mouth parts of the American lobster, *Homarus americanus*, collected from Maine and Cape Cod at the Northeast Atlantic coast of the United States (Obst et al. 2006). A phylogeographic study of the cycliophorans mentioned previously based on the mitochondrial gene COI indicated that the three species of cycliophorans were reproductively isolated on the three different hosts and that the free stages in the life cycle of cycliophorans have limited dispersal abilities (Obst et al. 2005). This study also showed a high genetic diversity of *Symbion americanus* and a low genetic diversity of *S. pandora*, and it was suggested that the latter species was of recent origin.

14.3 LIFE CYCLE

The life cycle of Cycliophora involves metagenesis with multiple stages and alternations between sessile stages that are permanently attached to a host and motile and free stages (Funch and Kristensen 1995; Funch 1996; Funch and Kristensen 1997) (Figure 14.3). The most prominent stage of the life cycle is the feeding stage, so named because it is the only stage in the life cycle with feeding structures and a digestive tract (Figure 14.4). Feeding stage individuals are often densely aggregated on the mouth parts of their decapod hosts and live on food particles collected by filter feeding (Figure 14.1). When a feeding stage individual grows, it continually forms internal new zooids with new feeding structures and gut, and these structures replace the structures associated with the old zooid (Figure 14.4). Larger and older feeding stage individuals also produce motile stages inside brood chambers (Funch and Kristensen 1997) (Figure 14.3). One feeding stage forms one motile stage in a brood chamber at a time, and it seems like asexual Pandora larva are produced first, then Prometheus larvae and finally females (Kristensen and Funch 2002). All motile stages are without a digestive tract.

The asexual part of the life cycle involves young feeding stage individuals that develop Pandora larvae in brood chambers (Figure 14.5). The Pandora larva is characterized by a ciliated locomotory disc and developing feeding structures inside. When mature, it escapes from the maternal feeding stage brood chamber and moves actively on the decapod host to seek a site on the mouth parts where it settles. This attached cyst-like stage then develops into a new small feeding stage when the internal feeding structures emerge (Funch and Kristensen 1997) (Figure 14.3).

The sexual part of the life cycle is initiated when smaller stages, the Prometheus larvae, are produced in the brood chambers of older feeding stages (Figures 14.3 and 14.6). Like the Pandora larva, the Prometheus larva uses a ciliated disc for locomotion, but contrary to the Pandora larva, it settles on the trunk of a cycliophoran feeding stage. Often, several Prometheus larvae are found on the same feeding stage individual. The preferred site for settlement is close to the cloacal opening of the feeding stage, and during settlement, a Prometheus larva typically orients itself with the posterior end as close to the cloacal opening as possible, directing the anterior end toward the attachment site of the feeding stage (Figures 14.4A and 14.6). Settlement involves secretion from gland cells that exits in the area of the ciliated disc and becomes an attachment disc. Dwarf males are produced inside the attached Prometheus larva, and one, two and three males have been observed developing simultaneously.

Females are produced inside the oldest feeding stages and are characterized by the presence of one oocyte (Figure 14.6). After escape from the maternal feeding stages, the females can be recognized by the presence of a single zygote (Figure 14.7). Females and Pandora larvae are almost similar in size, and females also use an anteroventral ciliated disc for locomotion when they are liberated from the brood chamber, and they settle on the mouth parts of the host. However, the preferred sites for settlement differ. Females prefer the lateral parts and articulations of the mouth parts of the host, while Pandora larvae prefer those medial segments of the mouth parts where availability of food particles is rich during host feeding (Obst and Funch 2006; Funch et al. 2008). When females settle, they degenerate and develop into chordoid cysts consisting of a female body cuticle containing a chordoid larva inside. These cysts and larvae are named after the presence of a characteristic longitudinal structure of similar vacuolated muscle cells (Figures 14.2 and 14.8). A chordoid larva has more locomotory ciliation compared to the other motile stages in the life cycle and has therefore been suggested to be a dispersal stage between hosts. This larval stage is capable of both crawling and swimming and is completely ciliated ventrally, including body ciliation separated from a ciliated foot. It has been suggested that the chordoid larva settles on the mouth parts of a host and develops into a small feeding stage, thereby completing the sexual life cycle (Figure 14.3) (Funch and Kristensen 1995).

14.4 EMBRYOGENESIS

In Cycliophora, the embryos are brooded inside females, but the type of cleavage is unknown, and polar bodies have never been observed. The zygote develops into a chordoid larva (Figure 14.8). The female develops one oocyte before it is liberated from the brood chamber of the feeding stage (Figure 14.6), and the first cleavage has been observed in a free-swimming female of *Symbion pandora* (Funch 1996), while an embryo consisting of four micromeres and four macromeres has been observed in a female after settlement (Neves et al. 2012). Based on these limited observations, it seems like cleavage is holoblastic. So, females that recently

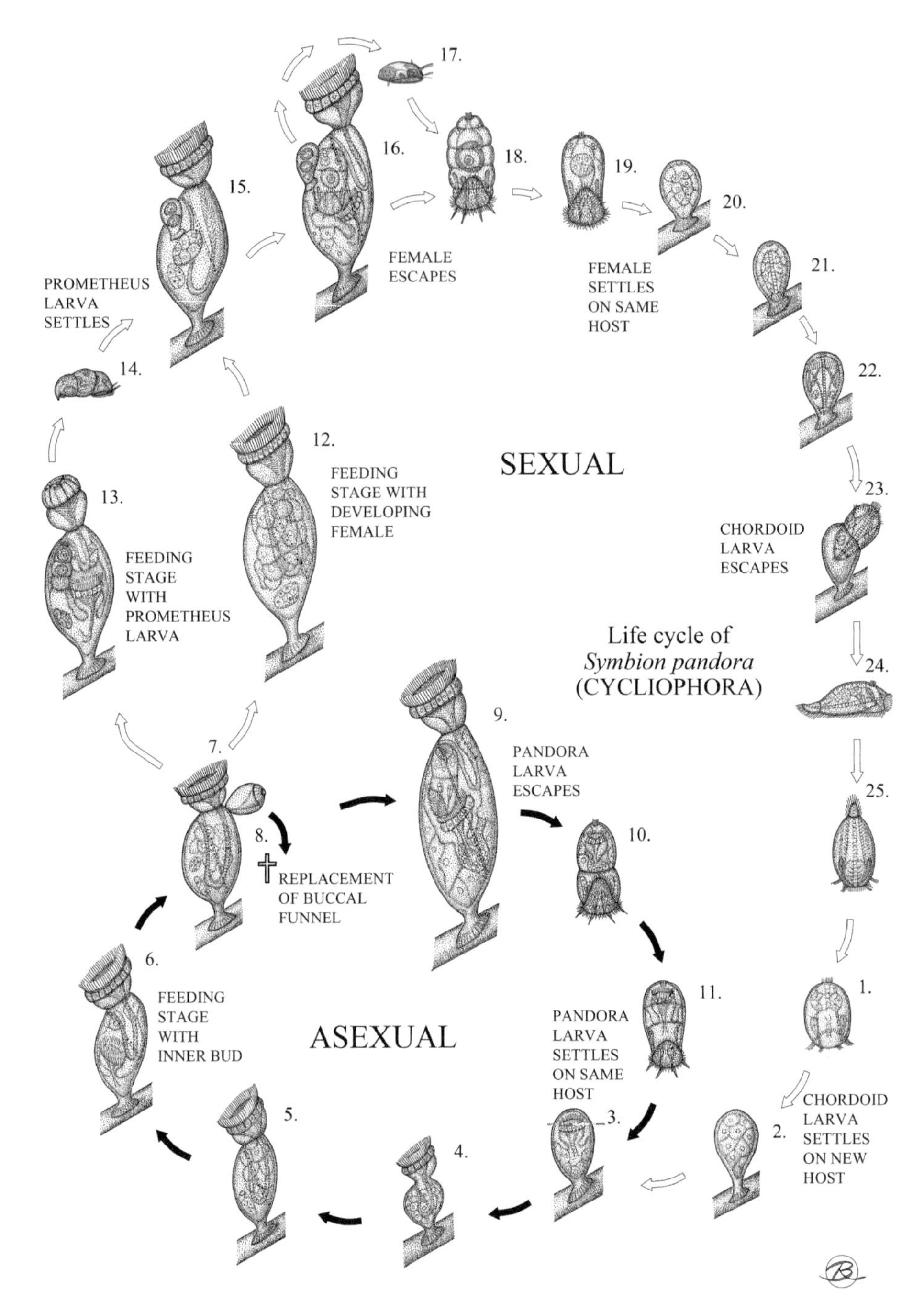

FIGURE 14.3 Proposed life cycle of *Symbion pandora*. The asexual cycle begins when a chordoid larva settles on the lobster host (1–2) and degenerates, while internal buds inside differentiate into feeding structures (3). The buccal funnel emerges, and filter feeding is enabled (4). The feeding stage then grows, and budding cells basally form a new zooid inside with a new buccal funnel, digestive tract and nervous system (5, 6). The new zooid replaces the old zooid (8). A larger and older feeding stage regenerates and replaces the feeding structures in a similar way but also forms a Pandora larva asexually inside a brood chamber (9). The fully developed Pandora larva then escapes the maternal feeding stage (9–10) and settles nearby on the host mouthparts (11–3). The larval structures degenerate, while the internal feeding structures matures, completing the asexual part of the life cycle (3). The factors involved when shifting to the sexual cycle are unknown, but the sexual part of the cycle involves older feeding stages that produce either one Prometheus larva (13) or one female (12) inside a brood chamber. When the Prometheus larva escapes, it settles (14) on the trunk of a feeding stage (15). Dwarf males develop inside the attached Prometheus larva from internal buds, while the female is produced inside the feeding stage (16). The fully mature dwarf male (17) might transfer the sperm during the release of the female or shortly afterward (18). Early cleavages have been observed before the female (19) settles on the mouthparts of the host (20). The female degenerates, while the internal embryo develops into a chordoid cyst (21). The chordoid larva escapes (23) and perhaps migrates (24, 25) to a new lobster host, where it settles on the mouthparts (1–2). Here budding cells inside develop into feeding structures while the larva degenerates (2–3), completing the sexual life cycle. (Material modified from: Peter Funch and Reinhardt Møbjerg Kristensen, Cycliophora is a new phylum with affinities to Entoprocta and Ectoprocta, Nature, published 1995, Nature Publishing Group.)

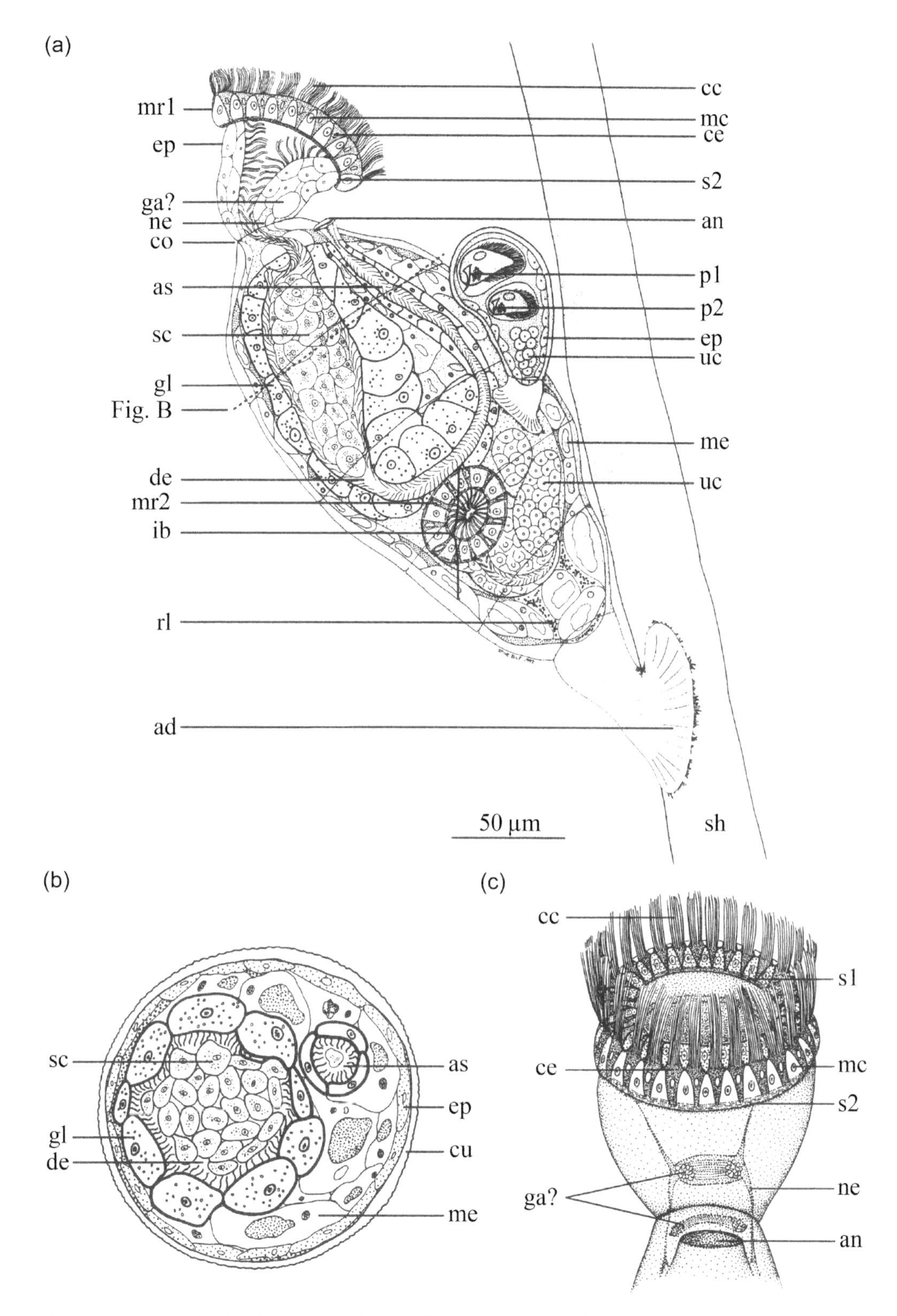

FIGURE 14.4 *Symbion pandora.* (a) Feeding stage individual (holotype) with attached Prometheus larva (allotype) attached to a seta of a mouthpart from *Nephrops norvegicus.* The mouth ring is in the everted feeding position. (b and c) Feeding stage individual. (b) Trunk in transverse section. (c) Buccal funnel with an everted mouth ring. Position of the nervous system is indicated. ad, adhesive disc; an, anus; as, ascending branch of the digestive tract; cc, compound cilia of the mouth ring; ce, ciliated epidermis; co, constriction (or "neck"); cu, cuticle; de, descending branch of digestive tract; ep, epidermis; ga, ganglion; gl, gut lining cell; ib, inner bud; mc, myoepithelial cell; me, mesenchyme; mr, mouth ring; ne, nerve; p, penis; rl, remnants of larval glands; sc, stomach cells; sh, seta from the host; s, sphincter; uc, undifferentiated cells. ([a] Material modified from Funch and Kristensen, Cycliophora is a new phylum with affinities to Entoprocta and Ectoprocta, Nature, published 1995, Nature Publishing Group; [b and c] Reproduced with permission from Funch and Kristensen, Cycliophora. In *Microscopic Anatomy of Invertebrates*, edited by F. W. Harrison and R. M. Woollacott, 409–474. New York etc.: Wiley-Liss Inc., published 1997.)

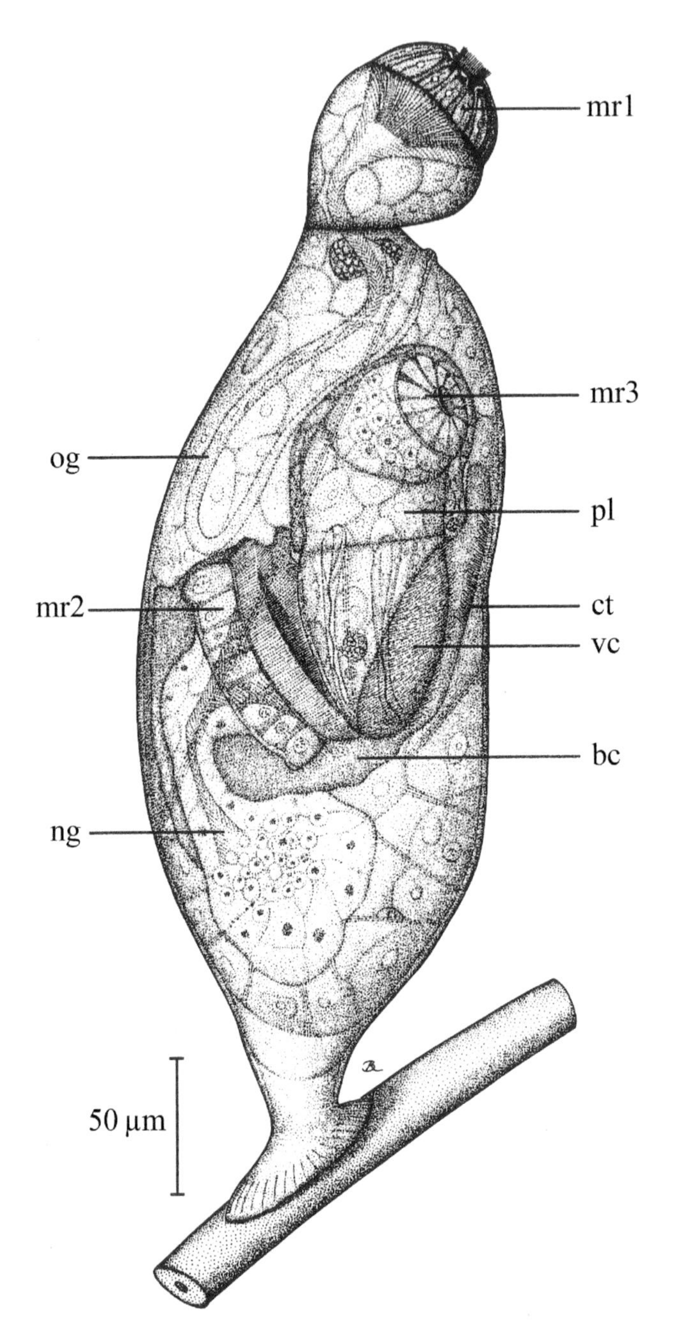

FIGURE 14.5 Young feeding stage of *Symbion pandora* attached to a host seta from the mouth parts of *Nephrops norvegicus*. Line drawing from whole mount. The feeding stage individual has a closed mouth ring (mrl) and an old gut (og) reduced in size that provides more space for the developing Pandora larva (pl) inside a brood chamber (bc). An inner bud is in the process of developing a new zooid with a mouth ring (mr2), a ciliated buccal funnel and an immature new gut (ng). The new buccal funnel and the Pandora larva develop inside the same brood chamber (bc) lined with a thin cuticle except at the anal side where cilia tufts (ct) are present. The anterior part of the Pandora larva has a ventral ciliated disc (vc), while the posterior part contains budding cells developing another new feeding stage, which is evident be the presence of a third mouth ring (mr3). (Reproduced with permission from Funch and Kristensen, Cycliophora. In *Microscopic Anatomy of Invertebrates*, edited by F. W. Harrison and R. M. Woollacott, 409–474. New York etc.: Wiley-Liss Inc., published 1997.)

attached to the mouth parts of the host contain the early developing embryo. Later the cells of the female degenerate, while the embryo inside develops a characteristic chordoid organ. This results in a stage named the chordoid cyst, which consists of a chordoid larva contained in an ovoid case originating from the body cuticle of the female (Figure 14.2). It has been suggested that chordoid larvae typically hatch stimulated by changes in external conditions such as host molting or death (Funch and Kristensen 1999; Kristensen and Funch 2002).

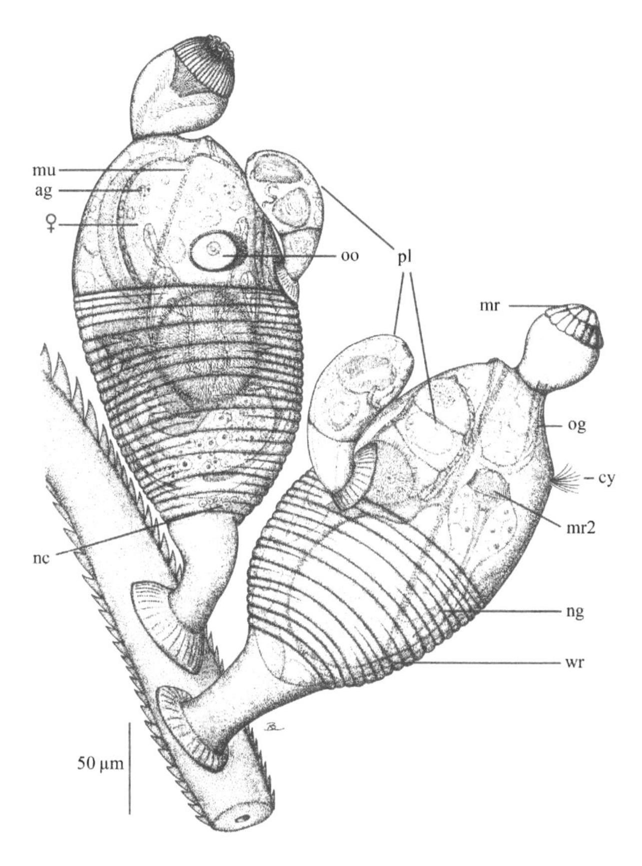

FIGURE 14.6 Two old feeding stages of *Symbion pandora* with numerous cuticular wrinkles (wr) and a Prometheus larva (pl) attached. Line drawing from whole mount. The feeding stages are attached to a host seta from the mouth parts of *Nephrops norvegicus*. Right feeding stage with degenerated gut and a Prometheus larva developing in the brood chamber. Left feeding stage with developing female inside the brood chamber. ag, accessory genital glands; cy, cyanobacteria; mr and mr2, mouth ring; mu, muscle; nc, necrotic cells; ng, new gut; og, old gut; oo, oocyte. (Reproduced with permission from Funch and Kristensen, Cycliophora. In *Microscopic Anatomy of Invertebrates*, edited by F. W. Harrison and R. M. Woollacott, 409–474. New York etc.: Wiley-Liss Inc., published 1997.)

14.5 ANATOMY

Cycliophorans are bilaterally symmetrical and acoelomate metazoans with a well-differentiated cuticle that apically has polygonal sculpturing. The feeding stages are sessile and vary in length from about 0.2 to 1 mm. The body of the feeding stage is divided into a distal buccal funnel, a short, slender neck, a trunk, a stalk and an adhesive disc basally that ensures a permanent attachment to the mouth parts of the crustacean host (Figure 14.4a). A few longitudinal muscles are present in the buccal funnel and trunk, but circular body wall muscles are absent (Neves, Kristensen et al. 2009; Neves, Cunha et al. 2010). The broader and distal part of the bell-shaped buccal funnel carries a radially symmetrical ciliated mouth ring that is used in filter feeding when it is everted (Figure 14.4c). Contraction of myoepithelial cells of the mouth ring results in inversion of the mouth ring that directs the cilia into the buccal cavity and closes the mouth opening.

The gut is U-shaped and lined with multiciliated cells. The anterior part of the digestive tract consists of a large mouth opening, the buccal funnel and a narrow S-shaped esophagus. The esophagus leads to an enlarged stomach containing secretory cells that reduce the stomach lumen to lacunae (Figure 14.4b). The tract narrows and bends into a U-turn that leads to an ascending intestine that opens distally in a slitlike transverse opening on the trunk close to the narrow neck (Figure 14.4c). An anal sphincter is present. This opening also serves as an exit for the brooded stages. The whole feeding apparatus including the buccal funnel is repeatedly regenerated from undifferentiated cells basal to the U-turn of the gut by internal budding. Each replacement of the old zooid with a new zooid leaves a wrinkle in the cuticle, and the number of cuticular scars indicates the age of the feeding stage individual. The youngest feeding stages have a smooth cuticle without wrinkles (Figure 14.5), while old feeding stages have many wrinkles (Figure 14.6). Brood chambers in feeding stages are lined with cuticle and contain fluid circulated by specific cilia (Figure 14.5). A brooded stage is fixed in the brood chamber with the anterior end directed toward the basal part of the maternal feeding stage, while the posterior part is connected to a placenta-like structure (Funch and Kristensen 1997).

The Pandora larva, the Prometheus larva and the female are smaller than the feeding stages and range in size between 80 and 200 µm. Their bodies are ovoid with presumed sensory organs consisting of bundles of paired long stiff ciliary organs anteriorly and a median ciliated pore posteriorly. The

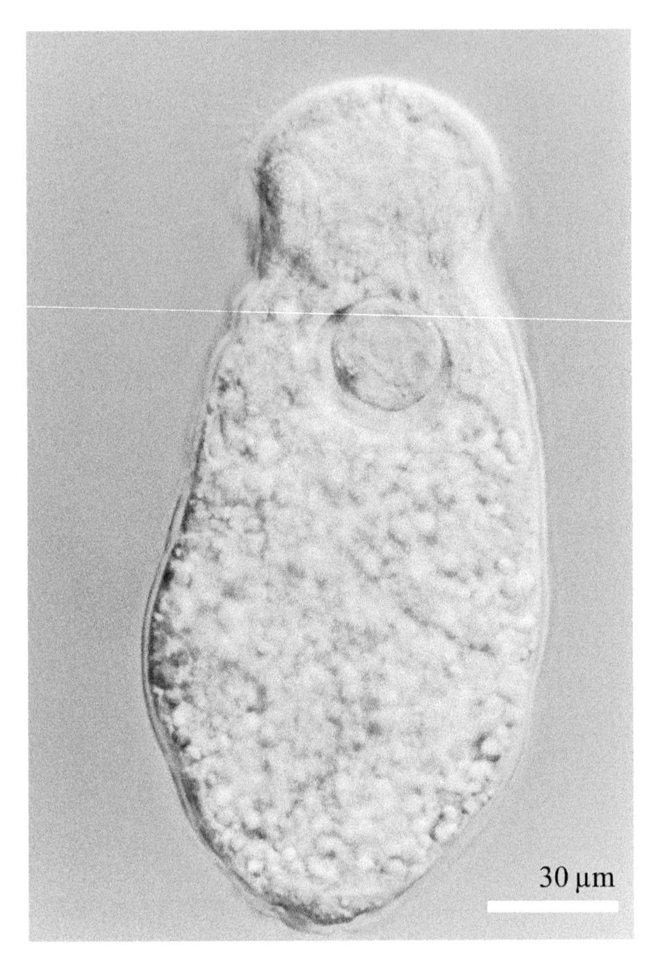

FIGURE 14.7 Female of *Symbion pandora* just released from the brood chamber of the maternal feeding stage. The upper part shows the anterior end characterized externally by the presence of motile longer stiff sensoria and shorter cilia that form the ventral ciliated disc. Gland cells are present laterally. A single oocyte is situated medially just posterior to the ciliated disc. DIC.

nervous system in these stages consists of a dorsal cerebral ganglion with a pair of lateral clusters of perikarya connected by a commissural neuropil and a pair of ventral longitudinal neurites (Neves, Kristensen et al. 2010). Anteriorly, dorsal and lateral gland cells with elongated gland necks with outlets in the area of the ventral ciliated sole are present. A digestive tract is lacking. After liberation from the feeding stage brood chamber, they have a brief motile phase. The locomotion is by ciliary gliding using an anteroventral ciliated sole (Funch and Kristensen 1997). Settlement and transition from free stages to the sessile stage involve secretion of the gland content over the ciliated sole that becomes the adhesive disc.

The males are the smallest life cycle stage, with an ovoid body, only around 30–40 μm long. They also possess an anteroventral sole for ciliary gliding, but in addition, they have two characteristic structures absent in the other life cycle stages. Their external ciliation includes a frontal ciliated field, and posteriorly a sickle-shaped penis is present. The penis is hidden in a ventral pouch but can be protruded (Obst and Funch 2003; Neves, da Cunha, Funch et al. 2010). They have a well-developed body wall musculature, a relatively large cerebral ganglion that occupies most of the anterior body and a pair of ventral neurites (Obst and Funch 2003; Neves, Kristensen et al. 2010).

The chordoid cysts and chordoid larvae are named after a characteristic longitudinal rod of 40–50 cylindrical muscle cells with a central vacuole surrounded by myofilaments—the chordoid organ (Funch 1996). The chordoid larvae are 150–210 μm long and have more external ciliation than any other cycliophoran life cycle stage (Figure 14.8). The ventral body is ciliated with two anterior ciliated bands followed by ciliated field separated from a foot with ventral ciliation. A free chordoid larva both swims and moves along the substrate by ciliary crawling. It has a pair of protonephridia, even though excretory organs are unknown in the other life cycle stages. The protonephridium consists of a single multiciliated terminal cell and at least one duct cell (Funch 1996). The nervous system consists of a dorsal bilobed cerebral ganglion and two paired longitudinal nerves (Neves, da Cunha, Kristensen et al. 2010). Presumed sensory organs include a pair of dorsal ciliated organs and a pair of lateral ciliated pits. A digestive tract is absent.

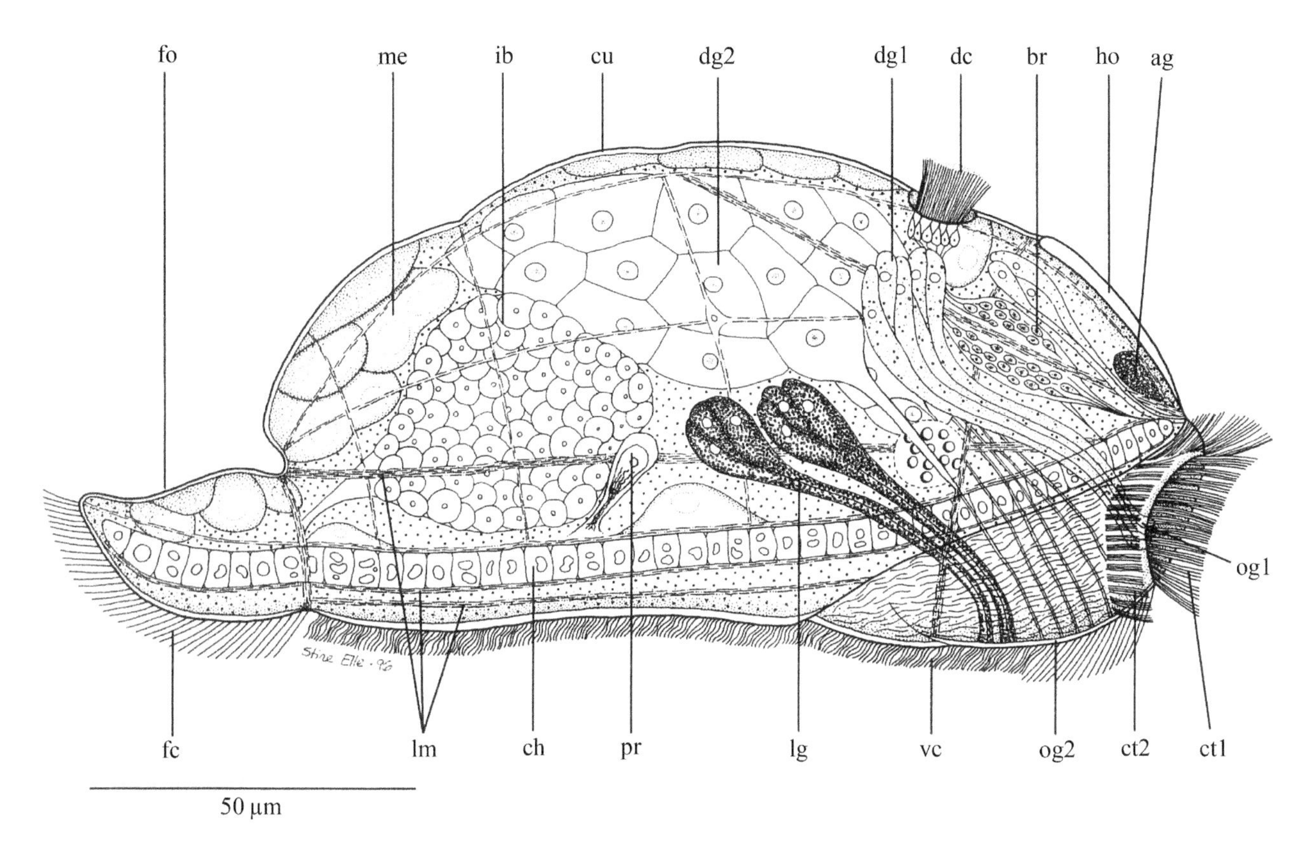

FIGURE 14.8 Chordoid larva of *Symbion pandora*, lateral view, line drawing from whole mount. The lateral and dorsal integument has an apical cuticle (cu) that dorsal to the brain forms a more rigid hood (ho). Posterior dorsal glands (dg2) with long gland necks extend into a ventral outlet complex (og2). Shorter dorsal glands (dgl), just posterior to the brain (br), extend into a smaller outlet complex (ogl) anteriorly. ag, anterior glands; ch, chordoid organ; ctl and ct2, ciliated band 1 and 2; dc, dorsal ciliated organ; fc, foot cilia; fo, foot; ib, inner bud; lg, lateral gland; lm, longitudinal muscles; me, mesenchyme; pr, protonephridium; vc, ventral cilia. (Reproduced with permission from Funch, the chordoid larva of *Symbion pandora* [Cycliophora] is a modified trochophore, Wiley-Liss Inc., published 1996.)

14.6 GENOMIC DATA

Genomic data on cycliophorans are scarce. However, a transcriptome is available for *Symbion americanus* generated from a single starved feeding stage individual (Laumer et al. 2015). For *S. pandora*, both transcriptomes and an EST library are available (Hejnol et al. 2009; Neves et al. 2017). Gene expression analysis showed that more than 10% of the genes were expressed differentially in *S. pandora*, when feeding stage individuals without attached Prometheus larvae (asexual phase) were compared with those with attached Prometheus larvae (sexual phase). Genes related to protein folding and RNA processing and splicing were upregulated in the asexual phase, while those involved in signal transduction and neurotransmission were upregulated in the sexual phase (Neves et al. 2017).

14.7 FUNCTIONAL APPROACHES: TOOLS FOR MOLECULAR AND CELLULAR ANALYSES

Ultrastructural studies of Cyliophora were applied and included in the first description of *Symbion pandora* (Funch and Kristensen 1995) and have been used to characterize various cell types (Funch 1996; Funch and Kristensen 1997). Cycliophoran cell types include multiciliated epidermal cells with compound cilia and erect microvilli, various types of unicellular glands especially in the free stages, different types of nerve cells and ciliated sensory organs, three types of cells in the protonephridia, strand-like cross-striated muscle cells, vacuolated cylindrical muscle cells of the chordoid organ, mesenchyme cells with large vacuoles with lipids and undifferentiated cells with large nuclei that divide and form the inner buds.

Immunoreactivity studies using fluorescence-coupled antibodies has given deeper insights into the anatomy and function of Cycliophora. The myoanatomy of all stages in the cycliophoran life cycle has been investigated using fluorescence-coupled phalloidin to label filamentous F-actin (Neves et al. 2008; Neves, Kristensen et al. 2009; Neves, Cunha et al. 2010), while the neuroanatomy of Cycliophora has been studied with antibodies directed for a number of markers such as serotonin, synapsin and FMRFamide (Wanninger 2005; Neves, da Cunha, Kristensen et al. 2010; Neves, Kristensen et al. 2010).

The standard fragment of the mitochondrial cytochrome c oxidase subunit I (COI) gene has been used for both species identification and phylogeographic analyses (Obst et al. 2005). Microsatellite loci have not been applied or characterized yet.

14.8 CHALLENGING QUESTIONS BOTH IN ACADEMIC AND APPLIED RESEARCH

One of the challenging questions that remain to be answered is the phylogenetic position of Cycliophora inside Spiralia. Phylogenetic affinities to Bryozoa and Entoprocta were suggested when Cycliophora was first described (Funch and Kristensen 1995), and later a sister group relationship between Entoprocta and Cycliophora was proposed (Funch and Kristensen 1997; Zrzavy et al. 1998; Sørensen et al. 2000). These suggestions were supported by limited and ambiguous morphological characters such as the presence of asexual reproduction by internal budding, complete nervous system degeneration during the transition and settlement from smaller free motile life cycle stages to sessile larger stages and mushroom-shaped extensions of the basal lamina into the epidermis. An alternative hypothesis was proposed based on only molecular data (18S rRNA), namely Syndermata (Rotifera + Acanthocephala) as sister group to Cycliophora (Winnepenninckx et al. 1998). This relationship to gnathiferan taxa was later supported in a number of phylogenetic analyses (Giribet et al. 2000; Peterson and Eernisse 2001; Zrzavy et al. 2001; Zrzavy 2003; Giribet et al. 2004). In the latter study based on four molecular loci, the phylogenetic position of Cycliophora was uncertain, but it tended to support a relationship to Syndermata (Rotifera + Acanthocephala), but the morphological data supporting this relationship were weak (Funch et al. 2005), although Wanninger (2005) suggested similarities in myoanatomy of the cycliophoran chordoid larva and certain rotifers. Phylogenetic analyses using more molecular data resurrected the cycliophoran affinity to entoprocts (Passamaneck and Halanych 2006; Paps et al. 2009), enforced by phylogenomic analyses based on expressed sequence tags (Hejnol et al. 2009; Nesnidal et al. 2013) and transcriptomes (Laumer et al. 2015; Kocot et al. 2017; Laumer et al. 2019). So, while the Cycliophora + Entoprocta clade seems to be well supported, its placement within Spiralia is still unsettled.

There are numerous remaining questions to clarify regarding the life cycle and reproduction in Cycliophora. First, fertilization has never been observed. It is known that females inside brood chambers have oocytes and that free females have embryos (Figures 14.6 and 14.7). It seems likely that fertilization could happen during escape or just after escape of the female, which could explain the preferred site for settlement of the Prometheus larva close to the cloaca opening of the feeding stage, which is also the site of escape of the female. Second, how do free motile stages select the right site for settlement and permanent attachment? Pandora larvae and chordoid larvae seem to prefer the same sites for settlement, namely the food-rich medial areas of the mouth parts of the host. Females prefer to settle upon areas of the mouth parts laterally, maybe because of less mechanical stress and risk of dislocation by the movements of the host, while Prometheus larvae settle upon feeding stages that develop females inside. In spite of these differences in preferred sites to settle, they are all equipped with morphologically similar long stiff ciliary sensory organs that are absent in the sessile stages in the life cycle. Most likely, these sensory organs are involved in sensing and testing if a given substrate is suitable for settlement, but nothing is known about the sensory physiology and type of mechanisms involved. The chordoid larva is equipped with more types of sensory organs, probably because it is a dispersal stage between hosts and uses some of these sensory organs for long-distance sensing. Third, the sex determination system in Cycliophora is unknown. Is haplodiploidy involved, and are cycliophoran dwarf males haploid like the males in, for example, monogonont rotifers? Probably not. In monogonont rotifers, haploid males develop from unfertilized meiotic eggs, while cycliophoran males seem to develop asexually from budding cells. Finally, the mechanism for shifting from asexual to sexual reproduction is unknown. It is unknown if a feeding stage produces a fixed number of Pandora larvae before the shift to sexual reproduction or if it depends on population density of cycliophorans on the host or food availability. Maybe starving of a feeding stage could induce formation of a Prometheus larva instead of a Pandora larva since the latter larva is large and requires more energy to produce.

Dwarf males of Cycliophora consist of less than 200 cells and have only few cell types (Obst and Funch 2003; Neves, Sørensen et al. 2009; Neves and Reichert 2015). Still, the body architecture is relatively complex, with well-developed nervous system, sensory organs, musculature and reproductive organs, which contradicts the general assumption about correlation of complexity of the body plan and the number of cells and cell types (Bell and Mooers 1997). Future exploration of the cycliophoran genome could provide new insights into how high body plan complexity can be achieved with few cells.

ACKNOWLEDGMENTS

I greatly acknowledge the collaboration with Stine Elle and Birgitte Rubæk and Reinhardt Møbjerg Kristensen producing the line drawings.

BIBLIOGRAPHY

Andersen, P. F. 1992. Beskrivelse af en ny klasse—Mesoprocta. Master thesis, University of Copenhagen. Copenhagen.

Baker, J. M., P. Funch, and G. Giribet. 2007. Cryptic speciation in the recently discovered American cycliophoran *Symbion americanus*: Genetic structure and population expansion. *Marine Biology* 151 (6):2183–2193.

Baker, J. M., and G. Giribet. 2007. A molecular phylogenetic approach to the phylum Cycliophora provides further evidence for cryptic speciation in *Symbion americanus*. *Zoologica Scripta* 36 (4):353–359.

Bell, G., and A. O. Mooers. 1997. Size and complexity among multicellular organisms. *Biological Journal of the Linnean Society* 60 (3):345–363.

Clément, P., and E. Wurdak. 1991. Rotifera. In *Microscopic Anatomy of Invertebrates*, edited by F. W. Harrison and R. M. Woollacott, 219–297. New York etc.: Wiley-Liss.

Funch, P. 1996. The chordoid larva of *Symbion pandora* (Cycliophora) is a modified trochophore. *Journal of Morphology* 230 (3): 231–263.

Funch, P., and R. M. Kristensen. 1995. Cycliophora is a new phylum with affinities to Entoprocta and Ectoprocta. *Nature* 378 (6558):711–714.

Funch, P., and R. M. Kristensen. 1997. Cycliophora. In *Microscopic Anatomy of Invertebrates*, edited by F. W. Harrison and R. M. Woollacott, 409–474. New York etc.: Wiley-Liss.

Funch, P., and R. M. Kristensen. 1999. Cycliophora. In *Encyclopedia of Reproduction*, edited by Ernst Knobil and Jimmy D. Neill, 800–808. San Diego: Academic Press.

Funch, P., and R. Neves. 2019. Cycliophora. In *Miscellaneous Invertebrates*, edited by A. Schmidt-Rhaesa, 87–110. Berlin/Boston: Walter de Gruyter GmbH & Co. KG.

Funch, P., M. V. Sørensen, and M. Obst. 2005. On the phylogenetic position of Rotifera: Have we come any further? *Hydrobiologia* 546:11–28.

Funch, P., P. Thor, and M. Obst. 2008. Symbiotic relations and feeding biology of *Symbion pandora* (Cycliophora) and *Triticella flava* (Bryozoa). *Vie et Milieu-Life and Environment* 58 (2):185–188.

Giribet, G., D. L. Distel, M. Polz, W. Sterrer, and W. C. Wheeler. 2000. Triploblastic relationships with emphasis on the acoelomates and the position of Gnathostomulida, Cycliophora, Plathelminthes, and Chaetognatha: A combined approach of 18S rDNA sequences and morphology. *Systematic Biology* 49 (3):539–562.

Giribet, G., M. V. Sørensen, P. Funch, R. M. Kristensen, and W. Sterrer. 2004. Investigations into the phylogenetic position of Micrognathozoa using four molecular loci. *Cladistics* 20 (1):1–13.

Hejnol, A., M. Obst, A. Stamatakis, M. Ott, G. W. Rouse, G. D. Edgecombe, P. Martinez, J. Baguna, X. Bailly, U. Jondelius, M. Wiens, W. E. G. Muller, E. Seaver, W. C. Wheeler, M. Q. Martindale, G. Giribet, and C. W. Dunn. 2009. Assessing the root of bilaterian animals with scalable phylogenomic methods. *Proceedings of the Royal Society B-Biological Sciences* 276 (1677):4261–4270.

Kocot, K. M., T. H. Struck, J. Merkel, D. S. Waits, C. Todt, P. M. Brannock, D. A. Weese, J. T. Cannon, L. L. Moroz, B. Lieb, and K. M. Halanych. 2017. Phylogenomics of Lophotrochozoa with consideration of systematic error. *Systematic Biology* 66 (2):256–282.

Kristensen, R. M. 2002. An introduction to Loricifera, Cycliophora, and Micrognathozoa. *Integrative and Comparative Biology* 42 (3):641–651.

Kristensen, R. M., and P. Funch. 2002. Phylum Cycliophora. In *Atlas of Marine Invertebrate Larvae*, edited by Craig M. Young, M. A. Sewell and M. Rice, 231–240. San Diego, CA: Academic Press.

Laumer, C. E., N. Bekkouche, A. Kerb, F. Goetz, R. C. Neves, M. V. Sørensen, R. M. Kristensen, A. Hejno, C. W. Dunn, G. Giribet, and K. Worsaae. 2015. Spiralian phylogeny informs the evolution of microscopic lineages. *Current Biology* 25 (15):2000–2006.

Laumer, C. E., R. Fernandez, S. Lemer, D. Combosch, K. M. Kocots, A. Riesgo, S. C. S. Andrade, W. Sterrer, M. V. Sørensen, and G. Giribet. 2019. Revisiting metazoan phylogeny with genomic sampling of all phyla. *Proceedings of the Royal Society B-Biological Sciences* 286 (1906).

Nesnidal, M. P., M. Helmkampf, A. Meyer, A. Witek, I. Bruchhaus, I. Ebersberger, T. Hankeln, B. Lieb, T. H. Struck, and B. Hausdorf. 2013. New phylogenomic data support the monophyly of Lophophorata and an ectoproct-phoronid clade and indicate that Polyzoa and Kryptrochozoa are caused by systematic bias. *BMC Evolutionary Biology* 13.

Neves, R. C., X. Bailly, and H. Reichert. 2014. Are copepods secondary hosts of Cycliophora? *Organisms Diversity & Evolution* 14 (4):363–367.

Neves, R. C., M. R. Cunha, P. Funch, R. M. Kristensen, and A. Wanninger. 2010. Comparative myoanatomy of cycliophoran life cycle stages. *Journal of Morphology* 271 (5):596–611.

Neves, R. C., M. R. da Cunha, P. Funch, A. Wanninger, and R. M. Kristensen. 2010. External morphology of the cycliophoran dwarf male: A comparative study of *Symbion pandora* and *S. americanus*. *Helgoland Marine Research* 64 (3):257–262.

Neves, R. C., M. R. da Cunha, R. M. Kristensen, and A. Wanninger. 2010. Expression of synapsin and co-localization with serotonin and RFamide-like immunoreactivity in the nervous system of the chordoid larva of *Symbion pandora* (Cycliophora). *Invertebrate Biology* 129 (1):17–26.

Neves, R. C., J. C. Guimaraes, S. Strempel, and H. Reichert. 2017. Transcriptome profiling of *Symbion pandora* (phylum Cycliophora): Insights from a differential gene expression analysis. *Organisms Diversity & Evolution* 17 (1):111–119.

Neves, R. C., R. M. Kristensen, and P. Funch. 2012. Ultrastructure and morphology of the cycliophoran female. *Journal of Morphology* 273 (8):850–869.

Neves, R., R. M. Kristensen, and A. Wanninger. 2008. New insights from the myoanatomy of *Symbion americanus* (phylum Cycliophora) revealed by confocal microscopy and 3D reconstruction software. *Journal of Morphology* 269 (12):1480–1480.

Neves, R. C., R. M. Kristensen, and A. Wanninger. 2009. Three-dimensional reconstruction of the musculature of various life cycle stages of the cycliophoran *Symbion americanus*. *Journal of Morphology* 270 (3):257–270.

Neves, R. C., R. M. Kristensen, and A. Wanninger. 2010. Serotonin immunoreactivity in the nervous system of the Pandora larva, the Prometheus larva, and the dwarf male of *Symbion americanus* (Cycliophora). *Zoologischer Anzeiger* 249 (1):1–12.

Neves, R. C., and H. Reichert. 2015. Microanatomy and development of the dwarf male of *Symbion pandora* (phylum Cycliophora): New insights from ultrastructural investigation based on serial section electron microscopy. *PLoS One* 10 (4).

Neves, R. C., K. J. K. Sørensen, R. M. Kristensen, and A. Wanninger. 2009. Cycliophoran dwarf males break the rule: High complexity with low cell numbers. *Biological Bulletin* 217 (1):2–5.

Obst, M., and P. Funch. 2003. Dwarf male of *Symbion pandora* (Cycliophora). *Journal of Morphology* 255 (3):261–278.

Obst, M., and P. Funch. 2006. The microhabitat of *Symbion pandora* (Cycliophora) on the mouthparts of its host *Nephrops norvegicus* (Decapoda: Nephropidae). *Marine Biology* 148 (5):945–951.

Obst, M., P. Funch, and G. Giribet. 2005. Hidden diversity and host specificity in cycliophorans: A phylogeographic analysis along the North Atlantic and Mediterranean Sea. *Molecular Ecology* 14 (14):4427–4440.

Obst, M., P. Funch, and R. M. Kristensen. 2006. A new species of Cycliophora from the mouthparts of the American lobster, *Homarus americanus* (Nephropidae, Decapoda). *Organisms Diversity & Evolution* 6 (2):83–97.

Paps, J., J. Baguna, and M. Riutort. 2009. Lophotrochozoa internal phylogeny: New insights from an up-to-date analysis of nuclear ribosomal genes. *Proceedings of the Royal Society B-Biological Sciences* 276 (1660):1245–1254.

Passamaneck, Y., and K. M. Halanych. 2006. Lophotrochozoan phylogeny assessed with LSU and SSU data: evidence of lophophorate polyphyly. *Molecular Phylogenetics and Evolution* 40 (1):20–28.

Peterson, K. J., and D. J. Eernisse. 2001. Animal phylogeny and the ancestry of bilaterians: Inferences from morphology and 18S rDNA gene sequences. *Evolution & Development* 3 (3):170–205.

Plaza, M. 2012. Cycliophoran host range. *Bulletin of the Peabody Museum of Natural History* 53 (2):389–395.

Sørensen, M. V., P. Funch, E. Willerslev, A. J. Hansen, and J. Olesen. 2000. On the phylogeny of the Metazoa in the light of Cycliophora and Micrognathozoa. *Zoologischer Anzeiger* 239 (3–4):297–318.

Wanninger, A. 2005. Immunocytochemistry of the nervous system and the musculature of the chordoid larva of *Symbion pandora* (Cycliophora). *Journal of Morphology* 265 (2):237–243.

Winnepenninckx, B. M. H., T. Backeljau, and R. M. Kristensen. 1998. Relations of the new phylum Cycliophora. *Nature* 393 (6686):636–638.

Zrzavy, J. 2003. Gastrotricha and metazoan phylogeny. *Zoologica Scripta* 32 (1):61–81.

Zrzavy, J., V. Hypsa, and D. F. Tietz. 2001. Myzostomida are not annelids: Molecular and morphological support for a clade of animals with anterior sperm flagella. *Cladistics* 17 (2):170–198.

Zrzavy, J., S. Mihulka, P. Kepka, A. Bezdek, and D. Tietz. 1998. Phylogeny of the Metazoa based on morphological and 18S ribosomal DNA evidence. *Cladistics* 14 (3):249–285.

15 Crustaceans

Nicolas Rabet

CONTENTS

15.1 HISTORY OF THE MODEL

The word crustacea is derived from the Latin *crusta*, which means that the body is covered with a hard shell. The name Crustacea was first proposed by Brünnich (1772). Nevertheless, it took decades for it to establish itself, and the boundaries of the group have also changed significantly.

Today, crustaceans are a paraphyletic group, representing approximately 70,000 currently valid species distributed in nearly 1,000 families and in 9 major lineages (Remipedia, Cephalocarida, Malacostraca, Copepoda, Thecostraca, Branchiopoda, Mystacocarida, Branchiura and Ostracoda) (Ahyong et al. 2011; Regier et al. 2010).

Large crustaceans (malacostracans and barnacles—Figure 15.1) have always been known to humanity because they have been eaten for thousands of years (Gutiérrez-Zugasti 2011; Zilhão et al. 2020). It is therefore quite logical that we can find crustaceans in old illustrations or in first classifications. In Aristotle's classification, some crustaceans were already listed under the name *μαλακόστρακα* (*malakostraka*), which means animals with soft (*malakós*) shell (*óstrakon*) (Zucker 2005). Even if the word Malacostraca evokes a classic name of the current classification, for a very long time, most crustaceans were integrated among the insects without a specific group. Others were ignored or sometimes classified with other organisms. For example, Linnaeus (1758) classified some crustaceans in the order of Aptera and recognized only three genera: *Cancer* with malacostracans and branchiopod anostracans; *Monoculus* with branchiura, other branchiopods, copepods, ostracods and two taxa including horseshoe crabs (which are now excluded from crustaceans); and *Oniscus*, regrouped malacostracan isopods. In addition, the cirripeds with genus *Lepas* was classified in the Vermes Testacea, while the parasitic copepods with the genus *Lernaea* were classified among the Vermes Mollusca.

Gradually, many species were described, and crustaceans were separated from insects on the basis of having a predominantly aquatic life, the presence of two pairs of antennae, biramate appendages and a nauplius larva. Like the morpho-anatomical diversity of the group, its classification has carried out numerous regroupings, and as such, many have been forgotten. The copepods, ostracods, branchiopods and cirripeds were gradually individualized and grouped in the entomostracans as opposed to the malacostracans (see Monod and Forest 1996). In the 20th century, new lineages of crustaceans were discovered, such as mystacocarids (Pennak and Zinn 1943), cephalocarids (Sanders 1955) and remipeds (Yager 1981). Bowman and Abele (1982) proposed a classification with six classes (Cephalocarida, Branchiopoda, Remipedia, Maxillopoda, Ostracoda and Malacostraca). The Maxillopoda grouped together the Mystacocarida, Cirripedia, Copepoda and Branchiura.

DOI: 10.1201/9781003217503-15

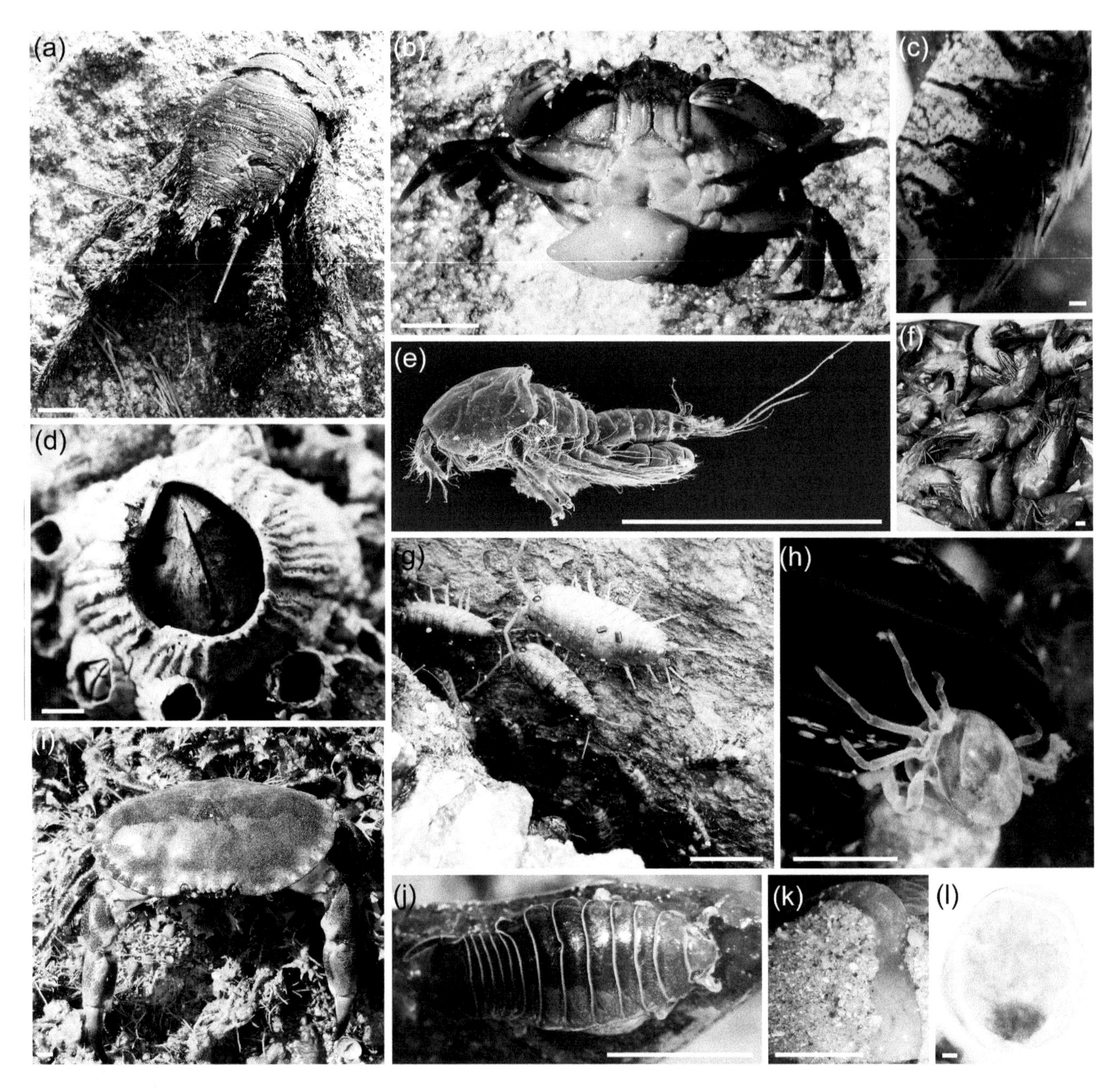

FIGURE 15.1 Marine crustacean (only Multicrustacea here) diversity illustrating morphological diversity, ecology and use. (a) *Galathea strigosa* (malacostracan); (b) *Carcinus maenas* (malacostracan) and *Sacculina carcini* (cirripeds); (c) *Palaemon elegans* eggs (malacostracan); (d) *Semibalanus balanoides* (barnacle); (e) *Tigriopus brevicornis* (copepod); (f) peneids in a market (malacostracan); (g) *Ligia oceanica* (malacostracan); (h) *Pinnotheres pisum* (malacostracan); (i) *Cancer pagurus* (malacostracan), Anilocra frontalis (malacostracan), *Processa edulis* (malacostraca), *caudal* gene expression in late embryo of *Sacculina carcini* (cirripeds). Scale bar: (a, b, f, g, h, i, j, k) = 1 cm; (c, d, e) = 1 mm; (l) = 10 μm.

Since then, molecular phylogenies have completely revolutionized this classification.

The pentastomides, which are respiratory parasites of vertebrates that were previously classified in many groups such as Tardigrada, Annelida, Platyhelminthes and Nematoda and have a strange, elongated, worm-like body ringed with two pairs of hooks, were finally integrated into the Branchiura thanks to the 18S gene sequencing comparison (Riley et al. 1978; Abele et al. 1989; Martin and Davis 2001; Lavrov et al. 2004).

Other analyses identified that the hexapods, previously believed to be close to crustaceans, were ultimately a lineage inside crustaceans (Regier et al. 2010) (Figure 15.2). As a result, crustaceans are not a monophyletic group but a paraphyletic group whose use remains practical to the extent that most animals are aquatic and share many ancestral characters. The name of the group incorporating hexapods among crustaceans is called the Pancrustacea, initially proposed by Zrzavý and Štys (1997), and some authors also use the name Tetraconata (Dohle 2001; Richter 2002). Several studies

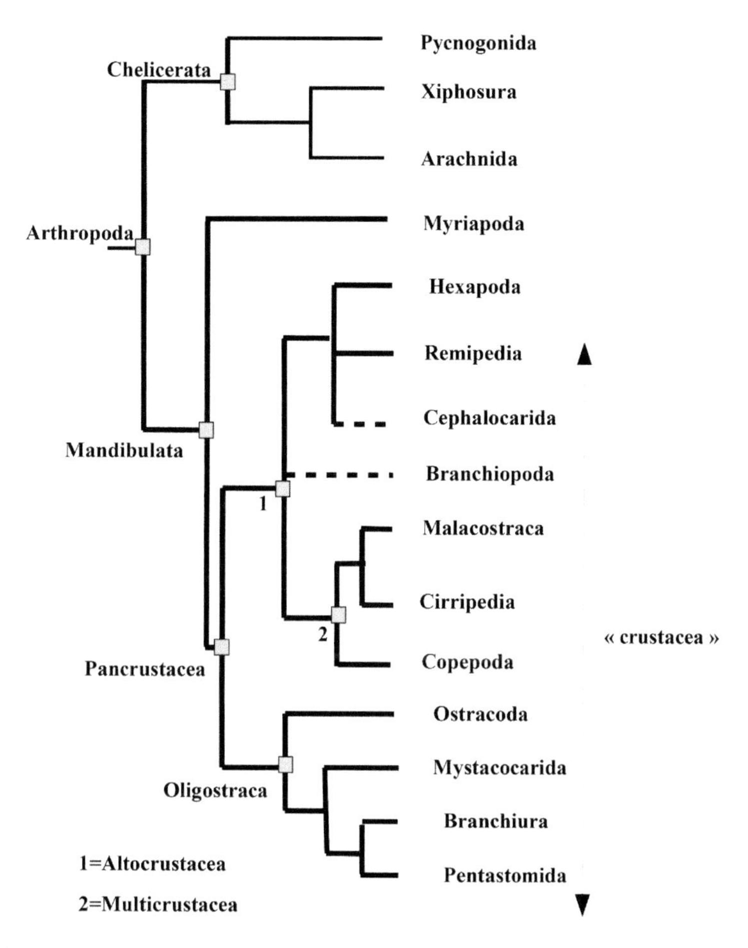

FIGURE 15.2 Phylogeny of Arthropoda. The dotted lines indicate that the position of these branches is uncertain. This figure clearly shows that crustaceans are paraphyletic. (From synthetic phylogeny built from Regier et al. 2010; Schwentner et al. 2017; Giribet and Edgecombe 2019.)

are now confirmed this important finding (Lee et al. 2013; Schwentner et al. 2017).

Another important change in crustacean phylogeny is that maxillopods are not monophyletic (Regier et al. 2005).

The relationships within the pancrustaceans are not entirely clear (Figure 15.1), mainly with respect to the position of branchiopods and cephalocarids (Schwentner et al. 2017; Giribet and Edgecombe 2019) (Figure 15.2). The earliest emergent group, called Oligostraca, contains Ostracoda, Branchiura, Tantulocarida, Mystacocarida and Pentastomida. It is the sister group of the rest of the Pancrustacea, called Altocrustacea and including Multicrustacea. The Multicrustacea contains the Malacostraca, Copepoda and Thecostraca (including cirripeds) (Figure 15.1). The position of Cephalocarida and Branchiopoda remains uncertain. All of these Pancrustacea lineages are very old, as evidenced by the fact that there were already malacostracans (Collette and Hagadorn 2010) and branchiopods (Waloszek 1993) present in the Cambrian era. Phylogenetic analysis has allowed scientists to confirm this (Regier et al. 2005), which implies that Pancrustacea has a truly ancient history with numerous lineages, a large part of which has probably disappeared.

In recent years, an important malacostracan amphipod model has been set up to study the development of crustaceans: *Parhyale hawaiensis* (Browne et al. 2005). This model is important enough to constitute the subject of an entire part of the next chapter, and, as such, it will not be included in this chapter. Furthermore, in this chapter, some continental aquatic organisms will be considered with strictly marine animals for reasons of phylogenetic coherence and usage.

15.2 GEOGRAPHICAL LOCATION

Crustaceans are extremely diverse and widely distributed all over the world in all climates. The place of the marine environment for crustaceans is considerable both in terms of the number of species and in the lineages represented. They also have considerable ecological functions. The whole will therefore be difficult to summarize, and we will focus on only some specific adaptations.

Some crustacean species inhabit the deepest marine environments, such as the malacostracan amphipod *Hirondellea gigas*, which lives in the Mariana Trench, sometimes at depths of more than 10,000 meters. It consumes sunken wood coming from the surface thanks to particular enzymatic activities detected in the animal's gut (Kobayashi et al. 2012) and has also developed an aluminum hydroxide gel that covers its exoskeleton and that may be linked to life at great depths (Kobayashi et al. 2019). In the deep sea, there are also many crustaceans that live around hydrothermal vents. Many of them use chemo-autotrophic bacteria that provide nutrients to animals. This is particularly the case with the malacostracan *Rimicaris exoculata* on the Mid-Atlantic Ridge, which harbors bacterial communities in its branchial cavities (Petersen et al. 2010; Zbinden et al. 2020).

The diversity of crustaceans is also considerable in the tidal zone, with some species able to survive conditions that vary according to the water level variations. Some, like the malacostracan *Carcinus maenas* (Figure 15.1), are able to temporarily acclimatize to the absence of water and resist consequent variations in the environment. Native to Europe, this particularly well-adapted species has colonized many temperate sites around the world (Jensen et al. 2007). In pools of the highest tidal levels, we can often observe copepods *Tigriopus* (Figure 15.1), which are also impacted by high temperatures and consequent variations in salinity (Fraser 1936; Raisuddin et al. 2007). As in many groups, underwater caves have also been colonized and can be the refuge of many specialized and original organisms. Among these are remipeds, a group of blind, predatory crustaceans that inhabit anchialine underwater caves (Yager 1981; Koenemann et al. 2007). These are also the only venomous crustaceans (von Reumont et al. 2014). Among the meiofauna, there are many species of crustaceans such as copepods and ostracods living in sediments. It is also in this type of biotope that we can find the odd cephalocarids (Sanders 1955; Neiber et al. 2011).

Many crustaceans such as ostracods, malacostracans, copepods and branchiopods have also colonized brackish or fresh water. The border between the two environments is not necessarily clear, and after passing through fresh water, some organisms then return to the marine environment, such as the marine cladocerans that represent few species but have a global distribution (Durbin et al. 2008). The hypersaline environments that form in coastal areas or sometimes in the middle of continents have also been colonized by crustaceans, in particular ostracods and copepods. However, the champion of resistance is unmistakably the branchiopod *Artemia*, which can survive in supersaturated salty environments up to 340 g/l (Gajardo and Beardmore 2012).

As it is sometimes difficult to dissociate marine crustaceans from freshwater or hypersaline crustaceans in an evolutionary way, they will be partially integrated in this chapter.

There have also been several colonizations by pancrustaceans of terrestrial environments such as hexapods or woodlice, but there are also terrestrial lineages in the adult state whose larvae are completely marine, as is the case for many terrestrial crabs or terrestrial hermit crabs. In this category, there are the largest land-living arthropods, like the coconut crab (*Birgus latro*) (Krieger et al. 2010). This hybrid lifestyle, which is also found in amphidromic crustaceans (living partially in freshwater and seawater), allows these animals to exploit the dispersive abilities of marine planktonic life and to colonize more or less isolated continental environments (Bauer 2013).

Crustaceans are also an essential component in the plankton of all seas. Some species live their entire life cycle as plankton and play a major ecological role (copepods, euphausiids). However, for many species, the passage through plankton is transient as part of a marine bentho-pelagic species or many terrestrial or freshwater crustaceans.

Crustaceans are so ubiquitous, it is almost impossible to study the aquatic environment without finding one!

15.3 LIFE CYCLE

In crustaceans, the life cycle presents extremely variable modalities. The majority of species are gonochoric with separate sexes, but there are cases of parthenogenesis in the brine shrimp *Artemia* (Bowen et al. 1978) and many freshwater and terrestrial species, probably due to the dispersive advantage (Scholtz et al. 2003; Kawai et al. 2009). There are cases of simultaneous hermaphroditism (both type of gonads are present simultaneously) in remipeds (Neiber et al. 2011), cephalocarids (Addis et al. 2012), cirripeds (Charnov 1987) and some branchiopods (Scanabissi and Mondini 2002; Weeks et al. 2014). Sequential hermaphrodism (change of sex during the life) is more observed in malacostracans (Benvenuto and Weeks 2020).

The mating modalities are also extremely varied in crustaceans and result in very different appendicular adaptations. The most original is undoubtedly the presence of a long penis in the barnacles which is always fixed and which compensates for the low mobility of the gametes (Barazandeh et al. 2013).

In most species, the mother will protect her offspring to allow the release of larvae. However, most calanoid copepods, euphausiids and dendrobranchiate decapods (Penaeoidea and Sergestoidea) shed their eggs into the water column (Lindley 1997).

In many crustaceans, the instability of trophic resources and living conditions has favored the development of a strategy of slowing down or stopping development during the deficit season (Alekseev and Starobogatov 1996). In this case, the eggs are laid and start diapausis. There are also resistance forms in anhydrobiosis or cryptobiosis (absence of metabolism with dehydration) (Fryer 1996; Alekseev and Starobogatov 1996). This innovation sometimes concerns the larvae, as in the copepod *Metacyclops minutus* (Maier 1992), but more often, it is the embryo that enters a state of suspended life. The embryo can be enveloped by different layers of varying natures and becomes resistant to drying out or freezing. In this form, we speak of a resting egg (also called a "duration egg" or "cyst"), and, when conditions are favorable, development resumes, leading to the release of a larva or an aquatic juvenile (Brendonck 2008).

In a group of malacostracan shrimps of the Alpheidae family, the existence of eusocial behavior has recently been reported, such as is found in insects and vertebrates (Duffy 1996).

In many species, the larvae released after hatching become planktonic. During this planktonic phase, the animals grow and disperse. At the end of the larval stages, there are animals whose adults remain in the plankton (many copepods, euphausiids) and others which emerge, most often becoming benthic. Sometimes the modifications are brutal and called metamorphosis for sessile animals, like in barnacles (Høeg and Møller 2006; Maruzzo et al. 2012). In this group, the transformation will result in a completely fixed animal. The choice of the fixation site is therefore essential for the survival of the individual, because it will subsequently have to withstand the conditions imposed by the environment. Recruitment is carried out by olfaction through antenna 1 of the substrate (Figure 15.4c). The bacterial film can be detected and, depending on its composition, induce the attachment of the cyprid larva (Rajitha et al. 2020). The presence of congeners due to the release of pheromones from adults that are not always necessarily from the same species is also an essential factor for fixation (Abramova et al. 2019). After an exploration phase using the attachment discs located at the end of the antenna 1, the final fixation is achieved by the deposition of a cement comprised of lipids and phosphoproteins (Liang et al. 2019).

In parasitic crustaceans, the life cycle is often highly modified. The most extensive parasitic life transformations are found in pentastomids, copepods and cirripeds. Adults are often very divergent from their non-parasitic parents, to the point that association with a taxonomic group has only been possible by studying the larval stages like for the cirriped Rhizocephala (Thompson 1836) (Figure 15.1b) or more recently by molecular data, like for the pentastomids (Abele et al. 1989). Rhizocephalic cirripeds are parasites characterized by considerable morphological transformations but also considerable modifications of their life cycle. The female larva will transform into a kentrogon, a kind of injection system that allows a few cells to invade the host, which will develop into a network resembling roots and allowing it to feed. The male larvae transform into trichogons and settle as hyperparasites on the females. The mature parasite profoundly modifies the physiology of the organism by feminizing it and blocking the molt (Delage 1884; Høeg and Lützen 1995).

15.4 EMBRYOGENESIS AND LARVAL DEVELOPMENT

In crustaceans, embryonic development is very variable depending on the groups or species. In the case of direct development, all of the ontogenetic stages lead to the release of a juvenile. In the case of the release of a larva, the steps missing to obtain a juvenile will be performed by larval development. The predominance of one or the other is therefore variable depending on phylogenetic history and ecological context, and both must be studied to understand the ontogeny of a species. In the case of the release of a larva, the essential difference with the equivalent embryonic stages in another species is at least the acquisition of mobility and sometimes early nutrition.

The modalities of embryonic development are extremely variable in crustaceans, and it is not possible to present them all here. We will use Chapter 16 as a reference for malacostracans, and here we will mainly develop the *Artemia* model, which is the organism with the best-studied anamorphic development.

15.4.1 Embryogenesis

The embryonic development of *Artemia* has been described by Benesch (1969) and Rosowski et al. (1997). After fertilization, the embryo forms a gastrula. Postgastrulean development until nauplius hatching occurs without any cell division (Olson and Clegg 1978). The 5,000 cells present in the gastrula organize and differentiate the head structures, including the three pairs of appendages and the salt gland. The rest of the head and the post-cephalic structures are formed from the remaining 2,000 cells. The posterior region of the embryo then takes the shape of a cone, and the ectoderm of this post-mandibular region takes on the appearance of a grid with long columns of cells arranged in parallel along the antero–posterior axis. The posterior region thus resembles that of other crustaceans, but in this case, it results from a phenomenon of reorganization. Upon hatching, the cells that compose the larva are small and diploid in the posterior region, while the cephalic elements (salt gland and appendages) are constituted by polyploid cells (Olson and Clegg 1978). At the gastrula stage, the embryo can go into cryptobiosis, and the dormant state is stabilized by the P26 protein (Malitan et al. 2019). In this case, the outer layers (shell) of the embryo are produced by the shell glands of the female (Morris and Afzelius 1967; Anderson 1970; Garreau de Loubresse 1974) and allow the protection of the embryo against variations in the environment. A shell gland specifically expressed gene (SGEG) has been found to be involved in egg shell formation. Lacking SGEG protein (by RNA interference) caused the eggs' shell to become translucent and induce a defective resting egg (Liu et al. 2009).

15.4.2 Larval Development

The emblematic larva of crustaceans is undoubtedly the nauplius larva (Figure 15.3). The first observation of nauplius dates back to the emergence of the first microscopes and was made by Antonie van Leeuwenhoek in 1699 on *Cyclops* copepods (Gurney 1942). Since then, it is found in many lineages of Pancrustacea and is probably one of the synapomorphies of this group (Regier et al. 2010). It is an externally unsegmented oligomeric head larva with three pairs of appendages and one pair of eyes corresponding to the most anterior part of the head (Figure 15.3a–c) (Dahms 2000). It shows similarities with the protonymphon larva of the pycnogonids, and the presence of homologous appendages (Figure 15.3d) suggests that this type of larva is possibly ancestral (Alexeeva et al. 2017). In crustaceans, the nauplius is the earliest larval stage observed.

The larval development of *Artemia* has been studied in detail (Anderson 1967; Benesch 1969; Schrehardt 1987). The development of the anterior structures leads to the replacement of structures composed of polyploid cells by the definitive adult organs, developed from diploid precursor

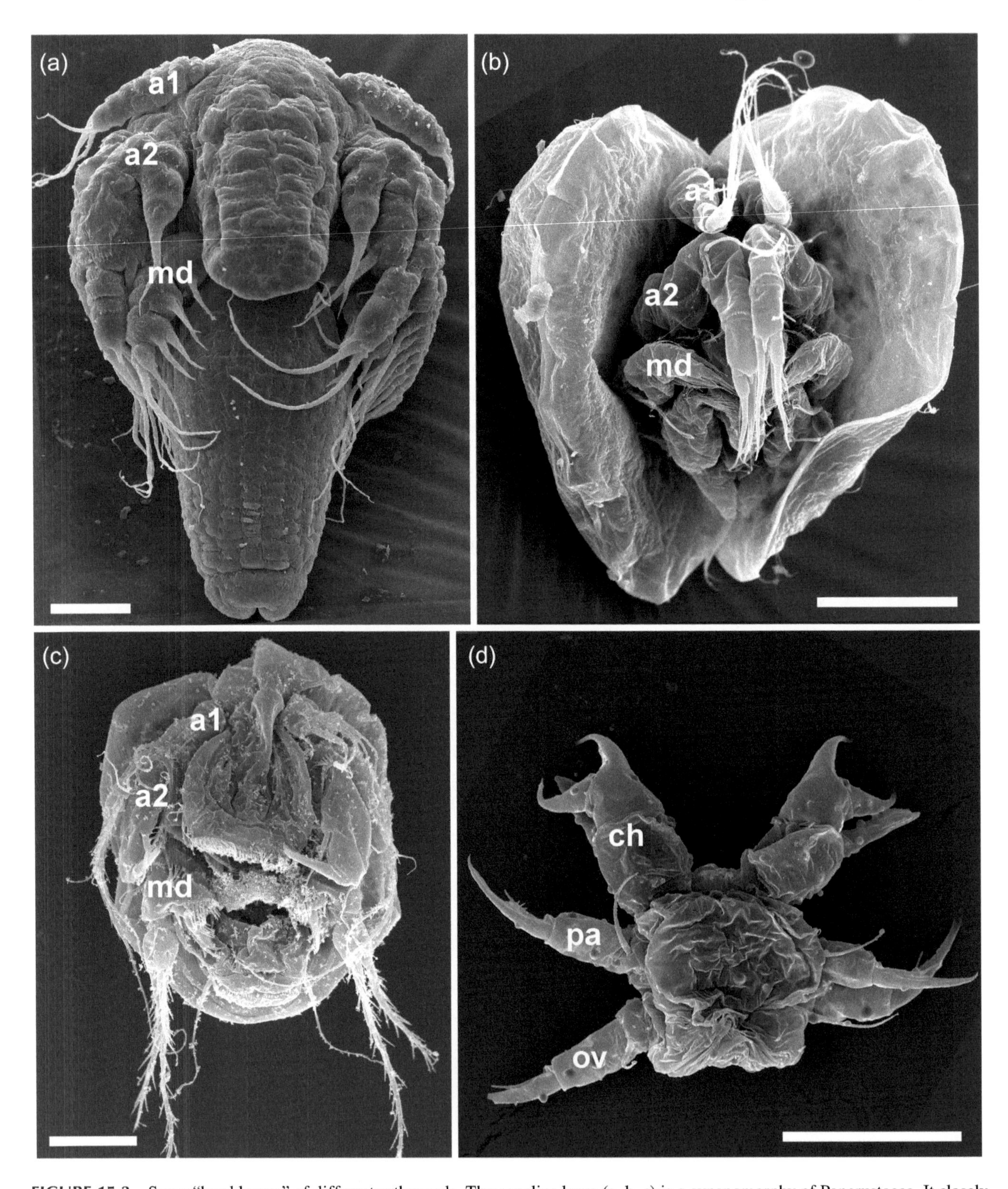

FIGURE 15.3 Some "head larvae" of different arthropods. The nauplius larva (a, b, c) is a synapomorphy of Pancrustacea. It closely resembles the protonymphon larva of sea spiders (d). (a) *Artemia franciscana* (branchiopods); (b) *Heterocypris incongruens* (ostracods); (c) *Tigriopus brevicornis* (copepods); (d) *Endeis sp.* (pycnogonids). a1: antenna 1, a2: antenna 2, md: mandible, ch: cheliphore, pa: palp, ov: oviger. The scale bar measures 50 μm for (a, b, c) and 10 μm for (d).

cells remaining within the cephalic structures (Olson and Clegg 1978).

In the posterior region of the larva, in front of the telson, a "morphogenetic differentiation area" is established. Along with this expression, the arrangement of the cells changes, forming rows of cells perpendicular to the anteroposterior axis (Figure 15.3a). In this same area, the intersegmental boundaries then appear by constriction of the ectoderm around the body, first creating the parasegments, then the final segments (Prpic 2008). We can therefore observe, in the same *Artemia* larva, a whole series of levels of development of the segments and their appendages (Figures 15.4, 15.5). When new appendages appear in the nauplius, these stages can be called metanauplius (Figure 15.4).

In the posterior region of the larva, in front of the telson, the segments appear and then gradually differentiate, making it possible to distinguish, at a given stage and in an arbitrary fashion, several levels of differentiation located from back to front as follows (Figures 15.4, 15.5):

- Initial cell proliferation;
- Cellular and genetic segmentation program;
- Segmental morphogenesis;
- Morphogenesis of the appendages.

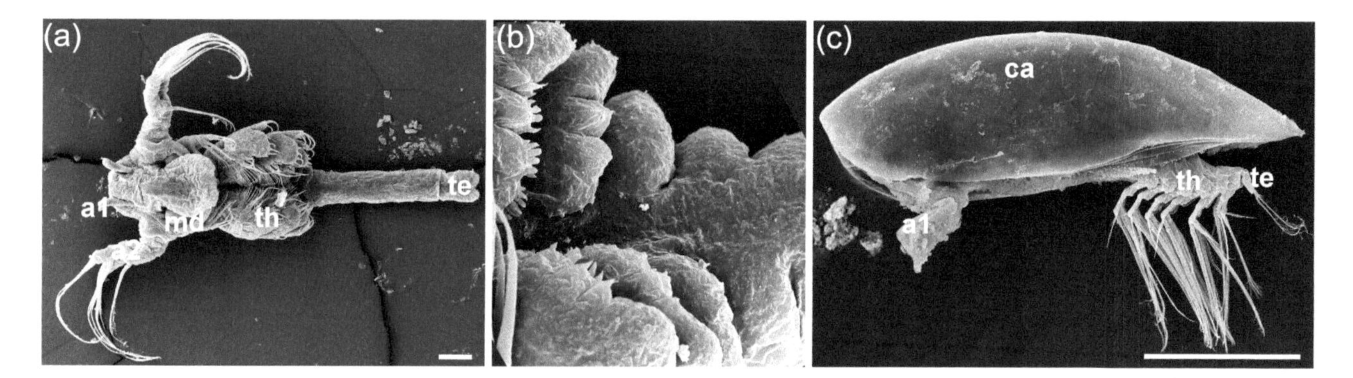

FIGURE 15.4 Larval development. (a, b) Metanauplius (late nauplius) stage of *Artemia franciscana* showing the levels of segment differentiation according to their position in the anteroposterior axis. (b) zoom of (a) at the level. (c) Cyprid stage of *Sacculina carcini*. This cyprid stage is a synapomorphy of Cirripeds and probably Thecostraceans. a1: antenna 1, a2: antenna 2, md: mandible, th: thorax, te: telson, ca: carapace. Scale bar: 100 μm.

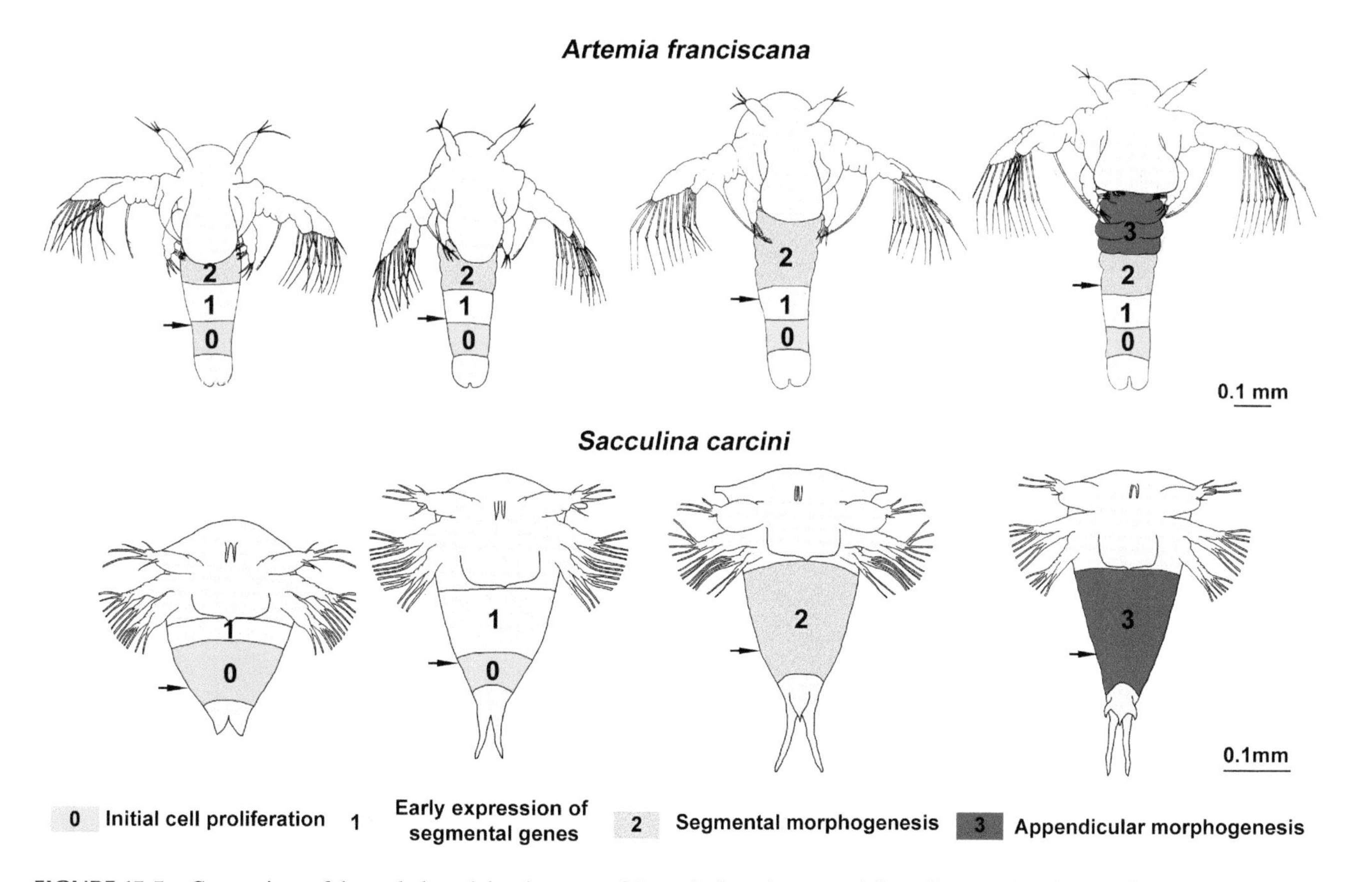

FIGURE 15.5 Comparison of the early larval development of *Artemia franciscana* and *Sacculina carcini*. *Artemia* has an anamorphous development with progressive elongation of the body. *Sacculina*, although producing nauplius, has an altered development showing synchronization of morphogenesis. The arrow indicates the position of a region of a specific thoracic segment during larval development. (The stage is redrawn after Collis and Walker 1994; Anderson 1967; Schrehardt 1987. The identifi cation of the territories is synthesized after Schrehardt 1987; Manzanares et al. 1993; Copf et al. 2003; Gibert et al. 2000; Rabet et al. 2001; Trédez 2016.)

This type of development is found in many lineages of crustaceans: Cephalocarida, Remipedia, Branchiopoda, Branchiura, Ostracoda, Copepoda, Mystacocarida, Malacostraca, Dendrobranchiata and Euphausiacea (Martin et al. 2014).

In cirripeds, nauplius are morphologically quite similar to the others, but larval development leads to a fairly synchronous intracuticular construction of thoracic segments that deviate clearly from the anamorphic model (see Figure 15.5) (Trédez et al. 2016). In addition, in this group, larval development leads to a typical stage called cypris, which precedes a metamorphosis for a fixed life (Høeg and Møller 2006; Maruzzo et al. 2012) (Figure 15.4).

In malacostracans, there are several direct or pseudo-direct developments, but in many groups, the hatching reaches a zoea-like larva stage (Jirikowski et al. 2015). This stage also appears in malacostracans producing a nauplius. The larva is characterized by a complete or nearly complete body segment number. It has functional thoracic appendages and most of the time has two eyes (Anger 2001). These generally planktonic larvae have specific names depending on their morphology and belong to different groups of malacostracans (protozoea, metazoea, mysis or phyllosoma) (Anger 2001)

In many malacostracans, an embryo with a nauplius-like form appears transiently in the embryo reminiscent of ancestral development (Scholtz 2002; Jirikowski et al. 2013; Jirikowski et al. 2015). Spawning at sea can be the subject of animal migration: Christmas Island has seen crab invasions due to a mass migration of animals during the egg-laying season (Adamczewska and Morris 2001).

15.5 ANATOMY

The morpho-anatomical diversity is quite exceptional (see Figure 15.1 only for Multicrustacea). The majority of animals have bilateral symmetry and a metameric organism. The head has an ocular region and appendages that are in sequence: two pairs of antennae (A1 and A2), the mandibles and two pairs of maxillae (M1 and M2). Both pairs of antennae and maxillae are characteristic of crustaceans (Scholtz and Edgecombe 2006). The head is made up of six segments (Zrzavý and Štys 1997). The posterior part of the body is terminated by the telson bearing the anus and sometimes with caudal furca (McLaughlin 1980). Between the head and the telson, the segments can be similar to each other and thus form a trunk in remipeds (Yager 1981; Neiber 2011), but more often, they are different and thus grouped into functional and morphological groups called tagmes. These body regions can therefore be specialized in locomotion, reproduction, respiration and nutrition functions and are generically called the thorax and the abdomen. In malacostracans, they can be called the pereion and the pleon (Mayrat and Saint Laurent 1996). It is quite possible that the tagmes are not homologous in the different groups and that the regroupings took place from an untagmatized ancestor (Averof and Akam 1995). Sometimes the head is fused with the thorax to form a cephalothorax or prosoma, with the addition of appendages associated with the function of food intake, the maxillipeds especially in copepods, some malacostracans and remipeds (Averof and Patel 1997; Yager 1981).

The number of body segments is often stable within a group, such as the hexapods. Thus, the Malacostraca has six cephalic segments, eight thoracic segments and six abdominal segments, with the exception of the leptostracans, which have seven. The different groups formerly classified in the Maxillopoda like the copepods, branchiurans, ostracods and the cirripeds have seven thoracic segments and four abdominal segments (Richter 2002).

On the other hand, in other lineages such as branchiopods or remipeds, the number of body segments can vary. For *Triops* (branchiopods), the number of segments changes within a population (Korn and Hundsdoerfer 2016).

The carapace is a structure that emerges from the posterior part of the head and covers part or all of the body. It is found in many groups of crustaceans with varying forms, and the hypothesis of its ancestrality in the line has been made (Calman 1909). The functions of the carapace are variable: in addition to a protective aspect of organisms, the carapace can have other functions such as having a role in hydrodynamics, protection of eggs, respiration and sometimes even in nutrition (Watling and Thiel 2013).

In cirripeds, the carapace turns into shell plates during metamorphosis (Watling and Thiel 2013). The cuticle of many crustaceans is associated with calcium carbonate, except in the plates of a small barnacle group, where it is composed of calcium phosphate (Lowenstam and Weiner 1992), a compound also found in the mandible of many malacostracans (Bentov et al. 2016).

In pancrustaceans, the appendages are ancestrally biramous. There is an outer branch called the expopodite and an inner branch called the endopodite. Additionally, there are expansions on the external (epipodite) or internal (endite) side. The function of these appendages is multiple and shows great flexibility with significant adaptive diversity (Boxshall 2004). In malacostraceans, there are appendages that can be transformed into a weapon, in particular in the form of a pincer. In some alpheid malacostraceans and stomatopods, the extreme speed of specialized appendages creates cavitation causing localized phenomena of extreme violence (Patek and Caldwell 2005; Lohse et al. 2001).

The appendages can even be leafy and have the functions of locomotion, nutrition and simultaneous respiration in branchiopods and in malacostracan leptostracans (Pabst and Scholtz 2009).

The morpho-anatomy of the body is particularly affected in the case of profound modification of the way of life and in particular when free life is abandoned. The fixed way of life in cirripeds leads to a profound modification of the animals, since the animal is fixed by the head and the locomotor appendages have been transformed into appendages used to capture prey (Høeg and Møller 2006) (Figures 15.1d, 15.7). Parasitic life also causes profound morpho-anatomical modification with the appearance of hooks or suction cups or even the introduction of ink or some sort of roots in some cases (Lavrov et al. 2004; Høeg and Lützen 1995) (Figure 15.1b).

15.6 GENOMIC DATA

New sequencing methods (NGS) make it possible to obtain DNA fragments at low cost to reconstruct genome fragments or complete genomes. With a bar-coding approach by PCR and transcriptome sequencing, we are able to obtain data for phylogenetic analyses essential to further understanding crustaceans and to proposing evolutionary scenarios. The mitochondrial genome has been obtained from many species, and there are rearrangements that may be useful in identifying or confirming delicate parts of the phylogeny. This is, for example, the case of a reorganization observed in the pentastomides that we found also in the branchiurans (Lavrov et al. 2004).

The first complete crustacean genome published is that of *Daphnia pulex* (Colbourne et al. 2011), but currently the number of sequenced genomes is increasing rapidly. However, the choice of crustacean models mainly concerns freshwater or brackish water models and few truly marine animals (Table 15.1).

There is a strong variation in the sizes of genomes in crustaceans. The smallest appears to be the branchiopod *Lepidurus*, with a little less than 0.11 Gb (Savojardo et al. 2019), and the largest, the arctic malacostracan *Ampelisca macrocephala*, seems to be the biggest with about 63.2 Gb (Rees et al. 2007), or almost 600 times bigger.

It would seem that crustaceans living in constant and cold environments would have genomes larger than others (Alfsnes et al. 2017). Similarly, the control region of the mitogenome in polar copepods of the genus *Calanus* is known to be the longest of the crustaceans (Weydmann et al. 2017).

15.7 FUNCTIONAL APPROACHES: TOOLS FOR MOLECULAR AND CELLULAR ANALYSES

Paryhale hawaiensis is arguably the richest and most tooled model today in crustaceans and will not be presented here (see Chapter 16). Historically, early work on larval gene expression used immunohisto-chemistry and *in situ* hybridization performed in *Artemia* through sonication processes to make the cuticle permeable (Manzanares et al. 1993; Averof and Akam 1995). This method has been improved by chemical permeabilization (Blin et al. 2003; Copf et al. 2003). RNAi has been successfully tested on *Artemia* (Copf et al. 2004) and on *Litopenaeus vannamei* (Robalino et al. 2004).

The intense development of crustacean cultures for food production was quickly accompanied by the proliferation of numerous studies on farming models. Studies have been conducted on genes related to biomineralization and genes related to RNAi machinery, but many of the studies are focused on reproductive mechanisms to optimize reproduction such as encoding genes for eyestalk neuropeptides, gene receptor-encoding genes and genes related to sexual differentiation (Sagi et al. 2013).

In addition, many diseases have developed due to the high concentrations of animals, the impact of which remains a major concern for aquaculture maintenance (Stentiford et al. 2012).

Thus, RNAi provides modern and promising tools to treat shrimp that can be affected by nearly 20 different viruses (Krishnan et al. 2009; Escobedo-Bonilla 2011; Gong and Zhang 2021).

TABLE 15.1
List of Complete Genomes Published

Species Name	Group	Habitat	Size in Gb	Publication
Acartia tonsa	Copepoda	Marine	2.5	Jørgensen et al. (2019b)
Amphibalanus amphitrite	Cirrepedia	Marine	0.481	Kim et al. (2019)
Apocyclops royi	Copepoda	Fresh to brackish water	0.45	Jørgensen et al. (2019a)
Armadillidium vulgare	Malacostraca	Terrestrial	1.72	Chebbi et al. (2019)
Daphnia pulex	Branchiopoda	Fresh water	0.2	Colbourne et al. (2011)
Daphnia magma	Branchiopoda	Fresh water	0.123	Lee et al. (2019)
Diaphanosoma celebensis	Branchiopoda	Brackish water	2.56	Kim et al. (2021)
Eriocheir sinensis	Malacostraca	Fresh water to marine	1.66	Song et al. (2016)
Eulimnadia texana	Branchiopoda	Fresh water	0.12	Baldwin-Brown et al. (2017)
Lepidurus apus	Branchiopoda	Fresh water	0.1075	Savojardo et al. (2019)
Lepidurus articus	Branchiopoda	Fresh water	0.1075	Savojardo et al. (2019)
Macrobrachium nipponense	Malacostraca	Fresh water	4.5	Jin et al. (2021)
Neocaridina denticulata	Malacostraca	Fresh water	3.2	Kenny et al. (2014)
Parhyale hawaiensis	Malacostraca	Marine	3.6	Kao et al. (2016)
Portunus trituberculatus	Malacostraca	Marine	1.0	Tang et al. (2020)
Procambarus clarkii	Malacostraca	Fresh water	8.5	Shi et al. (2018)
Procambarus virginalis	Malacostraca	Fresh water	3.5	Gutekunst et al. (2018)
Tigriopus californicus	Copepoda	Marine	0.190	Barreto et al. (2018)
Tigriopus japonicus	Copepoda	Marine	0.197	Jeong et al. (2020)
Tigriopus kingsejongensis	Copepoda	Marine	0.295	Kang et al. (2017)
Trinorchestia longiramus	Malacostraca	Semi-terrestrial	0.89	Patra et al. (2020)

Transgenesis was successfully performed on the freshwater branchiopod *Daphnia magma* (Kato et al. 2012).

15.8 CHALLENGING QUESTIONS BOTH IN ACADEMIC AND APPLIED RESEARCH

The diversity of crustaceans is such that we can ask many questions about the evolution of the development of these animals. We will start by discussing some aspects of research applied to the development of crustaceans, and then we will see some aspects of more fundamental research.

15.8.1 Crustaceans and Food

Crustaceans have always been a source of food for humanity, even concerning pre-modern human species, as evidence suggests Neanderthals ate them, too (Zilhão et al. 2020). Crustacean species consumed by humans are generally large in size and relatively abundant. The vast majority are malacostracans and among them mainly decapods. More occasionally, large barnacles are also consumed. In 2018, the marine capture production by fisheries was around 6 million tons per year in seawater and 0.45 million tons per year in freshwater. The farming of crustaceans in aquaculture represents 9.4 million tons per year (USD 69.3 billion) (FAO 2020). Crustacean farming is therefore an important source of food and is essentially based on controlling the development cycle of species, in particular the production of larvae or juveniles. The first breeding operations in Southeast Asia or America consisted of taking post-larvae and juveniles of malacostracan penaeid prawns in brackish water ponds in order to obtain extensive breeding. Indonesian "tambaks" are well-known examples of these traditional practices (Laubier and Laubier 1993; Escobedo-Bonilla 2011).

The development of the study of larval stages from the 19th century onward gradually made it possible to control the cycle of a species of interest, of which, in some cases, stocks were rapidly declining. The first step consisted of restocking, that is to say the release of larvae, which was practiced by the end of the 19th century. The results of the first lobster releases are not obvious (Laubier and Laubier 1993), but improvements in crustacean farming and behavioral testing may allow improving this practice (Carere et al. 2015).

Hudinaga (1942) completed the life cycle of *Penaeus japonicus* by identifying foods suitable for different stages. Panouse (1943) began to understand the hormonal regulation of *Leander serratus* reproduction allowing better control of shrimp reproduction. Hudinaga's work in the beginning of the 1960s enabled the first ton production of *Penaeus japonicus* reared in captivity. Production started to increase very significantly in the beginning of the 1980s (Laubier and Laubier 1993).

The resting eggs of the brine shrimp *Artemia* give aquaculture institutions the ability to obtain larvae at any desired time, since the cryptobiosis can be stopped by putting them back in water under appropriate conditions (Van Stappen et al. 2019) (Figure 15.6). This ability is combined with the fact that since Seale (1933), it is known that these larvae are good food for young fish. This organism is not strictly marine but lives and develops perfectly in sea water and can therefore serve as living food for many marine organisms at key stages of their development, forming a kind of artificial marine plankton. The production of *Artemia* larvae is suitable for 85% of the marine animals bred (Sorgeloos 1980).

It is therefore also essential for the aquarium hobbyists or the breeding of animals for scientific purposes, which is the case for many of our development models such as cnidarians (Lechable et al. 2020), many marine fishes (Madhu et al. 2012) or freshwater fish (Dabrowski and Miller 2018; Shima and Mitani 2004). *Artemia* are also used as food for other crustaceans like barnacles (Desai et al. 2006; Jonsson et al. 2018) or many malacostracans (Sorgeloos 1980).

15.8.2 Biofouling

Organism colonization called biofouling affects ships, buoys, pontoons, offshore structures and many other human marine constructions (Figure 15.7). Issues include increased costs, reduced speed, environmental concerns, corrosion and safety hazards (Bixler and Bhushan 2012). Antifouling methods currently employed, ranging from coatings to cleaning techniques, have a significant cost (Bixler and Bhushan 2012). Barnacles are among the most important fouling organisms in the marine environment (Abramova et al. 2019). Recruitment of these animals around the cyprid/juvenile

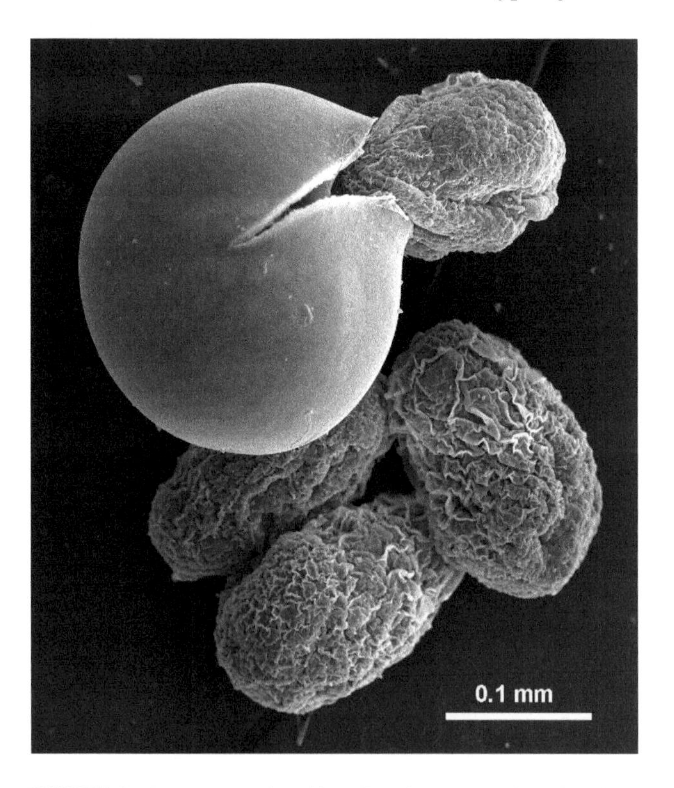

FIGURE 15.6 *Artemia* hatching. Resting egg and pre-hatching larvae of *Artemia franciscana* after re-filling. The nauplius larva still remains surrounded by the membrane and will soon swim. Hatch control is the basis of its success in marine aquaculture and fundamental research.

FIGURE 15.7 Biofouling by the barnacles *Amphibanalus amphitrite* and *Elminius modestus* in the port area of Saint Malo (North Brittany). (a) Tire used as port fender, (b) Underside of a boat needing cleaning.

stages is the key step in this problem, since fixation is definitive. Understanding the different stages of development from prospecting for the substrate to fixation through metamorphosis is therefore essential to prevent colonization. One strategy is to develop surfaces that are actively rejected by cyprids during the initial stages of the surface exploration, thus preventing attachment (Abramova et al. 2019). A more unexpected aspect of biofouling is that it can also serve as an indicator of the history of floating objects. Thus, the ambient temperature of the aircraft debris of the Boeing 777–200ER aircraft operated by Malaysian Airlines as MH370 was estimated from the biochemical analysis of the barnacles attached to the flaperon (Nesterov 2018).

15.8.3 Ecotoxicology

Small crustaceans are widely used in ecotoxicology because they represent an important link as a primary or even secondary consumer between primary producers and consumers of higher trophic levels, such as fish, for which they are an important food.

From the 1980s, *Artemia* was used very frequently as a standardized marine ecotoxicology test (Persoone and Wells 1987). Many new models have been added, such as calanoid copepods like *Acartia tonsa* or harpacticoid copepods like *Nitocra spinipes*, *Tisbe battagliai* and especially several species of *Tigriopus* (Figure 15.1e). The malacostracan amphipods of the genus *Corophium* are commonly used and more locally the malacostracan mysid *Mysidopsis bahia* (Pane et al. 2012).

15.8.4 Body Elongation and Segmentation

The anamorphosis that occurs in several groups of crustaceans is very reminiscent of the development that can be observed in other lineages of Metazoa, such as annelids (Chapter 13). The study of *Artemia* as an anamorphic organism has been initiated and has yielded interesting results (Averof and Akam 1995; Copf et al. 2003; Kontarakis et al. 2006; Copf et al. 2006; Prpic 2008). The thick cuticle and the lack of a functional tool are no doubt the reason studies on this model were abandoned at the expense of *Paryhale* (Chapter 16). In the years to come, however, it will be necessary to try to re-develop anamorphic models in order to be able to carry out the comparison with other Metazoa, because it is probable that larval retention modifies the ontogenetic sequences and can disrupt the comparisons.

15.8.5 Evolution of Ontogeny

In crustaceans, embryonic development can lead to the release of a juvenile resembling the adult, as in *Parhyale* (Chapter 16), but in many cases, embryonic development leads to the hatching of a larva whose development will often continue in plankton. Depending on the case, the released larva will have the number of body segments of the adult (zoe-like larva) or sometimes will be reduced to the most anterior region of the head (nauplius—Figure 15.3).

The body elongation processes will therefore be larval and/or embryonic in the different groups, with equivalent stages in both modes of development. Modalities of development largely remain to be studied. For a long time, it was believed that there was only a phenomenon of larval retention, but it seems possible that the limit of the passage between embryo and larva is more flexible and that, in particular, the nauplius larva has reappeared in malacostracans following a phenomenon of heterochrony (Jirikowski et al. 2015).

The same type of precise developmental comparison was initiated between a pseudo-direct and indirect development in branchiopods. It seems that the transition to direct development in cladocerans and cyclestherides has resulted in a modification of the ontogenetic stages with a compaction of certain stages of ancestrally anamorphic development (Fritsch et al. 2013). At the level of all crustaceans, this type of research still remains largely to be developed.

15.8.6 Terrestrialization and Origin of Insects

The transition from aquatic to aerial life requires profound physiological transformations, with the acquisition of important morpho-anatomical innovations affecting essential functions. This is a milestone in the history of the planet. There are several types of colonization of pancrustaceans in the aerial environment. In many decapod malacostracans, animals have retained the classic marine larval development, and therefore the adaptations to aerial life only concern juveniles and adults. There are also more colonizations with complete independence from the marine environment. The most important is undoubtedly that of the hexapods (Regier et al. 2010), but we can also cite the malacostracan amphipods and especially isopods. This last group would have colonized the mainland after the hexapods at the time of the Permian (Lins et al. 2017), but its phylogenetic history is still not understood (Dimitriou et al. 2019).

The research to be carried out concerns the acquisition of adaptations that are sometimes convergent between the lineages, such as the reduction of gill surface in different lines of land or intertidal crabs (O'Mahoney and Full 1984). The establishment of tracheae or pseudo-tracheae also appeared in a convergent manner in hexapods or malacostracan isopods (wood lice) and also elsewhere in arthropods (Cook et al. 2001; Csonka et al. 2013). Terrestrialization had other effects on the anatomical organization, such as the loss of lateral parts of the appendages and also the reduction of sensory structures. We can thus study the processes leading to the loss of antennas. In wood lice, terrestrialization has led to a strong reduction in A1 (Schmalfuss 1998), while in the hexapods, it is thought that it is the A2 that has entirely disappeared giving the intercalary segment. It is possible that developmental genes like *col* are involved in the appendage-less morphology of the intercalary segment of insects (Schaeper et al. 2010), but a comparative investigation must be carried out if similar mechanisms have been initiated following terrestrialization. To understand certain adaptations linked to terrestrial colonization, it is also possible to compare different lineages of aquatic pancrustaceans with insects to identify homologies between organs. This strategy made it possible to consider that the wings of insects could be derived from gills (Averof and Cohen 1997; Jockusch and Nagy 1997).

15.8.7 The Emergence of Parasitic Forms

The emergence of a parasitic lifestyle leads to profound changes in the life cycle and morpho-anatomy of organisms. In crustaceans, there are many parasitic forms, and the morpho-anatomical modifications are varied and more or less important. The case of cirripeds (Figure 15.1b) is particularly interesting because the larval stages are still very similar between the parasitic and non-parasitic forms. In this case, it is the metamorphosis from the cypris that is the key step in understanding the change in lifestyle (Høeg and Møller 2006) (Figure 15.4c). A detailed comparison of metamorphosis should make it possible to propose homologies between the post-metamorphosis stages and better understand the transformations in the lineages. It has already been identified that in *Sacculina carcini*, the naupliar stages are entirely lecitotrophic and synchronous, which is not the case in non-parasitic forms (Trédez et al. 2016) and suggests that there are therefore already modifications even before the cyprid stage.

15.8.8 Evolution of Cryptobiosis

Cryptobiosis is a very practical phenomenon for obtaining larvae at the right time (Figure 15.6), but the embryo in this suspended state of life is also a remarkable object of study. The brine shrimp is one of the three major models in this field, with nematodes and tardigrades (Hibshman et al. 2020). Several axes of research emerge from this problem: the formation of the shell of the resting eggs, the synthesis of trehalose, metabolic modifications with the synthesis of specific molecules such as Artemin, small Heat Shock proteins and late embryogeneisis abundant (LEA) proteins (Hibshman et al. 2020). Additionally, the structure of the particularly porous eggshell appears to be a carrier for nanocomposite material preparation and catalytic materials, opening up studies for new applied research (Wang et al. 2015; Zhao et al. 2019).

On the other hand, there is high variability in the shape and ornamentation of resting eggs among branchiopods (Figure 15.8). In particular, there are spherical, lenticular, tetrahedral or cylindrical shapes with a smooth, wrinkled or thorny surface (Figure 15.8) (Gilchrist 1978; Brendonck et al. 1992; Thiéry et al. 2007; Rabet 2010). A mathematical approach to these objects has already made it possible to understand that in *Tanymastix stagnalis*, the general shape is lenticular (Figure 15.8b) and corresponds to the intersection between two spheres. However, another shape can also be observed and would correspond to the intersection between two cylinders. In this case, the change in embryo shape would be due to an increase in volume (Thiéry et al. 2007). There are still many unanswered questions about the mechanisms allowing the construction of these shells and understanding how symmetry is acquired.

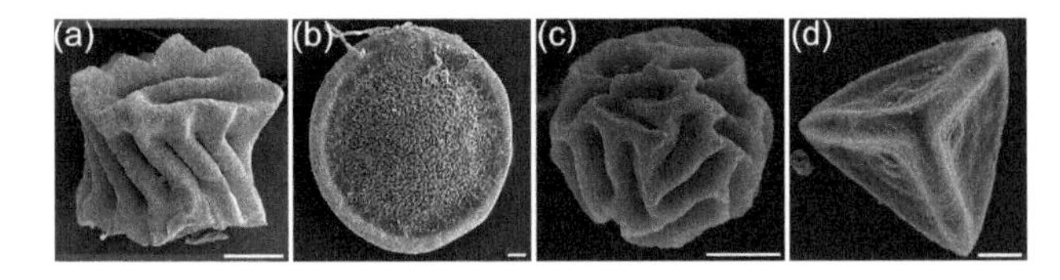

FIGURE 15.8 Variation of the resting egg shape in branchiopods. (a) Cylindrical, *Eulimnadia cylindrova*; (b) lenticular, *Tanymastix affinis*; (c) spherical, *Eulimnadia diversa*; (d) tetrahedral, *Streptocephalus archeri*.

BIBLIOGRAPHY

Abele, L.G., Kim, W. and B.E. Felgenhauer. 1989. Molecular evidence for inclusion of the phylum Pentastomida in the Crustacea. *Molecular Biology and Evolution*. 6:685–691.

Abramova, A., Lind, U., Blomberg, A. and M.A. Rosenblad. 2019. The complex barnacle perfume: Identification of waterborne pheromone homologues in *Balanus improvisus* and their differential expression during settlement. *Biofouling*. 35 (4):416–428.

Adamczewska, A.M. and S. Morris. 2001. Ecology and behavior of *Gecarcoidea natalis*, the Christmas Island Red Crab, during the annual breeding migration. *The Biological Bulletin*. 200:305–320.

Addis, A., Fabiano, F., Delogu, V. and M. Carcupino. 2012. Reproductive system morphology of *Lightiella magdalenina* (Crustacea, Cephalocarida): Functional and adaptive implications. *Invertebrate Reproduction and Development*. 57 (2):142–155.

Ahyong, S.T., Lowry, J.K. Alonso, M., Bamber, R.N., Boxshall, G.A., Castro, P., Gerken, S. et al. 2011. Subphylum Crustacea Brünnich, 1772. In: Zhang, Z.-Q. (Ed.), Animal biodiversity: An outline of higher-level classification and survey of taxonomic richness. *Zootaxa*. 3148:165–191.

Alekseev, V. and Y.I. Starobogatov. 1996. Types of diapause in Crustacea: Definitions, distribution, evolution. *Hydrobiologia*. 320 (1):15–26.

Alexeeva, N., Bogomolova, E., Tamberg, Y. and N. Shunatova. 2017. Oligomeric larvae of the pycnogonids revisited. *Journal of Morphology*. 278:1284–1304.

Alfsnes, K., Leinaas, H.P. and D.O. Hessen. 2017. Genome size in arthropods: Different roles of phylogeny, habitat and life history in insects and crustaceans. *Ecology and Evolution*. 7:5939–5947.

Anderson, D.T. 1967. Larval development and segment formation in the branchiopod crustaceans *Limnadia stanleyana* King (Conchostraca) and *Artemia salina* (L.)(Anostraca). *Australian Journal of Zoology*. 15:47–91.

Anderson, E., Lochhead, J.H., Lochhead, M.S., and E. Huebner. 1970. The origin and structure of the tertiary envelope in thick-shelled eggs of the brine shrimp, *Artemia*. *Journal of Ultrastructure and Molecular Structure Research*. 525:497–525.

Anger, K. 2001. *The Biology of Decapod Crustacean Larvae*. Publisher: A.A. Balkema. Boca Raton, FL: CRC Press.

Averof, M. and M. Akam. 1995. Hox genes and the diversification of insect and crustacean body plans. *Nature*. 376 (6539):420–423.

Averof, M. and S.M. Cohen. 1997. Evolutionary origin of insect wings from ancestral gills. *Nature*. 385:627–630.

Averof, M. and N.H. Patel. 1997. Crustacean appendage evolution associated with changes in Hox gene expression. *Nature*. 388 (6643):682–686.

Baldwin-Brown, J.G., Weeks S.C. and A.D. Long. 2017. A new standard for crustacean genomes: The highly contiguous, annotated genome assembly of the clam shrimp *Eulimnadia texana* reveals Hox gene order and identifies the sex chromosome. *Genome Biology and Evolution*. 10 (1):143–156.

Barazandeh, M., Davis, C.S., Neufeld, C.J., Coltman, D.W. and A.R. Palmer. 2013. Something Darwin didn't know about barnacles: Spermcast mating in a common stalked species. *Proceedings of the Royal Society B*. 280:20122919.

Barreto, F.S., Watson, E.T., Lima, T.G., Willett, C.S., Edmands, S., Li, W. and R.S. Burton. 2018. Genomic signatures of mitonuclear coevolution across populations of *Tigriopus californicus*. *Nature Ecology & Evolution*. 2 (8):1250–1257.

Bauer R.T. 2013. Amphidromy in shrimps: A life cycle between rivers and the sea. *Latin American Journal of Aquatic Research*. 41 (4):633–650.

Benesch, R. 1969. Zur Ontogenie und Morphologie von *Artemia salina* L. *Zoologische Jahrbücher. Abteilung für Anatomie und Ontogenie der Tiere Abteilung für Anatomie und Ontogenie der TiereS*.:307–458.

Bentov, S., Aflalo, E.D., Tynyakov, J., Glazer, L. and A. Sagi. 2016. Calcium phosphate mineralization is widely applied in crustacean mandibles. *Scientific Reports*. 6:22118.

Benvenuto, C. and S.C. Weeks. 2020. Hermaphroditism and gonochorism. In *"The Natural History of the Crustacea: Reproductive Biology": The Natural History of the Crustacea (6)*. Oxford: Oxford University Press.

Bixler, G.D. and B. Bhushan. 2012. Biofouling: Lessons from nature. *Philosophical Transactions of the Royal Society*. A370: 2381–2417.

Blin, M., Rabet, N., Deutsch, J., and Mouchel-Vielh, E. 2003. Possible implication of Hox genes *abdominal-B* and *abdominal-A* in the specification of genital and abdominal segments in cirripedes. *Development Genes and Evolution* 213:90–96.

Bowen, S.T., Durkin, J.P., Sterling, G., and L.S. Clark. 1978. *Artemia* hemoglobins: Genetic variation in parthenogenetic and zygogenetic populations. *Biological Bulletin*, 155 (2):273–287.

Bowman, T.E. and L.G. Abele. 1982. Classification of the recent Crustacea. In *Systematics, the Fossil Record, and Biogeography*, ed. Abele, L.G. (pp. 1–27). New York: Academic Press.

Boxshall, G.A. 2004. The evolution of arthropod limbs. *Biological Reviews*. 79 (2):253–300.

Brendonck, L., Hamer, M. and A. Thiery. 1992. Occurrence of tetrahedral eggs in the streptocephalidae daday (Branchiopoda: Anostraca) with descriptions of a new subgenus, *Parastreptocephalus*, and a new species, *Streptocephalus (Parastreptocephalus) zuluensis* Brendonck and Hamer. *Journal of Crustacean Biology*. 12 (2):282–297.

Brendonck, L., Rogers, D.C., Olesen, J., Weeks, S. and W.R. Hoeh. 2008. Global diversity of large branchiopods (Crustacea: Branchiopoda) in freshwater. *Hydrobiologia*. 595 (1):167–176.

Browne, W.E., Price, A.L., Gerberding, M. and N.H. Patel. 2005. Stages of embryonic development in the amphipod crustacean, *Parhyale hawaiensis*. *Genesis*. 42 (3):124–149.

Brünnich, M.T. 1772. Zoologiae fundamenta praelectionibus academicis accommodata: Grunde i Dyrelaeren. Hafniae et Lipsiae: Apud Frider. Christ. Pelt.

Calman, W.T. 1909. Crustacea. In *A Treatise on Zoology, Part 7, Fascicle 3*, ed. Lankester, E.R.·(pp. 1–346). London: Adam and Charles Black.

Carere, C., Nascetti, G., Carlini A., Santucci D. and E. Alleva. 2015. Actions for restocking of the European lobster (*Homarus gammarus*): A case study on the relevance of behaviour and welfare assessment of cultured juveniles. *Rendiconti Lincei. Scienze Fisiche e Naturali*. 26:59–64.

Charnov, E.L. 1987. Sexuality and hermaphroditism in barnacles: A natural selection approach. In *Barnacle Biology, Crustacean Issues 5*, ed. Southward, A.J. (pp. 89–103). Rotterdam: A. A. Belkema, Rotterdam.

Chebbi, M.A., Becking, T., Moumen, B., Giraud, I., Gilbert, C., Peccoud, J. and R. Cordaux. 2019. The genome of *Armadillidium vulgare* (Crustacea, Isopoda) provides insights into sex chromosome evolution in the context of cytoplasmic sex determination. *Molecular Biology and Evolution*. 36 (4): 727–741.

Colbourne, J.K., Pfrender, M.E., Gilbert, D., Thomas, W.K., Tucker, A., Oakley, T.H., Tokishita, S., Aerts, A., Arnold, G.J., Kumar Basu, M., Bauer, D.J., Cáceres, C.E., Carmel, L., Casola, C., Choi, J.-H., Detter, J.C., Dong, Q., Dusheyko, S., Eads, B.D., Fröhlich, T., Geiler-Samerotte, K.A., Gerlach, D., Hatcher, P., Jogdeo, S., Krijgsveld, J., Kriventseva, E.V., Kültz, D., Laforsch, C., Lindquist, E., Lopez, J., Manak, J.R., Muller, J., Pangilinan, J., Patwardhan, R.P., Pitluck S., Pritham, E.J., Rechtsteiner A., Rho, M., Rogozin, I.B., Sakarya, O., Salamov, A., Schaack, S., Shapiro, H., Shiga, Y., Skalitzky, C., Smith, Z., Souvorov, A., Sung, W., Tang, Z., Tsuchiya, D., Tu, H., Vos, H., Wang, M., Wolf, Y.I., Yamagata, H., Yamada T., Ye, Y., Shaw, J.R., Andrews, J., Crease, T.J., Tang, H., Lucas, S.M., Robertson, H.M., Bork, P., Koonin, E.V., Zdobnov, E.M., Grigoriev, I.V., Lynch, M. and J.L. Boore. 2011. The ecoresponsive genome of *Daphnia pulex*. *Science*. 331:555–561.

Collette, J.H. and J.W. Hagadorn. 2010. Early evolution of phyllocarid arthropods: Phylogeny and systematics of Cambrian-Devonian archaeostracans. *Journal of Paleontology*. 84 (5):795–820.

Collis, S.A. and G. Walker. 1994. The morphology of the naupliar stages of *Sacculina carcini* (Crustacea: Cirripedia: Rhizocephala). *Acta Zoologica*. 75 (4):297–303.

Cook, C.E., Smith, M.L., Telford, M.J., Bastianello, A. and M. Akam. 2001. Hox genes and the phylogeny of the arthropods. *Current Biology*. 11 (10):759–763.

Copf, T., Rabet, N. and M. Averof. 2006. Knockdown of spalt function by RNAi causes de-repression of Hox genes and homeotic transformations in the crustacean *Artemia franciscana*. *Developmental Biology*. 298 (1):87–94.

Copf, T., Rabet, N., Celniker, S.E. and M. Averof. 2003. Posterior patterning genes and the identification of a unique body region in the brine shrimp *Artemia franciscana*. *Development*. 130 (24):5915–5927.

Copf, T., Schröder, R. and M. Averof. 2004. Ancestral role of *caudal* genes in axis elongation and segmentation. *Proceedings of the National Academy of Sciences*. 101 (51):17711–17715.

Csonka, D., Halasy, K., Szabó, P., Mrak, P., Štrus, J. and E. Hornung. 2013. Eco-morphological studies on pleopodal lungs and cuticle in *Armadillidium* species (Crustacea, Isopoda, Oniscidea). *Arthropod Structure and Development*. 42:229–235.

Dabrowski, K. and M. Miller. 2018. Contested paradigm in raising zebrafish (*Danio rerio*). *Zebrafish*. 15 (3):295–309.

Dahms, H.-U. 2000. Phylogenetic implications of the Crustacean nauplius. *Hydrobiologia*. 417:91–99.

Delage, Y. 1884. Évolution de la Sacculine (*Sacculina carcini* Thomps.): crustacé endoparasite de l'ordre nouveau des Kentrogonides. *Archives de zoologie expérimentale et générale*. 2:417–736.

Desai, D.V., Anil, A.C. and K. Venkat. 2006. Reproduction in *Balanus amphitrite* Darwin (Cirripedia: Thoracica): Influence of temperature and food concentration. *Marine Biology*. 149:1431–1441.

Dimitriou, A.C., Taiti S. and S. Sfenthourakis. 2019. Genetic evidence against monophyly of Oniscidea implies a need to revise scenarios for the origin of terrestrial isopods. *Scientific Reports*. 9 (1):18508.

Dohle, W. 2001. Are the insects terrestrial crustaceans? A discussion of some new facts and arguments and the proposal of the proper name 'Tetraconata' for the monophyletic unit Crustacea + Hexapoda. In: Deuve, T. (ed.) Origin of the Hexapoda. *Annales de la Société entomologique de France (N.S.)*. 37:85–103.

Duffy, E. 1996. Eusociality in a coral-reef shrimp. *Nature*. 381 (6582):512–514.

Durbin, A.P., Hebert, D.N. and M.E.A. Cristescu. 2008. Comparative phylogeography of marine cladocerans. *Marine Biology*. 155:1–10.

Escobedo-Bonilla, C.M. 2011. Application of RNA interference (RNAi) against viral infections in shrimp. *Journal of Antivirals & Antiretrovirals*. S9.

FAO. 2020. *The State of World Fisheries and Aquaculture 2020: Sustainability in Action*. Rome. https://doi.org/10.4060/ca9229en

Fraser J.H. 1936. The occurrence, ecology and life history of *Tigriopus fulvus* (Fischer). *Journal of the Marine Biological Association of the United Kingdom*. 20 (3):523–536.

Fritsch, M., Bininda-Emonds, O.R.P. and S. Richter. 2013. Unraveling the origin of Cladocera by identifying heterochrony in the developmental sequences of Branchiopoda. *Frontiers in Zoology*. 10 (1):35.

Fryer, G.F.L.S. 1996. Diapause, a potent force in the evolution of freshwater crustaceans. *Hydrobiologia*. 320:1–14.

Gajardo, G. and J.A. Beardmore. 2012. The brine shrimp *Artemia*: Adapted to critical life conditions. *Frontiers in Zoology*. 3:185.

Garreau de Loubresse, N. 1974. Etude chronologique de la mise en place des enveloppes de l'œuf d'un crustacé phyllopode, *Tanymastix lacunae*. *Journal of Microscopy*. 20:21–38.

Gibert, J.M., Mouchel-Vielh, E., Quéinnec, E and J.S. Deutsch. 2000. Barnacle duplicate *engrailed* genes: Divergent expression patterns and evidence for a vestigial abdomen. *Evolution and Development*. 2:194–202.

Gilchrist, B.M. 1978. Scanning electron microscope studies of the egg shell in some anostraca (Crustacea: Branchiopoda). *Cell and Tissue Research*. 193:337–351.

Giribet, G. and G.D. Edgecombe. 2019. The phylogeny and evolutionary history of Arthropods. *Current Biology*. 29 (12):R592–R602.

Gong, Yi and Xiaobo Zhang. 2021. RNAi-based antiviral immunity of shrimp: Review. *Developmental and Comparative Immunology*. 115:103907.

Gurney, R. 1942. *Larvae of Decapod Crustacea*. London: Ray Society.

Gutekunst, J., Andriantsoa, R., Falckenhayn, C., Hanna, K., Stein, W., Rasamy, J. and F. Lyko. 2018. Clonal genome evolution and rapid invasive spread of the marbled crayfish. *Nature Ecology & Evolution*. 2:567–573.

Gutiérrez-Zugasti, I. 2011. The use of echinoids and crustaceans as food during the Pleistocene-Holocene transition in Northern Spain: Methodological contribution and dietary assessment. *The Journal of Island and Coastal Archaeology*. 6 (1):115–133.

Hibshman, J.D., Clegg, J.S. and B. Goldstein. 2020. Mechanisms of desiccation tolerance: Themes and variations in brine shrimp, roundworms, and tardigrades. *Frontiers in Physiology*. 11:592016.

Høeg, J.S. and J. Lützen. 1995. Life cycle and reproduction in the Cirripedia Rhizocephala. *Oceanography and Marine Biology: An Annual Review*. 33:427–485.

Høeg, J.T. and O.S. Møller. 2006. When similar beginnings lead to different ends: Constraints and diversity in cirripede larval development. *Invertebrate Reproduction and Development*. 49 (3):125–142.

Hudinaga, M. 1942. Reproduction, development and rearing of *Penaeus japonicus* Bate. *Japanese Journal of Environmental Entomology and Zoology*. 10:305–393.

Jensen, G.C., Sean McDonald, P. and D.A. Armstrong. 2007. Biotic resistance to green crab, *Carcinus maenas*, in California bays. *Marine Biology*. 151:2231–2243.

Jeong, C.B., Lee, B.Y., Choi, B.S., Kim, M.S., Park, J.C., Kim, D.-H., Wang M. et al. 2020. The genome of the harpacticoid copepod *Tigriopus japonicus*: Potential for its use in marine molecular ecotoxicology. *Aquatic Toxicology*. 222:105462.

Jin, S., Bian, C., Jiang, S., Han, K., Xiong, Y., Zhang, W., Shi, C. et al. 2021. A chromosome-level genome assembly of the oriental river prawn, *Macrobrachium nipponense*. *GigaScience*. 10:1–9.

Jirikowski, G.J., Richter, S. and C. Wolff. 2013. Myogenesis of Malacostraca: The "egg-nauplius" concept revisited. *Frontiers in Zoology*. 10:76.

Jirikowski, G.J., Wolff, C. and S. Richter. 2015. Evolution of eumalacostracan development: New insights into loss and reacquisition of larval stages revealed by heterochrony analysis. *EvoDevo*. 6:4.

Jockusch, E.L. and L.M. Nagy. 1997. Insect evolution: How did insect wings originate? *Current Biolology*. 7 (6):R358–R361.

Jonsson, P.R., Wrange, A.L., Lind, U., Abramova, A., Ogemark, M. and A. Blomberg. 2018. The barnacle *Balanus improvisus* as a marine model: Culturing and gene expression. *Journal of Visualized Experiments*. (138):e57825.

Jørgensen, S., B. Lykke, H. Nielsen, et al. 2019a. The whole genome sequence and mRNA transcriptome of the tropical cyclopoid copepod *Apocyclops royi*. *G3: Genes, Genomes, Genetics*. 9 (5):1295–1302.

Jørgensen, T.S., Petersen, B., Petersen, H.C.B., et al. 2019b. The genome and mRNA transcriptome of the cosmopolitan calanoid copepod *Acartia tonsa* Dana improve the understanding of copepod genome size evolution. *Genome Biology and Evolution*. 11 (5):1440–1450.

Kang, S., Ahn, D.-H., Lee, J.H., Lee, S.G., Shin, S.C., Lee, J., et al. 2017. The genome of the Antarctic-endemic copepod, *Tigriopus kingsejongensis*. *GigaScience*. 6:1–9.

Kao, D., Lai, A.G., Stamataki, E., Rosic, S., Konstantinides, N., Jarvis, E., Di Donfrancesco, et al. 2016. The genome of the crustacean *Parhyale hawaiensis*, a model for animal development, regeneration, immunity and lignocellulose digestion. *Elife*. 5:e20062.

Kato, Y., Matsuura, T. and H. Watanabe. 2012. Genomic integration and germline transmission of plasmid injected into crustacean *Daphnia magna* eggs. *PLoS One*. 7 (9):e45318.

Kawai, T., Scholtz, G., Morioka, S., Ramanamandimby, F., Lukhaup, C. and Y. Hanamura. 2009. Parthenogenetic alien crayfish (Decapoda: Cambaridae) spreading in Madagascar. *Journal of Crustacean Biology*. 29 (4):562–567.

Kenny, N.J., Sin, Y.W., Shen, X., Zhe, Q., Wang, W., Chan, T.F., Tobe, S.S., et al. 2014. Genomic sequence and experimental tractability of a new decapod shrimp model, *Neocaridina denticulate*. *Marine Drugs*. 12:1419–1437.

Kim, D.-H., Choi, B.-S., Kang, H.-M., Park, J.C., Kim, M.-S., Hagiwara, A. and J.-S. Lee. 2021. The genome of the marine water flea *Diaphanosoma celebensis*: Identification of phase I, II, and III detoxification genes and potential applications in marine molecular ecotoxicology. *Comparative Biochemistry and Physiology Part D: Genomics and Proteomics*. 37:100787.

Kim, J.-H., Kim, H.K., Chan, K.H., BKK, Kang, S. and W. Kim. 2019. Draft genome assembly of a fouling barnacle, *Amphibalanus amphitrite* (Darwin, 1854): The first reference genome for Thecostraca. *Frontiers in Ecology and Evolution*. 7:465.

Kobayashi, H., Hatada, Y., Tsubouchi, T., Nagahama, T. and H. Takami. 2012. The hadal Amphipod *Hirondellea gigas* possessing a unique cellulase for digesting wooden debris buried in the deepest seafloor. *PLoS One*. 7(8):e42727.

Kobayashi, H., Shimoshige, H., Nakajima, Y., Arai, W. and H. Takami. 2019. An aluminum shield enables the amphipod *Hirondellea gigas* to inhabit deep-sea environments. *PLoS One*. 14 (4):e0206710.

Koenemann, S., Schram, F.R., Iliffe, T.M., Hinderstein, L.M. and A. Bloechl. 2007. Behavior of remipedia in the laboratory, with supporting field observations. *Journal of Crustacean Biology*. 27:534–542.

Kontarakis, Z., Copf, T. and M. Averof. 2006. Expression of hunchback during trunk segmentation in the branchiopod crustacean *Artemia franciscana*. *Development Genes and Evolution*. 216 (2):89–93.

Korn, M. and A.K. Hundsdoerfer. 2016. Molecular phylogeny, morphology and taxonomy of Moroccan *Triops granarius* (Lucas, 1864) (Crustacea: Notostraca), with the description of two new species. *Zootaxa*. 4178 (3):328–346.

Krieger, J., Sandeman, R.E., Sandeman, D.C., Hansson, B.S. and S. Harzsch. 2010. Brain architecture of the largest living land arthropod, the giant robber crab *Birgus latro* (Crustacea, Anomura, Coenobitidae): Evidence for a prominent central olfactory pathway? *Frontiers in Zoology*. 7:25.

Krishnan, P., Gireesh-Babu, P., Rajendran, K.V. and A. Chaudhar. 2009. RNA interference-based therapeuticsfor shrimp viral diseases. *Aquatic Organisms*. 86:263–272.

Laubier, A. and L. Laubier. 1993. Marine crustacean farming: Present status and perspectives. *Aquatic Living Resources*. 6 (4):319–329.

Lavrov, D.V., Brown, W.M. and J.L. Boore. 2004. Phylogenetic position of the Pentastomida and (pan)crustacean relationships. *Proceedings of the Royal Society B: Biological Sciences*. 271:537–544.

Lechable, M., Jan, A., Duchene, A., Uveira, J., Weissbourd, B., Gissat, L., Collet, S., et al. 2020. An improved whole life cycle culture protocol for the hydrozoan genetic model *Clytia hemisphaerica*. *Open Biology*. 9 (11):bio051268.

Lee, B.-Y., Choi, B.-S., Kim, M.-S., Park, J.C., Jeong, C.-B., Han, J. and J.-S. Lee. 2019. The genome of the freshwater water flea *Daphnia magna*: A potential use for freshwater molecular ecotoxicology. *Aquatic Toxicology*. 210:69–84.

Lee, M.S.Y., Soubrier, J. and G.D. Edgecombe. 2013. Rates of phenotypic and genomic evolution during the Cambrian explosion. *Current Biology*. 23:1889–1895.

Liang, C., Strickland, J., Ye, Z., Wu, W., Hu, B. and D. Rittschof. 2019. Biochemistry of barnacle adhesion: An updated review. *Frontiers in Marine Science*. 6:2296–7745.

Lindley, J.A. 1997. Eggs and their incubation as factors in the ecology of planktonic crustacea. *Journal of Crustacean Biolology*. 17 (4):569–576.

Linnaeus, C. von. 1758. Caroli Linnaei Equitis De Stella Polari, Archiatri Regii, Med. & Botan. Profess. Upsal.; Systema Naturae Per Regna Tria Naturae, Secundum Classes, Ordines, Genera, Species, Cum Characteribus, Differentiis, Synonymis, Locis, Holmiae.

Lins, L.S.F., Ho, S.Y.W. and N. Lo. 2017. An evolutionary timescale for terrestrial isopods and a lack of molecular support for the monophyly of Oniscidea (Crustacea: Isopoda). *Organisms Diversity and Evolution*. 17:813–820.

Liu, Y.-L., Zhao, Y., Dai, Z.-M., Chen, H.-M., and W.-J. Yang. 2009. Formation of diapause cyst shell in brine shrimp, *Artemia parthenogenetica*, and its resistance role in environmental stresses. *The Journal of Biological Chemistry*. 284 (25):16931–16938.

Lohse, D., Schmitz, B. and M. Versluis. 2001. Snapping shrimp make flashing bubbles. *Nature*. 413 (6855):477–478.

Lowenstam, H.A. and S. Weiner. 1992. Phosphatic shell plate of the barnacle *Ibla* (Cirripedia): A bone-like structure. *Proceedings of the National Academy of Sciences*. 89 (22):10573–10577.

Madhu, R., Madhu, K. and T. Retheesh. 2012. Life history pathways in false clown *Amphiprion ocellaris* Cuvier, 1830: A journey from egg to adult under captive condition. *Journal of the Marine Biological Association of India*. 54 (1):77–90.

Maier, G. 1992. *Metacyclops minutus* (Claus, 1863): Population dynamics and life history characteristics of a rapidly developing copepod. *International Review of Hydrobiology*. 77: 455–466.

Malitan, H.S., Cohen, A.M. and T.H. MacRae. 2019. Knockdown of the small heat-shock protein p26 by RNA interference modifies the diapause proteome of *Artemia franciscana*. *Biochemistry and Cell Biology*. 97(4):471–479.

Manzanares, M., Marco, R. and R. Garesse. 1993. Genomic organization and developmental pattern of expression of the engrailed gene from the brine shrimp *Artemia*. *Development* 118:1209–1219.

Martin, J.W. and G.E. Davis. 2001. An updated classification of the recent Crustacea. *Natural History Museum of Los Angeles County Contributions in Science*. 39:1–124.

Martin, J.W., Olesen, J. and J.T. Høeg. 2014. *The Crustacean Nauplius in Atlas of Crustacean Larvae*, ed. Martin, J.W., Olesen, J. and J.T. Høeg. Baltimore, MD: Johns Hopkins University Press.

Maruzzo, D., Aldred, N., Clare, A.S. and J.T. Høeg. 2012. Metamorphosis in the cirripede crustacean *Balanus amphitrite*. *PLoS One*. 7 (5):e37408.

Mayrat, A. and M. de Saint Laurent. 1996. Considérations sur la Classe de Malacostracés. (Malacostraca Latreille, 1802). In *Traité de zoologie*, ed. GrasséÉ, P.-P. and J. Forest. (pp. 841–862). Masson, Paris: Crustacea.

McLaughlin, P.A. 1980. *Comparative Morphology of Recent Crustacea*. San Francisco: W. H. Freeman and Company. [General crustacean morphology].

Monod, T. and J. Forest. 1996. *Histoire de la classification des Crustacés, Traité de zoologie, Anatomie, systématique, biologie, Crustacés, Tome VII, fascicule II* (pp. 235–267). Paris, Milan, Barcelone: Masson.

Morris, J.E. and B.A. Afzelius. 1967. The structure of the shell and outer membranes in encysted *Artemia salina* embryos during cryptobiosis and development. *Journal of Ultrastructure Research*. 20 (3–4): 244–259.

Neiber, M.T., Hartke, T.R., Stemme, T., Bergmann, A., Rust, J., Iliffe, T.M. and S. Koenemann. 2011. Global biodiversity and phylogenetic evaluation of Remipedia (Crustacea). *PLoS One*. 6 (5):e19627.

Nesterov, O. 2018. Consideration of various aspects in a drift study of MH370 debris. *Ocean Science*. 14:387–402.

Olson, C.S. and J.S. Clegg. 1978. Cell division during the development of *Artemia salina*. *Wilhelm Roux's Archives of Developmental Biology*. 184:1–13.

O'Mahoney, P.M. and R.J. Full. 1984. Respiration of crabs in air and water. *Comparative Biochemistry and Physiology Part A: Physiology*. 79 (2):275–282.

Pabst, T. and G. Scholtz. 2009. The development of phyllopodous limbs in leptostraca and branchiopoda. *Journal of Crustacean Biology*. 29 (1):1–12.

Pane, L., Agrone, C., Giacco, E., Somà, A. and G.L. Mariottini. 2012. Utilization of marine crustaceans as study models: A new approach in marine ecotoxicology for European (REACH) regulation. In *Ecotoxicology*, ed. Begum, G. (pp. 91–106). Crotia: InTech.

Panouse, J.B. 1943. Influence de l'ablation du pédoncule oculaire sur la croissance de l'ovaire chez la crevette *Leander serratus*. *Comptes rendus hebdomadaires des séances de l'Académie des sciences*. 217:553–555.

Patek, S.N. and R.L Caldwell. 2005. Extreme impact and cavitation forces of a biological hammer: Strike forces of the peacock mantis shrimp *Odontodactylus scyllarus*. *Journal of Experimental Biology*. 208:3655–3664.

Patra, A.K., Chung, O., Yoo, J.Y., Kim, M.S., Yoon, M.G., Choi, J.-H. and Y. Yang. 2020. First draft genome for the sandhopper *Trinorchestia longiramus*. *Scientific Data*. 7 (1):85.

Pennak, R.W. and D.J. Zinn. 1943. Mystacocarida, a new order of Crustacea from intertidal beaches in Massachusetts and Connecticut. *Smithsonian Miscellaneous Collections*. 103 (9):1–11, pls.1–2.

Persoone, G. and P.G. Wells. 1987. *Artemia* in aquatic toxicology: A review. *Artemia Research and Its Applications*. 1:259–275.

Petersen, J., Ramette, M.A., Lott C., Cambon-Bonavita, M.A., Zbinden, M. and N. Dubilier. 2010. Dual symbiosis of the vent shrimp *Rimicaris exoculata* with filamentous gamma- and epsilonproteobacteria at four Mid-Atlantic Ridge hydrothermal vent fields. *Environmental Microbiology*. 12 (8): 2204–2218.

Prpic, N.-M. 2008. Parasegmental appendage allocation in annelids and arthropods and the homology of parapodia and arthropodia. *Frontiers in Zoology*. 5:17.

Rabet, N. 2010. Revision of the egg morphology of *Eulimnadia* (Crustacea, Branchiopoda, Spinicaudata). *Zoosystema* 32 (3):373–391.

Rabet, N., Gibert, J.M., Quéinnec, E., Deutsch, J.S. and E. Mouchel-Vielh. 2001. The *caudal* gene of the barnacle *Sacculina carcini* is not expressed in its vestigial abdomen. *Development Genes and Evolution*. 211:172–178.

Raisuddin, S., Kwok, K.W.H., Leung, K.M.Y., Schlenk, D. and J.-S. Lee. 2007. The copepod *Tigriopus*: A promising marine model organism for ecotoxicology and environmental genomics. *Aquatic Toxicology*. 83 (3):161–173.

Rajitha, K., Nancharaiaha, Y.V. and V.P. Venugopalan. 2020. Insight into bacterial biofilm-barnacle larvae interactions for environmentally benign antifouling strategies. *International Biodeterioration & Biodegradation*. 149:104937.

Rees, D.J., Dufresne, F., Glémet H. and C. Belzile. 2007. Amphipod genome sizes: First estimates for arctic species reveal genomic giants. *Genome*. 50 (2):151–158.

Regier, J.C., Jeffrey, W.S. and R.E. Kambic. 2005. Pancrustacean phylogeny: Hexapods are terrestrial crustaceans and maxillopods are not monophyletic. *Proceedings of the Royal Society B*. 272:395–401.

Regier, J.C., Shultz, J.W., Zwick, A., Hussey, A., Ball, B., Wetzer R., Martin, J.W. and C.W. Cunningham. 2010. Arthropod relationships revealed by phylogenomic analysis of nuclear protein-coding sequences. *Nature*. 463 (7284):1079–1083.

Richter, S. 2002. The Tetraconata concept: Hexapod-crustacean relationships and the phylogeny of Crustacea. *Organisms Diversity and Evolution*. 2:217–237.

Riley, J., Banaja, A.A. and J.L. James. 1978. The phylogenetic relationships of the Pentastomida: The case for their inclusion within the Crustacea. *International Journal for Parasitology*. 8 (4):245–254.

Robalino, J., Browdy, C.L., Prior, S., Metz, A., Parnell, P., Gross, P. and G. Warr. 2004. Induction of antiviral immunity by double-stranded RNA in a marine invertebrate. *Journal of Virology*. 78 (19):10442–10448.

Rosowski, J.R., Belk D., Gouthro, M.A. and K.W. Lee. 1997. Ultrastructure of the cyst shell and underlying membranes of the brine shrimp *Artemia franciscana* Kellogg (Anostraca) during postencystic development, emergence, and hatching. *Journal of Shellfish Research*. 16 (1):233–249.

Sagi, A., Manor, R. and T. Ventura. 2013. Gene silencing in crustaceans: From basic research to biotechnologies. *Genes*. 4: 620–645.

Sanders, H.L. 1955. The Cephalocarida, a new subclass of Crustacea from Long Island Sound. *Proceedings of National Academy of Sciences of the United States of America*. 41 (1):61–66.

Savojardo, C., Luchetti, A., Martelli, P.L., Casadio, R. and B. Mantovani. 2019. Draft genomes and genomic divergence of two *Lepidurus* tadpole shrimp species (Crustacea, Branchiopoda, Notostraca). *Molecular Ecology Resources*. 19 (1):235–244.

Scanabissi, F. and C. Mondini. 2002. A survey of the reproductive biology in Italian branchiopods. *Hydrobiologia*. 486 (1):263–272.

Schaeper, N.D., Pechmann, M., Damen, W.G.M., Prpic, N.M. and E.A. Wimmer. 2010. Evolutionary plasticity of collier function in head development of diverse arthropods. *Developmental Biology*. 344 (1):363–376.

Schmalfuss, H. 1998. Evolutionary strategies of the antennae in terrestrial Isopods. *Journal of Crustacean Biology*. 18 (1):10–24.

Scholtz, G. 2002. Evolution of the nauplius stage in malacostracan crustaceans. *Journal of Zoological Systematics and Evolutionary Research*. 38 (3):175–187.

Scholtz, G., Braband, A., Tolley, L., Reimann, A., Mittmann, B., Lukhaup, C., Steuerwald, F. and G. Vogt. 2003. Parthenogenesis in an outsider crayfish. *Nature*. 421:806.

Scholtz, G. and G.D. Edgecombe, 2006. The evolution of arthropod heads: reconciling morphological, developmental and palaeontological evidence. *Development Genes and Evolution*. 216:395–415.

Schrehardt, A. 1987. A scanning electron-microscope study of the post-embryonic development of Artemia. In *Artemia Research and Its Applications. Vol. 1: Morphology, Genetics, Strain Characterisation, Toxicology*, ed. Sorgeloos, P., Bengtson, D.A., Decleir, W. and E. Jaspers (pp. 5–32). Wetteren, Belgium: Universa Press.

Schwentner, M., Combosch, D.J., Pakes, J.N., and G. Giribet. 2017. A phylogenomic solution to the origin of insects by resolving

Crustacean-Hexapod relationships. *Current Biology*. 27 (12):1818–1824.

Seale, A. 1933. Brine shrimp (*Artemia*) as a satisfactory live food for fishes. *Transactions of the American Fisheries Society*. 63:129–130.

Shi, L., Yi, S., and Y. Li. 2018. Genome survey sequencing of red swamp crayfish *Procambarus clarkia*. *Molecular Biology Reports*. 45 (5):799–806.

Shima, A. and H. Mitani. 2004. Medaka as a research organism: Past, present and future. *Mechanisms of Development*. 121 (7–8):599–604.

Song, L., Bian, C., Luo, Y., Wang, L., You, X., Li, J., Qiu, Y., et al. 2016. Draft genome of the Chinese mitten crab, *Eriocheir sinensis*. *Gigascience*. 5:5.

Sorgeloos, P. 1980. The use of the brine shrimp *Artemia* in aquaculture. In *The Brine Shrimp Artemia. Vol. 3. Ecology, Culturing, Use in Aquaculture*, ed. Persoone, G., Sorgeloos, P., Roels, O. and E. Jaspers (p. 456). Wetteren, Belgium: Universa Press.

Stentiford, G.D., Neil, D.M., Peeler, E.J., Shields, J.D., Small, H.J., Flegel, T.W., Vlak, J.M., et al. 2012. Disease will limit future food supply from the global crustacean fishery and aquaculture sectors. *Journal of Invertebrate Pathology*. 110:141–157.

Tang, B., Zhang, D., Li, H., Jiang, S., Zhang, H., Xuan, F., Ge, B., et al. 2020. Chromosome-level genome assembly reveals the unique genome evolution of the swimming crab (*Portunus trituberculatus*). *GigaScience*. 9:1–10.

Thiéry, A., Rabet, N. and G. Nève. 2007. Models for intraspecific resting egg shape variation in a freshwater fairy shrimp *Tanymastix stagnalis* (L., 1758) (Crustacea, Branchiopoda). *Biological Journal of the Linnean Society*. 90 (1):55–60.

Thompson, J.V. 1836. Natural history and metamorphosis of an anomalous crustaceous parasite of *Carcinus maenas*, the *Sacculina carcini*. *Entomological Magazine*. 3:452–456.

Trédez, F., Rabet, N., Bellec, L. and F. Audebert. 2016. Synchronism of naupliar development of *Sacculina carcini* Thompson, 1836 (Pancrustacea, Rhizocephala) revealed by precise monitoring. *Helgoland Marine Research*. 70:26.

Van Stappen, G., Sui, L., Nguyen Hoa, V., Tamtin, M., Nyonje, B., de Medeiros Rocha, R., Sorgeloos, P. and G. Gajardo. 2019. Review on integrated production of the brine shrimp *Artemia* in solar salt ponds. *Reviews in Aquaculture*. 12:1054–1071.

von Reumont, B.M., Blanke, A., Richter, S. Alvarez, F., Bleidorn C. and R.A. Jenner. 2014. The first venomous crustacean revealed by transcriptomics and functional morphology: Remipede venom glands express a unique toxin cocktail dominated by enzymes and a neurotoxin. *Molecular Biology and Evolution*. 31 (1):48–58.

Waloszek, D. 1993. The upper Cambrian *Rehbachiella* and the phylogeny of Branchiopoda and Crustacea. *Fossils and Strata*. 32 (4):1–202.

Wang, S.F., Lv, F.J., Jiao, T.F., Ao, J.F., Zhang, X.C. and F.D. Jin. 2015. A novel porous carrier found in nature for nanocomposite materials preparation: A case study of *Artemia* egg shell-supported TiO_2 for formaldehyde removal. *Journal of Nanomaterials*. ID 963012.

Watling, L. and M. Thiel. 2013. The crustacean carapace: Morphology, function, development, and phylogenetic history. In *Book: The Natural History of the Crustacea, Vol. 1: Functional Morphology and Diversity* (pp. 103–127). Chapter: Publisher: Oxford: Oxford University Press Editors.

Weeks, S.C., Brantner, J.S., Astrop, T.I., Ott, D.W. and N. Rabet. 2014. The evolution of hermaphroditism from dioecy in crustaceans: Selfing hermaphroditism described in a fourth spinicaudatan genus. *Evolutionary Biology*. 41:251–261.

Weydmann, A., Przyłucka, A., Lubośny, M., Walczyńska, K.S., Serrão, E.A., Pearson, G.A. and A. Burzyński. 2017. Mitochondrial genomes of the key zooplankton copepods Arctic *Calanus glacialis* and North Atlantic *Calanus finmarchicus* with the longest crustacean non-coding regions. *Scientific Reports*. 7: 13702.

Yager, J. 1981. A new class of Crustacea from a marine cave in the Bahamas. *Journal of Crustacean Biology*. 1 (3):328–333.

Zbinden, M. and M.A. Cambon-Bonavita. 2020. *Rimicaris exoculata*: Biology and ecology of a shrimp from deep-sea hydrothermal vents associated with ectosymbiotic bacteria. *Journal Marine Ecology Progress Series*. 652:187–222.

Zhao, J., Yin, J., Zhong, J., Jiao, T., Bai, Z., Wang, S., Zhang, L., et al. 2019. Facile preparation of a self-assembled *Artemia* cyst shell-TiO_2-MoS_2 porous composite structure with highly efficient catalytic reduction of nitro compounds for wastewater treatment. *Nanotechnology*. 31 (8):085603.

Zilhão, J., Angelucci, D.E., Araújo Igreja, M., Arnold, L.J., Badal, E., Callapez, P., Cardoso, et al. 2020. Last interglacial Iberian Neandertals as fisher-hunter-gatherers. *Science*. 367 (6485):eaaz7943.

Zrzavý, J. and P. Štys. 1997. The basic body plan of arthropods: Insights from evolutionary morphology and developmental biology. *Journal of Evolutionary Biology*. 10 (3):353–367.

Zucker, A. 2005. *Aristote et les classifications zoologiques*. Louvain-La-Neuve, Paris; Dudley, MA: Peeters.

16 *Parhyale hawaiensis*, Crustacea

John Rallis, Gentian Kapai and Anastasios Pavlopoulos

CONTENTS

16.1 HISTORY OF THE MODEL

The marine crustacean species *Parhyale hawaiensis* (hereafter referred to as *Parhyale*) was first described by James D. Dana in 1853 from the Hawaiian island of Maui (Dana 1853; Shoemaker 1956; Myers 1985). It was first introduced in the laboratory of Prof. Nipam Patel in 1997 from a population that was collected from the filtration system of the Shedd Aquarium in Chicago (Rehm et al. 2009e). Since the early 2000s, it has emerged as an attractive experimental organism for modern biological and biomedical research. An increasing number of laboratories in America and Europe have embraced this model system for molecular, cellular, ecological, evolutionary, developmental genetic and functional genomic studies (Stamataki and Pavlopoulos 2016).

Parhyale is a member of the order Amphipoda, a diverse group of crustaceans with more than 10,000 identified species (Figure 16.1a) (Horton et al. 2020). Besides its biological and technical qualities described in the following sections, *Parhyale* was selected for its position in the arthropod phylogenetic tree. Amphipoda belong to the class Malacostraca that comprises well-known and nutritionally important crustaceans from the order Decapoda such as crabs, lobsters, shrimps and crayfish, as well as other familiar crustaceans such as mantis shrimps (Stomatopoda), woodlice (Isopoda), krill (Euphausiacea) and others (Figure 16.1b).

Although many high-level and low-level phylogenetic relationships still remain unresolved, several molecular phylogenetic and phylogenomic analyses have improved our knowledge on the relationships between malacostracans and the other crustacean and arthropod groups (Giribet and Edgecombe 2019). It is now almost universally accepted that insects (Hexapoda) represent a terrestrial lineage of crustaceans that together with the crustaceans constitute the monophyletic taxon Pancrustacea (Figure 16.1c). Within Pancrustacea, Remipedia are increasingly supported as the sister group to Hexapoda that together with Branchiopoda and Cephalocarida form a group called Allotriocarida (von Reumont et al. 2012; Schwentner et al. 2017). Malacostraca are more closely related to Copepoda and Thecostraca (with their exact relationships still unresolved) and form the sister group to Allotriocarida called Multicrustacea (Regier et al. 2010; Lozano-Fernandez et al. 2019). Finally, Oligostraca constitute the third major pancrustacean clade containing the Ostracoda, Mystacocarida, Branchiura and Pentastomida (Regier et al. 2010; Oakley et al. 2013). High-level arthropod relationships have been also adequately resolved, ending centuries of debates (Giribet and Edgecombe 2019). Myriapoda (centipedes, millipeds and allies) have been placed as the sister group to Pancrustacea, in a clade known as Mandibulata (jawed arthropods), and together with the Chelicerata (sea spiders, horseshoe crabs and arachnids)

DOI: 10.1201/9781003217503-16

they form the three main branches of extant Arthropoda (Figure 16.1c).

This improved phylogeny seeded the development of suitable crustacean species as experimental models for comparative studies to understand the conservation and divergence of developmental patterning mechanisms during pancrustacean and arthropod evolution. The insect *Drosophila melanogaster*, which is one of the premier animal models for developmental genetic and genomic research, has attracted disproportionately more attention compared to other emerging insect, crustacean, myriapod and chelicerate models. Acknowledging all the major contributions that *Drosophila* research has made in revealing many of the basic principles of animal development, its lineage represents only a tiny fraction of the morphological diversity and developmental strategies employed by arthropods alone. Over the last two decades, the availability of broadly applicable experimental approaches has bridged the technological gap between

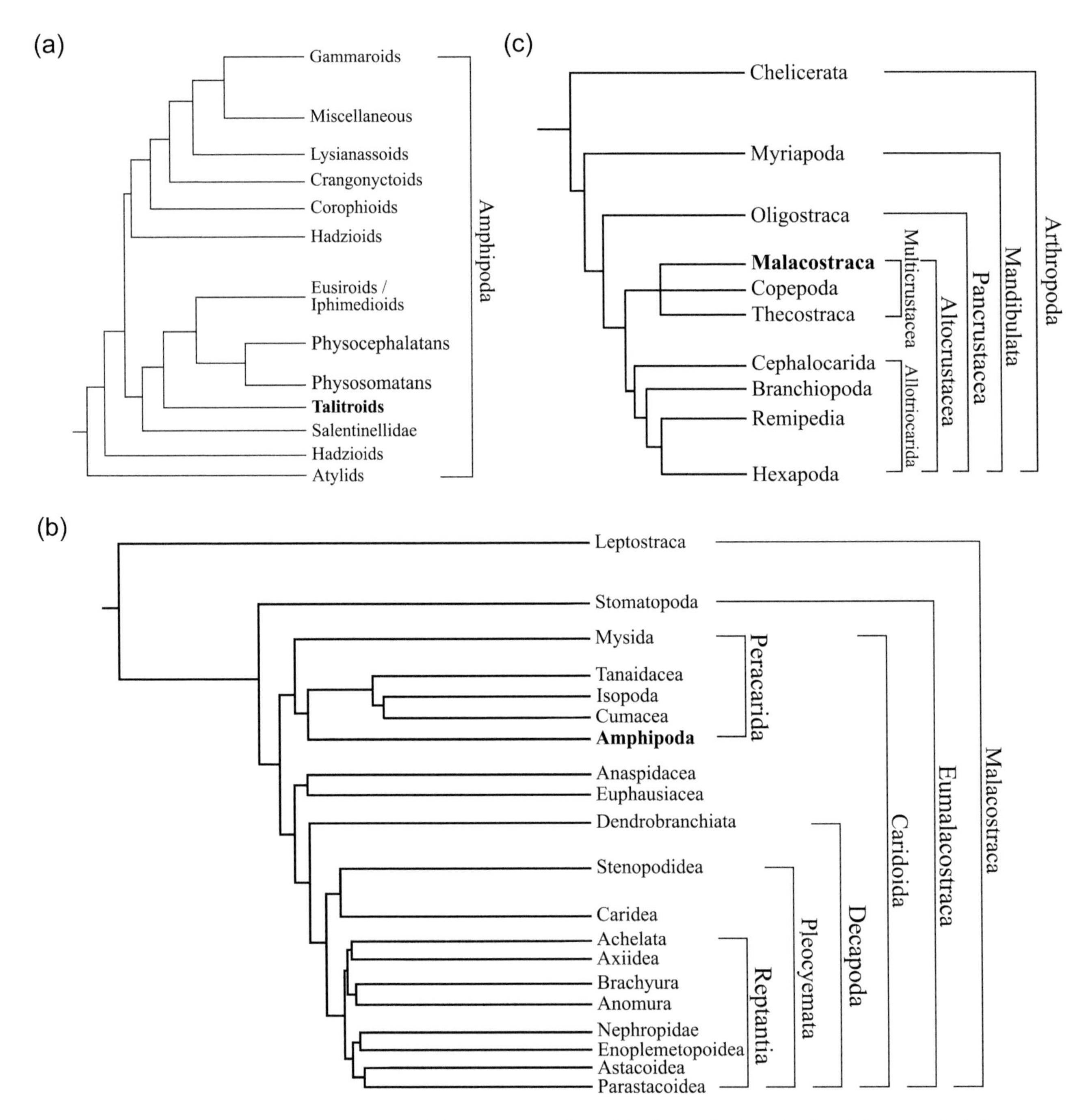

FIGURE 16.1 Phylogenetic affiliation of *Parhyale hawaiensis*. (a) One of the few available molecular phylogenies depicting the relationships between amphipod lineages, according to Copilaş-Ciocianu et al. 2020. *Parhyale* is a marine talitrid amphipod that belongs to the family Hyalidae. (b) Phylogenetic relationships within Malacostraca, according to (Schwentner et al. 2018). Note that many topologies are poorly supported and remain essentially unresolved. *Parhyale* is a peracarid amphipod. (c) Molecular tree of the arthropods, as reviewed by Giribet and Edgecombe 2019. *Parhyale* is a Malacostracan crustacean.

Drosophila and emergent arthropod models enabling both mechanistic insights into biological diversity, as well the study of unique traits and biological processes that are not accessible in standard model systems.

Parhyale is currently one of the very few available models representative of malacostracans, crustaceans and marine animals in general that is experimentally tractable and supported by a continuously expanding toolkit of techniques and resources (Kao et al. 2016). As a result, studies in *Parhyale* are increasing in scope and depth beyond the descriptive level, hypotheses can be tested functionally at a higher level of sophistication and novel discoveries are making research headlines (BBSRC Business Magazine 2017).

16.2 GEOGRAPHICAL LOCATION

Amphipods have inhabited almost all aquatic (marine, brackish and freshwater) environments, as well as moist terrestrial habitats, and play essential roles as detritovores or scavengers in nutrient recycling in these ecosystems (Copilaş-Ciocianu et al. 2020). *Parhyale* is an epibenthic detritovorous species with a worldwide, circumtropical distribution (Shoemaker 1956; Myers 1985). It lives in intertidal and shallow marine habitats, including bays, estuaries and mangrove litter; therefore, it can tolerate large changes in salinity, temperature and nutrient availability (Tararam et al. 1978; Poovachiranon et al. 1986).

Based on measurements of the population structure and dynamics in communities of intertidal shores, the *Parhyale* lifestyle is consistent with the opportunistic strategies adopted by epifaunal species inhabiting unpredictable environments (Alegretti et al. 2016). Population size varies during the year and grows rapidly during favorable environmental conditions. The rapid growth of *Parhyale* populations is attributed to their continuous reproductive capacity, a sex ratio biased toward females and multivoltinism (having several broods per season). The relatively low number of eggs per female (ranging between 5 and 30 per brood depending on the age and size of the female) is compensated for by the precocious sexual maturation of adults, as well as the low mortality of embryos and hatched juveniles that are kept by females in a ventral brood pouch. The average generation time of *Parhyale* in intertidal natural populations has been estimated at 3.5 months (Alegretti et al. 2016), but this is decreased to about 2 months in the laboratory. More broadly, this lifestyle enables *Parhyale* to thrive under controlled laboratory conditions, where the only major consideration is the continuous aeration of the cultures with air or water pumps due to their generally low tolerance to hypoxic conditions.

FIGURE 16.2 *Parhyale hawaiensis* as a laboratory experimental model. (a) Typical laboratory *Parhyale* culture in a plastic Tupperware (lid removed for the photo) containing artificial sea water, a layer of gravel (G), an air bubbler (AB) for aeration, a heating filament (HF) for a constant temperature at 26°C and a phosphate/nitrate remover (PNR) to keep the culture free of organic waste. (b) Petri dish with *Parhyale* mating pairs in precopulatory amplexus. (c) Adult male and (d) female *Parhyale*. Lateral views with anterior to the left and ventral to the bottom. The sexually dimorphic gnathopods are indicated with asterisks.

16.3 LIFE CYCLE

In the laboratory, *Parhyale* is cultured in large plastic containers on a bed of crushed coral gravel and covered in artificial sea water under continuous aeration (Figure 16.2a). Although they can tolerate a wide range of temperatures from at least 18°C to 30°C, they are routinely kept at 26°C to standardize developmental timing. *Parhyale* are omnivorous; therefore, different labs have adopted different diets ranging from plain carrots to rich mixes of larval shrimp and fish flakes supplemented with fatty acids and vitamins. Under these conditions and with frequent feeding and water change regimes, *Parhyale* has in the laboratory a life cycle of about two months. This relatively short generation time and the ease and cost effectiveness to grow this marine crustacean in dense cultures, as well as the daily availability of hundreds of individuals at any desired developmental stage throughout the year, make *Parhyale* a convenient model system for research purposes.

Parhyale is a sexually dimorphic species (Figure 16.2b–d). Adult males can be easily distinguished from females based on a pair of enlarged grasping appendages (the second pair of gnathopods) in their anterior thorax (Figure 16.2c, d). A sexually mature male uses the other first pair of unenlarged gnathopods to grasp and carry a female, guarding her against other males before copulation (Conlan 1991). The duration of this precopulatory amplexus varies from several hours to days, during which time the couple is capable of walking and swimming (Figure 16.2b). Shortly before copulation, the female molts, producing a new brood chamber (marsupium) under her ventral surface from flexible flaps (oostegites) extending medially from her thoracic appendages. The male then deposits sperm into the new marsupium, and the female ovulates, depositing her oocytes into the marsupium while the new exoskeleton is still flexible to allow their passage through the oviducts (Hyne 2011). The poorly understood process of fertilization takes place externally in the marsupium while the male and female separate. As noted earlier, *Parhyale* females lay about 5 to 30 eggs during each molting cycle depending on their age and size and can produce successive broods every few weeks during their lifetime. Considering also that females do not store sperm, this reproductive behavior is convenient for genetic research as single backcrosses and intercrosses can be set routinely to generate *Parhyale* inbred lines.

After fertilization, the embryos of each brood develop fairly synchronously inside the marsupium. Embryos at any stage of their development can be easily dissected or flashed out from the marsupial pouch of anesthetized gravid females (without sacrificing them) and cultured in Petri dishes in artificial seawater. Similar to the rest of amphipods, *Parhyale* are direct developers and lack intermediate larval stages (Figure 16.3). After about ten days of embryogenesis at 26°C, the juveniles that hatch and then are released from the marsupium resemble miniature versions of the adult form. Juveniles increase in size through successive molts and reach sexual maturations about six to seven weeks after hatching.

16.4 EMBRYOGENESIS

Parhyale was originally selected as a new crustacean model for comparative developmental studies (Rehm et al. 2009e). From the beginning, great effort has been invested in the detailed study of *Parhyale* embryogenesis that has been conveniently subdivided into well-defined stages based on morphological and molecular markers (Browne et al. 2005). Embryos have a number of useful properties for detailed microscopic inspection using brightfield or fluorescence imaging (Figure 16.3): the eggs are about 500 µm long, the eggshell is transparent, and early development takes place on the egg surface, resulting in a nice contrast between the embryo and the underlying opaque yolk that later on gets sequestered inside the developing midgut.

16.4.1 Early Cleavage Stages

Early cleavages of the *Parhyale* zygote (Figure 16.3, 3h) follow a holoblastic, radial, determinate and stereotyped pattern (Gerberding et al. 2002). The first cleavage occurs

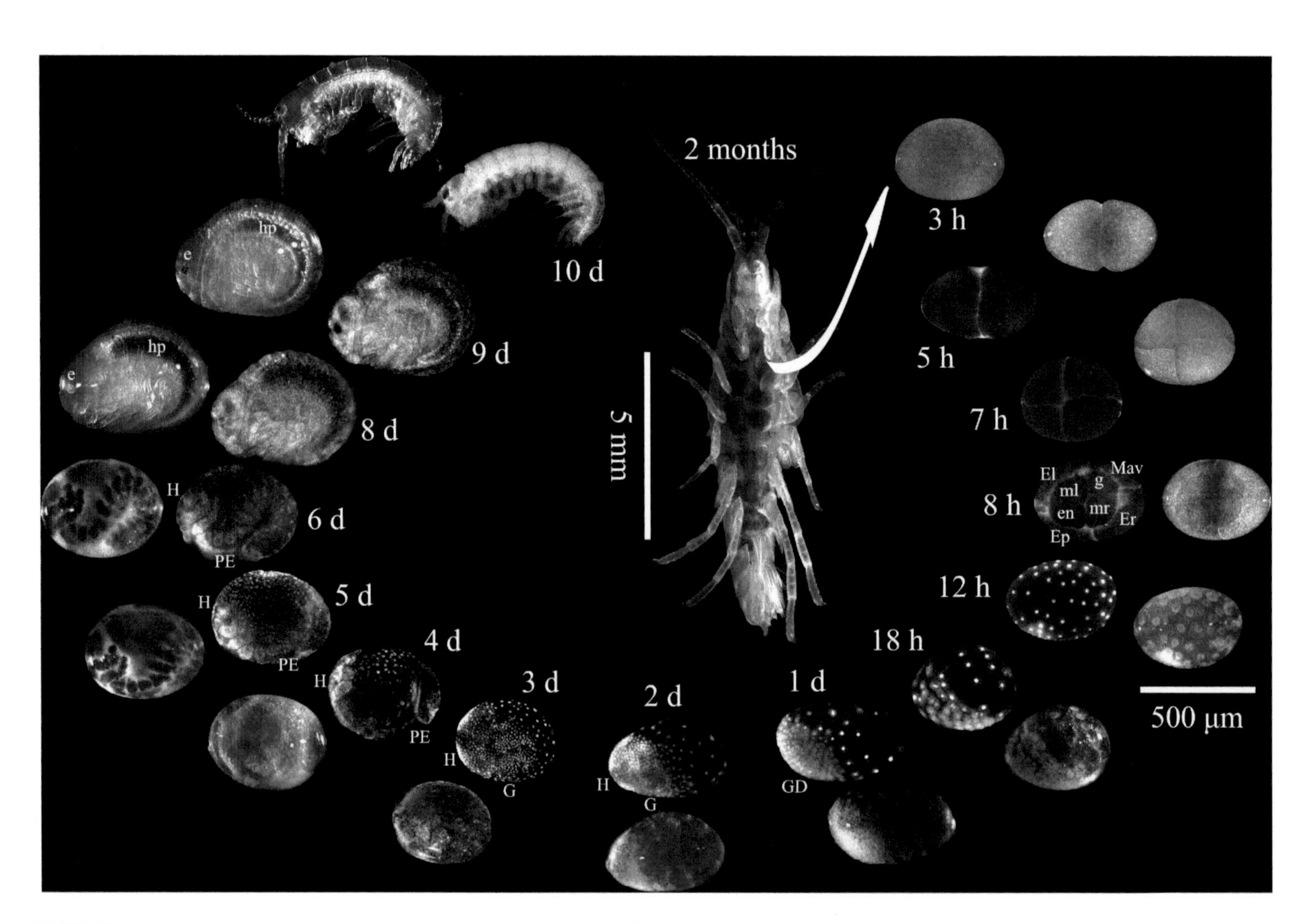

FIGURE 16.3 *Parhyale hawaiensis* embryogenesis. Brightfield images (aligned in the outer positions) and fluorescent images (aligned in the inner positions) of embryos at the indicated stages in hours (h) or days (d) after egg lay. Embryos can be removed from the marsupial pouch of anesthetized gravid females at any stage. The names of the macromeres and micromeres contributing to the different germ layers and the germ line are indicated in the eight-cell stage embryo (8 h). The juveniles that hatch from the eggs are miniature versions of the adults. All embryonic stages are shown to scale. Abbreviations: GD, germ disc; H, head; G, grid; PE, posterior end; hp, hepatopancreatic caecum; e, eye.

about four hours after egg lay (AEL) at 26°C (Figure 16.3, 5h). It is perpendicular to the long axis of the egg and slightly unequal, and the fate of each of the two blastomeres is already restricted to the left or right side of the animal with regard to a large fraction of the ectoderm and mesoderm. The second cleavage is parallel to the long axis of the egg and also slightly unequal (Figure 16.3, 7h), while the third cleavage (perpendicular to the other two) is highly unequal, producing a stereotypical arrangement of four macromeres and four micromeres uniquely identifiable based on their relative position and size (Figure 16.3, 8h). Each of these blastomeres has an invariant fate restricted to a single germ layer already at this early developmental stage (Gerberding et al. 2002; Browne et al. 2005; Price and Patel 2008; Hannibal et al. 2012). Three macromeres, termed El, Er and Ep, give rise to the ectoderm: El and Er contribute the left and right head ectoderm and parts of the left and right thoracic ectoderm, respectively, while Ep contributes the remaining thoracic and abdominal ectoderm, as well as a distinct column of cells marking the ventral midline of the embryo and separating its left and right sides. The fourth macromere, termed Mav, generates the visceral and somatic head mesoderm. Two micromeres, called mL and mr, form the left and right somatic trunk mesoderm, while the other two micromeres, called en and g, give rise to the endoderm and germ line, respectively. Despite these very early lineage restrictions, *Parhyale* embryos have the capacity to replace missing parts of the ectoderm and mesoderm after ablation of precursors during early development (Price et al. 2010). Similarly, although the germ line is normally specified in a cell-autonomous manner at the eight-cell stage (Extavour 2005; Ozhan-Kizil et al. 2009; Gupta and Extavour 2013), *Parhyale* has the astonishing flexibility to regenerate its germ line post-embryonically (presumably through reprogramming of somatic cells) after ablation of the g micromere (Modrell 2007; Kaczmarczyk 2014).

16.4.2 Gastrulation and Germ Disc Formation

Synchrony is gradually lost in later cleavages, and cells become yolk free as they extrude their yolk toward the center of the egg. The macromeres divide faster than the micromeres, forming a soccer ball-like embryo that consists of about 100 uniform cells around the egg surface at 12 hours AEL (Figure 16.3, 12h). Over the following 8 hours, gastrulation is effected by cell shape changes, neighbor exchange and cell migration (Figure 16.3, 18h) (Price and Patel 2008; Alwes et al. 2011; Chaw and Patel 2012). The group of Mav and g descendants (visible as a characteristic rosette) internalizes underneath a condensing epithelial monolayer formed by the El, Er and Ep descendants (ectoderm primordium), resulting in a multi-layered and bilaterally symmetric germ disc (embryo rudiment) at the anterior ventral side of the egg (Figure 16.3, 1d). The presumptive trunk somatic mesoderm (mL and mr descendants) and endoderm (en descendants) precursors internalize at the periphery of the germ disc. A few cells that do not contribute to the initial ventral germ disc remain widely distributed around the dorsal egg surface. The descendants of these cells contribute later on to the growing embryo proper, as well as to the adjoining extra-embryonic region.

16.4.3 Germ Band Extension and Segmentation

The germ disc grows by cell proliferation and recruitment of new cells laterally and posteriorly. About two days AEL, embryonic cells start organizing into an anterior pair of head lobes followed by a grid-like array that will give rise to the rest of the germ band (Figure 16.3, 2d–3d and Figure 16.4a, b). The ectodermal cells in this grid exhibit an ordered arrangement in transverse rows (perpendicular to the ventral midline) and longitudinal columns (parallel to the ventral midline) (Figure 16.4). The formation and growth of the ectodermal grid occur with an anterior-to-posterior progression, that is, the more anterior rows are formed first, and the more posterior rows are added sequentially at the posterior end of the grid (Figure 16.4b) (Browne et al. 2005). These rows will eventually give rise to most body units of *Parhyale* (called the post-naupliar region), and only the head region anterior to the mandibles (called the naupliar region) is formed from ectodermal cells outside the grid. Among all pancrustaceans and arthropods, this early patterning of the ectoderm by means of a highly ordered grid-like array of precursor cells is a unique common feature of Malacostracans (Dohle et al. 2003). Unlike most Malacostracans, though, that form this grid through the asymmetric repeated divisions of ectoderm stem cells called ectoteloblasts, amphipods like *Parhyale* lack ectoteloblasts and form the post-naupliar grid through the aforementioned progressive self-organization of scattered ectodermal cells into transverse rows of cells (Figure 16.4b).

Similar to *Drosophila* and the rest of the arthropods, the metameric organization of the early *Parhyale* embryo is parasegmental, with each transverse row of cells corresponding to one parasegment (Browne et al. 2005). Each row of cells undergoes two rounds of stereotyped and symmetric mitotic divisions, first producing a two-row and then a four-row parasegment (Figure 16.4c). These divisions are oriented parallel to the anterior–posterior axis, producing the ordered arrangement of daughter cells in well-defined longitudinal columns of cells. The geometric precision and invariance of the grid pattern enables to identify individual cells between the left and right side in each embryo and across embryos. A naming convention based on numbers and letters has been established by Prof. Wolfgang Dohle to indicate the position of cells in the one-, two- or four-row parasegments along the anterior–posterior axis and in the columns along the dorsal-ventral axis (Figure 16.4b, c) (Dohle et al. 2003; Browne et al. 2005).

The regularity of the grid dissolves during the following divisions that are not strictly longitudinal but have a more complex, yet still invariant, pattern. At the tissue

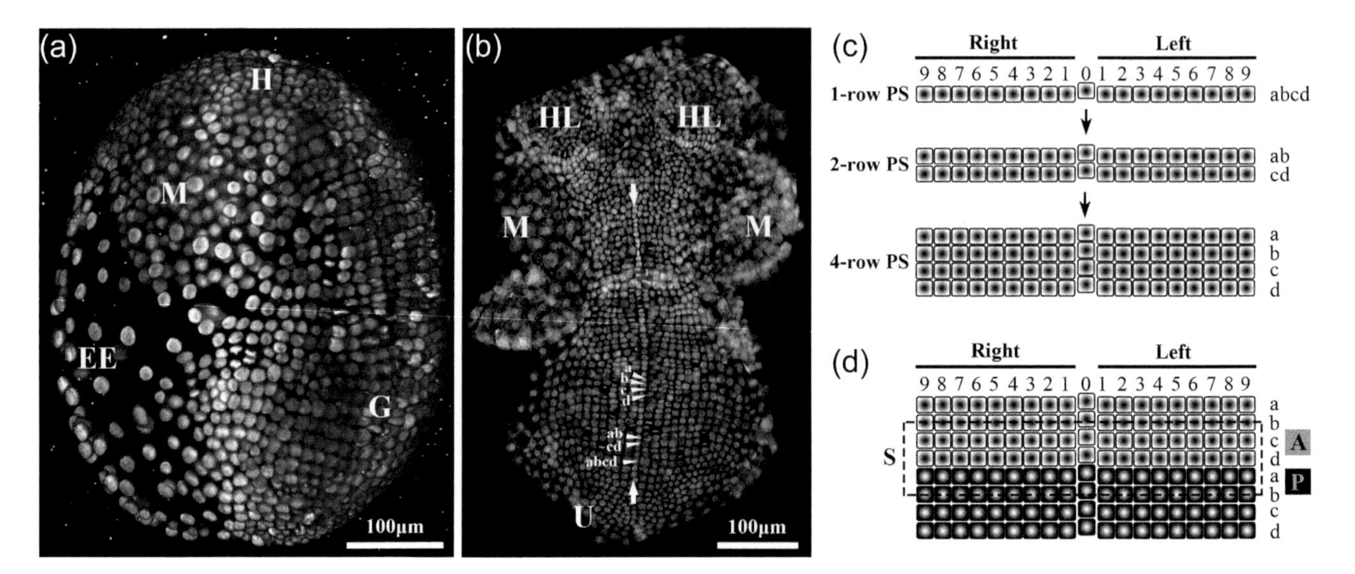

FIGURE 16.4 *Parhyale hawaiensis* ectoderm segmentation. (a) Right side of a live imaged *Parhyale* embryo with fluorescently labeled nuclei at the mid-germ band stage (anterior to the top and ventral to the right). Note the distinct organization and density of cells in the naupliar head region (H), the post-naupliar grid (G), the midgut primordium (M) and the extra-embryonic region (EE). (b) Ventral view of a similar staged fixed embryo with stained nuclei. From anterior (top) to posterior (bottom), the embryo is organized into the bilateral pairs of head lobes (HL) and midgut primordia (M), the conspicuous column of ectodermal cells marking the ventral midline (flanked by arrows) and the ectodermal grid with the constellation of parasegments that have undergone two rounds of mitotic cell divisions (four-row parasegment indicated with a, b, c and d), one round of cell divisions (two-row parasegment indicated with ab and cd), no cell division (one-row parasegment indicated with abcd) and unorganized cells before they become arranged in rows (U). (c) Schematic representation and naming convention of grid cells: one-row-parasegment (top) with abcd cells; two-row-parasegment (middle) with anterior ab and posterior cd cells; and four-row-parasegment (bottom) with a, b, c and d cells. Mediolateral columns are indexed by numbers with 0 denoting the ventral midline and 1, 2 . . . *n* the more lateral columns based on their distance from the midline. (d) Schematic representation of segmental organization. Cells from two neighboring parasegments (indicated with different patterns) contribute to each morphological segment (bounded by a rectangular line). Each segment is compartmentalized into anterior (A) and posterior (P) compartment cells derived from the anterior and posterior parasegment, respectively. Segmental boundaries run between progenies of the b cell rows.

level, transverse intersegmental furrows indicate the transition from the parasegmental to the segmental metameric organization of the embryo, and pairs of appendage buds start appearing ventrally, first in the anterior head segments and then more posteriorly (Figure 16.3, 4d). Like in other arthropods, each morphological segment and associated appendages are composed of cells from two neighboring parasegments without any cell mixing (Figure 16.4d): cells from the posterior rows of one parasegment contribute to the anterior compartment of the segment, while cells from the anterior rows of the following parasegment contribute to the posterior compartment of the segment (Browne et al. 2005; Wolff et al. 2018).

The mesoderm in *Parhyale* is derived from the mL and mr micromeres producing the left and right segmental mesoderm in the trunk, respectively, and the Mav macromere producing the head and visceral mesoderm (Gerberding et al. 2002; Price and Patel 2008; Vargas-Vila et al. 2010). The segmental trunk mesoderm develops in tight association with the overlying, growing ectodermal monolayer also with an anterior-to-posterior progression (Hannibal et al. 2012). In all Malacostracans, including *Parhyale*, the mesoderm in each trunk segment is formed from a row of eight founder cells, called mesoblasts, four in the left and four in the right hemisegment (Browne et al. 2005; Price and Patel 2008). The segmental rows of mesoblasts are the product of the asymmetric, repeated divisions of eight mesodermal stem cells, called mesoteloblasts, that are derived from the mL and mr lineages and are also uniquely identifiable based on their position and the use of a standardized nomenclature (Dohle et al. 2003). To summarize, axial elongation of the *Parhyale* germ band occurs by the sequential addition and division of new ectodermal and mesodermal rows. As the growing germ band reaches the posterior pole of the egg, it bends downward (Figure 16.3, 4d). During subsequent stages, the embryo acquires a comma shape, where the posterior abdominal trunk develops juxtaposed to the more anterior thoracic trunk.

16.4.4 Organogenesis

Ectodermal cells from the medial columns in the grid give rise to the nervous system and sternites, cells from the lateral columns give rise to the forming limbs and cells at the edge of the grid give rise to the dorsal body wall tergites (Vargas-Vila et al. 2010; Wolff et al. 2018). As the comma-shaped embryo continues to grow, the posterior terminus (telson) projects anteriorly until it reaches the anterior thoracic region (Figure 16.3, 5d–6d). Concurrent with axial elongation, the lateral edges of the ectoderm expand dorsally and the forming tergites from the two body halves fuse along the dorsal midline completing dorsal closure. Starting from the anterior head region backward and sequentially bulging out in the thorax and the abdomen, a total of 19 pairs of appendages develop along the *Parhyale* body (Figure 16.3, 4d–6d). Appendages increase in size and elongate along their respective proximal–distal axes (Browne et al. 2005; Wolff et al. 2018). As detailed in the next sections, the elaboration of the proximal–distal axis varies between different appendage types in terms of their pattern, size and shape, resulting in a remarkable morphological diversity along the anterior–posterior axis. Appendage growth, morphogenesis and differentiation continue until the late stages of embryogenesis, when the fully formed appendages occupy almost half of the egg space before hatching (Figure 16.3, 8d–9d).

The naupliar (anterior head) and post-naupliar somatic mesoderm are separated early on as they derive from the Mav macromere and the mL/mr micromeres at the eight-cell stage, respectively (Figure 16.3, 8h) (Gerberding et al. 2002; Browne et al. 2005). The micromere-derived rows of four mesoteloblasts (labeled M1 to M4 medial-to-lateral) under each side the ectodermal grid generate the segmental mesodermal founders (mesoblasts labeled m1 to m4) in the posterior head (second maxillary segment) and the thoracic and abdominal segments. Similar to the ectodermal structures, patterning of mesoderm occurs with an anterior (earlier developing) to posterior (later developing) progression. The origin and first division of mesoblasts has been described in *Parhyale* (Price and Patel 2008). The contribution of these mesoblasts to the different muscle groups along the dorsal–ventral body axis has been studied in the closely related amphipod *Orchestia cavimana* and is only briefly summarized here (Hunnekuhl and Wolff 2012). Descendant cells from the medial-most m1 mesoblasts give rise to the ventromedian muscles, cells from the central m2 and m3 mesoblasts generate the extrinsic and intrinsic musculature of the appendages and cells from the m3 and m4 mesoblasts give rise to the dorsolateral trunk musculature and the heart (Figure 16.7e).

The Mav macromere gives rise to the head musculature of the antennae and the mandibular and first maxillary segments (Price and Patel 2008; Price et al. 2010; Hunnekuhl and Wolff 2012), as well as to the visceral mesoderm. After gastrulation, a subset of the Mav progeny migrates under the developing head segments and becomes partitioned into the differentiating head segments in a less studied manner. The majority of Mav progeny, together with the descendants from the en micromere, give rise to the midgut tube that will eventually spread over and encapsulate the central yolk mass (Gerberding et al. 2002). During the germ band stages, the midgut primordium becomes visible as a bilateral pair of discs under the head lobes (Figure 16.4a, b). The discs increase in size, forming a continuous ventral layer that expands dorsally and posteriorly under the ectoderm and mesoderm to cover the yolk (Gerberding et al. 2002). The midgut develops a number of blind tubes (caeca) that function in food digestion and absorption (Schmitz and Scherrey 1983). The most conspicuous pair of anterior caeca, called hepatopancreatic caeca, extend in synchrony through peristaltic contractions from the anterior end of the midgut until the posterior abdomen of the embryo (Browne et al. 2005). The hepatopancreatic caeca flank and extend parallel to the midgut that is visible along the dorsal side (Figure 16.3, 8d–9d). The *Parhyale* heart develops as a muscular tube along the dorsal thoracic region with three pairs of lateral inflow valves and an anterior outflow valve, and it can be observed while beating on top of the midgut (Kontarakis et al. 2011b). At around the same stage when the heart starts beating, the bilaterally symmetric compound eyes become visible in the head capsule as small white clusters, each with about three ommatidia (Figure 16.3, 8d). During the last two days of embryogenesis, the eyes become dark pigmented, and *Parhyale* hatch with about eight to nine pigmented ommatidia per eye (Figure 16.3, 9d–10d), but this number increases gradually to about 50 in older adults (Ramos et al. 2019).

The smallest micromere g at the eight-cell stage is the source of germ line cells in the adult ovaries and testes (Figure 16.3, 8h) (Gerberding et al. 2002; Extavour 2005). There is strong evidence that germ cells in *Parhyale* are specified by a cell-autonomous mechanism (preformation) via the early asymmetric segregation of maternally provided germ line determinants (Extavour 2005; Modrell 2007; Gupta and Extavour 2013). The primordial germ cells (progeny of the g micromere) that have internalized and proliferated during the gastrulation and germ disc stages form a single medial cluster of about 15 cells under the posterior head ectoderm as the germ band elongates. During organogenesis stages, they split into two bilaterally opposed cell populations that migrate separately under the lateral ectoderm toward the dorsal side of the embryo (Extavour 2005; Browne et al. 2005). At the end of embryogenesis, when the eyes and the heart have formed, the primordial germ cells are aligned in two rows flanking the dorsal midline at the site of the future gonads (Extavour 2005).

16.5 ANATOMY

Parhyale displays the typical amphipod body plan that is laterally compressed and consists of a series of repeating segmental units along the anterior–posterior axis organized into three major tagmata: the head, the thorax and the abdomen (Figure 16.5a, b). The head (a.k.a. cephalon) is composed of six segments with five pairs of appendages. The most anterior limbless pre-antennal segment is followed by five segments bearing the first and second pair of antennae (An1 and An2; Figure 16.5a, b) and three pairs of medially fused gnathal appendages: the mandibles (Mn; Figure 16.5c) and the first and second maxillae (Mx1 and Mx2; Figure 16.5d). The thoracic region is composed of eight segments, each bearing a pair of jointed uniramous appendages (I-shaped limbs with a single proximal–distal axis) (Figure 16.5e–i). The abdominal region is composed of six segments, each bearing a pair of jointed biramous appendages (Y-shaped limbs with a bifurcated proximal–distal axis) (Figure 16.5J, K). Each thoracic and abdominal appendage consists of a proximal part and a distal part (Boxshall 2004; Pavlopoulos and Wolff 2020). The proximal part, called a protopod, is composed of two appendage articles (a.k.a. podomeres or limb segments), namely the proximal coxa and the distal basis (Figure 16.5g). The existence of a third proximal-most podomere, the precoxa, has been also proposed recently (Bruce and Patel 2020). In uniramous thoracic appendages (Figure 16.5e–i), a single branch extends distally from the protopod called the endopod (or telopod). In abdominal biramous appendages (Figure 16.5j, k), two branches extend distally from the protopod called the endopod (inner branch) and exopod (outer branch). As detailed in the following, different types of appendages develop also a variable number of ventral and/or dorsal outgrowths from their protopod called endites and exites, respectively,

The first thoracic segment (T1) is fused to the head that is also referred to as the cephalothorax. The T1 appendages, called maxillipeds (T1/Mxp; Figure 16.5e), are jointed, and uniramous like the more posterior thoracic appendages. However, unlike the other thoracic appendages and similar to the more anterior maxillae, maxillipeds are reduced in size, are medially fused at their base and have two prominent endites on their proximal segments (Figure 16.5e). Maxillipeds and gnathal appendages are specialized for feeding and have a compact arrangement around the mouth region (Figure 16.5b). The thoracic region behind T1, known as the pereon, is composed of seven segments (T2 to T8), each with a pair of uniramous appendages (a.k.a. pereopods or thoracopods) that articulate independently on

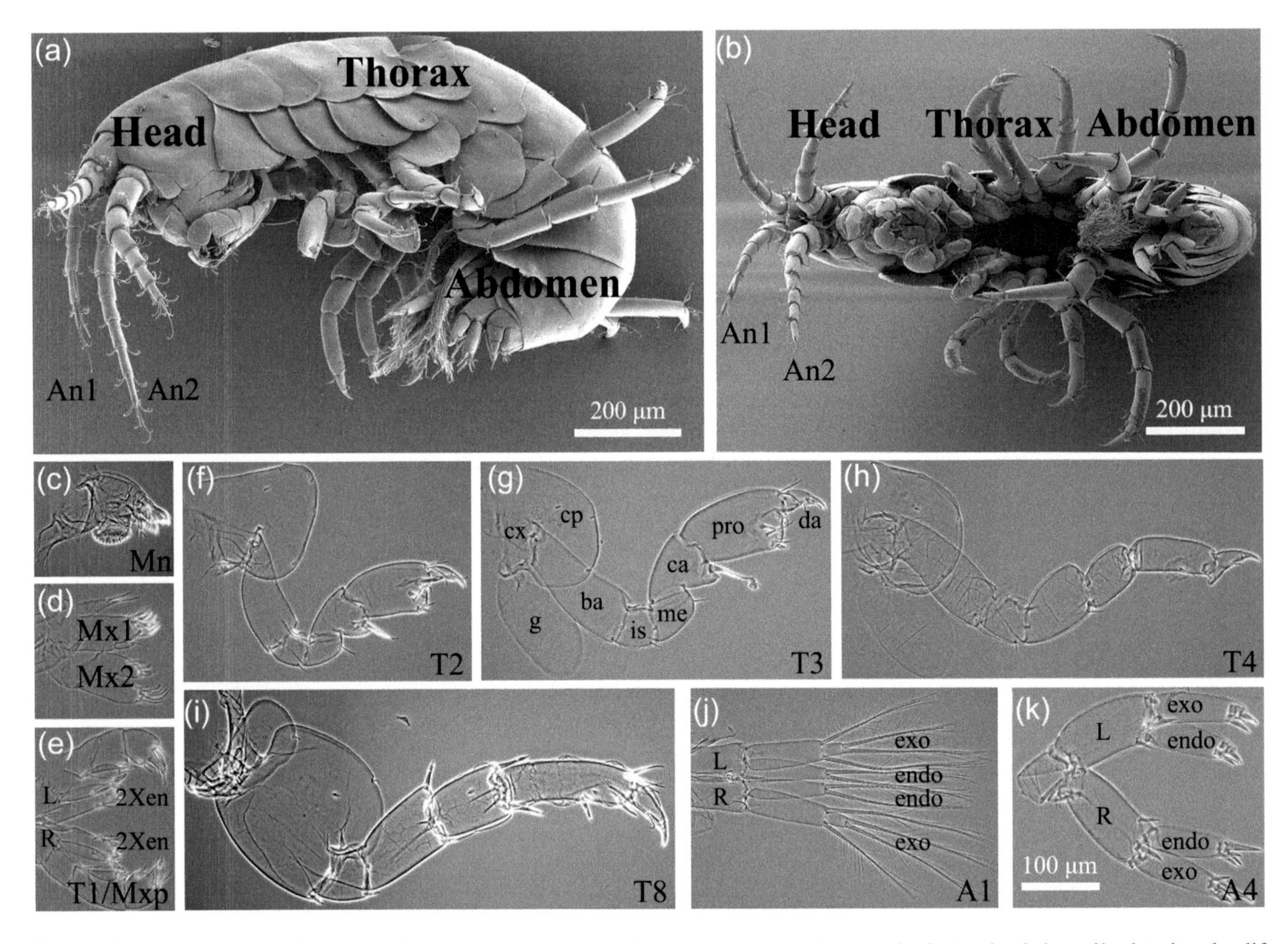

FIGURE 16.5 Appendage diversity in *Parhyale hawaiensis.* (a) Scanning electron micrograph of a *Parhyale* juvenile showing the different tagmata along the anterior–posterior body axis and the first and second pair of antennae (An1 and An2). Lateral view with anterior to the left and ventral to the bottom. (b) Similar to (a) from a ventral view. (c–k) Cuticle preparations of dissected appendages with their proximal side to left and their distal side to the right: (c) mandible (Mn); (d) Maxilla 1 (Mx1) and Maxilla 2 (Mx2); (e) bilateral pair of maxillipeds from the first thoracic segment (T1/Mxp) indicating the pair of endites (2Xen) on each side; (f) gnathopod from the second thoracic segment (T2); (g) gnathopod from the third thoracic segment (T3) indicating the seven segments, coxa (cx), basis (ba), ischium (is), merus (me), carpus (ca), propodus (pro) and dactylus (da), as well as the two exites, the coxal plate (cp) and the gill (g); (h) pereopod from the fourth thoracic segment (T4); (i) pereopod from the eighth thoracic segment (T8); (j) bilateral pair of pleopods from the first abdominal segment (A1) and (k) bilateral pair of uropods from the fourth abdominal segment (A4) indicating the endopod (endo) and exopod (exo) on each side. All appendages are shown to scale.

each side (Figure 16.5f–i). From proximal to distal, each jointed pereopod is made of seven segments: two protopodal segments (coxa and basis) and five endopodal segments (ischium, merus, carpus, propodus and dactylus) (Figure 16.5g). The T2 and T3 segments bear subchelate (clawed) grasping appendages, called gnathopods (Figure 16.5f, g), that are used for defense, grooming and as precopulatory organs (the T2 gnathopods) by males to carry the females (Holmquist 1982). The post-embryonic enlargement of the propodus and dactylus exclusively in the male T3 gnathopod is the most striking sexually dimorphic character in *Parhyale* (Figure 16.2c, d). The remaining five pereonic segments T4 to T8 bear elongated walking appendages (Figure 16.5h, i). Importantly, the opposite orientation between the T4/T5 pereopods that extend anteriorly and the T6/T7/T8 pereopods that extend posteriorly (Figure 16.5a, b) is what gives the group its name (from Greek words *αμφί* [amphi = both ways] and *πόδι* [podi = limb]). Besides their distinct function, podomere morphology and orientation, the T2–T8 pereopods are also distinguished by the presence or absence and the shape of exites attached on their protopodal coxa. Protective coxal plates of variable size and shapes are present on all pereopods, while respiratory gills are present on T3 to T7 appendages (Figure 16.5g). In the case of adult females, special endites (oostegites) forming the marsupium are attached on the pereopods T2 to T5.

The abdominal (pleonic) segments A1 to A6 develop two types of paired biramous appendages: pleopods on A1 to A3 (Figure 16.5j) and uropods on A4 to A6 (Figure 16.5k). Each of these biramous limbs has similar endopodal and exopodal branches. The A1–A3 pleopods (a.k.a. swimmerets) are highly setose and are coupled together for swimming and moving water over the thoracic gills. The A4–A6 uropods are thickened and spiky appendages used for jumping. The most posterior terminal structure is the telson, which is a small flap over the anus attached to segment A6. Overall, the morphological and functional specialization of body parts and associated appendages has been one of the main reasons for putting *Parhyale* forward as an attractive model organism for molecular, cellular, developmental and evolutionary studies described in Section 16.8.

Much less work has been invested in *Parhyale* to study the development, anatomy and physiology of the nervous system compared to other crustaceans (Wiese 2002). *Parhyale* neuroanatomy was recently described using a combination of histological, immuno-histochemical, optical and X-ray tomography methods (Wittfoth et al. 2019). The central nervous system consists of the brain and the ventral nerve cord. The ventral nerve cord is composed of the subesophageal ganglion, seven segmental ganglia of the pereon, three segmental ganglia of the pleosome and one fused ganglion of the urosome. The brain lies between the compound eyes in the dorsal part of the head capsule with its three neuromeres, the protocerebrum, deutocerebrum and tritocerebrum lining up from dorsal to ventral. The protocerebrum is equipped with the optic neuropils, the deutocerebrum with the antenna 1 neuropil and the olfactory lobe and the tritocerebrum with the antenna 2 neuropil. The three optic neuropils, the lamina, medulla and lobula, are in close proximity with each other, but only the lamina connects to the photoreceptors of the ommatidia in the compound eye (Wittfoth et al. 2019; Ramos et al. 2019). The architecture and neural connectivity of the *Parhyale* visual system have diverged from the typical organization exhibited by other malacostracan crustaceans and are associated with a shift to low spatial resolution and simple visual tasks (Ramos et al. 2019).

16.6 GENOMIC DATA

For many years, the high cost of next-generation sequencing technologies and the big size of malacostracan crustacean genomes have been prohibitive for amphipod genomics. Thanks to the decreasing sequencing costs, this limitation was overcome during the last five years, first with the sequencing, *de novo* assembly and annotation of the *Parhyale* genome in 2016, followed more recently by genome assemblies of variable quality for the amphipods *Hyalella azteca*, *Trinorchestia longiramous*, *Platorchestia hallaensis*, *Orchestia grillus* and *Gammarus roeselii* (Table 16.1) (Poynton et al. 2018; Patra et al. 2020a, 2020b; Cormier et al. 2021).

The *Parhyale* genome resembles and even exceeds in many respects the complexity of the human genome. The genome consists of 23 pairs of chromosomes ($2n = 46$; Figure 16.6a), and its size is estimated at 3.6 Gb. The huge genome size is associated with an expansion in repetitive and intronic sequences and exhibits very high levels of

TABLE 16.1
Sequenced Amphipod Genomes

Species	Size (Gb)	No. of Scaffolds	Scaffold N50 (Kb)	NCBI Link
Parhyale hawaiensis	2.75	278,189	20,229	www.ncbi.nlm.nih.gov/assembly/GCA_001587735.2
Hyalella azteca	0.55	18,000	215	www.ncbi.nlm.nih.gov/assembly/GCA_000764305.3
Trinorchestia longiramus	0.89	30,897	120	www.ncbi.nlm.nih.gov/assembly/GCA_006783055.1
Platorchestia hallaensis	1.18	39,873	87	www.ncbi.nlm.nih.gov/assembly/GCA_014220935.1
Orchestia grillus	0.81	143,039	17	www.ncbi.nlm.nih.gov/assembly/GCA_014899125.1
Gammarus roeselii	3.2	1,130,582	4.8	www.ncbi.nlm.nih.gov/assembly/GCA_016164225.1

heterozygosity and polymorphism (Kao et al. 2016). This published version of the genome called Phaw_3.0 (GenBank Accession number GCA_001587735.1) was sequenced to about 115x coverage from variable-sized shotgun and mate-pair Illumina libraries prepared from a single adult male from the Chicago-F iso-female line. The latest version of the genome, called Phaw_5.0 (GenBank Accession number GCA_001587735.2), was assembled from these reads supplemented with extra sequences to about 150x coverage from Dovetail Genomics proximity ligation libraries, which were generated from both in vitro reconstituted chromatin (so-called Chicago libraries prepared from the same genomic DNA used for the Illumina libraries) and native chromatin (so-called Hi-C libraries prepared from another adult male belonging to the same iso-female line) (Putnam et al. 2016). The resulting assembly with the Dovetail HiRise scaffolding pipeline has a total length of 2.75 Gb and consists of 278,189 scaffolds with an N50 of about 20 Mb and an L50 of 42 scaffolds (Table 16.1).

The availability of the high-quality reference genome has boosted functional studies of coding and non-coding sequences in *Parhyale*, as well as comparative genomic studies with other amphipods and animal taxa in general (Figure 16.6b–d) (Kao et al. 2016). The genome is accompanied and supported by an increasing number of other genome-wide resources, such as sex, stage and tissue-specific transcriptomes and proteomes, sequenced BAC clones, epigenetic marks and chromatin accessibility profiles (Parchem et al. 2010; Zeng et al. 2011; Zeng and Extavour 2012; Blythe et al. 2012; Nestorov et al. 2013; Trapp et al. 2016; Kao et al. 2016; Hunt et al. 2019; Artal et al. 2020). Annotation of the genome based on assembled *Parhyale* transcriptomes, homology with other model organisms and *ab initio* predictions has resulted in more than 28,000 protein-coding gene models (Kao et al. 2016). Most likely, this number is an overestimate of the actual protein-coding gene number (due to fragmented genes, different alleles or isoforms sorted as separate entries) that will be dropping as more genome-wide datasets become available. A much larger number of assembled transcripts with small predicted open reading frames have been classified as non-coding, bringing the total number of transcripts in the *Parhyale* transcriptome to over 280,000. These annotated non-coding RNAs include rRNAs, tRNAs, snRNAs, snoRNAs, eRNAs, ribozymes and lncRNAs, as well as non-coding RNAs and associated proteins of the siRNA, piRNA and miRNA pathways (Kao et al. 2016).

All common signaling pathways have been annotated in *Parhyale*, including components of the Wnt, TGF-β, Notch and FGF pathways. The genome encodes more than 1,100 transcription factors belonging to all major families, such as zinc-finger, helix-loop-helix, helix-turn-helix, ETS, Forkhead, homeobox-containing genes and others (Kao et al. 2016). As will be discussed in Section 16.8, particular efforts have been devoted to the analysis of transcription factors encoded by the nine *Parhyale* Hox genes that are organized in a cluster spanning more than 2 Mb (Serano et al. 2016; Kao et al. 2016; Pavlopoulos and Wolff 2020). Special attention has been given to the annotation of innate immunity genes and pathways as a resource for immunological studies relevant for crustacean food crop species (Kao et al. 2016; Lai and Aboobaker 2017). Another important discovery that emerged from comparative genomic and transcriptomic analyses is

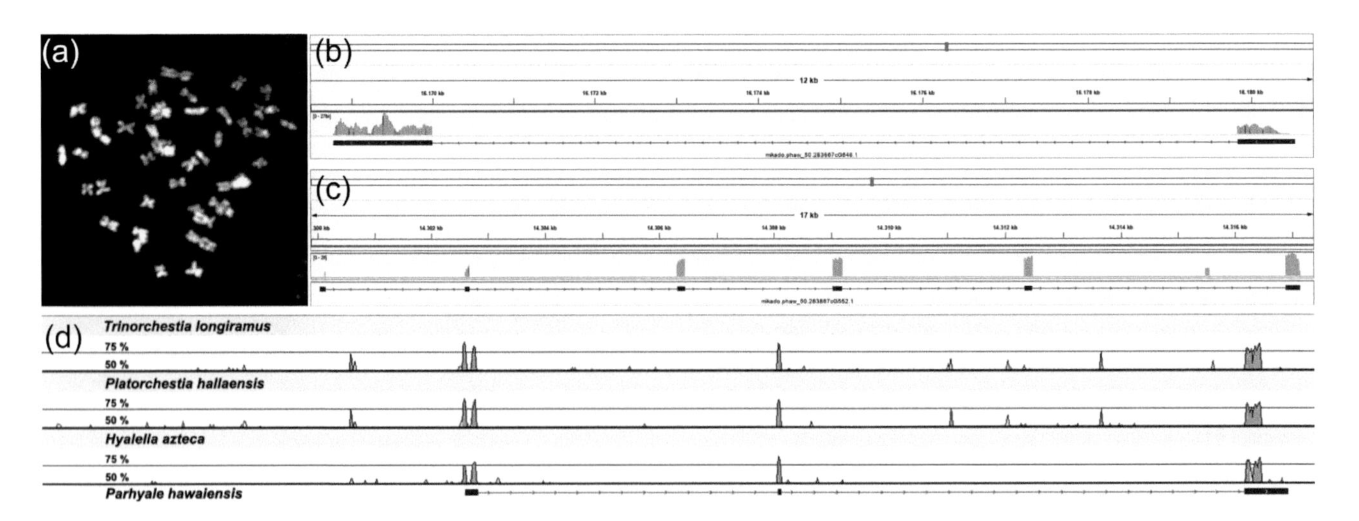

FIGURE 16.6 *Parhyale* genome-wide resources. (a) The karyotype of *Parhyale* consisting of 46 chromosomes. (b–c) Two examples of the *Parhyale* genome visualized with the Integrative Genomics Viewer. In each case, the small gray box at the top indicates the zoomed-in region of the scaffold that is displayed in detail. The span and the ruler underneath indicate the number of bases in display. Gene models are shown at the bottom, with filled boxes representing exons and thin lines representing introns. The track with the histograms above each gene model indicates the mapped reads from a transcriptomic data set. (d) Vista plots showing pairwise sequence comparisons for one locus between *Parhyale* and each of three other available amphipod genomes. High sequence similarity (above 50% indicated with histograms) is observed in exonic sequences (filled boxes) and in some non-exonic regions corresponding to putative conserved *cis*-regulatory sequences. ([b-c] Robinson et al. 2011.)

TABLE 16.2
Experimental Resources for *Parhyale* Research

Embryological manipulations	Cell microinjection Cell isolation Cell ablation (manual and photo-ablation)
Gene expression analysis	Colorimetric in situ hybridization Fluorescent hybridization chain reaction Colorimetric and fluorescent antibody staining
Transgenesis	Transposon-based (*Minos*) Integrase-based (*ΦC31*)
Gene trapping	Exon/enhancer trapping iTRAC (trap conversion)
Gain-of-function studies	Heat-inducible gene overexpression Binary systems (*UAS*/Gal4 under development)
Loss-of-function studies	CRISPR/Cas-based gene knock-out RNA interference-based gene knock-down Morpholino-based gene knock-down
Genome editing	CRISPR/Cas-based gene knock-in via homology-directed repair or non-homologous end joining
Imaging	Bright-field microscopy Laser scanning confocal microscopy Light-sheet microscopy Scanning and transmission electron microscopy

that the genomes of *Parhyale* and other marine crustaceans encode the full complement of enzymes required to extract metabolizable sugars from a lignocellulosic diet in the absence of symbiotic microorganisms (King et al. 2010; Kao et al. 2016). The capacity of marine crustaceans and *Parhyale* for autonomous wood digestion allows to harness the natural diversity in lignocellulose depolymerization mechanisms for green biofuel production and other biotechnological applications (Kern et al. 2013; Cragg et al. 2015; Chang and Lai 2018).

16.7 FUNCTIONAL APPROACHES: TOOLS FOR MOLECULAR AND CELLULAR ANALYSES

Parhyale has a set of biological and technical attributes that make it an attractive and powerful system for embryological and developmental genetic research (Rehm et al. 2009e; Stamataki and Pavlopoulos 2016). It is cultured easily and inexpensively in large numbers in the laboratory, it has a relatively fast life cycle, and a large number of transparent embryos are accessible at all stages of development and throughout the year. The arsenal of *Parhyale* tools and resources (Table 16.2) was built on a detailed description of the early embryo fate map and a comprehensive staging system for embryonic development (Gerberding et al. 2002; Browne et al. 2005). Robust protocols have been established for embryo dissection and fixation, as well as analysis of gene expression by colorimetric and fluorescent in situ hybridizations and immunohistochemistry (Rehm et al. 2009b, 2009c, 2009a; Choi et al. 2018). Likewise, a number of studies have demonstrated the amenability of *Parhyale* embryos to diverse embryological manipulations, including cell microinjection, labeling with lineage tracers, manual or photo-ablation, isolation and combinations thereof (Gerberding et al. 2002; Rehm et al. 2009d; Extavour 2005; Price et al. 2010; Hannibal et al. 2012; Nast and Extavour 2014; Kontarakis and Pavlopoulos 2014).

To facilitate functional genetic and genomic research in *Parhyale*, several efforts have been invested in developing an experimental toolkit of increasing scope and sophistication (Figure 16.7). Transgenesis in *Parhyale* was first achieved using the *Minos* transposon from *Drosophila hydei* that is active in a large variety of animal models (Pavlopoulos and Averof 2005; Pavlopoulos et al. 2007). Engineered transposons consist of the terminal inverted repeats of the *Minos* transposon flanking a transformation marker gene for detection of transgenic individuals (Figure 16.7d) and the desired transgene that is being tested (Figure 16.7e). Engineered transposons are mobilized from plasmids co-injected with a transient source of the *Minos* transposase into fertilized eggs and get randomly inserted into the genome (Kontarakis and Pavlopoulos 2014). Transposon-based transgenesis is used routinely to insert exogenous DNA into *Parhyale* (Pavlopoulos and Averof 2005; Pavlopoulos et al. 2009; Ramos et al. 2019) but has been also employed in unbiased gene trapping screens on a small scale to identify new gene functions (Kontarakis et al. 2011b). The characterization of endogenous heat-inducible promoters further allowed the development of conditional gene misexpression systems for gain-of-function studies in *Parhyale* (Pavlopoulos et al. 2009). The transgenic approaches in *Parhyale* have been expanded with the use of the bacteriophage *ΦC31* integrase for the site-specific insertion of transgenes into the genome (Kontarakis et al. 2011b). In addition, the combination of transposon with integrase-based transformation systems can increase the versatility of genetic manipulations in *Parhyale*, such as the redeployment of gene traps for creating cell and tissue markers for microscopy, drivers for ectopic gene expression, landing sites for inserting large cargos and other applications (Kontarakis et al. 2011a, 2011b).

Complementary loss-of-function studies in *Parhyale* were first conducted using RNA interference and morpholino-mediated gene knock-down approaches (Liubicich et al. 2009; Ozhan-Kizil et al. 2009). However, gene knock-down suffered a number of limitations, such as the incomplete and transient reduction in gene function. This problem was solved by employing targeted genome editing approaches based on the clustered regularly interspaced short palindromic repeats (CRISPR)/CRISPR-associated (Cas) system (Figure 16.7a–c). For reasons explained in the following, complete null phenotypes can be obtained with very high efficiency using CRISPR/Cas-based gene knock-out in *Parhyale* (Martin et al. 2016; Kao et al. 2016; Clark-Hachtel and Tomoyasu 2020; Bruce and Patel 2020). Moreover, the CRISPR/Cas system has been adapted to generate live fluorescent reporters of gene expression (Figure 16.7f) using both homology-dependent and homology-independent knock-in approaches in *Parhyale* (Serano et al. 2016; Kao et al. 2016).

It should be stressed that the effects of all aforementioned functional genetic manipulations are routinely analyzed first in treated embryos (in the G0 generation) and subsequently confirmed through the study of established transgenic or mutant lines (in the G1 or G2 generations) (Kontarakis and Pavlopoulos 2014; Kao et al. 2016). The early accessibility to fertilized eggs in *Parhyale*, together with their complete cleavage mode and slow tempo of development, results in high transgenesis rates and high CRISPR/Cas-mediated mutagenesis efficiencies in treated G0 embryos that exhibit very low levels of mosaicism and carry the genetic alterations both in their soma and in their germ line (Pavlopoulos and Averof 2005; Pavlopoulos et al. 2009; Martin et al. 2016; Kao et al. 2016; Clark-Hachtel and Tomoyasu 2020; Bruce and Patel 2020). Furthermore, the early and stereotyped lineage restrictions in the *Parhyale* embryo allow the comparison between the wild-type and the genetically altered conditions in the same embryo (Figure 16.7a)(Pavlopoulos and Averof 2005; Pavlopoulos et al. 2009; Martin et al. 2016), as well as the targeting of specific

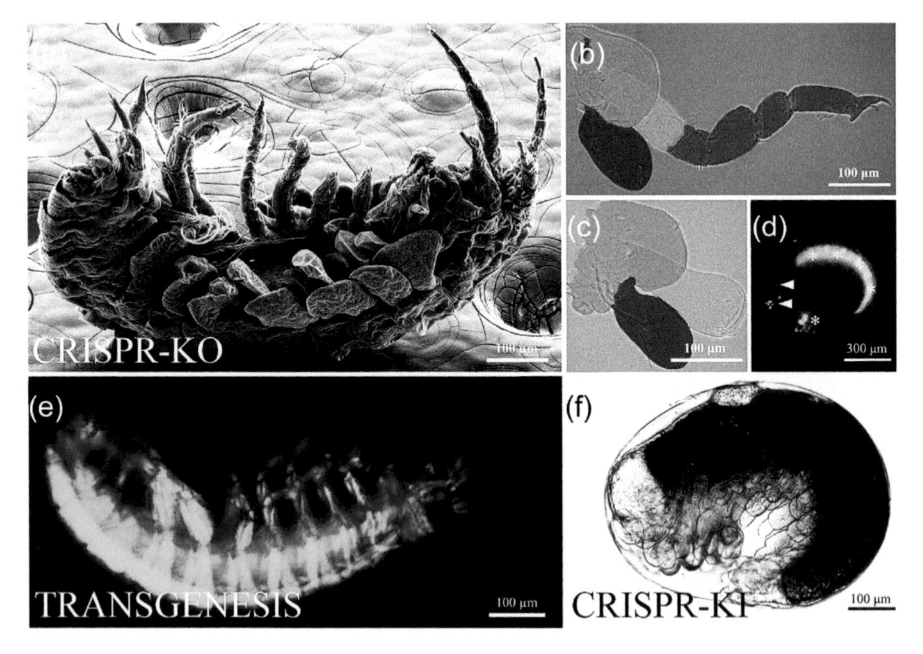

FIGURE 16.7 Functional approaches in *Parhyale*. (a) Phenotypic example of a CRISPR-based gene knock-out (CRISPR-KO) experiment. The image shows a scanning electron micrograph of a mosaic *Parhyale* juvenile with wild-type appendages on its right side and truncated appendages on its left side that are mutant for the limb patterning gene *Distal-less* (*Dll*). Lateral view with anterior to the right and ventral to the top. (b) Cuticle preparation of a wild-type and (c) a mutant thoracic T4 appendage after CRISPR-based *Dll* knock-out. The proximal side is to the left and the distal side to the right. Color masks in panels (a) to (c) indicate the distal appendage structures (magenta) that are missing after *Dll* knock-out, as well as the proximal appendage structures (coxal plates in orange, gills in red and basis in cyan) that are not affected. (d) Transgenic late-stage *Parhyale* embryo expressing two different fluorescent transgenesis markers in the head region (arrowheads): a *PhOpsin1*-driven expression in the compound eye shown in green and a *3xP3*-driven expression more dorsally, shown in magenta. Asterisks indicate non-specific autofluorescence detected in the gnathal appendages (green) and in the gut (magenta). Lateral view with anterior to the left and ventral to the bottom. (e) Transgenic *Parhyale* juvenile, oriented as in (a), expressing a muscle-specific fluorescent reporter construct shown in green. (f) CRISPR-mediated knock-in (CRISPR-KI) of a construct in the *Dll* locus driving expression of a fluorescent reporter in the appendages (shown in magenta) merged with the corresponding brightfield image. ([d] Ramos et al. 2019; Pavlopoulos and Averof 2005.)

lineages for labeling or ablation (Price et al. 2010; Alwes et al. 2011; Hannibal et al. 2012; Konstantinides and Averof 2014). All these features are very useful for experimentation in *Parhyale*, because they provide fast and reliable information about gene expression, regulation and function months before stable lines are available for analysis.

Parhyale is not only a genetically tractable but also an optically tractable experimental model, which is ideal to make the connection between the molecular and cellular basis of development. Light and electron microscopy analyses of fixed specimens have been used widely to characterize wild-type and mutant phenotypes in detail (Pavlopoulos et al. 2009; Serano et al. 2016; Martin et al. 2016; Ramos et al. 2019; Clark-Hachtel and Tomoyasu 2020; Bruce and Patel 2020). The increasing collection of genetic tools and transgenic lines for imaging, in combination with the transparency and low autofluorescence of embryos, have enabled the implementation of live microscopic inspections of cellular dynamics with exceptional spatial and temporal resolution. Different microscopy modalities, including bright-field, confocal and multi-view light-sheet microscopy, have been adapted successfully to image embryonic and post-embryonic processes over several days of development, such as *Parhyale* gastrulation and germ band formation, appendage development and regeneration (Price and Patel 2008; Alwes et al. 2011; Chaw and Patel 2012; Hannibal et al. 2012; Alwes et al. 2016; Wolff et al. 2018). Last but not least, thanks to a very productive collaboration between biologists, microscopists and computer scientists, a suite of sophisticated and open-source software is available for the visualization of image datasets, the manual and automated tracking of cells and the reconstruction and editing of cell lineages to understand the cellular behaviors contributing to tissue and organ development in *Parhyale* (Wolff et al. 2018; Salvador-Martínez et al. 2020; Sugawara et al. 2021).

16.8 CHALLENGING QUESTIONS BOTH IN ACADEMIC AND APPLIED RESEARCH

Parhyale lends itself to address several longstanding questions and problems in modern biological and biomedical research (Stamataki and Pavlopoulos 2016). Based on its phylogenetic position and its technical and biological attributes, it has increased the breadth and depth of comparative developmental studies with other pancrustacean, arthropod and animal groups. As a malacostracan crustacean, it is also closely related to shrimps, crabs and lobsters that have attracted research interest as commercially and nutritionally important crop species.

16.8.1 Developmental Basis of Morphological Evolution

Research in *Parhyale* was inspired by and has greatly contributed toward our understanding of the developmental mechanisms driving body plan evolution and specialization of body parts. Crustaceans exhibit a tremendous morphological diversity observed both within and between species. Seminal studies in crustaceans were among the first to implicate changes in the expression of *Hox* genes with the evolution of animal body plans and the diversification of developing appendages (Averof and Akam 1995; Averof and Patel 1997). Although expression studies of Hox genes have been carried out in all major crustacean lineages, the most comprehensive analysis of all nine *Hox* genes has been carried out in *Parhyale*, where they exhibit both spatial and temporal collinearity (Serano et al. 2016). Hox expression domains correspond to the subdivision of the body into morphologically and functionally distinct regions and correlate with the development of distinct appendages types. Importantly, systematic loss-of-function and gain-of-function studies of *Hox* genes in *Parhyale* have provided compelling evidence for the causal association between *Hox* genes and crustacean segmental organization and appendage diversification (Pavlopoulos et al. 2009; Liubicich et al. 2009; Martin et al. 2016). The homeotic transformations produced in these functional studies were recapitulating in *Parhyale* macroevolutionary changes observed in the body organization of other crustacean lineages, like the repeated evolution of feeding maxillipeds from locomotory appendages in the anterior thorax of many crustacean lineages or the change in the relative number of abdominal pleopods and uropods between malacostracan lineages (Averof et al. 2010; Martin et al. 2016; Pavlopoulos and Wolff 2020).

Along similar lines, expression and functional studies of developmental patterning genes in *Parhyale* have enabled to test century-old hypotheses about the homology and evolutionary novelty of arthropod appendages (McKenna et al. 2021). Considering that winged insects evolved from wingless crustaceans, different theories have been proposed to explain the origin of insect wings (Clark-Hachtel and Tomoyasu 2016): they are novel lateral outgrowths from the dorsal body wall (tergal origin or paranotal hypothesis), or they evolved from the exites of proximal leg segments (pleural origin hypothesis). By comparing the expression patterns and the loss-of-function phenotypes of leg, wing and body wall patterning genes between insects and *Parhyale*, it was proposed that the proximal exite-bearing leg segments present in the common ancestor of insects and crustaceans were incorporated into the insect body wall, giving rise to the insect wings (Clark-Hachtel and Tomoyasu 2020; Bruce and Patel 2020). Thus, these elegant studies in *Parhyale* have provided a fresh and unified model in favor of the evolution of insect wings from a pre-existing structure in their crustacean ancestor. A similar framework has been adopted to homologize pancrustacean, myriapod and chelicerate appendages, suggesting an eight-segment ground plan for the arthropod leg (Bruce 2021).

16.8.2 Molecular and Cellular Basis of Development

One of the biggest challenges in developmental biology is to understand how the genomic information encodes the morphogenetic cell behaviors, like cell proliferation and death, cell shape changes and cell movements, that produce the characteristic size and shape of developing tissues and organs in multicellular organisms (Heisenberg and Bellaïche 2013; Wan et al. 2019). The optical properties of the *Parhyale* egg and the embryonic development of its appendages as direct outgrowths from the body wall have enabled to advance beyond a gene-centric view of development and start integrating the molecular with the cellular aspects of appendage formation. In a tour-de-force study that involved advanced light-sheet fluorescence microscopy and image analysis tools, the complete lineage of developing *Parhyale* limbs was reconstructed with single-cell resolution (Wolff et al. 2018). The spatial coordinates for all constituent cells, their temporal dynamics and mother-daughter relationships were then analyzed to shed light on the cellular mechanisms driving appendage outgrowth, elongation and segmentation. These analyses revealed the cellular architecture and patterned cell activities operating at different stages of appendage development that were then correlated with the expression patterns of candidate patterning genes known from limb studies in *Drosophila* (Wolff et al. 2018). Interestingly, some of these cellular events were similar, but some were distinct compared to the textbook *Drosophila* paradigm, motivating future experiments to understand the conservation and divergence of appendage patterning mechanism during pancrustacean and arthropod evolution (Pavlopoulos and Wolff 2020).

In a broader perspective, all recent technical breakthroughs in *Parhyale* research provide the opportunity to study gene expression and function in the context of single-cell-resolution fate maps, both under wild-type and under genetically perturbed conditions. These multidisciplinary approaches will be employed by the community to advance our knowledge on longstanding questions in developmental biology, such as the identity and function of cell fate determinants (Nestorov et al. 2013; Gupta and Extavour 2013), the molecular and cellular mechanisms underlying embryo formation and healing (Alwes et al. 2011; Chaw and Patel 2012), the relative contributions of cell history and cell communication in development (Price et al. 2010; Hannibal et al. 2012) and the allometric growth of serially homologous appendages (Pavlopoulos et al. 2009; Martin et al. 2016).

16.8.3 Molecular and Cellular Basis of Regeneration

Besides studying embryonic development, *Parhyale* has emerged as an attractive model system for regenerative studies, as it has the capacity to replace lost tissues and entire body parts post-embryonically (Grillo et al. 2016). It has been demonstrated that *Parhyale* has the ability to regenerate missing limbs after amputation (Kontarakis et al. 2011b) and its germ line after ablation of the g micromere (Modrell 2007; Kaczmarczyk 2014). In principle, new cells for regeneration can be produced from the activation of pluripotent or lineage-restricted stem cells, as well as the de-differentiation or trans-differentiation of differentiated cells (Tanaka and Reddien 2011). Thanks to the early lineage restrictions in the *Parhyale* embryo, it has been possible to label and identify the source cells and examine their regenerative potential during regrowth of limbs (Konstantinides and Averof 2014). The sources for the new cells are restricted by their lineage and proximity to the regenerating appendage: the ectodermal and mesodermal lineages make distinct contributions to ectoderm-derived tissues (epidermis and neurons) and mesoderm-derived tissues (muscles and blood cells), respectively. Importantly, the availability of cell-specific markers led to the major discovery of invertebrate muscle stem cells in *Parhyale* that, similar to satellite cells in vertebrates, serve as progenitors for muscle repair during limb regeneration (Konstantinides and Averof 2014). It has been also possible to trace cell behaviors through live imaging of appendage regeneration in *Parhyale* with high resolution and over several days after amputation (Alwes et al. 2016). For example, the epidermis of the new limb is not formed by specialized stem cells but by the cell proliferation and redifferentiation of existing epidermal cells. Overall, crustaceans have a long history in regenerative research, albeit at the physiological and anatomical level. The addition of *Parhyale* as a new genetically and optically tractable regenerative model has opened new possibilities to dissect the molecular and cellular mechanisms that can be redeployed during its lifetime to replace missing limbs, germ cells and possibly other structures.

16.8.4 New Research Directions

We will conclude this chapter with some more exciting new research avenues that, like regeneration, were not conceivable when *Parhyale* was first introduced in the laboratory but have the potential to make big contributions to both basic and applied fields of research. The first steps have been taken already in establishing *Parhyale* as a model in the fields of chronobiology and ecotoxicology (Hunt et al. 2019; Artal et al. 2018, 2020; Diehl et al. 2021). Studies of the *Parhyale* innate immunity have been also proposed for disease control in crustacean aquaculture through a better understanding of infectious pathogens and host defense mechanisms (Kao et al. 2016; Lai and Aboobaker 2017). Last but not least, studies of lignocellulose digestion in *Parhyale* can offer novel insights into the ecologically important and understudied mechanisms of wood recycling in marine environments and can unleash their significant potential for biotechnological applications (Cragg et al. 2015; Kao et al. 2016; Chang and Lai 2018).

ACKNOWLEDGMENTS

This chapter was prepared amid the most challenging conditions imposed by the COVID-19 pandemic. It was only completed thanks to the unceasing patience and encouragement of our editors, Dr. Agnès Boutet and Dr. Bernd Schierwater, to whom we are deeply indebted. We would also like to thank Dr. Carsten Wolff and Suyash Kumar for providing the images in Figures 16.6a and 16.7a, and Dr. Evangelia Stamataki for comments on the manuscript. John Rallis and Gentian Kapai were supported by Fondation Santé graduate studentships and Anastasios Pavlopoulos by IMBB-FORTH intramural funds.

BIBLIOGRAPHY

Alegretti, L., de Aragão Umbuzeiro, G., and M.N. Flynn. 2016. "Population Dynamics of *Parhyale hawaiensis* (Dana, 1853) (Amphipoda: Hyalidae) Associated with an Intertidal Algal Belt in Southeastern Brazil." *Journal of Crustacean Biology* 36 (6): 785–791. doi:10.1163/1937240X-00002480.

Alwes, F., Enjolras, C., and M. Averof. 2016. "Live Imaging Reveals the Progenitors and Cell Dynamics of Limb Regeneration." *ELife* 5. doi:10.7554/eLife.19766.

Alwes, F., Hinchen, B., and C.G. Extavour. 2011. "Patterns of Cell Lineage, Movement, and Migration from Germ Layer Specification to Gastrulation in the Amphipod Crustacean *Parhyale hawaiensis*." *Developmental Biology* 359 (1): 110–123. doi:10.1016/j.ydbio.2011.07.029.

Artal, M.C., Dos Santos, A., Henry, T.B., and G. de Aragão Umbuzeiro. 2018. "Development of an Acute Toxicity Test with the Tropical Marine Amphipod *Parhyale hawaiensis*." *Ecotoxicology (London, England)* 27 (2): 103–108. doi:10.1007/s10646-017-1875-3.

Artal, M.C., Pereira, K.D., Luchessi, A.D., Okura, V.K., Henry, T.B., Marques-Souza, H., and G. de Aragão Umbuzeiro. 2020. "Transcriptome Analysis in *Parhyale hawaiensis* Reveal Sex-Specific Responses to AgNP and AgCl Exposure." *Environmental Pollution (Barking, Essex: 1987)* 260 (May): 113963. doi:10.1016/j.envpol.2020.113963.

Averof, M., and M. Akam. 1995. "Hox Genes and the Diversification of Insect and Crustacean Body Plans." *Nature* 376 (6539): 420–423. doi:10.1038/376420a0.

Averof, M., and N.H. Patel. 1997. "Crustacean Appendage Evolution Associated with Changes in Hox Gene Expression." *Nature* 388 (6643): 682–686. doi:10.1038/41786.

Averof, M., Pavlopoulos, A., and Z. Kontarakis. 2010. "Evolution of New Appendage Types by Gradual Changes in Hox Gene Expression: The Case of Crustacean Maxillipeds." 5.

"BBSRC Business Magazine." 2017. January 1. https://bbsrc.ukri.org/documents/bbsrc-business-winter-2017-pdf/.

Blythe, M.J., Malla, S., Everall, R., Shih, Y.H., Lemay, V., Moreton, J., Wilson, R., and A.A. Aboobaker. 2012. "High Through-Put Sequencing of the *Parhyale hawaiensis* MRNAs and microRNAs to Aid Comparative Developmental Studies." *PLoS One* 7 (3): e33784. doi:10.1371/journal.pone.0033784.

Boxshall, G.A. 2004. "The Evolution of Arthropod Limbs." *Biological Reviews of the Cambridge Philosophical Society* 79 (2): 253–300. doi:10.1017/s1464793103006274.

Browne, W.E., Price, A.L., Gerberding, M., and N.H. Patel. 2005. "Stages of Embryonic Development in the Amphipod Crustacean, *Parhyale hawaiensis*." *Genesis (New York, N.Y.: 2000)* 42 (3): 124–149. doi:10.1002/gene.20145.

Bruce, H.S. 2021. "How to Align Arthropod Leg Segments." *BioRxiv*, January. Cold Spring Harbor Laboratory, 2021.01.20.427514. doi:10.1101/2021.01.20.427514.

Bruce, H.S., and N.H. Patel. 2020. "Knockout of Crustacean Leg Patterning Genes Suggests That Insect Wings and Body Walls Evolved from Ancient Leg Segments." *Nature Ecology & Evolution* 4 (12): 1703–1712. doi:10.1038/s41559-020-01349-0.

Chang, W.H., and A.G. Lai. 2018. "Mixed Evolutionary Origins of Endogenous Biomass-Depolymerizing Enzymes in Animals." *BMC Genomics* 19 (1): 483. doi:10.1186/s12864-018-4861-0.

Chaw, R.C., and N.H. Patel. 2012. "Independent Migration of Cell Populations in the Early Gastrulation of the Amphipod Crustacean *Parhyale hawaiensis*." *Developmental Biology* 371 (1): 94–109. doi:10.1016/j.ydbio.2012.08.012.

Choi, H.M.T., Schwarzkopf, M., Fornace, M.E., Acharya, A., Artavanis, G., Stegmaier, J., Cunha, A., and N.A. Pierce. 2018. "Third-Generation In Situ Hybridization Chain Reaction: Multiplexed, Quantitative, Sensitive, Versatile, Robust." *Development (Cambridge, England)* 145 (12). doi:10.1242/dev.165753.

Clark-Hachtel, C.M., and Y. Tomoyasu. 2016. "Exploring the Origin of Insect Wings from an Evo-Devo Perspective." *Current Opinion in Insect Science* 13 (February): 77–85. doi:10.1016/j.cois.2015.12.005.

Clark-Hachtel, C.M., and Y. Tomoyasu. 2020. "Two Sets of Candidate Crustacean Wing Homologues and Their Implication for the Origin of Insect Wings." *Nature Ecology & Evolution*, August. doi:10.1038/s41559-020-1257-8.

Conlan, K.E. 1991. "Precopulatory Mating Behavior and Sexual Dimorphism in the Amphipod Crustacea." *Hydrobiologia* 223 (1): 255–282. doi:10.1007/BF00047644.

Copilaş-Ciocianu, D., Borko, Š., and C. Fišer. 2020. "The Late Blooming Amphipods: Global Change Promoted Post-Jurassic Ecological Radiation Despite Palaeozoic Origin." *Molecular Phylogenetics and Evolution* 143: 106664. doi:10.1016/j.ympev.2019.106664.

Cormier, A., Chebbi, M.A., Giraud, I., Wattier, R., Teixeira, M., Gilbert, C., Rigaud, T., and R. Cordaux. 2021. "Comparative Genomics of Strictly Vertically Transmitted, Feminizing Microsporidia Endosymbionts of Amphipod Crustaceans." *Genome Biology and Evolution* 13 (1). doi:10.1093/gbe/evaa245.

Cragg, S.M., Beckham, G.T., Bruce, N.C., Bugg, T.D.H., Distel, D.L., Dupree, P., Green Etxabe, A., Goodell, B.S., Jellison, J., McGeehan, J.E., McQueen-Mason, S.J., Schnorr, K., Walton, P.H., Watts, J.E.M., and M. Zimmer. 2015. "Lignocellulose Degradation Mechanisms across the Tree of Life." *Current Opinion in Chemical Biology, Energy • Mechanistic Biology* 29 (December): 108–119. doi:10.1016/j.cbpa.2015.10.018.

Dana, J.D. 1853. *United States Exploring Expedition: 14: Crustacea*. Sherman.

Diehl, O.J., Assano, P.K., da Costa, T.R.G., Oliveira, R., Marques-Souza, H., and G. de Aragão Umbuzeiro. 2021. "Antenna Regeneration as an Ecotoxicological Endpoint in a Marine Amphipod: A Proof of Concept Using Dimethyl Sulfoxide and Diflubenzuron." *Ecotoxicology (London, England)*, March. doi:10.1007/s10646-021-02395-5.

Dohle, W., Gerberding, M., Hejnol, A., and G. Scholtz. 2003. "Cell Lineage, Segment Differentiation, and Gene Expression in Crustaceans." In *Evolutionary Developmental Biology of Crustacea*. 1st Edition, 95–133. CRC Press. doi:10.13140/RG.2.1.4984.9841.

Extavour, C.G. 2005. "The Fate of Isolated Blastomeres with Respect to Germ Cell Formation in the Amphipod Crustacean *Parhyale hawaiensis*." *Developmental Biology* 277 (2): 387–402. doi:10.1016/j.ydbio.2004.09.030.

Gerberding, M., Browne, W.E., and N.H. Patel. 2002. "Cell Lineage Analysis of the Amphipod Crustacean *Parhyale hawaiensis* Reveals an Early Restriction of Cell Fates." *Development (Cambridge, England)* 129 (24): 5789–5801. doi:10.1242/dev.00155.

Giribet, G., and G.D. Edgecombe. 2019. "The Phylogeny and Evolutionary History of Arthropods." *Current Biology: CB* 29 (12): R592–602. doi:10.1016/j.cub.2019.04.057.

Grillo, M., Konstantinides, N., and M. Averof. 2016. "Old Questions, New Models: Unraveling Complex Organ Regeneration with New Experimental Approaches." *Current Opinion in Genetics & Development* 40 (October): 23–31. doi:10.1016/j.gde.2016.05.006.

Gupta, T., and C.G. Extavour. 2013. "Identification of a Putative Germ Plasm in the Amphipod *Parhyale hawaiensis*." *EvoDevo* 4 (1): 34. doi:10.1186/2041-9139-4-34.

Hannibal, R.L., Price, A.L., and N.H. Patel. 2012. "The Functional Relationship between Ectodermal and Mesodermal Segmentation in the Crustacean, *Parhyale hawaiensis*." *Developmental Biology* 361 (2): 427–438. doi:10.1016/j.ydbio.2011.09.033.

Heisenberg, C.P., and Y. Bellaïche. 2013. "Forces in Tissue Morphogenesis and Patterning." *Cell* 153 (5): 948–962. doi:10.1016/j.cell.2013.05.008.

Holmquist, J.G. 1982. "The Functional Morphology of Gnathopods: Importance in Grooming, and Variation with Regard to Habitat, in Talitroidean Amphipods." *Journal of Crustacean Biology* 2 (2): 159–179. doi:10.2307/1547997.

Horton, T., Lowry, J., De Broyer, C., Bellan-Santini, D., Coleman, C.O., Corbari, L., Costello, M.J., Daneliya, M., Dauvin, J.C., Fišer, C., Gasca, R., Grabowski, M., Guerra-García, J.M., Hendrycks, E., Hughes, L., Jaume, D., Jazdzewski, K., Kim, Y.H., King, R., Krapp-Schickel, T., LeCroy, S., Lörz, A.N., Mamos, T., Senna, A.R., Serejo, C., Sket, B., Souza-Filho, J.F., Tandberg, A.H., Thomas, J.D., Thurston, M., Vader, W., Väinölä, R., Vonk, R., White, K., and W. Zeidler. 2020. "World Amphipoda Database." www.marinespecies.org/amphipoda/.

Hunnekuhl, V.S., and C. Wolff. 2012. "Reconstruction of Cell Lineage and Spatiotemporal Pattern Formation of the Mesoderm in the Amphipod Crustacean *Orchestia cavimana*." *Developmental Dynamics: An Official Publication of the American Association of Anatomists* 241 (4): 697–717. doi:10.1002/dvdy.23758.

Hunt, B.J., Mallon, E.B., and E. Rosato. 2019. "In Silico Identification of a Molecular Circadian System with Novel Features in the Crustacean Model Organism *Parhyale hawaiensis*." *Frontiers in Physiology* 10. Frontiers. doi:10.3389/fphys.2019.01325.

Hyne, R.V. 2011. "Review of the Reproductive Biology of Amphipods and Their Endocrine Regulation: Identification of Mechanistic Pathways for Reproductive Toxicants." *Environmental Toxicology and Chemistry* 30 (12): 2647–2657. https://doi.org/10.1002/etc.673.

Kaczmarczyk, A.N. 2014. "Germline Maintenance and Regeneration in the Amphipod Crustacean, *Parhyale hawaiensis*." *UC Berkeley*. https://escholarship.org/uc/item/5h94b7kg.

Kao, D., Lai, A.G., Stamataki, E., Rosic, S., Konstantinides, N., Jarvis, E., Di Donfrancesco, A., Pouchkina-Stancheva, N., Sémon, M., Grillo, M., Bruce, H., Kumar, S., Siwanowicz, I., Le, A., Lemire, A., Eisen, M.B., Extavour, C., Browne, W.E., Wolff, C., Averof, M., Patel, N.H., Sarkies, P., Pavlopoulos, A., and A. Aboobaker. 2016. "The Genome of the Crustacean *Parhyale hawaiensis*, a Model for Animal Development, Regeneration, Immunity and Lignocellulose Digestion." *ELife* 5. doi:10.7554/eLife.20062.

Kern, M., McGeehan, J.E., Streeter, S.D., Martin, R.N.A., Besser, K., Elias, L., Eborall, W., Malyon, G.P., Payne, C.M., Himmel, M.E., Schnorr, K., Beckham, G.T., Cragg, S.M., Bruce, N.C., and S.J. McQueen-Mason. 2013. "Structural Characterization of a Unique Marine Animal Family 7 Cellobiohydrolase Suggests a Mechanism of Cellulase Salt Tolerance." *Proceedings of the National Academy of Sciences of the United States of America* 110 (25): 10189–10194. doi:10.1073/pnas.1301502110.

King, A.J., Cragg, S.M., Li, Y., Dymond, J., Guille, M.J., Bowles, D.J., Bruce, N.C., Graham, I.A., and S.J. McQueen-Mason. 2010. "Molecular Insight into Lignocellulose Digestion by a Marine Isopod in the Absence of Gut Microbes." *Proceedings of the National Academy of Sciences of the United States of America* 107 (12): 5345–5350. doi:10.1073/pnas.0914228107.

Konstantinides, N., and M. Averof. 2014. "A Common Cellular Basis for Muscle Regeneration in Arthropods and Vertebrates." *Science (New York, N.Y.)* 343 (6172): 788–791. doi:10.1126/science.1243529.

Kontarakis, Z., Konstantinides, N., Pavlopoulos, A., and M. Averof. 2011a. "Reconfiguring Gene Traps for New Tasks Using ITRAC." *Fly* 5 (4): 352–355. doi:10.4161/fly.5.4.18108.

Kontarakis, Z., and A. Pavlopoulos. 2014. "Transgenesis in Non-Model Organisms: The Case of Parhyale." *Methods in Molecular Biology (Clifton, N.J.)* 1196: 145–181. doi:10.1007/978-1-4939-1242-1_10.

Kontarakis, Z., Pavlopoulos, A., Kiupakis, A., Konstantinides, N., Douris, V., and M. Averof. 2011b. "A Versatile Strategy for Gene Trapping and Trap Conversion in Emerging Model Organisms." *Development (Cambridge, England)* 138 (12): 2625–2630. doi:10.1242/dev.066324.

Lai, A.G., and A.A. Aboobaker. 2017. "Comparative Genomic Analysis of Innate Immunity Reveals Novel and Conserved Components in Crustacean Food Crop Species." *BMC Genomics* 18 (1): 389. doi:10.1186/s12864-017-3769-4.

Liubicich, D.M., Serano, J.M., Pavlopoulos, A., Kontarakis, Z., Protas, M.E., Kwan, E., Chatterjee, S., Tran, K.D., Averof, M., and N.H. Patel. 2009. "Knockdown of Parhyale Ultrabithorax Recapitulates Evolutionary Changes in Crustacean Appendage Morphology." *Proceedings of the National Academy of Sciences of the United States of America* 106 (33): 13892–13896. doi:10.1073/pnas.0903105106.

Lozano-Fernandez, J., Giacomelli, M., Fleming, J.F., Chen, A., Vinther, J., Thomsen, P.F., Glenner, H., Palero, F., Legg, D.A., Iliffe, T.M., Pisani, D., and J. Olesen. 2019. "Pancrustacean Evolution Illuminated by Taxon-Rich Genomic-Scale Data Sets with an Expanded Remipede Sampling." *Genome Biology and Evolution* 11 (8). Oxford Academic: 2055–70. doi:10.1093/gbe/evz097.

Martin, A., Serano, J.M., Jarvis, E., Bruce, H.S., Wang, J., Ray, S., Barker, C.A., O'Connell, L.C., and N.H. Patel. 2016. "CRISPR/Cas9 Mutagenesis Reveals Versatile Roles of Hox Genes in Crustacean Limb Specification and Evolution." *Current Biology: CB* 26 (1): 14–26. doi:10.1016/j.cub.2015.11.021.

McKenna, K.Z., Wagner, G.P., and K.L. Cooper. 2021. "A Developmental Perspective of Homology and Evolutionary Novelty." *Current Topics in Developmental Biology* 141: 1–38. doi:10.1016/bs.ctdb.2020.12.001.

Modrell, M.S. 2007. "Early Cell Fate Specification in the Amphipod Crustacean, *Parhyale hawaiensis*." *UC Berkeley*.

Myers, A.A. 1985. "Shallow-Water, Coral Reef and Mangrove Amphipoda (Gammaridea) of Fiji." *Records of the Australian Museum, Supplement* 5 (December). The Australian Museum: 1–143. doi:10.3853/j.0812–7387.5.1985.99.

Nast, A.R., and C.G. Extavour. 2014. "Ablation of a Single Cell from Eight-Cell Embryos of the Amphipod Crustacean *Parhyale hawaiensis*." *Journal of Visualized Experiments: JoVE* (85) (March). doi:10.3791/51073.

Nestorov, P., Battke, F., Levesque, M.P., and M. Gerberding. 2013. "The Maternal Transcriptome of the Crustacean *Parhyale hawaiensis* Is Inherited Asymmetrically to Invariant Cell Lineages of the Ectoderm and Mesoderm." *PLoS One* 8 (2): e56049. doi:10.1371/journal.pone.0056049.

Oakley, T.H., Wolfe, J.M., Lindgren, A.R., and A.K. Zaharoff. 2013. "Phylotranscriptomics to Bring the Understudied into the Fold: Monophyletic Ostracoda, Fossil Placement, and Pancrustacean Phylogeny." *Molecular Biology and Evolution* 30 (1). Oxford Academic: 215–233. doi:10.1093/molbev/mss216.

Ozhan-Kizil, G., Havemann, J., and M. Gerberding. 2009. "Germ Cells in the Crustacean *Parhyale hawaiensis* Depend on Vasa Protein for Their Maintenance but Not for Their Formation." *Developmental Biology* 327 (1): 230–239. doi:10.1016/j.ydbio.2008.10.028.

Parchem, R.J., Poulin, F., Stuart, A.B., Amemiya, C.T., and N.H. Patel. 2010. "BAC Library for the Amphipod Crustacean, *Parhyale hawaiensis*." *Genomics* 95 (5): 261–267. doi:10.1016/j.ygeno.2010.03.005.

Patra, A.K., Chung, O., Yoo, J.Y., Baek, S.H., Jung, T.W., Kim, M.S., Yoon, M.G., Yang, Y., and J.H. Choi. 2020a. "The Draft Genome Sequence of a New Land-Hopper *Platorchestia hallaensis*." *Frontiers in Genetics* 11: 621301. doi:10.3389/fgene.2020.621301.

Patra, A.K., Chung, O., Yoo, J.Y., Kim, M.S., Yoon, M.G., Choi, J.H., and Y. Yang. 2020b. "First Draft Genome for the Sand-Hopper *Trinorchestia longiramus*." *Scientific Data* 7 (1): 85. doi:10.1038/s41597-020-0424-8.

Pavlopoulos, A., and M. Averof. 2005. "Establishing Genetic Transformation for Comparative Developmental Studies in the Crustacean *Parhyale hawaiensis*." *Proceedings of the National Academy of Sciences of the United States of America* 102 (22): 7888–7893. doi:10.1073/pnas.0501101102.

Pavlopoulos, A., Kontarakis, Z., Liubicich, D.M., Serano, J.M., Akam, M., Patel, N.H., and M. Averof. 2009. "Probing the Evolution of Appendage Specialization by Hox Gene Misexpression in an Emerging Model Crustacean." *Proceedings of the National Academy of Sciences of the United States of America* 106 (33): 13897–13902. doi:10.1073/pnas.0902804106.

Pavlopoulos, A., Oehler, S., Kapetanaki, M.G., and C. Savakis. 2007. "The DNA Transposon Minos as a Tool for Transgenesis and Functional Genomic Analysis in Vertebrates and Invertebrates." *Genome Biology* 8 (Suppl 1): S2. doi:10.1186/gb-2007-8-s1-s2.

Pavlopoulos, A., and C. Wolff. 2020. "Crustacean Limb Morphogenesis during Normal Development and Regeneration." In *The Natural History of the Crustacea: Developmental Biology and Larval Ecology*, edited by Klaus Anger, Steffen Harzsch, and Martin Thiel. Vol. 7. Oxford: Oxford University Press.

Poovachiranon, S., Boto, K., and N. Duke. 1986. "Food Preference Studies and Ingestion Rate Measurements of the Mangrove Amphipod *Parhyale hawaiensis* (Dana)." *Journal of Experimental Marine Biology and Ecology* 98 (1): 129–140. doi:10.1016/0022-0981(86)90078-X.

Poynton, H.C., Hasenbein, S., Benoit, J.B., Sepulveda, M.S., Poelchau, M.F., Hughes, D.S.T., Murali, S.C., Chen, S., Glastad, K.M., Goodisman, M.A.D., Werren, J.H., Vineis, J.H., Bowen, J.L., Friedrich, M., Jones, J., Robertson, H.M., Feyereisen, R., Mechler-Hickson, A., Mathers, N., Lee, C.E., Colbourne, J.K., Biales, A., Johnston, J.S., Wellborn, G.A., Rosendale, A.J., Cridge, A.G., Munoz-Torres, M.C., Bain, P.A., Manny, A.R., Major, K.M., Lambert, F.N., Vulpe, C.D., Tuck, P., Blalock, B.J., Lin, Y.Y., Smith, M.E., Ochoa-Acuña, H., Chen, M.J.M., Childers, C.P., Qu, J., Dugan, S., Lee, S.L., Chao, H., Dinh, H., Han, Y., Doddapaneni, H., Worley, K.C., Muzny, D.M., Gibbs, R.A., and S. Richards. 2018. "The Toxicogenome of *Hyalella azteca*: A Model for Sediment Ecotoxicology and Evolutionary Toxicology." *Environmental Science & Technology* 52 (10): 6009–6022. doi:10.1021/acs.est.8b00837.

Price, A.L., Modrell, M.S., Hannibal, R.L., and N.H. Patel. 2010. "Mesoderm and Ectoderm Lineages in the Crustacean *Parhyale hawaiensis* Display Intra-Germ Layer Compensation." *Developmental Biology* 341 (1): 256–266. doi:10.1016/j.ydbio.2009.12.006.

Price, A.L., and N.H. Patel. 2008. "Investigating Divergent Mechanisms of Mesoderm Development in Arthropods: The Expression of Ph-Twist and Ph-Mef2 in *Parhyale hawaiensis*." *Journal of Experimental Zoology: Part B, Molecular and Developmental Evolution* 310 (1): 24–40. doi:10.1002/jez.b.21135.

Putnam, N.H., O'Connell, B.L., Stites, J.C., Rice, B.J., Blanchette, M., Calef, R., Troll, C.J., Fields, A., Hartley, P.D., Sugnet, C.W., Haussler, D., Rokhsar, D.S., and R.E. Green. 2016. "Chromosome-Scale Shotgun Assembly Using an In Vitro Method for Long-Range Linkage." *Genome Research* 26 (3): 342–350. doi:10.1101/gr.193474.115.

Ramos, A.P., Gustafsson, O., Labert, N., Salecker, I., Nilsson, D.E., and M. Averof. 2019. "Analysis of the Genetically Tractable Crustacean *Parhyale hawaiensis* Reveals the Organisation of a Sensory System for Low-Resolution Vision." *BMC Biology* 17 (1): 67. doi:10.1186/s12915-019-0676-y.

Regier, J.C., Shultz, J.W., Zwick, A., Hussey, A., Ball, B., Wetzer, R., Martin, J.W., and C.W. Cunningham. 2010. "Arthropod Relationships Revealed by Phylogenomic Analysis of Nuclear Protein-Coding Sequences." *Nature* 463 (7284). Nature Publishing Group: 1079–1083. doi:10.1038/nature08742.

Rehm, J.E., Hannibal, R.L., Chaw, C.R., Vargas-Vila, M.A., and N.H. Patel. 2009a. "Antibody Staining of *Parhyale hawaiensis* Embryos." *Cold Spring Harbor Protocols* 2009 (1): pdb.prot5129. doi:10.1101/pdb.prot5129.

Rehm, J.E., Hannibal, R.L., Chaw, C.R., Vargas-Vila, M.A., and N.H. Patel. 2009b. "Fixation and Dissection of *Parhyale hawaiensis* Embryos." *Cold Spring Harbor Protocols* 2009 (1): pdb.prot5127. doi:10.1101/pdb.prot5127.

Rehm, J.E., Hannibal, R.L., Chaw, C.R., Vargas-Vila, M.A., and N.H. Patel. 2009c. "In Situ Hybridization of Labeled RNA Probes to Fixed *Parhyale hawaiensis* Embryos." *Cold Spring Harbor Protocols* 2009 (1): pdb.prot5130. doi:10.1101/pdb.prot5130.

Rehm, J.E., Hannibal, R.L., Chaw, C.R., Vargas-Vila, M.A., and N.H. Patel. 2009d. "Injection of *Parhyale hawaiensis* Blastomeres with Fluorescently Labeled Tracers." *Cold Spring Harbor Protocols* 2009 (1): pdb.prot5128. doi:10.1101/pdb.prot5128.

Rehm, J.E., Hannibal, R.L., Chaw, C.R., Vargas-Vila, M.A., and N.H. Patel. 2009e. "The Crustacean *Parhyale hawaiensis*: A New Model for Arthropod Development." *Cold Spring Harbor Protocols* 2009 (1): pdb.emo114. doi:10.1101/pdb.emo114.

Robinson, J.T., Thorvaldsdóttir, H., Winckler, W., Guttman, M., Lander, E.S., Getz, G., and J.P. Mesirov. 2011. "Integrative Genomics Viewer." *Nature Biotechnology* 29 (1): 24–26. doi:10.1038/nbt.1754.

Salvador-Martínez, I., Grillo, M., Averof, M., and M.J. Telford. 2020. "CeLaVi: An Interactive Cell Lineage Visualisation Tool." *BioRxiv*, December. Cold Spring Harbor Laboratory, 2020.12.14.422765. doi:10.1101/2020.12.14.422765.

Schmitz, E.H., and P.M. Scherrey. 1983. "Digestive Anatomy of *Halella azteca* (Crustacea, Amphipoda)." *Journal of Morphology* 175 (1): 91–100. doi:10.1002/jmor.1051750109.

Schwentner, M., Combosch, D.J., Pakes Nelson, J., and G. Giribet. 2017. "A Phylogenomic Solution to the Origin of Insects by Resolving Crustacean-Hexapod Relationships." *Current Biology: CB* 27 (12): 1818–1824.e5. doi:10.1016/j.cub.2017.05.040.

Schwentner, M., Richter, S., Rogers, C.D., and G. Giribet. 2018. "Tetraconatan Phylogeny with Special Focus on Malacostraca and Branchiopoda: Highlighting the Strength of Taxon-Specific Matrices in Phylogenomics." *Proceedings: Biological Sciences* 285 (1885). doi:10.1098/rspb.2018.1524.

Serano, J.M., Martin, A., Liubicich, D.M., Jarvis, E., Bruce, H.S., La, K., Browne, W.E., Grimwood, J., and N.H. Patel. 2016.

"Comprehensive Analysis of Hox Gene Expression in the Amphipod Crustacean *Parhyale hawaiensis*." *Developmental Biology* 409 (1): 297–309. doi:10.1016/j.ydbio.2015.10.029.

Shoemaker, C.R. 1956. "Observations on the Amphipod Genus *Parhyale*." http://repository.si.edu/xmlui/handle/10088/16629.

Stamataki, E., and A. Pavlopoulos. 2016. "Non-Insect Crustacean Models in Developmental Genetics Including an Encomium to *Parhyale hawaiensis*." *Current Opinion in Genetics & Development* 39: 149–156. doi:10.1016/j.gde.2016.07.004.

Sugawara, K., Cevrim, C., and M. Averof. 2021. "Tracking Cell Lineages in 3D by Incremental Deep Learning." *BioRxiv*, February. Cold Spring Harbor Laboratory, 2021. 02.26.432552. doi:10.1101/2021.02.26.432552.

Tanaka, E. M., and P. W. Reddien. 2011. "The Cellular Basis for Animal Regeneration." *Developmental Cell* 21 (1): 172–185. doi:10.1016/j.devcel.2011.06.016.

Tararam, A.S., Wakabara, Y., and F.P.P. Leite. 1978. "Notes on *Parhyale hawaiensis* (Dana), Crustacea-Amphipoda." *Bulletin of Marine Science* 28 (4): 782–786.

Trapp, J., Almunia, C., Gaillard, J.C., Pible, O., Chaumot, A., Geffard, O., and J. Armengaud. 2016. "Proteogenomic Insights into the Core-Proteome of Female Reproductive Tissues from Crustacean Amphipods." *Journal of Proteomics, Proteomics in Evolutionary Ecology*, 135 (March): 51–61. doi:10.1016/j.jprot.2015.06.017.

Vargas-Vila, M.A., Hannibal, R.L., Parchem, R.J., Liu, P.Z., and N.H. Patel. 2010. "A Prominent Requirement for Single-Minded and the Ventral Midline in Patterning the Dorsoventral Axis of the Crustacean *Parhyale hawaiensis*." *Development (Cambridge, England)* 137 (20): 3469–3476. doi:10.1242/dev.055160.

von Reumont, B.M., Jenner, R.A., Wills, M.A., Dell'Ampio, E., Pass, G., Ebersberger, I., Meyer, B., Koenemann, S., Iliffe, T.M., Stamatakis, A., Niehuis, O., Meusemann, K., and B. Misof. 2012. "Pancrustacean Phylogeny in the Light of New Phylogenomic Data: Support for Remipedia as the Possible Sister Group of Hexapoda." *Molecular Biology and Evolution* 29 (3). Oxford Academic: 1031–1045. doi:10.1093/molbev/msr270.

Wan, Y., McDole, K., and P.J. Keller. 2019. "Light-Sheet Microscopy and Its Potential for Understanding Developmental Processes." *Annual Review of Cell and Developmental Biology* 35: 655–681. doi:10.1146/annurev-cellbio-100818-125311.

Wiese, K. 2002. *The Crustacean Nervous System*. Berlin, Heidelberg: Springer-Verlag. www.springer.com/gp/book/9783540669005.

Wittfoth, C., Harzsch, S., Wolff, C., and A. Sombke. 2019. "The 'Amphi'-Brains of Amphipods: New Insights from the Neuroanatomy of *Parhyale hawaiensis* (Dana, 1853)." *Frontiers in Zoology* 16: 30. doi:10.1186/s12983-019-0330-0.

Wolff, C., Tinevez, J.Y., Pietzsch, T., Stamataki, E., Harich, B., Guignard, L., Preibisch, S., Shorte, S., Keller, P.J., Tomancak, P., and A. Pavlopoulos. 2018. "Multi-View Light-Sheet Imaging and Tracking with the MaMuT Software Reveals the Cell Lineage of a Direct Developing Arthropod Limb." *ELife* 7. doi:10.7554/eLife.34410.

Zeng, V., and C.G. Extavour. 2012. "ASGARD: An Open-Access Database of Annotated Transcriptomes for Emerging Model Arthropod Species." *Database: The Journal of Biological Databases and Curation* 2012: bas048. doi:10.1093/database/bas048.

Zeng, V., Villanueva, K.E., Ewen-Campen, B.S., Alwes, F., Browne, W.E., and C.G. Extavour. 2011. "De Novo Assembly and Characterization of a Maternal and Developmental Transcriptome for the Emerging Model Crustacean *Parhyale hawaiensis*." *BMC Genomics* 12 (November): 581. doi:10.1186/1471-2164-12-581.

17 Echinoderms

Focus on the Sea Urchin Model in Cellular and Developmental Biology

Florian Pontheaux, Fernando Roch, Julia Morales and Patrick Cormier

CONTENTS

17.1 HISTORICAL CONTRIBUTIONS OF SEA URCHIN GAMETES AND EMBRYOS

Sea urchins, and in particular their gametes, have been an important experimental model since the end of the 19th century and throughout the 20th century (reviewed in Monroy 1986; Briggs and Wessel 2006; Pederson 2006; Hamdoun et al. 2018). From Aristotle's description of sea urchins' feeding apparatus (350 BCE) to the genome sequencing of *Strongylocentrotus purpuratus* in the 21st century, echinoderms on many occasions have been the involuntary protagonists of the history of science (Sodergren et al. 2006; Pederson 2006). Indeed, as we will discuss in detail in the following sections, sea urchins have played a paramount role in the fields of embryology and cell biology (Pederson 2006; Briggs and Wessel 2006).

17.1.1 How Did the Optical Transparency of Sea Urchin Eggs Foster Significant Advances in the Understanding of Fertilization?

As early as 1840, Derbès, like Dufossé or von Baër, was probably seduced by the transparency of the sea urchin egg, which makes these animals an excellent experimental model system for the study of fertilization (Derbès 1847). Sea urchins are gonochoric, and their gametes can be easily obtained in large quantities: a single female can produce millions of eggs, and males release an even greater quantity of functional gametes. In addition, both eggs and sperm are immediately competent to accomplish fertilization without any complementary maturation. Consequently, the simple mixing of sperm and eggs initiates fertilization and development, which take place externally. Using this material, Derbès was able to produce accurate descriptions of fertilization, holoblastic radial cleavages and larval development. The size of the sea urchin eggs (≈100 μm diameter, see Table 17.1) and the optical characteristics of their oligolecithal cytoplasm make them a valuable system for manipulation, microinjection and observation under optical microscopy (Angione et al. 2015; Stepicheva and Song 2014). Derbès was the first scientist to hypothesize the existence of a transparent layer surrounding the unfertilized eggs: the egg-jelly. However, he did not properly grasp the importance of this protective coat, as he suggested that it was dispensable for fertilization (Briggs and Wessel 2006). We now know that this glycoprotein meshwork has several functions, which include attracting and activating the sperm and providing a carbohydrate-based mechanism to allow species-specific recognition (Vilela-silva et al. 2008). In fact, jelly layers

DOI: 10.1201/9781003217503-17

secrete chemo-attractants that drive sperm swimming, and more than 100 sperm-activating peptides have been identified in the egg-jelly of various sea urchin species (Darszon et al. 2005).

Two of these peptides are known as Resact, isolated from *Arbacia punctulata*, and Speract, purified from *Strongylocentrotus purpuratus*. They bind to their respective receptors that are placed on the sperm's outer membrane and trigger changes in sperm metabolism and motility by regulating its membrane potential (Darszon et al. 2005). Sperm swimming toward the egg is controlled by flagellar curvature modifications, which depend on oscillations in the intracellular Ca^{2+} concentration $[Ca^{2+}]_i$ (Böhmer et al. 2005). The egg-jelly is also responsible for the induction of the acrosome reaction of the sperm (Santella et al. 2012). In *S. purpuratus*, fucose-sulfate polymers are the jelly-coat specific components prompting acrosome reaction (SeGall and Lennarz 1979), a process that can be separated in consecutive phases (Vacquier 2012). First, the outer acrosomal membrane fuses with the plasma membrane of the sperm head, triggering actin polymerization. Then, the acrosomal vesicle releases its contents. Finally, the Bindin protein present on the acrosomal membrane is exposed to the egg surface. The acrosome reaction is essential for fertilization and ensures that it only occurs between gametes of homologous species. Fucose-sulfate polymers are central components of the egg-jelly, and their diversity seems to confer specificity to egg and sperm interactions (Pomin 2015).

The most remarkable and accurate observations of Derbès concern the establishment of the fertilization envelope. He documented the effects elicited by sperm on the egg for the first time, including the separation of the vitelline membrane from the egg plasma membrane. He interpreted the formation of this fertilization envelope as a landmark of fertilization (Derbès 1847), and high school, community college and university practical courses still use fertilization envelope elevation as the first visible sign of sperm-mediated egg activation (Vacquier 2011). Ernest Everett Just proposed that fertilization envelope elevation occurs within one minute after sperm-egg fusion and acts as a mechanical block to polyspermy (Just 1919) (reviewed in Byrnes and Newman 2014). E. E. Just was an African American cell biologist and embryologist of international renown who can be considered an early ecological developmental biologist (Just 1939) (reviewed in Byrnes and Eckberg 2006). Fertilization envelope elevation is accomplished by the cortical granule reaction occurring at the egg's surface. Several organelles are present on the cortex of unfertilized eggs, including cortical granules (Vacquier 1975), acidic vesicles (Sardet 1984; Morgan 2011) and endoplasmic reticulum (Sardet 1984). Cortical granules, about 1 μm in diameter, are especially abundant and are present immediately beneath the cytoplasmic membrane. Following sperm entry, their content is released into the space between the cell membrane and the structured mesh of proteins that forms the vitelline envelope. This exocytosis process releases several biological compounds. A trypsin-like protease called cortical granule serine-protease digests the proteins linking the cell membrane to the vitelline membrane and degrades the Bindin receptors, immediately removing any sperm (Haley and Wessel 1999). However, the cortical granules also release mucopolysaccharides, highly hydrophilic compounds generating an osmotic gradient that pumps water into the space between the cell membrane and the vitelline membrane, which swells and detaches from the egg (reviewed in the textbook Gilbert 2006). Finally, a peroxidase enzyme present in the cortical granules hardens the fertilization envelope by crosslinking the tyrosine residues of neighbouring proteins (Foerder and Shapiro 1977; Wong et al. 2004).

In 1876, Oskar Hertwig published the first observations indicating that only one sperm enters the egg during fertilization (Hertwig 1876; Fol 1879). Using the Mediterranean sea urchin *Paracentrotus lividus* (named at the time *Toxopneustes lividus*), he was also the first to observe the fusion of egg and sperm pronuclei (Hertwig 1876; Clift and Schuh 2013). Three years later, Hermann Fol further characterized the mechanism of sperm entry and made similar observations, primarily the gametes from the starfish *Marthasterias glacialis* (named at the time *Asterias glacialis*) and to a lesser extent *Paracentrotus lividus* (Fol 1879). Ever since, technical developments in optical microscopy have made it possible to refine these observations, and ultimately electron microscopy has enabled ultrastructural investigation of sea urchin fertilization. The surface of the fertilized egg changes abruptly during cortical granule exocytosis. Two minutes after insemination, actin filaments assemble and participate in the formation of the so-called fertilization cone (Tilney and Jaffe 1980). In *Arbacia punctulata*, the sperm passes through this structure, makes a 180° U-turn and comes to rest lateral to its penetration site (Longo and Anderson 1968). This process has been carefully documented using scanning electron microscopy (Schatten and Mazia 1976). The male pronucleus and its centriole separate from the mitochondria and flagellum, which then disassemble in the cytoplasm. According to Monroy (Monroy 1986), Friedrich Meves was the first to observe that sperm mitochondria do not proliferate in the egg, leading him to propose that embryonic mitochondria have a maternal origin (Meves 1912). After mitochondria and flagellum dissolution, the centriole localizes between the male pronucleus and the egg pronucleus. This centriole extends its microtubules to form an aster so that the two pronuclei migrate toward each other and occupy a central position in the egg, where karyogamy proceeds. DNA synthesis can occur during the migration of the two pronuclei or after their fusion into the zygote nucleus (Gilbert 2006). Centrosome inheritance in echinoderms is exclusively paternal (Zhang et al. 2004). The two sperm centrioles duplicate concomitantly with DNA synthesis and end up producing the centrosomes that will steer embryonic development (Longo and Plunkett 1973; Sluder 2016).

Embryonic development proceeds normally only if a single sperm enters the egg. Fertilization by two sperms leads to a triploid nucleus, where each sperm's centriole divides independently to form four centrosomes. Theodor

Boveri already observed in 1902 that dispermic sea urchin eggs develop abnormally or die, and based on these observations, he was the first scientist to speculate that malignant tumours could be the consequence of an abnormal chromosome constitution (Boveri 1902, translated in Boveri 2008; Maderspacher 2008; Scheer 2018). His contributions to the elucidation of the role played by chromosomes as the vectors of the genetic inheritance are widely acknowledged among cell biologists.

17.1.2 Sea Urchin's Contribution to Our Understanding of the Role Played by Calcium Signalling during Fertilization

The term "egg activation" designates the multiple changes—both biochemical and morphological—that transform the egg cytoplasm after sperm penetration and prepare the cell for mitosis. In sea urchins, egg meiotic maturation occurs in the female gonads before gamete spawning. Thus, the activation of the sea urchin egg, which already possesses a haploid pronucleus, is independent of meiotic maturation. The events triggered by the activation of the sea urchin egg can be classified as early responses, occurring within seconds, and late responses, taking place several minutes after fertilization (Allen and Griffin 1958; Gilbert 2006). The early responses include the fast block of polyspermy and the exocytosis of the cortical granules. Among the late responses, we can cite the activation of mRNA translation and the duplication of DNA. Strikingly, all these events can occur independently of fertilization and are also triggered by artificial activation or parthenogenesis, which was discovered in the sea urchin by Jacques Loeb (Loeb 1899; Monroy 1986). Analyzing the effect of ions on the sea urchin egg, he observed that a treatment with a hypertonic solution of $MgCl_2$ provokes the elevation of the fertilization envelope (Loeb 1899). As Monroy points out (Monroy 1986), Loeb's work prompted Otto Heinrich Warburg to use sea urchins to develop his work on oxygen consumption in living cells (Warburg 1908). He observed that the fertilization of sea urchin eggs resulted in a rapid and nearly six-fold increase in oxygen consumption. Refining this observation to the metabolic abnormalities of cancer cells, Warburg was awarded the Nobel Prize in 1931 for his discovery of the "nature and mode of action of the respiratory enzyme". In the context of fertilization, a specific NADPH oxidase of the egg's surface uses oxygen and produces a burst of hydrogen peroxide (Wong et al. 2004; Finkel 2011). Rather than damaging the egg, the hydrogen peroxide hardens the fertilization envelope and contributes to blocking polyspermy. H_2O_2 is produced by a Ca^{2+}-dependent mechanism that involves the reduction of one molecule of oxygen and the oxidation of two proton donors. Parthenogenetic activation by A23187 ionophore is sufficient to trigger this oxidative burst by using free cytosolic calcium (Wong et al. 2004).

The hypothesis of Ca^{2+} release following sea urchin fertilization was first proposed in the mid-20th century (Mazia 1937). The Ca^{2+} ion is essential for egg activation in all metazoans but more specifically in marine invertebrate deuterostomes, which has been extensively discussed (Runft et al. 2002; Whitaker 2006; Ramos and Wessel 2013; Costache et al. 2014; Swann and Lai 2016). Calcium release triggered by fertilization or ionophore treatment was first demonstrated in sea urchin eggs using the luminescent calcium sensor aequorin (Steinhardt and Epel 1974; Steinhardt et al. 1977). Two independent types of Ca^{2+} waves have been observed following fertilization in sea urchins. The first one, a small initial cortical flash, results from an action potential-mediated influx of extracellular Ca^{2+}. A second cytosolic wave, due to the release of Ca^{2+} from the intracellular stores, begins at the sperm entry point and travels throughout the cytoplasm to encompass the entire egg (Parrington et al. 2007; Whitaker and Steinhardt 1982). The initial cortical flash does not automatically provoke the second Ca^{2+} wave, which is a distinct process exclusively triggered by sperm arrival. Notably, fertilization elicits a single Ca^{2+} wave in the sea urchin, whereas it provokes multiple Ca^{2+} oscillations in ascidians and mammals (Whitaker 2006; Sardet et al. 1998; Dupont and Dumollard 2004).

Research into the mechanisms triggering the calcium wave in sea urchins has given rise to abundant literature (reviewed in Ramos and Wessel 2013). Just after fertilization, the Ca^{2+} rise occurs as a result of inositol 1,4,5-triphosphate (IP3)-mediated release of Ca^{2+} from the endoplasmic reticulum (Terasaki and Sardet 1991). Other intracellular second messengers, including nicotinic acid adenine dinucleotide phosphate (NAADP), cyclic guanosine monophosphate (cGMP), cyclic ADP-ribose (cADPR) and nitric oxide (NO), were shown to increase at fertilization and could trigger Ca^{2+} release (Kuroda et al. 2001). However, in contrast to IP3, none of these second messengers is indispensable to the fertilization wave in the sea urchin egg.

17.1.3 Sea Urchin Egg Abundance and Synchronous Early Embryonic Development Are Optimal for Biochemistry and Cell Biology Analyses

Unfertilized sea urchin eggs are physiologically blocked at the G1 stage of the cell cycle. Fertilization thus triggers entry into the S-phase and completion of the first mitotic division. Thanks to the large number of cells that can be recovered from a single female and their embryonic mitotic division synchronicity, these gametes have been crucial for the development of biochemical approaches studying cell cycle progression and protein translation (Evans et al. 1983; Humphreys 1969).

Unravelling the mechanisms controlling protein synthesis has been a central area of research in the 20th century (Thieffry and Burian 1996). In the 1940s, it was generally admitted that thymonucleic acid (DNA) existed only in animals and zymonucleic acid (RNA) in plants. However, Jean Brachet was the first biologist to localize both nucleic acids first in sea urchin and then in other animals (Brachet 1941). This critical observation led him to conclude that both nucleic acids could be present in all cells. To study this issue, Jean Brachet made several visits to the Biological Station of Roscoff, and he liked to describe the exciting atmosphere of this place in the early 30s (Brachet 1975). His results led him to suggest that there is a strong correlation between RNA levels and protein synthesis activity. Sea urchin eggs thus played a crucial role in demonstrating that RNAs are present in all cells and that they are implicated in the synthesis of proteins, as proposed by the central dogma of Francis Crick (DNA makes RNA, which in turn makes protein).

Sea urchin eggs permeability to radioactive precursors has helped elucidate the mechanisms controlling protein synthesis in relationships with the entry into mitosis in response to fertilization. Incorporation of exogenous amino acids into protein occurs only after fertilization in sea urchin. Indeed, RNA synthesis is negligible both before and after fertilization (Schmidt et al. 1948). Moreover, the inhibition of RNA transcription by actinomycin D alters neither protein synthesis rate nor the first mitotic divisions of early sea urchin embryos (Gross and Cousineau 1963), demonstrating that the zygotic genome activity is not required for early protein synthesis (Gross et al. 1964). These observations indicated for the first time that maternal mRNAs are already present in unfertilized eggs and strongly supported the notion that their translation is tightly controlled. Furthermore, the work of Hultin showed that the synthesis of specific proteins is required for mitosis entry (Hultin 1961), heralding the future discovery of Cyclins (Ernst 2011).

In late July 1982, once the teaching in the Woods Hole Marine Station was over and the sea urchin season was coming to an end, Tim Hunt performed the critical experiment that led to the discovery of Cyclins (Hunt 2002). Cyclins form complexes with Cyclin-dependent kinases (CDKs), a family of conserved serine/threonine kinases that phosphorylate substrates throughout the cell cycle (reviewed in Malumbres 2014). Before working on cell cycle control, Tim Hunt was interested in the regulation of protein synthesis. He wanted to compare the protein synthesis rates observed in normally fertilized and parthenogenetically activated eggs, using the calcium ionophore A23187. For this purpose, he studied the sea urchin *Arbacia punctulata*. Adding [^{35}S] methionine to an egg suspension and separating proteins by gel electrophoresis, he produced an autoradiogram where one specific band, which was later identified as a Cyclin, showed an unexpected behaviour (Evans et al. 1983). Whereas most bands became stronger and stronger with time, this protein accumulated after fertilization but disappeared rapidly just before blastomere cleavage. In 2001, Tim Hunt shared the Nobel Prize in physiology or medicine with Leland Hartwell and Paul Nurse for discovering the key regulators of the cell cycle.

17.1.4 Embryonic and Larval Development of the Sea Urchin in the Age of Molecular Biology

In the late 60s and early 70s, the rapid expansion of molecular biology was about to impact all the biology domains, including developmental biology. Notably, the first eukaryotic gene fragment isolated and introduced into the bacteria *E. coli* was obtained from unfractionated DNA from *Lytechinus pictus* and *Strongylocentrotus purpuratus* (Kedes et al. 1975). These fragments encoded for histones, making these sea urchin genes the first protein-coding eukaryotic genes ever cloned (Ernst 2011).

The study of sea urchin has provided many descriptions of developmental gene regulatory networks (dGRNs). These logic structures depict the sequential regulatory events determining cell fate in different tissues and embryonic layers. The genes involved in dGRNs encode for transcription factors and components of signalling pathways but also for effector genes acting downstream of cell fate determinants and for different cell state-specific markers. The configuration adopted by dGRNs, based on empirical data, provides a dynamic picture of the genetic interactions controlling spatial and temporal aspects of development (Martik et al. 2016). dGRNs are thus predictive and testable models which help in understanding why and when developmental functions take place. The dGRN controlling the specification of *S. purpuratus* endodermal and mesodermal layers was originally described before its genomic sequence was available (Davidson 2002). However, with the completion of the sea urchin genome (Sea Urchin Genome Sequencing Consortium et al. 2006), these original descriptions have never ceased to be enriched with new components and functional data (Davidson 2006; Oliveri et al. 2008; Su et al. 2009; Saudemont et al. 2010; Peter and Davidson 2010; Li et al. 2014). Different diagrams of dGRNs are available on the E. H. Davidson's laboratory webpage (http://grns.biotapestry.org/SpEndomes/). Eric Davidson was a US developmental biologist working at the California Institute of Technology and an inspiring figure for the community of developmental biologists, particularly those working with multicellular marine organisms (Ben-Tabou de-Leon 2017). He is renowned for his pioneering work on the characterization of regulatory networks and their roles in body plan evolution.

Here, we will summarize the mechanisms that initiate cell specification and the establishment of the main layers of the sea urchin embryo. More complete descriptions of these dGRNs are available in reviews (Arnone et al. 2015; Martik et al. 2016; Ben-Tabou de-Leon 2016). In sea urchins, the embryonic body plan is rapidly established after fertilization. At the 16-cell stage, maternal inputs plus zygotic transcription determine at least three distinctive dGRN states that control ectoderm, endoderm, and micromere determination (Martik et al. 2016). Ectoderm emanates from the animal pole, and endoderm and mesoderm derive from the vegetal pole. The canonical Wnt-β-catenin signalling pathway is involved in primary axis formation and endoderm specification (Wikramanayake et al. 1998; Logan et al. 1999; Wikramanayake et al. 2004). β-catenin is active in the vegetal pole and controls polarization along the animal–vegetal axis. When β-catenin enters the nucleus, it forms an active complex with the transcription factor Tcf, which initiates the specification of endoderm in the sea urchin vegetal half. At the 16-cell stage, the future endoderm and mesoderm are still assuming a common endomesodermic identity. A Delta-Notch signal controls the separation of these two embryonic territories. The Delta-ligand expression is activated indirectly by the β-catenin-Tcf input in the skeletogenic mesoderm (Oliveri et al. 2008). Cells receiving the Delta signal are specified as mesoderm, and the others acquire an endodermal fate (Sherwood and McClay 1997; Sherwood and McClay 1999; Sweet et al. 2002). The reception of Delta in the cells initiates the expression of the transcription factor GCM (*glial cell missing*) (Ransick and Davidson 2006; Croce and McClay 2010). Then, a triple positive feedback circuit involving GCM, GataE and Six1/2 is responsible for GCM expression maintenance in the mesoderm (Ransick and Davidson 2012; Ben-Tabou de-Leon 2016). Once the original endomesoderm and ectoderm GRN states are defined, further specification and signalling come into play to generate at least 15 different cell types, which are already distinguishable by early gastrulation (Peter and Davidson 2010; Peter and Davidson 2011; Martik et al. 2016).

For instance, the dorsal–ventral (DV) axis, also referred to as the oral–aboral axis, is morphologically distinguishable at the gastrula stage (Cavalieri and Spinelli 2015) and forms thanks to the activity of the Nodal and BMP ligands, which specify respectively ventral (oral) and dorsal (aboral) ectoderm (Duboc et al. 2004; reviewed in Molina et al. 2013). Nodal activates the expression of the BMP2/4 in the ventral ectoderm, but it also elicits the production of Chordin, which blocks the activation of the BMP receptors in this region. BMP2/4 of ventral origin can then diffuse to the dorsal side (Lapraz et al. 2009), where it specifies dorsal fates activating the phosphorylation of the transcription factor Smad1/5/8 (Floc'hlay et al. 2021).

In sea urchins, activation of the zygotic genome begins at the 16-cell stage. Thus, previous development is driven by maternal factors (reviewed in Kipryushina and Yakovlev 2020). Among the post-transcriptional processes involved in early embryonic development, mRNA translation regulation deserves particular attention (Morales et al. 2006; Cormier et al. 2016). By polysome profiling and RNA sequencing, the translatome, which gives a complete picture of the polysomal recruitment dynamics, has been investigated in the sea urchin *P. lividus* (Chassé et al. 2017). This translatome represents the first step to an inclusive analysis of the translational regulatory networks (TRNs) that control the egg-to-embryo transition as well as the early events patterning the sea urchin embryo (Chassé et al. 2018). Future challenges for sea urchin embryology will include deciphering the molecular mechanisms linking TRN and dGRN activities after fertilization.

17.2 ECHINODERM PHYLOGENY

The echinoderms are an ancient and successful taxon of marine animals grouping together more than 10,000 living species. The first representatives of this phylum, which has left behind an extensive fossil record, have been found in the Cambrian Stage 3 (520 Mya) (Zamora et al. 2013). The echinoderms are deuterostome organisms belonging to the Ambulacraria clade, which also includes the Hemichordata (Figure 17.1). Estimates based on molecular clocks indicate that these two taxa could have separated 580 Mya, during the Ediacaran (Erwin et al. 2011). The other branch of the deuterostomes, the Chordata, split even earlier and gave rise to the cephalochordates, the tunicates and the vertebrates (Lowe et al. 2015; Simakov et al. 2015).

The Paleozoic seas hosted at least 35 separate echinoderm clades presenting extremely diverse body plans. Most of them appeared during the Great Ordovician Biodiversification Event (GOBE), but only five of them made their way into the Mesozoic and have found a place in the modern faunas. These five clades correspond to the Crinoidea (sea lilies and feather stars), the Asteroidea (sea stars), the Ophiuroidea (brittle stars), the Holothuroidea (sea cucumbers) and the Echinoidea (sea urchins). Their representatives are all characterized by a typical pentaradial symmetry that is thought to have secondarily evolved from a bilateral ancestral form (Smith and Zamora 2013; Topper et al. 2019).

The ancient origin of the different echinoderm groups, which appeared during the Ordovician, has been a major obstacle to ascertaining their phylogenetic relationships. However, recent molecular phylogenies strongly support the so-called Asterozoan hypothesis that places the Ophiuroidea as the sister group of the Asteroidea (Reich et al. 2015; Telford et al. 2014; Cannon et al. 2014). According to these molecular phylogenies, Crinoidea appears as the basal branch of all the Echinoderms, with Holothuroidea being the closest relatives of Echinoidea (Figure 17.1). This last group underwent further diversification during the late Permian, the Mesozoic and the Cenozoic (Kroh and Smith 2010), producing a vast array of forms that have adopted remarkably different lifestyles and have adapted to all sorts of marine environments and climates.

The majority of the Echinoidea currently studied in the laboratory, including several edible species of commercial interest, belong to the order Camarodonta. The presence of this taxon in the fossil record has been dated back to the Miocene (Kroh and Smith 2010), but its different families may have originated earlier during the Middle Eocene and the Oligocene (45–23 Mya) (Láruson 2017). The recent characterization of the mitochondrial genomes and transcriptomes of several Camarodonta representatives (see Figure 17.2) have allowed researchers to establish the phylogeny of this group (Bronstein and Kroh 2019; Láruson 2017; Mongiardino Koch et al. 2018). At the same time, these molecular tools have provided an opportunity to develop comparative genomic approaches aimed at studying the molecular basis of the many anatomical, developmental, physiological and ecological specializations that characterize the different members of this taxon. Indeed, the density of available landmarks allows for the comparison of closely related species, such as the various Strongylocentrotidae representatives (divergence time estimated at 15–10 Mya) but also more distant species, such as the members of the Toxopneustidae, the Echinometridae, the Parechinidae or the Echinidae families (see Figure 17.2).

From a macroevolutionary perspective, these comparisons can nowadays be extended to other sea urchins

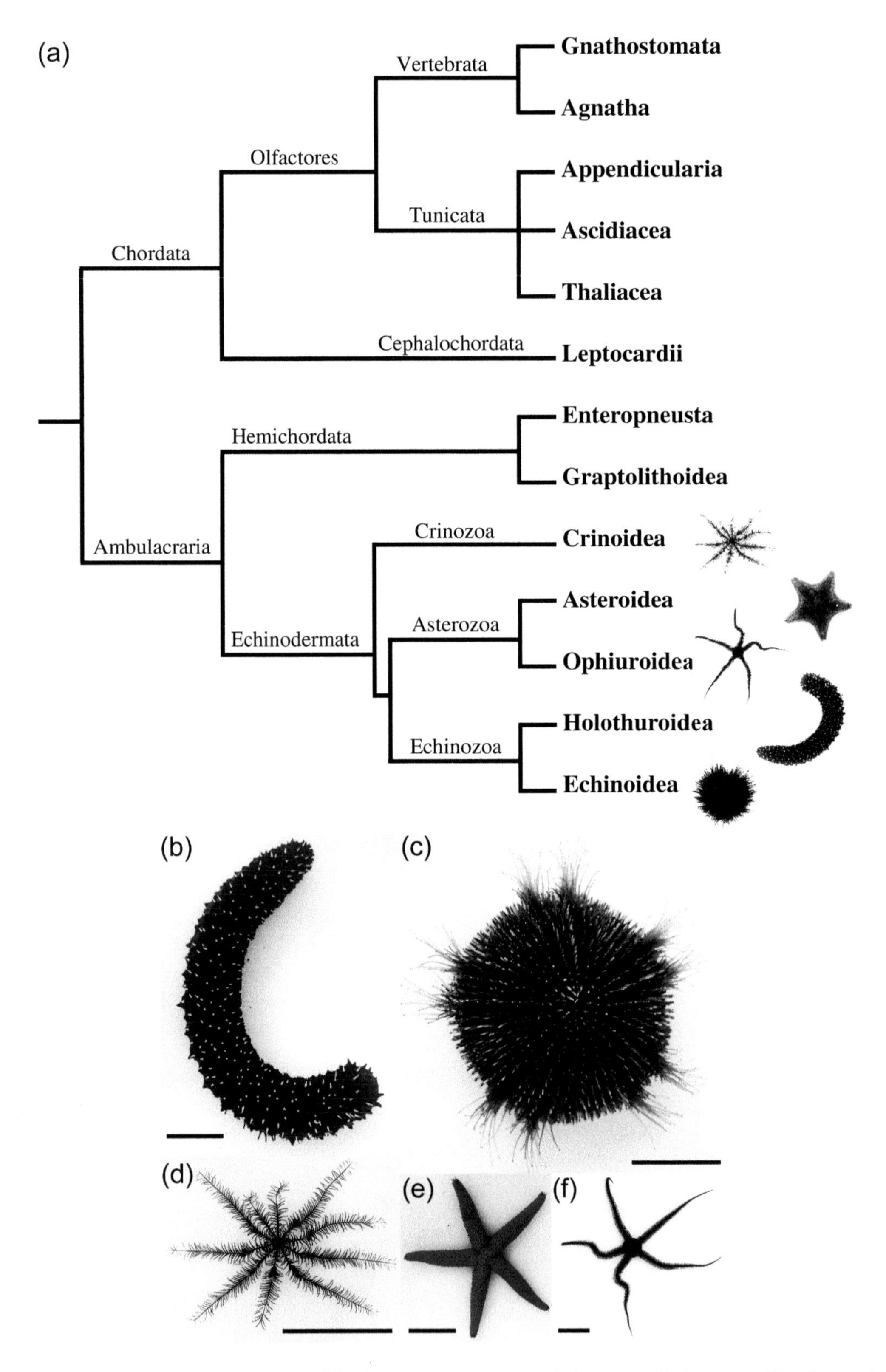

FIGURE 17.1 (a) Deuterostome group taxonomy: The deuterostome group includes two main lineages, Chordata and Ambulacraria. Chordata include cephalochordates, vertebrates and tunicates. Vertebrates are subdivided into Agnatha (e.g., myxines, lamprey) and Gnathostomata, which include Chondrichthyes (e.g., sharks, sawfish) and Osteichthyes (e.g., ray-finned fish, tetrapods). Tunicates are represented by ascidians, larvaceans (appendicularians) and thaliaceans. Ambulacraria include hemichordates and echinoderms, which are subdivided into five classes (crinoids, asteroids, ophiuroids, holothuroids and echinoids). Nodes and branches represent splits between taxons without any relative time reference. Each class of echinoderms is represented by black and white unscaled photographs. (b–f) Living adult representative echinoderms. (b) The holothuroid *Holothuria forskali*; (c) the echinoid *Sphaerechinus granularis*; (d) the crinoid *Antedon bifida*; (e) the asteroid *Echinaster sepositus*; (f) the ophiuroid *Ophiocomina nigra*. Animals were collected and maintained by the Roscoff Aquarium Service at the Roscoff Marine Station, France. Animals are shown at different scales and bars positioned at the bottom represent 5 cm.

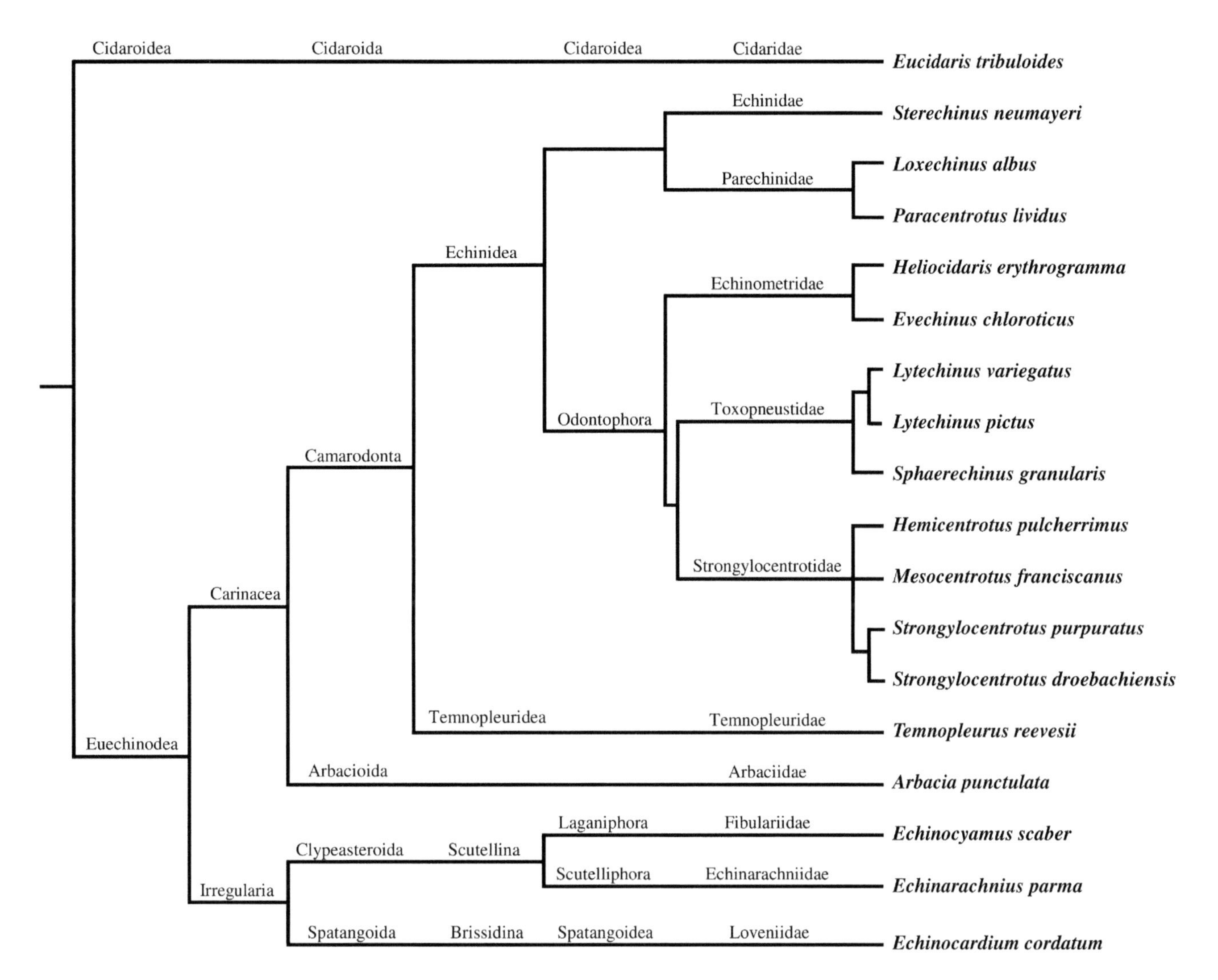

FIGURE 17.2 Echinoidea classification. The taxon of Echinoidea is mostly represented in laboratories by the Camarodonta order, but studies are also made on irregular sea urchins (Irregularia) and the most distant group Cidaroidea. Nodes and branches represent speciation without any relative time reference. Each column corresponds to the classification (subclass, infraclass, order, infraorder, superfamily, family, species) identified on the World Register of Marine Species (WoRMS). (From Schwentner et al. 2018.)

commonly used in the laboratory that belong to more distant orders, such as the different members of the genus *Arbacia*, the Scutellina *Echinarachnius parma* (sand burrowing sea urchin) and the primitive forms of the order Cidaroida, a basal group that separated from the rest of Echinoidea during the late Permian (250 Mya) (Kroh and Smith 2010). Notably, a draft of the genomic sequence of the Cidaroida *Eucidaris tribuloides* is also available, allowing comparisons with the two fully sequenced Camarodonta *Strongylocentrotus purpuratus* and *Lytechinus variegatus* (Kudtarkar and Cameron 2017).

17.3 GEOGRAPHICAL LOCATION OF ECHINODERMS

Echinoderms, with their large diversity of species (>2,000 Asteroidea, >2,000 Ophiuroidea, >600 Crinoidea, >4,000 Echinoidea and >1,700 Holothuroidea species) inhabit all the oceans and seas of the planet (see Figure 17.3). This group is exclusively marine and is absent from freshwater, although some species can be found in brackish waters (Pagett 1981).

Echinoderms are benthic, and some are considered subsoil species, since they can burrow a few tens of centimetres in the sand (e.g., *Echinocardium cordatum*). Echinoderms have managed to adapt to a wide variety of environments, ranging from the warm waters of the tropics to the coldest waters of the poles (McClintock et al. 2011). For instance, all five classes of echinoderms are present in the Arctic Ocean (Smirnov 1994), and the Antarctic Ocean hosts the sea urchin *Sterechinus neumayeri*, which is studied for its biological mechanisms adapted to sub-zero temperatures (Pace et al. 2010). Echinoderms are also found at all depths, with some sea urchins inhabiting environments as deep as 7,300 meters (Mironov 2008), starfish and brittle stars at 8,000 m (Mironov et al. 2016) and feather stars at 9,000 m (Oji et al.

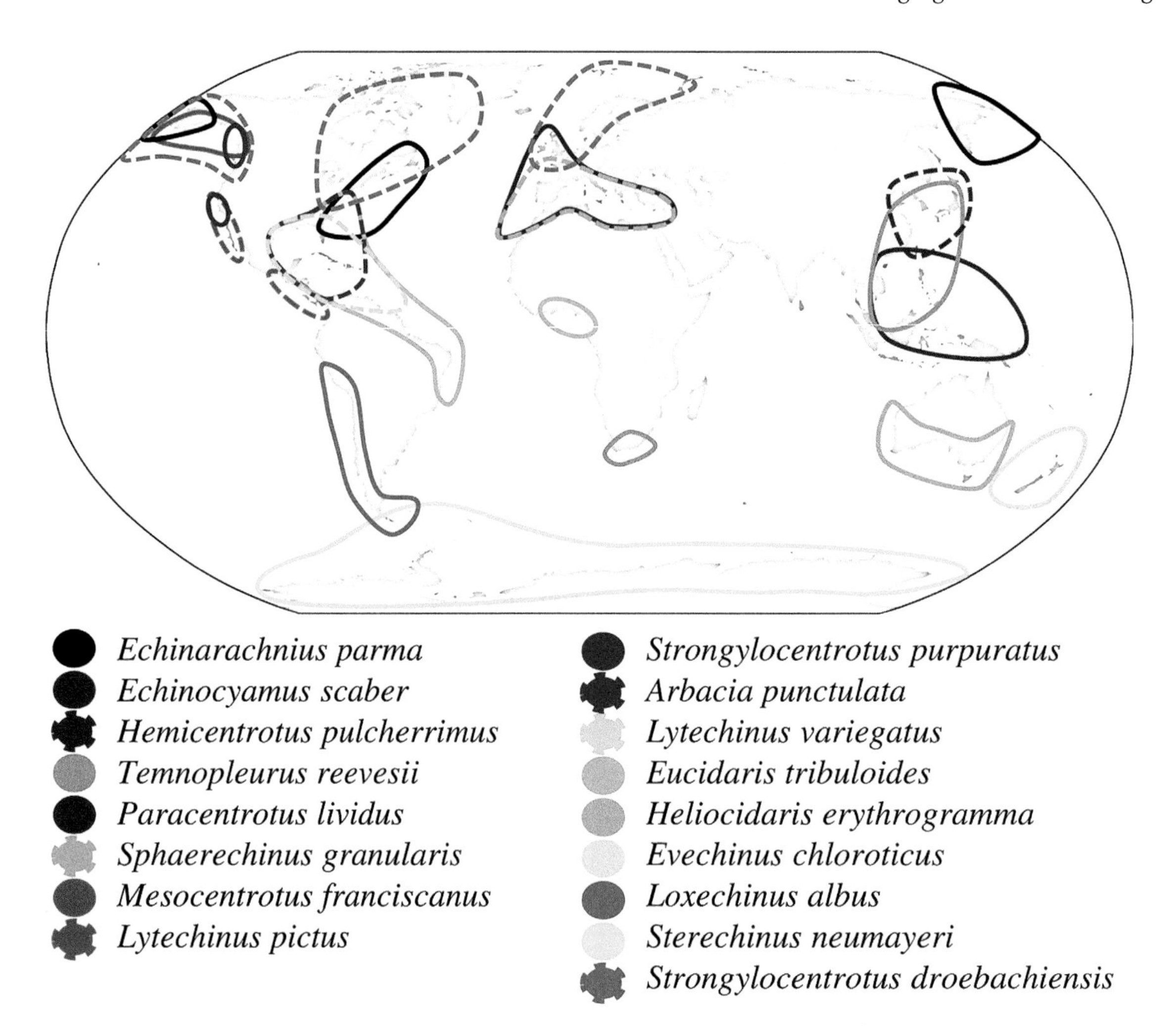

FIGURE 17.3 Geographic distribution of the main sea urchin species cited in this chapter. Geographical locations represent the major place where each species is found. Particular single occurrences in other areas can be found in the World Register of Marine Species (WoRMS). The cryptic species *Echinocardium cordatum* is widely distributed on the planet (not shown on the figure for clarity), divided into four distinct geographical lineages: one in the north-east and north-west Atlantic Ocean, one in the Mediterranean Sea and the north-east Atlantic Ocean, one in the Mediterranean Sea and one in the North and South Pacific Ocean avoiding the equatorial zone. Close colour shapes do not represent a taxonomic relationship between species but are here to help distinguish between species in the same region. (From Horton et al. 2020; Chenuil and Feral 2003; Egea et al. 2011.)

2009). Holothurians are though the record holders, as some specimens have been observed below 10,000 m (Mironov et al. 2019).

The most popular species in the laboratories (*A. punctulata*, *P. lividus*, *S. purpuratus*, *L. variegatus*, *L. pictus*, *H. pulcherrimus*) come mainly from the northern hemisphere. However, sea urchin species have been described in all oceans, including the Indian Ocean, the deep Pacific and the Arctic (Smirnov 1994; Price and Rowe 1996; Rowe and Richmond 2004; Filander and Griffiths 2017; Mironov et al. 2015; Mulochau et al. 2014). Other species cited in this chapter (e.g., *S. granularis*, *H. erythrogramma*, *S. neumayeri*, *M. franciscanus*, *S. droebachiensis*) illustrate different aspects of sea urchin diversity and facilitate the study of many biological questions, from phylogeny, adaptation and evolution to species conservation, community interactions and ecology.

17.4 SEA URCHIN LIFE CYCLE

Sea urchins are gonochoric with an average sex ratio of 1:1. Both sexes release their gametes (eggs or sperm) directly into the water column once a year, although in some species, a second period of spawning has also been reported, such as in *Paracentrotus lividus* (González-Irusta et al. 2010). In this animal, the mature season varies from January to June, and the spring equinox usually marks the height of the breeding season. In all echinoderms, the reproductive cycle and time of breeding (Table 17.1) may fluctuate, based on geographical (Figure 17.3) and local conditions. For example, *P. lividus* is found in the Mediterranean Sea and in the eastern Atlantic Ocean, from Scotland and Ireland to Southern Morocco and the Canary Islands. This species lives mainly in areas where winter water temperatures range from 10–15°C and summer temperatures from

TABLE 17.1
Breeding Season and Egg Diameter in Different Echinoidea Species

Echinoidea Species	Breeding Season	Egg Diameter	References
Eucidaris tribuloides	—	94 µm	(McAlister and Moran 2012; Lessios 1988; Lessios 1990)
Sterechinus neumayeri	September–November	180 µm	(Bosch et al. 1987; Stanwell-Smith and Peck 1998)
Paracentrotus lividus	May–September (AO) April–June and September–November (MS)	75 µm	(Hamdoun et al. 2018; Ouréns et al. 2011; Rocha et al. 2019; Byrne 1990)
Heliocidaris erythrogramma	November–February (SEA) February–May (SWA)	400–450 µm	(Binks et al. 2012; Foo et al. 2018; Raff 1987)
Evechinus chloroticus	November–February	—	(Delorme and Sewell 2016)
Hemicentrotus pulcherrimus	January–March	—	(Kiyomoto et al. 2014)
Mesocentrotus franciscanus	June–September	130 µm	(Bernard 1977; Bolton et al. 2000)
Strongylocentrotus purpuratus	November–March	80 µm	(Bolton et al. 2000; Hamdoun et al. 2018)
Strongylocentrotus droebachiensis	March–May	145 µm	(Himmelman 1978; Levitan 1993; Meidel and Scheibling 1998)
Lytechinus variegatus	May–September	100 µm	(Hamdoun et al. 2018; Lessios 1990, 1988; Schatten 1981)
Lytechinus pictus	May–September	120 µm	(Hamdoun et al. 2018)
Sphaerechinus granularis	April–June (Brittany) June–November (MS)	100 µm	(Guillou and Lumingas 1998; Guillou and Michel 1993; Vafidis et al. 2020)
Temnopleurus reevesii	July–January	100 µm	(Hamdoun et al. 2018)
Arbacia punctulata	June–August	69 µm	(Bolton et al. 2000; Gianguzza and Bonaviri 2013)
Echinarachnius parma	March–July (NWP)	110–135 µm	(Costello and Henley 1971; Drozdov and Vinnikova 2010; Summers and Hylander 1974)
Echinocardium cordatum	May–July (NWP) April–October (MS) May–October (AO)	110 µm	(Drozdov and Vinnikova 2010; Egea et al. 2011; Hibino et al. 2019)

AO: Atlantic Ocean, MS: Mediterranean Sea, NWP: North-West Pacific Ocean, SWA: South-West Australia, SEA: South-East Australia

18–25°C. Several factors, like temperature, photoperiod, resource availability and water turbulence contribute to the regulation of gametogenesis in these populations (Gago and Luís 2011). On the other hand, records from the North Pacific, Arctic and North Atlantic Oceans show that spawning of the sea urchin *Strongylocentrotus droebachiensis* may also be synchronized with the spring phytoplankton increase (Himmelman 1978; Starr et al. 1990). However, the main environmental factors triggering spawning and the molecular mechanisms that mediate this response are not yet known.

During the reproductive cycle, *P. lividus* gonads go through different development stages, which have been exhaustively characterized (Byrne 1990). Observation of gametogenesis in *P. lividus* through histological examinations allows us to classify the annual reproductive cycle of this species in six developmental stages: 1) recovery, 2) growing, 3) premature, 4) mature, 5) partly spawned and 6) spent. In turn, in *Strongylocentrotus droebachiensis*, four stages have been recognized by examining the activity of the two main cell populations composing the germinal epithelium (Walker et al. 2007, 2013). These populations are the germinal cells, which are either ova in the ovary or spermatogonia in the testis, and a group of somatic cells called nutritive phagocytes (NPs), which are functionally equivalent to the vertebrate Sertoli cells and are present in both sexes. Stage 1, called inter-gametogenesis, occurs directly after spring spawning and lasts for about three months. Residual reproductive cells are present, but otherwise, the gonads look empty. Toward the end of this stage, NP cells increase in number and resume nutrient storage, doubling their size by the end of this phase. In addition, reproductive cells begin to appear. NPs are involved in the phagocytosis of residual ova and spermatozoa, and thus participate in the recycling of derived nutrients. Stage 2 is called pre-gametogenesis and NP renewal. This stage begins in summer and lasts for approximately three to four months. Reproductive cells, present at the periphery of the gonad, increase both in number and size. Stage 3, gametogenesis and NP utilization, takes place during five winter months. The reproductive cells continue to develop and migrate into the centre of the gonad. Conversely, the NPs cells shrink, and their number decreases. Stage 4 corresponds to pre-spawning and spawning. This stage occurs in late winter and lasts around three months. The lumen of the gonad is packed with fully differentiated gametes, and the NP cells are barely observable. At the end of stage 4, spawning occurs, and gametes are released from the gonads by the gonopores.

Several holistic approaches have been generated to understanding the molecular mechanisms of gametogenesis and the events of the life cycle. Whole-genome and Q-PCR data have been obtained to identify genes expressed by *S. purpuratus* during oogenesis (Song et al. 2006). A general picture of protein abundance changes occurring during *P. lividus* gonad maturation has been generated by the proteomic approach (Ghisaura et al. 2016).

In Figure 17.4, we show the life cycle of the sea urchin *Sphaerechinus granularis*, which can be found at high densities in some locations of Brittany, such as the Glénan Islands and the Bay of Concarneau (Guillou and Michel 1993). However, captured adults maintained in appropriate conditions can release a large number of gametes from September to early July. Consequently, the availability of mature adults during most of the year makes this species a choice organism for cellular and biochemical studies (Feizbakhsh et al. 2020; Chassé et al. 2019). In the laboratory, gamete spawning may be induced artificially using several methods, such as intracoelomic injection of 0.1 M acetylcholine or of 0.5 M KCl. During the breeding season, when adults are mature, the expulsion of a small number of gametes may be obtained by a gentle shaking or by weak electrical stimulation, which facilitates the sexing of different individuals.

17.5 SEA URCHIN EMBRYOGENESIS

Sea urchins were one of the first animals to be used for embryological studies, that is, the development of a multicellular organism from a single cell (the fertilized egg) (reviewed in Ettensohn 2017). Therefore, the particular

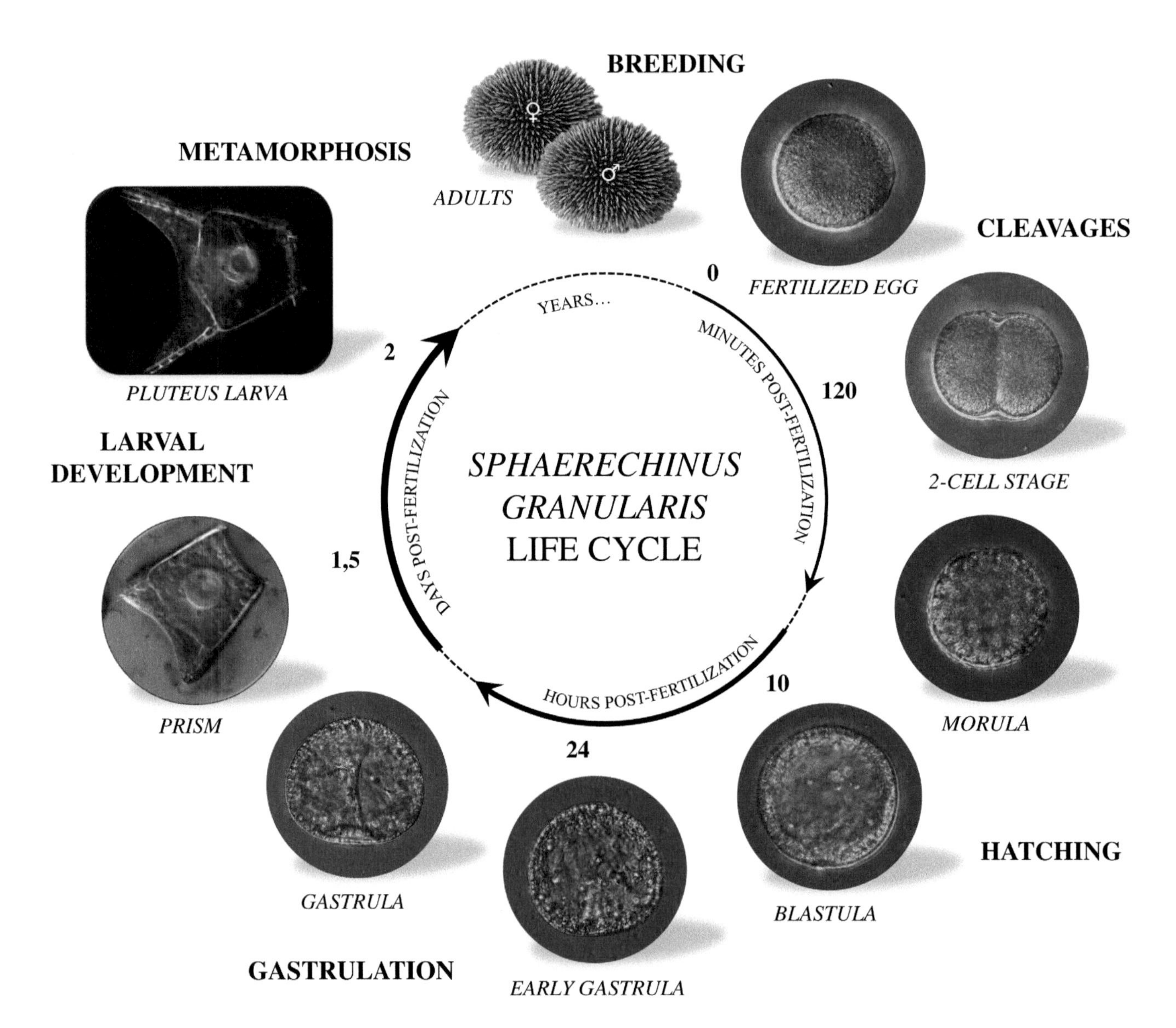

FIGURE 17.4 Life cycle of the sea urchin *Sphaerechinus granularis*. The sea urchin life cycle is composed of three periods of time with embryology (cleavages, hatching, gastrulation) taking minutes to hours, larval development taking days and growing individuals following the metamorphosis taking years. *Sphaerechinus granularis* development is synchronous, and times are noted. Microscopy pictures of *S. granularis* stages were taken with DIC filter on a Leica DMi8 microscope. Fertilized egg diameter is around 100 μm (×20 objectives) and slightly increases to prism and pluteus larva stage (×40 objectives). (From Delalande et al. 1998.)

development of a large set of species has been characterized in detail (see Table 17.2 and reviewed in Arnone et al. 2015; Hamdoun et al. 2018).

Sea urchin eggs are typically 80–200 µm in diameter and present an evenly distributed yolk (isolecithal; see Table 17.1). When released through the female gonoducts, unfertilized eggs are blocked at the G1 stage of the cell cycle, having completed their meiotic maturation in the ovary. Unfertilized eggs are polarized along a primordial axis, the animal–vegetal axis (A-V), which is specified during oogenesis and consequently is maternally established (Goldstein and Freeman 1997). Classically, the position of the animal pole corresponds to the extrusion site of the polar bodies. In some batches of *Paracentrotus lividus* eggs, a pigment band, initially described by Theodor Boveri (Schroeder 1980) and corresponding to a subequatorial accumulation of pigment granules, can be used as a visible marker of A-V polarity (Sardet and Chang 1985). A surface blister marking the animal pole has also been described in *Echinocardium cordatum* (Sardet and Chang 1985).

Bisection of an unfertilized egg through the equator, followed by independent fertilization of the two halves, results in an animal half that gives rise to an undifferentiated epithelial ball and a vegetal half that develops into a relatively normal pluteus (Horstadius 1939; Maruyama et al. 1985). The fates of the two halves are explained by the presence of genetic determinants in the vegetal pole and the subsequent participation of regulative interactions that implement the formation of the missing animal blastomeres in the vegetal half (reviewed in Angerer and Angerer 2000; Kipryushina and Yakovlev 2020).

Sea urchin embryos exhibit holoblastic cleavages; that is, they undergo a complete partition subdividing the whole egg into separate blastomeres. Cleavages are radial: the division planes form a right angle with respect to the previous division. The cleavage rate and the development speed usually depend on temperature. At 18°C, *Sphaerechinus granularis* zygotes reach the first division by 120 minutes, and each subsequent division occurs at regular intervals of nearly 60 minutes. In *Paracentrotus lividus*, the first cleavage is faster, occurring at 70–90 minutes post-fertilization. The first cleavage (Cl.1, 2 cells) is meridional (in the polar axis) and divides the egg into two equally sized blastomeres (Figure 17.5). The second cleavage (Cl.2, 4 cells) is perpendicular to the first but also

TABLE 17.2
Availability of Omics in Different Echinoidea Species and Their Main Research Thematics

	Omics Data	Main Research Thematics	References
Eucidaris tribuloides	G./T. available (Echinobase/NCBI)	Embryogenesis, Development, Global changing	(Erkenbrack et al. 2018)
Sterechinus neumayeri	T. available (NCBI)	Toxicity, Fertilization, Genetics, Global changing	(Dilly et al. 2015)
Loxechinus albus	T. available (NCBI)	Ecology, Genetics, Global changing	(Gaitán-Espitia et al. 2016)
Paracentrotus lividus	G. in progress (European consortium); T./Trl. available (NCBI)	Ecology, Toxicity, Fertilization, Embryogenesis, Development, Global changing, Economy	(Chassé et al. 2018; Gildor et al. 2016)
Heliocidaris erythrogramma	T. available (NCBI)	Fertilization, Embryogenesis, Development, Global changing	(Wygoda et al. 2014)
Evechinus chloroticus	T. available (NCBI)	Ecology, Toxicity, Global changing	(Gillard et al. 2014)
Hemicentrotus pulcherrimus	G./T. available (HpBase/NCBI)	Toxicity, Fertilization, Embryogenesis, Metabolism, Development, Genetics	(Kinjo et al. 2018)
Mesocentrotus franciscanus	T. available (NCBI)	Ecology, Fertilization, Genetics	(Wong et al. 2019)
Strongylocentrotus purpuratus	G./T. available (Echinobase/NCBI)	Toxicity, Fertilization, Embryogenesis, Development, Genetics, Global changing	(Kudtarkar and Cameron 2017; Sea Urchin Genome Sequencing Consortium et al. 2006; Tu et al. 2014)
Strongylocentrotus droebachiensis	Transcriptome available (NCBI)	Toxicity, Fertilization, Metabolism, Development, Global changing	(Runcie et al. 2017)
Lytechinus variegatus	G./T. available (Echinobase/NCBI)	Ecology, Toxicity, Fertilization, Development, Global changing	(Davidson et al. 2020; Hogan et al. 2020)
Lytechinus pictus	G. in progress; Transcriptomes available (Echinobase/NCBI)	Toxicity, Fertilization, Embryogenesis, Development	(Nesbit et al. 2019)
Sphaerechinus granularis	T. from ovaries available (Echinobase/NCBI)	Toxicity, Fertilization, Embryogenesis	(Reich et al. 2015)
Temnopleurus reevesii	See chapter 18	Genetics	(Suzuki and Yaguchi 2018)
Arbacia punctulata	T. available (NCBI)	Toxicity, Fertilization, Embryogenesis, Metabolism, Development	(Janies et al. 2016)
Echinarachnius parma	T. from ovaries available (Echinobase/NCBI)	Toxicity, Fertilization, Development	(Reich et al. 2015)
Echinocardium cordatum	T. available (NCBI)	Ecology, Toxicity, Development	(Romiguier et al. 2014)

G: Genome, T: Transcriptome, G./T: Genome and Transcriptome, Trl: Translatome

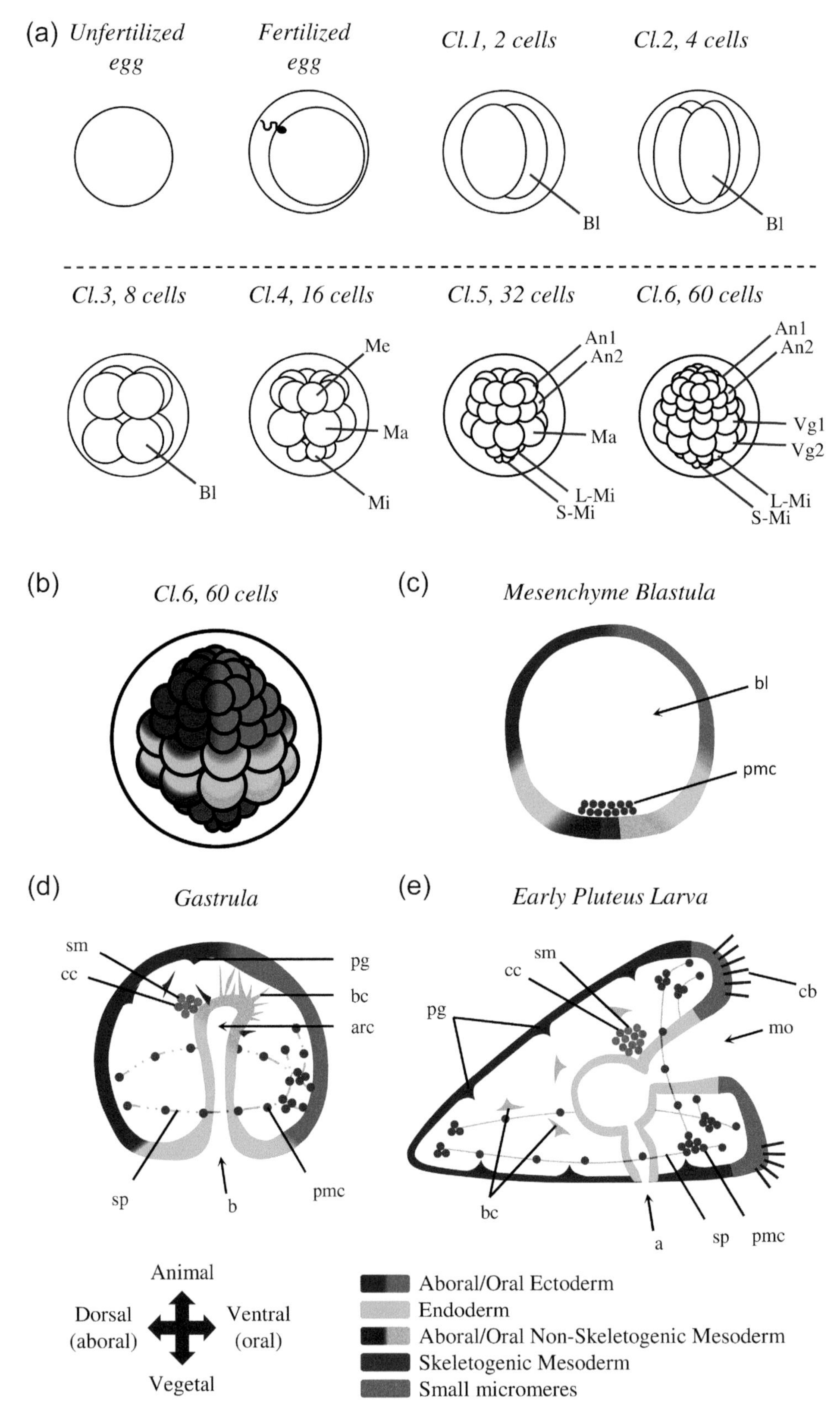

FIGURE 17.5 Diagrams of sea urchin embryo development. (a) The six first cleavages of a sea urchin embryo. (b) Diagram of the 60-cell stage, (c) mesenchyme blastula, (d) gastrula and (e) pluteus larva stage with the colouration of presumptive cell fates. See embryogenesis text part for more details on cleavage axis, cell fate and migration. (Cl: cleavage, Bl: blastomeres, Me: mesomeres, An: animal, Ma: macromeres, Vg: vegetal, Mi: micromeres, S/L-Mi: small/large micromeres, bl: blastocoel, pmc: primary mesenchyme cells, b: blastopore, arc: archenteron, cb: ciliary/ciliated bands, pg: red-pigmented cells, bc: blastocoel cells, cc: coelomic cells, sm: small micromeres, sp: larval spicules, mo: mouth, a: anus.)

occurs in a meridional plane, resulting in the production of four equally sized blastomeres. The third cleavage (Cl.3, 8 cells) is equatorial (at right angles of the polar axis), resulting in four upper and four lower blastomeres, all of equal sizes. The fourth cleavage (Cl.4, 16 cells) exhibits a complex and characteristic pattern that reveals the basic A-V polarity of the embryo. In the vegetal pole, the four blastomeres divide asymmetrically and horizontally, forming four small cells placed in the egg's pole (the micromeres) and four larger cells situated above (the macromeres). In the animal pole, the four blastomeres divide meridionally and symmetrically, resulting in eight equally sized cells (the mesomeres). At the fifth cleavage (Cl.5, 32 cells), the eight mesomeres at the animal half divide equatorially and symmetrically, resulting in two layers of cells called "an1" for the upper one and "an2" for the lower one. In the vegetal pole, the four macromeres instead divide meridionally, forming a tier of eight cells. The four micromeres divide horizontally and asymmetrically, resulting in four small micromeres at the extreme vegetal pole hemisphere and four large micromeres above. At the sixth cleavage (Cl.6, 60 cells), all the cleavage furrows are equatorial. The macromeres divide, giving rise to two eight-cell tiers called "veg1" and "veg2". The large micromeres divide as well, but not the small micromeres. In total, the 60-cell embryo shows, from top to bottom, 16 "an1" cells distributed in two layers of 8 cells each, 16 "an2" cells forming also two layers of 8 cells each, 8 "vg1", 8 "vg2", 8 large micromeres and 4 small micromeres.

The macromeres producing the "vg1" and "vg2" cells are the endomesoderm progenitors. The large micromeres contribute instead to the skeletogenic mesenchyme, and the small micromeres to the primordial germ cells (PGCs) (Okazaki 1975; Yajima and Wessel 2012). In the Echinoderm phylum, micromeres are only observed in echinoids and are thus considered a derived character. Asymmetric cell division is directed by the control of spindle and furrow cleavage position and by uneven repartition of molecules. Although the precise molecular mechanisms that orchestrate these asymmetric divisions are still poorly understood, it has been shown that the AGS/Pins proteins (activator of G-protein signalling/partner of Inscuteable) are required for normal asymmetrical division during micromere formation (Voronina and Wessel 2006; Poon et al. 2019).

As soon as the eight-cell stage is reached, a small central cavity forms in the centre of the embryo. As cleavage proceeds, this space enlarges and forms the blastocoel. A morula appears roughly six hours after fertilization, but at the 120-cell stage, the smooth-surfaced blastula becomes a continuous spherical monolayer surrounded by an outer hyaline layer. The epithelium sits on an inner basal membrane; cell adhesion is mediated by tight junctions. Cilia develop on the surface of the blastula, and their coordinated action triggers the rotation of the blastula within the fertilization envelope. Ten hours after fertilization, the blastula is composed of about 600 cells. Cell division rates decrease as the cell cycle lengthens. At the end of segmentation, the blastula is covered by cilia, presents a conspicuous apical ciliary tuft in the animal pole and starts secreting a hatching enzyme that digests the fertilization envelope. The synthesis of this hatching enzyme takes place in the animal-most two-thirds of the blastula and is likely to be restricted to the presumptive ectoderm territory (Lepage et al. 1992a, 1992b). Finally, a swimming blastula is released into the sea.

The blastula wall thickens at the vegetal pole, forming the vegetal plate. In the central region of this vegetal plate, the micromere descendants display pulsatile movements and start developing filopodia in their basal face. These cells lose their affinity for the outer hyaline structure and gain affinity for the fibronectin present in the basal lamina and the extracellular matrix lining the blastocoel (Fink and McClay 1985). Eventually, they detach from the epithelium and enter the blastocoel, forming the primary mesenchyme (Peterson and McClay 2003). As these cells are the first ones to ingress into the blastocoel, they are called primary mesenchyme cells (PMCs) (Burke et al. 1991). Adhering to the blastocoel matrix, these cells progress from the vegetal pole toward the animal pole and then reverse their trajectory. Finally, the PMCs reach an area located between the vegetal pole and the equator and form a ring pattern consisting of two ventrolateral cell clusters and dorsal and ventral interconnected chains of cells (Malinda and Ettensohn 1994). Then their filopodia coalesce, and the characteristic syncytial bridges of the larval skeleton appear. The primary mesenchyme cells of the sea urchin represent one of the best developmental models for studying mesodermal migration (Anstrom 1992; Ettensohn 1999; Ettensohn and Sweet 2000; Peterson and McClay 2003), and the cellular basis of skeletogenic cells has been characterized in detail (Okazaki 1975; Ettensohn and McClay 1988; Armstrong and McClay 1994). Moreover, the gene regulatory network (GRN) that controls their formation has also been described not only in species of the order Camarodonta (Oliveri and Davidson 2004; Oliveri et al. 2008) but also in other echinoid orders (Minokawa 2017).

In euechinoids, the ingression of the PMCs marks the onset of gastrulation. The invagination of the Vg2 territory in the blastocoel gives rise to the archenteron (primitive gut), opened to the outside by a circular blastopore (the future anus). Invagination of the vegetal plate, a universal feature of echinoderm gastrulation, is traditionally divided into "primary" and "secondary" invagination (Gustafson and Kinnander 1956). The "primary" invagination corresponds to an initial phase of gut extension that involves extensive extracellular matrix remodelling and cell shape changes (reviewed in Kominami and Takata 2004). Three hypotheses have been advanced to explain the "primary" invagination (reviewed in Ettensohn 2020). First, according to the so-called apical constriction hypothesis, a ring of vegetal plate cells become bottle shaped, compressing their apical ends (Kimberly and Hardin 1998). This cell shape modification causes the cells to pucker inward. However, bottle cells could be a specialized feature of euechinoids and not a general characteristic of all echinoderms (Ettensohn 2020). A second hypothesis proposes that invagination could be driven by changes in extracellular matrix composition (Lane et al. 1993). In fact, the hyaline layer is made up of two layers: an outer lamina composed of hyalin protein and glycoproteins and an inner lamina composed of fibropellin proteins (Hall and Vacquier 1982; Bisgrove et al. 1991). After PMC ingression, the vegetal plate cells secrete chondroitin sulfate proteoglycans into the inner lamina of the hyaline layer. As these chondroitin sulfate proteoglycans capture abundant water, the inner layer expands even if the outer layer remains stiff. The result is a force pushing the epithelium toward the blastocoel (Lane et al. 1993). A third hypothesis suggests another force arising from the concerted movement of cells toward the vegetal pole that may facilitate the invagination by drawing the buckled layer inward (Burke et al. 1991).

The "secondary" invagination ensues after a brief pause. During this stage, the archenteron extends and produces a long thin tube. The cells of the archenteron, which are organized as a monolayered epithelium, move over one another and flatten (Ettensohn 1985; Hardin 1989). In *Lytechinus variegatus*, gastrulation has been analyzed at a high resolution by live imaging and using transplantation techniques (Martik and McClay 2017). In this species, the process of archenteron elongation is mainly driven by the elongation of Vg2 endoderm cells. In fact, even if oriented cell divisions also contribute to gut elongation, cell proliferation inhibition does not preclude gastrulation, indicating that cell proliferation is not essential for this process (Stephens et al. 1986; Martik and McClay 2017).

The oral ectoderm of the gastrula flattens as the gastrula becomes roughly triangular, forming the prism larva. The embryonic radial symmetry is gradually replaced by a bilateral symmetry. An early sign of this transformation consists in the aggregation of primary mesenchyme cells into two clusters that develop in the opposite posterolateral–ventral angles of the prism larva. The cells of the primary mesenchyme form then a syncytium, in which two calcitic spicules develop. These spicules, flanking the primitive digestive tract, will constitute the endoskeleton of the pluteus larva. For this, the primary mesenchyme cells endocytose seawater from the larval internal body cavity and form a series of vacuoles where calcium can concentrate and precipitate as amorphous calcium carbonate (Kahil et al. 2020).

Once the archenteron reaches about two-thirds of its final length, the third and last stage of archenteron elongation begins (Hardin 1988). This phase is driven by the secondary mesenchyme cells, which extend filopodia through the blastocoel cavity to reach a specific area in the inner surface of the blastocoel roof (Hardin and McClay 1990). These filopodia pull the archenteron toward the animal pole and contact the region where the mouth will form. The mouth forms in the future ventral side of the larva after the fusion of the archenteron and the ectoderm epithelium. Typical of the deuterostomes, the mouth and the archenteron create a continuous digestive tube that joins the blastopore, which coincides with the anus.

During the processes of archenteron elongation, the secondary mesenchyme cells spread into the blastocoel fluid, where they form at least four non-skeletogenic mesoderm cells (Ettensohn and Ruffins 1993). Early in gastrulation, a population of red-pigmented cells forms (Gustafson and Wolpert 1967; Gibson and Burke 1985). It is interesting to note that independent knock out of the genes encoding for polyketide synthase, flavin monooxygenase family 3, and the glial cells missing (gcm) protein results in the disappearance of red-pigmented cells throughout the body of the larva (Wessel et al. 2020).

Later in gastrulation, a group of cells coming from the tip of the archenteron moves into the blastocoel and adopts a fibroblast-like morphology: they are the so-called basal cells (Cameron et al. 1991), or blastocoel cells (Tamboline and Burke 1992). At the end of gastrulation, two coelomic cavities appear as a bilateral out-pocketing of the foregut (Gustafson and Wolpert 1963). Afterwards, secondary mesenchymal cells move out of these coelomic cavities and produce the circumesophageal musculature of the pluteus larvae (Ishimoda-Takagi et al. 1984; Burke and Alvarez 1988; Wessel et al. 1990; Andrikou et al. 2013). While the right coelomic pouch remains rudimentary, the left coelomic pouch undergoes massive development to build many of the structures of the future adult sea urchin. The left side of the pluteus contributes to the formation of the future oral surface of the sea urchin adult (Aihara and Amemiya 2001). The left pouch splits into three smaller sacs. A duct-like structure, the hydroporic canal, extends from the anterior left coelomic pouch to the aboral ectoderm where the hydropore forms (Gustafson and Wolpert 1963). This hydroporic canal is covered by cilia and could be an excretory organ of the larvae (Hara et al. 2003) and later differentiates into a part of the adult water vascular system (Hyman 1955). The hydroporic canal formation constitutes the first morphological signature of left–right asymmetry in the pluteus larva (Luo and Su 2012). An invagination from the ectoderm fuses with the intermediate sac to form the imaginal rudiment, from which the pentaradial symmetry of the adult body plan is established (Smith et al. 2008). To facilitate the observation and the study of complex phases of development, a larval staging schematic of *Strongylocentrotus purpuratus* has been proposed (Smith et al. 2008). This schematic subdivides larval life into seven stages: 1) four-arm stage, 2) eight-arm stage, 3) vestibula invagination stage, 4) rudiment initiation stage, 5) pentagonal disc stage, 6) advanced rudiment stage and 7) tube-foot protrusion stage.

In the late gastrula, primary germ cells located in the archenteron tip incorporate into the imaginal rudiment. Skeletogenic mesenchyme cells penetrate the rudiment to produce the first skeletal plates of the future adult endoskeleton (Gilbert 2006). The rudiment separates from the rest of the larva during metamorphosis, reorganizes its digestive tract and then settles on the ocean floor, where the miniature sea urchin juvenile starts a benthic life.

This mode of development, however, is not universal among echinoids (reviewed in Raff 1987). Indeed, many sea urchins endowed with large eggs bypass the pluteus stage and directly form a non-feeding larva. For instance, *Peronella japonica*, a species that possesses 300-µm-diameter eggs, produces a partial pluteus with a variable skeleton but no larval gut. *Heliocidaris erythrogramma* produces from a 450-µm-diameter egg a free-floating larva but lacks any relic pluteus structure except for the vestibule. The sea urchin *Abatus cordatus*, with a 1,300-µm-diameter egg, undergoes direct development in a brood chamber placed inside the mother.

17.6 ANATOMY OF THE ADULT SEA URCHIN

A regular adult sea urchin resembles a sphere densely covered with spines. Animal size usually varies between 5 and 12 cm, but *Echinocyamus scaber*, an irregular echinoid, is the smallest known species (6 mm in size). The largest one is the red sea urchin *Mesocentrotus franciscanus* (syn. *Strongylocentrotus franciscanus*), with a body diameter of 15 to 17 cm and spines up to 30 cm.

Adult sea urchins exhibit a pentaradial symmetry with five equally sized parts radiating out of a central axis. The body is divided into radial (= ambulacral) and interradial (= interambulacral) alternate sectors. The mouth is present in the ventral side, and the anus appears in the dorsal—or aboral—region. The body plan is therefore organized around an oral–aboral axis, with no cephalic structures. Irregular echinoids, which include many species used for biological studies (Hibino et al. 2019), deviate from this regular pattern and belong to different clades such as the cidaroids (Order Cidaroida), the clypeasteroids (also known as sand dollars; Order Clypeasteroida) and the spatangoids (also known as heart urchins; Order Spatangoida). In these species, the anus and often also the mouth are no longer present in the two poles of the animal, generating a bilateral symmetry. Whereas regular sea urchins live often on rocky or sandy substrates, most of the irregular sea urchins are burrowing animals that dig in the sediment thanks to their specialized spines.

Sea urchins—like other echinoderms—have a dermaskeleton, which is a thin shell consisting of separate plates of hard calcite that is produced by mesenchyme cells of mesodermal origin. This dermaskeleton, called the test, is made of living cells surrounded by both organic and inorganic extracellular matrices. This calcium carbonate shell (mainly formed by $CaCO_3$) displays a specific three-dimensional organization known as stereom (an echinoderm synapomorphy). The cells constituting the stroma fill the open spaces of these stereomic structures with their mineral secretions. In echinoids, the plates forming the test are tightly apposed and bound together by connective tissue, generating a resistant armoured structure. A thin dermis and epidermis cover the dermaskeleton, which often bears protruding tubercles and rows of spines (in fact, the term echinoderm means in Greek "spiny skin"). The form and size of these spines are extremely variable. The base of the spine is attached by different sets of muscles capable of orienting the spine in different directions. The spine base contains a collagen matrix that can reversibly change its configuration and become flexible or rigid, which allows immobilizing the spine in one particular direction.

The ambulacra of most echinoderms, including echinoids, consist of longitudinal rows of tube feet (podia) protruding out of the test. Sea urchin adoral podia are highly specialized organs that have evolved to provide an efficient attachment to the substratum. These feet generally secrete in their tips a series of adhesive proteins sticking to different supports. Podia are the external appendages of the water vascular system and consequently can be hydraulically extended or contracted. This sophisticated hydraulic system consists of five radially arranged channels connected to a central ring channel surrounding the mouth. Water enters the system through the madreporite, a plate with a light-colored calcareous opening placed on the aboral side. The madreporite filters the seawater, which passes over a short stone channel and joins the ring channel. The tube feet are connected to five main radial channels by a network of lateral branches. Feet have two parts: the ampulla and the podium. The ampulla is a water-filled sac located inside the test and is flanked by circular and longitudinal muscles. The podium protrudes out of the test and is surrounded by a sheet of longitudinal muscles. When the muscles around the ampulla contract, water flows into the connected podium, inducing elongation. On the contrary, when the podia muscles contract, water returns to the ampulla and the podia retract. Differing from adoral podia, peristomal podia are not involved in adhesion and locomotion and have a sensory role. A large family of genes predicted to act in both chemo- and photoreception is expressed in tube feet or pedicellariae and reveals a complex sensory system in sea urchins (Sea Urchin Genome Sequencing Consortium et al. 2006). Pedicellariae are small claw-shaped structures found on the echinoderm endoskeleton, particularly in Asteroidea and Echinoidea. In some taxa, they are presented as cleaning appendages thought to keep the animal's surface free of parasites, debris and algae. Four primary forms of pedicellariae can be found in sea urchins: globiferous, triphyllous, ophicephalous and tridactylous. They typically present a claw shape consisting of three valves that have inspired the production of micro-actuated forceps (Leigh et al. 2012). Appendages, including tube feet, spines, pedicellariae and gills, are all present on the surface of the sea urchin. They present a broad diversity of shapes and offer a fantastic and strange spectacle under a simple dissecting microscope (for an excellent illustration of the different appendage types classified according to Hyman 1955, see Figure 4 of Burke et al. 2006).

These appendages are richly innervated sensory organs allowing sea urchins to interact with their environment (Yoshimura et al. 2012). Like other echinoderms, the sea urchin nervous system is dispersed, but it cannot be reduced to a loose neuron network. Although the adult is not cephalized, the radial nerve presents a segmental organization. The adult sea urchin nervous system is composed of five radial cords. They extend underneath the ambulacra and join their base by commissures that form the circumoral nerve ring, placed around the oesophagus next to the mouth (for a review, see Burke et al. 2006; Yoshimura et al. 2012). The radial nerve produces a series of extensions that pass through the test's pores and innervate the base of each appendage. Almost all tissues, including the viscera, are innervated (Burke et al. 2006), but the echinoderm nervous system is one of the least well studied among metazoans. Until the publication of a genomic view of the sea urchin nervous system (Burke et al. 2006), our knowledge about the echinoid nervous system relied exclusively on morphological studies. Now new lines of investigation have opened (Garcia-Arraras et al. 2001). The sea urchin genome encodes for all the regulatory proteins involved in neuronal specification, and many potential neuromodulators, neuropeptides and growth factors have been described, indicating that the echinoids use these modes of cell communication and regulation (Wood et al. 2018). While tube feet are non-ocular appendages, they do show localized expression of a set of retinal genes and many chemoreceptors, suggesting that they could be involved in light perception (Burke et al. 2006) and a wide range of other sensory modalities.

The digestive tract of echinoids is classically subdivided into different sections: mouth, buccal cavity, pharynx, oesophagus, stomach, intestine, rectum and anus (Hinman and Burke 2018). Sea urchins are benthic animals and eat organic matter that settles down from the column water, mainly preferring kelp, algae and sponges present in their habitat. Most of the irregular sea urchins, which live within the sediment, feed on its organic fraction. Sea urchins living in seaweed meadows graze and ingest macroalgae, including associated epibionts and microbiota (Burke et al. 2006). Echinoids possess a very sophisticated chewing apparatus, the lantern of Aristotle, which encircles the mouth opening and the pharynx. The lantern is composed of a pentamerous skeleton, including five teeth animated by a well-developed musculature (Ziegler et al. 2010). Sea urchins have an open circulatory system with an extensive body cavity filled with coelomic fluid. Passive gas exchange in the coelomic fluid takes place through gill-like appendages located around the mouth. Coelomocytes are free cells that are found in coelomic fluid and also among the tissue of various body parts. These cells are believed to play different functions, including nutrient transport and immune defence (Hakim et al. 2016).

In regular echinoids, sexes are separated, but the external morphology of males and females is indistinguishable. In the case of spatangoids, sexual dimorphism is apparent in the genital papillae; however, observing these structures is challenging, as they hide between the spines forming the apical system (Stauber 1993). The most prominent structures of the internal cavity of sea urchins are their five gonads (ovary or testis). These organs differentiate from a group of cells—the gonadal primordium—located in the dorsal mesentery of the newly metamorphosed juvenile (Chia and Xing 1996; Houk and Hinegardner 1980). The gonads are distinct organs delimited by a peritoneum; their innermost tissue layer contains the germinal epithelium. Each gonad forms a gonoduct joining the genital pore, an opening in the genital plates present on the aboral side of the animal. At the spawning period, eggs or sperm are released through these five genital pores. The group of Gary Wessel has extensively studied germ cell formation during echinoid development (for reviews, see Wessel et al. 2014; Swartz and Wessel 2015). The specification of these cells seems to be regulated by a conserved set of genes that include several classic germline markers such as Vasa, Nanos and Piwi. The germline cells derive from the small micromeres (Yajima and Wessel 2011), which appear early in embryogenesis during the fifth cleavage.

17.7 GENOMIC DATA OF ECHINODERMS

Strongylocentrotus purpuratus was the first fully sequenced echinoderm (Sea Urchin Genome Sequencing Consortium et al. 2006). It was also the first non-chordate deuterostome genome, allowing the characterization of gene family evolutionary dynamics within the Bilateria and Deuterostomia. The sea urchin genome contains roughly 23,300 genes representing nearly all vertebrate gene families without extensive redundancy. Some genes previously considered vertebrate exclusive were found in the sea urchin genome, tracing their origin back to the deuterostome lineage. Since its first release, the genome's assembly has been improved, and the latest release in 2019 is the v5.0 genome. Other echinoderm genomes have been sequenced following this pioneering work, including different representatives of each class. The echinoderm genomes available in the NCBI genome dataset are listed in Table 17.3.

The genome dataset is completed by a vast amount of RNA-Seq data that are accumulating at a steady pace. There are currently over 4,000 Echinodermata high-throughput datasets archived in the NCBI Sequence Read Archive (SRA) database. They are organized in 345 BioProjects (search in November 2020) concerning both nuclear and mitochondrial genomes and are useful for phylogenomic analysis, transcriptome analysis of developmental stages and adaptation to stress or climate change. Table 17.2 presents the omics availability in the different Echinoidea species listed in the biogeographic map and phylogenetic tree shown previously.

An important resource for biologists working on echinoderms is the Echinoderm genome database EchinoBase (www.echinobase.org, and its former version at legacy.echinobase.org; Kudtarkar and Cameron 2017). Originally set up for the annotation of the *S. purpuratus* genome, it has incorporated data for several other echinoderm species, and nowadays, it constitutes a crucial tool for studies on gene regulation, evolution and developmental and cellular biology.

Other useful databases are HpBase (devoted to the Asian sea urchin *H. pulcherrimus*; cell-innovation.nig.ac.jp/Hpul; Kinjo et al. 2018, and EchinoDB, comparative transcriptomics on 42 species of echinoderms; echinodb.uncc.edu; Janies et al. 2016).

TABLE 17.3
Echinodermata Genomes Available at the NCBI Genome Database (www.ncbi.nlm.nih.gov/datasets/)

	Genome Size (Mbp)	NCBI Latest Assembly	Year	Other Database
Echinoidea				
Strongylocentrotus purpuratus	921	Spur_5.0	2019	Echinobase: www.echinobase.org*
Lytechinus variegatus	1061	Lvar_3.0	2021	www.echinobase.org*
Eucidaris tribuloides	2187	Etri_1.0	2015	legacy.echinobase.org*
Hemicentrotus pulcherrimus	568	HpulGenome_v1	2018	HpBase: cell-innovation.nig.ac.jp/Hpul/**
Holothuroidea				
Actinopyga echinites	899	ASM1001598v1	2020	—
Apostichopus japonicus	804	ASM275485v1	2017	—
Apostichopus leukothele	480	ASM1001483v1	2020	—
Australostichopus mollis	1252	assembly_1.0	2020	—
Holothuria glaberrima	1128	ASM993650v1	2020	—
Paelopatides confundens	1128	ASM1131785v1	2020	—
Stichopus horrens	689	UKM_Sthorr_1.1	2019	—
Asteroidea				
Acanthaster planci	384	OKI-Apl_1.0	2016	Echinobase: www.echinobase.org*
Asterias rubens	417	eAstRub1.3	2020	www.echinobase.org*
Patiria miniata	811	Pmin_3.0	2020	www.echinobase.org*
Patiriella regularis	949	assembly_1.0	2017	—
Pisaster ochraceus	401	ASM1099431v1	2020	—
Ophuiroidea				
Ophionereis fasciata	1185	assembly_1.0	2017	—
Ophiothrix spiculata	2764	Ospi.un_1.0	2015	legacy.echinobase.org*
Crinoidea				
Anneissia japonica	589	ASM1163010v1	2020	Echinobase: www.echinobase.org*

*(Kudtarkar and Cameron 2017);
**(Kinjo et al. 2018)

17.8 FUNCTIONAL APPROACHES: TOOLS FOR MOLECULAR AND CELLULAR ANALYSES

Their external fertilization, the large number of gametes, the easy access to all stages of embryogenesis and the transparency of both eggs and embryos make echinoderms suitable organisms for different approaches in cellular biology, biochemistry and molecular biology. The availability of genome and transcriptome data (see genomic resources section) has facilitated gene expression analysis and manipulation in many sea urchin species and other echinoderms.

Spatial and temporal localization of mRNAs has been investigated by *in situ* hybridizations in several sea urchin species (Erkenbrack et al. 2019), as well as in other echinoderms (Fresques et al. 2014; Dylus et al. 2016; Yu et al. 2013). Localization of proteins at the cellular and embryonic levels by immunolocalization is often dependent on the availability of cross-reacting antibodies directed against vertebrate homologs of the protein of interest. Many commercial antibodies against mammalian proteins have indeed helped to decipher different molecular processes in sea urchins, such as microtubule dynamics and Cyclin B/CDK1 complex activity during embryonic divisions (see Figure 17.6). However, some specific antibodies directed against sea urchin proteins have also been developed in many laboratories (Venuti et al. 2004). The function of many molecular players and signalling pathways has also been investigated using different pharmacological inhibitors or activators (Mulner-Lorillon et al. 2017; Molina et al. 2017; Feizbakhsh et al. 2020). Finally, labelling of eggs and embryos with radioactive and non-radioactive precursors allows for the monitoring of metabolic activities (for example, protein synthesis; Chassé et al. 2019).

Manipulation of gene function and/or expression during embryogenesis is achieved by the microinjection of various reagents, such as exogenous mRNA coding for native proteins and dominant-negative forms, morpholinos that interfere with the translation or splicing of endogenous mRNAs and, more recently, CRISPR-Cas9 reagents permitting gene knock-out. Microinjection represents, thus far, the only way to efficiently introduce reagents into the sea urchin eggs or blastomeres. Several recently published methods have described microinjection techniques and applications (von Dassow et al. 2019; Molina et al. 2019; Chassé et al. 2019).

The genome-editing CRISPR/Cas9 technology has been successfully implemented in sea urchin to efficiently knockout developmental genes. So far, the genes targeted by CRISPR/Cas9 were selected because of a visible F0 phenotype: disruption of dorsoventral patterning for Nodal knockdown (Lin and Su 2016) or albinism as a visual readout for polyketide synthase 1 (Oulhen and Wessel 2016). Recently, the successful

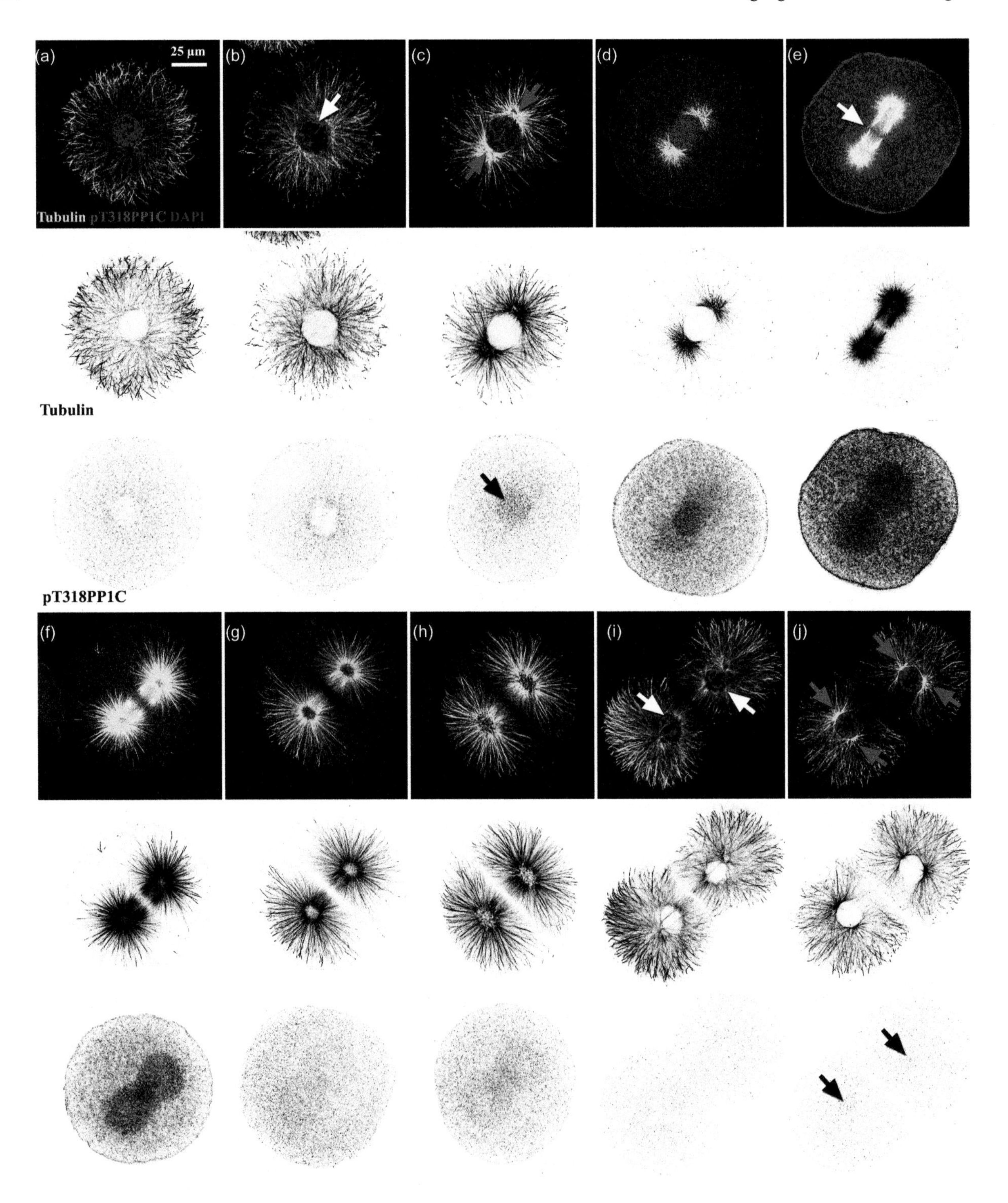

FIGURE 17.6 First mitotic division in *Sphaerechinus granularis* embryos. (a–j) Confocal micrographs describing progression through the first mitotic division in *S. granularis*. Embryos belonging to consecutive stages were labelled with anti-tubulin (shown in green, top panels; b/w, middle panels) and with an antibody against the T318 phosphorylated form of the phosphatase PP1C (red, top panels; b/w, bottom panels). The levels of this phospho-epitope reflect the activity of the Cyclin B/CDK1 complex (Chassé et al. 2016; Feizbakhsh et al. 2020). Nuclear DNA was labelled with DAPI (blue, top panels). (b) Chromatin condensation starts during early prophase (white arrow). (c) Later on, the phT318PP1C signal starts to accumulate in the nucleus (black arrow). The position of the MTOCs also becomes visible (red arrows). (d) Following the collapse of the microtubule radial network, the mitotic spindle begins to form. (e) During metaphase, the phT318PP1C levels reach their maximum, and the chromosomes align in the metaphasic plate (white arrow). The nuclear envelope has disappeared. (f–h) As sister chromatids separate during anaphase, the astral microtubules fill the entire cytoplasm. In parallel, the levels of phT318PP1C decrease dramatically. (i) Chromatin de-condensation begins in early telophase (white arrows). (j) By late telophase, the MTOCs of each daughter cell are apparent (red arrows). A faint phT318PP1C signal in the nuclei heralds the second mitotic division (black arrows).

production of a homozygous F2 mutant using the CRISPR-Cas9 system was obtained in *Temnopleurus reevesii*, which takes advantage of the relatively short life cycle of this species (Yaguchi et al. 2020; see also Chapter 18). This breakthrough gives us the possibility to implement genetic analyses in the sea urchin model (in species with short generation time) and study the function of many maternal factors and mRNAs.

17.9 CHALLENGING QUESTIONS

During their long evolutionary trajectory, echinoderms have adapted to all sorts of climatic conditions and have colonized most of the ocean floor, from the intertidal areas to the deep-sea benthos. The study of this adaptation capacity has just begun and should foster many exciting discoveries. Moreover, it has become evident that echinoderms constitute a valuable biological system to analyze the potential of marine species to adapt to anthropogenic disturbance. Their population densities are very sensitive to climate change, ocean acidification, eutrophication, overfishing, predatory removal and the introduction of alien species (Uthicke et al. 2009). In addition, this group of animals plays a crucial role in many marine habitats and food webs, and several members of this clade have been recognized as "keystone species" in different ecosystems (Power et al. 1996). Echinoderms have thus acquired an essential place in experimental marine ecology.

As detailed previously, many genomic resources are available nowadays for the researchers studying this clade, in particular for those interested in the analysis of the echinoids. These resources have greatly facilitated the development of comparative approaches aimed at understanding the genetic basis of adaptive traits. The density of available landmarks, including closely related species but also different groups separated by increasing phylogenetic distances, allows dissection at the molecular level of both micro- and macroevolutionary processes.

In echinoderms, many studies have focused on the acquisition of evolutionary novelties and the diversification of life strategies. For instance, it has been shown that several species have significantly accelerated their life cycles, reprogramming their ancestral planktotrophic larvae into non-feeding lecithotrophic forms (Raff and Byrne 2006). These evolutionary transitions have obvious adaptive roles. In lecithotrophic species, the life cycle becomes independent of fluctuations in plankton levels since their development relies on the nutrients supplied by their mothers. Indeed, it has been argued that the disturbance of planktonic food chains could contribute in the near future to the decline of planktotrophic species (Uthicke et al. 2009). At the same time, these developmental transitions can now be analyzed in great detail both at the cellular and molecular levels.

For instance, it has been shown in the *Heliocidaris* genus that the eggs of lecithotrophic species have undergone an outstanding increase in size, driven by a thorough remodelling of their oogenesis program (Byrne et al. 1999). Moreover, the comparison of gene regulatory networks controlling early development, like in the lecithotrophic *Heliocidaris erythrogramma* and the planktotrophic *Heliocidaris tuberculata*, provides important hints about the identity of the molecular players participating in evolutionary change (Israel et al. 2016). These approaches have greatly benefited from the deep knowledge of developmental networks acquired thanks to the study of early development in *Strongylocentrotus purpuratus* and other echinoderm species (Cary and Hinman 2017).

Echinoderm biology stands now at the intersection between ecology, cell and developmental and evolutionary biology and should greatly profit from this privileged position.

ACKNOWLEDGEMENTS AND FUNDING

We apologize to those whose work was not cited or discussed here because of the broad scope of this review and space limitations. We are indebted to the Marine and Diving Facility and the Aquarium Service of the Roscoff Marine Station for echinoderm collection and rearing. We are also grateful to the imaging platform of the FR2424 (Plateforme MerImage, SU/CNRS). We thank H. Flom for manuscript corrections. The authors acknowledge the support of "La Ligue contre le Cancer (coordination du Grand Ouest: comités Finistère, Côtes d'Armor, Deux-Sèvres, Morbihan, Ille-et-Villaine, Loire Atlantique, Charente, Sarthe)", the Brittany Regional Council (Région Bretagne), the Finistère Department Council (CG29). Florian Pontheaux is funded by a PhD fellowship from the French Education Ministry [SU; Doctoral School ED515].

BIBLIOGRAPHY

Aihara, M. and S. Amemiya. 2001. Left-right positioning of the adult rudiment in sea urchin larvae is directed by the right side. *Development* 128:4935–4948.

Allen, R.D. and J.L. Griffin. 1958. The time sequence of early events in the fertilization of sea urchin eggs. *Experimental Cell Research* 15:163–173.

Andrikou, C., Iovene, E., Rizzo, F., Oliveri, P. and M.I. Arnone. 2013. Myogenesis in the sea urchin embryo: The molecular fingerprint of the myoblast precursors. *EvoDevo* 4:33–48.

Angerer, L.M. and R.C. Angerer. 2000. Animal-vegetal axis patterning mechanisms in the early sea urchin embryo. *Developmental Biology* 218:1–12.

Angione, S.L., Oulhen, N., Brayboy, L.M., Tripathi, A. and G.M. Wessel. 2015. Simple perfusion apparatus for manipulation, tracking, and study of oocytes and embryos. *Fertility and Sterility* 103:281–290.

Anstrom, J.A. 1992. Microfilaments, cell shape changes, and the formation of primary mesenchyme in sea urchin embryos. *Journal of Experimental Zoology* 264:312–322.

Armstrong, N. and D.R. McClay. 1994. Skeletal pattern is specified autonomously by the primary mesenchyme cells in sea urchin embryos. *Developmental Biology* 162:329–338.

Arnone, M.I., Byrne, M. and P. Martinez. 2015. Echinodermata. In *Evolutionary Developmental Biology of Invertebrates*, ed. A. Wanninger, 6:1–58. Springer-Verlag, Wien.

Ben-Tabou de-Leon, S. 2017. The network remains. *History and Philosophy of the Life Sciences* 39:32.

Ben-Tabou de-Leon, S. 2016. Robustness and accuracy in sea urchin developmental gene regulatory networks. *Frontiers in Genetics* 7:16–21.

Bernard, F.R. 1977. Fishery and reproductive cycle of the red sea urchin, strongylocentrotus franciscanus, in British Columbia. *Journal of the Fisheries Board of Canada* 34:604–610.

Binks, R.M., Prince, J., Evans, J.P. and W.J. Kennington. 2012. More than bindin divergence: Reproductive isolation between sympatric subspecies of a sea urchin by asynchronous spawning. *Evolution* 66:3545–3557.

Bisgrove, B.W., Andrews, M.E. and R.A. Raff. 1991. Fibropellins, products of an EGF repeat-containing gene, form a unique extracellular matrix structure that surrounds the sea urchin embryo. *Developmental Biology* 146:89–99.

Böhmer, M., Van, Q., Weyand, I., Hagen, V., Beyermann, M., Matsumoto, M., Hoshi, M., Hildebrand, E. and U.B. Kaupp. 2005. Ca^{2+} spikes in the flagellum control chemotactic behavior of sperm. *EMBO J* 24:2741–2752.

Bolton, T.F., Thomas, F.I. and C.N. Leonard. 2000. Maternal energy investment in eggs and jelly coats surrounding eggs of the echinoid *Arbacia punctulata*. *The Biological Bulletin* 199:1–5.

Bosch, I., Beauchamp, K.A., Steele, M.E. and J.S. Pearse. 1987. Development, metamorphosis, and seasonal abundance of embryos and larvae of the Antarctic sea urchin *Sterechinus neumayeri*. *The Biological Bulletin* 173:126–135.

Boveri, T. 2008. Concerning the origin of malignant tumours by Theodor Boveri: Translated and annotated by Henry Harris. *Journal of Cell Science* 121:1–84.

Boveri, T. 1902. Über mehrpolige Mitosen als Mittel zur Analyse des Zellkerns. [Concerning multipolar mitoses as a means of analysing the cell nucleus]. *Verhandlungen der Physikalisch-Medizinischen Gesellschaft zu Würzburg* 35:67–90.

Brachet, J. 1975. From chemical to molecular sea urchin embryology. *American Zoologist* 15:485–491.

Brachet, J. 1941. La localisation des acides pentose-nucléiques dans les tissus animaux et les oeufs d'amphibiens en voie de développement. *Archive de Biologie* 53:207–257.

Briggs, E. and G.M. Wessel. 2006. In the beginning . . . animal fertilization and sea urchin development. *Developmental Biology* 300:15–26.

Bronstein, O. and A. Kroh. 2019. The first mitochondrial genome of the model echinoid *Lytechinus variegatus* and insights into Odontophoran phylogenetics. *Genomics* 111:710–718.

Burke, R.D. and C.M. Alvarez. 1988. Development of the esophageal muscles in embryos of the sea urchin *Strongylocentrotus purpuratus*. *Cell Tissue Research* 252:411–417.

Burke, R.D., Angerer, L.M., Elphick, M.R., Humphrey, G.W., Yaguchi, S., Kiyama, T., Liang, S., Mu, X., Agca, C., Klein, W.H., Brandhorst, B.P., Rowe, M., Wilson, K., Churcher, A.M., Taylor, J.S., Chen, N., Murray, G., Wang, D., Mellott, D., Olinski, R., Hallböök, F. and M.C. Thorndyke. 2006. A genomic view of the sea urchin nervous system. *Developmental Biology* 300:434–460.

Burke, R.D., Myers, R.L., Sexton, T.L. and C. Jackson. 1991. Cell movements during the initial phase of gastrulation in the sea urchin embryo. *Developmental Biology* 146:542–557.

Byrne, M. 1990. Annual reproductive cycles of the commercial sea urchin *Paracentrotus lividus* from an exposed intertidal and a sheltered subtidal habitat on the west coast of Ireland. *Marine Biology* 104:275–289.

Byrne, M., Villinski, J.T., Cisternas, P., Siegel, R.K., Popodi, E. and R.A. Raff. 1999. Maternal factors and the evolution of developmental mode: Evolution of oogenesis in *Heliocidaris erythrogramma*. *Developmental Genes and Evolution* 209:275–283.

Byrnes, W.M. and W.R. Eckberg. 2006. Ernest Everett Just (1883–1941): An early ecological developmental biologist. *Developmental Biology* 296:1–11.

Byrnes, W.M. and S.A. Newman. 2014. Ernest Everett Just: Egg and embryo as excitable systems. *Journal of Experimental Zoology. PartB, Molecular and Developmental Evolution* 322B:191–201.

Cameron, R.A., Fraser, S.E., Britten, R.J. and E.H. Davidson. 1991. Macromere cell fates during sea urchin development. *Development* 113:1085–1091.

Cannon, J.T., Kocot, K.M., Waits, D.S., Weese, D.A., Swalla, B.J., Santos, S.R. and K.M. Halanych. 2014. Phylogenomic resolution of the hemichordate and echinoderm clade. *Current Biology* 24:2827–2832.

Cary, G.A. and V.F. Hinman. 2017. Echinoderm development and evolution in the post-genomic era. *Developmental Biology* 427:203–211.

Cavalieri, V. and G. Spinelli. 2015. Symmetry breaking and establishment of dorsal/ventral polarity in the early sea urchin embryo. *Symmetry* 7:1721–1733.

Chassé, H., Aubert, J., Boulben, S., Le Corguille, G., Corre, E., Cormier, P. and J. Morales. 2018. Translatome analysis at the egg-to-embryo transition in sea urchin. *Nucleic Acids Research* 46:4607–4621.

Chassé, H., Boulben, S., Costache, V., Cormier, P. and J. Morales. 2017. Analysis of translation using polysome profiling. *Nucleic Acids Research* 45:e15. doi: 10.1093/nar/gkw907.

Chassé, H., Boulben, S., Glippa, V., Pontheaux, F., Cormier, P. and J. Morales. 2019. In vivo analysis of protein translation activity in sea urchin eggs and embryos. In *Methods in Cell Biology, Echinoderms, Part B*, ed. A. Hamdoun and K.R. Foltz, 150:335–352. Elsevier; Academic Press, London UK.

Chassé, H., Mulner-Lorillon, O., Boulben, S., Glippa, V., Morales, J. and P. Cormier. 2016. Cyclin B translation depends on mTOR activity after fertilization in sea urchin embryos. *PLoS One* 11:e0150318. https://doi.org/10.1371/journal.pone.0150318.

Chenuil, A. and J.-P. Féral. 2003. Sequences of mitochondrial DNA suggest that *Echinocardium cordatum* is a complex of several sympatric or hybridizing species: A pilot study. In *Echinoderm Research 2001*, ed. J.-P. Féral and B. David, 15–21. Proc. 6th Eur. Conf. Echinoderm, Banyuls-sur-mer, France. Swets & Zeitlinger Publishers, Lisse, NL.

Chia, F.S. and Xing, J. 1996. Echinoderm coelomocytes. *Zoological Studies* 35:231–254.

Clift, D. and Schuh, M. 2013. Restarting life: Fertilization and the transition from meiosis to mitosis. *Nature Reviews: Molecular Cell Biology* 14:549–562.

Cormier, P., Chassé, H., Cosson, B., Mulner-Lorillon, O. and J. Morales. 2016. Translational control in echinoderms: The calm before the storm. In *Evolution of the Protein Synthesis Machinery and Its Regulation*, ed. G. Hernandez and R. Jagus, 413–434. Springer International Publishing, Switzerland.

Costache, V., McDougall, A. and R. Dumollard. 2014. Cell cycle arrest and activation of development in marine invertebrate deuterostomes. *Biochemical and Biophysical Research Communications* 450:1175–1181.

Costello, D.P. and C. Henley. 1971. Methods for obtaining and handling marine eggs and embryos. *Marine Biological Laboratory*. https://hdl.handle.net/1912/295.

Croce, J.C. and D.R. McClay. 2010. Dynamics of Delta/Notch signaling on endomesoderm segregation in the sea urchin embryo. *Development* 137:83–91.

Darszon, A., Nishigaki, T., Wood, C., Treviño, C.L., Felix, R. and C. Beltrán. 2005. Calcium channels and Ca^{2+} fluctuations in sperm physiology. *International Review of Cytology* 243:79–172.

Davidson, E.H. 2006. Gene regulatory networks and the evolution of animal body plans. *Science* 311:796–800.

Davidson, E.H. 2002. A genomic regulatory network for development. *Science* 295:1669–1678.

Davidson, P.L., Guo, H., Wang, L., Berrio, A., Zhang, H., Chang, Y., Soborowski, A.L., McClay, D.R., Fan, G. and G.A. Wray. 2020. Chromosomal-level genome assembly of the sea urchin *Lytechinus variegatus* substantially improves functional genomic analyses. *Genome Biology and Evolution* 12:1080–1086.

Delalande, C., Monnier, A., Minella, O., Geneviere, A.M., Mulner-Lorillon, O., Belle, R. and P. Cormier. 1998. Developmental regulation of elongation factor-1 delta in sea urchin suggests appearance of a mechanism for alternative poly(A) site selection in gastrulae. *Experimental Cell Research* 242:228–234.

Delorme, N.J. and M.A. Sewell. 2016. Effects of warm acclimation on physiology and gonad development in the sea urchin *Evechinus chloroticus*. *Comparative Biochemistry and Physiology Part A: Molecular & Integrative Physiology* 198:33–40.

Derbès, A.A. 1847. Observations sur le méchanisme et les phenomènes qui accompagnent la formation de l'embryon chez l'oursin comestible. *Annales des Sciences Naturelles (Zool)* 8:80–98.

Dilly, G.F., Gaitán-Espitia, J.D. and G.E. Hofmann. 2015. Characterization of the Antarctic sea urchin (*Sterechinus neumayeri*) transcriptome and mitogenome: A molecular resource for phylogenetics, ecophysiology and global change biology. *Molecular Ecology Resources* 15:425–436.

Drozdov, A.L. and V.V. Vinnikova. 2010. Morphology of gametes in sea urchins from Peter the Great Bay, Sea of Japan. *Ontogenez* 41:47–57.

Duboc, V., Röttinger, E., Besnardeau, L. and T. Lepage. 2004. Nodal and BMP2/4 signaling organizes the oral-aboral axis of the sea urchin embryo. *Developmental Cell*. 6:397–410.

Dupont, G. and R. Dumollard. 2004. Simulation of calcium waves in ascidian eggs: Insights into the origin of the pacemaker sites and the possible nature of the sperm factor. *Journal of Cell Science* 117:4313–4323.

Dylus, D.V., Czarkwiani, A., Stångberg, J., Ortega-Martinez, O., Dupont, S. and P. Oliveri. 2016. Large-scale gene expression study in the ophiuroid *Amphiura filiformis* provides insights into evolution of gene regulatory networks. *EvoDevo* 7:2.

Egea, E., Mérigot, B., Mahé-Bézac, C., Féral, J.-P. and A. Chenuil. 2011a. Differential reproductive timing in *Echinocardium* spp.: The first Mediterranean survey allows interoceanic and interspecific comparisons. *Comptes Rendus Biologies* 334:13–23.

Erkenbrack, E.M., Croce, J.C., Miranda, E., Gautam, S., Martinez-Bartolome, M., Yaguchi, S. and R.C. Range. 2019. Whole mount in situ hybridization techniques for analysis of the spatial distribution of mRNAs in sea urchin embryos and early larvae. In *Methods in Cell Biology: Echinoderms, Part B*, ed. A. Hamdoun and K.R. Foltz, 150:177–196. Elsevier; Academic Press, London UK.

Erkenbrack, E.M., Davidson, E.H. and I.S. Peter. 2018. Conserved regulatory state expression controlled by divergent developmental gene regulatory networks in echinoids. *Development* 145:1–11.

Ernst, S.G. 2011. Offerings from an urchin. *Developmental Biology* 358:285–294.

Erwin, D.H., Laflamme, M., Tweedt, S.M., Sperling, E.A., Pisani, D. and K.J. Peterson. 2011. The Cambrian conundrum: Early divergence and later ecological success in the early history of animals. *Science* 334:1091–1097.

Ettensohn, C.A. 2020. The gene regulatory control of sea urchin gastrulation. *Mechanisms of Development* 162:103599.

Ettensohn, C.A. 2017. Sea urchins as a model system for studying embryonic development. In *Reference Module in Biomedical Sciences*, ed M.J. Caplan. Elsevier, Amsterdam.

Ettensohn, C.A. 1999. Cell movements in the sea urchin embryo. *Current Opinion in Genetics & Development* 9:461–465.

Ettensohn, C.A. 1985. Gastrulation in the sea urchin embryo is accompanied by the rearrangement of invaginating epithelial cells. *Developmental Biology* 112:383–390.

Ettensohn, C.A. and D.R. McClay. 1988. Cell lineage conversion in the sea urchin embryo. *Developmental Biology* 125:396–409.

Ettensohn, C.A. and S.W. Ruffins. 1993. Mesodermal cell interactions in the sea urchin embryo: Properties of skeletogenic secondary mesenchyme cells. *Development* 117:1275–1285.

Ettensohn, C.A. and H.C. Sweet. 2000. Patterning the early sea urchin embryo. *Current Topics in Developmental Biology* 50:1–44.

Evans, T., Rosenthal, E.T., Youngblom, J., Distel, D. and T. Hunt. 1983. Cyclin: A protein specified by maternal mRNA in sea urchin eggs that is destroyed at each cleavage division. *Cell* 33:389–396.

Feizbakhsh, O., Pontheaux, F., Glippa, V., Morales, J., Ruchaud, S., Cormier, P. and F. Roch. 2020. A peak of H3T3 phosphorylation occurs in synchrony with mitosis in sea urchin early embryos. *Cells* 9:898.

Filander, Z. and C. Griffiths. 2017. Illustrated guide to the echinoid (Echinodermata: Echinoidea) fauna of South Africa. *Zootaxa* 4296:1–72.

Fink, R.D. and D.R. McClay. 1985. Three cell recognition changes accompany the ingression of sea urchin primary mesenchyme cells. *Developmental Biology* 107:66–74.

Finkel, T. 2011. Signal transduction by reactive oxygen species. *The Journal of Cell Biology* 194:7–15.

Floc'hlay, S., Molina, M.D., Hernandez, C., Haillot, E., Thomas-Chollier, M., Lepage, T. and D. Thieffry. 2021. Deciphering and modelling the TGF-β signalling interplays specifying the dorsal-ventral axis of the sea urchin embryo. *Development* 148:dev189944.

Foerder, C.A. and B.M. Shapiro. 1977. Release of ovoperoxidase from sea urchin eggs hardens the fertilization membrane with tyrosine crosslinks. *Proceedings of the National Academy of Sciences of the United States of America* 74:4214–4218.

Fol, H. 1879. Recherches sur la fécondation et le commencement de l'hénogénie chez divers animaux. *Mémoires de la Société de Physique et d'Histoire Naturelle (Geneve)* 26:89–250.

Foo, S.A., Deaker, D. and M. Byrne. 2018. Cherchez la femme impact of ocean acidification on the egg jelly coat and attractants for sperm. *Journal of Experimental Biology* 221:jeb177188.

Fresques, T., Zazueta-Novoa, V., Reich, A. and G.M. Wessel. 2014. Selective accumulation of germ-line associated gene products in early development of the sea star and distinct differences from germ-line development in the sea urchin: Sea star germ-line gene expression. *Developmental Dynamics* 243:568–587.

Gago, J. and O.J. Luís. 2011. Comparison of spawning induction techniques on *Paracentrotus lividus* (Echinodermata: Echinoidea) broodstock. *Aquaculture International* 19:181–191.

Gaitán-Espitia, J.D., Sánchez, R., Bruning, P. and L. Cárdenas. 2016. Functional insights into the testis transcriptome of the edible sea urchin *Loxechinus albus*. *Scientific Reports* 6:36516.

Garcia-Arraras, J.E., Rojas-Soto, M., Jimenez, L.B. and L. Diaz-Miranda. 2001. The enteric nervous system of echinoderms: Unexpected complexity revealed by neurochemical analysis. *Journal of Experimental Biology* 204:865–873.

Ghisaura, S., Loi, B., Biosa, G., Baroli, M., Pagnozzi, D., Roggio, T., Uzzau, S., Anedda, R. and M.F. Addis. 2016. Proteomic changes occurring along gonad maturation in the edible sea urchin *Paracentrotus lividus*. *Journal of Proteomics* 144:63–72.

Gianguzza, P. and C. Bonaviri. 2013. Chapter 19: Arbacia. In *Developments in Aquaculture and Fisheries Science, Sea Urchins: Biology and Ecology*, ed. J.M. Lawrence, 275–283. Elsevier, London UK.

Gibson, A.W. and R.D. Burke. 1985. The origin of pigment cells in embryos of the sea urchin *Strongylocentrotus purpuratus*. *Developmental Biology* 107:414–419.

Gilbert, S.F. 2006. *Developmental Biology*, eighth ed. Sinauer Associates Inc., Sunderland, MA, USA.

Gildor, T., Malik, A., Sher, N., Avraham, L. and S. Ben-Tabou de-Leon. 2016. Quantitative developmental transcriptomes of the Mediterranean sea urchin *Paracentrotus lividus*. *Marine Genomics* 25:89–94.

Gillard, G.B., Garama, D.J. and C.M. Brown. 2014. The transcriptome of the NZ endemic sea urchin Kina (*Evechinus chloroticus*). *BMC Genomics* 15:45.

Goldstein, B. and G. Freeman. 1997. Axis specification in animal development. *Bioessays* 19:105–116.

González-Irusta, J.M., Goñi de Cerio, F. and J.C. Canteras. 2010. Reproductive cycle of the sea urchin *Paracentrotus lividus* in the Cantabrian Sea (Northern Spain): Environmental effects. *Journal of the Marine Biological Association of the United Kingdom* 90:699–709.

Gross, P.R. and G.H. Cousineau. 1963. Effects of actinomycin D on macromolecule synthesis and early development in sea urchin eggs. *Biochemical and Biophysical Research Communications* 10:321–326.

Gross, P.R., Malkin, L.I. and W.A. Moyer. 1964. Templates for the first proteins of embryonic development. *Proceedings of the National Academy of Sciences of the United States of America* 51:407–414.

Guillou, M. and L.J.L. Lumingas. 1998. The reproductive cycle of the 'blunt' sea urchin. *Aquaculture International* 6:147–160.

Guillou, M. and C. Michel. 1993. Reproduction and growth of *Sphaerechinus granularis* (Echinodermata: Echinoidea) in Southern Brittany. *Journal of the Marine Biological Association of the United Kingdom* 73:179–192.

Gustafson, T. and H. Kinnander. 1956. Microaquaria for time-lapse cinematographic studies of morphogenesis in swimming larvae and observations on sea urchin gastrulation. *Experimental Cell Research* 11:36–51.

Gustafson, T. and L. Wolpert. 1967. Cellular movement and contact in sea urchin morphogenesis. *Biological Reviews* 42:442–498.

Gustafson, T. and L. Wolpert. 1963. Studies on the cellular basis of morphogenesis in the sea urchin embryo: Formation of the coelom, the mouth, and the primary pore-canal. *Experimental Cell Research* 29:561–582.

Hakim, J.A., Koo, H., Kumar, R., Lefkowitz, E.J., Morrow, C.D., Powell, M.L., Watts, S.A. and A.K. Bej. 2016. The gut microbiome of the sea urchin, *Lytechinus variegatus*, from its natural habitat demonstrates selective attributes of microbial taxa and predictive metabolic profiles. *FEMS Microbiology Ecology* 92:146–157.

Haley, S.A. and G.M. Wessel. 1999. The cortical granule serine protease CGSP1 of the sea urchin, *Strongylocentrotus purpuratus*, is autocatalytic and contains a low-density lipoprotein receptor-like domain. *Developmental Biology* 211:1–10.

Hall, H.G. and V.D. Vacquier. 1982. The apical lamina of the sea urchin embryo: Major glycoproteins associated with the hyaline layer. *Developmental Biology* 89:168–178.

Hamdoun, A., Schrankel, C.S., Nesbit, K.T. and J.A. Espinoza. 2018. Sea urchins as lab animals for reproductive and developmental biology. *Encyclopedia of Reproduction* 6:696–703.

Hara, Y., Kuraishi, R., Uemura, I. and H. Katow. 2003. Asymmetric formation and possible function of the primary pore canal in plutei of *Temnopleurus hardwicki*. *Development, Growth & Differentiation* 45:295–308.

Hardin, J. 1989. Local shifts in position and polarized motility drive cell rearrangement during sea urchin gastrulation. *Developmental Biology* 136:430–445.

Hardin, J. 1988. The role of secondary mesenchyme cells during sea urchin gastrulation studied by laser ablation. *Development* 103:317–324.

Hardin, J. and D.R. McClay. 1990. Target recognition by the archenteron during sea urchin gastrulation. *Developmental Biology* 142:86–102.

Hertwig, O. 1876. Beitrage zur Erkenntnis der Bildung, Befruchtung, und Theilung des tierischen Eies. *Morphologisches Jahrbuch* 1:347–352.

Hibino, T., Minokawa, T. and A. Yamazaki. 2019. Chapter 4: Cidaroids, clypeasteroids, and spatangoids: Procurement, culture, and basic methods. In *Methods in Cell Biology: Echinoderms, Part B*, ed. A. Hamdoun and K.R. Foltz, 150:81–103. Elsevier; Academic Press, London UK.

Himmelman, J.H. 1978. Reproductive cycle of the green sea urchin, *Strongylocentrotus droebachiensis*. *Canadian Journal of Zoology* 56:1828–1836.

Hinman, V.F. and R.D. Burke. 2018. Embryonic neurogenesis in echinoderms. *Wiley Interdisciplinary Reviews. Developmental Biology* 7:e316.

Hogan, J.D., Keenan, J.L., Luo, L., Ibn-Salem, J., Lamba, A., Schatzberg, D., Piacentino, M.L., Zuch, D.T., Core, A.B., Blumberg, C., Timmermann, B., Grau, J.H., Speranza, E., Andrade-Navarro, M.A., Irie, N., Poustka, A.J. and C.A. Bradham. 2020. The developmental transcriptome for *Lytechinus variegatus* exhibits temporally punctuated gene expression changes. *Developmental Biology* 460:139–154.

Horstadius, S. 1939. The mechanics of sea urchin development, studied by operative methods. *Biological Reviews* 14:132–179.

Horton, T., Kroh, A., Ahyong, S., Bailly, N., Boyko, C.B., et al. 2020. *World Register of Marine Species (WoRMS)*. https://doi.org/10.14284/170.

Houk, M.S. and R.T. Hinegardner. 1980. The formation and early differentiation of sea urchin gonads. *The Biological Bulletin* 159:280–294.

Hultin, T. 1961. The effect of puromycin on protein metabolism and cell division in fertilized sea urchin eggs. *Experientia* 17:410–411.

Humphreys, T. 1969. Efficiency of translation of messenger-RNA before and after fertilization in sea urchins. *Developmental Biology* 20:435–458.

Hunt, T. 2002. Nobel lecture: Protein synthesis, proteolysis, and cell cycle transitions. *Bioscience Reports* 22:465–486.

Hyman, L.H. 1955. *The invertebrates, echinodermata (4)*. McGraw-Hill Education, New York, NY.

Ishimoda-Takagi, T., Chino, I. and H. Sato. 1984. Evidence for the involvement of muscle tropomyosin in the contractile

elements of the coelom-esophagus complex in sea urchin embryos. *Developmental Biology* 105:365–376.
Israel, J.W., Martik, M.L., Byrne, M., Raff, E.C., Raff, R.A., McClay, D.R. and G.A. Wray. 2016. Comparative developmental transcriptomics reveals rewiring of a highly conserved gene regulatory network during a major life history switch in the sea urchin genus *Heliocidaris*. *PLoS Biology* 14:e1002391.
Janies, D.A., Witter, Z., Linchangco, G.V., Foltz, D.W., Miller, A.K., Kerr, A.M., Jay, J., Reid, R.W. and G.A. Wray. 2016. EchinoDB, an application for comparative transcriptomics of deeply-sampled clades of echinoderms. *BMC Bioinformatics* 17:48.
Just, E.E. 1939. *The Biology of the Cell Surface*. The Technical Press LTD; Ave Maria Lane, Ludgate Hill, E.C., London.
Just, E.E. 1919. The fertilization reaction in *Echinarachnius parma*: I. Cortical response of the egg to insemination. *The Biological Bulletin* 36:110.
Kahil, K., Varsano, N., Sorrentino, A., Pereiro, E., Rez, P., Weiner, S. and L. Addadi. 2020. Cellular pathways of calcium transport and concentration toward mineral formation in sea urchin larvae. *Proceedings of the National Academy of Sciences of the United States of America* 117:30957–30965.
Kedes, L.H., Chang, A.C.Y., Houseman, D. and S.N. Cohen. 1975. Isolation of histone genes from unfractionated sea urchin DNA by subculture cloning in E. coli. *Nature* 255:533–538.
Kimberly, E.L. and J. Hardin. 1998. Bottle cells are required for the initiation of primary invagination in the sea urchin embryo. *Developmental Biology* 204:235–250.
Kinjo, S., Kiyomoto, M., Yamamoto, T., Ikeo, K. and S. Yaguchi. 2018. HpBase: A genome database of a sea urchin, *Hemicentrotus pulcherrimus*. *Development, Growth & Differentiation* 60:174–182.
Kipryushina, Y.O. and K.V. Yakovlev. 2020. Maternal control of early patterning in sea urchin embryos. *Differentiation* 113:28–37.
Kiyomoto, M., Hamanaka, G., Hirose, M. and M. Yamaguchi. 2014. Preserved echinoderm gametes as a useful and ready-to-use bioassay material. *Marine Environmental Research* 93:102–105.
Kominami, T. and Takata. 2004. Gastrulation in the sea urchin embryo: A model system for analyzing the morphogenesis of a monolayered epithelium. *Development, Growth & Differentiation* 46:309–326.
Kroh, A. and A.B. Smith. 2010. The phylogeny and classification of post-Palaeozoic echinoids. *Journal of Systematic Palaeontology* 8:147–212.
Kudtarkar, P. and R.A. Cameron. 2017. Echinobase: An expanding resource for echinoderm genomic information. *Database* 2017:bax074.
Kuroda, R., Kontani, K., Kanda, Y., Katada, T., Nakano, T., Satoh, Y., Suzuki, N. and H. Kuroda. 2001. Increase of cGMP, cADP-ribose and inositol 1,4,5-trisphosphate preceding Ca^{2+} transients in fertilization of sea urchin eggs. *Development* 128:4405–4414.
Lane, M.C., Koehl, M.A., Wilt, F. and R. Keller. 1993. A role for regulated secretion of apical extracellular matrix during epithelial invagination in the sea urchin. *Development* 117:1049–1060.
Lapraz, F., Besnardeau, L. and T. Lepage. 2009. Patterning of the dorsal-ventral axis in echinoderms: Insights into the evolution of the BMP-chordin signaling network. *PLoS Biology* 7:e1000248.
Láruson, Á.J. 2017. Rates and relations of mitochondrial genome evolution across the Echinoidea, with special focus on the superfamily Odontophora. *Ecology and Evolution* 7:4543–4551.
Leigh, S.J., Bowen, J., Purssell, C.P., Covington, J.A., Billson, D.R. and D.A. Hutchins. 2012. Rapid manufacture of monolithic micro-actuated forceps inspired by echinoderm pedicellariae. *Bioinspiration & Biomimetics* 7:044001.
Lepage, T., Ghiglione, C. and C. Gache. 1992a. Spatial and temporal expression pattern during sea urchin embryogenesis of a gene coding for a protease homologous to the human protein BMP-1 and to the product of the *Drosophila* dorsal-ventral patterning gene tolloid. *Development* 114:147–163.
Lepage, T., Sardet, C. and C. Gache, C. 1992b. Spatial expression of the hatching enzyme gene in the sea urchin embryo. *Developmental Biology* 150:23–32.
Lessios, H.A. 1990. Adaptation and phylogeny as determinants of egg size in echinoderms from the two sides of the isthmus of panama. *The American Naturalist* 135:1–13.
Lessios, H.A. 1988. Temporal and spatial variation in egg size of 13 Panamanian echinoids. *Journal of Experimental Marine Biology and Ecology* 114:217–239.
Levitan, D.R. 1993. The importance of sperm limitation to the evolution of egg size in marine invertebrates. *The American Naturalist* 141:517–536.
Li, E., Cui, M., Peter, I.S. and E.H. Davidson. 2014. Encoding regulatory state boundaries in the pregastrular oral ectoderm of the sea urchin embryo. *Proceedings of the National Academy of Sciences of the United States of America* 111:E906–E913.
Lin, C.-Y. and Y.-H. Su. 2016. Genome editing in sea urchin embryos by using a CRISPR/Cas9 system. *Developmental Biology* 409:420–428.
Loeb, J. 1899. On the nature of the process of fertilization and the artificial production of normal larvæ (plutei) from the unfertilized eggs of the sea urchin. *American Journal of Physiology* 3:135–138.
Logan, C.Y., Miller, J.R., Ferkowicz, M.J. and D.R. McClay. 1999. Nuclear beta-catenin is required to specify vegetal cell fates in the sea urchin embryo. *Development* 126:345–357.
Longo, F.J. and E. Anderson. 1968. The fine structure of pronuclear development and fusion in the sea urchin, *Arbacia punctulata*. *The Journal of Cell Biology* 39:339–368.
Longo, F.J. and W. Plunkett. 1973. The onset of DNA synthesis and its relation to morphogenetic events of the pronuclei in activated eggs of the sea urchin, *Arbacia punctulata*. *Developmental Biology* 30:56–67.
Lowe, C.J., Clarke, D.N., Medeiros, D.M., Rokhsar, D.S. and J. Gerhart. 2015. The deuterostome context of chordate origins. *Nature* 520:456–465.
Luo, Y.-J. and Y.-H. Su. 2012. Opposing nodal and BMP signals regulate left right asymmetry in the sea urchin larva. *PLoS Biol* 10:e1001402.
Maderspacher, F. 2008. Theodor Boveri and the natural experiment. *Current Biology* 18:R279–R286.
Malinda, K.M. and C.A Ettensohn. 1994. Primary mesenchyme cell migration in the sea urchin embryo: Distribution of directional cues. *Developmental Biology* 164:562–578.
Malumbres, M. 2014. Cyclin-dependent kinases. *Genome Biology* 15:122.
Martik, M.L., Lyons, D.C. and D.R. McClay. 2016. Developmental gene regulatory networks in sea urchins and what we can learn from them. *F1000Research* 5:203.
Martik, M.L. and D.R. McClay. 2017. New insights from a high-resolution look at gastrulation in the sea urchin, *Lytechinus variegatus*. *Mechanisms of Development* 148:3–10.
Maruyama, Y.K., Nakaseko, Y. and S. Yagi. 1985. Localization of cytoplasmic determinants responsible for primary

mesenchyme formation and gastrulation in the unfertilized egg of the sea urchin *Hemicentrotus pulcherrimus*. *Journal of Experimental Zoology* 236:155–163.

Mazia, D. 1937. The release of calcium in Arbacia eggs on fertilization. *Journal of Cellular and Comparative Physiology* 10:291–304.

McAlister, J.S. and A.L. Moran. 2012. Relationships among egg size, composition, and energy: A comparative study of geminate sea urchins. *PLoS One* 7:e41599.

McClintock, J.B., Amsler, M.O., Angus, R.A., Challener, R.C., Schram, J.B., Amsler, C.D., Mah, C.L., Cuce, J. and B.J. Baker. 2011. The Mg-calcite composition of Antarctic echinoderms: Important implications for predicting the impacts of ocean acidification. *The Journal of Geology* 119:457–466.

Meidel, S.K. and R.E. Scheibling. 1998. Annual reproductive cycle of the green sea urchin, *Strongylocentrotus droebachiensis*, in differing habitats in Nova Scotia, Canada. *Marine Biology* 131:461–478.

Meves, F. 1912. Verfolgung des sogenannten Mittelstückes des Echinidenspermiums im befruchteten Ei bis zum Ende der ersten Furchungstheilung. [Tracking of the so-called middle section of the echinid sperm in the fertilized egg until the end of the first cleavage division]. *Archiv für mikroskopische Anatomie* 80:811–823.

Minokawa, T. 2017. Comparative studies on the skeletogenic mesenchyme of echinoids. *Developmental Biology* 427:212–218.

Mironov, A.N. 2008. Pourtalesiid sea urchins (Echinodermata: Echinoidea) of the northern Mid-Atlantic Ridge. *Marine Biology Research* 4:3–24.

Mironov, A.N., Dilman, A.B., Vladychenskaya, I.P. and N.B. Petrov. 2016. Adaptive strategy of the Porcellanasterid sea stars. *Biology Bulletin Russian Academic Science* 43:503–516.

Mironov, A.N., Minin, K.V. and A.B. Dilman. 2015. Abyssal echinoid and asteroid fauna of the North Pacific. *Deep Sea Research Part II: Topical Studies in Oceanography* 111:357–375.

Mironov, A.N., Minin, K.V. and A.V. Kremenetskaia. 2019. Two new genera of the family Myriotrochidae (Echinodermata, Holothuroidea). *Progress in Oceanography* 178:102195.

Molina, M.D., de Crozé, N., Haillot, E. and T. Lepage. 2013. Nodal: Master and commander of the dorsalventral and left-right axes in the sea urchin embryo. *Current Opinion in Genetics & Development* 23:445–453.

Molina, M.D., Gache, C. and T. Lepage. 2019. Expression of exogenous mRNAs to study gene function in echinoderm embryos. In *Methods in Cell Biology: Echinoderms, Part B*, ed. A. Hamdoun and K.R. Foltz, 150:239–282. Elsevier; Academic Press, London UK.

Molina, M.D., Quirin, M., Haillot, E., Jimenez, F., Chessel, A. and T. Lepage. 2017. p38 MAPK as an essential regulator of dorsal-ventral axis specification and skeletogenesis during sea urchin development: A re-evaluation. *Development* 144:2270–2281.

Mongiardino Koch, N., Coppard, S.E., Lessios, H.A., Briggs, D.E.G., Mooi, R. and G.W. Rouse. 2018. A phylogenomic resolution of the sea urchin tree of life. *BMC Evolutionary Biology* 18:189.

Monroy, A. 1986. A centennial debt of developmental biology to the sea urchin. *The Biological Bulletin* 171:509–519.

Morales, J., Mulner-Lorillon, O., Cosson, B., Morin, E., Belle, R., Bradham, C.A., Beane, W.S. and P. Cormier. 2006. Translational control genes in the sea urchin genome. *Developmental Biology* 300:293–307.

Morgan, A.J. 2011. Sea urchin eggs in the acid reign. *Cell Calcium* 50:147–156.

Mulner-Lorillon, O., Chassé, H., Morales, J., Belle, R. and P. Cormier. 2017. MAPK/ERK activity is required for the successful progression of mitosis in sea urchin embryos. *Developmental Biology* 421:194–203.

Mulochau, T., Conand, C., Stöhr, S., Eléaume, M. and P. Chabanet. 2014. First inventory of the echinoderms from Juan de Nova (Iles Eparses, France) in the Mozambique Channel, south western Indian Ocean. *Western Indian Ocean Journal of Marine Science* 13:23–30.

Nesbit, K.T., Fleming, T., Batzel, G., Pouv, A., Rosenblatt, H.D., Pace, D.A., Hamdoun, A. and D.C. Lyons. 2019. The painted sea urchin, *Lytechinus pictus*, as a genetically-enabled developmental model. In *Methods in Cell Biology: Echinoderms, Part A*, ed. A. Hamdoun and K.R. Foltz, 150:105–123. Elsevier; Academic Press, London UK.

Oji, T., Ogawa, Y., Hunter, A.W. and K. Kitazawa, K. 2009. Discovery of dense aggregations of stalked crinoids in Izu-Ogasawara Trench, Japan. *Zoological Science* 26:406–408.

Okazaki, K. 1975. Spicule formation by isolated micromeres of the sea urchin embryo. *American Zoologist* 15:567–581.

Oliveri, P. and E.H. Davidson. 2004. Gene regulatory network controlling embryonic specification in the sea urchin. *Current Opinion in Genetics & Development* 14:351–360.

Oliveri, P., Tu, Q. and E.H. Davidson. 2008. Global regulatory logic for specification of an embryonic cell lineage. *Proceedings of the National Academy of Sciences of the United States of America* 105:5955–5962.

Oulhen, N. and G.M. Wessel. 2016. Albinism as a visual, in vivo guide for CRISPR/Cas9 functionality in the sea urchin embryo: Albinism as a visual metric for gene disruption in the sea urchin. *Molecular Reproduction and Development* 83:1046–1047.

Ouréns, R., Fernández, L. and J. Freire. 2011. Geographic, population, and seasonal patterns in the reproductive parameters of the sea urchin *Paracentrotus lividus*. *Marine Biology* 158:793–804.

Pace, D.A., Maxson, R. and D.T. Manahan. 2010. Ribosomal analysis of rapid rates of protein synthesis in the Antarctic sea urchin *Sterechinus neumayeri*. *The Biological Bulletin* 218:48–60.

Pagett, R.M. 1981. The penetration of brackish-water by the echinodermata. In *Feeding and Survival Srategies of Estuarine Organisms*, ed N. V. Jones and W. J. Wolff, 135–151. Marine Science. Springer US, Boston, MA.

Parrington, J., Davis, L.C., Galione, A. and G. Wessel. 2007. Flipping the switch: How a sperm activates the egg at fertilization. *Developmental Dynamics* 236:2027–2038.

Pederson, T. 2006. The sea urchin's siren. *Developmental Biology* 300:9–14.

Peter, I.S. and E.H. Davidson. 2011. A gene regulatory network controlling the embryonic specification of endoderm. *Nature* 474:635–639.

Peter, I.S. and E.H. Davidson. 2010. The endoderm gene regulatory network in sea urchin embryos up to mid-blastula stage. *Developmental Biology* 340:188–199.

Peterson, R.E. and D.R. McClay. 2003. Primary mesenchyme cell patterning during the early stages following ingression. *Developmental Biology* 254:68–78.

Pomin, V.H. 2015. Sulfated glycans in sea urchin fertilization. *Glycoconjugate Journal* 32:9–15.

Poon, J., Fries, A., Wessel, G.M. and M. Yajima. 2019. Evolutionary modification of AGS protein contributes to formation of micromeres in sea urchins. *Nature Communications* 10:3779.

Power, M.E., Tilman, D., Estes, J.A., Menge, B.A., Bond, W.J., Mills, L.S., Daily, G., Castilla, J.C., Lubchenco, J. and R.T. Paine. 1996. Challenges in the quest for keystones: Identifying keystone species is difficult: But essential to understanding how loss of species will affect ecosystems. *BioScience* 46:609–620.

Price, A.R.G. and F.W.E. Rowe. 1996. Indian Ocean echinoderms collected during the Sindbad Voyage (1980–81): 3. Ophiuroidea and echinoidea. *Bulletin of the Natural History Museum: Zoology Series* 62:71–82.

Raff, R.A. 1987. Constraint, flexibility, and phylogenetic history in the evolution of direct development in sea urchins. *Developmental Biology* 119:6–19.

Raff, R.A. and M. Byrne. 2006. The active evolutionary lives of echinoderm larvae. *Heredity* 97:244–252.

Ramos, I. and G.M. Wessel. 2013. Calcium pathway machinery at fertilization in echinoderms. *Cell Calcium* 53:16–23.

Ransick, A. and E.H. Davidson. 2012. Cis-regulatory logic driving glial cells missing: Self-sustaining circuitry in later embryogenesis. *Developmental Biology* 364:259–267.

Ransick, A. and E.H. Davidson. 2006. Cis-regulatory processing of Notch signaling input to the sea urchin glial cells missing gene during mesoderm specification. *Developmental Biology* 297:587–602.

Reich, A., Dunn, C., Akasaka, K. and G. Wessel. 2015. Phylogenomic analyses of echinodermata support the sister groups of asterozoa and echinozoa. *PLoS One* 10:e0119627.

Rocha, F., Rocha, A.C., Baião, L.F., Gadelha, J., Camacho, C., Carvalho, M.L., Arenas, F., Oliveira, A., Maia, M.R.G., Cabrita, A.R., Pintado, M., Nunes, M.L., Almeida, C.M.R. and L.M.P. Valente. 2019. Seasonal effect in nutritional quality and safety of the wild sea urchin *Paracentrotus lividus* harvested in the European Atlantic shores. *Food Chemistry* 282:84–94.

Romiguier, J., Gayral, P., Ballenghien, M., Bernard, A., Cahais, V., Chenuil, A., Chiari, Y., Dernat, R., Duret, L., Faivre, N., Loire, E., Lourenco, J.M., Nabholz, B., Roux, C., Tsagkogeorga, G., Weber, A.A.-T., Weinert, L.A., Belkhir, K., Bierne, N., Glémin, S. and N. Galtier. 2014. Comparative population genomics in animals uncovers the determinants of genetic diversity. *Nature* 515:261–263.

Rowe, F. and M. Richmond. 2004. A preliminary account of the shallow-water echinoderms of Rodrigues, Mauritius, Western Indian Ocean. *Journal of Natural History* 38:3273–3314.

Runcie, D.E., Dorey, N., Garfield, D.A., Stumpp, M., Dupont, S. and G.A. Wray. 2017. Genomic characterization of the evolutionary potential of the sea urchin *Strongylocentrotus droebachiensis* facing ocean acidification. *Genome Biology and Evolution* 8:3672–3684.

Runft, L.L., Jaffe, L.A. and L.M. Mehlmann. 2002. Egg activation at fertilization: Where it all begins. *Developmental Biology* 245:237–254.

Santella, L., Vasilev, F. and J.T. Chun. 2012. Fertilization in echinoderms. *Biochemical and Biophysical Research Communications* 425:588–594.

Sardet, C. 1984. The ultrastructure of the sea urchin egg cortex isolated before and after fertilization. *Developmental Biology* 105:196–210.

Sardet, C. and P. Chang. 1985. A marker of animal-vegetal polarity in the egg of the sea urchin *Paracentrotus lividus*. *Experimental Cell Research* 160:73–82.

Sardet, C., Roegiers, F., Dumollard, R., Rouviere, C. and A. McDougall. 1998. Calcium waves and oscillations in eggs. *Biophysical Chemistry* 72:131–140.

Saudemont, A., Haillot, E., Mekpoh, F., Bessodes, N., Quirin, M., Lapraz, F., Duboc, V., Röttinger, E., Range, R., Oisel, A., Besnardeau, L., Wincker, P. and T. Lepage. 2010. Ancestral regulatory circuits governing ectoderm patterning downstream of nodal and BMP2/4 revealed by gene regulatory network analysis in an echinoderm. *PLOS Genetics* 6:e1001259.

Schatten, G. 1981. The movements and fusion of the pronuclei at fertilization of the sea urchin *Lytechinus variegatus*: Time-lapse video microscopy. *Journal of Morphology* 167:231–247.

Schatten, G. and D. Mazia, D. 1976. The penetration of the spermatozoon through the sea urchin egg surface at fertilization: Observations from the outside on whole eggs and from the inside on isolated surfaces. *Experimental Cell Research* 98:325–337.

Scheer, U. 2018. Boveri's research at the Zoological Station Naples: Rediscovery of his original microscope slides at the University of Würzburg. *Marine Genomics* 40:1–8.

Schmidt, G., Hecht, L. and S.J. Thannhauser. 1948. The behavior of the nucleic acids during the early development of the sea urchin egg (Arbacia). *The Journal of General Physiology* 31:203–207.

Schroeder, T.E. 1980. Expressions of the prefertilization polar axis in sea urchin eggs. *Developmental Biology* 79:428–443.

Sea Urchin Genome Sequencing Consortium, Sodergren, E., Weinstock, G.M., Davidson, E.H., et al. The genome of the sea urchin *Strongylocentrotus purpuratus*. *Science* 314:941–952.

SeGall, G.K. and W.J. Lennarz. 1979. Chemical characterization of the component of the jelly coat from sea urchin eggs responsible for induction of the acrosome reaction. *Developmental Biology* 71:33–48.

Sherwood, D.R. and D.R. McClay. 1999. LvNotch signaling mediates secondary mesenchyme specification in the sea urchin embryo. *Development* 126:1703–1713.

Sherwood, D.R. and D.R. McClay. 1997. Identification and localization of a sea urchin Notch homologue: Insights into vegetal plate regionalization and Notch receptor regulation. *Development* 124:3363–3374.

Simakov, O., Kawashima, T., Marlétaz, F., et al. 2015. Hemichordate genomes and deuterostome origins. *Nature* 527:459–465.

Sluder, G. 2016. Using sea urchin gametes and zygotes to e centrosome duplication. *Cilia* 5:20.

Smirnov, A. 1994. Arctic echinoderms: Composition distribution and history of the fauna. In *Echinoderms through Time*, ed. B. David, A. Guille, J.P. Féral and M. Roux, 135–143. A.A. Balkema, Rotterdam, Brookfield.

Smith, A.B. and S. Zamora. 2013. Cambrian spiral-plated echinoderms from Gondwana reveal the earliest pentaradial body plan. *Proceedings of the Royal Society B* 280:20131197.

Smith, M.M., Smith, L.C., Cameron, R.A. and L.A. Urry. 2008. The larval stages of the sea urchin, *Strongylocentrotus purpuratus*. *Journal of Morphology* 269:713–733.

Sodergren, E., Shen, Y., Song, X., Zhang, L., Gibbs, R.A. and G.M. Weinstock. 2006. Shedding genomic light on Aristotle's lantern. *Developmental Biology* 300:2–8.

Song, J.L., Wong, J.L. and G.M. Wessel. 2006. Oogenesis: Single cell development and differentiation. *Developmental Biology* 300:385–405.

Stanwell-Smith, D. and L.S. Peck. 1998. Temperature and embryonic development in relation to spawning and field occurrence of larvae of three Antarctic echinoderms. *The Biological Bulletin* 194:44–52.

Starr, M., Himmelman, J.H. and J.-C. Therriault. 1990. Direct coupling of marine invertebrate spawning with phytoplankton blooms. *Science* 247:1071–1074.

Stauber, M. 1993. The lantern of Aristotle: Organization of its coelom and origin of its muscles (Echinodermata, Echinoida). *Zoomorphology* 113:137–151.

Steinhardt, R.A. and D. Epel. 1974. Activation of sea-urchin eggs by a calcium ionophore. *Proceedings of the National Academy of Sciences of the United States of America* 71:1915–1919.

Steinhardt, R.A., Zucker, R. and G. Schatten. 1977. Intracellular calcium release at fertilization in the sea urchin egg. *Developmental Biology* 58:185–196.

Stephens, L., Hardin, J., Keller, R. and F. Wilt. 1986. The effects of aphidicolin on morphogenesis and differentiation in the sea urchin embryo. *Developmental Biology* 118:64–69.

Stepicheva, N.A. and J.L. Song. 2014. High throughput microinjections of sea urchin zygotes. *Journal of Visualized Experiments JoVE* e50841.

Su, Y.-H., Li, E., Geiss, G.K., Longabaugh, W.J.R., Krämer, A. and E.H. Davidson. 2009. A perturbation model of the gene regulatory network for oral and aboral ectoderm specification in the sea urchin embryo. *Developmental Biology* 329:410–421.

Summers, R.G. and B.L. Hylander. 1974. An ultrastructural analysis of early fertilization in the sand dollar, *Echinarachnius parma*. *Cell and Tissue Research* 150:343–368.

Suzuki, H. and S. Yaguchi. 2018. Transforming growth factor-β signal regulates gut bending in the sea urchin embryo. *Development, Growth & Differenciation* 60:216–225.

Swann, K. and F.A. Lai. 2016. Egg activation at fertilization by a soluble sperm protein. *Physiologycal Reviews* 96:127–149.

Swartz, S.Z. and G.M. Wessel. 2015. Germ line versus soma in the transition from egg to embryo. *Current Topics in Developmental Biology* 113:149–190.

Sweet, H.C., Gehring, M. and C.A. Ettensohn. 2002. LvDelta is a mesoderm-inducing signal in the sea urchin embryo and can endow blastomeres with organizer-like properties. *Development* 129:1945–1955.

Tamboline, C.R. and R.D. Burke. 1992. Secondary mesenchyme of the sea urchin embryo: Ontogeny of blastocoelar cells. *Journal of Experimental Zoology* 262:51–60.

Telford, M.J., Lowe, C.J., Cameron, C.B., Ortega-Martinez, O., Aronowicz, J., Oliveri, P. and R.R. Copley. 2014. Phylogenomic analysis of echinoderm class relationships supports Asterozoa. *Proceedings of the Royal Society B* 281:20140479.

Terasaki, M. and C. Sardet. 1991. Demonstration of calcium uptake and release by sea urchin egg cortical endoplasmic reticulum. *Journal of Cell Biology* 115:1031–1037.

Thieffry, D. and R.M. Burian. 1996. Jean Brachet's alternative scheme for protein synthesis. *Trends in Biochemical Sciences* 21:114–117.

Tilney, L.G. and L.A. Jaffe. 1980. Actin, microvilli, and the fertilization cone of sea urchin eggs. *The Journal of Cell Biology* 87:771–782.

Topper, T.P., Guo, J., Clausen, S., Skovsted, C.B. and Z. Zhang. 2019. A stem group echinoderm from the basal Cambrian of China and the origins of Ambulacraria. *Nature Communications* 10:1366.

Tu, Q., Cameron, R.A. and E.H. Davidson. 2014. Quantitative developmental transcriptomes of the sea urchin *Strongylocentrotus purpuratus*. *Developmental Biology* 385:160–167.

Uthicke, S., Schaffelke, B. and M. Byrne. 2009. A boom: Bust phylum? Ecological and evolutionary consequences of density variations in echinoderms. *Ecological Monographs* 79:3–24.

Vacquier, V.D. 2012. The quest for the sea urchin egg receptor for sperm. *Biochemical and Biophysical Research Communications* 425:583–587.

Vacquier, V.D. 2011. Laboratory on sea urchin fertilization. *Molecular Reproduction and Development* 78:553–564.

Vacquier, V.D. 1975. The isolation of intact cortical granules from sea urchin eggs: Calcium lons trigger granule discharge. *Developmental Biology* 43:62–74.

Vafidis, D., Antoniadou, C. and V. Ioannidi. 2020. Population density, size structure, and reproductive cycle of the comestible sea urchin *Sphaerechinus granularis* (Echinodermata: Echinoidea) in the Pagasitikos Gulf (Aegean Sea). *Animals* 10:1506.

Venuti, J.M., Pepicelli, C. and Flowers, V.L. 2004. Analysis of sea urchin embryo gene expression by immunocytochemistry. In *Methods in Cell Biology: Development of Sea Urchins, Ascidians, and Other Invertebrate Deuterostomes: Experimental Approaches*, ed. C.A. Ettensohn, G.A. Wray and G.M. Wessel, 74:333–369. Elsevier; Academic Press, London UK.

Vilela-silva, A.-C.E.S., Hirohashi, N. and P.A.S. Mouro. 2008. The structure of sulfated polysaccharides ensures a carbohydrate-based mechanism for species recognition during sea urchin fertilization. *The International Journal of Developmental Biology* 52:551–559.

von Dassow, G., Valley, J. and K. Robbins. 2019. Microinjection of oocytes and embryos with synthetic mRNA encoding molecular probes. In *Methods in Cell Biology: Echinoderms, Part B*, ed. A. Hamdoun and K.R. Foltz, 150:189–222. Elsevier; Academic Press, London UK.

Voronina, E. and G.M. Wessel. 2006. Activator of G-protein signaling in asymmetric cell divisions of the sea urchin embryo. *Development, Growth & Differenciation* 48:549–557.

Walker, C.W., Lesser, M.P. and T. Unuma. 2013. Sea urchin gametogenesis: Structural, functional and molecular/genomic biology. In *Developments in Aquaculture and Fisheries Science: Sea Urchins: Biology and Ecology*, ed. J.M. Lawrence, 38:25–43. Elsevier, London UK.

Walker, C.W., Unuma, T. and M.P. Lesser. 2007. Chapter 2: Gametogenesis and reproduction of sea urchins. In *Developments in Aquaculture and Fisheries Science, Edible Sea Urchins: Biology and Ecology*, ed. J.M. Lawrence, 37:11–33. Elsevier, London, UK.

Warburg, O. 1908. Beobachtungen über die Oxydationsprozesse im Seeigelei. *Hoppe-Seyler's Zeitschrift für physiologische Chemie* 57:1–16.

Wessel, G.M., Brayboy, L., Fresques, T., Gustafson, E.A., Oulhen, N., Ramos, I., Reich, A., Swartz, S.Z., Yajima, M. and V. Zazueta. 2014. The biology of the germ line in echinoderms: The germ-cell lineage in echinoderms. *Molecular Reproduction and Development* 81:679–711.

Wessel, G.M., Kiyomoto, M., Shen, T.-L. and M. Yajima. 2020. Genetic manipulation of the pigment pathway in a sea urchin reveals distinct lineage commitment prior to metamorphosis in the bilateral to radial body plan transition. *Scientific Reports* 10:1973.

Wessel, G.M., Zhang, W. and W.H. Klein. 1990. Myosin heavy chain accumulates in dissimilar cell types of the macromere lineage in the sea urchin embryo. *Developmental Biology* 140:447–454.

Whitaker, M.J. 2006. Calcium at fertilization and in early development. *Physiological Reviews* 86:25–88.

Whitaker, M.J. and R.A. Steinhardt. 1982. Ionic regulation of egg activation. *Quaterly Reviews of Biophysics* 15:593–666.

Wikramanayake, A.H., Huang, L. and W.H. Klein. 1998. β-catenin is essential for patterning the maternally specified animal-vegetal axis in the sea urchin embryo. *Proceedings of the National Academy of Sciences of the United States of America* 95:9343–9348.

Wikramanayake, A.H., Peterson, R., Chen, J., Huang, L., Bince, J.M., McClay, D.R. and W.H. Klein. 2004. Nuclearß-catenin-dependent Wnt8 signaling in vegetal cells of the early sea urchin embryo regulates gastrulation and differentiation of endoderm and mesodermal cell lineages. *Genesis* 39:194–205.

Wong, J.L., Créton, R. and G.M. Wessel. 2004. The oxidative burst at fertilization is dependent upon activation of the dual oxidase udx1. *Developmental Cell* 7:801–814.

Wong, J.M., Gaitán-Espitia, J.D. and G.E. Hofmann. 2019. Transcriptional profiles of early stage red sea urchins (*Mesocentrotus franciscanus*) reveal differential regulation of gene expression across development. *Marine Genomics* 48:100692.

Wood, N.J., Mattiello, T., Rowe, M.L., Ward, L., Perillo, M., Arnone, M.I., Elphick, M.R. and P. Oliveri. 2018. Neuropeptidergic systems in pluteus larvae of the sea urchin *Strongylocentrotus purpuratus*: Neurochemical complexity in a "simple" nervous system. *Frontiers in Endocrinology* 9:628.

Wygoda, J.A., Yang, Y., Byrne, M. and G.A. Wray. 2014. Transcriptomic analysis of the highly derived radial body plan of a sea urchin. *Genome Biology and Evolution* 6:964–973.

Yaguchi, S., Yaguchi, J., Suzuki, H., Kinjo, S., Kiyomoto, M., Ikeo, K. and T. Yamamoto. 2020. Establishment of homozygous knock-out sea urchins. *Current Biology* 30:R427–R429.

Yajima, M. and G.M. Wessel. 2012. Autonomy in specification of primordial germ cells and their passive translocation in the sea urchin. *Development* 139:3786–3794.

Yajima, M. and G.M. Wessel. 2011. Small micromeres contribute to the germline in the sea urchin. *Development* 138:237–243.

Yoshimura, K., Iketani, T. and T. Motokawa. 2012. Do regular sea urchins show preference in which part of the body they orient forward in their walk? *Marine Biology* 159:959–965.

Yu, L., Yan, M., Sui, J., Sheng, W.-Q. and Z.-F. Zhang. 2013. Gonadogenesis and expression pattern of the vasa gene in the sea cucumber *Apostichopus japonicus* during early development. *Molecular Reproduction and Development* 80:744–752.

Zamora, S., Lefebvre, B., Javier Álvaro, J., Clausen, S., Elicki, O., Fatka, O., Jell, P., Kouchinsky, A., Lin, J.-P., Nardin, E., Parsley, R., Rozhnov, S., Sprinkle, J., Sumrall, C.D., Vizcaïno, D. and A.B. Smith. 2013. Chapter 13 Cambrian echinoderm diversity and palaeobiogeography. *Geological Society, London, Memoirs* 38:157–171.

Zhang, Q.Y., Tamura, M., Uetake, Y., Washitani-Nemoto, S. and S. Nemoto. 2004. Regulation of the paternal inheritance of centrosomes in starfish zygotes. *Developmental Biology* 266:190–200.

Ziegler, A., Mooi, R., Rolet, G. and C. De Ridder. 2010. Origin and evolutionary plasticity of the gastric caecum in sea urchins (Echinodermata: Echinoidea). *BMC Evolutionary Biology* 10:313.

18 Echinoderms

Temnopleurus reevesii

Shunsuke Yaguchi

CONTENTS

18.1 INTRODUCTION

Sea urchins have been used as model organisms in biological fields for more than a century. Their usefulness as such comes from certain aspects and characteristics: sea urchin adults are easily collectable from the oceans, their gametes are easily spawned by the simple intrablastocoelar injection of KCl and embryos and larvae develop synchronously in small containers like beakers. In addition, because their early development occurs outside of the adult bodies, scientists can routinely apply embryology techniques, such as microinjection and micromanipulation (Yaguchi 2019a; George et al. 2019), leading researchers to a number of high-impact achievements in various biological fields (Davidson 2010; Evans et al. 1983). On the other hand, because the life cycle of sea urchin is generally very long and it takes almost two years to obtain the next generation, it has been impossible to apply genetics to sea urchin studies in the laboratory. However, we have found that a sea urchin species, *Temnopleurus reevesii* (Figure 18.1a), can produce the next generation in a half-year, which is much shorter than the more commonly used species of sea urchins, such as *Strongylocentrotus purpuratus* and *Hemicentrotous pulcherrimus*, and has a potential to be applied to genetics. Therefore, in this chapter, I will introduce the biological characteristics of *T. reevesii* and its high potential to contribute to genetic studies of echinoderms.

18.2 HISTORY OF THE MODEL

Although most sea urchin species are attractive to human beings as tasty food ingredients, especially in Japan, *T. reevesii* is one of the exceptions due to its bitter taste. In addition, compared with other model sea urchins, such as *S. purpuratus*, *Lytechinus variegatus* in North America, *Paracentrotus lividus* in Europe and *H. pulcherrimus* in East Asia, *T. reevesii* has not been well studied in biology. Therefore, the presence of the species has been reported (Hegde et al. 2013), but there are only a handful of experimental biological data. As a comparative analysis of the developmental processes among the Temnopleurus group, the Kitazawa lab in Japan first described the development of *T. reevesii* (Kitazawa et al. 2010, 2014). Following this work, our group reported the high temperature tolerance and the neurogenesis of the embryos and larvae (Yaguchi et al. 2015). While culturing embryos/larvae/juveniles, we recognized that *T. reevesii* has a fast generation cycle, about half a year. By focusing on these characteristics, our group expected it would be possible to introduce the study of gene functions using genetics to this sea urchin and has started to prepare the genome and transcriptome resources, which will be published elsewhere soon. The genome information allowed us to use the CRISPR/Cas-9 system to knock out some genes, and in fact, we managed to obtain the first homozygous knock-out strain using this species (Yaguchi et al. 2020).

18.3 GEOGRAPHICAL LOCATION

It has been reported that *T. reevesii* is found in the western Pacific and Indian Oceans (Clark et al. 1971; Hegde et al. 2013). Since historically there have been few scientific groups using this species for research, there is a possibility that new habitats will be found elsewhere in the near future. In Japan, the Kitazawa group has reported that they used *T. reevesii* collected from the Seto Inland Sea (Kitazawa et al. 2010, 2014), whereas our group found the adults of this species in our research center's aquarium, into which seawater

DOI: 10.1201/9781003217503-18

FIGURE 18.1 The adult of *T. reevesii*. (a) *T. reevesii* is a regular sea urchin whose body has pentaradial symmetry. Bar = 5 mm. (b) The genital papilla from the gonopore of adult males (arrow). This is not observed in the gonopore of females (c).

is continuously pumped. It is expected that the larvae swim in the general area around the Shimoda Marine Research Center, University of Tsukuba, including the Sagami Bay and the Pacific Ocean, and were pumped into the aquarium overflow system, in which they metamorphosed. On the other hand, although we have tried to identify the habitat of *T. reevesii* around the Shimoda Marine Research Center, we have never succeeded in finding it through scuba diving or a remotely operated underwater vehicle (ROV). Some dredge investigations picked up young individuals of *T. reevesii* but never found mature adults. Some pictures on divers' private websites show the adults of *T. reevesii* in the Izu peninsula near Shimoda, suggesting that there is a suitable habitat around Shimoda Marine Research Center, but the population of these animals is not likely to be dense.

18.4 LIFE CYCLE

Like other model sea urchins, *T. reevesii* undergoes indirect development, in which the gametes spawned from the male and female are fertilized outside the adults' bodies and the early and late development proceed as plankton in the ocean. They swim in the ocean via the movement of cilia, which are located at the surface of each ectodermal cell. Because they sink in seawater if the ciliary beating stops, the embryos/larvae essentially keep afloat using their cilia. In addition, sea urchin larvae have anti-gravitaxis, prompting them to stay at the surface of the ocean (Mogami et al. 1988). Due to their benthic lives, the adults cannot migrate over a large area, suggesting that it is likely that they spread their geographical distribution during the planktonic embryo and larval stages. Larva consume micro-algae as a food source, and in the laboratory culture, we feed them a diatom species, *Chaetoceros calcitrans*, which is commercially available (SunCulture, Marinetech, Aichi). After 1 to 1.5 months, the adult rudiment appears on the left side of the eight-armed larval body, and it grows until metamorphosis. In our laboratory, the competent larvae of *H. pulcherrimus*, the major sea urchin model in Japan, rarely metamorphose without an inducer like biofilm, which is generally localized on rocks and/or the sea floor. However, the competent larvae of *T. reevesii* easily metamorphose in glass beakers by simply stopping the stirring of water (Yaguchi 2019b).

Juveniles eat the adhered diatoms until the shell diameter size is 1.5 mm, but their food preference changes to carnivorous when they become larger (Yaguchi 2019b). Therefore, they start to eat meat of fish, shellfish and even small sea urchins. It is surprising to note that they eat their same species but never other vegetarian species like *H. pulcherrimus*. The most prominent characteristic of *T. reevesii* as a model sea urchin in biology is that they grow very fast from juveniles to sexually mature adults. General model sea urchins like *S. purpuratus* or *H. pulcherrimus* take more than one to two years until they are stably producing gametes (Strathman 1987), but *T. reevesii* can reach the stage after a half-year by culturing above 20°C. Another advantage as a model sea urchin is the timing of producing eggs and sperm. In the general model sea urchins, they need a temperature stimulus from warm to cold (e.g. in *H. pulcherrimus*, the temperature change from 23°C to 13°C induces the maturation of gonads), but in *T. reevesii*, keeping the culturing seawater warm (above 20°C) is enough to induce the accumulation of sperm or eggs in the adult gonads. This characteristic allows scientists to repeatedly use the same individuals unless they become damaged due to spawning and to save a number of adult sea urchins for research purposes.

18.5 EMBRYOGENESIS

Because the adults that hold matured gonads were observed from May to December in the outside aquarium, it is expected that spawning and early embryogenesis occur during summer/fall in the wild, when the temperature of seawater is above 20°C. In addition, the fact that the embryos of this species have a wide range of temperature tolerance between 15 and 30°C has been described (Yaguchi et al. 2015). Therefore, in laboratory conditions, we generally culture them at room temperature (RT) (about 20°C) for long-term experiments like creating inbred strains and at 22°C for the purposes of developmental biology. The diameter of unfertilized and fertilized eggs of *T. reevesii* is about 80 μm (Figure 18.2a, b), which is smaller than that of *H. pulcherrimus*. When we culture them at RT, the first cleavage occurs between 1 and 1.5 hours, and the embryos reach the four-cell stage at about two hours. During these early cleavages, blastomeres do not attach to each other, unlike other model sea urchins. The blastomere strongly attaches to the hyaline layer (Figure 18.2c, d, arrow) (Yaguchi et al. 2015). These separated blastomeres group together around the 60-cell stage, an event called "compaction", and the development continues like other sea urchin embryos after that. At several hours after hatching (Figure 18.2e), primary mesenchyme cells (PMCs) ingress into the blastocoel from the posteriorly located vegetal plate, and gastrulation occurs from the same region. PMCs will be spiculogenic cells in prism/pluteus larval stages. As observed in other model sea urchin embryos, from the tip of the invaginating gut, the secondary mesenchyme cells (SMCs) ingress into the blastocoel (Figure

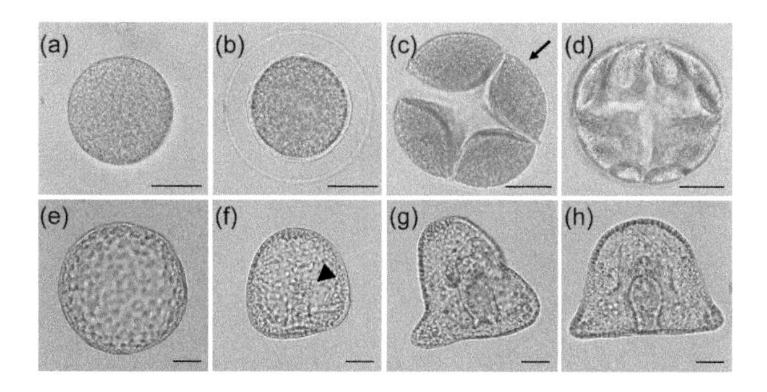

FIGURE 18.2 **Development of *T. reevesii* embryos/larvae.** (a) Unfertilized egg. (b) Fertilized egg with fertilization envelope and hyalin layer. (c) Four-cell stage. Arrow indicates the hyalin layer. (d) Sixteen-cell stage. (e) Hatched blastula. (f) Gastrula. Arrowhead indicates the ingress of secondary mesenchyme cells from the tip of invaginating gut. (g) Prism larva, lateral view. (h) Prism larva, ventral view. Bars = 40 μm.

18.2f, arrowhead). SMCs will be differentiated into muscles, pigment cells, the coelomic sac and blastocoel cells during the larval stages. After the tip of the gut fuses to the oral ectoderm in order to open the mouth, the endoderm starts to constrict to form the tripartite gut, which is composed of the esophagus, stomach and intestine (Figure 18.2g, h). The completion of gut differentiation allows the larvae to start food consumption (Yaguchi et al. 2015; Yaguchi 2019b). The number of larval arms increases during late pluteus stages from two to eight, as observed in other model sea urchins (Kitazawa et al. 2014). After 1 to 1.5 months after fertilization, the adult rudiment appears at the left side of the body, and it begins to metamorphose.

18.6 ANATOMY

Since *T. reevesii* is one of the regular sea urchins, the adult body has pentaradial symmetry covered with spines (Figure 18.1a). They move using tube feet, which are driven by the contractions of muscle and water force through the hydraulic system. All major anatomical characteristics are the same as those observed in the other regular sea urchins, but the genital papilla are notable in this species. The genital papilla clearly protrude from the gonopores in the male (Figure 18.1b) of *T. reevesii* but not from those of female (Figure 18.1c) (Yaguchi et al. 2015). This allows scientists to distinguish males and females when they obtain gametes, saving the time to collect eggs or sperm and saving the number of adults, because the researchers do not have to try multiple KCl injections on several individuals. The body shape and spine distribution of *T. reevesii* appear to be very similar to *Temnopleurus toreumaticus*. However, the spines of the former do not have a stripe pattern, while those of the latter do. The body color is essentially light brown, but it is variable; in fact, the strain kept in our laboratory is mutant, and its body color is highly pigmented and almost magenta. The size of the endoskeleton of adult *T. reevesii* is <5 cm in captivity in the laboratory, and the length of the spine is between about 1 to 3 cm.

18.7 GENOMIC DATA

In North America, Echinobase (Cary et al. 2018), a database for echinoderms (www.echinobase.org/entry/), publishes the genomic and transcriptomic data of several echinoderm species. In Europe, the genome and other genetic tools of the European model sea urchin, *P. lividus*, are in preparation (http://marimba.obs-vlfr.fr/organism/Paracentrotus/lividus) and will be made public soon. In Asia, we have the genome and transcriptome of *H. pulcherrimus* and have made a publicly available database for them, HpBase (Kinjo et al. 2018, 2021). The genome and transcriptome data of *T. reevesii* are in preparation, and the database is under construction and not yet publicly available but will be added to HpBase in near future. However, our laboratory used the information for gene knockout using the CRISPR/Cas-9 system (see Section 18.7), and it proved useful for these experiments. The genomic and transcriptome data will be available upon request to the author.

18.8 FUNCTIONAL APPROACHES: TOOLS FOR MOLECULAR AND CELLULAR ANALYSIS

As is the case for other model sea urchins, knockdown techniques using morpholino anti-sense oligonucleotides (MOs) and misexpression experiments using *in vitro* synthesized mRNA are available in *T. reevesii* (Suzuki and Yaguchi 2018). These reagents are introduced into unfertilized or fertilized eggs by microinjection. The microinjection techniques are common in any sea urchin species, and our laboratory uses an injection buffer that contains 22.5% glycerol for *H. pulcherrimus* eggs or blastomeres (40 mM HEPES, pH 8.0, 120 mM KCl, 22.5% glycerol). This buffer is also used for the North American *S. purpuratus*. On the other hand, glycerol-containing buffer kills the eggs of *T. reevesii*. Therefore, we use the injection buffer without glycerol. The details of the comparison and the methods of microinjection into sea urchin species are available elsewhere (Yaguchi 2019a).

To analyze the function of genes, *in situ* hybridization and immunohistochemistry are essential techniques and available to this species like other sea urchins. *T. reevesii* embryos/larvae have transparent bodies, which allow us to see the chromogenic and fluorescent signals very clearly (Figure 18.3a, b). In addition, almost all antibody reagents, which work against *H. pulcherrimus*, cross-react to *T. reevesii* embryos and larvae, but very few exceptions are present. For example, anti-phospho-Smad2/3 antibody (Abcam, Eugene, OR, USA) recognizes the phosphorylation site at the C-terminal of *H. pulcherrimus* Smad2/3 protein (. . . KQCSS*VS*; *phosphorylation site) but does not for *T. reevesii* because of its sequence difference (. . . KVCSS*MS*) (Suzuki and Yaguchi 2018).

One of the most prominent techniques in genetics is the knock-out. As mentioned, sea urchins have been considered not useful for genetics because of the length of their generation cycle. However, it takes about six months for *T. reevesii* to produce the next matured generation, which allows us

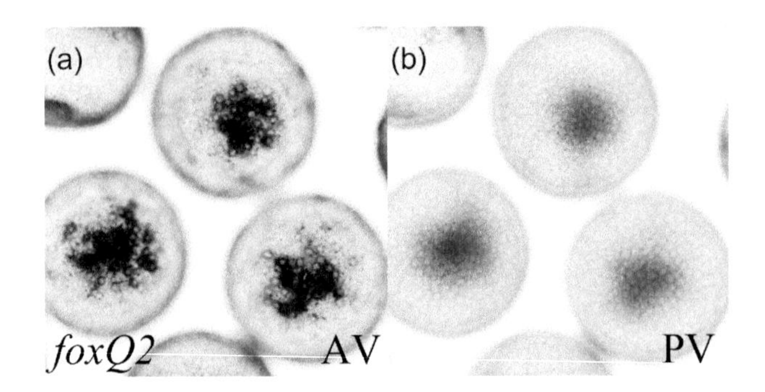

FIGURE 18.3 *In situ* hybridization using *T. reevesii*. (a) The expression of *foxQ2*, which is an essential transcription factor for the specification of anterior neuroectoderm. Anterior view (AV). (b) *foxQ2* does not express at the posterior end. Posterior view (PV).

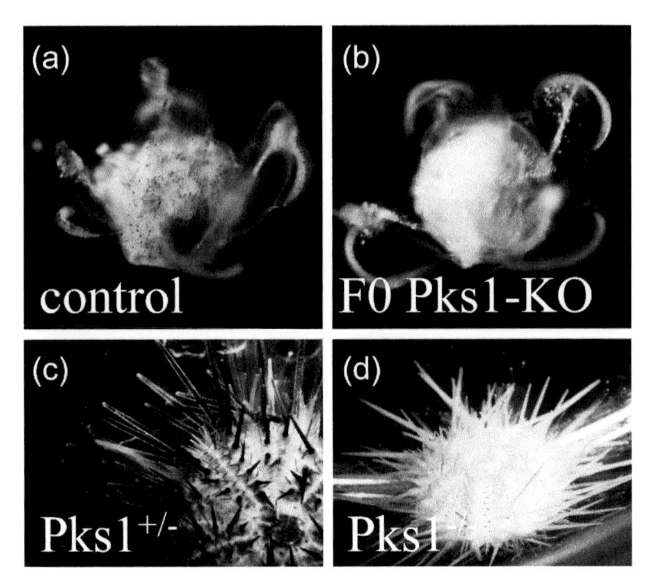

FIGURE 18.4 Pksl knock-out *T. reevesii*. (a) The late control (Cas-9 only injected) larva, which has an adult rudiment at the left side of the body. (b) Pksl knock-out F0 late pluteus larva, which loses pigmentation. (c) The heterogenous F2 adult (inbred magenta mutant is used as a control strain). (d) The homogenous Pksl knock-out F2 albino adult.

to challenge the status quo for sea urchins by introducing gene knock-out techniques to this species. In addition, the innovation of the CRISPR/Cas-9 system makes it easy for scientists to knock out genes in any organism, including sea urchins (Doudna and Charpentier 2014; Jao et al. 2013; Lin and Su 2016; Oulhen and Wessel 2016). The combination of the relatively short life cycle of *T. reevesii* and CRISPR/Cas-9 allowed us to produce the first homozygous knock-out strain of an albino sea urchin (Yaguchi et al. 2020). We focused on knocking out polyketide synthase 1 (Pksl), which plays the essential role in pigmentation (Akamatsu et al. 2010). We designed and synthesized five gRNAs against the second exon of the gene. Each gRNA was microinjected with hCas9 mRNA, which is synthesized from the plasmid (pCS2+hSpCas9; #51815 Addgene) *in vitro*. The efficiency of mutation was calculated with T7E1 assay (Vouillot et al. 2015), and #4 gRNA showed the highest efficiency. The injected embryos/larvae were cultured in 3L beakers with stirring until metamorphosis (Figure 18.4a, b), and the juveniles and young adults were cultured in a closed aquarium system (Yaguchi 2019b). Because the injected generation, that is, F0 generation, frequently contains mosaic genomic patterns even in one individual, the sperm or eggs are fertilized with wild type gametes and researchers obtain heterogeneous F1 generations. After confirming the genotype of individuals, we used the same types of sperm and eggs and then fertilized them to obtain the homozygous knock-out F2 mutant (Figure 18.4c, d). This research showed strong evidence for the availability of *T. reevesii* as a model organism in genetics, although the span of the life cycle is a little longer than those of other model organisms in this field, such as mice and fruit flies.

18.9 CHALLENGING QUESTIONS BOTH IN ACADEMIC AND APPLIED RESEARCH

Based on a number of previously published studies, gene regulatory analyses using sea urchin embryos have contributed much to biological fields and to an understanding of how gene expression is regulated. In fact, the most detailed and famous gene regulatory network in the world is about the specification of sea urchin endomesoderm (Davidson 2010; Cui et al. 2014). To investigate cis-regulatory elements, scientists utilized the microinjection of BAC-based reporter constructs into fertilized eggs (Nam et al. 2007; Sodergren et al. 2006; Buckley et al. 2019) and analyzed the data, which came from the mosaically integrated reporter constructs and variable patterns of individuals. A large number of experiments and the efforts of statistical processing helped scientists to confirm the results. Therefore, if people can analyze the endogenous gene expression pattern in embryos in which cis-regulatory elements were homozygously deleted by the CRISPR/Cas-9 system, the results will be more reliable and we can re-build more sophisticated gene regulatory networks.

Simple gene knock-outs are also available and efficient for analyzing gene functions in sea urchins. Although gene knock-downs using MO injection techniques can target only early embryogenesis, CRISPR/Cas-9-based knock-outs can target genes that function in later developmental stages and adults. This technique will help scientists understand the biology of sea urchins more thoroughly. However, metamorphosis during the sea urchin's life might be a barrier for genetics, because it is a really drastic event, and one expects that a number of genes function to create the adult body. In fact, when we knock out Smad2/3 with CRISPR/Cas-9, the mutants were all dead at the timing of metamorphosis (data not shown). It is also true that it is still not easy to obtain the next generation of sea urchins in the laboratory, even if *T. reevesii* is easier than other model sea urchins. Taken together, however, the combination of the CRISPR/Cas-9 system and *T. reevesii* promises to reveal numerous biological insights through sea urchin knock-out strains. Knock-in techniques have not yet been successful in sea urchins.

Although many sea urchin species are famous for being a source of tasty ingredients worldwide, *T. reevesii* is not suitable for food. The Japanese name of *T. reevesii* is "hari sanshou uni", and the meanings of "hari", "sanshou" and "uni" are "spined", "bitter/hot" and "sea urchins", respectively. Therefore, it is said that *T. reevesii* is not good as food, and, in fact, people do not find this species in seafood markets. However, in genetics, *T. reevesii* can be useful to understand gene functions related to the taste and the size of gonads. At the same time, when compared with other model sea urchins which are commonly used in food, it is a mystery why *T. reevesii* can grow faster. If this question can be answered using *T. reevesii*, sea urchin farmers in the fishery industries will obtain ideas for culturing sea urchins from the basic sciences.

BIBLIOGRAPHY

Akamatsu, H. O., Chilvers, M. I., Stewart, J. E. and Peever, T. L. 2010. Identification and Function of a Polyketide Synthase Gene Responsible for 1,8-Dihydroxynaphthalene-Melanin Pigment Biosynthesis in *Ascochyta rabiei*. *Current Genetics*. 56:349–360.

Buckley, K. M. and Ettensohn, C. A. 2019. Techniques for Analyzing Gene Expression Using BAC-Based Reporter Constructs. *Methods in Cell Biology*. 151:197–218.

Cary, G. A., Cameron, R. A. and Hinman, V. F. 2018. EchinoBase: Tools for Echinoderm Genome Analyses. *Methods in Cell Biology*. 151:349–369.

Clark, A. M. and Row, F. E. W. 1971. Monograph of Shallow-Water Indo-West Pacific Echinoderms. *Trustees of the British Museum of Natural History, London*. 238.

Cui, M., Siriwon, N., Li, E., Davidson, E. H. and Peter, I. S. 2014. Specific Functions of the Wnt Signaling System in Gene Regulatory Networks throughout the Early Sea Urchin Embryo. *Proceedings of the National Academy of Sciences of the United States of America*. 111:E5029–E5038.

Davidson, E. H. 2010. Emerging Properties of Animal Gene Regulatory Networks. *Nature*. 468:911–920.

Doudna, J. A. and Charpentier, E. 2014. The New Frontier of Genome Engineering with CRISPR-Cas9. *Science*. 346:6213.

Evans, T., Rosenthal, E. T., Youngblom, J., Distel, D. and Hunt, T. 1983. Cyclin: A Protein Specified by Maternal MRNA in Sea Urchin Eggs That Is Destroyed at Each Cleavage Division. *Cell*. 33:389–396.

George, A. N. and McClay, D. R. 2019. Methods for Transplantation of Sea Urchin Blastomeres. *Methods in Cell Biology*. 151:223–233.

Hegde, M. R. and Rivonker, C. U. 2013. A New Record of *Temnopleurus decipiens* (De Meijere, 1904) (Echinoidea, Temnopleuroida, Temnopleuridae) from Indian Waters. *Zoosystema*. 35:97–111.

Jao, L. E., Wente, S. R. and Chen, W. 2013. Efficient Multiplex Biallelic Zebrafish Genome Editing Using a CRISPR Nuclease System. *Proceedings of the National Academy of Sciences of the United States of America*. 110:13904–13909.

Kinjo, S., Kiyomoto, M., Yamamoto, T., Ikeo, K. and Yaguchi, S. 2018. HpBase: A Genome Database of a Sea Urchin, *Hemicentrotus pulcherrimus*. *Development Growth and Differentiation*. 60(3):174–182.

Kinjo, S., Kiyomoto, M., Yamamoto, T., Ikeo, K. and Yaguchi, S. 2021. Usage of Sea Urchin *Hemicentrotus pulcherrimus* Database, HpBase. *Methods in Molecular Biology*. 2219:267–275.

Kitazawa, C., Sakaguchi, C., Nishimura, H., Kobayashi, C., Baba, T. and Yamanaka, A. 2014. Development of the Sea Urchins *Temnopleurus toreumaticus* Leske, 1778 and *Temnopleurus reevesii* Gray, 1855 (Camarodonta: Temnopleuridae). *Zoological Studies*. 53:3.

Kitazawa, C., Tsuchihashi, Y., Egusa, Y., Genda, T. and Yamanaka, A. 2010. Morphogenesis during Early Development in Four Temnopleuridae Sea Urchins. *Information*. 13:1075–1089.

Lin, C. Y. and Su, Y. H. 2016. Genome Editing in Sea Urchin Embryos by Using a CRISPR/Cas9 System. *Developmental Biology*. 409:420–428.

Mogami, B. Y. Y., Oobayashi, C. and Baba, S. A. 1988. Negative Geotaxis in Sea Urchin Larvae: A Possible Role of Mechanoreception in the Late Stages of Development. *Journal of Experimental Biology*. 137:141–156.

Nam, J., Su, Y.-H., Lee, P. Y., Robertson, A. J., Coffman, J. A. and Davidson, E. H. 2007. Cis-Regulatory Control of the Nodal Gene, Initiator of the Sea Urchin Oral Ectoderm Gene Network. *Developmental Biology*. 306:860–869.

Oulhen, N. and Wessel, G. M. 2016. Albinism as a Visual, In Vivo Guide for CRISPR/Cas9 Functionality in the Sea Urchin Embryo. *Molecular Reproduction and Development*. 83:1046–1047.

Sodergren, E., Weinstock, G. M., Davidson, E. H., Cameron, R. A., Gibbs, R. A., Angerer, R. C., Angerer, L. M., et al. 2006. The Genome of the Sea Urchin *Strongylocentrotus purpuratus*. *Science*. 314:5801.

Strathman, M. F. 1987. *Reproduction and Development of Marine Invertebrates of the Northern Pacific Coast*. The University of Washington Press, Seatle, WA.

Suzuki, H. and Yaguchi, S. 2018. Transforming Growth Factor-β Signal Regulates Gut Bending in the Sea Urchin Embryo. *Development Growth and Differentiation*. 60:216–225.

Vouillot, L., Thélie, A. and Pollet, N. 2015. Comparison of T7E1 and Surveyor Mismatch Cleavage Assays to Detect Mutations Triggered by Engineered Nucleases. *G3: Genes, Genomes, Genetics*. 5:407–415.

Yaguchi, J. 2019a. Microinjection Methods for Sea Urchin Eggs and Blastomeres. *Methods in Cell Biology*. 150:173–188.

Yaguchi, S. 2019b. Temnopleurus as an Emerging Echinoderm Model. *Methods in Cell Biology*. 150:71–79.

Yaguchi, S., Yaguchi, J., Suzuki, H., Kinjo, S., Kiyomoto, M., Ikeo, K. and Yamamoto, T. 2020. Establishment of Homozygous Knock-Out Sea Urchins. *Current Biology*. 30:R427–R429.

Yaguchi, S., Yamazaki, A., Wada, W., Tsuchiya, Y., Sato, T., Shinagawa, H., Yamada, Y. and Yaguchi, J. 2015. Early Development and Neurogenesis of *Temnopleurus reevesii*. *Development Growth and Differentiation*. 57:242–250.

19 Cephalochordates

Salvatore D'Aniello and Stéphanie Bertrand

CONTENTS

19.1 HISTORY OF THE MODEL

Amphioxus are small, worm-like animals that resemble a fish without a head or a skeleton. They live burrowed in the sand of temperate and tropical costal areas, usually at shallow depths (1–50 m). Amphioxus, also called lancelets, is the common name for members of the cephalochordate clade. The first description of amphioxus came from a Chinese legend: Wenchang (or Wen Chang), the literature deity, was traveling around the world in search of new knowledge on the back of his pet crocodile. When the crocodile died in the Bay of Xiamen, larva emerged from its corpse. These "larva" were amphioxus, and even today the Chinese call amphioxus "Fish of the God of Literature" or "Wenchang fish" (Stokes and Holland 1998; Feng et al. 2016; Holland and Holland 2017). These animals are consumed as food in some Chinese regions, although the amphioxus population greatly decreased in the Bay of Xiamen during the second half of the 20th century.

While much more abundant in China than in Europe, the first scientific description of a cephalochordate came from the German zoologist and botanist Peter Simon Pallas in 1774, who named it *Limax lanceolatus* (Pallas 1774). He could only observe two fixed adult specimens from the Cornwall coast, UK, and classified amphioxus as a mollusk. In 1834, Gabriele Costa, a zoologist in Naples, Italy, described amphioxus as a fish and hypothesized it could represent the "missing link" between invertebrates and vertebrates (Costa 1834). He was able to observe live animals and described the oral cirri around the mouth as gills. For this reason, he gave the name *Branchiostoma* to the genus ("branchio" for "gills" and "stoma" for "mouth"). In 1836, William Yarrell, who was unfamiliar with Costa's work but knew about the description by Pallas, proposed "lancelet" as a common name for specimens from the Cornwall coast and changed the genus name *Limax*, given by Pallas, to *Amphioxus* ("amphi" for "both sides" and "oxus" for "pointed") (Yarrell 1836). Later on, the genus name became *Branchiostoma*. However, Yarrell is at the origin of the two common names of cephalochordate animals: amphioxus and lancelet. Thereafter, many zoologists developed an interest in amphioxus because of its proposed key evolutionary position as a close relative of vertebrates and made in-depth descriptions of its morphology; however, these zoologists were only working with adult specimens. The first researcher who described amphioxus embryos was the Russian embryologist Alexander Onufrievich Kowalevsky. After his studies

DOI: 10.1201/9781003217503-19

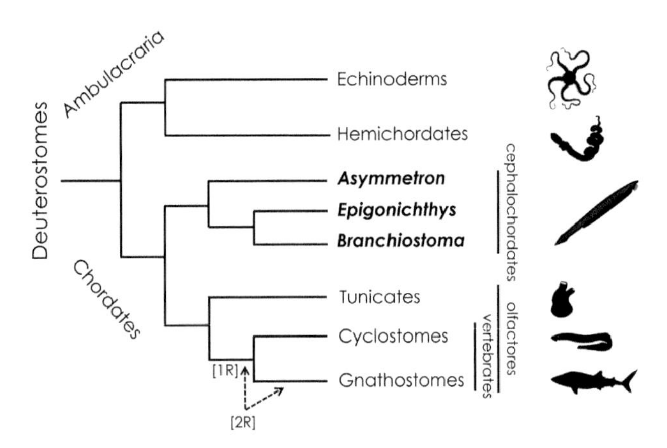

FIGURE 19.1 Deuterostome group classification. Deuterostomes are subdivided into Ambulacraria, composed of echinoderms and hemichordates, and chordates, which include cephalochordates and olfactores (tunicates and vertebrates). The three cephalochordate genera are represented in bold. The two whole genome duplications that occurred during vertebrate evolution are also indicated. The first took place before the divergence between gnathostomes (jawed vertebrates) and cyclostomes (lampreys and hagfish), whereas the position of the second is still debated.

in Russia and at the University of Heidelberg, Germany, he came to Naples in 1863 and 1864 in an attempt to obtain amphioxus embryos from local specimens (Davydoff 1960). Kowalevsky frequently collected amphioxus and kept them in his aquarium for months in hopes of the animals spawning. Finally, in May 1864, some adult animals spawned, and Kowalevsky was able, for the first time, to observe the development of amphioxus embryos (Kovalevskij 1867). He noticed that the blastula would flatten on one side that subsequently invaginated to create two embryonic layers through a process of gastrulation. His work was assembled in a manuscript thanks to which he obtained his Magister degree in St. Petersburg, Russia.

Many other zoologists became interested in amphioxus (Gans 1996), among whom were the famous Berthold Hatschek (Hatschek and Tuckey 1893) and Edwin Grant Conklin (Conklin 1932), who made many descriptions of amphioxus embryogenesis, as well as the German naturalist Ernst Haeckel, who wrote in the fifth edition of the book *The Evolution of Man*: "We begin with the lancelet—after man the most important and interesting of all animals. Man is at the highest summit, the lancelet at the lowest root, of the vertebrate stem" (Haeckel et al. 1905). However, being extant animals, cephalochordates cannot be at the root of vertebrates, but evolutionarily they are closely related; cephalochordates, together with vertebrates and their sister group the tunicates, form the chordate clade (Figure 19.1). This evolutionary proximity is one of the reasons many researchers use amphioxus as a model in research.

Therefore, the study of amphioxus development and its comparison with tunicate and vertebrate embryogenesis allows us to define ancestral traits of chordates and to understand the appearance of vertebrate-specific morphological characters.

During the second half of the 20th century, research on amphioxus slowed down in Europe and the United States while flourishing in China with the species *Branchiostoma belcheri* (Light 1923). Among Chinese researchers, Ti Chou Tung elegantly studied embryonic cell fate in amphioxus using vital staining and delicate micro-manipulations, providing the scientific community with important insights into cephalochordate development (Tung et al. 1958, 1960, 1962, 1965). Later, amphioxus entered the molecular biology era thanks to American researchers Dr. Linda and Prof. Nicholas Holland from the University of California, San Diego. They began to collect adults from the species *Branchiostoma floridae* in Tampa, Florida, during the summer of 1988 and were able to obtain embryos from *in vitro* fertilization and using gametes obtained by spawning induction of the adults through electric stimulation (Holland and Holland 1989). In collaboration with Prof. Peter Holland from Oxford University, they developed a protocol to analyze embryonic gene expression through whole mount *in situ* hybridization experiments, allowing the scientific community to renew its interest in amphioxus as a modern model to study the evolution of developmental mechanisms (Holland et al. 1992).

At the beginning of the 21st century, the development of new sequencing techniques accompanied the transition to whole-genome level studies for many organisms, including amphioxus. The first whole-genome sequence was obtained for the American species *B. floridae* (Putnam et al. 2008), followed by the genome of *B. belcheri* (Huang et al. 2012) and the genome and epigenome of the European species *B. lanceolatum* (Marletaz et al. 2018). These advances have made amphioxus a good model not only to understand morphological evolution in the chordate clade through developmental biology approaches but also to study the evolution of genome structure and function. Before any cephalochordate genome was published, multigene phylogenetic studies taking advantage of the whole genome sequencing of the tunicate *Oikopleura dioica* showed that, contrary to what was globally accepted in the community, tunicates, and not cephalochordates, are the sister group of vertebrates, with which they form the *Olfactores* clade (Delsuc et al. 2006). Comparing vertebrates and amphioxus thus gives us information on the chordate ancestor that probably had characters more closely related to those of vertebrates than previously thought!

19.2 GEOGRAPHICAL LOCATION

Cephalochordates include three genera—*Branchiostoma*, *Epigonichtys* and *Asymmetron*—with around 30–40 species described to date (Poss and Boschung 1996). All animals of this chordate group are very similar morphologically, the only major difference being that adults of the *Branchiostoma* genus species have two rows of gonads on both sides of the body, whereas *Asymmetron* and *Epigonichtys* species have only one row of gonads on the right side. Amphioxus live in the sand of the seafloor with the anterior part of their body sticking out of the sediment and feed by filtering the seawater. Cephalochordates are widely distributed, with species described along tropical and temperate coasts in sandy sediments all around the world (Poss and Boschung 1996). The precise distribution of each species is hard to define, as historically the identification of species was only based on morphological and meristic data, which, as stated before, are not sufficiently discriminant due to the high morphological resemblance among cephalochordates. Development of molecular identification is rising and recently allowed several research groups to suggest the existence of more species than previously described (Nishikawa 2004; Nohara et al. 2005; Kon et al. 2006; Kon et al. 2007; Igawa et al. 2017; Subirana et al. 2020). Moreover, regarding Asian species, recent studies showed that western Pacific lancelet populations that were for a long period recognized as belonging to one species, *B. belcheri*, belong instead to two distinct species, *B. belcheri* and *B. japonicum* (Zhang et al. 2006; Li et al. 2013). Molecular phylogenetic data also allowed the clarification of evolutionary relationships between species and showed that *Branchiostoma* and *Epigonichtys* are more closely related to each other than to the *Asymmetron* genus (Igawa et al. 2017). Interestingly, although *Asymmetron* and *Branchiostoma* diverged between 46 and 150 Mya (Igawa et al. 2017; Subirana et al. 2020), viable hybrid embryos from *A. lucayanum* and *B. floridae* can be obtained by *in vitro* fertilization (Holland et al. 2015).

19.3 LIFE CYCLE

19.3.1 Animals in the Field

Amphioxus are gonochoric animals presenting a typical bentho-pelagic life cycle. Males and females live burrowed in the sand, and during the breeding season, they swim into the water column just after sunset and release all their gametes into the environment: hundreds of oocytes are spawned by each female, whereas males release sperm full of spermatozoids. After external fertilization, the embryo continues its development protected by the fertilization envelope, also called the chorion. Hatching occurs at the end of the gastrulation process, and the ciliated embryo continues developing to form a planktonic larva that moves thanks to both the epidermal cilia and the newly formed trunk striated muscles. The larva then metamorphoses and becomes a juvenile that returns to a life in the sediment and reaches adulthood after sexual maturation (Stokes and Holland 1998).

The duration and timing of the breeding season depend on the species, as well as the speed of embryonic and post-embryonic development. In the *B. floridae* population of Tampa Bay, the breeding season starts in early May and ends at the beginning of September (Stokes and Holland 1996). During this period, animals might spawn several times and produce new gametes more or less every two weeks. In the Mediterranean *B. lanceolatum* population of Argelès-sur-Mer, France, the breeding season starts in May and ends in July, with animals capable of spawning at least twice during this period, although, contrary to observations made for *B. floridae*, animals from the same location do not always spawn synchronously (Fuentes et al. 2004; Fuentes et al. 2007). The two Asian species *B. belcheri* and *B. japonicum* can also spawn at least twice in the field during their reproductive seasons, which range from May to the end of July and from late April to late August, respectively (Zhang et al. 2007; Li et al. 2013). Finally, the *A. lucayanum* population from Bimini, the Bahamas, has two breeding periods during the year: in fall and spring, when the water temperature is moderate and the animals tend to spawn the same day, one or two days before the new moon (Holland and Holland 2010).

The length of the life cycle is variable from one species to the other: *B. floridae* can reach the adult stage several months after fertilization (Stokes and Holland 1998), whereas a whole year is needed for *B. belcheri* (Zhang et al. 2007) and more than two years for *B. lanceolatum* (Fuentes et al. 2007; Desdevises et al. 2011).

19.3.2 Animals in the Laboratory

For several years now, some research groups have tried to maintain live amphioxus in their laboratories. Two husbandry systems are mainly used for adults (Carvalho et al. 2017), which both consist of small tanks filled with seawater with or without sediment that are either placed in a water bath to stabilize the temperature or not. In both systems, the water is changed regularly by continuous flow or by big volume changes several times per day, and light is applied in order to get a day/night cycle of 24 hours. Less regular water changes have also been reported for inland laboratories without access to fresh seawater (Theodosiou et al. 2011; Benito-Gutierrez et al. 2013). Adult amphioxus in the field feed by filtering the sea water from which they ingest all the particles less than 100 μm in diameter (Ruppert et al. 2000). Studies of stable isotopes and feces showed that they consume a wide variety of organisms, from bacteria to zooplankton and phytoplankton (Chen et al. 2008; Pan et al. 2015). In the laboratory, a mixture of different algae can be efficiently used to feed adults, although they can survive for months without a food supply (Carvalho et al. 2017). Ripe adults of the four main species used for evo-devo studies—*B. floridae*, *B. belcheri*, *B. japonicum* and *B. lanceolatum*—can be induced to spawn in the laboratory in order to obtain gametes for *in vitro* fertilization (Garcia-Fernàndez et al. 2009). The artificial induction of gamete release was first achieved for *B. floridae* using an electric shock, undertaken at the time of the natural sunset on collected adults kept with a light on (Holland and Holland 1989). However, this method was shown to be efficient only on the days the animals collected would have spawned in the field. For *B. lanceolatum*, heat stimulation by increasing the temperature of the water by 4°C 24 to 36 hours before the desired spawning night can be efficiently used to induce spawning (Fuentes et al. 2007). This technique allows working with embryos at any desired day during the breeding season of this species. The same method has been successfully used in the other *Branchiostoma* species, although with apparently less efficiency. Interestingly, some rearing conditions allow us to obtain ripe animals all through the year for the Asian species *B. belcheri*, which has never been reliably achieved for any other species (Li et al. 2013; Holland et al. 2015).

Once embryos are obtained by *in vitro* fertilization, they can be cultivated easily in Petri dishes filled with seawater and placed in an incubator to control the temperature. The most delicate step in order to keep amphioxus in the laboratory during their whole life cycle is to raise the larva until they metamorphose to reach the juvenile stage. Larva can be raised in Petri dishes given unicellular algae as food until metamorphosis, but this system is time consuming, as the larva must be manually transferred into clean dishes every day under the binocular (Holland and Yu 2004). Another method, used for *B. belcheri* and *B. japonicum*, is to raise the larva in tanks, with or without sediment. Although by using biggest volume, water changes are less frequently required and easier to manage, the survival rate of larva is very low, at best 3–5% (Zhang et al. 2007). Finally, the only *Asymmetron* species for which laboratory rearing conditions have been reported is the *A. lucayanum* population of Bimini (Holland and Holland 2010; Holland et al. 2015). Adults can be kept in the laboratory in the overall same conditions as the *Branchiostoma* species and *in vitro* fertilization undertaken after spawning. However, the larva die after 10 days of culture with only one open pharyngeal slit, and later stages have yet to be obtained in the laboratory (Holland and Holland 2010; Holland et al. 2015).

19.4 EMBRYOGENESIS

Amphioxus embryogenesis was first described by Kowalevsky (Kovalevskij 1867) for the population of *B. lanceolatum* in the Gulf of Naples. After the zygote cell is formed by external fertilization in the water column, a fertilization envelope detaches from the plasmic membrane and grows, preventing polyspermy and protecting the embryo during its early developmental stages, as observed in other species, such as sea urchins (Holland and Holland 1989). Cephalochordates produce oligolecithal eggs (low amount of yolk evenly distributed in the oocyte) of around 80–100 μm diameter (depending on the species) that undergo a first holoblastic cleavage and produce two blastomeres. Each of these blastomeres is able to develop into a full normal embryo after separation (Tung et al. 1958), although it has been shown that at the larva stage, one of the twins develops an abnormal tail (Wu et al. 2011). The second cleavage is perpendicular to the first one, and the third cleavage is unequal, giving rise to the formation of four micromeres at the animal pole and four macromeres at the vegetal pole. After several additional synchronous divisions, the embryo reaches the blastula stage (Figure 19.2).

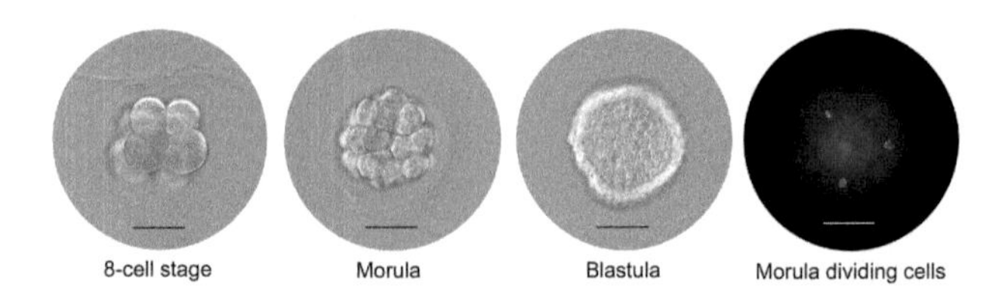

FIGURE 19.2 Cleavage stage. Pictures of *B. lanceolatum* embryos at the eight-cell, morula and blastula stages. During the cleavage period, divisions are synchronous, as shown by the anti-phospho-histone H3 immunostaining of chromosomes in all the cells at the morula stage. Scale bar = 50 μm.

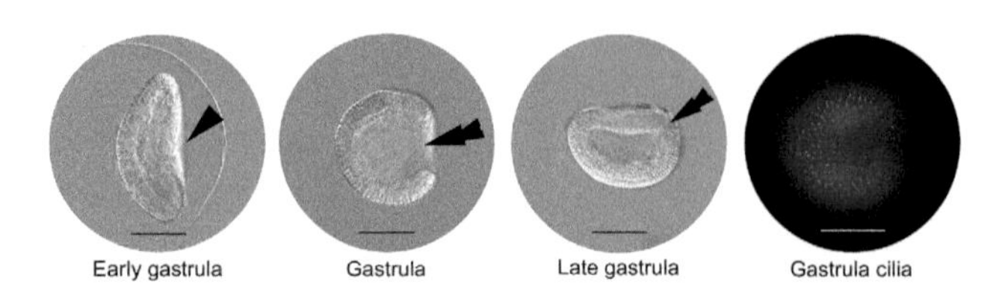

FIGURE 19.3 Gastrulation. Pictures of *B. lanceolatum* embryos during gastrulation. At the beginning of this developmental period, the vegetal plate invaginates (arrowhead) to form the internal layer called the mesendoderm. The opening that is formed is called the blastopore (double arrowheads), which will be completely covered by the epidermis at the end of gastrulation. During gastrulation, cilia grow as shown by anti-acetylated tubulin immunostaining, and the embryo starts to swim. Lateral views with anterior/animal to the left and dorsal to the top. Scale bar = 50 μm.

The blastula corresponds to a single cell layer surrounding a cavity called the blastocoel (Figure 19.2). At this stage, the vegetal region starts flattening and invaginates to form a gastrula with two touching germ layers: the ectoderm (external layer) and the mesendoderm (internal layer) (Figure 19.3). The cavity thus created corresponds to the archenteron, and its opening is called the blastopore. While gastrulation proceeds, cilia grow, and the embryo starts swimming inside the chorion (Figure 19.3).

During gastrulation, contrary to vertebrates, for example, few cells involute, and the two germ layers remain epithelial (Zhang et al. 1997). In the dorsal region, the ectoderm starts to flatten to form the neural plate. The rest of the ectoderm detaches and grows to cover the neural plate and close the blastopore. Before the neural plate is covered, the embryo hatches. Then neurulation proceeds with the neural plate rolling on itself, as observed in vertebrates, to become a hollow neural tube, enlarged in the anterior region, to form the cerebral vesicle. The epidermis that has covered the neural plate fuses in the midline, leaving an opening called the neuropore at the level of the cerebral vesicle (Figure 19.4). At the same time, the dorsal axial region of the mesendoderm starts to form the notochord, whereas in the dorsal paraxial region, pouches pinch off in a segmental manner to form the somites on both sides of the midline (Figure 19.4).

Somites form regularly from the anterior to the posterior region during embryo elongation, first by enterocoely and then by schizocoely from the tailbud. Somites in amphioxus are asymmetric, with the left somites shifted forward by half a somite. At the end of neurulation, the ventral mesendoderm has closed in the dorsal region and forms the future digestive tube. In its anterior region, two diverticula develop (called Hatchek's diverticula) on the right and left sides. The anterior ventral region of the endoderm enlarges to form the future

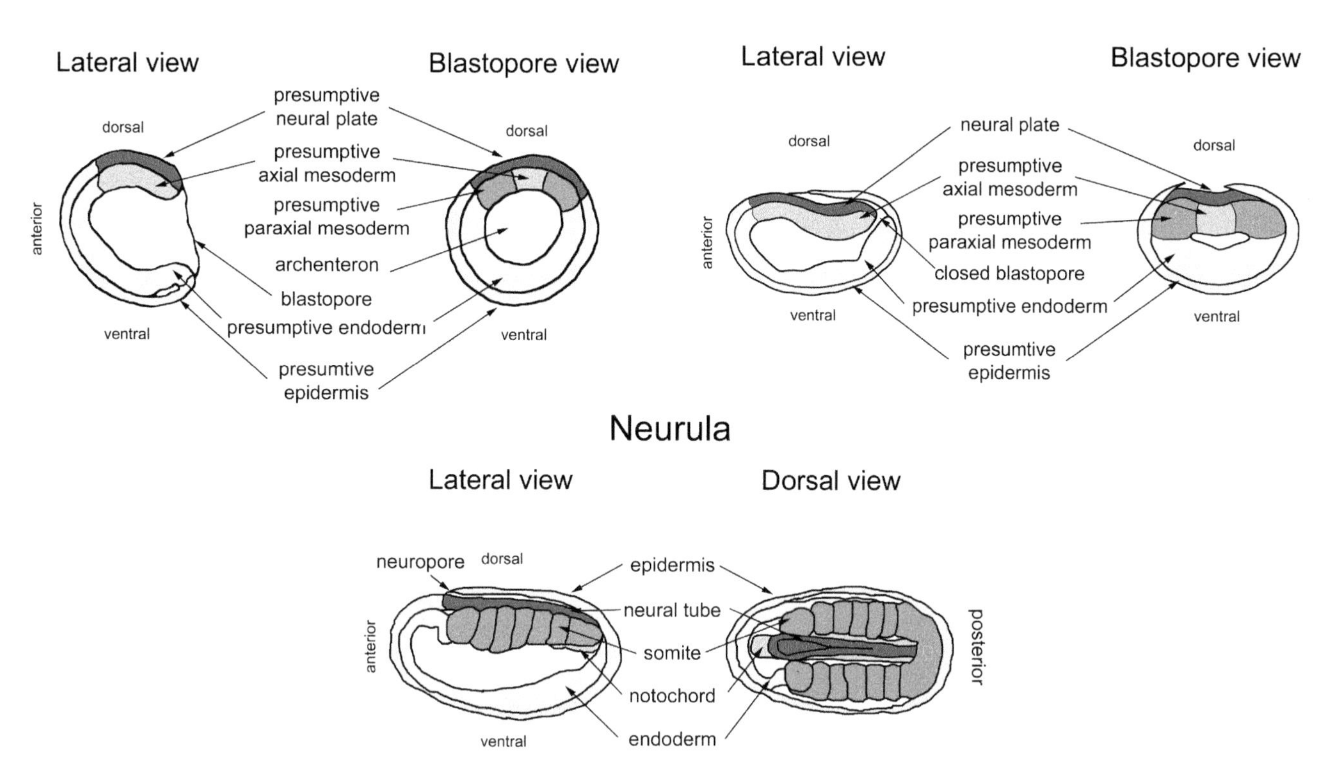

FIGURE 19.4 Diagram of embryos and presumptive fates from gastrula to neurula. Lateral views: dorsal to the top, anterior to the left. Blastopore views: dorsal to the top. Dorsal view: anterior to the left. The ectoderm-derived structures are in blue and light blue, the dorsal mesendoderm-derived structures are in red and orange and the ventral mesendoderm-derived structures are in green.

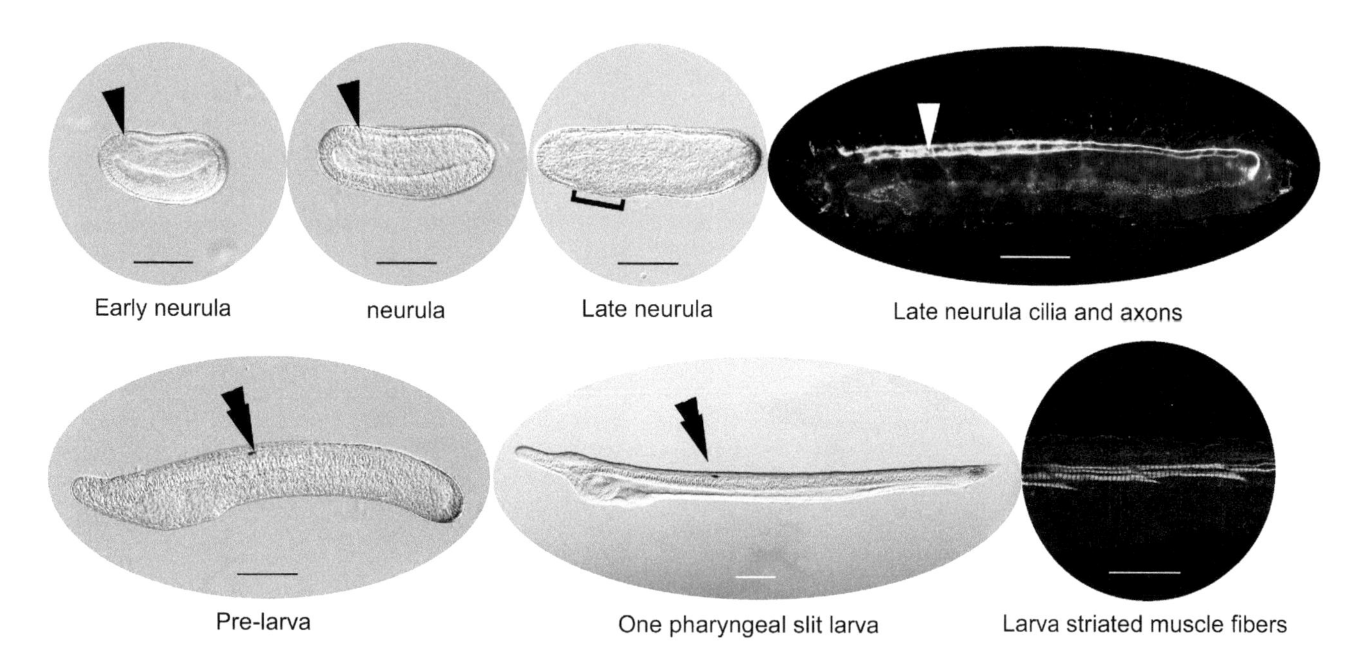

FIGURE 19.5 Neurulation. Pictures of *B. lanceolatum* neurula embryos and larva. At the beginning of the neurulation period, the epidermis has covered the rolling neural plate, leaving an anterior opening at the level of the cerebral vesicle called the neuropore (black arrowhead). In late neurula stage embryos, the pharyngeal region starts to enlarge (bracket) and neurons start to differentiate and grow axons (white arrowhead), as shown by the anti-acetylated tubulin immunostaining. Before the mouth opens, the pigment spot, which is associated with photoreceptor cells, is visible (double arrowhead). In the larva, striated muscle fibers are well developed, as shown by an enlarged picture of a larva after phalloidin-TexasRed labeling, allowing the animal to swim by both muscle contractions and cilia rotation. Lateral views with anterior to the left and dorsal to the top. Scale bar = 50 μm.

pharynx of the larva (Figure 19.5). The first pigment spot, which belongs to a photosensitive organ called the Hesse eyecup, appears. During neurulation, the formed somites elongate in the ventral region. The dorsal part, close to the notochord, forms striated muscle cells, whereas the ventral region participates in the formation of the circulatory system. The ventral region of the first left somite develops into the Hatschek's nephridium, the excretory organ of the larva, whereas the ventral part of the first right somite is considered a putative hematopoietic region. Finally, the left diverticulum becomes the preoral pit, or Hatschek's pit, and the right diverticulum becomes the rostral coelom, while the endostyle and the club-shaped gland (an organ specific to amphioxus) form from the wall of the pharyngeal endoderm. The mouth opens on the left side and the first pharyngeal slit on the ventral right side of the embryo that becomes a larva (Figure 19.5). At that time, the notochord has grown in the anterior region beyond the cerebral vesicle and segmented striated muscles

(a)

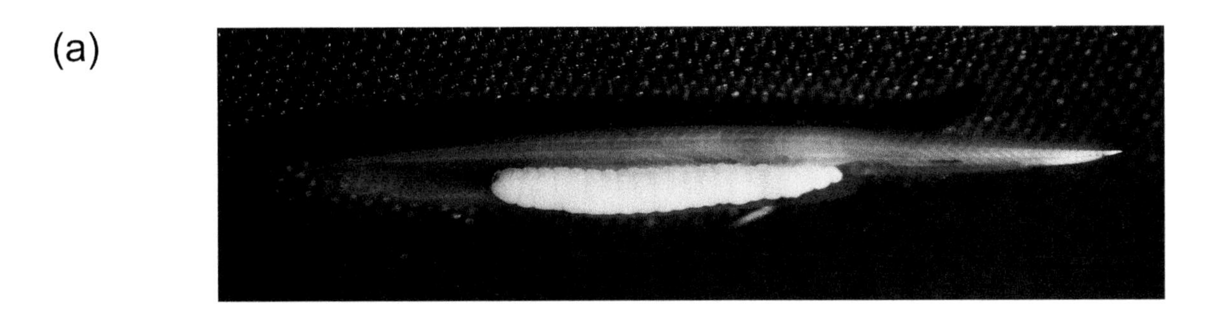

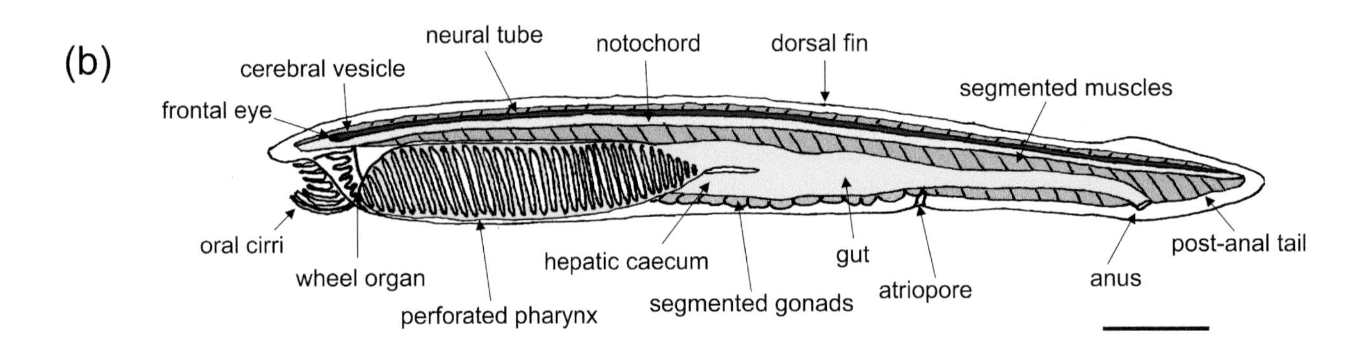

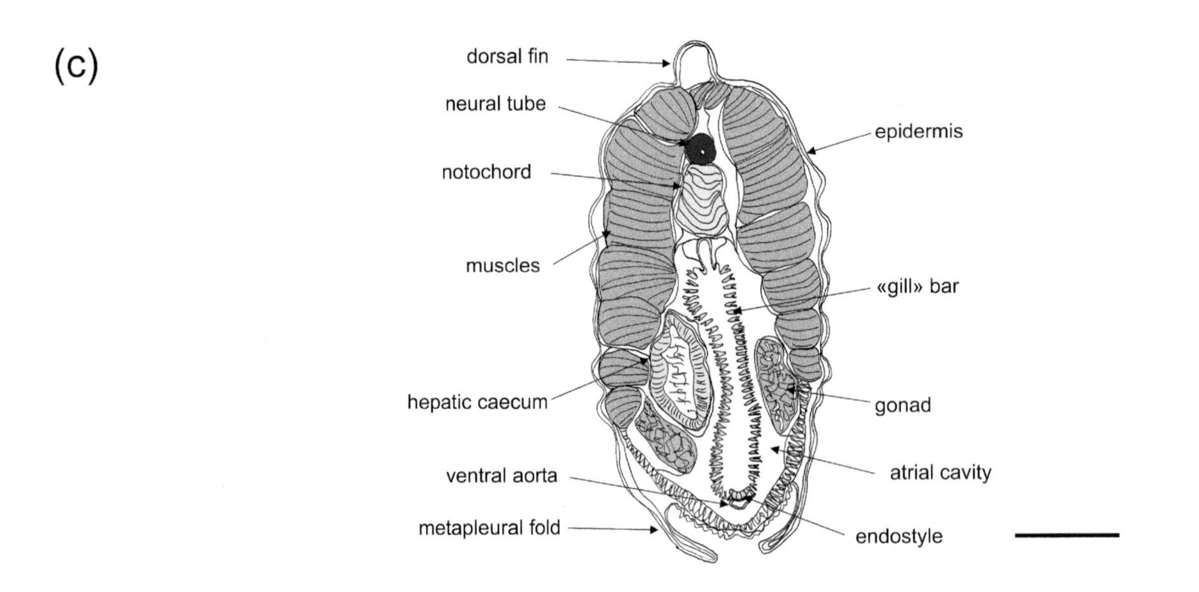

FIGURE 19.6 Morphology of cephalochordates. (a) Picture of an adult amphioxus of the *B. lanceolatum* species with visible gonads. Lateral view, anterior to the left and dorsal to the top, scale bar = 1 cm. (b) Diagram of the morphology of cephalochordates, lateral view with anterior to the left and dorsal to the top, scale bar = 1 cm. (c) Diagram of a cross-section at the level of the pharyngeal region. Dorsal to the top, scale bar = 0.5 cm. ([a] Courtesy of Guido Villani.)

have developed from the dorsal part of the somites, allowing the larva to swim by undulation in the plankton (Figure 19.5). The frontal eye, which is a photosensitive organ proposed to be homologous to the vertebrate retina, forms at the anterior tip of the cerebral vesicle. Finally, the anus opens and the larva starts to feed by filtering the seawater.

After this stage, the larva continues elongating, new somites are still forming in the posterior region and new pharyngeal slits open sequentially posterior to the first one. Once the number of slits has reached a threshold that depends on the species (between 9 and 18) (Holland and Yu 2004; Fuentes et al. 2007; Urata et al. 2007), the larva starts its metamorphosis. This post-embryonic process consists of many morphological modifications. The pharyngeal slits on the right side duplicate and form a second row that migrates toward the left region so that the juvenile possesses a row of slits on both sides of the body. The mouth migrates toward the ventral midline, as well as the endostyle, while the club-shaped gland disappears. Two membranes, called the metapleural folds, grow over the pharynx, cover it and fuse in the ventral midline, forming the atrial cavity that stays open in the posterior region at the level of the atriopore. At the same time, posterior to the pharynx, the hepatic caecum (a digestive gland) starts to bud from the digestive tract. Finally, the cilia of the epidermal cells are lost, and the juvenile migrates to the sediment.

19.5 ANATOMY

The anatomy of amphioxus has been extensively studied since its first scientific description, and a review of all the references can be found in Gans (1996). A diagram of amphioxus anatomy is presented in Figure 19.6. Amphioxus are elongated, almost transparent animals measuring just a few centimeters long at the adult stage. They are characterized by a prototypical chordate body plan and are considered vertebrate-like but simpler (Bertrand and Escriva 2011). As such, they possess a hollow nerve tube in the dorsal region, which forms a cerebral vesicle in the anterior part. Beneath the neural tube is a notochord, which is a rigid rod formed by aligned discoidal cells and which runs more anterior than the cerebral vesicle. This is why they are called cephalochordates ("cephalo" for "head", "chordate" for "notochord", name first proposed by Ernst Haeckel [Nielsen 2012]). The notochord is a shared character among chordates, with tunicates (or urochordates) presenting a notochord in the tail at the embryonic and larval stages at least and vertebrates having an embryonic notochord (except in their most anterior region) that disappears later on during the formation of the vertebral column in almost all species (Stemple 2004; Annona et al. 2015). Ventral to the notochord is the digestive tract: in the anterior region, the mouth is surrounded by oral cirri that form a net able to prevent the entry of big particles into the pharynx. The wheel organ, made of ciliated cells, borders the oral cavity. Posterior to it, the pharynx is windowed thanks to the pharyngeal slits present on both sides of the midline. Posterior to the pharynx are the gut and the hepatic caecum, the latter of which forms a tongue that is inserted between the pharynx and the wall of the atrium and that opens at the level of the junction between the intestine and the pharyngeal cavity. The ventral wall of the pharynx supports the endostyle, which produces mucus and has been proposed to be homologous to the vertebrate thyroid gland (Ogasawara 2000). Amphioxus swim by undulating their body thanks to the segmented V-shaped muscles that run all along their body on both sides. They also have segmented gonads whose gametes are first released into the atrial cavity and then into the sea water through the atriopore during spawning. The circulatory system consists of several contractile vessels and sinuses, and the vessels are formed by scattered endothelial cells embedded in a basal lamina (Moller and Philpott 1973a, 1973b). The proposed excretory system, although its function still needs to be clarified, corresponds to the Hatchek's nephridium derived from the ventral part of the first left somites and to other nephridia present as a succession of small paired structures associated with the pharyngeal slit clefts (Holland 2017).

19.6 GENOMIC DATA

Genomic and transcriptomic data are powerful resources to pose questions about genomic evolution and genetic control of development. Genomic and transcriptomic data are available for three *Branchiostoma* species (*B. floridae*, *B. belcheri* and *B. lanceolatum)* and transcriptomic data for one *Asymmetron* species *(Asymmetron lucayanum)* (see Table 19.1) (Putnam et al. 2008; Huang et al. 2012; Yue et al. 2014; Marletaz et al. 2018).

19.6.1 *Branchiostoma floridae*

This was the first genome to be sequenced and assembled in 2008. The project was supported by most of the research groups worldwide working with amphioxus (Holland et al. 2008; Putnam et al. 2008). The *B. floridae* genome was a key contribution to our understanding of chordate evolution and of the origin of vertebrates. It allowed for the reconstruction of the basic gene toolkit involved in development and cell signaling of the last common chordate ancestor. Although it was confirmed that amphioxus mostly contain a single-copy gene for each vertebrate paralogy group and that two rounds of whole-genome duplication predated the vertebrate lineage, it has also been assessed that the amphioxus genome has derived features represented by specific gene family expansion, such as the opsin one (Holland et al. 2008). Moreover, the *B. floridae* genome has allowed a reconstruction of the chromosomal organization of the chordate ancestor. (Access at https://mycocosm.jgi.doe.gov/Brafl1/Brafl1.home.html.)

19.6.2 *Branchiostoma belcheri*

The genome of this species was fully sequenced in 2012. The authors developed a novel automated pipeline named HaploMerger to create a better reference haploid assembly from the original diploid assembly, ensuring better sequence contiguity and continuity (Huang et al. 2012) (Access at http://genome.bucm.edu.cn/lancelet/gbrowser_wel.php.)

19.6.3 *Branchiostoma lanceolatum*

The genome of the Mediterranean amphioxus *B. lanceolatum* was published in 2018. Taking advantage of modern -omics approaches, the efforts of the genome sequencing consortium were focused on the analyses of this species' epigenome. For this purpose, DNA methylation, chromatin accessibility and histone modifications were characterized at the genomic scale. Additionally, transcriptomes across multiple developmental stages and adult tissues were produced. The main conclusion of this study (Acemel et al. 2016; Marletaz et al. 2018) (access at http://amphiencode.github.io), is that the genome of vertebrates has evolved by complexification at different levels, and we will detail this point in Section 19.8.

19.6.4 *Asymmetron lucayanum*

Transcriptomic data from larvae and adults have been generated for *A. lucayanum*, while the whole-genome sequence is not yet available. In their study, by comparing 430 orthologous gene groups among *A. lucayanum*, *B. floridae* and ten vertebrates, Yue and colleagues (2014) showed that cephalochordates are evolving, at the genetic scale, more slowly than any vertebrate, which is consistent with the substantial morphological similarities observed among extant cephalochordates that diverged more than 100 Mya.

19.7 FUNCTIONAL APPROACHES: TOOLS FOR MOLECULAR AND CELLULAR ANALYSES

Classical molecular biology approaches aimed at studying gene and protein localization are feasible in amphioxus, especially in embryos that are completely transparent. In particular, several protocols have been developed for *in situ* hybridization with labeled mRNA probes and for immunostaining approaches using antibodies against endogenous proteins. Moreover, the function of specific signaling pathways has been extensively studied using pharmacological treatments, since amphioxus embryos are particularly suitable for this kind of procedure. Examples on this topic are addressed in Section 19.8.

To understand the function of a given gene, it is necessary to interfere with its correct expression during development. This paradigm is at the base of the functional approaches used in developmental biology research. Classical tools to study gene function are overexpression (by mRNA injection or transient transgenesis), knock-down or knock-out (see Table 19.1). Microinjection is the tool of choice to introduce nucleic acids or proteins into the unfertilized amphioxus egg, rapidly followed by sperm fertilization (Holland and Yu 2004; Liu et al. 2013a; Hirsinger et al. 2015). The redistribution of the injected molecules in daughter cells after mitosis then guarantees gene repression or overexpression during embryonic development. Although there might not seem to be any specific reason for this kind of experiment to be difficult in comparison to similar models as ascidians and sea urchins, the hardness of the chorion and the fragility of the egg make the technique a bottleneck for functional analyses in amphioxus. Overexpression by mRNA injection of certain genes has been successfully achieved in all three main amphioxus species (*B. floridae, B. lanceolatum, B. belcheri*) (Onai et al. 2010; Li et al. 2017; Aldea et al. 2019; Zhang et al. 2019). Gene knock-down has been shown to be effective in *B. floridae* and *B. belcheri* by using gene-specific morpholinos that prevent the translation of mRNAs. Morpholino has been used to study the function of key transcription factors such as *Hox1* and *Pax1/9*, as well as the secreted protein *Dkk3* involved in head specification (Schubert et al. 2005, 2006; Holland and Onai 2011; Onai et al. 2012; Liu et al. 2013b; Liu et al. 2015).

Recently, a genomic mutagenesis approach has been developed in amphioxus by using the transcription activator-like effector nuclease (TALEN)-based technology. This knock-out application to amphioxus boosted the research in

TABLE 19.1
Availability of tools in different cephalochordate species

	B. floridae	B. belcheri	B. lanceolatum	A. lucayanum
Geographical location	Florida (USA), AO	Asia, PO	Europe, AO + MED	AO + IO + PO
Breeding season	May–September	May–July	May–July	Fall and Spring
Whole life cycle time	3 months	1 year	2 years	N/A
Whole life cycle in the lab	Yes	Yes	N/A	N/A, die at metamorphosis
Whole genome sequence	2008	2012	2018	N/A
Transcriptomes	Embryo larva & adult	Embryo larva & adult	Embryo larva & adult	Larva & adult
Overexpression	mRNA injection	mRNA injection	mRNA injection	N/A
Knock-down/knock-out	Morpholino injection, TALEN	TALEN	N/A	N/A
Transient transgenesis	Yes	Yes	Yes	N/A

AO: Atlantic Ocean, IO: Indian Ocean, PO: Pacific Ocean, MED: Mediterranean Sea the developmental biology field and filled the gaps with other chordate models (Li et al. 2014). Germ line mutagenesis has been used to study several important developmental genes, such as *Pax1/9*, *Pax3/7*, two ParaHox genes: *Pdx* and *Cdx*, *Hedgehog, Cerberus* and *Nodal* (Li et al. 2014; Wang et al. 2015; Hu et al. 2017; Li et al. 2017; Ren et al. 2020; Zhong et al. 2020; Zhu et al. 2020). Nevertheless, the long life cycle of amphioxus make these and other similar approaches very time consuming; this is the reason the tropical species *B. floridae* is more suitable than temperate species such as *B. lanceolatum*, which takes a few years to reach sexual maturity. It is foreseeable that in the next few years, gene function studies in amphioxus will also take advantage of the genome editing CRISPR/Cas9 (clustered regularly interspaced short palindromic repeats/Cas9) technique that represents the next-generation genome editing tool and provides high levels of gene-specific targeting and efficiency.

An efficient transgenic method to study enhancer activity has been recently developed for amphioxus: two transgenic amphioxus lines have been generated using the Tol2 transposon system, based on a hAT family transposon (Shi et al. 2018).

None of these functional approaches have been successfully developed in the *Asymmetron* genus, probably because only a few labs have access to live animals.

19.8 CHALLENGING QUESTIONS

Over the last decades, cephalochordates have become important animal models in the field of evo-devo. The phylogenetic position of amphioxus and its evolutionarily conserved morphology and genome organization make it an extremely useful organism for answering important evolutionary biology questions, in particular with respect to chordate evolutionary history. This section describes some important results obtained using amphioxus as a model as well as key questions for which the full answer is still to be found.

19.8.1 Chordate Genome and Evolution of Genomic Regulation

In the 1970s Susumu Ohno, a Japanese-American geneticist, proposed in his book *Evolution by Gene Duplication* that morphological novelties could result from gene duplications and that vertebrate genomes were built by one or probably two whole-genome duplications that took place during the invertebrate chordate to vertebrate transition (Ohno 1970). This hypothesis is named the 2R (for two rounds of duplication) hypothesis. Researchers have long tried to test this assumption using several arguments, such as the number of isozymes, the number of genes or the number of paralogues in vertebrates versus invertebrates. For example, it was shown that amphioxus has a single complete *Hox* gene cluster containing 15 genes, whereas mammals have four incomplete clusters (Amemiya et al. 2008; Putnam et al. 2008). The definitive argument for the 2R hypothesis came with the sequencing of the whole genome of the tunicate *Ciona intestinalis* and was confirmed by the sequencing of amphioxus's genome (Dehal and Boore 2005; Putnam et al. 2008). Cephalochordates, therefore, have an unduplicated genome compared to vertebrates, and it has been shown that, unlike tunicates, they have retained most of the genes present in the chordate ancestor genome, although some lineage-specific duplications occurred in several gene families (Holland et al. 2008). The cephalochordate genome thus represents the best proxy for the chordate ancestor genome, and analyses of *B. floridae* data allowed the reconstruction of the gene complement of the last common chordate ancestor and the partial reconstruction of its genomic organization (Holland et al. 2008).

Although the evolution of gene content during chordate evolution was probably crucial for their morphological diversification, the contribution of genome architecture and genome regulation is still to be finely studied. In this context, the recent description of the epigenome of the Mediterranean amphioxus, *B. lanceolatum*, already brought new insights. The characterization of the methylome, of chromatin accessibility and of histone modifications at different development stages and in several adult tissues allowed for the discovery of some functional changes that might have given rise to the greater complexity observed in vertebrates (Marletaz et al. 2018). For example, in vertebrates, there has been an increase in regulatory sequences, in particular those that regulate the expression of genes involved in the control of embryonic development. It was also shown that duplicate genes in vertebrates (after the 2R) have evolved mainly by subfunctionalization and specialization and that specialization of gene function was accompanied by an increase in regulatory complexity. Another study, focused on the Hox genomic region, showed that the complex regulation of Hox genes expression in vertebrate is in part due to the acquisition of a new three-dimensional organization of the chromatin around some of the Hox clusters (Acemel et al. 2016). Indeed, the amphioxus Hox gene cluster is contained in a single topologically associated domain (TAD), while in vertebrates, there are two TADs, one on each side of the cluster, and regulatory sequences present in these two TADs are responsible for the regulation of Hox genes expression in the limbs. This study of the *B. lanceolatum* genome also showed that although amphioxus presents a similar pattern of methylation to that of invertebrates (low methylation compared to vertebrates), the expression of some genes is regulated by demethylation in the same way as vertebrates (Marletaz et al. 2018). These recent data pave the way for a better understanding of the genomic regulation principles underlying the morphological and functional innovations of vertebrates. Nevertheless, further effort is necessary to overcome difficulties associated with enhancer element identification and understanding of their functional evolution throughout the last 500 million years.

19.8.2 Evolution of Vertebrate Morphological Traits

Although amphioxus share a typical chordate body plan with vertebrates, they lack key vertebrate characters such as the head, endoskeleton, migratory neural crest cells, placodes and paired appendages. Therefore, a comparative approach between invertebrate chordates and vertebrates should allow us to discover the main evolutionary innovations that led to the appearance of these complex structures, and amphioxus has been extensively used to answer such questions. In this section, as an example, the contribution of some studies using cephalochordates as a model for our understanding of the evolution of key vertebrate morphological features will be addressed.

19.8.2.1 Cartilage and Bones

One of the most iconic and specific structures of extant vertebrates is their endoskeleton made of cartilage and/or bone that is absent in tunicates and cephalochordates. However, in amphioxus, cartilage-like structures are found at the adult stage in the rods of the cirri that surround the mouth, which consist of cells embedded in a matrix, and in the "gill" bars of the pharynx, which were described as an acellular cartilage (Wright et al. 2001). Although it was proposed that both cartilage-like tissues were non-collagenous (Wright et al. 2001), it has been shown that fibrillar collagen, which is a major component of the cartilage matrix in vertebrates, is present in the pharyngeal "gill" bars (Rychel and Swalla 2007). In search of a conserved gene toolkit for cartilage formation, the expression pattern of amphioxus orthologues of genes controlling cartilage formation in vertebrates has been studied during embryogenesis. No co-expression could be observed, suggesting that cartilage did not appear by co-option of a pre-existing toolkit but probably by the appearance of new gene interactions (Meulemans and Bronner-Fraser 2007). However, these studies were carried out on embryos and not at later stages when the cartilage-like structures form (during metamorphosis). More recent studies using metamorphosing *B. floridae* larvae or regenerating oral cirri in adults have brought new insights on this issue (Kaneto and Wada 2011; Jandzik et al. 2015). It has been shown that during metamorphosis, *ColA*, coding for a collagen in amphioxus, is expressed in the forming oral cirri and in regenerating adult oral cirri as well as transcription factors required for cartilage formation in vertebrates (Kaneto and Wada 2011; Jandzik et al. 2015). The authors also showed that oral cirri formation is dependent upon FGF signaling, a signal which is required in vertebrates for cellular cartilage differentiation, and that adult regenerating cirri rods are expressing genes that are known to be required for osteogenesis in vertebrates (Kaneto and Wada 2011; Jandzik et al. 2015). All together, these data have shown that some elements of the chondrogenic and osteogenic programs of vertebrates were probably already required for the formation of cartilage-like structures in the chordate ancestor. However, more functional data, particularly focusing on amphioxus metamorphosis, are still required to understand the appearance of the vertebrate endoskeleton.

19.8.2.2 Neural Crest Cells

The neural crest cells (NCCs) are a specific transient population of cells specific to vertebrates that are sometimes referred to as the "fourth germ layer" of these animals (Gilbert 2000). They originate from the border of the neural plate at the time at which the neuroectoderm and the future epidermis separate during neurulation (Gilbert 2000). These cells undergo an epithelial-mesenchymal transition, delaminate and migrate all through the body where they differentiate into many different cell types such as melanocytes, adipocytes, neurons, smooth muscles, chondroblasts, odontoblasts and so on (Bronner and Simoes-Costa 2016). NCCs participate in the formation of structures that are vertebrate specific such as bones, cartilage and ganglia of the vertebrate head, and Gans and Northcutt even proposed that the vertebrates' "New Head" (an anterior structure with unsegmented muscles, well-developed brain and sensory organs) appearance was favored by the emergence of NCCs (Gans and Northcutt 1983). In amphioxus, there is no evidence of the existence of such cells, and it is considered that cephalochordates do not have migratory NCCs. However, neurulation occurs in a similar way as observed in vertebrates, and it has been shown that the neural plate border expresses genes that are orthologues of neural plate border specification genes in vertebrates (Yu et al. 2008). On the other hand, among the genes that are known to be required in vertebrates for the specification of NCC or among effector genes (that are downstream of the neural plate border specifying genes in the NCC gene regulatory network), only *Snail* is expressed in the neural plate border of amphioxus (Langeland et al. 1998). Concerning tunicates, the sister group of vertebrates, it has been shown in *Ciona intestinalis* that some cells expressing the NCC specification genes *Id*, *Snail*, *FoxD* and *Ets* differentiate into pigmented cells and that overexpression of *Twist* in these cells induces them to migrate (Abitua et al. 2012), suggesting that NCC would have appeared thanks to the recruitment of a "migratory" program at the neural plate border. However, tunicates have specific developmental modalities among chordates, and cephalochordates seem, at least during early embryogenesis, to develop most of their structures without any step of epithelial-mesenchymal transition, leaving the mystery of NCC emergence still incompletely resolved.

19.8.2.3 Eyes

Among the characters specific to vertebrates, the well-developed pair sensory organs are the most elaborate. The image-forming camera-type eye of vertebrates is a very complex structure composed of different tissues with various embryonic origins. Amphioxus, on the other hand, possess various photoreceptive organs: the lamellar body, Joseph cells, dorsal ocelli and the frontal eye, which is considered homologous to the vertebrate retina (Glardon et al. 1998; Pergner and Kozmik 2017). This very simple organ is formed at the larva stage at the tip of the cerebral vesicle, which is considered homologous to the vertebrate brain. The frontal eye consists of around six photoreceptor cells (Lacalli et al. 1994) of the ciliary type, like the cones and rods of the vertebrate retina, positioned posterior to nine pigment cells (Lacalli et al. 1994). The amphioxus photoreceptors and pigment cells express genes that are orthologous to genes known to be expressed in the photoreceptor cells and pigmented epithelium of the vertebrate retina, respectively (Vopalensky et al. 2012). Interestingly, other neurons positioned posterior to the row of photoreceptors were proposed to be homologous to the other cell types present in the vertebrate retina: interneurons and/or retinal ganglion cells (Lacalli et al. 1994; Lacalli 1996; Vopalensky et al. 2012). However, data are still missing in order to clearly answer this point. Another important aspect that would support the homology between the amphioxus frontal eye and vertebrate retina is the understanding of the developmental control of frontal eye formation. A recent study showed that, as in vertebrate embryos, inhibiting the Notch signaling pathway during amphioxus embryogenesis increases the number of photoreceptors formed (Pergner et al. 2020), but we are far from a complete understanding of the gene regulatory network underlying the formation of the frontal eye. Another key point that needs to be addressed is how vertebrate paired eyes evolved from a single, midline-positioned ancestral eye.

19.8.3 Evolution of Cell–Cell Signaling Pathways

Harmonious embryonic development relies on the capabilities of cells to communicate in order to construct the correct body plan. For this purpose, they use few signaling pathways, most of them being present in all metazoans (Barolo and Posakony 2002). One important question in the evo-devo field is therefore to understand how the evolution of these pathways (of their actors, roles and interactions) might have participated in the morphological diversification among animals. Amphioxus possess in their genome genes that code for the main actors of all the major signaling pathways, often with one orthologue for several paralogues in vertebrates that resulted from the two whole-genome duplications characterizing vertebrate early evolutionary history (Bertrand et al. 2017). One major issue that needs to be solved is how the multiplication of signaling pathway actors in vertebrates lead to the appearance of their morphological characters. There are still few data concerning this point, but we can cite the case of the retinoic acid receptors (RARs). This transcription factor, which is a nuclear receptor of retinoic acid, is encoded by a unique gene in amphioxus, whereas three paralogues, RARα, β and γ, are found in mammals. By comparing the expression pattern, the function and the binding capacity of vertebrate and amphioxus RARs, it has been proposed that RARβ kept chordate ancestral characteristics, whereas RARα and RARγ acquired new roles (i.e. neofunctionalization) during vertebrate evolution, which might explain the embryonic functions of retinoic acid that are specific to vertebrates (Escriva et al. 2006).

In cephalochordates, the developmental function of many cell–cell communication pathways has been studied mainly thanks to pharmacological treatments capable of inhibiting or activating these signals (for a review, see Bertrand et al. 2017). One of the advantages of using such an approach is the possibility to interfere with signaling pathways at different developmental time windows and therefore to study their implication in diverse developmental processes. Many data obtained in amphioxus have highlighted conservation in the use of different signals for the control of developmental processes with vertebrates, as might be expected given that chordates share a similar body plan. As an example, BMP and Nodal are opposing signals controlling the dorso–ventral patterning of the amphioxus embryo (Onai et al. 2010), the Wnt/β-catenin pathway regulates the formation of the dorsal organizer (Kozmikova and Kozmik 2020) and retinoic acid has been shown to act as a posteriorizing signal and to control the expression of Hox genes (Holland and Holland 1996; Escriva et al. 2002; Schubert et al. 2005), as is the case in vertebrates. However, we can point out some studies that reveal differences between amphioxus and vertebrates that might explain the emergence of some vertebrate novelties. In vertebrates, the somitogenesis process, which consists of the progressive segmentation of the paraxial mesoderm of the trunk during the embryo elongation (Pourquie 2001b), relies on the opposition of two main signals: the retinoic acid differentiating signal in the anterior region that acts in opposition to the fibroblast growth factor (FGF) and Wnt posterior proliferative signals (Pourquie 2001a). In amphioxus, the paraxial mesoderm gets segmented through a similar somitogenesis process, although it is also segmented in the anterior/head region, contrary to what happens in vertebrates. Interestingly, it has been shown in amphioxus that FGF controls only the formation of the anterior somites, that retinoic acid is not involved in this process and that FGF and retinoic acid do not seem to regulate each other during embryogenesis (Bertrand et al. 2011; Bertrand et al. 2015). These results might in part explain how the segmentation of the head mesoderm of vertebrates was lost during evolution and might indicate that the opposition between the FGF and retinoic acid signals, which controls the development of several vertebrate structures, would be a vertebrate novelty.

19.8.4 Evolution of the Immune System

The vertebrate immune system consists of two major components: innate and adaptive immunity. The former is common to all animals, while the latter was believed to be a vertebrate-specific system that relies on lymphocyte cells responsible for the so-called immune long-term memory. Amphioxus genomes possess homologs of most innate immune receptor genes found in vertebrates (Han et al. 2010; Dishaw et al. 2012), and many of these gene families have undergone large lineage-specific expansions, resulting in an extraordinary complexity and diversity of amphioxus innate immune gene complement (Huang et al. 2008). On the other hand, the identification of lymphocyte-like cells in the amphioxus pharynx and the finding of lymphoid proliferation and differentiation genes in cephalochordates indicate the presence of a kind of adaptive immunity system (Huang et al. 2007).

One of the most important events in the acquisition of adaptive immunity in vertebrates was the co-option of the RAG proteins for the antigen receptor gene assembly by V(D)

J recombination. It was long thought that RAG genes evolved from a transposon, and recent data in amphioxus support this hypothesis. Indeed, the amphioxus genome possesses a transposable element called *ProtoRAG* that codes for proteins showing sequence and function similarities with vertebrates RAG1 and RAG2 (Huang et al. 2016). These results highlight how amphioxus immune system studies might bring valuable insights into the evolution of vertebrate immunity.

19.8.5 Evolution of Regeneration

Regeneration is a variable feature in chordates, with some species capable of regenerating entire body parts, while others have only reduced abilities to do so. As a result, amphioxus has been shown to be a particularly relevant model organism for our understanding of the evolution and diversity of regeneration mechanisms in chordates. The first observations of this fascinating biological process go back to the beginning of the 20th century, but there has been a revival of interest in this topic in recent years. The latest pivotal studies have highlighted remarkable regenerative features of amphioxus both at the anatomical and molecular levels. In fact, similarities were found between tail regeneration in amphioxus and in vertebrates, although amphioxus can also rebuild the head region, a characteristic that vertebrates have lost (Kaneto and Wada 2011; Somorjai et al. 2012; Somorjai 2017; Liang et al. 2019). Moreover, the regeneration genetic toolkit seems in part to be conserved between amphioxus and vertebrates, as demonstrated by the key role of *Pax*, *Sox* and *Msx* genes (Somorjai et al. 2012; Somorjai 2017) and of the BMP signaling pathway (Liang et al. 2019). Nevertheless, since we are only beginning to dissect the regeneration process in cephalochordates, the potential of amphioxus as a non-vertebrate chordate regeneration model, and to what extent the progress made on understanding the regulation of amphioxus genome may highlight processes that are too complex in vertebrates, remains to be shown.

Importantly, in the last years, evidence of stem cell populations that could contribute to the regenerative process in amphioxus is opening new perspectives. Moreover, recent data suggest the possibility that cephalochordates possess an inherited mechanism for primordial germ cell (PGC) specification rather than an inductive one, as previously thought. PGCs are grouped posteriorly in the endoderm of the neurula tailbud and cluster near the anus at larval stages (Wu et al. 2011; Zhang et al. 2013; Dailey et al. 2016). It is thus very likely that what we will learn from cephalochordate research will complement and help further the study of regeneration and stem cells in vertebrates.

BIBLIOGRAPHY

Abitua, P. B., E. Wagner, I. A. Navarrete et al. 2012. Identification of a rudimentary neural crest in a non-vertebrate chordate. *Nature* 492 (7427):104–107.

Acemel, R. D., J. J. Tena, I. Irastorza-Azcarate et al. 2016. A single three-dimensional chromatin compartment in amphioxus indicates a stepwise evolution of vertebrate Hox bimodal regulation. *Nature Genetics* 48 (3):336–341.

Aldea, D., L. Subirana, C. Keime et al. 2019. Genetic regulation of amphioxus somitogenesis informs the evolution of the vertebrate head mesoderm. *Nature Ecology and Evolution* 3 (8):1233–1240.

Amemiya, C. T., S. J. Prohaska, A. Hill-Force et al. 2008. The amphioxus Hox cluster: Characterization, comparative genomics, and evolution. *Journal of Experimental Zoology Part B: Molecular and Developmental Evolution* 310 (5):465–477.

Annona, G., N. D. Holland and S. D'Aniello. 2015. Evolution of the notochord. *Evodevo* 6 (1):30.

Barolo, S. and J. W. Posakony. 2002. Three habits of highly effective signaling pathways: Principles of transcriptional control by developmental cell signaling. *Genes & Development* 16 (10):1167–1181.

Benito-Gutierrez, E., H. Weber, D. V. Bryant et al. 2013. Methods for generating year-round access to amphioxus in the laboratory. *PLoS One* 8 (8).

Bertrand, S., D. Aldea, S. Oulion et al. 2015. Evolution of the role of RA and FGF signals in the control of somitogenesis in chordates. *PLoS One* 10 (9):e0136587.

Bertrand, S., A. Camasses, I. Somorjai et al. 2011. Amphioxus FGF signaling predicts the acquisition of vertebrate morphological traits. *Proceedings of the National Academy of Sciences of the United States of America* 108 (22):9160–9165.

Bertrand, S. and H. Escriva. 2011. Evolutionary crossroads in developmental biology: Amphioxus. *Development* 138 (22): 4819–4830.

Bertrand, S., Y. Le Petillon, I. M. L. Somorjai et al. 2017. Developmental cell-cell communication pathways in the cephalochordate amphioxus: Actors and functions. *International Journal of Developmental Biology* 61 (10–11–12):697–722.

Bronner, M. E. and M. Simoes-Costa. 2016. The neural crest migrating into the twenty-first century. *Current Topics in Developmental Biology* 116:115–134.

Carvalho, J. E., F. Lahaye and M. Schubert. 2017. Keeping amphioxus in the laboratory: An update on available husbandry methods. *International Journal of Developmental Biology* 61 (10–11–12):773–783.

Chen, Y., S. G. Cheung and P. K. S. Shin. 2008. The diet of amphioxus in subtropical Hong Kong as indicated by fatty acid and stable isotopic analyses. *Journal of the Marine Biological Association of the United Kingdom* 88 (7):1487–1491.

Conklin, E. G. 1932. The embryology of amphioxus. *Journal of Morphology* 54 (1):69–151.

Costa, O. 1834. *Annuario zoologico. Cenni zoologici, ossia descrizione sommaria delle specie nuove di animali discoperti in diverse contrade del Regno nell'anno 1834.* Napoli: Azzolino.

Dailey, S. C., R. F. Planas, A. R. Espier et al. 2016. Asymmetric distribution of pl10 and bruno2, new members of a conserved core of early germline determinants in cephalochordates. *Frontiers in Ecology and Evolution* 3:156.

Davydoff, C. 1960. Alexandre Kovalevsky (1840–1901): Souvenirs d'un disciple. *Revue d'histoire des sciences* 325–348.

Dehal, P. and J. L. Boore. 2005. Two rounds of whole genome duplication in the ancestral vertebrate. *PLoS Biol.* 3 (10):e314.

Delsuc, F., H. Brinkmann, D. Chourrout et al. 2006. Tunicates and not cephalochordates are the closest living relatives of vertebrates. *Nature* 439 (7079):965–968.

Desdevises, Y., V. Maillet, M. Fuentes et al. 2011. A snapshot of the population structure of *Branchiostoma lanceolatum* in the Racou beach, France, during its spawning season. *PLoS One* 6 (4):e18520.

Dishaw, L. J., R. N. Haire and G. W. Litman. 2012. The amphioxus genome provides unique insight into the evolution of immunity. *Briefings in Functional Genomics* 11 (2):167–176.

Escriva, H., S. Bertrand, P. Germain et al. 2006. Neofunctionalization in vertebrates: The example of retinoic acid receptors. *PLoS Genetics* 2 (7):e102.

Escriva, H., N. D. Holland, H. Gronemeyer et al. 2002. The retinoic acid signaling pathway regulates anterior/posterior patterning in the nerve cord and pharynx of amphioxus, a chordate lacking neural crest. *Development* 129 (12):2905–2916.

Feng, Y., J. Li and A. Xu. 2016. Chapter 1: Amphioxus as a model for understanding the evolution of vertebrates. In *Amphioxus Immunity*, edited by Anlong Xu, 1–13. Beijing: Academic Press.

Fuentes, M., E. Benito, S. Bertrand et al. 2007. Insights into spawning behavior and development of the European amphioxus (*Branchiostoma lanceolatum*). *Journal of Experimental Zoology Part B: Molecular and Developmental Evolution* 308 (4):484–493.

Fuentes, M., M. Schubert, D. Dalfo et al. 2004. Preliminary observations on the spawning conditions of the European amphioxus (*Branchiostoma lanceolatum*) in captivity. *Journal* of *Experimental Zoology Part B*: *Molecular and Developmental Evolution* 302 (4):384–391.

Gans, C. 1996. Study of lancelets: The first 200 years. *Israel Journal of Zoology* 42:S3–S11.

Gans, C. and R. G. Northcutt. 1983. Neural crest and the origin of vertebrates: A new head. *Science* 220 (4594):268–273.

Garcia-Fernàndez, J., S. Jiménez-Delgado, J. Pascual-Anaya et al. 2009. From the American to the European amphioxus: Towards experimental Evo-Devo at the origin of chordates. *International Journal of Developmental Biology* 53 (8–10):1359–1366.

Gilbert, S. F. 2000. *Developmental Biology*. Sunderland, MA: Sinauer Associates.

Glardon, S., L. Z. Holland, W. J. Gehring et al. 1998. Isolation and developmental expression of the amphioxus Pax-6 gene (AmphiPax-6): Insights into eye and photoreceptor evolution. *Development* 125 (14):2701–2710.

Haeckel, E. 1905. *The Evolution of Man: A Popular Scientific Study*, vol. II. Joseph McCabe (trans. from the 5th ed.). New York: Putnam and Sons.

Han, Y., G. Huang, Q. Zhang et al. 2010. The primitive immune system of amphioxus provides insights into the ancestral structure of the vertebrate immune system. *Developmental & Comparative Immunology* 34 (8):791–796.

Hatschek, B. and J. Tuckey. 1893. *The Amphioxus and Its Development*. London: Swan, Sonnenschein & Co.

Hirsinger, E., J. E. Carvalho, C. Chevalier et al. 2015. Expression of fluorescent proteins in *Branchiostoma lanceolatum* by mRNA injection into unfertilized oocytes. *Journal of Visualized Experiments: JoVE* (95):52042.

Holland, L. Z., R. Albalat, K. Azumi et al. 2008. The amphioxus genome illuminates vertebrate origins and cephalochordate biology. *Genome Res*earch 18 (7):1100–1111.

Holland, L. Z. and N. D. Holland. 1996. Expression of AmphiHox-1 and AmphiPax-1 in amphioxus embryos treated with retinoic acid: Insights into evolution and patterning of the chordate nerve cord and pharynx. *Development* 122 (6): 1829–1838.

Holland, L. Z. and T. Onai. 2011. Analyses of gene function in amphioxus embryos by microinjection of mRNAs and morpholino oligonucleotides. *Methods in Molecular Biology* 770:423–438.

Holland, L. Z. and J. K. Yu. 2004. Cephalochordate (amphioxus) embryos: Procurement, culture, and basic methods. *Methods in Cell Biology* 74:195–215.

Holland, N. D. 2017. The long and winding path to understanding kidney structure in amphioxus: A review. *International Journal of Developmental Biology* 61 (10–12):683–688.

Holland, N. D. and L. Z. Holland. 1989. Fine-structural study of the cortical reaction and formation of the egg coats in a lancelet (= Amphioxus), *Branchiostoma floridae* (Phylum chordata, subphylum cephalochordata = acrania). *Biological Bulletin* 176 (2):111–122.

Holland, N. D. and L. Z. Holland. 2010. Laboratory spawning and development of the Bahama lancelet, *Asymmetron lucayanum* (cephalochordata): Fertilization through feeding larvae. *Biolical Bulletin* 219 (2):132–141.

Holland, N. D. and L. Z. Holland. 2017. The ups and downs of amphioxus biology: A history. *International Journal of Developmental Biology* 61 (10–12):575–583.

Holland, N. D., L. Z. Holland and A. Heimberg. 2015. Hybrids between the Florida amphioxus (*Branchiostoma floridae*) and the Bahamas lancelet (*Asymmetron lucayanum*): Developmental morphology and chromosome counts. *Biological Bulletin* 228 (1):13–24.

Holland, P. W., L. Z. Holland, N. A. Williams et al. 1992. An amphioxus homeobox gene: Sequence conservation, spatial expression during development and insights into vertebrate evolution. *Development* 116 (3):653–661.

Hu, G., G. Li, H. Wang et al. 2017. Hedgehog participates in the establishment of left-right asymmetry during amphioxus development by controlling Cerberus expression. *Development* 144 (24):4694–4703.

Huang, G., X. Xie, Y. Han et al. 2007. The identification of lymphocyte-like cells and lymphoid-related genes in amphioxus indicates the twilight for the emergence of adaptive immune system. *PLoS One* 2 (2):e206.

Huang, S., Z. Chen, G. Huang et al. 2012. HaploMerger: Reconstructing allelic relationships for polymorphic diploid genome assemblies. *Genome Res*earch 22 (8):1581–1588.

Huang, S., X. Tao, S. Yuan et al. 2016. Discovery of an active RAG transposon illuminates the origins of V(D)J recombination. *Cell* 166 (1):102–114.

Huang, S., S. Yuan, L. Guo et al. 2008. Genomic analysis of the immune gene repertoire of amphioxus reveals extraordinary innate complexity and diversity. *Genome Res*earch 18 (7):1112–1126.

Igawa, T., M. Nozawa, D. G. Suzuki et al. 2017. Evolutionary history of the extant amphioxus lineage with shallow-branching diversification. *Scientific Reports* 7 (1):1–14.

Jandzik, D., A. T. Garnett, T. A. Square et al. 2015. Evolution of the new vertebrate head by co-option of an ancient chordate skeletal tissue. *Nature* 518 (7540):534–537.

Kaneto, S. and H. Wada. 2011. Regeneration of amphioxus oral cirri and its skeletal rods: Implications for the origin of the vertebrate skeleton. *Journal of Experimental Zoology Part B: Molecular and Developmental Evolution* 316 (6):409–417.

Kon, T., M. Nohara, M. Nishida et al. 2006. Hidden ancient diversification in the circumtropical lancelet *Asymmetron lucayanum* complex. *Marine Biology* 149 (4):875–883.

Kon, T., M. Nohara, Y. Yamanoue et al. 2007. Phylogenetic position of a whale-fall lancelet (Cephalochordata) inferred from whole mitochondrial genome sequences. *BMC Evolutionary Biology* 7 (1).

Kovalevskij, A. O. 1867. *Entwickelungsgeschichte des Amphioxus lanceolatus*. St-Pétersboug: Eggers & Schmitzdorff.

Kozmikova, I. and Z. Kozmik. 2020. Wnt/beta-catenin signaling is an evolutionarily conserved determinant of chordate dorsal organizer. *Elife* 9:e56817.

Lacalli, T. C. 1996. Frontal eye circuitry, rostral sensory pathways and brain organization in amphioxus larvae: Evidence from 3D reconstructions. *Philosophical Transactions of the Royal Society of London: Series B: Biological Sciences* 351 (1337):243–263.

Lacalli, T. C., N. Holland and J. West. 1994. Landmarks in the anterior central nervous system of amphioxus larvae. *Philosophical Transactions of the Royal Society of London: Series B: Biological Sciences* 344 (1308):165–185.

Langeland, J. A., J. M. Tomsa, W. R. Jackman, Jr. et al. 1998. An amphioxus snail gene: Expression in paraxial mesoderm and neural plate suggests a conserved role in patterning the chordate embryo. *Development Genes and Evolution* 208 (10):569–577.

Li, G., J. Feng, Y. Lei et al. 2014. Mutagenesis at specific genomic loci of amphioxus *Branchiostoma belcheri* using TALEN method. *Journal of Genetics and Genomics* 41 (4):215–219.

Li, G., X. Liu, C. Xing et al. 2017. Cerberus-Nodal-Lefty-Pitx signaling cascade controls left-right asymmetry in amphioxus. *Proceedings of the National Academy of Sciences of the United States of America* 114 (14):3684–3689.

Li, G., Z. Shu and Y. Wang. 2013. Year-round reproduction and induced spawning of Chinese amphioxus, *Branchiostoma belcheri*, in laboratory. *PLoS One* 8 (9):e75461.

Liang, Y., D. Rathnayake, S. Huang et al. 2019. BMP signaling is required for amphioxus tail regeneration. *Development* 146 (4):dev166017.

Light, S. F. 1923. Amphioxus fisheries near the University of Amoy, China. *Science* 58 (1491):57–60.

Liu, X., G. Li, J. Feng et al. 2013a. An efficient microinjection method for unfertilized eggs of Asian amphioxus *Branchiostoma belcheri*. *Development Genes and Evolution* 223 (4):269–278.

Liu, X., G. Li, X. Liu et al. 2015. The role of the Pax1/9 gene in the early development of amphioxus pharyngeal gill slits. *Journal of Experimental Zoology Part B: Molecular and Developmental Evolution* 324 (1):30–40.

Liu, X., H. Wang, G. Li et al. 2013b. The function of DrPax1b gene in the embryonic development of zebrafish. *Genes & Genetic Systems* 88 (4):261–269.

Marletaz, F., P. N. Firbas, I. Maeso et al. 2018. Amphioxus functional genomics and the origins of vertebrate gene regulation. *Nature* 564 (7734):64–70.

Meulemans, D. and M. Bronner-Fraser. 2007. Insights from amphioxus into the evolution of vertebrate cartilage. *PLoS One* 2 (8):e787.

Moller, P. C. and C. W. Philpott. 1973a. Circulatory-system of amphioxus (*Branchiostoma floridae*). 1: Morphology of major vessels of pharyngeal area. *Journal of Morphology* 139 (4):389–406.

Moller, P. C. and C. W. Philpott. 1973b. Circulatory-system of amphioxus (*Branchiostoma floridae*). 2: Uptake of exogenous proteins by endothelial cells. *Zeitschrift Fur Zellforschung Und Mikroskopische Anatomie* 143 (1):135–141.

Nielsen, C. 2012. The authorship of higher chordate taxa. *Zoologica Scripta* 41 (4):435–436.

Nishikawa, T. 2004. A new deep-water lancelet (Cephalochordata) from off Cape Nomamisaki, SW Japan, with a proposal of the revised system recovering the genus *Asymmetron*. *Zoological Science* 21 (11):1131–1136.

Nohara, M., M. Nishida, M. Miya et al. 2005. Evolution of the mitochondrial genome in Cephalochordata as inferred from complete nucleotide sequences from two Epigonichthys species. *Journal of Molecular Evolution* 60 (4):526–537.

Ogasawara, M. 2000. Overlapping expression of amphioxus homologs of the thyroid transcription factor-1 gene and thyroid peroxidase gene in the endostyle: Insight into evolution of the thyroid gland. *Development Genes and Evolution* 210 (5):231–242.

Ohno, S. 1970. *Evolution by Gene Duplication*. Berlin, Heidelberg: Springer.

Onai, T., A. Takai, D. H. Setiamarga et al. 2012. Essential role of Dkk3 for head formation by inhibiting Wnt/beta-catenin and Nodal/Vg1 signaling pathways in the basal chordate amphioxus. *Evolution & Development* 14 (4):338–350.

Onai, T., J. K. Yu, I. L. Blitz et al. 2010. Opposing Nodal/Vg1 and BMP signals mediate axial patterning in embryos of the basal chordate amphioxus. *Developmental Biology* 344 (1):377–389.

Pallas, P. 1774. Limax lanceolatus: Descriptio limacis lanceolaris. *Spicilegia Zoologica, quibus novae imprimus et obscurae animalium species iconibus, descriptionibus. Gottlieb August Lange, Berlin* 10:19.

Pan, M. M., D. J. Yuan, S. W. Chen et al. 2015. Diversity and composition of the bacterial community in Amphioxus feces. *Journal of Basic Microbiology* 55 (11):1336–1342.

Pergner, J. and Z. Kozmik. 2017. Amphioxus photoreceptors: Insights into the evolution of vertebrate opsins, vision and circadian rhythmicity. *International Journal of Developmental Biology* 61 (10–11–12):665–681.

Pergner, J., A. Vavrova, I. Kozmikova et al. 2020. Molecular fingerprint of amphioxus frontal eye illuminates the evolution of homologous cell types in the chordate retina. *Frontiers in Cell and Developmental Biology* 8:705.

Poss, S. G. and H. T. Boschung. 1996. Lancelets (Cephalochordata: Branchiostomatidae): How many species are valid? *Israel Journal of Zoology* 42:S13–S66.

Pourquie, O. 2001a. The vertebrate segmentation clock. *Journal of Anatomy* 199 (Pt 1–2):169–175.

Pourquie, O. 2001b. Vertebrate somitogenesis. *Annual Review of Cell and Developmental Biology* 17:311–350.

Putnam, N. H., T. Butts, D. E. Ferrier et al. 2008. The amphioxus genome and the evolution of the chordate karyotype. *Nature* 453 (7198):1064–1071.

Ren, Q., Y. Zhong, X. Huang et al. 2020. Step-wise evolution of neural patterning by Hedgehog signalling in chordates. *Nature Ecology & Evolution* 4 (9):1247–1255.

Ruppert, E. E., T. R. Nash and A. J. Smith. 2000. The size range of suspended particles trapped and ingested by the filter-feeding lancelet *Branchiostoma floridae* (Cephalochordata: Acrania). *Journal of the Marine Biological Association of the United Kingdom* 80 (2):329–332.

Rychel, A. L. and B. J. Swalla. 2007. Development and evolution of chordate cartilage. *Journal of Experimental Zoology Part B: Molecular and Developmental Evolution* 308 (3):325–335.

Schubert, M., N. D. Holland, V. Laudet et al. 2006. A retinoic acid-Hox hierarchy controls both anterior/posterior patterning and neuronal specification in the developing central nervous system of the cephalochordate amphioxus. *Developmental Biology* 296 (1):190–202.

Schubert, M., J. K. Yu, N. D. Holland et al. 2005. Retinoic acid signaling acts via Hox1 to establish the posterior limit of the pharynx in the chordate amphioxus. *Development* 132 (1):61–73.

Shi, C., J. Huang, S. Chen et al. 2018. Generation of two transgenic amphioxus lines using the Tol2 transposon system. *Journal of Genetics and Genomics* 45 (9):513–516.

Somorjai, I. M. L. 2017. Amphioxus regeneration: Evolutionary and biomedical implications. *International Journal of Developmental Biology* 61 (10–11–12):689–696.

Somorjai, I. M. L., R. L. Somorjai, J. Garcia-Fernandez et al. 2012. Vertebrate-like regeneration in the invertebrate chordate amphioxus. *Proceedings of the National Academy of Sciences of the United States of America* 109 (2):517–522.

Stemple, D. L. 2004. The notochord. *Current Biology* 14 (20): R873–R874.

Stokes, M. D. and N. D. Holland. 1996. Reproduction of the Florida lancelet (*Branchiostoma floridae*): Spawning patterns and fluctuations in gonad indexes and nutritional reserves. *Invertebrate Biology* 115 (4):349–359.

Stokes, M. D. and N. D. Holland. 1998. The lancelet. *American Scientist* 86 (6):552–560.

Subirana, L., V. Farstey, S. Bertrand et al. 2020. Asymmetron lucayanum: How many species are valid? *PLoS One* 15 (3):e0229119.

Theodosiou, M., A. Colin, J. Schulz et al. 2011. Amphioxus spawning behavior in an artificial seawater facility. *Journal of Experimental Zoology Part B: Molecular and Developmental Evolution* 316 (4):263–275.

Tung, T. C., S. C. Wu and Y. F. Tung. 1958. The development of isolated blastomeres of Amphioxus. *Scientia Sinica* 7 (12):1280–1320.

Tung, T. C., S. C. Wu and Y. Y. Tung. 1960. The developmental potencies of the blastomere layers in Amphioxus egg at the 32-cell stage. *Scientia Sinica* 9:119–141.

Tung, T. C., S. C. Wu and Y. Y. F. Tung. 1962. Presumptive areas of egg of amphioxus. *Scientia Sinica* 11 (5):629–644.

Tung, T. C., S. C. Wu and Y. Y. F. Tung. 1965. Differentiation of prospective ectodermal and entodermal cells after transplantation to new surroundings in amphioxus. *Scientia Sinica* 14 (12):1785–1794.

Urata, M., N. Yamaguchi, Y. Henmi et al. 2007. Larval development of the oriental lancelet, *Branchiostoma belcheri*, in laboratory mass culture. *Zoological Science* 24 (8):787–797.

Vopalensky, P., J. Pergner, M. Liegertova et al. 2012. Molecular analysis of the amphioxus frontal eye unravels the evolutionary origin of the retina and pigment cells of the vertebrate eye. *Proceedings of the National Academy of Sciences of the United States of America* 109 (38):15383–15388.

Wang, H., G. Li and Y. Q. Wang. 2015. Generating amphioxus Hedgehog knockout mutants and phenotype analysis. *Yi Chuan* 37 (10):1036–1043.

Wright, G. M., F. W. Keeley and P. Robson. 2001. The unusual cartilaginous tissues of jawless craniates, cephalochordates and invertebrates. *Cell and Tissue Research* 304 (2):165–174.

Wu, H. R., Y. T. Chen, Y. H. Su et al. 2011. Asymmetric localization of germline markers Vasa and Nanos during early development in the amphioxus *Branchiostoma floridae*. *Developmental Biology* 353 (1):147–159.

Yarrell, W. 1836. *A History of British Fishes*. London: J. Van Voorst.

Yu, J. K., D. Meulemans, S. J. McKeown et al. 2008. Insights from the amphioxus genome on the origin of vertebrate neural crest. *Genome Research* 18 (7):1127–1132.

Yue, J. X., J. K. Yu, N. H. Putnam et al. 2014. The transcriptome of an amphioxus, *Asymmetron lucayanum*, from the Bahamas: A window into chordate evolution. *Genome Biology and Evolution* 6 (10):2681–2696.

Zhang, H., S. Chen, C. Shang et al. 2019. Interplay between Lefty and Nodal signaling is essential for the organizer and axial formation in amphioxus embryos. *Developmental Biology* 456 (1):63–73.

Zhang, Q. J., Y. J. Luo, H. R. Wu et al. 2013. Expression of germline markers in three species of amphioxus supports a preformation mechanism of germ cell development in cephalochordates. *Evodevo* 4 (1):17.

Zhang, Q. J., Y. Sun, J. Zhong et al. 2007. Continuous culture of two lancelets and production of the second filial generations in the laboratory. *The Journal of Experimental Zoology Part B: Molecular and Developmental Evolution* 308 (4):464–472.

Zhang, Q.-J., J. Zhong, S.-H. Fang et al. 2006. *Branchiostoma japonicum* and *B. belcheri* are distinct lancelets (Cephalochordata) in Xiamen waters in China. *Zoological Science* 23 (6):573–579.

Zhang, S. C., N. D. Holland and L. Z. Holland. 1997. Topographic changes in nascent and early mesoderm in amphioxus embryos studied by DiI labeling and by in situ hybridization for a Brachyury gene. *Development Genes and Evolution* 206 (8):532–535.

Zhong, Y., C. Herrera-Ubeda, J. Garcia-Fernandez et al. 2020. Mutation of amphioxus Pdx and Cdx demonstrates conserved roles for ParaHox genes in gut, anus and tail patterning. *BMC Biology* 18 (1):1–15.

Zhu, X., C. Shi, Y. Zhong et al. 2020. Cilia-driven asymmetric Hedgehog signalling determines the amphioxus left-right axis by controlling Dand5 expression. *Development* 147 (1): dev182469.

20 Solitary Ascidians

Gabriel Krasovec, Kilian Biasuz, Lisa M. Thomann and Jean-Philippe Chambon

CONTENTS

DOI: 10.1201/9781003217503-20

20.1 INTRODUCTION

The tunicates present various ecological behaviors comprising sessile or pelagic adult forms in addition to colonial or solitary animals. Solitary ascidians are present in several tunicate groups, meaning that both solitary and colonial ascidians are not restrictive or typical to a given clade. Whereas distribution of solitary ascidians is scattered in the urochordate tree, they share some common features, and these can be studied in a common specific chapter. Despite the large diversity of solitary ascidians, they can be characterized by typical features such as an individual sessile adult presenting two siphons (one inhalant and one exhalant, allowing the circulation of sea water), a pharynx supported by an endostyle and a large branchial basket structure. Usually hermaphrodites, fertilization takes place in sea water after the release of gametes and gives rise to a swimming pelagic larva which will have to settle in a definitive substrate. Unlike in colonial ascidians, asexual reproduction is not documented.

Solitary ascidians have had a noticeable historical contribution to developmental and cell biology studies and include several well-established models in marine biology such as *Ciona intestinalis*. As the sister group of vertebrates, ascidians genetics data and genomics tools have opened broad perspectives to understand the development and evolution of chordates. Moreover, the financial importance of some solitary ascidians species is notable as marine alimentary resources, like *Microcosmus sabatieri* (usually named "violet" or "sea fig") in the south of France; *Styela clava* in Korea; or *Halocyntia roretzi*, which has been popular in Japan. On the contrary, negative ecological consequences can result from invasive species, like *Styela clava*. The preceding succinct presentation of solitary ascidians highlights the necessity and relevance of an overview.

20.2 HISTORY OF THE MODEL

The evolutionary history of tunicates is documented by fossil records comprising organisms attested or suggested to be solitary ascidians. The fact that the first tunicate fossil evidence seems to correspond to solitary ascidians is probably due to a typical shape presenting two siphons in a "bag-shaped" morphology characterized by a pharynx and gill slit, making fossils of solitary ascidians easier to identify than other tunicates. The oldest attested representative is *Shankouclava shankpuense*, which has an estimated age of 524 million years corresponding to the second Turgenevian stage of the Cambrian, discovered in China (Chen et al. 2003). This discovery introduced tunicates, at least solitary ascidians, as part of the high diversity explosion of the Cambrian, witnessing the emergence of several major current groups of animals. Finally, some hypothetical identifications, such as *Yarnemia ascidiformis* (Chistyakov et al. 1984) or *Burykhia hunti* (Fedonkin et al. 2012) from the Russian Ediacaran (550 and 555 million years old, respectively), suggest an older appearance of ascidians.

The current species, *Ascidiella aspersa*, was the first experimental model in developmental biology, on which Laurent Chabry studied blastomere recombination at the end of the 19th century (Chabry 1887). Chabry destroyed one of the blastomeres of two-cell embryos and found that the surviving one was able to form a half-embryo (more precisely, a dwarf malformed larva). He obtained similar results with the same kind of experiment on four-cell embryos and deduced that an amputated early embryo is unable to compensate for deleted cells during the development. Consequently, pioneer experiments made by Chabry suggested that each part of the larva came from specific cells emerging during the first divisions. Next, Edwin Grant Conklin deepened our understanding of embryogenesis by working on the lineage of embryonic cells and the segregation of the egg cytoplasm of various species of solitary ascidians such as *Styela canopus* (Conklin 1905a, 1905b). He reconstructed the lineage of cells from the first divisions to the well-developed larva and confirmed the suggestions coming from Chabry's experiments; development is characterized by cell lineages, which give specific tissues in the future larva what was called "a development in mosaic". Conklin's studies on egg cytoplasm segregation, in addition to cell lineage characterization, led to the hypothesis that female determinants are present in the eggs to drive and participate in the cell fate establishment during development. Solitary ascidians consequently allowed the discovery of two fundamental points in developmental and cell biology: the existence of maternal determinants (now known as maternal RNA) and the existence of cell lineages. In the same period as Conklin's experiments, other biologists focused on tunicate reproduction biology, such as Thomas Morgan, who demonstrated in 1904, on *Ciona intestinalis*, that self-fertilization is blocked. We currently know that this kind of biological barrier has probably been selected to prevent consanguinity and facilitate genetic mixing and increasing variability (see embryogenesis section for details). From these pioneer studies by Chabry and Conklin, interest in solitary ascidian biology crossed time, and several biologists continued descriptive works. Throughout his career, Norman John Berrill developed ascidians as biological models (Berrill and Watson 1930; Berrill 1932a; Berrill and Watson 1936; Berrill and Sheldon 1964). He described various species (Berrill 1932b) and also focused on development and organ functionality, such as the gut and stomach (Berrill 1929). He particularly took advantage of solitary ascidians as an easy model to understand seminal functionality. Importantly, Berrill participated in the validation of the mosaic development theory, in opposition to regulative development, which considers that blastomere fate can be regulated during development to be able to form a normal embryo in case of cell destruction. Next, since the 70s, a new generation of researchers from several countries have expanded our understanding of ascidian biology. As one example among others, Guisseppina Ortolani worked on cell lineage differentiation or fertilization mechanisms on *Ciona*, *Phallusia* or *Ascidia*. She notably participated in the discovery of muscle cell lineages. Richard Whitteker validated Conklin's proposition in 1973 of the presence of maternal determinants in eggs driving cell lineage. In addition to research in Europe, strong expertise on solitary ascidians emerged in Japan, led by Noriyuki Satoh. One of Satoh's major contributions is his research on egg cytoplasmic factors establishing cell fate during embryogenesis. Thanks to horseradish peroxidase tracer techniques, he was able to follow cell lineage and identify maternal factors with monoclonal antibodies, an innovative approach at the beginning of its career. Next, he described the mechanism regulating expression of acetylcholinesterase in muscle differentiation. With his research and the formation of several future researchers, he actively participated in developing molecular techniques on ascidian species. For instance, he was at the origin of the first transcriptomic project but also on the sequencing of the *Ciona* genome. In addition, he provided, thanks to the ghost database, several molecular tools and data on *Ciona* development to the scientific community. Finally, in the 90s, the complementarity between developmental biology, genetics and incorporation of new molecular approaches opened new perspectives to discover maternal determinants, making mosaic development possible, but also on the importance of regulation between blastomeres. In 2002, the first ascidian genome, from *Ciona intestinalis*, was sequenced and annotated, opening an avenue of possibilities on embryogenesis, metamorphosis and molecular signaling pathway understanding. To date, several genomes and transcriptomes from different solitary ascidians such as *Phallusia mammilata*, *Ciona savigny*, *Molgula occulata* and *Halocynthia roretzi* have expanded the amount of molecular data on this group and contributed to easier molecular phylogenetic analysis, accessible molecular functionality comparison between chordates and experiment design. The International Tunicate Meeting (ITM), which occurs every two years, alternately in Japan, Europe and the United States, was initiated in 2001, illustrating the dynamism of research on ascidians where solitary species count as most of the biological models, in addition to a few colonial species such as *Botryllus genus* or Appendicularia such as *Oikopleura genus*.

This focus on solitary ascidian models' contribution to developmental biology is fundamental, but one must not ignore the debate on ascidian evolution and their position among the animals' phylogenetic tree in the 19th century. Ascidians have been considered close to molluscs for a long time because of their flask adult body devoid of hard structure. The first questioning of this belonging was made by Savigny in 1816, who recognized tunicates as distinct and separate from molluscs. Next, studies from Vadimir Kovalevsky during the 19th century questioned the relationship between tunicates and other animals. Indeed, Kovalevsky described the larval body plan of two species of solitary ascidians, *Ciona intestinalis* and *Phallusia mammilata*, and discovered an organization similar to chordate animals (1866). In particular, the presence of a dorsal chord in tadpole swimming larvae led to considering chordates as composed of three groups: tunicates (comprising solitary ascidians), cephalochordates (as genus *Amphioxus*) and vertebrates. Consequently, thanks to solitary ascidian larval descriptions, the phylogenetic position and evolutionary history of tunicates became better understood. From Kovalevsky's studies to the beginning of the 21st century, ascidians were considered the first divergent branch of chordates (making cephalochordates the sister-group of vertebrates). More recently, thanks to molecular phylogeny made possible by genome sequencing and statistical method development, it was established that tunicates are the sister-group of vertebrates, whereas cephalochordates are the first divergent chordate phylum, making tunicates the closest "invertebrates" to vertebrates (Delsuc et al. 2006). Consequently, ascidians became important in comparative studies from an evo-devo perspective to understand vertebrate evolution. Whereas the phylogenetic position of tunicates is now consensual and established, the relationship inside tunicates is more debated, and several phylogenies frequently emerge in the literature, although a consensus is currently appearing (Figure 20.1).

Tunicates are commonly considered to be composed of five major phyla: Appendicularia, Phlebobranchia, Aplousobranchia, Thaliacea and Stolidobranchia. Appendicularia are characterized by a pelagic lifestyle with a tadpole-shaped adult form, illustrated by the best-known species, *Oikopleura dioika*. Though Appendicularia are often positioned as the first branch separated from other tunicate groups, debate on the phylogenetic position of this group is not totally closed, and it could be the sister of the Stolidobranchia (Delsuc et al. 2006; Delsuc et al. 2018; Kocot et al. 2018; Tatián et al. 2011; Satoh 2013). The four other groups (Phlebobranchia, Aplousobranchia, Thaliacea, Stolidobranchia) are grouped together in recent phylogenetic analysis and form a monophyletic clade. Phylogeny inside this large group has been debated because of the difficulties of reconstructing the life history for several reasons: the convergent features, the secondary loss and the high evolution rate of DNA sequences, making molecular phylogeny difficult to perform. According to the current consensual phylogeny, Stolidobranchia was the first group to diverge from the others. Then, Phlebobranchia, Thaliacea and Aplousobranchia are considered monophyletic. Thaliacea diverged first, and Phlebobranchia grouped with Aplousobranchia to compose Enterogona.

Thaliacea, including salps, are pelagic only and form a planktonic colony made by the aggregation of multiple individuals. An important point to keep in mind is the presence of both solitary and colonial ascidians in Stolidobranchia and Phlebobranchia, whereas Aplousobranchia are only colonial and represent the group containing the highest number of species. In these three groups, adult forms are settled to the substrate, whereas Thaliacea are pelagic. Stolidobranchia, characterized by the presence of one gonad pair and an atrium formed from a unique indentation, is composed of colonial ascidians like *Botryllus schlosseri* as well as solitary ones such as *Molgula oculata*. Stolidobranchia are also characterized by a folded branchial sac. Phlebobranchia and Aplousobranchia, both usually grouped into Enterogona, possess an even number of gonads, and the atrium is formed by two indentations. Phlebobranchia present a branchial sac vascularized by longitudinal blood vessels, whereas Aplousobranchia have

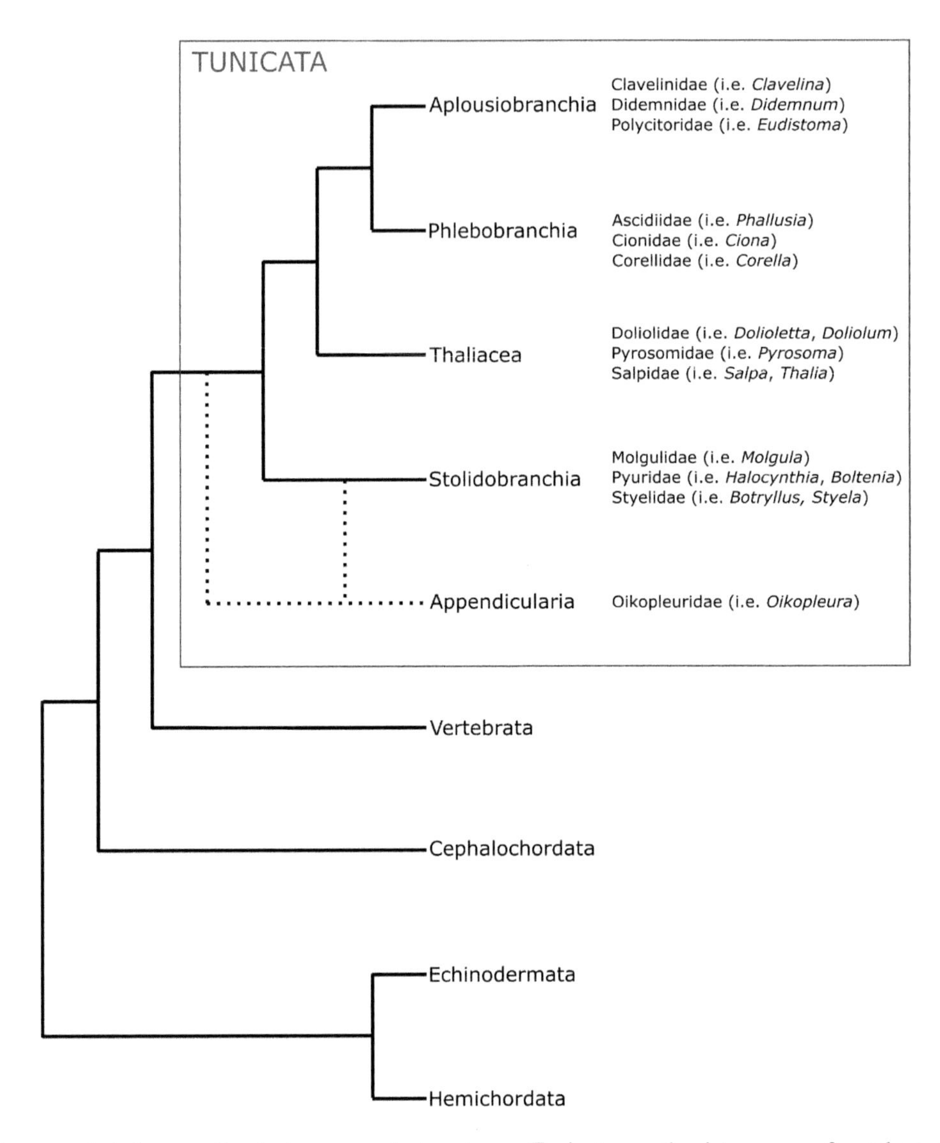

FIGURE 20.1 Consensual phylogeny of tunicates among deuterostomes. Tunicates are the sister-group of vertebrates. Among tunicates, Appendicularia are usually considered the basis of the phylogenetic tree. Solitary ascidian biological models belong mainly to the Stolidobranchia and Phlebobranchia groups.

a simple anatomy. The well-established biological models of solitary ascidians *Ciona intestinalis* and *Phallusia mammillata* belong to Phlebobranchia, a group also composed of a few colonial species such as *Perophora namei* with the particularity to present several individuals distributed along a long slender stolon. Aplousobranchia is composed of colonial species such as *Clavelina lepadiformis* or *Aplidium elegans*. Stolidobranchia and Phlebobranchia tunicates are both colonial and solitary, and this makes them ideal model animals to study in order to better understand evolution, convergence and the impact of environment to determine their lifestyle.

20.3 GEOGRAPHICAL DISTRIBUTION

Solitary ascidians are ubiquitously distributed across oceans and closed seas (Shenkar and Swalla 2011). The most-described species appear to originate from the Pacific region, possibly resulting from an artifact of sampling because taxonomists have been particularly active in this region. Solitary ascidians are marine, and no freshwater species have been reported. However, several species live in estuarine, and ascidians can usually support high variations of salinity (Lambert 2005; Shenkar and Swalla 2011). As an example, *Ciona intestinalis* can support a range of salinity from 12 to 40% and is able to survive a short bath in brackish water with a salinity less than 10% (Dybern 1967; Therriault and Herborg 2008). Solitary ascidians are also tolerant to temperatures lower than 1.9°C allowing, as we will see, survival at the poles (Primo and Vázquez 2009), but also to temperatures over 35°C, as reported in the Arabic Sea (Monniot and Monniot 1997). Resistance to variations could explain the ubiquitous repartition of ascidians. *Ciona intestinalis* is a perfect example showing the capacity of solitary ascidians to colonize various environments, leading to a ubiquitous distribution. It has been sampled in the Pacific Ocean (east and west), in the Atlantic on both American and European coasts and in the Mediterranean Sea.

In addition to the presence of several ubiquitous species, the capacity of larvae to settle in any substrate, such as soft sediments, rocks or coral reefs, facilitates colonization and expansion. Particularly, larvae can settle on several artificial substrates such as floating dock or ship hulls, leading to an artificial geographical spreading of some species at harbors around the world. Consequently, some solitary ascidians have a current ubiquitous repartition, but this does not seem natural as resulting from a secondary colonization mediated by human activities. For example, it has been reported in the port of Salvador, which receives cargo ships from several continents, that the ascidians species inventory presents a mix between possible endogenous ones (such as *Ascidia nordestina*), introduced ones (such as *Cnemidocarpa irene*) and ubiquitous ones. Importantly, for some solitary ascidians characterized by a wide/ubiquitous distribution, it can be difficult to evaluate if the geographical distribution is natural or artificial, resulting from centuries of spreading thanks to travels and maritime trades. It is thus assumed that some ascidians can have an unknown natural repartition. On the other hand, some cases of invasion are clearly documented. *Corella eumyota*, found natively in the southern hemisphere, is now established in the north Atlantic and Mediterranean Sea (Lambert et al. 1995; Collin et al. 2010). Moreover, *Styela genus* represents a relevant example of global repartition induced artificially. *Styela clava*, although coming from the northwest Pacific, was accidentally introduced in the East Pacific, Atlantic and European coasts. In Canada, this species has been described to disturb aquaculture, probably due to a overabundant population leading to the decrease of food availability for filter animal culture such as mussels or oysters, which suffer growth delay (Bourque et al. 2007; Arsenault et al. 2009). Coupled with dispersion driven by settlement on mobile artificial supports, some solitary ascidians can extend their life area by taking advantage of artificial waterways. This is the case of the Suez Canal, which has allowed to the endemic species *Herdmania momus* to disperse from the Red Sea toward the Mediterranean Sea (Shenkar and Loya 2008). Taken together, this high tolerance of ascidians to various environments, their capacity to spread thanks to artificial support and their potential impact on food availability for other filter animals make solitary ascidians a suitable model to understand the consequences of invasive species.

In opposition to species presenting a ubiquitous geographical repartition, some ascidians exhibit a specific distribution, making them endemic to a given area. The majority of ascidian species inventories reveal, in addition to new species description, a mixed composition with both ubiquitous and endemic species. This is typically the case in the Port of Salvador or more recently in the Gulf of Mexico. The Brazilian coast is also rich in endemic tunicates, such as the solitary ascidian *Eudistoma vannamei*. Relatively “closed” environments such as the Mediterranean Sea or the Red Sea present various endemic species, likely because of the reduced dispersal capacity compared to open environments. For example, 12 species are considered endemic to the Red Sea, representing 17% of the ascidian diversity (Shenkar and Loya 2008; Shenkar 2012).

Several solitary ascidians have been discovered in low-temperature environments in both the Arctic and Antarctic. *Styela rustica* can live in the north Atlantic in the Svalbard region, a colonization which seems recent (Demarchi et al. 2008). In the southern hemisphere, a number of species have been discovered in the South Shetland Islands such as *Styela wandeli* or *Molgula pedonculata* (Tatian et al. 1998). Antarctic species seem to be particularly adapted to survive in extreme conditions, such as *Cnemidocarpa verrucose*, known to be able to filter all ranges, particularly the finest, of organic particles to get enough nutrients in a poor environment (Tatián et al. 2004).

This large repartition shows also that the majority of solitary ascidians are shallow-water species and live on the continental shelf in harbors, reefs, and various coastal environments. In addition, abyssal species are also documented thanks to several sampling campaigns in the Pacific and other deep-sea regions. Abyssal species from the Pacific are represented by *Molgula sphaeroidea* or *Adagnesia bafida*, also discovered in the Atlantic at a depth of about 3,000 m. The deepest solitary ascidian discovered was in the Pacific at 7,000 m depth. Illustrating the ubiquitous presence of deep-sea species, we can also cite *Agnezia monnioti*, discovered in the Arabian Sea at 3,162 m depth. In *Styela gagetyleri*, localized in the same region but at 368 m depth (which is already considered a deep-sea conditions), the number of folds of the branchial sac is reduced, implying a decrease of cilia quantity and thus oxygen exchange surface. This could result from an adaptation to low oxygen levels and an optimization of the capacity to capture nutrients. Observations and species descriptions have led scientists to notice that abyssal species are in the high majority of solitary ascidians and not colonial ones. It has been proposed that the column shape of the body of solitary species allows a vertical elongation, creating a distance between the siphons and deep-sea soft and muddy sediments, whereas colonial ascidians are closed to the substrate and cover it in such a way that the siphon stays close to the mud, which could be problematic to capture food in a poor environment.

All studies made on Tunicate spatial distribution brought to light that solitary ascidians composed between 20% and 40% of the diversity (others are colonial ascidians) in tropical environments, whereas solitary ascidians represent most of the species at the two poles and temperate climates, with, for example, 58% and 70% of the diversity in the Antarctic and European coasts, respectively. This distribution is explained by the lifestyle of colonial ascidians presenting an indeterminate growth allowing colonization of most biological matter support in rich tropical environments.

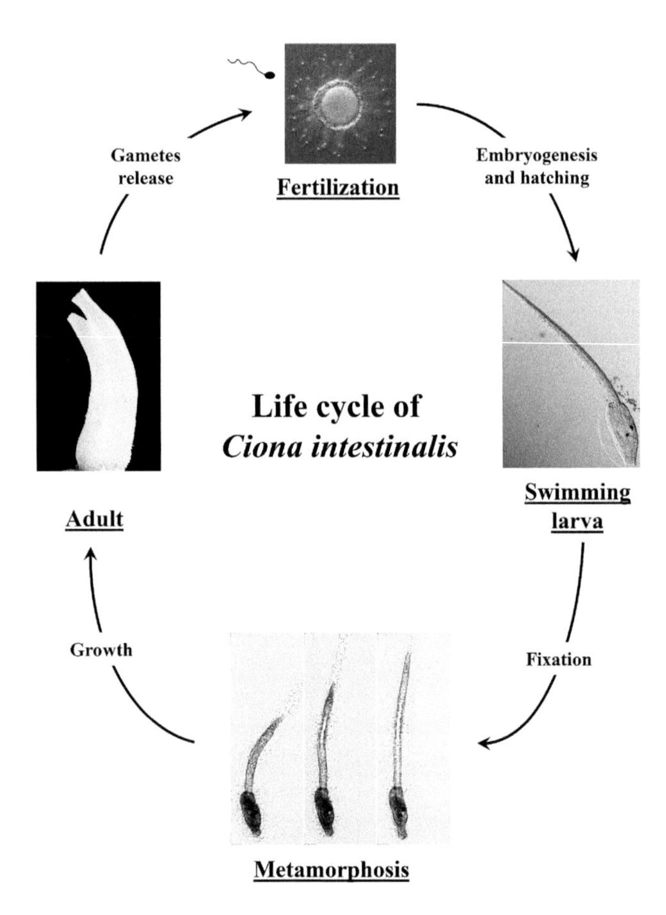

FIGURE 20.2 An example of solitary ascidian life cycle, *Ciona intestinalis*. After gametes are released, embryogenesis takes place in sea water and gives rise to a swimming larva in few hours. After a period of free swimming (four to eight hours in the case of *C. intestinalis*), the larva adheres to a substrate and starts metamorphosis, with the regression tail as the most dramatic event of this process. The pictures of the tail regression were captured from a time-lapse of *C. intestinalis* metamorphosis (Soulé and Chambon, unpublished data, photo credit Soulé and Chambon). After metamorphosis, the juvenile will give rise to a sexually mature adult in one to two months depending on the feeding conditions. (Adult picture photo courtesy of JP Chambon.)

20.4 LIFE CYCLE

Solitary ascidians are characterized by a bi-phasic life cycle (Figure 20.2), composed by a swimming larva and a sessile adult. Adults are usually hermaphrodites, producing both sperm and oocytes, accumulated in two separated gonoducts. Gamete production is controlled by a seasonal cycle and by light, and it can also be managed in culture. When gametes are mature, obscurity or light variations lead to their release in sea water, thereby inducing a synchronization of gamete release between individuals. Cross-fertilization (self-fertilization is usually blocked/sub-efficient) gives rise to a swimming tadpole larva after embryogenesis.

20.4.1 Hatching

At the end of embryogenesis, the fully formed larva is embedded in a chorion composed of a layer of maternal test cells (TCs) surrounded by a vitelline coat (VC) and at the most exterior part by follicular cells (FCs). The first tail movements appear before hatching, and these, coupled with apoptosis of test cells, contribute to the larva escaping from the chorion (Maury et al. 2006; Zega et al. 2006). Tail movements are due to muscle contractions under the control of the larval nervous system (reviewed in Meinertzhagen et al. 2004). From hatching, the larva adopts a pelagic behavior by swimming and dispersing in the environment.

20.4.2 Swimming and Pre-Metamorphic Phase

Using electrophysiological methods to record muscle tail contraction, the swimming behavior of *Ciona intestinalis* was characterized from hatching to the acquisition of metamorphic competence (Zega et al. 2006). Three different larval movements were observed: tail flicks, "spontaneous" swimming and shadow response. The *Ciona* larvae swim for longer periods and more frequently during the first hours after hatching. The swimming behavior changes during the free swimming phase and switches from photopositive to photonegative during the pre-metamorphic period. Using a Morpholino-knockdown approach against Ci-opsin1, the visual pigment expressed in the photoreceptor of the ocellus, it was observed that the *Ciona* larvae swimming behavior was affected (Inada et al. 2003), suggesting a photic control of the swimming phase. Recently, thanks to the recent completion of the *Ciona* larval central nervous system (CNS) connectome (Ryan et al. 2016), a group of photoreceptors that control the switch to the photonegative swimming behavior at the pre-metamorphic phase were identified (Salas et al. 2018). The competency for metamorphosis is acquired a few hours after hatching (8–12 hours in the case of *Ciona intestinalis*) and leads to the research of a substrate by the larvae. In its search for settlement, in addition to visual, geotactic and chemosensory inputs, the larva also exhibits strong thigmotactic behavior (Rudolf et al. 2019). These changes in behavior are probably correlated with the capacity of the larva to respond to a wide variety of external and endogenous signals (reviewed in Karaiskou et al. 2015). The settlement is the first step of metamorphosis and is mediated through the adhesive papilla, localized at the most anterior extremity of the larva. This is done preferentially on substrates (natural as well as artificial) presenting a bacterial film. The onset of metamorphosis is strictly associated with larva adhesion since papilla-cut larva are unable to fully metamorphose (Nakayama-Ishimura et al. 2009).

20.4.3 Metamorphosis

From settlement, the tadpole larva will undergo a metamorphosis characterized by a schematic sequence of events that transform a solitary ascidian larva to a juvenile one (Figure 20.3). Ascidian metamorphosis has been described by Cloney (1982), leading to characterization of ten successive steps globally shared between species despite a few variations: 1) secretion of adhesives by the anterior papilla,

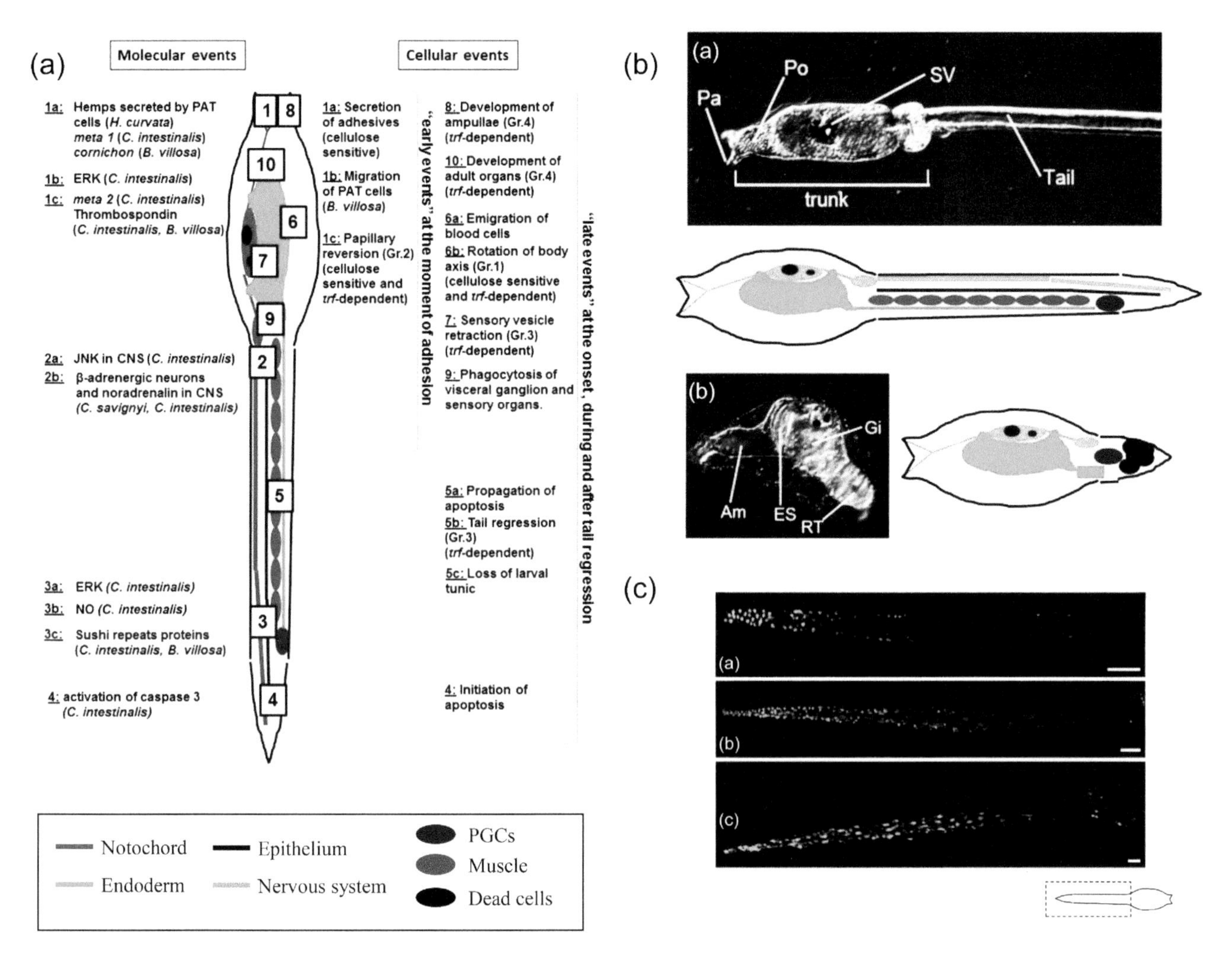

FIGURE 20.3 Metamorphosis of solitary ascidians. (a) Summary of molecular and cellular events that occur at the onset of the metamorphosis in solitary ascidians. Sequential numbers refer to the order of events. Gr.: Group according to classification in Nakayama-Ishimura *et al.* (2009). (b) Metamorphosis of the ascidian *Ciona intestinalis*. From the swimming larva and its schematic representation (a) to a juvenile soon after metamorphosis and its schematic representation (b). Pa, papilla; Po, preoral lobe; SV, sensory vesicle. The preoral lobe of larva is elongated and becomes transparent to be an ampulla (Am). Adult organs, such as endostyle (ES) and gills (Gi) start to develop in the trunk. The tail is retracted toward the trunk (RT). (c) TUNEL labeling of a metamorphic *Ciona intestinalis* larva tail at successive stages (a–c) of the tail regression. Schematic representation to show where the apoptotic cells are detected in the sequential TUNEL labeling. Apoptotic cells appear in green. Scale bars: 220 µm in (a); 140 µm in (b); 80 µm in (c). ([a] Adapted from Karaiskou et al. 2015; [b] adapted from Karaiskou et al. 2015; [c] adapted from Chambon et al. 2002.)

leading to larval settlement; 2) reversion and retraction of the papillae; 3) tail regression, also named tail resorption; 4) loss of the outer cuticle layer composing the tunic; 5) retraction of the sensory vesicles; 6) phagocytosis of sensory organs, visceral ganglion and cells of the axial complex and elimination of other specific larval structures (TLOs); 7) emigration of pigmented and blood cells from the epidermis to the external tunic; 8) digestive gut establishment by an expansion of the branchial basket in addition to visceral organ rotation through an arc of about 90°; 9) a global growth characterized by the expansion and elongation of the ampullae corresponding to the foot of the animals, allowing strong anchoring to the substrate concomitantly with tunic enlargement; and, finally, 10) total disappearance of larval rudiments, followed by the construction of adult tissues (PJOs). Next, the inhalant siphon opens first, and then the opening of the exhalant one allows the circulation of water in the pharynx, and the juvenile becomes ready to filter sea water to feed. In the past 20 years, many studies have allowed better comprehension at the molecular scale of these metamorphic events (reviewed in Karaiskou et al. 2015 and Figure 20.3).

Using gene profiling approaches, the secretion in the papillae of an EGF-like molecule named Hemps, which seems to control larva adhesion, was reported (Eri et al. 1999). The same approach in *Boltenia villosa* and *Ciona intestinalis* identified probable components of this potential adhesion regulated pathway (Davidson and Swalla. 2001; Nakayama et al. 2001). The activation of mitogen-activated protein kinase (MAPK) ERK was also reported in papillae around the time of adhesion and is a prerequisite for the subsequent tail regression event (Chambon et al. 2007).

Simultaneously, the JNK/MAPK pathway is also activated in the CNS, and similarly to the ERK pathway, it is essential for tail regression. The CNS seems to have a preponderant part in the onset of metamorphosis; expression of the β_1-adrenergic receptor was reported in this tissue in *Ciona intestinalis* and *Ciona savignyi* (Kimura et al. 2003). More recently, the neurotransmitter GABA was reported as a key regulator of *Ciona* metamorphosis (Hozumi et al. 2020), reinforcing the previous hypothesis of the preponderant role of the larval nervous system and sensory organs in selecting sites for adhesion and in the onset of metamorphosis (Cloney 1982). One of the most dramatic event of this process is the regression of the tail larva, which occurs a few hours after adhesion. Two not mutually exclusive mechanisms were reported during this event: the first involves the contractile properties of either the tail epithelial layer (observed in the solitary ascidian *Distaplia occidentalis, Aplidium constellatum, Diplosoma, Ecteinascidia turbiniata, C. intestinalis, Ascidia callosa, Corella willmeriana macdonaldi* and the colonial ascidian *Botryllus schlosseri*) or notochord cells (observed in *Boltenia villosa, Herdmania curvata, Styela gibbsii, Molgula mahattensis, Molgula occidentalis* and *Polycitor mutabilis*; reviewed by Cloney 1982); the second involves a massive apoptotic cell death of almost all of the cells that composed the tail and was observed in *C. intestinalis* (Chambon et al. 2002; Tarallo and Sordino 2004) and *Molgula oculata* (Jeffery 2002). Recently, using live microscopy, both mechanisms were observed during *Ciona intestinalis* tail regression, and they seem to be sequential, since initial contraction of the tip tail preceded apoptosis (Krasovec et al. 2019). Apoptosis appears to be the driving force of tail regression in solitary ascidians and affects almost all the cell types that compose the tail (the tunic, epidermal, notochord, tail muscle cells and the CNS), with two exceptions, the endodermal strand cells and the primordial germ cells (PGCs) (Figure 20.3). These two cell types escape apoptosis, the endodermal strand by migrating before the tail regression (Nakazawa et al. 2013), while the PGCs move toward the trunk at the time of tail regression in coordination with the progression of cell death (Krasovec et al. 2019). The most remarkable feature is that through sequential TUNEL pictures, it has been confirmed *in vivo* that apoptosis starts at the tail tip and continues up to the tail base by a perfect antero-posterior wave (Chambon et al. 2002; Krasovec et al. 2019). The same polarized propagation of apoptosis was reported in two other species of ascidians, *Molgula occidentalis* and *Asicidia ceratodes* (Jeffery 2002).

An arising and challenging question is the coordination mechanism of the metamorphic events. New insights were provided by the identification of the gene network downstream of the MAPK, ERK and JNK activation previously reported, respectively, in the papillae and the CNS. Among them is *Ci-sushi*, a gene under JNK control, with expression patterns at the tip of the tail, for which loss of function experiments lead to the inhibition of the initiation of apoptosis (Chambon et al. 2007). In addition, papilla and tail cut experiments on larva coupled with analyses of metamorphic mutants (*swimming juveniles* and *tail-regression fail* [*trf*]) allowed classification of metamorphic events in four groups (Nakayama-Ishimura et al. 2009). Group 1 includes a cellulose-sensitive and *trf*-independent event: body axis rotation; Group 2 encompasses a cellulose-sensitive and *trf*-dependent event: papillae retraction; Group 3 includes cellulose-independent and *trf*-dependent events, sensory vesicle retraction and tail regression; and Group 4 comprises cellulose-independent and *trf*-independent events, including ampullae formation and adult organ growth.

20.4.4 Juvenile and Adult

Metamorphosis in ascidians results in a dramatic modification of their body plan, transforming them in a few hours from swimming larva to sessile juvenile and after few months of growing to a sexually mature adult. Classically, juvenile growth timing depends on food availability and temperature. Consequently, the settled phase represents almost the entire life cycle, whereas the swimming phase is transitory and allows the dispersion of individuals.

20.5 EMBRYOGENESIS

20.5.1 Fertilization and Maternal Determinants

Ascidian embryogenesis is a rapid process involving a small number of cells (about 2,600 cells in *Ciona intestinalis*) and occurs within a chorion composed of test cells, a vitelline coat and follicular cells (Figure 20.4). It starts with fertilization, which, in solitary ascidians, occurs after the release of sperm and eggs into the surrounding seawater. To ensure fertilization, spermatozoids are activated and then attracted toward the eggs by a common factor released by mature oocytes (after germinal vesicle breakdown) called sperm-activating and sperm-attracting factor (SAAF) (Kondoh et al. 2008; Yoshida et al. 2002). The ascidians eggs are spawned embedded in a layer of follicular cells surrounding a vitelline coat, under which the test cells enclose the egg itself. In some species, such as *Styela plicata*, sperm and eggs are released at different times, while they are released simultaneously in *Ciona* and *Halocyntia*, allowing sperm to interact with self-eggs. In these latter species, which are known to be self-sterile, a self- and non-self recognition system was reported during fertilization, probably to promote outcrossing. In *Ciona*, this process is ensured by a couple of receptors expressed at the surface of the sperm (s-Themis A and B) and ligands expressed on the VC (v-Themis A and B). If a sperm containing s-Themis A and B interacts with an egg expressing both v-Themis A and B on the VC, its ability to bind the VC is reduced, and it is not able to fertilize the self-recognized egg (Harada et al. 2008). In addition to this self-recognition system, the polyspermy block involves a glycosidase enzyme released from the surface of FCs. It is interesting to notice that this enzyme activity release is not species specific, which means that sperm of a species could block the egg of an another (Lambert 2000). This sperm

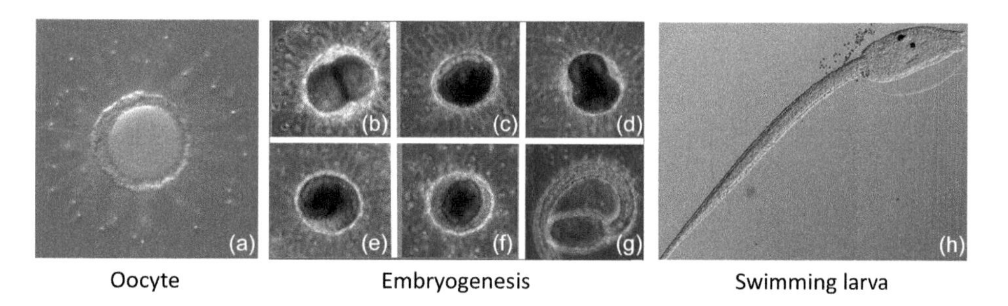

FIGURE 20.4 Embryogenesis of *Ciona intestinalis*. (a) Unfertilized oocyte in its chorion, FC (follicular cells), VC (vitelline coat), TC (test cells) (photo credit S. Darras); (b–g) capture from time-lapse microscopy of *Ciona intestinalis* embryogenesis in the chorion (photo credit J. Soule and JP Chambon); (b) two-cell stage; (c) mid-gastrula; (d) neurulae; (e) early tailbud; (f) tailbud; (g) hatching larva.; (h) swimming larva. (Photo courtesy of JP Chambon.)

competition may participate in the interspecific competition for space, leading to differential abundance of the ascidian community in natural environment.

Sperm entry into the egg results in a rise in calcium concentration through the egg, which initiates development, followed by a series of repetitive calcium waves. These waves are necessary for the completion of meiosis and initiate a signal-transduction cascade which brings about the remodeling of the male pronucleus and cytoskeletal rearrangements, as well as alterations in gene regulation at both the post-transcriptional and post-translational level (Tadros and Lipshitz 2009). The calcium waves are also responsible for the stimulation of ATP production necessary to match the energy demand associated with the onset of development (Dumollard and Sardet 2001). At this stage, the early embryo is dependent on maternal mRNAs and proteins, known as maternal factors, that are produced and stored in the egg during oogenesis to survive and develop prior to the full activation of the zygotic developmental program (Oda-Ishii et al. 2016). The transition from maternal products to zygotic factors occurs starting from the eight-cell stage and is called the maternal-to-zygotic transition (MZT) (Oda-Ishii et al. 2016; Treen et al. 2018). In ascidian embryos, four maternal factors are involved in the establishment of the first zygotic gene expression: ß-catenin, Tcf, Gata.a and Zic.r-a (also called Macho-1).

20.5.2 Ooplasmic Segregation and Establishment of Embryonic Axis

Following the completion of meiosis and the fusion of the male and female pronuclei, a series of synchronous and rapid cell divisions occur, called the cleavage stage. The first cleavage occurs 1 hr 45 min after fertilization in *Halocynthia roretzi* at 13°C and 1 hr in *Ciona intestinalis* at 18°C. Two synchronous and four asynchronous cleavages later, about 9 h later in *Halocynthia* and 5 h later in *Ciona*, the embryo will reach the 110-cell stage and the beginning of gastrulation (Figure 20.5a).

During this cleavage stage, establishment of the primary and secondary embryonic axis occurs. The primary axis, or animal–vegetal (AV) axis, of the embryo is set up during oogenesis. At fertilization, the sperm enters the egg in the animal hemisphere, defined by the position where the polar bodies form, and its nucleus is transported toward the vegetal pole by the actin-dependent contractions of the first ooplasmic segregation (Lemaire 2009; Satoh 1994). The secondary axis, or antero-posterior (AP) axis, is set up orthogonally to the AV axis following ooplasmic movements that localize asymmetric cleavage determinants to the posterior pole of the embryo. This asymmetric partitioning of determinants is responsible for the intrinsically different potentials of the anterior (so-called A- and a-line) and posterior (B- and b-line) blastomeres in response to induction (Feinberg et al. 2019).

20.5.3 Germ Layer Segregation

The 16-cell stage marks the onset of the mid-blastula transition, characterized by asynchronous cleavages, ß-catenin-dependent cell cycle asynchrony (Dumollard et al. 2013) and the appearance of the three germ layers of the embryo—endoderm, mesoderm and ectoderm. This process involves two binary fate choices coupled with the first two A-V-oriented rounds of cell divisions between the 8- and 32-cell stages. In both *Ciona* and *Halocynthia*, the first fate choice identifies the animal and vegetal destinies. It is driven by the transcriptional action of nuclear ß-catenin during the 8- and 16-cell stages, but as of today, the mechanisms responsible for the localization of ß-catenin are still unknown (Rothbächer et al. 2007; Hudson et al. 2013; Takatori et al. 2010).

In the A5.1 cell (Figure 20.5a) at the 16 cell-stage, nuclear localization of maternal ß-catenin controls the segregation of mesendoderm and ectoderm by forming a complex with TCF DNA-binding proteins to mediate the canonical Wnt signalling pathway. An active ß-catenin/TCF complex induces the mesendodermal fate by promoting the expression of notochord/neural/endodermal (NNE) factors *Foxa.a*, *Foxd* and *Fgf9/16/20* and by repressing ectoderm gene expression both directly and indirectly via NNE factors. Cells where the complex is inactive will acquire an ectodermal fate (Figure 20.5b) (Hudson et al. 2013; Hudson 2016).

The second binary fate choice takes place at the transition to the 32-cell stage and leads to the segregation of endoderm

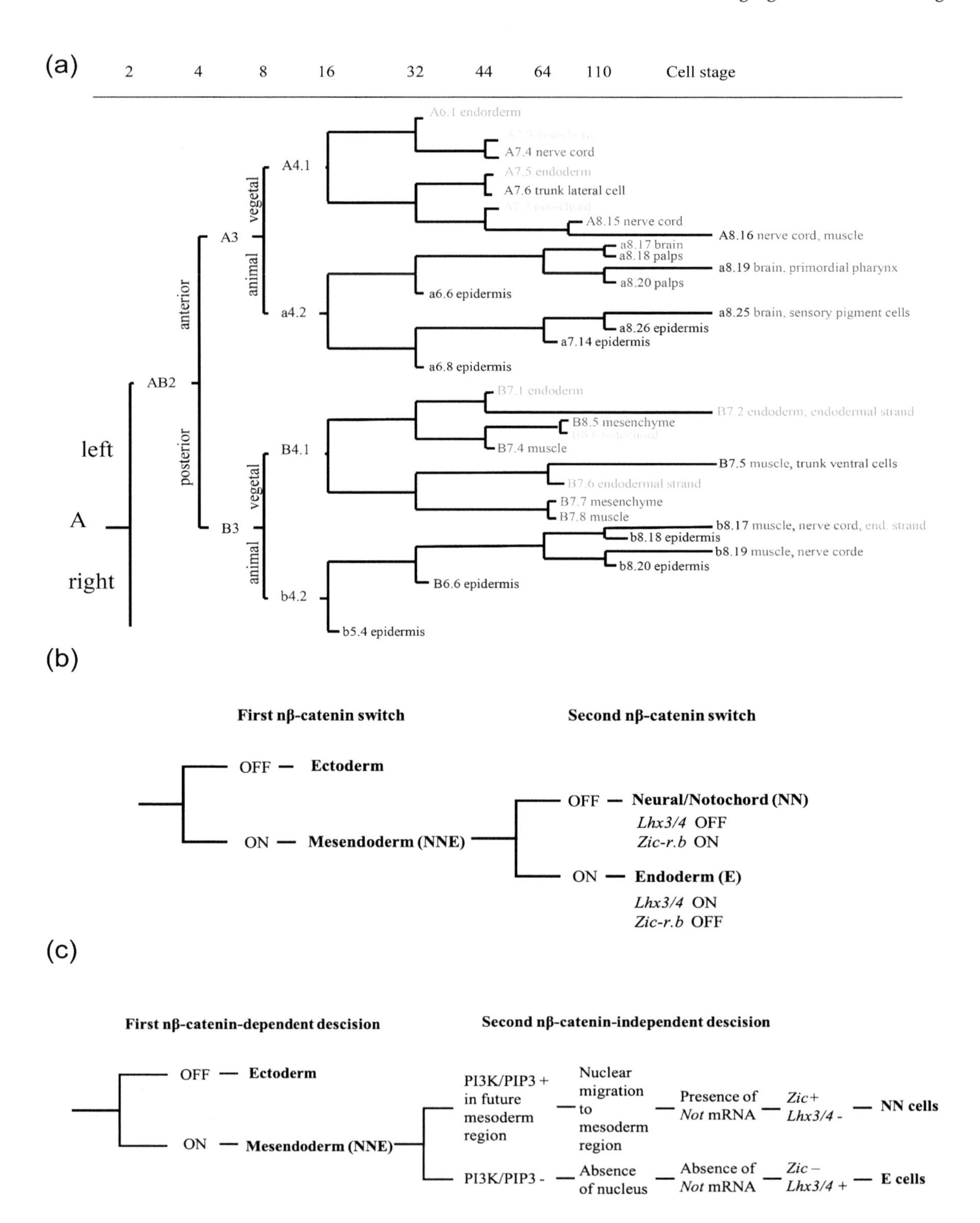

FIGURE 20.5 Cell lineage and developmental fate segregation in solitary ascidian embryos. (a) Cell lineage in ascidians. Lineage tree with the blastomere fate restriction at the successive cell divisions represented by color code (blue: nervous system, green: endoderm, red: muscle, orange: notochord, black: epidermal, gray: mesenchyme, purple: trunk lateral and ventral cells). Since ascidians are bilaterally symmetrical, only the left half of the embryo is shown. (b) Fate segregation in A-line mesendoderm lineages of *Ciona intestinalis*. Two successive rounds of nβ-catenin-driven binary fate decisions that segregate the mesendoderm lineages from the ectoderm lineages at the 16-cell stage and then the neural/notochord (NN) lineages from the endoderm (E) lineages at the 32-cell stage. (c) Fate segregation in the A-line mesendoderm lineage of *Halocynthia roretzi*. Two successive binary fate decisions that segregate the mesendoderm lineages from the ectoderm lineages at the 16-cell stage and then the neural/notochord lineages from the endoderm lineages at the 32-cell stage. The first is nβ-catenin-dependent. The second involves a ß-catenin-independent mechanism involving several Wnt pathway components, as Wnt5a and APC/GSK3 segregation of *not* mRNA transcripts. ([a] Modified from Kumano and Nishida 2007; [b] Hudson et al. 2016; [c] Takatori et al. 2010; Takatori et al. 2015.)

and notochord/neural (NN cells or mesoderm) from mesendoderm precursors. Two distinct regulatory processes have been discovered to achieve the same fate decision in the same A lineage in *Ciona* and *Halocynthia*.

In the case of *Ciona* embryos, this second fate choice involves a second ß-catenin-dependent process during the 32-cell stage. Continued activity of the ß-catenin/TCF complex in mesendodermal cells induces endoderm fate (E cells), whereas inactivation of the complex leads to the acquisition of the notochord/neural fate. During this second phase, ß-catenin/TCF works directly or indirectly in the E cells with the targets of the first phase of ß-catenin activity, *Foxa.a*, *FoxD* and *Fgf9/16/20*, to activate the E specifier *Lhx3/4* and to repress the NN specifier *Zic-r.b* (Figure 20.5B) (Hudson et al. 2016).

In *Halocynthia* embryos, a different mechanism exists. A possible explanation for this difference is the presence of nuclear ß-catenin in NN cells at the 32-cell stage (Hudson et al. 2013). Thus, *Halocynthia* NN specification depends on a Wnt-dependent but ß-catenin-independent mechanism involving *Not* mRNA transcripts. The asymmetrical partitioning of *Not* mRNA regulates the expression of transcription factors required for fate segregation. In endoderm cells, *Not* will be absent, and thus endoderm differentiation will occur. On the contrary, in NN cells, *Not* is present and will promote *Zic* expression as well as repressing *Lhx3/4* expression, thus promoting NN fate and repressing E fate (Figure 20.5c) (Hudson et al. 2016; Takatori et al. 2010; Takatori et al. 2015) (Figure 20.5c).

20.5.4 Larval Tail Muscle Formation

Muscle formation in ascidian is a well-known example of cell autonomous process first demonstrated by Conklin in 1905. However, recent studies have brought to light the importance of cell–cell interaction as another important factor.

At the larval stage, the only fully differentiated and functional muscles are those of the tail and most solitary species present between 18 and 21 muscle cells on either side of the tail. Muscle cells originated either from the primary muscle cell lineage and the B4.1 blastomeres or from the secondary lineage of A4.1 and b4.2 (Figure 20.5a) (Razy-Krajka and Stolfi 2019; Satoh 2013).

The primary lineage consists of 14 muscle cells located on either side of the tail specified following a cell autonomous specification and differentiation involving the Zic.r-a (Macho-1) maternal determinant. Zic.r-a will trigger the primary tail muscle specification regulatory network by activating the transcription of *Tbx6-related* (*Tbx6-r*) muscle determinants at the 16-cell stage and downstream factors at the 64-cell stage (Razy-Krajka and Stolfi 2019; Satoh 2013; Yagi et al. 2005). On the other hand, the secondary lineage gives rise to the muscle cells flanking the tip of the tail, whose numbers vary between species (ten cells of b4.2 origin in *Halocynthia* compared to four in *Ciona).* In the A-line, muscle potential is induced by intricate feed-forward signaling relay from the neighboring b6.5 lineage cells to A7.6 to A8.16. In *Ciona*, the Nodal and Delta/Notch signaling pathways are responsible for this, while in *Halocynthia*, a yet-unknown signal from the same b6.5 lineage induces the expression of Wnt5.a, which then promotes muscle fate in A8.16 (Figure 20.5a) (Tokuoka et al. 2007). Finally, the last muscle/neural cell fate decision in *Ciona* will see FGF/ERK signaling activating the muscle determinants *Tbx6-r.b* and *Mrf* expression. In *Halocynthia*, what regulates this final fate decision is yet another unknown parameter, but FGF/ERK signaling is not involved (Razy-Krajka and Stolfi 2019; Tokuoka et al. 2007).

20.5.5 Neural Plate Patterning

Similar to vertebrate neurulation, the ascidian neural plate is curled up dorsally to form a tube-like structure known as the neural tube. The neural plate emerges at the mid-gastrula stage and is composed of 40 cells at the neural plate stage, arranged in six rows and eight columns of cells along the A-P axis formed from posterior to anterior. The I and II rows compose the posterior neural plate and derive from the A-lineage. They will contribute to the caudal nerve cord, motor ganglions and posterior sensory vesicle. On the other hand, the a-lineage will give rise to the anterior four rows III to VI. Rows III and VI will contribute to the anterior part of the sensory vesicle, part of the oral siphon primordium and anterior brains. Finally, rows V and VI give rise to neurons of the peripheral neural system (PNS) (Hudson 2016; Imai et al. 2009; Wagner and Levine 2012). Once the neural tube is completely closed, the tail becomes distinguishable (Kumano and Nishida 2007).

Different signaling pathways are responsible for the patterning of the neural plate, such as Nodal, Nodal-dependent Snail, FGF/MEK/ERK and Delta/Notch (Hudson 2016; Hudson et al. 2007; Razy-Krajka and Stolfi 2019; Satoh 2013).

20.5.6 Neural Development

The ascidian nervous system is composed of the peripheral neural system and the central nervous system, and its development starts with neural induction at the 32-cell stage. CNS development starts in two blastomeres, pairs A6.2 and A6.4 (Figure 20.5a), which become neural fate restricted at the 64-cell stage under FGF induction (Hudson et al. 2016). It consists of approximately 330 cells and about 117 neurons and originates from three lineages: the A and a- and b-lines (Hudson et al. 2007). The CNS presents three morphologically distinct structures: the anterior-most sensory vesicle, the trunk ganglion (also called visceral ganglion) and the tail nerve cord (Hudson et al. 2016). The A-line blastomeres become fate restricted following a neuro-epidermal binary fate decision involving a β-catenin-driven binary fate switch. This lineage will give rise to the posterior part of the sensory vesicle as well as the ventral and lateral parts of both trunk ganglion and tail nerve cord (Hudson et al. 2013). The anterior part of the sensory vesicle and the dorsal part of the visceral ganglion and tail nerve cord respectively originate from the a-line (a6.5) and b-line (b6.5) blastomeres, which become restricted to neural fate at the 112-cell stage (Hudson et al. 2016; Roure et al. 2014).

PNS development starts with the birth of the a6.5 blastomere (Figure 20.5a). It is composed of different types of epidermal sensory neurons (ESNs): the papillary neurons of the adhesive papillae, the epidermal sensory neurons and the bipolar tail neurons (BTNs) distributed in the epidermis of the trunk and tail (Hudson 2016; Meinertzhagen and Okamura 2001).

20.5.7 Cardiac Development

The adult ascidian heart consists of a one-cell-layer single myocardial tube surrounded by a pericardium. It is formed of two distinct territories: the first heart field (FHF) and the second heart field (SHF) and originates from a single pair of blastomeres in the 64-cell stage embryos, the B7.5 cells (Figure 20.5a). The first division of the cardiac founder cells is symmetric and occurs during gastrulation. It leads to the appearance of two symmetrical pairs of pre-cardiac founder cells each consisting of a B8.9 and B8.10 blastomeres (Figure 20.5a) (Cooley et al. 2011). During neurulation, in each pre-cardiac lineage, founder cells divide a second time, asymmetrically this time, and each blastomere will give rise to four cells: two small anterior cells, which will migrate to form the heart, and two large posterior B7.5 granddaughter cells, which will differentiate as anterior tail muscles in both Halocynthia and *Ciona* (Figure 20.5a) (Christiaen et al. 2010; Davidson et al. 2006).

Two maternal determinants are responsible for the specification of the blastomeres: macho-1 and β-catenin. They activate the B7.5-specific expression of the transcription factor Mesp (Christiaen et al. 2009; Stolfi et al. 2010), which determines a competence domain facilitating either pre-cardiac or pre-vascular specification (Satou et al. 2004). Within the Mesp-expressing cells, subsequent inductive signals will induce specific identities. In the future cardioblasts, Mesp, in conjunction with FGF/MAPK signaling, will activate downstream components of the core cardiac regulatory (Davidson et al. 2006). BMP and FGF signalling will then either directly or indirectly regulate cardiac target gene expression of *FoxF* and the heart determinants *Nkx2.5*, *GATAa* and *Hand-like/NoTrlc* in the anterior the trunk ventral cells (Christiaen et al. 2010).

Following the second division, a first FGF-dependent migration of the trunk ventral cells (TVCs) to the ventral trunk region occurs. There they will undergo a series of successive asymmetric divisions along the mediolateral axis, followed by a second migration that will lead to a segregation of the heart cells from the lateral TVCs, precursors of the atrial siphon muscle (ASM) cells (Stolfi et al. 2010). The TVCs migrate dorsally toward each side of the trunk, where they will settle as a ring of cells at the base of the atrial siphon primordia (Stolfi et al. 2010).

20.5.8 Notochord

The ascidian larval notochord is composed of a single row of 40 cells that form through intercalation and originate from two of the four founder cell lineages. The anterior 32 notochord cells, termed the primary notochord, derive from the A-line founder lineage, whereas the posterior eight cells, termed the secondary notochord, are generated from the B-line founder lineage.

The anterior notochord precursors originate from A6.2 and A6.4 blastomeres, which are bipotential notochord/nerve cord precursors at the 32-cell stage. They are induced at the 32-cell stage and acquire developmental autonomy at the 64-cell stage (Jiang and Smith 2007).

In *Ciona* embryos, FGF and MAPK signaling are required at the 32–64-cell stage to polarize the blastomeres, which will divide asymmetrically into the induced notochord precursors and nerve cord precursors, which are the default fates (Hashimoto et al. 2011). In the secondary notochord lineages, which become fate restricted at the 110-cell stage (Jiang and Smith 2007), FGF signaling is necessary for two processes. It is first required at the 64-cell stage to suppress muscle fate in the mother cell of the notochord and mesenchyme precursors (Darras and Nishida 2001; Imai et al. 2002; Kim and Nishida 1999; Kim et al. 2000; Kim and Nishida 2001). Second, it is required to activate expression of *Ci-Nodal* in the b6.5 blastomere at the 32-cell stage, which is required for the specification of the secondary notochord precursor (Hudson and Yasuo 2005; Hudson and Yasuo 2006).

In the primary notochord precursors of *Halocynthia*, FGF is expressed in the notochord precursor and inhibited in the nerve cord precursor cells by the *Efna.d* signal coming from the animal hemisphere (Satou and Imai 2015). FGF expression leads to activation of Hr-Ets, which, coupled with Hr-FoxA and Hr-Zic.r-d, promotes the expression of the notochord-specific gene *Brachyury* (*Hr-Bra*) at the 64-cell stage. *Bra* then activates various downstream genes that are essential for notochord formation (Hashimoto et al. 2011). BMP2/4 is, on the other hand, implicated in the secondary notochord induction in *Halocynthia*. BMP2/4 is involved in the asymmetric cleavage of the B7.3 blastomeres as well as in the specification of secondary notochord cells (Darras and Nishida 2001).

20.5.9 Primordial Germ Cells

Primordial germ cells are the founders of gametes. It has been observed in several animals that the germ line is set aside early in embryogenesis and has to be "maintained" until differentiation of gametes in the mature gonads. PGCs can be specified by either inheritance of maternal determinant (pre-formation) or by induction (epigenesis). In ascidians, PGCs are specified during embryogenesis in posterior-vegetal blastomeres by the inheritance of postplasmic/PEM mRNAs in B7.6 blastomeres (reviewed in Kawamura et al.

2011), among them *Ci-Vasa*, an ATP-dependent DEAD-box RNA helicase, and *pem1*, which have been shown to repress mRNA transcription by inhibiting activating phosphorylations on the C-terminal domain (CTD) of the RNAPII (Shirae-Kurabayashi et al. 2006; Shirae-Kurabayashi et al. 2011; Kumano et al. 2011). During gastrulation, B7.6 divides asymmetrically, giving rise notably to B8.12, the founder of the eight PGCs localized at the tip of the larva tail at the end of the embryogenesis.

These cells will remain at this localization until the tail regression at metamorphosis, during which the PGCs will reach the trunk and the presumptive gonad.

20.6 ANATOMY

20.6.1 Larva

Anatomy of the larva is fundamental to the understanding of the phylogenetic affiliation of urochordates. The characteristic chordate body plan allowed Kovalevsky to discover that ascidians are closer to vertebrates and cephalochordates. The ascidian larva presents a morphology divided in two parts: the anterior trunk and the posterior tail. The larva is usually composed of a low number of cells, 2,600 cells in the case of *Ciona* intestinalis. A typical anatomy is common to the solitary ascidian larva with some tissues present all along the larva, whereas others are specific to the tail or to the trunk (Figure 20.6a).

The totality of the larval body is surrounded by the tunic, which is composed of a cellulose derivative, the tunicine. The epidermis, under the tunic, covers the entire animal body. Two internal tissues are distributed along the entire antero-posterior axis. The central nervous system is characterized by a dorsal neural tube as in a classical deuterostomian body plan organization. In the most anterior part of the trunk, neurones of the CNS compose the adhesive papilla, a sensitive structure which interacts with the environment to find a suitable substrate. These adhesive papillae allow the fixation of the larva. From the adhesive papilla to the tip of the tail, the CNS is then composed by the brain, in the trunk, the nerve ganglion which allows the junction between the posterior trunk and the most anterior part of the tail and finally the neural tube prolonged until the tip of the tail. Additional peripheral neurons are distributed along the tail epidermis. In the brain composing the CNS, an otolith and an ocellus are present and allow analyses of gravity and luminosity, respectively. The second tissue present both in the trunk and the tail is the endoderm. Endoderm is present in the postero-ventral part of the trunk and is prolonged in the ventral side of the tail by a line of cells named the endodermal strand.

The other tissues are specific either to the tail or of the trunk. In the tail, ventrally to the CNS but dorsally to other tissues, the notochord is present in almost the total length of the tail. Note that the presence of the notochord in the larva, absent in the adult, argues in favor of this model as suitable for the study of the anatomy and development of embryos and larvae to better understand animal evolution. The notochord plays the role of support structure for the muscles distributed laterally along the tail. These muscles allow the swimming movement of the larva after hatching and research on an adapted support for the settlement. Last, in the ventral side of the tip of the tail, in the posterior prolongation of the endodermal strand, are eight localized primordial germ cells, which will give rise to the gonads and the gametes in the adult. Finally, the larval trunk houses the heart in its ventral side and a sub-developed gut with a non-functional stomach. The outline of the pharynx is also present.

After hatching, the swimming phase and settlement lead to the metamorphosis phase, which will give rise to the adult animal. Tissues have been divided in three groups by Cloney according to their fate during metamorphosis, and this classification is still used. Group 1 correspond to tissues that exclusively function in the larval stage (transitory larval organs or TLOs) and can disappear during the metamorphosis; group 2 are tissues that function in both larval and adult stages (larval-juvenile organs/tissues or LJOs), conserved during the metamorphosis transition; and group 3 includes tissues emerging during the metamorphosis and consequently exclusively functioning in juvenile and the next adult stage (prospective juvenile organs or PJOs). Adult anatomy depends on LJOs, and PJO tissues compose a typical morphotype of solitary ascidians.

20.6.2 Juvenile and Adult

The adults of solitary ascidians are characterized by a bag-shaped morphology settled by a foot and distally to the point of fixation two siphons with sensory organs (usually a paired number) distributed around their opening (Figure 20.6b). The largest siphon, farthest from the foot, is the inhalant one, which allows the entry of the sea water in a large and sur-dimensioned pharynx upholstered with mucus and gill slits allowing respiration and filtration of nutriments. The pharynx is supported by a developed endostyle along its height on the side of the animal carrying the inhaling siphon. At the basis of the pharynx is the esophagus, driving aliments to the stomach, localized in the foot of the ascidian proximally to the substrate. From the stomach, the intestine climbs upward and the anus opens into the peribranchial cavity, opened on the outside by the exhaling siphon. Near the stomach, the heart surrounded by a pericarp manages the circulation through a few vessels carrying blood cells through the animal via a circuit organized around the gill sac. Around the stomach and the heart are localized the gonads, one for solitary ascidians belonging to Phlebobranchia and Aplousobranchia, two for those belonging to Stolidobranchia. Gonads produce both sperm and oocytes, which accumulate in two separated gonoducts alongside to the exhalant siphon parallel to the gut. Distally to the substrate and localized between the two siphons is the nerve ganglion from which the innervation is made toward the other organs of the animal. Finally, muscles are distributed all over the animal, participating in the maintenance body shape and fundamentally in pharynx contraction, thus allowing control of the water flow and its

(a)

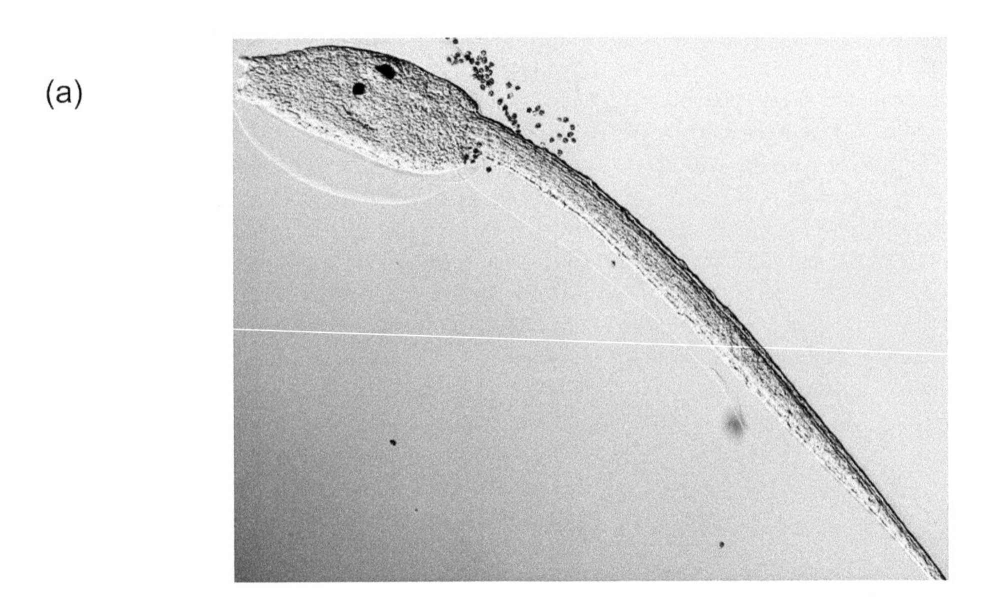

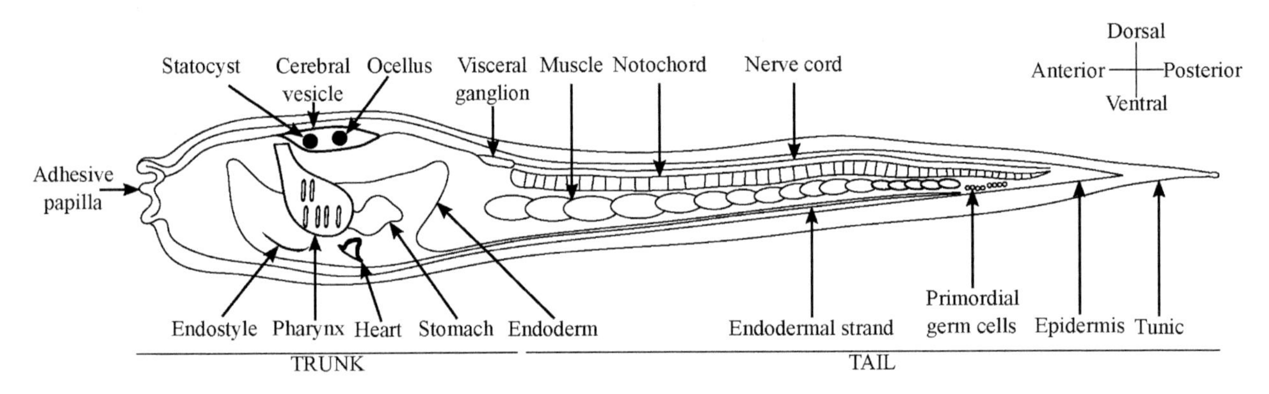

(b)

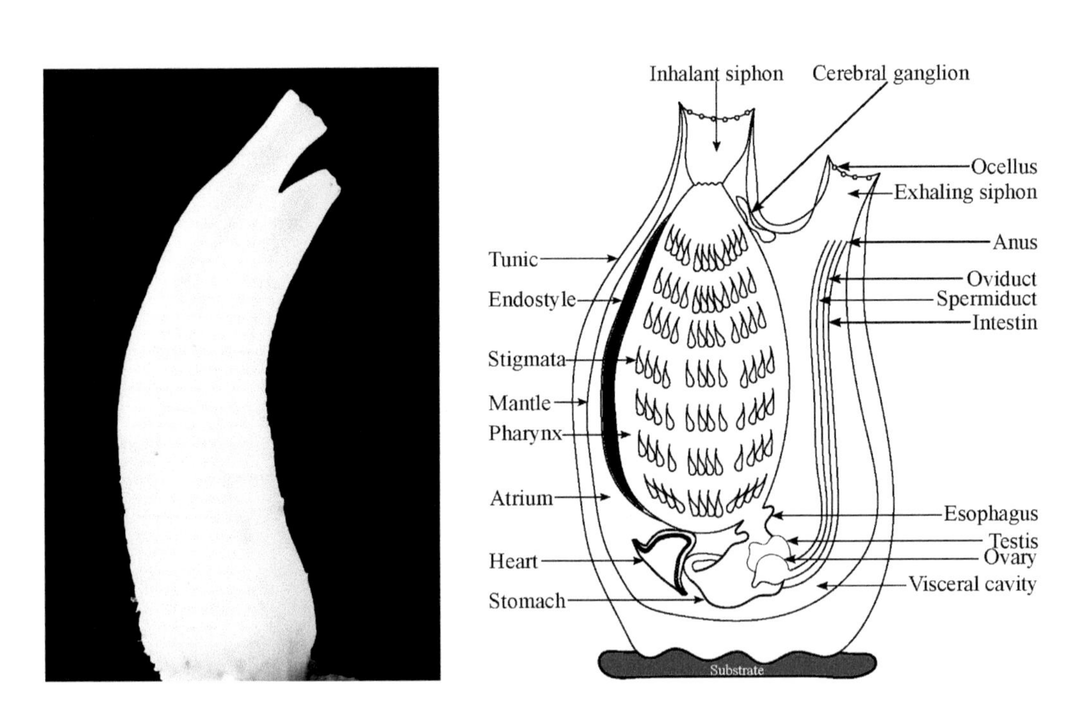

FIGURE 20.6 Classical anatomy representative of solitary ascidians. The larva, composed of a trunk and a tail, present a typical deuterostomian organization plan with a dorsal notochord. Adults are filtering individuals permanently settled to a substrate. Their body is organized around the pharynx and the two siphons, allowing circulation of water bringing food and oxygen.

brutal expulsion if necessary. In addition, muscles surround the siphons and allow them to open or close according to the animal's behavior.

20.7 GENOMIC, TRANSCRIPTOMIC, PROTEOMIC AND BIOINFORMATICS RESOURCES (DATABASES)

20.7.1 Genomics

The first solitary ascidian genome published was of *Ciona intestinalis* type A (now renamed *Ciona robusta*) in 2002, and most of the genomic DNA used for sequencing were isolated from the sperm of a single individual in Half Moon Bay, California (Dehal et al. 2002). The draft genome has been generated by the whole-genome shotgun method (WGS) with eight-fold coverage (Dehal et al. 2002). In this method, the whole genome of *Ciona* was fragmented (in around 3 kbp fragments) and cloned into plasmids (genomic library) for sequencing. In addition, two other libraries were made for this project, one with a mix of genomic DNA of three Japanese individuals cloned into bacterial artificial chromosomes (BACs) for BAC end sequencing and one from another Californian individual cloned into cosmids for cosmid library sequencing (Satoh 2004). Thanks to bioinformatic tools, all these reads were organized into overlapping contigs and then into scaffolds. The *Ciona* genome is approximatively ~159 Mb (comparable with *Drosophila*), rich in AT (65%; as a comparison, the human genome has 45%) and is composed of ~117 Mb of non-repetitive and euchromatic sequences, ~18 Mb of high-copy tandem repeats such as rRNA or tRNA and ~17 Mb of low-copy transposable elements (Satoh 2004). Like those of other invertebrates, the *Ciona* genome exhibits a very high level of allelic polymorphisms, with 1.2% of nucleotides differences between alleles. In 2008, the genome assembly was improved and led to the identification of 15,254 genes, 20% residing in operons, which contain a large majority of single-exon genes (Satou et al. 2008). Another particularity of the *Ciona* genome is its compaction, highlighted by the number of identified genes (15,254) in 117 Mb of euchromatic genome, which gives an average of a gene every 7.7 kb. Using the two-color fluorescent *in-situ* hybridization technique (FISH), a large part (around 82%) of the non-repetitive and euchromatic DNA has been mapped onto chromosomes but also a part of the rDNA and histones clusters (Shoguchi et al. 2006, 2008). *Ciona intestinalis* has 14 pairs of chromosomes, which are in majority telocentric. More recently, a new *Ciona intestinalis* type A assembled genome was published; this genome was sequenced by the Illumina technique and comes from an inbred line. This new genome suggests a previous overestimation on the genome size, since almost the entire genome was sequenced on ~123 Mb. This study also predicts a lower number of identified genes (14,072), which are all mapped on chromosomes (Satou et al. 2019).

From these genomes released, the genes involved in development are well characterized, among them transcription factors (~643), but also genes engaged in a variety of signaling and regulatory processes reported in vertebrate development, such as FGFs (Satou et al. 2002a), Smads (Yagi et al. 2003) and T-box genes (Takatori et al. 2004). Interestingly, developmental genes appear to be often a single copy in the *Ciona* genome, while they have been duplicated in vertebrates, simplifying functional studies, and they could help unravel complex developmental processes in vertebrates. In addition, some evolutionarily innovations were reported, such as a group of genes engaged in cellulose metabolism (Nakashima et al. 2004). There are also several lost genes in the *Ciona* genome, for example, several Hox genes (Hox7, 8, 9 and 11).

Taken together, these studies and the knowledge they brought (sequencing, annotation, physical map) make the *Ciona intestinalis* genome among the most useful to allow investigation at a global scale (chromosomal and genome-wide) of the regulation of gene regulatory networks during development. The *Ciona* genomic information is accessible at https://genome.jgi.doe.gov/portal/ but also in others databases (see the following for details).

Today, with the emergence of the high-throughput–next generation sequencing (reviewed in Pareek et al. 2011), genomes of several solitary but also colonial ascidians genomes have been performed. Interestingly, the choice of sequenced species is well distributed on ascidian phylogeny (Figure 20.1). Indeed, in addition to *Ciona intestinalis* type A (*Ciona robusta*), five Phlebobranchia were sequenced, two *Phallusia* (*Phallusia mammillata* and *Phallusia fumigata*), two additional *Ciona* (*Ciona savignyi, Ciona intestinalis type B*) and one *Corella* (*Corella inflata*); seven Stolidobranchia, three *Molgula* (*Molgula oculata, Molgula occulta, Molgula occidentalis*), one *Botrylloides* (*Botrylloides leachii*), one *Botryllus* (*Botryllus schlosseri*) and two *Halocyntia* (*Halocyntia roretzi, Halocyntia aurantum*). All these genomes and gene annotations are available in the ANISEED database (see the following for details). These genome decoding works allow comparative genomics of ascidians and promise very interesting insights into the A5.1 cell (Figure 20.5a) at the 16-cell stage for ascidian but also chordate evolution.

20.7.2 Transcriptomic

The first information about the ascidian transcriptome was obtained by express sequenced tag (EST) analyses (Satou et al. 2002b). This approach is based on the generation of cDNA clones from total mRNA purification in order to get gene expression information. The cDNA project conducted on *Ciona intestinalis* has generated gene expression information at different developmental stages of *Ciona*, such as fertilized egg, cleaving embryo, gastrulae/neurulae, tailbud embryo and tadpole larva but also in adult tissues corresponding to testis, ovary, endostyle, neural complex, heart and blood cells and whole young adults (Satoh 2013). This classification has also led to temporal and spatial information of gene expression; since the cDNA libraries used for EST analyses were not amplified or normalized, an abundance of

EST in each stage or tissue may reflect gene specific expression (EST count) (Satoh 2013).

These clones were sequenced and categorized (based on similarity to known proteins), and numbers of them were subjected to analysis by whole-mount in situ hybridization (ISH), revealing expression patterns of up to 1,000 genes during *Ciona* development and in adults (Satou et al. 2002b). Coupled with genomic information, cDNA analyses led to the identification and spatial expression profiles of almost all transcription factor genes, among them 46 basic helix-loop-helix, 26 basic leucine zipper domains, 15 E-twenty-six, 24 forkhead box, 21 high motility group, 83 homeobox family members and 17 nuclear receptor family members (Satou et al. 2003a; Wada et al. 2003; Yagi et al. 2003; Yamada et al. 2003) of genes encoding proteins involved in major signaling pathways (receptor tyrosine kinase, MAPK, Notch, Wnt, TGF-β, hedgehog, JAK/STAT) (Satou et al. 2003b, 2003c; Hino et al. 2003) but also gene encoding proteins involved in major cellular processes (cell polarity, actin dynamics, cell cycle, cell junction and extracellular matrix) (Sasakura et al. 2003b, 2003c; Kawashima et al. 2003; Chiba et al. 2003).

All the published and unpublished spatiotemporal data concerning EST included in the cDNA library and EST count are available in the GHOST and ANISEED databases.

A similar EST approach was conducted on five different developmental stages of the anural ascidian *Molgula tectiformis* and gives new insights on the molecular mechanisms of the tailless mode of development of this species (Gyoja et al. 2007).

From these initial works, different types of microarrays were prepared, coupled with cell sorting allowing the identification of the gene regulatory networks involved during heart precursor migration (Christiaen et al. 2008). Microarrays coupled with chemical inhibitors of either JNK or ERK/MAPK pathways also led to the identification of gene networks involved in the onset of metamorphosis (Chambon et al. 2007).

More recently, the recent emergence of single-cell RNA sequencing (scRNA-seq), coupled with previous genomic and transcriptomic data, revolutionized, as in other experimental models, the way to investigate cell specification during embryogenesis by allowing identification of novel cell types, or cell-state and dynamic. Applied to *Ciona* embryogenesis, from gastrulation to tadpole larva, scRNA-seq permitted the identification of 40 new cell types (40 neuronal subtypes) in the larva (Cao et al. 2019). In addition, this study also allowed a better comprehension of the evolution of vertebrate telencephalon by comparing *Ciona* larva gene expression data with other chordate animals.

In addition to EST data, new transcriptomic data coming from RNA-seq technologies and microarray are also integrated in ANISEED (see database section for details).

20.7.3 Proteomics

In addition to genomics and transcriptomics, proteomics completes the set of necessary data to address fundamental questions in developmental but also cell biology of solitary ascidians. These data were generated using the protein mass fingerprint-based method in which previous cleavage into smaller peptides of protein of interest is followed by mass spectrometry analysis (MALDI/TOF), eventually with a previous separation of proteins on 2D-gel electrophoresis.

Compared to genomics and transcriptomics, a few proteomics studies were reported, but recently this approach seems to be used as a tool to evaluate the environmental impact of ascidians. Using two conditions to rear *Ciona intestinalis*, at 18°C (the usual working temperature) and 22°C, a clear distinction in the protein expression pattern in ovaries was observed (Lopez et al. 2017). It was previously known that the reproductive capacity of this species is altered by temperature up to 20°C; in this study, a range of temperature-response proteins were identified, making proteomics on *Ciona* a good approach to evaluate the impact of global temperature change. More recently, a proteomics approach was performed on two solitary ascidians, *Microcosmus exasperatus* and *Polycarpa mytiligera*, both collected at different locations on the Mediterranean coast of Israel (five sites) and along the Red Sea coast (four sites) (Kuplik et al. 2019). Differentiated protein profiles were obtained in the two ascidians from different localities. Here again, proteomics analysis of ascidians may reflect the conditions in their environments and make this approach a potential good biomarker for monitoring coastal marine environment health.

Furthermore, proteomic methods in *Ciona* were used to investigate sperm cell components and to examine their functions (reviewed in Inaba 2007) but also to study the function and interactions of gametes (Satoh 2013). In addition, a proteomic analysis on three embryonic stages of *Ciona* intestinalis (unfertilized eggs 16-cell stage and tadpole larvae) allowed the creation of a protein expression profile and provided a dynamic overview of protein expression during embryogenesis. Interestingly, when a protein dataset was compared with mRNA levels at these same stages, nonparallel expression patterns of genes and proteins were observed (Nomura et al. 2009). In many cases, a change in protein network, protein expression, protein modification or localization is independent of gene expression or translation of new mRNA transcripts. A proteomic-based approach is capable of highlighting differential protein expression or modifications and will be essential to understand molecular mechanisms that sustain developmental process and/or cell behavior or cell fate in ascidians.

Ascidian proteomic datasets are available in the CIPRO database, which is an integrated *Ciona intestinalis* protein database (www.cipro.ibio.jp).

20.7.4 Databases

Several databases are available for the ascidian research community, and most of them emerged from ascidian laboratories. In this section, we provide a short description of the principal databases with a particular emphasis on GHOST and ANISEED, which are the main ascidian databases for the worldwide scientific community.

- The Ascidians Chemical Biology Database (ACBD) (created in 2010 in Japan) is a bibliographical database that compiles publications concerning the effect of chemical compounds on ascidian development and tends to promote ascidians as a model organism for whole-animal chemical screening.
- The Database of Tunicate Gene Regulation (DBTGR) (created in 2005 in Japan) focuses on tunicate gene regulation, including regulatory elements in the promoter region and the associated TF. In addition, it integrates a list of gene reporter constructs.
- The website of the Joint Genome Institute (JGI) (created in 1997 in United States) hosts the *Ciona intestinalis* type A genome and contains a genome browser.
- MAboya Gene Expression pattern and Sequence Tags (MAGEST) (created in 2000 in Japan) provides *Halocyntia roretzi* 3'- and 5'-tag sequences (20,000 clones) from the fertilized egg cDNA library, the amino acid fragment sequences predicted from the EST data set and the expression data from whole-mount in situ hybridization.
- *Ciona intestinalis* Adult *In Situ* hybridization Database (CiAID) (created in 2009 in Japan) gives access to gene expression patterns in adult juveniles with a body atlas.
- The *Ciona intestinalis* Protein (CIPRO) database (created in 2006 in Japan) is a *Ciona intestinalis* protein database that contains 3D expression profiling, 2D-PAGE and mass spectrometry-based large-scale analyses at various developmental stages, curated annotation data and various bioinformatic data.
- Four-Dimensional Ascidian Body Atlas (FABA) (created in 2010 in Japan) contains ascidian three-dimensional (3D) and cross-sectional images through the developmental time course (from fertilized egg to larva) to allow morphology comparison and provide a guideline for several functional studies of a body plan in chordate. Note that a second database called FABA2 (created in Japan) exists, focusing on later developmental stages, from hatching to seven-day-old juveniles.
- *Ciona intestinalis* Transgenic Line RESsources (CITRES) (created in 2012 in Japan) provides the ascidian research community with transgenic lines but also contains DNA constructs to perform transgenesis, image collections of *Ciona* GFP-expressing strains and publications.
- Ghost database (originally created in 2002 in Japan; http://ghost.zool.kyoto-u.ac.jp/) is one of the first ascidian databases available for the ascidian research community and the most useful from the beginning. This database provides all the data concerning the *Ciona intestinalis* EST project conducted by Satoh's lab (see transcriptomic section for details), such as EST count, that provide temporal expression information and published and unpublished ISH at several developmental stages. In addition to that, the database contains a genome browser, a search engine for specific expression or expression pattern of a given genes and gene annotation. At the beginning, this database represented an extraordinary source of molecular tools, since it provides a set of 13,464 unique cDNA clones available as the "*Ciona* intestinalis gene collection released" for the scientific community, ready for use in cDNA cloning, microarray analysis and other genome-wide analyses. Almost the entire database is now integrated in the ANISEED database.
- ANISEED (created in 2010 in France; www.aniseed.cnrs.fr) is the biggest and most complete database for the ascidian community (Dardaillon et al. 2019). There is a constant input of new data, and it provides functionally annotated gene and transcript models in both wild-type and experimentally manipulated conditions using formal anatomical ontologies. The advantages of this database are the extra information, going beyond genes by pointing out repeated elements and cis-regulatory modules and also providing orthology comparison within or even outside ascidians (tunicates, echinoderms, cephalochordates and vertebrates). There are enhanced functional annotations for each species, achieved by an improved orthology detection and manual curation of gene models. This database is user friendly, with three types of browsers, each offering a different but complementary point of view: a developmental browser which selects data based either on the gene expression or the territory of interest, an advanced genomic browser focusing on gene sets and gene regulation and a genomicus synteny browser that explores the conservation of local gene order across deuterostome. This later new release has a reference of the taxonomic range of 14 species, among them a non-ascidian species, the appendicularian *Oikopleura dioika*, which is a novelty. Finally, the new and powerful Morphonet morphogenetic browser enables a 4D exploration of gene expression profiles and territories.

20.8 FUNCTIONAL APPROACHES/TOOLS FOR MOLECULAR AND CELLULAR ANALYSES

In addition to classical over/ectopic expression of genes, several tools or technical approaches were developed by the ascidian community by taking advantage of biological particularities and/or experimental advantages offered by solitary ascidians.

20.8.1 Microinjection/Electroporation

To follow specific expression patterns of regulatory genes or to probe gene function, experimental biologists usually introduce reporter constructs or synthetic mRNA in fertilized eggs. In most animal models, these approaches are usually achieved by the microinjection technique. Solitary ascidians, essentially *Ciona*, allow an alternative technique, a simple electroporation method. This permits manipulation and screening of hundreds of synchronous developing embryos, either wild type or mutant, thus allowing greater confidence in functional screening, which is not possible with most of the other animal models.

20.8.2 Reporter Gene

The efficient introduction of reporter constructs by electroporation (Corbo et al. 1997), coupled with the facility (compared to the other animal models) to identify and clone the core promoter and associated enhancers of a given gene, made the solitary ascidian *Ciona intestinalis* an excellent model to study cis-regulation. Indeed, due to the *Ciona* compact genome, the *cis*-regulatory elements (CREs) are usually located within the first 1.5 kb upstream of the transcription start site, making it relatively easy to capture significant transcriptional units and clone them upstream of a reporter gene to drive its expression. Coupled with the electroporation technique, this allows a simple and rapid generation of hundreds of transient transgenic embryos expressing fluorescent proteins, which develop quickly to the larval stage (Zeller et al. 2006). These transient assays allowed rapid identification and characterization of up to 83 *Ciona cis*-regulatory elements, almost all enhancers, which activate transcription in a more or less tissue-specific manner (reviewed in Irvine 2013).

20.8.3 Loss-of-Function Approaches

To understand the molecular basis of development, experimental biologists expect to specifically inhibit the functions of a particular gene in particular cells at particular developmental stages. The basic technologies for examining gene functions by loss of function approaches have been established in *Ciona*, such as the knockdown of genes by antisense morpholino oligonucleotides (MOs) (Satou et al. 2001), transposon-mediated germ cell transformation and mutagenesis (Sasakura et al. 2003c, Sasakura et al. 2005), zinc-finger nucleases (ZFNs) (Kawai et al. 2012), transcriptional activator-like effector nucleases (TALENs) (Treen et al. 2014) and clustered regularly interspaced short palindromic repeats (CRISPR/Cas9) (Sasaki et al. 2014). These technologies have supported detailed and thorough analyses to reveal molecular and cellular mechanisms that underlie development of *Ciona*, since almost of them can be performed in a tissue-specific manner during embryogenesis.

20.8.3.1 MOs

The antisense morpholino oligonucleotide strategy consists of MOs that bind to the targeted mRNA and prevent translation. They were tested in a range of models, including ascidians, in which they were extensively used since they allow a rapid and high-throughput approach for functional studies. In addition, MOs are able to target maternal mRNA determinants as well as zygotic genes. The efficiency of this technique was first tested in *Ciona savignyi*, in which MOs were able to target the maternal pool of β-catenin mRNA and abolish endodermal differentiation (Satou et al. 2001). Since then, MOs were extensively used and allowed identification of key genes in tissue differentiation during embryogenesis, such as the maternal determinant Macho-1 for muscle differentiation in *Halocyntia roretzi* (Nishida and Sawada 2001) and in *Ciona savignyi* (Satou et al. 2002c); in tissue formation, for instance, chondroitin-6-O-sulfotransferase involved in *Ciona intestinalis* notochord morphogenesis (Nakamura et al. 2014); or even for cell fate, for example, Ci-Sushi, which controls the initiation of apoptosis at the onset of *Ciona intestinalis* metamorphosis (Chambon et al. 2007). However, there are several limitations to injecting MOs in solitary ascidians, notably the restricted numbers of mutants to analyze and the difficulty of interpreting some phenotypes due to off-target effects.

20.8.3.2 RNA Interference

Based on the introduction in the cells of double-strand RNA, which are converted in small interfering RNA (siRNA), causing the destruction of specific mRNA, this approach was successfully used in colonial ascidians but has had few successes with solitary ones, except the electroporation of short-hairpin RNA targeting tyrosinase-encoding gene in *Ciona* embryo leading to the absence of melanization of the tailbud pigmented cells (Nishiyama and Fujiwara 2008). To date, the use in solitary ascidians is very limited.

20.8.3.3 ZNFs and TALENs

The nuclease activity ZNFs and TALENs induces double-strand breaks (DSBs) at target sequences. In the case of ZNFs, mutations occur when DSBS are repaired by non-homologous end joining (NHEJ), which introduces insertional or deletional mutations at the target sequence. TALENs provoke mutations when the cellular DNA repair mechanisms fail. Both approaches were established in *Ciona intestinalis* by the Sasakura lab (Kawai et al. 2012; Treen et al. 2014) and are a very promising strategy to mutate endogenous genes during development. ZNFs were tested in a *Ciona* transgenic line expressing EGFP to introduce mutations in EGFP loci. When eggs were injected, it resulted in inheritable mutations with high frequency (about 100%), no toxic effect on embryogenesis and few off-target effects (Kawai et al. 2012). TALEN knockouts can be performed by electroporation and allow fast generation of mutants and a quick screening involving numbers of embryos not possible with other animals. Toxicity is a major concern with TALEN when ubiquitous knockouts are generated, but using tissue-specific promoters reduces this problem and allows mutations in a tissue-specific manner (Treen et al. 2014).

20.8.3.4 CRISPR/Cas 9

Since its discovery in 1987 (Ishino et al. 1987), CRISPR/Cas9 has become one of the most powerful tools for researchers to alter the genomes of a large range of organisms. CRISPR/Cas9 uses a short guide RNA (sgRNA) that binds to its target site; Cas9 protein is recruited to the binding site and induces a double-strand break at the target genomic region. In solitary ascidians, this technique was first successfully tested in *Ciona intestinalis* (Sasaki et al. 2014) and more recently in *Phallusia mammillata* (McDougall et al. 2021).

In *Ciona*, the most widely used application of CRISPR is for targeted mutagenesis in somatic cells of electroporated embryos. In this method established in 2014 by Sasakura lab (Sasaki et al. 2014) and recently improved by Stolfi (Gandhi et al. 2018), *in vitro* fertilized one-cell-stage embryos are electroporated with plasmids, allowing the zygotic expression of Cas9 protein and sgRNA. Interestingly, Cas9 can be expressed in a cell-specific manner, and the targeted mutations are a powerful means to dissect the tissue-specific functions of a gene during development.

20.8.4 Genetics, Mutagenesis and Transgenesis

Natural mutants often arise in wild populations, probably due to the high polymorphism between individuals within a given population (Satoh 2013). Moreover, the rapid life cycle and the possibility of self-fertilization (natural or induced with chemical or enzyme treatment), coupled with a rapid embryogenesis and a morphologically simple tadpole that allows simple phenotype detection, make both *Ciona* (*intestinalis* and *savignyi*) excellent models for mutagenesis. In addition to characterization of the *Ciona savignyi* natural mutant *frimousse* (Deschet and Smith 2004), Smith's lab took advantage of the self-fertility in this species to perform a mutagenesis screen notably using N-ethyl-N-nitrosourea (ENU)-induced mutations affecting early development. This random approach led to the isolation of a number of mutants with notochord defects such as *chongmague* and *chobi* (Nakatani et al. 1999). Since then, the transgenesis technique was established in *Ciona* using transposon-mediated transgenesis that allow creation of stable germ lines but also to use it for insertional mutagenesis and enhancer trapping.

The Tcl/mariner transposable element *Minos* (isolated from *Drosophila hydei*) is a small DNA transposon (2000 bp) activated by a "cut and paste" system in which a transposase is able to excise the transposon from the DNA and integrate it into a target sequence. When a plasmid containing *Minos* is microinjected or electroporated in *Ciona* eggs with transposase mRNA, *Minos* is excised from the vector DNA and integrated in the *Ciona* genome, and this event is observed in somatic and germ cells (Sasakura et al. 2003c). In the latter case, this insertion is inherited by the progeny, and its stability was reported over ten generations in several transgenic lines (Sasakura 2007). Insertions of *Minos* can disrupt gene function to create mutants, such as the *swimming juvenile*, which exhibits a cellulose synthesis defect and absence of tail regression during metamorphosis due to the integration of *Minos* at *Ci-CesA* promoter, a gene involved in cellulose synthesis in *Ciona intestinalis* (Sasakura et al. 2005). In addition to insertional mutagenesis, the transposon-based technique was also able to create stable marker lines when CRE of tissue-specific gene driving expression of fluorescent proteins were used with a *Minos*-based transposable element. Another potentiality of *Minos* transposons is the enhancer trapping technique. It consists of insertions using a reporter gene in a *Minos* transposons construct (GFP, for example), and if there is an enhancer close to the transposon insertions, the expression patterns of reporter genes are affected according to the enhancer. In *Ciona*, an intronic enhancer in the *Ci-Musashi* gene was identified by this approach (Awazu et al. 2004).

20.9 CHALLENGING QUESTIONS

Researchers in the ascidian field face many challenging questions. In this section, a brief overview of some of them will be given, followed by a detailed discussion of the unique opportunity provided by the ascidians to develop quantitative modeling of chordate embryos.

20.9.1 Evolution of Ascidians

As described in the genomic section, 11 ascidian genomes are now sequenced and annotated, some of them with transcriptomic data and identification of cis-regulatory modules. In addition, the compilation of these data in the ANISEED database will greatly facilitate comparative developmental genomics between ascidian species and allow new insights in ascidian evolution. Immediate application of this approach could lead to better understanding of the differences in gene-regulatory networks during embryogenesis observed between *Ciona intestinalis* and *Halocyntia roretzi* (see embryogenesis section for details). Indeed, these two species exhibit at least two differences for notochord and muscle secondary lineage which both require FGF but dependent on nodal and Delta/Notch for *Ciona* and independent of both of them for *Halocyntia*. Further analyses of the developmental genomics of these two species may allow evolutionary inference to better understand these changes.

Another example concerns the phenotypic change observed in several species that do not develop a tail during embryogenesis and do not develop notochord or tail muscles; instead, they give rise to non-motile tail-less larva without functional notochord or larval tail muscle or directly to a juvenile (Satoh 2013). Anural development occurred independently several times during ascidian evolution. Cross-fertilization approach of the tail-less *Molgula occulta*, and its close relative urodele species *Molgula oculata* gives rise to a hybrid embryo with a short tail containing a notochord. Swalla and Jeffery (1990) suggested an evolution of the anural mode of development by relatively simple genetic changes. Comparative genomics studies permitted by the release of the genome of these two species will certainly detect key genomics changes for these different modes of embryogenesis.

20.9.2 Ascidians for Therapeutic Advances

In the last few years, several studies have been conducted on the identification and characterization of chemical diversity produced from marine ascidians (Palanisamy et al. 2017). The essential part of these chemical compounds is used by ascidian species to prevent predatory fish, as an anti-fouling and anti-microbial mechanism and to control settlement (reviewed in Watters 2018). Ascidians, like several marine organisms, produce a rich variety of secondary metabolites with potential therapeutic properties in human medicine, with a range of biological activities such as cytotoxicity, antibiotic and immunosuppressive activities, inhibition of topoisomerases and cyclin-dependent kinases (Duran et al. 1998). Most of these compounds were identified by the liquid chromatography-mass spectrometry method. Among them, Ecteinascidin was isolated from *Ecteinascidia turbinata* and is currently used as a cancer drug to treat soft-tissue sarcoma and ovarian cancer (Gordon et al. 2016); Aplidin isolated from *Aplidium albicans* has given promising results in myeloma treatment (Delgado-Calle et al. 2019). In addition, anti-malarial effects were identified from extracts coming from three ascidians, *Microcosmus goanus, Ascidia sydneiensis* and *Phallusia nigra* (Mendiola et al. 2006). Between 1994 and 2014, up to 580 compounds were isolated from ascidians and offer a wide range of opportunities to identify molecules with therapeutic properties for human diseases.

In addition to screening for molecules with potential therapeutic effects, ascidian embryos have also started to be used as an experimental model to study the neurodevelopmental toxicity of different compounds (Dumollard et al. 2017).

20.9.3 When Developmental Biology Becomes Quantitative: A Big Step toward "Computable Embryos"

The transition from a single fertilized cell to a complex organism, with various cell types that compose its tissues in the correct numbers and their fine regulation in space and time, is the question at the heart of developmental biology. Decades of research in this field have designed a broad portrait of the fundamental processes involved during embryogenesis: from the description of the genetic programs of embryonic cells and the mechanisms regulating gene transcription to how cell fates and behaviors are coordinated by cell communication and the way this translates into morphogenesis.

Developmental mechanisms have traditionally been studied at the tissue level in a qualitative manner. For example, consider the current view of the classical chemical signaling during fate specification. A surprisingly small number of signaling pathways involving cell surface receptor and activating ligands act in widely different cellular contexts to produce the diversity of fate specification events occurring during embryogenesis (Perrimon et al. 2012). Despite this, many simple questions remain unanswered, such as: "What is the mechanism regulating the dose-response to increasing concentrations of ligands or receptors?" "How are ligand concentration and time of exposure integrated by cells?" To deepen the understanding of the principles which govern embryonic development, it is important to combine quantitative experimental approaches at the cellular scale with dynamic mathematical models including mechanistic details. For example, the recent development in quantitative imaging, sequencing, proteomics and physical measurements have allowed us to refine the historical morphogen concept, in which diffusible signaling molecules are proposed to coordinate cell fate specification and tissue formation using concentration-dependent mechanisms (a static readout), because it was insufficient to describe or model the complexities of patterning observed with these techniques in developing embryos (Garcia et al. 2020; Huang and Saunders 2020; Jaeger and Verd 2020; Rogers and Müller 2020; Schloop et al. 2020).

While physical modeling of life has a long history (Thompson 1917), it has remained a theoretical exercise for a long time: insufficient measurements of physical parameters for constraining models coupled with a largely qualitative and static description of phenotypes have rendered it difficult to apply physics to developing embryos and even to single cells. The recent technological breakthrough mentioned previously, however, reduced this difficulty while making "computable embryos" through a precise physical description of embryonic development more necessary than ever to capture key developmental concepts and bridge genomic information and dynamic phenotypes (Biasuz et al. 2018). First, our brains are simply unable to cope with the large amount of data generated, much of which are unrelated to the mechanism being studied. Second, biology involves several layers of feedback, resulting in unintuitive non-linear behaviors. Third, biology is a multiscale process in which macroscopic properties of cells and tissues arise from the mesoscopic properties of molecules or subcellular structures.

Ascidians definitively constitute a model of choice to build a global computational model of embryogenesis. Embryonic development is a continuous progression in time. The "computable embryo" is based on the idea that a mathematical description of the system can predict the future state of the embryo from the knowledge of its current state. This global computational model of embryogenesis at the single-cell, genome-wide and whole-embryo level is a challenging task and will only be achieved using the most appropriate developmental systems (Biasuz et al. 2018).

Solitary ascidian embryos seem to be good candidates for this breakthrough. At first glance, one would rather think of the *Drosophila melanogaster* or vertebrate embryos for this role. Indeed, thanks to decades of research, a deep understanding of core developmental mechanisms has been achieved, and powerful genetic and cell biology tools exist.

These embryos, while remaining a significant motor for defining new concepts, may, however, be too complicated to incorporate these concepts into a global model of embryogenesis. In contrast, ascidian embryos, as nematodes, are simpler and develop stereotypically with few cells and invariant cell lineages, so that each cell can be named and found at the same position in all embryos (Lemaire 2009). Unlike those of nematodes (Goldstein 2001), ascidian embryo geometries have even remained essentially unchanged since the emergence of the group, around 400 million years ago, despite extensive genomic divergence (Delsuc et al. 2018; Lemaire 2011). The development of ascidians is also characterized by earlier fate restriction than most animal embryos: 94 of the 112 early gastrula cells in the ascidian *Ciona* are fate restricted, each contributing to a single larval tissue type (Nishida 1987). Moreover, ascidians are closely related to vertebrates, as they belong to the vertebrate sister-group, but ascidians kept their genomic simplicity. Indeed, they diverged before the two rounds of whole-genome duplication events which occurred in the vertebrate lineage leading to the apparition of multiple paralogues for each gene (Dehal and Boore 2005), with potentially slightly divergent activities. Finally, ascidian embryos are small (~130 μm) and transparent, and they develop rapidly externally in sea water up to the larval stage (~12 h), making them very easy to image. Thus, ascidian embryos provide a rigid framework that allows combination of analyses at cellular resolution with mathematical modeling.

These advantageous properties of ascidian embryos, especially *Phallusia mammillata* embryos, which are fully transparent, combined with the breakthrough development of light sheet microscopy (Power and Huisken 2017), have enabled the production of the first digitized version of a metazoan embryo (Figure 20.7a) (Guignard et al. 2020). Based on automatic whole-cell segmentation and tracking over five cell generations of membrane-labeled cells with two-minute temporal resolution, this research offers a complete description of early ascidian embryo development, accounting for each cell in the ten embryos analyzed. Moreover, this quantitative and dynamic atlas of cell positions and geometries can be associated with the known cell fates and interactively explored through the MorphoNet online morphological browser (Leggio et al. 2019). These "digital embryos" show that ascidian development is reproducible down to the scale of cell–cell contacts and, combined with modeling and experimental manipulations, it allows us to establish contact area-dependent inductions as an alternative to classical morphogen gradients. This work opens the door to quantitative single-cell morphology and mechanical morphogenesis modeling.

In parallel with this work, another group combined high-resolution single-cell transcriptomics (single-cell RNA sequencing) and light-sheet imaging to build the first full comprehensive atlas which describes the genome-wide gene expression of every single cell of an embryo in the early stages of development, showing the evolution from a single cell up to gastrulation in the ascidian *Phallusia mammillata* (Sladitschek et al. 2020). By providing a complete representation of the gene expression programs, which instruct individual cells to form the different cell types necessary to build an embryo, and therefore by allowing us to know precisely cell-specific expression of transcription factors at the single-cell level, this study will significantly enhance current single-cell-based gene regulatory network inference algorithms (Aibar et al. 2017) and will help to further develop single-cell-based physical models of the different steps of transcriptional control during development. Moreover, these single-cell gene expression data will feed several layers of physical description of biological processes. For example, identification of cell-adhesion molecules will allow the refining of morphogenetic models, such as oriented cell divisions, cell shape changes or cell neighbor exchanges models (Etournay et al. 2015), thereby linking mechanical and genetic information at the cellular resolution.

In spite of the convenient properties and the recent advances that have been realized thanks to the ascidian embryo model, there is still a long way to go to be able to "compute the embryo".

Typically, studies at the single-cell level are in their early days, as can be illustrated by signal transduction studies. The MAPK/ERK signaling pathway is one of the important embryonic signaling pathways used by vertebrates and invertebrates, controlling many physiological processes (Lavoie et al. 2020), and is the main inducing pathway in early ascidian embryos (Lemaire 2009). The signaling cascade from the activation of the transmembrane receptor to the phosphorylation of the ERK nuclear targets is well described (Figure 20.7b) (Lavoie et al. 2020). Our current knowledge of this pathway is, however, mostly static, and an integrated understanding of its spatio-temporal dynamics is lacking (Patel and Shvartsman 2018). For example, it has been shown that the ERK pathway can trigger two qualitatively different types of ERK activity: pulsatile or continuous (Aikin et al. 2019). To understand these non-intuitive results, it is important to combine quantitative experimental approaches at the cellular resolution with dynamic mathematical models including mechanistic details. Genetically encoded fluorescent activity sensors that convert kinase activity into nucleocytoplasmic events have been recently developed (Durandau et al. 2015; Regot et al. 2014), and these tools can now be coupled with optogenetic systems in order to activate the ERK pathway with high spatiotemporal accuracy at different levels (Gagliardi and Pertz 2019). However, these techniques were only used to track a single pathway component at a time. Yet they suggest that multiplexing sensors at different levels of the cascade could reveal the dynamics of information flow through the cell. Such quantitative measures are required to more realistically model the catalogue of cell-signaling modalities (Biasuz et al. 2018).

The technological breakthroughs of the last quarter of the century have brought a whole new perspective to developmental biology, which is now seen through the combined lenses of mathematical modeling and experimental biology. A major challenge for the future will now be to integrate

(a)

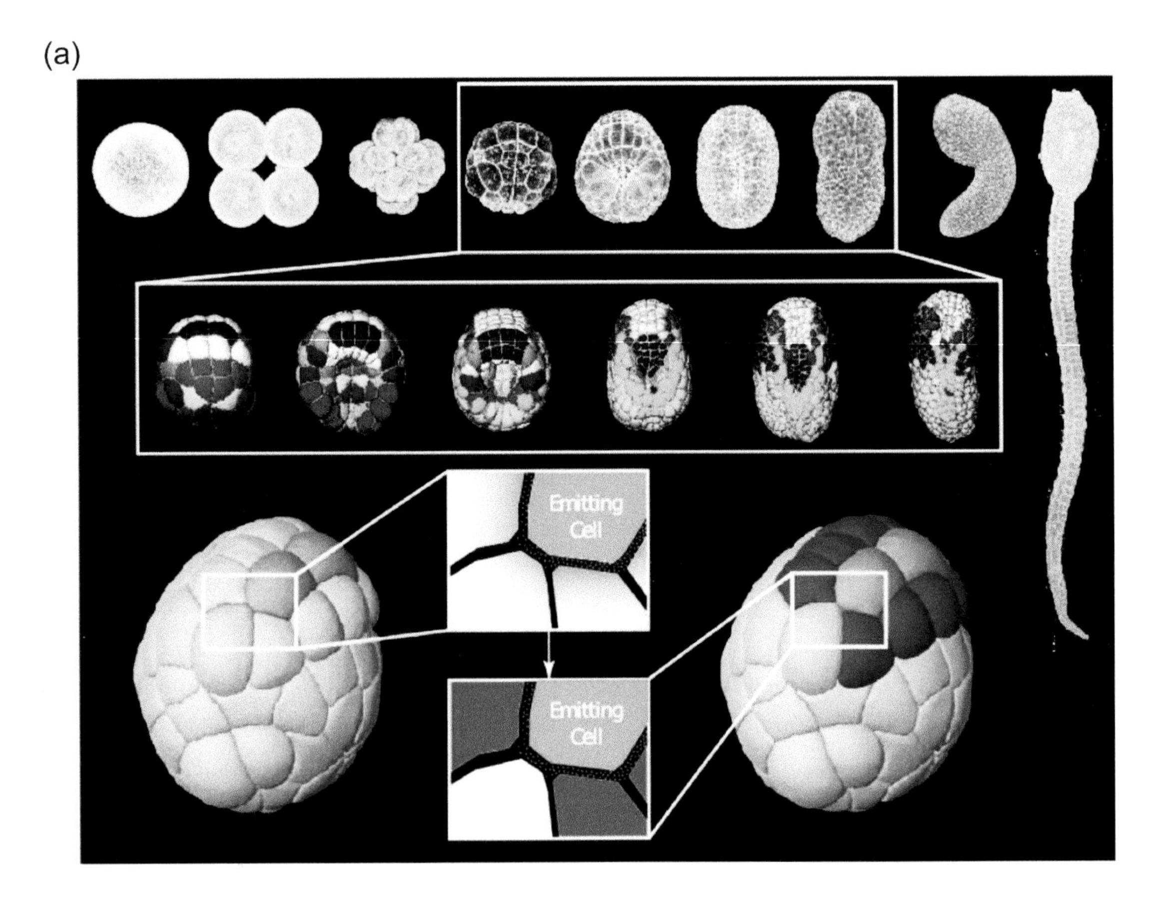

(b)

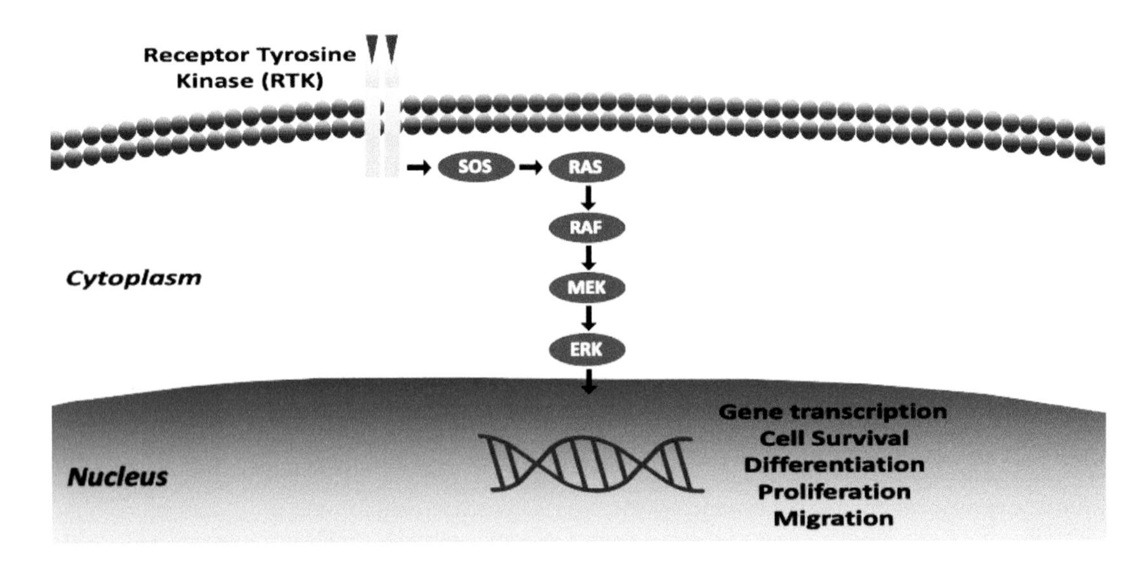

FIGURE 20.7 (a) Digitalization of *Phallusia mammillata* embryogenesis reveal contact area-dependent cell inductions. (Top) Light-sheet imaging of cell membranes (not shown) combined with automated cell segmentation and tracking allowed reconstruction of *Phallusia* embryogenesis between the 64-cell and initial tailbud stages. Digital embryos represented here are color-coded with cell fates. (Bottom) Illustration of the contact area-dependent mode of cell inductions. Light blue cells emit inducing extracellular signals (left). Among the neighbor cells which receive the signal, only the dark blue cells, which have the largest surface of contact with emitting cells, are induced (right). Digital embryos have been explored through the MorphoNet online morphological browser. (b) Simplified representation of the MAPK/ERK signaling pathway. ([a] Figure courtesty of Leo Guignard & Kilian Biasuz; [b] figure courtesy of Kilian Biasuz.)

partial models accounting for short-term activities into a global view of biological processes. Indeed, most of the modeling efforts were designed to shed light on specific processes over a short period of time. As a consequence, our physical knowledge of embryogenesis is reduced to a few unconnected kernels of insight. Increasing the number of kernels is imperative to "compute the embryo" but will not suffice: kernels will need to be incorporated into a bigger picture. The solitary ascidian embryos, which are simple and transparent and contain a relatively small number of cells and invariant cell lineages are perfect candidates to integrate these principles into a global model of embryogenesis.

20.10 GENERAL CONCLUSION

The last 20 years have been marked by extraordinary advances in the comprehensive biology of ascidians. Starting as the first experimental model organism in embryology, the ascidian embryo offers today an avenue of investigation in several biological research fields such as developmental biology, cell biology, comparative genomics, drug screening or evo-devo. The decoded genome of 13 ascidians, coupled with gene annotation, large transcriptomic data, proteomics, identification of cis-regulatory elements, large coverage of gene expression patterns by *in situ* hybridization, stereotyped and well-described cell lineages, physical maps of the genome onto chromosomes and routine generation of transgenic lines combined with cell line markers and single-cell transcriptomics (supported by FACS) render this "old" marine model one of the most promising for modern biology.

ACKNOWLEDGMENTS

The authors want to thank A. Karaiskou for critical reading of the manuscript and C. Dantec for her help with the database section.

BIBLIOGRAPHY

Aibar, S., González-Blas, C.B., Moerman, T., Huynh-Thu, V.A., Imrichova, H., Hulselmans, G., Rambow, F., Marine, J.C., Geurts, P., Aerts, J., van den Oord, J., Atak, Z.K., Wouters, J. & Aerts, S. 2017. SCENIC: Single-cell regulatory network inference and clustering. *Nature Methods* 14(11): 1083–1086.

Aikin, T.J., Peterson, A.F., Pokrass, M.J., Clark, H.R. & Regot, S. 2019. Collective MAPK signaling dynamics coordinates epithelial homeostasis. *bioRxiv* 826917.

Arsenault, G., Davidson, J. & Ramsay, A. 2009. Temporal and spatial development of an infestation of *Styela clava* on mussel farms in Malpeque Bay, Prince Edward Island, Canada. *Aquatic Invasions* 4: 189–194.

Awazu, S., Sasaki, A., Matsuoka, T., Satoh, N. & Sasakura, Y. 2004. An enhancer trap in the ascidian *Ciona intestinalis* identifies enhancers of its Musashi orthologous gene. *Developmental Biology* 275: 459–472.

Berrill, N.J. 1929. Digestion in ascidians and the influence of temperature. *Journal of Experimental Biology* 6: 275–292.

Berrill, N.J. 1932a. The mosaic development of the ascidian egg. *The Biological Bulletin* 63: 381–386.

Berrill, N.J. 1932b. Ascidians of the Bermudas. *The Biological Bulletin* 62: 77–88.

Berrill, N.J. & Sheldon, H. 1964. The fine structure of the connections between muscle cells in ascidian tadpole larva. *Journal of Cell Biology* 23: 664–669.

Berrill, N.J. & Watson, D.M.S. 1930. II: Studies in tunicate development. Part I: General physiology development of simple ascidians. *Philosophical Transactions of the Royal Society of London: Series B, Containing Papers of a Biological Character* 218: 37–78.

Berrill, N.J. & Watson, D.M.S. 1936. II: Studies in tunicate development. Part V: The evolution and classification of ascidians. *Philosophical Transactions of the Royal Society of London: Series B, Biological Sciences* 226: 43–70.

Biasuz, K., Leggio, B., Faure, E. & Lemaire, P. 2018. The "computable egg": Myth or useful concept? *Current Opinion in Systems Biology* 11: 91–97.

Bourque, D., Davidson, J., MacNair, N.G., Arsenault, G., LeBlanc, A.R., Landry, T. & Miron, G. 2007. Reproduction and early life history of an invasive ascidian *Styela clava* Herdman in Prince Edward Island, Canada. *Journal of Experimental Marine Biology and Ecology* 342: 78–84.

Cao, C., Lemaire, L.A., Wang, W., Yoon, P.H., Choi, Y.A., Parsons, L.R., Matese, J.C., Wang, W., Levine, M. & Chen, K. 2019. Comprehensive single-cell transcriptome lineages of a protovertebrate. *Nature* 571: 349–354.

Chabry, L. 1887. Contribution à l'embryologie normale et tératologique des ascidies simples. *Journal of Anatomy and Physiology* (Paris) 23: 167–319.

Chambon, J.-P., Nakayama, A., Takamura, K., McDougall, A. & Satoh, N. 2007. ERK- and JNK-signalling regulate gene networks that stimulate metamorphosis and apoptosis in tail tissues of ascidian tadpoles. *Development* 134: 1203–1219.

Chambon, J.-P., Soule, J., Pomies, P., Fort, P., Sahuquet, A., Alexandre, D., Mangeat, P.-H. & Baghdiguian. 2002. Tail regression in *Ciona intestinalis* (Prochordate) involves a Caspase-dependent apoptosis event associated with ERK activation. *Development* 129: 3105–3114.

Chen, J.-Y., Huang, D.-Y., Peng, Q.-Q., Chi, H.-M., Wang, X.-Q. & Feng, M. 2003. The first tunicate from the Early Cambrian of South China. *PNAS* 100: 8314–8318.

Chiba, S., Awazu, S., Itoh, M., Chin-Bow, S.T., Satoh, N., Satou, Y. & Hastings, K.E.M. 2003. A genomewide survey of developmentally relevant genes in *Ciona intestinalis*. *Development Genes and Evolution* 213: 291–302.

Chistyakov, V.G., Kalmykova, N.A., Nesov, L.A. & Suslov, G.A. 1984. О наличии вендских отложений в среднем течении р. Онеги и возможном существовании оболочечников (Tunicata: Chordata) в докембрии [On the presence of vendian deposits in the middle course of the Onega River and presumable existence of tunica (Tunicata: Chordata) in Precambrian]. *Vestnik Leningrad University (LGU)* 6: 11–19.

Christiaen, L., Davidson, B., Kawashima, T., Powell, W., Nolla, H., Vranizan, K. & Levine, M. 2008. The transcription/migration interface in heart precursors of *Ciona intestinalis*. *Science* 320: 1349–1352.

Christiaen, L., Stolfi, A., Davidson, B. & Levine, M. 2009. Spatio-temporal intersection of Lhx3 and Tbx6 defines the cardiac field through synergistic activation of Mesp. *Developmental Biology* 328: 552–560.

Christiaen, L., Stolfi, A. & Levine, M. 2010. BMP signaling coordinates gene expression and cell migration during precardiac

mesoderm development. *Developmental Biology* 340: 179–187.
Cloney, R.A. 1982. Ascidian larvae and the events of metamorphosis. *American Zoologist* 22: 817–826.
Collin, S., Oakley, J., Sewell, J. & Bishop, J. 2010. Widespread occurrence of the non-indigenous ascidian *Corella eumyota* Traustedt, 1882 on the shores of Plymouth Sound and Estuaries Special Area of Conservation, UK. *Aquatic Invasions* 5: 175–179.
Conklin, E.G. 1905a. *The Organization and Cell-Lineage of the Ascidian Egg*. Academy of Natural Sciences, Philadelphia.
Conklin, E.G. 1905b. Mosaic development in ascidian eggs. *Journal of Experimental Zoology* 2: 145–223.
Cooley, J., Whitaker, S., Sweeney, S., Fraser, S. & Davidson, B. 2011. Cytoskeletal polarity mediates localized induction of the heart progenitor lineage. *Nature Cell Biology* 13: 952–957.
Corbo, Joseph C., Levine, Michael & Zeller, Robert W. 1997. Characterization of a notochord-specific enhancer from the Brachyury promoter region of the ascidian, *Ciona intestinalis*. *Development* 124(3): 589–602.
Dardaillon, J., Dauga, D., Simion, P., Faure, E., Onuma, T.A., DeBiasse, M.B., Louis, A., Nitta, K.R., Naville, M., Besnardeau, L., Reeves, W., Wang, K., Fagotto, M., Guéroult-Bellone, M., Fujiwara, S., Dumollard, R., Veeman, M., Volff, J.-N., Roest Crollius, H., Douzery, E., Ryan, J.F., Davidson, B., Nishida, H., Dantec, C. & Lemaire, P. 2020. ANISEED 2019: 4D exploration of genetic data for an extended range of tunicates. *Nucleic Acids Research* 48: D668–D675.
Darras, S. & Nishida, H. 2001. The BMP signaling pathway is required together with the FGF pathway for notochord induction in the ascidian embryo. *Development* 128: 2629–2638.
Davidson, B., Shi, W., Beh, J., Christiaen, L. & Levine, M. 2006. FGF signaling delineates the cardiac progenitor field in the simple chordate, *Ciona intestinalis*. *Genes & Development* 20: 2728–2738.
Davidson, B. & Swalla, B.J. 2001. Isolation of genes involved in ascidian metamorphosis: Epidermal growth factor signaling and metamorphic competence. *Development Genes and Evolution* 211: 190–194.
Dehal, P. & Boore, J.L. 2005. Two rounds of whole genome duplication in the ancestral vertebrate. *PLoS Biology* 3: e314.
Dehal, P., Satou, Y., Campbell, R.K., Chapman, J., et al. 2002. The draft genome of *Ciona intestinalis*: Insights into chordate and vertebrate origins. *Science* 298: 2157–2167.
Delgado-Calle, J., Kurihara, N., Atkinson, E.G., Nelson, J., Miyagawa, K., Galmarini, C.M., Roodman, G.D. & Bellido, T. 2019. Aplidin (plitidepsin) is a novel anti-myeloma agent with potent anti-resorptive activity mediated by direct effects on osteoclasts. *Oncotarget* 10: 2709–2721.
Delsuc, F., Brinkmann, H., Chourrout, D. & Philippe, H. 2006. Tunicates and not cephalochordates are the closest living relatives of vertebrates. *Nature* 439: 965–968.
Delsuc, F., Philippe, H., Tsagkogeorga, G., Simion, P., Tilak, M.-K., Turon, X., López-Legentil, S., Piette, J., Lemaire, P. & Douzery, E.J.P. 2018. A phylogenomic framework and timescale for comparative studies of tunicates. *BMC Biology* 16: 39.
Demarchi, M., Chiappero, M., Laudien, J. & Sahade, R. 2008. Population genetic structure of the ascidian *Styela rustica* at Kongsfjorden, Svalbard, Arctic. *Journal of Experimental Marine Biology and Ecology* 364: 29–34.
Deschet, K. & Smith, W.C. 2004. Frimousse: A spontaneous ascidian mutant with anterior ectodermal fate transformation. *Current Biology* 14: R408–R410.
Dumollard, R., Gazo, I., Gomes, I.D.L., Besnardeau, L. & McDougall, A. 2017. Ascidians: An emerging marine model for drug discovery and screening. *Current Topics in Medicinal Chemistry* 17: 2056–2066.
Dumollard, R., Hebras, C., Besnardeau, L. & McDougall, A. 2013. Beta-catenin patterns the cell cycle during maternal-to-zygotic transition in urochordate embryos. *Developmental Biology* 384(2): 331–342.
Dumollard, R. & Sardet, C. 2001. Three different calcium wave pacemakers in ascidian eggs. *Journal of Cell Science* 114: 2471–2481.
Duran, Rosario, Zubia, Eva, Ortega, Maria J., Naranjo, Santiago & Salva, Javier. 1998. Phallusides, new glucosphingolipids from the ascidian *Phallusia fumigata*. *Tetrahedron* 54: 14597–14602.
Durandau, E., Aymoz, D. & Pelet, S. 2015. Dynamic single cell measurements of kinase activity by synthetic kinase activity relocation sensors. *BMC Biology* 13: 55.
Dybern, B.I. 1967. Settlement of sessile animals on eternite slabs in two polls near Bergen. *Sarsia* 29: 137–150.
Eri, R., Arnold, J.M., Hinman, V.F., Green, K.M., Jones, M.K., Degnan, B.M. & Lavin, M.F. 1999. Hemps, a novel EGF-like protein, plays a central role in ascidian metamorphosis. *Development* 126: 5809–5818.
Etournay, R., Popović, M., Merkel, M., Nandi, A., Blasse, C., Aigouy, B., Brandl, H., Myers, G., Salbreux, G., Jülicher, F. & Eaton, S. 2015. Interplay of cell dynamics and epithelial tension during morphogenesis of the *Drosophila* pupal wing. *eLife* 4: e07090.
Fedonkin, M.A., Vickers-Rich, P., Swalla, B.J., Trusler, P. & Hall, M. 2012. A new metazoan from the Vendian of the White Sea, Russia, with possible affinities to the ascidians. *Paleontological Journal* 46: 1–11.
Feinberg, S., Roure, A., Piron, J. & Darras, S. 2019. Antero-posterior ectoderm patterning by canonical Wnt signaling during ascidian development. *PLoS Genetics* 15: e1008054.
Gagliardi, P.A. & Pertz, O. 2019. Developmental ERK signaling illuminated. *Developmental Cell* 48: 289–290.
Gandhi, S., Razy-Krajka, F., Christiaen, L. & Stolfi, A. 2018. CRISPR knockouts in *Ciona* embryos. In: *Transgenic Ascidians* (Y. Sasakura, ed.), pp. 141–152. Springer, Singapore. *Advances in Experimental Medicine and Biology* 1029.
Garcia, H.G., Berrocal, A., Kim, Y.J., Martini, G. & Zhao, J. 2020. Lighting up the central dogma for predictive developmental biology. *Current Topics in Developmental Biology* 137: 1–35.
Goldstein, B. 2001. On the evolution of early development in the Nematoda. *Philosophical Transactions of the Royal Society B* 356: 1521–1531.
Gordon, E.M., Sankhala, K.K., Chawla, N. & Chawla, S.P. 2016. Trabectedin for soft tissue sarcoma: Current status and future perspectives. *Advances in Therapy* 33: 1055–1071.
Guignard, L., Fiúza, U.M., Leggio, B., Laussu, J., Faure, E., Michelin, G., Biasuz, K., Hufnagel, L., Malandain, G., Godin, C. & Lemaire, P. 2020. Contact area-dependent cell communication and the morphological invariance of ascidian embryogenesis. *Science* 369(6500): eaar5663.
Gyoja, F., Satou, Y., Shin-i, T., Kohara, Y., Swalla, B.J. & Satoh, N. 2007. Analysis of large scale expression sequenced tags (ESTs) from the anural ascidian, *Molgula tectiformis*. *Developmental Biology* 307: 460–482.
Harada, Y., Takagaki, Y., Sunagawa, M., Saito, T., Yamada, L., Taniguchi, H., Shoguchi, E. & Sawada, H. 2008. Mechanism of self-sterility in a hermaphroditic chordate. *Science* 320: 548–550.

Hashimoto, H., Enomoto, T., Kumano, G. & Nishida, H. 2011. The transcription factor FoxB mediates temporal loss of cellular competence for notochord induction in ascidian embryos. *Development* 138: 2591–2600.

Hino, K., Satou, Y., Yagi, K. & Satoh, N. 2003. A genomewide survey of developmentally relevant genes in *Ciona intestinalis*. *Development Genes and Evolution* 213: 264–272.

Hozumi, A., Matsunobu, S., Mita, K., Treen, N., Sugihara, T., Horie, T., Sakuma, T., Yamamoto, T., Shiraishi, A., Hamada, M., Satoh, N., Sakurai, K., Satake, H. & Sasakura, Y. 2020. GABA-induced GnRH release triggers chordate metamorphosis. *Current Biology* 30: 1555–1561.e4.

Huang, A. & Saunders, T.E. 2020. A matter of time: Formation and interpretation of the Bicoid morphogen gradient. *Current Topics in Developmental Biology* 137: 79–117. Elsevier.

Hudson, C. 2016. The central nervous system of ascidian larvae. *WIREs Developmental Biology* 5: 538–561.

Hudson, C., Kawai, N., Negishi, T. & Yasuo, H. 2013. β-catenin-driven binary fate specification segregates germ layers in ascidian embryos. *Current Biology* 23: 491–495.

Hudson, C., Lotito, S. & Yasuo, H. 2007. Sequential and combinatorial inputs from Nodal, Delta2/Notch and FGF/MEK/ERK signalling pathways establish a grid-like organisation of distinct cell identities in the ascidian neural plate. *Development* 134: 3527–3537.

Hudson, C., Sirour, C. & Yasuo, H. 2016. Co-expression of Foxa.a, Foxd and Fgf9/16/20 defines a transient mesendoderm regulatory state in ascidian embryos. *eLife* 5: e14692.

Hudson, C. & Yasuo, H. 2005. Patterning across the ascidian neural plate by lateral Nodal signalling sources. *Development* 132: 1199–1210.

Hudson, C. & Yasuo, H. 2006. A signalling relay involving Nodal and Delta ligands acts during secondary notochord induction in *Ciona* embryos. *Development* 133: 2855–2864.

Imai, K.S., Satoh, N. & Satou, Y. 2002. Early embryonic expression of FGF4/6/9 gene and its role in the induction of mesenchyme and notochord in *Ciona savignyi* embryos. *Development* 129: 1729–1738.

Imai, K.S., Stolfi, A., Levine, M. & Satou, Y. 2009. Gene regulatory networks underlying the compartmentalization of the *Ciona* central nervous system. *Development* 136: 285–293.

Inaba, K. 2007. Molecular basis of sperm flagellar axonemes: Structural and evolutionary aspects. *Annals of the New York Academy of Sciences* 1101: 506–526.

Inada, K., Horie, T., Kusakabe, T. & Tsuda, M. 2003. Targeted knockdown of an opsin gene inhibits the swimming behaviour photoresponse of ascidian larvae. *Neuroscience Letters* 347: 167–170.

Irvine, S.Q. 2013. Study of cis-regulatory elements in the ascidian *Ciona intestinalis*. *Current Genomics* 14: 56–67.

Ishino, Y., Shinagawa, H., Makino, K., Amemura, M. & Nakata, A. 1987. Nucleotide sequence of the iap gene, responsible for alkaline phosphatase isozyme conversion in *Escherichia coli*, and identification of the gene product. *Journal of Bacteriology* 169: 5429–5433.

Jaeger, J. & Verd, B. 2020. Dynamic positional information: Patterning mechanism versus precision in gradient-driven systems. *Current Topics in Developmental Biology* 137: 219–246. Elsevier.

Jeffery, W.R. 2002. Programmed cell death in the ascidian embryo: Modulation by FoxA5 and Manx and roles in the evolution of larval development. *Mechanisms of Development* 118: 111–124.

Jiang, D. & Smith, W.C. 2007. Ascidian notochord morphogenesis. *Developmental Dynamics* 236(7): 1748–1757.

Karaiskou, A., Swalla, B.J., Sasakura, Y. & Chambon, J.-P. 2015. Metamorphosis in solitary ascidians. *Genesis* 53: 34–47.

Kawai, N., Ochiai, H., Sakuma, T., Yamada, L., Sawada, H., Yamamoto, T. & Sasakura, Y. 2012. Efficient targeted mutagenesis of the chordate *Ciona intestinalis* genome with zinc-finger nucleases. *Development, Growth & Differentiation* 54: 535–545.

Kawamura, K., Tiozzo, S., Manni, L., Sunanaga, T., Burighel, P. & Tomaso, A.W.D. 2011. Germline cell formation and gonad regeneration in solitary and colonial ascidians. *Developmental Dynamics* 240: 299–308.

Kawashima, T., Tokuoka, M., Awazu, S., Satoh, N. & Satou, Y. 2003. A genomewide survey of developmentally relevant genes in *Ciona intestinalis*. *Development Genes and Evolution* 213: 284–290.

Kim, G.J. & Nishida, H. 1999. Suppression of muscle fate by cellular interaction is required for mesenchyme formation during ascidian embryogenesis. *Developmental Biology* 214: 9–22.

Kim, G.J. & Nishida, H. 2001. Role of the FGF and MEK signaling pathway in the ascidian embryo. *Development, Growth & Differentiation* 43: 521–533.

Kim, G.J., Yamada, A. & Nishida, H. 2000. An FGF signal from endoderm and localized factors in the posterior-vegetal egg cytoplasm pattern the mesodermal tissues in the ascidian embryo. *Development* 127: 2853–2862.

Kimura, Y., Yoshida, M. & Morisawa, M. 2003. Interaction between noradrenaline or adrenaline and the β1-adrenergic receptor in the nervous system triggers early metamorphosis of larvae in the ascidian, *Ciona savignyi*. *Developmental Biology* 258: 129–140.

Kocot, K.M., Tassia, M.G., Halanych, K.M. & Swalla, B.J. 2018. Phylogenomics offers resolution of major tunicate relationships. *Molecular Phylogenetics and Evolution* 121: 166–173.

Kondoh, E., Konno, A., Inaba, K., Oishi, T., Murata, M. & Yoshida, M. 2008. Valosin-containing protein/p97 interacts with sperm-activating and sperm-attracting factor (SAAF) in the ascidian egg and modulates sperm-attracting activity. *Development, Growth & Differentiation* 50: 665–673.

Kovalevsky, A.O. 1866. *Entwickelungsgeschichte der einfachen Ascidien*. St. Petersburg: Commissionäre der Kaiserlichen akademie der wissenschaften, Eggers et cie und H. Schmitzdorff.

Krasovec, G., Robine, K., Quéinnec, E., Karaiskou, A. & Chambon, J.P. 2019. Ci-hox12 tail gradient precedes and participates in the control of the apoptotic-dependent tail regression during *Ciona* larva metamorphosis. *Developmental Biology* 448: 237–246.

Kumano, G. & Nishida, H. 2007. Ascidian embryonic development: An emerging model system for the study of cell fate specification in chordates. *Developmental Dynamics* 236: 1732–1747.

Kumano, G., Takatori, N., Negishi, T., Takada, T. & Nishida, H. 2011. A maternal factor unique to ascidians silences the germline via binding to P-TEFb and RNAP II regulation. *Current Biology* 21: 1308–1313.

Kuplik, Z., Novak, L. & Shenkar, N. 2019. Proteomic profiling of ascidians as a tool for biomonitoring marine environments. *PLoS One* 14: e0215005.

Lambert, C.C. 2000. Germ-cell warfare in ascidians: Sperm from one species can interfere with the fertilization of a second species. *The Biological Bulletin* 198: 22–25.

Lambert, C.C., Lambert, I.M. & Lambert, G. 1995. Brooding strategies in solitary ascidians: Corella species from north and south temperate waters. *Canadian Journal of Zoology*. doi: 10.1139/z95-198.

Lambert, G. 2005. Ecology and natural history of the protochordates. *Canadian Journal of Zoology* 83: 34–50.

Lavoie, H., Gagnon, J. & Therrien, M. 2020. ERK signalling: A master regulator of cell behaviour, life and fate. *Nature Reviews Molecular Cell Biology* 21: 607–632.

Leggio, B., Laussu, J., Carlier, A., Godin, C., Lemaire, P. & Faure, E. 2019. MorphoNet: An interactive online morphological browser to explore complex multi-scale data. *Nature Communications* 10: 2812.

Lemaire, P. 2009. Unfolding a chordate developmental program, one cell at a time: Invariant cell lineages, short-range inductions and evolutionary plasticity in ascidians. *Developmental Biology* 332: 48–60.

Lemaire, P. 2011. Evolutionary crossroads in developmental biology: The tunicates. *Development* 138: 2143–2152.

Lopez, C.E., Sheehan, H.C., Vierra, D.A., Azzinaro, P.A., Meedel, T.H., Howlett, N.G. & Irvine, S.Q. 2017. Proteomic responses to elevated ocean temperature in ovaries of the ascidian *Ciona intestinalis*. *Biology Open* 6: 943–955.

Maury, B., Martinand-Mari, C., Chambon, J-P., Soulé, J., Degols, G., Sahuquet, A., Weill, M., Berthomieu, A., Fort, F., Mangeat, P. & Baghdiguian, S. 2006. Fertilization regulates apoptosis of *Ciona intestinalis* extra-embryonic cells through thyroxine (T4)-dependent NF-κB pathway activation during early embryonic development. *Developmental Biology* 289(1): 152–165.

McDougall, A., Hebras, C., Gomes, I. & Dumollard, R. 2021. Gene editing in the ascidian *Phallusia mammillata* and tail nerve cord formation. *Methods in Molecular Biology* 2219: 217–230.

Meinertzhagen, I.A., Lemaire, P. & Okamura, Y. 2004. The neurobiology of the ascidian tadpole larva: Recent developments in an ancient chordate. *Annual Review of Neuroscience* 27: 453–485.

Meinertzhagen, I.A. & Okamura, Y. 2001. The larval ascidian nervous system: The chordate brain from its small beginnings. *Trends in Neurosciences* 24: 401–410.

Mendiola, J., Hernández, H., Sariego, I., Rojas, L., Otero, A., Ramírez, A., de los Angeles Chávez, M., Payrol, J.A. & Hernández, A. 2006. Antimalarial activity from three ascidians: An exploration of different marine invertebrate phyla. *Transactions of the Royal Society of Tropical Medicine and Hygiene* 100: 909–916.

Monniot, C. & Monniot, F. 1997. Records of ascidians from Bahrain, Arabian Gulf with three new species. *Journal of Natural History* 31: 1623–1643.

Morgan, T.H. 1904. Self-fertilization induced by artificial means. *Journal of Experimental Zoology* 1: 135–178.

Nakamura, J., Yoshida, K., Sasakura, Y. & Fujiwara, S. 2014. Chondroitin 6-O-sulfotransferases are required for morphogenesis of the notochord in the ascidian embryo. *Developmental Dynamics* 243(12): 1637–1645.

Nakashima, K., Yamada, L., Satou, Y., Azuma, J. & Satoh, N. 2004. The evolutionary origin of animal cellulose synthase. *Development Genes and Evolution* 214: 81–88.

Nakatani, Y., Moody, R. & Smith, W.C. 1999. Mutations affecting tail and notochord development in the ascidian *Ciona savignyi*. *Development* 126: 3293–3301.

Nakayama, A., Satou, Y. & Satoh, N. 2001. Isolation and characterization of genes that are expressed during *Ciona intestinalis* metamorphosis. *Development Genes and Evolution* 211: 184–189.

Nakayama-Ishimura, A., Chambon, J., Horie, T., Satoh, N. & Sasakura, Y. 2009. Delineating metamorphic pathways in the ascidian *Ciona intestinalis*. *Developmental Biology* 326: 357–367.

Nakazawa, K., Yamazawa, T., Moriyama, Y., Ogura, Y., Kawai, N., Sasakura, Y. & Saiga, H. 2013. Formation of the digestive tract in *Ciona intestinalis* includes two distinct morphogenic processes between its anterior and posterior parts. *Developmental Dynamics* 242: 1172–1183.

Nishida, H. 1987. Cell lineage analysis in ascidian embryos by intracellular injection of a tracer enzyme: III: Up to the tissue restricted stage. *Developmental Biology* 121: 526–541.

Nishida, H. & Sawada, K. 2001. Macho-1 encodes a localized mRNA in ascidian eggs that specifies muscle fate during embryogenesis. *Nature* 409: 724–729.

Nishiyama, A. & Fujiwara, S. 2008. RNA interference by expressing short hairpin RNA in the *Ciona intestinalis* embryo. *Development, Growth & Differentiation* 50: 521–529.

Nomura, M., Nakajima, A. & Inaba, K. 2009. Proteomic profiles of embryonic development in the ascidian *Ciona intestinalis*. *Developmental Biology* 325 (2): 468–481.

Oda-Ishii, I., Kubo, A., Kari, W., Suzuki, N., Rothbächer, U. & Satou, Y. 2016. A maternal system initiating the zygotic developmental program through combinatorial repression in the ascidian embryo. *PLoS Genetics* 12: e1006045.

Palanisamy, S.K., Rajendran, N.M. & Marino, A. 2017. Natural products diversity of marine ascidians (Tunicates: Ascidiacea) and successful drugs in clinical development. *Natural Products and Bioprospecting* 7: 1–111.

Pareek, C.S., Smoczynski, R. & Tretyn, A. 2011. Sequencing technologies and genome sequencing. *Journal of Applied Genetics* 52: 413–435.

Patel, A.L. & Shvartsman, S.Y. 2018. Outstanding questions in developmental ERK signaling. *Development* 145.

Perrimon, N., Pitsouli, C. & Shilo, B.-Z. 2012. Signaling mechanisms controlling cell fate and embryonic patterning. *Cold Spring Harbor Perspectives in Biology* 4: a005975.

Power, R.M. & Huisken, J. 2017. A guide to light-sheet fluorescence microscopy for multiscale imaging. *Nature Methods* 14: 360–373.

Primo, C. & Vázquez, E. 2009. Antarctic ascidians: An isolated and homogeneous fauna. *Polar Research* 28: 403–414.

Razy-Krajka, F. & Stolfi, A. 2019. Regulation and evolution of muscle development in tunicates. *Evo Devo* 10: 13.

Regot, S., Hughey, J.J., Bajar, B.T., Carrasco, S. & Covert, M.W. 2014. High-sensitivity measurements of multiple kinase activities in live single cells. *Cell* 157: 1724–1734.

Rogers, K.W. & Müller, P. 2020. Optogenetic approaches to investigate spatiotemporal signaling during development. *Current Topics in Developmental Biology* 137: 37–77. Elsevier.

Rothbächer, U., Bertrand, V., Lamy, C. & Lemaire, P. 2007. A combinatorial code of maternal GATA, Ets and β-catenin-TCF transcription factors specifies and patterns the early ascidian ectoderm. *Development* 134(22): 4023–4032.

Roure, A., Lemaire, P. & Darras, S. 2014. An Otx/Nodal regulatory signature for posterior neural development in ascidians. *PLoS Genetics* 10: e1004548.

Rudolf, J., Dondorp, D., Canon, L., Tieo, S. & Chatzigeorgiou, M. 2019. Automated behavioural analysis reveals the basic behavioural repertoire of the urochordate *Ciona intestinalis*. *Scientific Reports* 9: 2416.

Ryan, Kerrianne, Lu, Zhiyuan & Meinertzhagen, Ian A. 2016. The CNS connectome of a tadpole larva of *Ciona intestinalis* (L.)

highlights sidedness in the brain of a chordate sibling. *eLife* 16962.001.

Salas, P., Vinaithirthan, V., Newman-Smith, E., Kourakis, M.J. & Smith, W.C. 2018. Photoreceptor specialization and the visuomotor repertoire of the primitive chordate *Ciona*. *Journal of Experimental Biology* 221.

Sasaki, H., Yoshida, K., Hozumi, A. & Sasakura, Y. 2014. CRISPR/Cas9-mediated gene knockout in the ascidian *Ciona intestinalis*. *Development, Growth & Differentiation* 56: 499–510.

Sasakura, Y. 2007. Germline transgenesis and insertional mutagenesis in the ascidian *Ciona intestinalis*. *Developmental Dynamics* 236: 1758–1767.

Sasakura, Y., Shoguchi, E., Takatori, N., Wada, S., Meinertzhagen, I.A., Satou, Y. & Satoh, N. 2003a. A genomewide survey of developmentally relevant genes in *Ciona intestinalis*. *Development Genes and Evolution* 213: 303–313.

Sasakura, Y., Yamada, L., Takatori, N., Satou, Y. & Satoh, N. 2003b. A genomewide survey of developmentally relevant genes in *Ciona intestinalis*. *Development Genes and Evolution* 213: 273–283.

Sasakura, Y., Awazu, S., Chiba, S. & Satoh, N. 2003c. Germ-line transgenesis of the Tc1/mariner superfamily transposon Minos in *Ciona intestinalis*. *Proceedings of the National Academy of Sciences* 100: 7726–7730.

Sasakura, Y., Nakashima, K., Awazu, S., Matsuoka, T., Nakayama, A., Azuma, J. & Satoh, N. 2005. Transposon-mediated insertional mutagenesis revealed the functions of animal cellulose synthase in the ascidian *Ciona intestinalis*. *Proceedings of the National Academy of Sciences* 102: 15134–15139.

Satoh, N. 1994. *Developmental Biology of Ascidians*. Cambridge University Press, Cambridge.

Satoh, N. 2004. Genomic resources for ascidians: Sequence/expression databases and genome projects. *Methods in Cell Biology* 74: 759–774. Academic Press.

Satoh, N. 2013. *Developmental Genomics of Ascidians*. Wiley-Blackwell.

Satou, Y. & Imai, K.S. 2015. Gene regulatory systems that control gene expression in the *Ciona* embryo. *Proceedings of the Japan Academy, Ser. B, Physical and Biological Sciences* 91: 33–51.

Satou, Y., Imai, K.S. & Satoh, N. 2001. Action of morpholinos in *Ciona* embryos. *Genesis* 30: 103–106.

Satou, Y., Imai, K.S. & Satoh, N. 2004. The ascidian Mesp gene specifies heart precursor cells. *Development* 131(11): 2533–2541.

Satou, Y., Imai, K.S. & Satoh, N. 2002a. FGF genes in the basal chordate *Ciona intestinalis*. *Development Genes and Evolution* 212: 432–438.

Satou, Y., Takatori, N., Fujiwara, S., Nishikata, T., Saiga, H., Kusakabe, T., Shin-i, T., Kohara, Y. & Satoh, N. 2002b. *Ciona intestinalis* cDNA projects: Expressed sequence tag analyses and gene expression profiles during embryogenesis. *Gene* 287: 83–96.

Satou, Y., Yagi, K., Imai, K.S., Yamada, L., Nishida, H. & Satoh, N. 2002c. Macho-1-related genes in *Ciona* embryos. *Development Genes and Evolution* 212: 87–92.

Satou, Y., Imai, K.S., Levine, M., Kohara, Y., Rokhsar, D. & Satoh, N. 2003a. A genomewide survey of developmentally relevant genes in *Ciona intestinalis*. *Development Genes and Evolution* 213: 213–221.

Satou, Y., Kawashima, T., Kohara, Y. & Satoh, N. 2003b. Large scale EST analyses in *Ciona intestinalis*. *Development Genes and Evolution* 213: 314–318.

Satou, Y., Sasakura, Y., Yamada, L., Imai, K.S., Satoh, N. & Degnan, B. 2003c. A genomewide survey of developmentally relevant genes in *Ciona intestinalis*. *Development Genes and Evolution* 213: 254–263.

Satou, Y., Mineta, K., Ogasawara, M., Sasakura, Y., Shoguchi, E., Ueno, K., Yamada, L., Matsumoto, J., Wasserscheid, J., Dewar, K., Wiley, G.B., Macmil, S.L., Roe, B.A., Zeller, R.W., Hastings, K.E., Lemaire, P., Lindquist, E., Endo, T., Hotta, K. & Inaba, K. 2008. Improved genome assembly and evidence-based global gene model set for the chordate *Ciona intestinalis*: New insight into intron and operon populations. *Genome Biology* 9: 1–11.

Satou, Y., Nakamura, R., Yu, D., Yoshida, R., Hamada, M., Fujie, M., Hisata, K., Takeda, H. & Satoh, N. 2019. A nearly complete genome of *Ciona intestinalis* type A (*C. robusta*) reveals the contribution of inversion to chromosomal evolution in the genus *Ciona*. *Genome Biology and Evolution* 11: 3144–3157.

Schloop, A.E., Bandodkar, P.U. & Reeves, G.T. 2020. Formation, interpretation, and regulation of the *Drosophila* dorsal/NF-κB gradient. *Current Topics in Developmental Biology* 137: 143–191. Elsevier.

Shenkar, N. 2012. Ascidian (Chordata, ascidiacea) diversity in the Red Sea. *Marine Biodiversity* 42: 459–469.

Shenkar, N. & Loya, Y. 2008. The solitary ascidian *Herdmania momus*: Native (Red Sea) versus non-indigenous (Mediterranean) populations. *Biological Invasions* 10: 1431–1439.

Shenkar, N. & Swalla, B.J. 2011. Global diversity of ascidiacea. *PLoS One* 6: e20657.

Shirae-Kurabayashi, M., Matsuda, K. & Nakamura, A. 2011. Ci-Pem-1 localizes to the nucleus and represses somatic gene transcription in the germline of *Ciona intestinalis* embryos. *Development* 138: 2871–2881.

Shirae-Kurabayashi, M., Nishikata, T., Takamura, K., Tanaka, K.J., Nakamoto, C. & Nakamura, A. 2006. Dynamic redistribution of vasa homolog and exclusion of somatic cell determinants during germ cell specification in *Ciona intestinalis*. *Development* 133: 2683–2693.

Shoguchi, E., Hamaguchi, M. & Satoh, N. 2008. Genome-wide network of regulatory genes for construction of a chordate embryo. *Developmental Biology* 316: 498–509.

Shoguchi, E., Kawashima, T., Satou, Y., Hamaguchi, M., Sin-I, T., Kohara, Y., Putnam, N., Rokhsar, D.S. & Satoh, N. 2006. Chromosomal mapping of 170 BAC clones in the ascidian *Ciona intestinalis*. *Genome Research* 16: 297–303.

Sladitschek, H.L., Fiuza, U.-M., Pavlinic, D., Benes, V., Hufnagel, L. & Neveu, P.A. 2020. MorphoSeq: Full single-cell transcriptome dynamics up to gastrulation in a chordate. *Cell* 181: 922–935.e21.

Stolfi, A., Gainous, T.B., Young, J.J., Mori, A., Levine, M. & Christiaen, L. 2010. Early chordate origins of the vertebrate second heart field. *Science* 329: 565–568.

Swalla, B.J. & Jeffery, W.R. 1990. Interspecific hybridization between an anural and urodele ascidian: Differential expression of urodele features suggests multiple mechanisms control anural development. *Developmental Biology* 142: 319–334.

Tadros, W. & Lipshitz, H.D. 2009. The maternal-to-zygotic transition: A play in two acts. *Development* 136: 3033–3042.

Takatori, N., Hotta, K., Mochizuki, Y., Satoh, G., Mitani, Y., Satoh, N., Satou, Y. & Takahashi, H. 2004. T-box genes in the ascidian *Ciona intestinalis*: Characterization of cDNAs and spatial expression. *Developmental Dynamics* 230: 743–753.

Takatori, N., Kumano, G., Saiga, H. & Nishida, H. 2010. Segregation of germ layer fates by nuclear migration-dependent localization of not mRNA. *Developmental Cell* 19: 589–598.

Takatori, N., Oonuma, K., Nishida, H. & Saiga, H. 2015. Polarization of PI3K activity initiated by ooplasmic segregation guides nuclear migration in the mesendoderm. *Developmental Cell* 35: 333–343.

Tarallo, R. & Sordino, P. 2004. Time course of programmed cell death in *Ciona intestinalis* in relation to mitotic activity and MAPK signaling. *Developmental Dynamics* 230: 251–262.

Tatian, M., Sahade, R.J., Doucet, M.E. & Esnal, G.B. 1998. Ascidians (Tunicata, ascidiacea) of Potter Cove, South Shetland Islands, Antarctica. *Antarctic Science* 10: 147–152.

Tatián, M., Lagger, C., Demarchi, M. & Mattoni, C. 2011. Molecular phylogeny endorses the relationship between carnivorous and filter-feeding tunicates (Tunicata, Ascidiacea). *Zoologica Scripta* 40: 603–612.

Tatián, M., Sahade, R.J. & Esnal, G.B. 2004. Diet components in the food of Antarctic ascidians living at low levels of primary production. *Antarctic Science* 16: 123–128.

Therriault, T.W. & Herborg, L. 2008. Predicting the potential distribution of the vase tunicate *Ciona intestinalis* in Canadian waters: Informing a risk assessment. *ICES Journal of Marine Science* 65(5): 788–794. doi: 10.1093/ICESJMS/FSN054.

Thompson, D.W. 1917. On growth and form. *Cambridge Univ. Press*. Available at: www.cabdirect.org/cabdirect/abstract/19431401837.

Tokuoka, M., Kumano, G. & Nishida, H. 2007. FGF9/16/20 and Wnt-5α signals are involved in specification of secondary muscle fate in embryos of the ascidian, *Halocynthia roretzi*. *Development Genes and Evolution* 217: 515–527.

Treen, N., Heist, T., Wang, W. & Levine, M. 2018. Depletion of maternal cyclin B3 contributes to zygotic genome activation in the *Ciona* embryo. *Current Biology* 28: 1150–1156.e4.

Treen, N., Yoshida, K., Sakula, T., Sasaki, H., Kawai, N. & Sasakura, Y. 2014. Tissue-specific and ubiquitous gene knockouts by TALEN electroporation provide new approaches to investigating gene function in *Ciona*. *Development* 141(2): 481–487.

Wada, S., Tokuoka, M., Shoguchi, E., Kobayashi, K., Di Gregorio, A., Spagnuolo, A., Branno, M., Kohara, Y., Rokhsar, D., Levine, M., Saiga, H., Satoh, N. & Satou, Y. 2003. A genomewide survey of developmentally relevant genes in *Ciona intestinalis*. *Development Genes and Evolution* 213: 222–234.

Wagner, E. & Levine, M. 2012. FGF signaling establishes the anterior border of the *Ciona* neural tube. *Development* 139: 2351–2359.

Watters, D.J. 2018. Ascidian toxins with potential for drug development. *Marine Drugs* 16: 162.

Yagi, K., Satou, Y., Mazet, F., Shimeld, S.M., Degnan, B., Rokhsar, D., Levine, M., Kohara, Y. & Satoh, N. 2003. A genomewide survey of developmentally relevant genes in *Ciona intestinalis*. *Development Genes and Evolution* 213: 235–244.

Yagi, K., Takatori, N., Satou, Y. & Satoh, N. 2005. Ci-Tbx6b and Ci-Tbx6c are key mediators of the maternal effect gene Ci-macho1 in muscle cell differentiation in *Ciona intestinalis* embryos. *Developmental Biology* 282: 535–549.

Yamada, L., Kobayashi, K., Degnan, B., Satoh, N. & Satou, Y. 2003. A genomewide survey of developmentally relevant genes in *Ciona intestinalis*. *Development Genes and Evolution* 213: 245–253.

Yoshida, M., Murata, M., Inaba, K. & Morisawa, M. 2002. A chemoattractant for ascidian spermatozoa is a sulfated steroid. *Proceedings of the National Academy of Sciences*. 99: 14831–14836.

Zega, G., Thorndyke, M.C. & Brown, E.R. 2006. Development of swimming behaviour in the larva of the ascidian *Ciona intestinalis*. *Journal of Experimental Biology* 209: 3405–3412.

Zeller, R.W., Virata, M.J. & Cone, A.C. 2006. Predictable mosaic transgene expression in ascidian embryos produced with a simple electroporation device. *Developmental Dynamics* 235: 1921–1932.

21 *Botryllus schlosseri*—A Model Colonial Species in Basic and Applied Studies

Oshrat Ben-Hamo and Baruch Rinkevich

CONTENTS

21.1 HISTORY OF THE MODEL

The apparent first description of *Botryllus schlosseri* colonies is attributed to Rondelet Guillaume (1555), under the name *uva marina*. With the increased interest in this species, about two centuries later, *Botryllus* was re-described by J.A. Schlosser and J. Ellis in a letter (1756) as: "I discovered a most extraordinary sea-production surrounding the stem of an old fucus teres [a brown algae]: it was of a hardish, but fleshy substance . . . of a light brown or ash colour, the whole surface covered over with bright yellow shining and star-like bodies". Later the animal was portrayed by Pallas (1766) as a zoophyte, that is to say, an animal-plant, and was named by Pallas *Alcyonium schlosseri* Pallas (1766). Linnaeus (1767) defined *Botryllus* as a soft coral from the family Alcyoniidae (Pallas 1766; Linnaeus 1767). Following these authors, Gärtner, Bruguière and Renier ascribed the animal as *Botryllus stellatus* (Gärtner 1774; Bruguière 1792; Renier 1793), and in 1816, the animal got its permanent name: *Botryllus schlosseri* (Savigny 1816). In a comprehensive review on this species, Manni et al. (2019) covered a list of authors who described *Botryllus* during the 19th century, and it will not be repeated here. Many of these papers were written in local languages (Italian, German, French) and the most comprehensive, as pioneering studies on *Botryllus*, are those published by Savigny (1816), Ganin (1870), Giard (1872) and Della Valle (1881). The famous biologist, zoologist and gifted painter Ernst Haeckel (1899) created a known drawing of *Botryllus* including anatomy.

The first biologist who successfully grew and bred *Botryllus schlosseri* colonies in the lab was Sabbadin in 1955. This opened a door for other laboratories to adopt *Botryllus* as a model in their studies. For the past decades, three main laboratories have been investigating and focusing on *Botryllus schlosseri*. These labs are located in California, United States (Weissman's lab); Italy (Sabbadin, Ballarin and Manni's labs); and Israel (Rinkevich's Lab). Several important milestones in the history of this species deal with the *Botryllus* palleal budding (asexual reproduction), whole body regeneration and allorecognition. The first study that described the complex weekly budding process in this species and the life and death cycles of *Botryllus* zooids (blastogenesis) was Spallanzani (1784). Important milestones in the study of bud development and life-and-death cycles were published by Metschnikow (1869), Hjort (1893) and Pizon (1893), Berrill (1941a, 1941b, 1951), Watterson (1945), Sabbadin (1955) and Izzard (1973). Also, the phenomenon of whole-body regeneration (vascular budding) in *Botryllus* was reported by Ganin (1870) and followed by Giard (1872) and Herdman (1924).

For allorecognition, the first documentation for fusion/rejection phenomena between contacting *Botryllus* colonies (self/non-self recognition) was made by Bancroft (1903). Only six decades thereafter, basic genetic studies and searches for allorecognition properties were followed by Sabbadin (1962) and Scofield et al. (1982) focusing on *Botryllus schlosseri*, while other studies evaluated allorecognition in *Botryllus primigenus* (Oka and Watanabe 1957, 1960; Taneda and Watanabe 1982a, 1982b; Taneda et al. 1985). Results were

DOI: 10.1201/9781003217503-21

intensified following the establishment of allorecognition assays (Rinkevich 1995) and animal breeding methodologies (Brunetti et al. 1984; Boyd et al. 1986; Rinkevich and Shapira 1998). Other studies on allorecognition contributed to the understanding of the initiation, the follow-up and the biology of chimerism; the involvement of stem cells in the process; and stem cell parasitism (e.g. Scofield et al. 1982; Rinkevich et al. 1993, 2013; Stoner and Weissman 1996; Stoner et al. 1999; Laird et al. 2005a; Corey et al. 2016).

Additional milestones in the research on *Botryllus schlosseri* are the publication of its draft genome, followed by the sequence of the histocompatibility locus (Voskoboynik et al. 2013a, 2013b).

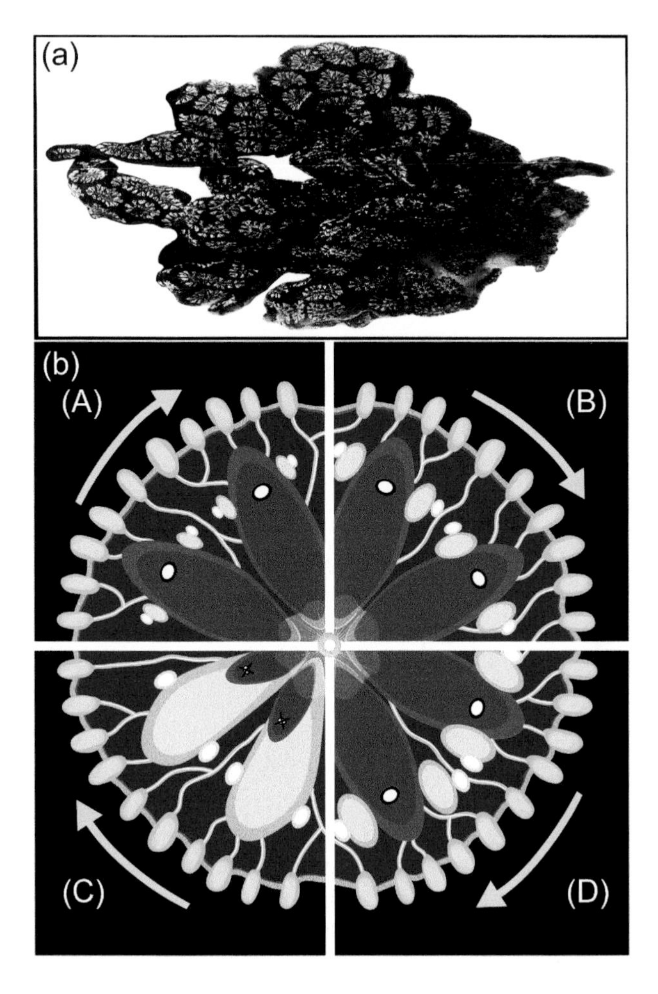

FIGURE 21.1 (a) A colony of *Botryllus schlosseri* (ca, 7 cm length) collected from a Chioggia, Italy, harbor, which naturally grow on algae as substrates. The colony is composed of hundreds of zooids arranged in colonial systems. (b) A diagram representing the four blastogenic stages that typify a weekly cycle (ca. seven days long). The green extensions represent the peripheral ampullae and their attached vasculature. Three generations of modules are shown in each stage: the mature zooids are colored in red, the primary buds in yellow and the secondary buds in white. Stage A, the beginning of a cycle, is signified by the opening of the oral and atrial siphons of the zooids. Open siphons enable the zooids to feed and breed. Secondary buds evaginate from the atrial wall of the primary buds. Primary buds are small and non-functional. Stage B is signified by visible heart-beats in the primary buds, while secondary buds develop as closed double-layered structures. In Stage C, primary buds almost complete development, while secondary buds commence organogenesis, primary subdivisions are completed and pigment cells accumulate in their outer epithelium. Stage D (takeover) starts by closing of the zooids' siphons and their continuous shrinkage until completely resorbed. At the same time the primary buds complete their development and are now fully grown, "waiting" for the takeover stage to conclude so that their oral siphons will be opened in the beginning of a new blastogenic cycle, enabling them to feed and breed. ([a-d] sensu Watanabe 1953 and Lauzon et al. 2002.)

21.2 GEOGRAPHICAL LOCATION

The colonial tunicate *Botryllus schlosseri* (Figure 21.1a, Figure 21.3) is a common shallow-water marine species, found from the intertidal zone to 200 m depth, above and under stones; on natural hard substrates; on algae and seaweeds; and on artificial substrates such as pilings, floats, pontoons, wharfs, ropes and ship bottoms (Rinkevich and Weissman 1991; Müller et al. 1994; Rinkevich et al. 1998a, 1998b), as well as on motile macroinvertebrates (Bernier et al. 2009) and on fish (Kayiş 2011). This species probably originated in the Atlantic European and Mediterranean seas (Van Name 1945; Berrill 1950; Paz et al. 2003; López-Legentil et al. 2006) and spread globally (Figure 21.2). Traits like fast adaptation to human-made environmental conditions (Lambert 2001; Lambert and Lambert 2003) and assumed high mutation rates acted as surrogates for the increase of genetic variability in just-established populations (Reem et al. 2013a). This further promotes the species invasiveness capacities by assisting pioneering colonies in quickly spreading in new sites and then their fast integration as common participants in assemblages of hard-bottom consortia (Lambert and Lambert 1998, 2003; Locke et al. 2009; Martin et al. 2011).

B. schlosseri is primarily recorded in marinas and harbors in the northern and southern hemispheres and has become a cosmopolitan alien species in marine human-made submerged hard substrates (Figure 21.2) (Rinkevich et al. 1998a, 1998b, 2001; Ben-Shlomo et al. 2001, 2006, 2010; Stoner et al. 2002; Paz et al. 2003; Bock et al. 2012; Reem et al. 2013a, 2013b, 2017; Yund et al. 2015; Karahan et al. 2016; Nydam et al. 2017). In the northern hemisphere, populations of *B. schlosseri* are distributed in all Atlantic coasts from the southern coast of India (Meenakshi and Senthamarai 2006; 8°22' N latitude, where sea water temperature ranges from 24 to 29.5°C), to the Norwegian sea ports (>62° N) with sea water temperatures ranging between 3 and 17°C, up to Alaska on the west coast of North America and British Columbia, Canada, and the east coast (Epelbaum et al. 2009), Japan (Rinkevich and Saito 1992; Rinkevich et al. 1992a), Korea and more (Figure 21.2). Populations of this species are further thriving under wide salinity ranges (18–34%; Epelbaum et al. 2009). In the southern hemisphere, this species is thriving in New Zealand (Ben-Shlomo et al. 2001), Australia and Tasmania (Kott 2005), South Africa (Millar 1955; Simon-Blecher 2003), Chile and Argentina (Figure 21.2) (Orensanz et al. 2002; Castilla et al. 2005; Ben-Shlomo

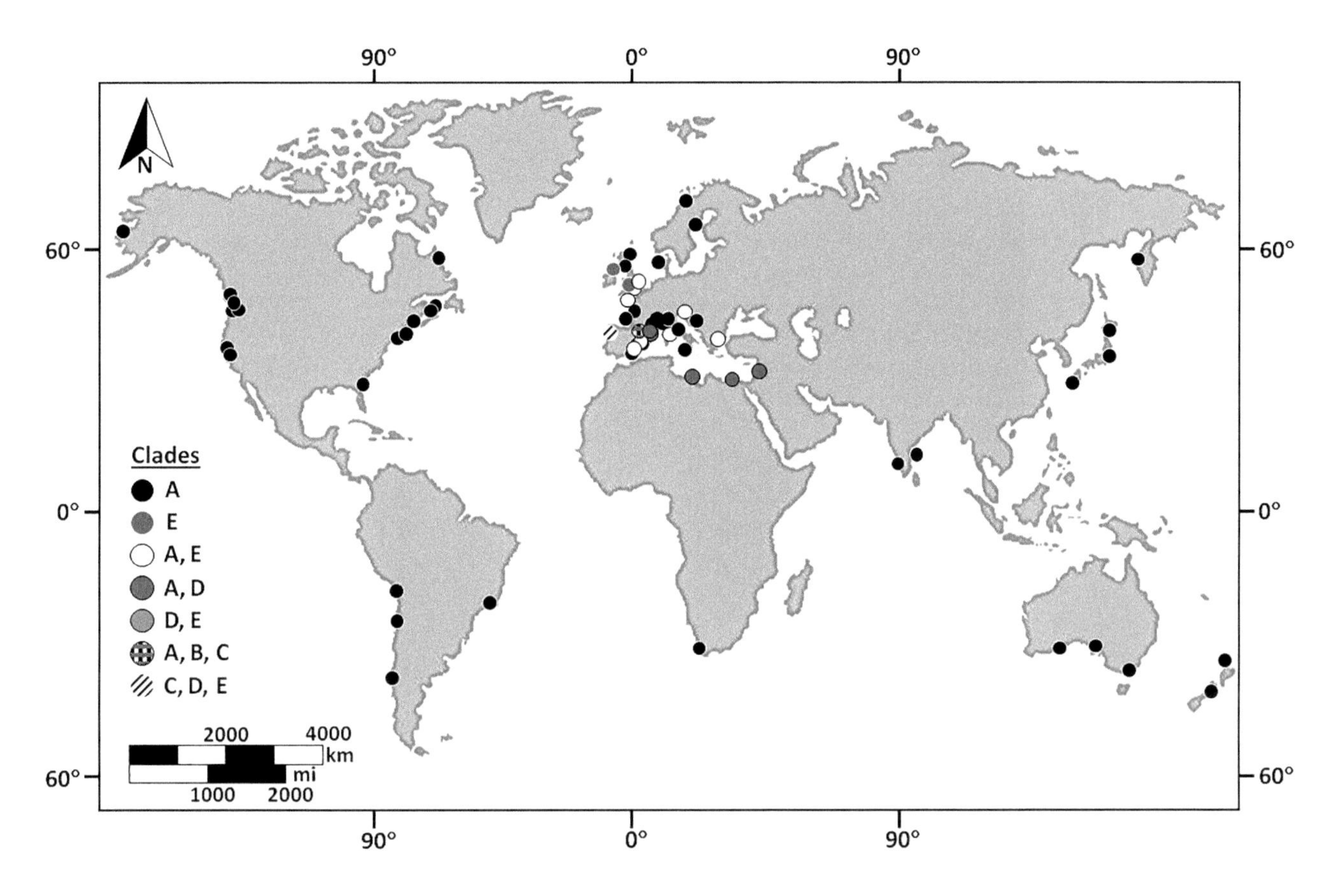

FIGURE 21.2 Global distribution of the *Botryllus schlosseri* five clades (a to e). The global distribution has been contributed by anthropogenic factors (see geographical location section). (Graphic assistance by Guy Paz.)

et al. 2010). Early suggestions (e.g. Van Name 1945) have implied that *B. schlosseri* originated in European waters, a proposal supported by Reem et al. (2017), while Yund et al. (2015) proposed that at least one haplotype in clade A (see the following) is native to the northwest Atlantic. Carlton (2005) proposed, albeit without supporting documentation, a possible Pacific origin. It is further assumed that this worldwide distribution pattern of *B. schlosseri* is primarily anthropogenic in nature, initiated during the last millennium with European travelers who sailed and explored the world, and further enhanced by aquaculture activities (Fitridge et al. 2012; Carman et al. 2016).

The use of the cytochrome oxidase subunit I (COI) marker for *B. schlosseri* population structures worldwide has resulted in the detection of five highly divergent *B. schlosseri* clades (termed A–E), leading to the assumption that *B. schlosseri* is a complex of five cryptic, and probably reproductively isolated, species (Bock et al. 2012). Yet Reem et al. (2017) revealed the possibility of admixture between individuals from clades A and E within two *B. schlosseri* Mediterranean populations, challenging this assumption. Clade A has emerged as a cosmopolitan, revealing significant differentiation patterns between native and invasive populations (Bock et al. 2012; Lin and Zhan 2016). The other four COI clades are restricted to the Mediterranean Sea and Atlantic European waters, with the wider distribution of clade E that is recorded from both sides of the La Manche channel and many coasts in the Mediterranean Sea, and clades B, C and D that are confined to a restricted few harbors (Figure 21.2). *B. schlosseri* clade B was found only in a single site, Vilanova, Spain, and in few samples (López-Legentil et al. 2006). *B. schlosseri* clade C was found in just three sites (López-Legentil et al. 2006; Pérez-Portela et al. 2009). López-Legentil et al. (2006) recorded few clade C specimens from Vilanova and Fornelos. Pérez-Portela et al. (2009) collected three samples from Ferrol, 7 km from Fornelos. This scarcity of data prevents the drawing of further conclusions.

21.3 LIFE CYCLE

The life cycle of the *Botryllus* colony reveals a complex astogeny (building of a colony body) where the continuous and synchronous exchange of asexual-derived generations of basic modules (the zooids in botryllid ascidians) takes place on a weekly basis, a phenomenon of cyclical death and rebirth that is called blastogenesis (Figure 21.1b) (Rinkevich 2002a, 2019; Manni et al. 2007, 2014, 2019; Rinkevich et al. 2013; Tiozzo et al. 2005). Upon accomplishing ontogeny, the first established basic module (oozooid) (Figure 21.3b) then commences astogeny, where similarly sized modules are continuously added in blastogenesis, a process also known as asexual reproduction (Figure 21.3c, d), dictated in *B. schlosseri* by synchronous and cyclical asexual multiplication processes; each lasts for about one week (Figure 21.1b) (Rinkevich 2002a, 2019; Manni et al. 2007, 2014, 2019). At the colony level, zooids are arranged in star-shaped systems, each with a common cloacal siphon in the center, and when the colony expands, each colonial system divides into two or more systems, each centered by a cloacal siphon (Figure 21.3d). The continuous developmental process of colonial growth is thus repeatedly interrupted by this phoenix-like (Rinkevich 2019) death and rebirth cycles of old and new modules, respectively (Figure 21.1b).

A mature *B. schlosseri* colony contemporaneously accommodates three successive generations of modules at any given time throughout the colony's lifespan, the zooids and two generations of buds, all arranged in a hierarchical subdivision within the colony (Figure 21.1b, Figure 21.4). The colony increases in size when more than one bud replaces each zooid of the old generation. The mature functioning modules are the zooids; the most-developed sets of buds but not yet active modules are the primary buds; and the youngest generation, the just-budded modules (the budlets), are the secondary buds (Figure 21.4). The development and growth of the three generations of modules are highly synchronized so that all modules of a certain cohort are exactly at the same differentiating state (Figure 21.1b) (Milkman 1967). Although a colony can live for several months to years, the colonial modules are transient, and the life span of each module, from onset of secondary bud to morphological resorption of the mature zooid, is about three weeks (three blastogenic cycles), whereas the functional-zooid status is for just one week/blastogenic cycle (Figure 21.1b) (under 20°C; Sabbadin 1955; Manni et al. 2007, 2019).

The budlets are formed and developed from the atrial wall (the peribranchial epithelium) of the primary buds as disc-shaped thickenings (Figure 21.4). The bud primordium curves perpendicularly to the primary bud wall and forms a small hemisphere and then tilts toward the anterior end of the primary bud, already establishing the anterior–posterior and dorsal–ventral axes (Sabbadin et al. 1975; Manni et al. 2007). At the end of the first week of the budlet's life, hearts are morphologically recognizable but do not function yet. Following the takeover stage (see the following) and along the second week of life, these modules become the primary buds, where additional organogenesis steps advance toward fully developed buds (Figure 21.4) (Berrill 1941a, 1941b; Izzard 1973). Following the next takeover stage and simultaneously at the beginning of the third and last week of the module's life, they become fully functional zooids, with open oral siphons, and are able to feed and breed (Figure 21.1b). All developmental stages of the three generations of modules are coordinated simultaneously, and the young zooids take over the colony from the older generation of zooids (morphologically illuminated by opening their oral siphons) simultaneously with the clearance and morphological absorption of the old zooids (Figure 21.1b) (Lauzon et al. 2000, 2002; Manni et al. 2007, 2014; Ballarin et al. 2010).

The takeover phase, 24–36 hours at the end of each blastogenic cycle, is the most dramatic astogenic process, where the old zooids gradually shrink and are absorbed into the colonial mass until completely disappearing (Lauzon et al. 1992, 2002; Manni et al. 2007; Ballarin et al. 2010). On the cellular level, the morphological clearance of the zooids is manifested by whole-zooid apoptosis and phagocytosis processes (Cima et al. 2003; Ballarin et al. 2010), and cell corpse clearance is assisted by hyaline amoebocytes and macrophage-like cells (Cima et al. 2003; Voskoboynik et al. 2002, 2004; Ballarin and Cima. 2005; Ballarin et al. 2008). Phagocyte digestion may lead to an oxidative stress, further enhancing zooidal senescence (Cima et al. 2010). Employing an anti-oxidant treatment (BHT) on the blastogenic cycle, Voskoboynik et al. (2004) have pointed to the importance of the macrophages in triggering apoptosis. The phagocyted materials are than recycled for other energy needs of the colony (Lauzon et al. 2002).

Two major staging methods associated with the complex development of the three module types within a single blastogenic cycle in botryllid ascidians were suggested (Watanabe 1953; Sabbadin 1955; modification of Sabbadin's method was suggested by Izzard 1973). The blastogenic cycle is either divided into four phases (Figure 21.1b) according to Watanabe (1953), or into 11 stages according to Sabbadin (1955). Each method has its pros and cons, and scientists use either method according to their research interests.

Few studies have searched for the molecular machinery controlling blastogenesis. One specific gene, *Athena*, was defined (Laird et al. 2005b) as differentially upregulated in the takeover stage as compared to the other blastogenic phases while being transcribed in the developing buds and absorbing zooids. Knockdown of the gene in *Botryllus* using RNAi and antisense morpholinos led to abnormal developmental syndromes of the buds. Further, the *Botryllus* homologue PL-10 also revealed a cyclical pattern associated with the blastogenic cycle with lower levels in old zooids as compared to young buds (Rosner et al. 2006). The same applies to 10 of the genes of the IAP family (a total of 25; Rosner et al. 2019) that were upregulated at late blastogenic stages C and D (Figure 21.1b) concurrent with increased expressions of apoptosis-inducing genes (AIF1, Bax, MCl1) and three caspases (caspase 2 and two orthologues of caspase 7), as in the reorganization of the colonial architecture (Rinkevich et al. 2013; Rosner et al. 2006, 2013, 2019).

When considering the yet-unknown cellular and molecular pathways which control astogeny in *B. schlosseri*, it is of interest to evaluate the operation of astogeny-associated gene families, as the same gene families may be used in ontogeny (e.g. Rosner et al. 2014). One of the first genes used for such comparisons is Pitx (Tiozzo and De Tomaso 2009; Tiozzo et al. 2005), a developmental regulator involved in organ development and in left-right asymmetry (Boorman and Shimeld 2002; Hamada et al. 2002). The *Botryllus* Pitx was present in earlier stages of bud development with similar expression patterns as in the developing embryos, suggesting a parallel role in module/embryo development (Tiozzo et al. 2005; Tiozzo and De Tomaso 2009). Other transcription factors involved in bud development are FoxA1, GATAa, GATAb, Otx, Gsc and Tbx2/3 (Ricci et al. 2016a).

Further research studied the expression along blastogenesis of three conserved signal transduction pathways, Wnt/β-catenin, TGF-β and MAPK/ERK (Rosner et al. 2014), by studying representative gene β-catenin (for Wnt/β-catenin pathway), p-Smad2 and p-Smad1/5/8 (for TGF-β pathway) and p-Mek1/2 (for MAPK/ERK pathway). Results revealed that while the same molecular machinery is functioning in *Botryllus schlosseri* astogeny and ontogeny, astogenic development is not an ontogenic replicate (Rosner et al. 2014).

Blastogenesis (Figure 21.1b) in *B. schlosseri* holds some unique characteristics for aging in colonial (modular) organisms that distinguishes this type of aging from aging in unitary organisms (Rinkevich 2017). Some characteristics that refer to non-random (genetic based) mortality were recorded in these organisms (Rinkevich et al. 1992b; Lauzon et al. 2000; Rabinowitz and Rinkevich 2004; Rinkevich 2017). The phenomena of budding, as well as module senescence, can be concurrently expressed at three hierarchical levels of colonial organization: the zooids, ramets and genets, including the weekly blastogenesis, the whole-genet programmed life span (Rinkevich et al. 1992b); ageing at the ramet level (Rabinowitz and Rinkevich 2004); and rejuvenilization after acute damage (Voskoboynik et al. 2002). To further understand *Botryllus* blastogenesis and aging at the molecular level, the aging-related heat-shock protein *mortalin* was studied (Ben-Hamo et al. 2018). RT-PCR and *in-situ* hybridization revealed significant upregulation of *mortalin* in colonies during the takeover phase as compared to other blastogenic phases. Quantitative PCR analyses of excised buds and zooids showed significantly higher levels of *mortalin* in buds as compared to functioning zooids. These findings are in line with literature that demonstrated lowering levels of mortalin in old organisms as compared to young organisms (Yokoyama et al. 2002; Kimura et al. 2007; Yaguchi et al. 2007), demonstrating a possible aging process that is restricted to the modules and associated with the blastogenic cycle (Ben-Hamo et al. 2018).

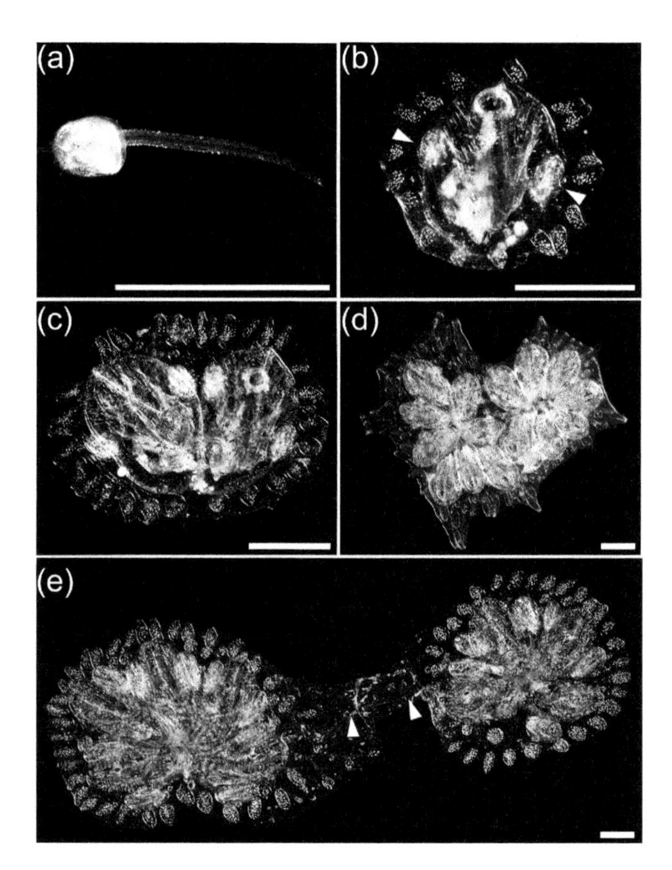

FIGURE 21.3 (a–d) Life stages of a *Botryllus schlosseri* colony. (a) *Botryllus* begins its life as a mobile larva, composed of a visceral trunk and locomotory tail. The mobility enables the larva to swim and find adequate substrate to settle on. (b) At metamorphosis, the attached larva becomes the first zooid, the oozooid, with open oral and atrial siphons. On both sides of the oozooid, the new generation of buds (white arrowheads) are formed and developed. (c) After a few days, the oozooid is resorbed and is replaced by two new zooids. (d) The numbers of zooids grows over time, forming a mature colony. The size of a colony differs between colonies and depends on the number of zooids (see colony in Figure 21.1a for comparison). (e) A chimera composed of two distinct colonies, connected via a blood vessel (two arrowheads). The chimera is formed after a physical contact between the ampullae of both of colonies (details in "Functional Approaches: Tools for Molecular and Cellular Analyses"). Scale bars = 0.5 mm.

21.4 EMBRYOGENESIS

Like the rest of the many colonial tunicates and unlike the other chordates, *Botryllus schlosseri* is an ovoviviparous species (Zaniolo et al. 1987) that reproduces both asexually and sexually, and colonies are simultaneous hermaphrodites (Berrill 1950; Rodriguez et al. 2014). The colonies can randomly switch between male and hermaphrodite states following physiological stress or become sterile (Rodriguez et al. 2016). Yet self-fertilization is eliminated, as male gonads mature two days following the eggs' fertilization by foreign sperm (Milkman and Borgmann 1963; Milkman 1967; Mukai 1977). Nonetheless, self-fertilization was successfully achieved under laboratory conditions (Milkman and Borgmann 1963; Milkman 1967; Rinkevich 1993) and in the field after surgical or natural separation of a colony into systems (Sabbadin 1971; Gasparini et al. 2014), documenting that in the absence of foreign sperm, self-fertilization may occur. Late fertilization is prone to embryo resorption at blastogenesis stage D (the takeover phase) due to larval delayed development (Milkman 1967, Stewart-Savage et al. 2001). Following metamorphosis of the larva (Figure 21.3a) into the first established zooid (the oozooid) (Figure 21.3b), it takes between 8 and 12 blastogenic cycles for the male gonads to develop and mature (Sabbadin 1971; Sabbadin and Zaniolo 1979). The female gonads mature afterward, establishing the hermaphrodite type of *B. schlosseri*'s sexual reproduction (Figure 21.4).

The gonads are first observed in the secondary buds. At onset, a bilateral gonad blastema appears where its medial part will give rise to the testis and the lateral part to the ovary. The oocytes and testes develop in the buds' blastema, and the oocytes move through several generations of buds (Sabbadin and Zaniolo 1979). Testes and ovaries are formed in the blastema and located in mesenchymal spaces between the epidermis and the peribranchial epithelium (Mukai 1977; Sabbadin and Zaniolo 1979). A study on the sister-species, *Botryllus primigenius*, showed that in cases where large oocytes inherited from former generations reach the blastema cell masses of the bud, part of the blastema is differentiated into the egg envelope, creating the egg follicle and a follicle stalk, while the other part of the blastema is differentiated into the testes. If ova are missing, the cell mass will differentiate into testes only (Mukai and Watanabe 1976). An ovary is composed of one to four oocytes (Zaniolo et al. 1987) of different sizes and developmental stages (Sabbadin and Zaniolo 1979) and contains a variable number of undifferentiated cells (Sabbadin and Zaniolo 1979). The globular egg that is enclosed within the ovum is layered by the chorion (acellular vitelline coat or egg-membrane) and the inner and outer follicle cell layers (Zaniolo et al. 1987; Manni et al. 1993) and is connected to the atrial epithelium by vesicular oviduct. Ovulation of eggs occurs inside the zooids at the onset of each blastogenic cycle, in blastogenic stage A (Milkman 1967; Rodriguez et al. 2016). During ovulation, the outer follicular layer is peeled off the egg, exposing the internal follicular layer. The outer layer then forms an ephemeral corpus luteum, and the egg ruptures and moves through the vesicular oviduct. Each egg hangs on the atrial wall and the epithelium of the oviduct, and together with the atrial epithelium, a cup-like "placenta" is formed. The inner follicular layer adheres to the placenta, forming junctional spots with the oviduct epithelium and the filamentous layer that anchors the layers, ensuring the attachment of the embryo to the parent. The corpus luteum is resorbed before gastrulation (Zaniolo et al. 1987) and then the outer follicular layer disintegrates and disappears (Mukai 1977). Oocytes, primordial germ cells (PGCs) and germ cells circulate freely in the blood system, temporarily occupy niches within colonial modules (zooids and buds) and move between generations of modules (Sabbadin and Zaniolo 1979; Magor et al. 1999; Voskoboynik et al. 2008). During their journey, PGCs present stemness genes such as BS-Vasa, BS-DDX1, γ-H2AX, BS-cadherin, phosphor-Smad1/5/8 and more (Rosner et al. 2013).

A testis (Figure 21.4) is a multilobe structure made of branched tubes ending in swollen follicles that host the undifferentiated germ stem cells and all daughter cells through spermatogenesis and where the most mature cells are located in the middle of the follicles and the least developed cells in the periphery (Burighel and Cloney 1997). Spermatogenesis initiates in blastogenic stage A1. During testis maturation (blastogenic stage B1), sperm is released from the atrial siphon, aided by the hydraulic force to be swept away far from the colony (Milkman 1967; Burighel and Cloney 1997), so along blastogenic stage C1, most of the sperm is already released (Rodriguez et al. 2016). Several associated gene expressions (e.g. tetraspanin-8, testis-specific serine/threonine protein kinase-1 and vitellogenin-1) typify spermatogenesis as Otoancorin, a marker for developing testes (Rodriguez et al. 2014).

The cleavage of the ascidian embryo is holoblastic and bilateral, and the gastrulation occurs by epiboly and invagination, while the large archenteron (where the notochord is formed) eliminates the blastocoel. The archenteron proliferates laterally, growing into a solid band of mesodermal cells in each side of the body, and, unlike other deuterostomes, in ascidians, the mesodermal bands do not arise by enterocoely and do not develop coelomic cavities. The differentiation of the ectoderm occurs along the mid-dorsal line into a neural tube where the ectoderm sinks inward and rolls upward, forming the neural tube. The embryo is developed to a lecithotropic, non-feeding larva (Figure 21.3a) that hatches and swims throughout the oral siphon into the outer world (Berrill 1950; Rodriguez et al. 2016), according to Mukai (1977).

The tadpole larva is divided into a visceral trunk and locomotory tail (Figure 21.3a). The trunk contains cerebral vesicle and viscera. The digestive system that exists in the larva does not function yet and will remain in the newly developed oozooid following metamorphosis (Figure 21.3b). The tail is propulsive and contains musculature, the notochord (a hollow tube that contains extracellular fluid), dorsal neural tube and an endodermal rudiment, while the dorsal and the ventral fins on the tail are folds of the larval tunic. The cerebral vesicle, which is located in the dilated anterior end of the neural tube, includes the ocellus and statocyst (Ruppert et al. 2004). The life-span of the swimming larva is short (less than one hour), following which the larva attaches to a substrate, aided by three anterior adhesive papillae and metamorphoses. The tail is retracted and absorbed, resulting in the loss of the notochord, dorsal hollow nerve cord, the musculature and the endodermal rudiment. The area between the adhesive papillae goes through a massive growth, resulting in a rotation of the body by 90°, positioning the siphons upward (opposite to the substrate). Then the atrium expands, enclosing the anus and the pharynx. The oozooid gives rise to the first zooid (Figure 21.3c), which, following several blastogenic cycles, will form a colonial entity (Figures 21.1a, 21.3d) (Ruppert et al. 2004).

21.5 ANATOMY

Botryllus schlosseri, like all other ascidians in the subphylum Tunicata, lacks the typical chordate features while possessing in the larval stage the essential chordate traits of a hollow dorsal nerve cord, a notochord, pharyngeal pouches and a tail (Berrill 1935; Ruppert et al. 2004).

Botryllus schlosseri colonies vary in color phenotypes, ranging from yellow, orange and brown to blue, green, gray and more. The intensity of colors and variation in coloration may be affected by age and environmental state of the colony (Milkman 1967; Lauzon et al. 2000), but the animal's basic color patterns are based on genetics (Sabbadin 1977). The sizes of *Botryllus* colonies are variable and can range from a few millimeters to several centimeters, depending on the number of zooids in a colony, from few to thousands (compare Figure 21.1a to Figure 21.3d) (Chadwick-Furman and Weissman 1995). From the anatomy point of view, a *Botryllus schlosseri* colony can be defined according to three levels of body organizations: the entire colony/genet level, the level of the system/ramet and the level of the modules (Rinkevich 2017). The following text considers the *Botryllus* anatomy at each level of organization.

The colonial mass (the genet, as well as each separated ramet) of *Botryllus schlosseri* is composed of a different number of modules (the zooids; in diverse developmental stages), which are embedded in the tunic and are connected to each other through a ramified blood system. The tunic is a gelatinous-like, fibrous, transparent extra-cellular matrix (Figure 21.1a, Figure 21.3) (Zaniolo 1981). It contains mainly carbohydrates and also proteins and motile cells (Smith and Dehnel 1971; Richmond 1991; Ruppert et al. 2004). A cellulose-like polymer named tunicin is abundant in the tunic. Tunicates are the only known animals that have a unique ability to produce cellulose-like materials using a cellulose-synthase (Nakashima et al. 2004, 2008; Inoue et al. 2019). The tunic envelops the zooids with a thin, dense cuticle layer that covers the entire tunic (Zaniolo 1981). Three types of test cells are found in the tunic. The first and most abundant cell type is the vacuolated motile cells, defined by filopodia that are homogeneously distributed in the tunic (Izzard 1974; Zaniolo 1981; Hirose et al. 1991; Hirose 2009). The other two test cell types are fusiform cells that are usually found adjacent to vessel walls and fibrocytes that have pseudopodia and are spread in the tunic (Zaniolo 1981; Hirose et al. 1991; Hirose 2009). In addition, diverse types of blood cells infiltrate and found in the tunic (Ruppert et al. 2004; Hirose 2009). The tunic and the test cells form together a complex connective-like tissue (Nakashima et al. 2008).

A ramified vasculature system is embedded in the tunic (Berrill 1950; Ruppert et al. 2004) and connects between all zooids. Each blood vessel in the network is made of an epithelium that connects to the zooids. The blood system of the zooids is open and contains lacunae across organs (Milkman 1967; Gasparini et al. 2007). The tunic vessels are uniquely lined by epidermis and epidermal basal lamina (Ruppert et al. 2004). The vessels are terminated in ampullae, numerous swollen thickening endings, sausage-like structures (Figure 21.4), located in the external boundaries of the tunic that help the colony to attach to or glide on the substrate (Katow and Watanabe 1978). The blood flows due to the contraction of both the zooids' hearts and the ampullae (Milkman 1967). The ampullae are the organs for primary physical contact sites in allorecognition and are the areas for self/non-self recognition between colonies (more details in section "Functional Approaches: Tools for Molecular and Cellular Analyses") (Katow and Watanabe 1980; Rinkevich and Weissman 1987a, 1987b).

The zooids are embedded within the tunic, in accordance with the *Botryllus*-specific pattern formation, as circular, star-like structures, each termed a system, ergo the epithet "star-ascidian" for *Botryllus* (a colony with two systems is shown in Figure 21.3d). Each system contains up to 10–12 zooids, and the numbers of systems/zooids in colonies vary, depending on free substrate space, environment conditions and colony vitality (Chadwick-Furman and Weissman 1995; Lauzon et al. 2000). The atrial siphons of the grouped zooids open into a common atrial chamber.

It is customary to separate (sub-clone) the colony into systems using a simple surgical procedure. When the separation is carried out properly, the separated systems, termed ramets, recover rapidly. Sub-cloning is a common procedure carried out in laboratories due to its experimental advantage in receiving a number of genetic-identical repeats (subcloning methodologies in Rinkevich and Weissman 1987a; Rinkevich 1995).

The zooids in *Botryllus* are divided into three groups according to their developmental stages, the zooids, the primary buds and the secondary buds (Figure 21.4; more on module development in the life cycle section). Here we will reveal the anatomy of the mature modules, since the buds are going through diverse stages of organogenesis. In a typical zooid, the soma is delineated by the body wall, the mantle, formed by the epidermis that contains connective tissue, blood vessels/lacunae and muscle strands. The zooid is oval, over 1 millimeter in length, and contains two openings: the oral (branchial or buccal) siphon, which is the mouth and is also used as the sperm/larvae doorway (Berrill 1950; Rodriguez et al. 2016), and the atrial siphon, which is an excretion site. The oral siphon is adorned with eight tentacles (four long and four short), leading to a pharynx, which is the branchial sac (Berrill 1950). Tunicates are filter-feeders, a process executed by the branchial sac (Figure 21.4), attaining their food from the seawater by intake of water through periodic contraction of the body wall. Food is filtered through the branchial sac by dedicated ciliatic cells arranged in slits named stigmata. This organ also participates in respiration process. Planktonic food is captured by the mucus in the branchial sac and then collected and transported via the cilia to the digestive system located in the visceral cavity, started from the pharynx, to the esophagus, the stomach, the U-shaped gut and last the atrium and outside the body through the atrial siphon (Berrill 1950). *Botryllus*, like other tunicates, lacks conventional nephridia. Instead, ammonia is released by diffusion, while other by-products such as uric acid and calcium oxalate are stored in specialized cells named nephrocytes that accumulate in various tissues (Ruppert et al. 2004)

The nervous system is composed of a cerebral ganglion and a neural (pyloric) gland. The cerebral ganglion is a rounded hollow "brain" located in a connective tissue, where the stemming nerves connect to the branchial siphon and to musculature (Ruppert et al. 2004). The pyloric gland is a hollow blind-sac stemming from the basal region of the stomach, branching over the wall of the intestine and ending in ampullae, and is involved in the evaluation of environmental signals (Burighel et al. 1998). A monoclonal antibody that is specific to the cells of the pyloric gland has been developed (Lapidot et al. 2003), a unique tool in research.

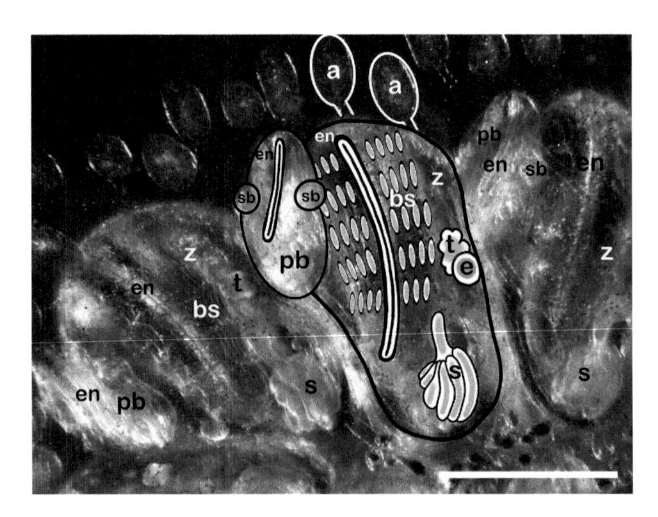

FIGURE 21.4 A close-up back-side photo of a colony growing on a glass slide, superimposed (in the center) with illustration describing the anatomy of the three generations of colonial modules. The zooid (z) is the mature module, and to the left of it is a primary bud (pb) marked with two secondary buds (sb) that appear as small round protrusions. The endostyle (en) is illustrated as elongated organ and is clearly seen in the zooid and in the primary bud. The branchial sac (bs) is composed of the endostyle and the stigmata, represented here by numerous oval-like thin structures. Zooids may contain testes (t) and an egg (e). "s" refers to the stomach. Blood vessel endings, the ampullae (a), are also marked as swollen structures at the periphery of the colony. Scale bar = 0.5 mm.

The heart is a long, tube-shaped structure made of a simple epithelium and striated muscle. The blood flows through the heart thanks to peristaltic movement waves. The zooids' heartbeats are synchronized for the rate and direction of flow. The hearts beat together so that the blood flows in the same direction for a few seconds and then stop and continue to beat in the opposite direction.

In the dorsal part of the zooid, a long, tube-like organ named the endostyle (Figure 21.4) is composed of eight zones, even and odd, where the odd zones have cilia and are in charge of mucus propelling, while the even zones manifest secretion (Burighel and Cloney 1997). The endostyle is a vertebrate thyroid homologue that further synthesizes and secretes thyroid hormones used for iodine metabolism (Ogasawara et al. 1999). In addition, it serves as a transient niche for hematopoietic stem cells (Voskoboynik et al. 2008) and is highly functional in feeding, secreting a mucus net aiding the branchial sac to capture food particles (Holley 1986; Burighel and Cloney 1997).

21.6 GENOMIC DATA

Using a novel high-throughput method for eukaryotic genome sequencing, a draft genome of *Botryllus schlosseri* from Monterey, California, possessing 27,000 estimated genes and 38,730 putative protein-coding loci, was published in 2013 (Voskoboynik et al. 2013a). Former genomic analyses using flow cytometry elucidated a genome size of 725 Mb (De Tomaso et al. 1998), based on 16 haploid chromosomes (according to Colombera 1963) or 13 (according to Voskoboynik et al. 2013a). About 65% of the *B. schlosseri* draft genome is composed of repeating segments, summing up to 6,601 repetitive families, each with three copies or more. A particular group of 1,400 large interspersed repeat gene families that are over 1 kb in length are located at dispersed genomic regions, with >10% that each possess >100 copies, and are found in several chromosomes. The average size of a gene is 3.6 kbp, and the average size of an exon is 170 bp. In order to estimate the protein-encoding genes, transcriptome sequence data were constructed from 19 *B. schlosseri* colonies. The transcriptome was compared to the list of the putative proteins, revealing at least 30% matches that support the sequenced genome validity (Voskoboynik et al. 2013a). Further, to evaluate the *Botryllus* phylogenetic relationships with other taxa, Voskoboynik et al. (2013a) compared 521 nuclear gene sequences (40,798 aligned amino acids) with homologous sequences from 14 other model species, including six vertebrates, a tunicate, a cephalochordate, an echinoderm, an insect, two cnidarians, a sponge and a choanoflagellate. Meta-analysis supported the prevailing notion that *Botryllus*, as a tunicate, belongs to the phylum Chordata. The predicted proteomics of *Botryllus*, which was compared with vertebrate proteomics, revealed high homologies: 77% with human, 85% with chicken and 86% with frog, suggesting a common ancestor. Also, *Botryllus* is the only protochordate that carries genes related to pregnancy-specific-glycoproteins (PSGs) (Voskoboynik et al. 2013a). The browser of the *Botryllus* genome is found at either of these links: http://botryllus.stanford.edu/botryllusgenome/ or http://hegemon.ucsd.edu/bot/.

Mitochondrial genome sequencing (Voskoboynik et al. 2013a) revealed 14,928-bp-long mtDNA that includes 24 tRNAs, 2 rRNAs and 13 proteins. This composition of proteins and nuclear acids is typical to tunicate mitochondrial genes. The sequences of the 13 putative proteins in the *Botryllus* mtDNA were further subjected to phylogenetic analyses and were compared to 66 organisms, including tunicates, vertebrates, cephalochordates, xenoturbellides, hemichordates, echinodermates and two outgroups (arthropods and mollusks), suggesting, as with the nuclear gene phylogenetics, a common ancestor with vertebrates. Results further demonstrated high substitution rates of nucleotides in tunicates and that the stolidobrancian tunicates (including *Botryllus*) create a monophyletic group (Voskoboynik et al. 2013a).

Gao et al. (2018) developed a large resource of *Botryllus* single-nucleotide polymorphism (SNP) using restriction site-associated DNA (RAD) tag sequencing, revealing 14,119 SNPs that are available for use. The SNPs served as markers to evaluate population genetic characteristics in *Botryllus.*

Studying *Botryllus* within diverse areas of interest such as astogeny of colonial organisms (blastogenesis; Manni et al. 2007; Ben-Hamo et al. 2018; Manni et al. 2019), regeneration (including whole-body regeneration; Rinkevich and Weissman 1990; Voskoboynik et al. 2007; Rinkevich and Rinkevich 2013; Rosner et al. 2019), allorecognition and population genetics gained informative genomic data, partially unveiling the cryptic biology underlying these phenomena. The following paragraphs summarize the central publications on the genomic data.

Blastogenesis has been well characterized in *Botryllus* both anatomically and ontogenically (Manni et al. 2019; Manni et al. 2014; Sabbadin et al. 1975; Izzard 1973; Berrill 1941a, 1941b). Recent years have yielded novel insights on the molecular processes underlying blastogenesis (Franchi et al. 2017; Campagna et al. 2016; Ricci et al. 2016a; Rodriguez et al. 2014; Rinkevich et al. 2013; Rosner et al. 2014, 2019; Qarri et al. 2020). Transcriptomes of three major stages along the *Botryllus* blastogenic cycle (mid-cycle, the pre-takeover and the takeover phases; Campagna et al. 2016; available at http://botryllus.cribi.unipd.it) have revealed 11,337 new genes, of which 581 transcripts were determined with complete open reading frames. Many sequences emerged as genes involved in apoptosis activation, de-activation and regulation (Campagna et al. 2016). Analyzing the differential expression for fertile vs. infertile *B. schlosseri* colonies, Rodriguez et al. (2014) revealed a set of genes that are differentially expressed in every blastogenesis stage analyzed. The highest numbers of differentially expressed genes were found in early stages, many of which are homologous to vertebrates. These genes have conserved roles in organism fertility (Rodriguez et al. 2014).

Ricci et al. (2016b) constructed transcriptomics libraries from epithelial tissues of developing buds and from non-developing buds and revealed differentially expressed gene expressions in the developing bud epithelial tissues that are associated with regeneration and stem cell functions and homologous to genes in other model organisms. Further sets of unknown genes were elucidated, indicating possible specific genes and functions associated with budding in *B. schlosseri* colonies (Ricci et al. 2016b), while in other cases, such as in response to reactive oxygen species (ROS) that emerge during the takeover stage (Cima et al. 1996; Voskoboynik et al. 2004), five transcripts for antioxidant defense enzymes [SOD (superoxide dismutase), GCLM (glutamyl-cysteine ligase modulatory subunit), GS (glutathione synthase), GPx3 and GPx5 (two glutathione peroxidases)] were identified (Franchi et al. 2017).

Allorecognition in botryllid ascidians is manifested when two or more genotypes come into physical contact with each other, resulting in either fusion (chimera formation) or rejection (see more in "Functional Approaches: Tools for Molecular and Cellular Analyses"). To assess the repertoire of differentially expressed genes during rejection, Oren et al. (2007) constructed expressed sequence tag (EST) libraries where allogeneic challenged colonies were compared to naïve counterparts and revealed dozens of specifically expressed genes homologous to genes involved in diverse immunological processes. The list includes stress proteins, pattern recognition receptors, complement proteins, proteases and protease inhibitors, cell adhesion and coagulation proteins, cytokine-related proteins, programmed cell death and proteasome-related proteins (Oren et al. 2007). Then Oren et al. (2010) elucidated transcriptional differences between the genotypes involved in the allogeneic rejection processes, the partner that displays the points of rejection (PORs; rejected partner) and the rejecting partner "causing" the PORs. Microarray and complementary qPCR assays revealed two distinct transcriptional landscapes for "rejected" vs. "rejecting" colonies in the same allogeneic assay. In the "rejected" colonies, 87% of the ESTs were downregulated as compared to the "rejecting" partner showing only minor changes (0.7%) in the allogeneic assay. In the "rejected" transcriptome, three functional groups were downregulated substantially: protein biosynthesis, cell structure and motility and immune-related genes, overall depicting the inhibition of response components rather than enhancement of immunologic responses (Oren et al. 2010).

Studies were further engaged with the *Botryllus* regeneration abilities and the roles of stem cells in this process (Braden et al. 2014; including whole-body regeneration; Rinkevich and Weissman 1990; Voskoboynik et al. 2007; Rinkevich and Rinkevich 2013; Rosner et al. 2019). According to these studies, stem cells circulate the blood system of the colonies and are confined to dedicated stem cell niches as the niches adjacent to the endostyle. Stem cells play a pivotal role in budding *de novo* of new generations of modules and in regeneration according to their genomic signatures. Three presumed stem cell populations were described in *Botryllus* (CP25, CP33 and CP34), and their expressed genes overlap with those of the mouse hematopoietic stem cells (Rosental et al. 2018).

21.7 FUNCTIONAL APPROACHES: TOOLS FOR MOLECULAR AND CELLULAR ANALYSES

Colonial tunicates such as *Botryllus schlosseri* express unique biological phenomena and are valuable models for variety of research fields, yielding novel discoveries and functional tools in the research. We detail an overview of three main tools that can be applied for diverse studies.

21.7.1 A Model for Chimerism

The first research tool is the use of *Botryllus schlosseri* as an accessible model system for allorecognition, primarily for chimera formation. Chimerism is the biological state where an organism is composed of cells originating from two genetically distinct conspecifics and is based on the capacity for morphological fusions between these organisms (Figure 21.3e). Artificial chimerism (performed in research institutions) is being achieved in model organisms such as frogs (Volpe and Earley 1970), rats (Fang 1971) and mice (Eichwald et al. 1959), established by uniting allogeneic cells during early embryonic stages or via surgical interventions in adults. These systems have proved an indispensible tool for a variety of research fields, such as hematology (Abkowitz et al. 2003), immunology (Liu et al. 2007), aging (Conboy et al. 2013) and more. Although parabiosis is an important system for studies, two main challenges keep it from being used on a wide scale in biology. First, growing public concern in recent years delegitimizes the use of adult parabionts in experimental settings, and second, the traumatic protocols cause enormous stress that may influence the results of the studies. *Botryllus* chimerism may alleviate these challenges.

Botryllid ascidians possess a unique type of immunity (allorecognition system) that may reveal the evolutionary routes for vertebrate immune systems (Magor et al. 1999; Weissman et al. 1990; Cooper et al. 1992; Rinkevich 2004, 2005a), as well as chimerism, revealing evolutionary and ecological aspects for this phenomenon (Rinkevich 2005b). Interest in *B. schlosseri* immunity has centered on allogeneic recognition and its consequences, as pairs of colonies that meet naturally (or in the laboratory) either anastomose contacting ampullae to form a vascular parabiont (Figure 21.3e) or develop cytotoxic lesions in the contact zones (termed points of rejection; Sabbadin and Astorri 1988; Teneda et al. 1985; Rinkevich and Weissman 1987a, 1987b, 1987c, 1991; Weissman et al. 1990; Rinkevich 1992, 1996, 1999a). In many cases, pairs of colonies that fused or rejected each other retreat, growing from their points/areas of contact (Rinkevich and Weissman 1988). *B. schlosseri* chimeras were widely recorded in the field (Ben-Shlomo et al. 2001), most likely the outcome of co-settlement aggregates of histocompatible kin colonies (Grosberg and Quinn 1986). Once colonies fuse, a second allorecognition phenomenon begins which leads to the morphological elimination (resorption) of one partner in the chimera (Rinkevich and Weissman 1987a, 1987b, 1987c, 1989; 1992a, 1992b; Sabbadin and Astorri 1988), termed allogeneic or chimeric resorption (Rinkevich 2005a) and based on a highly complex and polymorphic organization of histocompatibility alleles, revealing a clear hierarchy in the resorption phenomenon (Rinkevich 1993; Rinkevich et al. 1993). Yet a mild stress may change resorption directionality in *B. schlosseri* chimeras by expressing a non-genetic type of apoptotic pathways (Rinkevich et al. 1994).

One of the most interesting outcomes of chimerism in *B. schlosseri* are the phenomena of somatic/germ cell parasitism (Sabbadin and Zaniolo 1979; Pancer et al. 1995; Stoner and Weissman 1996; Magor et al. 1999; Stoner et al. 1999; Rinkevich and Yankelevich 2004; Simon-Blecher et al. 2004). Somatic and germ cell parasitism in chimeric *B. schlosseri* colonies are recognized when the soma and/or the gonads do not reflect equal contributions by the partners involved and are further recorded in "forced chimeras" established between allogeneic noncompatible partners (Rinkevich and Weissman 1998; Simon-Blecher et al. 2004). Germ cell parasitism in this system is fixed, reproducible, reveals hierarchical arrangements and, above all, is sexually inherited (Stoner et al. 1999; Rinkevich and Yankelevich 2004). In contrast, somatic cell parasitism, while reproducible and hierarchical, has not been characterized by the trait of sexual inheritance through a pedigree (Stoner et al. 1999). It may thus be concluded that somatic and germ cell parasitism are unlinked phenomena (Stoner et al. 1999; Magor et al. 1999; Rinkevich and Yankelevich 2004) and that for both types of cell parasitism, the chimeric entity enables foreign somatic and germ stem cells to hitchhike within the "winner" genotypes without being visible to natural selection forces that act on the winner genotypes (Rinkevich 2002a, 2002b, 2004a, 2004b, 2011a), part of the proposed "costs" for chimerism (Rinkevich 2002b, 2005b, 2011a). Yet several studies that evaluated "costs" and "benefits" predictions for chimerism in *B. schlosseri* revealed two major benefits, the shifts of the somatic constituents within chimeras in accordance with changes in environmental conditions and the expression of the heterosis phenomenon in chimeras, occurred via scrutinizing against genotypes that are less adapted to adverse environmental conditions (Rinkevich 1993, 2005b; Rinkevich et al. 1993; Rinkevich and Yankelevich 2004). This attests to the indispensable tool of *B. schlosseri* in the study on chimerism, allorecognition (see also Oren et al. 2010, 2013) and the evolution of immunity.

21.7.2 Accessible Regeneration/Aging Stem Cell-Mediated System

Scientific efforts that have been made over the years to study the biology of stem cells in vertebrates and have led to important understanding in the roles of stem cells in regeneration and aging (Conboy et al. 2015; Singer 2016; Bacakova et al. 2018; Busque et al. 2018; Keyes and Fuchs 2018). Since stem cells play a crucial role in regenerative abilities and aging of multi-cellular organisms, some consider these two phenomena opposite correlated and bounded by stem cell fitness (Conboy et al. 2015; Singer 2016; Keyes and Fuchs 2018). In comparison to the vast knowledge gained on stem cells in vertebrates, little is known on the function of stem cells in invertebrates (Vogt 2012; Ballarin et al. 2018). As opposed to vertebrates, invertebrates have impressive abilities to regenerate their bodies. Some hypotheses suggest reasoning for the gradual loss of regenerative abilities from invertebrates to vertebrates (Rinkevich and Rinkevich 2013; Luisetto et al. 2020). *Botryllus schlosseri* is an optimal model for studies of adult stem cells, regeneration and aging (Rosner et al. 2006, 2007, 2013, 2019; Voskoboynik et al. 2007, 2008, 2009; Rosner and Rinkevich 2007; Rinkevich 2011b; Rinkevich et al. 2013; Munday et al. 2015; Voskoboynik and Weissman 2015; Rinkevich 2017; Ben-Hamo et al. 2018; Qarri et al. 2020). Asexual budding cycles (blastogenesis) include *de novo* whole body regeneration every week throughout the life of colonies (more info in life-cycle section). In addition to the weekly death and growth cycles (Rinkevich 2019), *Botryllus* is able to perform vascular budding of new modules after amputating all existing modules except tunic and blood vessels (Sabbadin et al. 1975; Voskoboynik et al. 2007) and following major stress phenomena, including irradiation (Rinkevich and Weissman 1990; Voskoboynik et al. 2002, 2004; Qarri et al. 2020). Stem cells were further defined as units of selection of the species (Laird et al. 2005a; Rinkevich et al. 2009; Weissman 2015). Thus, *Botryllus* is a unique, omnipotent model organism for studies of regeneration, aging and stem cell biology.

21.7.3 Accessible *In Vitro* Invertebrate Cultures

In vitro approaches in research have advanced scientific disciplines, yet, in spite of significant efforts invested, they have not been successful in obtaining stable *in vitro* tissue cultures from any marine invertebrate, including from *Botryllus* (Rinkevich 1999b, 2005c, 2011b; Grasela et al. 2012). In spite of these failures, several primary cultures were developed successfully from embryos and larvae (Rinkevich and Rabinowitz 1994) and epithelial cell cultures from palleal buds (Rinkevich and Rabinowitz 1997; Rabinowitz and Rinkevich 2003, 2004, 2011; Rabinowitz et al. 2009). These *in vitro* approaches revealed that abrogating the *in vivo* colonial homeostasis resulted in extended life span and developmental features not recorded along blastogenesis. For example, extirpated buds (*in vitro* organ cultures) at blastogenesis stages B to D attached to the bottoms developed novel spheres (up to 1 mm diameter), and then they developed epithelial monolayers on substrates for the next ten days, about a fivefold increase in life expectancy under *in vitro* conditions. Further, instead of the apoptotic death of cells under normal blastogenesis (Lauzon et al. 2002), the *in vitro* death of epithelial monolayers was necrotic (Rabinowitz and Rinkevich 2004). Results revealed the unexpected regenerative power of isolated blastogenic stage D zooids (at the takeover phase process) under *in vitro* conditions that developed almost three times more epithelial monolayers than blastogenetic stages B and C buds, with a higher order of magnitude in monolayer-to-sphere ratio (Rabinowitz and Rinkevich 2004, 2011), and the vast majority of these stage D buds developed epithelial monolayers directly, without forming spheres. Generally speaking, Rabinowitz et al. (2009) showed enhanced expressions of actin, PL10, P-MEK, MAP-kinase, Piwi and cadherin in extirpated buds and monolayers, exhibiting *de novo* emergent stemness signatures.

21.8 CHALLENGING QUESTIONS BOTH IN ACADEMIC AND APPLIED RESEARCH

Botryllus schlosseri presents unique biological phenomena which are highly valuable to several fields in biology (Rinkevich 2002a; Manni et al. 2007; Voskoboynik and Weissman 2015; Manni et al. 2019). Yet studies on *Botryllus* are engaged with challenges that have not yet been solved. In the following, we will overview three major research challenges.

21.8.1 Breeding in the Laboratory

In spite of the growing scientific interest in using *Botryllus schlosseri* as a model organism in a wide range of scientific disciplines, only three laboratories worldwide hold colonies in captivity (in California, at Hopkins Marine Station, Stanford University; in Italy, at the University of Padova; and in Israel, at the National Institute of Oceanography, Haifa). In some other laboratories, such as in Japan (Shimoda Marine Station), some *B. schlosseri* colonies were held in the past. All these sites commonly have access to seawater facilities, while the methodologies of animal maintenance differ (e.g. in Israel, Rinkevich and Shapira 1998; in California, Boyd et al. 1986; in Italy, Brunetti et al. 1984). One of the challenges holding back development of brood stocks for research is therefore the development of methodologies and facilities for inland maintenance of the animals. For example, the use of artificial seawater has not yet been reported in the literature, and the current only way to hold stocks of breeding, healthy and fertile *Botryllus* colonies over time is the use of fresh seawater, in most cases using running seawater systems.

21.8.2 Lack of Sufficient Molecular Research Tools

For esoteric model organisms such as *Botryllus schlosseri*, one major obstacle is the lower efforts dedicated to developing adequate molecular tools by research laboratories and commercial companies, in contradiction to the investment in molecular tools for "popular" model organisms. Even basic tools, such as specific antibodies for *Botryllus*, cannot be commercially supplied and should be prepared in the lab, a time- and money-consuming process. Another struggle is the current failure to produce transgenic *Botryllus* or apply CRISPR gene editing on this species. These burdens slow the progress of research on *Botryllus* and can be eased if more laboratories will join the community of *Botryllus schlosseri* researchers.

21.8.3 Lack of Inbred Strains/Lines

In popular models, a variety of inbred lines and strains of animals are available, including strains that are being used as models for specific diseases and deficiencies. At the moment, there is no single inbred strain or line of *Botryllus*, and the diverse laboratories obtain the animals from their geographic marine locations, revealing high variations between animals. The lack of common strains for research may harm the ability to compare between studies due to variations between and within *Botryllus* ecotypes that stem from sampling different geographic locations and/or different *Botryllus* clades.

BIBLIOGRAPHY

Abkowitz, J.L., Abigail E.R., Sujata K., Michael W.L., and Jing, C. 2003. Mobilization of hematopoietic stem cells during homeostasis and after cytokine exposure. *Blood*. 102:1249–1253.

Bacakova, L., Zarubova, J., Travnickova, M., Musilkova, J., Pajorova, J., Slepicka, P., Kasalkova, N.S., Svorcik, V., Kolska, Z., Motarjemi, H., and Molitor, M. 2018. Stem cells: Their source, potency and use in regenerative therapies with focus on adipose-derived stem cells: A review. *Biotechnology Advances*. 36:1111–1126.

Ballarin, L., Burighel, P., and Cima, F. 2008. A tale of death and life: Natural apoptosis in the colonial ascidian *Botryllus schlosseri*

(Urochordata, Ascidiacea). *Current Pharmaceutical Design*. 14:138–147.

Ballarin, L., and Cima, F. 2005. Cytochemical properties of *Botryllus schlosseri* haemocytes: Indications for morpho-functional characterisation. *European Journal of Histochemistry*. 49:255–264.

Ballarin, L., Rinkevich, B., Bartscherer, K., Burzynski, A., Cambier, S., Cammarata, M., Domart-Coulon, I., Drobne, D., Encinas, J., Frank, U., Geneviere, A.M., Hobmayer, B., Löhelaid, H., Lyons, D., Martinez, P., Oliveri, P., Peric, L., Piraino, S., Ramšak, A., Rakers, S., Rentzsch, F., Rosner, A., Silva, T.H. da, Somorjai, I., Suleiman, S., and Coelho, A.V. 2018. Maristem: Stem cells of marine/aquatic invertebrates: From basic research to innovative applications. *Sustainability*. 10:1–21.

Ballarin, L., Schiavon, F., and Manni, L. 2010. Natural apoptosis during the blastogenetic cycle of the colonial ascidian *Botryllus schlosseri*: A morphological analysis. *Zoological Science*. 27: 96–102.

Bancroft, F.W. 1903. Variation and fusion of colonies in compound ascidians. *Proceedings of the California Academy of Sciences, 3rd Series*. 3:137–186.

Ben-Hamo, O., Rosner, A., Rabinowitz, C., Oren, M., and Rinkevich, B. 2018. Coupling astogenic aging in the colonial tunicate *Botryllus schlosseri* with the stress protein mortalin. *Developmental Biology*. 433:33–46.

Ben-Shlomo, R., Douek, J., and Rinkevich, B. 2001. Heterozygote deficiency and chimerism in remote populations of a colonial ascidian from New Zealand. *Marine Ecology Progress Series*. 209:109–117.

Ben-Shlomo, R., Paz, G., and Rinkevich, B. 2006. Postglacial-period and recent invasions shape the population genetics of botryllid ascidians along European Atlantic coasts. *Ecosystems*. 9:1118–1127.

Ben-Shlomo, R., Reem, E., Douek, J., and Rinkevich, B. 2010. Population genetics of the invasive ascidian *Botryllus schlosseri* from South American coasts. *Marine Ecology Progress Series*. 412:85–92.

Bernier, R.Y., Locke, A., and Hanson, J.M. 2009. Lobsters and crabs as potential vectors for tunicate dispersal in the southern Gulf of St. Lawrence, Canada. *Aquatic Invasions*. 4:105–110.

Berrill, N.J. 1935. VIII: Studies in tunicate development, Part III: Differential retardation and acceleration. *Philosophical Transactions of the Royal Society B: Biological Sciences*. 225:255–326.

Berrill, N.J. 1941a. The development of the bud in *Botryllus*. *The Biological Bulletin*. 80:169–184.

Berrill, N.J. 1941b. Size and morphogenesis in the bud of *Botryllus*. *The Biological Bulletin*. 80:185–193.

Berrill, N.J. 1950. *The Tunicata with an Account of the British Species*. Ray Society, London, UK. 3–56.

Berrill, N.J. 1951. Regeneration and budding in tunicates. *Biological Reviews*. 26:456–475.

Bock, D.G., Macisaac, H.J., and Cristescu, M.E. 2012. Multilocus genetic analyses differentiate between widespread and spatially restricted cryptic species in a model ascidian. *Proceedings of the Royal Society B: Biological Sciences*. 279:2377–2385.

Boorman, C.J., and Shimeld, S.M. 2002. The evolution of left-right asymmetry in chordates. *BioEssays*. 24:1004–1011.

Boyd, H.C., Brown, S.K., Harp, J.A., and Weissman, I.L. 1986. Growth and sexual maturation of laboratory-cultured Monterey *Botryllus schlosseri*. *Biological Bulletin*. 170:91–109.

Braden, B.P., Taketa, D.A., Pierce, J.D., Kassmer, S., Lewis, D.D., and De Tomaso, A.W. 2014. Vascular regeneration in a basal chordate is due to the presence of immobile, bi-functional cells. *PLoS One*. 9:e95460.

Bruguière, J.G. 1792. *Encyclopédie méthodique. Histoire naturelle des vers. Tome Premier*. Panckoucke, Paris. 1–757.

Brunetti, R., Marin, M.G., and Bressan, M. 1984. Combined effects of temperature and salinity on sexual reproduction and colonial growth of *Botryllus schlosseri* (Tunicata). *Bollettino di Zoologia*. 51:405–411.

Burighel, P., and Cloney, R.A. 1997. Urochordata: Ascidiacea. In *Microscopic Anatomy of Invertebrates*, eds. Harrison, F.W., and Ruppert, E.E. 221–347. Wiley-Liss Inc, New York.

Burighel, P., Lane, N.J., Zaniolo, G., and Manni, L. 1998. Neurogenic role of the neural gland in the development of the ascidian, *Botryllus schlosseri* (tunicata, urochordata). *Journal of Comparative Neurology*. 394:230–241.

Busque, L., Buscarlet, M., Mollica, L., and Levine, R.L. 2018. Concise review: Age-related clonal hematopoiesis: Stem cells tempting the devil. *Stem Cells*. 36:1287–1294.

Campagna, D., Gasparini, F., Franchi, N., Vitulo, N., Ballin, F., Manni, L., Valle, G., and Ballarin, L. 2016. Transcriptome dynamics in the asexual cycle of the chordate *Botryllus schlosseri*. *BMC Genomics*. 17:275.

Carlton J.T. 2005. Setting ascidian invasions on the global stage. In *Proc Int Invasive Sea Squirt Conf.* 21–22. Woods Hole Oceanographic Institution, Woods Hole, MA.

Carman, M.R., Lindell, S., Green-Beach, E., and Starczak, V.R. 2016. Treatments to eradicate invasive tunicate fouling from blue mussel seed and aquaculture socks. *Management of Biological Invasions*. 7:101–110.

Castilla, J.C., Uribe, M., Bahamonde, N., Clarke, M., Desqueyroux-Faúndez, R., Kong, I., Moyano, H., Rozbaczylo, N., Santelices, B., Valdovinos, C., and Zavala, P. 2005. Down under the Southeastern Pacific: Marine non-indigenous species in Chile. *Biological Invasions*. 7:213–232.

Chadwick-Furman, N.E., and Weissman, I.L. 1995. Life histories and senescence of *Botryllus schlosseri* (chordata, ascidiacea) in Monterey Bay. *Biological Bulletin*. 189:36–41.

Cima, F., Ballarin, L., and Sabbadin, A. 1996. New data on phagocytes and phagocytosis in the compound ascidian *Botryllus schlosseri* (tunicata, ascidiacea). *Italian Journal of Zoology*. 63:357–364.

Cima, F., Basso, G., and Ballarin, L. 2003. Apoptosis and phosphatidylserine-mediated recognition during the take-over phase of the colonial life-cycle in the ascidian *Botryllus schlosseri*. *Cell and Tissue Research*. 312:369–376.

Cima, F., Manni, L., Basso, G., Fortunato, E., Accordi, B., Schiavon, F., and Ballarin, L. 2010. Hovering between death and life: Natural apoptosis and phagocytes in the blastogenetic cycle of the colonial ascidian *Botryllus schlosseri*. *Developmental and Comparative Immunology*. 34:272–285.

Colombera, D. 1963. Chromosomes of *Botryllus schlosseri* (Ascidiacea). *La Ricerca scientifica. Parte 2, Rendiconti. Sezione B, Biologica*. 3:443–448.

Conboy, I.M., Conboy, M.J., and Rebo, J. 2015. Systemic problems: A perspective on stem cell aging and rejuvenation. *Aging*. 7:754–765.

Conboy, M.J., Conboy, I.M., and Rando, T.A. 2013. Heterochronic parabiosis: Historical perspective and methodological considerations for studies of aging and longevity. *Aging Cell*. 12:525–530.

Cooper, E.L., Rinkevich, B., Uhlenbruck, G., and Valembois, P. 1992. Invertebrate immunity: Another viewpoint. *Scandinavian Journal of Immunology*. 35:247–266.

Corey, D.M., Rosental, B., Kowarsky, M., Sinha, R., Ishizuka, K.J., Palmeri, K.J., Quake, S.R., Voskoboynik, A., and Weissman, I.L. 2016. Developmental cell death programs license cytotoxic cells to eliminate histocompatible partners. *Proceedings of the National Academy of Sciences of the United States of America*. 113:6520–6525.

Della Valle, A. 1881. Nuove contribuzioni alla storia naturale delle ascidie composte del Golfo di Napoli. *Atti della Accademia nazionale dei Lincei* (Ser. 3) 10:431–498.

De Tomaso, A.W., Saito, Y., Ishizuka, K.J., Palmeri, K.J., and Weissman, I.L. 1998. Mapping the genome of a model protochordate. i.a low resolution genetic map encompassing the fusion/histocompatibility (Fu/HC) locus of *Botryllus schlosseri*. *Genetics*. 149:277–287.

Eichwald, E.J., Lustgraaf, E.C., and Strainer, M. 1959. Genetic factors in parabiosis. *Journal of the National Cancer Institute*. 23:1193–1213.

Epelbaum, A., Herborg, L.M., Therriault, T.W., and Pearce, C.M. 2009. Temperature and salinity effects on growth, survival, reproduction, and potential distribution of two non-indigenous botryllid ascidians in British Columbia. *Journal of Experimental Marine Biology and Ecology*. 369:43–52.

Fang, C.H. 1971. Improvement of the method of rats parabiosis with aortic anastomoses. *Acta Medica Okayama*. 25:597–603.

Fitridge, I., Dempster, T., Guenther, J., and de Nys, R. 2012. The impact and control of biofouling in marine aquaculture: A review. *Biofouling*. 28:649–669.

Franchi, N., Ballin, F., and Ballarin, L. 2017. Protection from oxidative stress in immunocytes of the colonial ascidian *Botryllus schlosseri*: Transcript characterization and expression studies. *Biological Bulletin*. 232:45–57.

Ganin, M. 1870. Neue Tatsachen aus der Entwicklungsgeschichte der Ascidien. *Zeitschrift für wissenschaftliche Zoologie*. 20:512–518.

Gao, Y., Li, S., and Zhan, A. 2018. Genome-wide single nucleotide polymorphisms (SNPs) for a model invasive ascidian *Botryllus schlosseri*. *Genetica*. 146:227–234.

Gärtner, J. 1774. Zoophyta, quaedam minuta. In *Specilegia Zoologia fasc*, ed. Pallas, P.S., Lange, Berlin. 10:24–41.

Gasparini, F., Longo, F., Manni, L., Burighel, P., and Zaniolo, G. 2007. Tubular sprouting as a mode of vascular formation in a colonial ascidian (Tunicata). *Developmental Dynamics*. 236:719–731.

Gasparini, F., Manni, L., Cima, F., Zaniolo, G., Burighel, P., Caicci, F., Franchi, N., Schiavon, F., Rigon, F., Campagna, D., and Ballarin, L. 2014. Sexual and asexual reproduction in the colonial ascidian *Botryllus schlosseri*. *Genesis*. 16:1–16.

Giard, A.M. 1872. Recherches sur les ascidies composée ou synascidies. *Archives de zoologie expérimentale et générale*. 1:501–687.

Grasela, J.J., Pomponi, S.A., Rinkevich, B., and Grima, J. 2012. Efforts to develop a cultured sponge cell line: Revisiting an intractable problem. *In Vitro Cellular & Developmental Biology – Animal*. 48:12–20.

Grosberg, R.K., and Quinn, J.F. 1986. The genetic control and consequences of kin recognition by the larvae of a colonial marine invertebrate. *Nature*. 322:456–459.

Haeckel, E. 1899. *Kunstformen der Natur*. Verlag des Bibliograpischen Insitutts, Leipzig und Wien.

Hamada, H., Meno, C., Watanabe, D., and Saijoh, Y. 2002. Establishment of vertebrate left-right asymmetry. *Nature Reviews Genetics*. 3:103–113.

Herdman, E.C. 1924. *Botryllus*. Vol. 26. University Press.

Hirose, E. 2009. Ascidian tunic cells: Morphology and functional diversity of free cells outside the epidermis. *Invertebrate Biology*. 128:83–96.

Hirose, E., Saito, Y., and Watanabe, H. 1991. Tunic cell morphology and classification in botryllid ascidians. *Zoological Science*. 8:951–958.

Hjort, J. 1893. Über den Entwicklungscyclus der zusammengesetzen Ascidien. *Mittheilungen aus der Zoologischen Station zu Neapel*. 10:584–617.

Holley, M.C. 1986. Cell shape, spatial patterns of cilia, and mucus-net construction in the ascidian endostyle. *Tissue and Cell*. 18:667–684.

Inoue, J., Nakashima, K., and Satoh, N. 2019. ORTHOSCOPE analysis reveals the presence of the cellulose synthase gene in all tunicate genomes but not in other animal genomes. *Genes*. 10:1–9.

Izzard, C.S. 1973. Development of polarity and bilateral asymmetry in the palleal bud of *Botryllus schlosseri* (Pallas). *Journal of Morphology*. 139:1–26.

Izzard, C.S. 1974. Contractile filopodia and in vivo cell movement in the tunic of the ascidian, *Botryllus schlosseri*. *Journal of Cell Science*. 15:513–535.

Karahan, A., Douek, J., Paz, G., and Rinkevich, B. 2016. Population genetics features for persistent, but transient, *Botryllus schlosseri* (urochordata) congregations in a central Californian marina. *Molecular Phylogenetics and Evolution*. 101:19–31.

Katow, H., and Watanabe, H. 1978. Fine structure and possible role of ampullae on tunic supply and attachment in a compound ascidian, *Botryllus primigenus* Oka. *Journal of Ultrastructure Research*. 64:23–34.

Katow, H., and Watanabe, H. 1980. Fine structure of fusion reaction in compound ascidian *Botryllus primigenus* Oka. *Developmental Biology*. 76:1–14.

Kayiş, Ş. 2011. Ascidian tunicate, *Botryllus schlosseri* (Pallas, 1766) infestation on seahorse. *Bulletin of the European Association of Fish Pathologists*. 31:81–84.

Keyes, B.E., and Fuchs, E. 2018. Stem cells: Aging and transcriptional fingerprints. *Journal of Cell Biology*. 217:79–92.

Kimura, K., Tanaka, N., Nakamura, N., Takano, S., and Ohkuma, S. 2007. Knockdown of mitochondrial heat shock protein 70 promotes progeria-like phenotypes in *Caenorhabditis elegans*. *Journal of Biological Chemistry*. 282:5910–5918.

Kott, P. 2005. *Catalogue of Tunicata in Australian Waters*. Australian Biological Resources Study, Canberra.

Laird, D.J., Chang, W., Weissman, I.L., and Lauzon, R.J. 2005b. Identification of a novel gene involved in asexual organogenesis in the budding ascidian *Botryllus schlosseri*. *Developmental Dynamics*. 234:997–1005.

Laird, D.J., De Tomaso, A.W., and Weissman, I.L. 2005a. Stem cells are units of natural selection in a colonial ascidian. *Cell*. 123:1351–1360.

Lambert, C.C., and Lambert, G. 1998. Non-indigenous ascidians in Southern California harbors and marinas. *Marine Biology*. 130:675–688.

Lambert, C.C., and Lambert, G. 2003. Persistence and differential distribution of nonindigenous ascidians in harbors of the

Southern California Bight. *Marine Ecology Progress Series*. 259:145–161.

Lambert, G. 2001. A global overview of ascidian introductions and their possible impact on the endemic fauna. In *The Biology of Ascidians*, eds. Sawada, H., Yokosawa, H., and Lambert, C.C. 249–257. Springer, Tokyo.

Lapidot, Z., Paz, G., and Rinkevich, B. 2003. Monoclonal antibody specific to urochordate *Botryllus schlosseri* pyloric gland. *Marine Biotechnology*. 5:388–394.

Lauzon, R.J., Ishizuka, K.J., and Weissman, I.L. 1992. A cyclical, developmentally-regulated death phenomenon in a colonial urochordate. *Developmental Dynamics*. 194:71–83.

Lauzon, R.J., Ishizuka, K.J., and Weissman, I.L. 2002. Cyclical generation and degeneration of organs in a colonial urochordate involves crosstalk between old and new: A model for development and regeneration. *Developmental Biology*. 249:333–348.

Lauzon, R.J., Rinkevich, B., Patton, C.W., and Weissman, I.L. 2000. A morphological study of nonrandom senescence in a colonial urochordate. *Biological Bulletin*. 198:367–378.

Lin, Y., and Zhan, A. 2016. Population genetic structure and identification of loci under selection in the invasive tunicate, *Botryllus schlosseri*, using newly developed EST-SSRs. *Biochemical Systematics and Ecology*. 66:331–336.

Linnaeus, C. 1767. *Systema naturae per regna tria naturae, secundum classes, ordines, genera, species, cum characteribus, differentiis, synonymis, locis 12th Edition*. Salvius, Stockholm.

Liu, K., Waskow, C., Liu, X., Yao, K., Hoh, J., and Nussenzweig, M. 2007. Origin of dendritic cells in peripheral lymphoid organs of mice. *Nature Immunology*. 8:578–583.

Locke, A., Hanson, J.M., MacNair, N.G., and Smith, A.H. 2009. Rapid response to non-indigenous species. 2: Case studies of invasive tunicates in Prince Edward Island. *Aquatic Invasions*. 4:249–258.

López-Legentil, S., Turon, X., and Planes, S. 2006. Genetic structure of the star sea squirt, *Botryllus schlosseri*, introduced in southern European harbours. *Molecular Ecology*. 15:3957–3967.

Luisetto, M., Almukthar, N., and Hamid, G.A. 2020. Regeneration abilities of vertebrates and invertebrates and relationship with pharmacological research: Hypothesis of genetic evolution work and micro-environment inhibition role. *Journal of Cell Biology and Cell Metabolism*. 7:1–21.

Magor, B.G., De Tomaso, A.W., Rinkevich, B., and Weissman, I.L. 1999. Allorecognition in colonial tunicates: Protection against predatory cell lineages? *Immunological Revies*. 167:69–79.

Manni, L., Anselmi, C., Cima, F., Gasparini, F., Voskoboynik, A., Martini, M., Peronato, A., Burighel, P., Zaniolo, G., and Ballarin, L. 2019. Sixty years of experimental studies on the blastogenesis of the colonial tunicate *Botryllus schlosseri*. *Developmental Biology*. 448:293–308.

Manni, L., Gasparini, F., Hotta, K., Ishizuka, K.J., Ricci, L., Tiozzo, S., Voskoboynik, A., and Dauga, D. 2014. Ontology for the asexual development and anatomy of the colonial chordate *Botryllus schlosseri*. *PLoS One*. 9:1–15.

Manni, L., Zaniolo, G., and Burighel, P. 1993. Egg envelope cytodifferentiation in the colonial ascidian *Botryllus schlosseri* (tunicata). *Acta Zoologica Stockholm*. 74:103–113.

Manni, L., Zaniolo, G., Cima, F., Burighel, P., and Ballarin, L. 2007. *Botryllus schlosseri*: A model ascidian for the study of asexual reproduction. *Developmental Dynamics*. 236:335–352.

Martin, J.L., LeGresley, M.M., Thorpe, B., and McCurdy, P. 2011. Non-indigenous tunicates in the Bay of Fundy, eastern Canada (2006–2009). *Aquatic Invasions*. 6:405–412.

Meenakshi, V.K., and Senthamarai, S. 2006. First report on two species of ascidians to represent the genus *Botryllus* Gaertner, 1774 from Indian waters. *Journal of the Marine Biological Association of India*. 48:100–102.

Metschnikow, E. 1869. Entwicklungsgeschichtliche Beiträge. VII. Über die Larven und Knospen von *Botryllus*. *Bulletin de l'Académie impériale des sciences de St.-Pétersbourg*. 13:291–293.

Milkman, R. 1967. Genetic and developmental studies on *Botryllus schlosseri*. *Biological Bulletin*. 132:229–243.

Milkman, R., and Borgmann, M. 1963. External fertilization of *Botryllus schlosseri*. *Biological Bulletin*. 125:385.

Millar, R.H. 1955. On a collection of ascidians from South Africa. *Proceedings of the Zoological Society of London*. 125:169–221.

Mukai, H. 1977. Comparative studies on the structure of reproductive organs of four botryllid ascidians. *Journal of Morphology*. 152:363–379.

Mukai, H., and Watanabe, H. 1976. Studies on the formation of germ cells in a compound ascidian *Botryllus primigenus* Oka. *Journal of Morphology*. 148:337–362.

Müller, W.E.G., Pancer, Z., and Rinkevich, B. 1994. Molecular cloning and localization of a novel serine protease from the colonial tunicate *Botryllus schlosseri*. *Molecular Marine Biology and Biotechnology*. 3:70–77.

Munday, R., Rodriguez, D., Di Maio, A., Kassmer, S., Braden, B., Taketa, D.A., Langenbacher, A., and De Tomaso, A.W. 2015. Aging in the colonial chordate, *Botryllus schlosseri*. *Invertebrate Reproduction and Development*. 59:45–50.

Nakashima, K., Sugiyama, J., and Satoh, N. 2008. A spectroscopic assessment of cellulose and the molecular mechanisms of cellulose biosynthesis in the ascidian *Ciona intestinalis*. *Marine Genomics*. 1:9–14.

Nakashima, K., Yamada, L., Satou, Y., Azuma, J.I., and Satoh, N. 2004. The evolutionary origin of animal cellulose synthase. *Development Genes and Evolution*. 214:81–88.

Nydam, M.L., Giesbrecht, K.B., and Stephenson, E.E. 2017. Origin and dispersal history of two colonial ascidian clades in the *Botryllus schlosseri* species complex. *PLoS One*. 1:1–30.

Ogasawara, M., Lauro, R.D.I., and Satoh, N. 1999. Ascidian homologs of mammalian thyroid peroxidase genes are expressed in the thyroid-equivalent region of the endostyle. *Journal of Experimental Zoology Part B*. 285:158–169.

Oka, H., and Watanabe, H. 1957. Colony-specificity in compound ascidians as tested by fusion experiments. *Proceedings of the Japan Academy*. 33:657–659.

Oka, H., and Watanabe, H. 1960. Problems of colony-specificity in compound ascidians. *Bulletin of the Marine Biological Station of Asamushi*. 10:153–155.

Oren, M., Douek, J., Fishelson, Z., and Rinkevich, B. 2007. Identification of immune-relevant genes in histoincompatible rejecting colonies of the tunicate *Botryllus schlosseri*. *Developmental & Comparative Immunology*. 31:889–902.

Oren, M., Paz, G., Douek, J., Rosner, A., Amar, K.O., and Rinkevich, B. 2013. Marine invertebrates cross phyla comparisons reveal highly conserved immune machinery. *Immunobiology*. 218:484–495.

Oren, M., Paz, G., Douek, J., Rosner, A., Fishelson, Z., Goulet, T.L., Henckel, K., and Rinkevich, B. 2010. "Rejected" vs.

"rejecting" transcriptomes in allogeneic challenged colonial urochordates. *Molecular Immunology*. 47:2083–2093.

Orensanz, J.M., Schwindt, E., Pastorino, G., Bortolus, A., Casas, G., Darrigran, G., Elías, R., López Gappa, J.J., Obenat, S., Pascual, M., Penchaszadeh, P., Piriz, M.L., Scarabino, F., Spivak, E.D., and Vallarino, E.A. 2002. No longer the pristine confines of the world ocean: A survey of exotic marine species in the Southwestern Atlantic. *Biological Invasions*. 4:115–143.

Pallas, P.S. 1766. *Elenchus zoophytorum sistens generum adumbrationes generaliores et specierum cognitarum succinctas descriptiones cum selectis auctorum synonymis*. van Cleef, The Hague.

Pancer, Z., Gershon, H., and Rinkevich, B. 1995. Coexistence and possible parasitism of somatic and germ cell lines in chimeras of the colonial urochordate *Botryllus schlosseri*. *Biological Bulletin*. 189:106–112.

Paz, G., Douek, J., Mo, C., Goren, M., and Rinkevich, B. 2003. Genetic structure of *Botryllus schlosseri* (Tunicata) populations from the Mediterranean coast of Israel. *Marine Ecology Progress Series*. 250:153–162.

Pérez-Portela, R., Bishop, J.D.D., Davis, A.R., and Turon, X. 2009. Phylogeny of the families Pyuridae and Styelidae (Stolidobranchiata, Ascidiacea) inferred from mitochondrial and nuclear DNA sequences. *Molecular Phylogenetics and Evolution*. 50:560–570.

Pizon, A. 1893. Histoire de la blastogénèse chez les *Botryllides*. *Annales des sciences naturelles Series 7*. 14:1–386.

Qarri, A., Rosner, A., Rabinowitz, C., and Rinkevich, B. 2020. UV-B radiation bearings on ephemeral soma in the shallow water tunicate *Botryllus schlosseri*. *Ecotoxicology and Environmental Safety*. 196:110489.

Rabinowitz, C., Alfassi, G., and Rinkevich, B. 2009. Further portrayal of epithelial monolayers emergent de novo from extirpated ascidians palleal buds. *In Vitro Cellular & Developmental Biology – Animal*. 45:334–342.

Rabinowitz, C., and Rinkevich, B. 2003. Epithelial cell cultures from *Botryllus schlosseri* palleal buds: Accomplishments and challenges. *Methods in Cell Science*. 25:137–148.

Rabinowitz, C., and Rinkevich, B. 2004. *In vitro* delayed senescence of extirpated buds from zooids of the colonial tunicate *Botryllus schlosseri*. *Journal of Experimental Biology*. 207:1523–1532.

Rabinowitz, C., and Rinkevich, B. 2011. De novo emerged stemness signatures in epithelial monolayers developed from extirpated palleal buds. *In Vitro Cellular and Developmental Biology – Animal*. 47:26–31.

Reem, E., Douek, J., Katzir, G., and Rinkevich, B. 2013a. Long-term population genetic structure of an invasive urochordate: The ascidian *Botryllus schlosseri*. *Biological Invasions*. 15:225–241.

Reem, E., Douek, J., Paz, G., Katzir, G., and Rinkevich, B. 2017. Phylogenetics, biogeography and population genetics of the ascidian *Botryllus schlosseri* in the Mediterranean Sea and beyond. *Molecular Phylogenetics and Evolution*. 107:221–231.

Reem, E., Mohanty, I., Katzir, G., and Rinkevich, B. 2013b. Population genetic structure and modes of dispersal for the colonial ascidian *Botryllus schlosseri* along the Scandinavian Atlantic coasts. *Marine Ecology Progress Series*. 485:143–154.

Renier, S.A. 1793. Sopra il *Botrillo* piantanimale marino. Opuscoli scelti sulle scienze e sulle arti. *Milano*. 4:256–267.

Ricci, L., Cabrera, F., Lotito, S., and Tiozzo, S. 2016a. Redeployment of germ layers related TFs shows regionalized expression during two non-embryonic developments. *Developmental Biology*. 416:235–248.

Ricci, L., Chaurasia, A., Lapébie, P., Dru, P., Helm, R.R., Copley, R.R., and Tiozzo, S. 2016b. Identification of differentially expressed genes from multipotent epithelia at the onset of an asexual development. *Scientific Reports*. 6:1–10.

Richmond, P.A. 1991. Occurrence and functions of native cellulose. In *Biosynthesis and Biodegradation of Cellulose*, eds. Haigler, C.H., and Weimer, P.J. 5–23. Marcel Dekker, Inc., New York.

Rinkevich, B. 1992. Aspects of the incompatibility in botryllid ascidians. *Animal Biology*. 1:17–28.

Rinkevich, B. 1993. Immunological resorption in *Botryllus schlosseri* (Tunicata) chimeras is characterized by multilevel hierarchical organization of histocompatibility alleles: A speculative endeavor. *Biological Bulletin*. 184:342–345.

Rinkevich, B. 1995. Morphologically related allorecognition assays in botryllid ascidians. In *Techniques in Fish Immunology. 4: Immunology and Pathology of Aquatic Invertebrates*, eds. Stolen, J.S., Fletcher, T.C., Smith, S.A., Zelikoff, J.T., Kaattari, S.L., Anderson, R.S., Söderhäll, K., and Weeks-Perkins, B.A. 17–21. SOS Publications, Fair Haven, NJ.

Rinkevich, B. 1996. Bi-versus multichimerism in colonial urochordates: A hypothesis for links between natural tissue transplantation, allogenetics and evolutionary ecology. *Experimental and Clinical Immunogenetics*. 13:61–69.

Rinkevich, B. 1999a. Invertebrates versus vertebrates innate immunity: In the light of evolution. *Scandinavian Journal of Immunology*. 50:456–460.

Rinkevich, B. 1999b. Cell cultures from marine invertebrates: Obstacles, new approaches and recent improvements. *Journal of Biotechnology*. 70:133–153.

Rinkevich, B. 2002a. The colonial urochordate *Botryllus schlosseri*: From stem cells and natural tissue transplantation to issues in evolutionary ecology. *Bio Essays*. 24:730–740.

Rinkevich, B. 2002b. Germ cell parasitism as an ecological and evolutionary puzzle: Hitchhiking with positively selected genotypes. *Oikos*. 96:25–30.

Rinkevich, B. 2004a. Primitive immune systems: Are your ways my ways? *Immunological Reviews*. 198:25–35.

Rinkevich, B. 2004b. Will two walk together, except they have agreed? *Journal of Evolutionary Biology*. 17:1178–1179.

Rinkevich, B. 2005a. Rejection patterns in botryllid ascidian immunity: The first tier of allorecognition. *Canadian Journal of Zoology*. 83:101–121.

Rinkevich, B. 2005b. Natural chimerism in colonial urochordates. *Journal of Experimental Marine Biology and Ecology*. 322:93–109.

Rinkevich, B. 2005c. Marine invertebrate cell cultures: New millennium trends. *Marine Biotechnology*. 7:429–439.

Rinkevich, B. 2011a. Quo vadis chimerism? *Chimerism*. 2:1–5.

Rinkevich, B. 2011b. Cell cultures from marine invertebrates: New insights for capturing endless stemness. *Marine Biotechnology*. 13:345–354.

Rinkevich, B. 2017. Senescence in modular animals: Botryllid ascidians as a unique aging system. In *The Evolution of Senescence in the Tree of Life*, eds. Salguero-Gomez, R., Shefferson, R., and Jones, O. 220–237. Cambridge University Press, Cambridge.

Rinkevich, B. 2019. The tail of the underwater phoenix. *Developmental Biology*. 448:291–292.

Rinkevich, B., Lauzon, R.J., Brown, B.W.M., and Weissman, I.L. 1992b. Evidence for a programmed life span in a colonial protochordate. *Proceedings of the National Academy of Sciences of the United States of America*. 89:3546–3550.

Rinkevich, B., Porat, R., and Goren, M. 1998a. Ecological and life history characteristics of *Botryllus schlosseri* (Tunicata) populations inhabiting undersurface shallow-water stones. *Marine Ecology*. 19:129–145.

Rinkevich, B., Porat, R., and Goren, M. 1998b. On the development and reproduction of *Botryllus schlosseri* (Tunicata) colonies from the eastern Mediterranean Sea: Plasticity of life history traits. *Invertebrate Reproduction & Development*. 34:207–218.

Rinkevich, B., Paz, G., Douek, J., and Ben-Shlomo, R. 2001. Allorecognition and microsatellite allele polymorphism of *Botryllus schlosseri* from the Adriatic Sea. In *The Biology of Ascidians*, eds. Sawada H., Yokosawa H., Lambert C.C. 426–435. Springer, Tokyo.

Rinkevich, B., and Rabinowitz, C. 1994. Acquiring embryo-derived cell cultures and aseptic metamorphosis of larvae from the colonial protochordate *Botryllus schlosseri. Invertebrate Reproduction & Development*. 25:59–72.

Rinkevich, B., and Rabinowitz, C. 1997. Initiation of epithelial cell cultures from palleal buds of *Botryllus schlosseri*, a colonial tunicate. *In Vitro Cellular & Developmental Biology - Animal*. 33:422–424.

Rinkevich, B., and Rinkevich, Y. 2013. The "stars and stripes" metaphor for animal regeneration-elucidating two fundamental strategies along a continuum. *Cells*. 2:1–18.

Rinkevich, B., and Saito, Y. 1992. Self-nonself recognition in the colonial protochordate *Botryllus schlosseri* from Mutsu Bay, Japan. *Zoological Science*. 9:983–988.

Rinkevich, B., Saito, Y., and Weissman, I.L. 1993. A colonial invertebrate species that displays a hierarchy of allorecognition responses. *Biological Bulletin*. 184:79–86.

Rinkevich, B., and Shapira, M. 1998. An improved diet for inland broodstock and the establishment of an inbred line from *Botryllus schlosseri*, a colonial sea squirt (Ascidiacea). *Aquatic Living Resources*. 11:163–171.

Rinkevich, B., Shapira, M., Weissman, I.L., and Yasunory, S. 1992a. Allogeneic responses between three remote populations of the cosmopolitan ascidian *Botryllus schlosseri. Zoological Science*. 9:989–994.

Rinkevich, B., and Weissman, I.L. 1987a. A long-term study on fused subclones in the ascidian *Botryllus schlosseri*: The resorption phenomenon (Protochordata: Tunicata). *Journal of Zoology*. 213:717–733.

Rinkevich, B., and Weissman, I.L. 1987b. Chimeras in colonial invertebrates: A synergistic symbiosis or somatic- and germ-cell parasitism? *Symbiosis*. 4:117–134.

Rinkevich, B., and Weissman, I.L. 1987c. The fate of *Botryllus* (Ascidiacea) larvae cosettled with parental colonies: Beneficial or deleterious consequences? *Biological Bulletin*. 173:474–488.

Rinkevich, B., and Weissman, I.L. 1988. Retreat growth in the ascidian *Botryllus schlosseri*: The consequences of non-self recognition. In *Invertebrate Historecognition*, ed. Grosberg, R.K. 93–109. Plenum Press, New York.

Rinkevich, B., and Weissman, I.L. 1989. Variation in the outcomes following chimera formation in the colonial tunicate *Botryllus schlosseri. Bulletin of Marine Science*. 45:213–227.

Rinkevich, B., and Weissman, I.L. 1990. *Botryllus schlosseri* (Tunicata) whole colony irradiation: Do senescent zooid resorption and immunological resorption involve similar recognition events? *Journal of Experimental Zoology*. 253:189–201.

Rinkevich, B., and Weissman, I.L. 1991. Interpopulational allogeneic reactions in the colonial protochordate *Botryllus schlosseri. International Immunology*. 3:1265–1272.

Rinkevich, B., and Weissman, I.L. 1992a. Chimeras vs genetically homogeneous individuals: Potential fitness costs and benefits. *Oikos*. 63:119–124.

Rinkevich, B., and Weissman, I.L. 1992b. Allogeneic resorption in colonial protochordates: Consequences of nonself recognition. *Developmental & Comparative Immunology*. 16:275–286.

Rinkevich, B., and Weissman, I.L. 1998. Transplantation of Fu/HC-incompatible zooids in *Botryllus schlosseri* results in chimerism. *Biological Bulletin*. 195:98–106.

Rinkevich, B., Weissman, I.L., and Shapira, M. 1994. Alloimmune hierarchies and stress-induced reversals in the resorption of chimeric protochordate colonies. *Proceedings of the Royal Society of London Series B*. 258:215–220.

Rinkevich, B., and Yankelevich, I. 2004. Environmental split between germ cell parasitism and somatic cell synergism in chimeras of a colonial urochordate. *Journal of Experimental Biology*. 207:3531–3536.

Rinkevich, Y., Matranga, V., and Rinkevich. B. 2009. Stem cells in aquatic invertebrates: Common premises and emerging unique themes. In *Stem Cells in Marine Organisms*, eds. Rinkevich, B., and Matranga, B. 60–103. Springer, the Netherlands.

Rinkevich, Y., Voskoboynik, A., Rosner, A., Rabinowitz, C., Paz, G., Oren, M., Douek, J., Alfassi, G., Moiseeva, E., Ishizuka, K.J., Palmeri, K.J., and Weissman, I.L. 2013. Repeated, long-term cycling of putative stem cells between niches in a basal chordate. *Developmental Cell*. 24:76–88.

Rodriguez, D., Kassmer, S.H., and De Tomaso. A.W. 2016. Gonad development and hermaphroditism in the ascidian *Botryllus schlosseri. Molecular Reproduction & Development*. 84: 158–170.

Rodriguez, D., Sanders, E.N., Farell, K., Langenbacher, A.D., Taketa, D.A., Hopper, M.R., Kennedy, M., Gracey, A., and De Tomaso, A.W. 2014. Analysis of the basal chordate *Botryllus schlosseri* reveals a set of genes associated with fertility. *BMC Genomics*. 15:1183.

Rondelet, G. 1555. *Universae aquatilium historiae pars altera, cum veris ipsorum imaginibus*. M Bonhomme, Lugduni.

Rosental, B., Kowarsky, M., Seita, J., Corey, D.M., Ishizuka, K.J., Palmeri, K.J., Chen, S., Sinha, R., Okamoto, J., Mantalas, G., Manni, L., Raveh, T., Clarke, D.N., Quake, S.R., Weissman, I.L., Tsai, J.M., Newman, A.M., Neff, N.F., and Garry, P. 2018. Complex mammalian-like haematopoietic system found in a colonial chordate. *Nature*. 564:425–429.

Rosner, A., Alfassi, G., Moiseeva, E., Paz, G., Rabinowitz, C., Lapidot, Z., Douek, J., Haim, A., and Rinkevich, B. 2014. The involvement of three signal transduction pathways in botryllid ascidian astogeny, as revealed by expression patterns of representative genes. *International Journal of Developmental Biology*. 58:677–692.

Rosner, A., Kravchenko, O., and Rinkevich, B. 2019. IAP genes partake weighty roles in the astogeny and whole body regeneration in the colonial urochordate *Botryllus schlosseri. Developmental Biology*. 448:320–341.

Rosner, A., Moiseeva, E., Rabinowitz, C., and Rinkevich, B. 2013. Germ lineage properties in the urochordate *Botryllus schlosseri:* From markers to temporal niches. *Developmental Biology*. 384:356–374.

Rosner, A., Paz, G., and Rinkevich, B. 2006. Divergent roles of the DEAD-box protein BS-PL10, the urochordate homologue of human DDX3 and DDX3Y proteins, in colony astogeny and ontogeny. *Developmental Dynamics*. 235:1508–1521.

Rosner, A., Rabinowitz, C., Moiseeva, E., Voskoboynik, A., and Rinkevich, B. 2007. BS-cadherin in the colonial urochordate *Botryllus schlosseri*: One protein, many functions. *Developmental Biology*. 304:687–700.

Rosner, A., and Rinkevich, B. 2007. The DDX3 subfamily of the DEAD box helicases: Divergent roles as unveiled by studying different organisms and *in vitro* assays. *Current Medicinal Chemistry*. 14:2517–2525.

Ruppert, E.E., Fox, R.S., and Barnes, R.D. 2004. Chapter 29: Chordata. In *Invertebrate Zoology: A Functional Evolutionary Approach*. Vol. 6, 941–951. Saunders College Publishing, New York.

Sabbadin, A. 1955. Osservazioni sullo sviluppo, l'accrescimento e la riproduzione di *Botryllus schlosseri* (Pallas) in condizioni di laboratorio. *Bollettino di. Zoologia*. 22:243–263.

Sabbadin, A. 1962. Le basi genetiche della capacità di fusione fra colonie in *Botryllus schlosseri* (Ascidiacea). *Rendiconti Accademia Nazionale dei Lincei*. 32:1031–1035.

Sabbadin, A. 1971. Self- and cross-fertilization in the compound ascidian *Botryllus schlosseri*. *Developmental Biology*. 24:379–391.

Sabbadin, A. 1977.Linkage between two loci controlling colour polymorphism in the colonial ascidian, *Botryllus schlosseri*. *Experientia*. 33:876–877.

Sabbadin, A., and Astorri, C. 1988. Chimeras and histocompatibility in the colonial ascidian *Botryllus schlosseri*. *Developmental & Comparative Immunology*. 12:737–747.

Sabbadin, A., and Zaniolo, G. 1979. Sexual differentiation and germ cell transfer in the colonial ascidian *Botryllus schlosseri*. *Journal of Experimental Zoology*. 207:289–304.

Sabbadin, A., Zaniolo, G., and Majone, F. 1975. Determination of polarity and bilateral asymmetry in palleal and vascular buds of the ascidian *Botryllus schlosseri*. *Developmental Biology*. 46:79–87.

Savigny, J.C. 1816. *Mémoires sur les animaux sans vertèbres 2*. Doufour, Paris.

Schlosser, J.A., and Ellis, J. 1756. An account of a curious, fleshy, coral-like substance: In a letter to Mr. Peter Collinson, F. R. S. from Dr. John Albert Schlosser M. D. F. R. S. with some observations on it communicated to Mr. Collinson by Mr. John Ellis, F. R. S. *Philosophical Transections*. 49:449–452.

Scofield, V.L., Schlumpberger, J.M., West, L.A., and Weissman, I.L. 1982. Protochordate allorecognition is controlled by a MHC-like gene system. *Nature*. 295:499–502.

Simon-Blecher, N. 2003. Aspects of allorecognition in botryllid ascidians. PhD dissertation, Faculty of Life Sciences, Bar Ilan Univerity, Israel.

Simon-Blecher, N., Achituv, Y., and Rinkevich, B. 2004. Protochordate concordant xenotransplantation settings reveal outbreaks of donor cells and divergent life span traits. *Developmental & Comparative Immunology*. 28:983–991.

Singer, M.A. 2016. Stem cells and aging. *Stem Cell & Translational Investigation*. 3:1–8.

Smith, M.J., and Dehnel, P.A. 1971. The composition of tunic from four species of ascidians. *Comperative Biochemistry & Physiology Part B: Comparative Biochemistry*. 40:615–622.

Spallanzani, L. 1784. Giornale di esperienze sulla fauna marina della laguna di Chioggia. *Biblioteca Digitale Reggiana*. Mss. Reggio. B57.

Stewart-Savage, J., Phillippi, A., and Yund, P.O. 2001. Delayed insemination results in embryo mortality in a brooding ascidian. *Biological Bulletin*. 201:52–58.

Stoner, D.S., Ben-Shlomo, R., Rinkevich, B., and Weissman, I.L. 2002. Genetic variability of *Botryllus schlosseri* invasions to the east and west coasts of the USA. *Marine Ecology Progress Series*. 243:93–100.

Stoner, D.S., Rinkevich, B., and Weissman, I.L. 1999. Heritable germ and somatic cell lineage competitions in chimeric colonial protochordates. *Proceedings of the Nationall Academy of Sciences of the United States of America*. 96:9148–9153.

Stoner, D.S., and Weissman, I.L. 1996. Somatic and germ cell parasitism in a colonial ascidian: Possible role for a highly polymorphic allorecognition system. *Proceedings of the Nationall Academy of Sciences of the United States of America*. 93:15254–15259.

Taneda, Y., Saito, Y., and Watanabe, H. 1985. Self or non-self discrimination in ascidians. *Zoological Science*. 2:433–442.

Taneda, Y., and Watanabe, H. 1982a. Studies on colony specificity in the compound ascidian, *Botryllus primigenus* Oka. 1: Initiation of "nonfusion" reaction with special reference to blood cells infiltration. *Developmental & Comparative Immunology*. 6:43–52.

Taneda, Y., and Watanabe, H. 1982b. Effects of X-irradiation on colony specificity in the compound ascidian, *Botryllus primigenus* Oka. *Developmental & Comparative Immunology*. 6:665–673.

Tiozzo, S., Christiaen, L., Deyts, C., Manni, L., Joly, J.S., and Burighel, P. 2005. Embryonic versus blastogenetic development in the compound ascidian *Botryllus schlosseri*: Insights from Pitx expression patterns. *Developmental Dynamics*. 232:468–478.

Tiozzo, S., and De Tomaso, A.W. 2009. Functional analysis of Pitx during asexual regeneration in a basal chordate. *Evolutionary Development*. 11:152–162.

Van Name, W.G. 1945. The North and South American ascidians. *B Am Mus Mat Hist*. 84.

Vogt, G. 2012. Hidden treasures in stem cells of indeterminately growing bilaterian invertebrates. *Stem Cell Reviews and Reports*. 8:305–317.

Volpe, E.P., and Earley, E.M. 1970. Somatic cell mating and segregation in chimeric frogs. *Science*. 168:850–852.

Voskoboynik, A., Neff, N.F., Sahoo, D., Newman, A.M., Pushkarev, D., Koh, W., Passarelli, B., Fan, H.C., Mantalas, G.L., Palmeri, K.J., Ishizuka, K.J., Gissi, C., Griggio, F., Ben-Shlomo, R., Corey, D.M., Penland, L., White, R.A., Weissman, I.L., and Quake, S.R. 2013a. The genome sequence of the colonial chordate, *Botryllus schlosseri*. *eLife*. 2:1–24.

Voskoboynik, A., Newman, A.M., Corey, D.M., Sahoo, D., Pushkarev, D., Neff, N.F., Passarelli, B., Koh, W., Ishizuka, K.J., Palmeri, K.J., Dimov, I.K., Keasar, C., Fan, H.C., Mantalas, G.L., Sinha, R., Penland, L., Quake, S.R., and Weissman, I.L. 2013b. Identification of a colonial chordate histocompatibility gene. *Science*. 341:384–387.

Voskoboynik, A., Reznick, A.Z., and Rinkevich, B. 2002. Rejuvenescence and extension of an urochordate life span following a single, acute administration of an anti-oxidant, butylated hydroxytoluene. *Mechanisms of Ageing and Development*. 123:1203–1210.

Voskoboynik, A., Rinkevich, B., Weiss, A., Moiseeva, E., and Reznick, A.Z. 2004. Macrophage involvement for successful degeneration of apoptotic organs in the colonial urochordate *Botryllus schlosseri*. *Journal of Experimental Biology*. 207:2409–2416.

Voskoboynik, A., Rinkevich, B., and Weissman, I.L. 2009. Stem cells, chimerism and tolerance: Lessons from mammals and ascidians. In *Stem Cells in Marine Organisms*, eds. Rinkevich, B., and Matranga, B. 281–308. Springer, the Netherlands.

Voskoboynik, A., Simon-Blecher, N., Soen, Y., Rinkevich, B., De Tomaso, A.W., Ishizuka, K.J., and Weissman, I.L. 2007. Striving for normality: Whole body regeneration through a series of abnormal generations. *FASEB Journal*. 21:1335–1344.

Voskoboynik, A., Soen, Y., Rinkevich, Y., Rosner, A., Ueno, H., Reshef, R., Ishizuka, K.J., Palmeri, K.J., Moiseeva, E., Rinkevich, B., and Weissman, I.L. 2008. Identification of the endostyle as a stem cell niche in a colonial chordate. *Cell Stem Cell*. 3:456–464.

Voskoboynik, A., and Weissman, I.L. 2015. *Botryllus schlosseri*, an emerging model for the study of aging, stem cells, and mechanisms of regeneration. *Invertebrate Reproduction & Development*. 59:33–38.

Watanabe, H. 1953. Studies on the regulation in fused colonies in *Botryllus primigenus* (Ascidiae Compositae). *Science reports of the Tokyo Bunrika Daigaku. Section B*. 7:183–198.

Watterson, R.L. 1945. Asexual reproduction in the colonial tunicate, *Botryllus schlosseri* (pallas) savigny, with special reference to the developmental history of intersiphonal bands of pigment cells. *Biological Bulletin*. 88:71–103.

Weissman, I.L. 2015. Stem cells are units of natural selection for tissue formation, for germline development, and in cancer development. *Proceedings of the National Academy of Sciences of the United States of America*. 112:8922–8928.

Weissman, I.L., Saito, Y., and Rinkevich, B. 1990. Allorecognition histocompatibility in a protochordate species: Is the relationship to MHC somatic or structural? *Immunological Reviews*. 113:227–241.

Yaguchi, T., Aida, S., Kaul, S.C., and Wadhwa, R. 2007. Involvement of mortalin in cellular senescence from the perspective of its mitochondrial import, chaperone, and oxidative stress management functions. *Annals of the New York Academy of Sciences*. 1100:306–311.

Yokoyama, K., Fukumoto, K., Murakami, T., Harada, S., Hosono, R., Wadhwa, R., Mitsui, Y., and Ohkuma, S. 2002. Extended longevity of *Caenorhabditis elegans* by knocking in extra copies of hsp70F, a homolog of mot-2 (mortalin)/mthsp70/Grp75. *FEBS Letters*. 516:53–57.

Yund, P.O., Collins, C., and Johnson, S.L. 2015. Evidence of a native Northwest Atlantic COI Haplotype clade in the cryptogenic colonial ascidian *Botryllus schlosseri*. *Biological Bulletin*. 228:201–216.

Zaniolo, G. 1981. Histology of the ascidian *Botryllus schlosseri* tunic: In particular, the test cells. *Italian Journal of Zoology*. 48:169–178.

Zaniolo, G., Burighel, P., and Martinucci, G. 1987. Ovulation and placentation in *Botryllus schlosseri* (ascidiacea): An ultrastructural study. *Canadian Journal of Zoology*. 65:1181–1190.

22 Cyclostomes (Lamprey and Hagfish)

Fumiaki Sugahara

CONTENTS

22.1 INTRODUCTION

22.1.1 Cyclostomes for Evolutionary Research of Vertebrates

Living jawless fish diverged from a common vertebrate ancestor over 500 million years ago (mya). They comprise two groups, lampreys and hagfish, which form the monophyletic group Cyclostomata based on molecular phylogenetic analyses. Cyclostomes are important model organisms for understanding early vertebrate evolution because they retain many features that ancient jawless vertebrates had. However, it should be noted that since they are not "ancestral animals", cyclostomes lived independently from the jawed vertebrate (or gnathostome) lineages following divergence and thus possess independently evolved traits. Therefore, careful comparison of each trait among lampreys, hagfish and jawed vertebrates would allow us to determine which traits are primitive and which are derived and thus depict the ancestry of early vertebrates. Until recently, lampreys have been used as model organisms of jawless vertebrates, especially in developmental biology. Recently, however, it has become possible to obtain fertilized eggs from inshore

DOI: 10.1201/9781003217503-22

hagfish species and study their developmental mechanisms. In this chapter, the characteristics of both lampreys and hagfish are described as model organisms for the evolution of vertebrates, and challenging questions are suggested from genomic and developmental perspectives.

22.1.2 What Are Cyclostomes?

Cyclostomes comprise the extant lampreys and hagfish (Figure 22.1) as well as various extinct species. There are 38 extant lamprey species, of which 9 live in freshwater throughout their lifecycle, and 18 species feed parasitically as adults (Nelson et al. 2016). Adult lampreys have a sucker-shaped mouth with horny teeth instead of an articulated jaw with enameled teeth like gnathostomes (Figure 22.1b). Seven pairs of gill pores open behind the eyes. A single median nostril, called the nasohypophyseal duct, opens on the dorsal side of the head and ends in a blind sac. Lampreys do not have paired pectoral and pelvic fins, both of which are homologs of tetrapod limbs. All living lampreys have a larval stage called ammocoetes. During this stage, the eyes are undeveloped under the skin, and the mouth is not rounded but divided into upper and lower lips (Figure 22.1c). Ammocoetes larvae live at the bottom of rivers as filter feeders. After metamorphosis, some species live as parasites that feed by boring into the flesh of other fish to suck their blood, while others do not feed throughout the adult stage. Most parasitic species migrate from rivers to the sea after metamorphosis and return to the upstream of the river during the breeding season.

There are 29 extant hagfish species (Figure 22.1d). The vertebrae are almost absent. Similar to lampreys, a single nasohypophyseal duct opens at the rostral end of the head, but the internal duct does not end in a blind sac as it does in lampreys but rather opens into the pharynx. The eyes lack lenses, all extraocular muscles and nerve innervation (cranial nerves III, IV and VI). The 1–16 external gill openings are located relatively ventral and caudal compared with those of lampreys. The lateral line system is highly degenerate, and they have no paired fins. Hagfish are widely regarded as scavenger feeders and mostly eat dead animals using a tongue apparatus with a horny dental apparatus. When they encounter predators, they release mucous from 70 to 200 pores in the ventrolateral body that forms slime when coming into contact with seawater. Most hagfish species live in deep-sea habitats, but some species belonging to the genus *Eptatretus* live relatively inshore. For example, the Japanese inshore hagfish *Eptatretus burgeri* lives at depths of 10–270 m (Jørgensen et al. 2012). In contrast to lampreys, all hagfish species undergo direct development without the larval stage.

22.2 HISTORY OF THE MODEL

22.2.1 History of the Classification of Lampreys and Hagfish

Cyclostomes are important model organisms because they are the only extant jawless vertebrates, a characteristic that is shared with fossil Silurian and Devonian fish. Thus, they are in a unique phylogenetic position (Figure 22.2). However, the phylogenetic relationship between lampreys and hagfish has been the subject of controversy until recently. Carl Linneaus, the father of modern taxonomy, originally classified hagfish as *Vermes intestina*, since they lack vertebrae, which is the most important synapomorphy of vertebrates (Linnaeus 1758). In addition, the ammocoetes were initially thought to be a separate species from adult lampreys but were later revealed to be larval lampreys (Müller 1856). It was proposed that lampreys and hagfish be grouped together into "Cyclostome" based on their shared traits of a single nostril and lack of paired fins (Duméril 1806). However, Løvtrup (1977) stated that lampreys are more closely related to jawed vertebrates. Janvier (1996) supported Løvtrup's statement and proposed that hagfish should be placed as a sister group of the other vertebrates called "Craniata" and that lampreys and gnathostomes

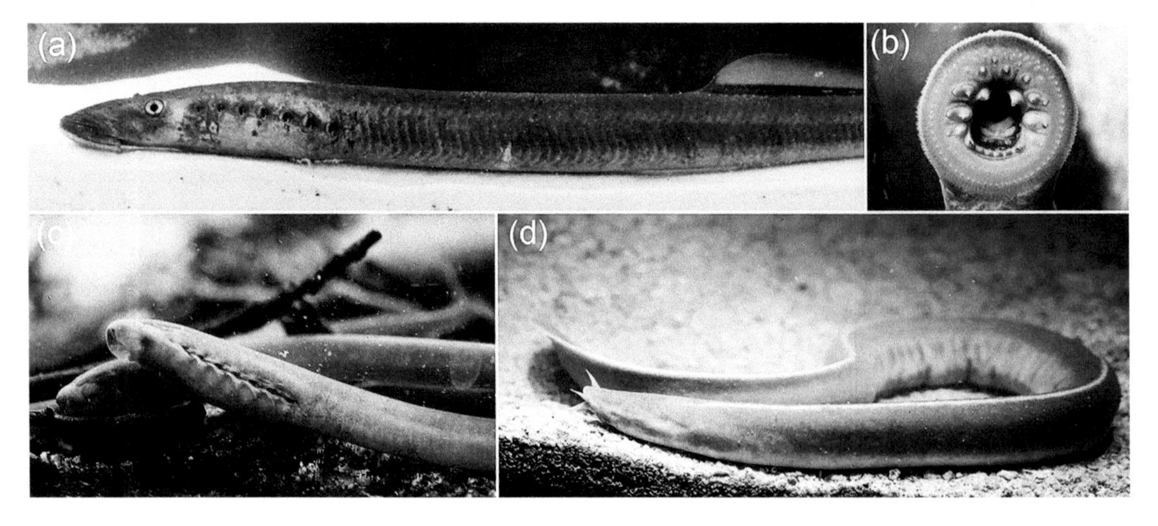

FIGURE 22.1 Lamprey and hagfish. (a–c) Arctic lamprey *Lethenteron camtschaticum*. (b) Oral funnel and horny teeth of adult lamprey. (c) Ammocoetes larvae of lamprey. Note that eyes are undeveloped under the skin, and upper and lower lips cover the mouth instead of the oral funnel. (d) Japanese inshore hagfish *Eptatretus burgeri*.

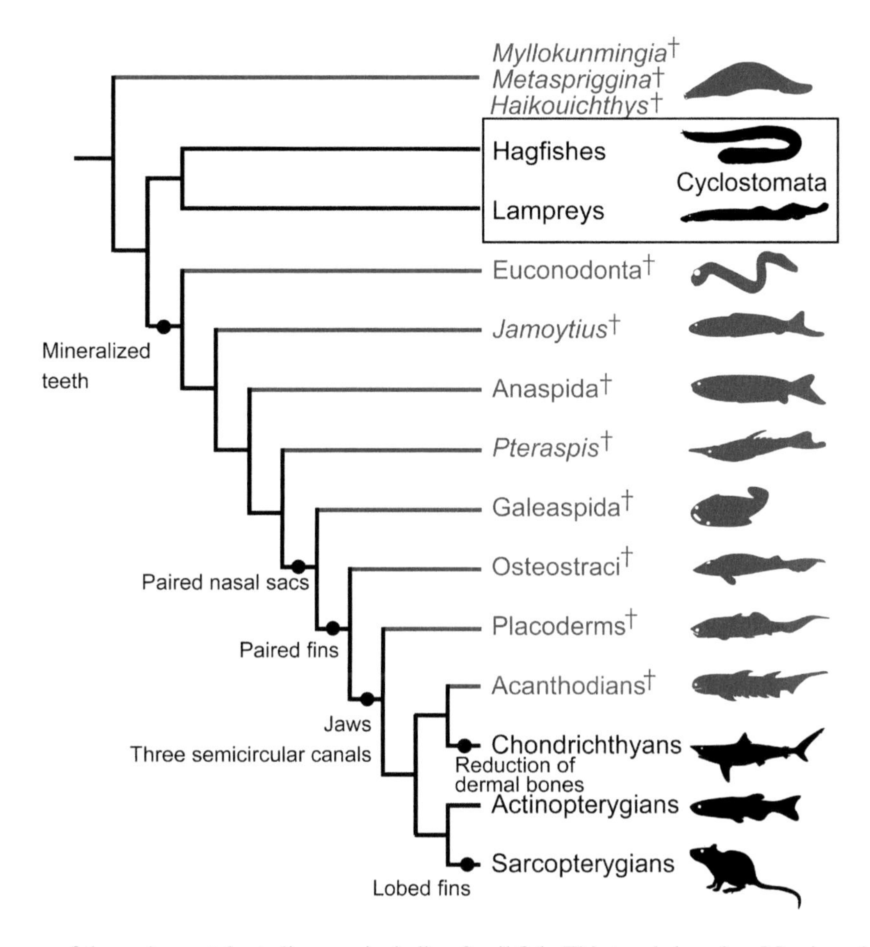

FIGURE 22.2 Phylogeny of the major vertebrate lineages including fossil fish. This tree is based on Morris and Caron (2014) for fossil jawless vertebrates and Zhu et al. (2013) for jawed vertebrates. Gray lines indicate extinct fossil lineages. Round spots indicate major changes toward crown gnathostomes. (From Janvier 1996; Gai et al. 2011).

be classified as "Vertebrata". This classification was widely accepted by paleontologists and morphologists until recently.

However, since the emergence of molecular phylogenetic analysis in the 1990s, lampreys and hagfish have been grouped as a monophyletic group (Kuraku et al. 1999; Mallatt and Sullivan 1998). This monophyletic theory has been repeatedly supported by the presence of cyclostome-specific miRNA (Heimberg et al. 2010), as well as the shared development of the head in lampreys and hagfish (Oisi et al. 2013). Thus, cyclostome monophyly has been widely supported (Figure 22.2).

22.2.2 Relationship with Fossil Vertebrates

The earliest vertebrates did not have an articulated jaw and are therefore called "Agnathan". Cambrian *Myllokunmingia*, *Metaspriggina* and *Haikouichthys* are thought to be early jawless vertebrates (Figure 22.2). Although the ancestors of cyclostomes might have diverged more than 500 million years ago, based on molecular phylogenetic studies (Kuraku and Kuratani 2006), no fossils have been found that can be identified as cyclostomes from this geological period. Later, the ancestor of cyclostomes split into two groups, the lampreys and hagfish, between 430 and 480 million years ago. The earliest lamprey fossil appears to be *Priscomyzon riniensis*, which lived during the Late Devonian (Gess et al. 2006). In addition, fossils of lamprey larvae have been found in the Lower Cretaceous, suggesting that the three-phased (larva–metamorphosis–adult) life cycle of the lamprey was established at least during this period (Chang et al. 2014). Conversely, hagfish fossils are rare, but *Myxinikela siroka* from the Carboniferous is a definite hagfish fossil (Bardack 1991). More ancient fossil fish have been found in the Devonian, and *Palaeospondylus gunni* is classified as a primitive hagfish (Hirasawa et al. 2016), but contrasting opinions have also been proposed based on the presence of three semicircular canals (Johanson et al. 2017).

After the divergence of cyclostomes, Conodonts, *Jamoytius*, Anaspida and *Pteraspis* are thought to have diverged (Figure 22.2). A Silurian osteocoderm (shell-skinned fish) group, Galeaspis, still did not have jaws but had two separated nasal sacs and a hypophyseal duct opening into the oral cavity as in gnathostomes. Therefore, they show intermediate head morphology between jawless and jawed vertebrates (Gai et al. 2011). Placoderms appear to have been the first group to acquire jaws (Figure 22.2), even though the head and brain morphology of primitive placoderms was similar to that of jawless vertebrates and cyclostomes (Dupret et al. 2014).

22.3 GEOGRAPHICAL LOCATION

22.3.1 Geographical Location of Lampreys

Most lamprey species live in the cool zone of the northern hemisphere, generally north of 30°N. The most cosmopolitan lamprey is the sea lamprey *Petromyzon marinus*, which is thus the species most commonly used as a model organism in North America and Europe. They live in the Great Lakes, the Atlantic and Pacific oceans and the Mediterranean Sea along the shores of Canada, the United States, Iceland, and Europe. They are mostly anadromous (seagoing), but the Great Lakes population is landlocked. This species is one of the largest lamprey species and can reach 1.2 m in length and 2.3 kg. The arctic lamprey *Lethenteron camtschaticum* is another important model organism for evolutionary developmental biology in the Far East. They are distributed throughout the Arctic extending south to Japan and Korea. Most of them are anadromous, but landlocked habitats have been observed in some areas (Yamazaki et al. 2011). The European river lamprey *Lampetra fluviatilis* (anadromous) and brook lamprey *Lampetra planeri* (freshwater) have been studied by European researchers. In the southern hemisphere, the pouched lamprey *Geotria australis* and the southern topeyed lamprey *Mordacia* are distributed in Australia (including Tasmania), New Zealand, Chile, Argentina, the Falkland Islands and South Georgia Island. Even though they are thought to have diverged from the northern lamprey 220–280 mya, there are fewer apparent morphological differences between them.

22.3.2 Geographical Location of Hagfish

Hagfish occur in all oceans except for the polar seas. All species prefer cool water (<15°C) and therefore live in deep water or locations where the water is cool. Extant hagfish can be divided into two major genera, *Myxine* and *Eptatretus*. The major morphological difference between them is the number of external gill apertures. That is, *Myxine* is defined as having one pair of common gill openings, whereas *Eptatretus* is characterized as having one duct as an exit from each gill pouch. The Atlantic hagfish *Myxine glutinosa* was first described by Linnaeus (1758) and is commonly found around the Atlantic Ocean in Europe and North America. Among the *Myxine* species, *M. glutinosa* lives in exceptionally shallow water (<40 m), but most *Myxine* species live in deep water where light does not reach. A relatively large number of studies have been reported on the behavior and embryonic development of *Eptatretus*, since they generally live in shallower seas than *Myxine*. The Pacific hagfish *Eptatretus stoutii* is distributed in the eastern north Pacific from Canada and the United States to Mexico in water of 16–633 m depth (Jørgensen et al. 2012). At the end of the 19th century, Bashford Dean collected fertilized eggs of *E. stoutii* (synonym: *Bdellostoma stoutii*) from Monterey Bay, California, and first described their embryonic development (Dean 1899). *E. burgeri* is distributed around Japan, Korea and Taiwan and has been used in developmental studies recently (Ota et al. 2007). As in lampreys, there are only a few genera in the southern hemisphere, such as *Notomyxine*, *Neomyxine* and *Nemamyxine*. It has been noted that these genera might have diverged early from the northern hagfishes based on 16S rDNA data (Fernholm et al. 2013). Further phylogenetic studies are needed to elucidate the phylogeny of extant hagfish.

22.4 LIFE CYCLE

22.4.1 Life Cycle of Lampreys

The life cycle of lampreys is highly complex, because they undergo three major morphological and physiological stages, ammocoetes larva, metamorphosis and adult. Mature adults spawn in nests of sand in the upper streams of rivers. Fertilized eggs hatch within two weeks and develop into ammocoetes larvae within about one month (see Section 22.5 for details). Ammocoetes larvae have undeveloped eyes under the skin, and their mouth is not rounded but divided into upper and lower lips (Figure 22.1c). They live as filter feeders, buried in mud, sand and organic detritus along rivers. The mucus secreted by the endostyle is used for this feeding behavior, as in amphioxi or ascidians. According to a study using stable isotope ratios ($\delta^{13}C$ and $\delta^{15}N$) in *P. marinus* larvae, they are primarily consumers of aquatic sediments, including macrophytes, algae and terrestrial plants (Evans and Bauer 2016). In an aquarium environment, dry yeast or the unicellular alga *Chlorogonium capillatum* (NIES-3374) can be used as a food source (Tetlock et al. 2012; Higuchi et al. 2019). The larval stage lasts for a number of years (e.g. *L. camtschaticum*: 2–5 years). The trigger for the transition to metamorphosis is probably not the length of the larval period but rather the larval size. Once larvae reach a certain length (e.g. *L. camtschaticum*: ~16 cm [Kataoka 1985]), they proceed to the metamorphic stage. Metamorphosis lasts for approximately one month. During this period, the oral apparatus changes into a round, sucker-like disc lined with horny teeth. The medial dorsal fin is higher, and the eyes are fully functional.

The adult life of lampreys varies considerably between parasitic and non-parasitic species. Many parasitic species are anadromous, migrating downstream to the sea and sucking on fish to feed on their blood. However, these species are not only parasites but also scavenge dead animals or prey on fresh fish as predators. Non-parasitic species spend their whole lives in freshwater and are sexually mature for less than a year. Usually, parasitic and non-parasitic behaviors are species specific, but the two types of behavior are sometimes found in the same species (Yamazaki et al. 2011).

Before the breeding season, parasitic species begin to migrate upstream. As they approach sexual maturity, males develop a urogenital papilla, a penis-like funnel-shaped organ elongated from the cloaca (Figure 22.4e). The abdomen of the female is visibly enlarged, and a post-cloacal finfold develops (Figure 22.4f). Mating behavior occurs in their nests, which are constructed by thrashing their bodies and mouths to remove stones. A male attaches itself to the female's head and wraps his tail around her trunk to assist in the extrusion of eggs. Finally, the couple vibrates vigorously for a few seconds to release eggs and sperm so they can be externally fertilized. All individuals die within a few days after spawning.

22.4.2 Life Cycle of Hagfish

In contrast to lampreys, the life cycle of hagfish might be relatively simple, because they undergo direct development with no larval and metamorphosis stages. However, many aspects of the hagfish life cycle remain unknown because they live in deep-sea habitats, and even their basic life history characteristics, such as growth rate, lifespan, sexual maturity and reproductive behavior, remain unclear. All of the described hagfish species prefer high salinity. For example, *M. glutinosa* dies rapidly in salinities of 20–25 ppt (Gustafson 1935). This could explain why hagfish do not occur in polar seas. Most species tend to live in deep waters. An unknown *Eptatretus* sp. was photographed at a depth of over 5,000 m (Sumich 1992). Although each species has a characteristic depth range, the range can be quite broad in some cases. For instance, *M. glutinosa* can be found at depths of 30 m in the northern Gulf of Maine, whereas this species has been collected at depths of 1,100 m in North America (Jørgensen et al. 2012).

E. burgeri is the only known species to show seasonal migration. On the Pacific side of mid-Japan, this species is found in quite shallow water (6–10 m depth) from mid-October to mid-July. Subsequently, these hagfish swim deeper than 50 m until September (Ichikawa et al. 2000). Although it is unknown whether this migration is related to water temperature or breeding behavior, researchers have failed to collect eggs by net sweeping at 40–110 m depth, suggesting that the spawning ground of this animal might be deeper than 100 m. Other studies have reported that differences in habitat depend on size and sex. Most *E. stoutii* are found at 100 m depth, where the ratio of males to females is 1:1, whereas larger females are predominant at 500 m (Jørgensen et al. 2012). Many species prefer to hide in the sand or mud on the sea floor, whereas others prefer the shade of rocks. Generally, hagfish are thought to be scavengers, eating dead fish and whales. However, many studies have showed that they are predators who attack and eat invertebrates and vertebrates, such as polychaetes, shrimp and fish. In addition, they are opportunistic scavengers on dead animals.

The most unique feature of hagfish is their ability to release large amounts of slime consisting of mucous and fibrous components from glands. This function is mainly defensive against predators. When they are physically attacked by predators, hagfish rapidly eject slime, which entrains large volumes of water and traps predators' head and gills. See Fudge et al. (2016) for further details of hagfish slime.

Little is known about hagfish reproduction, including the maturation mechanism, mating behavior, fertilization or embryonic development. This is because the location and timing of hagfish spawning remain unknown. The eggs seem to be fertilized externally, because hagfish do not have mating organs. However, mating behavior also remains unknown. Exceptionally, small numbers of fertilized eggs of the inshore hagfish *E. burgeri* have been collected every year since 2006 (Ota et al. 2007). In mid-August, pre-mature males and females can be caught at a depth of 100 m in the Sea of Japan. When they are kept on the bottom of the sea in cages, they lay eggs in late October (Oisi et al. 2015). Embryonic development is slow, with eggs taking approximately one year to hatch. Juveniles are almost identical to adults, except for carrying the yolk sac.

22.5 EMBRYOGENESIS

22.5.1 Development of Lamprey Embryos

Fertilized eggs of lampreys can be obtained by artificial fertilization during the breeding season once a year (Sugahara et al. 2015). *L. camtschaticum* eggs are approximately 1 mm in size, which is similar but slightly smaller than *Xenopus* eggs. Double-layered chorion surrounds the eggs. They are telolecithal eggs and show holoblastic cleavage. For staging, Tahara's developmental stages for *L. reisnneri* are widely used (Tahara 1988) (Figure 22.3a). At stage 13, gastrulation begins below the equator as in *Xenopus*. The blastopore is elliptical, while the yolk plug is not formed. At stage 17, the neural groove arises in the middle of the neural plate and changes to a neural fold. Both neural folds are almost parallel throughout the embryos, even in the head region, which is different from those in frogs and zebrafish. After neurulation, head protrusion is visible, and the cheek processes (mandibular arch and first pharyngeal pouch) appear on the lateral side of the protrusion. One of the unique features of lamprey embryos is that the nasal placode is single and fused with the hypophyseal placode at the anterior end to the mouth opening, forming a nasohypophyseal placode. Around stage 25 (approx. 10 dpf), eggs hatch and the heart starts beating. At stages 27 and 28, the eye spots are visible, and the velum starts pumping for ventilation. At stage 30 (approx. 30 dpf), ammocoetes larvae grow and dive into sand or mud to begin filter feeding.

22.5.2 Development of Hagfish Embryos

As mentioned, hagfish development remains unknown because there have been few published reports to describe hagfish embryology. Hagfish eggs are large (2 cm) compared with those of lampreys (Figure 22.3b) and are encased in a hard, orange eggshell that possesses anchor filaments (hook and loop tape-like structure) at both ends of the long axis to stick to each other, forming a cluster (Figure 22.3c). Little is known about early cleavage, but regarding the large amount of yolk, the cleavage style might be meroblastic. Embryonic development is slow. Surprisingly, the development can be observed from the outside of the eggshell four months after the eggs have been laid, and they appear to take approximately one year to hatch. So far, there are no normal stage tables for

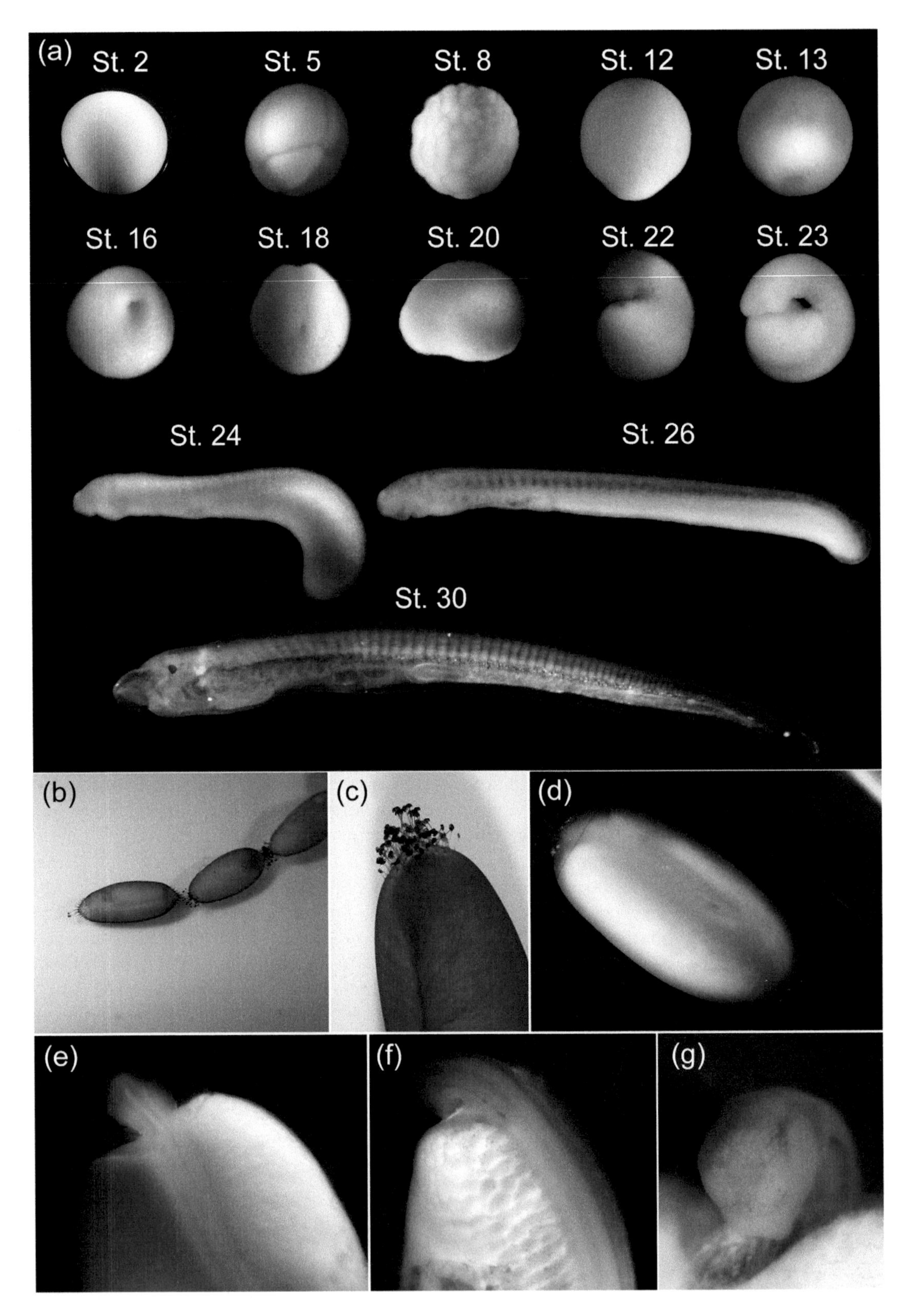

FIGURE 22.3 Embryonic development of the *L. camtschaticum* (a) and the *E. burgeri* embryos (b–g). (a) One-cell stage (St. 2), eight-cell stage (St.5), morula (St. 8), pre-gastrula (St. 12), early gastrula (St. 13), late gastrula (St. 16), early neurula (St. 18), late neurula (St. 20), head protrusion (St. 22), stomodaeum (St. 23), hatching (St. 24), melanophores (S. 26), ammocoete larva (St. 30). (b) Connected hagfish eggs. (c) "Hook and loop tape"-like structure at both ends of the long axis (d) External view of the hagfish embryo (pharyngula stage). Body axis can be seen, and head region is curved at the edge of the egg. (e) Mid-pharyngula embryo after removal of the eggshell. (f) Late-pharyngula embryo. (g) Anterior view of (f).

any hagfish species. However, researchers often refer to Dean's figure numbers as describing their developmental stages (Dean 1899; Oisi et al. 2013). The overall development is comparable to that of lampreys. For example, a single median nasohypophyseal (nasal, adenohypophysis) placode arises at the anterior ventral tip of the head. However, hagfish-specific developmental events can be also observed. The stomodeum is closed secondarily by the secondary oropharyngeal membrane. Subsequently, the primary oropharyngeal membrane disappears. This peculiar developmental event caused the endodermal origin of the adenohypophysis to be misidentified (Gorbman 1983). The nasohypophyseal duct opens into the pharynx in hagfish unlike in lampreys. The pharyngeal pouches and surrounding tissues are shifted caudally during the late developmental stage. Juveniles are almost identical to adults except for carrying the yolk sac.

22.6 ANATOMY

22.6.1 Lamprey Anatomy

The body of adult lampreys is cylindrical and covered with scaleless skin (Figure 22.4). On the head, the seven rounded external pharyngeal gill slits open just behind a pair of eyes. A single median nostril (or nasohypophyseal opening) lies on the dorsal midline between the eyes. This duct does not open into the pharynx and ends in a blind sac (Figure 22.4i). A pineal eye, which functions as a photoreceptor, is under the translucent skin, positioned just after the nostril. The oral funnel forms a sucking disk that enables attachment to other fish for feeding or rocks for holding their body in place. There are many horny teeth on the internal surface of the disk. Note that these are not homologous with the enameled teeth in other vertebrates. The dotted lateral lines are present around the head region to detect water flow.

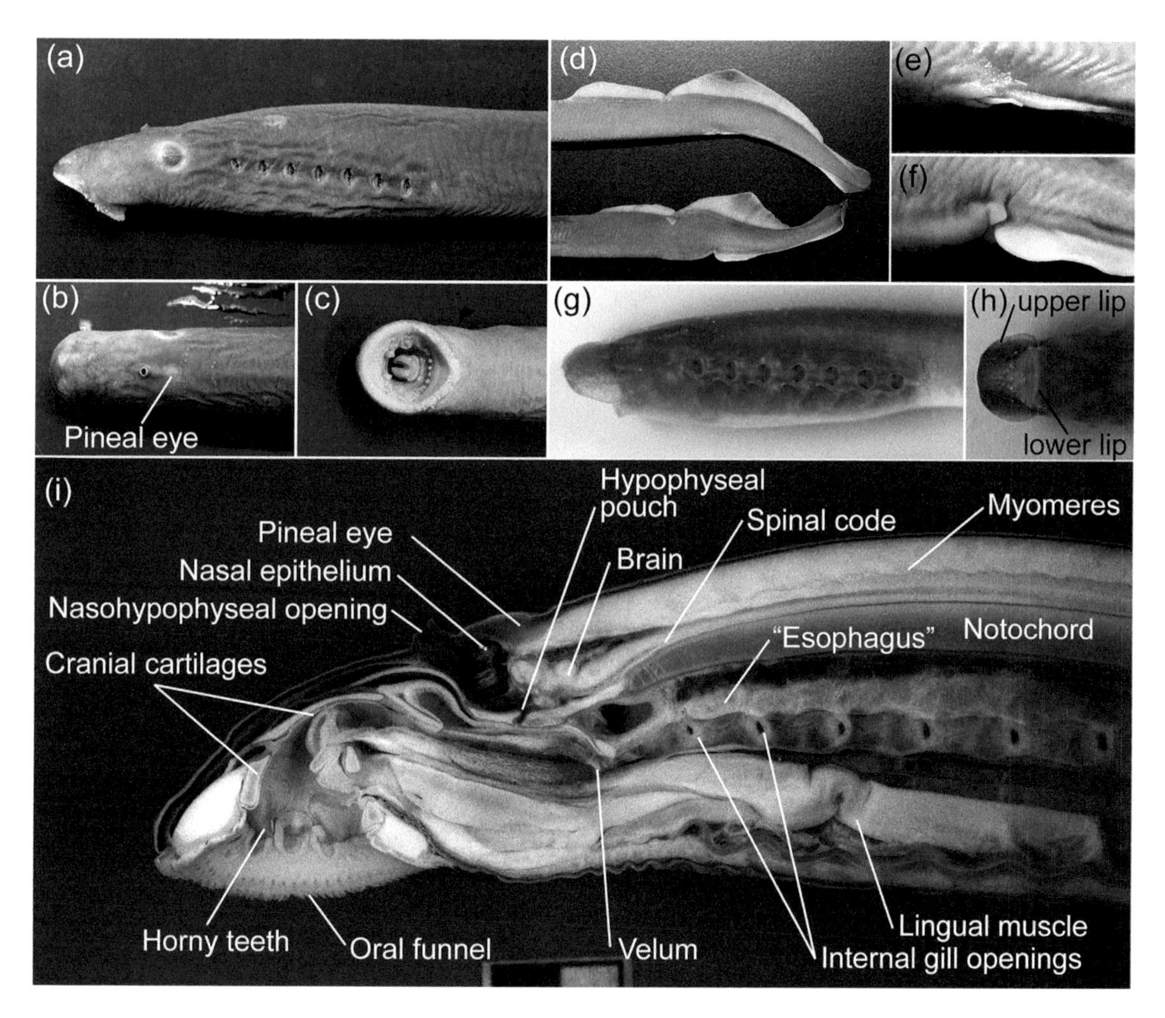

FIGURE 22.4 General anatomy of the lamprey, *L. camtschaticum*. (a–c) Lateral (a), Dorsal (b), and ventral (c) views of the head. (d) Abdomens of the mature male (above) and female (below). (e) Urogenital papilla of the mature male elongated from the cloaca. (f) Anal fin-like structure of the female. (g, h) Lateral (g) and ventral (h) views of the head of ammocoete larva. (i) Sagittal section of the adult lamprey. Note that the lamprey esophagus is termed the dorsal route of the pharynx (for respiration) and is not homologous with the esophagus in other vertebrates.

Lampreys do not possess paired fins, but two dorsal fins and caudal fins are present (Figure 22.4d). Usually, it is difficult to distinguish males from females based on external morphology. However, during the mating season, mature males can be distinguished by the presence of urogenital papilla (penis-like protrusion) anterior to the cloaca (Figure 22.4e). In contrast, an anal fin-like structure develops in mature females (Figure 22.4f).

Figure 22.4i shows a sagittal section of the anterior part of adult lampreys. The pharynx is subdivided dorsoventrally. The dorsal part is called the esophagus, and the ventral part is a respiratory tube connected with the gill openings. This subdivision develops during metamorphosis. The velum, positioned between the oral cavity and the pharynx, is a major pumping device during the larval stage but has no respiratory role in adults. True vertebrae are absent, and instead, dorsal cartilaginous arcualias protect the spinal cord. The notochord is fully functional as a supportive organ in the larval and adult body.

The gross anatomy of the lamprey brain is comparable to that of teleosts. The most significant difference between is that lamprey brains have a microscopic cerebellum (Figure 22.5a). In contrast, the pineal organ or epiphysis is well developed. In the inner ear, only two semicircular canals (anterior and posterior) are present, reminiscent of fossil osteostracans (Figure 22.5c; Higuchi et al. 2019).

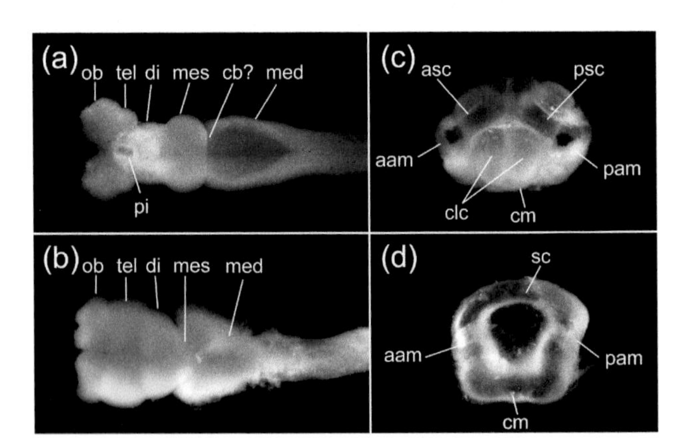

FIGURE 22.5 Brain and semicircular canals of the lamprey *L. camtschaticum* (a, c) and hagfish *E. burgeri* (b, d). asc, anterior semicircular canal; aam, anterior ampulla; cb, cerebellum; clc, ciliated chamber; cm, common macula; di, diencephalon; med, medulla oblongata, mes, mesencephalon; ob, olfactory bulb; pam, posterior ampulla; pi, pineal organ; psc, posterior semicircular canal; sc, semicircular canal; tel, telencephalon.

22.6.2 Hagfish Anatomy

The body of hagfish is eel-like, as in lampreys, and is covered with soft, scaleless skin (Figure 22.1). The rudimentary eyes lack lenses, extraocular muscles and innervating nerves [oculomotor (III), trochlear (IV) and abducens (VI)]. The pineal eye is absent. A few lateral line spots are found around the head surface as shallow grooves in *Eptatretus*, but these are absent in *Myxine* (Braun and Northcutt 1997). In contrast to lampreys, the hagfish mouth is normally occluded. They grasp food by protracting and retracting a pair of dental plates. Therefore, their retractor muscle is large (Figure 22.6f). There are six or eight barbels around the mouth innervated by the trigeminal nerve (V) with a sensory role. A single median nostril (or nasohypophyseal opening) opens at the anterior end of the head. Unlike lampreys, this duct does not end in a blind sac but rather opens into the pharynx (Figure 22.6f). This enables water to be taken from the nostril to the gill pouches while closing the mouth. The external gill openings are positioned relatively caudal-ventral compared with those in lampreys (Figure 22.6b). The number of gill openings varies among species, which reflects the number of gill pouches (5–16 pairs). Conversely, each branchial duct of *Myxine* tends to be fused and opened as a common external aperture on each side. Like lampreys, hagfish also do not possess paired fins, and only a continuous median fin is present on the posterior of the body. It is almost impossible to distinguish sex based on external morphology, but mature females are distinguishable by having large eggs in their abdomen. Velum movement generates a water current and acts as a ventilatory pump. Vertebral elements were traditionally considered to be absent from hagfish, whether cartilage or hard bone. However, recently, cartilaginous tissue (reminiscent of hemal arches in gnathostomes) has been found at the caudal–ventral part of the notochord (Ota et al. 2011). A unique feature is the presence of some accessory hearts in addition to the portal (true) heart. For example, *M. glutinosa* has five accessory hearts (a branchial, two cardinal and two caudal hearts) (Nishiguchi et al. 2016). These are not homologous to the portal heart in other vertebrates because of the lack of cardiac muscles. The accessory hearts are thought to play a role in assisting the portal heart.

The brain of hagfish show curious morphology in contrast to those of other vertebrates (Figure 22.5b). The olfactory bulb and cerebral hemisphere are strikingly larger, but the epiphysis and cerebellum are absent. Owing to this curious shape, it has been extremely difficult to homologize the subregions of the hagfish brain to those of other vertebrates (Conel 1929). In the inner ear, only a pair of single, donut-shaped semicircular canals are present (Figure 22.5d). Curiously, this single canal has two ampullae (the detector), whereas each canal has one ampulla in other vertebrates. Recent studies have suggested that the anterior and posterior halves of the canal are homologous to the anterior and posterior canals in lampreys, respectively (Higuchi et al. 2019).

As described, there are many specific features in lampreys and hagfish. It is important not to simply regard these traits as primitive, because they are not ancestral animals, but rather they diverged and lived independently from the jawed vertebrates for over 500 million years and so have traits that they acquired or lost independently. A careful comparison of each trait between the lampreys, hagfish and jawed vertebrates would allow us to depict the ancestry of early vertebrates (Sugahara et al. 2017).

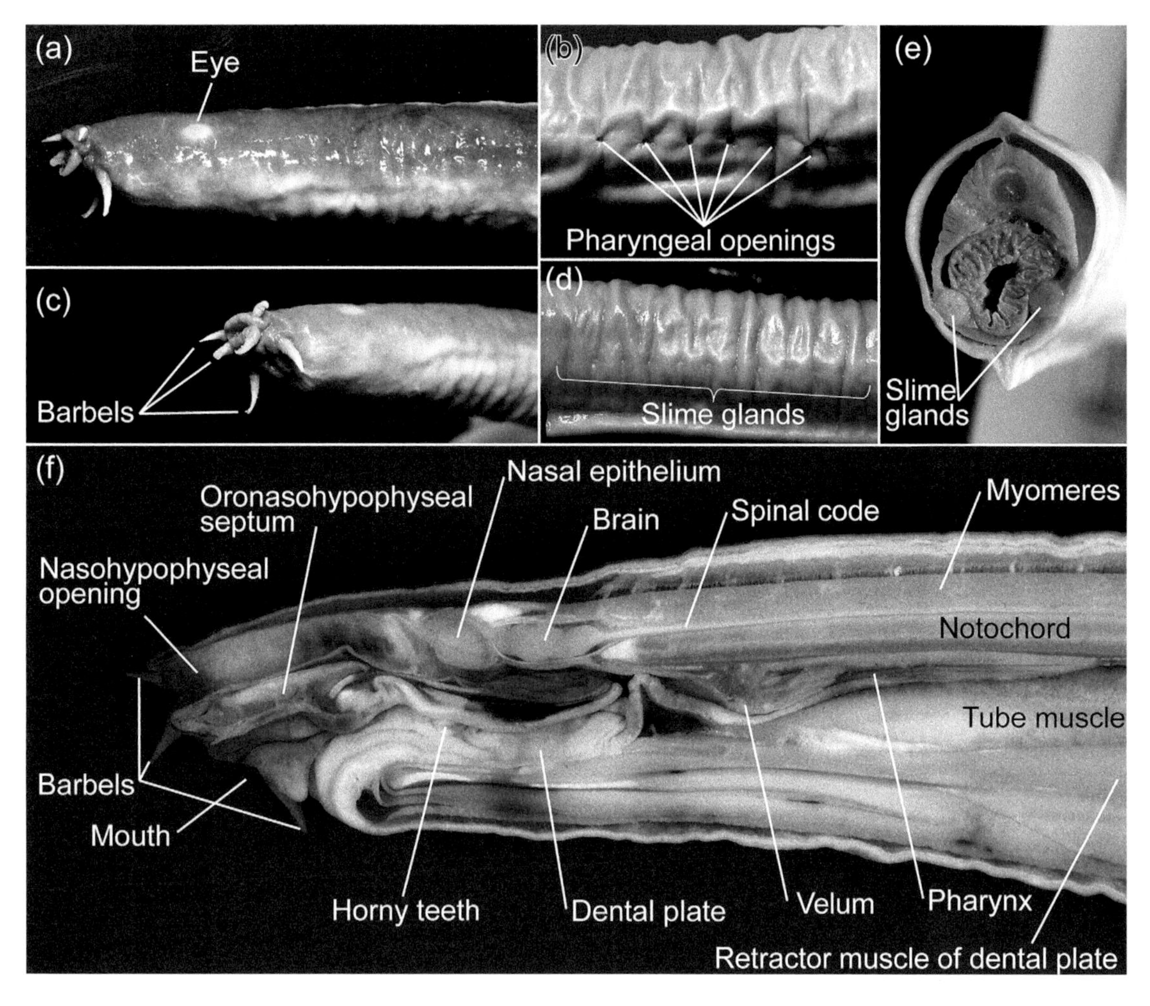

FIGURE 22.6 General anatomy of the hagfish, *E. burgeri*. The skin is artificially shrunk and shows bellow-like wrinkles by formalin fixation. (a, c) Lateral (a) and anterior (b) views of the head. (b) Pharyngeal openings on the ventral lateral body surfaces. Note that the last opening on the left side is slightly larger, called the pharyngocutaneous opening. This duct is directly connected to the pharynx and also fused with the common efferent gill duct on the left side. (d, e) Slime glands on the ventral lateral sides of the body. (f) Sagittal section.

22.7 GENOMIC DATA

22.7.1 Genomic Features of the Cyclostomes

All lamprey karyotypes are characterized by small, dot-shaped chromosomes (microchromosomes). In general, they have 100 or more chromosomes in somatic diploid cells. For example, germline diploid cells have 198 chromosomes in *P. marinus* and 168 chromosomes in *L. camtschaticum* (Ishijima et al. 2017). Males and females have the same number of chromosomes. The sex determination system is unclear but may be determined by the growth rate during the larval period (Johnson et al. 2017).

The genomic sequences of lampreys have been less well understood until recently because they contain high GC content in the coding region, which prevents sequencing by the traditional Sanger method and PCR-based gene cloning. Although the overall GC content is 46% in the *P. marinus* genome, the GC content in the coding regions is markedly higher (61%) than that in noncoding regions (Smith et al. 2013). Four-fold degenerate sites (GC_4) are especially high (around 70–90%) compared with those in hagfish (40–60%) (Kuraku et al. 2006). Another difficulty with sequencing is that the lamprey genome possesses highly repetitive elements that prevent the assembly of each fragment by next-generation sequencing. Recently, these difficulties have been overcome by optimizing the computational assembly that allowed us to assemble fragments from next-generation sequence data (Smith et al. 2018). Currently, the lamprey genome sequence is available from three species (*P. marinus, L. camtschaticum* and *Entosphenus tridentatus* (Table 22.1). Transcriptome data sets are also available for *P. marinus* and *L. camtschaticum*.

The chromosome number of hagfish is much lower than that of lampreys. For example, 52 are found in the diploid testis cells of *E. burgeri*, 48 in *E. stoutii* and 44 in *M. glutinosa*. Males and females have the same number of chromosomes. Sex determination is unknown. Recently, the genome sequence of the hagfish *E. burgeri* has been made available (Table 22.1).

TABLE 22.1
Major available genome resources in lampreys and hagfishes

Name	(Human)	Arctic lamprey	Sea lamprey	Inshore hagfish
Species	*Homo sapiens*	*Lethenteron camtschaticum*	*Petromyzon marinus*	*Eptatretus burgeri*
Source		testis	sperm	testis
Total sequence length (bp)	3,099,706,414	1,030,662,718	1,089,050,413	2,608,383,542
Scaffolds (bp)	67,794,873	86,125	1,434	10,846
N50 scaffold size	67,794,873	1,051,965	12,997,950	
Estimated genome size	3.1 Gb	1.6 Gb	N/A	2.9 Gb
Coverage		20.0x	62.36x	210x
Reference	https://www.ncbi.nlm.nih.gov/assembly/GCA_000466285.1/	https://www.ncbi.nlm.nih.gov/assembly/GCF_010993605.1	https://www.ncbi.nlm.nih.gov/assembly/GCA_900186335.2	https://www.ncbi.nlm.nih.gov/assembly/GCF_000001405.39

22.7.2 Chromosome Elimination and Programmed Sequence Loss in Cyclostomes

Chromosome elimination is a process in which some chromosomes are discarded during embryogenesis, whereas germline cells retain all chromosomes (Figure 22.7a). This process is widely seen in protostomes, such as nematodes and arthropods. In vertebrates, only hagfish were observed to expel some chromosomes from presumptive somatic cells. In *E. burgeri*, there are 36 chromosomes in somatic cells and 52 in the germline cells, suggesting that 16 chromosomes (20.9% DNA content) are eliminated during embryogenesis (Kohno et al. 1998; Figure 22.7b). These chromosomes contain highly repetitive DNA sequences and are highly heterochromatinized in germ cells (Kohno et al. 1998). Moreover, this event was recently observed in the lamprey *P. marinus* (Timoshevskiy et al. 2019), suggesting that this phenomenon is shared by both cyclostome lineages.

Another type of genome rearrangement is seen in cyclostomes, namely programmed sequence loss (Figure 22.7a). In *P. marinus*, the DNA content of haploid sperm is 2.31 pg, and that of blood cells is 1.82 pg (>20% of the genome, or 0.5 billion base pairs) (Smith et al. 2009); Figure 22.6c). Discarded sequences contain not only several different repetitive elements but also transcribed loci in the developmental stage. In hagfish, heterochromatinized regions that contain repetitive elements are widely eliminated (Kohno et al. 1998). Altogether, lampreys and hagfish undergo both genome rearrangement mechanisms and thus will provide critical insights into the evolution of genome rearrangement in the vertebrate lineage.

22.7.3 Hox Clusters and Whole-Genome Duplication

Ohno (1970) proposed that early vertebrates underwent two rounds of whole-genome duplication (2R WGD). This hypothesis has been supported by the number of Hox clusters. *Amphioxus* have a single Hox gene cluster in contrast to the four clusters in mammals (Figure 22.8). Moreover, teleosts might have experienced another WGD. This suggests that 2R WGD might have occurred in the ancestry of vertebrates. However, the timing of the WGD in the pre- or post-divergence of the cyclostomes remains unclear. Interestingly, recent genomic studies have revealed that both lampreys and hagfish have at least six Hox gene clusters in their genome (Mehta et al. 2013; Pascual-Anaya et al. 2018) (Figure 22.8). These results suggest that at least one independent (whole or partial) genome duplication event might have occurred in the cyclostome lineage, but it is still unclear whether cyclostomes share the gnathostome 2R, 1R or 0R of WGD (Figure 22.8).

22.8 FUNCTIONAL APPROACHES: TOOLS FOR MOLECULAR AND CELLULAR ANALYSES

22.8.1 Advantages of Lamprey Developmental Research

As it is difficult to obtain fertilized eggs, experimental embryology with hagfish has been limited to histological or gene and protein expression analyses (Oisi et al. 2015). Therefore, this topic focuses on the functional analysis of lamprey developmental biology (for normal histology, *in situ* hybridization, and immunohistochemical techniques on lamprey embryos, see Sugahara et al. 2015). Since the breeding season for lampreys occurs once a year, there are not many opportunities for experiments to be carried out compared with zebrafish and *Xenopus*. However, lampreys have some advantages over other model organisms. More than 10,000 eggs can be obtained from one female during a single artificial fertilization event. Hundreds of eggs and embryos can be incubated in a small plastic dish with fresh water (Figure 22.9a). Most

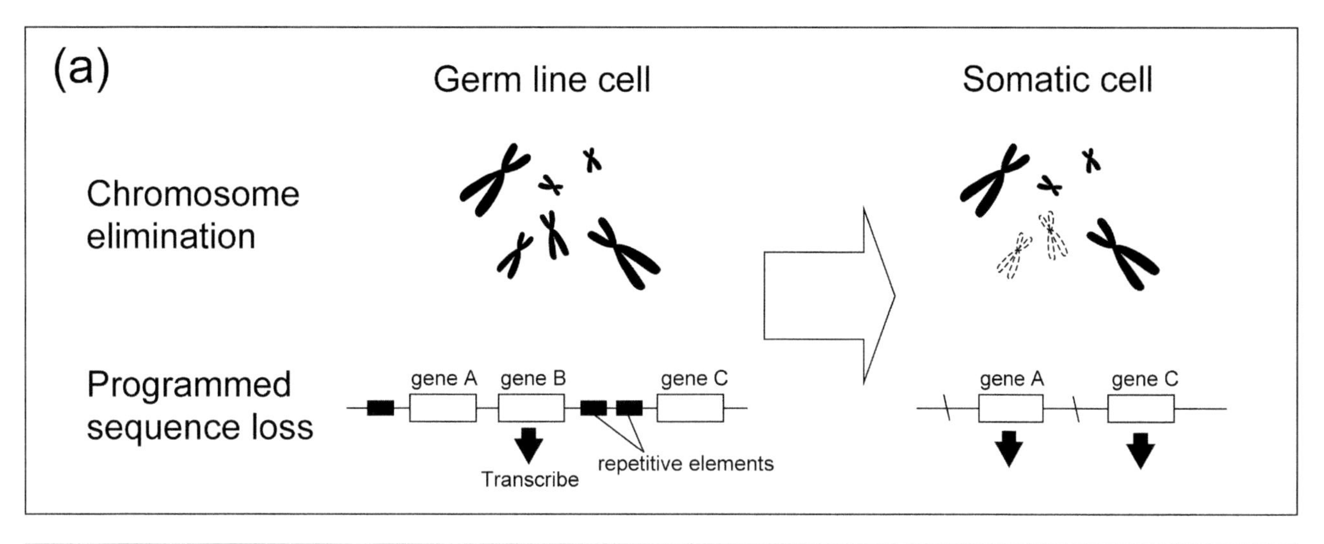

(b)

Species	Germ cell	Somatic cell	% of eliminated DNA amount
E. stoutii	48	34	52.8%
E. burgeri	52	36	20.9%
P. atami	48	34	40.0%
M. glutinosa	44	28	43.5%

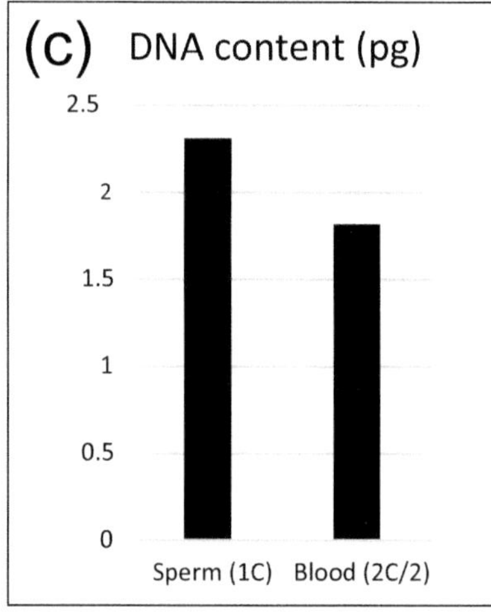

FIGURE 22.7 **Genome rearrangement in cyclostomes**. (a) Overview of the chromosome elimination and programmed sequence loss (b) Reduction of the chromosomes in hagfish species (c) Reduction of DNA content during development in the sea lamprey, *P. marinus*. 1C and 2C/2 indicates haploid genome size. Recent studies have revealed that both lampreys and hagfish undergo both reduction mechanisms. ([a] Modified from Semon et al. 2012; [b] based on Kohno et al. 1998; [c] adapted from Smith et al. 2009.)

experimental techniques developed for zebrafish or *Xenopus* can also be applied to lamprey embryos. In particular, lamprey eggs are particularly amenable to microinjection. They have a double chorion, which prevents them from exploding due to the water surface tension when eggs are removed from the water, and therefore, the eggs can be injected with liquids on dry mesh (Figure 22.9c, d). In addition, unlike fast-developing model organisms, the slow cleavage of lamprey embryos allows the injection of many eggs for a long time (over 5 h) during one or two cell stages.

22.8.2 Drug Application

Drug application in lamprey embryos is the easiest method for investigating certain gene functions or signaling pathways. Eggs or embryos can be exposed to an adequate concentration of the drug by immersion (Figure 22.9b). For instance, the following drugs have been used and showed certain effects on lamprey embryos: SU5402 for the blocking of FGF signaling (Tocris Bioscience; Sugahara et al. 2011), U0126 for the inhibition of MAP kinases (Tocris Bioscience; Jandzik et al. 2014), Cyclopamine for Hedgehog signaling (Calbiochem; Sugahara et al. 2011), DAPT for the Notch pathway inhibitor (Lara-Ramirez et al. 2019) and SB-505124 for the Nodal antagonist (Abcam; Lagadec et al. 2015). All-trans retinoic acid has also been used for enhancing retinoic acid signaling in a dose-dependent manner (Kuratani et al. 1998).

22.8.3 Morpholino Antisense Oligomers

Morpholino antisense oligomers (MOs) are useful tools for knocking down gene function in developmental biology research as conceived by Gene Tools LLC. The MOs are usually 25-mer nucleic acid analogs synthesized to bind to complementary target RNA. When MO binds to the 5′-UTR of mRNA, it can prevent translation of the coding region of the target gene by interfering with the progression of the ribosome. Once MO binds to the border of the

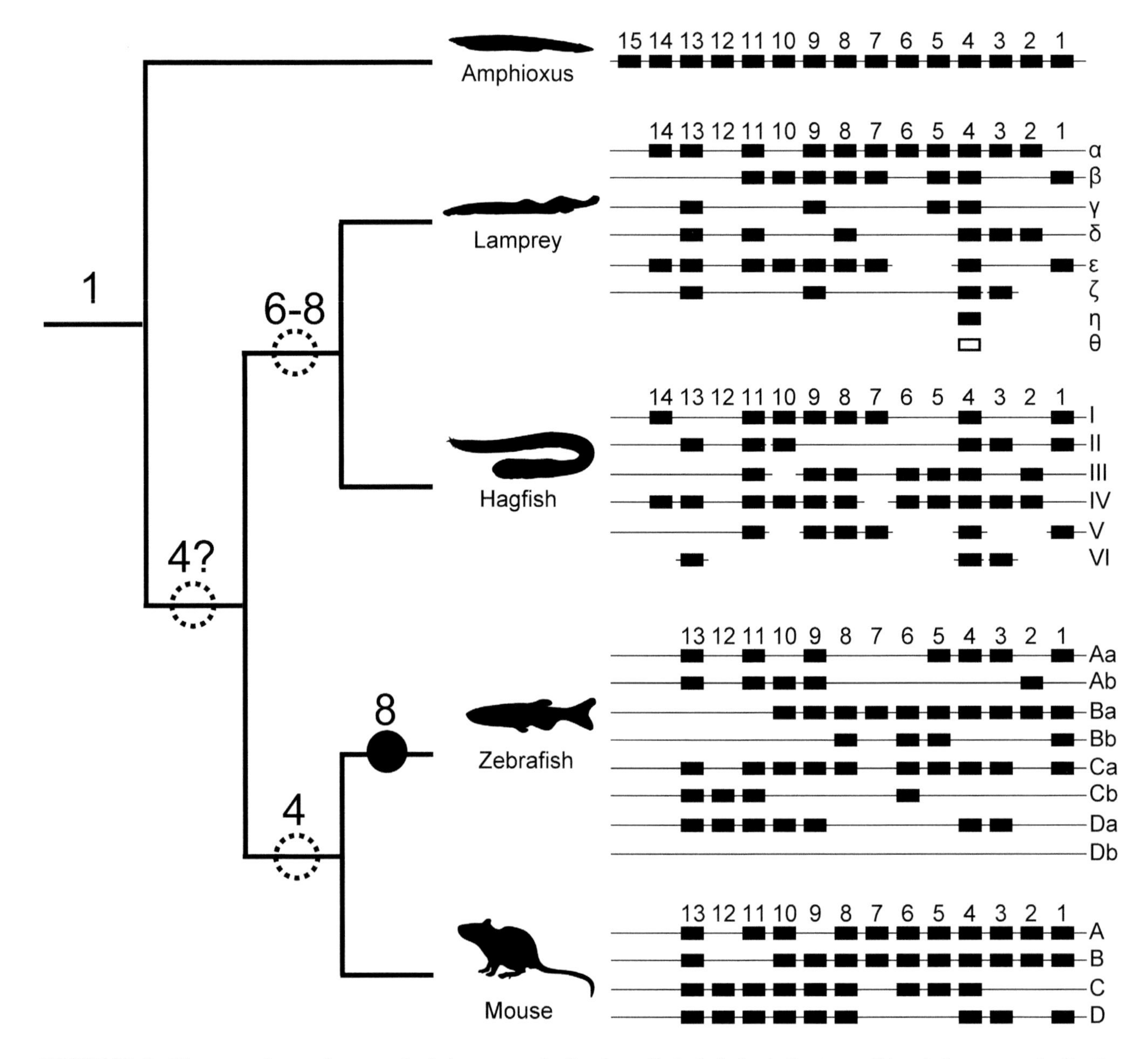

FIGURE 22.8 Hox genes in vertebrates and whole genome duplications. Dotted circles indicate possible whole genome duplication events. Black circle indicates teleost-specific whole genome duplication. Note that although zebrafish do not have HoxDb clusters, some teleost species (e.g. medaka and fugu) retain some genes belonging to HoxDb. (Adapted from Pascual-Anaya et al. 2018.)

introns on pre-mRNA, it can block splicing by interfering with a splice-directing small nuclear ribonucleoprotein (snRNP) complex. For investigating lamprey embryology, researchers can inject MOs by microinjection at the one- or two-cell stages. Five-mismatch MOs can be used as control experiments to distinguish side effects. When MOs are injected into one blastomere at the two-cell stage, the effect could be observed at only one side of the embryo. This enables easy comparison of morphological changes or gene expression (Nikitina et al. 2009).

22.8.4 CRISPR/Cas9 Gene Editing

CRISPR/Cas9 gene editing is a recently developed genetic engineering tool in molecular biology. The CRISPR/Cas9 system was originally a bacterial defense mechanism and was adapted to target mutagenesis in eukaryote genomes. In particular, it is a strong tool for producing knockout lines of animals, such as mice, flies, zebrafish and *Xenopus*. Mutations can be generated simply by injecting Cas9 (endonuclease) mRNA with a synthetic guide RNA into fertilized eggs. Once the Cas9-gRNA complex binds to the DNA target, Cas9 cleaves both strands. The resulting double-strand break is then repaired, but it frequently causes small insertions or deletions at the breaking sites, resulting in amino acid deletions, insertions or frameshift mutations of the target gene. Unfortunately, it is not practical to produce F_1 or F_2 generations of lampreys, and analysis has to be carried out at F_0. Usually, F_0 shows a mosaic for the mutation because CRISPR/Cas9 persists and functions beyond the one-cell

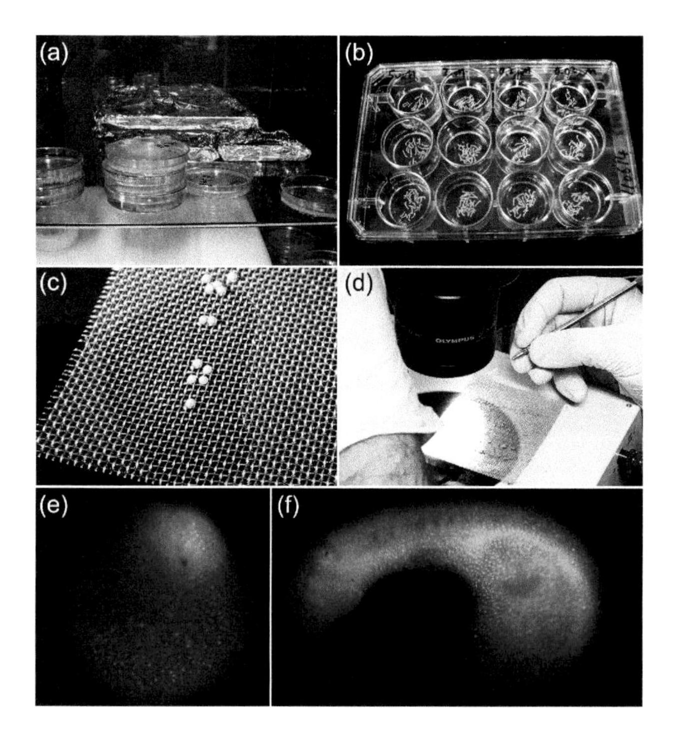

FIGURE 22.9 Embryonic manipulation of lamprey embryos. (a) Incubation of lamprey embryos in 9-cm dishes. Hundreds of embryos can be kept in one dish. Blue water is the 10% Steinberg's solution containing 0.6 ppm of methylene blue to prevent bacterial growth. (b) Drug application in lamprey embryos; 20–30 embryos can be exposed to a certain concentration of drugs in each 12-well dish. (c,d) Microinjection in lamprey embryos. The sieve mesh size is 0.61 mm, and wire diameter is 0.23 mm. (e,f) KAEDE (photoconvertible protein) expression in lamprey embryos (stages 18 and 23). KAEDE mRNA combined with nuclear localized signal injected in one-cell eggs after fertilization. The expression can be seen only in each cell nucleus and lasts at least until hatching stage.

stage. However, several reports have shown that CRISPR/Cas9-injected F_0 embryos effectively disrupted target genes, even though each cell was differentially mutated (Square et al. 2015).

22.9 CHALLENGING QUESTIONS

Finally, I suggest some challenging questions in the developmental and genomic fields from an evolutionary perspective.

22.9.1 Cerebellar Evolution

The cerebellum plays an essential role in controlling coordinated movements as well as cognitive and emotional functions in humans. All living gnathostomes have distinct, three-layered cerebella (granular, Purkinje and molecular layers). However, lampreys have an undifferentiated cerebellum, which is only visible as a dorsal lip at the anterior end of the rhombencephalon. They do not have a layered structure, but some cerebellum-specific neuron subtypes have been found. In contrast, the presence of the cerebellum in hagfish is uncertain. Recently, Sugahara et al. (2016) reported on the gene expression in lamprey and hagfish embryos that is essential for cerebellar development. When and how the cerebellum was established and acquired a three-layered structure during vertebrate evolution are intriguing questions. A comparison of cerebellar development between cyclostomes and gnathostomes would answer this question. See Sugahara et al. (2017) for detailed information.

22.9.2 Evolution of the Paired Nostrils

Most fossil jawless fish have a single median nostril, and cyclostomes might also retain this ancestral condition. During development, the gnathostome nasal placode is generated as paired and separated from the hypophyseal placode (Rathke's pouch). In contrast, the median nasal placode and hypophyseal placode arise as a single ectodermal thickening in lampreys and hagfish. The separation of the nasohypophyseal placode and subsequent changes in the migration of neural crest cells might be a key innovation for the acquisition of the jaw (Kuratani et al. 2001).

22.9.3 Origin of the Paired Appendages

Cyclostomes do not possess paired fins that are homologous to human arms and legs. So far, two major theories have been proposed to explain the origin of paired fins. The fin-fold theory posits that paired fins evolved from a longitudinal paired fin-fold. Anaspida, an early Silurian fish, might have had paired folds on the ventral side of the body (Janvier 1996). Another theory is the gill-arch theory that posits that the pectoral fins were the result of the transformation or co-option of the gill arches. It may well be the case that vertebrates acquired the pectoral fin first (see Osteostracans in Figure 22.2). It would be interesting to investigate whether cyclostomes have the potential to form paired appendages. In lamprey embryos, different distribution patterns of the lateral plate mesoderm, which contributes to limb growth, have been reported (Tulenko et al. 2013).

22.9.4 Evolution of the Thyroid Gland

The ammocoetes larvae of lampreys have an endostyle under the pharynx as a secreting organ for filter feeding. Non-vertebrate chordates, *Amphioxus* and ascidians also possess this organ. During metamorphosis, the lamprey endostyle changes into the thyroid gland. Therefore, it is thought that the chordate endostyle is homologous to the vertebrate thyroid gland and transitioned from an endostyle to the thyroid in lamprey evolutionary history (Ogasawara et al. 2001). This theory is based on the homology of the endostyle between lampreys and non-vertebrate chordates. However, the homology of the endostyle remains unclear. In addition, although the hagfish undergoes direct development and thus does not have an endostyle, there is only one much older study regarding the thyroid gland (Stockard 1906). Detailed analysis of hagfish thyroid gland development would shed light on this question.

22.9.5 Timing of Whole-Genome Duplication

As noted previously, hagfish and lampreys possess at least six Hox clusters (Mehta et al. 2013; Pascual-Anaya et al. 2018). Since the homology and relationship between each cluster and gnathostome clusters remain unclear, it is yet to be determined whether the two rounds of WGDs that gnathostome experienced occurred before or after the divergence of the cyclostomes. Deep, detailed comparative synteny analysis of cyclostome genomes would lead to a clearer understanding of the evolution of the vertebrate genome.

ACKNOWLEDGMENTS

I thank Shigeru Kuratani and his past and current laboratory members for cyclostome research, Noboru Sato and Hiroshi Nagashima for lamprey sampling and Osamu Kakitani for hagfish sampling.

BIBLIOGRAPHY

Bardack, D. 1991. First fossil hagfish (myxinoidea): A record from the Pennsylvanian of Illinois. *Science* 254 (5032):701–703.

Braun, C.B., and R.G. Northcutt. 1997. The lateral line system of hagfishes (Craniata: Myxinoidea). *Acta Zoologica* 78 (3):247–268.

Chang, M.M., Wu, F., Miao, D., and J. Zhang. 2014. Discovery of fossil lamprey larva from the Lower Cretaceous reveals its three-phased life cycle. *Proceedings of the National Academy of Sciences of the United States of America* 111 (43):15486–15490.

Conel, J.L. 1929. The development of the brain of *Bdellostoma stouti* 1: External growth changes. *Journal of Comparative Neurology* 47 (3):343–403.

Dean, B. 1899. On the embryology of *Bdellostoma stouti*: A general account of myxinoid development from the egg and segmentation to hatching. *Festschrift zum 70ten Geburststag Carl von Kupffer*:220–276.

Duméril, C. 1806. *Zoologie analytique, ou méthode naturelle de classification des animaux, Rendue plus facile a l'Aide de Tableaux Synoptiques*. Paris: Allais.

Dupret, V., Sanchez, S., Goujet, D., Tafforeau, P., and P.E. Ahlberg. 2014. A primitive placoderm sheds light on the origin of the jawed vertebrate face. *Nature* 507 (7493):500–3.

Evans, T.M., and J.E. Bauer. 2016. Identification of the nutritional resources of larval sea lamprey in two Great Lakes tributaries using stable isotopes. *Journal of Great Lakes Research* 42 (1):99–107.

Fernholm, B., Norén, M., Kullander, S.O., et al. 2013. Hagfish phylogeny and taxonomy, with description of the new genus *Rubicundus* (Craniata, Myxinidae). *Journal of Zoological Systematics Evolutionary Research* 51 (4):296–307.

Fudge, D.S., Herr, J.E., and T.M. Winegard. 2016. *Hagfish Slime, Hagfish Biology*. New York: CRC Press.

Gai, Z., Donoghue, P.C., Zhu, M., Janvier, P., and M. Stampanoni. 2011. Fossil jawless fish from China foreshadows early jawed vertebrate anatomy. *Nature* 476 (7360):324–327.

Gess, R.W., Coates M.I., and B.S. Rubidge. 2006. A lamprey from the Devonian period of South Africa. *Nature* 443 (7114):981–984.

Gorbman, A. 1983. Early development of the hagfish pituitary-gland: Evidence for the endodermal origin of the adenohypophysis. *American Zoologist* 23 (3):639–654.

Gustafson, G. 1935. On the biology of *Myxine glutinosa* L. *Arkiv Zool* 28:1–8.

Heimberg, A.M., Cowper-Sal-lari, R., Sémon, M., Donoghue, P.C., and K.J. Peterson. 2010. microRNAs reveal the interrelationships of hagfish, lampreys, and gnathostomes and the nature of the ancestral vertebrate. *Proceedings of the National Academy of Sciences of the United States of America* 107 (45):19379–19383.

Higuchi, S., Sugahara, F., Pascual-Anaya J., Takagi W., Oisi, Y., and S. Kuratani. 2019. Inner ear development in cyclostomes and evolution of the vertebrate semicircular canals. *Nature* 565 (7739):347–350.

Hirasawa, T., Oisi, Y., and S. Kuratani. 2016. Palaeospondylus as a primitive hagfish. *Zoological Letters* 2 (1):20.

Ichikawa, T., Kobayashi, H., and M. Nozaki. 2000. Seasonal migration of the hagfish, *Eptatretus burgeri*, Girard. *Zoological Science* 17 (2):217–223.

Ishijima, J., Uno, Y., Nunome, M. Nishida, C., Kuraku, S., and Y. Matsuda. 2017. Molecular cytogenetic characterization of chromosome site-specific repetitive sequences in the Arctic lamprey (*Lethenteron camtschaticum*, Petromyzontidae). *DNA Research* 24 (1):93–101.

Jandzik, D., Hawkins, M.B., Cattell, M.V., Cerny, R., Square, T.A., and D.M. Medeiros. 2014. Roles for FGF in lamprey pharyngeal pouch formation and skeletogenesis highlight ancestral functions in the vertebrate head. *Development* 141 (3):629–638.

Janvier, P. 1996. *Early Vertebrates*. Oxford: Oxford University Press.

Johnson, N.S., Swink, W.D., and T.O. Brenden. 2017. Field study suggests that sex determination in sea lamprey is directly influenced by larval growth rate. *Proceedings of the Royal Society*: *Biological Sciences* 284 (1851).

Johanson, Z., Smith, M., Sanchez, S., Senden, T., Trinajstic, K., and C. Pfaff. 2017. Questioning hagfish affinities of the enigmatic Devonian vertebrate Palaeospondylus. *Royal Society Open Science* 4 (7):170214.

Jørgensen, J.M., Lomholt, J.P., Weber R.E., and H. Malte. 2012. *The Biology of Hagfishes*. London: Springer Science & Business Media.

Kataoka, T. 1985. Studies on the propagation of *Lethenteron camtschaticum* (IV) (in Japanese). *Research Reports of Niigata Prefectural Freshwater Fisheries Experiment Station* 12:23–27.

Kohno, S., Kubota, S., and Y. Nakai. 1998. Chromatin diminution and chromosome elimination in hagfishes. In *The Biology of Hagfishes*. London: Springer.

Kuraku, S., Hoshiyama, D., Katoh, K., Suga, H., and T. Miyata. 1999. Monophyly of lampreys and hagfishes supported by nuclear DNA-coded genes. *Journal of Molecular Evolution* 49 (6):729–735.

Kuraku, S., Ishijima, J., Nishida-Umehara, C., Agata, K., Kuratani, S., and Y. Matsuda. 2006. cDNA-based gene mapping and GC3 profiling in the soft-shelled turtle suggest a chromosomal size-dependent GC bias shared by sauropsids. *Chromosome Research* 14 (2):187–202.

Kuraku, S., and S. Kuratani. 2006. Time scale for cyclostome evolution inferred with a phylogenetic diagnosis of hagfish and lamprey cDNA sequences. *Zoological Science* 23 (12):1053–1064.

Kuratani, S., Nobusada, Y., Horigome, N., and Y. Shigetani. 2001. Embryology of the lamprey and evolution of the vertebrate jaw: Insights from molecular and developmental perspectives. *Philosophical Transactions of the Royal Society B: Biological Sciences* 356 (1414):1615–1632.

Kuratani, S., Ueki, T., Hirano, S., and S. Aizawa. 1998. Rostral truncation of a cyclostome, *Lampetra japonica*, induced by all-trans retinoic acid defines the head/trunk interface of the vertebrate body. *Developmental Dynamics* 211 (1):35–51.

Lagadec, R., Laguerre, L., Menuet, A., Amara, A., Rocancourt, C., Péricard, P., Godard, B.G., Rodicio, M.C., Rodriguez-Moldes, I., Mayeur, H., Rougemont, Q., Mazan, S., and A. Boutet. 2015. The ancestral role of nodal signalling in breaking L/R symmetry in the vertebrate forebrain. *Nature Communications* 6.

Lara-Ramirez, R., Pérez-González, C., Anselmi, C., Patthey, C., and S.M. Shimeld. 2019. A Notch-regulated proliferative stem cell zone in the developing spinal cord is an ancestral vertebrate trait. *Development* 146 (1).

Linnaeus, C. 1758. *Systema naturae*. Vol. 1. Stockholm: Stockholm Laurentii Salvii.

Løvtrup, S. 1977. *The Phylogeny of Vertebrates*. New York: John Wiley.

Mallatt, J., and J. Sullivan. 1998. 28S and 18S rDNA sequences support the monophyly of lampreys and hagfishes. *Molecular Biology and Evolution* 15 (12):1706–1718.

Mehta, T.K., Ravi, V., Yamasaki, S., et al. 2013. Evidence for at least six Hox clusters in the Japanese lamprey (*Lethenteron japonicum*). *Proceedings of the National Academy of Sciences* 110 (40):16044–16049.

Morris, S.C., and J.B. Caron. 2014. A primitive fish from the Cambrian of North America. *Nature* 512 (7515):419–422.

Müller, A. 1856. XXVI: On the development of the lampreys. *Annals and Magazine of Natural History* 18 (106):298–301.

Nelson, J.S., Grande, T.C., and M.V.H. Wilson. 2016. *Fishes of the World*. Hoboken, NJ: John Wiley & Sons.

Nikitina, N., Bronner-Fraser, M., and T. Sauka-Spengler. 2009. The sea lamprey *Petromyzon marinus*: A model for evolutionary and developmental biology. *Cold Spring Harbor Protocols* 2009 (1):pdb.emo113.

Nishiguchi, Y., Tomita, T., Sato, et al. 2016. Examination of the hearts and blood vascular system of *Eptatretus okinoseanus* using computed tomography images, diagnostic sonography, and histology. *International Journal of Analytical Bio-Science* 4 (3):46–54.

Ogasawara, M., Shigetani, Y., Suzuki, S., Kuratani, S., and N. Satoh. 2001. Expression of thyroid transcription factor-1 (TTF-1) gene in the ventral forebrain and endostyle of the agnathan vertebrate, *Lampetra japonica*. *Genesis* 30 (2):51–58.

Ohno, S. 1970. *Evolution by Gene Duplication*. New York: Springer Science & Business Media.

Oisi, Y., Kakitani, O., Kuratani, S., and K.G. Ota. 2015. Analysis of embryonic gene expression patterns in the hagfish. In *In Situ Hybridization Methods*. New York: Springer.

Oisi, Y., Ota, K.G., Kuraku, S., Fujimoto, S., and S. Kuratani. 2013. Craniofacial development of hagfishes and the evolution of vertebrates. *Nature* 493 (7431):175–180.

Ota, K.G., Fujimoto, S., Oisi, Y., and S. Kuratani. 2011. Identification of vertebra-like elements and their possible differentiation from sclerotomes in the hagfish. *Nature Communications* 2:373.

Ota, K.G., Kuraku, S., and S. Kuratani. 2007. Hagfish embryology with reference to the evolution of the neural crest. *Nature* 446 (7136):672–675.

Pascual-Anaya, J., Sato, I., Sugahara, F., et al. 2018. Hagfish and lamprey Hox genes reveal conservation of temporal colinearity in vertebrates. *Nature Ecology Evolution* 2 (5):859–866.

Sémon, M., Schubert, M., and V. Laudet. 2012. Programmed genome rearrangements: In lampreys, all cells are not equal. *Current Biology* 22 (16):R641–643.

Smith, J.J., Antonacci, F., Eichler, E.E., and C.T. Amemiya. 2009. Programmed loss of millions of base pairs from a vertebrate genome. *Proceedings of the National Academy of Sciences of the United States of America* 106 (27):11212–11217.

Smith, J.J., Kuraku, S., Holt, C., et al. 2013. Sequencing of the sea lamprey (*Petromyzon marinus*) genome provides insights into vertebrate evolution. *Nature Genetics* 45 (4):415–421, 421e1–421e2.

Smith, J.J., Timoshevskaya, N., Ye, C., et al. 2018. The sea lamprey germline genome provides insights into programmed genome rearrangement and vertebrate evolution. *Nature Genetics* 50 (2):270–277.

Square, T., Romášek, M., Jandzik, D., Cattell, M.V., Klymkowsky, M., and D.M. Medeiros. 2015. CRISPR/Cas9-mediated mutagenesis in the sea lamprey *Petromyzon marinus*: A powerful tool for understanding ancestral gene functions in vertebrates. *Development* 142 (23):4180–4187.

Stockard, C.R. 1906. The development of the thyroid gland in *Bdellostoma stouti*. *Anatomischer. Anzeiger*. 29:91–99.

Sugahara, F., Aota, S., Kuraku, S., et al. 2011. Involvement of hedgehog and FGF signalling in the lamprey telencephalon: Evolution of regionalization and dorsoventral patterning of the vertebrate forebrain. *Development* 138 (6):1217–1226.

Sugahara, F., Murakami, Y., Pascual-Anaya, P., and S. Kuratani. 2017. Reconstructing the ancestral vertebrate brain. *Development, Growth & Differentiation* 59 (4):163–174.

Sugahara, F., Pascual-Anaya, J., Oisi, Y., et al. 2016. Evidence from cyclostomes for complex regionalization of the ancestral vertebrate brain. *Nature* 531 (7592):97–100.

Sugahara, F., Yasunori M., and S. Kuratani. 2015. Gene expression analysis of lamprey embryos. In *In Situ Hybridization Methods*. New York: Springer.

Sumich, J.L. 1992. Benthic communities. In *Introduction to the Biology of Marine Life*. Dubuque, Iowa: Wn. C. Brown.

Tahara, Y. 1988. Normal stages of development in the lamprey, *Lampetra reissneri* (Dybowski). *Zoological Science* 5 (1): 109–118.

Tetlock, A., Yost, C.K., Stavrinides, J., and R.G. Manzon. 2012. Changes in the gut microbiome of the sea lamprey during metamorphosis. *Applied Environmental Microbiology* 78 (21):7638–7644.

Timoshevskiy, V.A., Timoshevskaya, N.Y., and Smith, J.J. 2019. Germline-specific repetitive elements in programmatically eliminated chromosomes of the sea lamprey (*Petromyzon marinus*). *Genes (Basel)* 10 (10).

Tulenko, F.J., McCauley, D.J., Mackenzie, E.L., et al. 2013. Body wall development in lamprey and a new perspective on the origin of vertebrate paired fins. *Proceedings of the National Academy of Sciences of the United States of America* 110 (29):11899–11904.

Yamazaki, Y., Yokoyama, R., Nagai, T., and A. Goto. 2011. Formation of a fluvial non-parasitic population of *Lethenteron camtschaticum* as the first step in petromyzontid speciation. *Journal of Fish Biol* 79 (7):2043–2059.

Zhu, M., Yu, X., Ahlberg, P.E., et al. 2013. A Silurian placoderm with osteichthyan-like marginal jaw bones. *Nature* 502 (7470): 188–193.

23 Current Trends in *Chondrichthyes* Experimental Biology

Yasmine Lund-Ricard and Agnès Boutet

CONTENTS

DOI: 10.1201/9781003217503-23

23.1 INTRODUCTION TO *CHONDRICHTHYES* MODELS

23.1.1 Phylogeny

Chondrichthyes (cartilaginous fish) belong to gnathostomes (jawed vertebrates) and constitute the sister group of *Osteichthyes* (bony vertebrates). This monophyletic group diverged from a common ancestor with the *Osteichthyes* lineage about 420 million year ago (mya) (Brazeau and Friedman 2015) and occupies a pivotal position in gnathostomes. Within the *Chondrichthyes* class, there exists two sub-classes, *Elasmobranchii* (sharks, rays, skates and sawfish) and *Holocephali* (chimeras) (see Figure 23.1 to follow the description of *Chondrichthyes* phylogeny). The earliest trace of *Holocephali* can be found around 420 mya (Inoue et al. 2010). *Holocephali* include a single surviving order, Chimaeriformes (chimeras), with 39 extant species. One popular chimera is *Callorhinchus milii*, also known as the Australian ghost-shark. The elasmobranch subclass includes more than 1,000 species of sharks, skates and rays. Elasmobranchs are composed of eight orders of *Selachii* (modern sharks) and four orders of *Batoidea* (rays, skates, guitarfish and sawfish). Figure 23.1 recapitulates the main *Chondrichthyes* groups and mentions the species that will be discussed in this chapter. It is interesting to note that the *Chondrichthyes* group has survived the five mass extinctions over the last 400 million years.

Because of their phylogenetic position, *Chondrichthyes* have been used to shed light on the origin of gnathostomes. How the last common ancestor of all gnathostomes looked like is the subject of intense debate. Beside *Chondrichthyes* and *Osteichthyes*, jawed vertebrates comprise two paraphyletic groups of extinct animals, placoderms and acanthodians, whose fossils help specify the relationship of this common ancestor with cartilaginous and bony fish. Morphological data from fossil brain cases (Davis et al. 2012; Giles et al. 2015) and dermal skeletons (Zhu et al. 2013) have been used to build these hypotheses. In the study conducted by Davis et al. (2012), modern jawed vertebrates are proposed to be the result of the diversification of *Osteichthyes* away from an ancestral form similar to *Chondrichthyes*, to which acanthodians belonged. A study analyzing a shark-like

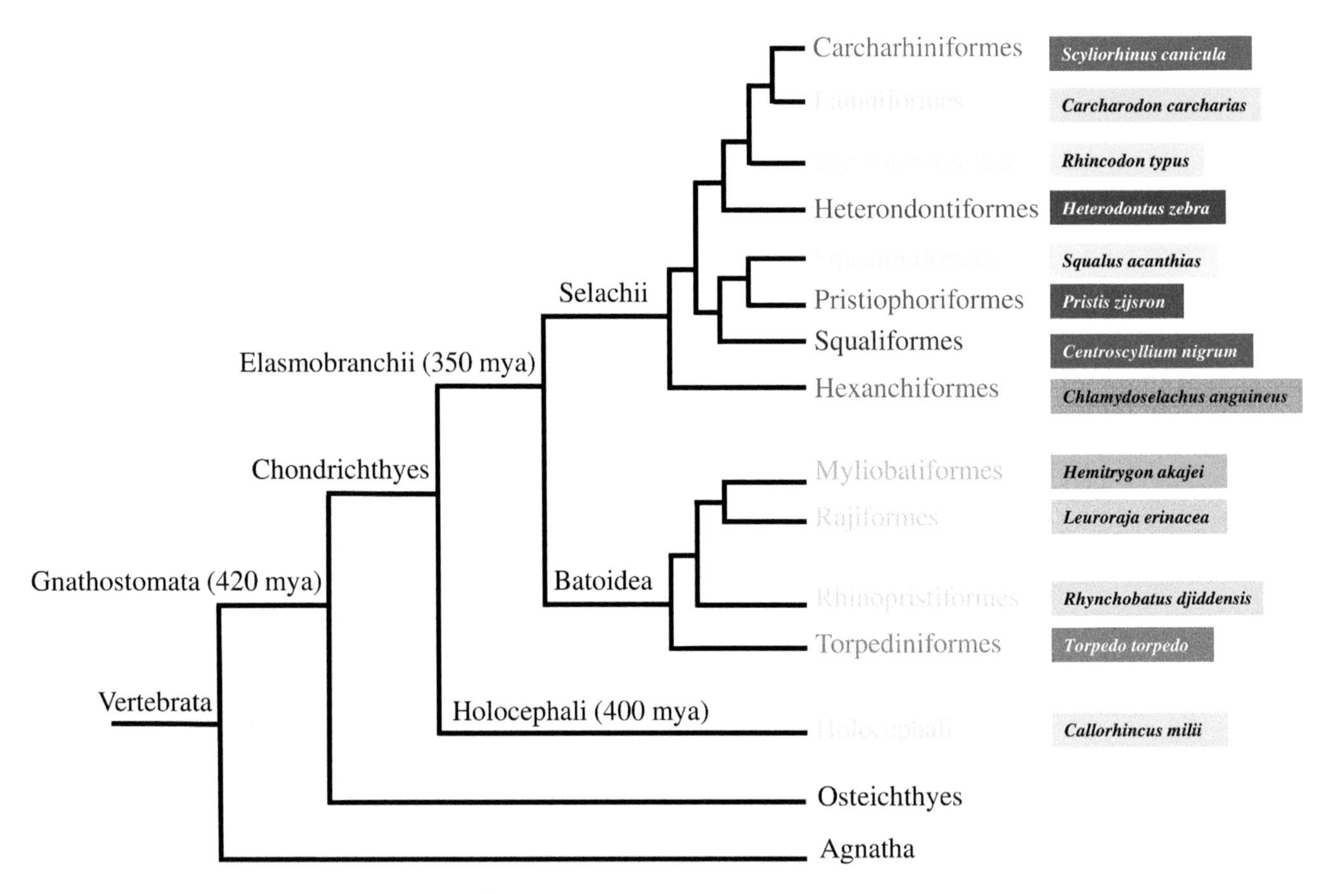

FIGURE 23.1 Phylogenic classification representing *Chondrichthyes* within vertebrates. Terminal clades are orders (Lamniformes, Rajiformes . . .), and each order is illustrated with an example species. *Chondrichthyes* comprise *Elasmobranchii* and *Holocephali*. *Elasmobranchii* include *Selachii* and *Batoidea*. The *Selachii* superorder encompasses eight orders: Carcharhiniformes (ground sharks), Heterodontiformes (bullhead sharks), Hexanchiformes (frilled and cow sharks), Lamniformes (mackerel sharks), Orectolobiformes (carpet sharks), Pristiophoriformes (sawsharks), Squaliformes (dogfish sharks) and Squatiniformes (angel sharks). The Batoidea superorder includes Myliobatiformes (stingrays and relatives), Rajiformes (skates and guitarfish), Torpediniformes (electric rays) and Rhinopristiformes (sawfish).

fossil concluded that the ancestral gnathostome condition for branchial arches was *Osteichthyes*-like (Pradel et al. 2014). Another study described an unexpected contrast between the endoskeletal structure in Janusiscus (an early Devonian gnathostome) and its superficially *Osteichthyes*-like dermal skeleton (Giles et al. 2015). The evolutionary history of jawed vertebrates is still debated, as newly uncovered fossils of early gnathostomes show unseen combinations of primitive and derived characters (Patterson 1981). For a detailed recent discussion about the evolution of jawed vertebrates, the reader can refer to the review from Brazeau and Friedman (2015).

23.2 *CHONDRICHTHYES* IN THE PAST AND PRESENT

Historically, scientific knowledge about *Chondrichthyes* remained limited compared to other vertebrates. Indeed, studying highly mobile animals in vast marine environments remained a challenge until the proper technologies were developed (Castro 2017). In 1868, Jonathan Couch reported descriptions and drawings of 35 *Chondrichthyes* species in the book *History of the Fishes of the British Islands* (1863, Figure 23.2), which constitutes one of the first atlases of the group. This diverse class contains some of the first animal models in experimental biology.

23.2.1 The Rise of *Chondrichthyes* as Models in Experimental Biology

The earliest mention of *Chondrichthyes* by scientists dates back to Aristotle (Demski and Wourms 2013). His observations include i) the distinction between oviparous and viviparous modes of reproduction in sharks, skates and rays; ii) description of the female and male reproductive system; iii) description of the shark and skate egg case structure and observations on embryonic development; and iv) notes on breeding seasons and migrations for "pupping" (Demski and Wourms 2013). Wourms (1997) extensively described the history of the rise of both *Osteichthyes* and *Chondrichthyes* embryology. He argues that the progressive development of knowledge of teleosts and *Chondrichthyes* embryology during the 19th century drove the birth of modern descriptive embryology. This led to the rise of comparative embryology associated with evolutionary studies and then to the experimental and physiological study of development (Wourms 1997). For example, Kastschenko (1888) used catshark embryos (*Scyliorhinus*

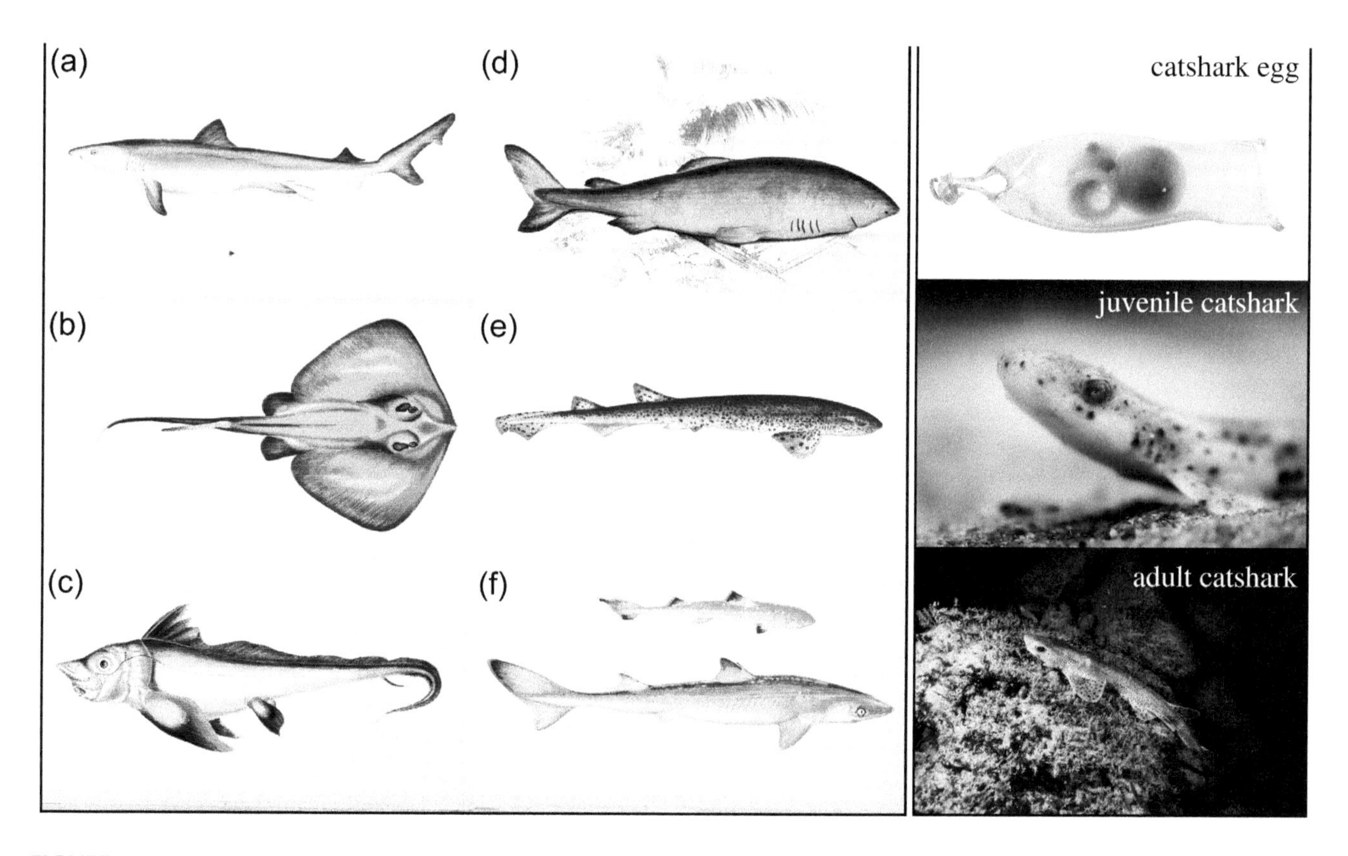

FIGURE 23.2 Drawings and pictures of *Chondrichthyes* species. (a–f) Drawings represent the white shark (a), sting ray (b), arctic chimera (c), Greenland shark (d), catshark (e) and picked dog (f). Right panel represents several steps of the catshark (*Scyliorhinus canicula*) life cycle: embryo, juvenile and adult stage. (*From A History of the Fishes of the British Islands* by Jonathan Couch, Vol I. 1868. Station Biologique de Roscoff (SBR) library collection and Biodiversity Heritage Library. Photos courtesy of © Station Biologique de Roscoff, Wilfried Thomas.)

canicula, Figure 23.2) as experimental models to test a developmental theory.

23.2.2 The Study of *Chondrichthyes* Behavior

The first reported studies on *Chondrichthyes* behavior emerged later. The initial studies on shark behavior include those carried out by Sheldon (1909, 1911) and by Parker (1914). The focus of these studies was the influence of the eyes, ears and other allied sense organs on the movements of the dogfish *Mustelus canis*. Remarkably, a military project entitled "Project Headgear" (1958–1971) conducted experiments in which sharks were trained to carry explosives. The details of this project have never been released. With increasingly sophisticated technology, the themes addressed in behavioral research have widened, and an array of studies can be found (Tricas and Gruber 2001; Sundström et al. 2001; Kelly et al. 2019; Gardiner 2012; Myrberg 2003; Gruber and Myrberg 1977; Hammerschlag 2016; Aidan et al. 2005). As the field of animal cognition expands, social learning in lemon sharks (Guttridge et al. 2013); tool use in batoids (Kuba et al. 2010); learning, habituations and memory in a benthic shark (Kimber et al. 2014); and spatial memory and orientation strategies in stingrays (Schluessel and Bleckmann 2005) have helped build a picture of *Chondrichthyes* cognitive functions. Schluessel (2015) reviewed the evidence for cognitive abilities in elasmobranchs.

23.2.3 Current Trends in *Chondrichthyes* Research

Current trends in *Chondrichthyes* research were analyzed in a recent review (Shiffman et al. 2020). This review depicts the trends in research efforts over three decades (1985–2016) by analyzing the content of all the abstracts presented at the annual conferences of the American Elasmobranch Society (AES), the oldest and largest professional society for the scientific study and management of these fish (Shiffman et al. 2020). AES research was most frequently on movement/telemetry, age and growth, population genetics, reproductive biology and diet/feeding ecology, with different areas of focus for different species or families. Certain biases exist in areas of investigations such as species "charisma" (e.g. white shark, *Carcharodon carcharias*), accessibility to long-term established field research programs (e.g. lemon shark, *Negaprion brevirostris*, and sandbar shark, *Carcharhinus plumbeus*) or ease of model maintenance for lab-based research (e.g. bonnethead shark, *Sphyrna tiburo*) (Shiffman et al. 2020). Nearly 90% of all described *Chondrichthyes* species have never been mentioned in an AES abstract, including some of the most threatened species in the Americas (Shiffman et al. 2020).

23.2.4 *Chondrichthyes* Conservation Status

Chondrichthyes are considered one of the most threatened vertebrate groups by the International Union for the Conservation of Nature (IUCN) Red List (McClenachan et al. 2012; Dulvy et al. 2014; White and Last 2012). The threats faced by *Chondrichthyes* can be grouped into the effects of various fishing activities and of habitat loss (Dulvy et al. 2014; Jennings et al. 2008) and environmental degradation such as pollution (Lyons and Wynne-Edwards 2018). Alarmingly, sharks are subject to a global slaughter; shark products such as dried fins have high commercial value and a high exposure to international trade (Gross 2019). Human exploitation of *Chondrichthyes* is aggravated by certain life history traits, like low fecundity, the production of small numbers of highly precocious young, slow growth rates and late sexual maturity (Collin 2012). In 2020, a study showed that fishing exploitation in the Mediterranean might exert an evolutionary pressure toward early maturation in the catshark, *Scyliorhinus canicula* (Ramírez-Amaro et al. 2020). Additionally, sharks are considered at a relatively high risk for climate change (Cavanagh et al. 2005; Rosa et al. 2014). Indeed, climate change is already affecting ocean temperatures, pH and oxygen levels. How ocean warming, acidification, deoxygenation and fishery exploitation may interact to impact *Chondrichthyes* populations is yet to be determined (Sims 2019; Rosa et al. 2017; Wheeler et al. 2020). The use of *Chondrichthyes* models in experimental biology must pay heed to conservation status.

23.2.5 The Science behind Conservation Efforts

Conservation efforts benefit from multidisciplinary approaches in assessing what conditions impact species survival. For example, quantifying distribution patterns and species-specific habitat associations in response to geographic and environmental drivers is critical to assessing risk of exposure to fishing, habitat degradation and the effects of climate change (Espinoza et al. 2014). *Chondrichthyes* extinction risk has been found to be determined by reproductive mode but not by body size (García et al. 2008). In this same study, extinction risk was highly correlated with phylogeny, and as such, the loss of species is predicted to be accompanied by a loss of phylogenetic diversity (García et al. 2008). Moreover, distribution patterns (Espinoza et al. 2014) ecosystem diversity (Boussarie et al. 2018), ecological context (Collin 2012) and behavior (Wheeler et al. 2020) are valuable for meaningful management and conservation. Behavioral differences within and between species, as well as the ecological context in which a species exists, can have important management implications. In an effort to combat the many threats *Chondrichthyes* face, several regions now have shark sanctuaries or have banned shark fishing—these regions include American Samoa, the Bahamas, Honduras, Dominican Republic, the Cook Islands, French Polynesia, Guam, the Maldives, Saba, St Marteen, New Caledonia, Bonaire, The Cayman Islands, the Marshall Islands, Micronesia, the Northern Mariana Islands and Palau (Bell 2018). These measures reveal that shark conservation has been understood as important.

23.3 BIOGEOGRAPHY

Chondrichthyes occupy a variety of ecological habitats all around the world. While some are restricted to relatively specific zones (as a function of temperature, osmolality or resources), other *Chondrichthyes* have wider distributions and migratory routes that lead them across the oceans. These habitats include;

- Benthic zones (e.g. the little skate *Leuroraja erinacea*)
- Coastal waters (e.g. the spiny dogfish *Squalus acanthias*)
- Cold waters (e.g. the Greenland shark *Somniosus microcephalus*)
- Deep sea (e.g. the Portuguese dogfish *Centroscymnus coelolepis*)
- Estuaries (e.g. the smalltooth sawfish *Pristis pectinate*)
- Lakes (e.g. the bull shark *Carcharhinus leucas*).
- Mangroves (e.g. the long comb sawfish *Pristis zijsron*)
- Open sea (e.g. pelagic sting ray *Pteroplatytrygon violacea*)
- Reefs (e.g. the blacktip reef shark *Carcharhinus melanopterus*)
- Rivers (e.g. the ocellate river stingray *Potamotrygon motor*)
- Tropical waters (e.g. the reef manta ray *Mobula alfredi*)

Depending on local availability, scientists have developed different models. In Europe, *Scyliorhinus canicula*, or the small-spotted catshark, can be described as a historical *Chondrichthyes* model in biology (Coolen et al. 2008) (Figure 23.2). Their spatial distribution spans from the Northeast and Eastern Central Atlantic, Norway and the Shetland Islands to Senegal (possibly along the Ivory Coast), as well as throughout the Mediterranean Sea. The IUCN defines the small-spotted catshark as one of the most abundant elasmobranchs in the Northeast Atlantic and Mediterranean Sea (IUCN SSC Shark Specialist Group et al. 2014). As such, the species is assessed as Least Concern.

23.4 *CHONDRICHTHYES* LIFE CYCLES

23.4.1 Reproductive Strategies

For all *Chondrichthyes*, fertilization is internal, and a paired pelvic male organ called claspers deliver sperm inside the female. Additionally to the pelvic claspers, *Holocephali* have a cephalic clasper (Tozer and Dagit 2004). Female elasmobranchs have been shown to store sperm (Pratt and Carrier 2001). Advantageously for science, *Chondrichthyes* are the vertebrates with the most diverse reproductive strategies; these include maternal investment, placental viviparity, ovoviviparity or strict lecithotrophic oviparity (yolk-dependent) (Dulvy and Reynolds 1997). These species-specific developmental specializations enable investigations on the evolution of reproductive strategies within a single clade (Mull et al. 2011). Ovoviviparous development, in which eggs hatch internally, is the norm in manta rays, the spiny dogfish, sawfish and whale sharks. The majority of *Chondrichthyes* species are oviparous (egg-laying): examples include the little skate and the small-spotted catshark. Viviparity or live birth is found in hammerhead sharks, bull sharks and blue sharks. Besides sexual reproduction, asexual parthenogenesis has been observed in captive *Chondrichthyes* such as the zebrashark (Dudgeon et al. 2017), the hammerhead shark (Chapman et al. 2007) and the sawfish (Fields et al. 2015). Fecundity is as few as 1 to 10 per litter in the electric ray, *Torpedo torpedo* (Diatta 2000), and as many as 300 per litter for the whale shark, *Rhincodon typus* (Joung et al. 1996).

Of these reproductive mechanisms, the most conducive to experimental manipulation is oviparity, as it facilitates handling. Importantly, oviparous species act as a steady sample bank for molecular and cellular investigations without needing to sacrifice the mothers. According to Compagno's review (1990) on *Chondrichthyes* life-history styles, approximately 43% of *Chondrichthyes* utilize oviparity, including all *Chimaeriformes* (chimeras), *Heterodontiformes* (bullhead sharks), *Rajoidae* (skates) and *Scyliorhinidae* (catsharks) (Compagno 1990). Many species can be maintained in captivity and will lay eggs throughout an annual season; embryos at various developmental stages can thus be obtained in the laboratory year-round. Artificial insemination has been reported for two oviparous species, the clearnose skate, *Raja eglanteria* (Luer et al. 2007), and the cloudy catshark, *Scyliorhinus torazame* (Motoyasu et al. 2003). Additionally, sperm storage allows wild-caught females to lay eggs for several months (*Scyliorhinus canicula*, Figure 23.2) without requiring males or captive mating events.

23.4.2 *Chondrichthyes* Species in Developmental Biology

Compared to other model species in genetics and development (such as *C. elegans* or *Drosophila*), the slow development of *Chondrichthyes* can be an advantage, as it confers a better spatial and temporal resolution. The choice of a *Chondrichthyes* model for developmental biology warrants knowledge on the species lifecycle; fecundity, sexual maturity and longevity. Estimated longevity can be as short as ten years for sharpnose sharks, *Rhizoprionodon spp.* (Cailliet et al. 2001), and as long as 272 years for Greenland sharks (Figure 23.2), *Somniosus microcephalus* (Nielsen et al. 2016).

A common *Chondrichthyes* shark model is the oviparous *S. canicula*. Detailed information on the small-spotted catshark such as maturity, fecundity and occurrence is described by Capapé (2008). This species deposits egg-cases protected by a horny capsule with long tendrils (Figure 23.2). Embryos, juveniles and adults (Figure 23.2) can be kept in lab facilities. Such is the case at the Station Biologique de Roscoff or at the Observatoire Océanologique de Banyuls-de-mer in France.

A *Chondrichthyes* skate model that is recurrent in developmental biology is the oviparous *Leucoraja erinacea*, or little skate (see details concerning the suitability of this animal as a lab model in Clifton et al. 2005). Little skates can be maintained in tanks, and egg-carrying females can be identified by palpation. Eggs are produced in pairs at intervals of about seven days, and hatching requires about six months at 15°C. Refrigerator temperatures can be used to hold embryonic development in stasis. Furthermore, the slow development of *Leucoraja erinacea* allows removal and *in vitro* culture of embryonic cells as well as transplantation of modified cells back into the embryo (Mattingly et al. 2004). Thanks to the reduced metabolic rates (ion transport and oxygen consumption) associated with cold-water habitats, the little skate exhibits an increased stability of cells, tissues and cellular macromolecules, including nucleic acids (Clifton et al. 2005).

Most holocephalans are found in the deep waters of the continental shelf and slope and as a result are unlikely candidates for captivity/lab use. The spotted ratfish (*Hydrolagus colliei*) is one notable exception occurring in near-shore waters (Tozer and Dagit 2004).

The small-spotted catshark and little skate are examples of how *Chondrichthyes* offer new perspectives for comparative studies of vertebrate development relative to the more traditional zebrafish, *Xenopus*, avian and mammalian developmental models. Table 23.1 compiles the existing papers on the development of specific *Chondrichthyes* species.

23.5 *CHONDRICHTHYES* EMBRYOGENESIS

23.5.1 Early Embryogenesis and Gastrulation

The main steps of early embryogenesis (*ovum* to gastrulation) of the elasmobranch embryo are documented for several oviparous species (see Table 23.1), and the following data are based on Balfour and Ballard's descriptions (Balfour 1878; Ballard et al. 1993). As in avian eggs, the cytoplasm

TABLE 23.1

Compilation of papers that describe a *Chondrichthyes*' embryogenesis (Conservation status of said species is detailed as reported by the IUCN Red List Status).

Picture	Species	Reference papers	Conservation status; IUCN Red List Status
	Chain Dogfish, *Scyliorhinus rotifer*	Castro et al., 1988	Least Concern
	Small Spotted Catshark, *Scyliorhinus canicula*	Ballard et al., 1993 Mellinger et al., 1994	Least Concern
	Elephant Shark, *Callorhincus milii*	Didier et al., 1998	Least Concern
	Clearnose Skate, *Raja eglanteria*	Luer et al., 2007	Least Concern
	Little Skate, *Leucoraja ocellata*	Maxwell et al., 2008	Near Threatened
	Greater Spotted Catshark, *Scyliorhinus stellaris*	Musa, Czachur, and Shiels 2018	Near Threatened
	Whitespotted Bamboo Shark, *Chiloscyllium punctatum*	Onimaru, Motone et al., 2018	Near Threatened
	Red Stingray, *Hemitrygon akajei*	Furumitsu, Wyffels and Yamaguchi et al., 2019	Near Threatened
	Frilled Shark, *Chlamydoselachus anguineus*	López-Romero et al., 2020	Least Concern

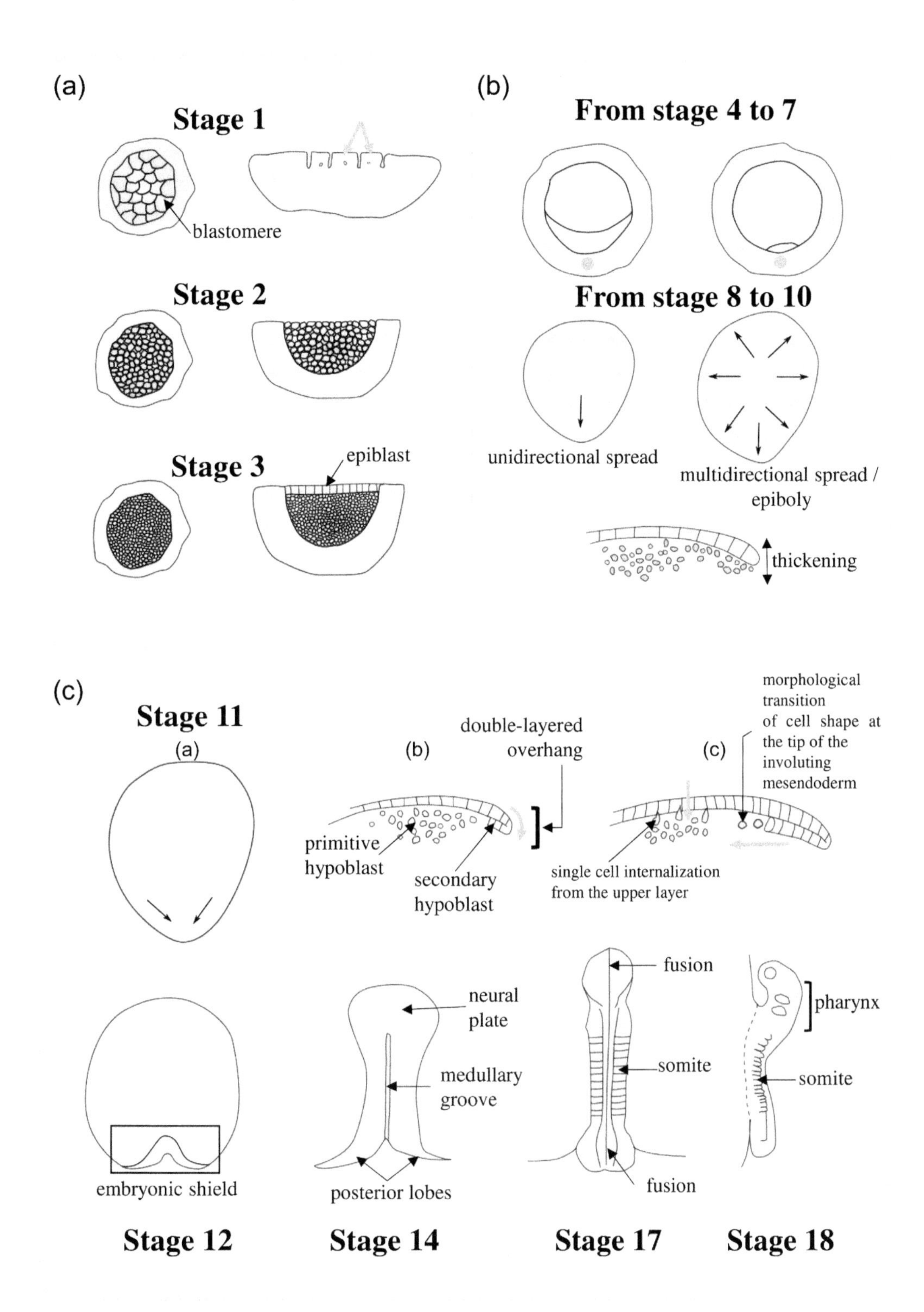

FIGURE 23.3 Early steps of catshark embryogenesis. (a) From stages 1 to 3. Right drawings: dorsal views of the embryo. Left drawings: cross-sections of the corresponding embryos. Pink arrows point to cells of the syncytial blastodisc. (b) From stages 4 to 10. Pink dots indicate the position of the posterior end of the embryo/blastodisc. Bottom drawing represents a cross-section of the posterior end of the embryo at stage 10. (c) Stages 11 to 18. (a) dorsal view of the embryo at stage 11. Arrows represent cells converging to the midline at the posterior end. (b, c) cross-sections of the embryo at the posterior end (over stage 11). In (b) the pink arrow represents mesendodermal cells involuting above the archenteron. (c) the horizontal pink arrow illustrates cell movements from the involuting mesendoderm. The vertical arrow illustrates the movement of single cells internalized from the upper layer. (Adapted from Balfour (1878), Vandebroek (1936), Ballard et al. (1993). Artwork: David Wahnoun, DigitalMarine.)

comprises large amounts of yolk, making segmentation possible in only a small portion of the telolecithal egg cell. These first cleavages start within the oviduct with the particularity of being incomplete. Cell membranes of the first blastomeres do not close up at their bases so that the cytoplasms and the underlying yolk are continuous and form a syncytial blastodisc (stage 1, Figure 23.3A). Around the 100-cell stage, the blastodisc is not syncytial anymore, and the loosely arranged blastomeres exhibit a spherical morphology (stage 2, Figure 23.3A). Later on, the density of inner blastomeres increases, and they are covered by an epithelium made of columnar cells, the epiblast (stage 3, Figure 23.3A). Dorsal views of the blastodisc will later display a crescent-like structure at the posterior end that will finally disappear (from stages 4 to 7, Figure 23.3B). From stages 8 to 10, the round blastodisc shifts to an oval shape due to a unidirectional posterior spread (Figure 23.3B). As epiboly proceeds, the spreading of the blastodisc becomes multidirectional, and a thickening starts to be observed at the posterior end. This cellular densification in the posterior area of the blastoderm tends to intensify at stage 10 (Figure 23.3B, cross-section).

An important feature of the stage 11 embryo is the folding of the epithelial upper layer over the yolk, generating a double-layered overhang (Figure 23.3C, a, b and c). The space created between this overhang and the yolk corresponds to the future archenteron of the embryo. Gastrulation properly starts at this stage, with the lower layer of the overhang representing the mesendodermal cells involuting above the archenteron (Figure 23.3C, b). This forming mesendodermal layer can be referred to as a secondary hypoblast, while the mass of inner blastomeres is called the primitive hypoblast (Ballard et al. 1993, Figure 23.3C, b). Several studies described cell movements accompanying mesendoderm and mesoderm formation during gastrulation. Cell tracking experiments showed that labeled cells within the upper layer of the overhang at the very beginning of stage 11 are later displaced onward within the involuting mesendoderm layer (Godard et al. 2014, Figure 23.3C, c). Similar experimental approaches revealed that single cells can be internalized from the upper layer of the blastoderm to take part in mesoderm formation (Godard et al. 2014, Figure 23.3C, c). On the other hand, Balfour (1878) observed that epithelial cells at the tip of the involuting mesendoderm undergo a morphological "transition", acquiring the shape of the inner rounded blastomeres (Figure 23.3C, c). Similar cell shape changes in this area have also been reported by Coolen et al. (2007). These observations suggest that epithelial cells both from the upper layer (epiblast) and from the tip of the involuting mesendoderm undergo an epithelial to mesenchymal transition (EMT) during gastrulation. *In situ* hybridization experiments performed with the mesoderm marker *Brachyury* at this stage in catshark embryos suggest other types of movements. In addition to being expressed at the site of the involuting mesendoderm, its expression pattern also describes a thin ring all around the blastoderm, which suggests that cells from the margin are converging to the midline at the posterior end of the embryo (Sauka-Spengler et al. 2003) (Figure 23.3C, a). The convergence of cells from the periphery of the blastoderm to the posterior end of the midline has been initially reported by Vandebroek (1936). In the future, development of live imaging approaches on elasmobranch embryos would definitively help to shed light on the spatial and temporal behaviors of their cells during gastrulation.

23.5.2 From Axis Formation to Pharynx Segmentation

At stage 12, the posterior end of the blastoderm exhibits a V-shaped structure referred to as the embryonic shield (Figure 23.3C). A slight depression is observed in the middle of the embryonic shield. It will give rise, by extension from posterior to anterior, to the medullary groove, stating the position of the embryonic axis (Figure 23.3C, stage 14). As the embryo increases in length, the anterior part will enlarge (neural plate, Figure 23.3C, stage 14) and rise to form the neural/medullary folds. In the posterior region, the two arms of the embryonic shield (posterior lobes, Figure 23.3C, stage 14) will progressively shrink to fuse and enclose the neural tube and the archenteric cavity (Figure 23.3C, stage 17). Similar fusion of the neural folds is observed in the anterior part (Figure 23.3c, stage 17). Several pairs of somites are formed during the process of neural tube closure (Figure 23.3C, stage 17). While the trunk pursues its segmentation through the formation of additional somite pairs, the pharynx area undergoes metamerization, too; several branchial clefts will appear (Figure 23.3C, stage 18).

23.6 *Chondrichthyes* Anatomy and Sensory Biology

23.6.1 External Features

In this section, typical *Chondrichthyes* body plans will be proposed for *Selachimorpha*, *Batoids* and *Holocephalans*, and general external features will be briefly discussed. For a more detailed account of *Chondrichthyes* anatomy, *The Dissection of Vertebrates, Second Edition* by Gerardo De Iuliis and Dino Pulerà is highly informative (2019).

All *Chondrichthyes* breathe through five to seven pairs of gills, depending on the species. As a general rule, pelagic (open sea) species have to keep swimming to ensure that oxygenated water is moving through their gills. Demersal species, which live in the water column near the sea floor, will actively pump water in through their spiracles and out through their gills (Salazar 2018). Spiracles are respiratory openings into the pharynx. For sharks, the gills are located on the sides of the body, while the gills are ventral for batoids (De Iuliis and Pulerà 2019). Elasmobranch gill structure and function are described by Wegner (2015). Holocephalans have a single gill opening, on each side, located just anterior to the base of the pectoral fin.

Most sharks, sawfish and chimeras have a heterocercal tail (with unequal upper and lower lobes). This particular structure has been showed to aid in locomotion (Wilga and Lauder 2002). For skates, however, tails range from a thick tail extending from the body to a whip to almost no tail. Stingrays (batoids) possess a venomous stinger located in the mid-area of the tail. This particularity has brought on studies on the chemistry of their venom (da Silva et al. 2015). In most holocephalans, the first dorsal fin is preceded by a venomous spine that can inflict a serious wound (Halstead and Bunker 1952).

Chondrichthyes have tough skin covered with dermal teeth, also called placoid scales (or dermal denticles). The dermal skeleton is the most ancestral mineralized skeleton (see Gillis et al. 2017 for more information) and dermal denticles in the skin of elasmobranchs as well as teeth in the head of all jawed vertebrates are remnants of this structure (Gillis et al. 2017). *Torpediniformes* (electric rays) form an exception, as they have a thick and flabby body, with smooth and loose skin. Notably, *Holocephali* lose their dermal denticles as adults to keep only those on the clasping organ seen on the caudal ventral surface of the male (Salazar 2018). Denticles usually provide protection and, in most cases, streamlining (Salazar 2018). On another level, denticles make the skin of the catshark and the common stingray a highly sought-after product for luxury lining and leatherwork. Called shagreen (or galuchat in French), the use of this skin to wrap travel cases and manufacture holders is mentioned by Buffon as early as 1789 in the second volume of *Histoire naturelle des poissons* (Buffon 1789).

In some shark species, such as the lantern shark, denticles even house bioluminescent bacteria that aid in intraspecific communication (Claes et al. 2015). In 2018, shark denticles were discovered to be laid out according to a Turing-like developmental mechanism explained by a reaction-diffusion system (Cooper et al. 2018).

As aforementioned, bioluminescence and biofluorescence can occur in certain *Chondrichthyes* species. Bioluminescence is the ability of living beings to radiate light on their own or with the help of certain symbiotes (e.g. bacteria). Biofluorescence is the process in which ambient light is absorbed via fluorescent compounds and reemitted at longer, lower-energy wavelengths. Examples of bioluminescent sharks include *Etmopterus spinax* (velvet belly lantern shark) (Claes et al. 2010), *Euprotomicrus bispinatus* (dwarf pelagic shark) (Hubbs et al. 1967) or *Squaliolus aliae* (smalleye pygmy shark) (Claes et al. 2012). They display light-emitting organs (photophores) on their undersides that form species-specific patterns over the flanks and abdomen. The ventral photophores are considered to participate in counter-illumination, a method of camouflage that uses light production to match background brightness and wavelength (Sparks et al. 2014). The bioluminescent flank markings may play a role in intraspecific communication (Gruber et al. 2016). The roles of biofluorescence are more elusive. The *Urotrygonidae* (American round stingrays), *Orectolobidae* (wobbegongs) and *Scyliorhinidae* (catsharks) families include fluorescent species. As these families are distantly related, biofluorescence is thought to have evolved at least three times in elasmobranchs (Gruber et al. 2016). The swell shark (*Cephaloscyllium ventriosum*), the chain catshark (*Scyliorhinus rotifer*) and round stingray (*Urobatis jamaicensis*) are known to exhibit bright green fluorescence (Sparks et al. 2014). The family of small molecules behind marine biofluorescence reviewed in Park et al. (2019) have been hypothesized to play a role in central nervous system signaling, resilience to microbial infections and photoprotection.

23.6.2 Internal Anatomy

This section is a selection of specific traits of *Chondrichthyes* anatomy deemed important to mention. As the etymology of the term *Chondrichthyes* indicates, they possess a cartilaginous skeleton.

For *Selachii*, the mouth is ventrally located. The upper and lower jaws are lined by multiple rows of serrated, triangular and pointed teeth that continuously grow and shed (De Iuliis and Pulerà 2019). Instead, batoids possess flattened plates for crushing bottom-dwelling prey (De Iuliis and Pulerà 2019). Gynandric heterodonty (sexual dimorphism in teeth) is very common in elasmobranchs, and Berio et al. (2020) described the intraspecific diversity of tooth morphology in the large-spotted catshark and revealed some of the ontogenic cues driving this sexual dimorphism. Holocephalans possess three pairs of tooth plates, two in the upper jaw and a single pair in the lower jaw (Tozer and Dagit 2004). Sawfish (Rhinopristiformes, Batoids, Elasmobranch) are characterized by a long, narrow and flattened rostrum (nose extension) lined with transversal teeth. This feature can also be found in sawsharks (Pristiophoriformes, Selachii, Elasmobranch).

Chondrichthyes have no swim bladders. Buoyancy is rather controlled with a large oil-filled liver, which reduces their specific density. An interesting feature of sharks is the valvular intestine, which bears a spiral valve, a corkscrew-shaped lower portion of the intestine that increases its effective length (De Iuliis and Pulerà 2019). Remarkably, chimeariformes lack stomachs (Salazar 2018).

Unlike mammals, *Chondrichthyes* do not have bone marrow, and red blood cells are produced in the spleen and the epigonal organ. The epigonal organ is a special tissue around the gonads that is only found in certain cartilaginous fish and thought to play a role in the immune system. Red blood cells are also produced in the Leydig's organ (nested along the top and bottom of the esophagus), which is also considered part of the immune system (Mattisson and Faänge 1982). The subclass *Holocephali* lacks both the Leydig's and epigonal organs.

Elasmobranch kidneys deserve a special mention, and the little skate and spotted catshark have been of particular interest for the study of kidney development. The functional unit of the kidney is the nephron, and the process of nephron formation is termed nephrogenesis. In mammals, nephrogenesis comes to a stop shortly after birth. This means nephron endowment is definitive in mammals at birth. Some elasmobranchs have been found to continually form nephrons even after embryonic development. Using kidney histological sections from a spotted catshark juvenile, Hentschel (1991) described nephrogenesis with similar morphological steps as found during mammalian nephrogenesis (Hentschel 1991). This unique capacity is a promising research area to better understand the orchestrating factors behind kidney morphogenesis.

Elasmobranch species possess a rectal (or salt) gland. This epithelial organ is located in the distal intestine and empties into the cloaca. It is composed of many tubules that serve a single function: the secretion of hypertonic NaCl solution (Forrest 2016). Initially discovered by Wendell Burger and Walter Hess (1960), this organ can be cannulated and perfused, and chloride secretion can be measured. As highlighted by Forrest (2016), this organ has helped in understanding the physiology of the mammalian thick ascending limb (TAL), an inaccessible portion of the kidney, which functions to filter sodium (Na^+), potassium (K^+) and chloride (Cl^-).

23.6.3 Sensory Biology

Chondrichthyes are gifted with a plethora of senses that are more or less developed depending on the species. The sensory biology of *Chondrichthyes* can be divided into visual, acoustic, mechanical, chemical, magnetic and electrical detection.

23.6.3.1 Photoreception

Studies that focus on visual function in *Chondrichthyes* have described differing sensitivities to light and colors (Douglas and Djamgoz 2012). Depending on the ecological niche they occupy, *Chondricthyes* have evolved different morphological adaptations to optimize photoreception. These include variation in eye size, eye positioning, mobile pupils, elaborate pupillary opercula and reflective retinal media (Walls 1942). The variety of pupil shapes (horizontal, oblique, U-crescent shaped slits) and pupillary opercula is striking. Usually, elasmobranchs benefit from large visual fields—a horizontal arc of up to 360° (McComb and Kajiura 2008)—while humans have a 210° horizontal arc. Elasmobranch retinas include both rod cells, which allow perception in dim-light conditions, and cone cells, which allow perception in bright-light conditions, higher acuity and possible color distinction (Jordan et al. 2013). Ecological factors seem to condition the proportion of rods and cones and the spectral sensitivity of cones. For example, species that inhabit the dysphotic and aphotic zone possess fewer to no cones (Collin et al. 2006). Concerning batoids, eyes are usually located dorsally, though lateral eye position can also be observed, and eyes can even be vestigial in some electric rays. Some batoids (skates, rays and guitarfish) exhibit several spectrally specific cone pigments that would entail the ability for color discrimination (Hart et al. 2004, Theiss et al. 2007). In 2016, the giant guitarfish (*Rhynchobatus djiddensis*) was discovered to possess the ability to retract its eyes, possibly as a means of protection during predation (Tomita et al. 2016).

23.6.3.2 Audition

Myrberg recounts the history of investigations concerning the hearing abilities of sharks in his *Acoustical Biological of Elasmobranch* review (2001). For sharks, the highest sensitivity has been demonstrated for low-frequency sounds (40 to 800 Hz). Specific sound characteristics attract free-ranging sharks: irregular pulses without sudden increases in intensity and frequencies below 80 Hz. Such characteristics are evocative of wounded or struggling prey (Myrberg 2001). This is an auditory explanation behind the role that sharks play in regulating the health of ocean populations. Recently, Parmentier et al. (2020) described the hearing abilities of the catshark, *Scyliorhinus canicula*, from early embryos to juveniles. Stage 31 embryos were able to detect sounds from 100 to 300 Hz, while juveniles were able to detect sounds from 100 to 600 Hz. As hearing development continues in the catshark, only the frequency range appears to widen, as sensitivity and thresholds were not found to improve with development (Parmentier et al. 2020). This last paper contains references to other studies on *Chondrichthyes* hearing abilities, namely hearing thresholds, frequency range and ear morphology.

23.6.3.3 Mechanosensory System

The mechanosensory systems of elasmobranchs include different tactile sense organs; receptor types and distribution depend on the species (Maruska 2001; Jordan 2008). These systems include lateral line canals, neuromasts and vesicules of Savi (types of sensory hair cells and their supporting cells) and spiracular organs. The lateral line marks the lateral line canals, which contain sensory nerve endings and open to the surface through tiny pores (De Iuliis and Pulerà 2019). These tactile sense organs respond to pressure variations induced by the velocity or acceleration of water flow. The electrosensory and lateral line systems of sawfish extend out along the rostrum. This allows them to sense and manipulate prey (Wueringer et al. 2011).

23.6.3.4 Chemoreception

Sensitivity to chemical signals through taste, chemical sense and olfaction constitutes another sense for *Chondrichthyes*. The underlying organs behind these functions include olfactory sacs (for olfaction) and taste papillae (gustation). Sharks have been found to locate potential food using the difference in bilateral odor arrival times (Gardiner and Atema 2010). Pharyngeal denticles and taste papillae possess receptors used for gustation. The morphological adaptations that are pharyngeal denticles could help sharks catch and direct food items and prevent injury of the mouth lining during food manipulation and consumption (Atkinson et al. 2016). Both dermal and oral denticles possess species-specific microstructural morphology that can be applied as a taxonomical tool (Bs et al. 2019). During odor source localization, combinatory signals will help locate potential prey. Gardiner and Atema (2007) looked into the contribution of different senses (olfaction, mechanoreception and vision) to odor perception in the smooth dogfish *Mustelus canis*. Interestingly, they found that the lateral line is required to locate odor sources (Gardiner and Atema 2007).

23.6.3.5 Magnetoreception

Fascinatingly, elasmobranchs have been observed to swim in straight lines for extended periods of time in a highly oriented manner and to navigate in relation to magnetic fields. These observations are true for tiger sharks (*Galeocerdo cuvier*; Holland et al. 1999), blue sharks (*Prionace glauca*; Carey et al. 1990) and scalloped hammerhead sharks (*Sphyrna lewini;* Klimley 1993). Meyer et al. (2005) showed experimentally that sharks can detect variations in the geomagnetic field. They performed condition experiments on captive sharks to determine how they detect magnetic fields and to measure detection thresholds. The anatomical modules underlying magnetoreception could be mediated directly via a magnetite-based sensory system or indirectly via the electrosensory system (Sundström et al. 2001). Indeed, the exact cells, molecules and receptors behind magnetoreception in elasmobranches remain unknown.

23.6.3.6 Electroreception

Electroreception is important in many *Chondrichthyes*. In 1678, Stefano Lorenzini first described pores dispersed on a shark's head without identifying their sensory role. It was only in the 1960s that their function began to be elucidated and identified as a modified part of the lateral line system. Named after Stefano Lorenzini, the ampullae of Lorenzini form a network of jelly-filled pores that act as sensing organs. These pores are connected to sensory cells by gel-filled canals and are highly sensitive to low-frequency electrical stimuli produced by both non-biological and biological sources. Ampullae of Lorenzini are mostly described in *Chondrichthyes*; however, they are also found in *Chondrostei. Chondrostei* are *Actinopterygii* in which the cartilaginous skeleton is a derived feature. They include reedfish, sturgeon and bichir. On the other hand, rays possess an electric organ that originates from modified nerve or muscle tissue. The electric field created by this organ is used for navigation, communication, mating (Feulner et al. 2009), defense and the incapacitation of prey.

Jordan et al. (2013) extensively reviewed both the current knowledge on elasmobranch sensory systems and the way in which these sensory systems could inspire methods for bycatch reduction. The following references will allow a deeper look into the sensory system anatomies of sharks (De Iuliis and Pulerà 2019), batoids (Bedore et al. 2014; Wueringer et al. 2011) and holocephalans (Tozer and Dagit 2004; Lisney 2010).

23.7 GENOMIC DATA

23.7.1 Genomes and Transcriptomes

With millions of species on earth, very few genomes or transcriptomes are in fact assembled, annotated and published. However, the availability of this data (genomes, transcriptomes or protein sequences) greatly accelerates studies on phylogeny (Li et al. 2012; Straube et al. 2015), species diversity and population structure (Boussarie et al. 2018), conservation (Corlett 2017), evolutionary history (Inoue et al. 2010; Renz et al. 2013) or human health research. More generally, studies that encompass diverse animal models to compare sequences have been critical for deciphering fundamental physiological mechanisms and conserved gene and protein functions. Another approach is to compare closely related genomes to identify divergent sequences that may underlie unique phenotypes (Stedman et al. 2004). Several studies have shown that non-coding sequences are more comparable between the genomes of humans and cartilaginous fish than between those of humans and zebrafish (Venkatesh et al. 2006; Lee et al. 2011). Both the slower molecular clock of cartilaginous fish relative to teleosts' (Venkatesh et al. 2014; Renz et al. 2013; Martin et al. 1992), as well as the extra whole-genome duplication specific of teleosts (Glasauer and Neuhauss 2014), can explain the comparability of human and *Chondrichthyes* genomes.

In 2013, a tissue-specific transcriptome was generated from the heart tissue of the great white shark (*Carcharodon carcharias*) (Richards et al. 2013). This represented the first transcriptome of any tissue for this species. Strikingly, this transcriptome revealed that the percentage of annotated transcripts involved in metabolic processes was more similar between the white shark and humans than between the white shark and a teleost (Richards et al. 2013). This finding is consistent with those of Venkatesh et al. (2006) who found genomic non-coding elements and the relative position of genes to be more similar between the elephant shark and humans than between the elephant shark and a teleost. In 2014, the first large-scale comparative transcriptomic survey of multiple cartilaginous fish tissues was analyzed: the pancreas, brain and liver of the lesser spotted catshark, *Scyliorhinus canicula* (Mulley et al. 2014). This study contributes to deciphering the molecular-level functions of pancreatic metabolic processes of *Chondrichthyes*. Uncommonly, *Chondrichthyes* possess the ability to both maintain stable blood glucose levels and tolerate extensive periods of hypoglycemia (Mulley et al. 2014). A high-coverage whole-genome sequencing project of *S. canicula* is underway (Génoscope, French National Sequencing Center and laboratory of Sylvie Mazan, Observatoire Océanologique de Banyuls sur Mer, France). A collection of catshark expressed sequence tags (ESTs) is also available in Mazan's lab. The first *Chondrichthyes* whole genome to be sequenced was of the holocephalan *Callorhincus milii*, published by Venkatesh et al. (2014). The genome size is approximately 1 Gbp. The same year, Wyffels et al. sequenced both the nuclear and mitochondrial genomes of the little skate (*Leucoraja erinacea*). The genome represents 3.42 Gbp across 49 chromosomes. Wyffels et al. (2014) introduced Skatebase (www.skatebase.org), a project for the collection of elasmobranch genomes to complete molecular resources for *Chondrichthyes* fish. Additionally, to the little skate genome, mitochondrion sequences from the ocellate spot skate (*Okamejei kenojei*) and thorny skate (*Amblyoraja radiata*) as well as transcriptomes from the spotted catshark and elephant shark can be found. Skatebase also regroups the *Chondrichthyes* sequence data found in NCBI databases, UniProtKB and the Protein Data Bank (PDB) of *Leucoraja erinacea, Callorhinchus milii* and *Scyliorhinus canicula*. Skateblast, hosted on Skatebase, provides a *Chondrichthyes*-specific blast platform with the previously mentioned data. Genomic contigs and features are available for download. In 2017, the draft sequencing and assembly of the genome of the whale shark, *Rhicodon typus*, was published by Read et al. (2017). The whale shark genome represents 3.44 Gbp. In 2018, the brown-banded bamboo shark, *Chiloscyllium punctatum*, and the cloudy catshark, *Scyliorhinus torazame*, *de novo* whole genomes as well as an improved assembly of the whale shark genome were presented by Hara et al. (2018). The genome size of the brownbanded bamboo shark is 4.7 Gbp and the cloudy catshark 6.7 Gbp. In 2018, both the zebra bull-head shark (Onimaru et al. 2018) and ocellate spot skate (Tanegashima et al. 2018) transcriptomes were published. Lastly, in 2019, the white-shark (*Carcharodon carcharias*) genome was published by Marra et al. (2019) with a size of 4.63 Gpb. Figure 23.4 represents a timeline of the *Chondrichthyes* genomes and transcriptomes with reference publications. Further information concerning gene repertoires, genome size variation, ploidy level, sequence composition can be found in a recent review dedicated to elasmobranch genomics (Kuraku 2021).

23.7.2 Gene Family Studies

A gene family is a set of several similar genes formed by duplication of a single original gene and generally with similar biochemical functions. The *Hox* family are well-known genes which act as major regulators of animal development. Developmental expression profiling and transcriptome analysis first described a lack of expression of the 11 *HoxC* genes in *S. canicula* and *L. erinacea* (Oulion et al. 2010, 2011; King et al. 2011). This finding was initially attributed to a genomic deletion of the entire *HoxC* cluster in these taxa (Oulion et al. 2010; King et al. 2011). A higher coverage sequencing has revealed that *HoxC* genes might in fact exist, but their genomic distributions and the elevated evolutionary rate of their sequences have rendered analysis difficult (Hara et al. 2018). Indeed, examination of several elasmobranch genome scaffolds comprising the presumed *HoxC* genes indicated that the cluster is far from as compact as the clusters of other vertebrate *Hox* genes (Hara et al. 2018). This type of situation highlights the importance of the quality of genomic databases that depends on sequencing depth and coverage. Furthermore, *Chondrichthyes* genomic databases can give insight on the evolution of vertebrate gene repertoires such as the gonadotropin-releasing hormone (GnRH) (Gaillard et al. 2018), *Fox* genes (Wotton et al. 2008) or detoxification gene modules (Fonseca et al. 2019).

23.8 TOOLS FOR MOLECULAR AND CELLULAR ANALYSES

23.8.1 Cell Lines

Cell lines are transformed cell populations with the ability to divide indefinitely. They are powerful tools in understanding physiological, pathophysiological and differentiation processes of specific cells under controlled environmental conditions. Until 2007, no *Chondrichthyes* cell line existed. Currently, two cell lines exist: the SAE cell line derived from *Squalus acanthias*, and LEE-1, derived from an early embryo of *Leucoraja erinacea*. The SAE cell line was the

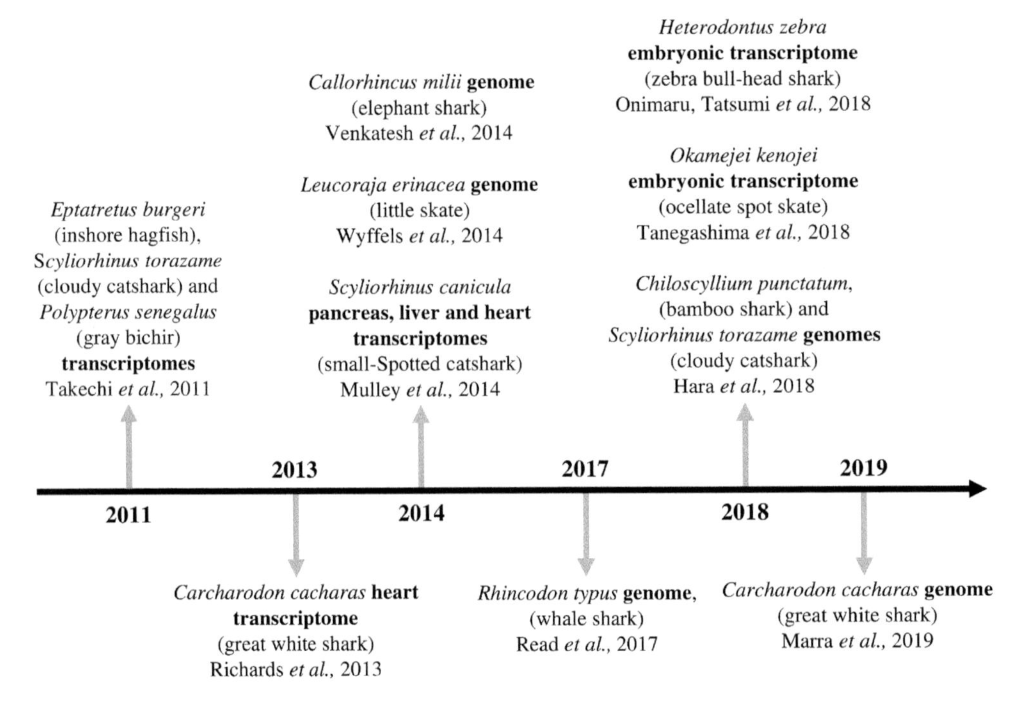

FIGURE 23.4 Timeline showing *Chondrichthyes* genome and transcriptome publications.

first multi-passage continuously proliferating cell line of a cartilaginous fish. Derived from *Squalus acanthias* mesenchymal cells, the primary culture was dispensed into several collagen-coated wells of a 48-well plate. This culture was maintained in a medium modified for fish species and supplemented with cell-type specific hormones, other proteins and sera and plated on a collagen substrate (Parton et al. 2007). SAE cells have been continuously proliferating for three years. For the LEE-1 cell line, isolation and culture were initiated with a stage 28 little skate embryo (Hwang et al. 2008). Similarly to the SAE cell line, cultures for the LEE-1 line were dispensed into collagen-coated wells of a 24-well plate with a basal nutrient medium supplemented with antibiotics and cell-type specific hormones, other proteins and sera.

23.8.2 Descriptive and Functional Approaches

Tools for molecular and cellular analyses have historically been developed with classical models (e.g., *Xenopus*, *Drosophila* or the mouse). The emergence of novel animal models has brought challenges in adapting these tools to varying frameworks. The value of *Chondrichthyes* models in experimental biology, which depends on the success of descriptive and functional approaches, is illustrated in Table 23.2. These approaches generate anatomical and structural data as well as valuable information on molecular mechanisms. The proposed methods can help deepen our understanding of the dynamics of developmental gene patterns, cell fate during morphogenesis, metabolic functions or the mechanisms of tissue regeneration.

TABLE 23.2
Compiled descriptive and functional approaches successfully performed on *Chondrichthyes* species with reference papers for protocol examples. The list of reference papers is not exhaustive

Descriptive approaches: Technique name	*Papers for reference*	*Functional approaches: Technique name*	*Papers for reference*
Alizarin red and Alcian blue clear staining	Eames BF et al. 2007 J Anat O'Shaughnessy KL et al. 2015 Nat Commun Onimaru K et al 2015 eLife Cooper RL et al. 2017 Evol Dev Gillis JA et al. 2017 PNAS	**Beads implantations**	O'Shaughnessy KL et al. 2015 Nat Commun
Cryostat sections **Cryo-scanning electron microscopy**	Sauka-Spengler T et al. 2001 Dev Genes EvolDean MN et al., 2008 Micro Today	**BrdU injection**	Vandenplas S et al. 2016 Dev Biol Lagadec R et al. 2018 Sci Rep
Electron microscopy (with sample coating)	Compagnucci C et al., 2013 Dev Biol	**Cell lines**	Parton A et al., 2007 Comp Biochem Physiol C Toxicol Pharmacol Hwang J-H et al., 2008 Comp Biochem Physiol C Toxicol Pharmacol
In situ hybridization on sections	O'Neill P et al., 2007 Dev Biol Jung H et al., 2018 Cell	**DiI Injection or DiI Cell labelling**	Godard BG et al. 2014 Biol Open illis JA et al. 2017 PNAS
Large-scale scan with high-resolution X-ray computed tomography	Coates MI et al., 2018 Proc Royal Soc B	**EdU injection**	Gillis JA et al., 2016 Dev
Micro-computerised tomography (MicroCT)	Dean MN et al. 2009 J Anat Rasch LJ et al., 2016 Dev Bio Cooper RL et al., 2017 Evol Dev	**Embryo cultures**	Onimaru K et al. 2015 eLife Onimaru K et al 2018. Dev Dyn
Shark MRI	3D Shark T1-Weighted MRI (Biomedical Research Imaging Center of the UNC school of Medicine)	**Extracellular recordings of the spinal cord**	Jung H et al., 2018 Cell
Paraffin embedding and sectioning for immunochemistry, histological coloration, in situ hybridization	Lagadec R et al., 2015 Nat Commun	***In ovo / Ex ovo* drug treatment**	Godard BG et al. 2014 Biol Open [1] Lagadec R et al. 2015 Nat Commun [1] Onimaru K et al., 2015 eLife [2] O'Shaughnessy KL et al., 2015 Nat Commun [3] Gillis JA et al., 2016 Dev [4] Cooper RL et al., 2017 Evol Dev [5] Jung H et al., 2018 Cell [6]
Retrograde labelling	Jung H et al., 2018 Cell	**TUNEL assays**	Debiais-Thibaud M et al., 2015 BMC Evol Biol
Vibratome sections	Jung H et al., 2018 Cell		
WISH (whole mount in situ hybridization)	Sauka-Spengler T et al. 2003 Dev Biol		

[1] Nodal inhibitor SB-505124 [2] Retinoic acid [3] Cyclopamine; 11-KT, SHH-N protein and flutamide [4] Cyclopamine [5] FGF-receptor inhibitor SU5402 [6] Electroporation of hox expression constructs.

23.9 CHALLENGING QUESTIONS

23.9.1 Endogenous *Chondrichthyes* Molecules for Biomedical Applications

23.9.1.1 Molecules Displaying Antibiotic Activity

The well-known squalamine is a cationic steroid isolated from stomach extracts of the spiny dogfish *Squalus acanthias*. It has been demonstrated to display antibacterial activity against Gram-negative and Gram-positive bacteria (Moore et al. 1993). Interestingly, the same study reported that squalamine induced osmotic lysis of *Paramecium caudatum* and had activity against *Candida albicans*, indicating that this shark molecule also holds antiprotozoal and fungicidal properties. As the research on squalamine progressed, it revealed that its chemical features extend beyond the antimicrobial field. This aspect will be presented in the following paragraph.

Microorganisms themselves can produce natural antimicrobial agents, meaning that bacterial symbionts in general can constitute an additional "tissue" to look for putative antibiotics. One specificity of *Chondrichthyes* is their considerable resistance to infection even when their skin is profoundly damaged due to events related to their lifestyle (mating, predation) or to anthropogenic activities. This observation strongly suggests that an innate immunity is operating through the mutualistic interactions taking place in the epidermal mucus layer between marine bacteria and shark epidermis. The most recent study that has investigated the property of these probiotic bacteria is the one from Ritchie et al. (2017). They analyzed the entire bacterial community of the epidermal mucus of three ray species (two marine and one freshwater) and of the clearnose skate, *Raja eglanteria*. They clearly identified particular strains displaying broad-spectrum antibiotic activity and activity against important nosocomial bacteria (Vancomycin-resistant *Enterococcus* [VRE] and Methicillin-resistant *S. aureus* [MRSA]). It goes without saying that interdisciplinary research, in this case intermingling marine microbiology and organism biology, has always sharpened our understanding of immune defense mechanisms. These data on shark epidermis might help medical research in seeking new antimicrobial compounds but also, more generally, in focusing on the preservation of symbiotic bacteria to prevent many types of human diseases and infections. As in *Chondrichthyes*, these bacteria play a fundamental role in our immunity.

23.9.1.2 The High Specificity of *Chondrichthyes* Antibodies

The *Chondrichthyes* adaptative immune system has many shared features with other gnathostomes (Flajnik 2018), except for their particular antibodies. These immunoglobulin (Ig)-like molecules, also called immunoglobulin new antigen receptors (IgNARs), are made of two heavy chains, lack light chains and bear a single variable region domain (V-NARs). In other words, they have one antigen recognition site instead of two, as is the case in the large majority of jawed vertebrate antibodies. Discovered in the 1990s (Greenberg et al. 1995; Roux et al. 1998), IgNARs rapidly raised important interest in the area of drug development. Indeed, the particular folding properties of V-NARs allow them to reach a large panel of protein sites, including hidden epitopes such as those found in the substrate pocket of enzymes that cannot be targeted by "classical" antibodies. Finally, V-NARs also present great solubility and stability, and their small size is another advantage within the field of antibody-based targeting strategy. Specific V-NARs from different elasmobranch species have already been developed to target viral proteins or toxins for medical applications such as anti-viral activity, immunodiagnostics or the development of biosensors. A list of these already-existing targeting V-NARs is available in the review from Kovaleva et al. (2014).

Within gnathostomes, camelids have also evolved such single-domain antibodies, from which the monomeric variable (V) antibody domain constitutes the VHH fragment. As they have been found only in sharks and camels so far, it is believed that these single-domain antibodies are the result of convergent evolution (Flajnik 2018). Nanobody is the name commonly used to indicate camelid VHH and shark V-NAR fragments. The important contribution that nanobodies can bring to the treatment of viral diseases has been spotlighted very recently, in the midst of the COVID-19 pandemic. Wrapp et al. (2020) managed to produce VHH fragments able to prevent the spike (S) glycoprotein of several coronavirus (SARS-CoV-1, SARS-CoV-2 and MERS-CoV) from interacting with their cellular receptors.

23.9.1.3 The Different Properties of Squalamine

As mentioned, squalamine is a polyvalent molecule that also displays antiviral activity, an ability linked to its biochemical properties. The positive charge on account of the spermidine moiety of squalamine (Moore et al. 1993) provides it with high affinity for negatively charged phospholipids of the membrane lipid bilayer (Selinsky et al. 2000). As anionic phospholipids are important to regulate surface charge and protein localization (Yeung et al. 2008), the neutralization of negative charges by squalamine may lead to the disruption of electrostatic potential and shuffle membrane-anchored proteins. This has been demonstrated for Rac1, a GTPase used by many viruses during the process of cell entry, which might impact the viral replication cycle (Zasloff et al. 2011). In the same study, they observed that a wide range of viral pathogens (such as those responsible for dengue, yellow fever, equine encephalitis and Hepatitis B) exhibit variable susceptibility to squalamine in both *in vitro* and *in vivo* tests (Zasloff et al. 2011).

The ability of squalamine to interact with the negatively charged lipids of the cell membrane also represents the underlying mechanism of α-synuclein aggregation impairment (Perni et al. 2017). These α-synuclein aggregates are part of pathogenesis hallmarks of several neurodegenerative disorders, and their destruction constitutes an important challenge to limiting toxicity within the brain parenchyma. Perni et al. (2017) also showed that squalamine exposure led to motility recovery in an animal model of Parkinson disease.

Finally, squalamine has also been demonstrated to impede tumor-associated angiogenesis and the growth of several solid neoplasms (reviewed in Luer and Walsh 2018; Márquez-Garbán et al. 2019). The mechanism of the angiostatic property of squalamine is not fully understood but might rely on, among other explanations, its ability to control endothelial cell shape/volume, as demonstrated by Sills et al. (1998) on embryonic vascular beds. More specifically, squalamine blocks the Na^+/H^+ exchanger (isoform NHE3) (Akhter et al. 1999). Such inhibition of the sodium-hydrogen antiporter will result in the modification of the hydrogen efflux out of the cell, which can explain volume change of endothelial cells.

23.9.1.4 Molecules Displaying Anti-Cancer Activity

Lacking bone marrow, lymphatic system and nodes, elasmobranchs have evolved two particular lymphomyeloïd structures: the epigonal organ associated with the gonads and the Leydig organ located around the esophageal wall, as previously mentioned (Honma et al. 1984). They are involved in the production of red blood cells and play an important role in immune system function.

With the aim to better characterize cell function of these tissues, Walsh and Luer (2018) first showed that cells from the Leydig and epigonal organs display phagocytic and pinocytic activities (Luer and Walsh 2018). Next, looking for more specific bioactive compounds, they tested epigonal conditioned medium (prepared from adult bonnethead shark) and found that it was able to inhibit growth of several mammalian tumor cell lines (Walsh et al. 2006). More specifically on Jurkat T-cell lines, this medium induced caspase-mediated apoptosis (Walsh et al. 2013), but the biochemical nature of this (or these) cell death inducer(s) released from shark epigonal conditioned medium still has/have to be discovered.

As previously mentioned, blocking the neovascularization that accompanies tumor growth is another way to restrain malignancy progression. Besides squalamine, Neovastat (AE-941), a shark cartilage extract, has been shown to inhibit matrix metallopeptidase and VEGF activity (Falardeau et al. 2001; Béliveau et al. 2002), which is consistent with antiangiogenic property. More specifically, Zheng et al. (2007) isolated from the cartilage of the blue shark *Prionace glauca*, a 15.5 kDa polypeptide (PG155) with the ability to reduce vessel formation in vertebrate embryos and tube formation of human umbilical vein endothelial cells (HUVECs). However, Neovastat hasn't gotten beyond phase II of clinical trials so far (Kang et al. 2019), meaning that the use of shark cartilage in the treatment of human malignancies is still exploratory.

23.9.2 Evo-Devo Studies in the Search for the Origin of Skeleton and Brain Asymmetries

23.9.2.1 Endoskeleton and Bone-Like Tissue in *Chondrichthyes*

Although skates, rays, sharks and chimeras are called cartilaginous fish, they possess mineralized structures in their endoskeleton and dermoskeleton (or exoskeleton). Their embryonic endoskeleton is made of a gel-like structure produced by chondrocytes: the hyaline cartilage, a cartilaginous matrix classically stained and observable using Alcian blue pigments. As development progresses, certain parts of the axial endoskeleton such as the vertebrae undergo mineralization, a process that can be visualized using Alizarin red staining.

Truncal vertebrae in elasmobranchs are made of i) a centrum that surrounds the notochord and ii) a dorsal neural arch delimiting the neural canal that contains the spinal cord. Caudal vertebrae also have, ventral to the centrum, a hemal arch that surrounds arteries and veins. Both the centrum and the neural arch of vertebrae of several elasmobranch species display Alcian blue staining at mid-embryogenesis, while Alizarin red coloration is observable in near-hatching embryos (Eames et al. 2007, Enault et al. 2015; Atake et al. 2019), meaning that a mineralization process is occurring on a cartilage-based matrix.

However, in contrast to the mineralization mechanism occurring in the long bones of *Osteichthyes* (the so-called endochondral ossification that also begins within cartilage but from the center to the periphery of the bone), the mineralization in elasmobranch vertebrae starts on the periphery of both the neural arch and the centrum. Interestingly, the expression pattern of type I and type II collagen in these elasmobranch structures is similar to that accompanying the shift from cartilage to mineralized cartilage during endochondral ossification of tetrapod long bones. Type II collagen (cartilage specific) is observed within the cartilaginous center of the neural arch, while type I collagen stains the outer surface of the neural arch (Eames et al. 2007; Enault et al. 2015). It is important, however, to outline that in several teleost species, some cartilages lack type II collagen expression, and bones can exhibit important immunostaining against type II collagen (Benjamin and Ralphs 1991). This indicates that the use of type II collagen as a pure cartilage marker must be considered cautiously.

Type X collagen is another collagen accompanying the process of endochondral ossification. Its expression was demonstrated in the mineralizing sites of catshark vertebrae but not in the type II collagen-expressing non-calcified elements (Debiais-Thibaud et al. 2019).

Another biochemical feature of mineralization is the presence of alkaline phosphatase (AP) activity that can be observed when the inner cartilage of tetrapod long bones is converted into a mineralized matrix. Such AP activity can also be detected in the mineralizing neural arches of near-hatching swell shark embryos (Eames et al. 2007).

Finally, Eames et al. (2007) described a specific cell population in the mineralizing sites of swell shark vertebrae that are morphologically different from chondrocytes, the rounded and well-separated cells embedded into the Alcian blue-positive matrix. These cells, located in the outer mineralizing layer of the neural arches, were surrounded by an Alizarin red-positive matrix and displayed an elongated shape (Eames et al. 2007). Similar flattened cells have been observed at the mineralizing sites of vertebrae in skates (Atake et al. 2019). The nature of these cells has not been investigated yet. Expression of signaling molecules (such as Ihh and several Wnt ligands) and transcription factors (mainly Sp7/Osterix and Runx2) known to be involved in the osteogenic program (Hartmann 2006) would be interesting to explore within the mineralizing elements of elasmobranchs. Such molecular studies would inform us about the mechanism underlying the calcification process in the elasmobranch axial endoskeleton and to what extent this mechanism shares genetic features with the one controlling endochondral ossification of the long bones in *Osteichthyes*. In the field of evo-devo, this last issue constitutes a fascinating question that can now be addressed, since functional experiments are possible (see Table 23.2) in several shark or skate species at different embryonic stages.

23.9.2.2 Exoskeleton (Teeth and Dermal Denticles) in *Chondrichthyes*

Teeth and dermal denticles (also named placoid or dermal scales) constitute the exoskeleton of *Chondrichthyes*. These mineralized appendages/structures made of enamel and dentine surrounding a pulp cavity are known under the general term of odontodes and can easily be observed in laboratories using Alizarin red staining. One important feature of elasmobranchs is that they are polyphyodont, meaning their teeth are continually replaced. The lower and upper jaws are lined by an initial row of mature individual teeth that can display several shapes (for example: needle-like, triangular or flattened). Posterior to this first line of teeth, multiple rows of developing teeth are present, intended to replace those that fall out. Unlike teeth, dermal denticles do not continuously regenerate throughout life. The dentition of holocephalans, the sister group of elasmobranchs, does not possess separate individual teeth but dental plates that grow continuously.

As with many other vertebrate embryonic structures, the development of teeth and dermal denticles involves reciprocal inductive interactions between an epithelium and its underlying mesenchyme, which are engineered by a set of signaling molecules and transcription factors. Using catshark models (*Scyliorhinus stellaris* and *S. canicula*), several works have demonstrated the expression of *Shh*, *Wnt/β-catenin*, *BMP* and *FGF* gene products in the developing dentition of these species, as reported in other bony vertebrates, which suggests that the dental gene regulatory network (GRN) is conserved within gnathostomes (Smith et al. 2009; Debiais-Thibaud et al. 2015; Martin et al. 2016; Rasch et al. 2016). However, the enamel knot, a transient signaling center present in the growing bud and controlling the morphogenesis of teeth cusps, seems to be missing in catshark teeth, indicating that the regulation point for cusp shape works differently in elasmobranchs (Debiais-Thibaud et al. 2015; Rasch et al. 2016).

BrdU pulse-chase experiments performed in embryonic and juvenile catsharks revealed the presence of slow cycling cells within the dental lamina, an epithelial tissue that interacts with the underlying mesenchyme and goes with tooth development (Martin et al. 2016; Vandenplas et al. 2016). These BrdU-positive cells that exhibit a low rate of mitosis constitute a stem cell population that expresses the Sox2 marker (Martin et al. 2016).

Questioning the homology between teeth and dermal denticles, Debiais-Thibaud et al. (2011) investigated the expression of several *Dlx* genes, a family of transcription factors involved in the early specification of dental epithelium and mesenchyme, and found that catshark teeth and caudal primary scales share common expression of *Dlx1*, *Dlx3*, *Dlx4* and *Dlx5* mRNAs. In addition, developing dermal scales in the catshark display the expression of signaling molecules such as BMP4, several FGFs and Shh (Debiais-Thibaud et al. 2015; Martin et al. 2016; Cooper et al. 2017). The conservation of the expression of this gene set supports the hypothesis that the appearance of additional odontodes on body surfaces or within cavities might be the result of a heterotopy, that is, of the dedicated gene regulatory network recruitment at this specific body part (Debiais-Thibaud et al. 2011; Martin et al. 2016).

In amniotes, integumentary structures such as feathers, hair and scales that also derive from epithelial placodes require FGF signaling for their development. FGF ligands are not only expressed in the developing dermal denticles in catsharks. *In ovo* injection of the FGF receptor inhibitor SU5402 leads to the perturbation of caudal dermal scale formation, indicating that this pathway is mandatory for their morphogenesis (Cooper et al. 2017). Such data also suggest that a common GRN might operate within the epithelial placodes of both amniote integumentary structures and elasmobranch dermal scales (Cooper et al. 2017).

Within the developing tooth or dermal denticle, enamel is produced by ameloblasts that differentiate from the epithelial compartment of the bud, while dentine is secreted by odontoblasts deriving from the mesenchymal compartment. Gillis et al. (2017) demonstrated that odontoblasts of the trunk denticles in the little skate (*Leucoraja erinacea*) are derived from trunk neural crest cells. This study constitutes one of the examples of successful cell-lineage tracing experiments in *Chondrichthyes* embryos (by means of DiI microinjection and staining). This work also shows that neural crest cells from the trunk can be skeletogenic, which is different from what has been reported in teleosts (Gillis et al. 2017).

An exhaustive discussion about the origin of teeth in vertebrates and their evolutionary relationship with odontodes in extinct or living species can be found in the recent review from Donoghue and Rücklin (2016).

23.9.2.3 Evolution of Brain Asymmetries in Vertebrates

The position of *Chondrichthyes* as the sister group of all bony vertebrates (*Osteichthyes*, Figure 23.1) undoubtedly makes cartilaginous fish species valuable to study the evolution of a biological structure or process. A recent example is the mechanism underlying asymmetry of the epithalamus, whose evolutionary history in gnathostomes has been brought to light thanks to an elasmobranch model.

The epithalamus arises from the dorsal part of the diencephalon and is composed of two habenular nuclei and a pineal complex (pineal and parapineal glands). In a great majority of vertebrate species, the habenular nuclei display left/right (L/R) asymmetries in size, neurotransmitter and developmental gene expression and in neuronal organization (Concha and Wilson 2001). In addition, while the pineal gland, involved in melatonin secretion, is generally located on the midline, the parapineal gland is found to be connected to the left habenulae and, in rare cases, to the right one (Boutet 2017). Finally, during zebrafish embryogenesis, the dorsal diencephalon displays a left-sided activity of the Nodal pathway known to be involved in L/R asymmetry of internal organs (Signore et al. 2009).

In zebrafish, the connection of the parapineal gland to the left habenula is important, as its experimental removal restores the symmetry of the two habenulae. In contrast, Nodal abrogation leads to randomized connection of the parapineal gland: it is either associated with the left habenula (50% of the time) or the right (50%). In other terms, asymmetry is still present, but laterality is lost (Signore et al. 2009).

The absence of data concerning Nodal expression outside the *Osteichthyes* group and the fact that the left-sided expression of Nodal in the diencephalon had been reported only in teleosts led to the hypothesis that L/R laterality of the epithalamus might have been stochastic at the base of the vertebrate lineage. Experiments performed with the catshark indicated, however, that Nodal is asymmetrically expressed in the dorsal diencephalon as in zebrafish and that it controls habenular asymmetries, including neurogenic asymmetry (Lagadec et al. 2015; Lagadec et al. 2018). Similar results are obtained using lamprey embryos (cyclostome/agnatha; Lagadec et al. 2015). These findings obtained from jawless vertebrates and from *Chondrichthyes* demonstrate that epithalamic asymmetry was not random in the last common ancestor of vertebrates and that diencephalic left-sided Nodal expression was already present in this ancestor. *Chondrichthyes*, and also cyclostomes, thus allowed to understand the evolution of the mechanisms driving this particular brain asymmetry (Boutet 2017).

Note that evolutionary scenarios dealing with brain asymmetry or other processes are never set in stone and can be redrafted later on in light of data collected from additional species. This last point highlights the importance for experimental biology to diversify its models as much as possible. Much more than bringing complexity, embryonic and molecular results raised from a wide range of models, scattered over several taxa, contribute to broadening our view related to evolutionary mechanisms. Data obtained from fossil records are also very useful in such a kind of study.

23.9.3 The Elasmobranch Properties of Kidney Regeneration

As previously mentioned, elasmobranch fish have been found to possess a regenerative kidney. In 2003, Elger et al. described a nephrogenic zone in the adult kidneys of the little skate, *Leucoraja erinacea*. This nephrogenic zone represents a niche within the kidney where stem cell-like cells could reside. The tissue responds to partial reduction of renal mass with the formation of new nephrons. The morphogenic process of neonephrogenesis appears to be an important mechanism for renal growth, as well as for repair of injured kidneys. Renal hypertrophy (a common response to renal mass reduction in humans) contributed only slightly to the reconstitution of the little skate renal mass following the renal reduction experiment (Elger et al. 2003). The morphological analyses demonstrated that a zone of embryonic renal tissue persists in adult skates (Elger et al. 2003). *S. canicula, S. acanthias* and *L. erinacea* have been powerful models for the description of kidney morphogenesis, and multiple studies have detailed renal morphogenesis and architecture using sections (Hentschel 1987; Hentschel 1991; Elger et al. 2003; Cutler et al. 2012). This neonephrogenetic ability found in *Chondrichthyes* is a valuable framework which warrants studies on stem cell homeostasis during nephron ontogeny or repair.

As a conclusion, it appears that *Chondrichthyes* have accompanied experimental biology for a long time. The place they occupy in the vertebrate phylogenetic tree and their particular physiological and biological properties, such as the possibility to regenerate the adult kidney, to replace teeth continually or the unique structure of their antibodies make cartilaginous fish metazoans of great interest.

Human impact on Earth's ecosystems remains, however, overwhelming and a great threat to hundreds of *Chondrichthyes* species. Conservation status has to be taken into account when choosing a model for experimental studies if we want cartilaginous fish to continue to reveal new secrets for the next decades and beyond.

ACKNOWLEDGMENTS

We thank Nicole Guyard from the SBR library, Wilfried Thomas from the marine diving facility of the SBR and all the people at the Roscoff Aquarium Service (RAS) for their valuable help. We are also grateful to David Wahnoun and Haley Flom from the Erasmus+ funded project, DigitalMarine. Research in the laboratory is funded by Sorbonne Universités Emergence Grant [SU-16-R-EMR610_Seakidstem] (IDEX SUPER), and the "Ligue contre le Cancer" (Grand Ouest)]. Y.L.-R. is a student funded by Sorbonne Université (Ecole doctorale Complexité du Vivant ED515).

BIBLIOGRAPHY

Aidan, M.R., Hammerschalg, N., Collier, R.S. and Fallows, C. 2005. Predatory behaviour of white sharks (*Carcharodon carcharias*) at Seal Island, South Africa. *J. Mar. Biol. UK* 85: 1121–1135.

Akhter, S., Nath, S.K., Tse, C.M., Williams, J., Zasloff, M. and Donowitz, M. 1999. Squalamine, a novel cationic steroid, specifically inhibits the brush-border Na^+/H^+ exchanger isoform NHE3. *Am. J. Physiol.* 276: C136–144.

Atake, O., Cooper, D.M.L. and Eames, B.F. 2019. Bone-like features in skate suggest a novel elasmobranch synapomorphy and deep homology of trabecular mineralization patterns. *Acta Biomater.* 84: 424–436.

Atkinson, C.J.L., Martin, K.J., Fraser, G.J. and Collin, S.P. 2016. Morphology and distribution of taste papillae and oral denticles in the developing oropharyngeal cavity of the bamboo shark, *Chiloscyllium punctatum*. *Biol. Open* 5: 1759–1769.

Balfour, F.M. 1878. *A Monograph on the Development of Elasmobranch Fishes*. Hardpress, Miami.

Ballard, W.W., Mellinger, J. and Lechenault, H. 1993. A series of normal stages for development of *Scyliorhinus canicula*, the lesser spotted dogfish (Chondrichthyes: Scyliorhinidae). *Journal of Experimental Zoology* 267: 318–336

Bedore, C.N., Harris, L.L. and Kajiura, S.M. 2014. Behavioral responses of batoid elasmobranchs to prey-simulating electric fields are correlated to peripheral sensory morphology and ecology. *Zoology* 117: 95–103.

Béliveau, R., Gingras, D., Kruger, E.A., Lamy, S., Sirois, P., Simard, B., Sirois, M.G., Tranqui, L., Baffert, F., Beaulieu, E., Dimitriadou, V., Pépin, M.-C., Courjal, F., Ricard, I., Poyet, P., Falardeau, P., Figg, W.D. and Dupont, E. 2002. The antiangiogenic agent neovastat (AE-941) inhibits vascular endothelial growth factor-mediated biological effects. *Clin. Cancer Res. Off. J. Am. Assoc. Cancer Res.* 8: 1242–1250.

Bell, G. 2018. *Shark Sanctuaries around the World*. The Pew Charitable Trust, Philadelphia, PA.

Benjamin, M. and Ralphs, J.R. 1991. Extracellular matrix of connective tissues in the heads of teleosts. *J. Anat.* 179: 137–148.

Berio, F., Evin, A., Goudemand, N. and Debiais-Thibaud, M. 2020. The intraspecific diversity of tooth morphology in the large-spotted catshark *Scyliorhinus stellaris*: Insights into the ontogenetic cues driving sexual dimorphism. *J Anat.* 237:960–978.

Boussarie, G., Bakker, J., Wangensteen, O.S., Mariani, S., Bonnin, L., Juhel, J.-B., Kiszka, J.J., Kulbicki, M., Manel, S., Robbins, W.D., Vigliola, L. and Mouillot, D. 2018. Environmental DNA illuminates the dark diversity of sharks. *Sci. Adv.* 4.

Boutet, A. 2017. The evolution of asymmetric photosensitive structures in metazoans and the Nodal connection. *Mech. Dev.* 147: 49–60.

Brazeau, M.D. and Friedman, M. 2015. The origin and early phylogenetic history of jawed vertebrates. *Nature* 520: 490–497.

Bs, R., Af, A., Jr, K.J. and Reg, R. 2019. Microstructural morphology of dermal and oral denticles of the sharpnose sevengill shark *Heptranchias perlo* (Elasmobranchii: Hexanchidae), a deep-water species. *Microsc. Res. Tech.* 82: 1243–1248.

Buffon, G.-L.L. 1789. *Histoire Naturelles de Poissons*. L'imprimerie Royale, Paris.

Burger, J.W. and Hess, W.N. 1960. Function of the rectal gland in the spiny dogfish. *Science* 131: 670–671.

Cailliet, G.M., Andrews, A.H., Burton, E.J., Watters, D.L., Kline, D.E. and Ferry-Graham, L.A. 2001. Age determination and validation studies of marine fishes: Do deep-dwellers live longer? *Exp. Gerontol.* 36: 739–764.

Capapé, C., Vergne, Y., Guélorguet, O. and Quignard, J.P. 2008. Maturity, fecundity and occurrence of the smallspotted catshark *Scyliorhinus canicula* (Chondrichthyes: Scyliorhinidae) off the Languedocian caost (Southern France, north-western Mediterranean). *Life Environ.* 58: 47–55.

Carey, F.G., Scharold, J.V. and Kalmijn, A.J. 1990. Movements of blue sharks (*Prionace glauca*) in depth and course. *Mar. Biol.* 106: 329–342.

Castro, J.I. 2017. The origins and rise of shark biology in the 20th century. *Mar. Fish. Rev.* 78: 14–33.

Castro, J.I., Bubucis P.M. and Overstrom, N. 1988. The reproductive biology of the chain dogfish. *Copeia*. 3: 740–746.

Cavanagh, R.D., Camhi, M., Burgess, G.H., Cailliet, G.M., Fordham, S.V., Simpfendorfer, C.A. and Musick, J.A. 2005. *Sharks, rays and chimaeras: The status of the chondrichthyan fishes* (S.L. Folwer, ed.). IUCN. Gland, Switzerland and Cambridge, UK.

Chapman, D.D., Shivji, M.S., Louis, E., Sommer, J., Fletcher, H. and Prodöhl, P.A. 2007. Virgin birth in a hammerhead shark. *Biol. Lett.* 3: 425–427.

Claes, J.M., Aksnes, D.L. and Mallefet, J. 2010. Phantom hunter of the fjords: Camouflage by counterillumination in a shark (*Etmopterus spinax*). *J. Exp. Mar. Biol. Ecol.* 388: 28–32.

Claes, J.M., Ho, H.-C. and Mallefet, J. 2012. Control of luminescence from pygmy shark (*Squaliolus aliae*) photophores. *J. Exp. Biol.* 215: 1691–1699.

Claes, J.M., Nilsson, D.-E., Mallefet, J. and Straube, N. 2015. The presence of lateral photophores correlates with increased speciation in deep-sea bioluminescent sharks. *R. Soc. Open Sci.* 2: 150219.

Clifton, S., Amemiya, C.T., Barnes, D., Forrest, J.N., Mattingly, C., Mardis, E., Minx, P., Waren, W. and Wilson, R.K. 2005. Proposal to sequence the genomes of the spiny dogfish shark (*Squalus acanthias*) and the little skate (*Raja erinacea*). Washington University, Genome Sequencing Center. Available at https://www.genome.gov/Pages/Research/Sequencing/SeqProposals/SharkSkateSeq.pdf

Coates, M.I., Finarelli, J.A., Sansom, I.J., Andreev, P.S., Criswell, K.E., Tietjen, K., Rivers, M.L. and La Riviere, P.J. 2018. An early chondrichthyan and the evolutionary assembly of a shark body plan. *Proc. R. Soc. B* 285: 20172418.

Collin, S.P. 2012. The neuroecology of cartilaginous fishes: Sensory strategies for survival. *Brain. Behav. Evol.* 80: 80–96.

Collin, S.P., Lisney, T.J. and Hart, N.S. 2006. Visual communication in elasmobranchs. In: *Communication in Fishes*. Science Publishers Inc., Enfield, NH and Plymouth, UK.

Compagno, L.J.V. 1990. Alternative life-history styles of cartilaginous fishes in time and space. *Environ. Biol. Fishes* 28: 33–75.

Compagnucci, C., Debiais-Thibaud, M., Coolen, M., Fish, J., Griffin, J.N., Bertocchini, F., Minoux, M., Rijli, F.M., Borday-Birraux, V., Casane, D., Mazan, S. and Depew, M.J. 2013. Pattern and polarity in the development and evolution of the gnathostome jaw: Both conservation and heterotopy in the branchial arches of the shark, *Scyliorhinus canicula*. *Dev. Biol.* 377: 428–448.

Concha, M.L. and Wilson, S.W. 2001. Asymmetry in the epithalamus of vertebrates. *J. Anat.* 199: 63–84.

Coolen, M., Menuet, A., Chassoux, D., Compagnucci, C., Henry, S., Lévèque, L., Silva, C.D., Gavory, F., Samain, S., Wincker, P., Thermes, C., D'Aubenton-Carafa, Y., Rodriguez-Moldes, I., Naylor, G., Depew, M., Sourdaine, P. and Mazan, S. 2008. The dogfish *Scyliorhinus canicula*: A reference in jawed vertebrates. *Cold Spring Harb. Protoc.* 2008: pdb.emo111.

Coolen, M., Sauka-Spengler, T., Nicolle, D., Le-Mentec, C., Lallemand, Y., Silva, C.D., Plouhinec, J.-L., Robert, B., Wincker, P., SHI, D.-L. and Mazan, S. 2007. Evolution of axis specification mechanisms in jawed vertebrates: Insights from a chondrichthyan. *PLoS One* 2: e374.

Cooper, R.L., Martin, K.J., Rasch, L.J. and Fraser, G.J. 2017. Developing an ancient epithelial appendage: FGF signalling regulates early tail denticle formation in sharks. *Evo Devo* 8.

Cooper, R.L., Thiery, A.P., Fletcher, A.G., Delbarre, D.J., Rasch, L.J. and Fraser, G.J. 2018. An ancient Turing-like patterning

mechanism regulates skin denticle development in sharks. *Sci. Adv*. 4: eaau5484.
Corlett, R.T. 2017. A bigger toolbox: Biotechnology in biodiversity conservation. *Trends Biotechnol*. 35: 55–65.
Couch, J. 1863. *A History of the Fishes of the British Islands*. Groombridge and sons. London, UK.
Cutler, C.P., Harmon, S., Walsh, J. and Burch, K. 2012. Characterization of aquaporin 4 protein expression and localization in tissues of the dogfish (*Squalus acanthias*). *Front. Physiol*. 3.
da Silva, N.J., Clementino Ferreira, K.R., Leite Pinto, R.N. and Aird, S.D. 2015. A severe accident caused by an ocellate river stingray (*Potamotrygon motoro*) in Central Brazil: How well do we really understand stingray venom chemistry, envenomation, and therapeutics? *Toxins* 7: 2272–2288.
Davis, S.P., Finarelli, J.A. and Coates, M.I. 2012. Acanthodes and shark-like conditions in the last common ancestor of modern gnathostomes. *Nature* 486: 247–250.
Dean, M.N., Mull, C.G., Gorb, S.N. and Summers, A.P. 2009. Ontogeny of the tessellated skeleton: Insight from the skeletal growth of the round stingray *Urobatis halleri*. *J Anat*. 215: 227–239.
Debiais-Thibaud, M., Chiori, R., Enault, S., Oulion, S., Germon, I., Martinand-Mari, C., Casane, D. and Borday-Birraux, V. 2015. Tooth and scale morphogenesis in shark: An alternative process to the mammalian enamel knot system. *BMC Evol. Biol*. 15.
Debiais-Thibaud, M., Oulion, S., Bourrat, F., Laurenti, P., Casane, D. and Borday-Birraux, V. 2011. The homology of odontodes in gnathostomes: Insights from Dlx gene expression in the dogfish, *Scyliorhinus canicula*. *BMC Evol. Biol*. 11: 307.
Debiais-Thibaud, M., Simion, P., Ventéo, S., Muñoz, D., Marcellini, S., Mazan, S. and Haitina, T. 2019. Skeletal mineralization in association with type X collagen expression is an ancestral feature for jawed vertebrates. *Mol. Biol. Evol*. 36: 2265–2276.
De Iuliis, G. and Pulerà, D. 2019. Chapter 3: The Shark. In: *The Dissection of Vertebrates* (Third Edition) (G. De Iuliis and D. Pulerà, eds.), pp. 53–109. Academic Press, Boston.
Demski, L.S. and Wourms, J.P. 2013. *The Reproduction and Development of Sharks, Skates, Rays and Ratfishes*. Springer Science & Business Media, New York, NY.
Diatta, Y. 2000. Reproductive biology of the common torpedo, *Torpedo torpedo* (Linnaeus, 1758) (Pisces, Torpedinidae) from the coast of Senegal (Eastern Tropical Atlantic). Available at: https://core.ac.uk/reader/39078552.
Didier, D.A. 1998. Embryonic staging and external features of development of the Chimaroid fish, *Callorhincus milii* (Holocephali, Callorhinchidae). *J. Morphol*. 236: 25–47.
Donoghue, P.C.J. and Rücklin, M. 2016. The ins and outs of the evolutionary origin of teeth. *Evol. Dev*. 18: 19–30.
Douglas, R. and Djamgoz, M. 2012. *The Visual System of Fish*. Springer Science & Business Media, New York, NY.
Dudgeon, C.L., Coulton, L., Bone, R., Ovenden, J.R. and Thomas, S. 2017. Switch from sexual to parthenogenetic reproduction in a zebra shark. *Sci. Rep*. 7: 40537.
Dulvy, N.K., Fowler, S.L., Musick, J.A., Cavanagh, R.D., Kyne, P.M., Harrison, L.R., Carlson, J.K., Davidson, L.N., Fordham, S.V., Francis, M.P., Pollock, C.M., Simpfendorfer, C.A., Burgess, G.H., Carpenter, K.E., Compagno, L.J., Ebert, D.A., Gibson, C., Heupel, M.R., Livingstone, S.R., Sanciangco, J.C., Stevens, J.D., Valenti, S. and White, W.T. 2014. Extinction risk and conservation of the world's sharks and rays. *eLife* 3: e00590.
Dulvy, N.K. and Reynolds, J.D. 1997. Evolutionary transitions among egg-laying, live-bearing and maternal inputs in sharks and rays. *Proc. R. Soc. B Biol. Sci*. 264: 1309–1315.
Eames, B.F., Allen, N., Young, J., Kaplan, A., Helms, J.A. and Schneider, R.A. 2007. Skeletogenesis in the swell shark *Cephaloscyllium ventriosum*. *J. Anat*. 210: 542–554.
Elger, M., Hentschel, H., Litteral, J., Wellner, M., Kirsch, T., Luft, F.C. and Haller, H. 2003. Nephrogenesis is induced by partial nephrectomy in the elasmobranch *Leucoraja erinacea*. *J. Am. Soc. Nephrol*. 14: 1506–1518.
Enault, S., Muñoz, D.N., Silva, W.T.A.F., Borday-Birraux, V., Bonade, M., Oulion, S., Ventéo, S., Marcellini, S. and Debiais-Thibaud, M. 2015. Molecular footprinting of skeletal tissues in the catshark *Scyliorhinus canicula* and the clawed frog *Xenopus tropicalis* identifies conserved and derived features of vertebrate calcification. *Front. Genet*. 6.
Espinoza, M., Cappo, M., Heupel, M.R., Tobin, A.J. and Simpfendorfer, C.A. 2014. Quantifying shark distribution patterns and species-habitat associations: Implications of marine park zoning. *PLoS One* 9: e106885.
Falardeau, P., Champagne, P., Poyet, P., Hariton, C. and Dupont, E. 2001. Neovastat, a naturally occurring multifunctional antiangiogenic drug, in phase III clinical trials. *Semin. Oncol*. 28: 620–625.
Feulner, P.G.D., Plath, M., Engelmann, J., Kirschbaum, F. and Tiedemann, R. 2009. Electrifying love: Electric fish use species-specific discharge for mate recognition. *Biol. Lett*. 5: 225–228.
Fields, A.T., Feldheim, K.A., Poulakis, G.R. and Chapman, D.D. 2015. Facultative parthenogenesis in a critically endangered wild vertebrate. *Curr. Biol*. 25: R446–R447.
Flajnik, M.F. 2018. A cold-blooded view of adaptive immunity. *Nat. Rev. Immunol*. 18: 438–453.
Fonseca, E.S.S., Ruivo, R., Machado, A.M., Conrado, F., Tay, B.-H., Venkatesh, B., Santos, M.M. and Castro, L.F.C. 2019. Evolutionary plasticity in detoxification gene modules: The preservation and loss of the pregnane X receptor in *Chondrichthyes* lineages. *Int. J. Mol. Sci*. 20: 2331.
Forrest, J.N. 2016. The shark rectal gland model: A champion of receptor mediated chloride secretion through CFTR. *Trans. Am. Clin. Climatol. Assoc*. 127: 162–175.
Furumitsu, K., Wyffels, J.T. and Yamaguchi, A. 2019. Reproduction and embryonic development of the red stingray *Hemitrygon akajei* from Ariake Bay, Japan. *Ichtyol. Res*. 66: 419–436.
Gaillard, A.-L., Tay, B.-H., Pérez Sirkin, D.I., Lafont, A.-G., De Flori, C., Vissio, P.G., Mazan, S., Dufour, S., Venkatesh, B. and Tostivint, H. 2018. Characterization of gonadotropin-releasing hormone (GnRH) genes from cartilaginous fish: Evolutionary perspectives. *Front. Neurosci*. 12: 607.
García, V.B., Lucifora, L.O. and Myers, R.A. 2008. The importance of habitat and life history to extinction risk in sharks, skates, rays and chimaeras. *Proc. R. Soc. B Biol. Sci*. 275: 83–89.
Gardiner, J.M. 2012. Sensory physiology and behavior of elasmobranchs. In: *Biology of Sharks and Their Relatives*. CRC Press, Boca Raton, FL.
Gardiner, J.M. and Atema, J. 2007. Sharks need the lateral line to locate odor sources: Rheotaxis and eddy chemotaxis. *J. Exp. Biol*. 210: 1925–1934.
Gardiner, J.M. and Atema, J. 2010. The function of bilateral odor arrival time differences in olfactory orientation of sharks. *Curr. Biol. CB* 20: 1187–1191.
Giles, S., Friedman, M. and Brazeau, M.D. 2015. Osteichthyan-like cranial conditions in an Early Devonian stem gnathostome. *Nature* 520: 82–85.

Gillis, J.A., Alsema, E.C. and Criswell, K.E. 2017. Trunk neural crest origin of dermal denticles in a cartilaginous fish. *Proc. Natl. Acad. Sci.* 114: 13200–13205.

Gillis J.A. and Hall, B.K. 2016. A shared role for Sonic Hedgehog signaling in patterning chondrichthyan gill arch appendages and tetrapod limbs. *Dev.* 143: 1313–1317.

Glasauer, S.M.K. and Neuhauss, S.C.F. 2014. Whole-genome duplication in TELEOST fishes and its evolutionary consequences. *Mol. Genet. Genomics MGG* 289: 1045–1060.

Godard, B.G., Coolen, M., Le Panse, S., Gombault, A., Ferreiro-Galve, S., Laguerre, L., Lagadec, R., Wincker, P., Poulain, J., Da Silva, C., Kuraku, S., Carre, W., Boutet, A. and Mazan, S. 2014. Mechanisms of endoderm formation in a cartilaginous fish reveal ancestral and homoplastic traits in jawed vertebrates. *Biol. Open* 3: 1098–1107.

Greenberg, A.S., Avila, D., Hughes, M., Hughes, A., McKinney, E.C. and Flajnik, M.F. 1995. A new antigen receptor gene family that undergoes rearrangement and extensive somatic diversification in sharks. *Nature* 374: 168–173.

Gross, M. 2019. Stop the global slaughter of sharks. *Curr. Biol.* 29: R819–R822.

Gruber, D.F., Loew, E.R., Deheyn, D.D., Akkaynak, D., Gaffney, J.P., Smith, W.L., Davis, M.P., Stern, J.H., Pieribone, V.A. and Sparks, J.S. 2016. Biofluorescence in catsharks (Scyliorhinidae): Fundamental description and relevance for elasmobranch visual ecology. *Sci. Rep.* 6: 1–16.

Gruber, S.H. and Myrberg, A.A. 1977. Approaches to the study of the behavior of sharks. *Am. Zool.* 17: 471–486.

Guttridge, T.L., van Dijk, S., Stamhuis, E.J., Krause, J., Gruber, S.H. and Brown, C. 2013. Social learning in juvenile lemon sharks, *Negaprion brevirostris*. *Anim. Cogn.* 16: 55–64.

Halstead, B.W. and Bunker, N.C. 1952. The venom apparatus of the ratfish, *Hydrolagus colliei*. *Copeia* 1952: 128–138.

Hammerschlag, N. 2016. Nocturnal and crepuscular behavior in elasmobranchs: A review of movement, habitat use, foraging, and reproduction in the dark. *Bull. Mar. Sci.* doi: 10.5343/bms.2016.1046.

Hara, Y., Yamaguchi, K., Onimaru, K., Kadota, M., Koyanagi, M., Keeley, S.D., Tatsumi, K., Tanaka, K., Motone, F., Kageyama, Y., Nozu, R., Adachi, N., Nishimura, O., Nakagawa, R., Tanegashima, C., Kiyatake, I., Matsumoto, R., Murakumo, K., Nishida, K., Terakita, A., Kuratani, S., Sato, K., Hyodo, S. and Kuraku, S. 2018. Shark genomes provide insights into elasmobranch evolution and the origin of vertebrates. *Nat. Ecol. Evol.* 2: 1761–1771.

Hart, N.S., Lisney, T.J., Marshall, N.J. and Collin, S.P. 2004. Multiple cone visual pigments and the potential for trichromatic colour vision in two species of elasmobranch. *J. Exp. Biol.* 207: 4587–4594.

Hartmann, C. 2006. A Wnt canon orchestrating osteoblastogenesis. *Trends Cell Biol.* 16: 151–158.

Hentschel, H. 1987. Renal architecture of the dogfish *Scyliorhinus caniculus* (Chondrichthyes, Elasmobranchii). *Zoomorphology* 107: 115–125.

Hentschel, H. 1991. Developing nephrons in adolescent dogfish, *Scyliorhinus caniculus* (L.), with reference to ultrastructure of early stages, histogenesis of the renal countercurrent system, and nephron segmentation in marine elasmobranchs. *Am. J. Anat.* 190: 309–333.

Holland, K.N., Wetherbee, B.M., Lowe, C.G. and Meyer, C.G. 1999. Movements of tiger sharks (*Galeocerdo cuvier*) in coastal Hawaiian waters. *Mar. Biol.* 134: 665–673.

Honma, Y., Okabe, K. and Chiba, A. 1984. Comparative histology of the Leydig and epigonal organs in some elasmobranchs. *Jpn. J. Ichthyol.* 31: 47–54.

Hubbs, C.L., Iwai, T. and Matsubara, K. 1967. *External and Internal Characters, Horizontal and Vertical Distribution, Luminescence, and Food of the Dwarf Pelagic Shark, Euprotomicrus Bispinatus*. University of California Press, Berkeley and Los Angeles, LA.

Hwang, J.-H., Parton, A., Czechanski, A., Ballatori, N. and Barnes, D. 2008. Arachidonic acid-induced expression of the organic solute and steroid transporter-beta (Ost-beta) in a cartilaginous fish cell line. *Comp. Biochem. Physiol. Toxicol. Pharmacol. CBP* 148: 39–47.

Inoue, J.G., Miya, M., Lam, K., Tay, B.-H., Danks, J.A., Bell, J., Walker, T.I. and Venkatesh, B. 2010. Evolutionary origin and phylogeny of the modern holocephalans (Chondrichthyes: Chimaeriformes): A mitogenomic perspective. *Mol. Biol. Evol.* 27: 2576–2586.

IUCN SSC Shark Specialist Group, C.M. (IUCN S.S.S.G., Javier Guallart (IUCN SSC Shark Specialist Group), Ferid Haka (IUCN SSC Shark Specialist Group), Titian Schembri (IUCN SSC Shark Specialist Group), Nicola Ungaro (IUCN SSC Shark Specialist Group), Jim Ellis (IUCN SSC Shark Specialist Group), Abella, A. and Kirsteen, M. 2014. IUCN red list of threatened species: Small spotted catshark. In: *IUCN Red List of Threatened Species*. Gland, Switzerland and Cambridge, UK.

Jennings, D.E., Gruber, S.H., Franks, B.R., Kessel, S.T. and Robertson, A.L. 2008. Effects of large-scale anthropogenic development on juvenile lemon shark (*Negaprion brevirostris*) populations of Bimini, Bahamas. *Environ. Biol. Fishes* 83: 369–377.

Jordan, L.K. 2008. Comparative morphology of stingray lateral line canal and electrosensory systems. *J. Morphol.* 269: 1325–1339.

Jordan, L.K., Mandelman, J.W., McComb, D.M., Fordham, S.V., Carlson, J.K. and Werner, T.B. 2013. Linking sensory biology and fisheries bycatch reduction in elasmobranch fishes: A review with new directions for research. *Conserv. Physiol.* 1.

Joung, S.-J., Chen, C.-T., Clark, E., Uchida, S. and Huang, W.Y.P. 1996. The whale shark, *Rhincodon typus*, is a livebearer: 300 embryos found in one 'megamamma' supreme. *Environ. Biol. Fishes* 46: 219–223.

Jung H., Baek M., D'Elia, K.P., Boisvert C., Currie P.D., Tay B.H., Venkatesh B., Brown S.M., Heguy A., Schoppik D. and Dasen, J.S. 2018. The ancient origins of neural substrates for land walking. *Cell.* 172: 667–682.

Kang, B., Park, H. and Kim, B. 2019. Anticancer activity and underlying mechanism of phytochemicals against multiple myeloma. *Int. J. Mol. Sci.* 20.

Kastschenko, N. 1888. Zur Entwicklungsgeschichte des Selachierembryos. *Anat Anz* 3: 445–467.

Kelly, M.L., Collin, S.P., Hemmi, J.M. and Lesku, J.A. 2019. Evidence for sleep in sharks and rays: Behavioural, physiological, and evolutionary considerations. *Brain. Behav. Evol.* 94: 37–50.

Kimber, J.A., Sims, D.W., Bellamy, P.H. and Gill, A.B. 2014. Elasmobranch cognitive ability: Using electroreceptive foraging behaviour to demonstrate learning, habituation and memory in a benthic shark. *Anim. Cogn.* 17: 55–65.

King, B.L., Gillis, J.A., Carlisle, H.R. and Dahn, R.D. 2011. A natural deletion of the HoxC cluster in elasmobranch fishes. *Science* 334: 1517.

Klimley, A.P. 1993. Highly directional swimming by scalloped hammerhead sharks, *Sphyrna lewini*, and subsurface irradiance, temperature, bathymetry, and geomagnetic field. *Mar. Biol.* 117: 1–22.

Kovaleva, M., Ferguson, L., Steven, J., Porter, A. and Barelle, C. 2014. Shark variable new antigen receptor biologics: A novel technology platform for therapeutic drug development. *Expert Opin. Biol. Ther*. 14: 1527–1539.

Kuba, M.J., Byrne, R.A. and Burghardt, G.M. 2010. A new method for studying problem solving and tool use in stingrays (*Potamotrygon castexi*). *Anim. Cogn*. 13: 507–513.

Kuraku, S. 2021. Shark and ray genomics for disentangling their morphological diversity and vertebrate evolution. *Dev Biol*. 477: 262–272

Lagadec, R., Laguerre, L., Menuet, A., Amara, A., Rocancourt, C., Péricard, P., Godard, B.G., Celina Rodicio, M., Rodriguez-Moldes, I., Mayeur, H., Rougemont, Q., Mazan, S. and Boutet, A. 2015. The ancestral role of nodal signalling in breaking L/R symmetry in the vertebrate forebrain. *Nat. Commun*. 6: 6686.

Lagadec, R., Lanoizelet, M., Sánchez-Farías, N., Hérard, F., Menuet, A., Mayeur, H., Billoud, B., Rodriguez-Moldes, I., Candal, E. and Mazan, S. 2018. Neurogenetic asymmetries in the catshark developing habenulae: Mechanistic and evolutionary implications. *Sci. Rep*. 8: 4616.

Lee, A.P., Kerk, S.Y., Tan, Y.Y., Brenner, S. and Venkatesh, B. 2011. Ancient vertebrate conserved noncoding elements have been evolving rapidly in teleost fishes. *Mol. Biol. Evol*. 28: 1205–1215.

Li, C., Matthes-Rosana, K.A., Garcia, M. and Naylor, G.J.P. 2012. Phylogenetics of chondrichthyes and the problem of rooting phylogenies with distant outgroups. *Mol. Phylogenet. Evol*. 63: 365–373.

Lisney, T.J. 2010. A review of the sensory biology of chimaeroid fishes (Chondrichthyes: Holocephali). *Rev. Fish Biol. Fish*. 20: 571–590.

López-Romero, F.A., Klimpfinger, C., Tanaka, S. and Kriwet, J. 2020. Growth trajectories of prenatal embryos of the deep-sea shark *Chlamydoselachus anguineus* (Chondrichthyes). *J. Fish. Biol*. 97: 212–224.

Luer, C.A. and Walsh, C.J. 2018. Potential human health applications from marine biomedical research with elasmobranch fishes. *Fishes* 3: 47.

Luer, C.A., Walsh, C.J., Bodine, A.B. and Wyffels, J.T. 2007. Normal embryonic development in the clearnose skate, *Raja eglanteria*, with experimental observations on artificial insemination. *Environ. Biol. Fishes* 80: 239.

Lyons, K. and Wynne-Edwards, K.E. 2018. Legacy polychlorinated biphenyl contamination impairs male embryonic development in an elasmobranch with matrotrophic histotrophy, the round stingray (*Urobatis halleri*). *Environ. Toxicol. Chem*. 37: 2904–2911.

Márquez-Garbán, D.C., Gorrín-Rivas, M., Chen, H.-W., Sterling, C., Elashoff, D., Hamilton, N. and Pietras, R.J. 2019. Squalamine blocks tumor-associated angiogenesis and growth of human breast cancer cells with or without HER-2/neu overexpression. *Cancer Lett*. 449: 66–75.

Marra, N.J., Stanhope, M.J., Jue, N.K., Wang, M., Sun, Q., Pavinski Bitar, P., Richards, V.P., Komissarov, A., Rayko, M., Kliver, S., Stanhope, B.J., Winkler, C., O'Brien, S.J., Antunes, A., Jorgensen, S. and Shivji, M.S. 2019. White shark genome reveals ancient elasmobranch adaptations associated with wound healing and the maintenance of genome stability. *Proc. Natl. Acad. Sci. U. S. A*. 116: 4446–4455.

Martin, A.P., Naylor, G.J.P. and Palumbi, S.R. 1992. Rates of mitochondrial DNA evolution in sharks are slow compared with mammals. *Nature* 357: 153–155.

Martin, K.J., Rasch, L.J., Cooper, R.L., Metscher, B.D., Johanson, Z. and Fraser, G.J. 2016. Sox2+ progenitors in sharks link taste development with the evolution of regenerative teeth from denticles. *Proc. Natl. Acad. Sci*. 113: 14769–14774.

Maruska, K.P. 2001. Morphology of the mechanosensory lateral line system in elasmobranch fishes: Ecological and behavioral considerations. *Environ. Biol. Fishes* 60: 47–75.

Mattingly, C., Parton, A., Dowell, L., Rafferty, J. and Barnes, D. 2004. Cell and molecular biology of marine elasmobranchs: *Squalus acanthias* and *Raja erinacea*. *Zebrafish* 1: 111–120.

Mattisson, A. and Faänge, R. 1982. The cellular structure of the Leydig organ in the shark, *Etmopterus spinax* (l.). *Biol. Bull*. 162: 182–194.

Maxwell, E.E., Fröbisch, N.B. and Heppleston, A.C. 2008. Variability and conservation in late chondrichthyan development: Ontogeny of the winter skate (*Leucoraja ocellata*). *Anat. Rec*. 291: 1079–1087.

McClenachan, L., Cooper, A.B., Carpenter, K.E. and Dulvy, N.K. 2012. Extinction risk and bottlenecks in the conservation of charismatic marine species. *Conserv. Lett*. 5: 73–80.

McComb, D.M. and Kajiura, S.M. 2008. Visual fields of four batoid fishes: A comparative study. *J. Exp. Biol*. 211: 482–490.

Mellinger, J. 1994. L'œuf de roussette (*Scyliorhinus canicula*) incubé au labortatoire: un matériel de recherche pour l'embryologiste, l'éthologiste, le physiologiste. *Ichtyophysiol. Acta*. 17: 9–27.

Meyer, C.G., Holland, K.N. and Papastamatiou, Y.P. 2005. Sharks can detect changes in the geomagnetic field. *J. R. Soc. Interface* 2: 129–130.

Moore, K.S., Wehrli, S., Roder, H., Rogers, M., Forrest, J.N., McCrimmon, D. and Zasloff, M. 1993. Squalamine: An aminosterol antibiotic from the shark. *Proc. Natl. Acad. Sci. U. S. A*. 90: 1354–1358.

Motoyasu, M., Yoshiyuki, I., Shigenori, K., Haruyuki, I. and Tooru, I. 2003. Artificial insemination of the cloudy catshark. *JAZA* 44: 39–43.

Mull, C.G., Yopak, K.E. and Dulvy, N.K. 2011. Does more maternal investment mean a larger brain? Evolutionary relationships between reproductive mode and brain size in chondrichthyans. *Mar. Freshw. Res*. 62: 567.

Mulley, J.F., Hargreaves, A.D., Hegarty, M.J., Heller, R.S. and Swain, M.T. 2014. Transcriptomic analysis of the lesser spotted catshark (*Scyliorhinus canicula*) pancreas, liver and brain reveals molecular level conservation of vertebrate pancreas function. *BMC Genomics* 15.

Musa, S.M., Czachur M.V. and Shiels, H.A. 2018. Oviparous elasmobranch inside the egg case in 7 key stages. *PLoS One*. 13.

Myrberg, A.A. 2003. Cognition in elasmobranch fishes, a likely possibility. In: *ResearchGate*. Max Planck Institute, Tutzing, Germany.

Myrberg, A.A. 2001. The acoustical biology of elasmobranchs. *Environ. Biol. Fishes* 60: 31–46.

Nielsen, J., Hedeholm, R.B., Heinemeier, J., Bushnell, P.G., Christiansen, J.S., Olsen, J., Ramsey, C.B., Brill, R.W., Simon, M., Steffensen, K.F. and Steffensen, J.F. 2016. Eye lens radiocarbon reveals centuries of longevity in the Greenland shark (*Somniosus microcephalus*). *Science* 353: 702–704.

Onimaru, K., Tatsumi, K., Shibagaki, K. and Kuraku, S. 2018. A de novo transcriptome assembly of the zebra bullhead shark, *Heterodontus zebra*. *Sci. Data* 5: 1–5.

Oulion, S., Borday-Birraux, V., Debiais-Thibaud, M., Mazan, S., Laurenti, P. and Casane, D. 2011. Evolution of repeated structures along the body axis of jawed vertebrates, insights from the *Scyliorhinus canicula* Hox code. *Evol. Dev*. 13: 247–259.

Oulion, S., Debiais-Thibaud, M., d'Aubenton-Carafa, Y., Thermes, C., Da Silva, C., Bernard-Samain, S., Gavory, F., Wincker, P., Mazan, S. and Casane, D. 2010. Evolution of Hox gene clusters in gnathostomes: Insights from a survey of a shark

(*Scyliorhinus canicula*) transcriptome. *Mol. Biol. Evol.* 27: 2829–2838.

Park, H.B., Lam, Y.C., Gaffney, J.P., Weaver, J.C., Krivoshik, S.R., Hamchand, R., Pieribone, V., Gruber, D.F. and Crawford, J.M. 2019. Bright green biofluorescence in sharks derives from bromo-kynurenine metabolism. *iScience* 19: 1291–1336.

Parker, G.H. 1914. The directive influence of the sense of smell in the dogfish. *Bull. US. Bureau. Fish.* 33: 61–68.

Parmentier, E., Banse, M., Boistel, R., Compère, P., Bertucci, F. and Colleye, O. 2020. The development of hearing abilities in the shark *Scyliorhinus canicula*. *J. Anat.* n/a.

Parton, A., Forest, D., Kobayashi, H., Dowell, L., Bayne, C. and Barnes, D. 2007. Cell and molecular biology of SAE, a cell line from the spiny dogfish shark, *Squalus acanthias*. *Comp. Biochem. Physiol. Toxicol. Pharmacol. CBP* 145: 111–119.

Patterson, C. 1981. Significance of fossils in determining evolutionary relationships. *Annu. Rev. Ecol. Syst.* 12: 195–223.

Perni, M., Galvagnion, C., Maltsev, A., Meisl, G., Müller, M.B.D., Challa, P.K., Kirkegaard, J.B., Flagmeier, P., Cohen, S.I.A., Cascella, R., Chen, S.W., Limbocker, R., Sormanni, P., Heller, G.T., Aprile, F.A., Cremades, N., Cecchi, C., Chiti, F., Nollen, E.A.A., Knowles, T.P.J., Vendruscolo, M., Bax, A., Zasloff, M. and Dobson, C.M. 2017. A natural product inhibits the initiation of α-synuclein aggregation and suppresses its toxicity. *Proc. Natl. Acad. Sci. U. S. A.* 114: E1009–E1017.

Pradel, A., Maisey, J.G., Tafforeau, P., Mapes, R.H. and Mallatt, J. 2014. A Palaeozoic shark with osteichthyan-like branchial arches. *Nature* 509: 608–611.

Pratt, H.L. and Carrier, J.C. 2001. A review of elasmobranch reproductive behavior with a case study on the nurse shark, *Ginglymostoma cirratum*. *Environ. Biol. Fishes* 60: 157–188.

Ramírez-Amaro, S., Ordines, F., Esteban, A., García, C., Guijarro, B., Salmerón, F., Terrasa, B. and Massutí, E. 2020. The diversity of recent trends for chondrichthyans in the Mediterranean reflects fishing exploitation and a potential evolutionary pressure towards early maturation. *Sci. Rep.* 10: 1–18.

Rasch, L.J., Martin, K.J., Cooper, R.L., Metscher, B.D., Underwood, C.J. and Fraser, G.J. 2016. An ancient dental gene set governs development and continuous regeneration of teeth in sharks. *Dev. Biol.* 415: 347–370.

Read, T.D., Petit, R.A., Joseph, S.J., Alam, Md.T., Weil, M.R., Ahmad, M., Bhimani, R., Vuong, J.S., Haase, C.P., Webb, D.H., Tan, M. and Dove, A.D.M. 2017. Draft sequencing and assembly of the genome of the world's largest fish, the whale shark: *Rhincodon typus* Smith 1828. *BMC Genomics* 18: 532.

Renz, A.J., Meyer, A. and Kuraku, S. 2013. Revealing less derived nature of cartilaginous fish genomes with their evolutionary time scale inferred with nuclear genes. *PLoS One* 8: e66400.

Richards, V.P., Suzuki, H., Stanhope, M.J. and Shivji, M.S. 2013. Characterization of the heart transcriptome of the white shark (*Carcharodon carcharias*). *BMC Genomics* 14: 697.

Ritchie, K.B., Schwarz, M., Mueller, J., Lapacek, V.A., Merselis, D., Walsh, C.J. and Luer, C.A. 2017. Survey of antibiotic-producing bacteria associated with the epidermal mucus layers of rays and skates. *Front. Microbiol.* 8.

Rosa, R., Baptista, M., Lopes, V.M., Pegado, M.R., Paula, J.R., Trübenbach, K., Leal, M.C., Calado, R. and Repolho, T. 2014. Early-life exposure to climate change impairs tropical shark survival. *Proc. Biol. Sci.* 281.

Rosa, R., Rummer, J.L. and Munday, P.L. 2017. Biological responses of sharks to ocean acidification. *Biol. Lett.* 13: 20160796.

Roux, K.H., Greenberg, A.S., Greene, L., Strelets, L., Avila, D., McKinney, E.C. and Flajnik, M.F. 1998. Structural analysis of the nurse shark (new) antigen receptor (NAR): Molecular convergence of NAR and unusual mammalian immunoglobulins. *Proc. Natl. Acad. Sci.* 95: 11804–11809.

Salazar, A. 2018. *Advanced Chordate Zoology* (First Edition). Ed-Tech Press, Waltham Abbey, UK.

Sauka-Spengler, T., Baratte, B., Lepage, M. and Mazan, S. 2003. Characterization of Brachyury genes in the dogfish *S. canicula* and the lamprey *L. fluviatilis*: Insights into gastrulation in a chondrichthyan. *Dev. Biol.* 263: 296–307.

Schluessel, V. 2015. Who would have thought that 'Jaws' also has brains? Cognitive functions in elasmobranchs. *Anim. Cogn.* 18: 19–37.

Schluessel, V. and Bleckmann, H. 2005. Spatial memory and orientation strategies in the elasmobranch *Potamotrygon motoro*. *J. Comp. Physiol. A* 191: 695–706.

Selinsky, B.S., Smith, R., Frangiosi, A., Vonbaur, B. and Pedersen, L. 2000. Squalamine is not a proton ionophore. *Biochim. Biophys. Acta* 1464: 135–141.

Sheldon, R.E. 1909. The reactions of the dogfish to chemical stimuli. *J. Comp. Neurol. Psychol.* 19: 273–311.

Sheldon, R.E. 1911. The sense of smell in Selachians. *J. Exp. Zool.* 10: 51–62.

Shiffman, D.S., Ajemian, M.J., Carrier, J.C., Daly-Engel, T.S., Davis, M.M., Dulvy, N.K., Grubbs, R.D., Hinojosa, N.A., Imhoff, J., Kolmann, M.A., Nash, C.S., Paig-Tran, E.W.M., Peele, E.E., Skubel, R.A., Wetherbee, B.M., Whitenack, L.B. and Wyffels, J.T. 2020. Trends in chondrichthyan research: An analysis of three decades of conference abstracts. *Copeia* 108: 122–131.

Signore, I.A., Guerrero, N., Loosli, F., Colombo, A., Villalón, A., Wittbrodt, J. and Concha, M.L. 2009. Zebrafish and medaka: Model organisms for a comparative developmental approach of brain asymmetry. *Philos. Trans. R. Soc. B Biol. Sci.* 364: 991–1003.

Sills, A. K., Williams, J. I., Tyler, B. M., Epstein, D. S., Sipos, E. P., Davis, J. D., McLane, M. P., Pitchford, S., Chesire, K., Gannon, F. H., Kinney, W. A., hao, T. L., Donowitz, M., Laterra, J., Zasloff, M. and Brem, H. 1998. Squalamine inhibits angiogenesis and solid tumor growth in vivo and perturbs embryonic vasculature. *Cancer Res.* 58:2784–2792.

Sims, D.W. 2019. *8.9 the significance of ocean deoxygenation for Elasmobranchs. Ocean deoxygenation: Everyone's problem; Causes, impacts, consequences and solution.* Edited by D. Laffolry and JM Baxter. IUCN. Gland, Switzerland and Cambridge, UK.

Smith, M.M., Fraser, G.J., Chaplin, N., Hobbs, C. and Graham, A. 2009. Reiterative pattern of Sonic Hedgehog expression in the catshark dentition reveals a phylogenetic template for jawed vertebrates. *Proc. R. Soc. B Biol. Sci.* 276: 1225–1233.

Sparks, J.S., Schelly, R.C., Smith, W.L., Davis, M.P., Tchernov, D., Pieribone, V.A. and Gruber, D.F. 2014. The covert world of fish biofluorescence: A phylogenetically widespread and phenotypically variable phenomenon. *PLoS One* 9.

Stedman, H.H., Kozyak, B.W., Nelson, A., Thesier, D.M., Su, L.T., Low, D.W., Bridges, C.R., Shrager, J.B., Minugh-Purvis, N. and Mitchell, M.A. 2004. Myosin gene mutation correlates with anatomical changes in the human lineage. *Nature* 428: 415–418.

Straube, N., Li, C., Claes, J.M., Corrigan, S. and Naylor, G.J.P. 2015. Molecular phylogeny of squaliformes and first occurrence of bioluminescence in sharks. *BMC Evol. Biol.* 15.

Sundström, L.F., Gruber, S.H., Clermont, S.M., Correia, J.P.S., de Marignac, J.R.C., Morrissey, J.F., Lowrance, C.R., Thomassen,

L. and Oliveira, M.T. 2001. Review of elasmobranch behavioral studies using ultrasonic telemetry with special reference to the lemon shark, *Negaprion brevirostris*, around Bimini Islands, Bahamas. *Environ. Biol. Fishes* 60: 225–250.

Tanegashima, C., Nishimura, O., Motone, F., Tatsumi, K., Kadota, M. and Kuraku, S. 2018. Embryonic transcriptome sequencing of the ocellate spot skate *Okamejei kenojei*. *Sci. Data* 5: 1–6.

Theiss, S.M., Lisney, T.J., Collin, S.P. and Hart, N.S. 2007. Colour vision and visual ecology of the blue-spotted maskray, *Dasyatis kuhlii* Müller & Henle, 1814. *J. Comp. Physiol. A* 193: 67–79.

Tomita, T., Murakumo, K., Miyamoto, K., Sato, K., Oka, S., Kamisako, H. and Toda, M. 2016. Eye retraction in the giant guitarfish, *Rhynchobatus djiddensis* (Elasmobranchii: Batoidea): A novel mechanism for eye protection in batoid fishes. *Zoology* 119: 30–35.

Tozer, H. and Dagit, D.D. 2004. Husbandry of spotted ratfish, *Hydrolagus colliei*. In: *The Elasmobranch Husbandry Manual: Captive Care of Sharks, Rays and their Relatives*, pp. 488–491. Ohio Biological Survey, Inc. Columbus, OH.

Tricas, T.C and Gruber, S.H. 2001. *The Behavior and Sensory Biology of Elasmobranch Fishes: An Anthology in Memory of Donald Richard Nelson: Developments in Environmental Biology of Fishes*. Kluwer Academic Publishers, Dordrecht, Netherlands.

Vandebroek, G. 1936. Les mouvements morphogénétiques au cours de la gastrulation chez *Scyllium canicula* Cuv. In: *Archives de Biologie*, pp. 499–582. Bruxelles.

Vandenplas, S., Vandeghinste, R., Boutet, A., Mazan, S. and Huysseune, A. 2016. Slow cycling cells in the continuous dental lamina of *Scyliorhinus canicula*: New evidence for stem cells in sharks. *Dev. Biol.* 413: 39–49.

Venkatesh, B., Kirkness, E.F., Loh, Y.-H., Halpern, A.L., Lee, A.P., Johnson, J., Dandona, N., Viswanathan, L.D., Tay, A., Venter, J.C., Strausberg, R.L. and Brenner, S. 2006. Ancient noncoding elements conserved in the human genome. *Science* 314: 1892–1892.

Venkatesh, B., Lee, A.P., Ravi, V., Maurya, A.K., Lian, M.M., Swann, J.B., Ohta, Y., Flajnik, M.F., Sutoh, Y., Kasahara, M., Hoon, S., Gangu, V., Roy, S.W., Irimia, M., Korzh, V., Kondrychyn, I., Lim, Z.W., Tay, B.-H., Tohari, S., Kong, K.W., Ho, S., Lorente-Galdos, B., Quilez, J., Marques-Bonet, T., Raney, B.J., Ingham, P.W., Tay, A., Hillier, L.W., Minx, P., Boehm, T., Wilson, R.K., Brenner, S. and Warren, W.C. 2014. Elephant shark genome provides unique insights into gnathostome evolution. *Nature* 505. 174–179.

Walls, G.L. 1942. The vertebrate eye and its adaptive radiation. *Anat. Rec.* 88: 411–413.

Walsh, C.J., Luer, C.A., Bodine, A.B., Smith, C.A., Cox, H.L., Noyes, D.R. and Maura, G. 2006. Elasmobranch immune cells as a source of novel tumor cell inhibitors: Implications for public health. *Integr. Comp. Biol.* 46: 1072–1081.

Walsh, C.J., Luer, C.A., Yordy, J.E., Cantu, T., Miedema, J., Leggett, S.R., Leigh, B., Adams, P., Ciesla, M., Bennett, C. and Bodine, A.B. 2013. Epigonal conditioned media from bonnethead shark, *Sphyrna tiburo*, induces apoptosis in a T-cell leukemia cell line, jurkat E6–1. *Mar. Drugs* 11: 3224–3257.

Wegner, N.C. 2015. 3: Elasmobranch gill structure. In: *Fish Physiology* (R.E. Shadwick, A.P. Farrell and C.J. Brauner, eds.), pp. 101–151. Academic Press. Elsevier, Amsterdam, the Netherlands.

Wheeler, C.R., Gervais, C.R., Johnson, M.S., Vance, S., Rosa, R., Mandelman, J.W. and Rummer, J.L. 2020. Anthropogenic stressors influence reproduction and development in elasmobranch fishes. *Rev. Fish Biol. Fish.* 30: 373–386.

White, W.T. and Last, P.R. 2012. A review of the taxonomy of chondrichthyan fishes: A modern perspective. *J. Fish Biol.* 80: 901–917.

Wilga, C.D. and Lauder, G.V. 2002. Function of the heterocercal tail in sharks: Quantitative wake dynamics during steady horizontal swimming and vertical maneuvering. *J. Exp. Biol.* 205: 2365–2374.

Wotton, K.R., Mazet, F. and Shimeld, S.M. 2008. Expression of FoxC, FoxF, FoxL1, and FoxQ1 genes in the dogfish *Scyliorhinus canicula* defines ancient and derived roles for fox genes in vertebrate development. *Dev. Dyn.* 237: 1590–1603.

Wourms, J.P. 1997. The rise of fish embryology in the nineteenth century. *Am. Zool.* 37: 269–310.

Wrapp, D., De Vlieger, D., Corbett, K.S., Torres, G.M., Wang, N., Van Breedam, W., Roose, K., van Schie, L., Hoffmann, M., Pöhlmann, S., Graham, B.S., Callewaert, N., Schepens, B., Saelens, X. and McLellan, J.S. 2020. Structural basis for potent neutralization of betacoronaviruses by single-domain camelid antibodies. *Cell* 181: 1004–1015.e15.

Wueringer, B.E., Peverell, S.C., Seymour, J., L. Squire, J., Kajiura, S.M. and Collin, S.P. 2011. Sensory systems in sawfishes. 1: The ampullae of Lorenzini. *Brain. Behav. Evol.* 78: 139–149.

Wyffels, J., L. King, B., Vincent, J., Chen, C., Wu, C.H. and Polson, S.W. 2014. SkateBase, an elasmobranch genome project and collection of molecular resources for chondrichthyan fishes. *F1000Research* 3.

Yeung, T., Gilbert, G.E., Shi, J., Silvius, J., Kapus, A. and Grinstein, S. 2008. Membrane phosphatidylserine regulates surface charge and protein localization. *Science* 319: 210–213.

Zasloff, M., Adams, A.P., Beckerman, B., Campbell, A., Han, Z., Luijten, E., Meza, I., Julander, J., Mishra, A., Qu, W., Taylor, J.M., Weaver, S.C. and Wong, G.C.L. 2011. Squalamine as a broad-spectrum systemic antiviral agent with therapeutic potential. *Proc. Natl. Acad. Sci.* 108: 15978–15983.

Zheng, L., Ling, P., Wang, Z., Niu, R., Hu, C., Zhang, T. and Lin, X. 2007. A novel polypeptide from shark cartilage with potent anti-angiogenic activity. *Cancer Biol. Ther.* 6: 775–780.

Zhu, M., Yu, X., Ahlberg, P.E., Choo, B., Lu, J., Qiao, T., Qu, Q., Zhao, W., Jia, L., Blom, H. and Zhu, Y. 2013. A Silurian placoderm with osteichthyan-like marginal jaw bones. *Nature* 502: 188–193.

24 Anemonefishes

Marleen Klann, Manon Mercader, Pauline Salis, Mathieu Reynaud, Natacha Roux, Vincent Laudet and Laurence Besseau

CONTENTS

24.1 HISTORY OF THE MODEL

I noticed a very pretty little fish which hovered in the water close by, and nearly over the anemone. This fish was six inches long, the head bright orange, and the body vertically banded with broad rings of opaque white and orange alternately, three bands of each. As the fish remained stationary, and did not appear to be alarmed at my movements, I made several attempts to catch it; but it always eluded my efforts, not darting away, however, as might be expected, but always returning presently to the same spot. . . . I visited from time to time the place where the anemone was fixed, and each time, in spite of all my disturbance of it, I found the little fish there also. This singular persistence of the fish to the same spot, and to the close vicinity of

DOI: 10.1201/9781003217503-24

FIGURE 24.1 Colony of *A. clarkii* (a) and cohabitation of *A. clarkii* and *A. sandaracinos* (b) in Okinawa, Japan. ([a] Photo courtesy of Manon Mercader; [b] photo courtesy of Kina Hayashi.)

> the great anemone, aroused in me strong suspicions of the existence of some connection between them.
>
> **(Collingwood 1868)**

This is the first written description of an anemonefish* (Figure 24.1) and its peculiar lifestyle, observed by English naturalist Cuthbert Collingwood in 1866 at Fiery Cross Reef off the coast of Borneo. The remarkable symbiosis between anemonefishes and giant sea anemones has since then received a lot of attention, becoming one of the main examples of mutualistic interactions (Apprill 2020). It is actually the keen interest for this interaction that first drove scientists to study these fish (Mariscal 1970; Lubbock and Smith 1980; Fautin 1991), but, as scuba diving became popular, rending shallow environments easily accessible, multiple aspects of their biology and ecology soon started to be investigated (Mariscal 1970; Allen 1974; Moyer 1980; Ochi 1985; Murata et al. 1986). Indeed, anemonefishes are unthought-of models for marine ecologists as, unlike many marine fishes, they can be easily located at a given site as well as followed through time. Besides, they are also relatively easy to capture and, being one of the most iconic tropical reef fish species, they quickly became a must-have for aquarium hobbyists. They were one of the first captive-bred marine fish back in the 1970s, and now, many species as well as a variety of fancy mutants can easily be found in pet shops. This combination of efficient rearing and convenient sampling possibilities makes anemonefishes excellent model organisms not only for marine ecologists but also for a multitude of biological fields (reviewed in Roux et al. 2020). Until now, studies on behavior (Buston 2003a; Rueger et al. 2018), physiology (Park et al. 2011; Miura et al. 2013), development (Salis et al. 2018b; Roux et al. 2019b), evolution (Litsios et al. 2012a; Rolland et al. 2018) and population dynamics (Nanninga et al. 2015; Salles et al. 2015), just to mention a few, have been conducted using anemonefishes.

24.2 GEOGRAPHICAL LOCATION AND PHYLOGENY

Anemonefishes form a clade of at least 30 species in genera *Premnas* and *Amphiprion*, including two species that are natural hybrids (*A. leukokranos* [*A. sandaracinos* X *A. chrysopterus*] and *A. thiellei* [*A. sandaracinos* X *A. ocellaris*]) within the Pomacentridae family (Frédérich and Parmentier 2016). All are living as symbionts with ten sea anemone species that belong to three distantly related families (*Thalassianthidae, Actinidae, Stichodactilidae*) (Allen 1974; Fautin and Allen 1997; Ollerton et al. 2007; Allen et al. 2008, 2010). This mutualistic relationship is the driving force of their diversification through adaptive radiation (Litsios et al. 2012b). However, diversification of giant sea anemones occurred before the establishment of this symbiotic relationship. Since their taxonomy is still unclear, the specificity between anemonefishes and their hosts will likely be revisited (Titus et al. 2019; Nguyen et al. 2020).

Historically, anemonefishes were categorized into six morphology-based groups; genus *Premnas* formed a group on its own, and *Amphiprion* was divided into four subgenera: *Actinicola, Paramphiprion, Phalerebus* and *Amphiprion* (the last one sub-divided into two species complex: *ephippium*-complex and *clarkii*-complex) (Allen 1974; Allen et al. 2008, 2010). It was also believed that the ancestral anemonefish was able to live in association with multiple sea anemone species (i.e. generalist) that later radiated into various more specialized species (Elliott et al. 1999). This process is commonly used to explain the evolution of symbiotic organisms (Futuyma and Moreno 1988). *A. clarkii* was then believed to be at the base of the anemonefish phylogenetic tree, as it is the most widespread and generalist species of the tribe. It is also less dependent on its host sea anemone due to its good swimming performance and its morphology, which resembles that of other free-living pomacentrids. However, the latest molecular phylogenetic studies do not support those hypotheses based on morphological traits. They support the monophyletic origin of anemonefish species, but the topologies found are inconsistent with the grouping into the six complexes mentioned previously. They also place *A. percula* and *A. ocellaris*, both specialists and poor swimmers, at the basal node of the tree (Santini and Polacco 2006; Litsios et al. 2012a, 2014b) (Figure 24.2a).

All 30 species of anemonefish inhabit coral reef environments in the warm, tropical waters of the Indo-Pacific Ocean, from Australia to the Ryukyu archipelago and from Thailand to the Marshall Islands (Figure 24.2 B) (Allen 1974; Fautin and Allen 1992, 1997; Allen et al. 2008, 2010). Distribution varies greatly from one species to another, with some being widespread (e.g. *A. clarkii, P. biaculeatus*) (Figure 24.2c), while others have a restricted regional distribution (e.g. *A. bicinctus, A. percula*) (Figure 24.2d) or are even confined to a few islands (e.g. *A. chagosensis, A. fuscocaudatus*) (Figure 24.2e). The highest diversity is found in the Coral Triangle (Fautin 1988; Elliott & Mariscal 2001; Camp et al. 2016), which is probably their center of origin (Santini and Polacco 2006; Litsios et al. 2014b). In the Madang region (Papua New Guinea), nine species of anemonefish can be found in sympatry. Such coexistence is explained by niche differentiation, species coexisting through resource partitioning by using different host anemone species and/or habitat (e.g. depth, localization in the reef). They can even

* The term anemonefi shes, rather than clownfi shes, is used in this chapter to refer to *Amphiprion* and *Premnas* even though other fi shes (pomacentrid and also non-pomacentrid; Randall & Fautin 2002) can eventually live in sea anemones. This choice was made to avoid confusion due to the variety of common names employed for the different species of this clade.

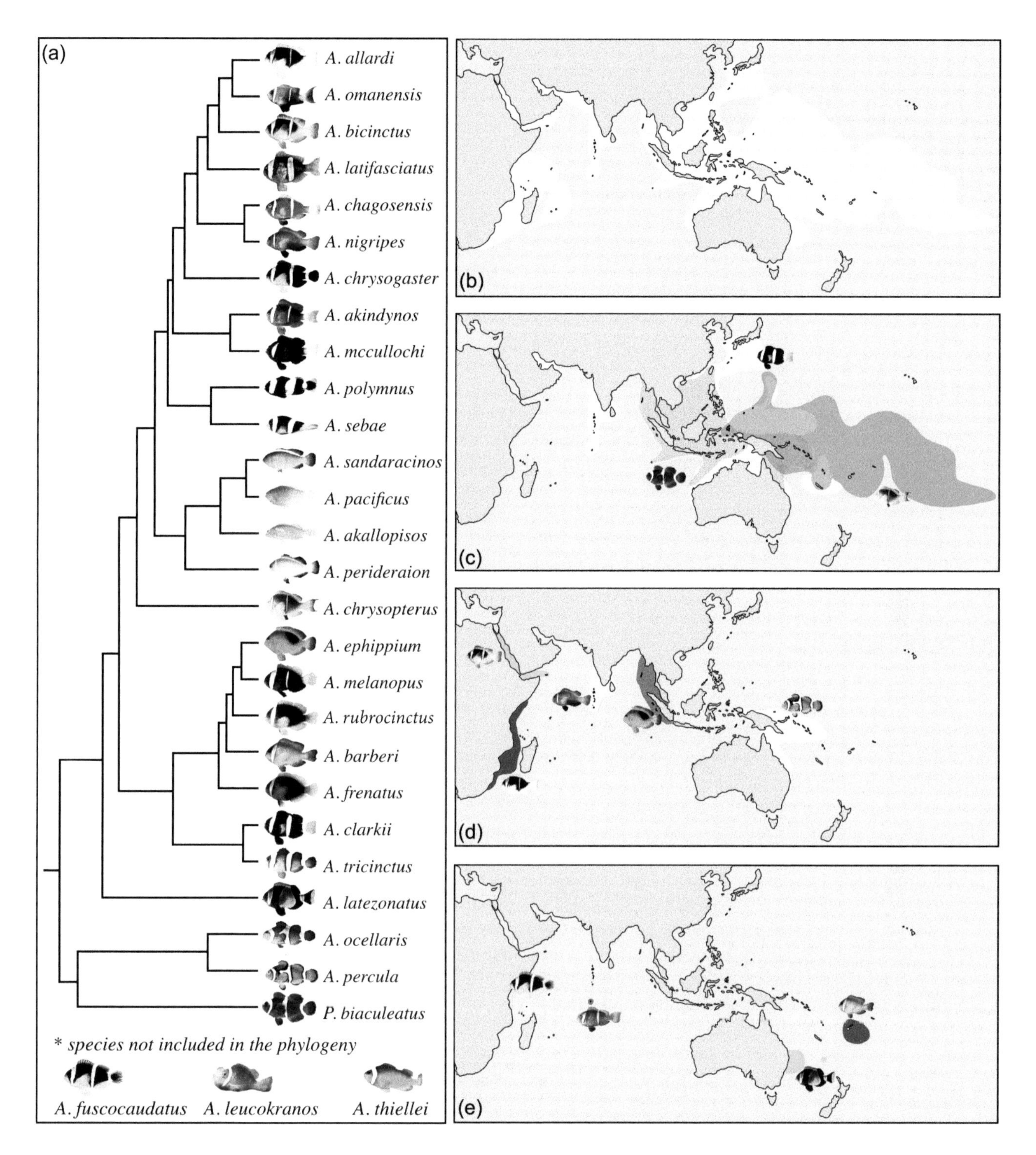

FIGURE 24.2 Phylogenetic relationship and geographic distribution of anemonefishes. Phylogenetic tree of 27 anemonefish species. Three species could not be included in the tree because they are either rare (*A. fuscocaudatus*) or hybrid species (*A. leucokranos* and *A. thiellei*) (a) Anemonefishes are distributed across the Indo-Pacific Ocean (b), with some species being widespread, such as *A. chryosopterus, A. clarkii* and *P. biaculeatus* (c); regional, such as *A. allardi, A. bicinctus, A. ephippium, A. nigripes* and *A. percula* (d); or restricted to specific areas, such as *A. barberi, A. chagosensis, A. fuscocaudatus* and *A. latezonatus* (e). (Adapted from the published work of Litsios et al. 2014b; Rolland et al. 2018.)

coexist in the same anemone (Figure 24.1b) by partitioning space in it (Elliott and Mariscal 2001; Camp et al. 2016; Hayashi et al. 2018). Anemonefishes can also be found in the Red Sea, the southwest coasts of Africa, the Maldives, French Polynesia and as far north as the southeast coast of Japan, where the warm Kuroshio current carrying tropical waters provide them adequate conditions (Moyer 1976; Fautin and Allen 1992; Fautin and Allen 1997). According to their evolutionary history, anemonefishes first spread from the Coral Triangle and then colonized the Indian and central Pacific Oceans, where they diversified around four million years ago (Mya), leading to their present distribution and diversity (Litsios et al. 2014b). In accordance with this model, farther from the coral triangle, species richness declines (Camp et al. 2016). While six species can still be found in sympatry in Okinawa (Japan) (Hayashi et al. 2018) or Lizard Island (Great Barrier Reef), only one is living in the Red Sea or French Polynesia (Allen 1974; Fautin 1988; Elliott and Mariscal 2001). Anemonefishes are not found in some Pacific islands such as the Hawaiian Islands, Johnston Atoll and the Marquesas (Randall 1955), nor on the coast of Central and South America or the Atlantic. This pattern of distribution is common to many Indo-Pacific species, which are unable to disperse past the East Pacific Barrier (Briggs 1961; Robertson et al. 2004). Since anemonefishes are obligate symbionts, their distribution is strictly dependent on their Actinian host's distribution and specific habitat requirements. Due to their endosymbiotic zooxanthellae host, sea anemones are restricted to the photic zone (≤200 m), and therefore anemonefishes are mainly found in clear shallow waters, usually no deeper than 50 m.

24.3 LIFE CYCLE

Anemonefishes exhibit the classical bi-partite life cycle of most reef fish, which is composed of a pelagic dispersive larval phase followed by a demersal juvenile and adult phase (Leis 1991) (Figure 24.3). However, their peculiar lifestyle distinguishes them from other species.

Anemonefishes live in socially well-structured colonies composed of a dominant breeding pair and several immature individuals (Figure 24.1a). A sized-based dominance hierarchy structures each colony; the largest fish is a dominant female, which defends the colony, and the second largest is a sub-dominant male taking care of the demersal eggs (Olivotto and Geffroy 2017). This monogamous pair is surrounded by smaller, sexually immature individuals, ranked by size, the smallest (youngest recruit) being at the bottom of the hierarchy (Fautin and Allen 1992; Buston 2003a; Iwata et al. 2012; Casas et al. 2016; Olivotto and Geffroy 2017). Anemonefishes have been described as protandrous sequential hermaphrodites, and the sex change from functional male to female is size dependent and/or socially mediated (Fricke and Fricke 1977). When the female disappears from the group, the male changes sex, and the third-ranked fish inherits the male breeding position and territory, thus forming a new monogamous pair (Buston 2004b; Mitchell 2005). Therefore, the size hierarchy represents a queue to attain dominant status and reproduction, individuals only ascending in rank when a higher-ranked individual disappears (Rueger et al. 2018).

Reproduction occurs all year around (except in extreme parts of their distribution range, where reproduction stops during winter), every two to three weeks, usually a week before or after a full moon (Seymour et al. 2018). The breeding couple adopts a specific behavior, which varies among species but generally includes male and female swimming close to each other and touching bellies. This "parade" is initiated by the female, which subsequently lays between 100 and 1,000 eggs, depending on species and conditions, in a roughly circular patch that are immediately fertilized by the male (Allen 1974; Buston and Elith 2011). Eggs are attached to a rock in the direct vicinity of the host sea anemone. This makes anemonefish benthic spawners, unlike most coral reef fish that spawn in the open ocean.

Embryonic development lasts between seven and ten days, during which mainly the male takes care of the eggs by fanning and mouthing them, removing dead ones (which are eaten) and keeping the nest clean (Allen 1974). Hatching occurs just after dusk, and larvae disperse in the open ocean for up to 15 days. The embryonic phase of anemonefish development is rather long compared to other fish species even when compared to other Pomacentridae (e.g. one day for the night sergeant *Abudefduf taurus*, three days for the threespot dascyllus *D. trimaculatus*) (Kavanagh and Alford 2003). Therefore, hatching larvae already have the ability to swim, feed and catch prey merely hours after hatching (Putra et al. 2012). This makes anemonefish larval development one of the shortest known for coral reef fishes (for instance, most pomacentrids have a pelagic larval duration [PLD] that lasts approximately 25 days) (Victor and Wellington 2000; Berumen et al. 2010).

After this dispersive pelagic phase, larvae metamorphose into juvenile individuals. Metamorphosis is a crucial developmental step mediated by thyroid hormones, during which morphological, physiological, behavioral and ecological changes lead to the loss of larval attributes (Laudet 2011). At this time, juveniles look like small adults and leave the open ocean to enter the reef, a process known as recruitment (Figure 24.3). More details on embryonic and larval development as well as on metamorphosis are provided in Section 24.4. Once recruited to the reef, juveniles actively search for an adequate sea anemone using environmental cues and their sensory abilities (Leis et al. 2011; Paris et al. 2013; Barth et al. 2015) to settle and establish the fascinating symbiosis that is so typical of anemonefishes.

The long-term association between anemonefishes and their sea anemones is considered a mutualistic relationship, as the sea anemone provides protection to the anemonefishes, which in turn provide nitrogen and carbon to their host and its endosymbiotic zooxanthellae (playing an important role in their nutrition) (Cleveland et al. 2011), provide protection against predators (mainly butterflyfishes) (Fautin 1991) and reduce hypoxia through aeration-like behavior (Herbert et al. 2017).

This association has always intrigued scientists for two main reasons. First, there is a complex species specificity of this mutualistic relationship, probably related to the toxicity levels of the hosts (Litsios et al. 2012b; Nedosyko et al. 2014; Marcionetti et al. 2019). A few anemonefish species live only in one sea anemone species, such as *A. sebae* and *P. biaculatus* (i.e. specialists). On the contrary, other species may have two or even ten possible hosts such as *A. ocellaris*,

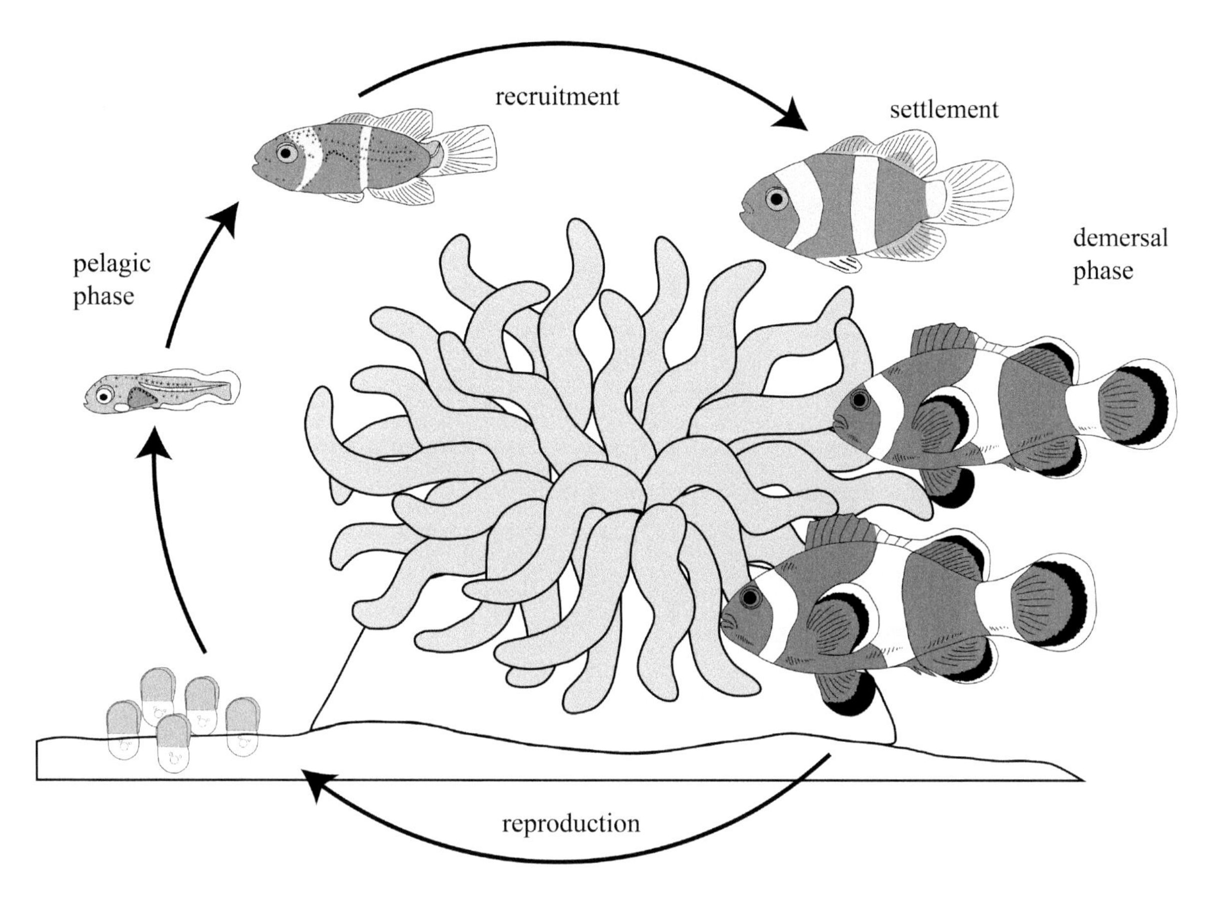

FIGURE 24.3 Anemonefish life cycle. Eggs are laid on the substrate close to the host sea anemone. After hatching, the pelagic larvae disperse in the open ocean. Recruitment to the reef coincides with metamorphosis from larvae to juveniles, which then settle into a sea anemone.

A. bicinctus, A clarkii and *A. perideraion* (i.e. generalists) (Fautin and Allen 1997) (Table 24.1).

Second, anemonefishes are able to live unharmed inside the tentacles of their host, which are known to discharge stinging cells called nematocysts (Mebs 2009). Two main hypotheses have been formulated to explain this ability. The first one suggests that anemonefishes coat themselves with sea anemone mucus, which is therefore used as a chemical camouflage (Fautin 1991; Scott 2008). This is achieved during an acclimation process that corresponds to a series of behaviors during which anemonefishes carefully enter their hosts (Schlichter 1968). First, they kiss the tentacles, then touch them with their pectoral fins and finally scrub their entire body against the tentacles. This behavior has been observed in several species, but not all, and it also seems different depending on the sea anemone species. Surprisingly, *A. clarkii* needs to acclimate when entering in *Entacmea quadricolor* but not when entering the more toxic *Stichodactyla haddoni* (Lubbock 1981; Elliott and Mariscal 1997; Mebs 2009). The second hypothesis suggests that anemonefishes are protected from sea anemone stinging by their own mucus that either prevents nematocyst discharge or protects the fish from the consequence of the discharge. Indeed, it has been shown that *A. ocellaris* lacks *N*-acetylneuraminic acid in its mucus, which is normally detected by sea anemone tentacles to discharge stinging cells (Abdullah and Saad 2015). All these studies suggest that the mucus of both partners is the key to understanding how anemonefishes are able to live in sea anemones without being harmed. Moreover, it has recently been demonstrated that changes in the microbial composition are occurring in both partners during initiation of the symbiosis, suggesting a potential role of bacterial communities in the establishment of this relationship (Pratte et al. 2018; Roux et al. 2019a).

After settlement, anemonefishes integrate into the colony hierarchy, queuing for breeding positions. Why and how anemonefishes engage in such a social system is starting to be understood thanks to extensive work on *A. percula* colonies and may have a great contribution to the understanding of complex societies. Buston and collaborators have shown that members of a colony are not composed of close relatives (2007) and that non-breeders don't provide alloparental care, their presence having neither a positive or negative effect on the dominant pair's breeding success (Buston 2004a). Non-breeders can adjust their size and growth rate in order to maintain a clear size difference with respect to individuals of higher social rank so that conflicts are limited, thereby reducing the risk of eviction and the potential cost to the breeding dominant pair (Buston 2003a). Consequently, there seem to be no direct benefits of living in such social groups. However, withholding reproduction by staying small and not contesting to remain part of the colony might represent a better option than either leaving the host anemone to breed elsewhere (because of predation risk) or contesting for breeding

TABLE 24.1
Summary of host anemone specificity among all 30 members of the clade (*A.* – *Amphiprion, P.* – *Premnas*).

	C. adh	*E. qua*	*H. aur*	*H. cri*	*H. mag*	*H. mal*	*M. dor*	*S. gig*	*S. had*	*S. mer*
A. akallopisos										
A. akindynos										
A. allardi										
A. barberi										
A. bicinctus										
A. chagosensis										
A. chrysogaster										
A. chrysopterus										
A. clarkii										
A. ephippium										
A. frenatus										
A. fuscocaudatus										
A. latezonatus										
A. latifasciatus										
A. leucokranos										
A. mccullochi										
A. melanopus										
A. nigripes										
A. ocellaris										
A. omanensis										
A. pacificus										
A. percula										
A. perideraion										
A. polymnus										
A. rubrocinctus										
A. sandaracinos										
A. sebae										
A. thiellei										
A. tricinctus										
P. biaculeatus										

* *C. adh – Cryptodendrum adhaesivum, E. qua – Entacmaea quadricolor, H. aur – Heteractis aurora, H. cri – Heteractis crispa, H. mag - Heteractis magnifica, H. mal – Heteractis malu, M. dor – Macrodactyla doreensis, S. gig – Stichodactyla gigantea, S. had – Stichodactyla haddoni, S. mer – Stichodactyla mertensii*

(because of the risk of being evicted or even killed; Buston 2003b; Rueger et al. 2018). Moreover, long-term benefits can come from staying in the colony, as subordinates will inherit the territory in which they reside after the death of breeding individuals (Buston 2004b).

Once they are finally able to reach the highest hierarchical rank, anemonefishes have to undergo a protandrous sex change (from functional male to functional female). Hermaphroditism is widely found in at least 27 teleost families, including Pomacentridae. Indeed, among vertebrates, teleost fish exhibit the greatest diversity in sex determination in relation to a remarkable plasticity of gonadal development and sexual expression (Munday et al. 2006; Liu et al. 2017; Ortega-Recalde et al. 2020).

However, even though the social hierarchy of anemonefishes has been well described for several species, the internal mechanisms at play during protandrous sex change are still poorly understood. Nonetheless, one of the main advantages of anemonefishes as model organisms is that sex change can be experimentally induced, both in field and laboratory conditions, by simply removing the dominant female. It is thus possible to study the molecular and physiological mechanisms governing sex change by following the dominant male during its transition into a functional female.

Histological analysis of gonads revealed that juveniles develop bisexual gonads, otherwise known as ovotestis, possessing both male and female tissues which are topographically distinct but not separated (Kobayashi et al. 2013; Todd et al. 2016; Gemmell et al. 2019). Once sexual maturity is reached, the ovotestis of the reproducing male exhibits a functional male territory, where spermatogenesis occurs, and an immature female territory (Kobayashi et al. 2010). During protandrous sex change, oogenesis occurs in the developing female area of the ovotestis, while the male territory progressively disappears (Casas et al. 2016). This histological scenario of gonadal protandrous transition is the same for all species of anemonefish studied so far (Godwin 1994; Kobayashi et al. 2013; Casas et al. 2016). Studies have reported that cellular changes within the ovotestis are subjected to endocrine control during sex change (Kobayashi et al. 2010; Miura et al. 2013). Like in other sequential hermaphroditic fish, the gonadal sex change is accompanied by major shifts in plasma levels of sex steroid hormones, mainly characterized by a decrease of 11-ketotestosterone levels and a subsequent 17β-estradiol increase (Godwin and Thomas 1993; Miura et al. 2013). Even though observed experimentally, the upstream mechanisms controlling the shift in sex steroid secretion still remain poorly understood. It has been suggested that the crosstalk between the hypotholamo-pituitary-gonadal (HPG) and hypothalamo-pituitary-interenal (HPI) axes plays a central role in the neuroendocrine regulation of protandrous sex change in anemonefishes (Godwin et al. 1996; Lamm et al. 2015). The association between stress and hermaphroditism was first described in *A. melanopus*, in which a peak of serum cortisol levels were observed during later sex change stages (Godwin and Thomas 1993; Goikoetxea et al. 2017; Geffroy and Douhard 2019).

Natural mortality of adult anemonefishes is very low compared to other coral reef fishes, which is most probably due to them being protected from predators by living within their host anemone. Mortality rate is not affected by environmental (e.g. reef, depth, anemone diameter) or demographic (e.g. number of individuals, density and standard length) parameters (Buston 2003b). However, it differs according to the hierarchical rank occupied by the fish. Since low-ranked individuals can be evicted from the anemone and thus undergo greater predatory pressure, juveniles suffer higher mortality than dominant individuals (Buston 2003b; Salles et al. 2015). Standard evolutionary theories of aging (i.e. mutation accumulation, antagonistic pleiotropy and disposable soma theory) predict that low extrinsic mortality leads to the evolution of slow senescence and an extended lifespan (Medawar 1952; Williams 1957; Kirkwood 1977). Anemonefishes are a great example confirming these theories, with some species having been observed to live over 20 years (Sahm et al. 2019), while predictions estimate a lifespan of up to 30 years (Buston and García 2007). Such longevity is exceptional for small fishes and at least twice the estimated longevity for other pomacentrids (Buston and García 2007; Sahm et al. 2019).

24.4 DEVELOPMENT

Anemonefish eggs are capsule shaped, and their size varies depending on the species, with a length from 1.3–1.5 mm (*A. ephippium*) to 2.4–2.6 mm (*A. nigripes*) and a width from 0.53–0.72 mm (*A. ephippium*) to 1.0–1.2 mm (*A. percula*) (Dhaneesh et al. 2009; Anil et al. 2012; Krishna 2018). The developing embryo is separated from a large amount of yolk (i.e. polylecithal, telolecithal egg), which is colored yellow to orange or even red (due to the presence of carotenoids), similar to the parent coloration. The side of the egg that is attached to the substrate (via a glutinous substance and/or threads) has consistently been recognized as the animal pole. Fertilization activates the egg and is characterized by cytoplasmic movements, which result in the formation of a dome-shaped blastodisc (Yasir and Qin 2007; Thomas et al. 2015; Krishna 2018). The chorion is transparent and leaves a narrow perivitelline space. Embryonic development usually lasts between six and eight days, depending on species and temperature. Major developmental changes will be described for all species, as they are very similar to each other, only differing in the exact timing. The following species and literature were compared for this: *A. akallopisos* (Dhaneesh et al. 2012), *A. bicinctus* (Shabana and Helal 2006), *A. ephippium* (Krishna 2018), *A. frenatus* (Ghosh et al. 2009), *A. melanopus* (Green 2004), *A. nigripes* (Anil et al. 2012), *A. ocellaris* (Liew et al. 2006, Yasir and Qin 2007, Madhu et al. 2012, Salis et al.), *A. percula* (Dhaneesh et al. 2009), *A. polymnus* (Rattanayuvakorn et al. 2005) and *A. sebae* (Thomas et al. 2015; Gunasekaran et al. 2017). To avoid disruption, these studies will not be cited again in the following descriptions.

24.4.1 Embryonic Stage 1: Early Cleavages (Figure 24.4a)

This stage comprises four synchronous division cycles that lead from a zygote to a 16-cell stage. All blastomeres of a given cell stage are of equal size. Cleavages are meroblastic (partial cleavage) and discoidal (cleavage furrows do not penetrate the yolk). The yolk exhibits prominent fat/oil globules throughout these cleavages.

24.4.2 Embryonic Stage 2: Late Cleavages (Figure 24.4b)

This stage comprises the division of the 16-cell stage until the start of gastrulation. All blastomeres are of equal size, partially overlapping each other as they arrange themselves into several layers (sphere shape) before they start to spread. The fat/oil globules decrease in number and size and are typically located toward the vegetal pole.

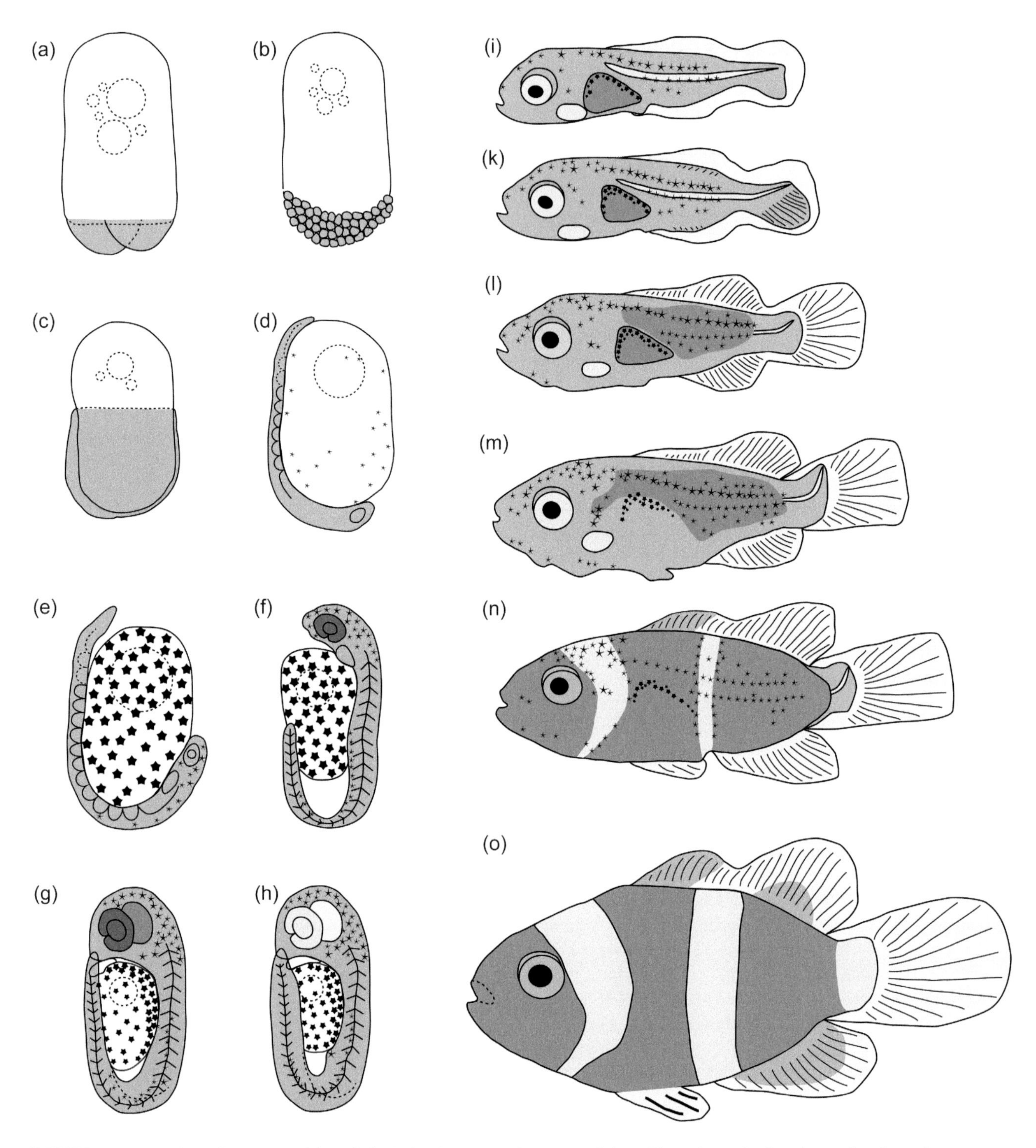

FIGURE 24.4 Embryonic (a–h) and larval (i–o) development of anemonefishes. The schematic drawings of embryonic stages are representative for all anemonefish species and do not refer to a single species, whereas *A. ocellaris* was used as representative for larval schematics (according to Roux et al. 2019b).

24.4.3 Embryonic Stage 3: Gastrulation (Figure 24.4c)

This stage comprises gastrulation, the formation of the three germ layers: ectoderm, mesoderm and endoderm. During the first step, epiboly, blastomeres flatten, move and extend toward the vegetal pole, covering the underlying yolk. Terms like 50% or 75% epiboly describe how much yolk has been covered by the blastoderm (i.e. the connective sheet of blastomeres). Formation of the embryonic shield, the future embryo, is achieved by a local thickening of blastomeres during 30–75% epiboly.

24.4.4 Embryonic Stage 4: Cephalization and Somite Development (Figure 24.4d)

The head, including optic buds (located at the animal pole), as well as neural ectoderm, is formed. The tail bud begins to develop later on. Overall, this stage marks the beginning of organogenesis and metamerization. The first appearance of paired somites occurs before 100% epiboly is reached (around 60–80% epiboly). Stellate melanophores begin to cover the yolk.

24.4.5 Embryonic Stage 5: Turn-Over (Figure 24.4e)

The entire body of the embryo is covered with few melanophores, particularly abundant in the head region. The head is clearly distinguishable, and the brain has differentiated into three parts: the prosencephalon, mesencephalon and rhombencephalon. Primitive optic buds/vesicles have formed, with subsequent induction of eye formation (eye cup, lens and cornea). Somitogenesis (trunk segmentation) is finished at the end of this stage. The body is transparent due to the absence of muscular structure at beginning, but later on, myotomes are recognizable. The embryo completely turns itself (body reversal by positioning the head toward the vegetal pole) while the tip of the tail is still attached to the yolk sac. This is a critical step for further development to proceed. The body is attached to the yolk sac, while the tail detaches from the yolk toward the end of this stage and exhibits increasing tail movements. A tubular, pink-colored heart has been differentiated and begins to beat.

24.4.6 Embryonic Stage 6: Blood Formation (Figure 24.4f)

The head and tail of the embryo have distinctly separated from the yolk, which is reduced in its volume. The body length has increased distinctly. Transparent (later a light shade of pink) spherical blood cells and subsequently blood circulation can be observed. Pigmentation is prominent in the head, especially in the large eyes displaying brownish pigments, but less in the tail region. Skeletal muscles and myotomes become clearly visible.

24.4.7 Embryonic Stage 7: Remaining Organ and Fin Development (Figure 24.4g)

The head occupies one-third of the capsule space and has salient eyes with brown melanin pigmentation. The size of the entire embryo has increased substantially, with the tail reaching the posterior part of the eyes, and it displays continuous movement. The yolk sac becomes quite small, and yellow pigments start to appear on the trunk. Branchial arches with ventilating gills and opercula, a looped alimentary tract and jaws have developed. The fin folds have developed and are clearly visible.

24.4.8 Embryonic Stage 8: Hatching (Figure 24.4h)

A hindgut has formed, and the embryo fully occupies the capsule. The spinal cord is not flexed. The eyes are turning and silver shining (eyeshine from the tapetum). The embryo tries to hatch out: vigorous movements of the tail rupture an area close to the base of the eggshell (where the egg is attached to the substrate). The hatchlings emerge tail first, which usually takes place after sunset in complete darkness.

A relatively short larval development follows hatching and precedes metamorphosis. Even though developmental time frames for larvae are more variable than for embryos, the following studies have been combined to describe larval development and metamorphosis for anemonefishes in general: *A. ephippium* (Krishna 2018), *A. frenatus* (Putra et al. 2012), *A. nigripes* (Anil et al. 2012), *A. ocellaris* (Madhu et al. 2012; Roux et al. 2019b), *A. perideraion* (Salis et al. 2018a) and *A. sebae* (Gunasekaran et al. 2017).

24.4.9 Larval Stage 1: Preflexion of the Notochord (Figure 24.4i)

The larvae are mainly transparent, with some melanophores and xanthophores scattered over the head and body. Additionally, one or two horizontal lines of melanophores are present on the trunk, along the ventral midline. The embryonic fin folds remain undifferentiated and transparent. The notochord is still straight, in preflexion. Larvae are able to feed on live prey soon after hatching and process the food in a short, straight alimentary canal with the anus located in the middle of the body length. Stomach, midgut and hindgut are distinct, and the liver and pancreas are differentiated. The larvae display phototropic behavior and swim at the top of the water column.

24.4.10 Larval Stage 2: Flexion of the Notochord (Figure 24.4k)

The embryonic fin folds start to differentiate into the caudal, dorsal and anal fins, which exhibit first signs of soft rays. The notochord begins to flex by bending dorsally.

24.4.11 Larval Stage 3: Postflexion of the Notochord (Figure 24.4l)

The embryonic fin folds have completely differentiated into caudal, dorsal and anal fins. Both anal and dorsal fins exhibit the complete set of soft rays and spines that start to appear in a posterior–anterior gradient. The pelvic fins begin to differentiate. The notochord is in postflexion, resulting in a vertical position of the hypural bones. There are no major changes in pigmentation pattern or swimming behavior.

24.4.12 Larval Stage 4: Pelvic Spine (Figure 24.4m)

All fins, including the pelvic fins, are fully developed and possess all soft rays and spines. The numbers of melanophores and xanthophores scattered over the body are increasing. There is also a marked change in behavior, as larvae are not attracted to light anymore but swim close to the bottom. This can be considered the beginning of metamorphosis, which is accompanied by a shift from a pelagic to an epibenthic lifestyle.

24.4.13 Larval Stage 5: Appearance of White Bands (Figure 24.4n)

During this stage, pigmentation patterns changes drastically. On one hand, chromatophores (bearing pigments, which shift from yellow to orange/red) are beginning to spread into the dorsal and anal fins as well as the caudal peduncle and head. On the other hand, the horizontal lines of melanophores start to disappear. Instead, the vertical white bands on the head and, depending on the species, on the body (*A. ephippium*, *A. frenatus*, *A. ocellaris*) start to emerge. They are transparent at the beginning but will adopt white color subsequently. Melanophores align at the border of the white bands. During metamorphosis, anemonefish larvae also undergo a rapid and extensive cranial remodeling that is linked with a change in preferred food items (Cooper et al. 2020). Furthermore, the shape of the body changes, and the width of the dorso-ventral axis increases, resulting in a more oval shape.

24.4.14 Larval Stage 6: Maturation of Adult Color Pattern (Figure 24.4o)

Although the final maturation of the adult pigmentation is highly dependent on the anemonefish species, it is generally characterized by an increase in the thickness of the white bands. Pigmentation of the fins is completed during this stage in all species, with the caudal fin being the last to gain color. In *A. ocellaris*, for example, a third white band appears on the caudal peduncle after approximately 20 dph (days post-hatching), resulting in an adult that possesses three white bands. In *A. ephippium*, on the other hand, both the head and body white bands increase in thickness before they start to disappear. It has been described that this process starts with the middle portion of the body band at 50–55 dph and then slowly regresses toward the dorsal and ventral sites (completion by 160 dph). After that, the head band starts to disappear at approximately 240 dph and is completely gone by 300–310 dph. Similarly, larvae of *A. frenatus* exhibit a transient white band on the body at 20 dph, which subsequently disappears.

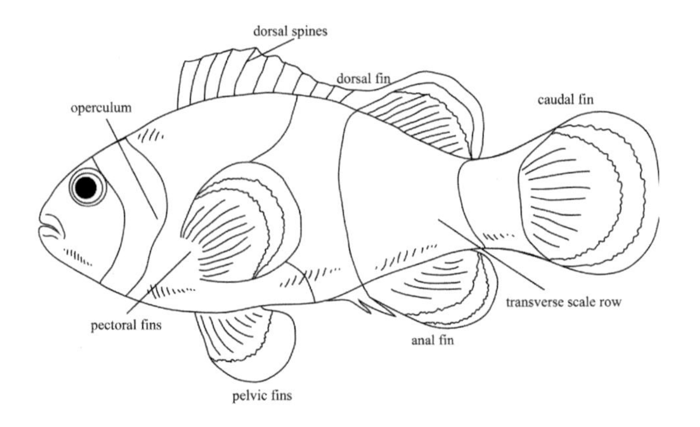

FIGURE 24.5 Schematic drawing of *A. ocellaris* showing external anatomical features.

24.5 ANATOMY

The following anatomical features can be used to distinguish members of the Amphiprioninae (Figure 24.5) from the remainder of the pomacentrids (Allen 1974; Nelson et al. 2016):

1. Nine to 11 dorsal spines
2. Suborbital, preopercle, opercle and interopercle bones with serrated or spinous margins and/or sculptured with radiating striae
3. Usually more than 50 transverse scale rows

Many tribe members also share the following features:

1. Teeth are uniserial and usually conical
2. Snout is mostly naked
3. Color pattern consists of one to three whitish bands on a darker background, which can be of various shades of orange, red, brown or black [exceptions are (i) *A. akallopisos*, *A. ephippium*, and *A. pacificus*, which do not have any bands, and (ii) *A. perideraion* and *A. sandaracinos*, which exhibit a dorsal stripe]

Anemonefishes are small sized (5–15 cm), and their body is oval and compressed (laterally thin) with a well-defined head and tail. As vertebrates, they possess all the characteristic organs and organ systems that specify this clade, such as a notochord, which develops into a vertebral column, gill arches, and neural crest cells. As representatives of the ray-finned fishes (Actinopterygii), the external anatomy is characterized by the presence of fin rays in the paired and unpaired fins, an operculum, a lateral line system and overlapping scales (Figure 24.5). Furthermore, they have specialized internal organs, such as three pairs of gill arches and a swim bladder.

The brains of anemonefishes exhibit typical features of teleostean brains; among others, these are: (i) large rhombencephalon; (ii) large unpaired cerebellum; (iii) two pronounced tectal halves located dorsal to the midbrain tegmentum and diencephalon; (iv) large, paired hypothalamic inferior lobe bulging out in the ventral brain surface; and (v) relatively small, everted telencephalon and relatively large olfactory bulbs (Nieuwenhuys et al. 1998). Furthermore, the visual system of *A. akindynos* was studied in high detail by

Sieb and colleagues (2019), who showed that retinal cones are arranged in a repetitive pattern, with four double cones surrounding a single cone.

All species of anemonefishes can produce and hear sounds, mainly composed of chirps and short and long pops (Parmentier et al. 2005; Parmentier et al. 2009). Pops are usually displayed as an aggressive, agonistic behavior against both conspecifics and heterospecifics. On the other hand, courtship sounds are more complex and differ in the number of pulses, pulse duration and dominant frequency. Sounds convey information about the size of the individual producing it, therefore implying the social rank of the emitter (Colleye et al. 2009). Sounds are produced by a series of cranial-focal interactions (Parmentier et al. 2007). First, the hyoid bar is lowered rapidly. Second, the sonic ligament, which connects the hyoid bar and internal parts of the mandible, is stretched and therefore forces the mandible to turn around its articulation, which in turn is closing the mouth. Third, the sound itself is made by collisions of the jaw teeth, with the jaw potentially acting as an amplifier. The sonic ligament represents a novel adaptation of the skeletal repertoire of anemonefish and other damselfish.

24.6 GENOMIC DATA

Actinopterygian fishes have a complex genomic history, and anemonefishes are of course no exception. In the 1970s, Susumu Ohno highlighted the importance of gene duplications as an important evolutionary mechanism that allows the creation of novelties during evolution (Ohno 1970). He further hypothesized that two rounds (2R) of whole genome duplications (WGDs) occurred early during vertebrate evolution. This was a controversial claim at the time, but it is now clear that there were effectively two genome duplications at the base of vertebrates. This is the famous "2R hypothesis", which is now largely accepted even if there are still many discussions about the precise timing and even magnitude of these duplications (reviewed in Onimaru and Kuraku 2018).

In actinopterygians, the situation is even more complex, as a third genome duplication occurred at the base of the group (Meyer and Schartl 1999; Jaillon et al. 2004). This WGD is estimated to have taken place ca. 300 Mya and is often called the "teleost-specific genome duplication" or "Ts3R" (reviewed in Glasauer and Neuhauss 2014). Within teleosts, there were several more recent lineage-specific events, such as a fourth round of WGD in salmonids ca. 100 Mya (Berthelot et al. 2014) or in the lineage of carps within cyprinids ca. 5–10 Mya (Li et al. 2015). Anemonefishes are at the typical level of teleost fishes for which three WGDs have occurred: the two at the base of vertebrates, plus the one at the base of teleost fishes.

These events provide a higher complexity in terms of gene numbers in teleost fishes than in other vertebrate lineages such as birds or mammals. This may also be linked to the great number of species in teleosts as well as their extraordinary phenotypic diversity, although the link between WGDs and species diversity is still a matter of debate (Glasauer and Neuhauss 2014; Onimaru and Kuraku 2018).

The so-called DDC model (duplication-degeneration-complementation) predicts three possible outcomes following duplication of a gene: (i) non-functionalization (i.e. the loss of one of the duplicates), (ii) neo-functionalization (i.e. one of the copies retains the ancestral role, while the other duplicate assumes a novel functionality) or (iii) sub-functionalization (i.e. both duplicates assume a part of the function of the single ancestral gene). While the model predicts that the most likely outcome following duplication of a gene is the loss of one of the duplicates (i.e. non-functionalization), there are now several examples of neo-functionalization and sub-functionalization of duplicated genes (e.g. Kawaguchi et al. 2013 for stickleback hatching enzymes or Bertrand et al. 2004 for nuclear receptors in zebrafish).

This complex evolutionary history must be taken into account when the genome data of anemonefishes is analyzed. The genomic era of anemonefish research started in 2018 with the first complete genome, that of *A. ocellaris*, which was generated using a mix of nanopore and Illumina sequencing (Tan et al. 2018). The coverage of this genome was low (11X), but this allowed the prediction of around 27,000 genes and a genome size of 800 to 900 million base pairs (Mbp). Then, the genomes of *A. frenatus* (Marcionetti et al. 2018) and *A. percula* (Lehmann et al. 2019) followed, as well as a high-density genetic map of *A. bicinctus* (Casas et al. 2018). Genome size and gene number have been estimated to be of ca. 850 Mbp and 26,900 genes for *A. frenatus* and 908 Mb and 26,600 genes for *A. percula*. The *A. percula* genome, determined by using single molecule real-time Pacific Bioscience technology, was of exceptional quality, as the authors also performed Hi-C-based chromosome contact mapping, resulting in a genome assembly into 24 chromosomes (reviewed in Hotaling and Kelley 2019). This was in accordance with previous karyotypic studies done on *A. perideraion* (Supiwong et al. 2015). This *A. percula* genome is now a unique resource for the whole community. Another major achievement was the genome assembly and annotation of nine species of anemonefish (*A. akallopisos, A. bicinctus, A. melanopus, A. nigripes, A. ocellaris, A. perideraion, A. polymnus, A. sebae* and *P. biaculeatus*) and a related damselfish outgroup, allowing for the first time insights into the genomics of anemonefish radiation and identification of genes that may be implicated in the symbiosis with sea anemones (Marcionetti et al. 2019). These datasets have already been used by independent authors to analyze specific gene sets such as peptidic hormones (Southey et al. 2020). Certainly, this is only the beginning of the anemonefish genomic era. We can anticipate that soon the genomes of all 30 known species of anemonefish will be available. Several genomes of distinct populations of anemonefishes are currently being sequenced, thus opening the way to population genomic analysis of these iconic fishes.

Complete genome sequences have been complemented by several transcriptomic data sets that started to tackle specific questions. A transcriptome of *A. ocellaris* post-embryonic development, spanning newly hatched larvae until settled juveniles, has been determined (Roux et al. in preparation). Another area of interest is the identification of genes related to the differently colored areas (white, orange and black) of *A. ocellaris* (Maytin et al. 2018; Salis et al. 2019a). This, combined with detailed pharmacological and microscopic analysis, has allowed researchers to determine that iridophores are responsible for the white color in this species but also to identify new iridophore and xanthophore genes in fish (Salis et al. 2019a, reviewed in Irion and Nüsslein-Volhard 2019; Patterson and Parichy 2019). Transcriptomic analysis has also been applied to the spectacular sex change abilities of anemonefishes. For example, a study of *A. bicinctus* from the Red Sea has revealed a complex genomic response in the brain and subsequently in the gonads with a prominent effect on genes implicated in steroidogenesis (Casas et al. 2016). Genes implicated in reproduction have also been studied in *A. ocellaris* (Yang et al. 2019).

Last, transcriptome analysis was used in the context of aging, as anemonefishes are known to have a long lifespan (Sahm et al. 2019). The authors have detected positively selected genes in *A. clarkii* and *A. percula* and tested if these genes were similar to those found in other models of aging such as mole rats or short-lived killifishes. They concluded that molecular convergence is likely to occur in the evolution of lifespan.

These examples are in fact the exhaustive list of genomic and transcriptomic studies done so far on anemonefishes. Due to low-cost high-throughput sequencing, it is likely that this will increase exponentially in the coming years as these fishes will be used more and more as experimental models which allow to link ecological, evolutionary and developmental studies.

24.7 FUNCTIONAL APPROACHES: TOOLS FOR MOLECULAR AND CELLULAR ANALYSIS

24.7.1 Husbandry

Generally, the success of an emerging model species is linked to a feasible husbandry as well as the ease of obtaining samples. For marine teleosts, this can pose difficulties, as it might be difficult to achieve reproduction in captivity or to reliably locate them in the natural environment. Anemonefishes provide an excellent model for both scenarios. On the one hand, due to their close association with sea anemones, researchers are able to locate and re-locate anemonefishes with relative ease in the wild, enabling them to conduct long-term experiments with the same individuals. On the other hand, they are very well adapted for captive life, having been in the hobbyist trade for decades. For tropical marine fishes, anemonefishes are relatively tolerant to temperature (24°C to 28°C) and salinity variations (25 to 40‰) (Dhaneesh et al. 2012). Smaller species, like *A. ocellaris*, *A. percula* and *A. sandaracinos*, can be kept in 60-L tanks, while bigger species, such as *A. clarkii*, *A. frenatus* and *P. biaculetatus*, will need up to 200-L tanks. In captivity, anemonefishes thrive without the addition of sea anemones and establish breeding pairs, which usually reproduce all year around. Both partners will participate in selection of an appropriate substrate and its cleaning, usually a terra cotta pot, ceramic tiles or even the glass walls. Egg clutch sizes vary greatly between and within species and depend on previous reproductive experience, nutrition and body size. A sufficient amount of eggs can be obtained for experimental purposes (up to 700–1,000 eggs) every 14–21 days. For experiments that require embryonic stages (such as micro-injection), the eggs can be scraped off substrate (for example, with a razor blade) and can be transferred to an egg tumbler or petri dishes for incubation. For experiments that require larval stages, the eggs remain with the parents until they are supposed to hatch (night of hatching). For hatching, they can be transferred into a separate aquarium by replacing the substrate with the attached eggs. Alternatively, if external water circulation can be interrupted, the larvae can hatch in the parent's aquarium and subsequently be transferred to a different aquarium by attracting them with a light source. This, however, is only advisable if there is no sea anemone in the same aquarium. Larvae can either be raised in small aquaria (20–30 L) or in 500–1,000-mL beakers (containing 1–20 larvae per beaker; Roux et al.). They are first fed with a mixture of micro algae and rotifers and later on *Artemia nauplii*. Juveniles are also fed with *Artemia nauplii* and either powdered food or food pellets (depending on size). The diet of adult fish is diverse and can be adjusted easily: *Artemia*, food pellets, chopped mussels, squid, shrimp and egg yolk, as well as vitamin supplements (Anil et al. 2012).

Several standard approaches have been successfully established in anemonefishes, and only a few will be highlighted here.

24.7.2 *In Situ Hybridization*

In situ hybridization is a very powerful tool to study temporal and spatial requirements of specific genes in their cellular context. In *A. frenatus*, embryonic mesodermal and neuroectodermal development has been followed by gene expression analysis of *no tail* (*ntl*) and *sox3*, respectively (Ghosh et al. 2009). Further, a comparative expression analysis of *orthodenticle homeobox 2* (*otx2*) in the olfactory placode of larval *A. percula* indicates that this gene is required for olfactory responses to settlement cues (Veilleux et al. 2013). Moreover, *in situ* hybridization can validate results acquired employing alternative approaches, such as transcriptomics. For example, a recent study revealed several upregulated genes in the white skin of *A. ocellaris*, some of which could be confirmed via *in situ* hybridization on juvenile skin sections (Salis et al. 2019a). Fluorescent *in situ* hybridization (FISH) has also been successfully established in anemonefishes. In *A. akindynos*, it has been shown that long wavelength-sensitive (LWS)-related opsin genes are exclusively expressed in double cones, while short wavelength-sensitive (SWS)-related opsins are only expressed in the interspaced single cones (Stieb et al. 2019).

24.7.3 Immunoassay

Commercial enzyme immunoassay (EIA) kits are available to analyze biochemical aspects of cells, such as hormones, neurotransmitters and second messenger molecules (such as cAMP). In 2010, Mills and colleagues validated two such kits for measuring 11-ketotestosterone and cortisol concentration, respectively, using blood plasma from *A. chrysopterus* and *A. percula*. They found that a minimum of 5–7 μL blood plasma is sufficient to confidently estimate steroid hormone concentrations, which is especially valuable when working in the field. Other hormones, such as thyroid hormones, can be routinely measured using phenobarbital extraction and ELISA detection according to the method developed by Kawakami et al. (2008) and Holzer et al. (2017).

24.7.4 Use of Drugs for Functional Experiments

Pharmacological reagents/small molecules have been used widely in zebrafish, *Danio rerio*, and helped to broaden our understanding of zebrafish biology. To date, only few of them have been tested in anemonefishes, but they pose a great potential in a variety of fields. For example, it has been shown that the small molecule TAE 684 inhibits *Alk* and *Ltk* dependent iridophores in zebrafish (Rodrigues et al. 2012). In *A. ocellaris*, TAE 684 treatment of larvae results in juveniles without white bands, thus providing evidence that iridiophores are responsible for the white color of anemonefishes (Salis et al. 2019a). Furthermore, treatment with BMP inhibitors, such as dorsomorphin or DMH1, in early embryonic stages can result in dorsalization in zebrafish (Yu et al. 2008) and *A. ocellaris* (M. Klann personal observations) alike.

24.7.5 Cell Culture

So far, there is only one report on cell culture from anemonefish explants, even though this technique is extremely valuable for research projects focusing, for example, on virology, cytobiology and oncology/disease, but also for environmental toxicology/ecotoxicology or genetics/genomics. Patkaew and colleagues (2014) used *A. ocellaris* vertebrae explants to establish a corresponding primary culture. Four days after the initial implantation, fibroblastic cells could be seen, which then multiplied rapidly, reaching 70–80% confluence within four to five days. The fifth passage was preserved in liquid nitrogen for one month and subsequently assessed. The average viability after thawing and seeding has been reported with 80%, with a 57% cell recovery and no obvious changes in cell morphology or growth pattern. Even though they do not give details, the authors also state that the employed explant method (without the use of enzymes) resulted in successful primary cultures from gills, skin and vertebrae from other anemonefishes.

24.7.6 Genetic Markers

Genetic markers, particularly microsatellites, have been developed and are now available for several anemonefish species. They are widely used to study population genetics and have been used for example to investigate phylogeographic connectivity (Dohna et al. 2015), detect and monitor hybridization events (He et al. 2019; Gainsford et al. 2020), elucidate self-recruitment of larval dispersal (Jones et al. 2005), estimate connectivity between marine protected areas (MPAs) (Planes et al. 2009) and even to determine the composition of social groups (Buston et al. 2007). A substantial number of population genetic and dynamic studies have been done on *A. percula* populations of Kimbe Bay (Papua New Guinea), with the notable construction of the first multigenerational pedigree for a marine fish population (Salles et al. 2016). Such genealogy provides an opportunity to investigate how maternal effect, environment or even philopatry can shape wild fish populations (Salles et al. 2020). Probably due to its localization in the diversity center of anemonefishes, Kimbe Bay represents a privileged study site for the investigation and testing of numerous ecological and evolutionary theories and mechanisms. For example, a recent study demonstrated that the combination of ecological and social pressure promotes the evolution of non-breeding strategies (Branconi et al. 2020). The integration of the generated data provides an invaluable cornerstone for future studies in the general field of ecology and evolution.

24.8 CHALLENGING QUESTIONS, BOTH IN ACADEMIC AND APPLIED RESEARCH

Anemonefishes are ideal emerging model systems to answer a wide range of questions in biology, including but not limited to conservation, host recognition, evolutionary mechanisms and biomedical research. Missing functional approaches are also discussed at the end of this section.

24.8.1 Human Impact and Conservation

Anemonefishes live in coral reefs, which are among the most threatened ecosystems. Many anthropogenic stressors act either globally or at a local scale: global warming, pollution, ocean acidification and deoxygenation, to name just a few (Altieri et al. 2017; Albright et al. 2018; Hughes et al. 2018; Porter et al. 2018). The effects of stressors on coral reef fishes can be studied at different levels, including growth, physiology, development, genetics, bioaccumulation and behavior. Information gained in any of these fields will provide a better understanding of the coral reef ecosystem and ultimately, its conservation. A few exploratory studies investigating the effect of anthropogenic stressors on anemonefishes have already been conducted, and some will be introduced subsequently. A chemical compound found in sunscreens acting as a UV filter (benzophenone-3) perturbed feeding and swimming behavior and led to a decrease of body weight even at small concentrations of 1 mg/l (Chen et al. 2018; Barone et al. 2019), whereas higher concentrations of 100 mg/l resulted in 25% increased mortality rate (Barone et al. 2019). The direct impact of global warming (increased water temperature) on the physiology of anemonefishes has been investigated. The cellular stress responses (quantification of molecular biomarkers) of adults raised for one month at 26°C (control) or 30°C (elevated temperature) have been compared, and tissue-specific differences could be found, with muscles, gills and liver being the most reactive tissues (Madeira et al. 2016). The authors concluded that if individuals are not able to adapt to elevated temperatures, lower reproductive success, reduced growth and disease resistance would most likely occur (Madeira et al. 2016). Sea anemone bleaching (loss of symbiotic zooxanthellae) poses an important indirect effect of global warming for anemonefishes. It has been shown that juveniles of *A. chrysopterus* living in bleached sea anemones (*H. magnifica*) had an increased standard metabolic rate (up to 8%) when compared to juveniles from unbleached sea anemones (Norin et al. 2018). The authors suggested that this increased minimum cost of living might result in reduced fitness (revised energy allocation) such as reduced growth rate, spawning frequency or lower fecundity. In the same species, it has been shown that fish living in bleached hosts experienced changes in stress and reproductive hormones (cortisol and 11-KT and 17β-estradiol, respectively) (Beldade et al. 2017). Spawning frequency and clutch sizes were lower than in unbleached hosts (respectively, 51% and 64%), while egg mortality was higher (38%), leading to an overall fecundity decrease of 73%. However, after host recovery, all hormonal and reproductive parameters went back to their pre-bleaching levels. This strongly suggests a key role of hormonal response plasticity in fish acclimation to climate changes (Beldade et al. 2017). Similarly, a decrease in egg production in bleached anemone has been reported for *A. polymnus* (Saenz-Agudelo et al. 2011). None of the previously mentioned studies reported mortality of adult fish subsequent to a bleaching event. However, by following two consecutive bleaching events, Hayashi and Reimer (2020) showed that host anemones took longer to recover after the second bleaching and that one individual even completely disappeared, together with the anemonefish pair living in it. This study indicates that if temperature abnormalities are to happen regularly, sea anemone resilience to bleaching might be impaired, which can have direct consequences for anemonefishes. Another indirect effect of global warming is ocean acidification. Indeed, when reared under simulated ocean acidification conditions, olfactory and auditory abilities of anemonefish larvae were disrupted, which usually provide important cues to locate the reef and their hosts (Munday et al. 2008; Dixson et al. 2010; Simpson et al. 2011; Holmberg et al. 2019). Noise induced by humans is classified as a form of pollution. Indeed, a study showed that embryos of *A. melanopus* reared under the influence of playback boat noise exhibited faster heart rates (about 10% increase of cardiovascular activity) than ambient reef controls (Fakan and McCormick 2019). Although survival rates of embryos subjected to noise did not change, it is possible that embryogenesis is nevertheless negatively affected, leading to larvae and juveniles with reduced fitness (Fakan and McCormick 2019). Besides boat noise, anemonefishes can also be directly affected by other recreational activities such as scuba diving. Indeed, divers tend to approach these iconic fishes as closely as possible, but this human attitude could induce changes in the behavior and stress level of the fish (Hayashi et al. 2019a). In the long run, repeated human presence could affect anemonefish fitness by impairing essential behaviors such as courtship, egg care and feeding (Nanninga et al. 2017). Another drawback of their popularity is that anemonefishes are highly targeted by the aquarium trade. Indeed, the same attributes that make them good model organisms attract aquarists (longevity and exotic symbiosis) and permit easy harvesting in their natural environment (Shuman et al. 2005). Pomacentrids represent around 76% of wild-caught ornamental fish imported in the United States, with *A. percula* and *A ocellaris* in fifth place (after four species of damselfish) (Rhyne et al. 2012), even though they can be captive-bred easily. Anemonefishes represent up to 57% of all collected organisms in the Philippines (Shuman et al. 2005). There, exploited sites exhibit lower anemonefish biomass than protected sites, and fish size distribution tends to be skewed toward small fish. For *A. clarkii*, even the number of individuals present in exploited sites was lower, and similar results were observed for the anemone *H. crispa* (Shuman et al. 2005). Those results reflect the non-negligible impact of aquarium trade on anemonefishes and host anemone populations.

Another human impact that has been studied is coastline anthropization. Recent studies showed that it could not only lead to low replenishment rates but also affect community structures and diversity of anemonefishes (Hayashi et al. 2019b; Hayashi et al. 2020).

While many aspects of anemonefishes biology and ecology have been studied, very little has been done to integrate those findings in applied fields such as conservation biology (but see Planes et al. 2009; Hayashi et al. 2019b, 2020), which, in the actual context of ever-growing human pressures, should be one of the priorities of the research community.

24.8.2 Host Recognition and Settlement Clues

Numerous studies have focused on the symbiotic relationship between anemonefishes and their host anemones, with the aim to understand how juvenile recruitment occurs. Although it is well documented that anemonefishes can distinguish different host anemones and their health status (bleached vs. unbleached) using chemical cues (Murata et al. 1986; Arvedlund and Nielsen 1996; Arvedlund et al. 1999; Miyagawa-Kohshima et al. 2014; Scott and Dixson 2016), composition and structure of these chemicals still remain unknown. A study found an upregulation of *otx2* expression, a transcription factor frequently associated with olfactory imprinting, in larvae which were exposed to settlement odors compared with no-odor control larvae of *A. percula* (Veilleux et al. 2013). This chemical imprinting is believed to occur during late embryonic development and the first hours after hatching and is sufficient to recognize all species-specific partner host anemones regardless of the parents' host anemone (Arvedlund et al. 2000; Miyagawa-Kohshima et al. 2014). However, it has also been shown that anemonefishes possess a limited innate recognition

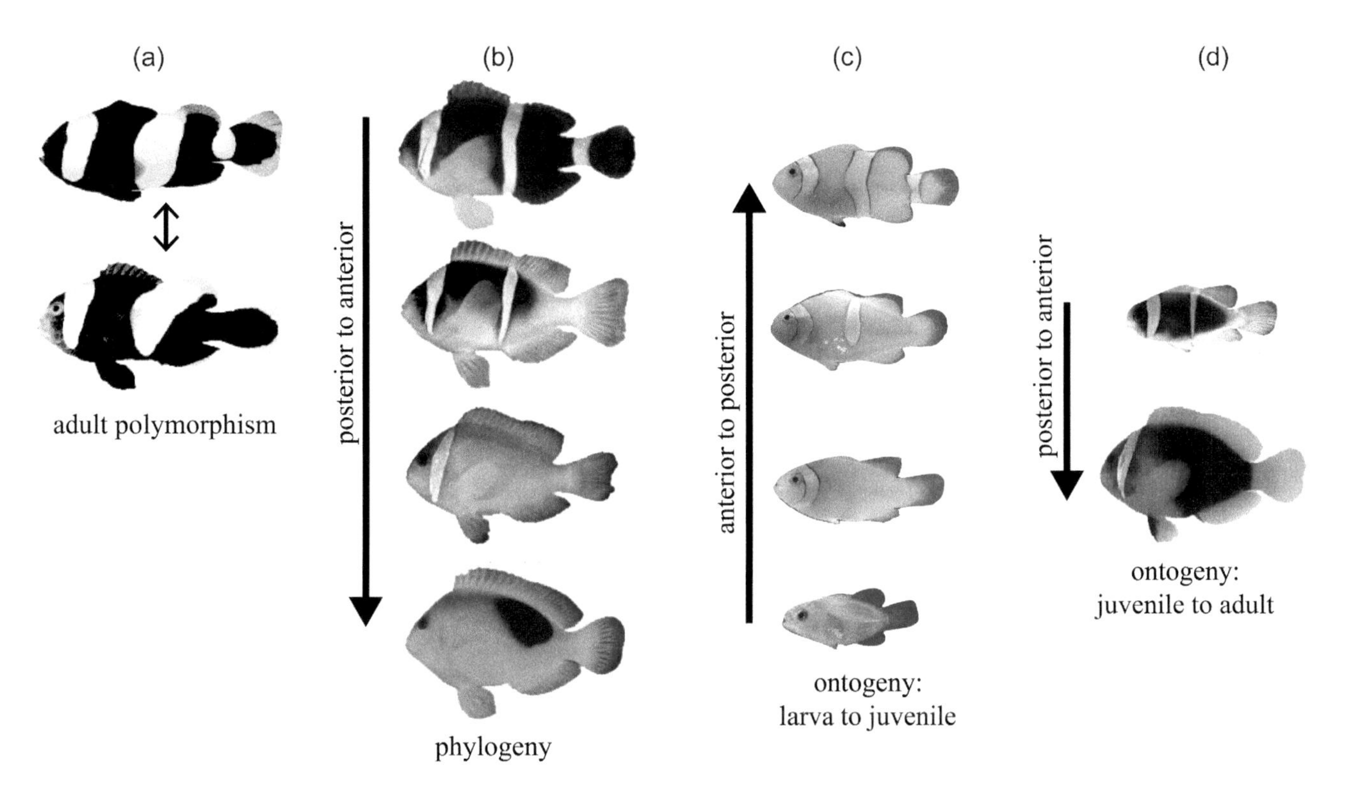

FIGURE 24.6 Evolutionary and developmental white band acquisition. Opposing trends have been described, but the underling mechanisms remain unsolved.

of partner and non-partner host anemones (Miyagawa-Kohshima et al. 2014). Field experiments further showed that new recruits do not discriminate between occupied and unoccupied host anemones (Elliott et al. 1995) but did encounter highly aggressive behavior from the resident fish (especially resident juveniles). Usually the new recruit would cease approaching an inhabited host after several aggressive interactions and try to locate a different host (Elliott et al. 1995). This eviction of juvenile anemonefishes has been widely noted and is believed to be the reason for the formation of sub-symbiotic partnerships if symbiotic partnership cannot be established (i.e. use of a sea anemone species that is not preferred) (Miyagawa-Kohshima et al. 2014). Most studies on anemonefish settlement have focused on the cues involved when selecting a host anemone, but cues to settle out of the plankton into the benthic reef habitat are less well investigated. They are unlikely to be the same, as it has been shown that chemical cues from anemones can only guide juveniles if they are relatively close to and downstream of an anemone (typically 2 m, with a maximum around 8 m) (Elliott et al. 1995). Due to the relative ease of obtaining naive larvae (i.e. aquarium-raised without sea anemone contact), field experiments can be conducted to validate experimental hypotheses. Once we have a better understanding of anemonefish settlement, we will be able to investigate how other coral reef fish larvae select nurseries and/or microhabitats. Selection of an appropriate substrate is of great importance for young fish, as it will ultimately determine their survival and breeding success.

24.8.3 Evolutionary Mechanisms

Anemonefish phylogeny has been used to investigate how hybridization and species diversification are linked (Litsios and Salamin 2014). This phylogeny was also used to compared the evolution rate of anemonefishes at both intra- and interspecific scales (i.e. micro- and macro- evolution) (Rolland et al. 2018). Other new approaches, such as quantitative genetics, might also provide a better understanding of evolutionary mechanisms. This kind of approach assesses how phenotypes are shaped given the relatedness between individuals sharing similar traits and the environment in which they are living (Thomson et al. 2018). For example, Salles et al. (2020) estimated the proportion of variance in lifetime reproductive success (LRS) explained by genetic and environmental factors. When compared to environment, genetics play a minor role, resulting in low heritability and evolvability. This suggests that in its current state, the population potential for evolutionary change is very limited, highlighting the importance of plasticity to enable rapid adaptive responses. Another complex feature observed in anemonefishes is color polymorphism, which has been noted to occur at multiple scales, with melanization being the predominant one (see Figure 24.1 for an example in *A. clarkii*). Geographical variation in coloration is common among widely distributed species, but sympatric variations have also been reported in populations in which sexual dichromatism and ontogenetic differences govern pigmentation (Moyer 1976; Fautin and Allen 1997). A suite of interacting and conditional ecological factors encompassing social rank, host anemone species and location had been identified as the primary factors predicting distribution of melanistic morphs (Militz et al. 2016). However, phylogenetic studies on melanistic *A. clarkii* showed that specimens cluster by color rather than geographical origin: a melanistic specimen from Bali is more closely related to another melanistic individual originating from Papua New Guinea than to a syntopic orange *A. clarkii* (Litsios et al. 2014a). Another common polymorphic feature of anemonefish color pattern is the variation of band number, regularly observed in *A. clarkii, A. melanopus* and *A. plolymnus* (Figure 24.6a). This suggests complex mechanisms might be involved in anemonefish polymorphism. Salis and colleagues (2018b) mapped the occurrence and number of bands on the phylogeny to reconstruct the ancestral state and could show that the diversification of anemonefish color pattern results from successive caudal to rostral losses of bands during evolution (Figure 24.6b). This is in contrast with the developmental acquisition of bands, which appear in an anterior to posterior gradient (Figure 24.6c). Interestingly, juveniles of some species have supplementary bands that disappear later caudo-rostrally (Figure 24.6d). The reduction of band number during ontogeny matches the sequence of band loss during evolution, demonstrating that diversification in color pattern among anemonefish lineages resulted from changes in developmental processes. The functional aspect of anemonefish skin color and pattern remains unclear. However, it has been suggested that color patterns may (i) be used in advertising social rank (Fautin and Allen 1997; Militz et al. 2016), (ii) signal individual identity (Fricke 1973; Buston 2003a), (iii) provide disruptive coloration (Salis et al. 2018b) and (iv) be used for species recognition (Salis et al. 2018b; Salis et al. 2019b). Yet developmental mechanisms underlying the color pattern formation have still not been identified. However, a Turing-like model (that patterns zebrafish or angelfish, for example) cannot explain the appearance and/or disappearance of bands during ontogeny, thus suggesting that band formation is controlled by specific patterning mechanisms that remain to be analyzed. The dorsal fin might act as a spatial reference, since its size and geometry have been significantly correlated with the number of white bands (Salis et al. 2018b). Given the increase in interdisciplinary studies, considerable improvement in the understanding of evolutionary mechanisms should be expected in the coming years.

24.8.4 Biomedical Research

Anemonefishes are a promising model system for biomedical research, even though studies in this field are limited so far. On one hand, they have a relatively long life span and, on the other hand, their ability to avoid nematocyst discharge is rare among vertebrates. Anemonefishes are one of a few species that offer the opportunity to study longevity and aging. Indeed, they have a long life expectancy, which is approximately six times longer than that predicted for other small fish (Buston and García 2007; Sahm et al. 2019), and they reproduce monthly all year around. Using anemonefish, a recent study (Sahm et al. 2019) suggested that the mito-nuclear balance (i.e. balance between expression of nuclear and mitochondrially encoded mitochondrial proteins) plays a key role in aging, which opens the gate to explore those genetic pathways involved.

Although many studies have attempted to unveil how anemonefishes avoid the negative effects of nematocyst stinging, there are still many open questions and various competing hypotheses (see Section 24.3). Indeed, a field study with several species of anemonefish showed that new naive recruits (around 20 dph) are able to enter their host anemones without being harmed on the first attempt (Elliott et al. 1995). Occasionally, the new recruits adhered to the tentacle but usually could break free and, after a short acclimation process, could enter unharmed. From a biomedical standpoint, it is of great interest, as understanding how anemonefishes avoid being stung by the hosts' nematocysts might lay a foundation for possible prevention and therapy of negative human interactions with jellyfish, for example. Additionally and rather unexpectedly, the anemonefish queuing system has been used to serve as the basis of a novel brain tumor segmentation algorithm (Mc and Subramanian 2016).

24.8.5 Missing Functional Approaches

Casas et al. (2016) performed the first *de novo* transcriptome analysis of wild *A. bicinctus* and highlighted the rapid and complex genomic responses of the brain during sex change, which is subsequently transmitted to the gonads. This transcriptomic data (Casas et al. 2016; Yang et al. 2019) will broaden our understanding not only of the physiological mechanisms involved but also of the perception and processing of external cues into a coordinated response that characterizes sex change (Lamm et al. 2015; Liu et al. 2017). Advances in molecular endocrinology, genomic and transcriptomic data in anemonefishes will allow opening new avenues in our understanding of sex change and sex determination in fishes and more widely in vertebrates. Moreover, extensive efforts have been put in by several research groups to establish micro-injection (Roux et al. 2020) and associated genome editing, such as CRISPR/Cas9 in anemonefishes (Mitchell et al. 2020). This is a much-needed toolkit to gain functional data and will be applicable to a range of research areas. Micro-injection is possible, yet mortality rates are still high, and obtaining larvae remains difficult (Mitchell et al. 2020; Roux et al. 2020). However, once established, the possibility of modifying specific genetic aspects will advance the field of anemonefish research, as well as research on coral reef fish, immensely. Although there are several pet shop mutants available with diverse color patterns, the underlying mutations and exact mechanisms have not been studied in detail.

24.9 CONCLUSION

This chapter summarizes the past and most recent research finding as well as future perspectives, revealing the great potential anemonefishes offer as emerging marine fish models. Future research on anemonefishes will complement studies on traditional model organisms in a wide variety of biological areas, from pigmentation to neurobiology. Their unique biological attributes open perspectives to tackle new questions related to aging, sexual differentiation, symbiosis, growth or even social organization. Anemonefishes have and will always remain prominent models for ecological studies, but now those can be linked with lab based evo-devo approaches, which is hardly possible with other model organisms. As there is a lack of convenient experimental models for marine fishes, we hope and strongly believe that this model will find its place in the vast array of new models available for the biologists of tomorrow.

BIBLIOGRAPHY

Abdullah NS, Saad S. 2015. Rapid detection of N-acetylneuraminic acid from false clownfish using HPLC-FLD for symbiosis to host sea anemone. *Asian Journal of Applied Sciences* 3:858–864.

Albright R, Takeshita Y, Koweek DA, Ninokawa A, Wolfe K, Rivlin T, Nebuchina Y, Young J, Caldeira K. 2018. Carbon dioxide addition to coral reef waters suppresses net community calcification. *Nature* 555:516–519.

Allen GR. 1974. *The anemonefish: Their classification and biology*. Second edition. T.F.H. Publications Inc., Neptune City, NJ.

Allen GR, Drew J, Fenner D. 2010. *Amphiprion pacificus*, a new species of anemonefish (Pomacentridae) from Fiji, Tonga, Samoa, and Wallis Island. *Aqua, International Journal of Ichthyology* 16:10.

Allen GR, Drew J, Kaufman L. 2008. *Amphiprion barberi*, a new species of anemonefish (Pomacentridae) from Fiji, Tonga, and Samoa. *Aqua, International Journal of Ichthyology* 14:10.

Altieri AH, Harrison SB, Seemann J, Collin R, Diaz RJ, Knowlton N. 2017. Tropical dead zones and mass mortalities on coral reefs. *Proceedings of the National Academy of Sciences* 114:3660–3665.

Anil MK, Santhosh B, Prasad BO, George RM. 2012. Broodstock development and breeding of black-finned anemone fish *Amphiprion nigripes* Regan, 1908 under captive conditions. *Indian Journal of Fisheries* 59:77–82.

Apprill A. 2020. The role of symbioses in the adaptation and stress responses of marine organisms. *Annual Review of Marine Science* 12:291–314.

Arvedlund M, Larsen K, Winsor H. 2000. The embryonic development of the olfactory system in *Amphiprion melanopus* (Perciformes: Pomacentridae) related to the host imprinting hypothesis. *Journal of the Marine Biological Association of the United Kingdom* 80:1103–1109.

Arvedlund M, McCormick M, Fautin D, Bildsøe M. 1999. Host recognition and possible imprinting in the anemonefish *Amphiprion melanopus* (Pisces: Pomacentridae). *Marine Ecology Progress Series* 188:207–218.

Arvedlund M, Nielsen LE. 1996. Do the anemonefish *Amphiprion ocellaris* (Pisces: Pomacentridae) imprint themselves to their host sea anemone heteractis magnifica (Anthozoa: Actinidae)? *Ethology* 102:197–211.

Barone AN, Hayes CE, Kerr JJ, Lee RC, Flaherty DB. 2019. Acute toxicity testing of TiO_2-based vs. oxybenzone-based sunscreens on clownfish (*Amphiprion ocellaris*). *Environmental Science and Pollution Research* 26:14513–14520.

Barth P, Berenshtein I, Besson M, Roux N, Parmentier E, Banaigs B, Lecchini D. 2015. From the ocean to a reef habitat: How do the larvae of coral reef fishes find their way home? A state of art on the latest advances. *Vie et milieu* 65:91–100.

Beldade R, Blandin A, O'Donnell R, Mills SC. 2017. Cascading effects of thermally-induced anemone bleaching on associated anemonefish hormonal stress response and reproduction. *Nature Communications* 8:716.

Berthelot C et al. 2014. The rainbow trout genome provides novel insights into evolution after whole-genome duplication in vertebrates. *Nature Communications* 5:3657.

Bertrand S, Brunet FG, Escriva H, Parmentier G, Laudet V, Robinson-Rechavi M. 2004. Evolutionary genomics of nuclear receptors: From twenty-five ancestral genes to derived endocrine systems. *Molecular Biology and Evolution* 21:1923–1937.

Berumen ML, Walsh HJ, Raventos N, Planes S, Jones GP, Starczak V, Thorrold SR. 2010. Otolith geochemistry does not reflect dispersal history of clownfish larvae. *Coral Reefs* 29:883–891.

Branconi R, Barbasch TA, Francis RK, Srinivasan M, Jones GP, Buston PM. 2020. Ecological and social constraints combine to promote evolution of non-breeding strategies in clownfish. *Communications Biology* 3:649.

Briggs JC. 1961. The East Pacific Barrier and the distribution of marine shore fishes. *Evolution* 15:545–554.

Buston PM. 2003a. Size and growth modification in clownfish. *Nature* 424:145–146.

Buston PM. 2003b. Mortality is associated with social rank in the clown anemonefish (*Amphiprion percula*). *Marine Biology* 143:811–815.

Buston, Buston PM. 2004a. Does the presence of non-breeders enhance the fitness of breeders? An experimental analysis in the clown anemonefish *Amphiprion percula*. *Behavioral Ecology and Sociobiology* 57:23–31.

Buston PM. 2004b. Territory inheritance in clownfish. *Proceedings of the Royal Society of London*. Series B: Biological Sciences 271. Available from https://royalsocietypublishing.org/doi/10.1098/rsbl.2003.0156 (accessed November 8, 2020).

Buston PM, Bogdanowicz SM, Wong A, Harrison RG. 2007. Are clownfish groups composed of close relatives? An analysis of microsatellite DNA variation in *Amphiprion percula*. *Molecular Ecology* 16:3671–3678.

Buston PM, Elith J. 2011. Determinants of reproductive success in dominant pairs of clownfish: A boosted regression tree analysis: Determinants of reproductive success. *Journal of Animal Ecology* 80:528–538.

Buston PM, García MB. 2007. An extraordinary life span estimate for the clown anemonefish *Amphiprion percula*. *Journal of Fish Biology* 70:1710–1719.

Camp EF, Hobbs J-PA, De Brauwer M, Dumbrell AJ, Smith DJ. 2016. Cohabitation promotes high diversity of clownfishes in the Coral Triangle. *Proceedings of the Royal Society B: Biological Sciences* 283:20160277.

Casas L, Saborido-Rey F, Ryu T, Michell C, Ravasi T, Irigoien X. 2016. Sex change in clownfish: Molecular insights from transcriptome analysis. *Scientific Reports* 6:35461.

Casas L, Saenz-Agudelo P, Irigoien X. 2018. High-throughput sequencing and linkage mapping of a clownfish genome provide insights on the distribution of molecular players involved in sex change. *Scientific Reports* 8:4073.

Chen T-H, Hsieh C-Y, Ko F-C, Cheng J-O. 2018. Effect of the UV-filter benzophenone-3 on intra-colonial social behaviors of the false clown anemonefish (*Amphiprion ocellaris*). *Science of The Total Environment* 644:1625–1629.

Christ MCJ, Subramanian R. 2016. Clown fish queuing and switching optimization algorithm for brain tumor segmentation. *Biomedical Research* 27:5.

Cleveland A, Verde EA, Lee RW. 2011. Nutritional exchange in a tropical tripartite symbiosis: Direct evidence for the transfer of nutrients from anemonefish to host anemone and zooxanthellae. *Marine Biology* 158:589–602.

Colleye O, Frederich B, Vandewalle P, Casadevall M, Parmentier E. 2009. Agonistic sounds in the skunk clownfish *Amphiprion akallopisos*: Size-related variation in acoustic features. *Journal of Fish Biology* 75:908–916.

Collingwood C. 1868. *Rambles of a naturalist on the shores and waters of the China Sea*. John Murray, London.

Cooper W, Van Hall R, Sweet E, Milewski H, DeLeon Z, Verderber A, DeLeon A, Galindo D, Lazono O. 2020. Functional morphogenesis from embryos to adults: Late development shapes trophic niche in coral reef damselfishes. *Evolution & Development* 22:221–240.

Dhaneesh KV, Kumar TTA, Shunmugaraj T. 2009. Embryonic development of percula clownfish, *Amphiprion percula* (Lacepede, 1802). *Middle-East Journal of Scientific Research* 4:84–89.

Dhaneesh KV, Nanthini Devi K, Ajith Kumar TT, Balasubramanian T, Tissera K. 2012. Breeding, embryonic development and salinity tolerance of skunk clownfish *Amphiprion akallopisos*. *Journal of King Saud University-Science* 24:201–209.

Dixson DL, Munday PL, Jones GP. 2010. Ocean acidification disrupts the innate ability of fish to detect predator olfactory cues. *Ecology Letters* 13:68–75.

Dohna TA, Timm J, Hamid L, Kochzius M. 2015. Limited connectivity and a phylogeographic break characterize populations of the pink anemonefish, *Amphiprion perideraion*, in the Indo-Malay Archipelago: Inferences from a mitochondrial and microsatellite loci. *Ecology and Evolution* 5:1717–1733.

Elliott JK, Elliott JM, Mariscal RN. 1995. Host selection, location, and association behaviors of anemonefishes in field settlement experiments. *Marine Biology* 122:377–389.

Elliott JK, Lougheed SC, Bateman B, McPhee LK, Boag PT. 1999. Molecular phylogenetic evidence for the evolution of specialization in anemonefishes. *Proceedings of the Royal Society of London*. Series B: Biological Sciences 266:677–685.

Elliott JK, Mariscal RN. 1997. Acclimation or innate protection of anemonefishes from sea anemones? *Copeia*:284–289.

Elliott JK, Mariscal RN. 2001. Coexistence of nine anemonefish species: Differential host and habitat utilization, size and recruitment. *Marine Biology* 138:23–36.

Fakan EP, McCormick MI. 2019. Boat noise affects the early life history of two damselfishes. *Marine Pollution Bulletin* 141:493–500.

Fautin DG. 1988. Sea anemones of Madang Province. *Science in New Guinea* 14:22–29.

Fautin DG. 1991. The anemonefish symbiosis: What is known and what is not? *Symbiosis* 10:23, 46.

Fautin DG, Allen GR. 1992. Field guide to anemonefishes and their host sea anemones. *Western Australian Museum*. Available from https://books.google.fr/books?id=WRLGjwEACAAJ.

Fautin DG, Allen GR. 1997. *Anemone fishes and their host sea anemones: A guide for aquarists and divers*. Rev. edition. Western Australian Museum, Perth, WA.

Frédérich B, Parmentier E, editors. 2016. *Biology of damselfishes*. CRC Press, Taylor & Francis Group, Boca Raton.

Fricke HW. 1973. Individual partner recognition in fish: Field studies on *Amphiprion bicinctus*. *Die Naturwissenschaften* 60:204–205.

Fricke HW, Fricke S. 1977. Monogamy and sex change by aggressive dominance in coral reef fish. *Nature* 266:830–832.

Futuyma DJ, Moreno G. 1988. The evolution of ecological specialization. *Annual Review of Ecology and Systematics* 19:207–233.

Gainsford A, Jones GP, Gardner MG, van Herwerden L. 2020. Characterisation and cross-amplification of 42 microsatellite markers in two Amphiprion species (Pomacentridae) and a natural hybrid anemonefish to inform genetic structure within a hybrid zone. *Molecular Biology Reports* 47:1521–1525.

Geffroy B, Douhard M. 2019. The adaptive sex in stressful environments. *Trends in Ecology & Evolution* 34:628–640.

Gemmell NJ, Todd EV, Goikoetxea A, Ortega-Recalde O, Hore TA. 2019. Natural sex change in fish. Pages 71–117 in *Current topics in developmental biology book series: Sex determination in vertebrates*. Elsevier, New Zealand.

Ghosh J, Wilson RW, Kudoh T. 2009. Normal development of the tomato clownfish *Amphiprion frenatus*: Live imaging and *in situ* hybridization analyses of mesodermal and neurectodermal development. *Journal of Fish Biology* 75:2287–2298.

Glasauer SMK, Neuhauss SCF. 2014. Whole-genome duplication in teleost fishes and its evolutionary consequences. *Molelucar Genetics and Genomics 289*:1045–1060.

Godwin JR. 1994. Behavioural aspects of protandrous sex change in the anemonefish, *Amphiprion melanopus*, and endocrine correlates. *Animal Behaviour* 48:551–567.

Godwin JR, Crews D, Warner RR. 1996. Behavioural sex change in the absence of gonads in a coral reef fish. *Proceedings of the Royal Society of London*. Series B: Biological Sciences 263:1683–1688.

Godwin JR, Thomas P. 1993. Sex change and steroid profiles in the protandrous anemonefish *Amphiprion melanopus* (Pomacentridae, teleostei). *General and Comparative Endocrinology* 91:144–157.

Goikoetxea A, Todd EV, Gemmell NJ. 2017. Stress and sex: Does cortisol mediate sex change in fish? *Reproduction* 154:R149–R160.

Green BS. 2004. Embryogenesis and oxygen consumption in benthic egg clutches of a tropical clownfish, *Amphiprion melanopus* (Pomacentridae). *Comparative Biochemistry and Physiology Part A: Molecular & Integrative Physiology* 138:33–38.

Gunasekaran K, Sarvanakumar A, Selvam D, Mahesh R. 2017. Embryonic and larval developmental stages of sebae clownfish *Amphiprion sebae* (Bleeker 1853) in captive condition. *Indian Journal of Marine Sciences* 46:8.

Hayashi K, Reimer JD. 2020. Five-year study on the bleaching of anemonefish-hosting anemones (Cnidaria: Anthozoa: Actiniaria) in subtropical Okinawajima Island. *Regional Studies in Marine Science* 35:101240.

Hayashi K, Tachihara K, Reimer JD. 2018. Patterns of coexistence of six anemonefish species around subtropical Okinawa-jima Island, Japan. *Coral Reefs* 37:1027–1038.

Hayashi K, Tachihara K, Reimer JD. 2019a. Species and sexual differences in human-oriented behavior of anemonefish at Okinawa Island, Japan. *Marine Ecology Progress Series* 616:219–224.

Hayashi K, Tachihara K, Reimer JD. 2019b. Low density populations of anemonefish with low replenishment rates on a reef edge with anthropogenic impacts. *Environmental Biology of Fishes* 102:41–54.

Hayashi K, Tachihara K, Reimer JD. 2020. Loss of natural coastline influences species diversity of anemonefish and host anemones in the Ryukyu Archipelago. *Aquatic Conservation: Marine and Freshwater Ecosystems*. doi:10.1002/aqc.3435.

He S, Planes S, Sinclair-Taylor TH, Berumen ML. 2019. Diagnostic nuclear markers for hybrid Nemos in Kimbe Bay,

PNG-*Amphiprion chrysopterus* x *Amphiprion sandaracinos* hybrids. *Marine Biodiversity* 49:1261–1269.

Herbert NA, Bröhl S, Springer K, Kunzmann A. 2017. Clownfish in hypoxic anemones replenish host O_2 at only localised scales. *Scientific Reports* 7:6547.

Holmberg RJ et al. 2019. Ocean acidification alters morphology of all otolith types in Clark's anemonefish (*Amphiprion clarkii*). *Peer Journal of Life and Environment* 7:e6152.

Holzer G, Besson M, Lambert A, Barth P, Gillet B, Hughes S, Leulier F, Viriot L, Lecchini D, Laudet V. 2017. Fish larval recruitment to reefs is a thyroid hormone-mediated metamorphosis sensitive to the pesticide chlorpyrifos. *Elife* e27595.

Hotaling S, Kelley JL. 2019. The rising tide of high-quality genomic resources. *Molecular Ecology Resources* 19:567–569.

Hughes TP et al. 2018. Global warming transforms coral reef assemblages. *Nature* 556:492–496.

Irion U, Nüsslein-Volhard C. 2019. The identification of genes involved in the evolution of color patterns in fish. *Current Opinion in Genetics & Development* 57:31–38.

Iwata E, Mikami K, Manbo J, Moriya-Ito K, Sasaki H. 2012. Social interaction influences blood cortisol values and brain aromatase genes in the protandrous false clown anemonefish, *Amphiprion ocellaris*. *Zoological Science* 29:849–855.

Jaillon O et al. 2004. Genome duplication in the teleost fish *Tetraodon nigroviridis* reveals the early vertebrate protokaryotype. *Nature* 431:946–957.

Jones GP, Planes S, Thorrold SR. 2005. Coral reef fish larvae settle close to home. *Current Biology* 15:1314–1318.

Kavanagh KD, Alford RA. 2003. Sensory and skeletal development and growth in relation to the duration of the embryonic and larval stages in damselfishes (Pomacentridae): Development and growth in damselfishes. *Biological Journal of the Linnean Society* 80:187–206.

Kawaguchi M, Takahashi H, Takehana Y, Naruse K, Nishida M, Yasumasu S. 2013. Sub-functionalization of duplicated genes in the evolution of nine-spined stickleback hatching enzyme: Sub-functionalization of hatching enzyme. *Journal of Experimental Zoology Part B: Molecular and Developmental Evolution* 320:140–150.

Kawakami Y, Nozaki J, Seoka M, Kumai H, Ohta H. 2008. Characterization of thyroid hormones and thyroid hormone receptors during the early development of Pacific bluefin tuna (*Thunnus orientalis*). *General and Comparative Endocrinology* 155:597–606.

Kirkwood T. 1977. Evolution of aging. *Nature* 270:301–304.

Kobayashi Y, Horiguchi R, Miura S, Nakamura M. 2010. Sex- and tissue-specific expression of P450 aromatase (cyp19a1a) in the yellowtail clownfish, *Amphiprion clarkii*. *Comparative Biochemistry and Physiology Part A: Molecular & Integrative Physiology* 155:237–244.

Kobayashi Y, Nagahama Y, Nakamura M. 2013. Diversity and plasticity of sex determination and differentiation in fishes. *Sexual Development* 7:115–125.

Krishna R. 2018. Larval development and growth of red saddleback anemonefish, *Amphiprion ephippium* (Bloch, 1790) under captive conditions. *Indian Journal of Geo-Marine Sciences* 47:2421–2428.

Lamm MS, Liu H, Gemmell NJ, Godwin JR. 2015. The need for speed: Neuroendocrine regulation of socially-controlled sex change. *Integrative and Comparative Biology* 55:307–322.

Laudet V. 2011. The origins and evolution of vertebrate metamorphosis. *Current Biology* 21:R726–R737.

Lehmann R et al. 2019. Finding Nemo's genes: A chromosome-scale reference assembly of the genome of the orange clownfish *Amphiprion percula*. *Molecular Ecology Resources* 19:570–585.

Leis JM. 1991. The pelagic stage of reef fishes: The larval biology of coral reef fishes. Pages 183–230 in Sale PF, editor. *The ecology of fishes on coral reefs*. Academic Press, San Diego.

Leis JM, Siebeck U, Dixson DL. 2011. How Nemo finds home: The neuroecology of dispersal and of population connectivity in larvae of marine fishes. *Integrative and Comparative Biology* 51:826–843.

Li J-T, Hou G-Y, Kong X-F, Li C-Y, Zeng J-M, Li H-D, Xiao G-B, Li X-M, Sun X-W. 2015. The fate of recent duplicated genes following a fourth-round whole genome duplication in a tetraploid fish, common carp (*Cyprinus carpio*). *Scientific Reports* 5:8199.

Liew HJ, Ambak MA, Abol-Munafi AB, Chuah TS. 2006. Embryonic development of clownfish *Amphiprion ocellaris* under laboratory conditions. *Journal of Sustainability Science and Management* 1:64–73.

Litsios G, Pearman PB, Lanterbecq D, Tolou N, Salamin N. 2014b. The radiation of the clownfishes has two geographical replicates. *Journal of Biogeography* 41:2140–2149.

Litsios G, Salamin N. 2014. Hybridisation and diversification in the adaptive radiation of clownfishes. *BMC Evolutionary Biology* 14:245.

Litsios G, Sims CA, Wüest RO, Pearman PB, Zimmermann NE, Salamin N. 2012a. Mutualism with sea anemones triggered the adaptive radiation of clownfishes. *BMC Evolutionary Biology* 12:212.

Litsios G, Sims CA, Wüest RO, Pearman PB, Zimmermann NE, Salamin N. 2012b. Mutualism with sea anemones triggered the adaptive radiation of clownfishes. *BMC Evolutionary Biology* 12:1.

Liu H, Todd EV, Lokman PM, Lamm MS, Godwin JR, Gemmell NJ. 2017. Sexual plasticity: A fishy tale. *Molecular Reproduction and Development* 84:171–194.

Lubbock R. 1981. The clownfish/anemone symbiosis: A problem of cell recognition. *Parasitology* 82(159):173.

Lubbock R, Smith DC. 1980. Why are clownfishes not stung by sea anemones? *Proceedings of the Royal Society of London*. Series B: Biological Sciences 207:35–61. Royal Society.

Madeira C, Madeira D, Diniz MS, Cabral HN, Vinagre C. 2016. Thermal acclimation in clownfish: An integrated biomarker response and multi-tissue experimental approach. *Ecological Indicators* 71:280–292.

Madhu R, Madhu K, Retheesh T. 2012. Life history pathways in false clown *Amphiprion ocellaris* Cuvier, 1830: A journey from egg to adult under captive condition. *Journal of Marine Biological Association of India* 54:77–90.

Marcionetti A, Rossier V, Bertrand JAM, Litsios G, Salamin N. 2018. First draft genome of an iconic clownfish species (*Amphiprion frenatus*). *Molecular Ecology Resources* 18:1092–1101.

Marcionetti A, Rossier V, Roux N, Salis P, Laudet V, Salamin N. 2019. Insights into the genomics of clownfish adaptive radiation: Genetic basis of the mutualism with sea anemones. *Genome Biology and Evolution* 11:869–882.

Mariscal RN. 1970. The nature of the symbiosis between Indo-Pacific anemone fishes and sea anemones. *Marine Biology* 6:58–65.

Maytin AK, Davies SW, Smith GE, Mullen SP, Buston PM. 2018. De novo transcriptome assembly of the clown anemonefish (*Amphiprion percula*): A new resource to study the evolution of fish color. *Frontiers in Marine Science* 5:284.

Mebs D. 2009. Chemical biology of the mutualistic relationships of sea anemones with fish and crustaceans. *Toxicon* 54:1071–1074.

Medawar P. 1952. *An unsolved problem of biology: Printed lecture*. University College London, London.

Meyer A, Schartl M. 1999. Gene and genome duplications in vertebrates: The one-to-four (-to-eight in fish) rule and the

evolution of novel gene functions. *Current Opinion in Cell Biology* 11:699–704.
Militz TA, McCormick MI, Schoeman DS, Kinch J, Southgate PC. 2016. Frequency and distribution of melanistic morphs in coexisting population of nine clownfish species in Papua New Guinea. *Marine Biology* 163:200–210.
Mills SC, Mourier J, Galzin R. 2010. Plasma cortisol and 11-ketotestosterone enzyme immunoassay (EIA) kit validation for three fish species: The orange clownfish *Amphiprion percula*, the orangefin anemonefish *Amphiprion chrysopterus* and the blacktip reef shark *Carcharhinus melanopterus*. *Journal of Fish Biology* 77:769–777.
Mitchell J. 2005. Queue selection and switching by false clown anemonefish, *Amphiprion ocellaris*. *Animal Behaviour* 69:643–652.
Mitchell LJ, Tettamanti V, Marshall JN, Cheney KL, Cortesi F. 2020. CRISPR/Cas9-mediated generation of biallelic G0 anemonefish (*Amphiprion ocellaris*) mutants. preprint. *Molecular Biology*. Available from http://biorxiv.org/lookup/doi/10.1101/2020.10.07.330746 (accessed November 15, 2020).
Miura S, Kobayashi Y, Bhandari RK, Nakamura M. 2013. Estrogen favors the differentiation of ovarian tissues in the ambisexual gonads of anemonefish *Amphiprion clarkii*: The role of estrogen for gonad in anemonefish. *Journal of Experimental Zoology Part A: Ecological Genetics and Physiology* 319:560–568.
Miyagawa-Kohshima K et al. 2014. Embryonic learning of chemical cues via the parents' host in anemonefish (*Amphiprion ocellaris*). *Journal of Experimental Marine Biology and Ecology* 457:160–172.
Moyer JT. 1976. Geographical variation and social dominance in Japanese populations of the anemonefish *Amphiprion clarkii*. *Japane Journal of Ichthyology* 23:12–22.
Moyer JT. 1980. Influence of temperate waters on the behavior of the tropical anemonefish *Amphiprion clarkii* at Miyake-jima, Japan. *Bulletin of Marine Science*:261–272.
Munday PL, Buston P, Warner R. 2006. Diversity and flexibility of sex-change strategies in animals. *Trends in Ecology & Evolution* 21:89–95.
Munday PL, Jones GP, Pratchett MS, Williams AJ. 2008. Climate change and the future for coral reef fishes. *Fish and Fisheries* 9:261–285.
Murata M, Miyagawa-Kohshima K, Nakanishi K, Naya Y. 1986. Characterization of compounds that induce symbiosis between sea anemone and anemone fish. *Science* 234:585–587.
Nanninga GB, Côté IM, Beldade R, Mills SC. 2017. Behavioural acclimation to cameras and observers in coral reef fishes. *Ethology* 123:705–711.
Nanninga GB, Saenz-Agudelo P, Zhan P, Hoteit I, Berumen ML. 2015. Not finding Nemo: Limited reef-scale retention in a coral reef fish. *Coral Reefs* 34:383–392.
Nedosyko AM, Young JE, Edwards JW, Burke da Silva K. 2014. Searching for a toxic key to unlock the mystery of anemonefish and anemone symbiosis. *PLoS One* 9:e98449.
Nelson JS, Grande T, Wilson MVH. 2016. *Fishes of the world*. Fifth edition. John Wiley & Sons, Hoboken, NJ.
Nguyen H-TT, Dang BT, Glenner H, Geffen AJ. 2020. Cophylogenetic analysis of the relationship between anemonefish *Amphiprion* (Perciformes: Pomacentridae) and their symbiotic host anemones (Anthozoa: Actiniaria). *Marine Biology Research* 16:117–133.
Nieuwenhuys R, ten Donkelaar HJ, Nicholson C. 1998. *The central nervous system of vertebrates*. Springer Berlin Heidelberg, Berlin, Heidelberg. Available from http://link.springer.com/10.1007/978-3-642-18262-4 (accessed November 13, 2020).
Norin T, Mills SC, Crespel A, Cortese D, Killen SS, Beldade R. 2018. Anemone bleaching increases the metabolic demands of symbiont anemonefish. *Proceedings of the Royal Society B: Biological Sciences* 285:20180282.
Ochi H. 1985. Temporal patterns of breeding and larval settlement in a temperate population of the tropical anemonefish, *Amphiprion clarkii*. *Japanese Journal of Ichthyology* 32:248–257.
Ohno S. 1970. *Evolution by gene duplication*. Springer Berlin Heidelberg, Berlin, Heidelberg.
Olivotto I, Geffroy B. 2017. Clownfish. Pages 177–199 in Calado R, Olivotto I, Oliver MP, Holt GJ, editors. *Marine ornamental species aquaculture*. First edition. Wiley Online Books, Chichester, West Sussex, UK.
Ollerton J, McCollin D, Fautin DG, Allen GR. 2007. Finding NEMO: Nestedness engendered by mutualistic organization in anemonefish and their hosts. *Proceedings of the Royal Society B: Biological Sciences* 274:591–598.
Onimaru K, Kuraku S. 2018. Inference of the ancestral vertebrate phenotype through vestiges of the whole-genome duplications. *Briefings in Functional Genomics* 17:352–361.
Ortega-Recalde O, Goikoetxea A, Hore TA, Todd EV, Gemmell NJ. 2020. The genetics and epigenetics of sex change in fish. *Annual Review of Animal Biosciences* 8:47–69.
Paris CB, Atema J, Irisson J-O, Kingsford M, Gerlach G, Guigand CM. 2013. Reef odor: A wake up call for navigation in reef fish larvae. *PLoS One* 8:e72808.
Park MS, Shin HS, Kil G-S, Lee J, Choi CY. 2011. Monitoring of Na+/K+-ATPase mRNA expression in the cinnamon clownfish, *Amphiprion melanopus*, exposed to an osmotic stress environment: Profiles on the effects of exogenous hormone. *Ichthyological Research* 58:195–201.
Parmentier E, Colleye O, Fine ML, Frederich B, Vandewalle P, Herrel A. 2007. Sound production in the clownfish *Amphiprion clarkii*. *Science* 316:1006.
Parmentier E, Colleye O, Mann D. 2009. Hearing ability in three clownfish species. *Journal of Experimental Biology* 212:2023–2026.
Parmentier E, Lagardère JP, Vandewalle P, Fine ML. 2005. Geographical variation in sound production in the anemonefish *Amphiprion akallopisos*. *Proceedings of the Royal Society B: Biological Sciences* 272:1697–1703.
Patkaew S, Direkbusarakom S, Tantithakura O. 2014. A simple method for cell culture of 'Nemo' ocellaris clownfish (*Amphiprion ocellaris*, Cuvier 1830). *Cell Biology International Reports* 7.
Patterson LB, Parichy DM. 2019. Zebrafish pigment pattern formation: Insights into the development and evolution of adult form. *Annual Review of Genetics* 53:505–530.
Planes S, Jones GP, Thorrold SR. 2009. Larval dispersal connects fish populations in a network of marine protected areas. *Proceedings of the National Academy of Sciences* 106:5693–5697.
Porter SN, Humphries MS, Buah-Kwofie A, Schleyer MH. 2018. Accumulation of organochlorine pesticides in reef organisms from marginal coral reefs in South Africa and links with coastal groundwater. *Marine Pollution Bulletin* 137:295–305.
Pratte ZA, Patin NV, McWhirt ME, Caughman AM, Parris DJ, Stewart FJ. 2018. Association with a sea anemone alters the skin microbiome of clownfish. *Coral Reefs* 37:1119–1125.
Putra DF, Abol-Munafi AB, Muchlisin ZA, Chen J-C. 2012. Preliminary studies on morphology and digestive tract development of tomato clownfish, *Amphiprion frenatus* under captive condition. *International Journal of the Bioflux Society AACL Bioflux* 5:8.
Randall JE. 1955. Fishes of the Gilbert Islands. *Atoll Research Bulletin* 47:1–243.
Randall JE, Fautin D. 2002. Fishes other than anemonefishes that associate with sea anemones. *Coral Reefs* 21:188–190.

Rattanayuvakorn S, Mungkornkarn P, Thongpan A, Chatchavalvanich K. 2005. Embryonic development of saddleback anemonefish, *Amphiprion polymnus*, Linnaeus (1758). *Natural Science* 39:455–463.

Rhyne AL, Tlusty MF, Schofield PJ, Kaufman L, Morris JA, Bruckner AW. 2012. Revealing the appetite of the marine aquarium fish trade: The volume and biodiversity of fish imported into the United States. *PLoS One* 7:e35808.

Robertson DR, Grove JS, McCosker JE. 2004. Tropical transpacific shore fishes. *Pacific Science* 58:507–565.

Rodrigues FSLM, Yang X, Nikaido M, Liu Q, Kelsh RN. 2012. A simple, highly visual *in vivo* screen for anaplastic lymphoma kinase inhibitors. *ACS Chemical Biology* 7:1968–1974.

Rolland J, Silvestro D, Litsios G, Faye L, Salamin N. 2018. Clownfishes evolution below and above the species level. *Proceedings of the Royal Society B: Biological Sciences* 285:20171796.

Roux N, Lami R, Salis P, Magré K, Romans P, Masanet P, Lecchini D, Laudet V. 2019a. Sea anemone and clownfish microbiota diversity and variation during the initial steps of symbiosis. *Scientific Reports* 9:19491.

Roux N, Salis P, Lambert A, Logeux V, Soulat O, Romans P, Frédérich B, Lecchini D, Laudet V. 2019b. Staging and normal table of postembryonic development of the clownfish (*Amphiprion ocellaris*). *Developmental Dynamics* 248:545–568.

Roux N, Salis P, Lee S-H, Besseau L, Laudet V. 2020. Anemonefish, a model for eco-evo-devo. *EvoDevo* 11:20.

Roux N, Logeux V, Trouillard N., Pillot R Magré K, Salis P, Lecchini D, Besseau L, Laudet V, Romans P. 2021. A star is born again: Methods for larval rearing of an emerging model organism, the False clownfish Amphiprion ocellaris. *Journal of Experimental Zoology Part B: Molecular and Developmental Evolution*, Jun: 336(4): 376–85.

Rueger T, Barbasch TA, Wong MYL, Srinivasan M, Jones GP, Buston PM. 2018. Reproductive control via the threat of eviction in the clown anemonefish. *Proceedings of the Royal Society B: Biological Sciences* 285:20181295.

Saenz-Agudelo P, Jones GP, Thorrold SR, Planes S. 2011. Detrimental effects of host anemone bleaching on anemonefish populations. *Coral Reefs* 30:497–506.

Sahm A, Almaida-Pagán P, Bens M, Mutalipassi M, Lucas-Sánchez A, de Costa Ruiz J, Görlach M, Cellerino A. 2019. Analysis of the coding sequences of clownfish reveals molecular convergence in the evolution of lifespan. *BMC Evolutionary Biology* 19:89.

Salis P et al. 2019a. Developmental and comparative transcriptomic identification of iridophore contribution to white barring in clownfish. *Pigment Cell & Melanoma Research* 32:391–402.

Salis P, Lorin T, Laudet V, Frédérich B. 2019b. Magic traits in magic fish: Understanding color pattern evolution using reef fish. *Trends in Genetics* 35:265–278.

Salis P, Roux N, Lecchini D, Laudet V. 2018a. The post-embryonic development of *Amphiprion perideraion* reveals a decoupling between morphological and pigmentation change. *Société Française d'Ichtyologie*. Available from http://sfi-cybium.fr/fr/post-embryonic-development-amphiprion-perideraion-reveals-decoupling-between-morphological-and (accessed November 13, 2020).

Salis P, Roux N, Soulat O, Lecchini D, Laudet V, Frédérich B. 2018b. Ontogenetic and phylogenetic simplification during white stripe evolution in clownfishes. *BMC Biology* 16(1):90.

Salis P, Lee SH, Roux N, Lecchini D, Laudet V. 2021. The real Nemo movie: Description of embryonic development in Amphiprion ocellaris from first division to hatching. *Developmental Dynamics*, May 7.

Salles OC, Almany GR, Berumen ML, Jones GP, Saenz-Agudelo P, Srinivasan M, Thorrold SR, Pujol B, Planes S. 2020. Strong habitat and weak genetic effects shape the lifetime reproductive success in a wild clownfish population. *Ecology Letters* 23:265–273.

Salles OC, Maynard JA, Joannides M, Barbu CM, Saenz-Agudelo P, Almany GR, Berumen ML, Thorrold SR, Jones GP, Planes S. 2015. Coral reef fish populations can persist without immigration. *Proceedings of the Royal Society B: Biological Sciences* 282:20151311.

Salles OC, Pujol B, Maynard JA, Almany GR, Berumen ML, Jones GP, Saenz-Agudelo P, Srinivasan M, Thorrold SR, Planes S. 2016. First genealogy for a wild marine fish population reveals multigenerational philopatry. *Proceedings of the National Academy of Sciences* 113:13245–13250.

Santini S, Polacco G. 2006. Finding Nemo: Molecular phylogeny and evolution of the unusual life style of anemonefish. *Gene* 385:19–27.

Schlichter D. 1968. Das Zusammenleben von Riffanemonen und Anemonenfischen. *Zeitschrift für Tierpsychologie* 25:933–954.

Scott A, Dixson DL. 2016. Reef fishes can recognize bleached habitat during settlement: Sea anemone bleaching alters anemonefish host selection. *Proceedings of the Royal Society B: Biological Sciences* 283:20152694.

Scott MW. 2008. *Damsefishes and anemonefishes: The complete illustrated guide to their identification, behaviors and captive care*. T. F. H. Publications, Neptune City.

Seymour J, Barbasch T, Buston P. 2018. Lunar cycles of reproduction in the clown anemonefish *Amphiprion percula*: Individual-level strategies and population-level patterns. *Marine Ecology Progress Series* 594:193–201.

Shabana NMA, Helal AM. 2006. Reproduction in captivity, broodstock rearing and embryology of the anemone fish *Amphiprion bicintus* inhabiting the Red Sea. *Egyptian Journal of Aquatic Research* 32:438–446.

Shuman CS, Hodgson G, Ambrose RF. 2005. Population impacts of collecting sea anemones and anemonefish for the marine aquarium trade in the Philippines. *Coral Reefs* 24:564–573.

Simpson SD, Munday PL, Wittenrich ML, Manassa R, Dixson DL, Gagliano M, Yan HY. 2011. Ocean acidification erodes crucial auditory behaviour in a marine fish. *Biology Letters* 7:917–920.

Southey BR, Rodriguez-Zas SL, Rhodes JS, Sweedler JV. 2020. Characterization of the prohormone complement in Amphiprion and related fish species integrating genome and transcriptome assemblies. *PLoS One* 15:e0228562.

Stieb SM, de Busserolles F, Carleton KL, Cortesi F, Chung W-S, Dalton BE, Hammond LA, Marshall NJ. 2019. A detailed investigation of the visual system and visual ecology of the Barrier Reef anemonefish, *Amphiprion akindynos*. *Scientific Reports* 9:16459.

Supiwong W, Tanomtong A, Pinthong K, Kaewmad P, Poungnak P, Jangsuwan N. 2015. The first chromosomal characteristics of nucleolar organizer regions and karyological analysis of pink anemonefish, *Amphiprion perideraion* (Perciformes, Amphiprioninae). *Cytologia* 80:271–278.

Tan MH, Austin CM, Hammer MP, Lee YP, Croft LJ, Gan HM. 2018. Finding Nemo: Hybrid assembly with Oxford Nanopore and Illumina reads greatly improves the clownfish (*Amphiprion ocellaris*) genome assembly. *GigaScience* 7. Available from https://academic.oup.com/gigascience/article/doi/10.1093/gigascience/gix137/4803946 (accessed September 25, 2020).

Thomas D, Prakashand C, Gopakumar G. 2015. Spawning behaviour and embryonic development in the sebae anemonefish *Amphiprion sebae* (Bleeker, 1853). *Indian Journal of Fisheries* 62:58–65.

Thomson CE, Winney IS, Salles OC, Pujol B. 2018. A guide to using a multiple-matrix animal model to disentangle genetic and nongenetic causes of phenotypic variance. *PLoS One* 13:e0197720.

Titus BM et al. 2019. Phylogenetic relationships among the clownfish-hosting sea anemones. *Molecular Phylogenetics and Evolution* 139:106526.

Todd EV, Liu H, Muncaster S, Gemmell NJ. 2016. Bending genders: The biology of natural sex change in fish. *Sexual Development* 10:223–241.

Veilleux HD, Van Herwerden L, Cole NJ, Don EK, De Santis C, Dixson DL, Wenger AS, Munday PL. 2013. Otx2 expression and implications for olfactory imprinting in the anemonefish, *Amphiprion percula*. *Biology Open* 2:907–915.

Victor B, Wellington G. 2000. Endemism and the pelagic larval duration of reef fishes in the eastern Pacific Ocean. *Marine Ecology Progress Series* 205:241–248.

Williams GC. 1957. Pleiotropy, natural selection, and the evolution of senescence. *Evolution* 11:398–411.

Yang W, Lin B, Li G, Chen H, Liu M. 2019. Sequencing and transcriptome analysis for reproduction-related genes identification and SSRs discovery in sequential hermaphrodite *Amphiprion ocellaris*. *Turkish Journal of Fisheries and Aquatic Sciences* 19. Available from www.trjfas.org/pdf/issue_19_12/1207.pdf (accessed November 11, 2020).

Yasir I, Qin JG. 2007. Embryology and early ontogeny of an anemonefish *Amphiprion ocellaris*. *Journal of the Marine Biological Association of the United Kingdom* 87:1025–1033.

Yu PB, Hong CC, Sachidanandan C, Babitt JL, Deng DY, Hoyng SA, Lin HY, Bloch KD, Peterson RT. 2008. Dorsomorphin inhibits BMP signals required for embryogenesis and iron metabolism. *Nature Chemical Biology* 4:33–41.

Index

Note: Page numbers in *italic* indicate a figure and page numbers in **bold** indicate a table on the corresponding page.

D

E

F